Processes on the Early Earth

edited by

Wolf Uwe Reimold
Museum für Naturkunde (Mineralogy)
Humboldt University
Invalidenstrasse 43
10115 Berlin
Germany

Roger L. Gibson
Impact Cratering Research Group
School of Geosciences
University of the Witwatersrand
Private Bag 3
2050 Wits
Johannesburg
South Africa

THE
GEOLOGICAL
SOCIETY
OF AMERICA

Special Paper 405

3300 Penrose Place, P.O. Box 9140 ▪ Boulder, Colorado 80301-9140 USA

2006

Published by The Geological Society of America, Inc.
3300 Penrose Place, P.O. Box 9140, Boulder, Colorado 80301-9140, USA
www.geosociety.org

Printed in U.S.A.

GSA Books Science Editor; Abhijit Basu

Library of Congress Cataloging-in-Publication Data

Processes on the early Earth / edited by Wolf Uwe Reimold, Roger L. Gibson.
p. cm. (Special paper ; 405)
Includes bibliographical references and index.
ISBN-10 0-8137-2405-8 (pbk.)
ISBN-13 978-0-8137-2405-8 (pbk.)
1. Geology, Stratigraphic—Archean—Congresses. 2. Sediments (Geology)—Congresses. 3. Cryptoexplosion structures—Congresses. 4. Life—Origin—Congresses. 5. Earth—Crust—Congresses. I. Reimold, W.U. II. Gibson, Roger L. (Roger Lawrence), 1963-. Special papers (Geological Society of America) ; 405.
QE653.3 .P76 2006
551.7'12 22

2006040772

Cover: This painting shows a possible view of the early Earth, some time between 4.4 and 4 billion years ago. In contrast to earlier speculations, there is now evidence of liquid water starting on the order of 100 million years after the Moon-forming impact event. The young Moon was much closer to Earth than today and did not yet show any of the mare that dominate its Earth-facing side today. A significantly higher impact rate than today, as the tail end of the accretion of the planets, led to a barrage of both Earth and Moon. One medium-size impact event can be seen in the right side of the image, whereas smaller events happen on the Moon. The ocean dominates the surface of Earth, but a number of island-arc formations, with intense volcanism, indicate the presence of early continental crust. Earth is covered by a dense, CO_2-dominated atmosphere. Painting by Dona Jalufka (Vienna, 2005.)

10 9 8 7 6 5 4 3 2 1

Contents

Preface

The first-ever Field Forum hosted jointly by the Geological Society of America and the Geological Society of South Africa was held in South Africa 4–9 July, 2004. It attracted 43 participants from 13 countries who gathered to investigate and debate recent developments dealing with "Processes on the Early Earth." Themes within the forum included: accretionary processes; the age of Earth and the oldest rocks on Earth; the early impact history of the Earth and Moon; Archean crust formation and geodynamics; impact flux through time; traces of impact events in Archean strata; significance of large impact events for development of early life; earliest formation of life on Earth; early sedimentary processes; and Archean oceans.

Discussion was facilitated by two superlative Archean field localities in the Kaapvaal Craton, namely the Barberton Mountain Land (one of the oldest and best-preserved volcano-sedimentary greenstone belts in the world) and the Vredefort Dome (the central part of the erosional remnant of the world's largest and oldest known impact structure, recently declared a World Heritage Site). Together, these locations provided an opportunity to examine events spanning more than 1000 million years of Archean history and crustal environments ranging from the surface to the deep lithosphere.

In the past few decades, new geochronological and isotope geological tools have helped the investigation of Earth's oldest rocks and environments, and provided new insight into the workings on and in our planet during Archean and even Hadean times. In addition, impact cratering has become accepted as one of the fundamental processes in the solar system, and its inherent physical and chemical processes have become increasingly well understood. Current debates in this discipline are centered on the role this process played during the initial stage of our planet's evolution as well as in the course of the development of life on Earth. Zircon geochronology has indicated evidence for the existence of the earliest continental crust back to ca. 4.4 Ga. The degree to which modern-day tectonic processes were active in Archean times remains a topic of considerable interest. The discovery of microfossils in a range of Archean environments has opened up new possibilities for multidisciplinary cooperation and has captured the attention of scientists and laypeople alike. Some of the world's most important ore deposits were developed during Archean times, making the study and understanding of these rocks important for the detection of new mineral resources.

The southern African Kaapvaal Craton provides a unique, large Archean terrane that allows investigation of all the aforementioned major aspects of Archean geology. Cooperation between the GSA and GSSA on the Field Forum, as well as substantial sponsorship obtained for this event, provided an opportunity to bring together established experts in several of these fields, but also a number of newcomers and postgraduate students, thus providing an opportunity for multilevel exchange. Invited lectures set the scene, and field outcrop examinations involved both outcrop-specific introductory talks, as well as short reviews by experts on specific subjects, including—among others—impact spherule layers and the record of the early atmosphere. A more detailed review of the Field Forum proceedings was published in *GSA Today* (Reimold et al., 2004).

The Field Forum was made possible through sizable sponsorship from the National Research Foundation of South Africa, as well as from the Council for Geoscience (Pretoria). The editors of this volume also would like to thank their co-organizers of the Field Forum: Carl R. Anhaeusser, Economic Geology Research Institute, School of Geosciences, University of the Witwatersrand, Johannesburg, South Africa; Ken A. Eriksson, Department of Geosciences, Virginia Polytechnic Institute and State University, Blacksburg, Virginia, USA; Axel Hofmann, School of Geological Sciences, Howard College Campus, University of KwaZulu-Natal, Durban, South Africa; Christian Koeberl, Department of Geological Sciences, University of Vienna, Vienna, Austria; Bruce M. Simonson, Geology Department, Oberlin College, Oberlin, Ohio, USA; and Frances Westall, Centre de Biophysique Moléculaire, CNRS, Orléans, France.

The contents of this proceedings volume have a strong focus on the southern African Archean, but there are also several contributions dealing with Canadian and Australian Archean terranes. The sequence chosen for this volume comprises a first section dealing with impact processes and evidence thereof in the Archean record. This is followed by a variety of contributions dealing with the development of early life and tectonic processes and evolution during the Archean. The volume commences with an in-depth review by Christian Koeberl of the evidence for impact in the earliest stages of the evolution of the solar system, including a review of the earliest impact evidence on the Earth's surface. Richard Grieve and coworkers examine the results of modeling of differential scaling of impact melt and transient cavity volumes on the Earth, with a comparison of lunar evidence. They investigate the formation of giant impact melt pools on the early Earth and study the possibility that these melt volumes could have differentiated to form felsic rocks which, in turn, could have been possible sources for pre–3.9 Ga zircons.

The earliest direct evidence for impact on the early Earth in the terrestrial impact record is represented by several, ca. 3.4 Ga, spherule layers in Barberton and Western Australian Archean sequences. The contributions by Axel Hofmann and co-workers, and by Issaku Kohl and co-workers, examine aspects of these Paleoarchean spherule layers in the Barberton Mountain Land as well as of the Neoarchean Monteville spherule layer in the Transvaal Supergroup. Hofmann et al. review the somewhat controversial stratigraphic, petrographic, and geochemical evidence regarding the Barberton spherule layers and draw attention to some phenomenological and petrographic aspects of accretionary lapilli beds in the Barberton stratigraphy. The latter is of particular interest with regard to the recent discoveries of accretionary lapilli in several impactite sequences. Kohl et al. report the results of a detailed petrographic study of the Monteville spherules, with the aim of providing a basis for intercontinental correlation with the Neoarchean spherule layer in the Jeerinah Formation of the Hamersley Basin of Western Australia.

Russell Hartlaub and co-workers examine the U-Pb and Lu-Hf chronological records for Paleoarchean (>3.7 Ga) crust. TIMS (thermal ionization mass spectrometry) and SIMS chronology for early Archean crust in Southern Greenland is compared and critically evaluated by Krogh and Kamo, who call for caution in the interpretation of complex zircon age distributions for very early terrestrial rocks.

Frances Westall and her group of collaborators contribute a very detailed review of the current knowledge about, and the careful and multidisciplinary methodology to be applied to the study of, traces of early life in Archean environments. This is followed by an exemplary study by Beate Orberger and co-workers on the >3.4 Ga occurrence of trace micro-fossils in the Kitty's Gap area of the Pilbara Craton, which also investigates in detail the relevant sedimentary environment.

Annika Dziggel and her team investigate the oldest formations in the Barberton greenstone belt and present evidence for the existence of an evolved granitoid crust older than ca. 3.52 Ga, with early sediments, older than 3.43 Ga, formed within a possible island-arc environment. John Shervais reports on his investigation of potential Archean accretionary mélange complexes, finding that there may well be evidence for Phanerozoic-style plate tectonics as old as 3.8 Ga. Carl Anhaeusser proposes that discontinuous laterally extensive ultramafic belts within the Kaapvaal Craton could represent paleosutures marking ocean closures during the Early and Mesoarchean that led to consolidation of the craton. He arrives at a similar model of convergent plate tectonics ("ophiolite tectonics") in the Archean as Shervais. A field-based structural analysis of 3.1 Ga granite batholiths in the southern part of the Barberton Mountain Land by Richard Belcher and Alex Kisters provides new constraints on the tectonics operating during the final phase of consolidation of the Kaapvaal Craton. The process proposed involves granite emplacement in a contractional plate margin environment followed by gravitational collapse of the tectonically and magmatically thickened crust.

Richard Armstrong and his team have obtained extensive new U-Pb single-zircon SHRIMP (sensitive high-resolution ion microprobe) data for lithologically and structurally well constrained samples from the Archean basement of the Vredefort Dome, a large window into the crust of the Kaapvaal craton. These new results are discussed against the various tectonic models for the formation and evolution of this craton, concluding that the new results favor that the oldest sialic rocks (tonalite-trondhjemite-granodiorite lithologies) in the dome were formed in an island-arc tectonic setting at ca. 3.12 Ga, which was accreted onto the craton at ca. 3.1 Ga. Allan Wilson and Claire Grant report on a very detailed physical volcanological and geochemical study of the volcanic strata of the Mesoarchean Pongola Supergroup, thereby providing one of the earliest detailed records of volcanic processes in the Archean.

Two papers examine the nature of craton boundaries: Paul Dirks and Hielke Jelsma provide a synthesis of the polyphase evolution of the northeastern margin of the Zimbabwe craton. Exhumation of the craton margin occurred as the result of oblique collision between the craton and an unknown crustal block to the north at ca. 2.6 Ga. Subsequent tectonic events at 1.1–0.5 Ga formed the Zambezi Belt, but did not markedly affect the craton margin. Jackson Barton and colleagues suggest an alternative geodynamic model involving Turkic-style accretion of terranes during the Late Archean and Paleoproterozoic between the Zimbabwe and Kaapvaal cratons, to form the Central Zone of the Limpopo Belt.

The final trilogy of papers deals with aspects of one of the world's foremost ore deposits, namely the gold and uranium deposits of the Witwatersrand Basin. A longstanding controversy about this deposit was centered on the question whether the Witwatersrand gold originates from Archean placer sedimentation or is the result of later hydrothermal processes. Baojin Zhao and co-workers report new stable isotope results for various phases in the Ventersdorp Contact Reef, and conclude that the metamorphic fluids involved in the formation of sulfide mineralization were probably derived from the basin itself. By implication, this supports the conclusion that gold was locally remobilized from preexisting placer concentrations—providing further support for the modified-placer theory of Witwatersrand gold mineralization. Gillian Drennan and Lawrence Robb contribute an extensive review of the hydrocarbon phases and fluids related to Witwatersrand gold mineralization. They find that the circulation of hydrocarbon-rich fluids and subsequent precipitation of bitumen and gold-sulfide mineralization, late in the evolution of the basin, provide an explanation for the apparent hydrothermal component of Witwatersrand mineralization. Rudy Boer and Uwe Reimold contribute new analytical results on fluid inclusions in various lithologies associated with the Ventersdorp Contact Reef. They conclude that evidence for fluid movement associated with the 2.02 Ga Vredefort impact event exists in various parts of the basin, but that fluid flow was relatively minor and localized.

We would like to express our gratitude to all contributors to this Special Paper and our appreciation for the support received from the reviewers: Carl Anhaeusser, Nick Arndt, Jay Barton,

Alex Bevan, Guy Charlesworth, Marthinus Cloete, Alex Deutsch, Paul Dirks, Burkhard Dressler, Maarten de Wit, Gillian Drennan, Ken Eriksson, Richard Grieve, Russell Hartlaub, Joerg Herrmann, Axel Hofmann, Keith Kenyon, Christian Koeberl, Tim Kusky, Jonathan Law, J. Marsh, Alan Nutman, Marc Poujol, Steve Prevec, David Prieur, Ulrich Riller, Bruce Simonson, Andre Smit, Dirk van Reenen, Torsten Vennemann, Maud Walsh, Malcolm Waters, and several colleagues who prefer to remain anonymous. GSA Books Science Editor Abhijit Basu and the GSA publications team are thanked for their encouragement and technical assistance.

REFERENCE CITED

Reimold, W.U., Anhaeusser, C.R., Eriksson, K.A., Hofmann, A., Gibson, R.L., Koeberl, C., Simonson, B.M., and Westall, F., 2004, Field Forum Report: Processes on the Early Earth: GSA Today, v. 14, no. 11, p. 28–29.

Wolf Uwe Reimold
Roger L. Gibson

Participants in the Geological Society of America–Geological Society of South Africa Field Forum "Processes on the Early Earth," Kaapvaal Craton, 4–9 July 2004. Photograph courtesy of Christian Koeberl, University of Vienna.

Geological Society of America
Special Paper 405
2006

The record of impact processes on the early Earth: A review of the first 2.5 billion years

Christian Koeberl[†]
Department of Geological Sciences, University of Vienna, Althanstrasse 14, A-1090 Vienna, Austria

ABSTRACT

Collisions and impact processes have been important throughout the history of the solar system, including that of the Earth. Small bodies in the early solar system, the planetesimals, grew through collisions, ultimately forming the planets. The Earth started growing ca. 4.56 Ga in this way. Its early history was dominated by violent impacts and collisions, of which we only have circumstantial evidence. The Earth was still growing and had reached ~70%–80% of its present mass when at ca. 4.5 Ga a Mars-sized protoplanet collided with Earth, leading to the formation of the moon—at least according to the currently most popular hypothesis of lunar origin. After its formation, the moon was subjected to intense post-accretionary bombardment between ca. 4.5 and 3.9 Ga. In addition, there is convincing evidence that the Moon experienced an interval of intense bombardment with a maximum at ca. 3.85 ± 0.05 Ga; subsequent mare plains as old as 3.7 or 3.8 Ga are preserved. It is evident that if a late heavy bombardment occurred on the Moon, the Earth must have been subjected to an impact flux at least as intense as that recorded on the Moon. The consequences for the Earth must have been devastating, although the exact consequences are the subject of debate (total remelting of the crust versus minimal effects on possibly emerging life forms). So far, no unequivocal record of a late heavy bombardment on the early Earth has been found. The earliest rocks on Earth date back to slightly after the end of the heavy bombardment, although there are relict zircons that have ages of up to 4.4 Ga (in which, however, no impact-characteristic shock features were found so far). In terms of evidence for impact on Earth, the first solid evidence exists in the form of various spherule layers found in South Africa and Australia with ages between ca. 3.4 and 2.5 Ga; these layers represent several (the exact number is still unknown) large-scale impact events. The oldest documented (and preserved) impact craters on Earth have ages of 2.02 and 1.86 Ga. Thus, the impact record for more than half of the geological history of the Earth is incomplete and not well preserved, and we mostly have only indirect evidence regarding the impact record and its effects during the first 2.5 b.y. of Earth history.

Keywords: early Earth, Hadean, impact processes, origin of the Moon, impact spherules, late heavy bombardment.

[†]E-mail: christian.koeberl@univie.ac.at.

Koeberl, C., 2006, The record of impact processes on the early Earth: A review of the first 2.5 billion years, *in* Reimold, W.U., and Gibson, R.L., Processes on the Early Earth: Geological Society of America Special Paper 405, p. 1–22, doi: 10.1130/2006.2405(01). For permission to copy, contact editing@geosociety.org.

INTRODUCTION: CRATERING PROCESSES

It has only recently been recognized in the geological sciences how important the process of impact cratering is on a planetary scale (e.g., Wilhelms, 1987; Taylor, 1992b; Wetherill, 1994). During the last few decades, planetary scientists and astronomers (e.g., Wilhelms, 1987; Shoemaker et al., 1990; Taylor, 1992b; Wetherill, 1994) have demonstrated that the surfaces of the Moon, Mercury, Venus, Mars, the asteroids, and the moons of the outer gas planets are all covered (some surfaces to saturation) with meteorite impact craters (Fig. 1). Impact cratering is a high-energy event that occurs at more or less irregular intervals (although over long periods of time, an average cratering rate can be established; e.g., Shoemaker et al., 1990). Part of the problem regarding recognition of the remnants of impact events is the fact that terrestrial processes (sedimentation, erosion, plate tectonics) either cover or erase the surface expression of impact structures on Earth. Many impact structures are covered by younger (i.e., post-impact) sediments and are not visible on the surface. Others were mostly (or entirely) destroyed by erosion. In some cases, ejecta have been found far from any possible impact structure. The study of these ejecta has led, in some cases, to the discovery of the source impact craters. Important witnesses for the characteristics of the impact process are the affected rocks. As described in more detail below, crater structures are filled with melted, shocked, and brecciated rocks. Some of these are in situ, and others have been transported, in some cases considerable distances, from the source crater, representing so-called ejecta. Some of the ejecta material can fall back directly into the crater, and most of the ejecta end up close to the crater (<5 crater radii; these are called *proximal* ejecta), but a small fraction may travel much greater distances and is then considered *distal* ejecta. The recent reviews by Montanari and Koeberl (2000), Koeberl (2001), and Simonson and Glass (2004) describe impact ejecta in some detail.

Figure 1. Densely cratered part of the surface of the moon. This Apollo 16 metric camera image (AS16–3021) shows the Moon's eastern limb and far side. The lower left part of the image shows a portion of the Moon visible from Earth. The dark area at the 8 o'clock position on the edge is Mare Crisium (~900 km diameter of outermost ring). To the right of that is Mare Smythii. The upper right area shows the heavily cratered lunar far side.

To determine whether specific rocks have been subjected to impact, it is necessary to identify criteria that allow us to distinguish the effects of such processes from those resulting from normal terrestrial geological processes. Only the presence of diagnostic shock metamorphic effects, and in some cases the discovery of meteorites or traces thereof, are generally accepted as providing unambiguous evidence for an impact origin. Shock deformation can be present in macroscopic form (shatter cones) or in microscopic form (see below). The same two criteria apply to distal impact ejecta layers and allow to confirm that material found in such layers originated in an impact event at a possibly still unknown location. As of mid 2005, ~170 impact structures have been identified on Earth based on these criteria. With one exception (Vredefort, see below), all of these are younger than 2 Ga.

In nature, shock metamorphic effects are uniquely characteristic of shock levels associated with hypervelocity impact, and the only natural process known to date associated with such effects is the hypervelocity impact of extraterrestrial bodies. Shock metamorphic effects are best expressed in various breccias that are found within and around the crater structure. During impact, shock pressures of ≥100 GPa and temperatures ≥3000 °C are produced in large volumes of target rock. Such values are significantly different from conditions for endogenic metamorphism of crustal rocks, and shock compression is not a thermodynamically reversible process; thus, most of the structural and phase changes in minerals and rocks are characteristic of the high pressures (5 to >50 GPa) and extreme strain rates (10^6–10^8 s^{-1}) associated with impact. As described in various review papers (e.g., Stöffler and Langenhorst, 1994; Grieve et al., 1996; Montanari and Koeberl, 2000; Koeberl, 2002), a wide variety of shock metamorphic effects has been identified. The best diagnostic indicators for shock metamorphism are features visible in the polarizing microscope. They include planar microdeformation features (especially the well-known planar deformation features, PDFs); optical mosaicism; changes in refractive index, birefringence, and optical axis angle; isotropization (e.g., formation of diaplectic glasses); and phase changes (high pressure phases; melting).

Another important line of evidence for impact processes comes from geochemical indications of the presence of an extraterrestrial component (cf. Koeberl, 1998). In this respect the study during the past 25 yr of rocks from the Cretaceous-Tertiary

(K-T) boundary has been very important (Alvarez et al., 1980). Only elements that have high abundances in meteorites but low abundances in terrestrial crustal rocks are useful for such studies, for example, the siderophile platinum-group elements (PGEs: Ru, Rh, Pd, Os, Ir, and Pt) and other siderophile elements (e.g., Co, Ni). Elevated abundances of siderophile elements in impact melt rocks or breccias (and impact ejecta) compared to target rock abundances can be indicative of the presence of either a chondritic or an iron meteoritic component, provided that this elevated abundance can be unambiguously shown not to be of terrestrial (e.g., mantle) origin. Contamination from achondritic projectiles (stony meteorites that underwent magmatic differentiation) is more difficult to recognize, because they have significantly lower abundances of the key siderophile elements. It is also necessary to sample all possible target rocks to determine the so-called indigenous component (i.e., the contribution to the siderophile element content of the impact melt rocks from the target). Iridium is most often determined as a proxy for all PGEs (Fig. 2), because it can be measured with the best detection limit of all PGEs by neutron activation analysis (which was, for a long time, the only more-or-less-routine method for Ir measurements at sub-ppb abundance levels in small samples). Studies of impact glasses from some small craters for which the meteorite has been partly preserved indicated that the siderophile elements can show differences in their interelement ratios (fractionation)

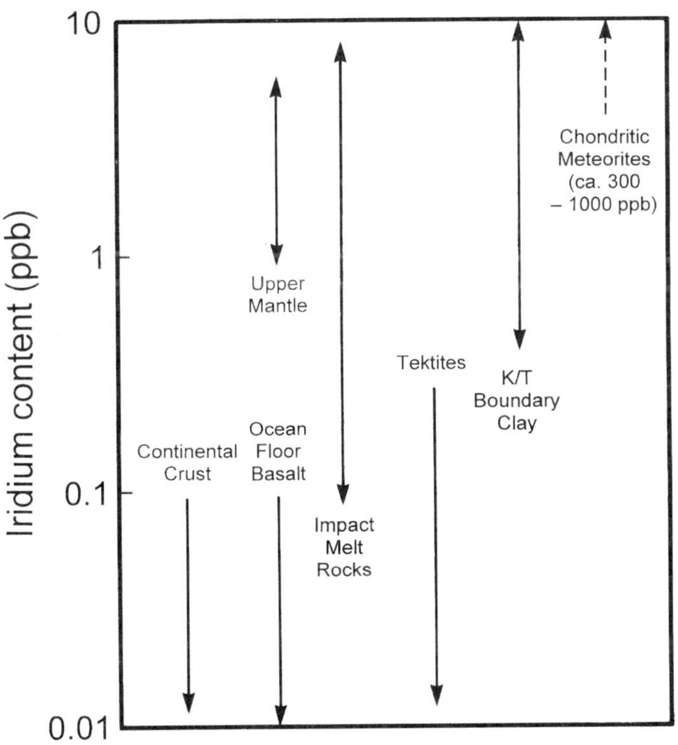

Figure 2. Siderophile elements, especially the platinum group elements, typically have high contents in extraterrestrial materials and significantly lower contents in terrestrial rocks. Shown here is the range of the contents of the platinum group element iridium in typical terrestrial and extraterrestrial rocks (diagram after Koeberl, 1998).

compared to the initial ratios in the impacting meteorite. Other types of fractionation have also been observed; for example, impact ejecta from the Acraman structure in Australia show deviations from chondritic PGE patterns due to low-temperature hydrothermal alteration (see summary in Koeberl, 1998).

These problems can, in part, be overcome by the use of isotopic tracers for extraterrestrial components. Most prominent among those are the Os and Cr isotopic methods. In the first case, ^{187}Re undergoes β^- decay to ^{187}Os. In meteorites (and mantle rocks), the Os/Re elemental ratio is ~10, whereas in crustal rocks it is ~0.1. This leads to significant differences in the ^{187}Os content of meteorites and crustal rocks relative to other Os isotopes (the nonradiogenic ^{188}Os is commonly used for normalization purposes), so that the ^{187}Os/^{188}Os ratio of meteorites is ~0.1, whereas that of crustal rocks is much higher. In addition, chondritic meteorites typically contain several hundred ppb of Os, and crustal rocks contain on the order of 0.02 ppb Os. Thus, even a small (<1 wt%) chondritic component completely dominates the Os isotopic composition of a breccia or melt that is predominantly composed of crustal rocks (see the review by Koeberl and Shirey, 1997, for more details). With this method it is possible to determine small (subpercent) contributions of extraterrestrial material to breccias and melt rocks, but it is not possible to determine the meteorite type.

In contrast, the Cr isotopic method (cf. Shukolyukov and Lugmair, 1998) relies on the fact that all terrestrial rocks have a uniform Cr isotopic composition, whereas different meteorite types have different isotopic anomalies (the difference between carbonaceous chondrites and other meteorites being particularly significant). The Cr isotopic method has, thus, the advantage over both the use of Os isotopes or PGE abundance ratios of being selective not only regarding the Cr source (terrestrial versus extraterrestrial) but also regarding the meteorite type. The complicated and time-consuming analytical procedure is, however, problematic. Also, a significant proportion of the Cr in an impactite, compared to the abundance in the target, has to be of extraterrestrial origin. As discussed in detail by Koeberl et al. (2002), the detection limit of this method is a function of the Cr content in the (terrestrial) target rocks that were involved in the formation of the impact breccias or melt rocks. For example, if the average Cr concentration in the target is ~185 ppm (i.e., the average Cr concentration in the bulk continental crust), only an extraterrestrial component of more than 1.2% can be detected. In contrast, the Os isotopic method can detect meteoritic components at about one tenth of these abundance levels. Thus, while being more selective than the Os isotopic method, the Cr isotopic method is less sensitive.

After this short discussion of the methods for the recognition of the products of impact events, I will now briefly review the impact history of the Earth from its origin to ca. 2 Ga. This interval covers more than half of the history of our planet but includes only one impact structure plus limited evidence in the form of several ejecta layers for a few other impact events between 3.4 and 2.5 Ga.

ACCRETION OF THE EARTH AND THE MOON-FORMING IMPACT

As noted above, collisions between planetary bodies have been of fundamental importance in the evolution of the solar system. Studies of the past several decades have shown fairly convincingly that the planets formed by collision and hierarchical growth starting from small objects, i.e., from dust to planetesimals to planets (e.g., Wetherill, 1994; Taylor, 1992a, 1992b). It is currently assumed that the Earth and other planets formed by accretion of smaller objects (cf. Canup and Agnor, 2000). Late during the accretion of the Earth (some time around or after 4.5 Ga), when the Earth had ~70%–90% of its final mass, it was probably impacted by a Mars-sized body, which has become the prevailing hypothesis for the origin of the Moon (e.g., Canup and Asphaug, 2001; Canup, 2004a, 2004b; and papers in Canup and Righter, 2000). The consequences of such an impact event for the proto-Earth would have been severe and would have included almost complete remelting of the Earth, loss of any primary atmosphere, and admixture of material from the impactor. The material remaining in orbit after accretion of the Moon would have continued to impact onto the Earth (and the Moon) for millions of years or more. Core formation was coeval with the accretion, and the core of the Mars-sized impactor is likely to have merged with the core of the proto-Earth almost instantaneously.

It has been known for some time that the age of the Earth is close to that of most meteorites at 4.55 Ga (see review in, e.g., Taylor, 1992b). More recent studies, using a variety of isotope systems, have yielded somewhat younger ages for the Earth and have provided information on the accretion history of the Earth (e.g., references in Canup and Righter, 2000; Zhang, 2002). Of course, it has to be defined what "age of the Earth" really means: when the meteorites formed, which are dated at ca.4.57–4.54 Ga (see Zhang, 2002), no large planetesimals or planetlike objects existed. The question that has to be asked is, how long did it take for the Earth to accrete? Can we set the "age" of the Earth to when it was accreted to 50% of its current mass, or, as more commonly done, 90%? Was the core formation completed at the time? Was it fast or slow? All these ages are difficult to impossible to determine at this time; it is likely that any "reset age" would be the age of the last giant impact (most likely the moon-forming impact event, see below).

A combination of the U-Pb, Hf-W, I-Pu-U-Xe, Sm-Nd, and Nb-Zr isotopic systems provides some information. New Sm-Nd isotopic data, using the ^{142}Nd/^{144}Nd ratio, indicate that global differentiation of the Earth already seems to have occurred at 4.53 Ga, while the accretion must still have been in progress (Boyet and Carlson, 2005). As pointed out by Zhang (2002), information from the U-Pb and Hf-W isotopic systems rules out rapid accretion (at ca. 4.55 Ga) followed by smooth and continuous core formation but allows at least two different models, as illustrated in Figure 3. I-Pu-U-Xe data indicate a relatively young age of 4.45 ± 0.02 Ga for Xe closure, and the earliest possible formation age of the crust, as constrained by U-Pb, Sm-Nd, and Nb-Zr isotope data, is ca. 4.45 ± 0.05 Ga. Zhang (2002) summarized these data and discussed that they allow for two scenarios for the accretion and differentiation of the Earth: (1) continuous Earth accretion and simultaneous core formation with a mean age of 4.53 Ga (representing a mean accretion time of 30 m.y. from the age of the meteorites) (Fig. 3A); or (2) a single age of 4.45 ± 0.02 Ga for all events related to instantaneous differentiation (e.g., core formation), probably representing the time of the last giant impact; at this time Earth had accreted ~80%–90% of its present mass (Fig. 3B). In the first scenario, there were numerous large impacts, but they were not powerful enough to rehomogenize the whole Earth. The last of such impacts (by a body the size of the Moon or greater) occurred at ca. 4.45 Ga; it stripped the atmosphere from the Earth and remelted the crust of the Earth. In contrast to the second option, in this version some history of the proto-Earth before 4.45 Ga is preserved in the isotope data. However, other data (see below) indicate that the impacting body was larger

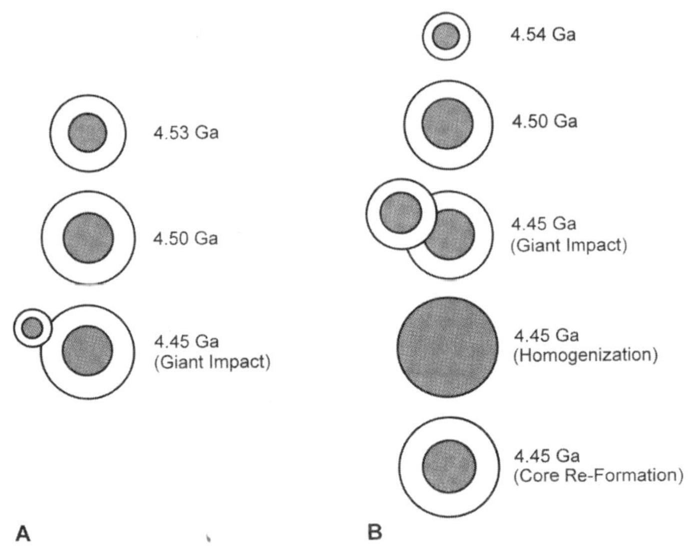

Figure 3. Two possible scenarios for the accretion of the Earth, core formation, late impact, and Xe retention. Both versions agree with the available noble gas evidence. Diagrams after Zhang (2002). (A) During continuous accretion that leads to the formation of the Earth, the core also continually forms and increases in size, until a giant impact of a lunar-sized body stripped the Earth of its early atmosphere and removed any Xe that had formed previously. After the giant impact Xe is retained, and thus the Xe retention age indicates the time of the impact. (B) In a more likely scenario (see text), the Earth grew, from 4.54–4.45 Ga, to ~70% of its present mass (the accretion was probably faster early on and then slowed down exponentially). The core grew as well. At 4.45 Ga, a Mars-sized impactor hit the Earth, causing stripping off of the atmosphere (and resetting the Xe isotope system) and total rehomogenization of the core and mantle, with some contribution to them from the impactor. The addition of mass from the impactor led to a sudden increase in total Earth mass. The core reformed fairly quickly, probably within just a few million years, and this was followed, after 4.45 Ga, by accretion of the remaining Earth mass in the form of late veneers.

than Moon sized (e.g., Canup, 2004a, 2004b). The essential accretion and core formation of the Earth appears to have been completed ca. 100–50 Ma after the initial collapse of the solar nebula (e.g., Lee and Halliday, 1995, 1996; Halliday et al., 1996), in a time scale apparently more stretched out than that for smaller planetesimals and Mars (e.g., Wetherill, 1994).

Much of the material impacting on the Earth during its growth (pre–4.45 and post–4.45 Ga) is thought to have been in the form of differentiated planetesimals, although undifferentiated volatile-rich objects probably also played a part. As pointed out by Halliday (2004), this complex accretion history is not well simulated by existing isotopic models. Customarily, two-stage U-Pb and Hf-W model ages have been used to define an artificial time, at which the Earth's core instantaneously formed or last equilibrated with the silicate Earth. Such considerations might have been useful for understanding the broad picture, but calculations based especially on short-lived nuclides are of limited importance for open systems like the Earth, which grew over a significant time interval. Even the initial accretion of the embryo of the Earth took (possibly considerably) longer than 5 m.y., depending on the amount of gas present in the solar nebula (cf. Taylor, 1992b), rendering two-stage model ages meaningless in terms of precise rates or absolute time. In particular, this is the case when considering a short-lived isotope like ^{182}Hf, which has a half-life of 9 m.y., because late core formation would have no effect on the W isotopic composition of the Earth. The two-stage Hf-W model age for the Earth's core would be 30 Ma (e.g., Lee and Halliday, 1995, 1996; Halliday, 2004), which defines the last time the Earth could have globally re-equilibrated. However, Halliday (2004) notes that there is no basis for believing that such an event occurred, and as far as W isotopes are concerned, roughly half the Earth's core could have formed much later.

As mentioned above, it is currently believed that the Moon most likely formed in the collision of a Mars-sized planetary embryo with the proto-Earth. The exact mass of the Earth at the time of this impact is not certain but was probably on the order of 80%–90% of its current mass (e.g., Canup, 2004a, 2004b; Kleine et al., 2004). Hydrodynamic simulations (see Canup, 2004a, 2004b) of potential Moon-forming impacts revealed that the most favorable conditions for producing a sufficiently massive and iron-depleted (Taylor, 1982) protolunar disk involve collisions with impact angles of ~45° and impact velocities of less than 4 km/s. Such impacts place about one lunar mass of material into Earth orbit outside the Roche limit, with ~10%–30% of the material consisting of vapor. Most of the orbiting material, which eventually became the Moon, was derived from the impactor, agreeing with geochemical considerations (cf. Taylor, 1982, 1992b, 1993; papers in Canup and Righter, 2000).

Accretion of the Moon from the protolunar disk followed very rapidly. This impact event is thought to have been the last of the major collisions that were involved in the accretion of the Earth, and the enormous energy deposited onto the Earth during the impact event was thought to have been responsible for leaving the Earth covered with a global magma ocean of debated duration (e.g., papers in Canup and Righter, 2000), leading to the establishment of the equilibrium distribution of the moderately siderophile elements between core and mantle toward the end of the Earth's accretion. Tungsten isotopic considerations are important for understanding the early history of the Earth. The small excess of ^{182}W in the bulk silicate Earth (BSE) compared to the isotopic composition of chondrites indicates that core formation occurred during the lifetime of the now-extinct (9 m.y. half-life) parent isotope ^{182}Hf. The potential of tungsten isotope geochemistry is illustrated in Figure 4. Figure 4A shows a case in which the Moon-forming impact occurred 55 m.y. after T = 0 (= origin of the solar system), with a bulk silicate Hf/W ratio of the impactor mantle (BSI; bulk silicate impactor) of 15, the same as that of the BSE before impact, and 26% of the impactor core equilibrated with the BSE; the giant impact would drastically increase the ε^{182}W value of the BSE. Only with these assumptions can a present-day ε^{182}W value of zero be accomplished. However, the Hf/W ratios in planetary mantles were unlikely constant (Halliday, 2004). A second possibility, shown in Figure 4B, involves a Hf/W ratio of 5 in both BSI and BSE (similar to that of Mars) and a giant impact at 50 Ma; in this case a present-day value of ε^{182}W of zero is achieved if only 4% of the tungsten in the mantle of the impactor equilibrates with the BSE during the giant impact. The exact timing of these events is still debated, however (e.g., Kleine et al., 2004). The results of these considerations (Halliday, 2004) show that the initial Hf/W ratios of the BSI (and possibly the BSE) were less than those in the Moon or silicate Earth today, maybe similar to values inferred for the Martian mantle.

The Moon-forming impactor planet, called Theia (e.g., Halliday, 2004), contributed ~10% of the (then) mass of the Earth, thus contributing a further 9% of the current Earth mass. After this event, there was probably no significant amount of accretion (Halliday, 2004), because significant further accretion would have disrupted the identical oxygen isotopic compositions of the Earth and the Moon. Calculations related to the formation of the Moon (Canup, 2004a, 2004b) indicate that after the Moon-forming impact the Earth was more than 95% accreted. This is not inconsistent with a so-called "late veneer" of extraterrestrial material that is required to explain, for example, the relatively high PGE contents of the Earth's mantle (see, e.g., papers in Canup and Righter, 2000), but in terms of total mass of the Earth it was not important, having probably contributed not more than 1% of additional Earth mass. Independent evidence for a late veneer comes from Os isotopic studies (Morgan, 1985) and a comparison of mantle rocks with the composition of the bodies that bombarded the ancient lunar crust (Morgan, 1986).

The age of the Moon itself is not well defined, depending on which isotope system is used. For example, U-Pb and Sm-Nd ages of ca. 50–120 Ma (relative to the age of the solar system) have been quoted (see review by Halliday, 2004, for references). Initial Hf-W isotope data indicated a model age of ca. 55 Ma

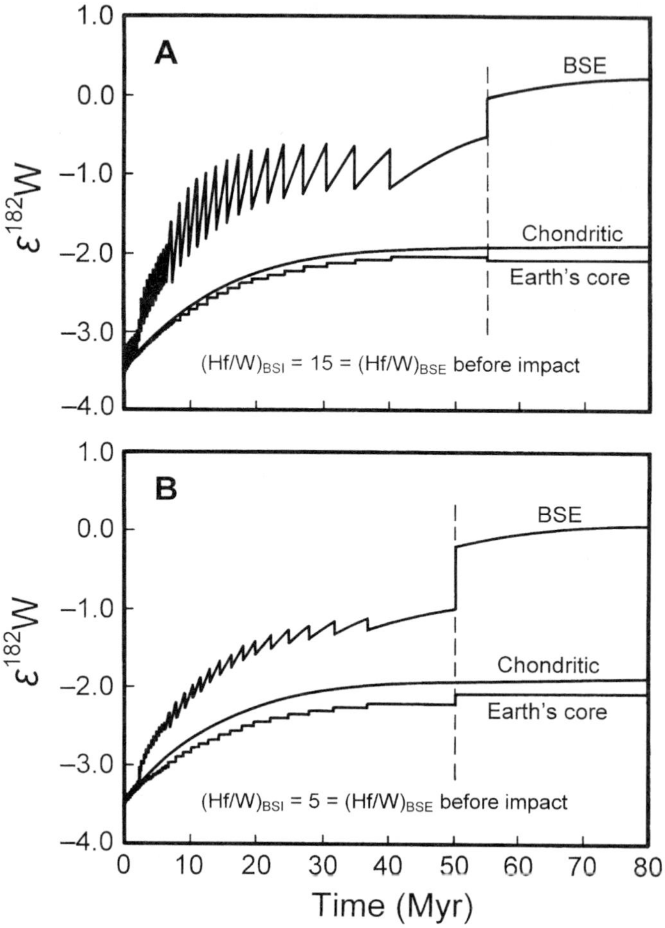

Figure 4. Two possible scenarios for the evolution of the tungsten isotopic composition of the Earth, after Halliday (2004). Both are tuned to reproduce the present-day $\varepsilon^{182}W$ value (near zero) for the bulk silicate Earth (BSE). There is a small, but resolvable excess of $\varepsilon^{182}W$ of the BSE compared to chondrites, which is explained by core formation during the lifetime of ^{182}Hf, the precursor of ^{182}W. The step-pattern in the values for the Earth's core and the BSE indicate accretion of large planetesimals followed by equilibration between mantle and core. The curve showing the evolution of the total (bulk) Earth is identical to the one for chondritic meteorites, which are believed to be compositionally similar to the source of the planetesimals that formed the Earth. In case (A), a (Moon-forming) giant impact happens at 55 Ma after the beginning of the formation of the solar system. The ratio of Hf/W of 15 in the BSE is assumed to be identical to that in the mantle of the impactor (BSI; bulk silicate impactor). This model requires that 26% of the core of the impactor, called Theia, would equilibrate with the BSE. In case (B), the giant impact occurs already at t = 50 Ma, and a ratio of Hf/W of only 5 in the BSI and BSE is assumed. In this model only 4% of the core of Theia would need to equilibrate with the BSE, but even such a number is difficult to reconcile with the W isotopic composition of the Moon. These diagrams demonstrate how model-dependent the implications regarding timing and amount of core formation and giant impact are.

(Lee et al., 1997), but a major portion of this was produced from cosmogenic ^{182}Ta, leading to a revised Hf-W model age of ca. 30 Ma (Halliday, 2004) to 40 Ma (Kleine et al., 2004). However, such a number assumes that the Moon formed from undifferentiated (with respect to Hf-W) chondritic material, which is unlikely. A variety of considerations, as described by Halliday (2004), lead to the current best estimate of the Hf-W age of the Moon between 44 and 54 Ma, which agrees with an independent estimate, based on a different W isotope data set by Kleine et al. (2004). After the giant impact with the proto-Earth, the Moon seems to have accreted rapidly (e.g., Canup, 2004b).

The consequences of such an impact event for the proto-Earth would have been severe, ranging from almost complete melting and formation of a magma ocean, thermal loss of pre-existing atmosphere, changes in spin rate and spin-axis orientation, to accretion of material from the impactor directly or through rapid fallout from orbital debris below the Roche limit. Much of the material blasted off in the impact eventually reimpacted the Earth; some of the ensuing Earth-orbiting debris would have rapidly coalesced to form most of the Moon and probably some smaller moonlets. Some of this geocentrically orbiting material would have continued to impact both bodies for perhaps tens of millions of years (e.g., Canup, 2004b) after the Moon-forming impact.

In terms of the evolution of a magma ocean, Kleine et al. (2004) found from their data that formation of a magma ocean as a result of the (late) Moon-forming impact event alone is not sufficient to remove the radiogenic ^{182}W from the Earth's mantle that would have accumulated during early core formation. They suggest that more than half of the metal-silicate equilibration was already achieved before the formation of the Moon, indicating core formation by metal-silicate separation had already occurred in earlier magma oceans. Kleine et al. (2004), thus, concluded that the Moon-forming impact was only the latest in a series of large-scale impact events that led to several episodes of magma ocean formation or that a magma ocean existed for a prolonged time period, probably during most of the early period of Earth formation (however, see below for considerations on the early crust of the Earth). New trace element fractionation data and ^{146}Sm-^{142}Nd geochronology indicated that an episode of mantle fractionation took place at around 4.46 ± 0.11 Ga, very soon after the end of the main accretion period and the formation of the Moon (Caro et al., 2005).

The period on Earth between the end of accretion and the production of the oldest known crustal rocks is commonly referred to as the Hadean Eon (e.g., Harland et al., 1989), which is a chronostratic division. Its terminal boundary is actually not defined on the Earth; Harland et al. (1989) equated it with the Orientale impact on the Moon. As explained by Ryder et al. (2000), Orientale is the last of the largest impact basins on the Moon, but it is difficult to date absolutely, as Orientale is far removed from any sites sampled so far. Based on crater counts, the abundance of secondary craters, and stratigraphic considerations, it is older than the oldest affected mare plains and, thus, construed as older than ~3.80 Ga. In this review the term Hadean is used to represent the time period between the formation of the Moon at ca. 4.5 Ga and the beginning of the continuous terrestrial rock record at 3.8 Ga.

Walter and Trønnes (2004) described the physical and chemical processes of the differentiation of the early Earth as a consequence of the impact-induced magma oceans. The depth and duration of a magma ocean depends on a variety of parameters that include the mass and energy deposited onto the proto-Earth and the initial temperature and the presence or absence of an atmosphere. Without an atmosphere, heat radiation into space is very efficient, and the deeper parts of the magma ocean could cool and crystallize within thousands of years (with the upper part of a shallow upper mantle staying molten longer). With a rock vapor or steam (H_2O and/or CO_2) atmosphere present, the cooling times of even the lower parts of a magma ocean could be prolonged considerably, up to several tens of millions of years (e.g., Abe, 1997; Matsui and Abe, 1986; Walter and Trønnes, 2004). Following the differentiation and solidification of the magma ocean, the first crust formed.

WHAT HAPPENED TO THE EARTH'S OLDEST CRUST?

The terrestrial rock record, in form of crustal rocks, extends back to only ~89% of the history of the planet to ca. 3.90 Ga (with the oldest rocks showing very limited exposures at Isua and Akilia, Greenland, and Acasta, Canada; see, e.g., papers in Fowler et al., 2003; Hartlaub et al., this volume). The record is improved somewhat if we include the information that can be obtained from rare detrital zircon grains found in rocks at Mount Narryer and Jack Hills, Western Australia (e.g., Compston and Pidgeon, 1986), the oldest one of which was dated at 4.40 Ga (Wilde et al., 2001). So the question is, what happened to the crust that must have existed before 3.9 Ga?

It is likely that the Moon-forming impact led to large-scale melting of the Earth and the existence of an early magma ocean. Mantle temperatures in the Hadean were much higher (probably by a factor of up to five) than today as a result of a higher heat flow (about half of all heat produced by ^{235}U decay to ^{207}Pb was released during the Hadean, adding several hundred degrees to the internal temperature of the Earth) and thermal energy released during the impact of late accretionary bodies (e.g., Davies, 1985). The terrestrial heat flow at the surface at 4.45 Ga was probably on the order of 1.3×10^{14} W, compared to a present-day value of 4×10^{13} W (Smith, 1981). Due to the hotter mantle and the temperature dependency of the dehydration reactions that control how much water is recycled to the mantle, the Hadean mantle was drier than its modern counterpart. The ocean, which was already in existence probably as early as 4.4 Ga (Valley et al., 2002), contained most of the Earth's water and its volume may have been up to twice that of today's oceans.

The nature and amount of the earliest crust on Earth has been debated, but comparison with other planets suggests that the earliest crust on Earth was basaltic (e.g., Taylor, 1989; Arndt and Chauvel, 1991). As noted by Russell and Arndt (2005), high-degree melting of hot dry Hadean mantle at ocean ridges and plumes resulted in a crust ~30 km thick, overlain in places by extensive and thick mafic volcanic plateaus, but continental crust, by contrast, was relatively thin and mostly submarine. Morphological, mineralogical, and geochemical characteristics of the 4.4–4.1 Ga zircons (see below) indicate a composite granitoid source of continental provenance for these zircons. Thus, there is evidence for at least minor amounts of felsic igneous rocks in the Hadean, which may have formed in small amounts from remelting of basaltic crust that sank back into the mantle (e.g., Taylor, 1989). Granitic crust requires multistep derivation from the primitive mantle by recycling of subducted basaltic crust through a "wet" mantle, which will slowly lead to an increasing amount of granitic crust through time (Taylor and McLennan, 1995). Thus, the Hadean Earth was most probably characterized by a thick basaltic crust covered by an ocean, with little dry land and minor amounts of felsic rocks (granitoids).

Russell and Arndt (2005) give an excellent description of the nature of the early crust, its probable composition, and plate tectonic processes operating in the Archean, and also give a vivid account of the topography of the Hadean Earth, which differed significantly from that of the modern Earth. Because of the higher concentrations of radioactive elements in the Hadean granitic crust, it was hotter, less viscous, and thinner than modern continental crust, and, due to the hotter mantle, the continental lithosphere was thinner. The Hadean oceanic crust was much thicker than modern oceanic crust, as thick or thicker than the early Archean continental crust. Due to the larger ocean mass, a greater area of the Earth's surface was flooded. In contrast to today, most of the continents were submerged, and only mountain ranges at convergent margins and vast volcanic plateaus occasionally breached the ocean surface (Russell and Arndt, 2005), resulting in little subaerial weathering and erosion of the mainly submerged continents and comparatively little clastic sedimentation.

Grieve et al. (this volume) considered the possibility that early large impact events led to the formation of felsic crust on the Hadean Earth. From computer modeling of such large-scale impacts, they found that impact melt volumes exceed transient cavity volumes at transient cavity diameters greater than ~500 km on the Earth. Impact basins on the Moon would have to be much larger—on the order of 300 km in diameter—before such effects would occur, and the form of such large basins on the Earth would be unlike Mare Orientale (a common analog) but more like the Valhalla multi-ring basin on the Jovian moon Callisto. These structures would, according to Grieve et al. (this volume), be characterized by extensive mafic impact melt pools that would differentiate, leading to the crystallization of more felsic rocks. Thus Hadean impacts (between ca. 4.45 and 3.9 Ga) provide a mechanism to reprocess the initial basaltic Hadean crust to produce pockets of felsic crust, as evidenced by pre–3.9 Ga relict zircon crystals (see below), without necessitating crustal recycling mechanisms. And scaling of lunar impact basins yields a cumulative melt production on the Hadean Earth by such basins alone to the order of 10^{11}–10^{12} km^3. Grieve et al. (this volume) suggest that the cumulative felsic rocks produced through differentiation of such impact melt pools could approach volumes comparable to 50% of the current continental crust.

Very few rocks on Earth with ages approaching 3.9 Ga have been found, but some rare older detrital zircon grains up to ca. 4.4 Ga are known (e.g., Compston and Pidgeon, 1986; Maas et al., 1992; Amelin et al., 1999). In contrast to the suggestion by Grieve et al. (this volume) that crustal recycling is not necessary in the Hadean, the data obtained by Amelin et al. (1999) do indicate that such recycling did occur, however. These authors studied the Lu-Hf isotopic signatures of single zircons from the Narryer Gneiss Complex (Australia) with ages of up to 4.14 Ga. They found that about half the grains they analyzed seemed to have formed by remelting of significantly older crust, indicating that crust production and reworking seem to have been important processes during the Hadean.

In addition, there is some evidence that the Earth's upper mantle had already undergone some differentiation at the time of formation of the oldest rocks preserved on the Earth's surface (e.g., Harper and Jacobsen, 1992; McCulloch and Bennett, 1993; Bowring and Housh, 1995; Kamber and Moorbath, 1998; Kamber et al., 1998; Boyet et al., 2003). In particular, there is evidence from ^{142}Nd isotopic studies (involving the short-lived ^{146}Sm-^{142}Nd decay, half-life 103 m.y.) that the Earth's upper mantle had already undergone some differentiation at the time of formation of the oldest rocks preserved on the Earth's surface, but until recently there was disagreement about the reality of the observed isotopic anomaly. However, the existence of ^{142}Nd anomalies in Isua rocks has been confirmed by Caro et al. (2003). These authors found a statistically significant ^{142}Nd anomaly of 15 ± 4 ppm, which, together with initial ε^{143}Nd values, indicates an age of mantle differentiation of 4.46 ± 0.12 Ga. This age could well be related to the Moon-forming impact discussed above, which led to large-scale melting of the Earth and the existence of an early magma ocean, in which the mantle differentiation occurred. Early crust formation, which is indicated by the isotopic data, may have been aided by the higher than present heat flow. There has been a lot of debate regarding the existence of plate tectonics in the Hadean (or even in the Archean; e.g., Kröner and Layer, 1992), but the evidence for recycling of early crust (e.g., Kamber et al., 2003; Harper and Jacobsen, 1992; McCulloch and Bennett, 1993; Bowring and Housh, 1995; Kamber and Moorbath, 1998; Kamber et al., 1998; Boyet et al., 2003) coupled with the higher heat flow (see discussion above on mantle temperatures) seems to demand that plate tectonics, even in limited form, as well as accretion of felsic crust, were operating shortly after the solidification of the magma ocean. Any late accretionary bodies would have added further thermal energy to the already elevated budget of heat flow in the early Earth (e.g., Smith, 1981; Davies, 1985; Taylor, 1993). Alternatively, the mechanism suggested by Grieve et al. (this volume), through production of large amounts of felsic crust in differentiated impact melt sheets, may also account for the production and recycling of differentiated early crust.

Grieve (1980), Frey (1980), and Grieve and Parmentier (1984) discussed the effect of large impact events resulting in impact structures with diameters exceeding 100 km on the ancient Earth, with ages larger than 3.8 Ga. By scaling the lunar impact record to the Earth these authors concluded that ~2500–3000 impact structures with diameters larger than ~100 km could have formed. Their simulations yielded close to 1000 craters with diameters exceeding 200 km, and possibly some ten structures with diameters larger than that of the Imbrium basin on the Moon (~1300 km diameter). This crater population would have covered ~40% of the surface of the Earth. Using the minimum estimate for the cratering frequency, Grieve (1980) derived a cumulative energy of ~10^{29} J added to the Hadean Earth due to impact events and concluded that the net effect of large impact events was to localize and accelerate a variety of endogenic geological activities.

Recent data indicate that the magma ocean may have been shorter lived than previously thought and that differentiation and recycling could have started shortly after the Moon-forming impact. Work by Peck et al. (2001) and Wilde et al. (2001) showed the existence of detrital zircon cores from the Narryer Gneiss Terrane in Western Australia with ages of up to 4404 ± 8 Ma, very close in age to the inferred age of the Moon-forming impact event. These authors found variations in the oxygen isotopic composition of these zircons that indicate that these zircons have grown in evolved granitic magmas and show evidence for low-temperature surficial processes, including diagenesis, weathering, and low-temperature alteration (see also Sleep et al., 2001). In other words, the data seem to indicate the presence of liquid water on the surface of the Earth ~50 m.y. after the Moon-forming giant impact and the presence of granitic (not just basaltic) pockets of continental crust. Granitic crust requires multistep derivation from the primitive mantle by recycling of subducted basaltic crust through a "wet" mantle, which will slowly lead to an increasing amount of granitic crust through time (Taylor and McLennan, 1995). Valley et al. (2002) called, thus, for a cool early Earth and took these data to indicate that the frequency of meteorite impacts during the time span between 4.4 and 4.0 Ga was less than previously thought.

Recently, a study by Watson and Harrison (2005) provided independent confirmation of the cool early Earth with water concept by determining the crystallization temperatures of the Western Australian Hadean zircons, using a geothermometer based on the titanium content of these zircons. Their data showed that the crystallization temperatures of these zircons cluster at ~700 °C, which is indistinguishable from temperatures of granitoid zircon growth today and strongly suggests a regulated mechanism producing zircon-bearing rocks during the Hadean. The low temperatures found by Watson and Harrison (2005) substantiate the existence of wet, minimum-melting conditions within 200 m.y. of solar system formation and further confirm the earlier suggestion (see above) that Earth had settled into a pattern of crust formation, erosion, and sediment recycling as early as 4.35 Ga.

These zircons are the only record that persisted from Hadean times. It would be interesting to study these Hadean zircons (or at least their cores, as most of them have been overgrown later) for any shock effects, although this could be difficult because of later annealing. The question has been raised why there is no preserved crust of >4 Ga (e.g., Bowring and Housh, 1995), and it has been suggested (e.g., Koeberl and Sharpton, 1988; Koeberl et al., 2000) that any early Hadean crust was destroyed at around the same time that massive impacts reshaped the surface of the Moon (see next section). Any existing crust would have been mixed back into the upper mantle. Any sedimentological record, which would host information specific to surface environments such as the rate and violence of meteorite impact and the presence of life, has been lost from Hadean times.

A LATE HEAVY BOMBARDMENT AT 3.9 GA?

In contrast to the youthful age for most of the crust of the Earth, the surface of the Moon displays abundant evidence of an intense bombardment at some time between its original crustal formation and the outpourings of lava that form the dark mare plains. Even prior to the Apollo missions, these plains were calculated to be ca. 3.6 Ga based on crater counting statistics and realistic flux estimates, and, thus, the heavy bombardment was inferred to be ancient (Hartmann, 1966). Isotope analysis of lunar highland samples showed isotopic resetting from thermal heating, for which there is abundant evidence in terms of heating and melting by impact events with ages of around 3.9–3.8 Ga. The most ancient volcanic rocks from mare plains have ages of ca. 3.8 Ga (see, e.g., Taylor, 1982; Wilhelms, 1987). The highland ages have been interpreted to either represent a short and intense "late" heavy bombardment period at 3.85 ± 0.05 Ga (e.g., Tera et al., 1974; Wetherill, 1975; Ryder, 1990) or the tail end of an extended post-accretionary bombardment (e.g., Baldwin, 1974; Hartmann, 1975), as discussed in detail by Hartmann et al. (2000). It should be noted that the term "late heavy bombardment" (LHB) has been used by different authors to denote different things. On the one hand, it is used to describe the long-term (but decreasing) heavy bombardment between the formation of the Earth (and Moon) at ca. 3.8 Ga, and on the other hand it has been used to describe an intense spike in bombardment around 3.85 Ga. Herein, I use the term LHB in the latter sense.

As mentioned above, the accretion of the Earth appears to have been completed ~50–100 m.y. after the initial formation of the solar nebula, defining the initiation of the so-called Hadean Eon. Whereas there is almost no evidence of terrestrial witnesses to the Hadean Eon, the pre-Nectarian and Nectarian periods cover this time interval on the Moon. Soon after the formation of the Moon, the feldspathic highland crust formed (e.g., Taylor, 1982). On Earth such a crust did not form, probably because feldspar-rich anorthosite melts are buoyant only in the absence of water, and the Moon, having originated in the volatile-depleting giant impact event, had not even traces of water left. The morphology of the highlands of the Moon records numerous impacts that occurred prior to the extrusion of the volcanic flows that form the mare that so clearly mark the near side of the Moon today (e.g., Wilhelms, 1987). Geochronological studies of brecciated highland samples showed that impact-related thermal events are concentrated at ca. 3.9–3.8 Ga.

The best way to date an impact event is from a clast-free or clast-poor impact melt (cf. Deutsch and Schärer, 1994), and Ryder (1990) discussed the lack of impact melts in the sample collections that are older than ca. 3.95–3.92 Ga (Fig. 5); this is unlikely to be the result of resetting of all older ages, given the difficulties of such resetting without complete melting in an impact event (Deutsch and Schärer, 1994). Thus, the lack of such ages in lunar impact melts can be taken as evidence that there were no significant impact events prior to that time, in agreement with the interpretations of Valley et al. (2002) and Watson and Harrison (2005) regarding the "cool early Earth" between ca. 4.4 and 4 Ga. There is not enough space here to review the arguments for the ages of the large impact basins (now Maria) on the Moon, and the reader is referred to Ryder et al. (2000) and Stöffler and Ryder (2001) for details. It should

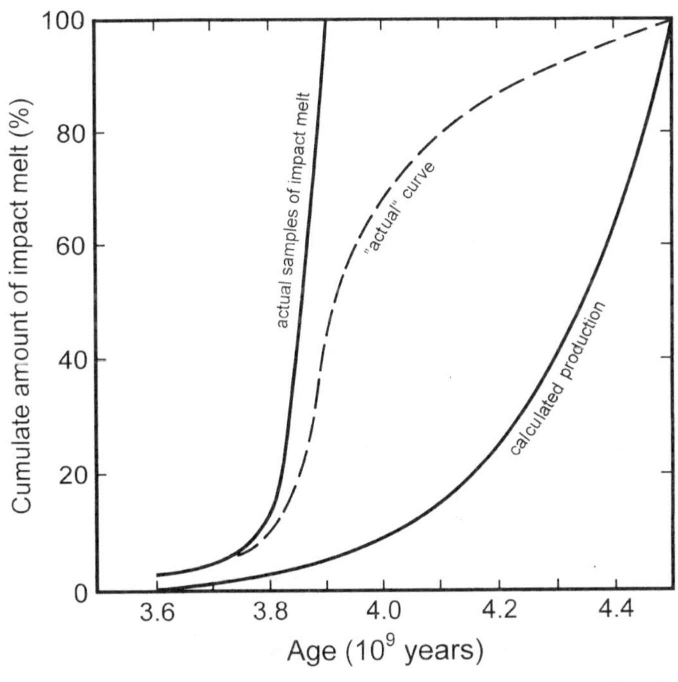

Figure 5. Cumulative amount of impact melt in the lunar highlands, after Ryder (1990). The curve labeled "actual" refers to earlier data that did not take into account the lack of impact melt rock samples with ages greater than 3.9 Ga; if corrected for this effect, as explained by Ryder (1990), the curve is shifted toward the left and results in the curve labeled "actual samples of impact melt." Both curves deviate significantly from the one that was calculated for a continuous heavy bombardment as proposed by Hartmann (1966, 1975) or Baldwin (1974).

be noted, though, that all large impact basins have formation ages of ca. 4–3.8 Ga (the magma that filled these impact basins to form the Mare is, of course, younger, at ca. 3 Ga).

The most likely ages are for the Imbrium Basin, 3.85 ± 0.02 Ga; Nectaris, ca. 3.90 Ga; Serenitatis, 3.893 ± 0.009 Ga; and Crisium, ca. 3.89 Ga, with several other basins, e.g., Herzsprung and Humorum, having formed after Nectaris (Ryder et al., 2000). Thus, there was considerable bombardment of the Moon in the 60 m.y. between 3.90 Ga and ca. 3.84 Ga. Recent work by Cohen et al. (2000) and Kring and Cohen (2002) confirmed that the ages of impact melts from the lunar meteorites, which constitute random collections of lunar surface material, also cluster in the same age bracket (Fig. 6); these observations support the interpretations of Ryder (1990, 2002) and Ryder et al. (2000). This period of bombardment ended at 3.85 Ga with the near simultaneous creation of the Imbrium, Schrödinger, and Orientale basins.

The data summarized by Ryder et al. (2000) indicate a massive decline in the flux of bombardment on the Moon over a short period of time. The cratering rate derived from the surface of the Nectaris ejecta, which have an age of 3.90 Ga, is a factor of four higher than the rate calculated from Imbrium ejecta, which, in turn, is a factor of two to four higher than that for the oldest mare plains (=3.80 Ga). Thus, during the period 3.90 Ga to ca. 3.85 Ga the flux was ~1000 times greater than it is now and still a few hundred times greater in the subsequent 50 m.y. It is possible that the decline took place over only the first 10 of these 50 m.y., so that by 3.84 Ga or 3.83 Ga the flux was decreasing to levels similar to present-day levels.

An independent argument in favor of an LHB can also be made using the masses of basin-forming projectiles on the Moon. Ryder (2001, 2002) provided a detailed discussion of this topic, so a short summary will suffice. The masses of the Imbrium and Orientale projectiles, for instance, have been estimated at between 8×10^{20} to 2×10^{21} g and 4×10^{20} to 1.5×10^{21} g, respectively. Thus, the aggregate mass of the ~15 Nectarian and early Imbrian basin-forming projectiles would be on the order of 10^{21} to 10^{22} g. Given that the formation ages of these basins are within ~80 m.y. (see Fig. 7), this leads to a lower limit of the accretion rate on the Moon of $\sim 1.2 \times 10^{13}$ g/year. This is about one to two orders of magnitude above the smoothly declining curve of lunar accretion and three orders of magnitude above a back-extrapolated current accretion rate. If we assume a much longer age spread for these basins (Nectaris age at 4.12 Ga), the average mass accretion rate of at least 3×10^{12} g/year for 300 m.y. is still about an order of magnitude larger than the back-extrapolated curve (solid line in Fig. 7). If one assumes that the mass accretion in the first case is the tail end of the early lunar accretion and not a spike, this means that back-extrapolating the masses

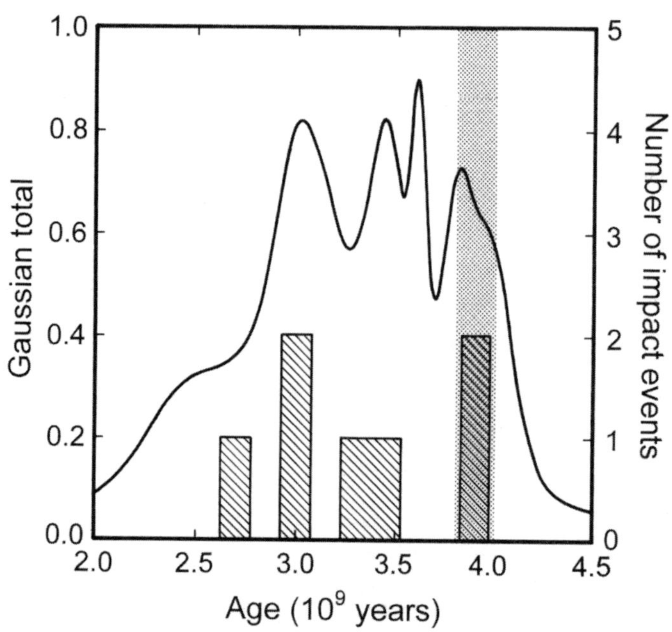

Figure 6. Idealized histogram of ages of 31 impact melt clasts from lunar meteorites, after Cohen et al. (2000). The shaded bar marks the age constraints of the LHB (late heavy bombardment) as derived by Ryder (1990) and Ryder et al. (2000). These measurements provide an independent confirmation of the lack of impact melt ages of >3.9 Ga, as noted by Ryder (1990). In this case there cannot be any bias from one or two large impacts that might have dominated the Apollo rock collections.

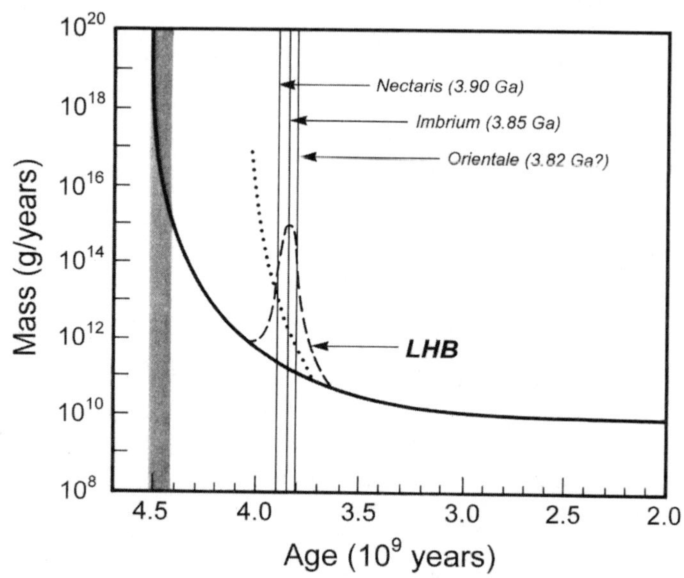

Figure 7. Various interpretations of the mass flux (accretion rate) on the Moon (simplified after Koeberl, 2004; modified after Ryder, 2001, 2002). Ages of the most important major impact basins are indicated. The solid line is the present-day background flux (constrained by present-day accretion rates and lunar impact data; e.g., Culler et al., 2000) extrapolated back in time toward the origin of the solar system. The curve marked LHB indicates the spike in the accretion rate that is suggested to be the LHB. The dotted line indicates an accretion curve that includes the masses of the basin-forming projectiles; this curve is unlikely, because it leads to the accretion of the Moon at 4.1 Ga instead of 4.5–4.4 Ga (indicated by the gray band, which marks the age of the Moon obtained from isotopic constraints).

of basin-forming projectiles accreting onto the Moon between 3.8 and 4 Ga indicates that the current mass of the Moon was exceeded already at ca. 4.1 Ga instead of 4.45 Ga. An alternative (but similar) interpretation of the early impact record by Arrhenius and Lepland (2000) is shown in Figure 8. Thus, the evidence marshaled by Ryder (1990, 2001, 2002), Ryder et al. (2000), and Kring and Cohen (2002) makes it difficult not to accept that a cataclysmic LHB occurred in the inner solar system at ca. 3.85 Ga. Such a peak in impact flux in the inner solar system was also noted by Bogard (1995) on the basis of meteorite ages.

Norman et al. (2002) recognized that chemical evidence for multiple impact events and the clear signatures of specific types of meteoritic impactors in the Apollo 17 melt breccias show that the lunar crust was not comprehensively reworked by prior impacts from 4.5–3.9 Ga, an observation more consistent with a late cataclysm than a smoothly declining accretionary flux. Late accretion of chondritic material during a 4.0–3.8 Ga cataclysm may have contributed to siderophile element heterogeneity on the Earth but would not have made a significant contribution to the volatile budget of the Earth or oxidation of the terrestrial mantle.

Thus, while there is only limited evidence of the Hadean impact history on the Earth, the evidence from the Moon, which preserves the early impact history of the solar system so much better than the Earth, is convincing. Therefore, it seems quite likely that an LHB really did happen in the inner solar system (cf. also Bogard, 1995; Kring and Cohen, 2002). This raises the question regarding the source of the impactors. It needs to be emphasized that the spike in the impact flux in the inner solar system could have been relatively steep, i.e., it could be that the time window was less than 100 m.y., because

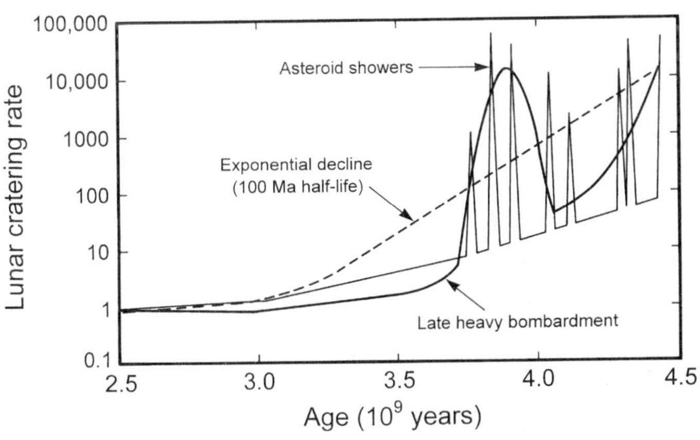

Figure 8. An alternative depiction of the possibilities of the early impact record, after Arrhenius and Lepland (2000). The dashed line shows the back-extrapolated exponential decline curve, assuming a continuously declining heavy bombardment similar to that envisioned by Hartmann et al. (2000), whereas repeated asteroid impact showers could also "wipe the slate clean" in terms of an earlier impact record. The solid line shows the preferred interpretation of a late heavy bombardment spike centered around 3.85 Ga.

the current limits of our knowledge of the lunar impact ages are given by the precision of the geochronologic methods. The spike could have had a half-life of only 20 or 30 m.y. In addition, it is necessary to have a supply of large bodies, as the impactors that formed the large lunar basins must have been several tens of kilometers (maybe up to 50 km) in diameter, and there must have been many of them. This makes a short-term disturbance of the asteroid belt a rather unlikely source. Recent studies in celestial mechanics indicate a possible mechanism that could supply a short-term spike in an otherwise steady or decreasing background flux of impactors (Zappalà et al., 1998) from asteroidal sources, but, as discussed by Morbidelli et al. (2001), asteroids from the main belt are a rather unlikely source of impactors for the LHB, because it is difficult to envision a mechanism that would deliver a short-term flux increase into the inner solar system.

A new idea was suggested by Levison et al. (2001), who noted from calculations of the accretion of planets that Uranus and Neptune did not form at the same time as the terrestrial planets and Jupiter and Saturn, but that their formation was delayed until ca. 3.9 Ga. Once these planets started accreting, which would have happened very rapidly (within 10–20 m.y.), abundant planetesimals would be scattered throughout the solar system and could have been responsible for the LHB. The possibility that (proto)–Kuiper-Belt objects are somehow relevant for the LHB is an intriguing one. The mass flux into the inner solar system at the time of the LHB must have been significant. Values ranging from ~6×10^{21} to 10^{23} g of material accreted each by the Moon, Mars, and Earth have been calculated (e.g., Ryder, 2001, 2002; Levison et al., 2001). It can be assumed that only a fraction of the objects entering the inner solar system would be accreted by the planets and the Moon. Thus, the total amount of matter injected into the inner solar system could amount to a sizeable fraction of the total mass of the asteroid belt. Mechanisms to destabilize several percent of the mass of the asteroid belt are difficult to quantify, although the model of outer planet migration of Levison et al. (2001) also causes instability in the asteroid belt. In contrast, the mass of objects in the paleo–Kuiper Belt was several orders of magnitude larger than the mass of the asteroid belt. Destabilization of these objects would provide a much more plausible source of projectiles for the LHB. As summarized by Koeberl (2004), it is conceivable that within a few 100 m.y. after formation of Neptune this body had slowly migrated into a planetesimal-rich zone, whereupon it started to accrete more mass and rapidly migrated outward toward 40 AU, in the process scattering the proto–Kuiper Belt objects all over the solar system, with the LHB in the inner solar system being a consequence of this process. Very recently, Gomes et al. (2005) provided detailed numerical calculations that confirm that the formation of the outer planets and the delayed migration of Uranus and Neptune outward can provide a qualitative and quantitative explanation of the LHB. This is the first time that a viable and internally consistent mechanism for the LHB has been found.

EVIDENCE FOR A LATE HEAVY BOMBARDMENT ON THE EARTH?

In any given time span, the Earth has been subjected to a higher impact flux than the Moon, as it has a larger diameter and a much larger gravitational cross section, thus making it an easier target to hit (e.g., Maher and Stevenson, 1988; Oberbeck and Fogleman, 1989; Gogarten-Boekels et al., 1995; Zahnle and Sleep, 1997). If an LHB occurred on the Moon, the Earth was subject to scaling of the same impact flux due to the ratio of the cross sections (Sleep et al., 1989), which would have resulted in an impact rate ≥20 times greater than the lunar one, comprising both more and larger impact events. The consequences for the hydrosphere, atmosphere, and even the lithosphere of Earth at that time must have been devastating (Zahnle and Sleep, 1997; Grieve, 1980; Frey, 1980).

Given the evidence for an LHB on the Moon and elsewhere as discussed above, it is clear that during that time the Earth must have been subjected to an even more intense bombardment. Thus, the question arises if there is any evidence of this bombardment preserved on Earth. The oldest rocks on Earth, from Acasta in Canada and the Isua/Akilia area in Greenland (e.g., papers in Fowler et al., 2003), are the obvious places to conduct such a search. In an early work, Appel (1979) reported finding chromite grains in Isua rocks with sizes of 10–30 μm and of compositions unlike those found in other similar rocks, which he interpreted as being of extraterrestrial origin. It is noticeable that chromite grains are also all that remains in more or less unaltered form of the late Ordovician fossil meteorites reported by, e.g., Schmitz et al. (2003). Thus, it is conceivable that the chromite grains found by Appel (1979) in Isua rocks could be remnants of the earliest documented flux of extraterrestrial material. However, no detailed follow-up studies of this material have been made. Cr isotope studies could be helpful if ever sufficient numbers of these chromites could be extracted.

In a more recent attempt to find a possible earliest Archean extraterrestrial component, samples of some of the oldest rocks on Earth from Isua, Greenland, were analyzed by Koeberl et al. (2000) for their chemical composition, including the PGE abundances. The idea was to search for a signature similar to what is observed in some ejecta layers (e.g., Koeberl, 1998). Unfortunately, the results were ambiguous, and no clear meteoritic signature was found, as discussed in detail by Ryder et al. (2000). A similar study by Anbar et al. (2001) on rocks from the nearby Akilia Island, southwestern Greenland, also failed to detect any significant amount of Ir that would indicate an extraterrestrial component. In addition, zircon extracted from the rocks at Isua were studied by Koeberl et al. (2000) for the possible presence of shock metamorphic effects similar to those found in more recently shocked zircons from impact deposits (e.g., Bohor et al., 1993). Zircon is a very resistant mineral and would be expected to exhibit impact-related shock features if it had been subjected to any impact events in the past. The petrographical studies of Isua zircons failed to show evidence for shock metamorphism (Koeberl et al., 2000).

Several possible reasons could explain the absence of any evidence from these samples for an LHB on Earth. First, the number of samples was probably too small. Second, calculations that correlate impact melt production with crater dimension (Melosh, 1989; Cintala and Grieve, 1994, 1998; Grieve et al., this volume) show a breakdown of such a correlation for very large impact structures. As the magnitude of the impact increases, the melt volume relative to the transient crater size increases, with a larger proportion of melt being retained inside the crater and the depth of melting for large impact structures exceeding the depth of excavation. Thus, the large-scale melting associated with such large impacts will actually reduce the amount of shocked rocks that are formed and preserved. In such impact events, which produce impact structures of several hundred kilometers in diameter, thermal metamorphism could be more important than shock metamorphism. However, craters smaller than a few hundred kilometers in diameter would still produce abundant shocked rocks.

Furthermore, while the Isua rocks are usually considered to have an age of ca. 3.85 Ga (e.g., Mojzsis et al., 1996), it is only the cores of zircons in these rocks that have such high ages. It has been argued that while the zircons are indeed >3.8 Ga, the rocks themselves may only be ca. 3.65 Ga (e.g., Moorbath and Kamber, 1998; Moorbath et al., 1997; Moorbath, 2005). In this case, the LHB would have long ceased and no direct evidence for an extraterrestrial component could be obtained. Indeed, the data and arguments of Mojzsis et al. (1996), who used carbon isotopic compositions of graphite in apatites from these Isua rocks to suggest that life could have evolved as early as 3.85 Ga, have been shown to be problematic for a variety of other reasons as well (e.g., van Zuilen et al., 2002; Fedo and Whitehouse, 2002; Lepland et al., 2005; and see below). Thus, it is possible that the influence of impact events on early life could have been severe (e.g., Sleep et al., 1989; Zahnle and Sleep, 1997), although Ryder (2002) suggested that this influence did not necessarily lead to sterilization of the planet.

Aside from the intriguing chromite grains reported by Appel (1979), the only evidence for an LHB on the Earth seems to come from recent tungsten (W) isotopic studies of Schoenberg et al. (2002), who found W anomalies in ca. 3.85 Ga metasedimentary rocks from Greenland (Fig. 9). Such an anomaly, if real, is difficult to explain by anything but an extraterrestrial component (although Bolhar et al., 2005, consider another mechanism, e.g., increased neutron irradiation effects on the early Earth). However, it is difficult to understand why a similar meteoritic signal would not show up in the PGE abundances, as high W abundances are not common in meteorites. In fact, Frei and Rosing (2005) searched for such a meteoritic signature in similar samples from the Isua Supracrustal Belt in Greenland using the chromium isotopic method and found no trace of an extraterrestrial component; they, thus, comment that they cannot confirm the findings of Schoenberg et al. (2002). Thus, little if any evi-

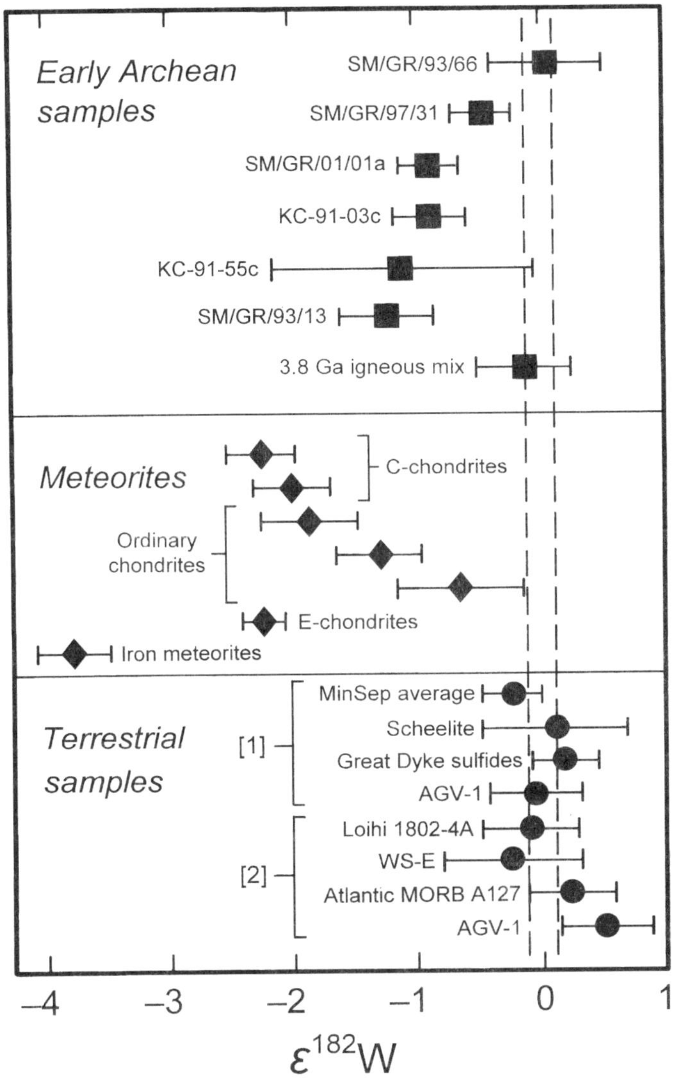

Figure 9. Summary of tungsten isotopic compositions of terrestrial rocks, meteorites, and early Archean samples, after Schoenberg et al. (2002). Terrestrial samples and chondritic meteorites (especially carbonaceous chondrites) show a small but measurable difference of ~2 epsilon units, which indicates that core formation started while ^{182}Hf was still present in the solar system. The deviation of the tungsten isotopic composition in the Archean samples was interpreted by Schoenberg et al. (2002) to indicate the traces of a meteoritic component derived from the LHB. See Schoenberg et al. (2002) for sample details.

dence of an LHB phase at 3.85 Ga is preserved on Earth, but the reason may well be that the LHB destroyed any previous impact record with the production of enormous amounts of melts on the Earth at that time due to large-scale impact events (Cintala and Grieve, 1994, 1998), erasing any traces even of the LHB.

IMPACTS AND THE ORIGIN OF LIFE ON EARTH

The accretionary flux in the inner solar system from ca. 4.4–4.0 Ga seems to have been more benign (Valley et al., 2002) than assumed earlier (e.g., Sleep et al., 1989). When scaled to Earth, Ryder (2002) calculated that even the impact energies associated with the bombardment between ca. 4.4 and 3.9 Ga may not have produced ocean-evaporating, globally sterilizing events, in agreement with the conclusions of Valley et al. (2002). Earlier discussions that such events took place are based on the extrapolation of a nonexistent lunar record to the Hadean (Ryder, 2002). The impact situation on the Earth from ca. 4.4–3.8 Ga may have been comparatively (to the LHB) peaceful, and the impacts themselves during that time could have been thermally and hydrothermally beneficial. Ryder (2002) surmised that the origin of life could have taken place at any time between 4.4 and 3.85 Ga, given the current impact constraints, and that there is no justification for the claim that life originated (or reoriginated) as late as 3.85 Ga in response to the end of hostile impact conditions.

However, accepting the LHB as a reality implies that severe environmental changes occurred at ca. 3.85 Ga. During this cataclysm the Earth would have undergone events of an order of magnitude larger than that of the Moon and experienced many more of them than the Moon had to endure. According to Zahnle and Sleep (1997), hundreds of objects with sizes similar to those of the Imbrium and Orientale impactors must have struck the Earth during the basin-forming era. Nonetheless, an Imbrium-sized impactor scaled to terrestrial collision energy had only 1% of the energy needed to evaporate Earth's oceans, assuming an ocean mass similar to today's and that 25% of the energy of such an impact was partitioned into the ocean (Zahnle and Sleep, 1997). Such an impact would have vaporized only the upper few tens of meters of ocean by heating from above by an impact-produced atmosphere of hot silicate vapor. Even impacts orders of magnitude larger would have been far from sufficient, boiling off only a few hundred meters of ocean water. Ocean-vaporizing impacts were considered rare after 4.4 Ga in the estimation of Zahnle and Sleep (1997). The lunar record is in agreement with this and suggests that impacts of such magnitude may have been rare to absent (Ryder, 2002).

It is quite possible that life started on Earth as soon as the environment became safe enough, and the existence of liquid water as early as 4.4 Ga (e.g., Wilde et al., 2001) or, as more recent data show, around 4.2 Ga (Cavosie et al., 2005) may have provided the necessary boundary conditions. Nevertheless, the details of the development of early life (and its record) remain relatively unconstrained, although a rapid development has been suggested (e.g., Shock et al., 2000).

Several studies have considered the effects of impact on the atmosphere and hydrosphere, again, particularly for very large events (Maher and Stevenson, 1988; Oberbeck and Fogleman, 1989; Sleep et al., 1989; Chyba, 1993; Zahnle and Sleep, 1997). These studies have largely been expressed in the context of the early evolution of life and impact-induced sterilization. An Imbrium-scale impact onto the early Earth would have had the ultimate effect of boiling off ~40 m of sea water, with a subsequent hot surface layer and annihilation of any surface ecosystems (Zahnle and Sleep, 1997); possible larger events would have caused correspondingly more devastating effects. It requires the impact of an asteroid several hundred kilometers in diameter to totally vaporize one present-day ocean mass of water (Fig. 10), which is not very likely, given the oxygen isotope evidence from Hadean zircons discussed above. Also, the scale of such events would probably be too destructive to allow preservation of evidence. It is the probability for these vaporizing impacts on the early Earth that has led to the general impression that impact events were a negative forcing function for the development and evolution of emergent life (e.g., Grieve, 1998). Such large early impact events also must have led to the processing and reprocessing of the Earth's early atmosphere, producing molecules that are precursors for the abiotic synthesis of complex organic molecules (Fegley et al., 1986).

The time of the actual beginning of life on Earth is a highly contentious topic (cf. Bada, 2004) that cannot be discussed in any great detail here; a short summary has to suffice. It is possible, but not proven in any way, that the clement conditions following the solidification of the magma ocean ca. 4.4–4.2 Ga

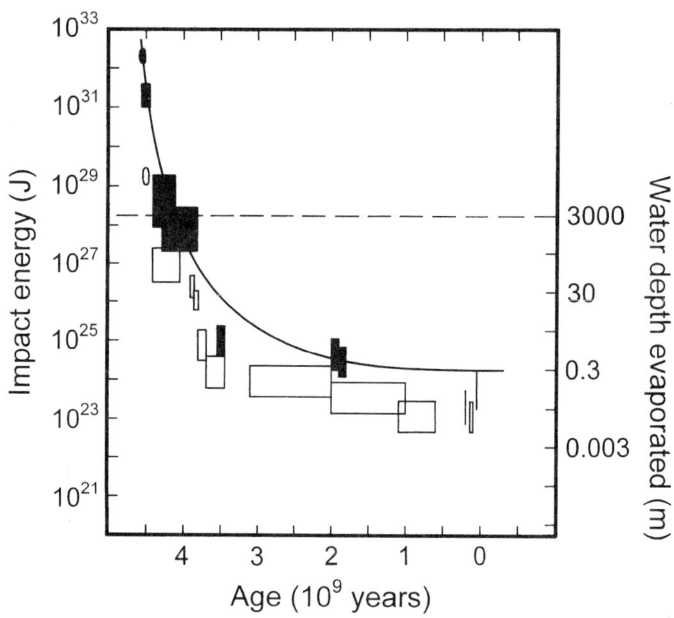

Figure 10. Impact energies of the largest impacts documented on the Earth (filled boxes) and on the Moon (open boxes), after Sleep et al. (1989). The curve indicates the exponential decay of the original accretion flux, similar to what has been shown in Figures 7 or 8. All boxes to the right of ~3.95 Ga and at or below an impact energy of 10^{26} J are documented; the lunar ones are known impact basins. The terrestrial box at ca. 3.5–3.4 Ga indicates the impact events indicated by spherule layers at Barberton, South Africa, and the boxes at around 2–1.9 Ga mark the two oldest impact structures, Vredefort and Sudbury. The line shortly before the zero age point marks the Cretaceous-Tertiary boundary impact event. The dashed line corresponds to total vaporization of the ocean. However, it is debatable (Ryder, 2002; Valley et al., 2002) whether or not the impact rate before 3.9 Ga was really as high as assumed by Sleep et al. (1989; their large boxes to the left of 4 Ga). If the LHB (late heavy bombardment) model applies (as shown in Fig. 7), then ocean-vaporizing events would have been rare to absent, and maybe only one such event might have happened during the LHB.

with liquid water present may have been sufficient to provide the basis for life on Earth. The absence of a rock record does not allow for the testing of this hypothesis. However, reports of isotopically light graphite in early Archean Isua and Akilia supracrustal rocks from southwestern Greenland have been suggested to indicate remains of the earliest life on Earth (e.g., Schidlowski, 1988; Mojzsis et al., 1996; Rosing, 1999). In particular, Mojzsis et al. (1996) concluded that isotopically light graphite occurring as inclusions in apatite crystals in an Isua sample supported a biogenic origin. This apatite-graphite association was proposed to be indicative of biogenic sedimentary apatite, and such an association was considered to indicate an independent biomarker in sedimentary rocks of early Archean age. Mojzsis et al. (1996) reported that these geochemical, petrographic, and isotopic criteria, which they interpreted to indicate a biogenic origin for apatite-hosted graphite inclusions in amphibolite facies Isua rocks, also applied to a sample of granulite facies quartz-pyroxene rock from the island of Akilia (southwest of Isua). Some researchers (e.g., Nutman et al., 1997) consider the Akilia rocks to be even older, by several hundred m.y., than those of Isua (which would bring them close to 3.85 Ga). However, the validity of the original arguments was questioned, by e.g. Rosing et al. (1996), Myers and Crowley (2000), Fedo and Whitehouse (2002), and Moorbath (2005), and more recent petrographic and geochemical studies of graphite, including graphite inclusions in apatite from Isua (e.g., van Zuilen et al., 2002; Lepland et al., 2005; Moorbath, 2005), have challenged the biogenic interpretations, as the graphite-apatite association is found only in secondary carbonate veins and consequently lacks biologic relevance (references in Lepland et al., 2005). The interpretation of isotopically light graphite in Isua rocks thought to represent graded sedimentary beds (Rosing, 1999) is not affected by these arguments.

The problems of using the apatite-graphite criterion as a biosignature in amphibolite facies rocks at Isua raise concerns about applying it to even higher-grade rocks from Akilia, as noted by Lepland et al. (2005). Most importantly, Lepland et al. (2005) reported results of petrographic analysis of several samples from Akilia, including the actual sample used in the original study of Mojzsis et al. (1996), and found that none of the apatite crystals in these samples actually contain any graphite inclusions. Thus, no good evidence currently exists for life on Earth in the earliest preserved rocks on our planet.

LATER (EARLY ARCHEAN) IMPACT EVENTS

In the absence of any conclusive impact record in the oldest rocks on Earth, we need to look at "younger" rocks. The first "real" rock record of impact events dates ~400–500 m.y. after the end of the LHB, in the form of (distal?) ejecta layers (see reviews by Simonson and Glass, 2004, and Simonson et al., 2004). Four distinct spherule horizons in the Barberton Greenstone belt, South Africa (designated S1 to S4), with ages between ca. 3.5 and 3.2 Ga, have been proposed as being of impact origin (e.g., Lowe et al., 2003). The spherules are mostly spherical particles, up to a few mm across, of quenched melt droplets that supposedly formed by condensation from vapor clouds. The spherule layers are coarse grained and have been interpreted to reflect high-energy depositional events in otherwise low-energy, quiet-water environments. The original mineralogical and chemical composition of the spherules has been almost completely changed by alteration. The stratigraphic positions of these layers at different geographic locations are difficult to correlate, and the possibility exists that some of the layers represent duplication (e.g., Hofmann et al., this volume). These spherule layers show extreme enrichments in the PGEs (in some cases exceeding the PGE abundances found in chondritic meteorites), unlike modern impact ejecta deposits (for example, those at the Cretaceous-Tertiary boundary, or in the late Eocene, see, e.g., Montanari and Koeberl, 2000, for a review), which caused some questions regarding the initial impact interpretation (see, for example, Hofmann et al., this volume).

An example of a sample from a Barberton spherule layer (3.48 Ma; Byerly et al., 2002) is shown in Figure 11. These spherule layers from the Barberton Greenstone Belt have been interpreted as the result of large asteroid or comet impacts onto the early Earth (see review by Lowe et al., 2003). The extreme enrichments in the PGEs (e.g., Kyte et al., 1992; Koeberl and Reimold, 1995; Reimold et al., 2000), among other problems, caused Koeberl and Reimold (1995) to question the impact interpretation. Figure 12 shows a correlation between the abundances of Ir and As, a very mobile element, in samples from the Barberton spherule layers, all of which have been subject to pervasive transformation into secondary mineral assemblages. This relationship between Ir and As, thus, indicates remobilization of both elements; this means that the PGE signature in these samples is not primary (e.g., Koeberl and Reimold, 1995; Reimold et al., 2000).

More recently, though, Shukolyukov et al. (2000) and Kyte et al. (2003) reported Cr isotopic anomalies (using methods and arguments of Shukolyukov and Lugmair, 1998) in samples from several of these layers that seem to support the presence of an extraterrestrial component.

Other occurrences of unusual spherule layers were reported by Simonson (1992) from the Hamersley Basin in Western Australia. On the basis of similarities to microtektites and mikrokrystites, Simonson (1992) interpreted the spherules as having formed in an impact event and having been redeposited in a sediment gravity flow. Later, three additional spherule-bearing layers were found in the Hamersley Basin sequence, which were also interpreted to be of impact origin (e.g., Simonson et al., 1998). Simonson et al. (2000a, 2000b) also reported the discovery of a similar spherule layer (ca. 2.6 Ma) in the Monteville Formation of the Transvaal Supergroup in South Africa, which might be correlated with one of the Australian layers (e.g., Simonson and Hassler, 1997; Simonson et al., 1999; Rasmussen and Koeberl, 2004).

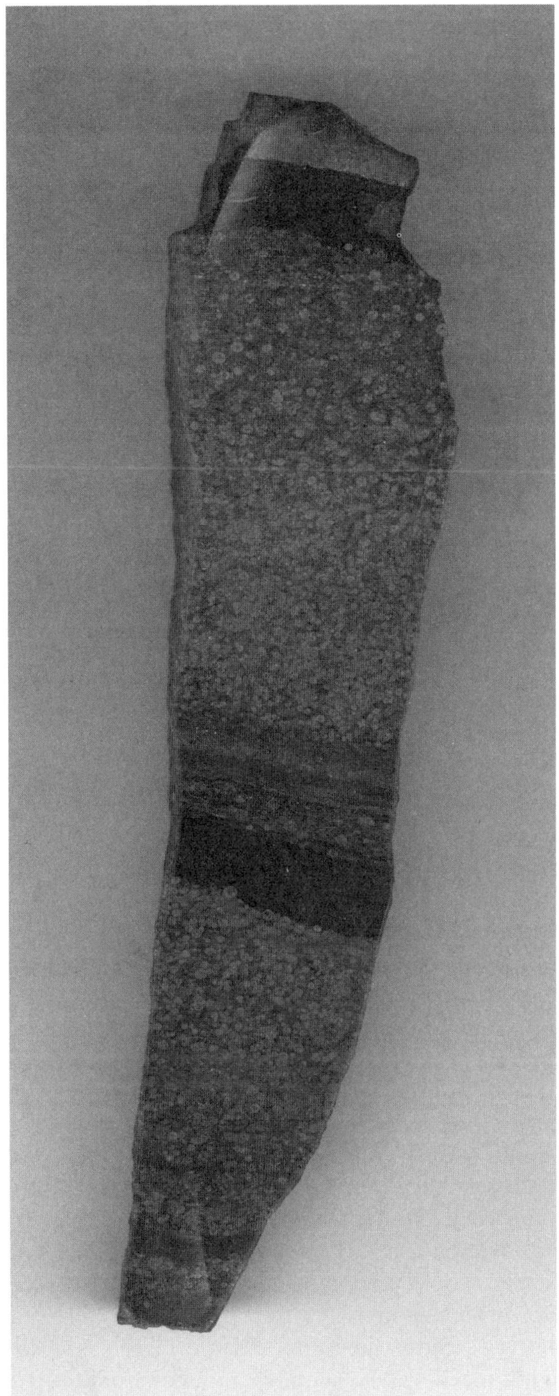

Figure 11. Spherule layer sample from one of the impact-derived layers of 3.48 Ga age (S2 or S3; see discussion of stratigraphic complexities in Koeberl and Reimold, 1995, and Hofmann et al., this volume), from Sheba Mine, Barberton, South Africa; the sample shows mm-sized, clearly sorted spherules in bands. It is thought that the repetition of spherule bands in samples such as this one was caused by tectonic duplication (folding; Reimold et al., 2000). Length of sample: 16 cm.

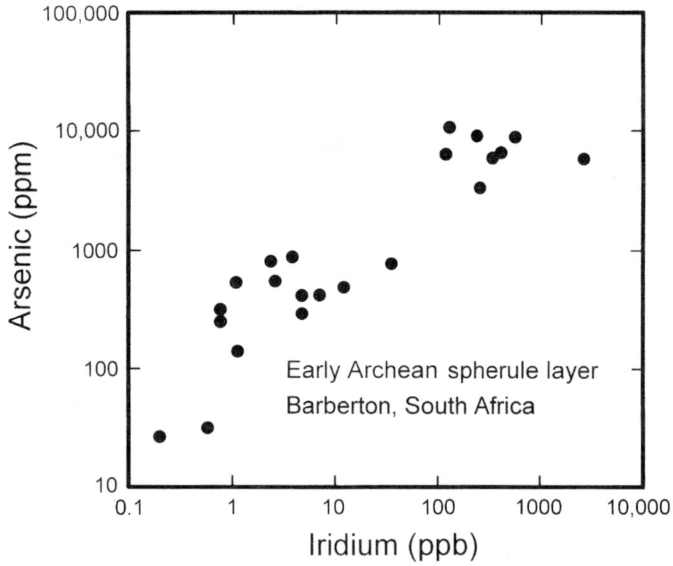

Figure 12. Correlation between As and Ir in samples from the S2 or S3 (see previous caption regarding uncertainties) Barberton spherule layers; data from Koeberl and Reimold (1995) and Reimold et al. (2000). The correlation indicates that the PGE (platinum group element) concentrations in these samples—which exceed even pure chondritic concentrations anyway—must be the result of secondary hydrothermal activity.

Until recently, no shocked minerals have been reported to be associated with any of these spherule layers. It was suggested that this is because the impacts occurred into oceanic crust, which has little or no quartz, and whatever else there was in terms of shocked minerals had long been destroyed by alteration (Simonson et al., 1998). Rasmussen and Koeberl (2004), however, were able to describe one shocked quartz grain in a sample from the 2.63 Ga Jeerinah impact layer in Australia; this is so far the only evidence of diagnostic shock features. Unfortunately, so far no definitive criteria for the identification of Archean impact deposits are known. For none of the South African (Barberton and Monteville) or Australian spherule layers have source craters been found, and given the scarcity of the early Archean geological record, it is unlikely they will ever be found. It is not clear why impact events in the Archean would predominantly produce large volumes of spherules, which are mostly absent from post-Archean impact deposits (i.e., those for which source craters are known), although a recent report about ejecta beds from the 1.85 Ga Sudbury impact structure also included observation of spherule horizons (Addison et al., 2005). The question regarding how to identify Archean impact deposits remains open and will hopefully be addressed in future studies (but see Simonson and Harnik, 2000, and Simonson, 2003, for discussions on this subject).

Nevertheless, the discovery of these various spherule layers provides interesting material for the discussion about the importance of impact events in the Earth's history. Several lines of geological and geochemical evidence indicate that the level of atmospheric oxygen was extremely low before 2.45 Ga and that it had reached considerable levels by 2.22 Ga (cf. Bekker et al., 2004). Bekker et al. (2004) presented evidence that the abundance of oxygen in the terrestrial atmosphere underwent a dramatic increase by 2.32 Ga. This could have influenced the preservation of early impact signals due to the change in oxidation parameters during the impact event (e.g., production of oxidized versus reduced impact deposits, such as spinels) and/or altered the nature of the deposits due to different weathering in oxidized and reduced atmospheres.

THE OLDEST IMPACT STRUCTURES PRESERVED ON EARTH

The earliest remnants on Earth of actual impact structures—following the more controversial and difficult to quantify evidence from the early Archean spherule beds—date to ca. 2 Ga. Thus, for more than half of the "life" of our planet we do not have any good preserved remnants of the many impact structures that must have formed and for which we have indirect evidence (also on the Moon) throughout the first 2.5 b.y. of Earth evolution. The oldest known terrestrial impact structures are the Vredefort and Sudbury structures with ages of 2023 ± 4 Ma and 1850 ± 3 Ma, respectively (see Reimold and Gibson, 1996; Gibson and Reimold, 2001), which represent the complete documented pre–1.85 Ga terrestrial impact record.

As there have been several good reviews of the geology and importance of Vredefort (e.g., Gibson and Reimold, 2001) and Sudbury (e.g., Grieve and Therriault, 2000), as well as on the subsequent impact events from 2 Ga to today, this is the place to end the current review. I will briefly summarize some information on Vredefort and end at 2 Ga. The Vredefort Dome is an ~80-km-wide feature centered ~120 km southwest of Johannesburg, South Africa, and is widely accepted to be the deeply eroded central uplift of a large (initial diameter on the order of 200–300 km), complex impact structure that formed 2023 ± 4 Ma (Kamo et al., 1996). Therefore, it represents the remnant of the oldest currently known impact structure on Earth. The dome comprises an ~40-km-wide core of early Archean high-grade gneisses that is surrounded by a subvertically dipping sequence of late Archean to Paleoproterozoic supracrustal rocks and intrusions. The rocks exposed there display a variety of impact-related features as described in the first chapter of this review, including shatter cones, coesite and stishovite, planar deformation features in quartz, dykes of impact melt breccia (the so-called "Vredefort Granophyre," which contains traces of an extraterrestrial component), and abundant pseudotachylitic breccias (Reimold and Gibson, 2005). The absence of both a definitive crater morphology and a coherent impact melt body and fallback breccias indicates that considerable erosion has taken place after the impact event, removing the top 5–10 km of the post-impact surface. In much of the core of the dome, the Archean gneisses preserve pre-impact mineralogy and fabrics, which formed during upper amphibolite to granulite facies regional metamorphism at 3.2–3.1 Ga (Armstrong et al., this volume). Gibson and Reimold (2005) investigated the shock metamorphic zonation around the center of the dome.

In contrast to the 1.85 Ga Sudbury Structure, for which impact ejecta (including an up-to-10-cm-thick spherule layer) have recently been identified in Minnesota and western Ontario 650–875 km from the impact site (Addison et al., 2005), no confirmed ejecta have to date been recovered for the Vredefort impact. Vredefort deposits would have formed some 200 m.y. prior to the earliest sedimentary deposits known from southern Africa, the Waterberg Group occurring widespread in the northern parts of South Africa (e.g., Simpson et al., 2004). To date, no evidence for shocked materials has been reported from Waterberg sediments. Interestingly, Chadwick et al. (2001) found a spherule layer in Greenland that is in the same age bracket as Vredefort and Sudbury, at 2.1–1.9 Ga based on stratigraphic considerations, and could represent distal ejecta from either of these two impact events.

SUMMARY AND OUTLOOK

Following its formation by impact-driven accretion, the Earth has been subjected early in its history to a significant number of impact events. This early impact record is better documented in lunar rocks than in terrestrial rocks. As a consequence of the accretion of the Earth and possibly also related to the giant moon-forming impact around 4.45 Ga, a large-scale magma ocean is postulated to have existed on Earth, which may have persisted for up to several hundred m.y. Related thermal effects would have effectively obliterated any petrographic or geochemical remnants of the early intense impact bombardment. Following the Moon-forming giant impact event, there is some evidence that the Earth's upper mantle underwent differentiation to form an early crust, some of which might have involved granitic rocks; remnants of these would be the earliest terrestrial zircons dated at 4.4–4.3 Ga. Geochemical data show that shortly after the giant impact, clement conditions seem to have dominated on Earth, with liquid water present. Some isotopic evidence exists that plate tectonics and crustal differentiation could have operated already in the Hadean period. Very few rocks on Earth with ages approaching 3.9 Ga have been found; some rare older detrital zircon grains up to ca. 4.4 Ga are known. It is possible that the absence of any rocks older than ca. 3.9 Ga is the result of an intense episode of impacts, the so-called late heavy bombardment (LHB) at around 3.8–4 Ga, during which impact-induced mixing may have recycled early crustal fragments back into the upper mantle. Any sedimentological record, which would host information specific to surface environments such as the rate and violence of meteorite impact and the presence of life, has been lost from Hadean times. The oldest rocks on Earth that could be searched for an impact record are the ca. 4 Ga Acasta Gneiss Complex in Canada and the =3.8 Ga rocks from Isua, Greenland. However, the results of such studies have been ambiguous to date, and no clear impact signature has been found in the Earth's oldest rocks.

Another interesting topic is the influence of impact events on the development and evolution of life (cf. Cockell and Bland, 2005). Early interpretations of the impact record suggested a "dead" time span before ca. 3.9 Ga because of the deleterious effects of impacts, but more recent data show that water must have existed on the Earth as far back as ca. 4.4 Ga, and only during the LHB were ocean-vaporizing impacts events possible (but by no means proven). Unfortunately the record of early life on Earth remains quite controversial, suffering from the limited rock record and bad preservation, similar factors that are hampering our understanding of early impact events (cf. papers in Gilmour and Koeberl, 2000).

Possible impact debris layers have been documented in 3.2–3.47 Ga Archean successions in the Barberton Greenstone Belt (South Africa) and Pilbara Craton (Australia). These spherule layers show extreme enrichments in the PGEs, unlike modern ejecta deposits, but Cr isotopic anomalies in samples from some of these layers seem to support the presence of an extraterrestrial component. Similar spherule layers (of ca. 2.5–2.63 Ma) were found in the Hamersley Basin in Australia as well as in the Monteville Formation of the Transvaal Supergroup in South Africa; these South African and Australian layers may be correlated with each other. Until recently, no shocked quartz, commonly regarded as unambiguous evidence for a hypervelocity impact event, has been found in distal ejecta horizons older than ca. 600 Ma; an exception is a single shocked quartz grain from the 2.63 Ga Jeerinah impact layer in Australia. The exact number of impact spherule layers is not known, as it is not clear which ones might correlate with each other, but a minimum seven different events between 3.4 and 2.5 Ga and at 2.1–1.8 Ga has been suggested.

The oldest impact crater on Earth dates to 2 Ga (Vredefort in South Africa), and for the next billion years the impact record on Earth is quite sparse in terms of both craters and ejecta layers. The early record of impact on the Earth is rather limited and mostly circumstantial. Thus, the impact record on Earth is so far quite limited: nothing for the first billion years, then some spherule layers until ca. 2.5 Ga, and then less than a handful of impact craters prior to ca. 750 Ma. Nevertheless, the discovery of these spherule layers aids in the discussion of the importance of impact events in the early parts of the Earth's history.

Figure 13 gives a summary of the impact record of the Earth, starting with the preserved rock record. The number of documented events are few—seven or eight impact layers in the time span between 3.5 and 2.4 Ga, three impact structures between ca. 700 Ma and 2.1 Ga, and the rest of the ~170 presently known impact structures all being younger than that. Clearly the "early" impact record on Earth, which spans more than half of the age of our planet, is still a wide-open field of research.

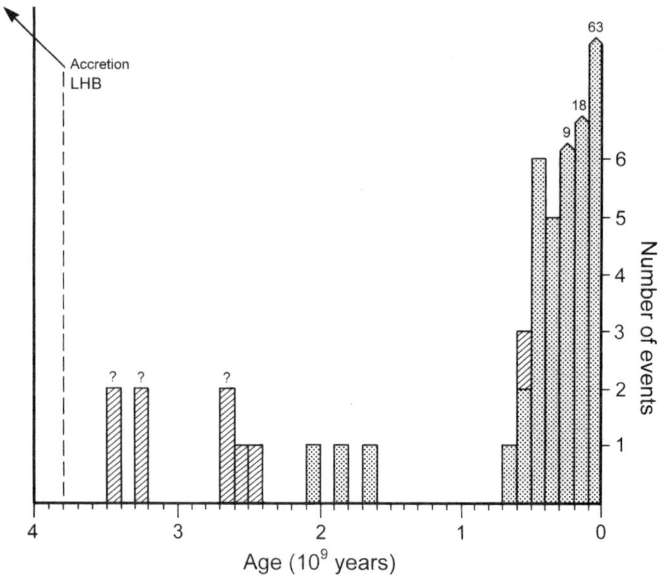

Figure 13. Summary of the impact history (status 2005) documented on Earth. Note that the number of impact structures within the past 300 m.y. by far exceeds the scale used here, and actual numbers are written on top of the histogram bars. The dot-pattern indicates impact structures, whereas the hatched pattern signifies ejecta and/or spherule layers. The accretion and LHB (late heavy bombardment) flux is indicated by the arrow on the upper left. The diagram demonstrates how limited the impact record over most of the history of the Earth really is.

ACKNOWLEDGMENTS

This research was supported by the Fonds zur Förderung der wissenschaftlichen Forschung in Austria. I am grateful to D. Jalufka (University of Vienna) for expert drafting of all the figures, to S. Moorbath (Oxford University) for literature and discussion, and to R. Grieve (Natural Resources Canada) and W.U. Reimold (University of the Witwatersrand) for helpful reviews.

REFERENCES CITED

Abe, Y., 1997, Thermal and chemical evolution of the terrestrial magma ocean: Physics of the Earth and Planetary Interiors, v. 100, p. 27–39, doi: 10.1016/S0031-9201(96)03229-3.

Addison, A.D., Brumpton, G.R., Vallini, D.A., McNaughton, N.J., Davis, D.W., Kissin, S.A., Fralick, P.W., and Hammond, A.L., 2005, Discovery of distal ejecta from the 1850 Ma Sudbury impact event: Geology, v. 33, p. 193–196, doi: 10.1130/G21048.1.

Alvarez, L.W., Alvarez, W., Asaro, F., and Michel, H.V., 1980, Extraterrestrial cause for the Cretaceous-Tertiary extinction: Science, v. 208, p. 1095–1108.

Amelin, Y., Lee, D.-C., Halliday, A.N., and Pidgeon, R.T., 1999, Nature of the Earth's oldest crust from hafnium isotopes in single detrital zircons: Nature, v. 399, p. 252–255, doi: 10.1038/20426.

Anbar, A.D., Zahnle, K.J., Arnold, G.L., and Mojzsis, S.J., 2001, Extraterrestrial iridium, sediment accumulation and the habitability of the early Earth's surface: Journal of Geophysical Research, v. 106, p. 3219–3236, doi: 10.1029/2000JE001272.

Appel, P.W.U., 1979, Cosmic grains in an iron-formation from the early Precambrian Isua supercrustal belt: West Greenland: Journal of Geology, v. 87, p. 573–578.

Armstrong, R.A., Lana, C., Reimold, W.U., and Gibson, R.L., 2006, this volume, SHRIMP zircon age constraints on Mesoarchean crustal development in the Vredefort dome, central Kaapvaal craton, South Africa, *in* Reimold, W.U., and Gibson, R.L., Processes on the early Earth: Geological Society of America Special Paper 405, doi: 10.1130/2006.2405(13).

Arndt, N., and Chauvel, C., 1991, Crust of the Hadean Earth: Bulletin of the Geological Society of Denmark, v. 39, p. 145–151.

Arrhenius, G., and Lepland, A., 2000, Accretion of moon and earth and the emergence of life: Chemical Geology, v. 169, p. 69–82, doi: 10.1016/S0009-2541(00)00333-8.

Bada, J.L., 2004, How life began on Earth: A status report: Earth and Planetary Science Letters, v. 226, p. 1–15, doi: 10.1016/j.epsl.2004.07.036.

Baldwin, R.B., 1974, Was there a "Terminal Lunar Cataclysm" 3.9–4.0 × 10^9 years ago?: Icarus, v. 23, p. 157–166, doi: 10.1016/0019-1035(74)90003-7.

Bekker, A., Holland, H.D., Wang, P.-L., Rumble, D., III, Stein, H.J., Hannah, J.L., Coetzee, L.L., and Beukes, N.J., 2004, Dating the rise of atmospheric oxygen: Nature, v. 427, p. 117–120, doi: 10.1038/nature02260.

Bogard, D.D., 1995, Impact ages of meteorites: Meteoritics & Planetary Science, v. 30, p. 244–268.

Bohor, B.F., Betterton, W.J., and Krogh, T.E., 1993, Impact-shocked zircons: Discovery of shock-induced textures reflecting increasing degrees of shock metamorphism: Earth and Planetary Science Letters, v. 119, p. 419–424, doi: 10.1016/0012-821X(93)90149-4.

Bolhar, R., Kamber, B.S., Moorbath, S., Whitehouse, M.J., and Collerson, K.D., 2005, Chemical characterization of the Earth's most ancient clastic metasediments from Isua Greenstone Belt, southern West Greenland: Geochimica et Cosmochimica Acta, v. 69, p. 1555–1573, doi: 10.1016/j.gca.2004.09.023.

Bowring, S.A., and Housh, T., 1995, The Earth's early evolution: Science, v. 269, p. 1535–1540.

Boyet, M., and Carlson, R.W., 2005, ^{142}Nd evidence for early (>4.53 Ga) global differentiation of the silicate Earth: Science, v. 309, p. 576–581, doi: 10.1126/science.1113634.

Boyet, M., Rosing, M., Blichert-Toft, J., Storey, M., and Albarede, F., 2003, ^{142}Nd evidence for early earth differentiation: Lunar and Planetary Science, v. 34, abs. 1945 (CD-ROM).

Byerly, G.R., Lowe, D.R., Wooden, J.L., and Xie, X., 2002, An Archean impact layer from the Pilbara and Kaapvaal Cratons: Science, v. 297, p. 1325–1327, doi: 10.1126/science.1073934.

Canup, R.M., 2004a, Dynamics of lunar formation: Annual Review of Astronomy and Astrophysics, v. 42, p. 441–475, doi: 10.1146/annurev.astro.41.082201.113457.

Canup, R.M., 2004b, Simulations of a late lunar-forming impact: Icarus, v. 168, p. 433–456, doi: 10.1016/j.icarus.2003.09.028.

Canup, R.M., and Agnor, C.B., 2000, Accretion of the terrestrial planets and the Earth-Moon system, *in* Canup, R., and Righter, K., eds., Origin of the Earth and Moon: Tucson, University of Arizona Press, p. 113–129.

Canup, R.M., and Asphaug, E., 2001, Origin of the Moon in a giant impact near the end of the Earth's formation: Nature, v. 412, p. 708–712, doi: 10.1038/35089010.

Canup, R., and Righter, K., eds., 2000, Origin of the Earth and Moon: Tucson, University of Arizona Press, 555 p.

Caro, G., Bourdon, B., Birck, J.-L., and Moorbath, S., 2003, ^{146}Sm-^{142}Nd evidence from Isua metamorphosed sediments for early differentiation of the Earth's mantle: Nature, v. 423, p. 428–432, doi: 10.1038/nature01668.

Caro, G., Bourdon, B., Wood, B.J., and Corgne, A., 2005, Trace-element fractionation in Hadean mantle generated by melt-segregation from a magma ocean: Nature, v. 436, p. 246–249, doi: 10.1038/nature03827.

Cavosie A.J., Valley J.W., and Wilde, S.A., 2005, Magmatic $\delta^{18}O$ in 4400–3900 Ma detrital zircons: A record of the alteration and recycling of crust in the Early Archean: Earth and Planetary Science Letters, v. 235, p. 663–681.

Chadwick, B., Claeys, P., and Simonson, B.M., 2001, New evidence for a large Palaeoproterozoic impact: Spherules in a dolomite layer in the Ketilidian orogen, South Greenland: Journal of the Geological Society of London, v. 158, p. 331–340.

Chyba, C.F., 1993, The violent environment for the origins of life: Progress and uncertainties: Geochimica et Cosmochimica Acta, v. 57, p. 3351–3358, doi: 10.1016/0016-7037(93)90543-6.

Cintala, M.J., and Grieve, R.A.F., 1994, The effects of differential scaling of impact melt and crater dimensions on lunar and terrestrial craters: Some brief examples, *in* Dressler, B.O., Grieve, R.A.F., and Sharpton, V.L., eds., Large Meteorite Impacts and Planetary Evolution: Geological Society of America Special Paper 293, p. 51–59.

Cintala, M.J., and Grieve, R.A.F., 1998, Scaling impact melting and crater dimensions: Implications for the lunar cratering record: Meteoritics & Planetary Science, v. 33, p. 889–912.

Cockell, C.S., and Bland, P.A., 2005, The evolutionary and ecological benefit of asteroid and comet impacts: Trends in Ecology & Evolution, v. 20, p. 175–179, doi: 10.1016/j.tree.2005.01.009.

Cohen, B.A., Swindle, T.D., and Kring, D.A., 2000, Support for the lunar cataclysm hypothesis from lunar meteorite melt ages: Science, v. 290, p. 1754–1756, doi: 10.1126/science.290.5497.1754.

Compston, W., and Pidgeon, R.T., 1986, Jack Hills, evidence of more very old zircons in Western Australia: Nature, v. 291, p. 193–196.

Culler, T.S., Becker, T.A., Muller, R.A., and Renne, P.R., 2000, Lunar impact history from $^{40}Ar/^{39}Ar$ dating of glass spherules: Science, v. 287, p. 1785–1788, doi: 10.1126/science.287.5459.1785.

Davies, G.F., 1985, Heat deposition and retention in a solid planet growing by impacts: Icarus, v. 63, p. 45–68, doi: 10.1016/0019-1035(85)90020-X.

Deutsch, A., and Schärer, U., 1994, Dating terrestrial impact events: Meteoritics, v. 29, p. 301–322.

Fedo, C.M., and Whitehouse, M.J., 2002, Metasomatic origin of quartz-pyroxene rock, Akilia, Greenland, and implications for Earth's earliest life: Science, v. 296, p. 1448–1452, doi: 10.1126/science.1070336.

Fegley, B., Jr., Prinn, R.G., Hartman, H., and Watkins, G.H., 1986, Chemical effects of large impacts on the Earth's primitive atmosphere: Nature, v. 319, p. 305–308, doi: 10.1038/319305a0.

Fowler, C.M.R., Ebinger, C., and Hawkesworth, C.J., eds., 2003, The early Earth: Physical, Chemical, and Biological Development: Geological Society [London] Special Publication 199, 352 p.

Frei, R., and Rosing, M.T., 2005, Search for traces of the late heavy bombardment on Earth—Results from high precision chromium isotopes: Earth and Planetary Science Letters, v. 236, p. 28–40, doi: 10.1016/j.epsl.2005.05.024.

Frey, H., 1980, Crustal evolution of the early Earth: The role of major impacts: Precambrian Research, v. 10, p. 195–216, doi: 10.1016/0301-9268(80)90012-1.

Gibson, R.L., and Reimold, W.U., 2001, The Vredefort Impact Structure, South Africa—The scientific evidence and a two-day excursion guide: Pretoria, South Africa, Council for Geoscience Memoir 92, 111 p.

Gibson, R.L, and Reimold, W.U., 2005, Shock pressure distribution in the Vredefort impact structure, South Africa, *in* Kenkmann, T., Hörz, F., and Deutsch, A., eds., Large Meteorite Impacts and Planetary Evolution III: Geological Society of America Special Paper 384, p. 329–349.

Gilmour, I., and Koeberl, C., eds., 2000, Impacts and the Early Earth: Lecture Notes in Earth Sciences 91: Heidelberg, Springer-Verlag, 445 p.

Gogarten-Boekels, M., Hilario, E., and Gogarten, P., 1995, The effects of heavy meteorite bombardment on the early evolution—The emergence of the three domains of life: Origins of Life and Evolution of the Biosphere, v. 25, p. 251–264, doi: 10.1007/BF01581588.

Gomes, R., Levison, H.F., Tsiganis, K., and Morbidelli, A., 2005, Origin of the cataclysmic late heavy bombardment period of the terrestrial planets: Nature, v. 435, p. 466–469, doi: 10.1038/nature03676.

Grieve, R.A.F., 1980, Impact bombardment and its role in proto-continental growth on the early Earth: Precambrian Research, v. 10, p. 217–247, doi: 10.1016/0301-9268(80)90013-3.

Grieve, R.A.F., 1998, Extraterrestrial impacts on Earth: The evidence and its consequences, *in* Grady, M.M., Hutchinson, R., McCall, G.J.H., and Rothery, D.A., eds., Meteorites: Flux with Time and Impact Effects: Geological Society [London] Special Publication 140, p. 105–131.

Grieve, R.A.F., and Parmentier, E.M., 1984, Impact phenomena as factors in the evolution of the Earth, *in* Comparative Planetology, Proceedings of the 27th International Geological Congress: VNU Science Press, v. 19, p. 99–114.

Grieve, R.A.F., and Therriault, A., 2000, Vredefort, Sudbury, Chicxulub; three of a kind: Annual Review of Earth and Planetary Sciences, v. 28, p. 305–338, doi: 10.1146/annurev.earth.28.1.305.

Grieve, R.A.F., Langenhorst, F., and Stöffler, D., 1996, Shock metamorphism of quartz in nature and experiment: II. Significance in geoscience: Meteoritics & Planetary Science, v. 31, p. 6–35.

Grieve, R.A.F., Cintala, M.J., and Therriault, A.M., 2006, this volume, Large-scale impacts and the evolution of the Earth's crust: The early years, *in* Reimold, W.U., and Gibson, R.L., Processes on the early Earth: Geological Society of America Special Paper 405, doi: 10.1130/2006.2405(02).

Halliday, A.N., 2004, Mixing, volatile loss and compositional change during impact-driven accretion of the Earth: Nature, v. 427, p. 505–509, doi: 10.1038/nature02275.

Halliday, A., Rehkämper, M., Le, D.-C., and Yi, W., 1996, Early evolution of the Earth and Moon: New constraints on Hf-W geochemistry: Earth and Planetary Science Letters, v. 142, p. 75–89, doi: 10.1016/0012-821X(96)00096-9.

Harland, W.B., Armstrong, R.L., Cox, A.V., Craig, L.E., Smith, A.G., and Smith, D.G., 1989, A Geologic Time Scale 1989: Cambridge, Cambridge University Press, 263 p.

Harper, C.L., and Jacobsen, S.B., 1992, Evidence from coupled ^{147}Sm-^{143}Nd and ^{146}Sm-^{142}Nd systematics for very early (4.5-Gyr) differentiation of the Earth's mantle: Nature, v. 360, p. 728–732, doi: 10.1038/360728a0.

Hartlaub, R.P., Heaman, L.M., Simonetti, A., and Böhm, C.O., 2006, this volume, Relics of Earth's earliest crust: U-Pb, Lu-Hf, and morphological characteristics of >3.7 Ga detrital zircon of the western Canadian Shield, *in* Reimold, W.U., and Gibson, R.L., Processes on the early Earth: Geological Society of America Special Paper 405, doi: 10.1130/2006.2405(05).

Hartmann, W.K., 1966, Early lunar cratering: Icarus, v. 5, p. 406–418, doi: 10.1016/0019-1035(66)90054-6.

Hartmann, W.K., 1975, Lunar "cataclysm": A misconception?: Icarus, v. 24, p. 181–187, doi: 10.1016/0019-1035(75)90095-0.

Hartmann, W.K., Ryder, G., Dones, L., and Grinspoon, D.H., 2000, The time-dependent intense bombardment of the primordial Earth-Moon system, *in* Canup, R., and Righter, K., eds., Origin of the Earth and Moon: Tucson, University of Arizona Press, p. 493–512.

Hofmann, A., Reimold, W.U., and Koeberl, C., 2006, this volume, Archean spherule layers in the Barberton Greenstone Belt, South Africa: A discussion of problematics related to the impact interpretation, *in* Reimold, W.U., and Gibson, R.L., Processes on the early Earth: Geological Society of America Special Paper 405, doi: 10.1130/2006.2405(03).

Kamber, B.S., and Moorbath, S., 1998, Initial Pb of the Amitsoq gneiss revisited: Implication for the timing of the early Archaean crustal evolution in West Greenland: Chemical Geology, v. 150, p. 19–41, doi: 10.1016/S0009-2541(98)00059-X.

Kamber, B.S., Moorbath, S., and Whitehouse, M.T., 1998, Extreme Nd-isotope heterogeneity in the early Archean—Fact or fiction? Case histories from northern Canada and West Greenland—Reply: Chemical Geology, v. 148, p. 219–224, doi: 10.1016/S0009-2541(98)00032-1.

Kamber, B.S., Collerson, K.D., Moorbath, S., and Whitehouse, M.J., 2003, Inheritance of early Archaean Pb-isotope variability from long-lived Hadean protocrust: Contributions to Mineralogy and Petrology, v. 145, p. 25–46.

Kamo, S.L., Reimold, W.U., Krogh, T.E., and Colliston, W.P., 1996, A 2.023 Ga age for the Vredefort impact event and a first report of shock metamorphosed zircons in pseudotachylite breccias and Granophyre: Earth and Planetary Science Letters, v. 144, p. 369–388, doi: 10.1016/S0012-821X(96)00180-X.

Kleine, T., Mezger, K., Palme, H., and Münker, C., 2004, The W isotope evolution of the bulk silicate Earth: Constraints on the timing and mechanisms of core formation and accretion: Earth and Planetary Science Letters, v. 228, p. 109–123, doi: 10.1016/j.epsl.2004.09.023.

Koeberl, C., 1998, Identification of meteoritical components in impactites, *in* Grady, M.M., Hutchison, R., McCall, G.J.H., and Rothery, D.A., eds., Meteorites: Flux with Time and Impact Effects: Geological Society of [London] Special Publication 140, p. 133–152.

Koeberl, C., 2001, The sedimentary record of impact events, *in* Peucker-Ehrenbrink, B., and Schmitz, B., eds., Accretion of Extraterrestrial Matter throughout Earth's History: New York, Kluwer Academic/Plenum Publishers, p. 333–378.

Koeberl, C., 2002, Mineralogical and geochemical aspects of impact craters: Mineralogical Magazine, v. 66, p. 745–768, doi: 10.1180/0026461026650059.

Koeberl, C., 2004, The late heavy bombardment in the inner solar system: Is there a connection to Kuiper Belt objects?: Earth, Moon, and Planets, v. 92, p. 79–87, doi: 10.1023/B:MOON.0000031927.85937.9e.

Koeberl, C., and Reimold, W.U., 1995, Early Archean spherule beds in the Barberton Mountain Land, South Africa: No evidence for impact origin: Precambrian Research, v. 74, p. 1–33, doi: 10.1016/0301-9268(94)00099-D.

Koeberl, C., and Sharpton, V.L., 1988, Giant impacts and their influence on the early Earth, *in* Papers presented to the Conference on "Origin of the Earth": Houston, Lunar Planetary Institute, p. 47–48.

Koeberl, C., and Shirey, S.B., 1997, Re-Os systematics as a diagnostic tool for the study of impact craters and distal ejecta: Palaeogeography, Palaeoclimatology, Palaeoecology, v. 132, p. 25–46, doi: 10.1016/S0031-0182(97)00045-X.

Koeberl, C., Reimold, W.U., McDonald, I., and Rosing, M., 2000, Search for petrographical and geochemical evidence for the late heavy bombardment on Earth in early Archean rocks from Isua, Greenland, *in* Gilmour, I., and Koeberl, C., eds., Impacts and the early Earth: Lecture Notes in Earth Sciences 91: Heidelberg, Springer-Verlag, p. 73–97.

Koeberl, C., Peucker-Ehrenbrink, B., Reimold, W.U., Shukolyukov, A., and Lugmair, G.W., 2002, A comparison of the osmium and chromium isotopic methods for the detection of meteoritic components in impactites: Examples from the Morokweng and Vredefort impact structures, South Africa, *in* Koeberl, C., and MacLeod, K.G., Catastrophic Events and Mass Extinctions: Impacts and Beyond: Geological Society of America Special Paper 356, p. 607–617.

Kring, D.A., and Cohen, B.A., 2002, Cataclysmic bombardment throughout the inner solar system 3.9–4.0 Ga: Journal of Geophysical Research, v. 107, no. 2, p. 4-1–4-5, doi 10.1029/2001JE001529.

Kröner, A., and Layer, P.W., 1992, Crust formation and plate motion in the early Archean: Science, v. 256, p. 1405–1411.

Kyte, F.T., Zhou, L., and Lowe, D.R., 1992, Noble metal abundances in an early Archean impact deposit: Geochimica et Cosmochimica Acta, v. 56, p. 1365–1372, doi: 10.1016/0016-7037(92)90067-S.

Kyte, F.T., Shukolyukov, A., Lugmair, G.W., Lowe, D.R., and Byerly, G.R., 2003, Early Archean spherule beds: Chromium isotopes confirm origin through multiple impacts of projectiles of carbonaceous chondrite type: Geology, v. 31, p. 283–286, doi: 10.1130/0091-7613(2003)031<0283:EASBCI>2.0.CO;2.

Lee, D.-C., and Halliday, A.N., 1995, Hafnium-tungsten chronometry and the timing of terrestrial core formation: Nature, v. 378, p. 771–774, doi: 10.1038/378771a0.

Lee, D.-C., and Halliday, A.N., 1996, Hf-W isotopic evidence from rapid accretion and differentiation in the solar system: Science, v. 274, p. 1876–1879, doi: 10.1126/science.274.5294.1876.

Lee, D.-C., Halliday, A.N., Snyder, G.A., and Taylor, L.A., 1997, Age and origin of the Moon: Science, v. 278, p. 1098–1103, doi: 10.1126/science.278.5340.1098.

Lepland, A., van Zuilen, M.A., Arrhenius, G., Whitehouse, M.J., and Fedo, C.M., 2005, Questioning the evidence for Earth's earliest life—Akilia revisited: Geology, v. 33, p. 77–79, doi: 10.1130/G20890.1.

Levison, H.F., Dones, L., Chapman, C.R., Stern, S.A., Duncan, M.J., and Zahnle, K., 2001, Could the lunar "Late Heavy Bombardment" have been triggered by the formation of Uranus and Neptune?: Icarus, v. 151, p. 286–306, doi: 10.1006/icar.2001.6608.

Lowe, D.R., Byerly, G.R., Kyte, F.T., Shukolyukov, A., Asaro, F., and Krull, A., 2003, Spherule beds 3.47–3.24 billion years old in the Barberton Greenstone Belt, South Africa: A record of large meteorite impacts and their influence on early crustal and biological evolution: Astrobiology, v. 3, p. 7–48, doi: 10.1089/153110703321632408.

Maas, R., Kinny, P.D., Williams, I.S., Froude, D.O., and Compston, W., 1992, The Earth's oldest known crust: A geochronological and geochemical study of 3900–4200 Ma old zircons from Mt. Narryer and Jack Hills, Western Australia: Geochimica et Cosmochimica Acta, v. 56, p. 1281–1300, doi: 10.1016/0016-7037(92)90062-N.

Maher, K.A., and Stevenson, D.J., 1988, Impact frustration of the origin of life: Nature, v. 331, p. 612–614, doi: 10.1038/331612a0.

Matsui, T., and Abe, Y., 1986, Evolution of an impact-induced atmosphere and magma ocean on the accreting Earth: Nature, v. 319, p. 303–305, doi: 10.1038/319303a0.

McCulloch, M.T., and Bennett, V.C., 1993, Evolution of the early Earth: Constraints from ^{143}Nd-^{142}Nd isotopic systematics: Lithos, v. 30, p. 237–255, doi: 10.1016/0024-4937(93)90038-E.

Melosh, H.J., 1989, Impact Cratering: A Geologic Process: New York, Oxford University Press, 245 p.

Mojzsis, S.J., Arrhenius, G., McKeegan, K.D., Harrison, T.M., Nutman, A.P., and Friend, C.R.L., 1996, Evidence for life on Earth before 3800 million years ago: Nature, v. 384, p. 55–59, doi: 10.1038/384055a0.

Montanari, A., and Koeberl, C., 2000, Impact Stratigraphy: The Italian Record: Lecture Notes in Earth Sciences 93: Heidelberg, Springer-Verlag, 364 p.

Moorbath, S., 2005, Oldest rocks, earliest life, and the Hadean-Archaean transition: Applied Geochemistry, v. 20, p. 819–824, doi: 10.1016/j.apgeochem.2005.01.001.

Moorbath, S., and Kamber, B.S., 1998, A reassessment of the timing of early Archaean crustal evolution in West Greenland: Geology of Greenland Survey Bulletin, v. 180, p. 88–93.

Moorbath, S., Whitehouse, M.J., and Kamber, B.S., 1997, Extreme Nd-isotope heterogeneity in the early Archaean—Fact or fiction? Case histories from northern Canada and West Greenland: Chemical Geology, v. 135, p. 213–231, doi: 10.1016/S0009-2541(96)00117-9.

Morbidelli, A., Petit, J.-M., Gladman, B., and Chambers, J., 2001, A plausible cause of the late heavy bombardment: Meteoritics & Planetary Science, v. 36, p. 371–380.

Morgan, J.W., 1985, Osmium isotope constraints on Earth's late accretionary history: Nature, v. 317, p. 703–705, doi: 10.1038/317703a0.

Morgan, J.W., 1986, Ultramafic xenoliths: Clues to the Earth's late accretionary history: Journal of Geophysical Research, v. 91, p. 12,375–12,387.

Myers, J.S., and Crowley, J.L., 2000, Vestiges of life in the oldest Greenland rocks? A review of early Archean geology in the Godthåbsfjord region, and reappraisal of life evidence for >3850 Ma life on Akilia: Precambrian Research, v. 103, p. 101–124, doi: 10.1016/S0301-9268(00)00089-9.

Norman, M.D., Bennett, V.C., and Ryder, G., 2002, Targeting the impactors: Siderophile element signatures of lunar impact melts from Serenitatis: Earth and Planetary Science Letters, v. 202, p. 217–228, doi: 10.1016/S0012-821X(02)00780-X.

Nutman, A.P., Mojzsis, S.J., and Friend, C.R.L., 1997, Recognition of >3850 Ma water-lain sediments in West Greenland and their significance for the early Archean Earth: Geochimica et Cosmochimica Acta, v. 61, p. 2475–2484, doi: 10.1016/S0016-7037(97)00097-5.

Oberbeck, V., and Fogleman, G., 1989, Impacts and the origin of life: Nature, v. 339, p. 434, doi: 10.1038/339434b0.

Peck, W.H., Valley, J.W., Wilde, S.A., and Graham, C.M., 2001, Oxygen isotope ratios and rare earth elements in 3.3 to 4.4 Ga zircons: Ion microprobe evidence for high δ^{16}O continental crust and oceans in the early Archean: Geochimica et Cosmochimica Acta, v. 65, p. 4215–4229, doi: 10.1016/S0016-7037(01)00711-6.

Rasmussen, B., and Koeberl, C., 2004, Iridium anomalies and shocked quartz in a late Archean spherule layer from the Pilbara craton: New evidence for a major asteroid impact at 2.63 Ga: Geology, v. 32, p. 1029–1032, doi: 10.1130/G20825.1.

Reimold, W.U., and Gibson, R.L., 1996, Geology and evolution of the Vredefort impact structure, South Africa: Journal of African Earth Sciences, v. 23, p. 125–162, doi: 10.1016/S0899-5362(96)00059-0.

Reimold, W.U., and Gibson, R.L., 2005, "Pseudotachylites" in large impact structures, *in* Koeberl, C., and Henkel, H., eds., Impact Tectonics: Impact Studies vol. 8: Heidelberg, Springer, p. 1–53.

Reimold, W.U., Koeberl, C., Johnson, J., and McDonald, I., 2000, Early Archean spherule beds in the Barberton Mountain Land, South Africa: Impact or terrestrial origin?, *in* Gilmour, I., and Koeberl, C., eds., Impacts and the early Earth: Lecture Notes in Earth Sciences 91: Heidelberg, Springer-Verlag, p. 117–180.

Rosing, M.T., 1999, ^{13}C-depleted carbon microparticles in >3700-Ma sea-floor sedimentary rocks from western Greenland: Science, v. 283, p. 674–676, doi: 10.1126/science.283.5402.674.

Rosing, M.T., Rose, N.M., Bridgwater, D., and Thomsen, H.S., 1996, Earliest part of Earth's stratigraphic record: A reappraisal of the >3.7 Ga Isua (Greenland) supracrustal sequence: Geology, v. 24, p. 43–46, doi: 10.1130/0091-7613(1996)024<0043:EPOESS>2.3.CO;2.

Russell, M.J., and Arndt, N.T., 2005, Geodynamic and metabolic cycles in the Hadean: Biogeosciences, v. 2, p. 97–111.

Ryder, G., 1990, Lunar samples, lunar accretion, and the early bombardment history of the Moon: Eos (Transactions, American Geophysical Union), v. 71, p. 313–323.

Ryder, G., 2001, Mass flux during the ancient lunar bombardment: The cataclysm: Lunar and Planetary Science, v. 32, abs. 1326 (CD-ROM).

Ryder, G., 2002, Mass flux in the ancient Earth-Moon system and benign implications for the origin of life on Earth: Journal of Geophysical Research, v. 107, p. 6-1–6-14, doi: 10.1029/2001JE001583,

Ryder, G., Koeberl, C., and Mojzsis, S.J., 2000, Heavy bombardment on the Earth ~3.85 Ga: The search for petrographic and geochemical evidence, *in* Canup, R.M., and Righter, K., eds., Origin of the Earth and Moon: Tucson, University of Arizona Press, p. 475–492.

Schidlowski, M., 1988, A 3,800-million-year isotopic record of life from carbon in sedimentary rocks: Nature, v. 333, p. 313–318, doi: 10.1038/333313a0.

Schmitz, B., Haggstrom, T., and Tassinari, M., 2003, Sediment-dispersed extraterrestrial chromite traces a major asteroid disruption event: Science, v. 300, p. 961–964, doi: 10.1126/science.1082182.

Schoenberg, R., Kamber, B.S., Collerson, K.D., and Moorbath, S., 2002, Tungsten isotope evidence from ~3.8-Gyr metamorphosed sediments for early meteorite bombardment of the Earth: Nature, v. 418, p. 403–405, doi: 10.1038/nature00923.

Shock, E.L., Amend, J.P., and Zolotov, M.Y., 2000, The early Earth vs. the origin of life, *in* Canup, R.M., and Righter, K., eds., Origin of the Earth and Moon: Tucson, University of Arizona Press, p. 527–543.

Shoemaker, E.M., Wolfe, R.F., and Shoemaker, C.S., 1990, Asteroid and comet flux in the neighborhood of Earth, *in* Sharpton, V.L., and Ward, P.D., eds., Global Catastrophes in Earth History: Geological Society of America Special Paper 247, p. 155–170.

Shukolyukov, A., and Lugmair, G.W., 1998, Isotopic evidence for the Cretaceous-Tertiary impactor and its type: Science, v. 282, p. 927–929, doi: 10.1126/science.282.5390.927.

Shukolyukov, A., Kyte, F.T., Lugmair, G.W., Lowe, D.R., and Byerly, G.R., 2000, The oldest impact deposits on earth—First confirmation of an extraterrestrial component, *in* Gilmour, I., and Koeberl, C., eds., Impacts and the Early Earth: Lecture Notes in Earth Sciences 91: Heidelberg, Springer-Verlag, p. 99–115.

Simonson, B.M., 1992, Geological evidence for a strewn field of impact spherules in the early Precambrian Hamersley Basin of Western Australia: Geological Society of America Bulletin, v. 104, p. 829–839, doi: 10.1130/0016-7606(1992)104<0829:GEFASF>2.3.CO;2.

Simonson, B.M., 2003, Petrographic criteria for recognizing certain types of impact spherules in well-preserved Precambrian successions: Astrobiology, v. 3, p. 49–65, doi: 10.1089/153110703321632417.

Simonson, B.M., and Glass, B.P., 2004, Spherule layers—Records of ancient impacts: Annual Review of Earth and Planetary Sciences, v. 32, p. 329–361, doi: 10.1146/annurev.earth.32.101802.120458.

Simonson, B.M., and Harnik, P., 2000, Have distal impact ejecta changed through geologic time?: Geology, v. 28, p. 975–978, doi: 10.1130/0091-7613(2000)028<0975:HDIECT>2.3.CO;2.

Simonson, B.M., and Hassler, S.W., 1997, Revised correlations in the early Precambrian Hamersley Basin based on a horizon of resedimented impact spherules: Australian Journal of Earth Sciences, v. 44, p. 37–48.

Simonson, B.M., Davies, D., Wallace, M., Reeves, S., and Hassler, S.W., 1998, Iridium anomaly but no shocked quartz from late Archean microkrystite layer: Oceanic impact ejecta?: Geology, v. 26, p. 195–198, doi: 10.1130/0091-7613(1998)026<0195:IABNSQ>2.3.CO;2.

Simonson, B.M., Hassler, S.W., and Beukes, N.J., 1999, Late Archean impact spherule layer in South Africa that may correlate with a Western Australian layer, *in* Dressler, B.O., and Sharpton, V.L., eds., Large Meteorite Impacts and Planetary Evolution II: Geological Society of America Special Paper 339, p. 249–261.

Simonson, B.M., Davies, D., and Hassler, S.W., 2000a, Discovery of a layer of probable impact melt spherules in the late Archean Jeerinah Formation, Fortescue Group, Western Australia: Australian Journal of Earth Sciences, v. 47, p. 315–325, doi: 10.1046/j.1440-0952.2000.00784.x.

Simonson, B.M., Koeberl, C., McDonald, I., and Reimold, W.U., 2000b, Geochemical evidence for an impact origin for a late Archean spherule layer, Transvaal Supergroup, South Africa: Geology, v. 28, p. 1103–1106, doi: 10.1130/0091-7613(2000)028<1103:GEFAIO>2.3.CO;2.

Simonson, B.M., Byerly, G.R., and Lowe, D.R., 2004, The early Precambrian stratigraphic record of large extraterrestrial impacts, *in* Eriksson, P.G., Altermann, W., Catuneanu, O., Mueller, W.U., and Nelson, D.R., eds., The Precambrian Earth: Tempos and Events: Amsterdam, Elsevier, p. 27–45.

Simpson, E.L., Kuklis, K.A., Eriksson, P.G., Bumby, A.J., and van Jaarsveld, C.F., 2004, Description and interpretation of 1.8 Ga playa deposits, Makgabeng Formation, Waterberg group, South Africa: Sedimentary Geology, v. 163, p. 279–292, doi: 10.1016/j.sedgeo.2003.07.002.

Sleep, N.H., Zahnle, K.J., Kasting, J.F., and Morowitz, H.J., 1989, Annihilation of ecosystems by large asteroid impacts on the early Earth: Nature, v. 342, p. 139–142, doi: 10.1038/342139a0.

Sleep, N.H., Zahnle, K.J., and Neuhoff, P.S., 2001, Initiation of clement surface conditions on the earliest Earth: Proceedings of the National Academy of Sciences of the United States of America, v. 98, p. 3666–3672, doi: 10.1073/pnas.071045698.

Smith, J.V., 1981, The first 800 million years of Earth's history: Philosophical Transactions of the Royal Society, London, v. A301, p. 401–422.

Stöffler, D., and Langenhorst, F., 1994, Shock metamorphism of quartz in nature and experiment: I: Basic observation and theory: Meteoritics, v. 29, p. 155–181.

Stöffler, D., and Ryder, G., 2001, Stratigraphy and isotope ages of lunar geologic units: Chronological standard for the inner solar system: Space Science Reviews, v. 96, p. 7–53, doi: 10.1023/A:1017343517892.

Taylor, S.R., 1982, Planetary Science: A Lunar Perspective: Houston, Lunar and Planetary Institute, 481 p.

Taylor, S.R., 1989, Growth of planetary crusts: Tectonophysics, v. 161, p. 147–156, doi: 10.1016/0040-1951(89)90151-0.

Taylor, S.R., 1992a, The origin of the Earth, *in* Brown G., Hawkesworth C., and Wilson C., eds., Understanding the Earth: Cambridge, UK, Cambridge University Press, p. 25–43.

Taylor, S.R., 1992b, Solar System Evolution: A New Perspective: Cambridge, UK, Cambridge University Press, 307 p.

Taylor, S.R., 1993, Early accretion history of the Earth and the Moon-forming event: Lithos, v. 30, p. 207–221, doi: 10.1016/0024-4937(93)90036-C.

Taylor, S.R., and McLenna, S.M., 1995, The geochemical evolution of the continental crust: Reviews of Geophysics, v. 33, p. 241–265.

Tera, F., Papanastassiou, D.A., and Wasserburg, G.J., 1974, Isotopic evidence for a terminal lunar cataclysm: Earth and Planetary Science Letters, v. 22, p. 1–21, doi: 10.1016/0012-821X(74)90059-4.

Valley, J.W., Peck, W.H., King, E.M., and Wilde, S.A., 2002, A cool early Earth: Geology, v. 30, p. 351–354, doi: 10.1130/0091-7613(2002)030<0351:ACEE>2.0.CO;2.

van Zuilen, M.A., Lepland, A., and Arrhenius, G., 2002, Reassessing the evidence for the earliest traces of life: Nature, v. 418, p. 627–630, doi: 10.1038/nature00934.

Walter, M.J., and Trønnes, R.G., 2004, Early Earth differentiation: Earth and Planetary Science Letters, v. 225, p. 253–269, doi: 10.1016/j.epsl.2004.07.008.

Watson, E.B., and Harrison, T.M., 2005, Zircon thermometer reveals minimum melting conditions on earliest Earth: Science, v. 308, p. 841–844, doi: 10.1126/science.1110873.

Wetherill, G.W., 1975, Late heavy bombardment of the Moon and terrestrial planets: Proceedings of the 6th Lunar Science Conference, p. 1539–1561.

Wetherill, G.W., 1994, Provenance of the terrestrial planets: Geochimica et Cosmochimica Acta, v. 58, p. 4513–4520, doi: 10.1016/0016-7037(94)90352-2.

Wilde, S.A., Valley, J.W., Peck, W.H., and Graham, C.M., 2001, Evidence from detrital zircons for the existence of continental crust and oceans on the Earth 4.4 Gyr ago: Nature, v. 409, p. 175–181, doi: 10.1038/35051550.

Wilhelms, D.E., 1987, The geologic history of the Moon: U.S. Geological Survey Professional Paper 1348, 302 p.

Zahnle, K.J., and Sleep, N.H., 1997, Impacts and the early evolution of life, *in* Thomas, P.J., Chyba, C.F., and McKay, C.P., eds., Comets and the Origin and Evolution of Life: New York, Springer-Verlag, p. 175–208.

Zappalà, V., Cellino, A., Gladman, B.J., Manley, S., and Migliorini, F., 1998, Asteroid showers on Earth after family breakup events: Icarus, v. 134, p. 176–179, doi: 10.1006/icar.1998.5946.

Zhang, Y., 2002, The age and accretion of the Earth: Earth-Science Reviews, v. 59, p. 235–263, doi: 10.1016/S0012-8252(02)00077-6.

Manuscript Accepted by the Society 4 October 2005

Printed in the USA

Geological Society of America
Special Paper 405
2006

Large-scale impacts and the evolution of the Earth's crust: The early years

Richard A.F. Grieve
Earth Sciences Sector, Natural Resources Canada, Ottawa, Ontario K1A 0E4, Canada
Mark J. Cintala
NASA Johnson Space Center, Houston, Texas 77055, USA
Ann M. Therriault
Earth Sciences Sector, Natural Resources Canada, Ottawa, Ontario K1A 0E4, Canada

ABSTRACT

Modeling the effects of differential scaling of impact melt and transient cavity volumes indicates that impact melt volumes exceed transient cavity volumes at transient cavity diameters greater than ~500 km on the Earth. This condition is not realized on the Moon until transient cavity diameters are greater than ~3000 km. A reasoned case is made that because of this "differential scaling," the large impact "basins" comparable in size to such lunar basins as Orientale, which must have been formed on the Hadean Earth, did not have an Orientale-like form. While their exact form is unknown, they were likely shallow structures, and they would have been characterized by voluminous central impact pools. These melt pools with their closed-system environment would likely differentiate, leading to the crystallization of more felsic rocks. This reprocessing of the Hadean crust by large-scale impacts provides a mechanism to produce pre–3.9 Ga zircons, without calling for plate-tectonic–related or other mechanisms of crustal recycling to produce felsic rocks. While the impact melting at a single one of such Hadean impact "basins" would be impressive, the cumulative effects would be potentially staggering, particularly if impact velocities during Hadean time were, as believed, lower than current velocities. Based on the number of large multi-ring basins on the Moon scaled to terrestrial conditions, cumulative melt production on the Hadean Earth by such basins alone would reach ~10^{11}–10^{12} km^3. With the assumption that the impact melt pools were basaltic in composition, modeling with the crystal-melt fractionation software MELTS suggests the cumulative volume of felsic rocks potentially produced through the evolution of such impact melt pools could be significant.

Keywords: impact melting, multi-ring basin, differentiation, zircons, continental crust.

INTRODUCTION

It is a widely held view that the terrestrial planets were formed in an equivalent manner through the rapid accretion of planetismals, with the conversion of much of their kinetic energy to heat resulting in melting and the formation of magma oceans. In the case of the Earth, accretion is believed to have been completed ca. 30 Ma after the condensation of solids in the solar nebula (e.g., Kleine et al., 2002). Following accretion and separation of the initial crust and mantle through differentiation,

Grieve, R.A.F., Cintala, M.J., and Therriault, A.M., 2006, Large-scale impacts and the evolution of the Earth's crust: The early years, *in* Reimold, W.U., and Gibson, R.L., Processes on the Early Earth: Geological Society of America Special Paper 405, p. 23–31, doi: 10.1130/2006.2405(02). For permission to copy, contact editing@geosociety.org.

the evolutionary paths of the terrestrial planets diverged depending on their thermal regime, size, gravitation acceleration, and the presence or not of water. It is also widely held that there is another commonality in the early evolution of the terrestrial planets, namely that following the formation of their initial crusts, they were subjected to continued intense impact at a variety of scales. There is, however, continued debate based on interpretations of the lunar impact record as to whether this flux of impacting bodies rapidly declined from 4.4 to 3.5 Ga (Hartmann et al., 2000; Neukum and Ivanov, 1994) or was punctuated by a period of intense bombardment at ca. 3.9 Ga (e.g., Tera et al., 1974; Kring and Cohen, 2002; Ryder, 2002; Koeberl, 2003).

No direct evidence of this period in Earth's evolutionary history, sometimes referred to as the Hadean, is preserved beyond some zircons with U-Pb ages greater than 3.9 Ga (e.g., Amelin et al., 2000). There are also some Lu-Hf isotopic data suggesting the formation of some felsic crustal rocks by ca. 4.3 Ga (Bizzarro et al., 2003), with some evidence also for possibly an earlier mafic protocrust (e.g., Chase and Patchett, 1988). Similarly, Sm-Nd isotopic data result in model ages compatible with large-scale, initial differentiation of the Earth's mantle relatively soon after planetary formation (Harper and Jacobsen, 1992). They also suggest that the protocrust may have been isolated from its mantle source and that crustal recycling in the early Hadean was relatively inefficient (Caro et al., 2003).

If reliance on relatively limited isotopic data to address issues of the evolution of the earliest crust of the Earth can be regarded as having a degree of uncertainty, then the effects of impact on the Hadean Earth must be considered highly speculative at best. Most discussions consider the evidence from the Moon as a proxy for the contemporaneous bombardment of Earth after scaling upward the numbers of impacts that result from the greater gravitational cross section of the Earth and for the higher impact velocity, due to the effect of the higher escape velocity of the Earth. Reasoned speculations as to the effects of such a bombardment on the Hadean Earth have included the frustration of the development and evolution of surface life (e.g., Maher and Stevenson, 1988; Chyba, 1993), the erosion of the atmosphere (e.g., Melosh and Vickery, 1989), and the evaporation of portions, if not all, of the hydrosphere (e.g., Sleep et al., 1989; Zahnle and Sleep, 1997). Models of the effects on terrestrial crustal evolution have generally made analogies with the effects of the formation of the large, multi-ring basins on the Moon (Fig. 1), albeit allowing for a thinner lithosphere and the presence of water on the Earth (e.g., Green, 1972; Frey, 1980; Grieve, 1980).

In this contribution, we present the case that the effects of very large impacts on the crusts of Hadean Earth and the Moon were not equivalent. This arises from the fact that in an impact event, the size of the transient cavity produced directly by the cratering flow field is a function of projectile size, impact velocity, the physical properties of the projectile and target, and planetary gravity. Planetary gravity acts to limit transient cavity growth and is thus relatively more effective in inhibiting growth on planetary bodies with higher gravity and in larger impacts (longer event times) on a given planet. The absolute rate of shock-wave attenuation and thus the distance from the point of impact at which the target rocks experience a specific shock pressure is a function of the same variables, with the exception that it is not a function of planetary gravity. The concept of this differential scaling with respect to the relative sizes and geometries of the transient cavity and the portion of the target that is melted is illustrated schematically in Figure 2. Here, we summarize the consequences of this differential scaling for the relative amount of impact melting and the nature of large terrestrial impact "basins" in the Hadean, compared to the lunar case, and offer some reasoned speculations on the evolution of such "basins" and their potential role in early crustal evolution on Earth.

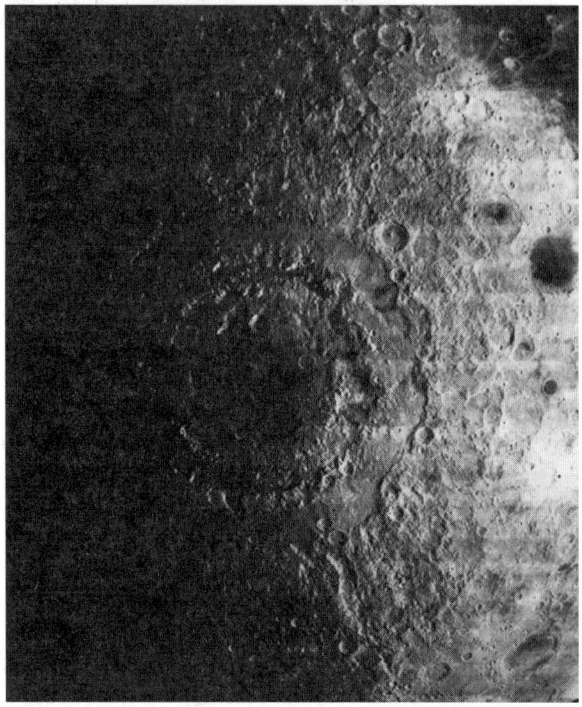

Figure 1. Multi-ring basin Orientale, which has a diameter of ~930 km as defined by the outer ring, the Cordillera mountains. Orientale is the youngest and best-preserved multi-ring basin on the Moon (Lunar Orbiter photograph).

DIFFERENCES BETWEEN THE EARTH AND THE MOON

In determining the first principles of the differences between very large impact events on the Earth and Moon, we have used a previously described thermodynamic model to estimate impact melt volumes (Grieve and Cintala, 1992, 1997; Cintala and Grieve, 1994, 1998). In this model, phase changes are determined by calculating the change in entropy, as a function of pressure, temperature, and density on the Hugoniots of

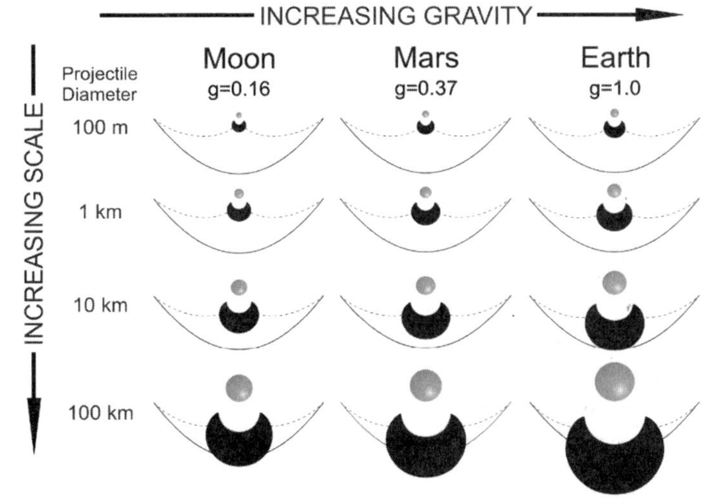

Figure 2. Schematic representation, scaled to transient cavity size, of the concept of differential scaling. Indicated are the relative sizes and geometries of impactor (gray), melt (black), transient cavity (solid line), and ejected volume (between dashed line and the ground surface), with increasing gravity (g) relative to the Earth (left to right) and/or increasing event size (top to bottom), all other impact parameters remaining fixed.

the projectile and target materials. The calculated impact melt and vapor volumes and decay of shock stress in the model are similar to those of more complex, finite-difference models (e.g., O'Keefe and Ahrens, 1977). The calculated melt volumes also compare favorably with empirical estimates of the volumes of impact melt rocks over seven orders of magnitude at terrestrial impact structures (Grieve and Cintala, 1992) and with more recent hydrocode calculations (Pierazzo et al., 1997).

Various materials can be used for the projectile and target through use of their modified Murnaghan equations of state (e.g., Duvall, 1958). To exemplify the first principles and "average" behavior, we have confined the model calculations to chondritic projectiles, which were approximated by the equation of state of basalt with a chondritic density of 3.58 g cm^{-3} (Grieve and Cintala, 1992). Other types of impacting bodies can be considered but for specific impact conditions, melt volumes produced by "chondritic" bodies are intermediate to those produced by ice and iron bodies and can be considered to represent an "average" volume (Cintala, 1992; Grieve and Cintala, 1992).

The target materials were assumed to be anorthosite for the Moon and basalt for the Earth, and all impacts were assumed to be vertical. In reality, the impact angle would vary and average 45°. Hydrocode modeling indicates that vertical impacts are the most efficient in producing impact melt, with the volume of impact melt decreasing by somewhat less than 20% at impact angles of 45° (Pierazzo and Melosh, 2000). While impact angle and impacting body composition are variables in determining melt production, their net average effect is small in comparison to melt production variations due to impact velocity variations and do not affect the basic conclusions and first-order principles outlined here. The model formulations and the values of the specific thermodynamic variables used have been given previously (Cintala, 1992; Grieve and Cintala, 1992; Cintala and Grieve, 1994) and are not repeated.

Transient cavity dimensions resulting from the impact conditions that yield specific melt volumes were modeled with the scaling relationships of Schmidt and Housen (1987). For the purposes of a standard geometric comparison and discussion, it was assumed that the transient cavity at its maximum growth in all directions defines an idealized paraboloid of revolution, with a depth-to-diameter ratio of 1:3, regardless of event size (Melosh, 1989; Turtle et al., 2005). Transient cavity diameters were scaled to final crater diameters, using the empirical "modification-scaling" relationship of Croft (1985). As this empirical scaling relationship was developed from observations of lunar impact structures with the diameter range from 30 km to 300 km, it likely degrades on extrapolation to very large impact structures such as multi-ring basins. There is also some debate as to what constitutes the rim-crest diameter and the relative location of the transient cavity at multi-ring basins (e.g., Spudis, 1993; Turtle et al., 2005). Nevertheless, for the purposes of comparison and facilitating visualization with respect to the sizes and appearance of known lunar multi-ring basins, the modification-scaling relationship is assumed to be valid, regardless of event size.

The exact details of the geometry of the volume in the target from which impact melts are initially generated are not known. The conditions imposed by the model, however, result in nearly spherical geometries, which are similar to those in other works (e.g., Croft, 1982; Melosh, 1989). The relative fraction of melt ejected from the transient cavity was not explicitly calculated for specific impact events. As gravity acts against the ejection process, the relative fraction of melt ejected decreases with higher planetary gravity and with increasing event size, i.e., longer times (Fig. 2; Melosh, 1989). While these details may affect the exact values of particular parameters, they do not negate the first-order principles of the relative changes in volumes and comparative initial spatial geometries of the impact melt and transient cavity, with the volume of melt increasing as approximately the fourth power of the transient cavity diameter and occupying an ever-increasing fraction of the transient cavity volume, as event size increases (Fig. 2).

Figure 3 illustrates the modeled changes in relative melt and transient cavity volumes for large impact events, with chondritic impactors, under lunar and terrestrial gravity at an impact velocity of 20 km s^{-1}. In the case of the Earth (target-basalt), the impact melt volume exceeds that of the transient cavity at transient cavity diameters >500 km (Fig. 3). This condition is not reached on the Moon (target-anorthosite) until transient cavity diameters are >3000 km (Fig. 3), which is essentially equivalent to whole-Moon melting (radius 1738 km). The melt volumes in these very large impact events are minimum estimates for the specified model impact conditions, as at these

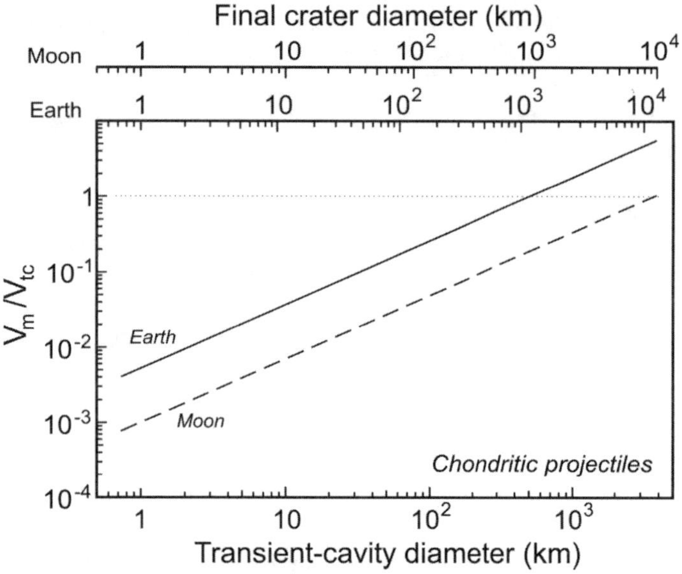

Figure 3. Logarithmic plot of the variation in the ratio of the volume of melt (V_m) to the volume of transient cavity (V_{tc}) with transient cavity (lower horizontal scale) and/or final crater (upper horizontal scale) diameter for the Earth (target-basalt) and the Moon (target-anorthosite), for chondritic impactors at an impact velocity of 20 km s^{-1}. See text for details and implications.

sizes the increase in temperature of the target rocks with depth, due to the geotherm, will increase overall melt production (Ivanov and Melosh, 2003).

Also shown in Figure 3 are the equivalent final crater diameters for structures on the Earth and Moon, using the modification-scaling relationship of Croft (1985). The canonical lunar multi-ring basin is Orientale, with a generally quoted rim diameter of ~930 km (Fig. 1; Spudis, 1993). A potentially equivalent-sized "basin" on the Earth would correspond to the situation in which the transient cavity and impact melt volumes were approximately equivalent (Fig. 3). Given that such a transient cavity on Earth is largely within material with no strength (i.e., melt; Fig. 2), it is not likely that Croft's modification-scaling relationship is applicable, as noted earlier. Nevertheless, it is clear that an impact event of such a size on the Hadean Earth would not result in an impact basin with the form of Orientale (Fig. 1) even with the appropriately scaled reduction in topography due to the higher terrestrial gravity. If the early terrestrial lithosphere was relatively thin, tectonic ring theory (McKinnon and Melosh, 1980; Melosh, 1989) could favor Valhalla-like structures as observed on Callisto, with multiple external graben features as the final form. This, however, is speculative. While the detailed final form of impact structures at such scales on the Hadean Earth may be unknown, an essential feature of such structures would have been a very voluminous central melt pool.

While impacts sufficient to produce an Orientale-size, multi-ring basin on the Moon did not produce similar multi-ring basin morphologies on the Earth, this is not equivalent to stating that no lunarlike, multi-ring basins were formed on the Hadean Earth. Smaller impact events are likely to have produced the more classic multi-ring basin form. For example, the three largest known terrestrial impact structures, Chicxulub, Sudbury, and Vredefort, with estimated original diameters in the 200–300 km range, have been ascribed some attributes of multi-ring basins (e.g., Morgan and Warner, 1999; Stöffler et al., 1994; Therriault et al., 1997). It is not clear, however, how their various attributed annuli correspond, in detail, to the obvious topographic rings observed in lunar multi-ring basins (Grieve and Therriault, 2000). Lunarlike, multi-ring basin forms (Alexopolous and McKinnon, 1994) for the four largest impact structures on Venus (i.e., Klenova, Meitner, Mead, and Isabella; ~150–270 km in diameter) are probably the best evidence for the case of lunarlike, multi-ring basins on the Hadean Earth, given the similar planetary gravities. At these size ranges, the amount of impact melting within the transient cavity approaches or exceeds the depth of the cavity but is only on the order of 30%–40% of the volume of the transient cavity (Fig. 3).

While no real case can be made for impact-induced volcanism by decompression melting in the Archean and younger Earth (e.g., Ivanov and Melosh, 2003), this may not be the case for the Hadean Earth. In addition to direct impact melting, recent model calculations indicate that subsolidus shock heating from post-accretion impacts could raise the temperature of the lithosphere by 200–400 °C, relative to the geotherm (Ivanov, 2004). Depending on the lithospheric thickness, there may be a contribution from shock heating to the underlying asthenosphere sufficient to overcome the latent heat of fusion. It would be possible to evaluate the extent of this contribution to impact-related melting if the thermal structure of the Hadean Earth and, more importantly, the post-impact, geometric configuration of the subsurface isotherms beneath such Hadean impact "basins" were known, as they determine the post-impact pressure and temperature (impact plus geotherm) relations (Ivanov and Melosh, 2003). As the Hadean "basins" are unlike known large, terrestrial complex impact structures or the traditional multi-ring basins of the Moon, this post-impact subsurface geometry is unknown. Accordingly, and as we are illustrating first-order principles, no attempt has been made to model this potential effect, and it is offered only as a possibility for the production of additional melted material, as a result of these very large impacts on the Hadean Earth.

Cumulative Effects

We have assumed that the earliest crust of the Earth was a partial melt of the mantle, i.e., basaltic in composition. Therefore, the composition of the massive impact melt pools, which were a principal feature of these large Hadean impact "basins," would have ranged from basaltic to more mafic compositions for the largest events, where the melt volume could have reached into the mantle. The depth of the melt pools within these "basins" is unknown, as the final post-modification configuration and form of the "basins" is unknown. The depths would have been variable and in the range of many kilometers to many tens of kilometers. What seems clear is that the surface and crust of the Hadean Earth would have had extensive and voluminous impact-produced melt pools of mafic composition (Fig. 4). Given the appropriate cooling times, bodies of basaltic melt >300 m thick differentiate in the terrestrial environment, with the potential degree of differentiation being a function of the thickness of the body (Jaupart and Tait, 1995). It is therefore expected that these thick, closed-system melt pools would have differentiated into an ultramafic-mafic base and a more felsic top, much in the same manner as the ~2.5-km-thick impact melt sheet, now manifested as the Sudbury Igneous Complex, at the originally ~200–250 km diameter Sudbury impact structure, Canada (Therriault et al., 2002). The results of individual impacts on the Hadean Earth would have been impressive. For example, a terrestrial impact event with the magnitude of the one that resulted in the Orientale basin on the Moon would have generated in excess of 10^7 km^3 of impact melt. If only 10% of the initial melt volume took the form of felsic differentiates, they would have been comparable in volume to the Columbia River basalts (Swanson et al., 1989).

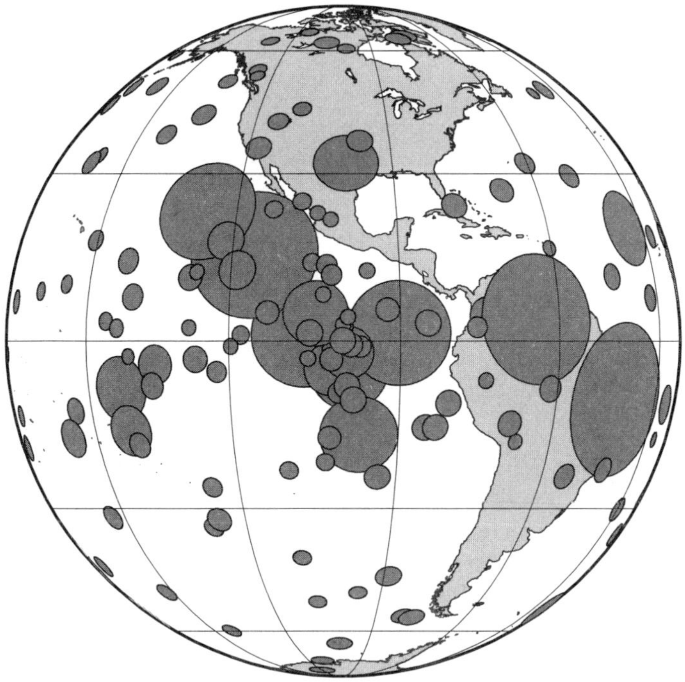

Figure 4. Cartoon of the number of impact melt pools (approximated by transient cavity diameters), in which melt volumes exceed transient cavity volumes, on the Hadean Earth for the minimizing condition of an impact velocity of 22.4 km s^{-1} (Table 2). The number of potential terrestrial impact basins (melt pools) was based on the lunar record of multi-ring basins formed at ca. 4.5–3.8 Ga (see text for details of the transformation to terrestrial conditions). Locations of melt pools were generated by random numbers, and current land mass is shown for comparison.

In order to derive some sense of the cumulative effects of these impacts on Hadean crustal evolution, we used the number of well-known impact basins on the Moon as a starting point. Spudis (1993) provided an inventory of multi-ring basins on the Moon, and the 10 identified basins with diameters (main topographic ring) >700 km were used to calculate the numbers of similar-sized impacts on the Hadean Earth. The relative numbers of impacts on the Earth and Moon are a function of their relative gravitational cross sections, which are, in turn, a function of impact velocity. The result of the calculation of the relative gravitational cross sections for impact velocities of 7.5, 10, and 15 km s^{-1} on the Moon and the equivalent terrestrial impact velocities are given in Table 1. At the higher impact velocities, which are similar to today's average asteroidal impact velocities, the Earth would receive ~25 impacts of comparable magnitude for every lunar impact. It is generally held, however, that impact velocities in the solar system have increased with time (Wetherill, 1975), due to the action of gravitational forces. The lower impact velocities and considerably larger ratios for the gravitational cross sections are likely more appropriate for Hadean times on Earth.

To calculate the appropriate cumulative amount of impact melt produced by the scaled numbers of terrestrial impacts, it was necessary to estimate the size distribution of impacting bodies for the various specified impact velocities as follows. The transient cavity diameter for each of the 10 lunar basins was estimated from the "modification-scaling" relationship of Croft (1985). The scaling relationships of Schmidt and Housen (1987) were then applied to determine the diameter of the assumed spherical chondritic impactor that made each of those transient cavities, for each assumed lunar impact velocity (Table 1). The intrinsic approach velocity, v_∞, corresponding to each impact velocity, was then extracted by taking into account the lunar escape velocity and the escape velocity of the Earth (at the current lunar distance).

As each assumed impact velocity generates a different size distribution of impactors, a least-squares fit was applied to each set of 10 lunar impactors. Those fits were then used in calculating the impacting population for the Earth for each value of the impact velocity corresponding to the different v_∞ and resulting

TABLE 1. RELATIVE GRAVITATIONAL CROSS SECTIONS AND TERRESTRIAL IMPACT VELOCITIES FOR SPECIFIC LUNAR IMPACT VELOCITIES

V_i, Impact velocity, Moon (km s^{-1})	V_i, Impact velocity, Earth (km s^{-1})	Gravitational cross section, Moon (km^2)	Gravitational cross section, Earth (km^2)	Ratio cross-sectional areas, Earth/Moon
7.5	14.9	1.34×10^7	1.29×10^9	96.14
10	17.4	1.09×10^7	5.41×10^8	49.76
15	22.4	9.91×10^6	2.54×10^8	25.76

different capture cross sections. For instance, 10 km s^{-1} lunar impacts correspond to terrestrial impacts at 17.4 km s^{-1}, because of the greater terrestrial mass, and the resulting terrestrial capture cross section would then be almost 50 times greater than that of the Moon. The least-squares fit to the cumulative number of impactors as a function of impactor diameter gives a diameter of 66.5 km for the body that, at 10 km s^{-1}, created the smallest lunar basin, Tsiolkovsky-Stark, on the list of Spudis (1993). As there were 10 lunar impactors of this size or larger, it is assumed from the ratio of cross-sectional areas that there were 498 terrestrial impactors of this size or larger. It is then a straightforward process to apply the least-squares fit incrementally to each impactor to calculate its size.

Recently, Petro and Pieters (2004) estimated the transient cavity diameters of 42 lunar basins, ranging in rim diameter from ~365 km to 1100 km, based on a mean trend derived from the geophysical constraints of Wieczorek and Phillips (1999). At basins the size of Tsiolkovsky-Stark (~700 km), the transient cavity diameter estimates of Petro and Pieters (2004) are similar to those used here, based on the "modification scaling" relationship of Croft (1985). At diameters equivalent to that of Imbrium (~1160 km), the transient cavity diameter estimates of Petro and Pieters (2004) are ~25% larger than those used here and would therefore require larger impactors for a given impact velocity. The size of the impactors used here, therefore, could be considered a set of minimum estimates.

Having the size distributions and impact velocities for the impactors, the sizes of the transient cavities and the accompanying volumes of impact melt can be calculated for the different terrestrial impacting populations. The cumulative melt volumes for those terrestrial impact events, in which the impact melt volumes equal or exceed the transient cavity volume, are given in Table 2 for the various impact velocities. Although higher impact velocities in a single event result in the production of relatively more impact melt, the net effect of the increase in the relative gravitational cross-section ratio with lower impact velocities (Table 1) results in a larger number of individual impact events and a considerably larger cumulative volume of impact melt having been produced on the Hadean Earth (Fig. 5; Table 2).

TABLE 2. CUMULATIVE IMPACT MELT VOLUMES FOR IMPACTS ON THE HADEAN EARTH, IN WHICH MELT VOLUMES EQUAL OR EXCEED TRANSIENT CAVITY VOLUMES

V_i (km s^{-1})	Cumulative impact melt volumes (km^3)
14.9	1.31×10^{12}
17.4	3.25×10^{11}
22.4	7.87×10^{10}

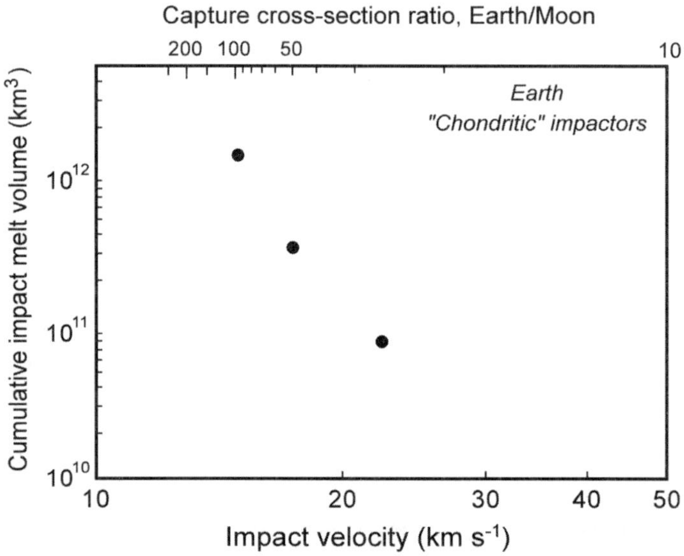

Figure 5. Logarithmic plot of cumulative volumes of impact melt produced on the Hadean Earth, in which melt volumes exceed transient cavity volumes, as a function of impact velocity (lower horizontal scale) and/or capture cross-section ratio (upper horizontal scale) of the Earth/Moon (Table 2).

POST-IMPACT EVOLUTION

It is a widely held view that impact melts are a compositionally homogenous mixture of the target lithologies that were shock-melted. While this is true for impacts that result in impact structures in the size range of tens of kilometers (e.g., Grieve et al., 1977; Koeberl and Reimold, 2003), it does not necessarily hold for very large impact structures. For example, small compositional variations in impact melt rocks are evident at some terrestrial impact structures 100 km in diameter (Grieve and Floran, 1978; Kettrup et al., 2003). Some of this compositional variation may be due to the assimilation of different populations of lithic and mineral clasts. In the very large impact events under consideration here, however, the absolute distances are so large that differential particle velocities within the melted volume were likely insufficient to result in complete homogenization. Accordingly, and in keeping with the spirit of dealing only with first-order principles, the post-impact evolution of the melt pools within these Hadean impact "basins" was modeled as if the melts had the composition of basalt.

The use of the crystal-melt fractional modeling procedures in the software MELTS (Ghiorso and Sack, 1995) indicates that if differentiation were 100% efficient, crystallizing basaltic liquid at a pressure of 0.1 GPa would produce a residual liquid of tonalitic composition at 1200–900 °C on the quartz-fayalite-magnetite (QFM-1) buffer, with a volume of ~20%–30% of that of the original basaltic liquid. For the purposes of illustration here, it is assumed that differentiation is only 10% efficient at physically separating crystallizing phases and residual liquid. For the end-member estimates of the cumulative volumes of impact melt produced in these large Hadean "basins" (Table 2), this corresponds to a potential volume of ~1.5×10^8 km^3 to 4×10^9 km^3 of tonalitic magma. This can be compared to the volume of the present-day continental crust of ~8×10^9 km^3. These estimates for the model conditions used here are best regarded as upper limits. In reality, the largest Hadean impact events would result in impact melts more mafic than basalt in terms of average initial composition, while others would overlap spatially (Fig. 4), and some of the Hadean crust would be impact-melted and subsequently differentiated more than once.

Whatever the case, it is apparent that the model considerations here raise the possibility of considerable volumes of felsic rock being generated in a near-surface environment, even at 1% differentiation efficiency of an initial basaltic composition for these Hadean impact melts. These felsic differentiates could provide a source for pre–3.9 Ga zircons, without necessitating the recycling of an initial basaltic crust. Impact melt systems are also closed systems and preserve the isotopic signatures of the preexisting target rocks in such isotopic systems as Nd-Sm and Re-Os, while resetting isotopic clocks such as U-Pb (e.g., Deutsch, 1994; Dickin et al., 1999; Morgan et al., 2002).

Neither the effects of these large-scale impacts nor the evolution of the Earth's crust by internally driven processes existed in isolation during the Hadean. The Earth differs from the other terrestrial planets in having both oceanic and continental crust with subduction zones and plate tectonics, providing the physical environment for the production of continental crust throughout most of Earth's post-Hadean history. The presence of water is essential in the production of continental crust by the subduction of hydrated oceanic crust (Stern, 2002) and appears critical in reducing the strength of the lithosphere to permit plate tectonics to operate in the first place (Solomatov, 2003). The onset and character of the earliest generation of continental crust through the regime of plate tectonics is speculative, but the partial melting of subducted or delaminated, hydrated, and metamorphosed oceanic crust beneath more buoyant crust is the common theme for the production of early tonalitic and trondhjemetic rocks (e.g., Foley et al., 2002, 2003; Rapp et al., 2003). While the crystallized impact melt pools with their upper felsic differentiates would not inherently have had a lower density, delamination of their mafic lower regions could have produced residual blocks of more buoyant, felsic rocks. If this occurred, one could speculate that such buoyant crustal blocks could also possibly have played a role in the onset and development of protocontinental crust, through some form of regime more akin to traditional plate tectonics.

CONCLUDING REMARKS

It has been suggested that the lack of preservation of pre–3.9 Ga rocks on Earth was due to destruction by impact bombardment (e.g., Nutman et al., 2001). While the experience of the preserved lithologies from the lunar highlands in the Apollo collection would suggest that this would be unlikely, the sense of the massive amount of impact melting in the Hadean suggested here calls for a decidedly nonlunar case for the Earth. If the impact bombardment was punctuated at ca. 3.9 Ga (Tera et al., 1974; Kring and Cohen, 2002; Ryder, 2002), it conjures the scenario of a period of massive crustal remelting or reprocessing at ca. 3.9 Ga on the Hadean Earth. The lack of preservation, however, of terrestrial rocks from this period, including any that can be attributed directly to impact [assuming they could be recognized as such (Grieve et al., 1991)], is a testament to the powerful endogenic geologic processes that have shaped subsequent terrestrial crustal evolution. While this work has focused on the effects of large impacts on the Hadean crust, these endogenic forces must have been operating in the Hadean, and what is presented here would be incremental to the effects of endogenic terrestrial geologic processes.

ACKNOWLEDGMENTS

We would like to acknowledge reviews of an earlier version of this work by H. J. Melosh and W. B. McKinnon and reviews of this updated and expanded version by A. Deutsch, C. Koeberl, and M. Pilkington. Receipt of a German Science Foundation Mercator Guest Professorship at the von Humboldt University, Berlin, to R.A.F.G. is gratefully acknowledged. Contribution from the Geological Survey of Canada 2004310.

REFERENCES CITED

Alexopolous, J.S., and McKinnon, W.B., 1994, Large impact craters and basins on Venus, with implications for ring mechanics on the terrestrial planets, *in* Dressler, B.O., Grieve, R.A.F., and Sharpton, V.L., eds., Large Meteorite Impacts and Planetary Evolution: Geological Society of America Special Paper 293, p. 29–51.

Amelin, Y., Lee, D.C., and Halliday, A.N., 2000, Early-middle Archaean crustal evolution deduced from Lu-Hf and U-Pb isotopic studies of single grain zircons: Geochimica et Cosmochimica Acta, v. 64, p. 4205–4225, doi: 10.1016/S0016-7037(00)00493-2.

Bizzarro, M., Baker, J.A., Haack, H., Ulfbeck, D., and Rosing, M., 2003, Early history of Earth's crust-mantle system inferred from hafnium isotopes in chondrites: Nature, v. 421, p. 931–933, doi: 10.1038/nature01421.

Caro, G., Bourdon, B., Birck, J.-L., and Moorbath, S., 2003, ^{146}Sm-^{142}Nd evidence from Isua metamorphosed sediments for early differentiation of the Earth's mantle: Nature, v. 423, p. 428–431, doi: 10.1038/nature01668.

Chase, C.G., and Patchett, P.J., 1988, Stored mafic/ultramafic crust and early Archean mantle depletion: Earth and Planetary Science Letters, v. 91, p. 66–72, doi: 10.1016/0012-821X(88)90151-3.

Chyba, C.F., 1993, The violent environment of the origin of life: Progress and uncertainties: Geochimica et Cosmochimica Acta, v. 57, p. 3351–3358, doi: 10.1016/0016-7037(93)90543-6.

Cintala, M.J., 1992, Impact-induced thermal effects in the lunar and mercurian regolith: Journal of Geophysical Research, v. 97, p. 947–974.

Cintala, M.J., and Grieve, R.A.F., 1994, The effects of differential scaling of impact melt and crater dimensions on lunar and terrestrial craters: Some brief examples, *in* Dressler, B.O., Grieve, R.A.F., and Sharpton, V.L., eds., Large Meteorite Impacts and Planetary Evolution: Geological Society of America Special Paper 293, p. 51–59.

Cintala, M.J., and Grieve, R.A.F., 1998, Scaling impact melting and crater dimensions: Implications for the lunar cratering record: Meteoritics & Planetary Science, v. 33, p. 889–912.

Croft, S.K., 1982, First order estimate of shock and vaporization in oceanic impacts, *in* Silver, L.T., and Schultz, P.H., eds., Geological Implications of Impacts of Large Asteroids and Comets on Earth: Geological Society of America Special Paper 190, p. 143–152.

Croft, S.K., 1985, The scaling of complex craters: Journal of Geophysical Research, v. 90, p. C828–C842.

Deutsch, A., 1994, Isotope systematics support the impact origin of the Sudbury Structure (Ontario, Canada), *in* Dressler, B.O., Grieve, R.A.F., and Sharpton, V.L., eds., Large Meteorite Impacts and Planetary Evolution: Geological Society of America Special Paper 293, p. 289–302.

Dickin, A.P., Nguyen, T., and Crockett, J.H., 1999, Isotopic evidence for a single impact melt origin for the Sudbury Igneous Complex, *in* Dressler, B.O., and Sharpton, V.L., eds., Large Meteorite Impacts and Planetary Evolution II: Colorado: Geological Society of America Special Paper 339, p. 361–371.

Duvall, G.E., 1958, Pressure-volume relations in solids: American Journal of Physics, v. 26, p. 235–238, doi: 10.1119/1.1996112.

Foley, S., Tiepolo, M., and Vannucci, R., 2002, Growth of early continental crust controlled by melting of amphibolite in subduction zones: Nature, v. 417, p. 837–840, doi: 10.1038/nature00799.

Foley, S., Buhre, S., and Jacob, D.E., 2003, Evolution of the Archaean crust by delamination and shallow subduction: Nature, v. 421, p. 249–252.

Frey, H., 1980, Crustal evolution of the early Earth: The role of major impacts: Precambrian Research, v. 10, p. 195–216, doi: 10.1016/0301-9268(80) 90012-1.

Ghiorso, M.S., and Sack, R.O., 1995, Chemical mass transfer in magmatic processes: IV: A revised and internally consistent thermodynamic model for the interpolation and extrapolation of liquid-solid equilibria in magmatic systems at elevated temperatures and pressures: Contributions to Mineralogy and Petrology, v. 119, p. 197–212.

Green, D.H., 1972, Archean greenstone belts may include terrestrial equivalents of lunar maria?: Earth and Planetary Science Letters, v. 15, p. 263–270, doi: 10.1016/0012-821X(72)90172-0.

Grieve, R.A.F., 1980, Impact bombardment and its role in proto-continental growth on the early Earth: Precambrian Research, v. 10, p. 217–247, doi: 10.1016/0301-9268(80)90013-3.

Grieve, R.A.F., and Cintala, M.J., 1992, An analysis of differential impact melt-crater scaling and implications for the terrestrial impact record: Meteoritics, v. 27, p. 526–539.

Grieve, R.A.F., and Cintala, M.J., 1997, Planetary differences in impact melting: Advances in Space Research, v. 20, p. 1551–1560, doi: 10.1016/S0273-1177(97)00877-6.

Grieve, R.A.F., and Floran, R.J., 1978, Manicouagan impact melt, Quebec 2: Chemical interrelations with basement and formational processes: Journal of Geophysical Research, v. 83, p. 2761–2771.

Grieve, R.A.F., and Therriault, A.M., 2000, Vredefort, Sudbury, Chicxulub: Three of a kind?: Annual Review of Earth and Planetary Sciences, v. 28, p. 305–338, doi: 10.1146/annurev.earth.28.1.305.

Grieve, R.A.F., Dence, M.R., and Robertson, P.B., 1977, Cratering processes: As interpreted from the occurrence of impact melts, *in* Roddy, D.J., Pepin, R.O., and Merrill, R.B., eds., Impact and Explosion Cratering: New York, Pergamon, p. 791–814.

Grieve, R.A.F., Stöffler, D., and Deutsch, A., 1991, The Sudbury Structure: Controversial or misunderstood?: Journal of Geophysical Research, v. 96, p. 22,753–22,764.

Harper, C.L., Jr., and Jacobsen, S.B., 1992, Evidence from coupled ^{147}Sm-^{143}Nd and ^{146}Sm-^{142}Nd systematic for very early (4.5 Gyr) differentiation of the Earth's mantle: Nature, v. 360, p. 728–732, doi: 10.1038/360728a0.

Hartmann, W.K., Ryder, G., Dones, L., and Grinspoon, D., 2000, The time-dependent intense bombardment of the Earth/Moon system, *in* Canup, R.M., and Righter, K., eds., Origin of the Earth and Moon: Tucson, University of Arizona Press, p. 493–512.

Ivanov, B.A., 2004, Heating of the lithosphere during meteorite cratering: Solar System Research, v. 38, p. 304–318.

Ivanov, B.A., and Melosh, H.J., 2003, Impacts do not initiate volcanic eruptions I: Eruptions close to the crater: Geology, v. 31, p. 869–872, doi: 10.1130/G19669.1.

Jaupart, C., and Tait, S., 1995, Dynamics of differentiation in magma reservoirs: Journal of Geophysical Research, v. 100, p. 17,615–17,636, doi: 10.1029/95JB01239.

Kettrup, B., Deutsch, A., and Masaitis, V.L., 2003, Homogeneous impact melts produced by a heterogeneous target? Sr-Nd isotopic evidence from the Popigai crater, Russia: Geochimica et Cosmochimica Acta, v. 67, p. 733–750, doi: 10.1016/S0016-7037(02)01143-2.

Kleine, T., Münker, C., Mezger, K., and Palme, H., 2002, Rapid accretion and core formation on asteroids and terrestrial planets from Hf-W chronometry: Nature, v. 418, p. 952–955, doi: 10.1038/nature00982.

Koeberl, C., 2003, The late heavy bombardment in the inner Solar System: Is there any connection to Kuiper Belt objects?: Earth, Moon, and Planets, v. 92, p. 79–87, doi: 10.1023/B:MOON.0000031927.85937.9e.

Koeberl, C., and Reimold, W.U., 2003, Geochemistry and petrology of impact breccias and target rocks from the 145 Ma Morokweng impact structure, South Africa: Geochimica et Cosmochimica Acta, v. 67, p. 1837–1862, doi: 10.1016/S0016-7037(02)00994-8.

Kring, D.A., and Cohen, B.A., 2002, Cataclysmic bombardment throughout the inner solar system 3.9–4.0 Ga: Journal of Geophysical Research, v. 107, E2, p. 5009, doi: 10.1029/2001JE001529.

Maher, K.A., and Stevenson, D.J., 1988, Impact frustration and the origin of life: Nature, v. 331, p. 612–614, doi: 10.1038/331612a0.

McKinnon, W.B., and Melosh, H.J., 1980, Evolution of planetary lithospheres: Evidence from multiringed structures on Ganymede and Callisto: Icarus, v. 44, p. 454–471, doi: 10.1016/0019-1035(80)90037-8.

Melosh, H.J., 1989, Impact Cratering: A Geologic Process: New York, Oxford University Press, 245 p.

Melosh, H.J., and Vickery, A.M., 1989, Impact erosion of the primordial atmosphere of Mars: Nature, v. 338, p. 487–489, doi: 10.1038/338487a0.

Morgan, J., and Warner, M., 1999, Chicxulub: The third dimension of a multi-ring impact basin: Geology, v. 27, p. 407–410, doi: 10.1130/0091-7613(1999)027<0407:CTTDOA>2.3.CO;2.

Morgan, J.W., Walker, R.J., Horan, M.F., Beary, E.J., and Naldrett, A.J., 2002, ^{190}Pt-^{186}Os and ^{187}Re-^{187}Os systematics of the Sudbury Igneous Complex: Geochimica et Cosmochimica Acta, v. 66, p. 273–290, doi: 10.1016/S0016-7037(01)00768-2.

Neukum, G., and Ivanov, B., 1994, Crater size distributions and impact probabilities on Earth from lunar, terrestrial planet and asteroid cratering data, *in* Gehrels, T., ed., Hazards due to Comets and Asteroids: Tucson, University of Arizona Press, p. 359–416.

Nutman, A.P., Friend, C.R.L., and Bennett, V.C., 2001, Review of the oldest (4400–3600 Ma) geological and mineralogical record: Glimpses of the beginning: Episodes, v. 24, p. 93–101.

O'Keefe, J.D., and Ahrens, T.J., 1977, Impact-induced energy partitioning, melting and vaporization on terrestrial planets, *in* Proceedings, Lunar and Planetary Science Conference 8th, p. 3357–3374.

Petro, N.E., and Pieters, C.M., 2004, Surviving the heavy bombardment: Ancient material at the surface of South Pole–Aitken Basin: Journal of Geophysical Research, v. 109, E6, p. 6004, doi: 10.1029/2003JE002182.

Pierazzo, E., and Melosh, H.J., 2000, Melt production in oblique impacts: Icarus, v. 145, p. 252–261, doi: 10.1006/icar.1999.6332.

Pierazzo, E., Vickery, A.M., and Melosh, H.J., 1997, A re-evaluation of impact melt production: Icarus, v. 127, p. 408–432, doi: 10.1006/icar.1997.5713.

Rapp, R.P., Shimizu, N., and Norman, M.D., 2003, Growth of early continental crust by partial melting of eclogite: Nature, v. 425, p. 605–609, doi: 10.1038/nature02031.

Ryder, G., 2002, Mass flux in the ancient Earth-Moon system and benign implications for the origin of life on Earth: Journal of Geophysical Research, v. 107, E4, doi: 10.1029/2001JE1583.

Schmidt, R.M., and Housen, K.R., 1987, Some recent advances in the scaling of impact and explosion cratering: International Journal of Impact Engineering, v. 5, p. 543–560.

Sleep, N.H., Zahnle, K.J., Kasting, J.F., and Morowitz, H.J., 1989, Annihilation of ecosystems by large asteroidal impacts on the early Earth: Nature, v. 342, p. 139–142, doi: 10.1038/342139a0.

Solomatov, V.S., 2003, Towards understanding the conditions required for plate tectonics to occur on Earth-like planets: 34th Lunar and Planetary Science Conference, abstract 1406, (CD-ROM).

Spudis, P.D., 1993, The Geology of Multi-ring Basins: The Moon and Other Planets: Cambridge, UK, Cambridge University Press, 263 p.

Stern, R.J., 2002, Subduction zones: Reviews of Geophysics, v. 40, no. 4, p. 1012, doi: 10.129/2001RG000108.

Stöffler, D., Deutsch, A., Avermann, M., Bischoff, L., Brockmeyer, P., Buhl, D., Lakomy, R., and Müller-Mohr, V., 1994, The formation of the Sudbury Structure, Canada: Toward a unified impact model, *in* Dressler, B.O., Grieve, R.A.F., and Sharpton, V.L., eds., Large Meteorite Impacts and Planetary Evolution: Geological Society of America Special Paper 293, p. 303–318.

Swanson, D.A., Cameron, K.A., Evarts, R.C., Pringle, P.T., and Vance, J.A., 1989, Cenozoic volcanism in the Cascade range and Columbia Plateau, southwestern Washington and northernmost Oregon: Field excursions to volcanic terranes in the western United States, Volume II: Cascades and Intermountain West: New Mexico Bureau of Mines and Minerals Memoir, v. 47, p. 1–43.

Tera, F., Papanastassiou, D.M., and Wasserburg, G.J., 1974, Isotopic evidence for a terminal lunar cataclysm: Earth and Planetary Science Letters, v. 22, p. 1–21, doi: 10.1016/0012-821X(74)90059-4.

Therriault, A.M., Grieve, R.A.F., and Reimold, W.U., 1997, Original size of the Vredefort Structure: Implications for the geological evolution of Witwatersrand Basin: Meteoritics & Planetary Science, v. 32, p. 71–77.

Therriault, A.M., Fowler, A.D., and Grieve, R.A.F., 2002, The Sudbury Igneous Complex: A differentiated impact melt sheet: Economic Geology and the Bulletin of the Society of Economic Geologists, v. 97, p. 1521–1540.

Turtle, E.D., Pierazzo, E., Collins, G.S., Osinski, G.R., Melosh, H.J., Morgan, J.V., and Reimold, W.U., 2005, Impact structures: What does crater diameter mean?, *in* Kenkmann, T., Hörz, F., and Deutsch, A., eds., Large Meteorite Impacts III: Geological Society of America Special Paper 384, p. 1–24.

Wetherill, G.W., 1975, Late heavy bombardment of the Moon and terrestrial planets, *in* Proceedings of the Lunar Science Conference 6th, p. 1539–1561.

Wieczorek, M.A., and Phillips, R.J., 1999, Lunar multiring basins and the cratering process: Icarus, v. 139, p. 246–259, doi: 10.1006/icar.1999.6102.

Zahnle, K.J., and Sleep, N.H., 1997, Impacts and the early evolution of life, *in* Thomas, P.J., Chyba, C.F., and McKay, D.P., eds., Comets and the Origin and Evolution of Life: New York, Springer-Verlag, p. 175–208.

Manuscript Accepted by the Society 4 October 2005

Printed in the USA

Geological Society of America
Special Paper 405
2006

Archean spherule layers in the Barberton greenstone belt, South Africa: A discussion of problems related to the impact interpretation

Axel Hofmann†
Impact Cratering Research Group, School of Geosciences, University of the Witwatersrand, Private Bag 3, 2050 Wits, Johannesburg, South Africa
Wolf Uwe Reimold
Impact Cratering Research Group, School of Geosciences, University of the Witwatersrand, Private Bag 3, 2050 Wits, Johannesburg, South Africa, and
Museum für Naturkunde (Mineralogy), Humboldt University, Invalidenstrasse 43, 10115 Berlin, Germany
Christian Koeberl
Department of Geological Sciences, University of Vienna, Althanstrasse 14, A-1090 Vienna, Austria

ABSTRACT

This review of the current knowledge of impact spherule layers in the Barberton greenstone belt, together with new petrographic, geochemical, and field data on a number of impact and volcanic spherical particle horizons, highlight a number of problems with the proposition of frequent and large meteorite impacts during mid-Archean times. Field data indicate that some of the four previously proposed impact spherule layers may be laterally correlative units that may have formed from the same impact event. Petrographic work reveals the presence of volcaniclastic particles associated with some spherule layers, while other layers not regarded as impact deposits contain clasts commonly observed in the spherule beds. Major and trace element compositions of spherule layers reflect the composition of the immediate host rocks. The existing platinum group element and chromium isotope data are difficult to reconcile with the current knowledge of the composition of meteoritic debris. A thorough discussion of these problems is necessary before meaningful estimates of Archean impact flux and bolide diameters should be attempted.

Keywords: Archean, impact process, Barberton greenstone belt, South Africa, spherule layer, petrography, accretionary lapilli, geochemistry.

INTRODUCTION

Spherule-bearing layers of purported meteorite impact origin are known from the mid-Archean Barberton greenstone belt of South Africa (Lowe et al., 2003) and the late Archean to Paleoproterozoic Transvaal and Mount Bruce Supergroups of South Africa and Western Australia, respectively (Simonson and Glass, 2004). Layers containing impact spherules as well as microtektites and microkrystites have also been described from the Eocene, K/T boundary, Triassic, and Devonian (review by Simonson and Glass, 2004), although these layers are generally much thinner than those in the Precambrian, and the spherules show different textures with regard to the more ancient ones (Simonson and Harnik, 2000). A spherule horizon

†Present address: School of Geological Sciences, Howard College Campus, University of KwaZulu-Natal, 4041 Durban, South Africa; e-mail: hofmann@ukzn.ac.za.

Hofmann, A., Reimold, W.U., and Koeberl, C., 2006, Archean spherule layers in the Barberton greenstone belt, South Africa: A discussion of problems related to the impact interpretation, *in* Reimold, W.U., and Gibson, R.L., Processes on the Early Earth: Geological Society of America Special Paper 405, p. 33–56, doi: 10.1130/2006.2405(03). For permission to copy, contact editing@geosociety.org.

ca. 1.8–2.1 Ga has been discovered in southern Greenland by Chadwick et al. (2001) and tentatively related to either the Sudbury or Vredefort impact structures, while distal impact ejecta derived from the Sudbury event has recently been described ~700 km away from the impact site (Addison et al., 2005).

In the Barberton greenstone belt, four specific spherule horizons named S1 to S4 have been proposed as being of impact origin (Lowe et al., 1989, 2003). Zircon dating yielded ages of 3.47–3.24 Ga for these strata, with the S1 horizon seemingly having a coeval counterpart in the Pilbara craton (Byerly et al., 2002). Spherules in the Barberton belt are interpreted to have formed by condensation of global clouds of impact-generated rock vapor and to represent distal impact deposits far from the original crater sites (Lowe et al., 1989; Kyte et al., 1992; Byerly and Lowe, 1994; Lowe et al., 2003). There has been debate about whether some or all of these horizons ought to be considered of impact origin (Koeberl et al., 1993; Koeberl and Reimold, 1995; Reimold et al., 2000), mostly on stratigraphic and geochemical grounds.

Apart from the S1–S4 spherule horizons, a number of other layers with spheroidal particles of volcanic origin, such as accretionary lapilli, have been identified in the Barberton stratigraphy (Heinrichs, 1984; Lowe, 1999a). A careful study of these layers is necessary, because accretionary lapilli have been observed as an integral part of recently discovered ejecta from the Sudbury impact (Addison et al., 2005).

Because of widespread silica metasomatism in the Barberton greenstone belt (Paris et al., 1985; Duchac and Hanor, 1987; Lowe, 1999b), unequivocal distinction between spherical particles of volcanic or impact origin may not always be easy. Although recent chromium isotope analyses clearly indicate an extraterrestrial component in some of the spherule layers (Shukolyukov et al., 2000; Kyte et al., 2003), many problems concerning mainly their geochemical characteristics still remain unanswered.

However, a better understanding of the Barberton spherule layers is significant for a better understanding of early Earth evolution. Impacts by large extraterrestrial objects are undoubtedly capable of causing major changes in Earth's atmosphere, hydrosphere, biosphere, and lithosphere (e.g., Sleep and Zahnle, 1998; Glikson, 1999). On the other hand, ancient impact ejecta may also help in elucidating secular variations in the evolution of the Earth (Simonson and Harnik, 2000; Simonson, 2004). In this contribution we present a critical review of the current knowledge about the Barberton spherule layers and point to problematic aspects associated with the impact origin of these strata. This discussion is augmented by new petrographic, geochemical, and field data on impact and volcanic spherical particles from the Barberton greenstone belt.

SPHERULE LAYERS FROM THE BARBERTON GREENSTONE BELT

Four seemingly impact-derived spherule layers have so far been described from the Barberton greenstone belt: S1 in the upper part of the Hooggenoeg Formation and S2 to S4 in the Fig Tree Group (for their geographic and stratigraphic positions, see Figs. 1–3; also Lowe et al., 1989; Lowe and Byerly, 1999; Lowe et al., 2003). While S4 has been reported from only a single locality, the other layers in the Fig Tree Group crop out at several localities in the central part of the greenstone belt (Fig. 2).

Difficulties have been encountered with the correlation and numbering of these spherule layers (Lowe et al., 2003). This is because the central part of the Barberton greenstone belt consists of several fault-bounded tectono-stratigraphic units that structurally duplicate sections of the uppermost Onverwacht volcanic rocks and overlying Fig Tree strata (e.g., Lowe et al., 1999). Zircon dating has suggested that the Onverwacht–Fig Tree contact in different tectono-stratigraphic units could be diachronous (Kröner et al., 1991; Byerly et al., 1996). These complexities resulted in changes of the numbering scheme for the S2 and S3 beds at some locations in the past literature (Lowe et al., 1989; Lowe et al., 2003). On the basis of geochronological data, spinel character and abundance, Ir contents, and spherule types, Lowe et al. (2003) suggested that, in southern outcrops, S2 is present at the Onverwacht–Fig Tree contact and S3 some 50–150 m higher up in the sequence. In contrast, S3 occurs along the Onverwacht–Fig Tree contact in northern areas, mostly in those areas north of the Inyoka Fault, and in some parts of the Barite Valley Syncline (Fig. 1). Lowe et al. (2003) also mentioned "at least two," so far undescribed additional spherule layers (e.g., at the contact between the Sheba and Belvue Road Formations, Fig. 3) and suggested that some layers currently correlated with S3 may represent S2 or other spherule layers.

Apart from possible impact spherules, a variety of other spherical particles occur in sedimentary event deposits in the Barberton greenstone belt. These particles include accretionary lapilli and quenched liquid silicate droplets of possible volcanic origin. In the following, we discuss some salient aspects of these respective horizons, with new observations reported from a suite of new samples, some from new localities.

Spherule Layer S1

S1 is a laterally continuous (>25 km) layer of spherule-bearing sandstone that is interbedded with a sequence of green and black banded chert several meters thick (Lowe et al., 2003). This chert horizon (H4c of Lowe and Byerly, 1999) occurs in the upper part of the Hooggenoeg Formation where it overlies a komatiite sequence (Figs. 1 and 3). The spherule layer consists of two normally graded beds 10–20 cm thick that consist of spherules mixed with chert intraclasts and possible impact-produced debris (Lowe et al., 2003). Ir contents are up to 3 ppb and are similar to background values (generally <5 ppb for Barberton komatiites; Lowe et al., 1989, 2003). Chromium isotope analysis has not been performed on this bed. S1 has been interpreted to represent a current- or wave-deposited layer in an otherwise low-energy, relatively deep-water environment (Lowe et al., 2003).

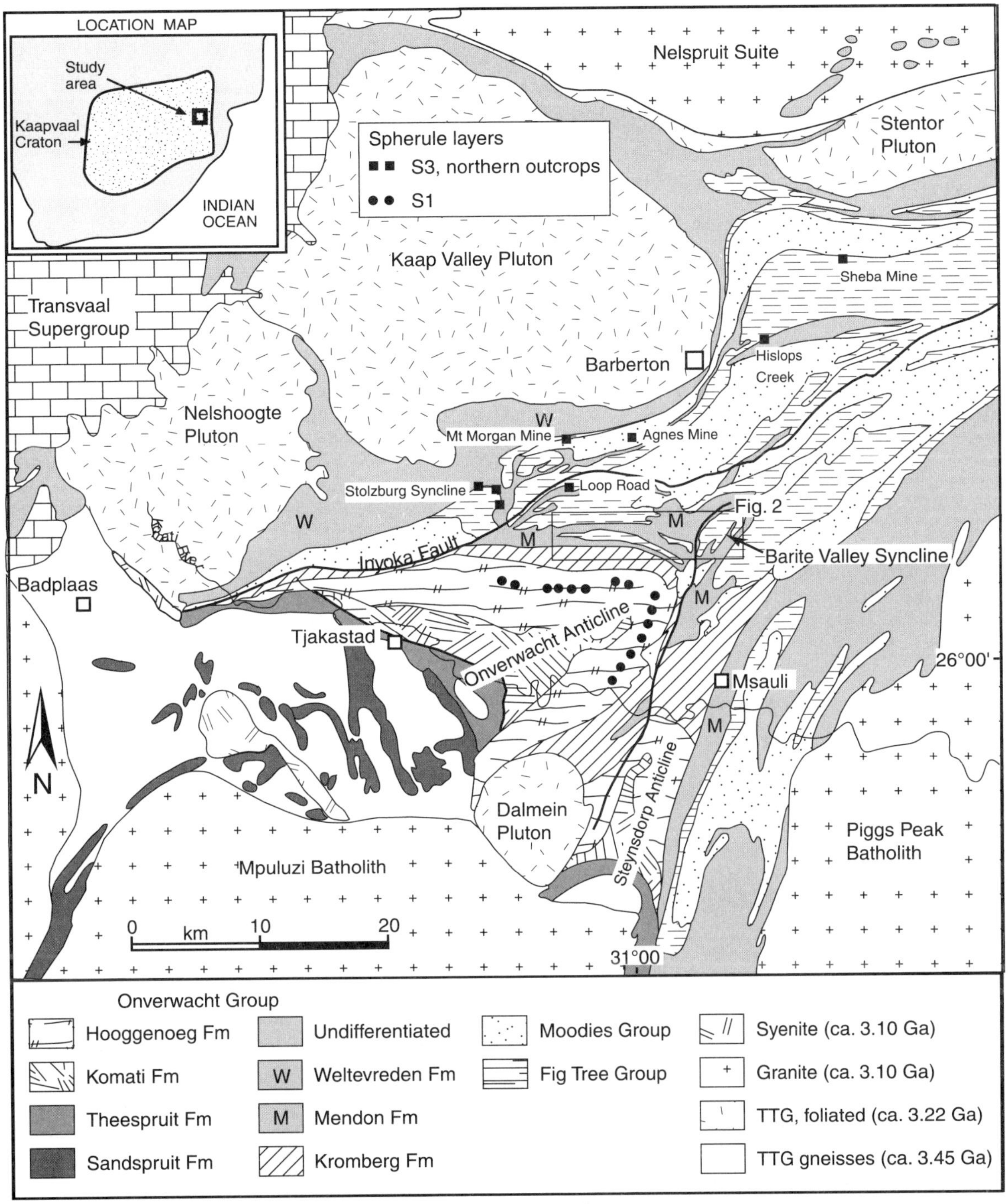

Figure 1. Geological map of the western part of the Barberton greenstone belt and immediately surrounding granitic terrane, showing the distribution of spherule layers S1 and S3 of the northern areas (modified from Kamo and Davis, 1994; Lowe et al., 2003). TTG—tonalite-trondhjemite-granodiorite.

Samples of silicified sandstone from H4c have been studied in thin section. One sample (H9, Table 1) contains abundant spherules similar to those described by Lowe et al. (2003). These are characterized by spherical, dumbbell, and ovoid shapes, contain acicular crystallites that have been replaced by sericite, and have central or off-centered nuclei composed of somewhat coarser-grained silica (Fig. 4A). Other clasts are more variable in shape and include fragments probably derived from underlying chert and volcanic rocks. Possible pyroclastic grains also occur and include altered glass shards of the type shown in Figure 4B, grains of pumiceous appearance, and porphyritic-vesicular grains (Fig. 4C).

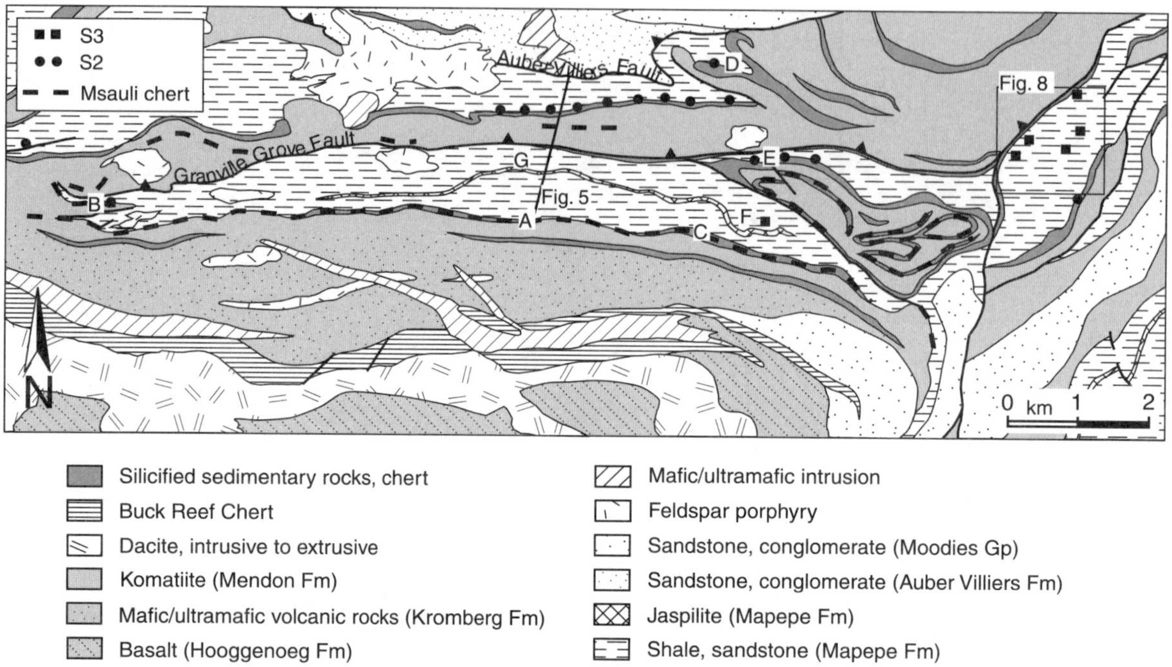

Figure 2. Geological map of the central part of the Barberton greenstone belt, showing the distribution of spherule layer occurrences. A–G are localities discussed in text (see Fig. 1 for location of area; modified from Ward, 2000; Lowe and Byerly, 1999; Lowe et al., 2003).

Spherule Layer S2 and Other Event Deposits

A spherule layer is locally present along the Onverwacht–Fig Tree contact near the top of a sequence of banded black cherts, which are assigned to the Mendon Formation of the Onverwacht Group by Lowe and Byerly (1999), and 0–5 m below siltstones, fine-grained sandstones, and, locally, ferruginous siltstones and cherts of the Fig Tree Group (Figs. 2 and 5). The Onverwacht–Fig Tree contact in the southernmost outcrop belt of Fig Tree strata is characterized by altered ultramafic rocks of the Mendon Formation that are overlain by a thick unit of bedded accretionary lapillistone (Msauli Chert; Lowe, 1999a; also see section below, Accretionary Lapilli). The Msauli Chert is overlain by banded black and white chert, followed by siltstone and shale of the Fig Tree Group (Fig. 5). The Onverwacht–Fig Tree contact in this southern section is, according to Lowe and Byerly (1999), devoid of a spherule layer. Instead, a characteristic horizon of sandstone and chert pebble conglomerate is preserved at the stratigraphic level where one would expect to find S2 (Figs. 5 and 6A). A similar conglomerate commonly occurs at the base of the banded chert sequence (Fig. 5). All of these horizons formed during high-energy current events in an otherwise low-energy, probably deep-water environment and could record volcanic-, seismic-, or impact-related events. Scattered, unusually large accretionary lapilli (Fig. 11A *in* Lowe and Byerly, 1999) are associated with some of the sandstone beds.

Samples of the sandstone and conglomerate event deposits (Table 1) are compositionally diverse and contain a wide variety of different clast types. Large, sedimentary chert clasts are common and most likely represent reworked material derived from underlying or laterally equivalent beds, although carbonaceous matter is much less common in the clasts compared to the host rock. Apparent volcanic clasts are common, as indicated by porphyritic, vesicular, and acicular textures (Fig. 6B). Some of these may be reworked volcanic rocks derived from the erosion of underlying or laterally equivalent units, but most of them appear to represent pyroclastic fragments that were also affected by reworking. These grains sometimes form distinct monomict volcanic sandstone beds (Fig. 6B). Clasts of chert-cemented, probably volcaniclastic, sandstone are also common in this interval. One sample (MC10, Table 1) contains 2 vol% of spherical, ovoid, dumbbell, and teardrop-shaped clasts, probably of volcanic origin, that resemble impact spherules (Figs. 6D, 6E, and 6F) with regard to petrographic criteria for spherule recognition (Simonson, 2003; also see below).

Fig Tree strata of the previous (southern) section are tectonically overlain by a second unit of Onverwacht and Fig Tree rocks along a thrust fault (Figs. 2 and 5; Granville Grove Fault of Lowe and Byerly, 1999). S2 is well developed at many localities along the Onverwacht–Fig Tree contact in this stratigraphic unit where the S2 type section is situated (Fig. 2, locality D). S2 is a medium to very thick (up to 3 m thick) bed of fine chert pebble conglomerate containing disseminated spherules. In many sections, S2 consists of two spherule layers separated by an erosion surface or thin chert beds (Lowe et al., 2003). S2 marks a high-energy wave or current event in an otherwise low-energy,

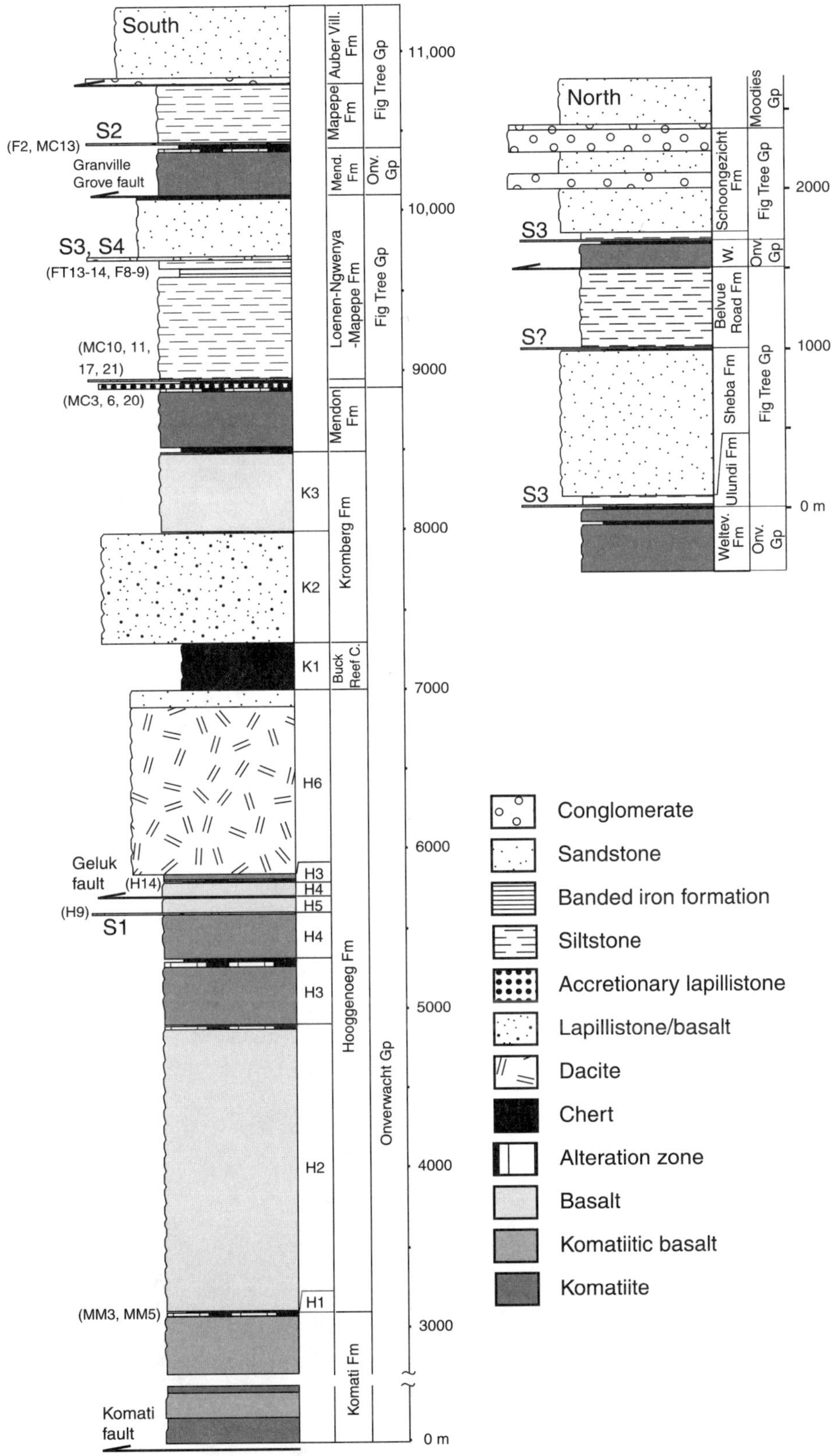

Figure 3. Stratigraphic logs of the Barberton greenstone belt from the western limb of the Onverwacht anticline (Fig. 1), showing the stratigraphic position of spherule layers (and samples used in this study) in the southern area and in areas north of the Inyoka Fault, as exemplified by the stratigraphy of the Stolzburg Syncline (modified from Lowe and Byerly, 1999; Lowe et al., 1999, 2003).

possibly deep-water environment, in which deposition of intercalated banded cherts took place. Iridium values of this layer are up to 3.9 ppb, similar to komatiite values (Lowe et al., 1989, 2003). However, one sample of S2 yielded a small deviation from the terrestrial chromium isotope value, which was calculated into a 1% extraterrestrial component. Values of this order are typical for known impact deposits (Kyte et al., 2003).

Sample F2 (Table 1) is from the S2 type locality, and sample MC13 (Table 1) was collected south of it (Fig. 2, locality E). Spherules are spherical or ovoid or show irregular shapes as a result of compaction. They are either zoned, with coarse-grained quartz in the center that is surrounded by a rim of chert and phyllosilicate, or they are texturally massive, in which case they consist mainly of finely intergrown chert and phyllosilicate (Figs. 7A and 7B). A detailed description of S2 spherules is presented by Lowe et al. (2003). The spherules are admixed with silicified sedimentary and volcanic fragments and quartz lithoclasts that may have been derived from reworking of underlying rocks. Unlike the S1 layer, unequivocal pyroclastic grains were not observed.

Spherule Layer S3

In the southern parts of the Barberton belt, S3 occurs ~800 m above the lower contact of the Fig Tree Group (Fig. 5). In contrast, in the northern areas, the spherule layer at the Onverwacht–Fig Tree contact, previously correlated with S2 (Lowe et al., 1989), is now correlated with S3 (Lowe et al., 2003). Because of these uncertainties, individual layers have to be considered separately.

In its type section (Fig. 2, locality F), S3 occurs in the middle of an ~11-m-thick section of silicified sandstone and chert pebble conglomerate, termed Jay's Chert (Lowe et al., 2003), which occurs several tens of meters above a thick succession of jaspilitic banded iron formation (Fig. 2). S3 at this locality is a spherule-bearing sandstone to fine chert pebble conglomerate that contains abundant pyrite. Iridium contents are very high (426–519 ppb, Lowe et al., 2003). Jay's Chert is part of a unit of pebbly sandstone and conglomerate horizons ~50 m thick within finely laminated siltstones that can be traced for several kilometers to the west; spherules have been described from the type locality (Lowe et al., 2003) and from an ~50-cm-thick unit of two graded beds to the west (Fig. 2, locality G; Fig. 5; Hofmann, 2005a).

Other S3 sites described by Lowe et al. (2003) include a thin bed of almost pure spherules on the east limb of the Barite Valley Syncline (Figs. 8 and 9) ~140 m above the base of the Fig Tree Group and a layer near the base of the Fig Tree Group on the west limb of the Barite Valley Syncline. Spherules also occur as individual grains or spherule-bearing rock fragments ~10 m below the latter bed in up to 150 m long and 2 m wide dykes of carbonaceous chert. These spherules were regarded to have been derived from the overlying bed by downward movement into fissures, now represented by chert dykes (Lowe et al., 2003).

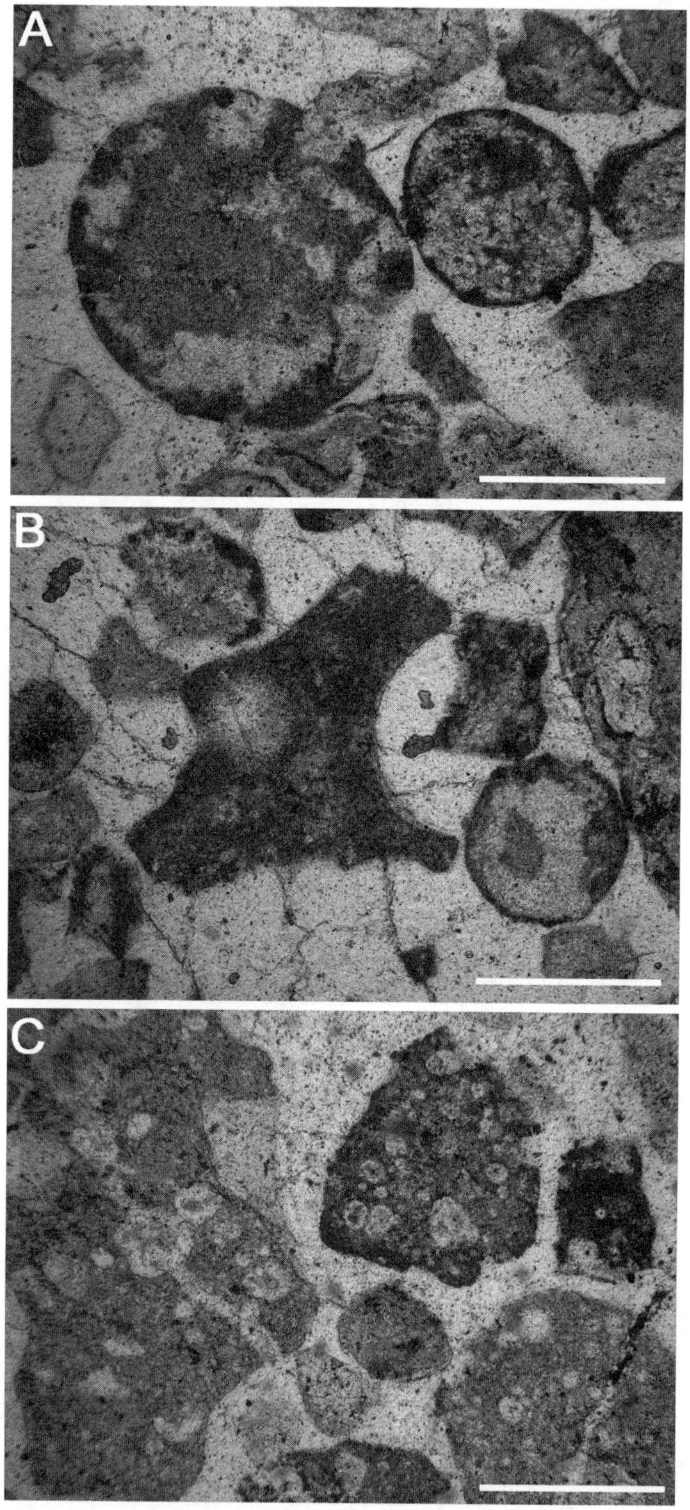

Figure 4. Particles from the S1 spherule layer. Scale bars are 500 µm long. (A) Pair of spherules showing typical dark rim. (B) Irregular clast probably representing an altered glass shard as indicated by broken vesicle wall. (C) Porphyritic-vesicular scoriaceous grain.

TABLE 1. DESCRIPTION OF SAMPLES OF SPHERULE LAYERS AND VOLCANOGENIC STRATA

Sample	Stratigraphy, rock type	Grain size, packing	Constituents
Spherule layers			
H9	S1 spherule layer (Fig. 3), spherule-bearing sandstone (upper layer)	30 vol% spherules (300–800 µm), other clasts 100 µm–1.2 mm, well sorted, moderate to loose packing, ~30 vol% clear chert matrix.	Spherical to irregular and angular clasts (including broken spherical grains) of chert and phyllosilicate micromosaics. Spherules of possible impact origin are mostly spherical (minor dumbbell and ovoid shapes) and typically have a very thin, dark outer rim; broken spherules lack this rim along fresh surfaces. Some spherules contain sericite-replaced, acicular crystallites <100 µm long, whereas others have central or off-centered particles of coarser-grained silica. Other clasts are more variable in shape. Some clasts (mostly not spherical) consist of clear, massive chert and probably represent reworked silicified sediments. Others (including spherical clasts) have a porphyritic texture. Rare particles represent volcanic glass shards, as indicated by broken bubble walls; pumicious grains also occur.
F2	S2 spherule layer (Fig. 2, loc. D), spherule-bearing fine chert pebble conglomerate	25 vol% spherules (200 µm–1 mm), other clasts 200 µm–2 cm, loose packing, 40 vol% chert matrix.	Many spherules are zoned, with central zones of coarse-grained, typically elongate quartz crystals, surrounded by finer-grained outer zones of chert and phyllosilicate. Some spherules are massive and consist of finely intergrown chert and phyllosilicate. Many spherules are flattened and squashed. Other clasts are mostly angular and consist of pure chert or a chert-sericite micromosaic. Rare clasts consist of small, irregular, possibly volcaniclastic grains disseminated in a chert matrix, but unequivocal pyroclastic grains were not observed. Dark rims are common around many clasts.
MC13	S2 spherule layer (Fig. 2, loc. E), spherule-bearing fine chert pebble conglomerate	10 vol% spherules (100–700 µm), other clasts 100 µm–1 cm, 15 vol% chert matrix, strongly silicified.	Spherules are spherical, ovoid, and irregular as a result of compaction. They consist of coarse quartz or massive, finely intergrown chert and phyllosilicate and are characterized by opaque rims. 5 vol% of disseminated, angular, detrital, 100–400 µm mono-quartz grains. Other particles include angular clasts of pure chert, carbonaceous chert, and a number of replaced volcanic rock fragments.
F8, F9	Jay's chert (host lithology of S3/4; Fig. 2, loc. F), fine chert pebble conglomerate	100 µm–1 cm clasts, tight packing, extensive silicification, 5 vol% coarse dolomite as matrix or replacement.	Poorly sorted, angular, mostly sand-sized mono-quartz and massive chert grains. Quartz and other clasts are partly replaced by chert. Carbonaceous chert clasts are common; some show detrital carbonaceous grains and are clearly sedimentary, others show a mylonite-like fabric and are of unknown origin. Compared to intercalated spherule beds, the rock is more coarse-grained and contains abundant chert clasts, but no chlorite-rich clasts; quartz clasts occur in both units, however.
FT13	S3 spherule layer (Fig. 2, loc. F), spherule-bearing sandstone	15 vol% spherules (500–1000 µm); fine- to medium-grained sandstone, moderate packing, 15 vol% coarse chert matrix.	Poorly sorted, angular, mostly 20–200 µm clasts of mono-quartz, minor chert, and chlorite-rich clasts; many chert and chlorite clasts are elongate, angular, splintery, shardlike fragments. Spherical to ellipsoidal (compacted) spherules and spherule fragments are disseminated. Many spherules consist of chlorite and are either massive or contain a spherical, partly off-centered core of coarse quartz. Others consist of massive chlorite in the center and grade into clear chert toward the margin; this texture is probably a replacement feature. Disseminated idiomorphic grains of py and cpy are present.
FT14	S4 spherule layer (Fig. 2, loc. F), spherule-bearing sand- and claystone	70 vol% spherules, two sandstone layers intercalated with a lamina of graded silty claystone, tight-packing, fine-grained clastic material as matrix.	Spherules consist of massive chlorite and some subspherical nuclei of chert; the spherules are strongly compacted and squashed; the matrix consists of 30–300 µm, angular clasts of mono-quartz, chert, and phyllosilicate-rich clasts. The claystone layer contains disseminated, ellipsoidal (i.e., less compacted), large (up to 2 cm) chlorite spherules with some off-centered chert nuclei. A few spherules are replaced by microcrystalline quartz along their margins.
F14	Spherule layer (BVS, west limb; Fig. 9), spherule-bearing fine chert pebble conglomerate	15 vol% spherules (400 µm–1.5 mm), other clasts 200 µm–1 cm, loose packing, 40 vol% chert matrix.	Isolated spherical, ellipsoidal, and dumbbell spherules, together with angular and subrounded to rounded massive chert clasts. Spherules are composed of mosaic barite and massive chert and phyllosilicate micromosaics. They often display distinct narrow phyllosilicate-rich rims with radial growth of opaques (oxides, chromite?). Centered and off-centered nuclei, forming a large part of the clast and generally replaced by barite, are common. Minor finely disseminated sulfides (py, cpy) are present.
F13	Spherules in chert dike (BVS, west limb; Fig. 9), spherule-bearing sandstone	30 vol% spherules (200 µm–1.2 mm), 30 vol% phyllosilicate-chert matrix, weathered sample.	Spherical to ovoid (probably due to compaction) spherules that are locally densely packed and entirely replaced by quartz and sericite micromosaics. Cryptocrystalline chert fills the interstices between spherules, but angular to subrounded fragments of chert and quartz also occur.
F12	Spherule layer (BVS, east limb; Fig. 9), spherule sandstone	90 vol% spherules, 700 µm–1.5 mm, mostly 1 mm, well-sorted, 10 vol% sericite-chert matrix.	Spherules are replaced by chert and sericite and are variably compacted. Opaques (oxides, chromite?) occur preferentially along the spherule rim. Traces of disseminated sulfides are present. Many spherules display outer zones with radial mineral growth of quartz and sericite. Compared to spherules in F14, these zones are substantially wider, making up 1/3 of the spherules. Some spherules have nuclei that are not always in the center and that consist of coarse quartz or massive, finely intergrown chert and sericite.

(*continued*)

TABLE 1. DESCRIPTION OF SAMPLES OF SPHERULE LAYERS AND VOLCANOGENIC STRATA (*continued*)

Sample	Stratigraphy, rock type	Grain size, packing	Constituents
Event layers, Onverwacht–Fig Tree contact			
MC10	Onverwacht–Fig Tree event deposit (Fig. 2, loc. A), very coarse-grained sandstone	300 µm–2 mm clasts, loose packing, 35 vol% matrix, some grains are rimmed by botryoidal chert.	Amazing diversity of different globular particles. Smooth, well-rounded, spherical to ellipsoidal and irregularly shaped, angular clasts. All are replaced by clear chert or quartz or by chert-phyllosilicate micromosaics. Thin, dark outer rims are common. Many clasts, including spherical ones, have a porphyritic (volcanic?) texture due to the presence of idiomorphic domains (50–150 µm) of clear chert. Some contain replaced acicular crystallites. 2 vol% of the clasts are spherical, ovoid, dumbbell, or teardrop-shaped and resemble impact spherules. They consist of massive chert-phyllosilicate micromosaics and some have an off-centered chert core.
MC11	Onverwacht–Fig Tree event deposit (Fig. 2, loc. A), fine chert pebble conglomerate	200 µm–2 cm clasts, 30 vol% chert matrix, all grains are rimmed by botryoidal chert.	All clasts are very well rounded, except for subordinate angular clasts of pure chert. Most clasts are massive silicified siltstone and claystone clasts. Many clasts have a porphyritic texture, as indicated by idiomorphic, equant chert domains, although some of these domains are spherical and may represent former vesicles; some clasts contain acicular crystallites.
MC17	Onverwacht–Fig Tree event deposit (Fig. 2, loc. C), poorly sorted chert pebble conglomerate	Poorly sorted, 100 µm–5 cm, 30 vol% clear chert matrix.	Clasts > ~1 mm are mostly well rounded, tabular clasts of massive to laminated, relatively pure chert with minor sericite and some carbonaceous matter. Some pebbles consist of clasts disseminated in a 50 vol% chert matrix. These clasts are 500 µm-2 mm, irregular and angular, and consist of mixtures of chert, phyllosilicate, and opaques, some with fuchsite. They have an opaque, oxide-rich rim and some show an acicular or porphyritic texture; the clasts probably represent reworked volcanic rock or pyroclastic fragments. A distinct variety of clasts is mostly <800 µm, is rich in opaques and phyllosilicates, and has a common black rim (absent along fresh edges); the clasts are angular to well rounded, and several show porphyritic and acicular textures.
MC21	Onverwacht–Fig Tree event deposit (Fig. 2, loc. B), coarse-grained volcanic sandstone	Thinly bedded, coarse- (500 µm–2 mm) and fine-grained (50–300 µm) layers, 30 vol% clear chert matrix.	Monomict composition, all clasts are irregular-shaped, subrounded to very angular, and consist of chert, sericite, chlorite, and minor opaques; all clasts contain equant to tabular, partly idiomorphic domains of clear chert >200 µm, possibly pseudomorphs after olivine, and chert-filled vesicles, making up ~30 vol% of the clasts.
Accretionary lapillistone, volcaniclastic sandstone			
MM3	Middle Marker (Fig. 3), volcaniclastic sandstone	200–500 µm clasts, mostly chert with minor but varying amounts of intergrown sericite, loose packing, 40 vol% matrix.	Mostly equant, sometimes spherical, but also many irregular, angular clasts of massive, mostly fine-grained chert in slightly coarser-grained silica matrix. Many particles have a very thin, dark outer rim. Many clasts represent shards of only sparsely vesicular glass, as indicated by the presence of broken bubble walls. The shards have a slightly coarser-grained outer rim of silica lacking phyllosilicate. Spherical particles may represent quenched lava droplets, although teardrop and dumbbell-shaped particles are absent; an accretionary origin for the spherules cannot be discounted.
MM5	Middle Marker (Fig. 3), accretionary lapillistone	500 µm–2 mm lapilli, 30 vol% of relatively coarse-grained chert matrix.	Accretionary lapilli consist of intergrown chert and phyllosilicate (sericite and yellowish-brown phyllosilicate) and are generally spherical to ovoid, the latter probably a result of compaction. Broken and rare dumbbell-shaped lapilli also occur. Clasts (<200 µm) of coarsely crystalline quartz (sometimes with a phyllosilicate rim) or phyllosilicate, many of which are distinctly shard-shaped (angular, broken bubble walls), are common in the lapilli.
H14	Hooggenoeg Formation (H3c?, Fig. 3), accretionary lapillistone	100–800 µm lapilli, mostly 150 µm, ~30 vol% of matrix, extreme silicification makes clast identification difficult.	Uncommon spherical, but abundant equant and many angular clasts. Many, mostly angular, phyllosilicate-rich clasts that may be replaced volcanic rock fragments. Spherical particles are relatively coarse (>500 µm) and are internally coarse-grained and heterogeneous (presence of angular inclusions) typical for accretionary lapilli. The clasts are probably a mixture of reworked sedimentary material and pyroclastic (lapilli, shards) material.
H7	Hooggenoeg Formation (H5c), accretionary lapillistone	300 µm–1 mm lapilli, 30 vol% of relatively coarse chert matrix.	Laminated chert with laminae rich in accretionary lapilli. Equant, spherical to irregular lapilli of relatively coarse chert and intergrown brownish phyllosilicate. Rare compound, dumbbell-shaped lapilli. Silt- to fine sand-sized, angular, shardlike particles with broken bubble walls are common.
MC3	Msauli Chert (Fig. 3), accretionary lapillistone	500 µm–1.2 mm lapilli, mostly 800 µm, ~20 vol% matrix of clear chert.	Well-rounded lapilli with a characteristic "rough" outline, replaced by silica and brown phyllosilicate. Lapilli consist of silt- to sand-sized grains that are dominated by phyllosilicates and rare clear chert. There are abundant angular grains (100–200 µm) mostly in the matrix, but also in the lapilli, that consist of chlorite and may represent nonvesicular volcanic glass shards.
MC16	Msauli Chert (Komati Gorge), accretionary lapillistone	1.5–3 mm lapilli, 20 vol% matrix of coarse quartz and dolomite.	Completely chert-replaced, mostly spherical to ovoid particles. Lapilli outlines only discernible as a result of a sericite-rich rim. No zonation recognizable.
MC20	Msauli Chert (type locality), accretionary lapillistone	1.5–3 mm lapilli, loosely packed, supported by matrix (~40 vol%) of chlorite, silica, and carbonate.	The lapilli are subspherical and consist of finely intergrown chert and minor sericite. Several lapilli contain a nucleus of chert or chlorite-replaced material. The lapilli vary in texture, are fine to coarse grained, and are massive to concentrically layered.

Note: BVS, Barite Valley Syncline; py, Pyrite; cpy, Chalcopyrite.

In the northern part of the Barberton greenstone belt (Fig. 1), thin spherule beds up to ~20 cm thick occur at the Onverwacht–Fig Tree contact at Sheba Mine and Hislops Creek, Agnes Mine, Mount Morgan Mine, Stolzburg Syncline, and the Loop Road locality of Lowe et al. (2003). Detailed mineralogical descriptions and geochemical information of Sheba and Agnes samples were published by Koeberl et al. (1993), Koeberl and Reimold (1995), and Reimold et al. (2000). Iridium and other platinum-group element contents of spherule-rich samples are high at Sheba Mine (104–725 ppb Ir; Lowe et al., 2003), variable on the west limb of the Barite Valley Syncline (2–145 ppb Ir, although it is unclear which bed was analyzed), but generally close to komatiite background values at most other localities (<5 ppb). However, several beds consisting of almost pure spherules contain remarkably low Ir values, such as S3 on the east limb of the Barite Valley Syncline (5.8 ppb) and S3 at Mount Morgan Mine (1–13 ppb, Lowe et al., 2003). High Ir values in the latter bed only occur in the uppermost part where spherules are admixed with other detritus (Lowe et al., 2003). Koeberl et al. (1993), Koeberl and Reimold (1995), and Reimold et al. (2000) observed extremely high Ir concentrations, up to 2700 ppb, for spherule-bearing samples from the Sheba and Agnes Mine localities (Fig. 1).

A distinct variety of spherules containing dendritic and skeletal Cr- and Ni-rich spinels are common in spherule layers now regarded as S3. They are absent at some localities, especially where spherules were reworked and admixed with intrabasinal detritus (Lowe et al., 2003). The chemical composition of the spinels has been described to be unique and distinct from spinels in komatiites and other volcanic rocks (Byerly and Lowe, 1994). Detailed mineralogical and geochemical studies of this material have also been reported by Reimold et al. (2000). They observed that spinel grains are zoned, with a core enriched in Ni, a middle Ni-rich zone, and a comparatively Ni-depleted but Zn-rich rim zone. The Zn enrichment of outer zones was taken as clear evidence for secondary, hydrothermal processes having affected these grains. Many S3 samples studied to date have been characterized as sulfide rich. Most of this sulfide is pyrite, but chalcopyrite, sphalerite, galena, and As-bearing sulfides such as gersdorffite (a Ni-As sulfide) have been described (Koeberl et al., 1993; Koeberl and Reimold, 1995; Reimold et al., 2000). Matrices of such samples are composed of secondary quartz, sericite, barite, and siderite, with other carbonates occurring sporadically.

Many so-called S3 occurrences in the southern and eastern sections, i.e., mostly those beds that occur within the Fig Tree Group, are intercalated with shallow-water deposits and commonly preserve evidence for current and/or wave reworking and mixing with intrabasinal detritus. The nearly pure spherule bed on the east limb of the Barite Valley Syncline was reported to contain evidence for tidal current activity (Lowe et al., 2003). Spherule layer exposures in more northern areas consist of a thin, rather undiluted bed, which has been interpreted to represent a fall-deposited layer in a deeper-water setting (Lowe et al., 2003).

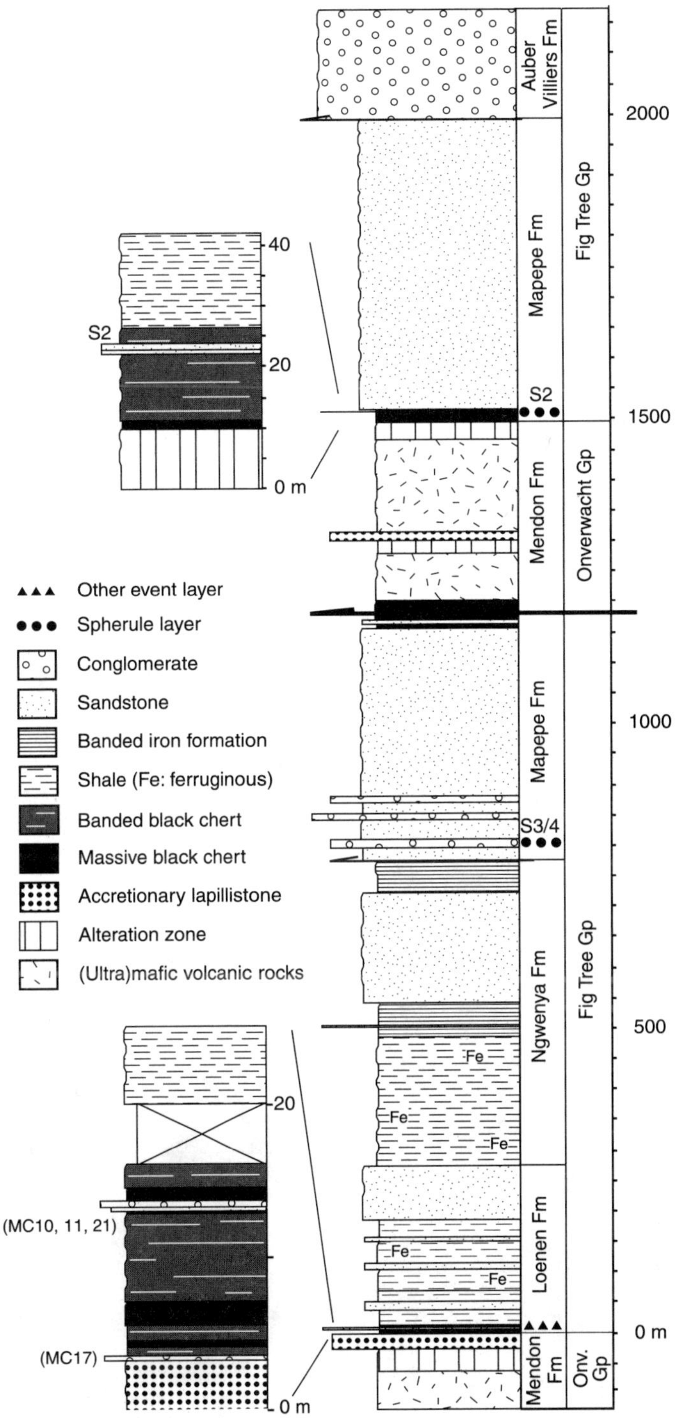

Figure 5. Stratigraphic log of the Fig Tree Group (see Fig. 2 for location of section; after Hofmann, 2005a).

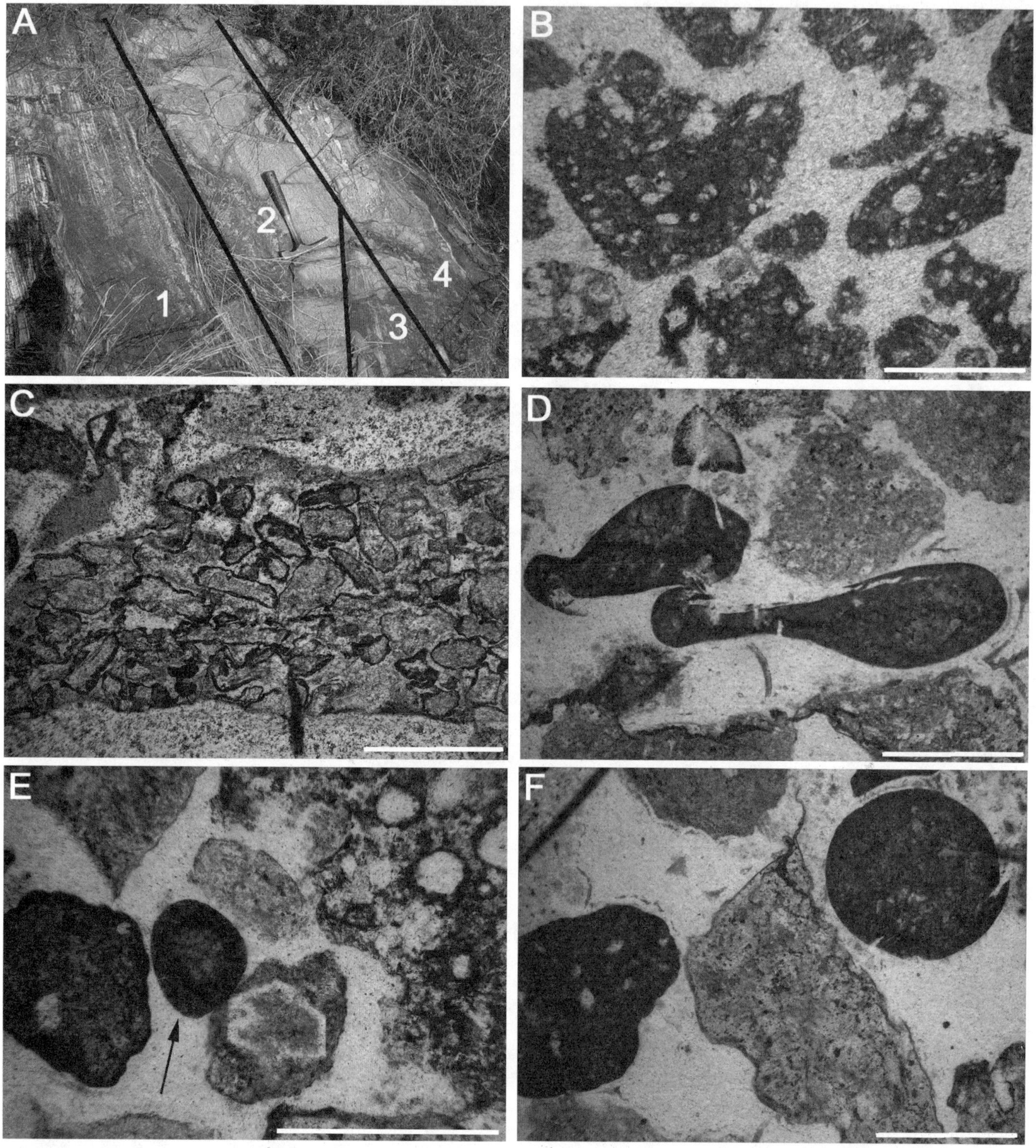

Figure 6. (A) Event deposit at the Onverwacht–Fig Tree contact a few meters below Fig Tree shales (Fig. 5). Banded black and white chert (1) is overlain by normally graded pebbly sandstone (2). An erosive scour is filled with banded chert (3) and overlain by chert pebble conglomerate (4). (B) Porphyritic-vesicular scoriaceous grains constituting a monomict volcaniclastic sandstone bed. (C) Clast of chert-cemented, probably volcaniclastic sandstone. (D) Teardrop-shaped chlorite-rich clast surrounded by porphyritic fragments. (E) Subspherical chlorite-rich grain (arrow) with central chlorite-replaced core together with vesicular and porphyritic clasts. (F) Spherical chlorite-rich grain (upper left) together with texturally different clasts. Scale bars are 500 μm long.

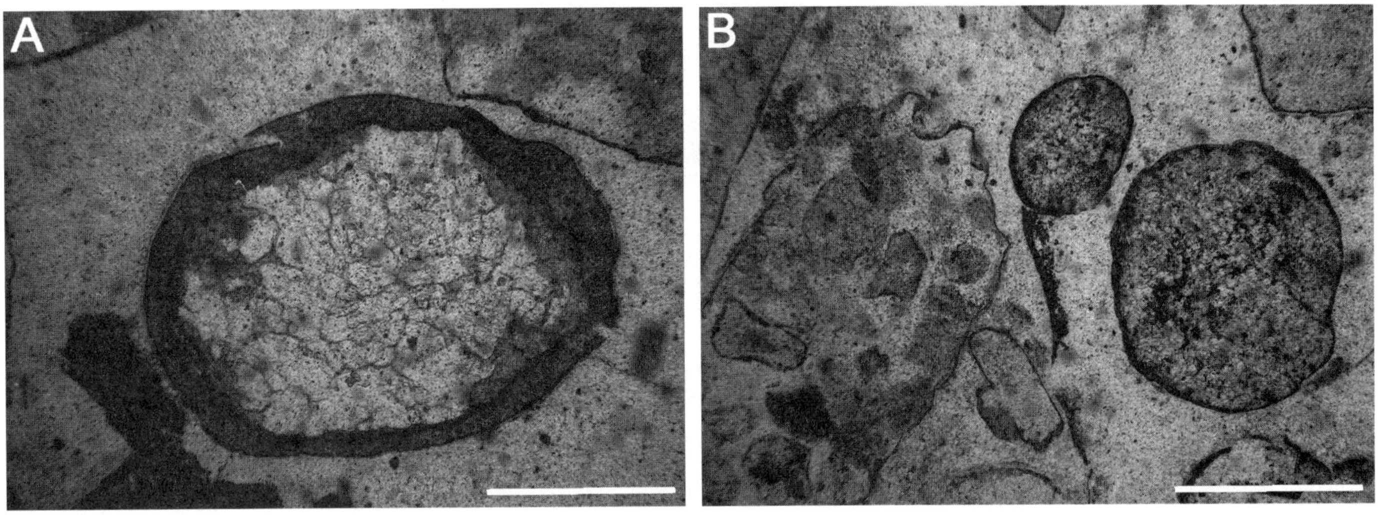

Figure 7. Particles from the S2 spherule layer. Scale bars are 500 µm long. (A) Zoned spherule with coarse-grained quartz in the center and a thin chert-phyllosilicate rim. (B) Two massive spherules of chert-phyllosilicate micromosaics and a clast (left) of chert-cemented, probably volcaniclastic, sandstone. Similar clasts are common in other event beds (Fig. 6C).

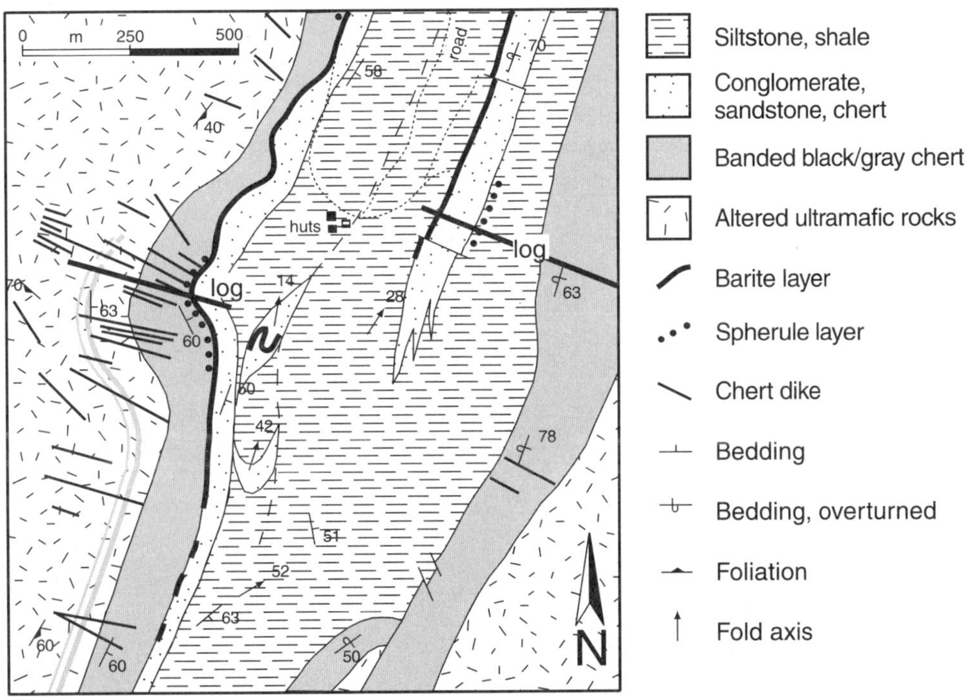

Figure 8. Geological map of the southern part of the Barite Valley Syncline (see Fig. 2 for location of section; modified from Heinrichs and Reimer, 1977; Lowe and Byerly, 2003). Positions of logs shown in Figure 9 are indicated.

One sample of S3 from the type locality and one sample from the Sheba Mine spherule bed have been analyzed for chromium isotopes and both yielded isotopic anomalies close to the value for carbonaceous chondrites (Kyte et al., 2003). This result was taken by these authors to imply an unusually high extraterrestrial component of 50%–60% and that all the Cr in these samples is of extraterrestrial origin.

A sample from the S3 type locality (FT13, Table 1) is a fine- to medium-grained sandstone that has no more than 15 vol% of round to ellipsoidal spherules. Many spherules consist of chlorite and are either massive or contain a spherical, partly off-centered nucleus of coarse quartz. They are disseminated in a groundmass of poorly sorted, angular clasts of quartz, minor chert, and chlorite-rich clasts surrounded by a chert

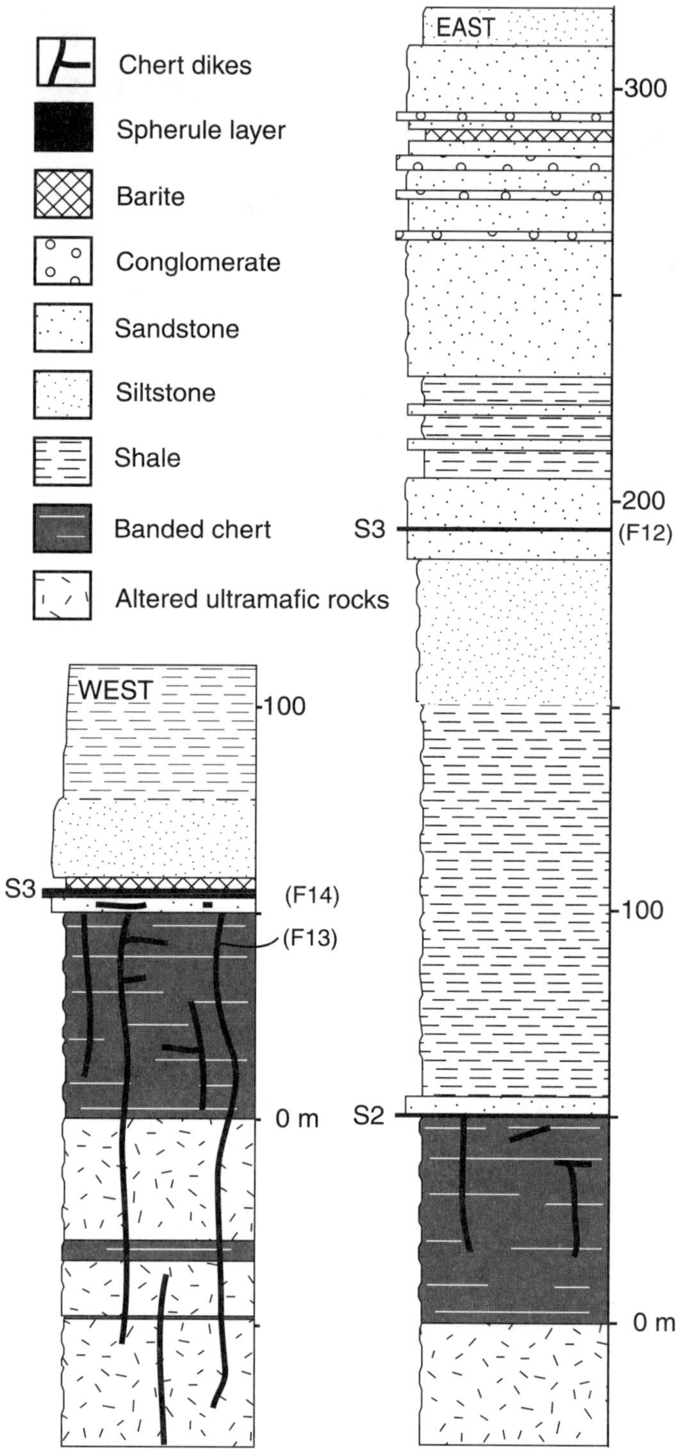

Figure 9. Stratigraphic logs from the west and east limbs of the Barite Valley Syncline (modified from Lowe and Nocita, 1999; Lowe and Byerly, 2003). See Figure 8 for position of measured sections. Different stratigraphies suggest that the Barite Valley Syncline is not a simple, unfaulted syncline.

matrix; many chert and chlorite clasts are elongate, angular, splintery fragments similar to glass shards (Fig. 10A). Disseminated, commonly idiomorphic sulfide grains are common. Samples from Jay's Chert (F8, F9, Table 1), the host lithology, are compositionally distinct in that they contain abundant sedimentary chert clasts but no chlorite-rich clasts. A sample of either S3 or S4 (Fig. 2, locality G) shows well-developed botryoidal rims typical for many late Archean spherule occurrences (Simonson, 2003).

The spherule layer from the west limb of the Barite Valley Syncline (F14, Table 1) comprises 15 vol% of spherical, ellipsoidal, and dumbbell-shaped spherules that are partly replaced by barite and are mixed with angular to rounded, massive chert clasts probably derived from the underlying sedimentary chert beds. Pyroclastic material was not observed. Spherules in underlying (i.e., stratigraphically lower but crosscutting) chert dikes (F13, Table 1) are more tightly packed and are entirely replaced by quartz and sericite micromosaics. The spherule layer from the east limb of the Barite Valley Syncline (F12, Table 1) is almost totally composed of well sorted, commonly zoned spherules that consist of finely intergrown chert and sericite (Fig. 10C).

Spherule Layer S4

Spherule layer S4 has been reported from a single locality, 6.5 m above S3 at the S3 type section (Fig. 2, locality F). It crops out for only 1 m along strike because of erosion by overlying strata. It is a thin, current-deposited bed containing chlorite-replaced spherules, spherule debris, and terrigenous clastic grains. Iridium contents are high and range from 8 to 450 ppb. Spinels have not been observed, and the platinum group element carrier phase(s) remains unidentified (Lowe et al., 2003). The chromium isotopic compositions of two samples from this locality are nonterrestrial and indicate a high extraterrestrial component of 15%–30% (Kyte et al., 2003). The Ir data indicate an even higher extraterrestrial component.

A sample from the S4 type locality (FT14, Table 1) comprises 70 vol% of strongly compacted and squashed spherules, which are composed of massive chlorite and have some subspherical nuclei of chert (Fig. 10D). They are, thus, very similar to spherules from the S3 layer. The matrix consists of angular clasts of quartz, chert, and phyllosilicate-rich clasts that may represent replaced lithoclasts.

Accretionary Lapilli

Bedded cherts containing layers composed of silicified accretionary lapilli are widespread in the Barberton greenstone belt and have been studied by a number of workers (Lowe and Knauth, 1977; Reimer, 1983; Heinrichs, 1980, 1984; Stanistreet et al., 1981; Lowe, 1999a). Accretionary lapilli range from sand to fine pebble size (typically 0.5–5 mm, i.e., they are strictly speaking not all lapilli but also include particles of coarse ash

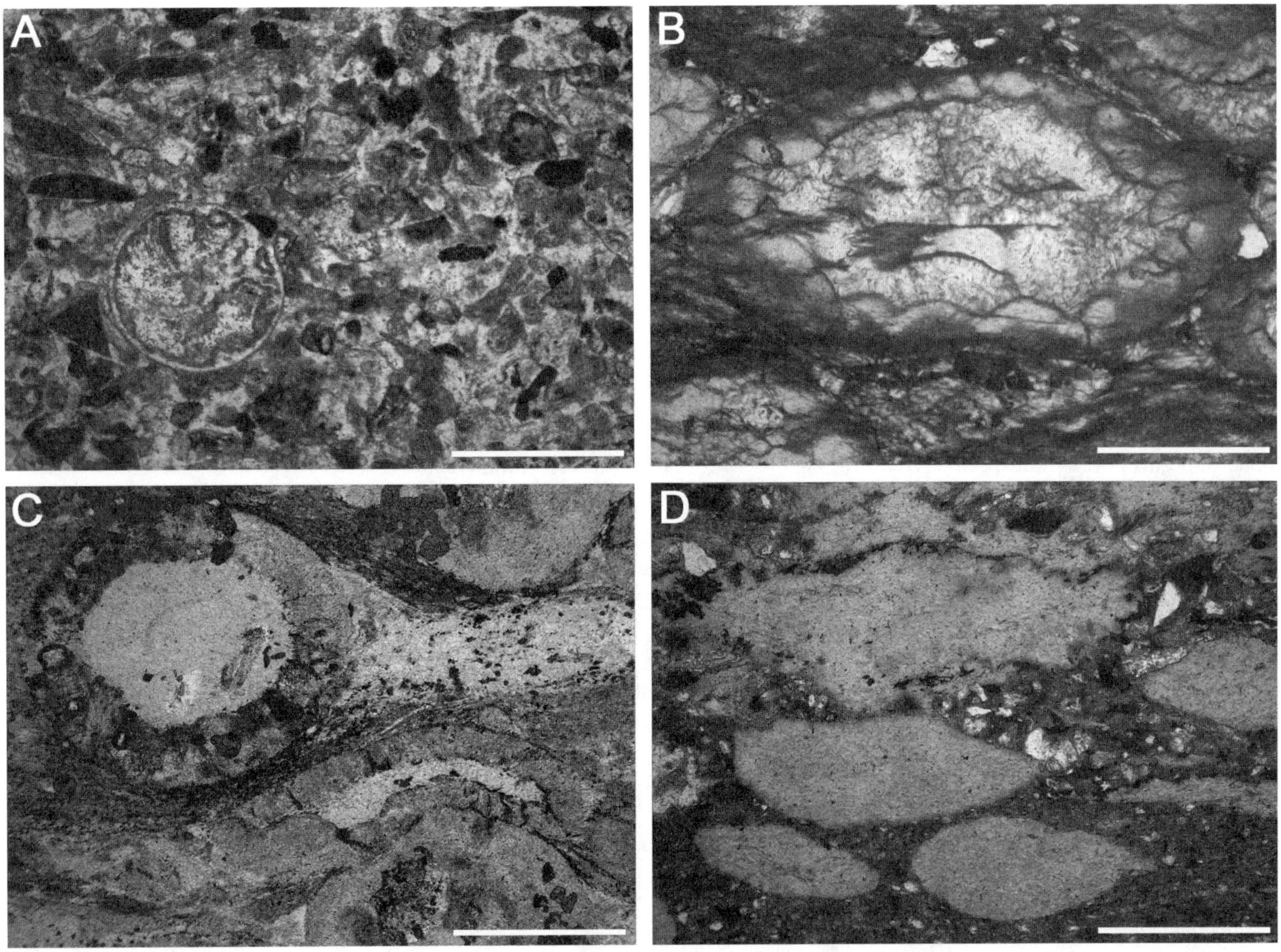

Figure 10. Particles from the S3 and S4 spherule layers. Scale bars are 500 μm long. (A) Chert-replaced spherule within a matrix of angular fragments of chert and chlorite (S3 type section). (B) Compacted chert-sericite spherule with botryoidal rims growing inward. (C) Variably compacted chert-sericite spherules with botryoidal rims. (D) Flattened chlorite spherules in a matrix of quartz-bearing silty claystone (S4 type section).

size), are massive or concentrically laminated, and may have a nucleus. They are abundant most notably in bedded chert horizons overlying ultramafic volcanic rock sequences (Lowe, 1999a) and are particularly well developed in the Msauli Chert at the Onverwacht–Fig Tree contact, which consists of a unit of lapillistone ~20 m thick (Fig. 5). According to Heinrichs (1984), Msauli Chert lapilli consist of microcrystalline quartz and white mica with minor but varying amounts of chlorite, ferroan dolomite, authigenic K-feldspar, rutile, and opaques. The presence of vesicular and pumiceous ash particles within the accretionary lapilli and in the groundmass supports a volcanogenic origin. Ash particles are mostly vitric fragments; crystal and lithic fragments occur in subordinate amounts. Based on the absence of quartz and feldspar phenocrysts coupled with low Al_2O_3 and Zr but high Cr contents (Lowe, 1999b), the Msauli Chert is regarded as comprising pyroclastic debris of komatiitic parentage.

Accretionary lapilli of the Msauli Chert are found in normally graded beds that consist of lapillistone at the base and komatiitic tuff, now green chert, at the top (Heinrichs, 1984). Current and climbing-ripple lamination as well as cross-bedding are common in the middle to upper parts of many beds. The graded beds show partial or complete Bouma sequences and have been interpreted as the deposits of turbidity currents that formed upon the entering of pyroclastic falls and base surges into water (Stanistreet et al., 1981; Heinrichs, 1984). The absence of pelagic intervals and the similar composition throughout the section coupled with a uniform paleocurrent pattern were regarded to reflect a major phreatoplinian eruption cycle (Heinrichs, 1984). On the other hand, Lowe (1999a) interpreted the graded beds as shallow-water deposits and suggested that the Bouma sequence reflected the declining rate of pyroclastic fall into flowing water rather than declining velocities of turbidity currents. The accretionary lapilli were regarded as having formed during hydromagmatic explosions in the closing stages of major episodes of komatiitic volcanism, when broad shield volcanoes became subaerial.

Accretionary lapilli, although now found as reworked material in subaquatic deposits, are generally thought to form in phreatomagmatic eruption columns and ash clouds, suggesting shallow-water to subaerial volcanism (e.g., Fisher and Schmincke, 1984). Interestingly, accretionary lapilli in the Barberton greenstone belt are mostly associated with ultramafic volcanism (Lowe, 1999a). This relationship would suggest local subaerial conditions only during times of komatiite volcanism. It seems clear that the mechanisms related to the formation of ultramafic accretionary lapilli in the Archean are not yet understood.

For comparative purposes and in order to delineate the similarities and differences between accretionary lapilli and impact spherule horizons, respectively, several samples of pyroclastic deposits containing accretionary lapilli were obtained: two from the Middle Marker (Fig. 3), one sample each from chert horizons H3c and H5c at the middle and top of the Hooggenoeg Formation (Fig. 3), and three samples from the Msauli Chert (Fig. 5). Accretionary lapilli from the Middle Marker (MM5, Table 1) are mostly spherical. Angular shard-shaped clasts with broken bubble walls are common in the lapilli (Fig. 11A). A volcaniclastic sandstone intercalated with lapillistone (MM3, Table 1) consists of a variety of spherical to angular clasts, some of which clearly represent shards of sparsely vesicular glass. Spherical particles may represent quenched lava droplets, although teardrop and dumbbell-shaped particles are absent, and some show agglutinated grains typical for accretionary lapilli (Fig. 11B).

Accretionary lapilli from the top of the Hooggenoeg Formation (H7, Table 1) are generally coarse grained and spherical,

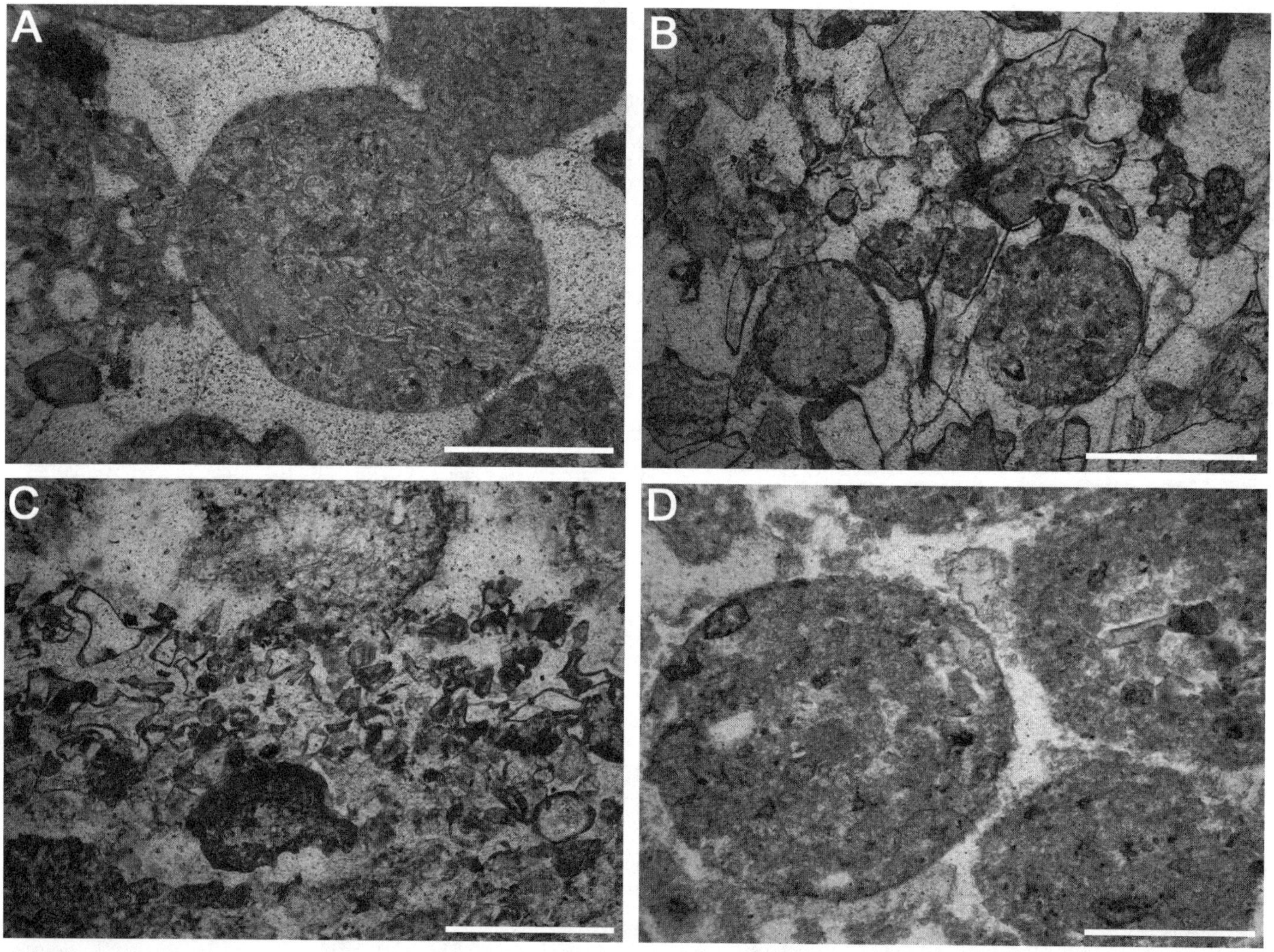

Figure 11. Accretionary lapilli and associated particles. Scale bars are 500 μm long. (A) Accretionary lapillus comprising replaced glass shards (Middle Marker). (B) Volcaniclastic sandstone consisting of vesicular glass shards and spherical grains that probably represent accretionary lapilli (Middle Marker). (C) Irregular accretionary lapilli embedded in volcaniclastic sandstone made up of altered glass shards (Hooggenoeg Formation). (D) Accretionary lapilli showing indistinct concentric lamination (Msauli Chert).

but rare dumbbell particles occur apart from abundant shardlike grains (Fig. 11C). Lapilli from the Msauli Chert are quite variable in texture, and fine- to coarse-grained and massive to concentrically layered types occur (Fig. 11D). Clasts resembling nonvesicular volcanic glass shards are common in this layer.

GEOCHEMISTRY

Most of the samples described here (for sample detail, refer to Table 1) were analyzed for selected major and trace element contents by instrumental neutron activation analysis. Samples were reduced by jaw crusher into small chips. After cleaning, vein-free chips were handpicked and crushed in an agate mortar. This material was then pulverized using an agate mill. About 150 mg of sample powder of each sample was sealed into polyethelene vials and irradiated together with international geological standard reference materials (for quality control) at the Triga reactor of the Atomic Institute of the Austrian Universities in Vienna. Details on the counting procedure, instrumentation, standards, calculations, accuracy, and precision are given in Koeberl (1993). The results are listed in Table 2.

All samples have been affected by silica metasomatism. The change in composition as a result of silicification has been documented for volcanic rocks of the Barberton belt (Duchac and Hanor, 1987; Hofmann, 2005b). Silicification is generally associated with a depletion of most elements except for K_2O, Rb, and Ba, which are enriched due to alteration overprint. Sc, V, and Cr were relatively immobile, whereas Ni, Co, Cu, and Zn were more mobile and became depleted in volcanic rocks but enriched in some sedimentary rocks. LREE (light rare earth elements) became enriched relative to the HREE (heavy rare earth elements) during progressive alteration of volcanic rocks (Hofmann, 2005b).

Samples of accretionary lapillistone, including volcaniclastic sandstone from the Middle Marker, are characterized by relatively high Cr/Th and low Th/Sc ratios similar to komatiites and komatiitic basalts (see Hofmann, 2005a for a summary of geochemical characteristics of Barberton rocks), suggesting that they represent ultramafic pyroclastic deposits. Relatively unfractionated REE (rare earth element) patterns support this view (Fig. 12A). Cr/Ni ratios are relatively high, which may reflect Ni depletion during silicification.

The geochemistry of chert horizons in the Hooggenoeg Formation, one of which (H4c) represents the host rock for spherule layer S1, generally reflects the composition of the volcanic rocks with which they are interbedded (Hofmann, 2005b). This may reflect the presence of detrital material eroded from underlying rocks or admixtures of tuffaceous material. Spherule layer S1 contains high Sc and Cr contents, resulting in trace element ratios similar to rocks of komatiitic basalt composition. Such rocks underlie chert horizon H4c.

Two samples of spherule layer S2 from different localities are chemically very similar except for variable Cr content, indicating that Cr is heterogeneously distributed in the samples. The samples are characterized by enriched LREE and unfractionated HREE identical to underlying chert and overlying shale (Fig. 12B). They also have similar trace element contents to the host rocks, suggesting that their chemical compositions reflect the composition of eroded and admixed host rock material. The Cr content is within the range of Cr concentrations of the host rocks (few ppm in Fig Tree cherts versus up to 1260 ppm in Fig Tree shales, Hofmann, 2005a).

Spherule samples derived from the east and west limbs of the Barite Valley Syncline are compositionally distinct. Those derived from the west limb show LREE–enriched patterns, while the sample from the east limb is characterized by an LREE–depleted pattern (Fig. 12C). Ba contents are exceptionally high for all samples.

Samples from the S3 and S4 spherule layers and from the immediate host rock (Jay's Chert) at the type localities show very similar trace element ratios. They are characterized by enriched LREE and unfractionated HREE patterns similar to underlying shale (Fig. 12D). Cr and Ni are enriched in the spherule layers relative to the immediate host rock (Table 2).

Samples from the event beds along the Onverwacht–Fig Tree contact (Table 1) have trace element ratios similar to basaltic rocks from the Barberton greenstone belt. REE patterns are characterized by LREE enrichment and are thus similar to some of the spherule samples analyzed. Iridium levels are only elevated in layers regarded to consist of impact spherules. The highest Ir values obtained are associated with the S3 and S4 spherule layers (Table 2).

PROBLEMS WITH AN IMPACT ORIGIN FOR THE SPHERULE LAYERS

Problems associated with stratigraphic relationships and the impact origin of specific spherule beds have been emphasized in the past (Koeberl et al., 1993; Koeberl and Reimold, 1995; Reimold et al., 2000). Lowe et al. (2003) presented a detailed discussion on the origin and positioning of the Barberton spherule layers. In the following section, we present a discussion that roughly follows the points raised by these authors. Discussion of the difficulties of impact spherule recognition is also included.

Spherule Recognition

According to the study of late Archean and Paleoproterozoic spherule layers by Simonson (2003; also Kohl et al., this volume), the best criteria to recognize spherule beds of likely impact origin include (1) a predominance of highly spherical grains in the sand-sized fraction; (2) the presence of grains with unusual shapes, such as teardrops and dumbbells; (3) fibroradial aggregates of K-feldspar crystals nucleated on the edges of spherules and growing inward; and (4) well-defined internal spots representing both cement-filled vesicles and replaced glass cores not always located in the centers of such grains. The

TABLE 2. MAJOR AND TRACE ELEMENT DATA AND SELECTED ELEMENT RATIOS

Sample	H-9	F-2	MC-13	F-8	F-9	FT-13	FT-14	F-14	F-13	F-12	MC-10	MC-21	MM-3	MM-5	H-14	H-3	MC-3	MC-16	MC-20
Stratigraphy	H4c			Jay's ch.	Jay's ch.	Jay's ch.	Jay's ch.	BVS-W	BVS-W	BVS-E	Onv-FT	Onv-FT	M. Marker	M. Marker	H3c?	H5c	Msauli ch.	Msauli ch.	Msauli ch.
Rock type	S1	S2	S2			S3	S4	S2-3?	S2-3?	S2-3?	event bed	event bed	volc. sst	acc. lapilli	acc. lapilli	acc. lapilli	acc. lapilli	acc. lapilli	acc. lapilli
Na	1180	162	1200	77	97	175	34	60.3	213	278	367	391	115	978	81	139	66	453	419
Sc	19.9	4.41	2.97	2.6	2.29	10.6	4.55	4.19	3.87	41.8	12.5	15.9	14.8	42.3	7.43	13.5	24.6	18.6	34
Cr	728	2.24	290	331.0	133	980	1667	145	450	525	980	1265	635	477	1030	140	328	258	977
Fe (wt%)	0.26	0.30	0.12	3.38	1.23	4.58	11.9	0.15	0.15	2.00	1.04	0.14	0.69	1.07	0.31	0.55	3.79	0.75	3.56
Co	0.69	0.69	3.19	17.9	5.61	70.7	57.5	8.53	1.49	5.26	60.8	67.3	5.91	5.28	12.2	4.31	9.09	9.95	23.2
Ni	10	16	22.9	170	60	519	484	71.5	29.1	147	220	379	58	30	115	22	90	56.7	134
Zn	28	6.3	3.89	10	6.5	15.4	30.7	3.69	6.63	38.9	18	42.3	38	4	2	80	14	6.94	18.5
As	10.2	44.1	70.5	14.6	5.68	54	16.3	13.9	20.3	1.79	32.7	490	1.81	1.40	0.72	4.79	0.62	0.52	38.6
Se	0.65	0.30	0.51	0.43	0.29	2.80	0.58	29.9	0.23	2.57	0.35	<1.3	0.11	0.17	0.08	0.20	0.32	0.01	<0.7
Br	0.21	0.54	0.3	0.47	0.22	<0.4	<0.6	0.4	<0.4	0.7	0.43	<0.8	0.30	0.2	0.32	0.28	0.67	0.5	1.0
Rb	67.8	30.0	8.18	1.7	2.2	14.3	2.29	6.99	63.6	335	20.1	23.9	31.1	92.1	23.6	32.8	12.1	108	89
Sr	<10	12	26	8	8	<20	<26	1374	10.8	<24	10	<56	6	8	10	14	10	5.47	2.65
Zr	30	30	22.7	30	35	72.3	126	13.9	94.7	61.9	40	104	15	13	3	20	9	19.1	63.9
Sb	0.43	2.72	2.26	0.5	0.27	1.12	0.35	2.96	1.36	0.66	0.17	8.31	0.12	0.10	0.024	0.066	0.082	0.17	0.82
Cs	1.32	0.33	0.39	0.1	0.17	0.71	28.2	0.36	1.21	8.84	1.01	0.6	0.17	0.49	0.53	0.83	1.05	0.61	2.09
Ba	205	450	118	495	1150	2730	60.2	15%	13400	5604	100	64.3	50	110	100	60	30	14.5	116
La	1.29	4.23	2.74	4.1	8.14	4.97	28.5	1.82	27.0	1.69	8.07	5.82	0.11	0.34	0.24	0.41	0.19	0.35	2.17
Ce	2.59	7.30	4.84	6.41	11.9	11.1	47.4	2.28	26.0	2.84	19.1	13.9	0.27	0.43	0.31	0.76	0.47	1.1	5.28
Nd	1.4	3.83	2.47	3.43	5.72	5.86	20.3	1.2	16.9	1.9	9.28	8	0.18	0.41	0.29	0.51	0.45	1.23	3.57
Sm	0.35	0.78	0.49	0.40	0.92	1.38	2.50	0.19	3.19	0.58	1.55	1.78	0.057	0.10	0.059	0.15	0.16	0.48	1.09
Eu	0.13	0.24	0.20	0.15	0.26	0.52	0.70	0.09	0.85	0.26	0.43	0.57	0.042	0.054	0.028	0.045	0.075	0.18	0.57
Gd	0.19	0.77	0.53	0.44	0.97	1.87	1.41	0.15	2.27	1.07	0.90	1.11	0.07	0.15	0.10	0.18	0.23	0.51	1.68
Tb	0.035	0.12	0.09	0.075	0.11	0.31	0.26	0.03	0.30	0.28	0.11	0.16	0.015	0.025	0.016	0.024	0.051	0.09	0.28
Tm	0.02	0.07	0.05	0.053	0.06	0.12	0.22	0.02	0.12	0.26	0.056	0.07	0.01	0.02	<0.01	<0.02	0.04	0.04	0.12
Yb	0.12	0.49	0.32	0.41	0.43	0.88	1.52	0.10	0.66	2.38	0.41	0.48	0.055	0.17	0.04	0.087	0.30	0.31	1.15
Lu	0.018	0.068	0.05	0.058	0.063	0.12	0.25	0.01	0.09	0.36	0.055	0.07	0.0079	0.024	0.0064	0.010	0.048	0.04	0.16
Hf	1.02	0.59	0.73	0.76	0.65	1.48	2.93	0.32	1.10	2.83	1.05	1.59	0.22	0.81	0.13	0.73	0.30	0.37	0.6
Ta	0.21	0.073	0.11	0.069	0.11	0.18	0.36	0.04	0.15	0.32	0.29	0.4	0.017	0.052	0.023	0.11	0.034	0.02	0.03
Ir (ppb)	<0.8	1.3	3.43	<0.3	<0.4	77.8	102	0.66	<0.7	5.06	<1	<0.5	<0.8	<0.8	0.3	<0.7	<0.9	0.06	0.78
Au (ppb)	0.6	<1	0.7	0.4	0.4	4.8	1.6	1.6	0.8	2.2	0.6	1.0	0.1	0.5	0.1	0.5	<1	2.4	1.5
Th	0.19	0.47	0.63	0.70	0.97	1.42	1.91	0.30	0.44	0.87	0.70	0.56	0.02	0.076	0.035	0.11	0.062	0.06	0.07
U	0.48	0.49	0.37	0.31	0.15	0.39	3.24	0.57	0.80	0.49	0.47	0.45	0.085	0.13	0.13	0.20	0.075	<0.3	<0.5
Th/Sc	0.010	0.107	0.212	0.268	0.424	0.134	0.420	0.072	0.114	0.021	0.056	0.035	0.001	0.002	0.005	0.008	0.003	0.003	0.002
Cr/Th	3832	5	460	473	137	690	873	483	1023	603	1400	2259	31750	6276	29429	1273	5290	4300	13957
La/Sc	0.065	0.959	0.923	1.559	3.555	0.469	6.264	0.434	6.977	0.040	0.646	0.366	0.007	0.008	0.032	0.030	0.008	0.019	0.064
Cr/Ni	72.8	0.1	12.7	1.9	2.2	1.9	3.4	2.0	15.5	3.6	4.5	3.3	10.9	15.9	9.0	6.4	3.6	4.6	7.3
LaN/SmN	2.32	3.41	3.52	6.40	5.57	2.27	7.17	6.03	5.32	1.83	3.28	2.06	1.21	2.14	2.56	1.72	0.75	0.46	1.25
GdN/YbN	1.28	1.27	1.34	0.87	1.82	1.71	0.75	1.21	2.78	0.36	1.77	1.87	1.03	0.71	2.02	1.67	0.62	1.33	1.18

Note: See Table 1 for sample description. Data are in ppm, except as noted.

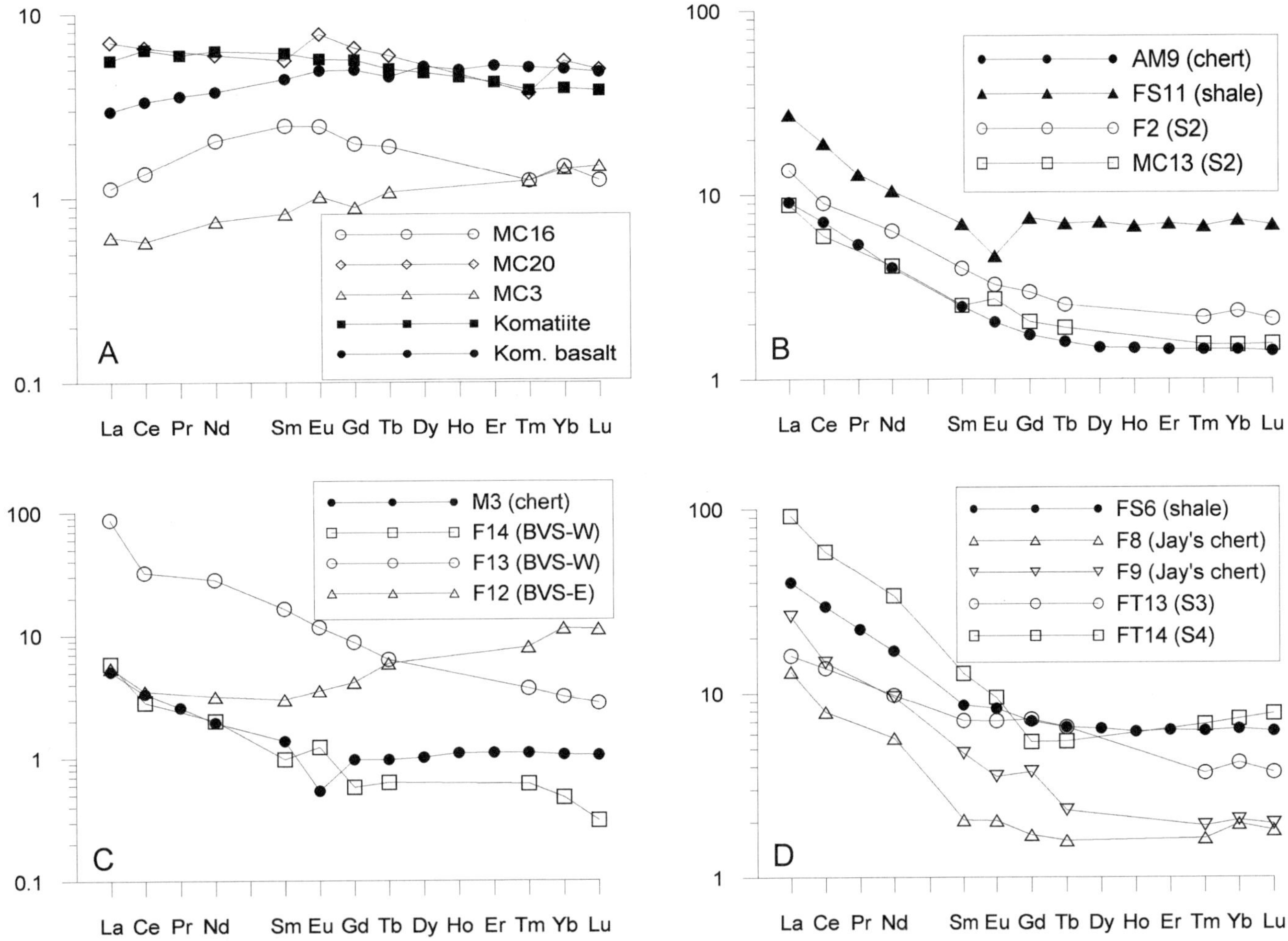

Figure 12. Chondrite-normalized (Boynton, 1984) rare earth element plots. (A) Accretionary lapillistone of the Msauli Chert in comparison with average compositions of komatiite (Komati Formation, n = 7, Lahaye et al., 1995) and komatiitic basalt (Weltevreden Formation, n = 2, Lahaye et al., 1995) from the Barberton greenstone belt. (B) Composition of spherule layer S2 in comparison with chert below and shale above the spherule layer. (C) Composition of spherule samples from the Barite Valley Syncline (BVS); the composition of chert below the spherule layer from the west limb of the syncline is also shown. (D) Composition of spherule layers S3 and S4 in comparison to the host rock (Jay's Chert) and underlying shale. See Hofmann (2005a) for chert and shale data and sample localities.

problem with using some of these criteria for Barberton spherules is that the spherule layers together with the interbedded host rocks have been subjected to intense alteration and metasomatism (de Wit et al., 1982; Paris et al., 1985; Lowe and Byerly, 1986; Duchac and Hanor, 1987; Hanor and Duchac, 1990; Lowe, 1999b) as well as tectonic deformation in many cases. The original mineral assemblages of all our samples have been replaced by chert, sometimes together with barite, siderite, and various phyllosilicate phases. Sericite and chlorite are minerals observed widely. However, the presence of botryoidal rims probably originally showing a radial-fibrous texture in some samples (Fig. 10B), the common presence of off-centered spherical "nuclei" in many samples, and characteristic shapes are consistent with their origin as former melt droplets, although such textural indicators fail to differentiate between impact and volcanic melt droplets.

Spheroidal particles that can be confused with impact spherules include volcanic spherules formed during lava fountaining, accretionary lapilli, and sedimentary ooids (Simonson, 2003). No ooids with their characteristic radial-concentric structures and central nuclei have yet been observed in Barberton sedimentary rocks. Accretionary lapilli are common in the stratigraphy; they generally form highly spherical grains, including dumbbell shapes in rare cases. Because accretionary lapilli are mechanical accretions of volcanic ash, they can generally be identified by the presence of vitric tuff particles, both within the lapilli and in the matrix, together with a frequently observed concentric lamination (Fig. 11D). Spherules representing quenched melt droplets of either volcanic or impact origin cannot be distinguished using textural characteristics, but tuffs that formed by lava fountaining are believed to occur over much smaller areas (<10 km lateral extent) than impact deposits (Simonson, 2003).

Thin layers consisting mostly of former melt droplets and which are laterally very extensive are, thus, more likely to be of impact origin. However, layers consisting of only a small proportion of spherules but much reworked material and showing evidence that they formed by density currents could well have formed by volcanic processes and then have been redistributed over much larger areas by such currents, covering distances well exceeding tens of kilometers. This holds especially true for layers that contain unequivocal pyroclastic material. Most of the Barberton spherule layers and similar beds reflect short-lived, high-energy events and show evidence for waning current activity. Although preferably attributed to tsunami wave action (Lowe et al., 2003), high-energy density currents could have been equally capable of depositing many of these layers.

Correlation of Spherule Beds

The main reason for the current uncertainties with respect to the correlation of spherule layers situated along the Onverwacht–Fig Tree contact is that the contact in different tectono-stratigraphic units is apparently diachronous (Kröner et al., 1991; Byerly et al., 1996). Near the S2 type section, a dacitic tuff bed intercalated with ferruginous cherts occurring ~20 m above the spherule layer has yielded a SHRIMP (sensitive high-resolution ion microprobe) zircon age of 3258 ± 3 Ma (Byerly et al., 1996), suggesting that the spherule bed is slightly older. This age is corroborated by zircon ages for a dacitic breccia and a tuffaceous sandstone of the Auber Villiers Formation of ca. 3255 Ma (Byerly et al., 1996). This unit tectonically overlies the Fig Tree strata in the vicinity of the S2 type locality along the Auber Villiers Fault (Fig. 2).

On the other hand, a dacitic tuff underlying a spherule bed at the Onverwacht–Fig Tree contact near the Inyoka Fault and close to the Maid of the Mountain Mine (Loop Road locality of Lowe et al., 2003; Fig. 1) has been dated at 3243 ± 4 Ma (Kröner et al., 1991), suggesting that this spherule layer is younger than the previous layer. This 3243 Ma age is in contrast to zircon ages of ca. 3255 Ma obtained from felsic siliciclastic rocks of the Bien Venue Formation (Kröner et al., 1991; Kohler, 2003). The Bien Venue Formation locally overlies the Sheba and Belvue Road Formations in the northern part of the greenstone belt and has been correlated with the Auber Villiers Formation (Kohler and Anhaeusser, 2002), suggesting that the Onverwacht–Fig Tree contact cannot be younger than 3255 Ma in the northern part of the greenstone belt as well. It seems likely that the Onverwacht–Fig Tree contact is *broadly* synchronous throughout the Barberton greenstone belt; i.e., it must be located within the time interval of ca. 3260–3255 Ma. This is supported by the lack of evidence for continued mafic magmatism in the southern part of the greenstone belt after ca. 3255 Ma.

The age of 3243 Ma is difficult to interpret. According to Lowe et al. (2003), the sample locality lies within a few meters of a major fault and associated rocks are strongly altered, possibly affecting the clarity of the field relationships. The age could also reflect intrusive felsic magmatism, as indicated by an identical age from a felsic porphyry in the Kromberg Formation (Kröner et al., 1991). If the above assumptions are correct, the spherule layer developed along the Onverwacht–Fig Tree contact could be a single layer. If this is the case, spinel character and abundance, Ir contents, and spherule types cannot be used for correlation purposes as has been suggested by Lowe et al. (2003). This could also mean that the event deposit located along the Onverwacht–Fig Tree contact formed during meteorite impact; the presence of accretionary lapilli in this unit would argue for relatively proximal ejecta deposits.

On the other hand, there is geochemical evidence (Hofmann, 2005a) that the Fig Tree Group consists of different stratigraphic packages in different tectono-stratigraphic units. Based on lithological and geochemical data (Heinrichs, 1980; Hofmann, 2005a), the Fig Tree Group in the lower tectono-stratigraphic unit consists of the Loenen, Ngwenya, and Mapepe Formations. Spherule layers S3 and S4 occur at the base of the Mapepe Formation (Fig. 5). In the upper tectono-stratigraphic unit, the Loenen and possibly the Ngwenya Formation are missing, and S2 occurs at the base of the Mapepe Formation. If the proposed correlations are correct, then S2 and S3/S4 could represent laterally correlative beds.

The stratigraphy of the Fig Tree Group in the Barite Valley Syncline is even more complex. Tuffaceous rocks ~300 m above the Onverwacht–Fig Tree contact and ~150 m above a spherule layer along the eastern limb of the Barite Valley Syncline (Fig. 9) have been dated at 3227 ± 4 Ma (Kröner et al., 1991). This age is identical to ages obtained for the Schoongezicht Formation (Kröner et al., 1991; Kamo and Davis, 1994), which is the stratigraphically uppermost unit of the Fig Tree Group. This age indicates that the section of Fig Tree Group rocks in the Barite Valley Syncline must be condensed and could contain unconformities, as suggested by Heinrichs and Reimer (1977) and Lowe and Nocita (1999), making stratigraphic correlations between spherule layers at this locality difficult.

A further point worth mentioning in this context is the regional correlation of spherule bed S1. Zircons in this spherule layer have been dated at 3470 ± 2 Ma (Byerly et al., 2002). The same age was reported by these authors for a spherule layer in the Pilbara craton and used as evidence that these two deposits could represent a single fallout layer of global extent. The zircons were interpreted as locally derived detritus eroded by impact-generated tsunami waves and representing coeval or slightly older felsic igneous activity, thus providing an approximate depositional age for the spherule layer (Byerly et al., 2002). It must be noted, however, that zircons from the Middle Marker chert, which is situated ~2600 m stratigraphically below S1, yielded, within error, the same age (3472 ± 5 Ma; Armstrong et al., 1990), casting doubts on the usefulness of these zircon ages for correlation purposes. Considering the fact that possible impact spherules in S1 samples only constitute a minor component, the correlation between the zircons and the impact spherules also remains tenuous.

How Many Spherule Layers?

We have shown that layer S2 and layer S3 of the northern facies could be correlative. It is also possible that layer S2 and layer S3/S4 of the southern facies are the same. We acknowledge the fact that stratigraphic correlations in the Fig Tree Group are difficult, but these difficulties need to be addressed before inferences on the impact flux during the Archean should be made. A further problem is that many spherule layers occur as pairs of normally graded beds (e.g., the S2 layer) that can be separated by at least several decimeters of strata. Could this mean that the S4 layer, which occurs ~7 m above S3, consists of reworked spherules that originally formed during the S3 event and were only later introduced to the depositional site? Spherules in both layers are compositionally and texturally similar, and both layers have a large calculated extraterrestrial component (Kyte et al., 2003). The number of spherule layers in the Barberton greenstone belt may thus be less than currently believed. However, apparently additional spherule layers have since been discovered, for example at the contact between the Sheba and Belvue Road Formations (Fig. 3; Lowe et al., 2003), but have not yet been described in detail.

Siderophile Element (e.g., Iridium) Enrichment

Layers S3 and S4 show Ir enrichments that for a considerable number of samples are similar to, and even exceed, the chondritic level by factors of up to 4 (Koeberl et al., 1993; Koeberl and Reimold, 1995; Reimold et al., 2000; Kyte et al., 2003). Several explanations have been offered by Lowe et al. (2003), including element fractionation in impact-produced rock vapor clouds that resulted in the formation of compositionally heterogeneous spherules, fractionation of Ir-enriched spinel-bearing spherules during fall and reworking due to density contrasts, pressure solution of spherule beds with Ir enrichment in the residual material, and hydrothermal alteration that resulted in element redistribution. The question remains, though, why these processes would have operated only in the case of the Barberton spherule layers. PGE abundances in more recent impact deposits (e.g., K/T boundary layer) or even in the 2.6 Ga impact spherule layers in South Africa (e.g., the Monteville layer) and in Australia (e.g., the Jeerinah layer) are much more in line with expectations for impact deposits (i.e., they contain only up to a few percent of extraterrestrial material; Simonson and Glass, 2004; Rasmussen and Koeberl, 2004).

The observations by Koeberl et al. (1993), Koeberl and Reimold (1995), and Reimold et al. (2000) included extremely high PGE abundances in some spherule layer samples, which also showed near-chondritic abundance ratios. Furthermore, they described the association of high PGE abundances with secondary sulfide mineralization (such as gersdorffite) and a positive correlation between chalcophile elements (e.g., As, Sb, Se) and the siderophile element contents. This latter finding suggested that the hydrothermal alteration that pervasively affected the Barberton spherule layers did not affect the siderophile element contents, unless the alteration was actually responsible for these strong enrichments of siderophile elements. This latter possibility seriously questions the validity of the widely held belief that the Ir contents are a meteoritic component and can be used to determine sizes of bolides that caused the impacts responsible for the formation of these spherule layers. In addition, it is difficult to understand why the PGE interelement ratios would not change significantly during any post-formational hydrothermal redistribution and sulfide mineralization.

The earlier geochemical work by Koeberl et al. (1993) and Koeberl and Reimold (1995) has been criticized (Lowe et al., 2003) as having focused on sulfide- and gold-mineralized—and, thus, hydrothermally affected—areas at Sheba gold mine. In order to address the issue of element redistribution during hydrothermal alteration, the material investigated by Reimold et al. (2000) was derived from outside the immediate sulfide/gold-mineralized zone mined at Agnes gold mine. However, again, good correlation between siderophile element abundances and those for highly mobile, likely hydrothermally introduced elements was observed. In addition, our current examination of samples from all spherule beds so far described in the Barberton belt by Lowe and coworkers has shown that at least trace sulfide mineralization is a feature of all these beds. On the other hand, it has been repeatedly noted that Ir should not be considered in isolation. Koeberl and Reimold (1995) and Reimold et al. (2000) demonstrated that Ir is well correlated with other volatile and mobile elements, such as Zn or As. The point made was that Ir concentrations cannot be attributed, per se, to a meteoritic component and applied to calculate a meteoritic component or even infer sizes of projectiles.

Lowe et al. (2003) noted that Ir abundances as high as, and even higher than, those of the spherule layers have been reported from the K/T boundary layer and meteorites. Although extremely high (or low) Ir concentrations in individual samples of impact deposits can be a result of heterogeneous distribution of ejecta phases or chemical redistribution by post-depositional diagenetic or metasomatic processes (Shukolyukov et al., 2000), there is no K/T boundary sample in the world that has Ir (or other PGE) contents that exceed the chondritic abundances. Ir contents at various terrestrial K/T boundary sites vary from ~1–26 ppb; in marine layers values up to ~50 ppb have been measured, and the most organic-rich sublayers at marine sites (e.g., Stevns Klint; Caravaca) have ~60–180 ppb Ir (e.g., Schmitz, 1988). There are some meteorites (iron meteorites) that have Ir contents up to 20 ppm (which is indeed higher than chondrites at 0.7 ppm or the highest values in the Barberton spherule layers at ~2.4 ppm), but this is irrelevant because the impactor certainly was not an iron meteorite for a variety of reasons, such as the chromium isotope compositions that indicate a carbonaceous chondritic impactor (Shukolyukov et al., 2000; Kyte et al., 2003).

Volcanic versus Impact Origin for Spherules

Quenched silicate melt droplets form from either lava fountaining during volcanic eruptions or from meteorite impacts and show identical petrographic characteristics (Heiken, 1972; Simonson, 2003). Because of this, unequivocal petrographic distinction between volcanic and impact spherules is hardly possible. Lowe et al. (2003) emphasized a couple of points as being in favor of a meteorite impact origin for the Barberton spherules and spherule layers, such as the regional extent of the spherule beds, the compositional diversity of the spherules, their distinctiveness as compared to other volcanic particles, and their restricted presence in unique beds. However, all these points also apply to accretionary lapilli that form distinct, laterally traceable horizons in the Barberton greenstone belt. Possibly the strongest point raised by Lowe et al. (2003) for an impact origin in this context may be the absence of pyroclastic material, such as shards, vesicular grains, or phenocrysts, in the spherule layers. Although this may hold true for many of the spherule layers, it is definitely not the case for the S1 layer, which contains glass shards (Fig. 4B) and scoriaceous particles (Fig. 4C). Shardlike particles were also observed associated with S3 (Fig. 10A). The presence of pyroclastic grains associated with the S1 spherule bed thus raises the possibility that the spherules could be either of volcanic origin rather than of impact origin or that materials of different origin were mixed prior to deposition or in the depositional environment.

On the other hand, certain layers not regarded as impact deposits and containing abundant pyroclastic material also contain spherical particles that resemble impact spherules (Figs. 6D–6F). Furthermore, some of these layers, such as the layer at the top of banded cherts along the stratigraphically lowermost Onverwacht–Fig Tree contact (Fig. 5), are laterally as extensive as the spherule layers. This and similar layers contain conspicuous clasts that are characterized by a porphyritic texture, as indicated by idiomorphic, equant chert domains (possibly after olivine), although some of these domains are spherical and may represent chert-filled vesicles (Fig. 6B). The same clast type occurs in S1 (see above) and is very common in 3.4 Ga silicified arenitic sedimentary rocks of the Pilbara craton (Dunlop and Buick, 1981). Interestingly, the impact layer formed by the ca. 2 Ma Eltanin asteroid (Gersonde et al., 1997) contains abundant olivine-phyric vesicular particles, very similar to the ones described here, that have been interpreted as shock-melted debris derived from this oceanic impact (Kyte and Brownlee, 1985). This discussion shows that certain clasts that resemble pyroclastic material may have originated during impact events. This holds true even for accretionary lapilli (Graup, 1981), which seem to be more common in impact ejecta than previously thought, as indicated by a number of recent discoveries (e.g., Addison et al., 2005; Schulte and Kontny, 2005). Unequivocal petrographic distinction between volcanic and impact-derived particles is thus very difficult, if not impossible.

The geochemical evidence presented in favor of a meteorite impact origin, such as Ir anomalies (Lowe et al., 1989; Kyte et al., 1992), near-chondritic PGE ratios (Kyte et al., 1992), spinel compositions unlike those in associated komatiites (Byerly and Lowe, 1994), and Cr/Ir ratios similar to those in carbonaceous chondrites (Lowe et al., 1989, 2003) cannot rule out other processes completely, as discussed in various sections of this contribution. Chromium isotope ratios from the Barberton layers have been stated to indicate an extraterrestrial Cr component, suggesting carbonaceous chondrite projectiles (Shukolyukov et al., 2000; Kyte et al., 2003). With respect to spinel composition, it may be true that the spinels in the spherule layers are unlike igneous spinels in interbedded komatiites (Byerly and Lowe, 1994). However, the spinels in these layers are also compositionally different compared to similar spinels from known impact layers (e.g., late Eocene; K/T boundary). Reimold et al. (2000) determined that at least some Cr-spinel in Barberton spherule samples is distinctly zoned, with outer zonation likely a result of hydrothermal overprint. These authors raised the questions as to whether cosmic chromium remains in a regionally open chemical system without significant fractionation and whether chromium isotopic signatures in the various Barberton spherule beds are similar. This latter question still requires further work. Clearly, the chemical characteristics of the spherule-bearing layers in the Barberton stratigraphy that, to date, have not been directly linked to impact must be established in order to be able to better constrain the impact signature.

Spherule Formation

Spherules in the Barberton greenstone belt are interpreted to have formed by condensation of global clouds of impact-generated rock vapor rather than ballistic particles thrown out from the impact sites (Lowe et al., 1989; Kyte et al., 1992; Byerly and Lowe, 1994; Lowe et al., 2003). This interpretation is based on (1) the absence of ballistic particles in the spherule beds (except for possible detritus in S1), indicating that the beds represent distal impact deposits far from the original crater sites; (2) an assumed high proportion of bolide material in the layers (Kyte et al., 1992; Byerly and Lowe, 1994); and (3) the similarity of the Barberton spherules to younger microkrystites (Lowe et al., 2003).

The interpretation of these layers as distal impact ejecta is clearly problematic. The high proportion of spherule material in some of these strata is in contrast to distal impact ejecta from more recent impact events that have highly diluted spherules, if any. There is also the excessively high proportion of meteoritic component in these layers, with more modern microkrystite layers having a much lower meteoritic component (Glass et al., 2004). Clearly, our understanding of the distribution of meteoritic material in and outside the crater region is far from good. In addition, the distribution of impact ejecta itself, especially that of impact melt that would likely contain the bulk of the nonvaporized meteoritic projectile mass, is also not well constrained. Our knowledge is based mainly on work on quite small impact events, such as the Ries or Bosumtwi events that produced crater structures of 24 and 10 km, respectively, but that produced substantial amounts of impact melt that was dispersed widely in the form of tektites, in addition to the sizeable amounts of melt incorporated into suevitic breccias. Furthermore, there are the K/T boundary and late Eocene ejecta layers (for a recent review, see Simonson and Glass, 2004). However, a direct comparison between these more recent impact events and those recorded in the Archean impact layers is naturally hindered by the different target assemblages—more mafic in the case of the Archean impacts—as well as the comprehensive secondary alteration of the Archean impact deposits. Overall, there are no definite constraints on the size or the velocity of the bolides, and these factors would obviously play a major role in the production of kinetic energy and, consequently, melt volume generated. In the absence of much detail regarding the formation of large impact structures in mafic volcanic terrains (the terrestrial impact record is very limited in this regard), it is not well known how melt formation scales with event magnitude and how melt accumulation and/or dispersal behave in such events of impact into oceanic, basaltic crust. Another critical factor with regard to scaling of impact events from levels of meteoritic contamination in ejecta is the fact that distal impact ejecta cannot be taken a priori as representing target-derived ejecta alone. In contrast, it is mandatory to consider these materials, at least in the closer environs of impact structures, as mixtures of ejecta plus ground-derived debris.

Bolide Size Estimate

Various parameters have been used to estimate the size of the bolides that are thought to be responsible for the Barberton spherule layers, including spherule bed thickness (Sleep et al., 1989), spherule size (Lowe et al., 1989; Melosh and Vickery, 1991; Byerly and Lowe, 1994), Ir concentration (Kyte et al., 1992; Byerly and Lowe, 1994), and extraterrestrial Cr content (Shukolyukov et al., 2000). Most yielded estimates of 20–50 km for the bolides. It is clear that some of these estimates are problematic for several reasons. For example, it is unknown if the beds were deposited globally and, even if they were deposited over vast areas, how their thicknesses varied spatially. The zircon evidence (Byerly et al., 2002) that suggests that spherule layer S1 can be correlated with a similar bed in the Warrawoona Group of the Pilbara craton is ambiguous (as discussed in detail above). In addition, Ir values are highly erratic and can exceed chondritic values, probably as a result of secondary alteration and mineralization processes. Many other variables, such as target composition and water depth, are equally unknown, making any bolide size estimates highly speculative at best.

CONCLUSIONS

The identification and significance of Precambrian impact deposits remains an interesting problem for future studies. Many geologic and petrographic criteria for the identification of Archean impact deposits cannot clearly differentiate between a volcanic and an impact origin. Geochemical studies face problems associated with hydrothermal alteration or insufficient data, as the Cr isotope method has not yet been applied widely. In addition, some classic criteria for the identification of impacts (observation of source craters and shocked minerals) may not be applicable to these ancient rocks.

No source craters have yet been identified for the Archean and Paleoproterozoic spherule layers from South Africa and Australia. Given the scarcity of the Archean geological record, it is most likely that they will never be found, especially if the impacts took place on oceanic crust. The oldest-known continental impact structure on Earth is the originally 250–300-km-large Vredefort structure in South Africa (Gibson and Reimold, 2001, for a recent review; also Koeberl, this volume). The Vredefort impact structure is deeply eroded, with 7–10 km erosion depth. No unequivocal trace of ejecta from this event has been determined to date. Only the putative proposition by Chadwick et al. (2001) that the ca. 1.8–2.1 Ga Greenland spherule occurrence could be related to either the Vredefort or Sudbury impact event has been noted, although it may not be related to either one of these impact events.

None of the Archean or Proterozoic spherule layers known to date is associated with shocked minerals, which are the hallmark for all confirmed impact structures and ejecta. The notable exception is a single grain of quartz with planar deformation features from the 2.63 Ga Jeerinah layer (Rasmussen and Koeberl, 2004). It has been suggested that the Archean impacts involved oceanic crust, resulting in the absence of shocked quartz or zircon, and that alteration of the spherule beds overprinted potential shock features associated with other minerals (Byerly and Lowe, 1994; Simonson et al., 1998; Lowe et al., 2003). Unshocked quartz and zircon grains are not uncommon in the Barberton spherule layers, although they have been regarded to represent locally derived detritus (Byerly et al., 2002).

It is not clear why proposed impact events in the Archean and Paleoproterozoic predominantly produced large volumes of spherules (aggregate thickness of spherules in spherule layers of up to several decimeters; Simonson and Harnik, 2000), which are mostly absent from post-Archean impact deposits, i.e., those for which source craters are known. If this is purely the result of a higher impact flux and of the impact of much larger bolides at that time, we should expect to find many other, and generally thinner, impact-induced spherules or other meteoritic debris-bearing layers in these ancient strata. No evidence for this has so far been presented. Although high-energy event deposits are common in the Barberton greenstone belt throughout the stratigraphy, it is more reasonable to regard these beds as deposits of tsunami waves and gravity flows, triggered by volcanic or earthquake activity rather than by impacts. The impact flux during Fig Tree times was probably slightly higher than today, maybe by as much as a factor of two from back-extrapolating the known cratering rates, but this would not account for the large number of large impacts proposed by Lowe and coworkers half a billion years after the late heavy bombardment (Ryder et al., 2000; Ryder, 2002; Koeberl, 2004).

Despite the discovery by Shukolyukov et al. (2000) that the S4 layer and by Kyte et al. (2003) that also the S3 (and possibly the S2) layers have chromium isotope anomalies that point toward a carbonaceous chondritic component, many questions remain to be addressed. We accept that there is an extraterrestrial component present in some layers, but this does not invalidate the many questions noted above, such as the unusually high calculated extraterrestrial Cr content and extremely high Ir values found in some of the spherule layers, up to 2700 ppb Ir (Koeberl and Reimold, 1995), which is four times the Ir concentration in chondrites (including carbonaceous chondrites). Clearly, the spherule layers do not consist of 100% of meteoritic material. No impact layer ever does. Even coherent impact melt rocks in the central parts of large impact structures only rarely carry more than 1%–2% meteoritic component (e.g., Koeberl, 1998). Many of the Barberton spherule layers, including those that yielded high Ir values, are not even composed of 100% spherules and instead are strongly diluted by locally derived detritus (or a mixture of impact-ejected target material plus local detritus). Dilution by early diagenetic chert cement, which can constitute up to 40 vol% of spherule samples (Table 1), also needs to be taken into account. Apart from elevated PGE and some transition metal contents, the trace element contents of the spherule layers are similar to the composition of the host rocks because of the incorporation of locally derived detritus.

The problem of excessive amounts of extraterrestrial component was observed by Kyte et al. (2003, p. 286), who concluded "the Cr isotope data require extraterrestrial accretion as a source of materials in these beds, and we must find models that fit the data." Consequently, there must have been specific concentration mechanisms at work. Kyte et al. (2003, p. 286) mentioned "interplanetary dust storms" but were themselves critical of this possibility as it "is much less likely and inconsistent with large tsunamis proposed by Lowe et al. (1989, 2003)." It is again noted that the spherule beds consist of a mix of possible impact, volcanogenic, and reworked sedimentary material plus an early diagenetic and hydrothermally added component.

Another problem is why the PGE ratios did not change during hydrothermal alteration that resulted in sulfide mineralization of some layers. And when exactly did this secondary mineralization happen? It has definitely taken place much later than the deposition of the spherule layers, most likely later than 3.1 Ga (de Ronde et al., 1991). The detection of extraterrestrial Cr does not answer any of these questions raised. Further research on the Barberton spherule beds is required before inferences on the Archean impact flux can be attempted. The stratigraphic problems reviewed here can only be addressed by further lateral mapping in association with detailed analyses of the clast and spherule populations in samples from all known occurrences. A study of other high-energy event layers in the Barberton greenstone belt is also necessary in order to establish compositional similarities and differences in the spherule layers. Notably, the Ir carrier is still elusive (Reimold et al., 2000), and further microchemical work on the unique spinel mineralogy is required.

ACKNOWLEDGMENTS

AH acknowledges funding by the Deutsche Forschungsgemeinschaft (Ho 2507/1-1/2) and from the University of the Witwatersrand Research Committee. WUR is funded by the National Research Foundation of South Africa. CK has been supported by the Austrian Science Foundation, project P17194-N10. We are grateful to Johan Eksteen, Mpumalanga Parks Board, for access and Colin Wille, Taurus Estate, for access and hospitality. Reviews by Bruce Simonson and Alex Bevan and editorial comments by Roger Gibson are gratefully acknowledged. This is University of the Witwatersrand Impact Cratering Research Group Contribution No. 97.

REFERENCES CITED

Addison, W.D., Brumpton, G.R., Vallini, D.A., McNaughton, N.J., Davis, D.W., Kissin, S.A., Fralick, P.W., and Hammond, A.L., 2005, Discovery of distal ejecta from the 1850 Ma Sudbury impact event: Geology, v. 33, p. 193–196, doi: 10.1130/G21048.1.

Armstrong, R.A., Compston, W., de Wit, M.J., and Williams, I.S., 1990, The stratigraphy of the 3.5–3.2 Ga Barberton greenstone belt revisited: A single zircon ion microprobe study: Earth and Planetary Science Letters, v. 101, p. 90–106, doi: 10.1016/0012-821X(90)90127-J.

Boynton, W.V., 1984, Geochemistry of the rare earth elements: Meteorite studies, *in* Henderson, P., ed., Rare Earth Element Geochemistry: Amsterdam, Elsevier, p. 63–114.

Byerly, G.R., and Lowe, D.R., 1994, Spinel from Archean impact spherules: Geochimica et Cosmochimica Acta, v. 58, p. 3469–3486, doi: 10.1016/0016-7037(94)90099-X.

Byerly, G.R., Kröner, A., Lowe, D.R., Todt, W., and Walsh, M.M., 1996, Prolonged magmatism and time constraints for sediment deposition in the early Archean Barberton greenstone belt: Evidence from the Upper Onverwacht and Fig Tree groups: Precambrian Research, v. 78, p. 125–138, doi: 10.1016/0301-9268(95)00073-9.

Byerly, G.R., Lowe, D.R., Wooden, J.L., and Xie, X., 2002, An Archean impact layer from the Pilbara and Kaapvaal cratons: Science, v. 297, p. 1325–1327, doi: 10.1126/science.1073934.

Chadwick, B., Claeys, P., and Simonson, B.M., 2001, New evidence for a large Palaeoproterozoic impact: Spherules in a dolomite layer in the Ketilidian orogen, South Greenland: Journal of the Geological Society of London, v. 158, p. 331–340.

de Ronde, C.E.J., Kamo, S., Davis, D.W., de Wit, M.J., and Spooner, E.T.C., 1991, Field, geochemical and U-Pb isotopic constraints from hypabyssal felsic intrusions within the Barberton greenstone belt, South Africa: Implications for tectonics and the timing of gold mineralization: Precambrian Research, v. 49, p. 261–280, doi: 10.1016/0301-9268(91)90037-B.

de Wit, M.J., Hart, R.J., Martin, A., and Abbott, P., 1982, Archaean abiogenic and probable biogenic structures associated with mineralized hydrothermal vent systems and regional metasomatism, with implications for greenstone belt studies: Economic Geology and the Bulletin of the Society of Economic Geologists, v. 77, p. 1783–1802.

Duchac, K.C., and Hanor, J.S., 1987, Origin and timing of metasomatic silicification of an early Archean komatiite sequence, Barberton Mountain Land, South Africa: Precambrian Research, v. 37, p. 125–146, doi: 10.1016/0301-9268(87)90075-1.

Dunlop, J.S.R., and Buick, R., 1981, Archaean epiclastic sediments derived from mafic volcanics, North Pole, Pilbara Block, Western Australia: Geological Society of Australia Special Publication 7, p. 225–233.

Fisher, R.V., and Schmincke, H.U., 1984, Pyroclastic Rocks: Berlin, Springer-Verlag, 472 p.

Gersonde, R., Kyte, F.T., Bleil, U., Diekmann, B., Flores, J.A., Gohl, K., Grahl, G., Hagen, R., Kuhn, G., Sierro, F.J., Völker, D., Abelmann, A., and Bostwick, J.A., 1997, Geological record and reconstruction of the late Pliocene impact of the Eltanin asteroid in the Southern Ocean: Nature, v. 390, p. 357–363, doi: 10.1038/37044.

Gibson, R.L., and Reimold, W.U., 2001, The Vredefort Impact Structure, South Africa—The Scientific Evidence and a Two-Day Excursion Guide: Pretoria, Council for Geoscience, Memoir 92, 111 p.

Glass, B.P., Huber, H., and Koeberl, C., 2004, Geochemistry of Cenozoic microtektites and clinopyroxene-bearing spherules: Geochimica et Cosmochimica Acta, v. 68, p. 3971–4006, doi: 10.1016/j.gca.2004.02.026.

Glikson, A.Y., 1999, Oceanic mega-impacts and crustal evolution: Geology, v. 27, p. 387–390, doi: 10.1130/0091-7613(1999)027<0387:OMIACE>2.3.CO;2.

Graup, G., 1981, Terrestrial chondrules, glass spherules and accretionary lapilli from the suevite, Ries Crater, Germany: Earth and Planetary Science Letters, v. 55, p. 407–418, doi: 10.1016/0012-821X(81)90168-0.

Hanor, J.S., and Duchac, K.C., 1990, Isovolumetric silicification of early Archean komatiites: Geochemical mass balances and constraints on origin: The Journal of Geology, v. 98, p. 863–877.

Heiken, G., 1972, Morphology and petrography of volcanic ashes: Geological Society of America Bulletin, v. 83, p. 1961–1988.

Heinrichs, T., 1980, Lithostratigraphische Untersuchungen in der Fig Tree Gruppe des Barberton Greenstone Belt zwischen Umsoli und Lomati (Südafrika): Göttinger Arbeiten zur Geologie und Paläontologie 22, 118 p.

Heinrichs, T., 1984, The Umsoli chert, turbidite testament for a major phreatoplinian event at the Onverwacht/Fig Tree transition (Swaziland Supergroup, Archaean, South Africa): Precambrian Research, v. 24, p. 237–283, doi: 10.1016/0301-9268(84)90061-5.

Heinrichs, T.K., and Reimer, T.O., 1977, A sedimentary barite deposit from the Archean Fig Tree Group of the Barberton Mountain Land (South Africa): Economic Geology and the Bulletin of the Society of Economic Geologists, v. 72, p. 1426–1441.

Hofmann, A., 2005a, The geochemistry of sedimentary rocks from the fig Tree Group, Barberton greenstone belt: Implications for tectonic, hydrothermal and surface processes during the mid-Archaean times: Precambrian Research, v. 143, p. 23–49.

Hofmann, A., 2005b, Silica alteration zones in the Barberton greenstone belt: A window into subseafloor processes 3.5–3.3 Ga ago: Durban, South Africa, GEO2005, Abstracts Volume, University of KwaZulu-Natal, p. 111–112.

Kamo, S.L., and Davis, D.W., 1994, Reassessment of Archean crustal development in the Barberton Mountain Land, South Africa, based on U-Pb dating: Tectonics, v. 13, p. 167–192, doi: 10.1029/93TC02254.

Koeberl, C., 1993, Instrumental neutron activation analysis of geochemical and cosmochemical samples: A fast and reliable method for small sample analysis: Journal of Radioanalytical and Nuclear Chemistry, v. 168, p. 47–60, doi: 10.1007/BF02040877.

Koeberl, C., 1998, Identification of meteoritic components in impactites, *in* Grady, M.M., Hutchison, R., McCall, G.J.H., and Rothery, D.A., eds., Meteorites: Flux with Time and Impact Effects: Geological Society [London] Special Publication 140, p. 133–153.

Koeberl, C., 2004, The late heavy bombardment in the inner solar system: Is there a connection to Kuiper Belt objects?: Earth, Moon, and Planets, v. 92, p. 79–87.

Koeberl, C., 2006, this volume, The record of impact processes on the early Earth: A review of the first 2.5 billion years, *in* Reimold, W.U., and Gibson, R.L., Processes on the early Earth: Geological Society of America Special Paper 405, 10.1130/2006.2405(01).

Koeberl, C., and Reimold, W.U., 1995, Early Archaean spherule beds in the Barberton Mountain Land, South Africa: No evidence for impact origin: Precambrian Research, v. 74, p. 1–33, doi: 10.1016/0301-9268(94)00099-D.

Koeberl, C., Reimold, W.U., and Boer, R.H., 1993, Geochemistry and mineralogy of early Archean spherule beds, Barberton Mountain Land, South Africa: Evidence for origin by impact doubtful: Earth and Planetary Science Letters, v. 119, p. 441–452, doi: 10.1016/0012-821X(93)90152-Y.

Kohl, I., Simonson, B.M., and Berke, M., 2006, this volume, Diagenetic alteration of impact spherules in the Neoarchean Monteville layer, South Africa, *in* Reimold, W.U., and Gibson, R.L., Processes on the early Earth: Geological Society of America Special Paper 405, 10.1130/2006.2405(04).

Kohler, E.A., 2003, The geology of the Archaean granitoid-greenstone terrain in the vicinity of Three Sisters, Barberton Greenstone Belt: Geological Survey of South Africa Bulletin 133, 150 p.

Kohler, E.A., and Anhaeusser, C.R., 2002, Geology and geodynamic setting of Archaean silicic metavolcaniclastic rocks of the Bien Venue Formation, Fig Tree Group, northeast Barberton greenstone belt, South Africa: Precambrian Research, v. 116, p. 199–235, doi: 10.1016/S0301-9268(02)00021-9.

Kröner, A., Byerly, G.R., and Lowe, D.R., 1991, Chronology of early Archean granite-greenstone evolution in the Barberton Mountain Land, South Africa, based on precise dating by single zircon evaporation: Earth and Planetary Science Letters, v. 103, p. 41–54, doi: 10.1016/0012-821X(91)90148-B.

Kyte, F.T., and Brownlee, D.E., 1985, Unmelted meteoritic debris in the late Pliocene iridium anomaly: Evidence for the ocean impact of a nonchondritic asteroid: Geochimica et Cosmochimica Acta, v. 49, p. 1095–1108, doi: 10.1016/0016-7037(85)90001-8.

Kyte, F.T., Zhou, L., and Lowe, D.R., 1992, Noble metal abundances in an early Archean impact deposit: Geochimica et Cosmochimica Acta, v. 56, p. 1365–1372, doi: 10.1016/0016-7037(92)90067-S.

Kyte, F.T., Shukolyukov, A., Lugmair, G.W., Lowe, D.R., and Byerly, G.R., 2003, Early Archean spherule beds: Chromium isotopes confirm origin through multiple impacts of projectiles of carbonaceous chondrite type: Geology, v. 31, p. 283–286, doi: 10.1130/0091-7613(2003)031<0283:EASBCI>2.0.CO;2.

Lahaye, Y., Arndt, N., Byerly, G., Chauvel, C., Foucade, S., and Gruau, G., 1995, The influence of alteration on the trace-element and Nd-isotopic compositions of komatiites: Chemical Geology, v. 126, p. 43–64, doi: 10.1016/0009-2541(95)00102-1.

Lowe, D.R., 1999a, Shallow-water sedimentation of accretionary lapilli-bearing strata of the Msauli Chert: Evidence of explosive hydromagmatic komatiitic volcanism, *in* Lowe, D.R., and Byerly, G.R., eds., Geologic Evolution of the Barberton Greenstone Belt, South Africa: Geological Society of America Special Paper 329, p. 213–232.

Lowe, D.R., 1999b, Petrology and sedimentology of cherts and related silicified sedimentary rocks in the Swaziland Supergroup, *in* Lowe, D.R., and Byerly, G.R., eds., Geologic Evolution of the Barberton Greenstone Belt, South Africa: Geological Society of America Special Paper 329, p. 83–114.

Lowe, D.R., and Byerly, G.R., 1986, Archean flow-top alteration zones formed initially in a low-temperature sulphate-rich environment: Nature, v. 324, p. 245–248, doi: 10.1038/324245a0.

Lowe, D.R., and Byerly, G.R., 1999, Stratigraphy of the west-central part of the Barberton Greenstone Belt, South Africa, *in* Lowe, D.R., and Byerly,

G.R., eds., Geologic Evolution of the Barberton Greenstone Belt, South Africa: Geological Society of America Special Paper 329, p. 1–36.

Lowe, D.R., and Byerly, G.R., 2003, Field Guide to the Geology of the 3.5–3.2 Ga Barberton Greenstone Belt, South Africa: Guidebook prepared for Field Conference, Archean Surface Processes, June 23–July 2, 2003, 184 p.

Lowe, D.R., and Knauth, L.P., 1977, Sedimentology of the Onverwacht Group (3.4 billion years), Transvaal, South Africa, and its bearing on the characteristics and evolution of the early Earth: The Journal of Geology, v. 85, p. 699–723.

Lowe, D.R., and Nocita, B.W., 1999, Foreland basin sedimentation in the Mapepe Formation, southern-facies Fig Tree Group, *in* Lowe, D.R., and Byerly, G.R., eds., Geologic Evolution of the Barberton Greenstone Belt, South Africa: Geological Society of America Special Paper 329, p. 233–258.

Lowe, D.R., Byerly, G.R., Asaro, F., and Kyte, F.T., 1989, Geological and geochemical record of 3 400-million-year-old terrestrial meteorite impacts: Science, v. 245, p. 959–962.

Lowe, D.R., Byerly, G.R., and Heubeck, C., 1999, Structural divisions and development of the west-central part of the Barberton greenstone belt, *in* Lowe, D.R., and Byerly, G.R., eds., Geologic Evolution of the Barberton Greenstone Belt, South Africa: Geological Society of America Special Paper 329, p. 37–82.

Lowe, D.R., Byerly, G.R., Kyte, F.T., Shukolyukov, A., Asaro, F., and Krull, A., 2003, Spherule beds 3.47–3.24 billion years old in the Barberton Greenstone Belt, South Africa: A record of large meteorite impacts and their influence on early crustal and biological evolution: Astrobiology, v. 3, p. 7–47, doi: 10.1089/153110703321632408.

Melosh, H.J., and Vickery, A.M., 1991, Melt droplet formation in energetic impact events: Nature, v. 350, p. 494–497, doi: 10.1038/350494a0.

Paris, I., Stanistreet, I.G., and Hughes, M.J., 1985, Cherts of the Barberton greenstone belt interpreted as products of submarine exhalative activity: The Journal of Geology, v. 93, p. 111–129.

Rasmussen, B., and Koeberl, C., 2004, Iridium anomalies and shocked quartz in a late Archean spherule layer from the Pilbara craton: New evidence for a major asteroid impact at 2.63 Ga: Geology, v. 32, p. 1029–1032, doi: 10.1130/G20825.1.

Reimer, T.O., 1983, Accretionary lapilli and other spheroidal rocks from the Archaean Swaziland Supergroup, Barberton Mountain Land, South Africa, *in* Peryt, T.M., ed., Coated Grains: New York, Springer-Verlag, p. 619–634.

Reimold, W.U., Koeberl, C., Johnson, S., and McDonald, I., 2000, Early Archean spherule beds in the Barberton Mountain Land, South Africa: Impact or terrestrial origin, *in* Gilmour, I., and Koeberl, C., eds., Impacts and the Early Earth: Berlin-Heidelberg, Springer-Verlag, p. 117–180.

Ryder, G., 2002, Mass flux in the ancient Earth-Moon system and benign implications for the origin of life on Earth: Journal of Geophysical Research, v. 107, p. 6-1–6-14.

Ryder, G., Koeberl, C., and Mojzsis, S.J., 2000, Heavy bombardment on the Earth ~3.85 Ga: The search for petrographic and geochemical evidence, *in* Canup, R., and Righter, K., eds., Origin of the Earth and Moon: Tucson, University of Arizona Press, p. 475–492.

Schmitz, B., 1988, Origin of microlayering in worldwide distributed Ir-rich Cretaceous/Tertiary boundary clays: Geology, v. 16, p. 1068–1072, doi: 10.1130/0091-7613(1988)016<1068:OOMIWD>2.3.CO;2.

Schulte, P., and Kontny, A., 2005, Chicxulub impact ejecta from the Cretaceous-Paleogene (K-P) boundary in northeastern Mexico, *in* Kenkmann, T., Hörz, F., and Deutsch, A., eds., Large Meteorite Impacts III: Geological Society of America Special Paper 384, p. 191–221.

Shukolyukov, A., Kyte, F.T., Lugmair, G.W., Lowe, D.R., and Byerly, G.R., 2000, The oldest impact deposits on earth—First confirmation of an extraterrestrial component, *in* Gilmour, I., and Koeberl, C., eds., Impacts and the Early Earth: Berlin-Heidelberg, Springer-Verlag, p. 99–116.

Simonson, B.M., 2003, Petrographic criteria for recognizing certain types of impact spherules in well-preserved Precambrian successions: Astrobiology, v. 3, p. 49–65, doi: 10.1089/153110703321632417.

Simonson, B.M., 2004, Spherule layers as archives of both impact and earth history, *in* Reimold, W.U., and Hofmann, A., comp., Abstract volume: Field Forum on Processes on the Early Earth, Kaapvaal Craton, 4–9 July 2004: Johannesburg, University of the Witwatersrand, p. 90–92.

Simonson, B.M., and Harnik, P., 2000, Have distal impact ejecta changed through geologic time?: Geology, v. 28, p. 975–978, doi: 10.1130/0091-7613(2000)028<0975:HDIECT>2.3.CO;2.

Simonson, B.M., and Glass, B.P., 2004, Spherule layers—Records of ancient impacts: Annual Reviews Earth and Planetary Science, v. 32, p. 329–361, doi: 10.1146/annurev.earth.32.101802.120458.

Simonson, B.M., Davies, D., Wallace, M., Reeves, S., and Hassler, S.W., 1998, Iridium anomaly but no shocked quartz from late Archean microkrystite layer: Oceanic impact ejecta?: Geology, v. 26, p. 195–198, doi: 10.1130/0091-7613(1998)026<0195:IABNSQ>2.3.CO;2.

Sleep, N.H., and Zahnle, K.J., 1998, Refugia from asteroid impacts on early Mars and the early Earth: Journal of Geophysical Research, v. 103, p. 28,529–28,544, doi: 10.1029/98JE01809.

Sleep, N.H., Zahnle, K.J., Kasting, J.F., and Morowitz, H.J., 1989, Annihilation of ecosystems by large asteroid impacts on the early Earth: Nature, v. 342, p. 139–142, doi: 10.1038/342139a0.

Stanistreet, I.G., de Wit, M.J., and Fripp, R.E.P., 1981, Do graded units of accretionary spheroids in the Barberton Greenstone Belt indicate Archaean deep water environment?: Nature, v. 293, p. 280–284, doi: 10.1038/293280a0.

Ward, J.H.W., 2000, Metallogenic map of the Barberton greenstone belt: South Africa and Swaziland: Pretoria Council for Geoscience, scale 1:100,000, 1 sheet.

Manuscript Accepted by the Society 4 October 2005

Printed in the USA

Geological Society of America
Special Paper 405
2006

Diagenetic alteration of impact spherules in the Neoarchean Monteville layer, South Africa

Issaku Kohl
Bruce M. Simonson
Geology Department, Oberlin College, Oberlin, Ohio 44074-1052, USA
Melissa Berke
U.S. Geological Survey, 12201 Sunrise Valley Drive, MS 926A, Reston, Virginia 20192, USA

ABSTRACT

Intercontinental correlation of distal Archean impact ejecta layers can be used to help create a global time-stratigraphic framework for early Earth events. For example, an impact spherule layer in the Neoarchean Monteville Formation (Griqualand West Basin, South Africa) may be correlated with layers in one or more formations in Western Australia. To help assess the degree to which diagenetic alteration would hinder such correlations, we performed a petrographic study of spherules in the Monteville layer. Most of the spherules in the Monteville layer have botryoidal rims composed of radial-fibrous K-feldspar, but compaction and replacement have greatly altered their appearance and mineralogy. Moreover, the Monteville spherule layer consists of three main subunits, and spherule compaction varies between subunits as well as across the Griqualand West region. Compaction is about three times greater in a medial spherule-rich subunit as compared to a basal subunit rich in large intraclasts, resulting in better preservation of the shapes of melt particles in the latter. However, spherule rims have comparable numbers of fractures in both subunits, indicating the melt particles were fractured prior to compaction. Some spherules contain mica ribbons with a septarian geometry. Fracturing via rapid thermal quenching could help explain all of these features. If hot spherules possessing crystalline rims were thermally shocked when they hit the ocean, fractures would have the observed geometries and provide pathways for fluid infiltration and local replacement of glass by mica. Although heavily distorted, impact spherules in the Monteville layer are very similar to those in the Hesta occurrence of the Neoarchean Jeerinah spherule layer of the Hamersley Basin, even showing similar diagenetic histories. In this instance, diagenetic alteration may actually help rather than hinder intercontinental correlation of impact spherule layers.

Keywords: Precambrian, ejecta, replacement, compaction, stratigraphic correlation.

Kohl, I., Simonson, B.M., and Berke, M., 2006, Diagenetic alteration of impact spherules in the Neoarchean Monteville layer, South Africa, *in* Reimold, W.U., and Gibson, R.L., Processes on the Early Earth: Geological Society of America Special Paper 405, p. 57–73, doi: 10.1130/2006.2405(04). For permission to copy, contact editing@geosociety.org.

INTRODUCTION

Asteroid impacts were instrumental in shaping the early Earth's environment. These powerful events are recorded in the craters that they leave behind, as well as their proximal and distal ejecta deposits. Oceans cover most of the current Earth's surface and probably covered an even larger proportion of it early in Earth's history (Eriksson, 1995). This suggests that a majority of these early impacts occurred in the open ocean or on continental shelves instead of on dry land. However, due to the lack of pre-Mesozoic oceanic crust, workers can only make inferences from distal ejecta deposits that date back to 3.47 Ga (Byerly et al., 2002). Distal ejecta layers also have the potential to serve as important stratigraphic markers due to their relatively instantaneous emplacement and regional or even global distribution (e.g., Montanari and Koeberl, 2000; Koeberl and Martinez-Ruiz, 2003). In order for these layers to be effective time-stratigraphic markers, workers must be able to correlate distal ejecta deposits reliably on an intercontinental scale. This has proven relatively effective in the case of the K/T impact (e.g., Smit, 1999), but Archean successions lack the rich fossil record that made this global correlation possible. Strong diagenetic overprinting is also a potential impediment hindering the correlation of spherule deposits in different geographic areas.

The goal of this study was to investigate the textures of spherules from a Neoarchean impact layer in the Monteville Formation of the Griqualand West Basin, South Africa, which is potentially contemporaneous with late Archean layers found in the Hamersley Basin, Western Australia (Simonson et al., 1999, 2000a), to see whether diagenetic alteration would render this correlation unworkable. Although there is clear textural evidence that the spherule population in the Monteville layer was seriously modified by diagenetic processes, especially via compaction and replacement, we found parallels between these spherules and those of certain layers in the Hamersley Basin that support the proposed correlation. We hope our results will assist future workers as new layers are discovered and new correlations are attempted.

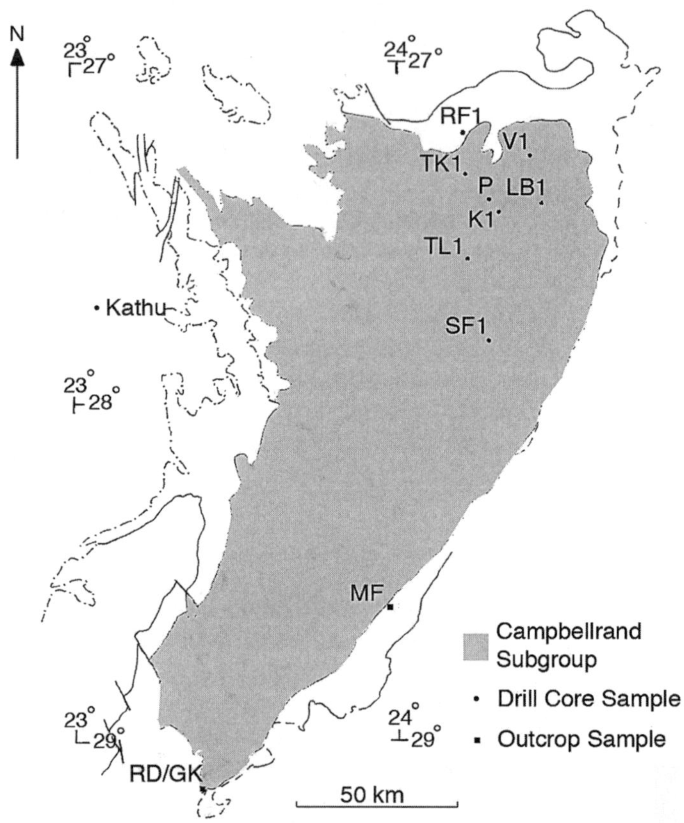

Figure 1. Map of Griqualand West Basin showing extent of near-surface occurrences of the Campbellrand Subgroup (see Beukes, 1987, for general location and stratigraphic nomenclature) and locations of cores and surface exposures studied. Abbreviations for surface exposures are MF—Monteville Farm, GK—Goudkop, and RD—Reid's Drift; the other abbreviations are for cores from the Pering Mine Region and are listed in Simonson et al. (1999). Solid lines represent depositional contacts with the underlying Lokommona Formation and small-displacement faults offsetting same. Dashed and dot-dashed lines represent nondepositional contacts with other formations. Location of map area indicated by 4 latitude/longitude markers. Adapted from Keyser (1997).

Geologic Setting

The Monteville Formation is the lowermost unit of the Campbellrand Subgroup of the Transvaal Supergroup in the Griqualand West Basin, South Africa (Fig. 1). The Campbellrand Subgroup is a major carbonate buildup located along the western margin of the Kaapvaal Craton. Most of the preserved Campbellrand underlies the Ghaap Plateau, which occupies a large part of the Griqualand West Basin (Beukes, 1987). The Monteville Formation is best exposed on the northeastern edge of the basin along the Campbellrand escarpment, but it extends throughout most of the basin in the subsurface. Stratigraphically, the Monteville Formation occupies a transitional position between the underlying shale-rich Lokammona Formation, which was deposited in relatively deep water, and the overlying Reivilo Formation consisting of shallow-water carbonate deposited on a platform directly above the Monteville Formation. The lithologies and sedimentary structures of the Monteville Formation make it clear that it contains the last deep-basinal strata deposited in a deeper shelf to slope environment before the transition to an extended period of platformal deposition known as the Ghaap Facies (Beukes, 1987).

The Monteville Formation contains a single layer rich in impact spherules that has been identified in three distinct geographic regions (Fig. 1; Simonson et al., 1999). Isotopic age dates of tuffs located within the Transvaal Supergroup constrain the spherule layer's age between 2640 Ma and 2520 Ma (Simonson et al., 2000a). Known surface exposures of the spherule layer are restricted to the Campbellrand escarpment, the best ones occurring on the Monteville Farm at the type section of the formation. Here we define the Monteville Farm region as including the Monteville Farm site proper (MF) and two closely spaced sites 50 km to the southwest of Monteville Farm at locations informally designated Reid's Drift (RD) and Goudkop (GK) (Fig. 1). Reid's Drift and Goudkop were not mentioned in Simonson et al. (1999) because they were discovered only after this paper was published. The Monteville spherule layer was also intersected in a suite of drill cores (RF1, V1, TK1, P, LB1, K1, TL1, SF1) in a trapezoidal region centered on the recently closed Pering Mine 145 km north-northeast of Monteville Farm (Fig. 1). Samples from most but not all of these cores were studied in detail. The Monteville spherule layer was also intersected in a third distinct area in the deep subsurface by a single research core drilled near Kathu, which is located 130 km west along the edge of the Griqualand West Basin (Fig. 1). We did not include samples from the Kathu core in our study because spherules and related particles are present in low concentrations and have been extensively replaced by carbonates.

Monteville Farm Region

In the Monteville Farm region, the Monteville layer's measured thickness is 29–80 cm, and it is divided into two distinct subunits (Simonson et al., 1999). The lower subunit of the spherule layer consists of gravel-size intraclasts mixed with well-sorted, sand-size interstitial material. The intraclasts consist of dolomite, shale, and pyrite and have textures suggesting they were derived from the underlying strata (Simonson et al., 1999). There is a wide size range within the intraclasts and, like all Monteville strata in surface exposures, they show obvious signs of recent surface weathering, including pyrite oxidation. Carbonate intraclasts can be as large as 180 cm long and 13 cm thick, and shale intraclasts are up to 13 × 18 cm in cross section. Pyrite intraclasts are rarely longer than a few centimeters (Fig. 2).

The interstitial sand ranges down to fine sand size and is composed in part of the same constituents as the intraclasts, but there is also an abundance of spherules. The spherules and other interstitial sand are concentrated in discrete pockets, which we interpret as large pores between intraclasts (Fig. 2). The intraclast-rich subunit is overlain by an upper subunit that is much finer and less spherule rich, consisting mostly of fine- to medium-sand–size carbonate detritus, with sparse gravel-size intraclasts. The carbonate is coarsely crystalline diagenetic dolomite, and the few spherules present exhibit textures similar to those contained in the lower subunit of the layer, although they are commonly fragmented. Although detrital textures are largely obscured, the presence of probable hummocky cross-stratification indicates that the upper subunit originally consisted of well-sorted sand (Simonson et al., 1999; Hassler and Simonson, 2001). We did not apply our techniques to samples from the upper subunit in the Monteville Farm region, because spherules and related particles occur in relatively low concentrations and are extensively replaced. In addition, we were only able to identify the upper subunit at Monteville Farm but neither at Reid's Drift nor at Goudkop, and we were unable to identify the medial spherule-rich subunit described below at any of the surface exposures (Fig. 3).

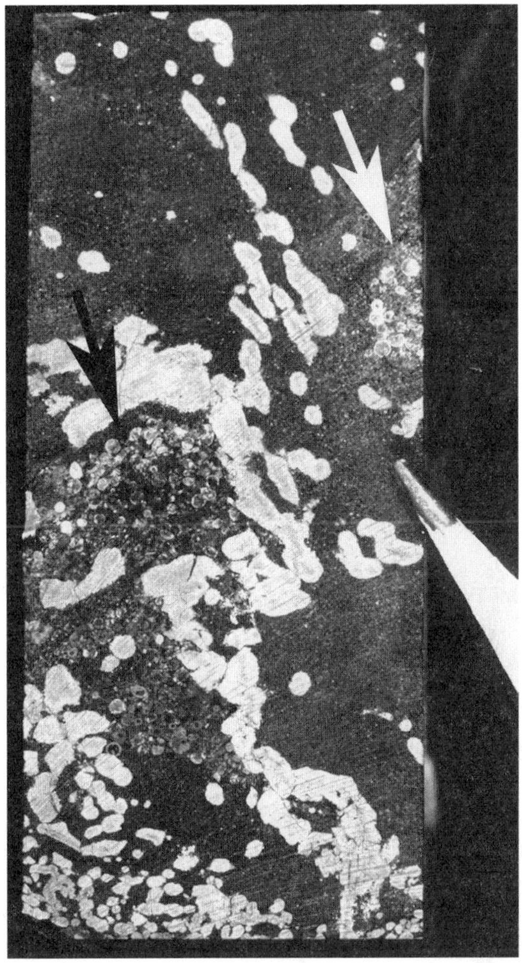

Figure 2. Unpolished surface of typical sample from intraclast-rich subunit in core SF1 shown in original orientation (top of image is up). Dark gray areas are carbonate that mostly belongs to large intraclasts. White to light gray solid bodies with oblong to subcircular cross sections are early diagenetic pyrite concretions; some are still lodged inside carbonate intraclasts, and others are liberated to form discrete intraclasts. Black arrow points to top of pocket rich in well-sorted spherules (white circles, many with dark central spots) interpreted as sand interstitial to large intraclasts; white arrow points to smaller, similar pocket of spherules on edge of core. Diagonal lines across face are marks left by rock saw. Core is 2.8 cm wide right to left, pencil point on right.

Pering Mine Region

The Monteville spherule layer was identified in nine cores drilled in a trapezoidal area 65 km north to south, 20 km east to west, and centered around Pering Mine (Simonson et al., 1999) ("P" in Fig. 1). In most of these cores, the lowest subunit of the layer contains a high concentration of spherules in an ~5-cm-thick bed that possesses a sharp erosional base. Again in most cores, this spherule-rich basal subunit is overlain by a distinctly different subunit that contains a much lower abundance of spherules. This upper subunit consists of a thin layer of quartzose sandstone in the southernmost cores but expands to a 2-m-thick

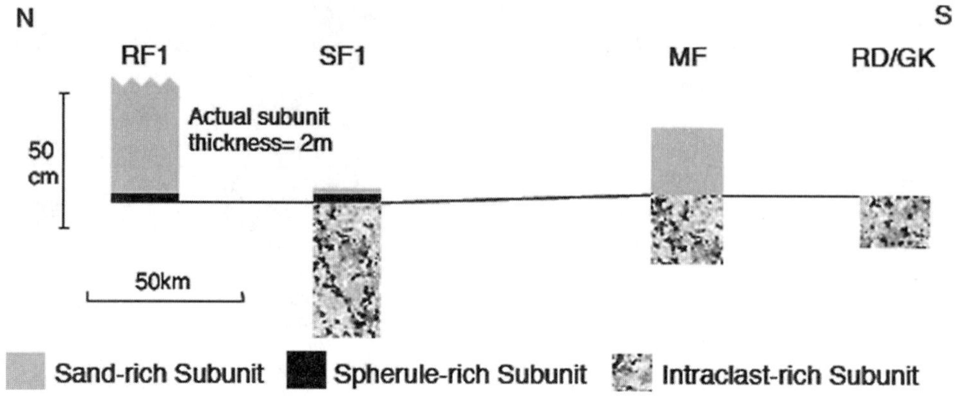

Figure 3. Schematic stratigraphic columns of the Monteville spherule layer showing stratigraphic relationships and variations in thicknesses of the three subunits along a roughly north-south transect within the study area; location abbreviations are the same as in Figure 1. Note that upper subunit of RF1 is artificially truncated.

layer rich in carbonate in the northernmost cores (Fig. 5 *in* Simonson et al., 1999). Spherules tend to be finer, more fragmented, and widely dispersed in this upper subunit, which we correlate with the upper subunit at Monteville Farm (Fig. 3). Small, unaltered pyrite intraclasts are found in these cores but are much finer than their counterparts at the Monteville Farm locality. In addition to these two subunits, the SF1 core from the southern edge of the Pering Mine Region (Fig. 1) contains a basal subunit beneath the spherule-rich subunit. The basal subunit is rich in gravel-size intraclasts of carbonate and pyrite that are similar in size and abundance to these phases in the lower subunit of the layer at Monteville Farm (Simonson et al., 1999). The basal intraclast-rich subunit is ~50 cm thick, and the material between the intraclasts consists of fine- to medium-sand–size dolomite detritus, with high concentrations of spherules (Fig. 2). We did not apply our techniques to samples from the upper subunit in the cores of the Pering Mine region, again because spherules and related particles occur in relatively low concentrations.

Petrographic Methods

Samples were first screened by general petrographic examination, including comparison to published photomicrographs and descriptions of impact spherules from various layers including the K/T boundary layer (Smit et al., 1992; Smit, 1999), late Eocene microtektite and microkrystite strewn fields (Glass, 2002), other microtektite strewn fields (Simonson and Glass, 2004, and references therein), and late Archean layers in the Hamersley Basin of Western Australia (Simonson et al., 1999, 2000a, 2000b; Simonson, 2003). We also examined spherule-bearing thin sections from the K/T boundary layer at the Mimbral site in northern Mexico (Smit et al., 1992) and ODP (Ocean Drilling Program) site 1049C drilled on Blake Nose (Norris et al., 1999) for comparison. Eight samples from three surface sites in the Monteville Farm region and ten samples from nine cores in the Pering Mine region that were rich in well-preserved spherules were then chosen for quantification of the characteristics of the spherules via the methods described in the following paragraphs.

Over 400 points were counted for each of 18 samples to determine the relative abundances of 8 textural categories (Table 1) encompassing all the former melt particles in the Monteville layer. The relative abundances of these categories were then tabulated independent of the presence of other types of detritus (Table 2). In addition to spherule types, the abundance of interstitial material, fine-grained sheet silicates, sparry carbonate crystals, and particles with internal textures completely obscured by replacement were also quantified by point counting. The replacement noted in surface samples was mainly by carbonate and pyrite (the latter weathered subsequently to iron oxyhydroxides). In core samples, replacement is dominated by sheet silicate phases, most likely sericite and a chlorite phase, as well as minor amounts of carbonate in certain samples. None of the carbonate minerals were analyzed quantitatively, but the majority appear to be dolomite rather than calcite based on their weak response to dilute HCl.

Average aspect ratios (long dimension/short dimension) of spherules were determined by measuring long and short axis lengths for 35–40 former melt particles with botryoidal rims (categories 1, 2, 4, and 6) in each sample. Restricted areas judged to be typical were chosen in a given sample, and then all spherules in that area were measured. The thickness and number of fractures showing offsets in the rim of each particle were measured at the same time (Table 3). The average number of contacts per grain was determined by counting the actual number of contacts in each of 20 successive grains along a horizontal transect within a given sample in an area chosen to be typical of each core or surface exposure.

In addition, electron microprobe analyses of a typical spherule from sample SF1–1 were obtained (Table 4). The spherule was in a polished thin section that was carbon-coated prior to analysis. The electron beam was set for all analyses at 15 kV acceleration

TABLE 1. SUMMARY OF MAIN CHARACTERISTICS OF THE EIGHT TEXTURAL CATEGORIES INTO WHICH ALL SPHERULES AND RELATED MELT PARTICLES WERE PLACED DURING POINT COUNTS FOR THIS STUDY

Category number	Cross-sectional shape	Aspect ratio	Botryoidal rim?	Characteristics of rim	Central spot?	Additional characteristics
1	Highly elongate	≥4	yes	Brittle deformation	yes	May appear like a rubber band lying on a flat surface or a haphazard cluster of fragments
2	Oval	1.5–4	yes	Brittle deformation	yes	Less elongated than type 1
3	Oval	1.5–4	no		no	Hololathic internally = entirely filled with lath-shaped crystals of K-feldspar
4	Oval	1.5–4	yes	Unfractured	yes	
5	Circular	1–1.5	no		no	Hololathic
6	Circular	1–1.5	yes	Unfractured	yes	
7	Angular to subrounded	N/A	no		N/A	Shows possible flow banding, abundant fine vesicles, and widespread internal replacement by sheet silicate
8	Fragmented, e.g., <half of orig. circular perimeter	N/A	yes	Present but not whole	Also fragmented	May appear as a piece of pie or a section of a circular rim; flow textures truncated on broken edges, but not original outlines

TABLE 2. RELATIVE ABUNDANCES (IN PERCENT) OF DIFFERENT TYPES OF SPHERULES IN VARIOUS SAMPLES BASED ON POINT COUNTS

Slide no.	Location	Outcrop	Core	Subunit†	Spherule types (same as categories in Table 1)									n
					1	2	3	4	5	6	7	8	Total (%)	
RF1–1DE	RF1		X	S	54.9	25.7	7.4	3.2	0.3	0.7	7.1	0.7	100.0	403
V1–1A	V1		X	S	30.3	29.5	3.4	1.4	10.5	17.3	5.7	1.8	99.9	402
TK1–1Ac	TK1		X	S	36.1	28.8	6.4	2.2	7.0	7.8	10.2	1.4	99.9	406
LB1–1Da	LB1		X	S	20.5	21.7	13.2	2.4	6.4	11.1	18.8	5.9	100.0	401
P11–1a	P		X	S	37.9	22.3	1.6	1.6	9.5	11.5	12.9	2.6	99.9	403
P9–1a	P		X	S	27.1	24.2	2.0	4.7	10.4	11.2	13.1	7.2	99.9	404
K1–12a	K1		X	S	31.0	34.0	2.5	1.3	7.8	12.8	7.8	2.8	100.0	403
TL1–1A2	TL1		X	S	39.7	28.9	2.5	2.5	6.1	10.3	7.5	2.5	100.0	401
SF1–1	SF1		X	S	36.4	22.9	6.2	5.5	8.1	12.1	7.0	1.8	100.0	403
SF1–2	SF1		X	I	10.4	35.6	2.6	3.5	7.2	25.9	9.2	5.7	100.1	970
U63–1	MF	X		I	22.6	25.8	3.5	1.6	7.1	8.6	16.3	14.4	99.9	1981
V110–1A	RD	X		I	13.4	44.0	0.7	3.0	2.1	11.1	18.5	7.3	100.1	882
V110–1B	RD	X		I	22.0	41.7	0.7	0.3	2.1	12.4	15.4	5.4	100.0	706
U126–1a	RD	X		I	19.1	41.0	1.4	0.6	3.7	8.8	19.5	5.8	99.9	1076
U126–1b	RD	X		I	17.4	40.0	1.1	1.4	1.5	10.2	21.9	6.5	100.0	857
U98–1A	GK	X		I	10.1	25.2	2.3	0.5	2.3	9.4	34.0	16.3	100.1	1654
U98–1B	GK	X		I	15.7	31.4	1.5	2.5	1.9	11.5	23.7	11.8	100.0	1157
U98–2	GK	X		I	13.1	32.9	0.3	2.2	2.2	13.4	23.6	12.4	100.1	1303
W104–3A	Hesta	X		N/A	23.9	21.4	4.2	1.8	2.9	6.4	36.7	2.6	99.9	737

Notes: Types 1 through 8 are categories whose characteristics are listed in Table 1. All spherule counts are normalized to 100% by volume, and the number of points counted per sample (n) is listed in column on right. Abbreviations of locations are same as in Figure 1 save for Hesta, which is a site in the Hamersley Basin of Western Australia where the Jeerinah layer is exposed (see Hassler et al., 2005 for more details).

†I—intraclast-rich subunit, S—spherule-rich subunit.

TABLE 3. ADDITIONAL DATA ON DEGREE OF ELONGATION AND RIM CHARACTERISTICS OF SPHERULES IN SAMPLES THROUGHOUT STUDY AREA

Slide no.	Location	Outcrop	Core	Subunit†	Aspect ratio (l/w)		Rim thickness (mm)		Number of fractures in rim	
					ave.	st. dev.	ave.	st. dev.	ave.	st. dev.
RF1–1DE	RF1		X	S	5.9	4.1	0.04	0.036	2.1	2.8
V1–1A	V1		X	S	4.8	3.2	0.06	0.038	5.2	2.8
TK1–1Ac	TK1		X	S	5.3	4.3	0.036	0.028	4.8	3.5
LB1–1Da	LB1		X	S	3.8	2.5	0.058	0.038	5.6	3.6
P11–1a	P		X	S	4.7	2.4	0.051	0.033	5.6	3.2
P9–1a	P		X	S	3.2	2.3	0.065	0.039	5.8	4.5
K1–12a	K1		X	S	4.6	3	0.058	0.04	8.2	6.2
TL1–1A2	TL1		X	S	3.5	2	0.069	0.029	6.9	5.1
SF1–1	SF1		X	S	3.8	2.8	0.066	0.043	6.7	5.4
SF1–2	SF1		X	I	1.7	1	0.067	0.04	5.2	5.1
U61–1A	MF	X		I	2.4	1.6	0.085	0.054	9.5	5.6
V110–1A	RD	X		I	1.6	0.72	0.058	0.046	5.1	4.8
V110–1B	RD	X		I	1.7	0.93	0.056	0.043	7.4	3.9
U126–1a	RD	X		I	1.8	0.95	0.064	0.05	6.1	3.6
U126–1b	RD	X		I	1.8	1.3	0.056	0.029	7.5	5.4
U98–1A	GK	X		I	1.4	0.71	0.062	0.042	5.4	3.7
U98–1B	GK	X		I	1.6	0.94	0.059	0.039	4.7	4.3
U98–2	GK	X		I	1.6	0.93	0.072	0.05	4.9	2.9

Notes: Abbreviations of locations are same as in Figure 1.

†I—intraclast-rich subunit, S—spherule-rich subunit.

TABLE 4. REPLICATE ELECTRON MICROPROBE ANALYSES OF DIFFERENT PARTS OF A SINGLE SPHERULE FROM THE SPHERULE-RICH SUBUNIT IN CORE SF1

	Radial-fibrous rim—minerals optically identified as K-feldspar					Low-biref. parts of clear central spot—mins. optically identified as K-feldspar					Hi-biref. parts of clear central spot—mins. optically identified as muscovite				
	1	2	3	4	5	1	2	3	4	5	1	2	3	4	5
SiO_2	65.1	64.3	64.3	64.3	64.7	62.6	64.5	65.1	64	64.8	47.7	48.4	48.2	48.5	47.9
TiO_2	N.D.	N.D.	N.D.	N.D.	N.D.	N.D.	N.D.	N.D.	N.D.	N.D.	0.13	0.12	- -	- -	- -
Al_2O_3	18.2	18.5	18	18.1	18.6	19.4	18.6	18.7	18.3	18.6	36	35.7	34.7	35.6	35.5
FeO	0.12	- -	- -	- -	- -	0.1	- -	0.1	- -	- -	0.16	0.18	0.15	0.4	- -
MnO	- -	- -	- -	0.1	- -	- -	- -	- -	- -	0.1	- -	0.12	0.1	- -	- -
MgO	- -	- -	- -	- -	- -	0.48	- -	- -	- -	- -	0.55	0.53	0.69	0.79	0.79
CaO	- -	- -	- -	- -	- -	- -	- -	- -	- -	- -	- -	- -	- -	- -	- -
Na_2O	0.14	0.1	0.25	0.23	0.13	0.35	0.1	0.26	0.24	0.1	0.1	0.15	0.1	0.28	0.1
K_2O	16.7	16.8	16.7	16.8	16.1	15.7	16.8	16.7	16.7	16.7	10.3	10.4	10.3	10.4	10.6
Cr_2O_3	N.D.	N.D.	N.D.	N.D.	N.D.	N.D.	N.D.	N.D.	N.D.	N.D.	- -	0.1	- -	- -	0.1
Totals	100.26	99.7	99.25	99.53	99.53	96.7	100	100.86	99.24	100.3	94.94	95.7	94.24	95.97	94.99

Notes: Analyses performed at Rand Afrikaans University (now University of Johannesburg) by Dr. Jens Gutzmer. N.D.—not determined; - - —below detection limits (0.3 wt%); units—wt%.

voltage and 10 nA absorbed beam current on brass. Spectra for each spot analysis were acquired for 80 seconds. Analyses were performed at Rand Afrikaans University (now called University of Johannesburg) on a Cameca CAMEBAX 355 electron microprobe system equipped with 3 wavelength-dispersive spectrometers (WDS) and an additional Link energy-dispersive spectrometer (EDS), carrying a Li-drifted Si-detector. The system was controlled by the Link ExL II and Lemas software packages. Only the EDS system was utilized for the analyses. Routine calibration of the energy-dispersive spectrometer was performed using a pure Co metal standard. The raw EDS spectra were automatically ZAF-corrected by the ExL II software.

PETROGRAPHIC DESCRIPTIONS

Monteville Farm Region

General Petrography

The eight samples from the spherule layer at all three sites in the Monteville Farm region are fairly uniform in composition. Their major constituents are gravel-size shale and pyrite intraclasts mixed with fine to very coarse sand consisting of a mixture of spherules, irregular melt particles, carbonate grains, and shale and pyrite fragments. The shale fragments possess tabular to angular shapes with occasional relict bedding. The large intraclasts appear to form a grain-supported framework, indicated by the local presence of sparry carbonate fillings in what appear to be sheltered pores above geopetally distributed carbonate sand and spherules. Original clast outlines within the sand fraction are often obscure due to replacement by coarse carbonate crystals (Fig. 4). Sparry carbonate crystals between intraclasts and former melt particles probably represent void-filling cement and/or replacements of earlier generation cement. Other carbonate crystals occur in pyrite-free, cross-cutting veins that can be up to 2 mm wide. Some of the vein crystals show deformation twins indicative of temperatures as high as 300 °C (Passchier and Trouw, 1996). These veins cut across a petrographically distinct type of carbonate consisting of crystals 50–100 μm large, with intercrystalline pyrite.

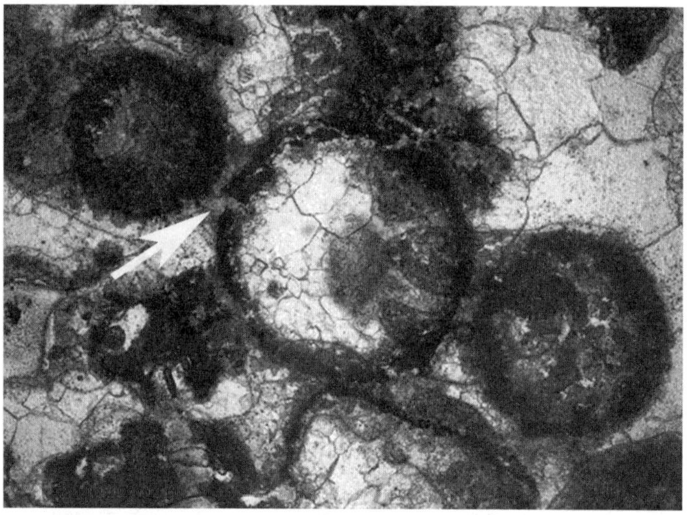

Figure 4. Photomicrograph (plane polarized light) of uncompacted spherules largely replaced by coarsely crystalline carbonate in sample U63–1 from Monteville Farm. Note nice circular cross sections, fractures in rim in spherule in middle (e.g., one indicated by white arrow), and high minus-cement porosity. Diameter of spherule in middle is 0.9 mm.

All of the samples from the Monteville Farm region are from surface exposures and show extensive evidence of weathering, most prominently in the form of iron-oxide staining. The sources of this iron are most likely the pyrite intraclasts (which are almost entirely oxidized) and possibly ferroan carbonate phases. In reflected light, silvery metallic cores are visible within red intraclasts; they are thought to be pyrite weathered to varying degrees, whereby the earthy red material is the most altered phase and possibly consists of oxy-hydroxides. The samples also show signs of minor compaction, such as in situ grain fracturing and pressure-solved grain contacts. A selvage composed of sericite, chlorite, and possibly pyrite is sometimes present along elongated grain contacts. Sand grains in the samples we studied from the Monteville Farm region, all from the lower subunit of the Monteville layer, average 1.3 contacts per grain, indicating that they have not been compacted very much. Uncompacted sands consisting of equidimensional, well-sorted clasts average 0.9–1.6 contacts per grain (Pettijohn, 1975), and the sand in the lower Monteville subunit appears to have originally consisted of well-sorted clasts with low aspect ratios (Table 3).

Petrography of Impact Ejecta

Spherical splash forms. These particles show a range of internal textures, with much of this variation reflecting different degrees of replacement. Some are composed of K-feldspar with sericite and, much less commonly, quartz, whereas others have been completely replaced by carbonate save for thin opaque rims with relict, inward-radiating, botryoidal terminations (Fig. 4). Different amounts of sparry carbonate crystals are present within some of these particles. Still other spherules show indiscriminant replacement by large angular crystals of an opaque phase, most likely hematite formed from authigenic, now surface-weathered pyrite.

Although textures within the spherules that are not obscured by replacement are highly variable, one or both of two textural elements appear, to some degree, in almost every spherule, as is typical for late Archean to early Paleoproterozoic impact spherules in general (Simonson, 2003). Acicular to lath-shaped crystals of what is now K-feldspar constitute one of these elements. Most of the K-feldspar crystals are arranged in fan-shaped radial arrays. Much less commonly, spherules contain lath-shaped crystals that are randomly oriented. A complete continuum exists between spherules in which the K-feldspar laths form rims and those in which the laths fill the entire cross section; this latter texture will be referred to hereafter as hololathic.

The other common textural element involves clear internal spots that are highly variable and filled by sheet silicate, quartz, K-feldspar, and/or carbonate. Two optically distinct varieties of sheet silicates are present, one appearing clear and the other pale green in plane polarized light. The clear sheet silicate is most likely sericite, whereas the pale green one is composed of smaller crystals and appears to be chlorite, as it exhibits lower birefringence than the sericite (although this could be an impression caused by the smaller grain size). Some of the clear spots possess well-defined circular to oval cross sections; rare teardrop-shaped clear spots are also present. We interpret these as filled vesicles within melt droplets by analogy to similar textures in Phanerozoic impact spherules, especially those found in the K/T boundary layer (see Fig. 5 *in* Smit et al., 1992; Smit, 1999). In contrast to the clear spots with well-defined geometric shapes in cross section, other clear spots have more irregular boundaries defined by the uneven termination of inwardly radiating K-feldspar (see Fig. 7b *in* Simonson, 2003). These are more likely to have originated as relict glass cores that were dissolved and/or replaced during later diagenesis and are again analogous to observations of certain K/T spherules (Bohor and Glass, 1995). Finally, many spherules have clear central spots that are so extensive that their origin is unclear. Quite a few spherules have been so extensively replaced by carbonate that all internal textures have been obliterated and all that remains are thin opaque botryoidal rims. Replacement carbonate occurs as microcrystalline to blocky crystals, the latter filling spherules with only one to three crystals. It is possible that bubbles actually occupied most of the interiors of some of these spherules.

Nonspherical splash forms. A minority of the particles with spherule-type internal textures have shapes that are highly elongate and similar to those of peanuts, tear drops, and other more exotic splash forms. Some have thin opaque botryoidal rims and are completely replaced by blocky carbonate crystals. Such particles are often oriented perpendicular or oblique—instead of parallel—to bedding, indicating they were not produced by the compaction of spheres. Some elongate splash forms exhibit hololathic textures. K-feldspar crystals within these nonspherical particles tend to be at the larger end of the K-feldspar size spectrum, resulting in relatively more coarsely crystalline particles.

Fragmented splash forms. Fragments of broken spherules and other particles are abundant and have two distinct modes of occurrence. The most easily recognizable fragments exhibit radial-fibrous or botryoidal textures typical of spherule rims, and some occur as isolated fragments, whereas others occur in clusters. Most isolated fragments consist of K-feldspar arranged in radial fibrous fans with occasional opaque rims exhibiting botryoidal terminations on one side versus smoothly curved circular boundaries that represent arcs of a larger circle on the opposite side. This indicates that these particles are segments of a once-continuous rim, and the fact that the individual fragments do not resemble or fit into any clasts in close proximity suggests that they were fragmented prior to deposition. The fragments range in character from "pie slices," with just one or two adjacent fans on the former rim, to more extensive rim fragments that represent up to half the circumference of an original spherule.

In contrast to isolated occurrences, other fragments occur in clusters of slightly curved to angular fragments that closely resemble one another and in many cases clearly fit into one another. We observed two types of clusters: (1) groups of randomly oriented fragments, where rims appear to have been completely disarticulated and displaced (Fig. 5; lower portion), and (2) groups of particles displaying ovoid cross sections (Fig. 5; central grain). Some of the most elongate clusters show little displacement of the broken pieces and preserve, to a degree, a continuous rim (Fig. 6). In the ovoid clusters, the botryoidal sides of the fragments generally face inward, whereas the smoother, curved edges face outward. Both elongated individual fragments and the long axes of the clusters themselves are usually parallel to bedding. In intraclast-rich samples, fragment clusters often contain sericite internal to the broken rim, although carbonate usually occupies most of the area inside the rim. Some clusters define a circular outline that is nearly undisturbed save for thin carbonate-filled gaps that cut the rims and are optically continuous with the carbonate crystals on one or both sides of the rim (Fig. 4). The fragments in circular clusters are geometrically similar to sectors of intact spherules with botryoidal rims.

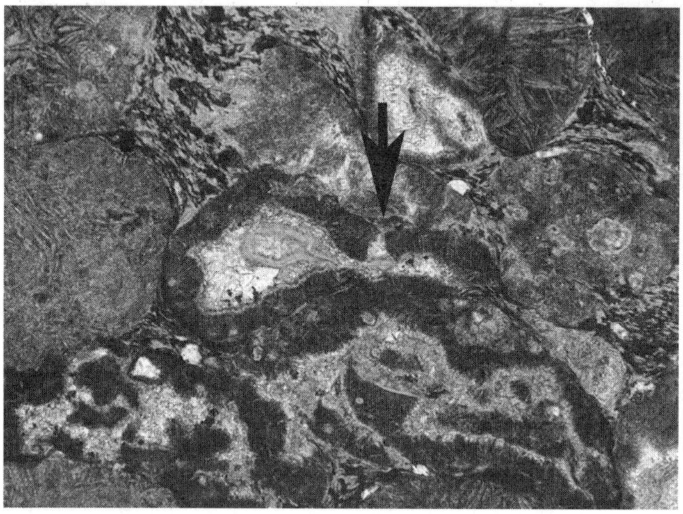

Figure 5. Photomicrograph (plane polarized light) of compacted spherules in sample from core K1. Spherule in middle exhibits "eggshell"-type geometry, including wedge-shaped fracture opening in middle of upper rim (directly beneath black arrow). Light gray core of middle spherule is composed of sericite that exhibits distinct mica ribbons on left, but individual ribbons cannot be distinguished on right where most compression has occurred. Long dimension of spherule in middle is 0.60 mm.

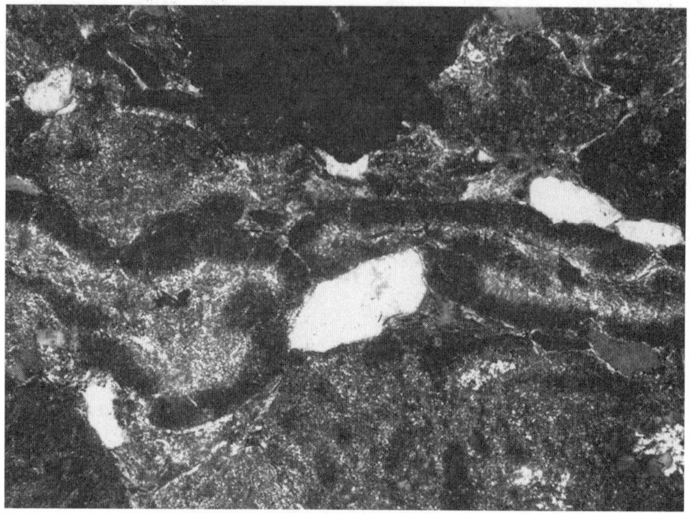

Figure 6. Photomicrograph (crossed polarizers) of oblong detrital quartz grain (center) coated with birefringent mica selvage from core LB1. Note how quartz grain is pressure-solving into rims of adjacent compacted spherules, especially along upper right edge. Light gray material between and within grains is mostly fine mica, rims and other dark material is mostly K-feldspar, and long dimension of central quartz grain is 0.25 mm.

Irregular particles. The Monteville layer in the Monteville Farm region also contains irregular particles of former melt, textures of which are almost identical to those in the Carawine layer of the Hamersley Basin (Simonson et al., 2000b). Irregular particles are generally larger and more angular than the spherules and show an array of internal textures. Almost all of the large irregular particles have two distinct textural aspects. The first consists of light bands composed of K-feldspar alternating with dark to light bands of sheet silicate. The bands form patterns that range from planar to wavy to completely swirled and strongly resemble flow bands in molten materials such as tektites and obsidian. Infilled bubbles are the second major textural element. They range in abundance from 0% to around 20% by area and are filled by sericite of varying coarseness. They are similar to vesicles in the spherules, although they tend to be smaller. Even particles that were extensively replaced by sheet silicate are still identifiable as "irregulars" by their abundant vesicles.

Mica ribbons. In six of the surface samples, a small percentage of spherules contain one or more isopachous sericite ribbons that appear to consist of fibers, as they display coherent length–slow extinction (Fig. 7). The ribbons occur in three distinct textural positions: (1) coating botryoidal terminations along the inner edges of rims, (2) lining the edges of vesicles in a concentric orientation, and (3) forming a radially branching network throughout the irregular (i.e., nonvesicular) clear spots. The ribbons forming branching networks resemble the "mesh texture" of serpentinites (e.g., Fig. 5.99 *in* Vernon, 2004) in that original material appears to have been replaced by parallel fibers nucleated on and growing away from a network of microfractures. Where ribbons occupy two or more of the three textural positions mentioned above within a given spherule, they are connected to form a network that appears to have a septarian geometry (Fig. 7). The remainder of the material filling clear spots in spherules containing mica ribbons consists of either radial fibrous K-feldspar or coarser crystalline muscovite.

Figure 7. Photomicrograph (crossed polarizers) of spherule containing mica ribbons from core TL1. Note how mica ribbons pinch together in right half of spherule, which is more compacted than left half, although the contrast is not as drastic as in the spherule in Figure 5. Clear core is composed of sericite; radial-fibrous rim consists of K-feldspar. Field of view is 1.40 mm wide from right to left.

Ribbons with the same geometries that lack coherent extinction patterns also occur in some spherules. These ribbons are composed of the chloritic sheet silicate instead of sericite, tend to be slightly wider, and have a massive texture. A given spherule may contain ribbons with either one or both compositions, but no sericite ribbons were seen in spherules that are dominantly chloritic. In spherules with chloritic ribbons, clear spots are filled with relatively coarser muscovite with or without the presence of any other chloritic sheet silicate. Spherules from the K/T boundary layer at both ODP site 1049 (Fig. 8; see Norris et al., 1999, for location information) and DSDP (Deep Sea Drilling Project) hole 603B (see Fig. 6 *in* Smit et al., 1992) show textures that appear quite similar to these mica ribbons.

Pering Mine Region, Spherule-Rich Subunit

General Petrography

Spherules from the nine cores in the Pering Mine region average 0.65 mm in diameter (Simonson et al., 1999) and consist almost entirely of K-feldspar and sericite. The space interstitial to spherules and other sand grains is almost entirely occupied by the optically chloritic sheet silicate (Fig. 5). Its uniform appearance and optical orientation suggest that it is authigenic in nature, but it could represent a fine-grained detrital matrix that was replaced and/or oriented by compaction. Minor detrital quartz is also present (Fig. 6), as are opaque minerals that are evenly dispersed and probably authigenic. The fine sizes of the interstitial sheet silicate and opaque fractions support an authigenic origin, as it is highly unlikely that silt- and mud-size particles would be deposited together with well-sorted coarse sand-size particles. In addition, coarsely crystalline carbonate has indiscriminately replaced former melt particles in three samples, leaving traces of melt-derived textures.

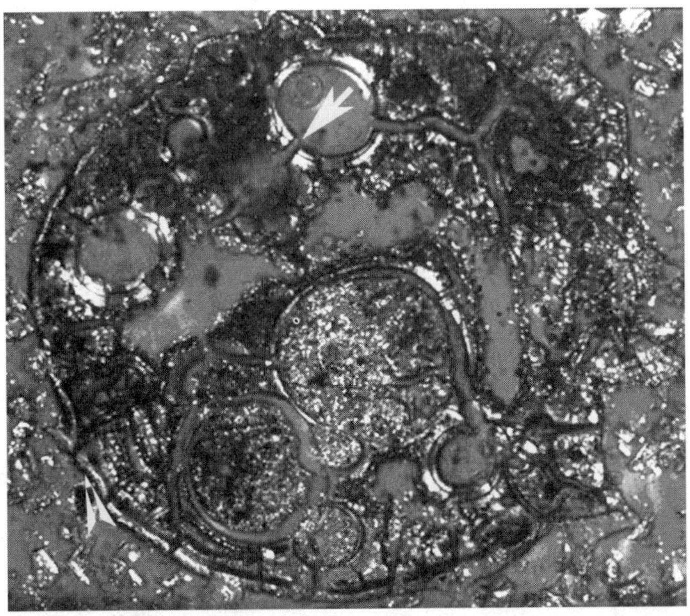

Figure 8. Photomicrograph (crossed polarizers) of spherule from K/T boundary layer in drill core from ODP (Ocean Drilling Program) Hole 1049C drilled on Blake Nose in the western North Atlantic (Norris et al., 1999). Edge of spherule and internal vesicles have optically oriented linings of an undetermined sheet silicate mineral. Linings are fragmented in situ (e.g., one indicated by white arrow), and some piles of completely disarticulated rim fragments were observed nearby. Diameter of spherule is 0.77 mm.

There is clear textural evidence that the spherule-rich subunit experienced a high degree of compaction with individual grains having been deformed via both plastic and brittle mechanisms. One line of evidence is the number of contacts per grain, 7.1 on average. An average of 5.2 contacts per grain is typical for well-sorted sands buried to a depth of 2570 m (Pettijohn, 1975). Strong compaction is also evidenced by the fact that almost all grains are in full contact with surrounding grains along their entire perimeters. This is consistent with a high abundance of long and concavoconvex contacts between grains (Fig. 6). Thin selvages of sheet silicates dominated by muscovite but also containing the chloritic phase are found along many of these contacts. Detrital quartz grains appear to have acted as battering rams that indent into former melt particles along concavoconvex grain boundaries (Fig. 6). Since quartz is normally more susceptible to pressure solution than feldspar (Trurnit, 1968), we attribute this behavior to the presence of a thin selvage of insoluble sheet silicate armoring the surfaces of the quartz grains. Simonson et al. (1999) observed similar textures within the overlying quartz-sand–rich subunit, where pressure solution between quartz grains is also evident.

Petrography of Impact Ejecta

Spherical splash forms. Almost every former melt particle in the Pering Mine region shows the same two textural elements noted for the spherules in samples from the Monteville region, namely acicular to lath-shaped crystals of K-feldspar and the presence of circular to irregular, clear, central spots. Commonly radial fibrous aggregates of K-feldspar possess a multitude of submicroscopic inclusions, appearing tan in plane-polarized light, and generally are most abundant near the outer edge of the spherule (Figs. 5 and 6). Acicular crystals vary in size; their widths range from submicroscopic to 10 µm and their lengths from 20 to 200 µm. The laths also vary in size, but they are thicker, ranging from 10 to 20 µm in width, and they are 20–300 µm long. These sizes are comparable to those found in samples from the Monteville Farm region. Aggregates of acicular crystals and/or laths may be arranged to form a rim composed of inward-facing botryoidal terminations or, less commonly when no clear core is present, are hololathic.

As in the Monteville Farm area, the other major textural elements in spherules are clear spots composed of K-feldspar, sericite, and/or a chloritic sheet silicate, with either well-defined circular to oval cross sections or irregular boundaries. Sericite displays one of several habits, appearing in little fans, as vermiform crystal packets, or as larger crystals up to 20 µm long. The habits of the microcrystalline chloritic sheet silicate are similar to the clear sericite, yet individual crystals tend to be smaller. The spots with regular boundaries are often defined by a circle of opaque dust or a band of fine white mica. The irregular-edged clear spots begin where the radial fibrous or lath-shaped K-feldspar crystals terminate and are usually composed of sericite, although a small number consist of K-feldspar. Clear spots with irregular edges have been interpreted as replaced glass cores, in part because filled vesicles generally possess circular or ovoid cross sections (Simonson, 2003). Botryoidal textures have been replaced by sericite in some spherules, yet they are still recognizable due to a thin opaque band preserving the telltale scalloped shape of the final growth surface within the sericite (Fig. 9). There are also intermediate stages of replacement where sericite pods can be seen on K-feldspar fans (Fig. 10).

Nonspherical splash forms. This type of particle is rare in the Pering Mine region, but a few teardrop- and peanut-shaped splash forms have been observed. When they are present, they almost always show signs of brittle deformation acting upon the botryoidal rim of the particles. Nonspherical splash forms display both clear spots and hololathic K-feldspar textures internally.

Fragmented splash forms. Just as in the Monteville Farm region, spherule fragments occur both as solitary clasts and in clusters. The isolated fragments are compositionally similar to those found in the Monteville Farm region and, as in those samples, appear to have cracked and to have been separated prior to deposition. The fragment clusters likewise appear to be similar to those in the Monteville Farm region in that we observed two types: (1) randomly oriented fragment collections where rims appear to have been completely disarticulated, and (2) highly elongate clusters in which fragments still line up to define the shape of a once continuous rim. Internally, when a rim can be defined, clusters can be filled with either pure sericite or a combination of sericite and finer-grained chloritic sheet silicate, with minor pyrite (Fig. 11). The sheet silicates within the clear spots in some spherules with relatively continuous rims show a preferred length–slow orientation parallel to bedding. In a few cases, detrital quartz grains are located inside semicontinuous fragmented rims.

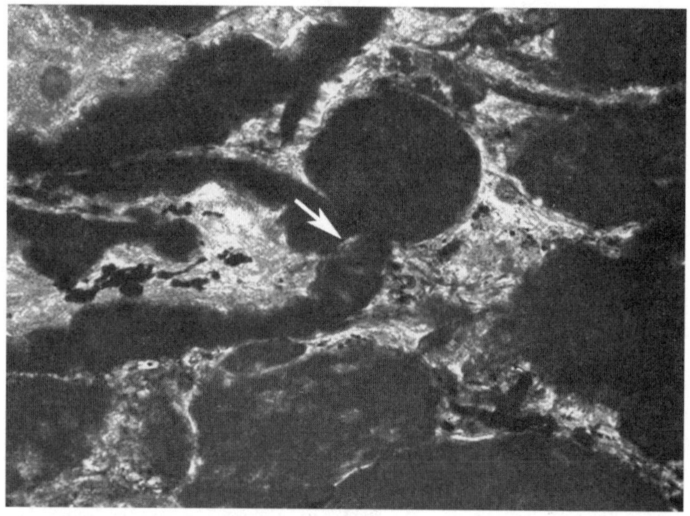

Figure 10. Photomicrograph (crossed polarizers) of compacted spherules from core RF1. Note three minute ovoid zones of recrystallization by light, high-birefringence sericite (top one indicated by white arrow) in the hinge of an opaque rim on the most highly compacted grain. Field of view is 1.05 mm wide from right to left.

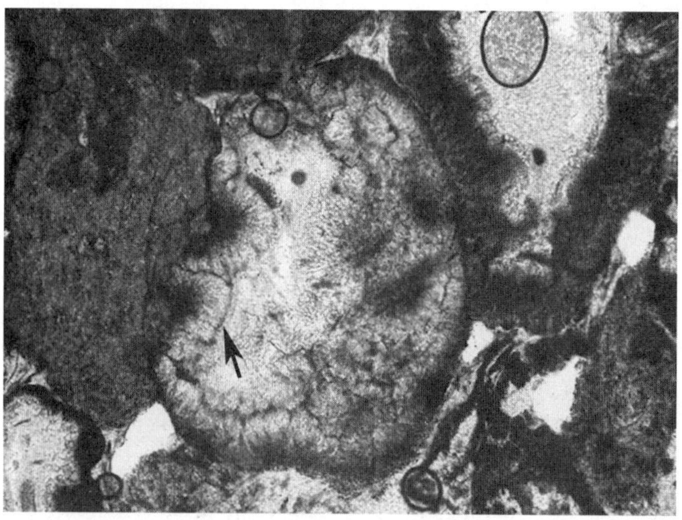

Figure 9. Photomicrograph (plane polarized light) of tightly compacted spherules from core P11. Note well-developed concavoconvex and long contacts between adjacent spherules. Botryoidal K-feldspar rim on spherule in middle has been replaced by sericite, but its shape is still outlined by opaque dust (indicated by black arrow). Spherule in middle is 0.55 mm across.

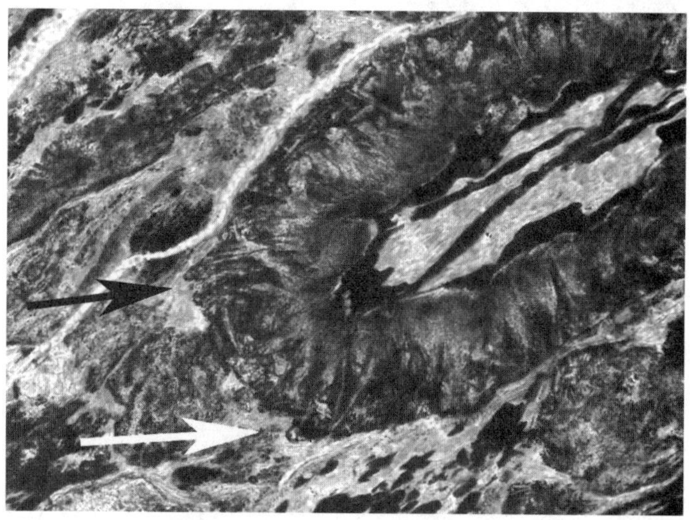

Figure 11. Photomicrograph (plane polarized light) of highly compacted spherules from core RF1. Shortening in the most tightly curved part of the rim of the largest, darkest spherule has been accommodated by microfaults that have imparted a toothed geometry to the outer edge of the rim (indicated by black and white arrows). The dark rim is 0.2 mm wide, and displacement along microfractures is on the order of 0.01 mm.

Irregular particles. Spherule-rich samples from the Pering Mine region also contain irregular particles with flow-banded and vesicular textures similar to those found in the surface samples. No consistent textural differences were noted between the irregular particles in the Monteville Farm and Pering Mine regions.

Mica ribbons. In the Pering Mine region, a large number of grains contain mica ribbons. The ribbons have the same textures as described previously, again occurring in three distinct varieties. The textural positions of these varieties are (1) lining former vesicles, (2) at the boundary between the rim and the central clear spot, and (3) branching in a septarian geometry within the clear core. Overall, the ribbons are more abundant in samples from the Pering Mine region compared to samples from the Monteville Farm region. Within some elongate melt particles that show brittle fracturing and flattening, a preferred orientation of sericite crystals is sometimes developed that is length-slow parallel to bedding.

Pering Mine Region, Intraclast-Rich Subunit

There is only one known occurrence of the intraclast-rich subunit in the Pering Mine region (Table 1), where it is directly beneath a spherule-rich subunit that is 2.5 cm thick (Fig. 3). The spherule-rich subunit shows textures similar to those described above; the following descriptions apply only to this single known core occurrence of the intraclast-rich subunit.

As with the samples from the Monteville Farm region, the intraclast-rich subunit found in the Pering Mine region contains widely scattered pockets rich in spherules (Fig. 2) that are believed to be interstitial to large intraclasts consisting predominantly of dolomite. The former melt particles in these pockets show textures similar to those in the overlying spherule-rich units, but some textures are also present that were not observed in any other samples. Foremost among these is growth-banded quartz filling both vesicles and central clear spots within spherules. There is also a high abundance of carbonate cement comparable to the cement of intraclast-rich samples from the Monteville Farm region.

Sheet silicate replacement like that seen throughout the spherule-rich subunit in the Pering Mine region is also present but generally less common in the intraclast-rich subunit. A moderate amount of microcrystalline quartz is present and restricted to intergranular material. Some spherules show evidence of brittle deformation in the form of fractures in radial-fibrous spherule rims and shortening in the vertical direction. We attribute this to compaction, although an average of three contacts per grain indicates that this sand is much less compacted than the spherule-rich subunit. In addition to brittle shortening, long grain contacts and very thin intergranular selvages composed of sheet silicate phases provide evidence of compaction. Many spherules with blocky carbonate crystals at their centers exhibit carbonate-filled fractures in their rims that display optical continuity with crystals on one or both sides of the rim. Although the former melt particles in the intraclast-rich subunit from Pering Mine exhibit textures found both in the spherule-rich layer and in the outcropping intraclast-rich layer, the population of melt particles is more similar to the latter. No mica ribbons were observed in samples from the single core occurrence of the intraclast-rich subunit.

DISCUSSION

Original Composition

Based on both optical characteristics and some microprobe analyses of sample SF1–1 (Table 4), the spherules in the Monteville formation are now composed mainly of K-feldspar and sericite except where they are replaced by carbonate. Various lines of evidence indicate that these are all authigenic phases formed largely by replacement of both glass and crystalline phases formed from a melt. For example, the splash-form shapes of the spherules indicate they were originally molten and consisted largely of glass at the time of deposition, yet they now consist entirely of crystals. Another line of evidence is the fact that the carbonate-free spherule-rich samples consist almost entirely of K-feldspar and sericite. None of the rock types likely to make up the country rock in the target area (e.g., granite or basalt) consist of only these two minerals, indicating it is very unlikely that the spherules are isochemical with the original impact melt.

Moreover, microprobe analysis of spherules from sample SF1–1 indicates the K-feldspar has almost no Na_2O (Table 4). Such a composition is typical of authigenic K-feldspar formed under low-temperature diagenetic conditions, e.g., via replacement (Simonson, 1992), but not via eutectic melt crystallization. This is not unusual as few if any original phases have survived to the present in Precambrian spherule layers, with the exception of skeletal spinels in a few layers (e.g., Byerly and Lowe, 1994). Even though they have been replaced wholesale and now consist largely of K-rich phases, it is likely that the spherules in the Monteville Formation were originally basaltic in composition. This is strongly implied by the fact that the internal textures in the lath-rich spherules bear a strong resemblance to partially crystallized basaltic melts formed both experimentally and naturally (Lofgren, 1977). Hassler and Simonson (2001) also noted basaltic tuffs in the Griqualand West Basin that showed similar replacement by K-feldspar.

Variation Between Intraclast-Rich and Spherule-Rich Subunits

The most noticeable contrast between the intraclast-rich and spherule-rich subunits is in the degree of compaction. Elongate, severely compacted grains showing evidence of brittle deformation (Category 1, Fig. 5) are much more abundant in the spherule-rich bed, whereas moderately compacted to spherical grains (Categories 2, 4, and 6) are more abundant within the intraclast-rich bed, although some also show brittle fracturing within their rims (Category 2). The fact that the number of contacts per grain is three times greater in the spherule-rich subunit than it is in the intraclast-rich bed also indicates the spherule-rich bed has undergone more compaction, despite the fact that the intraclast-rich subunit was buried to the same depth and experienced similar amounts of vertical compressive stress as the spherule-rich subunit. Higher average aspect ratios also indicate the spherule-rich subunit is more highly compacted than the intraclast-rich one (Fig. 12). Aspect ratios in samples from the intraclast-rich subunit fall between 1.4 and 2.4, whereas those from the spherule-rich subunit fall between 3.2 and 5.9. This also suggests around three times more vertical shortening in the spherule-rich layer.

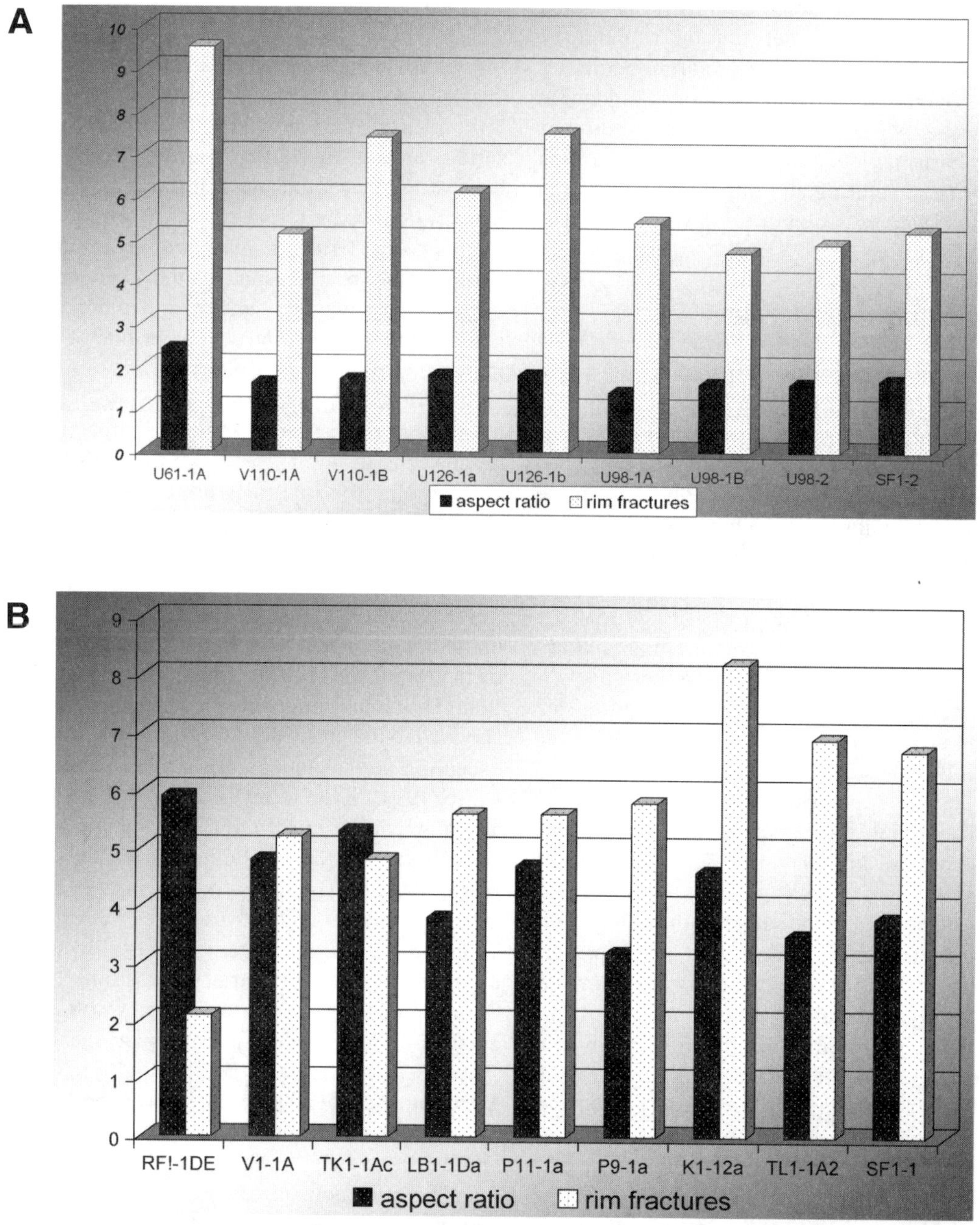

Figure 12. Bar graphs showing average values for aspect ratios (black bars) and number of rim fractures (white bars) for spherules in (A) samples from the intraclast-rich subunit (all from the Pering Mine region) and (B) the spherule-rich subunit (all from the Monteville Farm region save for SF1–1, which is from the Pering Mine region). See Table 2 for sample locations and Table 3 for values of averages plus standard deviations.

The presence of localized shelter porosity in the intraclast-rich subunit indicates that the larger intraclasts formed a framework that could have protected the sand grains from some of the compressive stress. This, in turn, could have helped preserve intergranular porosity long enough to allow the emplacement of carbonate cement early enough in diagenesis for the sands in the intraclast-rich subunit to experience much less compactional deformation than those in the spherule-rich subunit (Fig. 4). The presence of abundant carbonate clasts within the intraclast-rich subunit could have aided in the early cementation of this subunit, as the clasts themselves may have provided a ready cement source. The absence of fine-grained matrix in this subunit is evidence for its deposition by a high-energy traction mechanism, such as a wave or unidirectional current, rather than via a debris-flow–type mechanism. The highest matrix percentage counted in the spherule-rich subunit was 23% by volume. There is a very small percentage of possible cement in the spherule-rich subunit and what there is consists mainly of sheet silicate. Even if the fine-grained matrix is detrital in origin, which we think is unlikely, it still would do little to help prevent pore volume loss during compaction, as sheet silicates are generally much easier to compact physically than sparry carbonate cement.

In contrast to the differences in compaction, the thicknesses of spherule rims are very similar in samples from the spherule-rich and the intraclast-rich subunits. Rim thicknesses in all samples range from 40 to 85 μm, with most falling into the range of 50 to 70 μm (Table 3). Likewise, the number of fractures per rim is rather consistent (Table 3) and does not vary systematically as a function of aspect ratio, contacts per grain, or any other measured quantity. This lack of correlation suggests the rim fractures formed via a mechanism unrelated to burial stress, presumably prior to compaction.

Variations Within Intraclast-Rich Samples

Within the Monteville Farm region, the relative abundances of different types of former melt particles are relatively consistent (Table 2), and the amount of compaction is generally low. There is a moderate decrease in the number of rim fractures per spherule from northeast (Monteville Farm) to southwest (Reid's Drift and Goudkop), but their aspect ratio stays relatively constant between 1.4 and 2.4 (Table 3). The subtle differences in former melt particle abundance may simply reflect varying degrees of cementation and replacement, but the diagenetic history is such that it is difficult to work out a detailed history of cementation. One problem is the abundance of pyrite, which not only obscures the size of carbonate crystals but obscures textures even more where pyrite is oxidized. The abundance of former melt particles within the material interstitial to the large intraclasts does vary within the region from a minimum of 20% to a maximum of 59% by volume. It is worth noting that the relative abundances of the different particle types were not affected to any great degree by this high level of disparity, which strongly suggests that former melt particles were not differentially replaced by carbonate to any significant degree during diagenesis. The fact that the intraclast-rich subunit samples from both core (SF1) and surface exposures (the Monteville Farm region) show similar textural character, including those features related to compaction (Fig. 13), further suggests that diagnostic ratios of different types of particles were similar over wide geographic areas and have not been significantly altered by surface weathering processes.

Within the Pering Mine intraclast-rich sample (SF1-1), a smaller amount of brittle grain deformation is seen than in the other samples, as it contains 7% fewer severely compacted

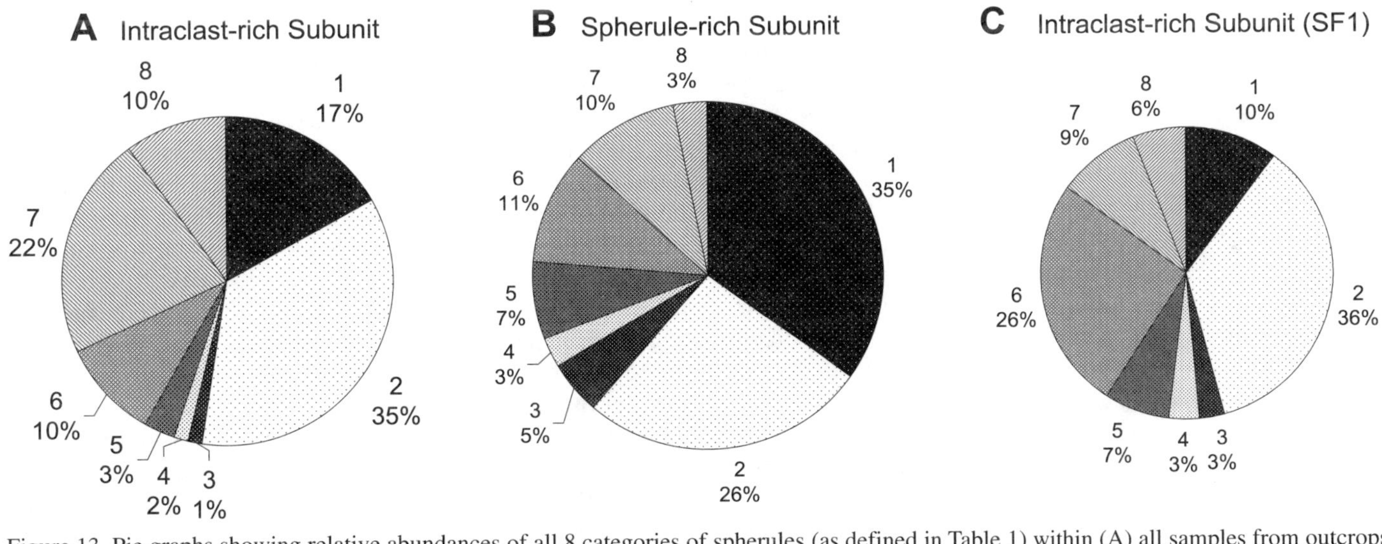

Figure 13. Pie graphs showing relative abundances of all 8 categories of spherules (as defined in Table 1) within (A) all samples from outcrops of the intraclast-rich subunit, (B) all samples from the spherule-rich subunit, which was seen exclusively in core, and (C) samples from the intraclast-rich subunit in core SF1; based on point count data presented in Table 2.

elongate grains (Category 1) and 16% more spherical particles with botryoidal rims (Category 6) than any of the other samples. In general, if severely compacted elongate grains were originally spherical or ovoid, then the higher number of spherical grains in sample SF1-1 indicates a lesser degree of compaction. We interpret individual former melt particles in the SF1-1 core sample as having experienced a lesser degree of brittle deformation, on average, than those of the intraclast-rich subunit exposed at the surface. Nevertheless, the higher average of contacts per grain in the one core sample of the intraclast-rich subunit as compared to samples from surface occurrences indicates overall compaction was more extreme in the Pering Mine region.

The population of former melt particles in the intraclast-rich subunit in the SF1 core differs slightly from that of the Monteville Farm region in that irregular shapes increase by 13% in abundance in the Monteville Farm region. This could reflect a decrease in depositional energy from south to north inasmuch as irregular particles tend to be larger and more crystalline than spherules. This, in turn, would be consistent with evidence presented by Simonson et al. (1999) that the Monteville layer was deposited by high-energy waves and currents, possibly impact induced, directed roughly normal to the edge of the Kaapvaal Craton, which locally is roughly east-west. It is worth noting that in core SF1, the percentage of irregular particles in the intraclast-rich subunit is equal to that of the spherule-rich subunit (Table 2).

Variations Within Spherule-Rich Samples

Samples from the spherule-rich subunit all come from drill cores from the Pering Mine region, and they display a relatively uniform population of melt particles (Table 2), despite the degree of compaction generally increasing to the north (Table 3). The latter cannot be attributed to a progressive increase in load-induced stress, because the sediment package thins to the north (see Fig. 14 *in* Beukes, 1987). The average aspect ratio generally increases to the north, whereas rim fractures (Fig. 11) decrease slightly in abundance in that direction. Both these phenomena are thought to result from an increase in the thermal regime in the northern part of the study region where more dolerite dikes were observed in the cores than in those from the southern region. Generally, grains obscured by replacement are seen to decrease in abundance as the number of severely compacted grains increases. However, the spherule layer in the northernmost core (RF1) was only 4 m below a large dike or sill and displays both anomalously high numbers of elongate grains and one of the highest values for obscured textures (Table 2). The reason for this anomalous relationship could be that the highly elongate particles flattened by brittle deformation of their rims were replaced most readily due to the additional surface area resulting from fracturing. In contrast to the above trend, centrally located samples (e.g., those from P11, P9, and K1 cores) all behave in a similar fashion, showing moderate amounts of compaction and replacement. Transitionally located cores (V1 and TK1) show degrees of compaction intermediate between that of RF1 and that found in the centrally located cores. In summary, varying degrees of replacement and compaction were observed even within the small Pering Mine region, but they show subtle trends that may be the result of lateral thermal gradients.

Mica Ribbons

Particles displaying mica ribbons (Fig. 7) are more abundant in samples from the spherule-rich subunit than in those from the intraclast-rich subunit (Fig. 14). Within a given sample, mica ribbons are less abundant in more highly compacted grains. Most mica ribbons occur in spherules possessing radial fibrous K-feldspar rims, the only exception being where they line vesicles. We believe vesicle linings formed as void-filling cements, and they occur in all types of grains. In spherules with K-feldspar rims, the ribbons not lining vesicles occur along contacts between rims and irregular clear spots and/or as radially branching networks inside irregular spots. These irregular clear spots are thought to represent glass cores that were devitrified or replaced by sericite and/or acicular K-feldspar during diagenesis (Simonson, 2003). We interpret the origin of the nonvesicular mica ribbons as follows.

It is likely that impact glass, including the glass cores of spherules, would still be hot enough from shock melting and/or atmospheric reheating to crack via the quick drop in temperature and resulting contractional stress once they came into contact with seawater. Given their crystallinity and radial crystal orientation, rims would be unlikely to contract as much, resulting in: (1) cores being pulled away from rims, and (2) cracks forming inside cores with a geometry similar to cracks seen in septarian nodules (Pettijohn, 1975). It has been proposed that thermal contraction cracking is a mechanism for microtektite fragmentation in Cenozoic strewn fields (Glass et al., 1997). As microtektites consist entirely of glass and possess no rim, complete disarticulation of a microtektite is more likely than in the case of a microkrystite. If spherules possessing internal microfractures are reworked by a high-energy event as those in the Monteville layer clearly were, the possibility of fragmentation would increase. Perhaps the presence of unfractured crystalline rims helped many spherule cores to avoid fragmentation during deposition. Preexisting internal fractures should also make it easier for spherule cores to disaggregate under burial-induced pure shear stress. Microcracks could also serve as conduits for fluid entry and help guide parallel fibrous replacement via nucleation along cracks in a manner analogous to axiolitic replacement of volcanic glass shards or the formation of mesh texture in serpentinites. A similar effect could take place along the edge of the glass core where it is in contact with the crystalline rim. This replacement is likely to have resulted in the parallel fibrous textures we describe as isopachous mica ribbons.

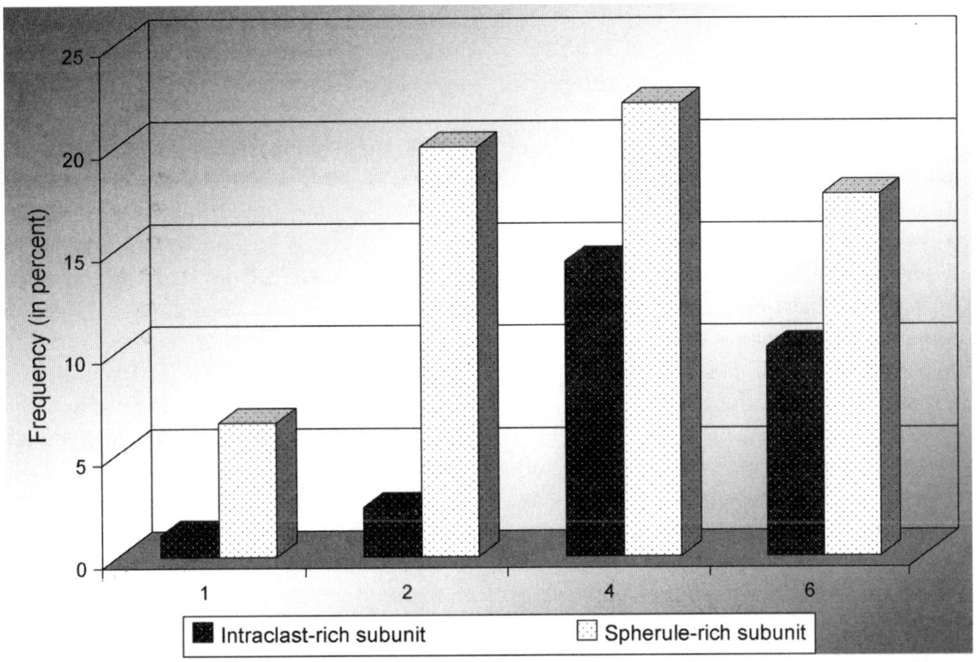

Figure 14. Bar graph showing average percentages of spherules in categories (defined in Table 1) that contain mica ribbons: categories 1, 2, 4, and 6. Black bars represent all samples in a given category from the intraclast-rich subunit, whereas white bars represent all samples in a given category from the spherule-rich subunit.

Deformation Textures and Modes of Vertical Shortening

About half of the splash-form population at each site experienced brittle compaction, as evidenced by rim fractures showing varying degrees of displacement, often resulting in disaggregation into fragment clusters. The small groups of fragments are oriented in a way that, along with their rim structures, suggests fragmentation of spheroidal to elongate splash forms with relatively thin rims. In the more highly compacted spherule-rich subunit, we have documented a direct correlation between thickness of spherule rim and degree of compaction; specifically, individual spherules with thicker rims are likely to have lower aspect ratios in all samples (Table 3). Spherule rims show brittle fractures that indicate they are acting as rigid shells enclosing a much weaker central core of sheet silicate or possibly a void in some cases.

Another dominant mechanism resulting in long and concavo-convex grain boundaries as well as bent but not fractured rim segments is diffusional mass transfer (DMT) of minerals such as K-feldspar and quartz (Passchier and Trouw, 1996). The difference in behavior between central clear spots now composed of sericite versus fracturing of rims made of feldspar presumably reflects different responses to stress diagnostic of different minerals. Sheet silicate crystals oriented length-slow parallel to bedding suggest that reorientation or oriented growth are shortening mechanisms that occurred within the cores of spherules either through DMT or simple mechanical compaction. Sheet silicates deform by slip limited to discrete surfaces. Therefore, they show abundant evidence of accommodation mechanisms other than DMT (Passchier and Trouw, 1996). So, reorientation and slip-dominated mechanical compaction are more likely to have occurred in the sheet silicate cores.

Rim fractures show different geometries on different grains, and two geometries are dominant, with spherules containing either one or both types. The first geometry resembles the "eggshell" fracturing of a hollow spheroidal particle (Wilkinson and Landing, 1978) and is characterized by wedge-shaped fractures that can taper either into or out of the center of a spherule (Fig. 6). "Eggshell" fractures only form where a central void is present into which the fracturing rim can be displaced. Detrital quartz grains located inside rims showing the "eggshell" geometry confirm the fact that voids were present in such spherules; these quartz grains are usually accompanied by the chloritic sheet silicate. The second geometry consists of slip-dominated planar fractures at high angles to the rim. In such particles, offset is almost entirely parallel to the fracture, whereas extension perpendicular to the fracture is insignificant (Fig. 11). In this case, internal volume loss via DMT or slip-dominated reorientation of sheet silicate crystals could accommodate shortening without the need for voids, thereby avoiding an "eggshell"-type geometry. These rim microfaults are often accompanied by ovoid zones of recrystallized sheet silicates within the rim in the areas experiencing the most shortening (Fig. 10), indicating DMT aided migration of sheet silicate material, most likely from within the core toward a location in the rim. These two different shortening mechanisms could have worked in proportionally variable combinations.

SUMMARY AND CONCLUSIONS

Spherules in the Monteville Formation are now composed primarily of authigenic K-feldspar, muscovite (sericite) and chloritic phases, and carbonate. All of these phases are thought to be the result of replacement and therefore do not reflect the original mineralogy of the melt particles. However, pseudomorphed crystal textures within the spherules suggest that they were originally basaltic in composition. The fact that regionally associated tuffs thought to be basaltic in composition are replaced in a very similar fashion supports this interpretation. Since the bulk of the mass in impact ejecta is derived from the target rock, this in turn suggests formation of the Monteville spherules by an impact into oceanic crust.

Various lines of evidence indicate the intraclast-rich subunit experienced less compaction than the spherule-rich subunit, e.g., its higher content of interstitial carbonate spar is interpreted as cement or a replacement of cement. This lower degree of compaction within the intraclast-rich subunit leads us to believe it is the least altered and most closely representative of the original population of former melt particles. However, the fact that the number of rim fractures shows no increase as the degree of compaction increases between the two subunits suggests that the particle rims experienced structural weakening in equal measure throughout the study area prior to deposition. The formation of cracks via thermal shrinkage in glass cores and differential shrinkage between radial fibrous rims and glass cores could provide the mechanism for this structural weakening and resulting fracturing early in diagenesis. The microcracks in some spherules then served as conduits for fluid entry and localized replacement by parallel fibrous sheet silicates to form mica ribbons. In other spherules, the microcracks were apparently conduits for dissolution instead, helping to create voids that enabled "eggshell"-type compaction. In still other spherules, diffusional mass transfer (DMT) and/or crystal reorientation enabled grains to shorten via microfaulting without obvious fragmentation of the rim. Fracturing of rims and reorientation or DMT of centrally located sheet silicate minerals indicates some compaction either postdates or was contemporaneous with the later stages of replacement. Despite all these modifications, the former melt particle populations within both the intraclast-rich and the spherule-rich subunits show only slight disparities over distances on the order of 100 km, both within and between subunits.

Implications for Intercontinental Correlation

Simonson et al. (1999) discussed the possible correlation of the Monteville layer with spherule layers in the Jeerinah, Wittenoom, and/or Carawine formations in the Hamersley Basin based on age constraints available at the time, as well as sedimentologic and gross petrographic similarities. Since this paper was published, another outcrop of the Jeerinah layer was found at the Hesta locality in the Hamersley Basin (Simonson et al., 2001; Hassler et al., 2005). Additional occurrences in drill cores are described by Rasmussen and Koeberl (2004). This layer shows some striking similarities to the Carawine layer, particularly an abundance of irregular melt particles. To further test the proposed correlations, a sample from the Hesta occurrence was point counted using the methods described above, and it shows striking similarities to the Monteville spherule layer (Table 2). Not only are the relict textures that the spherules inherited from their molten state very similar, they seem to have experienced similar diagenetic histories as well, most notably the formation of isopachous mica ribbons in spherules. Our studies of the Monteville layer suggest that the melt particles in spherule layers can be relatively homogeneous on a scale of a few hundred kilometers and retain a significant amount of textural information despite widespread diagenetic alteration. The similarities with the Jeerinah layer in Australia indicate this may also be true over much larger distances and include similarities in diagenetic as well as primary features. It may provide additional evidence that the Pilbara and Kaapvaal cratons were both part of a single depositional basin at the time the Monteville layer formed, as some workers have suggested (e.g., Cheney, 1996).

ACKNOWLEDGMENTS

Field work was supported by grants from the National Geographic Society and Oberlin College, Oberlin, Ohio, United States. We thank Pering Mine (Pty) Ltd. and Gold Fields of South Africa Ltd. for permission to examine cores through the Monteville Formation. Former Oberlin students Paul Jambor and Brooke Wilkerson provided field assistance, and Michael Cardiff and Trista Thornberry helped in the laboratory. We also thank Jan Smit and the Smithsonian Institution for the loan of samples from the K/T boundary layer. We are especially indebted to Dr. Nic Beukes of Rand Afrikaans University for introducing us to the field area and making the arrangements that made this study possible. Dr. Jens Gutzmer of the University of Johannesburg performed the electron microprobe analyses given in Table 4. The manuscript benefited greatly from careful reviews by Drs. Burkhard Dressler and Christian Koeberl.

REFERENCES CITED

Beukes, N.J., 1987, Facies relations, depositional environments and diagenesis in a major early Proterozoic stromatolitic carbonate platform to basinal sequence, Campbellrand Subgroup, Transvaal Supergroup, southern Africa: Sedimentary Geology, v. 54, p. 1–46, doi: 10.1016/0037-0738(87)90002-9.

Bohor, B.F., and Glass, B.P., 1995, Origin and diagenesis of K/T impact spherules—From Haiti to Wyoming and beyond: Meteoritics, v. 30, p. 182–198.

Byerly, G.R., and Lowe, D.R., 1994, Spinel from Archean impact spherules: Geochimica et Cosmochimica Acta, v. 58, p. 3469–3486, doi: 10.1016/0016-7037(94)90099-X.

Byerly, G.R., Lowe, D.R., Wooden, J.L., and Xie, X., 2002, An Archean impact layer from the Pilbara and Kaapvaal Cratons: Science, v. 297, p. 1325–1327, doi: 10.1126/science.1073934.

Cheney, E.S., 1996, Sequence stratigraphy and plate tectonic significance of the Transvaal succession of southern Africa and its equivalent in Western Australia: Precambrian Research, v. 79, p. 3–24, doi: 10.1016/0301-9268(95)00085-2.

Eriksson, K.A., 1995, Crustal growth, surface processes and atmospheric evolution of the early Earth, *in* Coward, M.P., and Ries, A.C., eds., Early Precambrian Processes: Geological Society [London] Special Publication 95, p. 11–25.

Glass, B.P., 2002, Upper Eocene impact ejecta/spherule layers in marine sediments: Chemie der Erde, v. 62, p. 173–196, doi: 10.1078/0009-2819-00017.

Glass, B.P., Muenow, D.W., Bohor, B.F., and Meeker, G.P., 1997, Fragmentation and hydration of tektites and microtektites: Meteoritics & Planetary Science, v. 32, p. 333–341.

Hassler, S.W., and Simonson, B.M., 2001, The sedimentary record of extraterrestrial impacts in deep-shelf environments: Evidence from the early Precambrian: The Journal of Geology, v. 109, p. 1–19, doi: 10.1086/317958.

Hassler, S.W., Simonson, B.M., Sumner, D.Y., and Murphy, M., 2005, Neoarchaean impact spherule layers in the Fortescue and Hamersley Groups, Western Australia: Stratigraphic and depositional implications of recorrelation: Australian Journal of Earth Sciences, v. 52, p. 759–771.

Keyser, N., compiler, 1997, Geological Map of the Republic of South Africa and the Kingdoms of Lesotho and Swaziland: Pretoria, Council for Geoscience, scale 1:1,000,000, 4 sheets.

Koeberl, C., and Martinez-Ruiz, F., eds., 2003, Impact Markers in the Stratigraphic Record: Berlin, Springer, Impact Studies 3, 347 p.

Lofgren, G.E., 1977, Dynamic crystallization experiments bearing on the origin of textures in impact-generated liquids: Proceedings of the Eighth Lunar Science Conference, p. 2079–2095.

Montanari, A., and Koeberl, C., 2000, Impact Stratigraphy: The Italian Record: Berlin, Springer, Lecture Notes in Earth Sciences 93, 364 p.

Norris, R.D., Huber, B.T., and Self-Trail, J., 1999, Synchroneity of the K-T oceanic mass extinction and meteorite impact: Blake Nose, western North Atlantic: Geology, v. 27, p. 419–422, doi: 10.1130/0091-7613(1999)027<0419:SOTKTO>2.3.CO;2.

Passchier, C.W., and Trouw, R.A.J., 1996, Microtectonics: Berlin, Springer, 289 p.

Pettijohn, F.J., 1975, Sedimentary Rocks, third edition: New York, Harper and Row, 628 p.

Rasmussen, B., and Koeberl, C., 2004, Iridium anomalies and shocked quartz in a late Archean spherule layer from the Pilbara craton: New evidence for a major asteroid impact at 2.63 Ga: Geology, v. 32, p. 1029–1032, doi: 10.1130/G20825.1.

Simonson, B.M., 1992, Geological evidence for a strewn field of impact spherules in the early Precambrian Hamersley Basin of Western Australia: Geological Society of America Bulletin, v. 104, p. 829–839, doi: 10.1130/0016-7606(1992)104<0829:GEFASF>2.3.CO;2.

Simonson, B.M., 2003, Petrographic criteria for recognizing certain types of impact spherules in well-preserved Precambrian successions: Astrobiology, v. 3, p. 49–65, doi: 10.1089/153110703321632417.

Simonson, B.M., and Glass, B.P., 2004, Spherule layers—Records of ancient impacts: Annual Review of Earth and Planetary Sciences, v. 32, p. 329–361, doi: 10.1146/annurev.earth.32.101802.120458.

Simonson, B.M., Hassler, S.W., and Beukes, N., 1999, Late Archean impact spherule layer in South Africa that may correlate with a Western Australian layer, *in* Dressler, B.O., and Sharpton, V.L., eds., Large Meteorite Impacts and Planetary Evolution II: Geological Society of America Special Paper 339, p. 249–261.

Simonson, B.M., Davies, D., and Hassler, S.W., 2000a, Discovery of a layer of probable impact melt spherules in the late Archean Jeerinah Formation (Fortescue Group, Western Australia): Australian Journal of Earth Sciences, v. 47, p. 315–325, doi: 10.1046/j.1440-0952.2000.00784.x.

Simonson, B.M., Hornstein, M., and Hassler, S.W., 2000b, Particles in late Archean Carawine Dolomite (Western Australia) resemble Muong Nong-type tektites, *in* Gilmour, I., and Koeberl, C., eds., Impacts and the Early Earth: Berlin, Springer, Lecture Notes in Earth Sciences 92, p. 181–213.

Simonson, B.M., Cardiff, M., and Schubel, K.A., 2001, New evidence that a spherule layer in the late Archean Jeerinah Formation was produced by a major impact: Lunar and Planetary Sciences Conference XXXII [abs. 1141]: Houston, Lunar and Planetary Institute, CD-ROM.

Smit, J., 1999, The global stratigraphy of the Cretaceous-Tertiary boundary impact ejecta: Annual Review of Earth and Planetary Sciences, v. 27, p. 75–113, doi: 10.1146/annurev.earth.27.1.75.

Smit, J., Alvarez, W., Montanari, A., Swinburne, N., Van Kempen, T.M., Klaver, G.T., and Lustenhouwer, W.J., 1992, "Tektites" and microkrystites at the Cretaceous Tertiary boundary: Two strewn fields, one crater?: Proceedings of Lunar and Planetary Science, v. 22, p. 87–100.

Trurnit, P., 1968, Pressure solution phenomena in detrital rocks: Sedimentary Geology, v. 2, p. 89–114, doi: 10.1016/0037-0738(68)90030-4.

Vernon, R.H., 2004, A Practical Guide to Rock Microstructure: Cambridge, U.K., Cambridge University Press, 594 p.

Wilkinson, B.H., and Landing, E., 1978, "Eggshell diagenesis" and primary radial fabric in calcite ooids: Journal of Sedimentary Petrology, v. 48, p. 1129–1138.

Manuscript Accepted by the Society 4 October 2005

Geological Society of America
Special Paper 405
2006

Relicts of Earth's earliest crust: U-Pb, Lu-Hf, and morphological characteristics of >3.7 Ga detrital zircon of the western Canadian Shield

R.P. Hartlaub[†]
L.M. Heaman
A. Simonetti
Department of Earth and Atmospheric Sciences, 1-26 Earth Sciences Building, University of Alberta, Edmonton, Alberta T6G 2E3, Canada
C.O. Böhm
Manitoba Geological Survey, Manitoba Industry, Economic Development and Mines, 360-1395 Ellice Ave., Winnipeg, Manitoba R3G 3P2, Canada

ABSTRACT

Ancient (>3.7 Ga) detrital zircons represent some of the few remaining relicts of Earth's earliest evolution. A metagreywacke from the northwestern margin of the Superior Province, Canada, has abundant Paleoarchean detrital zircon with peaks in age distribution at 3.86, 3.79, 3.74, and 3.32 Ga. A fuchsitic quartzite from the western margin of the Rae Province, Canada, contains entirely Paleoarchean detritus with peaks in age distribution at 3.86, 3.76, and 3.72 Ga. Both samples contain a small (2%–4%) proportion of zircon grains that are ≥3.9 Ga. Hf isotopic analysis indicates that a large proportion of the Paleoarchean zircon from both samples was derived by reworking of significantly older crust, consistent with previously published evidence for scarce pre–4.0 Ga continental crust from Jack Hills, Australia, and the Acasta Gneiss, Canada. When comparing the detrital zircon age distributions obtained in this study with known terranes with intact Paleoarchean rocks, most similarity is observed with the Itsaq Gneiss Complex of western Greenland. A lack of ca. 3.6–3.3 Ga igneous crystallization and detrital zircon ages is apparent in the data from the western Rae Province and from the northwestern Superior Province. Many terranes with significant evidence for 3.7–3.9 Ga crust also have a well-defined lack of crust formation at this time, possibly indicating a hitherto undetected relationship between these Paleoarchean terranes.

Keywords: geochronology, zircon, Archean, Hafnium, Superior Province.

[†]Current address: British Columbia Institute of Technology, 3700 Willingdon Avenue, Burnaby, British Columbia V5G 3H2, Canada; hartlaub@ualberta.net.

Hartlaub, R.P., Heaman, L.M., Simonetti, A., and Böhm, C.O., 2006, Relicts of Earth's earliest crust: U-Pb, Lu-Hf, and morphological characteristics of >3.7 Ga detrital zircon of the western Canadian Shield, *in* Reimold, W.U., and Gibson, R.L., Processes on the Early Earth: Geological Society of America Special Paper 405, p. 75–89, doi: 10.1130/2006.2405(05). For permission to copy, contact editing@geosociety.org.

INTRODUCTION

Our knowledge of Earth's earliest evolution is directly linked to our understanding of the few existing fragments of the planet's oldest crust. Although direct study of these ancient fragments is important, the limited number of identified ancient terranes has restricted our ability to make global comparisons and conclusions. Detritus from this ancient crust is, however, a good tracer because much of the oldest crust may have been recycled back into the mantle through erosion and subduction, or is now deeply buried within the cratons. Study of the proportion, internal structure, and mineral inclusion record of ancient detrital zircon has helped to elucidate the nature of early continental crust (Maas et al., 1992), the presence of oceans (Wilde et al., 2001), and the possibility of crustal growth versus recycling (Nutman, 2001). The hafnium isotopic composition of ancient zircon has also been utilized in several studies of early Earth crust-mantle evolution (Patchett et al., 1981; Vervoort et al., 1996; Amelin et al., 1999). Utilizing zircon, rather than whole-rock samples, offers the significant advantage of having a direct U-Pb age on the same material from which Hf isotopes are recorded. The high Hf content (~10,000 ppm) in zircon enables measurement of single grains on laser ablation multicollector inductively coupled plasma mass spectrometry (MC-ICPMS) instruments (Amelin et al., 1999), and the low Lu/Hf ratios (typically less than 0.0015) imply that in situ radiogenic growth will make a minimal contribution to the $^{176}Hf/^{177}Hf$ ratio. Recent developments in laser ablation MC-ICPMS techniques and instrumentation now enable both U-Pb and Lu-Hf isotopic compositions to be measured on individual parts of zircon grains (e.g., Woodhead et al., 2004). This development is important for zircon with metamorphic overgrowths; a common characteristic of zircons found within ancient gneisses.

In this paper we first review some of the oldest exposed crustal fragments and then describe the morphological, U-Pb, and Lu-Hf characteristics of >3.7 Ga detrital zircon grains that were separated from metasedimentary rocks of the Archean Superior and Rae Provinces of the western Canadian Shield. Our new results expand both the number and global distribution of well-studied 3.7 to >3.9 Ga zircons. We compare the characteristics of these ancient detrital grains to known ancient crustal terranes and discuss the implications of this new detrital zircon data set to the understanding of early-Earth crustal processes.

GLOBAL EXAMPLES OF 3.7–3.9 GA CRUST

Although there is rare ≥4.0 Ga crust exposed in the western Slave Province of northwestern Canada (Acasta gneiss of Bowring and Williams, 1999), and rare >4.1 Ga detrital zircons have been detected in Western Australia (e.g., Froude et al., 1983; Wilde et al., 2001), there appears to be much more evidence, globally, for the existence of sialic crust between 3.7 and 3.9 Ga. The largest and best studied examples of ca. 3.7–3.9 Ga crust are located within the Precambrian shields of Canada and

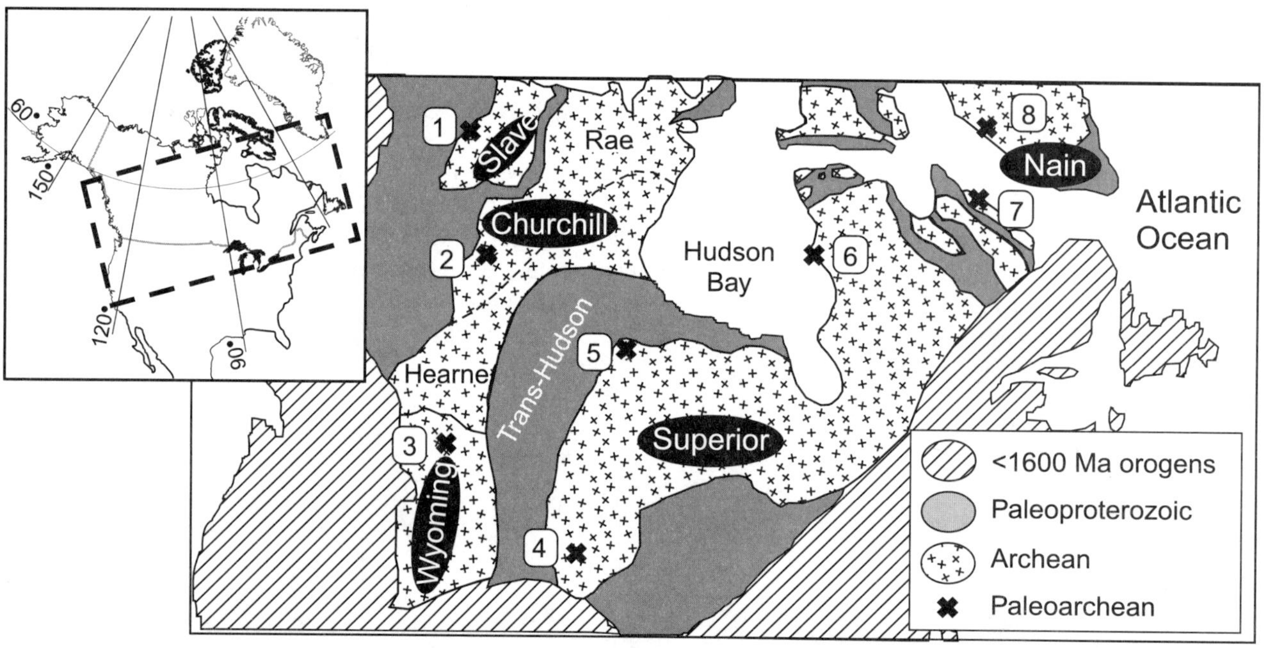

Figure 1. Simplified geological map of North America highlighting the major Archean cratons and locations of identified and suspected ancient (≥3.5 Ga) crust. Upper left inset indicates the region of North America examined in Figure 1. Overlying Proterozoic and younger sedimentary basins have been removed. 1—Acasta Gneiss, 2—western Rae Province, 3—northern Wyoming Province, 4—Minnesota River Valley, 5—Assean Lake ancient crustal complex, 6—Nuvvuagittuq Supracrustal Sequence, 7—Saglek Block, 8—Itsaq Gneiss Complex.

Greenland (Fig. 1). Evidence for such ancient crust also exists in several other locations around the world. This section will highlight some of the most important examples, but readers are directed to other reviews (e.g., Moorbath et al., 1986; Bowring and Housh, 1995), as well as the manuscripts cited herein, for additional details.

Itsaq Gneiss Complex, Western Greenland

The largest and best-studied area of Paleoarchean crust occurs in southwestern Greenland (Fig. 1) near the capital city of Nuuk. Paleoarchean crust in this region was referred to as the Itsaq Gneiss Complex and contains the oldest known supracrustal rocks in the world (Nutman et al., 1996). A review of these Paleoarchean terranes and their accretion history with new zircon U-Pb dates was provided by Nutman et al. (2004). The Isua greenstone belt, the oldest known greenstone belt in the world, is an amphibolite facies, highly deformed belt of 3.7 to ≥3.8 Ga sedimentary and volcanic rocks (Moorbath et al., 1973; Nutman et al., 1996; Appel et al., 1998; Myers, 2001). Although the supracrustal rocks have received abundant study, 3.6 to ca. 3.86 Ga tonalitic gneisses are the dominant lithology throughout the Itsaq Gneiss Complex (Nutman et al., 1996, 1999; Crowley, 2003). In the gneiss belts surrounding the Isua greenstone belt these tonalites range in age from 3810 to 3795 Ma and are typified by oscillatory zoned zircon (Nutman et al., 1999; Crowley, 2003). Together, the above gneiss belts make up the Isukasia terrane of Nutman et al. (2004). To the south of the Isukasia Terrane, tonalites of the Faeringehaven terrane (Nutman et al., 2004) range in age from ca. 3.85–3.66 Ga, and volumetrically minor granites are ca. 3.66–3.60 Ga old (Nutman et al., 1993, 1996); these ages are not, however, without controversy (e.g., Whitehouse et al., 1999; Nutman et al., 2001; Whitehouse et al., 2001). North of the Isukasia terrane, complex gneisses of the Quarliit Taserssuat Assemblage and Aasivik terrane yield highly disturbed U-Pb zircon ages in the range of ca. 3600–3780 Ma (Rosing et al., 2001; Nutman et al. 2004).

Saglek Block, Labrador, Canada

The Saglek Block of northern Labrador, Canada, contains Paleoarchean crust exposed at surface (Baadsgaard et al., 1979) and in drill hole (Wasteneys et al., 1996). The recognition of ancient crust in this area is consistent with the interpretation that the Precambrian shields of Labrador and Greenland were joined as the North Atlantic Craton prior to rifting that formed the modern Labrador Sea (Wasteneys et al., 1996). The oldest component of the Saglek Block consists of 3732 ± 6 Ma felsic orthogneisses (Schiøtte et al., 1989) and 3742 ± 12 amphibolitic gneisses (Wasteneys et al., 1996). These rocks, broadly termed the Uivak gneisses (Baadsgaard et al., 1979), were migmatized at ca. 3.62 Ga and locally contain xenocrysts as old as 3863 ± 12 Ma (Schiøtte et al., 1989). The Lister Gneiss, a volumetrically less important component of the Saglek Block, has much younger crystallization ages of 3213 +21/–3 Ma, 3171 ± 30 Ma (Wasteneys et al., 1996), and 3235 ± 8 Ma (Schiøtte et al., 1989).

Nuvvuagittuq Supracrustal Sequence, Northern Superior Province

The Nuvvuagittuq Supracrustal Sequence is a recently discovered sliver of >3.8 Ga volcanic and sedimentary rock that lies along the northern margin of the Superior Province in Quebec, Canada (Fig. 1). Zircon with primary magmatic zoning from a felsic tuff horizon within the supracrustal package yielded a U-Pb crystallization age of 3823 ± 18 Ma (David et al., 2004). A tonalite from the area has yielded a younger U-Pb age of 3650 ± 5 Ma (David et al., 2004).

The Minnesota River Valley, United States

Initial interest in the Minnesota River Valley terrane, located along the southwestern margin of the Superior Province (Fig. 1), commenced when Goldich and Hedge (1974) proposed rocks were present with ages of ca. 3.8 Ga. Although these ages were disputed (Farhat and Wetherill, 1975), the antiquity of rocks in the region was still suspected and was later confirmed (Michard-Vitrac et al., 1977). New U-Pb SHRIMP (sensitive high-resolution ion microprobe) results by Bickford et al. (2004) are helping to further unravel the complex geological history of the region. These data indicate that some of the early gneisses formed at ca. 3.5 Ga and were metamorphosed and injected by tonalites at ca. 3.3–3.4 Ga. Younger sediments in the area have well-defined detrital zircon age distribution peaks at 3520, 3380, 3140, and 2600 Ma (Bickford et al., 2004). Therefore, although initial results suggested >3.7 Ga crust in the Minnesota River Valley, recent work has yet to define a history commencing much before ca. 3.5 Ga.

North China Craton

Liu et al. (1992) presented U-Pb ion microprobe evidence for ≥3800 Ma crust in the northern portion of the North China craton. At one location they found that all detrital zircons from a fuchsitic metaquartzite were older than 3550 Ma. At a second location 400 km away they identified sheared gneisses with a 3804 ± 5 Ma protolith age with evidence for ca. 3300 Ma metamorphism and perhaps migmatization. Other granites in the region have 3306 ± 13 Ma and 2962 ± 4 Ma crystallization ages (Liu et al., 1992). Additional evidence for ca. 3.8 Ga crust, migmatized at ca. 3.3 Ga, from the northern portion of the craton was presented by Song et al. (1996). In the southern portion of the craton, felsic granulite xenoliths from Mesozoic volcanics have U-Pb and Hf ages that indicate that a portion of the lower crust there is ≥3.6 Ga old (Zheng et al., 2004).

Napier Complex, Antarctica

The Napier Complex of Enderby Land, Antarctica, contains ancient high-grade rocks with a multistage history of deformation and magmatism (Harley and Black, 1997). The oldest rocks in this complex are found near Mount Sones, but their complex history has made interpreting their protolith ages difficult (e.g., Williams et al., 1984). Initial U-Pb zircon SHRIMP analyses on the Mount Sones orthogneiss indicated a protolith age of 3927 +/–10 Ma. A recent reexamination of this gneiss indicates that 3800 +50/–100 Ma is a better estimate of its age (Harley and Black, 1997). Other orthogneisses in the same area have U-Pb zircon ages of 3773 +13/–11 and 3840 +30/–20 Ma (Harley and Black, 1997).

Narryer Gneiss Complex, Australia

The Narryer Gneiss Complex of Western Australia is dominated by ca. 3.65–3.60 Ga migmatites, with minor ca. 3.4–3.5 and 3.3 Ga components (Kinny et al., 1988; Nutman et al., 1991). Poor exposure, combined with the high metamorphic grade and highly strained nature of much of the complex, makes interpretation of the early Archean history difficult (Nutman et al., 1991), but zircon cores in some of the rocks have U-Pb ages approaching 3.70–3.75 Ga (Kinny et al., 1988; Kinny and Nutman, 1996; Pidgeon and Wilde, 1998). Metasedimentary rocks in the region are known to contain ancient ≥3.9 Ga detrital zircons (e.g., Maas et al., 1992).

SAMPLE SELECTION

Western Rae Province (Murmac Bay Area)

A Paleoarchean component of the Rae Province, Canada (Fig. 1), was first identified in Hartlaub et al. (2004a) and corroborated by Hartlaub (2004). In these studies Paleoarchean detrital zircons were identified in sedimentary rocks of the Murmac Bay Group, and Paleoarchean xenocrystic zircons were identified in ca. 1.9–2.0 Ga granites. Locally, metaquartzite of the Murmac Bay Group unconformably overlies 3.0 Ga granitoid basement, and a Neoarchean depositional age was inferred (Hartlaub et al., 2004a). The discovery of 2.3 Ga volcaniclastic rocks within the Murmac Bay Group indicates, however, that the majority of sediment was deposited at ca. 2.3 Ga (Hartlaub et al., 2004b). A middle-amphibolite facies fuchsitic quartzite from the area contains detrital zircons with entirely Paleoarchean U-Pb ages (Hartlaub et al., 2004a), but this ID-TIMS (isotope dissolution thermal ionization mass spectrometry) study was hampered by significant discordance and the lack of a statistically meaningful sample population. In order to better identify and describe these ancient detrital grains, zircons from the fuchsitic quartzite were mounted in epoxy, imaged by backscattered electron microscopy, and analyzed by laser ablation MC-ICPMS.

Northwestern Superior Province (Assean Lake Area)

The outline of a Paleoarchean crustal block in the Assean Lake area at the northwestern margin of the Superior Province (Fig. 1) was identified by Böhm et al. (2000) based on Sm-Nd and initial U-Pb zircon results. Combined SHRIMP (sensitive high resolution ion microbe) and ID-TIMS U-Pb data from detrital zircons in metasediments from this block indicated that the source of this detritus had a history spanning 3.2–3.9 Ga (Böhm et al., 2000, 2003). Felsic orthogneisses with ca. 3.1–3.2 Ga crystallization ages are exposed in the area, and all units have undergone upper amphibolite facies Neoarchean metamorphism (Böhm et al., 2003). One sample of metagreywacke from the Assean Lake area (CB97–12) contains abundant 3.7–3.9 Ga detrital zircons. Although some description of the detritus from this sample was provided in Böhm et al. (2003), a second sample (CB00–56) from the same outcrop was collected for additional description and imaging, as well as U-Pb and Lu-Hf analysis. Approximately 50 detrital zircon grains from this second sample were mounted in an epoxy puck and imaged by backscattered electron microscopy prior to analysis by laser ablation MC-ICPMS.

METHODOLOGY AND RESULTS

U-Pb Laser Ablation MC-ICPMS Geochronology

In order to obtain ages for a large population of detrital zircon, a rapid and precise method of analysis is required. Laser ablation multi-collector inductively coupled plasma mass spectrometry (LA-MC-ICPMS) offers this ability with the advantage that Lu-Hf isotope ratios can be measured from the same zircon grains due to the relatively nondestructive nature of the analysis. The new LA-MC-ICPMS laboratory at the University of Alberta utilizes an innovative Nu Plasma MC-ICPMS collector design that has three ion multipliers, enabling simultaneous collection of ^{207}Pb, ^{206}Pb, and ^{204}Pb (Simonetti et al., 2005). ^{235}U, ^{238}U, ^{205}Tl, and ^{203}Tl were determined on Faraday collectors. Following the methods of Taylor et al. (2003), the Faraday-ion counter bias and ion counter calibration were determined using a mixed standard solution prior to each analytical session. In this case, the standard solution consists of Pb (NIST SRM 981) and Tl (NIST SRM 997). Prior to each individual analysis a 30-second blank measurement was undertaken. The measured ^{204}Hg/^{202}Hg value was found to be statistically identical during both background and laser ablation analyses, enabling blank measurements to correct for both ^{204}Pb background and ^{204}Hg interference. A New Wave Research 213 nanometer Nd:YAG (neodimum-doped yttrium aluminum garnet) laser unit was employed to ablate selected 40 μm spots on zircon grains mounted in epoxy. Low energy densities (2–3 J/cm^2) and repetition rates (4 Hz) were utilized due to the high sensitivity of the ion counters. These conditions resulted in ablation pits that were typically 15 μm deep after a 30-second run time. A He carrier gas was utilized to reduce the

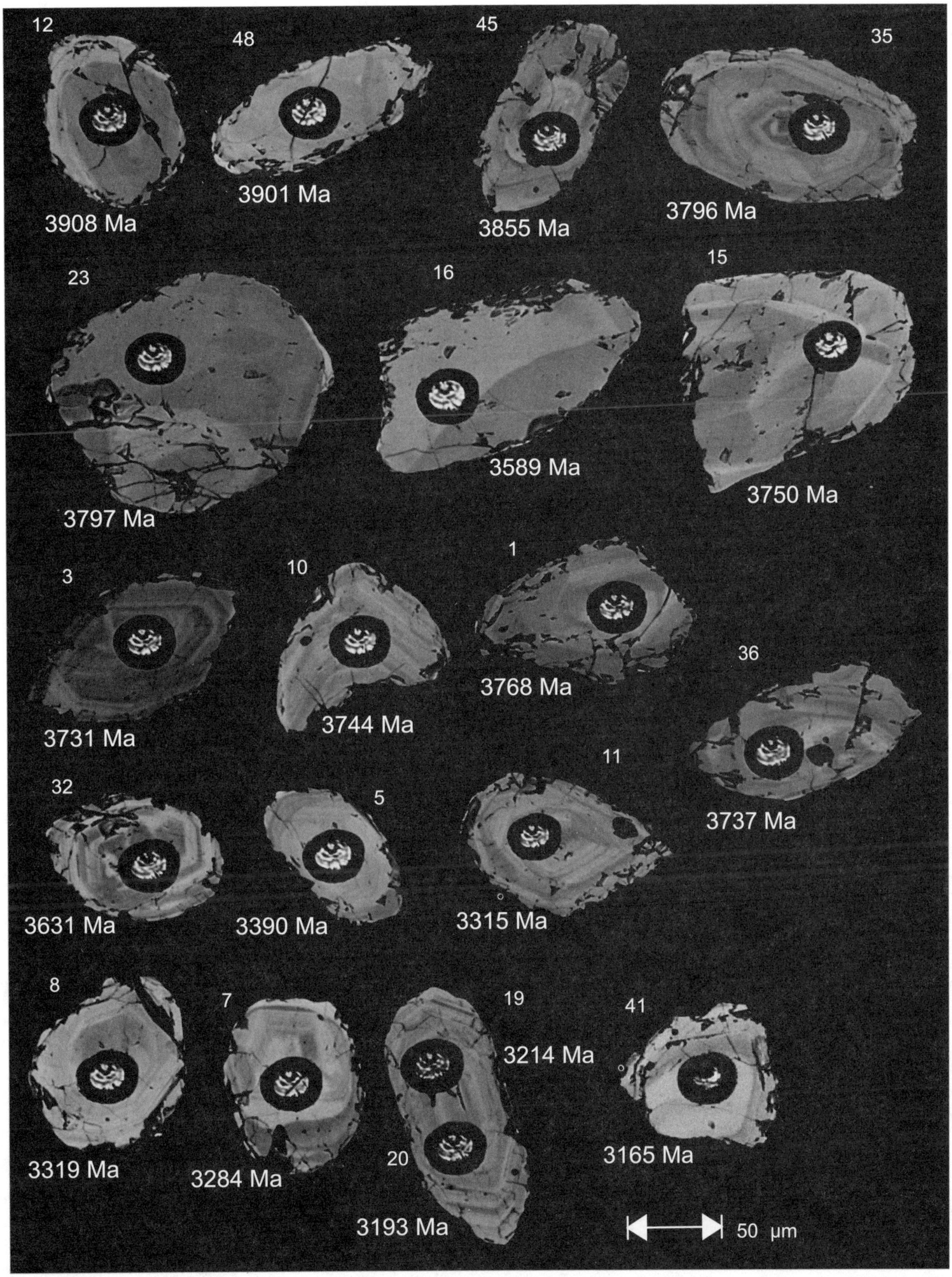

Figure 2. Electron microprobe, back-scattered electron images of detrital zircons from metagreywacke at Assean Lake, northwestern Superior Province (CB00–56). Laser spots (38–40 μm in diameter) with their associated U-Pb ages are prominently visible.

amount of Pb-U within-run fractionation and to increase sensitivity (e.g., Eggins et al., 1998). The integration time is one second, with a total dwell time of 30 seconds at each spot. During each analysis, $^{205}Tl/^{203}Tl$ is aspirated into the sample nebulizer to correct for mass bias on Pb isotopes (e.g., Machado and Simonetti, 2001). Measured ratios were corrected for machine drift by bracketing each dozen analyses of unknowns with one or more standard analyses. Gas flow rates through a DSN-100 (desolvation nebuliser) desolvating nebulizer were adjusted during standard analyses at the beginning of each analytical session. The internal precision on the MC-ICPMS analyses is largely dependent on signal intensity but is typically ≥0.2% (2σ) on $^{207}Pb/^{206}Pb$ in this study. External reproducibility (2σ) of <1% for $^{207}Pb/^{206}Pb$ and <3% for $^{206}Pb/^{238}U$ was determined by analysis of the homogeneous, international SRM NIST (standard reference material, National Institute of Standards and Technology) 612 glass wafer, and the international BR266 and 91500 and in-house LH94–15 zircon standards. Concordia and relative probability plots were prepared with Isoplot version 3.0 (Ludwig, 2003). For plotting purposes, error correlation coefficients were calculated utilizing the method of Ludwig (1980); however those analyses with low ^{204}Pb cps and low precision (≥2% error on $^{206}Pb/^{238}U$) were calculated following Horstwood et al. (2003).

Assean Lake Metagreywacke (CB00–56, UTM 659050E/6236250N NAD27, Zone 14)

The Assean Lake metagreywacke sample, from the northwestern margin of the Superior Province, contains an abundance of moderately to well-rounded zircons that have numerous fractures (Fig. 2). As zircon from different source rocks can have identical external morphologies (Hoskin and Schaltegger, 2003), we shall describe both external and internal grain morphologies for the main zircon populations of this rock. The relative probability plot of zircon ages (Fig. 3) has prominent peaks at 3.86, 3.79, 3.74, and 3.32 Ga. The two oldest zircons (Nos. 12 and 48, Table 1, Fig. 2) are concordant with ages of 3901 and 3908 Ma; proportionally, they make up ~4% of the detrital population. These two grains appear structureless, but both show extremely faint zoning in the core; Number 12 has a bright outer rim (Fig. 2). Recrystallization of zircon that was originally igneous in nature (Hoskin and Black, 2000) may be responsible for the massive internal appearance of these grains. The prominent ca. 3.86 Ga peak of zircon ages found in this sample (Fig. 3) was also identified in a SHRIMP study (Böhm et al., 2003) of a sample collected at the same outcrop. The ca. 3.86 Ga grains are commonly prismatic with visible occillatory zoning (e.g., No. 45, Fig. 2), consistent with an igneous derivation (Hoskin and Schaltegger, 2003). Ca. 3.79 Ga detrital zircons are the most abundant population in the sample (Fig. 3), including both prismatic, oscillatory zoned (e.g., No. 35, Fig. 2) and homogeneous grains with no igneous zoning (e.g., No. 23, Fig. 2). Ca. 3.73–3.75 Ga zircon grains are also an important constituent of the sample; many of these grains display excellent oscillatory zoning (e.g., Nos. 3 and 10, Fig. 2). Numerous, concordant, ca. 3.32 Ga old grains are moderately to well rounded and have good growth zoning (Nos. 7, 8, and 11, Fig. 2). The youngest concordant age of 3165 ± 27 Ma was recorded from grain 41 (Fig. 2, Table 1). Although the age of this highly fractured grain fragment may represent a detrital age, we suspect that it has undergone significant metamorphic recrystallization (e.g., Hoskin and Black, 2000), coeval with the widespread injection of ca. 3.17–3.18 Ga (Böhm et al., 2003; Hartlaub, 2005) tonalites and granodiorites.

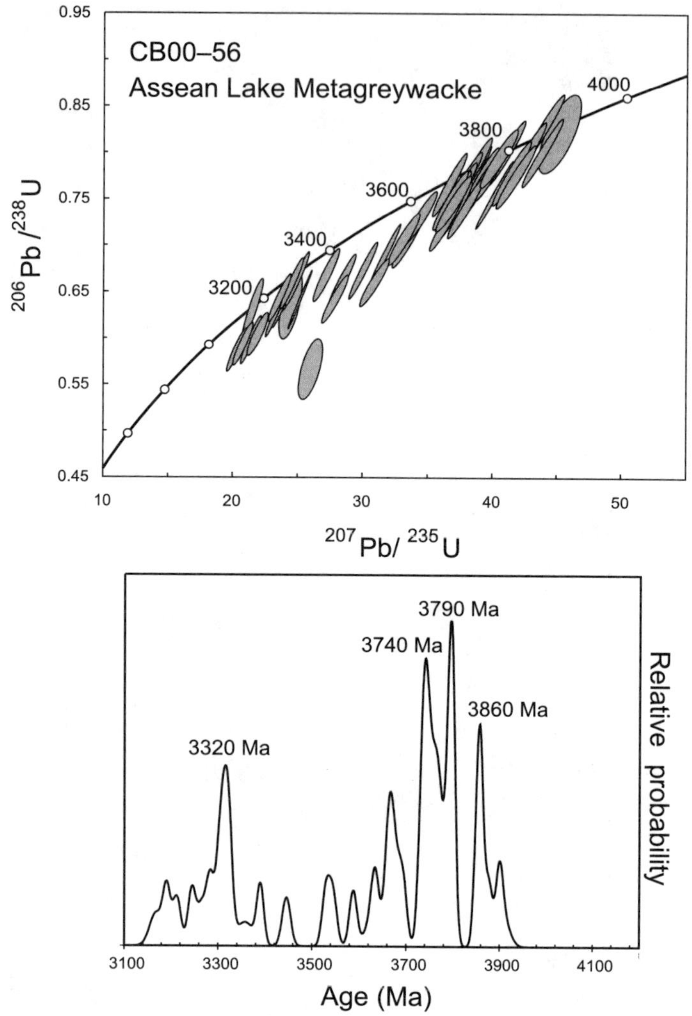

Figure 3. U-Pb concordia and relative probability plots for detrital zircons from a sample of metagreywacke from Assean Lake, northwestern Superior Province (CB00–56). Data-point error ellipses are 2σ.

Murmac Bay Group Quartzite (4700–6106, UTM 644765E/6604545N NAD27, Zone 12)

Fuchsitic quartzite from the Murmac Bay Area, western Rae Province contains entirely Paleoarchean detrital zircons (Fig. 4). Three main peaks of detrital grain ages occur at 3.86,

TABLE 1. LA-MC-ICPMS U-Pb RESULTS FOR DETRITAL ZIRCONS

	$^{206}Pb/^{238}U$	±1σ	$^{207}Pb/^{235}U^{\dagger}$	±1σ	$^{207}Pb/^{206}Pb$	±1σ	$^{207}Pb/^{206}Pb$ Age (Ma)	± Ma (2σ)	%Disc
Assean Lake Pelite (CB00-56)									
1	0.73763	0.01296	37.063	0.556	0.36438	0.0019	3768	±16	7.1
2	0.75531	0.01481	40.158	0.602	0.38547	0.0020	3853	±15	7.7
3	0.74923	0.01309	36.757	0.551	0.35561	0.0019	3731	±16	4.4
4	0.72643	0.01293	36.469	0.547	0.36379	0.0020	3766	±16	8.5
5	0.64228	0.01274	25.233	0.379	0.28486	0.0014	3390	±16	7.2
6	0.74091	0.01333	37.914	0.569	0.37099	0.0019	3795	±16	7.6
7	0.64408	0.01042	23.645	0.355	0.26619	0.0013	3284	±16	3.0
8	0.65897	0.01209	24.748	0.371	0.27231	0.0014	3319	±16	2.2
9	0.79413	0.01282	40.485	0.607	0.36995	0.0020	3791	±16	0.8
10	0.75714	0.01337	37.406	0.561	0.35856	0.0018	3744	±15	3.8
11	0.64068	0.01181	23.969	0.360	0.27158	0.0014	3315	±16	4.7
12	0.82106	0.01717	45.229	0.679	0.39983	0.0020	3908	±15	1.5
13	0.63611	0.01503	24.471	0.367	0.27925	0.0014	3359	±16	7.0
14	0.76098	0.01351	38.850	0.583	0.37065	0.0019	3794	±15	5.0
15	0.74582	0.01215	37.004	0.555	0.36011	0.0018	3750	±15	5.5
16	0.67357	0.01243	30.063	0.451	0.32405	0.0017	3589	±16	9.6
17	0.63005	0.01154	23.386	0.351	0.26942	0.0014	3303	±16	5.8
18	0.66055	0.01038	30.952	0.464	0.34022	0.0018	3664	±16	13.7
19	0.59439	0.00953	20.858	0.313	0.25471	0.0015	3214	±18	8.1
20	0.58820	0.00952	20.361	0.305	0.25124	0.0013	3193	±16	8.2
21	0.77239	0.01376	39.062	0.586	0.36697	0.0019	3779	±15	3.1
22	0.76528	0.01414	38.108	0.572	0.36080	0.0019	3753	±16	3.1
23	0.78107	0.01330	39.971	0.600	0.37134	0.0019	3797	±15	2.6
24	0.80002	0.01427	41.045	0.616	0.37228	0.0019	3801	±15	0.4
25	0.59868	0.01054	21.443	0.322	0.25997	0.0014	3247	±17	8.6
26	0.76832	0.01354	37.800	0.567	0.35703	0.0018	3737	±16	2.2
27	0.77860	0.01406	41.497	0.623	0.38671	0.0020	3858	±15	5.0
28	0.64268	0.01052	27.905	0.419	0.31428	0.0017	3542	±17	12.2
29	0.79807	0.01445	42.609	0.639	0.38714	0.0020	3860	±16	2.7
30	0.77466	0.01381	41.828	0.627	0.39182	0.0020	3878	±16	6.1
31	0.69029	0.01340	31.766	0.477	0.33386	0.0018	3635	±16	8.9
32	0.56718	0.01340	26.051	0.391	0.33303	0.0018	3631	±17	25.0
33	0.70730	0.01140	33.195	0.498	0.34144	0.0019	3669	±17	7.8
34	0.70629	0.01226	33.552	0.503	0.34527	0.0018	3686	±16	8.5
35	0.78424	0.01414	40.105	0.602	0.37103	0.0019	3796	±15	2.2
36	0.78296	0.01374	38.534	0.578	0.35697	0.0018	3737	±16	0.3
37	0.66871	0.01156	27.239	0.409	0.29446	0.0018	3441	±19	5.2
38	0.66417	0.01193	25.005	0.375	0.27313	0.0014	3324	±16	1.6
39	0.65953	0.01122	24.577	0.369	0.27038	0.0014	3308	±16	1.7
40	0.76625	0.01394	36.686	0.550	0.34752	0.0018	3696	±16	1.0
41	0.63276	0.01287	21.580	0.324	0.24686	0.0022	3165	±28	0.2
42	0.60453	0.00942	21.958	0.329	0.26332	0.0018	3267	±22	8.4
43	0.73080	0.01139	34.463	0.517	0.34208	0.0017	3672	±16	4.8
44	0.66228	0.01168	28.467	0.427	0.31191	0.0017	3530	±16	9.2
45	0.82656	0.01505	43.950	0.659	0.38578	0.0019	3855	±15	–1.0
46	0.77053	0.01527	38.472	0.577	0.36227	0.0020	3759	±17	2.7
47	0.74851	0.01336	36.967	0.555	0.35852	0.0019	3744	±16	4.9
48	0.80051	0.01493	43.896	0.658	0.39774	0.0020	3901	±15	3.7

(continued)

3.76, and 3.72 Ga. The single grain older than 3.9 Ga is small, well rounded, and has extremely faint growth zoning near its rim (No. 30, Fig. 5); this morphology is similar to that of the ≥3.9 Ga zircons of the Assean Lake metagreywacke (Sample CB00–56). Ca. 3.86 Ga zircons have a diversity of shapes and internal structures (Nos. 2, 4, 7, Fig. 5). Grain 4 has numerous well-defined growth zones, with several of the later zones transecting the earlier ones. Grain 7 has a partially recrystallized core and a zoned rim, whereas grain 2 has a relatively thick massive rim that cuts right across the axis of the prismatic and massive core. Together, these ca. 3.86 Ga grains have a complex history of igneous growth and later Paleoarchean high-grade metamorphism. The majority of zircon grains with ages between 3.70 and 3.77 Ga are prismatic and have well-defined internal oscillatory zoning (Nos. 5, 9, 28, 37, and 40, Fig. 5). Since growth zoning is a typical feature of magmatic zircon (Corfu et al., 2003), the source of this detritus was very likely felsic to intermediate igneous rocks. Some of the ca. 3.70–3.77 Ga detrital zircons from this sample have, however, a massive structureless core and a faint patchy appearance (e.g., No. 29, Fig. 5). This patchy appearance is likely the result of solid-state recrystallization of the zircon during high-grade metamorphism (e.g., Hoskin and Black, 2000; Corfu et al., 2003). Well-defined "ghost" zoning and transgressive recrystallization fronts (Hoskin and Black, 2000) occur in the cores of several grains (e.g., Nos. 7 and 10, Fig. 5). Although grain 17 (Fig. 5) has the youngest $^{207}Pb/^{206}Pb$ age (3412 ± 11 Ma), it may have undergone significant lead loss and isotopic mixing during Proterozoic regional metamorphism (e.g., Hartlaub et al., 2004a), as it has numerous cracks, a well-defined metamorphic overgrowth, and is slightly discordant.

TABLE 1. LA-MC-ICPMS U-Pb RESULTS FOR DETRITAL ZIRCONS (*continued*)

	$^{206}Pb/^{238}U$	±1σ	$^{207}Pb/^{235}U^{\dagger}$	±1σ	$^{207}Pb/^{206}Pb$	±1σ	$^{207}Pb/^{206}Pb$ Age (Ma)	± Ma (2σ)	%Disc
Fookes Lake Quartzite (4700–6106)									
1	0.78345	0.01356	39.019	0.585	0.36128	0.0018	3755	±15	0.9
2	0.83557	0.01586	44.399	0.666	0.38552	0.0019	3854	±15	–2.1
3	0.76863	0.01260	38.396	0.576	0.36236	0.0018	3760	±15	2.9
4	0.82196	0.01478	43.814	0.657	0.38676	0.0019	3858	±15	–0.3
5	0.67443	0.01082	32.938	0.494	0.35437	0.0018	3726	±16	13.8
6	0.78879	0.01306	41.382	0.621	0.38074	0.0019	3835	±15	3.0
7	0.82555	0.01376	43.987	0.660	0.38651	0.0019	3857	±15	–0.8
8	0.77222	0.01554	38.490	0.577	0.36158	0.0018	3757	±15	2.4
9	0.79438	0.01506	39.680	0.595	0.36237	0.0018	3760	±15	–0.3
10	0.70285	0.01297	31.599	0.474	0.32618	0.0017	3599	±16	6.0
11	0.84449	0.01432	41.793	0.627	0.35901	0.0018	3746	±15	–7.2
12	0.81840	0.01408	41.805	0.627	0.37050	0.0019	3793	±15	–2.1
13	0.79136	0.01413	38.346	0.575	0.35132	0.0018	3713	±16	–1.6
14	0.69953	0.01114	34.115	0.512	0.35418	0.0021	3725	±18	10.6
15	0.85931	0.01437	45.274	0.679	0.38232	0.0019	3841	±15	–5.5
16	0.78443	0.01435	39.659	0.595	0.36677	0.0019	3778	±16	1.6
17	0.67632	0.01292	26.958	0.404	0.28888	0.0020	3412	±21	3.1
18	0.82240	0.01295	39.856	0.598	0.35167	0.0018	3714	±16	–5.5
19	0.76573	0.01344	39.627	0.594	0.37544	0.0019	3814	±16	5.1
20	0.73278	0.01170	36.080	0.541	0.35710	0.0018	3738	±15	6.7
21	0.74912	0.01201	36.988	0.555	0.35817	0.0019	3742	±16	4.8
22	0.78429	0.01353	37.819	0.567	0.34970	0.0018	3706	±15	–1.0
23	0.79568	0.01448	39.537	0.593	0.36038	0.0018	3751	±15	–0.8
24	0.85608	0.01424	45.699	0.686	0.38722	0.0019	3860	±15	–4.4
25	0.73914	0.01330	35.702	0.536	0.35035	0.0018	3709	±15	5.0
26	0.79501	0.01580	41.263	0.619	0.37648	0.0019	3818	±15	1.6
27	0.82596	0.01290	43.565	0.653	0.38259	0.0020	3842	±15	–1.4
28	0.81619	0.01280	39.639	0.595	0.35223	0.0018	3717	±16	–4.7
29	0.79268	0.01459	39.404	0.591	0.36058	0.0018	3752	±15	–0.4
30	0.85450	0.01366	47.826	0.717	0.40595	0.0020	3931	±15	–1.7
31	0.76722	0.01434	37.424	0.561	0.35369	0.0018	3723	±16	1.8
32	0.70317	0.01145	32.601	0.489	0.33620	0.0018	3646	±16	7.5
33	0.78546	0.01387	40.244	0.604	0.37157	0.0019	3798	±15	2.1
34	0.82522	0.01408	41.048	0.616	0.36087	0.0018	3754	±15	–4.4
35	0.67142	0.01101	30.353	0.455	0.32781	0.0018	3607	±16	10.5
36	0.67066	0.01086	30.804	0.462	0.33306	0.0019	3631	±17	11.3
37	0.75563	0.01341	37.877	0.568	0.36346	0.0018	3764	±15	4.7
38	0.73062	0.01221	35.128	0.527	0.34869	0.0018	3701	±15	5.8
39	0.76321	0.01290	36.874	0.553	0.35039	0.0018	3709	±15	1.9
40	0.73593	0.01329	36.324	0.545	0.35792	0.0018	3741	±15	6.4
41	0.79249	0.01263	39.269	0.589	0.35936	0.0018	3747	±15	–0.5
42	0.74970	0.01422	36.991	0.555	0.35778	0.0018	3740	±15	4.7

Notes: †Calculated value based on $^{238}U/^{235}U$ = 137.88; Disc—discordance.

Lu-Hf Isotopes

Laser ablation analyses for in situ Hf isotope compositions were performed using the same instrumentation as for U-Pb analysis. Ablation runs utilized spot sizes of 40–60 μm at a repetition rate of 5 Hz corresponding to an energy density of 9 J/cm^2 for a duration of ~30–60 seconds. The analytical protocol for Yb- and Lu-isobaric interference corrections during laser ablation runs is described in Machado and Simonetti (2001). External reproducibility and accuracy of the analytical protocol was verified by repeated analysis of the international SHRIMP zircon standard BR266, which yields an average $^{176}Hf/^{177}Hf$ value of 0.281625 ± 0.000056 (n = 25; 2σ error) in excellent agreement with the $^{176}Hf/^{177}Hf$ value of 0.281621 ± 0.000024 (n = 95; 2σ) for BR266 (Woodhead et al., 2004). The error for our laser ablation $^{176}Hf/^{177}Hf$ measurements (Table 2) are slightly larger than those for single-grain solution-mode MC-ICPMS analysis; however, this difference is compensated by the large number of grains that can be analyzed in a short period of time and by our ability to analyze specific zones within the zircons. The capacity to analyze specific regions within zircons allows us to avoid mixing core and overgrowth Hf compositions. The choice of decay constant for ^{176}Lu significantly affects the chondrite uniform reservoir (CHUR) and DM (depleted mantle) evolution trends. As suggested by Patchett et al. (2004), we utilize a decay constant for ^{176}Lu of 1.867×10^{-11} year^{-1} in our calculations of initial Hf values and for determining CHUR evolution trends. In addition, this ^{176}Lu λ value is consistent with Lu-Hf isochron work from igneous rocks of known U-Pb age (Scherer et al., 2001; Söderlund et al., 2004), counting experiments (e.g., Nir-El and Lavi, 1998), and data from chondritic meteorites (Patchett et al., 2004).

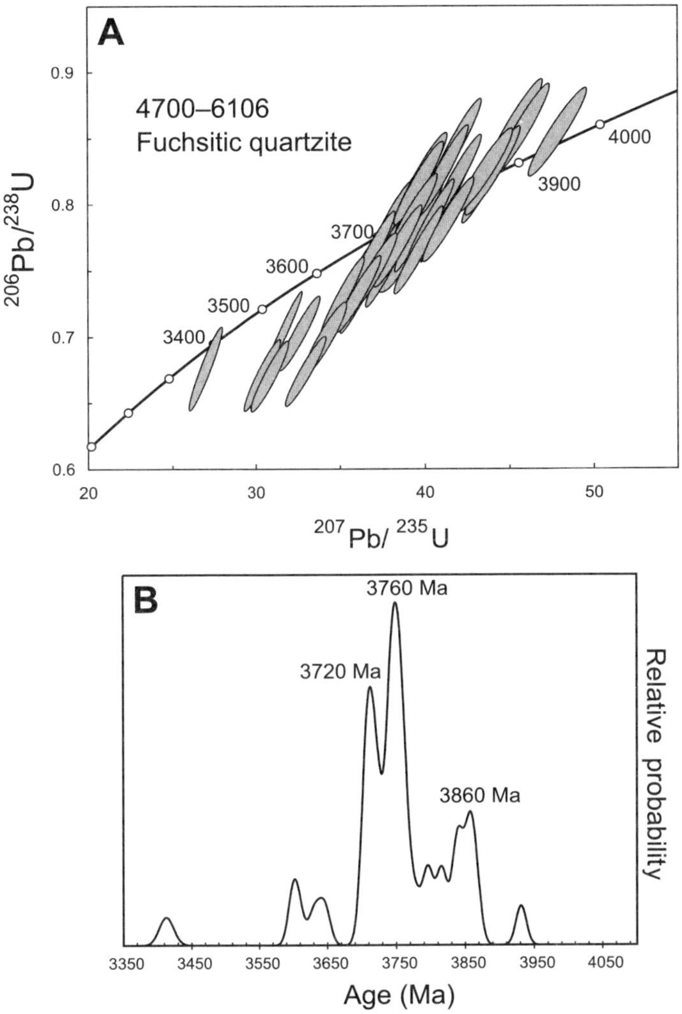

Figure 4. U-Pb concordia and relative probability plots for detrital zircons from a sample of fuchsitic quartzite from the Uranium City area, Rae Province (4700–6106). Data-point error ellipses are 2σ.

Results

Hf and Nd isotopic results are commonly presented as epsilon values that compare measured isotopic ratios with those from chondrites (CHUR). The CHUR value is thought to roughly estimate the initial bulk silicate earth composition for refractory elements like Lu and Hf (Patchett et al., 2004). Therefore, the choice of a CHUR composition will influence both the resulting epsilon units and subsequent interpretation of crust-mantle evolution. To avoid complication, we have cited εHf values in Table 2 based solely on the CHUR composition of Blichert-Toft and Albarède (1997; $^{176}Lu/^{177}Hf = 0.0332$ and $^{176}Hf/^{177}Hf = 0.282772$), although these values are almost identical to those based on Patchett et al. (2004; $^{176}Lu/^{177}Hf = 0.0342$ and $^{176}Hf/^{177}Hf = 0.282843$). Use of CHUR values from Bizzarro et al. (2003) would yield substantially lower $^{176}Hf/^{177}Hf$ values and, therefore, more positive εHf values. Due to these differences, we have elected to display our results on a plot of initial $^{176}Hf/^{177}Hf$ versus time rather than εHf versus time (Fig. 6). A DM evolution curve utilizing modern mid-oceanic ridge basalt (MORB) values from Salters and White (1998; $^{176}Hf/^{177}Hf = 0.28319$ and $^{176}Lu/^{177}Hf = 0.0378$) is also included for comparative purposes.

Ancient detrital zircons from the western Rae and the northwestern Superior Provinces have similar Hf isotopic ratios (Fig. 6, Table 2). The majority of the 3.6–3.9 Ga zircon grains have negative εHf values between –2 and –10 (Table 2) when compared to the CHUR values of Blichert-Toft and Albarède (1997). Our Hf results are broadly similar to the results of Amelin et al. (1999, 2000) for zircons from Jack Hills, Australia, and from the Acasta gneisses, Canada (Fig. 6). Some of our detrital zircons with 3.6–3.7 Ga ages have significantly lower $^{176}Hf/^{177}Hf$ values than the zircons from the Isua area of Western Greenland (Isukasia terrane of Nutman et al., 2004); we note, however, that some of the grains of this age from the Assean Lake area are moderately discordant with regard to their U-Pb isotope systematics (Fig. 3) and, therefore, may record >3.7 Ga Hf values. The εHf values indicate that the Assean Lake and Murmac Bay detrital zircons in this study are derived from evolved, reworked crust. A similar interpretation for the Jack Hills Metaconglomerate and Acasta Gneiss data was presented by Amelin et al. (1999, 2000). Massive cores are found in some of the ancient grains (Figs. 2 and 5). Similar ca. 3.86 Ga massive cores were reported from the Uivak Gneisses of Labrador; a metamorphic origin for these cores was thought likely (Schiøtte et al., 1989). The negative εHf composition of these ancient grains would be consistent with metamorphism of older evolved crust. The oldest (ca. 3.93 Ga) zircon grain in our study (9704–6106–37, Table 2) is also derived from a highly evolved crustal source. Although this grain has near concordant U-Pb systematics, we suspect that the grain has undergone some Paleoarchean lead loss and may record an older Hf composition than its $^{207}Pb/^{206}Pb$ age would indicate (see Amelin et al. [2000] for a description of the effects of ancient Pb loss on $^{176}Hf/^{177}Hf$ values).

DISCUSSION AND CONCLUSIONS

Hf Evolution of the Crust and Mantle

The upper limit of $^{176}Hf/^{177}Hf$ values versus time (Fig. 6) for zircon from Jack Hills, the Acasta Gneiss, the Assean Lake Area, and the Western Rae Province is roughly linear and plots just below the CHUR evolution line of Patchett et al. (2004). Unfortunately, the choice of CHUR and bulk silicate earth (BSE) Lu-Hf values can only be chosen on a subjective basis (Patchett et al., 2004), making interpretation of our data difficult. Utilizing the CHUR values of Blichert-Toft and Albarède (1997) or Patchett et al. (2004), the majority of Paleoarchean zircon data in this study are consistent with early (>4.0 Ga) crust formation but do not indicate an early (>3.9 Ga) mantle depletion event. Utilizing Bizzarro et al.'s (2003) CHUR values

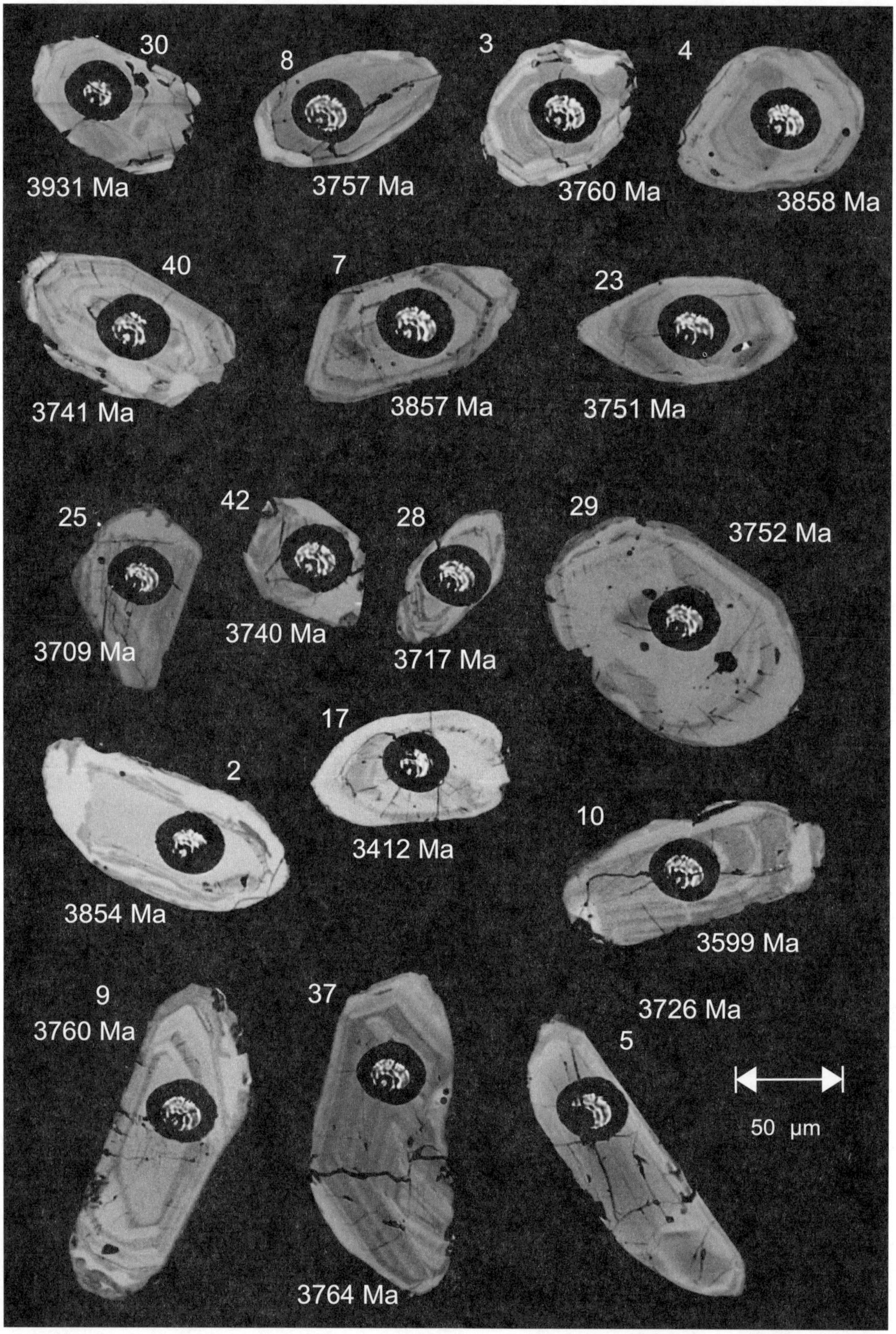

Figure 5. Electron microprobe, back-scattered electron images of detrital zircons from fuchsitic quartzite, western Rae Province (4700–6106). Laser spots (38–40 μm in diameter) with their associated U-Pb ages are prominently visible.

TABLE 2. Lu-Hf DATA FOR DETRITAL ZIRCONS

Grain no.	Measured $^{176}Lu/^{177}Hf$	Measured $^{176}Hf/^{177}Hf$	±	t (Ma)	Initial $^{176}Hf/^{177}Hf$	εHf (t)[a]
CB00–56–1	0.00080	0.280240	59	3768	0.280181	–6
CB00–56–2	0.00113	0.280127	28	3853	0.280042	–9
CB00–56–3	0.00079	0.280238	33	3731	0.280180	–6
CB00–56–4	0.00039	0.280354	30	3766	0.280328	–10
CB00–56–6	0.00060	0.280382	49	3795	0.280338	0
CB00–56–7	0.00096	0.280501	51	3284	0.280440	–8
CB00–56–8	0.00096	0.280588	29	3319	0.280527	–4
CB00–56–10	0.00056	0.280318	22	3744	0.280277	–3
CB00–56–11	0.00117	0.280329	11	3315	0.280254	–14
CB00–56–12	0.00060	0.280149	57	3908	0.280104	–6
CB00–56–13	0.00059	0.280585	67	3359	0.280547	–3
CB00–56–14	0.00084	0.280197	38	3794	0.280135	–7
CB00–56–15	0.00093	0.280224	17	3750	0.280156	–7
CB00–56–16	0.00113	0.280414	34	3589	0.280335	–5
CB00–56–18	0.00089	0.280219	65	3664	0.280156	–10
CB00–56–19	0.00085	0.280625	37	3214	0.280572	–5
CB00–56–21	0.00041	0.280215	34	3779	0.280184	–6
CB00–56–22	0.00045	0.280264	58	3753	0.280231	–5
CB00–56–23	0.00049	0.280307	20	3797	0.280271	–2
CB00–56–24	0.00110	0.280221	93	3801	0.280140	–7
CB00–56–25	0.00121	0.280523	34	3247	0.280448	–9
CB00–56–29	0.00107	0.280143	52	3860	0.280063	–8
CB00–56–30	0.00038	0.280220	29	3878	0.280191	–3
CB00–56–31	0.00062	0.280444	93	3635	0.280400	–1
CB00–56–32	0.00085	0.280453	53	3631	0.280394	–2
CB00–56–34	0.00036	0.280457	58	3686	0.280431	1
CB00–56–35	0.00091	0.280175	25	3796	0.280108	–8
CB00–56–36	0.00059	0.280177	45	3737	0.280134	–9
CB00–56–37	0.00138	0.280217	53	3441	0.280125	–16
CB00–56–38	0.00081	0.280567	26	3324	0.280515	–5
CB00–56–39	0.00110	0.280525	34	3308	0.280455	–7
CB00–56–40	0.00074	0.280310	37	3696	0.280258	–5
CB00–56–41	0.00126	0.280606	46	3165	0.280530	–7
CB00–56–42	0.00105	0.280394	19	3267	0.280328	–13
CB00–56–44	0.00078	0.280320	45	3530	0.280267	–9
CB00–56–45	0.00131	0.280090	24	3855	0.279993	–11
CB00–56–46	0.00055	0.280245	28	3759	0.280205	–5
CB00–56–47	0.00057	0.280158	22	3744	0.280116	–9
4700–6106–3	0.00058	0.280341	33	3760	0.280299	–2
4700–6106–4	0.00100	0.280264	22	3858	0.280190	–4
4700–6106–5	0.00088	0.280309	25	3726	0.280246	–5
4700–6106–6	0.00139	0.280189	22	3835	0.280086	–8
4700–6106–7	0.00146	0.280146	47	3857	0.280037	–9
4700–6106–8	0.00108	0.280247	32	3757	0.280169	–7
4700–6106–9	0.00166	0.280334	24	3760	0.280213	–5
4700–6106–10	0.00099	0.280269	14	3599	0.280201	–9
4700–6106–11	0.00119	0.280238	17	3746	0.280152	–8
4700–6106–12	0.00158	0.280257	49	3793	0.280141	–7
4700–6106–13	0.00100	0.280275	23	3713	0.280203	–7
4700–6106–15	0.00114	0.280127	58	3841	0.280042	–9
4700–6106–16	0.00059	0.280278	18	3778	0.280235	–4
4700–6106–18	0.00092	0.280303	44	3714	0.280237	–5
4700–6106–19	0.00101	0.280242	16	3814	0.280168	–5
4700–6106–20	0.00154	0.280169	22	3738	0.280058	–11
4700–6106–21	0.00103	0.280175	21	3742	0.280100	–10
4700–6106–22	0.00044	0.280350	15	3706	0.280318	–3
4700–6106–23	0.00164	0.280218	32	3751	0.280099	–9
4700–6106–24	0.00127	0.280189	25	3860	0.280094	–7
4700–6106–25	0.00138	0.280208	19	3709	0.280109	–10
4700–6106–28	0.00162	0.280293	38	3717	0.280177	–8
4700–6106–30	0.00120	0.279998	56	3931	0.279906	–12
4700–6106–31	0.00109	0.280168	43	3723	0.280089	–10
4700–6106–32	0.00088	0.280371	31	3646	0.280309	–4
4700–6106–34	0.00135	0.280211	17	3754	0.280112	–9
4700–6106–35	0.00114	0.280202	29	3607	0.280122	–12
4700–6106–36	0.00120	0.280301	41	3631	0.280217	–8
4700–6106–37	0.00146	0.280295	55	3764	0.280188	–6
4700–6106–38	0.00092	0.280340	53	3701	0.280274	–4
4700–6106–39	0.00104	0.280416	33	3709	0.280341	–2
4700–6106–40	0.00098	0.280301	18	3741	0.280230	–5
4700–6106–41	0.00138	0.280304	25	3747	0.280204	–6
4700–6106–42	0.00118	0.280269	13	3740	0.280184	–7

Notes: [a]εHf(t) value is determined utilizing the CHUR values of Blichert-Toft and Albarède (1997) and a ^{176}Lu λ value of 1.867 × 10^{-11} $year^{-1}$.

would indicate, however, a juvenile depleted mantle source for the majority of ≥3.7 Ga zircon in this study. Resolution of the major discrepancy in Hf interpretations will require additional study and assessment of Lu-Hf CHUR and BSE parameters.

Correlation between Ancient Terranes or Just Coincidence?

The relative probability plots of ancient detrital zircons from the Rae and Superior Province metasediments (Figs. 2 and 4) highlight several similarities. The majority of zircons in both samples have U-Pb ages between 3.72 and 3.79 Ga, with a gap between 3.80 and 3.84 Ga. Another important similarity between the two samples is a peak in ages at 3.86 Ga (Figs. 3 and 4). Finally, both samples were derived from source terranes that have a history that extends back to at least 3.9 Ga. The only known exposures of rocks this old are the Acasta Gneisses of the Slave Province (Bowring and Williams, 1999), but we do not consider this to be a likely matching source terrane for either of the sediments, as 3.72–3.86 Ga crust has not been recorded in that area. Ancient >3.7 Ga detrital zircons have been discovered in the Beartooth Mountains of the Wyoming Province (Mueller et al., 1992; Fig. 1). The preponderance of detrital zircons in the region have $^{207}Pb/^{206}Pb$ ages between 3.2 and 3.4 Ga but a significant number of 3.7–3.8 Ga grains occurring in some samples (Mueller et al., 1998). Although there are no known ancient (>3.3 Ga) rocks exposed in this region, the detrital zircon record does extend back to ca. 4.0 Ga, consistent with the oldest zircons found in our study. The Narryer Gneiss Complex of Western Australia also has rocks that contain ancient ≥3.9 Ga detrital zircons (e.g., Maas et al., 1992). Gneissic igneous rocks in this region have a complex metamorphic history, with some zircon cores having maximum ages of ca. 3.6–3.75 Ga (Kinny and Nutman, 1996; Pidgeon and Wilde, 1998); however, there are no ≥3.8 Ga crystallization ages in this complex, whereas they represent a significant component of the detritus in our study.

Any terrains with a history matching the source rocks to the Assean Lake and Murmac Bay area sediments should have significant evidence for 3.7 to ≥3.9 Ga crust and little evidence for 3.3–3.6 Ga crust. The age compilation in Figure 7 compares these two areas to ancient terranes that have well-constrained crust formation between 3.7 and 3.9 Ga. The Itsaq Gneiss Complex of western Greenland shows the most similarities to the ancient detrital provenance in the Superior and Rae Provinces. The ca. 3.86 Ga peaks as well as the 3.79–3.72 Ga peaks in detrital zircon from the Assean Lake and Murmac Bay areas are well matched by the crystallization ages of the tonalite-trondjemite-granodiorite (TTG) suites in this complex (Fig. 7). Derivation of these detrital grains from TTG source rocks would be consistent with their predominantly prismatic and oscillatory zoned morphologies. One characteristic of our ancient detritus that appears missing from western Greenland is exposed ≥3.9 Ga crust. However, Nutman et al. (1996) suggested that ca. 3.87 Ga quartz-dioritic gneiss cuts even older

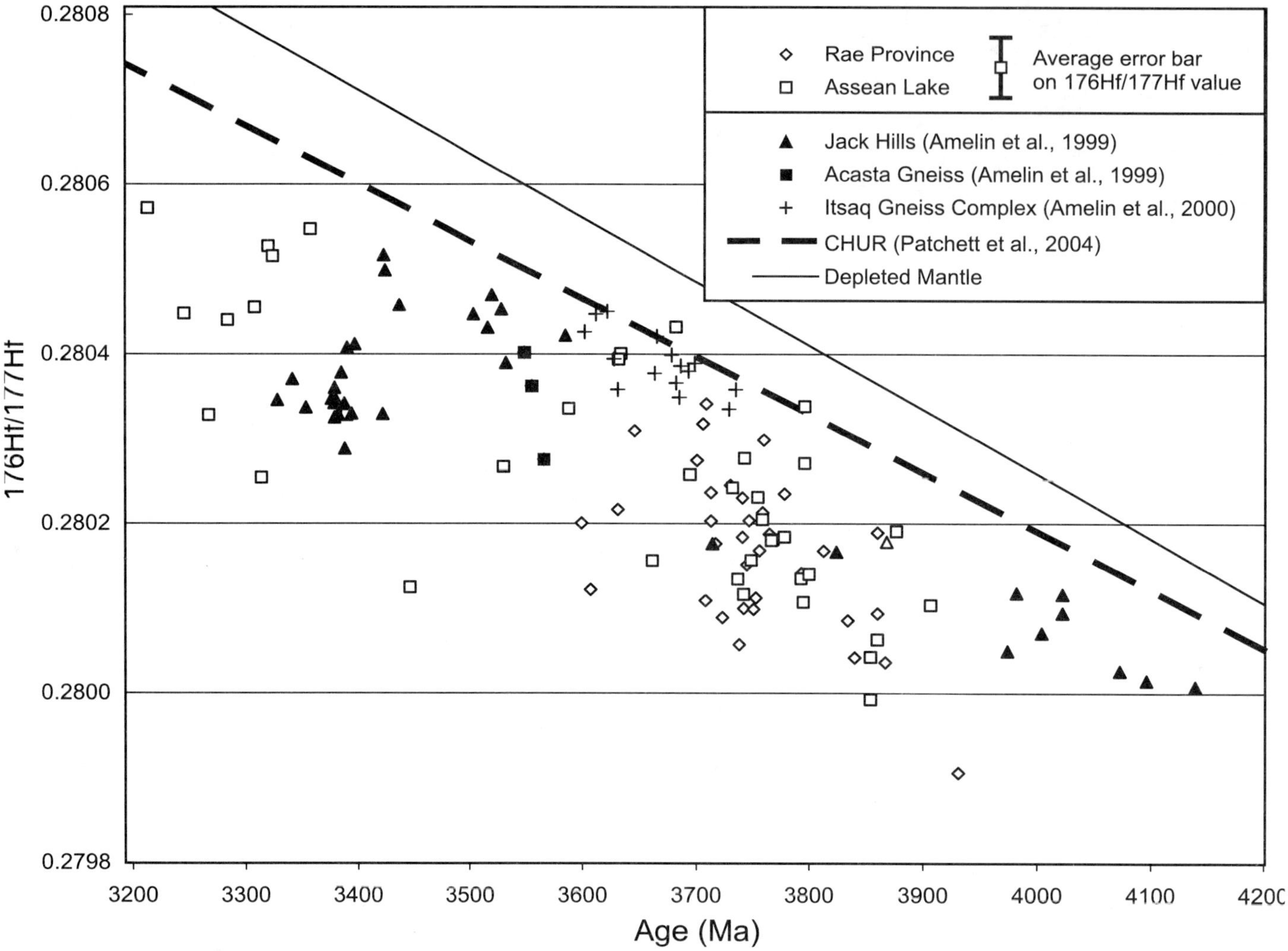

Figure 6. $^{176}Hf/^{177}Hf$ versus time plot for zircon from this study (open symbols) and previous studies. Zircon analyses with >10% U-Pb discordance have been excluded from this plot. Cited $^{176}Hf/^{177}Hf$ ratios have been recalculated using a decay constant for ^{176}Lu of 1.867×10^{-11} year^{-1} (Patchett et al., 2004).

supracrustal rocks in the area and a single ca. 3.9 Ga detrital zircon was found in a sediment dominated by 3.84–3.86 Ga detritus (Nutman et al., 1997). Could these ancient fragments have formed or been welded together during early Paleoarchean time and then dispersed by ca 3.3 Ga?

The large number of Neoarchean cratonic fragments has made supercontinent reconstructions difficult (e.g., Bleeker, 2003), but the quantity of ca. 3.7–3.9 Ga crustal fragments is much smaller. Paleoarchean continental reconstruction may, therefore, be possible if these remaining fragments represent a significant proportion of the early Archean crust. At this stage there are insufficient data to make definitive correlations, but additional study of the igneous, metamorphic, sedimentological, and isotopic characteristics of these terranes will enable discussion on global Paleoarchean continental formation and dispersion. With the spread of new LA-MC-ICPMS technology, we expect the acquisition of these important data to occur at a much more rapid pace. In particular, as our database of well-studied detrital and igneous zircon from ancient terranes expands, matching age, morphological characteristics, Lu-Hf isotope, and geochemical characteristics will become increasingly viable.

ACKNOWLEDGMENTS

Major funding for this project was provided by the Manitoba Geological Survey, Manitoba Hydro, and a Natural Sciences and Engineering Research Council (NSERC) Collaborative Research and Development Grant. Discussions with Tom Chacko, Ken Ashton, and Tim Corkery aided our advancement of this manuscript. Allen Nutman and Marc Poujol are thanked for providing insightful, constructive, and timely reviews.

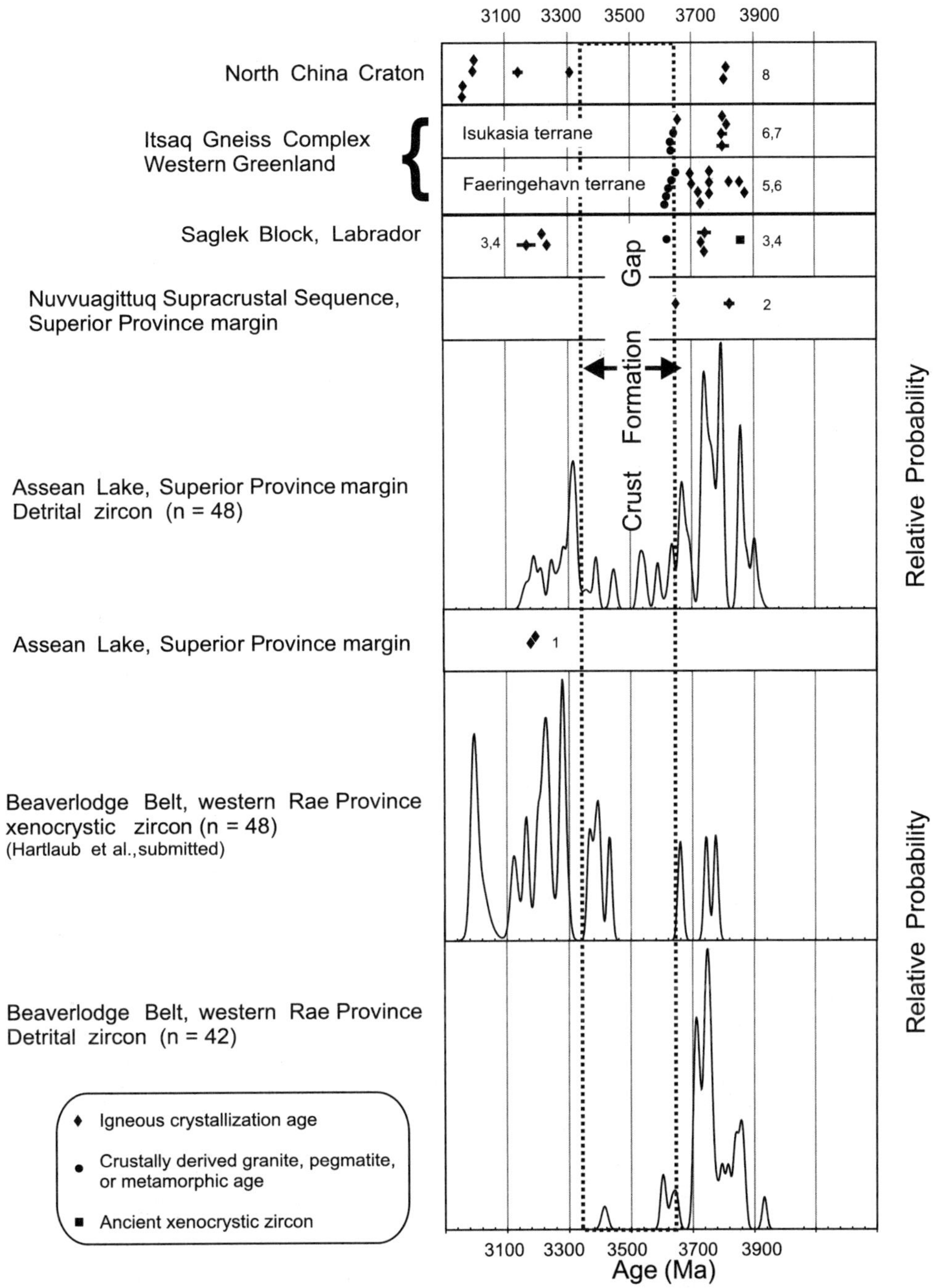

Figure 7. Comparative age chart for terranes that have evidence of 3.7–3.86 Ga crustal growth. This chart explores only the igneous record of these terranes; for a detailed examination of metamorphic overprints see associated references. Only representative analyses are presented for the Itsaq Gneiss Complex; for a more detailed comparison of the terranes of Western Greenland see Nutman et al. (2004). Cited ages are restricted to U-Pb zircon dates. References: 1—Böhm et al., 2003; 2—David et al., 2004; 3—Wasteneys et al., 1996; 4—Schiøtte et al., 1989; 5—Nutman et al., 1993; 6—Nutman et al., 1996; 7—Crowley, 2003; 8—Song et al., 1996.

REFERENCES CITED

Amelin, Y., Lee, D.C., Halliday, A.N., and Pidgeon, R.T., 1999, Nature of the Earth's earliest crust from hafnium isotopes in single detrital zircons: Nature, v. 399, no. 6733, p. 252–255, doi: 10.1038/20426.

Amelin, Y., Lee, D.C., and Halliday, A.N., 2000, Early-middle Archaean crustal evolution deduced from Lu-Hf and U-Pb isotopic studies of single zircon grains: Geochimica et Cosmochimica Acta, v. 64, no. 24, p. 4205–4225, doi: 10.1016/S0016-7037(00)00493-2.

Appel, P.W.U., Fedo, C.M., Moorbath, S., and Myers, J.S., 1998, Recognizable primary volcanic and sedimentary features in a low-strain domain of the highly deformed, oldest known (approximate to 3.7–3.8 Gyr) Greenstone Belt, Isua: West Greenland: Terra Nova, v. 10, no. 2, p. 57–62, doi: 10.1046/j.1365-3121.1998.00162.x.

Baadsgaard, H., Collerson, K.D., and Bridgwater, D., 1979, Archean gneiss complex of Northern Labrador, 1: Preliminary U-Th-Pb geochronology: Canadian Journal of Earth Sciences, v. 16, no. 4, p. 951–961.

Bickford, M.E., Wooden, J.L., and Bauer, R.L., 2004, New SHRIMP U-Pb zircon ages for the Paleoarchean to Mesoarchean rocks of the Minnesota River Valley: Geological Society of America Programs with Abstracts, v. 36, no. 5, p. 458.

Bizzarro, M., Baker, J.A., Haack, H., Ulfbeck, D., and Rosing, M., 2003, Early history of Earth's crust-mantle system inferred from hafnium isotopes in chondrites: Nature, v. 421, no. 6926, p. 931–933, doi: 10.1038/nature01421.

Blichert-Toft, J., and Albarède, F., 1997, The Lu-Hf isotope geochemistry of chondrites and the evolution of the mantle-crust system: Earth and Planetary Science Letters, v. 148, no. 1-2, p. 243–258, doi: 10.1016/S0012-821X(97)00040-X.

Bleeker, W., 2003, The late Archean record: A puzzle in ca. 35 pieces: Lithos, v. 71, no. 2-4, p. 99–134, doi: 10.1016/j.lithos.2003.07.003.

Böhm, C.O., Heaman, L.M., Creaser, R.A., and Corkery, M.T., 2000, Discovery of pre–3.5 Ga exotic crust at the northwestern Superior Province margin, Manitoba: Geology, v. 28, no. 1, p. 75–78, doi: 10.1130/0091-7613(2000)028<0075:DOPGEC>2.3.CO;2.

Böhm, C.O., Heaman, L.M., Stern, R.A., Corkery, M.T., and Creaser, R.A., 2003, Nature of Assean Lake ancient crust, Manitoba: A combined SHRIMP-ID-TIMS U-Pb geochronology and Sm-Nd isotope study: Precambrian Research, v. 126, no. 1-2, p. 55–94, doi: 10.1016/S0301-9268(03)00127-X.

Bowring, S.A., and Housh, T., 1995, The Earth's early evolution: Science, v. 269, no. 5230, p. 1535–1540.

Bowring, S.A., and Williams, I.S., 1999, Priscoan (4.00–4.03 Ga) orthogneisses from northwestern Canada: Contributions to Mineralogy and Petrology, v. 134, no. 1, p. 3–16, doi: 10.1007/s004100050465.

Corfu, F., Hanchar, J.M., Hoskin, P.W.O., and Kinny, P., 2003, Atlas of zircon textures, *in* Hanchar, J.M., and Hoskin, P.W.O., eds., Zircon: Reviews in Mineralogy and Geochemistry: Washington, D.C., The Mineralogical Society of America, p. 469–500.

Crowley, J.L., 2003, U-Pb geochronology of 3810–3630 Ma granitoid rocks south of the Isua greenstone belt, southern West Greenland: Precambrian Research, v. 126, no. 3-4, p. 235–257, doi: 10.1016/S0301-9268(03)00097-4.

David, J., Godin, L., Berclaz, A., Maurice, C., Parent, M., Francis, D., and Stevenson, R.K., 2004, Geology and geochronology of the Nuvvuagittuq Supracrustal Sequence: An example of paleoarchean crust (ca. 3.8 Ga) in the northeastern Superior Province: Eos (Transactions, American Geophysical Union), v. 85(17), Joint Assembly Supplement, Abstract V23D-03.

Eggins, S.M., Kinsley, L.P.J., and Shelley, J.M.G., 1998, Deposition and element fractionation processes during atmospheric pressure laser sampling for analysis by ICP-MS: Applied Surface Science, v. 127–129, p. 278–286, doi: 10.1016/S0169-4332(97)00643-0.

Farhat, J.S., and Wetherill, G.W., 1975, Interpretation of apparent ages in Minnesota: Nature, v. 257, no. 5528, p. 721–722, doi: 10.1038/257721a0.

Froude, D.O., Ireland, T.R., Kinny, P.D., Williams, I.S., Compston, W., Williams, I.R., and Myers, J.S., 1983, Ion microprobe identification of 4,100–4,200-myr-old terrestrial zircons: Nature, v. 304, no. 5927, p. 616–618, doi: 10.1038/304616a0.

Goldich, S.S., and Hedge, C.E., 1974, 3,800 Myr granitic gneiss in southwestern Minnesota: Nature, v. 252, no. 5483, p. 467–468, doi: 10.1038/252467a0.

Harley, S.L., and Black, L.P., 1997, A revised Archaean chronology for the Napier Complex, Enderby Land, from SHRIMP ion-microprobe studies: Antarctic Science, v. 9, no. 1, p. 74–91.

Hartlaub, R.P., 2004, Archean and Proterozoic evolution of the Beaverlodge Belt, Churchill craton, Canada [Ph.D. thesis]: Edmonton, University of Alberta, 189 p.

Hartlaub, R.P., Heaman, L.M., Ashton, K.E., and Chacko, T., 2004a, The Archean Murmac Bay Group: Evidence for a giant Archean rift in the Rae Province, Canada: Precambrian Research, v. 131, no. 3-4, p. 345–372, doi: 10.1016/j.precamres.2004.01.001.

Hartlaub, R.P., Heaman, L.M., Ashton, K.E., Chacko, T., Simonetti, A., and Creaser, R.A., 2004b, Two billion years of crustal reworking recorded in the Beaverlodge Belt of northwestern Saskatchewan: Summary of results and implications for mineral exploration: Saskatoon, Saskatchewan Geological Survey Open House, Abstract Vol., p. 29.

Hartlaub, R.P., Böhm, C.O., Heaman, L.M., and Simonetti, A., 2005, Northwestern Superior craton margin, Manitoba: An overview of Archean and Proterozoic episodes of crustal growth, erosion, and orogenesis (parts of NTS 54D and 64A), *in* Report of Activities 2005, Manitoba Industry, Economic Developments and Mines: Manitoba Geological Survey, p. 54–60.

Horstwood, M.S.A., Foster, G.L., Parrish, R.R., Noble, S.R., and Nowell, G.M., 2003, Common-Pb corrected in situ U-Pb accessory mineral geochronology by LA-MC-ICP-MS: Journal of Analytical Atomic Spectrometry, v. 18, no. 8, p. 837–846, doi: 10.1039/b304365g.

Hoskin, P.W.O., and Black, L.P., 2000, Metamorphic zircon formation by solid-state recrystallization of protolith igneous zircon: Journal of Metamorphic Geology, v. 18, no. 4, p. 423–439, doi: 10.1046/j.1525-1314.2000.00266.x.

Hoskin, P.W.O., and Schaltegger, U., 2003, The composition of zircon and igneous and metamorphic petrogenesis, *in* Hanchar, J.M., and Hoskin, P.W.O., eds., Zircon: Reviews in Mineralogy and Geochemistry: Washington, D.C., The Mineralogical Society of America, p. 27–62.

Kinny, P.D., Williams, I.S., Froude, D.O., Ireland, T.R., and Compston, W., 1988, Early Archaean zircon ages from orthogneisses and anorthosites at Mount Narryer, Western Australia: Precambrian Research, v. 38, p. 325–341.

Kinny, P.D., and Nutman, A.P., 1996, Zirconology of the Meeberrie Gneiss, Yilgarn Craton, Western Australia: An early Archaean migmatite: Precambrian Research, v. 78, no. 1-3, p. 165–178, doi: 10.1016/0301-9268(95)00076-3.

Liu, D.Y., Nutman, A.P., Compston, W., Wu, J.S., and Shen, Q.H., 1992, Remnants of greater-than-or-equal-to 3800 Ma crust in the Chinese part of the Sino-Korean Craton: Geology, v. 20, no. 4, p. 339–342, doi: 10.1130/0091-7613(1992)020<0339:ROMCIT>2.3.CO;2.

Ludwig, K.R., 1980, Calculation of uncertainties of U-Pb isotope data: Earth and Planetary Science Letters, v. 46, no. 2, p. 212–220, doi: 10.1016/0012-821X(80)90007-2.

Ludwig, K.R., 2003, Isoplot/Ex, a geochronological toolkit for Microsoft Excel, Version 3.0: Berkeley, California, Berkeley Geochronology Center Special Publication 4.

Maas, R., Kinny, P.D., Williams, I.S., Froude, D.O., and Compston, W., 1992, The Earth's oldest known crust—A geochronological and geochemical study of 3900–4200 Ma old detrital zircons from Mt Narryer and Jack Hills, Western Australia: Geochimica et Cosmochimica Acta, v. 56, no. 3, p. 1281–1300, doi: 10.1016/0016-7037(92)90062-N.

Machado, N., and Simonetti, A., 2001, U-Pb dating and Hf isotopic composition of zircon by laser ablation-MC-ICP-MS, *in* Sylvester, P., ed., Laser Ablation-ICPMS in the Earth Sciences: Mineralogical Association of Canada Short Course, v. 29, p. 121–146.

Michard-Vitrac, A., Lancelot, J., Allegre, C.J., and Moorbath, S., 1977, U-Pb ages on single zircons from early Precambrian rocks of West Greenland and Minnesota River Valley: Earth and Planetary Science Letters, v. 35, no. 3, p. 449–453, doi: 10.1016/0012-821X(77)90077-2.

Moorbath, S., O'Nions, R.K., and Pankhurst, R.J., 1973, Early Archaean age for Isua Iron Formation, West Greenland: Nature, v. 245, no. 5421, p. 138–139.

Moorbath, S., Taylor, P.N., and Jones, N.W., 1986, Dating the oldest terrestrial rocks—Fact and fiction: Chemical Geology, v. 57, no. 1-2, p. 63–86, doi: 10.1016/0009-2541(86)90094-X.

Mueller, P.A., Wooden, J.L., and Nutman, A.P., 1992, 3.96 Ga zircons from an Archean Quartzite, Beartooth Mountains, Montana: Geology, v. 20, no. 4, p. 327–330, doi: 10.1130/0091-7613(1992)020<0327:GZFAAQ>2.3.CO;2.

Mueller, P.A., Wooden, J.L., Nutman, A.P., and Mogk, D.W., 1998, Early Archean crust in the northern Wyoming province—Evidence from U-Pb ages of detrital zircons: Precambrian Research, v. 91, no. 3-4, p. 295–307, doi: 10.1016/S0301-9268(98)00055-2.

Myers, J.S., 2001, Protoliths of the 3.8–3.7 Ga Isua greenstone belt: West Greenland: Precambrian Research, v. 105, no. 2-4, p. 129–141, doi: 10.1016/S0301-9268(00)00108-X.

Nir-El, Y., and Lavi, N., 1998, Measurement of the half-life of Lu-176: Applied Radiation and Isotopes, v. 49, no. 12, p. 1653–1655, doi: 10.1016/S0969-8043(97)10007-0.

Nutman, A.P., 2001, On the scarcity of >3900 Ma detrital zircons in ≥3500 Ma metasediments: Precambrian Research, v. 105, no. 2-4, p. 93–114, doi: 10.1016/S0301-9268(00)00106-6.

Nutman, A.P., Kinny, P.D., Compston, W., and Williams, I.S., 1991, Shrimp U-Pb zircon geochronology of the Narryer Gneiss Complex, Western Australia: Precambrian Research, v. 52, no. 3-4, p. 275–300, doi: 10.1016/0301-9268(91)90084-N.

Nutman, A.P., Friend, C.R.L., Kinny, P.D., and McGregor, V.R., 1993, Anatomy of an early Archean gneiss complex—3900 to 3600 Ma crustal evolution in southern West Greenland: Geology, v. 21, no. 5, p. 415–418, doi: 10.1130/0091-7613(1993)021<0415:AOAEAG>2.3.CO;2.

Nutman, A.P., McGregor, V.R., Friend, C.R.L., Bennett, V.C., and Kinny, P.D., 1996, The Itsaq Gneiss Complex of southern west Greenland; the world's most extensive record of early crustal evolution (3900–3600 Ma): Precambrian Research, v. 78, no. 1-3, p. 1–39, doi: 10.1016/0301-9268(95)00066-6.

Nutman, A.P., Bennett, V.C., Friend, C.R.L., and Rosing, M.T., 1997, ~3710 and ≥3790 Ma volcanic sequences in the Isua (Greenland) supracrustal belt; structural and Nd isotope implications: Chemical Geology, v. 141, no. 3-4, p. 271–287, doi: 10.1016/S0009-2541(97)00084-3.

Nutman, A.P., Bennett, V.C., Friend, C.R.L., and Norman, M.D., 1999, Meta-igneous (non-gneissic) tonalites and quartz-diorites from an extensive ca. 3800 Ma terrain south of the Isua supracrustal belt, southern West Greenland: Constraints on early crust formation: Contributions to Mineralogy and Petrology, v. 137, no. 4, p. 364–388, doi: 10.1007/s004100050556.

Nutman, A.P., McGregor, V.R., Bennett, V.C., and Friend, C.R.L., 2001, Age significance of U-Th-Pb zircon data from early Archaean rocks of west Greenland—A reassessment based on combined ion-microprobe and imaging studies—Comment: Chemical Geology, v. 175, no. 3-4, p. 191–199, doi: 10.1016/S0009-2541(00)00297-7.

Nutman, A.P., Friend, C.R.L., Barker, S.L.L., and McGregor, V.R., 2004, Inventory and assessment of Palaeoarchaean gneiss terrains and detrital zircons in southern West Greenland: Precambrian Research, v. 135, no. 4, p. 281–314, doi: 10.1016/j.precamres.2004.09.002.

Patchett, P.J., Kouvo, O., Hedge, C.E., and Tatsumoto, M., 1981, Evolution of continental-crust and mantle heterogeneity—Evidence from Hf isotopes: Contributions to Mineralogy and Petrology, v. 78, no. 3, p. 279–297.

Patchett, P.J., Vervoort, J.D., Soderlund, U., and Salters, V.J.M., 2004, Lu-Hf and Sm-Nd isotopic systematics in chondrites and their constraints on the Lu-Hf properties of the Earth: Earth and Planetary Science Letters, v. 222, no. 1, p. 29–41, doi: 10.1016/j.epsl.2004.02.030.

Pidgeon, R.T., and Wilde, S.A., 1998, The interpretation of complex zircon U-Pb systems in Archaean granitoids and gneisses from the Jack Hills, Narryer gneiss Terrane, Western Australia: Precambrian Research, v. 91, no. 3-4, p. 309–332, doi: 10.1016/S0301-9268(98)00056-4.

Rosing, M.T., Nutman, A.P., and Lofqvist, L., 2001, A new fragment of the early Earth crust: The Aasivik terrane of West Greenland: Precambrian Research, v. 105, no. 2-4, p. 115–128, doi: 10.1016/S0301-9268(00)00107-8.

Salters, V.J.M., and White, W.M., 1998, Hf isotope constraints on mantle evolution: Chemical Geology, v. 145, no. 3-4, p. 447–460, doi: 10.1016/S0009-2541(97)00154-X.

Scherer, E., Munker, C., and Mezger, K., 2001, Calibration of the lutetium-hafnium clock: Science, v. 293, no. 5530, p. 683–687.

Schiøtte, L., Compston, W., and Bridgwater, D., 1989, Ion Probe U-Th-Pb zircon dating of polymetamorphic orthogneisses from Northern Labrador, Canada: Canadian Journal of Earth Sciences, v. 26, no. 8, p. 1533–1556.

Simonetti, A., Heaman, L.M., Hartlaub, R.P., Creaser, R.A., MacHattie, T., and Böhm, C., 2005, U-Pb zircon dating by laser ablation-MC-ICP-MC using a new multiple ion counting faraday collector array: Journal of Analytical Atomic Spectrometry, v. 20, p. 677–686.

Söderlund, U., Patchett, J.P., Vervoort, J.D., and Isachsen, C.E., 2004, The Lu-176 decay constant determined by Lu-Hf and U-Pb isotope systematics of Precambrian mafic intrusions: Earth and Planetary Science Letters, v. 219, no. 3-4, p. 311–324, doi: 10.1016/S0012-821X(04)00012-3.

Song, B.A., Nutman, A.P., Liu, D.Y., and Wu, J.S., 1996, 3800 to 2500 Ma crustal evolution in the Anshan area of Liaoning Province, northeastern China: Precambrian Research, v. 78, no. 1-3, p. 79–94, doi: 10.1016/0301-9268(95)00070-4.

Taylor, R.N., Warneke, T., Milton, J.A., Croudace, I.W., Warwick, P.E., and Nesbitt, R.W., 2003, Multiple ion counting determination of plutonium isotope ratios using multi-collector ICP-MS: Journal of Analytical Atomic Spectrometry, v. 18, no. 5, p. 480–484, doi: 10.1039/b300432e.

Vervoort, J.D., Patchett, P.J., Gehrels, G.E., and Nutman, A.P., 1996, Constraints on early Earth differentiation from hafnium and neodymium isotopes: Nature, v. 379, no. 6566, p. 624–627, doi: 10.1038/379624a0.

Wasteneys, H.A., Wardle, R.J., and Krogh, T.E., 1996, Extrapolation of tectonic boundaries across the Labrador shelf: U-Pb geochronology of well samples: Canadian Journal of Earth Sciences, v. 33, no. 9, p. 1308–1324.

Whitehouse, M.J., Kamber, B.S., and Moorbath, S., 1999, Age significance of U-Th-Pb zircon data from early Archaean rocks of west Greenland—A reassessment based on combined ion-microprobe and imaging studies: Chemical Geology, v. 160, no. 3, p. 201–224, doi: 10.1016/S0009-2541(99)00066-2.

Whitehouse, M.J., Kamber, B.S., and Moorbath, S., 2001, Age significance of U-Th-Pb zircon data from early Archaean rocks of west Greenland—A reassessment based on combined ion-microprobe and imaging studies—Reply: Chemical Geology, v. 175, no. 3-4, p. 201–208, doi: 10.1016/S0009-2541(00)00301-6.

Wilde, S.A., Valley, J.W., Peck, W.H., and Graham, C.M., 2001, Evidence from detrital zircons for the existence of continental crust and oceans on the Earth 4.4 Gyr ago: Nature, v. 409, no. 6817, p. 175–178, doi: 10.1038/35051550.

Williams, I.S., Compston, W., Black, L.P., Ireland, T.R., and Foster, J.J., 1984, Unsupported radiogenic Pb in zircon—A Cause of anomalously high Pb-Pb, U-Pb and Th-Pb ages: Contributions to Mineralogy and Petrology, v. 88, no. 4, p. 322–327, doi: 10.1007/BF00376756.

Woodhead, J., Hergt, J., Shelley, M., Eggins, S., and Kemp, R., 2004, Zircon Hf-isotope analysis with an excimer laser, depth profiling, ablation of complex geometries, and concomitant age estimation: Chemical Geology, v. 209, no. 1-2, p. 121–135, doi: 10.1016/j.chemgeo.2004.04.026.

Zheng, J.P., Griffin, W.L., O'Reilly, S.Y., Lu, F.X., Wang, C.Y., and Zhang, M., 2004, 3.6 Ga lower crust in central China: New evidence on the assembly of the North China craton: Geology, v. 32, no. 3, p. 229–232, doi: 10.1130/G20133.1.

Manuscript Accepted by the Society 4 October 2005

Printed in the USA

Geological Society of America
Special Paper 405
2006

Precise U-Pb zircon ID-TIMS ages provide an alternative interpretation to early ion microprobe ages and new insights into Archean crustal processes, northern Labrador

Thomas E. Krogh
Sandra L. Kamo†
Jack Satterly Geochronology Laboratory, Department of Geology, University of Toronto, Ontario, M5S 3B1, Canada

ABSTRACT

Geochronological investigations of early Archean gneiss complexes by U-Pb dating methods are complicated by the presence of zircon populations that contain multiple ages of magmatic and metamorphic growth, and inherited components from premagmatic crust formation processes. In an early ion-microprobe study in the Saglek block, Nain Province, northern Labrador, identical ages for magmatism of 3.73 Ga were reported for both amphibolite- and granulite-grade orthogneisses, based on the oldest $^{207}Pb/^{206}Pb$ ages in two and three grains, respectively (Schiøtte et al., 1989). Here we present zircon ages for the same rock units as in the previous study of 3348 ± 7 Ma and 3634 ± 31 Ma, respectively, obtained by isotope-dilution thermal ionization mass spectrometry (ID-TIMS) methods, and based on the best quality grains from the dominant population of each rock. Zircon cores in the amphibolite-facies orthogneiss were separated from igneous overgrowths using a brief HF acid treatment and give ages ranging from 3700 to 3760 Ma, the age reported for magmatism in Schiøtte et al. (1989). The time of major metamorphism in the amphibolite-facies orthogneiss is recorded by both Pb loss in the magmatic zircons and by new zircon growth within remnants of cross-cutting metadiabase at 2702 ± 2 Ma. The age for the early stage of granulite-facies metamorphism was reported by Schiøtte et al. (1989) as 2766 ± 17 Ma compared to the age reported here at 2740 ± 4 Ma.

The results of this comparative study are instructive for the interpretation of sequential inherited, magmatic, and metamorphic zircon components in a complex population and for devising sampling strategies based on simple geological relations.

Our new results demonstrate the way different ages can be obtained for the same rock unit due to contrasting approaches to dating geological events, and it is suggested that the earlier published data require reinterpretation.

The new data reported here add to the database on igneous activity in the region so that the formation of crustal components is documented at 3850–3800, 3750–3700, ca. 3630, 3348, and 3200 Ma. This crustal instability may reflect the lack of an early mantle keel that was not fully established until the 2740–2702 Ma episode of meta-

†Corresponding author e-mail: skamo@geology.utoronto.ca.

Krogh, T.E., and Kamo, S.L., 2006, Precise U-Pb zircon ID-TIMS ages provide an alternative interpretation to early ion microprobe ages and new insights into Archean crustal processes, northern Labrador, *in* Reimold, W.U., and Gibson, R.L., Processes on the Early Earth: Geological Society of America Special Paper 405, p. 91–103, doi: 10.1130/2006.2405(06). For permission to copy, contact editing@geosociety.org.

morphism and ductile flow. A thin-skinned protocrust with multiple ages of intrusion and metamorphism, possibly driven by plume-related heat from below, i.e., multistage crustal recycling rather than competent plate interactions, is suggested for the formation of these Archean gneisses.

Keywords: zircon, U-Pb geochronology, early Archean, northern Labrador.

INTRODUCTION

A field trip to the Saglek Fjord area in northern Labrador in 1989 with Dr. David Bridgwater of the Copenhagen Museum, Denmark, provided a unique opportunity to sample rocks from two sites that were part of a then recently published paper reporting ion-microprobe (SHRIMP) ages by Schiøtte et al. (1989). The basis of our new study was to do a direct dating comparison by ID-TIMS (isotope-dilution thermal ionization mass spectrometry) in order to evaluate the relative merits of the two U-Pb dating methods.

Both of these studies and several others on the North Atlantic Craton of southwestern Greenland (e.g., Nutman et al., 1997; Whitehouse et al., 1999) indicate that the presence of inherited zircons and multiple ages for magmatism are widespread. In this paper, we discuss these observations in comparison to the development of the short-lived, extensive 2.70–2.75 Ga volcano-plutonic terrane of the Superior Province (e.g., Corfu and Davis, 1992). The contrast in tectonic regime appears to reflect different crustal processes operating in the Paleoarchean to Neoarchean. Our new ages add to the Archean knowledge base and have implications for correlations between Labrador and Greenland and for models of the tectonic evolution of the North Atlantic Craton and other mainly pre–3.0 Ga terranes.

Since Schiøtte et al.'s pioneering SHRIMP study (1989), ion-microprobe methods have advanced greatly (e.g., Ireland and Williams, 2003). Nevertheless, an examination of the contrasting approaches and results and the rationale behind how best to unravel complex, multiepisodic geological histories by zircon dating remains instructive.

GEOLOGICAL SETTING AND SAMPLE LOCATIONS

South of Saglek Fjord on the Labrador coast is a vast region of gray quartzofeldspathic gneiss (Uivak 1 gneiss), most of which underwent migmatization both before and after intrusion of the basic to ultrabasic Saglek dykes (Bridgwater et al., 1975). Folded, subhorizontal, tectonically interleaved supracrustal units are seen on the walls of the Saglek Fjord and elsewhere in the region (Ryan et al., 1983; Collerson et al., 1989) and are thought to represent tectonic intercalations that occurred before and during the late Archean high-grade metamorphic event (Schiøtte et al., 1989). A major, subvertical, north-south–trending fault (the "Handy fault," Bridgwater et al., 1975) separates different crustal levels of the Nain Province so that orthogneisses of granulite grade are present to the west and orthogneisses of amphibolite grade are present to the east of the fault (Fig. 1; Bridgwater et al., 1990). It was the intent of Schiøtte et al.'s original study (1989) to compare the ages of emplacement and metamorphism on either side of this fault. Schiøtte et al. selected and we resampled an amphibolite-facies orthogneiss on Little Island situated in the mouth of the Saglek Fjord (Fig. 1). The island comprises amphibolite-facies, gray and pink orthogneiss (TK89–5) containing remnants of meter-wide, amphibolitized Saglek dykes (TK89–6). The dykes are enclosed by highly strained orthogneiss that is locally ruptured and subsequently infilled by late, locally derived, unstrained pegmatite (TK89–7; Fig. 2). We sampled these to date the time of post-dyke emplacement metamorphism and the end stage of ductile flow. A second orthogneiss sample, a porphyritic granodiorite (DB82.12), was also collected on the east side of Little Island, where it appears to locally crosscut the foliation of the gray gneissic host rock but is itself cut by mafic dykes, and where both are variably deformed, likely during late metamorphism.

The second sampling locality is 14 km to the southwest of Little Island on Torr Bay. Here, granulite orthogneiss (TK89–10; Fig. 3) was sampled. At this site, Schiøtte et al. (1989) also collected a local orthopyroxene-bearing pegmatitic melt from the ruptured orthogneiss that was believed to be the product of metamorphic melting of the LILE–depleted (large ion lithophile element) host. Trains of ultramafic inclusions are common (Collerson et al., 1989), and where one of these had ruptured during the late stages of ductile flow, we collected an unstrained orthopyroxene-bearing pegmatitic infill (TK89–8) shown in Figure 3.

Petrographic descriptions and a more detailed geological setting can be found in Schiøtte et al. (1989) and Ryan and Martineau (1992), respectively.

ANALYTICAL PROCEDURES

Zircon and titanite U-Pb analysis followed modified procedures of Krogh (1973) and Corfu (1988), respectively. All zircons were air abraded to remove surfaces affected by secondary Pb loss (Krogh, 1982). U-Pb isotopic data are presented in Table 1, and analytical details can be found in the footnotes. Errors on ratios and ages are quoted at 95% confidence levels and plotted at 2σ. The programs of Davis (1982) and Ludwig (2000) were used for regression calculations.

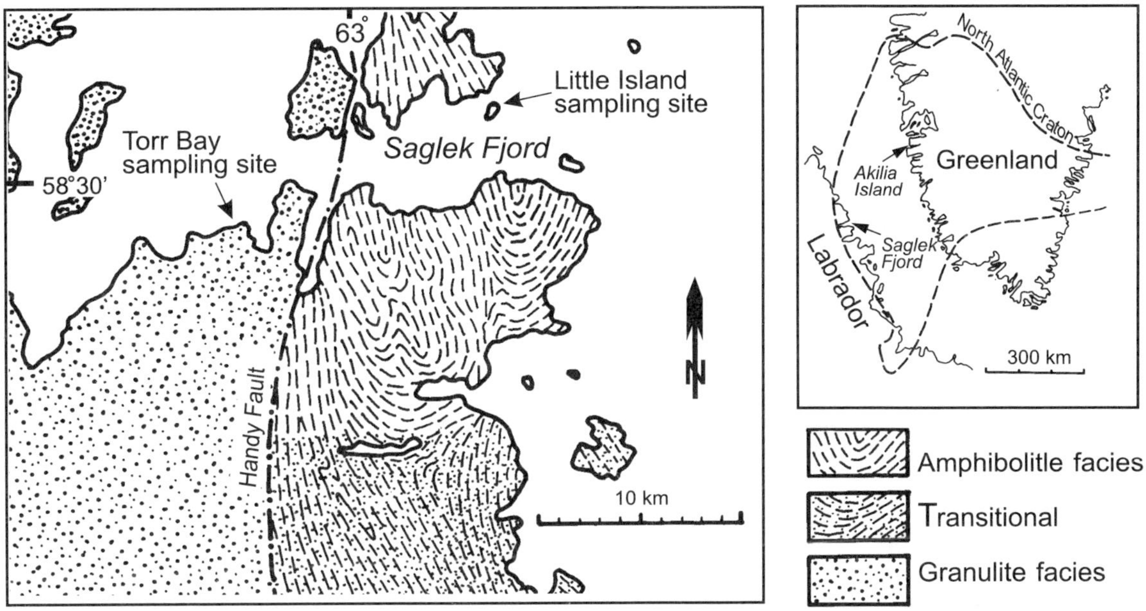

Figure 1. Simplified map of the Saglek Fjord area in northern Labrador showing the distribution of metamorphic facies in the region of the two sampling sites (Little Island and Torr Bay). These are the same sampling sites as reported in Schiøtte et al. (1989). Figure from Schiøtte et al. (1989).

ZIRCON POPULATION DESCRIPTIONS, ION-MICROPROBE AND ID-TIMS RESULTS, INTERPRETATIONS, AND COMPARISONS

In this section, sample descriptions and U-Pb data obtained by the two methods are presented back-to-back for each sampling site, followed by a comparison of the results from both methods. Table 2 is a summary of ages obtained by both methods. The general dating strategies of the two studies, especially for conventional ID-TIMS, are given in the description of results for the first site.

Results of Previously Published Ion-Microprobe Study of Amphibolite Facies Uivak 1 Gneiss (Sample DB82.74.1)

In their study, Schiøtte et al. (1989) mounted and polished relatively large zircon grains (200–600 µm, Fig. 2 *in* Schiøtte et al., 1989) and sought to assign the origin of different zones to inheritance and magmatic or metamorphic growth. The grains were imaged using both transmitted and reflected light, with the latter enhanced by etching with HF vapor (Krogh and Davis, 1974) that was done after the 20–40 µm spot analyses. This method darkens intense radiation-damaged zones, but leaves zones below a certain threshold untouched, and is especially useful in geologically old, radiation-damaged zircons of the type encountered in this study.

Zircons from the amphibolite-grade orthogneiss were described as purplish brown with minor metamorphic rounding and pronounced igneous zoning. The largest grains contain massive (i.e., unzoned) low U interiors that give the oldest ^{207}Pb-^{206}Pb ages. A magmatic emplacement age of 3733 ± 9 Ma was obtained by averaging data from 3 spot analyses on one grain and one spot analysis on another grain. Use of this low U central zone as the magmatic emplacement age was justified by the observation that the highly zoned outer portion, interpreted as igneous, showed a conformable rather than resorbed interface with the interior. One rounded, low U interior domain, overgrown by magmatic zoning, gave a minimum age of 3863 ± 12 Ma, which was attributed to inheritance. A total of 31 spot analyses on 10 grains were reported for this sample, but spot locations were not specified.

The average U concentration is 566 ppm, Th/U ratios for magmatic grains range from 0.01 to 0.62, and 9 of 31 data fall within the range of 0.3–0.5 (i.e., the range of Th/U found in 8 of 9 magmatic grains in our study). An age for the time of metamorphism for this site was not reported.

Results From Conventional Isotope Dilution Methods for Amphibolite Facies Uivak 1 Gneiss (Samples TK89–5 and DB82.12) and Associated Amphibolite (TK89–6) and Pegmatite (TK89–7)

Our investigation was unavoidably delayed so that the data presented here were obtained ~10 yr ago. Isotope dilution analyses take longer to perform but give much higher analytical precision and therefore require fewer age determinations. The data are generally reproducible to within measurement uncertainty (±1–2 Ma), and replicate analyses provide a test to show that

Figure 2. (A) Photograph of the outcrop on Little Island where samples were collected. Amphibolite-grade orthogneiss (TK89–5) was collected from the least deformed material away from the contact with metadiabase. Sample TK89–6 is a metadiabase derived from a Saglek dyke. Pegmatite sample TK89–7 was emplaced during the late stages of ductile deformation. (B) Photograph taken near where samples were collected showing all 3 rock types. Note the ruptured amphibolitized Saglek dyke surrounded by orthogneiss and the locally developed pegmatite hosted within the metadiabase.

only a single growth stage has been sampled. Rigorous zircon selection criteria and strict preparation procedures are employed to limit, or eliminate, the effects of recent Pb loss that the body of historic analytical data demonstrates is present in most Precambrian zircons. All grains are selected from the least paramagnetic fraction (Silver and Deutsch, 1963) and are chosen based on high luster, lack of color (i.e., low in U), and having no cracks. Exterior, geologically leached surfaces are removed by air abrasion as a final step that improves concordance and also improves precision by reducing the common Pb correction (Krogh, 1982). Each of these steps and considerations will generally result in the analysis of lower U grains with reduced radiation damage and discordance. Identification of low U grains can be further achieved by exposing the grains to NaOH or HF at 220 °C for ~2 h. This procedure causes radiation-damaged

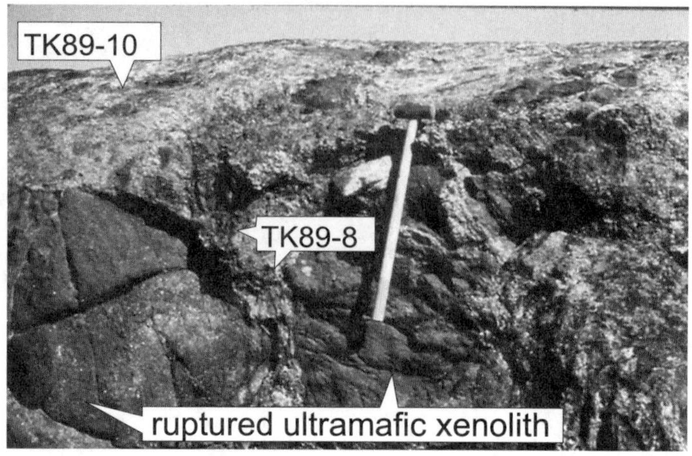

Figure 3. Photograph of sampling locality at Torr Bay. Granulite-grade orthogneiss (TK89–10) was sampled from the least deformed part of the outcrop, and pegmatite (TK89–8) was taken from within a ruptured ultramafic xenolith.

zones, i.e., zones that generally show weak or no cathodoluminescence (CL), to become visible and permits the discrimination of low U grains or low U tips from high U cores (Krogh, 1998). A very minor HF treatment (i.e., 1–5 min at room temperature) played a vital role in our investigation as this allowed the separation of low U cores from highly zoned, partly altered overgrowths (Fig. 4). After a HF treatment such grains fall apart when touched, or they separate along the core-overgrowth interface due to differential radiation-induced expansion.

We generally interpret components of a zircon population using simple principles: (1) for most suitable igneous rocks a primary zircon population must be present and be dominant, (2) zircons that formed prior to an intensive metamorphism generally show resorbed surfaces (e.g., Vavra et al., 1996), whereas those that formed late during a high-grade metamorphic event are generally sharply facetted and euhedral (e.g., van Breemen et al., 1986; Schaltegger et al., 1999), (3) most igneous zircons show oscillatory zoning (e.g., Vavra, 1990), (4) fine-grained metamorphosed mafic rocks are generally devoid of primary zircons but yield unique, high-quality, multifaceted spherical grains in amphibolite or higher-grade metamorphic rocks (e.g., van Breemen et al., 1986; Krogh, 1993), (5) precise replicate analyses of tips of grains formed in pegmatites during the late stages of a metamorphic event can prove that mixed ages have been avoided (Krogh, 1994), and (6) independently determined ages for the time of metamorphism can facilitate accurate upper intercept ages, even when no concordant domains are present (e.g., Krogh, 1994).

Visual examination of our amphibolite-facies orthogneiss zircon population (TK89–5) indicates a predominance of small (50–150 μm) prismatic zircons and minor subpopulations of large grains with cores (200–600 μm) and equant yellow grains (100–200 μm). Each of these distinct grain types has been dated, and the results are described below according to type.

TABLE 1. U-Pb DATA FOR ORTHOGNEISSES AND ASSOCIATED ROCKS FROM THE SAGLEK FJORD AREA, NORTHERN LABRADOR, CANADA

No.	Fraction description	Weight (mg)	U (ppm)	Th/U	Pbtot (pg)	PbCom (pg)	$^{207}Pb/^{204}Pb$	$^{206}Pb/^{238}U$	2 Sig	$^{207}Pb/^{235}U$	2 Sig	$^{207}Pb/^{206}Pb$	2 Sig	$^{206}Pb/^{238}U$ Age (Ma)	2 Sig	$^{207}Pb/^{235}U$ Age (Ma)	2 Sig	$^{207}Pb/^{206}Pb$ Age (Ma)	2 Sig	Disc (%)	Corr Coeff
Little Island																					
TK89–5 Amphibolite-Facies Orthogneiss																					
Isolated Zircon Cores																					
1	1 pink round	0.006	141	0.31	817	8.5	1790	0.7805	0.0027	38.472	0.135	0.35749	0.00052	3718.9	9.7	3732.1	3.5	3739.2	2.2	0.7	0.913
2	1 purple, 5 min HF	0.014	159	0.42	2143	1.8	21078	0.7689	0.0016	37.012	0.090	0.34911	0.00030	3676.9	5.8	3693.9	2.4	3703.1	1.3	0.9	0.939
3	1 purple, 5 min. HF	0.003	231	0.47	658	1.0	11052	0.7556	0.0022	35.432	0.099	0.34009	0.00061	3628.1	8.2	3650.7	2.8	3663.2	2.7	1.3	0.806
4	1/2 core, 1 min. HF	0.001	156	0.40	158	1.7	1752	0.7967	0.0040	41.630	0.217	0.37898	0.00042	3777.3	14.5	3810.3	5.2	3827.7	1.7	1.7	0.977
5	1 pink 4:1 clear euhedral flawless	0.002	159	1.64	359	0.8	6836	0.7429	0.0032	34.524	0.154	0.33702	0.00038	3581.5	11.9	3625.1	4.4	3649.3	1.7	2.4	0.968
Magmatic Zircons (Euhedral Pink)																					
6	1 pale pink euhedral	0.004	101	0.49	314	1.4	3239	0.6473	0.0016	23.415	0.063	0.26235	0.00029	3217.7	6.3	3244.4	2.6	3260.9	1.8	1.7	0.910
7	1 pale pink clear subhedral tip	0.001	56	0.51	45	0.7	985	0.6590	0.0055	24.096	0.199	0.26519	0.00051	3263.2	21.3	3272.3	8.1	3277.8	3.0	0.6	0.973
8	1 clear euhedral pink	0.002	145	0.42	227	0.8	4081	0.6593	0.0022	24.336	0.085	0.26769	0.00028	3264.6	8.5	3281.9	3.4	3292.5	1.6	1.1	0.954
9	1 pink clear largest	0.009	74	0.52	529	1.4	5450	0.6574	0.0022	24.076	0.081	0.26562	0.00038	3257.0	8.5	3271.5	3.3	3280.4	2.3	0.9	0.908
10	1 pink clear smallest	0.001	98	0.40	77	2.5	475	0.6638	0.0115	24.763	0.432	0.27054	0.00055	3282.0	44.8	3298.9	17.1	3309.2	3.2	1.0	0.993
11	1 pink to pale brown, darkest in group	0.002	265	0.28	443	1.0	7336	0.7016	0.0022	30.273	0.097	0.31295	0.00043	3426.6	8.3	3495.7	3.1	3535.5	2.1	4.0	0.908
Yellow Zircons																					
12	1 round clear yellow	0.002	2712	0.01	2783	31.8	920	0.4978	0.0012	12.854	0.032	0.18728	0.00027	2604.3	5.0	2669.0	2.4	2718.4	2.4	5.1	0.824
13	1 round clear yellow	0.006	3578	0.02	10864	49.1	2176	0.4929	0.0012	12.529	0.034	0.18437	0.00021	2583.1	5.3	2644.9	2.6	2692.6	1.9	4.9	0.910
14	1 round clear yellow	0.003	2161	0.02	3313	19.5	1682	0.4970	0.0014	12.678	0.032	0.18502	0.00036	2600.7	5.9	2656.0	2.4	2698.4	3.2	4.4	0.728
Titanite																					
15	multititanite clear brown fragments unabraded	0.159	142	0.78	13824	987	115	0.4790	0.0012	10.989	0.071	0.16639	0.00100	2522.9	5.2	2522.2	6.0	2521.6	10.1	–0.1	0.358
DB82–12 Amphibolite-Facies Orthogneiss																					
16	1 euhedral pinkish brown	0.007	61	0.58	356	1.4	3626	0.6746	0.0016	25.510	0.067	0.27425	0.00031	3323.7	6.3	3327.9	2.6	3330.5	1.8	0.3	0.906
17	1 euhedral pale pink best	0.001	331	0.50	266	0.5	7494	0.6665	0.0024	24.914	0.095	0.27112	0.00026	3292.2	9.4	3304.9	3.7	3312.5	1.5	0.8	0.967
18	1 euhedral clear pink	0.003	78	0.51	185	1.4	1876	0.6558	0.0022	24.189	0.093	0.26753	0.00060	3250.7	8.5	3276.0	3.7	3291.6	3.5	1.6	0.816
19	3 euhedral pale pink flawless	0.003	205	0.43	459	0.9	7211	0.6330	0.0015	22.227	0.059	0.25468	0.00025	3161.3	5.9	3193.7	2.6	3214.1	1.6	2.1	0.928
TK89–6 Metadiabase (hosted in TK89-5)																					
20	1 brown resorbed some faces	0.002	243	0.26	273	1.4	2098	0.5185	0.0014	13.259	0.039	0.18546	0.00025	2692.8	5.8	2698.2	2.8	2702.3	2.2	0.4	0.891
21	1 clear colorless	0.002	27	0.11	29	1.7	209	0.5136	0.0023	13.111	0.068	0.18513	0.00064	2672.1	9.6	2687.7	4.9	2699.4	5.7	1.2	0.756
22	5 titanite fragments light brown unabraded	0.048	248	1.0	7194	150	364	0.4739	0.0013	10.759	0.036	0.16467	0.00030	2500.5	5.6	2502.6	3.1	2504.2	3.1	0.2	0.836
TK89–7 Pegmatite (hosted in TK89–6)																					
23	1 brown euhedral fragment 5 min. 5% HF	0.001	296	0.13	150	0.9	1749	0.4879	0.0019	11.482	0.047	0.17068	0.00024	2561.7	8.4	2563.1	3.8	2564.3	2.4	0.1	0.200
Torr Bay																					
TK89–10 Granulite-Facies Orthogneiss																					
Magmatic Zircons																					
24	1 dark pink subhedral clear	0.002	472	0.41	852	1.3	10870	0.7348	0.0020	32.549	0.097	0.32128	0.00029	3551.1	7.4	3567.0	2.9	3576.0	1.4	0.9	0.954
25	1 equant subrounded surficial cracks	0.001	1962	0.10	1665	1.6	19046	0.7330	0.0028	32.406	0.130	0.32063	0.00035	3544.7	10.6	3562.7	4.0	3572.8	1.7	1.0	0.962
26	1 dark pink subhedral	0.002	726	0.35	1284	0.7	29056	0.7297	0.0020	31.932	0.094	0.31737	0.00030	3532.4	7.4	3548.2	2.9	3557.2	1.5	0.9	0.946
27	1 pink round 3 hours NaOH	0.016	73	0.40	1036	0.8	22403	0.7240	0.0017	31.422	0.081	0.31475	0.00033	3511.2	6.2	3532.4	2.5	3544.4	1.6	1.2	0.915
28	1 pale pink clear 3 hours NaOH no chemistry	0.002	56	0.28	96	0.5	3363	0.7225	0.0025	31.221	0.111	0.31340	0.00040	3505.5	9.3	3526.0	3.5	3537.7	2.0	1.2	0.933
29	1 pale pink clear 3 hours NaOH no chemistry	0.001	42	0.29	36	3.4	189	0.7150	0.0038	30.566	0.200	0.31005	0.00089	3477.3	14.1	3505.2	6.4	3521.2	4.4	1.6	0.905
30	1 dark pink 1/2 grain	0.001	644	0.70	581	1.1	8207	0.7054	0.0038	29.519	0.163	0.30353	0.00030	3440.9	14.4	3470.9	5.4	3488.3	1.5	1.8	0.984
Colorless Grains and Tips																					
31	10 clear colorless	0.033	13	0.39	247	0.7	3809	0.5344	0.0013	14.371	0.040	0.19503	0.00017	2760.0	5.5	2774.5	2.6	2785.1	1.4	1.1	0.954
32	1 slightly grey clear tip no chemistry	0.004	48	0.35	113	0.5	2407	0.5339	0.0021	14.188	0.058	0.19274	0.00021	2757.8	8.7	2762.4	3.9	2765.7	1.8	0.3	0.963
33	1 euhedral well facetted clear equant	0.006	11	0.09	35	1.0	440	0.5290	0.0022	13.813	0.065	0.18940	0.00041	2737.0	9.2	2737.0	4.4	2737.0	3.6	0	0.887
34	1 pale pink round clear	0.002	152	0.59	187	2.3	875	0.5286	0.0026	13.861	0.062	0.19017	0.00048	2735.6	10.8	2740.2	4.3	2743.6	4.2	0.4	0.855
35	1 colourless clear tip	0.001	12	0.05	7	3.3	38	0.5224	0.0110	13.314	0.614	0.18483	0.00655	2709.5	46.5	2702.2	43.5	2696.7	58.4	–0.6	0.677
TK89-8 Pegmatite (hosted in TK89–10)																					
36	1 pink clear euhedral tip	0.048	19	2.3	727	1.8	3019	0.5215	0.0014	13.364	0.039	0.18587	0.00018	2705.5	5.9	2705.8	2.8	2705.9	1.6	0	0.941
37	1 euhedral purple	0.103	256	1.2	17757	2.5	65488	0.5199	0.0012	13.225	0.035	0.18448	0.00018	2698.9	5.2	2695.9	2.5	2693.5	1.6	–0.2	0.931
38	1 euhedral pink tip	0.015	16	2.2	188	3.3	461	0.5212	0.0012	13.365	0.040	0.18597	0.00026	2704.4	4.9	2705.8	2.8	2706.9	2.3	0.1	0.888

Notes: All zircons have been air abraded; zircons have had anion exchange chemistry unless noted otherwise. Weights are approximate but estimated to be within ±50% (this does not affect the isotopic data, only the concentrations of U and Pb). Pbtot—Total amount of radiogenic Pb; PbCom—total common Pb. Th/U calculated from radiogenic $^{208}Pb/^{206}Pb$ ratio and $^{207}Pb/^{206}Pb$ age assuming concordance. $^{207}Pb/^{204}Pb$ corrected for fractionation and common Pb in the spike. All other isotopic ratios corrected for blank (up to 10 pg blank isotopic composition $^{206}Pb/^{204}Pb = 18.221$; $^{207}Pb/^{204}Pb = 15.612$; $^{208}Pb/^{204}Pb = 39.36$), initial common Pb above 10 pg corrected using Stacey and Kramers (1975), common Pb in the spike, and fractionation. %Disc—percent discordance for the given $^{207}Pb/^{206}Pb$ age. Uranium decay constants are from Jaffey et al. (1971).

TABLE 2. COMPARATIVE CHRONOLOGY OF EVENTS RECORDED IN GNEISSES OF THE SAGLEK FJORD AREA OF NORTHERN LABRADOR, CANADA

Granulite-Facies Area		Amphibolite-Facies Area	
ID-TIMS	Ion-microprobe (Ma)	ID-TIMS	Ion-microprobe (Ma)
Premagmatic History			
n.d.	>ca. 3840	3988 + 27/–25 Ma 3813 + 26/–22 Ma	>3863 ± 12
Primary Magmatism			
3634 ± 31 Ma (3593 ± 11 Ma minimum age)	3730 ± 8	3348 ± 7 Ma	3733 ± 9
Metamorphism			
New zircons in orthogneiss: 2740 ± 4 Ma	3630 ± 8; 2766 ± 17	New zircons in metadiabase: 2702 ± 2 Ma New zircons in orthogneiss: 2.7 Ga	n.d.
Late Stages of Ductile Flow			
New zircons in pegmatite hosted in ruptured ultramafic xenolith: 2694 ± 2 Ma	Pegmatitic infill in ruptured orthogneiss: 2744 ± 4; 2711 ± 20	New zircons in pegmatite: 2564 ± 2 Ma	n.d.
Titanite crystallization			
		In orthogneiss: 2522 ± 10 Ma In amphibolite: 2504 ± 3 Ma	n.d.

Note: n.d.—not determined.

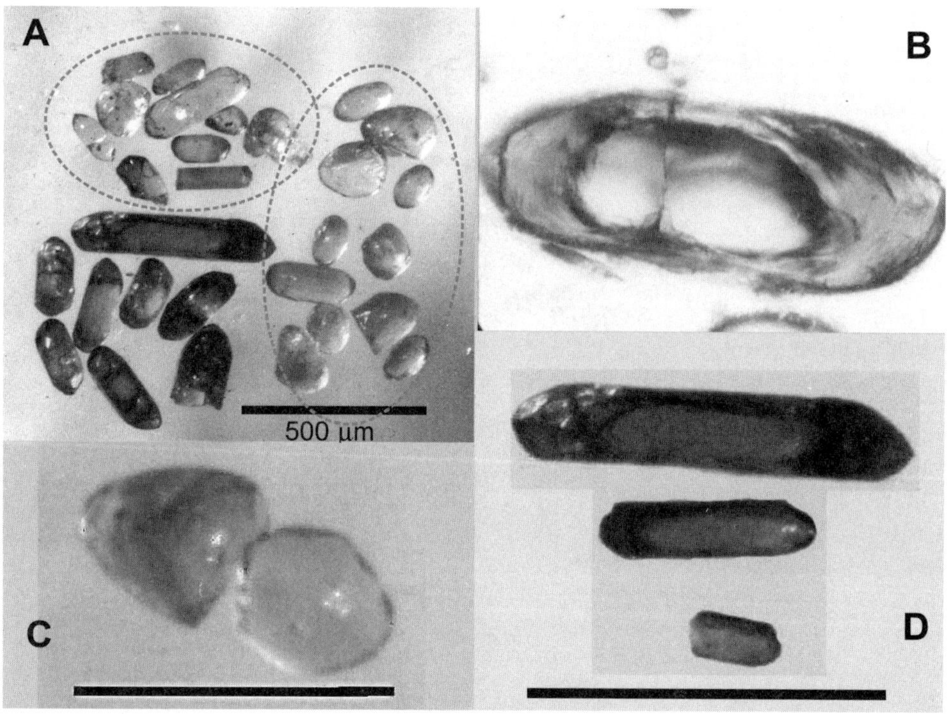

Figure 4. Photos of zircons from orthogneiss TK89–5. (A) 3 distinct groups of zircons: grains in the top left oval are pink, euhedral magmatic grains; the group in the oval on the right are high-U, yellow equant grains, and the third group in the lower left are grains that have visible core-overgrowth relations; scale bar, 500 µm. (B) Zircon core mantled by overgrowth that has an apparent reentrant angle along the crack from premagmatic resorption; photo taken under transmitted light. (C) Zircon core and overgrowth separated after a brief (<5 min) exposure to HF; scale bar, 250 µm. (D) Partially isolated cores after brief HF exposure; the larger zircon core was separated from the elongate grain shown in (A); scale bar, 400 µm. Note the rounded terminations due to premagmatic resorption.

Magmatic Zircons

There is a predominance of highly zoned, brown to purple, altered, 2:1 to 4:1 prismatic grains. A primary magmatic component is virtually certain in a rock with this granitoid composition, and we interpret this dominant grain type as magmatic in origin. These grains show a continuous variation toward pale pink, crack-free zircons, and five of these were selected for analysis. The second orthogneiss sample from Little Island (DB82.12), a porphyritic granodiorite, contains pale pink, prismatic grains with metamorphic rounding from resorption and a greater proportion of crack-free grains with no visible evidence of core-overgrowth relations.

Data for the five single grains from TK89–5 (analyses 6–10) give an upper intercept age of 3347 ± 19 Ma (12% probability of fit) when regressed with the age for metamorphism of 2702 ± 2 Ma (described below). Four data for the nearby porphyry (DB 82.12, analyses 16–19) give an identical age of 3351 ± 21 Ma if anchored at 2702 Ma (6% probability of fit). The combined data for 9 grains (Table 1, Fig. 5) yield an age of 3348 ± 7 Ma (42% probability of fit) using the 2702 Ma independent age for metamorphism of a rock from the same outcrop. Six of these nine grains have U concentrations of 100 ppm or less, with the others at 160, 200, and 330 ppm, and Th/U ratios for eight of nine grains are between 0.3 and 0.5, with one at 0.6. One grain (analysis 11) gave data that plots with the ca. 3.7 Ga array discussed below, indicating that both core and overgrowth are present in this grain and that precise analyses are more valuable in defining growth stages than visual appearance.

Zircon Core/Mantle Components

Relatively large grains of this subpopulation of TK89–5 (200–600 μm) typically show low U central zones overgrown by oscillatory zoned zircon that we interpret as magmatic material (Fig. 4A). We consider these grains to have been incorporated prior to the main stage of magmatic crystallization. In Figure 4B note the re-entrant angle along the crack of the grain in the photo; in Figure 4C there is a rounded interface between core and overgrowth; and in Figure 4D note the well-rounded

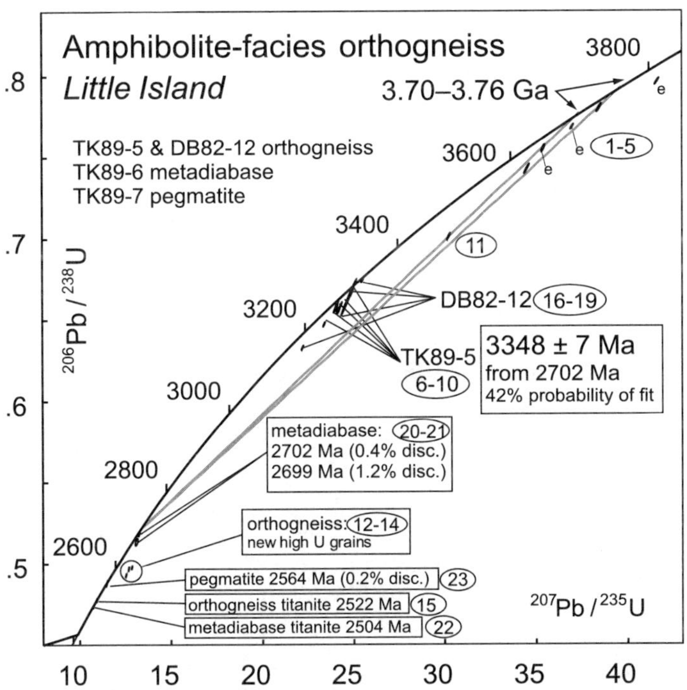

Figure 5. Concordia figure showing U-Pb data for the amphibolite-facies orthogneiss (TK89–5 and DB82–12) and associated metadiabase (TK89–6) and pegmatite (TK89–7). Numbers in ellipses are the analysis numbers keyed to Table 1. The "e" denotes the zircon cores that were liberated from overgrowth after a 5 min treatment in 5% HF acid.

terminations exposed in the partially de-clad cores. The longest grain present in Figure 4A is shown again in Figure 4D above the same grain at an intermediate stage with the overgrowth partially removed, prior to abrasion. Isotopic data for two cores that were separated from their overgrowths (analyses 2 and 3), one round grain with no overgrowth (analysis 1), one 4:1 prismatic, clear, euhedral grain (analysis 5), and a single, pinkish brown grain (analysis 11) form an array of upper intercept ages from 3.70 to 3.76 Ga when anchored by the 2702 Ma time of metamorphism described below. For analyses 1–5 these are: 3760 ± 12 Ma, 3732 ± 6 Ma, 3704 ± 13 Ma, 3876 ± 14 Ma, and 3731 ± 15 Ma, respectively. If the effects of recent Pb loss in these grains have not been completely removed, the projected ages would be too old, and any Pb loss that occurred during the 3348 Ma magmatic event would make them too young. The oldest $^{207}Pb/^{206}Pb$ minimum age for a rounded core type in the 3.7–3.76 Ga group is 3739 ± 2 Ma, which is analytically identical to the age noted as magmatic by Schiøtte et al. (1989). We also note that one of our grains in this array is a 4:1 pink, clear, euhedral grain of the sort analyzed by Schiøtte et al. (1989); hence inheritance of igneous type grains is probable. Both morphology and the age difference define a xenocrystic origin for the ca. 3.7 Ga component.

Curiously, the data for the five analyses of inherited grains in Figure 5 also form two mixing lines that are each collinear from the 3348 Ma magmatic age and project onto the concordia curve at 3813+26/–22 Ma and 3988 +27/–25 Ma, which can be interpreted as possible primary ages for incorporated protoliths. However, this interpretation is considered less probable given the pervasive overprinting caused by the 2.7 Ga event and recent Pb loss. A single analysis of one-half of a rounded released core (analysis 4) has a distinctly older minimum age of 3828 ± 2 Ma, which supports the concept that the low U central zones within zoned magmatic grains are inherited.

Metamorphic Zircon

A minor subgroup in the amphibolite-facies orthogneiss population comprises equant, rounded, yellow grains (Fig. 4A). These give U-Pb data that are discordant (4%–5%) and have $^{207}Pb/^{206}Pb$ ages of 2718 ± 2 Ma, 2698 ± 3 Ma, and 2693 ± 2 Ma (analyses 12–14, Fig. 5). The ages, high U concentrations (2200–3600 ppm), and morphological characteristics differentiate these grains from the older populations. Also, low Th/U ratios of 0.01–0.02 are consistent with amphibolite-facies metamorphic zircon growth (van Breemen et al., 1986; our unpublished data).

Metadiabase (TK89–6) that cuts the host orthogneiss contains zircon that gives ages of 2702 ± 2 Ma and 2699 ± 6 Ma (0.4% and 1.3% discordant, analyses 20 and 21, respectively). Uranium concentrations are 240 and 30 ppm, and Th/U ratios are 0.26 and 0.11, respectively. The 2702 Ma age is considered the best estimate for the time of metamorphism and is in general agreement with the ages of the high U zircons in the orthogneiss. However, due to discordance and minor inheritance in one analysis, the high U zircons are considered less useful for pinpointing a precise time of metamorphism.

A single, concordant analysis (analysis 23) of a brown, sharply facetted, euhedral grain from an unstrained pegmatite in a ruptured Saglek dyke (TK89–7; Fig. 2) records a late stage of ductile flow at 2564 ± 2 Ma (0.1% discordant, 296 ppm U, Th/U 0.12).

Titanite in the orthogneiss gives a concordant age of 2522 ± 10 Ma and titanite from the metadiabase is concordant at 2504 ± 3 Ma (analyses 15 and 23, respectively), as if they formed or cooled through their blocking temperatures long after the zircon forming events. Crustal heating during emplacement of late Archean granites (Baadsgaard et al., 1979) may have played a role in this regard.

Comparison of Results for Amphibolite Gneiss

In both the SHRIMP and ID-TIMS studies, central domains of large zircons overgrown by altered, zoned magmatic material were analyzed. Schiøtte et al. (1989) interpreted the minimum age for two of these as indicating the time of magmatic emplacement at 3733 ± 9 Ma (2σ), based on the presence of magmatic zoning in one case and a conformable interface to the outer, magmatically zoned material in another. In our study, we found these interior domains to have rounded forms when released from altered overgrowths after the HF treatment and to yield a ca. 3.7–2.7 Ga array (Fig. 5). One of the cores gave a minimum age of 3739 ± 2 Ma, which is analytically identical to their proposed magmatic age, and we note also that one of our grains is a 4:1 prismatic, clear, euhedral "magmatic" type; hence, inheritance of igneous grains does occur. In contrast, data for five high-quality, relatively small, euhedral grains, when combined with those from a coeval associated intrusion, gave an age of 3348 ± 7 Ma (42% probability of fit) when anchored at 2702 Ma, the age for metamorphism at this outcrop. These grains are much smaller (i.e., 50–100 μm, 2:1 to 4:1 prisms) than those illustrated in Figure 2 of the Schiøtte et al. (1989) study, and it is possible that this grain type was not sampled in their study. Also, U concentrations of 100 ppm or less are much lower than an average of 566 ppm U in the grains analyzed by SHRIMP. Thus, it is possible that selection of the larger, high U grains that contain an abundance of inherited components affected the outcome of their study.

The simplest explanation of our morphological and isotopic observations is that the ca. 3.7 Ga component is inherited along with the >3.8 Ga rounded core type component reported in both studies. The fact that our data for the magmatic grains is collinear, whereas that for the cores is not, may imply variable primary ages for the source, but the effects of minor, recent Pb loss cannot be ruled out. The fact that the ion-microprobe data is not collinear but is displaced toward a ca. 2.0 Ga intercept is explained as being due to "a combination of late Archean and recent Pb loss" (p. 1536, line 16; Schiøtte et al., 1989). Conformity of our data to the 2702 Ma age for metamorphism reflects our strategy of selecting the highest-quality grains designed to eliminate recent Pb loss.

Results of a Previously Published Ion-Microprobe Study for Granulite Grade Uivak 1 Gneiss (DB82.21c) and Related Local Pegmatitic Melt (DB82.21a)

The dominant zircon type in this population was described by Schiøtte et al. (1989) as purplish-brown grains of generally high quality with "blurred euhedral zoning" and rounded terminations. Visual distinction between newly crystallized zircon and recrystallized portions is noted as being not clear in some cases, and pre-igneous rounded cores that have U levels and Th/U ratios that overlap with those for the igneous zircons were also noted. Relatively large grains (~200–600 μm) were mounted and imaged as in the previous sample.

The age for the granulite-grade orthogneiss was derived by averaging the oldest minimum ages for single spot analyses on 3 grains that were selected from 28 spots on 22 grains to give an age of 3730 ± 8 Ma. The authors also noted a strong concentration of ages in the range 3605–3630 Ma and assigned this to an early metamorphic remelting event that produced a second generation of igneous zircons. Average U levels were noted as being generally low (<100–200 ppm), with Th/U ratios of 0.1–0.8 and 5 of 28 spots giving values between 0.3 and 0.4 (i.e., the range of Th/U we observed in 5 of 7 grains of our igneous population). Ages for rounded inclusions are as high as ca. 3840 Ma, whereas others are close to their igneous zircon age. A subpopulation of low U, medium Th grains gave near concordant ages that agree within error and have an average value of 2766 ± 17 Ma. In this case, 2 sigma uncertainties of ±60–90 Ma are noted due to the relatively low U levels (100–150 ppm). Zircons from a local pegmatitic melt, described as euhedrally zoned, with U levels in excess of 1000 ppm, yielded data that plot below concordia with a mean $^{207}Pb/^{206}Pb$ age of 2744 ± 4 Ma and a ca. 2.5 Ga $^{206}Pb/^{238}U$ age.

ID-TIMS Results for Granulite-Grade Uivak 1 Orthogneiss (TK89–10) and Related Late Metamorphic Pegmatitic Infill (TK89–8)

Zircons from our sample are generally high quality, pink to brown to colorless, with variably resorbed surfaces and only traces of igneous zoning. A regression line calculated through the data for 7 euhedral grains gives an upper intercept age of 3634 ± 31 Ma and a lower intercept at 3054 +142/–130 Ma (Fig. 6; analyses 24–30) with a 70% probability of fit. If the data are anchored at 2740 Ma, our independent age of metamorphism for this locality (see below), the top 6 analyses yield a minimum upper intercept age of 3597 Ma, with 0% probability of fit. Three of these six grains were selected from a group of several hundred after a 3 h etch treatment in NaOH at 220 °C that reveals the level of radiation damage. In this case, grains with U concentrations of 42, 56, and 73 ppm U were selected

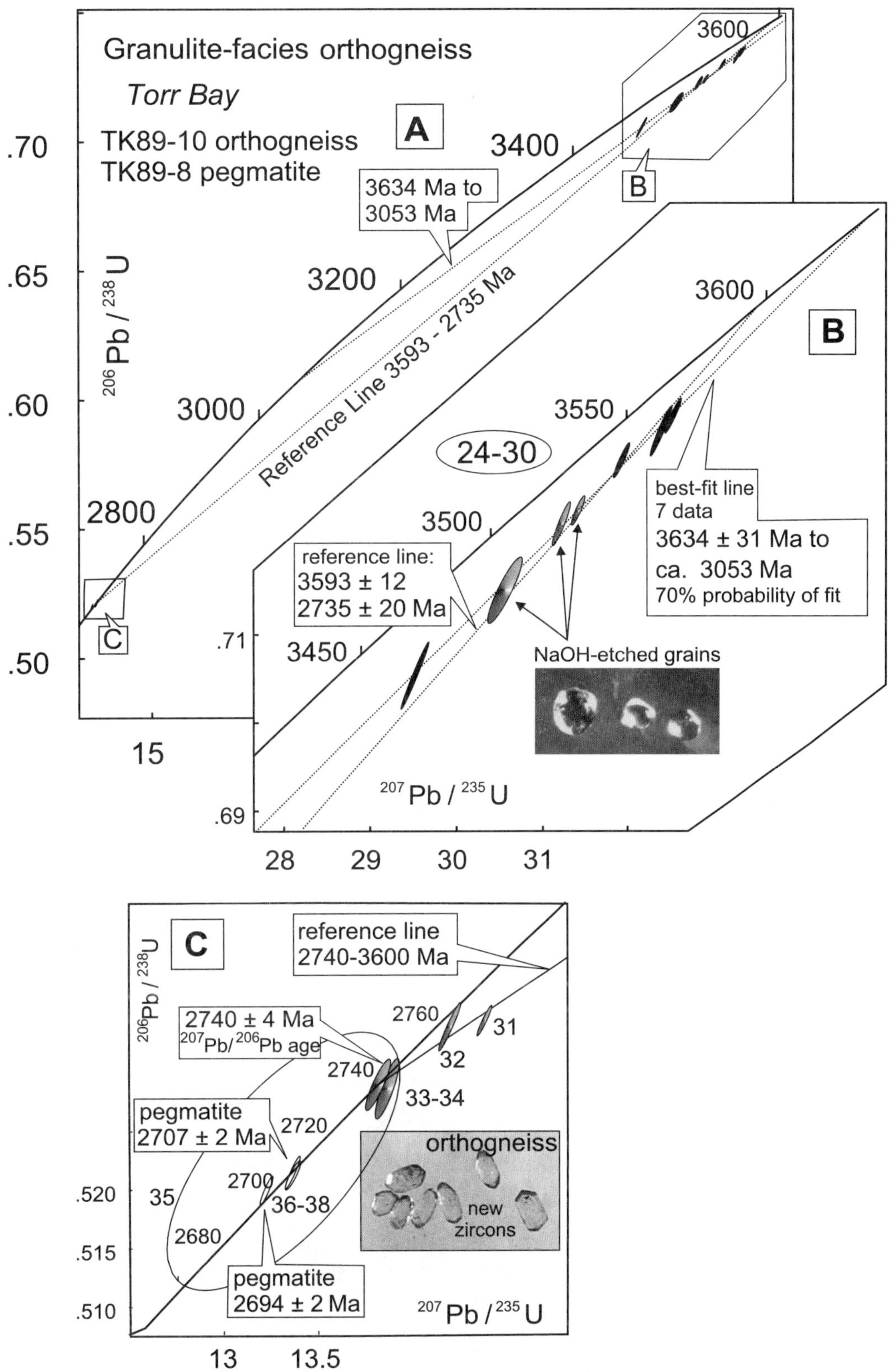

Figure 6. Concordia figure showing U-Pb data for (A) the granulite-facies orthogneiss (TK89–10) and reference line to the age of the pegmatite, (B) enlargement of the top of the mixing line showing the data for the orthogneiss, and (C) enlargement of the lower end of the reference line showing data from the pegmatite (TK89–8). Numbers in ellipses refer to the analysis number keyed to Table 1. The NaOH–etched grains were heated to 200 °C for 3 h in a saturated solution.

compared to 472, 1962, 726, and 644 ppm for selected high-quality grains that were not etched (Fig. 6B). It is unusual for etched grains not to produce more concordant data; nevertheless, because etched grains rarely show recent Pb loss, their conformity to the data produced from unetched grains enhances the reliability of the array. Given the 2740 Ma age for metamorphic zircons in this rock (see below), the 3054 Ma lower intercept was unexpected.

From previous experience it is known that data plotting furthest down a discordia line are more susceptible to a second stage of Pb loss, and therefore plot below, not above, the mixing line toward the time of metamorphism, as observed here. If this deviation from the discordia line to 2744 Ma is real, it would require that the zircons in the granulite orthogneiss underwent Pb loss at some time before the 2740 Ma event, perhaps at the 3350 Ma magmatic event that produced the amphibolite-facies orthogneiss or possibly minor new growth occurred during a ca. 3600 Ma event. Th/U ratios for 5 of 7 grains are between 0.3 and 0.4, with two others at 0.1 and 0.7. To resolve these alternative interpretations, we would need to analyze multiple, resorbed tips, but this is beyond the scope of the present study.

A distinctly different minor subpopulation of sharply facetted, euhedral, equant, pink grains is also present, and data for two of these are concordant with $^{207}Pb/^{206}Pb$ ages of 2737 ± 4 Ma (0% discordant; analysis 33) and 2744 ± 4 Ma (0.4% discordant; analysis 34) for a combined age of 2740 ± 4 Ma. Data for a third colorless clear tip (analysis 35) gave a similar result with a very large error (Fig. 6C) due to a combination of small sample size, low U, and a slightly enlarged common Pb content (3 pg versus ~1 pg). A replicate analysis using 10 grains (analysis 31) confirms the ca. 2740 Ma age with a much smaller error but shows a trace of inheritance, as does data for a fourth single grain (analysis 32). Note that even with minor inheritance, however, analysis 31 verifies the time of metamorphism, given that the magmatic age is known. The ability to combine low U grains into one analysis permits precise isotope dilution results for grains that would yield unacceptably large uncertainties by ion microprobe analysis, even if a regular (~10–20) spot analysis pooling approach were used in the latter.

Zircons from the pegmatite that occurs within a ruptured ultramafic inclusion hosted in the granulite-facies orthogneiss are sharply facetted, euhedral, and brown to pink to colorless. Data for two euhedral tips that were selected to avoid older components give analytically identical results of 2706 ± 2 and 2707 ± 2 Ma, and are 0% and 0.1% discordant, respectively (analyses 36 and 38). Data for a dark pink to purple euhedral grain (analysis 37) gave a slightly younger age of 2694 ± 2 Ma. Th/U ratios in these zircons are high (2.3, 1.2, 2.2), and U concentrations in 2 of 3 grains are relatively low at 16 and 19 ppm, perhaps reflecting their U-depleted ultramafic host. Note that our 2 sigma errors are ±2 Ma for such low U grains, which is generally unachievable for data obtained by ion microprobe.

Comparison of Results for Granulite Facies Uivak 1 Orthogneiss

Our data for seven selected high-quality grains from the main magmatic population give no indication of the previously reported older 3730 and 3840 Ma ages but rather define a collinear array with an upper intercept at 3634 ± 31 Ma. Th/U ratios for 5 of 7 of these grains give typical igneous values of 0.3–0.4 compared to 5 of 28 in the ion-microprobe study. A more coherent, alteration-free zircon population in our study is suggested.

The simplest explanation of these age differences is that the ca. 3605–3630 Ma "second-stage" magmatic ages of the previous study and our 3634 Ma age represent the time of magmatic emplacement of this orthogneiss. The older component that gave the 3730 ± 8 Ma ion-microprobe age would then represent inherited, weakly zoned xenocrysts, which they note have similar ages, Th/U, and U concentrations to some of their rounded core types. The fact that the early results plot along a 3.6 Ga to a 2.5–2.0 Ga array and not to the ca. 2.7 Ga event indicates metamorphic resetting as well as recent Pb loss in most grains, as was noted by Schiøtte et al. (1989).

The timing of metamorphism from new zircon growth in the orthogneiss is 2740 ± 4 Ma compared to the ion-microprobe age of 2766 ± 17 Ma. Discordant data for zircons from the late-tectonic, granulite pegmatite were previously published at 2744 ± 4 Ma, whereas our new concordant data for the same event indicates the latest stages of metamorphic growth at 2706 ± 2 Ma, 2707 ± 2 Ma, and 2693 ± 2 Ma. The difference in age for the two metamorphic grain types by the two methods may result from the limitation of the ion microprobe to detect small amounts of inheritance due to a large analytical uncertainty. In contrast, ID-TIMS analyses permit pooling of grains without the danger of mixed ages, which can usually be detected due to the higher analytical precision (e.g., analysis 31). The age difference between new growth in the host (our 2740 Ma age) and the late orthopyroxene-bearing melt (our 2707 Ma and 2694 Ma) reflects the protracted duration of this high-grade event also noted by Schiøtte et al. (1989). Precise ID analyses of low U grains tend to give us the most reliable results, but these have the highest analytical uncertainty in the ion microprobe results. The simple field choice to select pegmatitic infill from an ultramafic host rather than the orthogneiss for determining the time of late metamorphism probably accounts for the low U concentrations (~20 ppm versus 1000 ppm) and concordant versus discordant data.

GEOLOGICAL IMPLICATIONS

Our new ages for the two Saglek Fjord sampling sites indicate magmatism at both ca. 3630 Ma and ca. 3350 Ma, with ages for older crust in the region of 3700–3750 Ma and 3800–3850 Ma, represented by inherited grains. The latter ages were also reported for tonalitic gneiss on Akilia Island (see

Fig. 1), a southwestern Greenland correlative of the Saglek area (e.g., Nutman and Collerson, 1991; Nutman et al., 1997). Our 3630 Ma age is also similar to the 3.65 Ga tonalite gneiss age published by Whitehouse et al. (1999). Data for detrital grains from quartzite interlayers in tonalite in the Saglek area also indicate the presence of 3800–3850 Ma crust (Nutman and Collerson, 1991).

Combined with the previously published results from Greenland, our new ages indicate a remarkably long period for magmatism in the North Atlantic Craton: 3850–3800 Ma, 3750–3700 Ma, ca. 3650 Ma, 3348 Ma, and 3235 Ma (the latter from Schiøtte et al., 1989). Granulite-facies metamorphism at Akilia Island occurred at 3650–3670 Ma, with a second high-grade metamorphic event that produced ductile flow and boudinage at 2710 Ma (our unpublished data; Myers and Crowley, 2000). In addition, our data indicate granulite-facies metamorphism in the interval 2740 Ma to 2707 Ma and amphibolite-facies metamorphism at 2702 Ma in the Saglek Fjord area. Trains of ultramafic inclusions imply proximal mantle and disruption associated with the 2740–2707 Ma ductile flow. Disrupted rafts of granulite-grade metasediments that are intercalated with tonalite result from the same process (Nutman and Collerson, 1991). The timing of the tectonic interleaving is debatable. Nutman and Collerson (1991) suggest that it occurred during the emplacement of the 3700–3730 Ma tonalite, whereas our data for ductile flow-related pegmatite suggest 2740–2707 Ma. Perhaps both episodes and others in between are possible.

The results for the North Atlantic Craton are in stark contrast to those for the volcano-plutonic belts in the Superior Province. These formed mainly over a remarkably short interval between 2750 and 2700 Ma at high structural levels with multiple coeval components occurring over a distance of ~1000 km (Corfu and Davis, 1992). Unlike in the North Atlantic Craton, inherited zircons are extremely rare and regionwide reheating and remobilization, long after emplacement, is lacking. Such crustal stability is commonly attributed to the formation of a subcrustal keel. We then attribute the lack of crustal stability, i.e., the multiple ages of magmatism and metamorphism between ca. 3850 and 3200 Ma in the North Atlantic Craton, to the lack of such a keel until the final 2740–2702 Ma remobilization. Progressive development before this time due to magma extraction is probable.

In general, the lithological assemblage and fabric in the North Atlantic Craton resemble those exposed in the Kapuskasing uplift in the Superior Province, which is a midcrustal cross section from below the world's largest greenstone belt (Krogh, 1994). Here, granulite-grade orthogneiss is interlayered with supracrustal metasedimentary and metavolcanic rocks that are laminated with a subhorizontal gneissic fabric. Metamorphic ages young with depth, and the presence of granitic clasts with formation ages younger than arc accretion at higher levels requires hot, ductile, post-accretion, shallow underplating. Heat from below is required because units on either side of the uplift have low-grade assemblages, which rules out continental overthrusting. Nutman and Collerson (1991) proposed a similar mechanism for their ca. 3730–3700 Ma tectonic interleaving in the North Atlantic Craton. We concur but note that at least some of the interleaving at Saglek occurred in the 2740–2702 Ma interval. Here, the material assembled had been extracted during multiple episodes between 3850 Ma and 3200 Ma with abundant crustal recycling indicated by inherited zircons. Heat from the proximal mantle indicated by ultramafic inclusions, perhaps plume related, rather than due to plate margin collisions or overthrusting as indicated by Friend et al. (1988) is suggested. This thin-skinned, multistage tectonic regime would then represent an intermediate stage between the ca. 2750–2700 Ma episode of massive crust production and the pre–4.0 Ga processes that recycled and dispersed any crustal components created. Perhaps it is no coincidence that our proposed episode of stabilizing crustal underplating in the North Atlantic Craton occurred at the same time as the 2750–2700 Ma Superior Province event, as both would be driven by mantle processes.

CONCLUSIONS

Since the early work of Schiøtte et al. (1989) ion-microprobe instrumentation has undergone major advancements and data treatment methods have evolved, as have other techniques. Nevertheless, there remains an interpretative aspect to all isotopic data and our approach is to minimize this by using available geological relations in selecting rock samples, followed by strict zircon selection criteria. The present study was initiated some years ago as a comparative investigation of the two emerging and rapidly evolving methods. However, due to a significant amount of elapsed time, the main purpose of this contribution is to address the accuracy of the ages in the database of the Saglek Fjord area in order to best reconstruct its tectonic history.

1. The primary ages of two Saglek Fjord area orthogneisses (Uivak gneiss) that occur on either side of the north-south Handy fault are younger than previously reported and are different (3348 ± 7 Ma and 3634 ± 31 Ma), i.e., not analytically identical, at 3733 ± 9 Ma and 3730 ± 8 Ma as previously reported.

2. We attribute the difference in ages to the practice in the Schiøtte et al. (1989) study of averaging the oldest minimum ages for zoned zircon grains without recognizing that some of these may be inherited.

3. Crustal components in the Saglek Fjord area including those represented by xenocrysts span a remarkably long period: 3850–3800 Ma, 3750–3700 Ma, ca. 3650 Ma, 3348 Ma, and 3200 Ma.

4. Granulite-facies metamorphism, associated in its late stages with rupturing of ultramafic xenoliths and regional ductile flow, recorded west of the Handy fault, occurred in the interval 2740 ± 4 Ma to 2707 ± 2 Ma. This contrasts with the previously published age for metamorphism in the orthogneiss from the same locality at 2766 ± 7 Ma. Amphibolite-facies metamorphism east of the Handy fault is dated at 2702 ± 2 Ma.

5. Simple, geologically based rock sampling strategies on the outcrop, such as collecting rocks that contain only metamorphic zircons, combined with precise ID-TIMS analyses of selected high-quality grains that represent the main population can lead to assumption-free unambiguous age data.

6. Intrusions of multiple ages that yielded many zircon grains with cores in the interval 3850–3200 Ma contrast with the Superior Province short 2750–2700 Ma interval of formation and almost a complete lack of cores.

7. Intercalation of metasedimentary, tonalitic, and ultramafic units in the Saglek Fjord area imply proximal mantle, crustal softening and ductile flow suggestive of heat from below rather than plate collisions or overthrusting. Much of this occurred during the 2740–2702 Ma metamorphism, although earlier similar episodes are probable.

8. The multiple episodes of magmatism and metamorphism in the North Atlantic Craton imply a lack of an early stable mantle keel in this region that reached its final stage of development during the interval when most of the Superior Province was created.

9. Crustal components formed at 3850–3800 Ma, 3760–3700 Ma, 3670–3600 Ma, and a 2740–2700 Ma metamorphism are present at both Saglek and Akilia Island in southwestern Greenland.

ACKNOWLEDGMENTS

This study would not have been possible without the kindness of Dr. David Bridgwater, who provided guidance in collecting the rock samples. The Natural Sciences and Engineering Research Council (NSERC) of Canada is gratefully acknowledged for financial support. Comments and criticisms from R. Hartlaub and an anonymous reviewer helped to improve the manuscript and are appreciated.

REFERENCES CITED

Baadsgaard, H., Collerson, K.D., and Bridgwater, D., 1979, The Archean Gneiss complex of northern Labrador, 1, Preliminary U-Th-Pb geochronology: Canadian Journal of Earth Sciences, v. 16, p. 951–961.

Bridgwater, D., Collerson, K.D., Hurst, R.W., and Jesseau, C.W., 1975, Field characters of the early Precambrian rocks from Saglek, coast of Labrador, *in* Report of Activities Part A, April to October 1974: Geological Survey of Canada Paper 75-01A, p. 287–296.

Bridgwater, D., Mengel, F., Schiøtte, L., and Winter, J., 1990, Research on the Archean rocks of northern Labrador, *in* Pereira, C.P.G., Walsh, D.G., and Blackwood, R.F., eds., Current Research: Newfoundland and Labrador Geological Survey Report 90-1, Progress Report 1989, p. 227–236.

Collerson, K.D., McCulloch, M.T., and Nutman, A., 1989, Sr and Nd isotope systematics of polymetamorphic Archean gneisses from southern West Greenland and northern Labrador: Canadian Journal of Earth Sciences, v. 26, p. 446–466.

Corfu, F., 1988, Differential response of U-Pb systems in coexisting accessory minerals, Winnipeg River Subprovince, Canadian Shield: Implications for Archean crustal growth and stabilization: Contributions to Mineralogy and Petrology, v. 98, p. 312–325, doi: 10.1007/BF00375182.

Corfu, F., and Davis, D., 1992, A U-Pb geochronological framework for the western Superior Province, *in* Thurston, P.C., Sutcliffe, R.H., Stott, G.M., and Williams, H.R., eds., The Geology of Ontario: Ontario Geological Survey Special Volume 4, Part 2, p. 1335–1346.

Davis, D.W., 1982, Optimum linear regression and error estimation applied to U-Pb data: Canadian Journal of Earth Sciences, v. 19, p. 2141–2149.

Friend, C.R.L., Nutman, A.P., and McGregor, V.R., 1988, Late Archean terrane accretion in the Godthab region, southern West Greenland: Nature, v. 335, p. 535–538, doi: 10.1038/335535a0.

Ireland, T., and Williams, I.S., 2003, Considerations in zircon geochronology by SIMS, *in* Hanchar, J.M., and Hoskin, P.W.O., eds., Zircon: Reviews in Mineralogy and Geochemistry, v. 53, p. 215–241.

Jaffey, A.H., Flynn, K.F., Glendenin, L.E., Bentley, W.C., and Essling, A.M., 1971, Precision measurement of half-lives and specific activities of ^{235}U and ^{238}U: Physical Review, v. 4, p. 1889–1906, doi: 10.1103/PhysRevC.4.1889.

Krogh, T.E., 1973, A low contamination method for hydrothermal decomposition of zircon and extraction of U and Pb for isotopic age determinations: Geochimica et Cosmochimica Acta, v. 37, p. 485–494, doi: 10.1016/0016-7037(73)90213-5.

Krogh, T.E., 1982, Improved accuracy of U-Pb zircon ages by the creation of more concordant systems using an air abrasion technique: Geochimica et Cosmochimica Acta, v. 46, p. 637–649, doi: 10.1016/0016-7037(82)90165-X.

Krogh, T.E., 1993, High precision U-Pb ages for granulite metamorphism and deformation in the Archean Kapuskasing structural zone, Ontario: Implications for structure and development of the lower crust: Earth and Planetary Science Letters, v. 119, p. 1–18, doi: 10.1016/0012-821X(93)90002-Q.

Krogh, T.E., 1994, Identification of concordant zircons using etch techniques: Berkeley, California, Abstracts of the Eighth International Conference on Geochronology, Cosmochronology, and Isotope Geology, June 5–11, p. 180.

Krogh, T.E., 1998, Etch selection and abrasion techniques applied to complex populations and revealed spurious ion probe ages: Goldschmidt Conference Abstracts Volume, Part 2, Mineralogical Magazine, v. A 62, p. 816–817.

Krogh, T.E., and Davis, G.L., 1974, Alteration in zircons with discordant U-Pb ages: Carnegie Institution Washington Yearbook, v. 73, p. 560–567.

Ludwig, K., 2000, Isoplot/Ex version 2.2: A geochronological toolkit for Microsoft Excel: Berkeley, California, Berkeley Geochronology Center Special Publication 1a, 55 p.

Myers, J.S., and Crowley, J., 2000, Vestiges of life in the oldest Greenland rocks? A review of early Archean geology in the Godthabsfjord region, and reappraisal of field evidence for >3850 Ma life on Akilia: Precambrian Research, v. 103, p. 101–124, doi: 10.1016/S0301-9268(00)00089-9.

Nutman, A.P., and Collerson, K.D., 1991, Very early Archean crustal-accretion complexes preserved in the North Atlantic Craton: Geology, v. 19, p. 791–794, doi: 10.1130/0091-7613(1991)019<0791:VEACAC>2.3.CO;2.

Nutman, A.P., Mojzsis, S.J., and Friend, C.R.L., 1997, Recognition of ≥3850 Ma water-lain sediments in West Greenland and their significance for the early Archean Earth: Geochimica et Cosmochimica Acta, v. 61, p. 2475–2484, doi: 10.1016/S0016-7037(97)00097-5.

Ryan, B., and Martineau, Y., 1992, Geology of the Saglek Fiord–Hebron Fiord area, Labrador (NTS 14L/2, 3,6, 7): Newfoundland Department of Mines and Energy, Geological Survey Branch, Open File Map 92-18 and Geological Survey of Canada, Open File 2466, scale 1:100,000, 2 sheets.

Ryan, B., Martineau, Y., Bridgwater, D., Schiøtte, L., and Lewry, J., 1983, The Archean-Proterozoic boundary in the Saglek Fiord Area, Labrador: Report 1, *in* Current Research, Part A: Geological Survey of Canada paper 83-01A, p. 297–304.

Schaltegger, U., Fanning, C.M., Gunther, D., Maurin, J.C., Schulman, K., and Gebauer, D., 1999, Growth, annealing and recrystallization of zircon and preservation of monazite in high-grade metamorphism: Conventional and in-situ U-Pb isotope, cathodoluminescence and microchemical evidence: Contributions to Mineralogy and Petrology, v. 134, p. 186–201, doi: 10.1007/s004100050478.

Schiøtte, L., Compston, W., and Bridgwater, D., 1989, Ion probe U-Th-Pb zircon dating of polymetamorphic orthogneisses from northern Labrador, Canada: Canadian Journal of Earth Sciences, v. 26, p. 1533–1556.

Silver, L., and Deutsch, S., 1963, Uranium-lead isotopic variations in zircons: A case study: The Journal of Geology, v. 71, p. 721–758.

Stacey, J.S., and Kramers, J.D., 1975, Approximation of terrestrial lead isotope evolution by a two-stage model: Earth and Planetary Science Letters, v. 34, p. 207–226.

van Breemen, O., Davidson, A., Loveridge, W.D., and Sullivan, R.W., 1986, U-Pb zircon geochronology of Grenville tectonites, granulites and igneous precursors, Parry Sound, Ontario: Geological Association of Canada Special Paper 31, p. 191–208.

Vavra, G., 1990, On the kinematics of zircon growth and its petrogenetic significance: A cathodoluminescence study: Contributions to Mineralogy and Petrology, v. 106, p. 90–99, doi: 10.1007/BF00306410.

Vavra, G., Gebauer, D., Schmid, R., and Compston, W., 1996, Multiple zircon growth and recrystallization during polyphase Late Carboniferous to Triassic metamorphism in granulites of the Ivrea Zone (Southern Alps): An ion microprobe (SHRIMP) study: Contributions to Mineralogy and Petrology, v. 122, p. 337–358, doi: 10.1007/s004100050132.

Whitehouse, M.J., Kamber, B.S., and Moorbath, S., 1999, Age significance of U-Th-Pb zircon data from early Archean rocks of west Greenland—A reassessment based on combined ion-microprobe and imaging studies: Chemical Geology, v. 160, p. 201–224, doi: 10.1016/S0009-2541(99)00066-2.

Manuscript Accepted by the Society 4 October 2005

Printed in the USA

Geological Society of America
Special Paper 405
2006

The 3.466 Ga "Kitty's Gap Chert," an early Archean microbial ecosystem

Frances Westall[†]
Centre de Biophysique Moléculaire, CNRS, Rue Charles Sadron, 45071 Orléans cedex 2, France
Sjoukje T. de Vries[‡]
Wouter Nijman[§]
Faculty of Geosciences, Utrecht University, Budapestlaan 4, 3584 CD Utrecht, The Netherlands
Virgile Rouchon[#]
Département des Sciences de la Terre, UMR CNRS-UPS Interaction and Dynamics of Surface Environments, Université Paris Sud, Bât. 504, 91405 Orsay Cedex, France and
Planetary Science Laboratory, Osaka University, 1-1 Machikaneyama, Toyonaka, 560-0043 Osaka, Japan
Beate Orberger[††]
Département des Sciences de la Terre, UMR CNRS-UPS Interaction and Dynamics of Surface Environments, Université Paris Sud, Bât. 504, 91405 Orsay Cedex, France
Victoria Pearson[‡‡]
Jon Watson[§§]
Alexander Verchovsky[##]
Ian Wright[†††]
Planetary and Space Sciences Research Institute (PSSRI), The Open University, Walton Hall, Milton Keynes, K7 6AA, UK
Jean-Noël Rouzaud[‡‡‡]
Centre de la Matière Divisée, CNRS, 1B rue de la Férollerie, 45071 Orléans cedex 2, France
Daniele Marchesini[§§§]
via Torino 10, 40139 Bologna, Italy
Anne Severine[###]
Department of Geological Sciences, University of Orléans, 45071 Orléans cedex 2, France

[†]E-mail: westall@cnrs-orleans.fr.
[‡]E-mail: stdevries@geo.uu.nl.
[§]E-mail: wnijman@terra.geo.uu.nl.
[#]E-mail: rouchon@geol.u-psud.fr.
[††]E-mail: orberger@geol.u-psud.fr.
[‡‡]E-mail: v.k.pearson@open.acuk.
[§§]E-mail: j.watson@open.ac.uk.
[##]E-mail: a.verchovsky@open.ac.uk.
[†††]E-mail: i.p.wright@open.ac.uk.
[‡‡‡]Present address: Laboratoire de Géologie, Ecole Normale Supérieure de Paris, 24 rue Lhomond, 75231 Paris Cedex 5, France; e-mail: rouzaud@mailhost.geologie.ens.fr.
[§§§]E-mail: danilel@tin.it.
[###]E-mail: anne.severine@free.fr.

Westall, F., de Vries, S.T., Nijman, W., Rouchon, V., Orberger, B., Pearson, V., Watson, J., Verchovsky, A., Wright, I., Rouzaud, J.-N., Marchesini, D., and Severine, A., 2006, The 3.466 Ga "Kitty's Gap Chert," an early Archean microbial ecosystem, *in* Reimold, W.U., and Gibson, R.L., Processes on the Early Earth: Geological Society of America Special Paper 405, p. 105–131, doi: 10.1130/2006.2405(07). For permission to copy, contact editing@geosociety.org.

ABSTRACT

A multidisciplinary study of silicified volcanoclastic, near-shore deposits from the 3.446 Ga "Kitty's Gap Chert," Warrawoona Group, Pilbara, reveals that they contain a wealth of carbonaceous microbial fossil remains. The volcanoclastic sediments host predominantly colonies of coccoidal microorganisms that occur in two modal size ranges, 0.4–0.5 μm and 0.75–0.8 μm. These microbial colonies coat the surfaces of the volcanic particles and form either dense, carpetlike associations up to tens of micrometers in diameter comprising hundreds of individuals. They also form less dense concentrations that include many chainlike associations of coccoids. All colonies are associated with a polymer film (extracellular polymeric substances—EPS) that coats both the organisms and their substrate. Multispecies biofilms formed at a boundary representing a short period of nondeposition. They consisted predominantly of coccoids and EPS but also included common, small filaments tens of micrometers in length and 0.25 μm in width and rare, short rods 1 μm in length. Carbon isotopic compositions of about –26‰ to –30‰, measured on individual layers, are compatible with microbial fractionation. The biofilms include possible anoxygenic-photosynthesizing organisms (the filaments), whereas the colonies coating the volcanic clasts probably represent chemolithotrophic organisms. The interaction between the microbes, their colonies and biofilms, and their environment is intimate and complex. The environment provided the substrate and the nutrient, energy, and carbon sources, whereas the metabolic activity of the microbes contributed to the early diagenetic alteration of the volcanic particles, to the binding of the sediment, and to their silicification. The microorganisms were preserved by rapid silicification, with the silica coming partly from hydrothermal sources and partly from pore water enrichment in Si due to the devitrification of the volcanic protoliths (partially mediated by microbial activity).

Our multidisciplinary approach to the study of this sample demonstrates the importance of using complementary methods in order to understand the complex microbe/sediment interactions and to be able to relate different types of microbial colonies/biofilms to different microenvironments. The observations and conclusions from this study have important consequences for the methods that need to be used in the search for traces of past life in general and especially in the search for past life on other planets such as Mars.

Keywords: Archean, "Kitty's Gap Chert," near-shore environment, microfossil, anoxia, chemolithotroph, anoxygenic photosynthesizer.

INTRODUCTION

Given the complex interactions of microorganisms with their environment, we have approached the study of fossil microorganisms and biofilms using a variety of methods (Westall, 2005a), ranging from (1) macroscopic to microscopic interpretation of the ecologic environment, (2) optical microscope and geochemical characterization of the composition of the primary sedimentary material, as well as of the early diagenetic and metamorphic changes that took place, (3) optical and high-resolution scanning electron microscope characterization of the microfossil remains, (4) high-resolution transmission electron microscope study, Raman spectroscopy, and isotopic investigations of the organic carbon remains, and (5) nuclear microprobe and isotopic investigation of nitrogen, trapped as a chemical fossil in mica minerals (Rouchon et al., 2005). This multidisciplinary approach has allowed us to (1) make detailed interpretations of the macro- and microenvironment, (2) conclude that the microstructures observed have a high probability of biogenicity, and (3) understand the early diagenetic and metamorphic history of the rock.

The variety of methods that we have adopted is essential in providing an acceptable level of confidence for the interpretation of potentially biogenic structures in these ancient sediments, especially given the recent discussions concerning the identification of microbial fossils and the origin (biogenic or abiogenic) of the organic carbon in early Archean cherts (Brasier et al., 2002; van Zuilen et al., 2002, 2003; Westall, 2004) or in Martian meteorites (McKay et al., 1996; Cady et al., 2003). The silicified volcano-sedimentary sequences of the early Archean terrains in the Pilbara in northwestern Australia and also in Barberton in South Africa are the oldest, well-preserved sediments known to date. Although older sediments (3.65–3.85 Ga) occur in southwestern Greenland (McGregor and Mason, 1977; Fedo, 2000; Rosing, 1999), amphibolite to granulite facies metamorphism has so altered the original texture and mineralogy of these rocks that they cannot be used for microfossil study. In fact, the microfossils that had been described, e.g., by Pflug (1979, 2001), Pflug and Jaeschke-Boyer (1979), and Robbins (1987), have been reinterpreted as recent endolithic contaminants (Westall and Folk, 2003). The sediments from the Pilbara and Barberton regions are therefore the oldest rocks that can be used in the search for fossilized traces of ancient life. A number of investigations of microfossils in Pilbara rocks have been previously made (see Schopf and Walter, 1983 and references therein; Schopf and Packer, 1987; Schopf, 1993; Hofmann et al., 1999; Schopf et al., 2002; Allwood et al., 2005a, 2005b). However, many of the biogenic interpretations of individual structures as fossilized cyanobacteria, of domical stromatoloids as microbial stromatolites, and even of the negative values of carbon isotopes that are usually taken as indicators of biogenic processes, have been questioned (e.g., Buick, 1990; Lowe, 1994; Brasier et al., 2002; van Zuilen et al., 2002). In the light of these discussions and given the submicrometer- to micrometer-scale nature of the individual microbial structures preserved within the early Archean cherts from the Pilbara, we believe that an investigation integrating observations and analyses on a meter to submicrometer scale is both necessary and realistic in order to reliably study potential microfossils and their biofilms in situ in ancient sediments. In this study we therefore make special use of high-resolution scanning electron microscopy, coupled with light element analysis (EDX), to reveal the wealth of detail and information that is trapped within the cherts and that cannot be obtained by the traditional, optical microscope methods that have previously been used in the search for Archean microfossils.

Microorganisms and Biofilms

Microorganisms in nature do not occur in isolation; they form colonies, often consisting of multispecies consortia as well as biofilms and mats. The term "biofilm" usually refers to a laterally extensive development of microbial colonies encased in a polymer matrix of extracellular polymeric substances (EPS), whereas "microbial mat" refers to a generally well-developed biofilm that includes both trapped detrital particles and authigenically precipitated minerals. The terms are, however, used interchangeably (Levitt and Krumbein, 2003). Individual microbial colonies can form on any kind of substrate and do not necessarily form biofilms or mats. The relationship between the microorganisms and their substrate is complex; the substrate provides a support for the microbial colony and also serves as a source of energy and nutrients (Madigan et al., 2000). Microbes can take advantage of redox reactions occurring at the surface of reactive particles, such as minerals and volcanoclastic material, that provide the electrons necessary for driving their energetic metabolic processes and, consequently, biosynthesis. Such microorganisms are known as chemolithotrophs; they obtain both their energy and their carbon from inorganic sources, whereby the carbon comes from CO_2 (autotrophy), organic compounds that are readily biodegradable, or organic molecules (mixotrophy). Alteration of the particles (often through the corrosive action of excreted microbial metabolites) liberates the essential elements (macronutrients) that the microorganisms require, such as N, P, S, Mg, K, Ca, Na, and Fe, as well as metallic micronutrients (including Fe, Cr, Co, Cu, Mn, Mo, Ni, Si, W, V, and Zn). Thus, diagenetic alteration of mineral or volcanic surfaces exposed to seawater, including the pore waters of volcanic sediments, can be regarded as a combination of microbial and physicochemical processes acting on the particles.

Biofilms and microbial mats are architecturally complex structures (Charaklis and Wilderer, 1989) that can also be compositionally heterogeneous, consisting of a variety of cohabiting microorganisms that form a consortium. The biofilm structure is advantageous to the organisms because (1) the water-rich polymer film provides them with protection, especially in severe environmental situations, (2) within the biofilm, microorganisms can create their own microenvironment with controlled redox and pH conditions and nutrient concentrations, and (3) the different organisms within a biofilm are complementary and not necessarily in competition with each other; each organism is specialized such that what one organism excretes as a metabolite, another can use as a nutrient. Moreover, the different types of organisms obtain their energy and carbon sources in different ways that are not necessarily in competition with each other.

Biofilms interact with their substrate in a variety of ways such as sediment stabilization and/or lithification. The exopolymeric substances (EPS) are cohesive polymers that bind or cement sediment particles together (Krumbein et al., 1994; Westall and Rincé, 1994). The physical expression of the films can be delicate or they can be very thick and robust, depending on the time available for development or the environment of development. These films can form very rapidly in ephemeral environments, such as tidal mud flats (Stal, 1994). Colonies of microbes, biofilms, and mats are heavily implicated in the lithification of sediments (Krumbein et al., 1994). The EPS as well as the envelopes around the individual microorganisms are rich in reactive functional groups, such as hydroxyls, phosphoryls, and carboxyls, that readily complex metal ions (Monty et al., 1991; Westall et al., 1995). Depending on the availability of metal ions and suitable physical conditions, ions such as Si, Fe, or Ca (among others) form a complex with the functional groups of the organic substrate, then polymerize, thus acting as a chemical fixative and fossilizing agent. Therefore, although microbes and their biofilms may be ephemeral on geological time scales, they have the potential of being rapidly preserved and, thus, the potential to pass into the geological record.

Apart from the differences in temporal scales between microbial and geological phenomena, microbial spatial scales are also orders of magnitude smaller than geological spatial scales. Individual microbes are of the order of some hundreds of nanometers to several micrometers in size, depending on the species (and some filaments are >100 µm in length). [*N.B.* Most microbial organisms in nature are smaller in size than those in a laboratory culture.] Microbial colonies may consist of a few to very many ($\geq 10^3$) individuals and cover areas ranging from a few to many hundreds of micrometers in diameter. Microbial biofilms or mats, on the other hand, are generally much larger, of the order of hundreds of μm^2 to m^2 and even up to km^2 in dimension.

When studying the fossilized remains of microbes, colonies of microbes, and their biofilms and mats in rocks, account needs to be taken of the effects of the differences in spatial and time scales on quantification of the microbes in a rock (and therefore on elemental concentrations such as organic C, N, and S, as well as the isotopic ratios of these elements). Although microbes and their biofilms may appear to be relatively common in sediments, their volumetric importance may be insignificant with respect to that of the bulk sediment. Biomass development depends upon variables such as environmental conditions including nutrient availability, pH, and salt concentrations, as well as on metabolic pathway. In the shallow water environment studied, the organisms and their biofilms are surface specific: they may be common in one microenvironment (on a scale of 10^{1-3} µm) but not in the adjacent environment. Thus, although a rock may be locally enriched in microbial remains (and C, N, and S) on a scale of 10^{1-3} µm, the bulk sample may contain only small quantities of these components.

Geological Setting of the "Kitty's Gap Chert"

The object of this study is a chert formation, here called the "Kitty's Gap Chert," that is part of the Panorama Formation in the Warrawoona Group and outcrops 50 km northeast of the town of Marble Bar in the Coppin Gap Greenstone Belt, Pilbara, northwestern Australia (Fig. 1) (Hickman, 1983, 1990; DiMarco and Lowe, 1989; Eriksson et al., 1994; Van Kranendonk et al., 2002). Dating of a felsic unit in the formation shows it to be 3.445 ± 4 Ga (de Vries, 2004) and therefore the lateral equivalent of the Panorama Formation in the North Pole Dome area (cf. also Williams, 1990). It consists of a series of packets of silicified volanoclastic sediments (Fig. 2) that were deposited in a near-shore, channel-and-flat environment. Each packet of sediments is separated by material of more mafic nature (de Vries, 2004). The cherts are underlain by rhyolitic and dacitic-rhyolitic volcanics and overlain by basalt. The whole volcano-sedimentary complex of the Panorama Formation was deposited in a large basinal structure characterized by growth faults (Nijman et al., 1999; de Vries, 2004). Silica-rich hydrothermal fluids used this fracture system as a conduit to the surface (Fig. 1). At the sample location, a hydrothermal silica vein penetrates into the lowermost layer of the silicified volcanoclastic sediments, fracturing them and dissipating laterally into

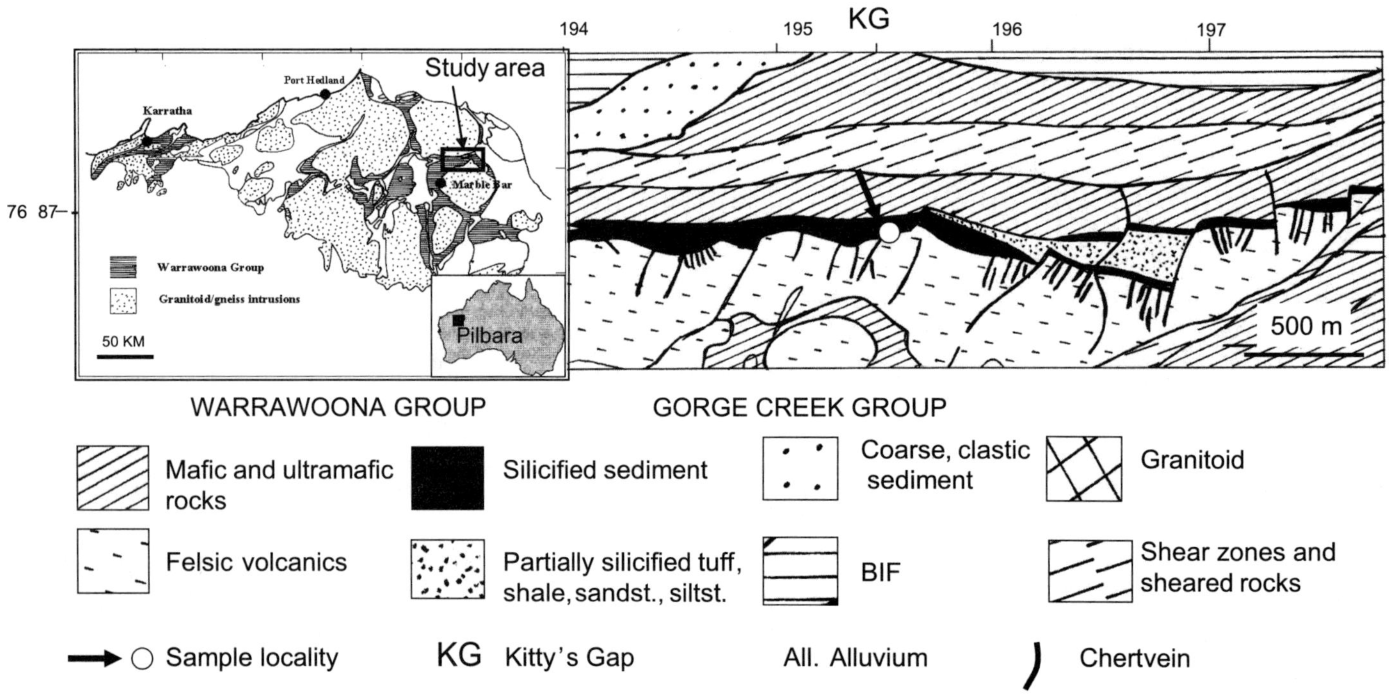

Figure 1. Pilbara study area location and detailed geological map of Kitty's Gap volcano-sedimentary sequence (after de Vries, 2004). Coordinates refer to Australian map grid zone 51.

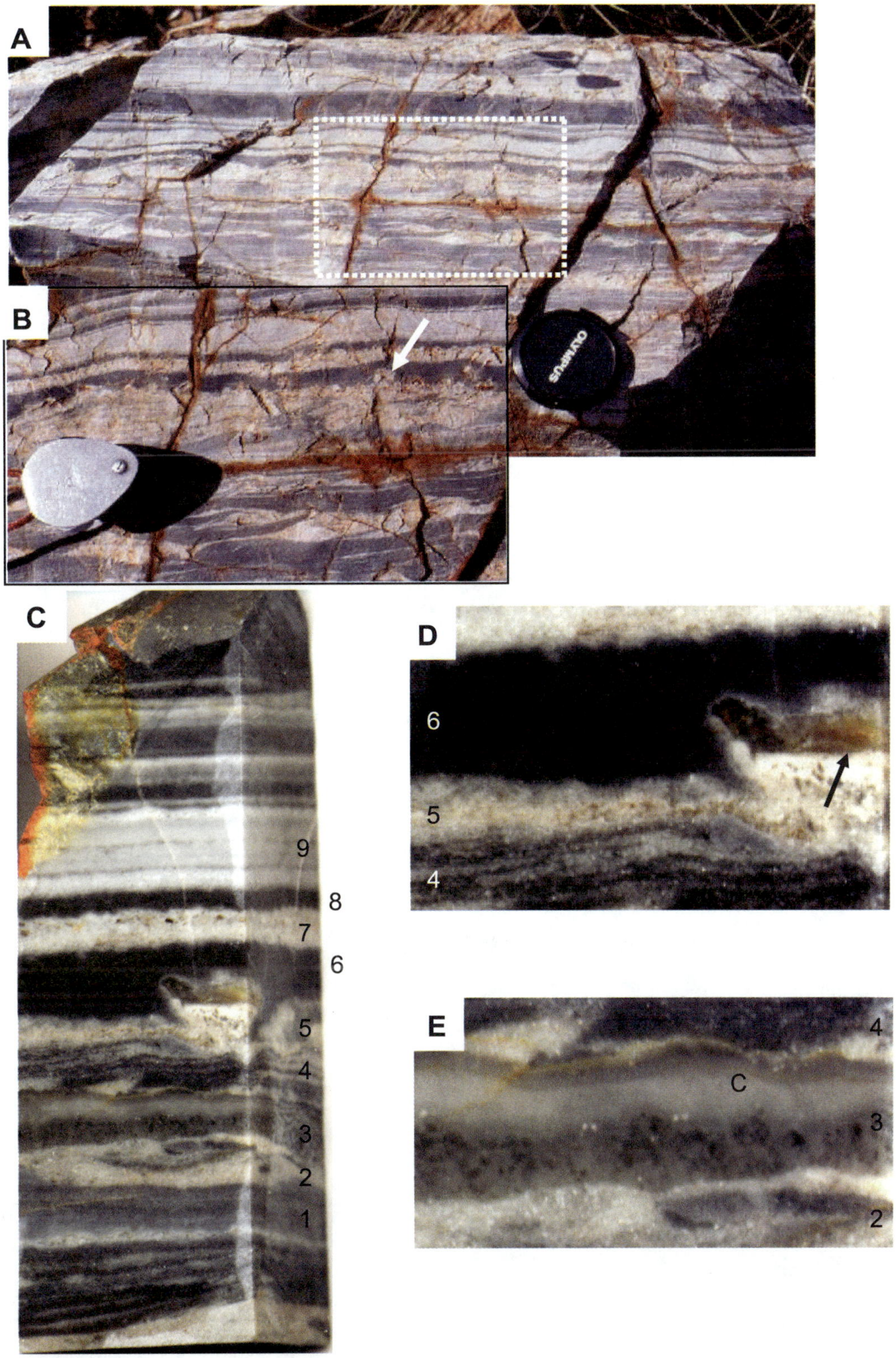

Figure 2. (A, B) Field view of silicified, near-shore channel-and-flat, volcanoclastic sediments that are the subject of this study. Scale—lens cap 5.4 cm. (B). Detail of flaser-linsen and cross-bedding structures, as well as the layer (Layer 5) embedded pumice fragments (arrow). Scale—hand lens 3.5 cm long. (C, D, E) Cut surface of rock showing different layers, 1–9, described in the text. (D) Detail of Layers 5 and 6. Note clear horizontal distinction between lower, more silicified part of pumice fragment and upper, translucent, hydromuscovite-rich part (arrow). This boundary is in lateral continuation with the top of Layer 5. (E) Detail of Layer 3 highlighting a lower, speckled stratum consisting of silicified, Ti oxide-rich volcanic particles, and an upper, translucent stratum of chalcedony (C) that precipitated probably in an open crack during early diagenesis.

them. The influence of the disruptive pressure of the hydrothermal fluids decreases laterally along the sediment layers with distance from the chert vein. The silicified sediments now outcrop as an east-west oriented ridge, in which the volcano-sedimentary layers are dipping vertically or are slightly overturned. The rocks underwent burial metamorphism up to the upper stages of the prehnite-pumpellyite facies (Kisch and Nijman, 2004).

MATERIALS AND METHODS

Sample

The sample studied was obtained from the lowermost layer of the "Kitty's Gap Chert," immediately overlying the felsic volcanics (sample location coordinates: 0195366E; 7686751S (the grid numbers refer to Australian map grid zone 51) [120°04.53′E; 20°53.62′S] (Fig. 1). This sample was chosen because the field exposure of the silicified volcanoclastics exhibited an intriguing, laterally continuous horizon (over a scale of several meters) characterized by small lumps of the order of 0.5 cm in size on the top of ripple-marked sediments (Fig. 2). Subsequent analysis showed that the "lumps" were, in fact, pieces of silicified pumice and that a delicate microbial mat had formed at the top of the surface of ripple-marked sediments and pumice fragments. Two samples of this horizon were taken, one ~2 m east of the chert vein and one further away from the vein at a distance of 1 m from the previous sample (i.e., 3 m from the vein).

Sedimentologic and Mineralogic Characterization

Careful field characterization at the outcrop level of the formation was undertaken in order to understand the geological and environmental context of the deposit (de Vries, 2004). Further information concerning the fine-scale sedimentological environment and composition of the particles came from optical and scanning electron microscope study of the cut rock surface and thin sections. The microscopes used included an Olympus BX51 optical microscope and a Jeol FEG [field emission scanning electron microscope] SEM 6400 (NASA–Johnson Space Center, Houston) and a Hitachi FEG SEM S4200 (University of Orléans) as scanning electron microscopes.

Geochemical Characterization

Two Cameca SX50 electron microprobes (at the BRGM [Bureau de la Recherché Géologique et Minéralogique], Orléans and Centre Camparis, Université Paris VI) were used for major, minor, and trace element analyses in minerals. Working conditions at the BRGM–Orléans were: acceleration voltage 15 kV, beam current 12nA, and beam width 1 µm^2. The standards used included: K—orthoclase ($KAlSi_3O_8$), Ti—MnTiO3 (synthetic), Fe—hematite (Fe_2O_3), Na—albite ($NaAlSi_3O_8$), Ca—andradite ($Ca_3Fe_2Si_3O_{12}$), Cl—vanadinite ($Pb_5(VO_4)_3Cl$), Cu—Cu metal, Si—albite ($NaAlSi_3O_8$), S—barytes ($BaSO_4$), U—UO_2, Cr—Cr_2O_3 (synthetic), Al—corundum (Al_2O_3), V—vanadinite ($Pb_5(VO_4)_3Cl$), Mg—olivine $(Mg,Fe)_2SiO_4$. Working conditions for the Cameca SX50 at the Centre Camparis, Université Paris VI, are described in Orberger et al. (this volume). The nuclear microprobe (nuclear reaction analyses (NRA) of the Laboratoire Pierre Süe (CEA–CNRS [Commisariat à l'Enérgie Atomique]) was used to locate C and N in minerals (Rouchon et al., 2005). Further details related to the in-depth geochemical investigations (electron and nuclear microprobes, ICP-AEX [inductively-coupled plasma atomic emission spectrometry] and ICP-MS [inductively-coupled plasma mass spectrometry]) are given in Orberger et al. (this volume) and Rouchon et al. (2005).

Organic Carbon/Nitrogen Investigations

High-Resolution Transmission Electron Microscopy (TEM)

For the high-resolution transmission electron microscope study of the carbonaceous component of the sample, a small portion of the pumice layer horizon with the biofilm and a small portion of the underlying layer of ripple-marked volcanoclastic sediments were digested in HF and HCl to eliminate the mineral fraction. A small drop of the residue in suspension in alcohol was deposited on a prepared lacey TEM grid and observed with a Philips CM20 instrument fitted with an energy-dispersive spectrometer (EDX system). The (002) lattice fringe mode of observation allows the direct imaging of the polyaromatic layers forming the skeleton of carbon materials (Boulmier et al., 1982; Oberlin, 1989; Rouzaud and Clinard, 2002).

Raman Spectroscopy

Raman spectroscopy (DILOR XY800) was performed on the same layers in the cut surface of the sample to obtain information on the maturity of the carbonaceous material. The working conditions were as follows: excitation provided by a green beam at 514.5 nm by an Ar Coherent Radiation laser; multichannel detector DDd WRIGHT; objective: MDPLAN Olympus ×100; acquisition rates of 50–300 s with a final power value of 0.4–1mW.

Bulk Carbon Analysis and Carbon Isotope Analysis

Carbon concentrations and isotopic composition were measured on bulk samples of individual layers at the Open University by stepped combustion (from 200 to 1400 °C with 100 increments) using a static mode machine MS-86 (Wright and Pillinger, 1989; Verchovsky et al., 2002). The system blank was ~10 ng and the precision for $\delta^{13}C$ is ± 1‰. The samples were crushed to ~0.1 mm grain size before analysis, and aliquots with weights from 2 to 22 mg were used. In one case (Layer 7) we also analyzed an uncrushed sample. In the experiment, the carbon yield was also measured. The system blank was ~5 ng of C.

Microfossil Investigations

Sample Preparation

Preliminary identification of a potential microbial biofilm at the pumice layer surface was made in thin section. For high-resolution scanning electron microscope study, a number of preparations of the sample taken furthest away from the intruding chert vein were made to ensure reproducibility and to ensure that the microstructures studied were in situ and formed contemporaneously with sediment deposition (i.e., endogenous). These preparations included a raw, freshly broken surface, a polished thin-section surface, and a simple cut surface. The preparations were cleaned in milliQ water in an ultrasonic bath, dried, and then dipped in alcohol and flamed to eliminate surface contamination. Some of the specimens were delicately etched in the fumes of HF for 1 h. The etched preparations were then abundantly rinsed in milliQ water before being dried and coated with either Au or Au/Pt for SEM observation. The SEMs used are those listed above (a Jeol FEG SEM 6400 and a Hitachi FEG SEM S4200). All the instruments were equipped with a light element EDX detector that was used to make qualitative spot analyses of the elemental composition, as well as maps of various elements, especially C, associated with specific microstructures. Observations were made in secondary as well as backscattered electron mode. Most of the microfossil observations were made on the etched cut sample surface, and others were made on the thin-section surface.

Microfossil Identification

The identification of the microfossils was based on two main aspects: (1) compatibility of the environment of deposition with the existence of living organisms, and (2) demonstration of characteristics indicative of biogenicity. We recognize that there is no one observation or measurement that, on its own, is proof of biogenicity. Rather, many lines of evidence need to be assembled to arrive at a probability of biogenicity. The probability of biogenicity can be expressed in many ways, for instance as a number (e.g., from one to five with increasing confidence of biogenicity) or in terms of adjectives (e.g., strong, moderate, weak).

The criteria indicative of biogenicity used in this study are related to the structural and biogeochemical aspects of an organism. These criteria will vary depending on the species of microorganism. For example, the morphological criteria used for investigating most prokaryote microorganisms are different in many respects from those used for studying one well-known group of prokaryotes, the cyanobacteria, e.g., size, shape, cell structure, and differentiation. Moreover, the organic chemical signals (different organic macromolecules associated with the sheath and membranes) are different. The criteria we use also include features related to the interrelationship between the microorganisms and their immediate environment (see also Westall, 1999; Westall et al., 2000).

Structural characteristics. The structural characteristics concern the organisms as physical entities as individuals as well as their colonies, biofilms, or mats. They include the following: (1) morphological phenomena such as size, shape, evidence for life and death; for example, as in cell division (i.e., evidence for a living process), cell collapse (i.e., evidence for cell death or lysis; this is the process by which an enzyme, released automatically when a cell dies and breaks down the outer cell envelope so that the internal cytoplasm can escape, thus leading to collapse of the cell. This process leads to variations in the surface texture of the cell, such as wrinkling as opposed to the turgid structures and textures typical of living cells), and flexibility; and (2) colony and biofilm/mat characteristics, such as association with a number of other organisms of the same species, association with different species in the same colony (consortium), association with microbially produced polymer (extracellular polymeric substances, EPS), and formation of physical entities, i.e. as individual cells as well as colonies, biofilms, or mats. We also used statistical analysis to evaluate the size-shape range of individual microstructures within individual colonies and for a comparison of different colonies forming in different layers (microenvironments) of the sediment. In this exercise, ~2000 individual structures were measured (diameter and length/width ratios with standard deviation [SD] values) from 20 different colonies in Layers 3, 4, 5, and 6 to obtain the modal diameters of the cells.

Biogeochemical characteristics. The biogeochemical characteristics concern the metabolism of the organisms, such as (1) an organic carbon composition (including the degradation products of specific molecules, in the case of fossil microorganisms; *N.B.* in certain conditions such as hot springs, the microbial structure may be fossilized but all trace of organic carbon may have been removed by oxidation, e.g., Cady and Farmer, 1996); (2) association of other important macro- or micronutrients, such as N, P, S, Mg, K, Ca, Na, Fe, Cr, Co, Cu, Mn, Mo, Ni, Si, W, V, and Zn; (3) the metabolic fractionation of the isotopes of elements such as C, S, N, and Fe; and (4) complexation of other elements, such as heavy metals and rare earth elements (REEs) to organic matter.

Other characteristics related to the metabolic activity of microorganisms can be used as indications of biogenicity, such as biocorrosion. A recent study found corrosion pits in the glassy rinds of early Archean pillow basalts from Barberton, South Africa, that were attributed to microbial activity (Furnes et al., 2005).

Microbe-environment interactions. Microorganisms are surface specific and generally need a substrate for reproduction (except when they are in the planktonic form). An understanding of the interactions between individual microorganisms, their colonies, biofilms, and mats and their microenvironment is therefore very important. This means that study of the fossil microorganisms should go hand-in-hand with study of their immediate microenvironment (hence the multidisciplinary approach that we have used). The information thus obtained provides important clues to potential sources of energy, carbon, and nutrients. Study of direct interactions of the microbes with their substrates provides information on substrate corrosion, biofilm/mat formation, and sediment stabilization including fossilization. As noted above, the potential rapidity of the fossilization process (in hours to days, depending on the species availability of fossilizing cations and physical environmental conditions; Westall et al., 1995; Westall, 1997; Toporski et al., 2002; Orange, 2004) means that even extremely ephemeral microbial developments have the potential of being preserved. Thus, fossilized microbial communities also contribute to sediment stabilization and, in the case of the early Archean sediments, to their early silicification.

RESULTS

Sedimentology and Mineralogy

The studied sample (Fig. 2) is a chert, composed presently of ~80–99 vol% of cryptocrystalline quartz. The rock texture is layered and shows characteristic features, such as microripples and flaser-linsen bedding, which are typical for sediments deposited in an aqueous, near-shore environment. On the basis of microscopic investigations, these sediments can be classified as volcanoclastic in origin.

The sample consists of a number of micrometer- to millimeter-thick layers, the colors of which alternate between light gray and dark gray. The light gray layers consist of coarser material, rich in large pseudomorphosed volcanic mineral particles (Ti-bearing mica, K-feldspar, amphiboles, glass shards, and mineral debris), glassy precursors, and a higher abundance of well-formed, detrital zircon, secondary monazite, and fluorencite. The volcanic clasts have sizes ranging from 30 to 500 μm. The darker gray, fine-grained material contains almost no large clasts and is formed of intergrown microcrystalline quartz and hydromuscovite. The individual layers of this sample have been numbered for ease of description (Figs. 2C–2E). Each of the layers is traceable laterally on a scale of meters. Further mineralogical details are given in Rouchon et al. (2004a, 2004b, 2005) and Orberger et al. (this volume).

Layer 1 (light colored) consists of flat-lying to slightly channel-bedded (on a centimeter scale) laminae, containing fine-grained (30–100 μm) angular volcanic clasts.

Layer 2 is light colored in hand specimen but in thin section consists of a dense concentration of brownish, very fine-grained clasts outlined by microcrystallites (1–2 μm) of Ti-oxide. Its brownish color is given by the large amount of hydromuscovite pseudomorphs replacing primary volcanic clasts.

Layer 3 is light colored and translucent and consists of two sublayers. The lower part has a speckled appearance in hand specimen due to concentrations of Ti-oxide crystallites (Figs. 2E and 3A–3C). It consists of a 2-mm-thick accumulation of ghost structures having subhedral to euhedral shapes. These structures, now consisting of microcrystalline quartz, range in size from 200 to 500 μm. The microprobe data indicate that the ghost particles represent hydromuscovite replacing K-feldspars and volcanic glass shards. Micrometer-sized Ti-oxide crystals outline the protominerals or protoglass shards, indicating that Ti was liberated during silification of, probably, proto-biotite. Scarce, well-formed crystals of monazite are also present. In texture, this sublayer is well sorted and matrix supported (Figs. 3A and 3B). It is overlain abruptly by a fine, crinkly laminated, millimeter-thin layer of chalcedonic quartz (Fig. 2E). The laminations parallel to the top and bottom of this layer indicate that the quartz precipitated in an open crack.

Layer 4 consists of a 1–2 cm layer of poorly sorted (10–300 μm sized), densely packed, and almost grain-supported, angular volcanic shards (Fig. 3D) exhibiting rippled, flaser-linsen bedding (Figs. 2A and 2C). Ti-oxides invariably outline pseudomorphs of hydromuscovite after volcanic clasts. The finer layers comprising the linsen bedding consist of very fine, clay-sized material.

The bedding laminae on top of Layer 4 appear to be truncated by an overlying, 1-mm-thick, light-colored (in hand specimen) layer (Layer 5) (Figs. 2C–2D and 3E–3F). Although it has the same sedimentological characteristics as Layer 4 in terms of grain size, composition, and sorting, this layer has a cloudy, brownish appearance in optical microscopy (Fig. 3E). The ghost particles are still coated by Ti-oxides, but their outlines are diffuse and the whole layer is richer in Ti-oxide crystallites than the underlying layer. There is a sharp contact between the top of Layer 5 and the base of Layer 6 (Figs. 2D and 3E–3G).

Partially embedded in Layer 5 are large (up to 0.5 cm), irregularly shaped fragments of angular to subrounded, scoriaceous particles (Figs. 2B–2D). The electron microprobe analyses suggest a composition similar to that of the silicified, hydromuscovite-replaced volcanic clasts in Layer 4. Many of these large volcanic fragments exhibit distinct optical and compositional differences between the lower portion embedded in Layer 5 and the upper portion protruding into the overlying Layer 6 (Fig. 2D) that is laterally contiguous with the sharp boundary between Layers 5 and 6. The lower halves of the fragments are heterogeneous in color (brownish gray in thin section but white in hand specimen) and texture. The upper halves often have a translucent, yellow-brown color (compared to the matte, white color of the lower part). In some cases, there is a significant concentration of Ti-oxide crystals at the boundary between the upper and lower parts of these large volcanic fragments. Ti-oxide crystallites also outline the edges of primary structures that we interpret to be quartz-filled vesicles, according to their larger quartz grain sizes and SiO_2 enrichment compared to the groundmass of the fragments. Individual pseudomorphs after volcanic minerals, together with the presence of vesicles, indicate that these fragments may have been pumice. In hand specimen, a millimeter-thick, whitish (in hand specimen) surface continuously coats the top of Layer 5 and the pumice fragments (Fig. 2D). Stylolites are sometimes formed below and/or around the pumice fragments and along bedding-parallel planes within Layer 5, especially closer to the intruding chert vein.

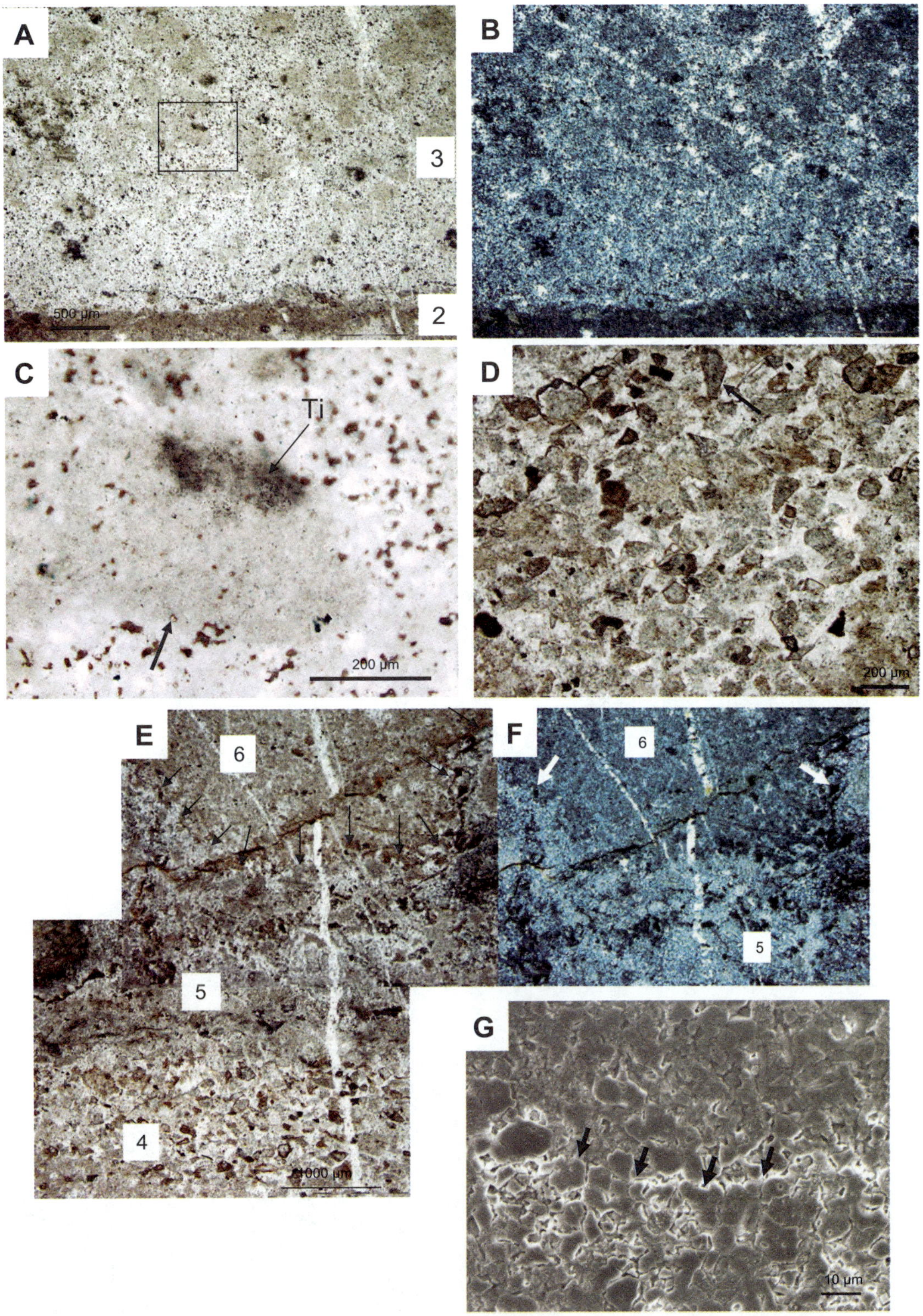

Figure 3. Thin section photomicrographs. (A, B) Normal light and polarized light images of highly silicified, ghost volcanic particles in Layer 3. Layers shown marked by numbers. Box marks location of (C). (C) Detail of one of the ghost particles (lower edge marked by an arrow) with a concentration of minute Ti oxide crystallites (Ti) at its upper edge. (D). Detail of the angular volcanic detrital particles whose sharp outlines (e.g., arrow) are caused by presence of Ti oxide crystallites on their surfaces (cf. Orberger et al., this volume). (E, F) Top of Layer 4, Layer 5, and base of Layer 6. Layer 5 is compositionally and texturally the same as Layer 4, but in thin section appears more diffuse. The top of Layer 5 is shown by black arrows (E), whereas the edges of two pumice fragments are indicated by white arrows (F). (G) SEM photomicrograph of a delicately HF-etched, cut surface showing clear boundary between Layers 5 and 6 (arrows).

Layer 6 is dark in hand specimen and ranges in thickness between 1 and 2 cm along strike (Fig. 2D). It is very fine grained and homogeneous, containing rare, small (20–100 µm) ghost particles (Figs. 2D and 3E–3F). It has a similar composition to that of the underlying volcanic layers, suggesting that it probably consisted of a volcanic dust protolith replaced by very fine-grained quartz (hence the black color in hand specimen) and hydromuscovite. The lower part of this layer is characterized by a fine, brownish horizon (in optical microscopy) that conformably follows the top of Layer 5, including the pumice fragments (Fig. 12A). Blocky, carbonaceous particles with a subrounded shape and sizes ranging from 6 to 20 µm occur in Layers 5 and 6 (Fig. 4). In terms of texture and structure, these particles can be divided into two groups. One type has a rather homogeneous, granular texture and a composition that includes traces of S and Cl, as well as SiO_2. The other type has a filamentous texture and has filaments similar to those described below (in the Microfossils section) attached to it. Sulfur and Cl do not appear in the EDX spectra of this type of carbonaceous clast (Fig. 4C).

Layer 7 is light colored and consists of flat-lying laminae of relatively coarser (50–300 µm), volcanic particles (Fig. 2C). The grains in the lowermost horizon of Layer 7 have been pressed into the underlying sediments in a texture suggesting

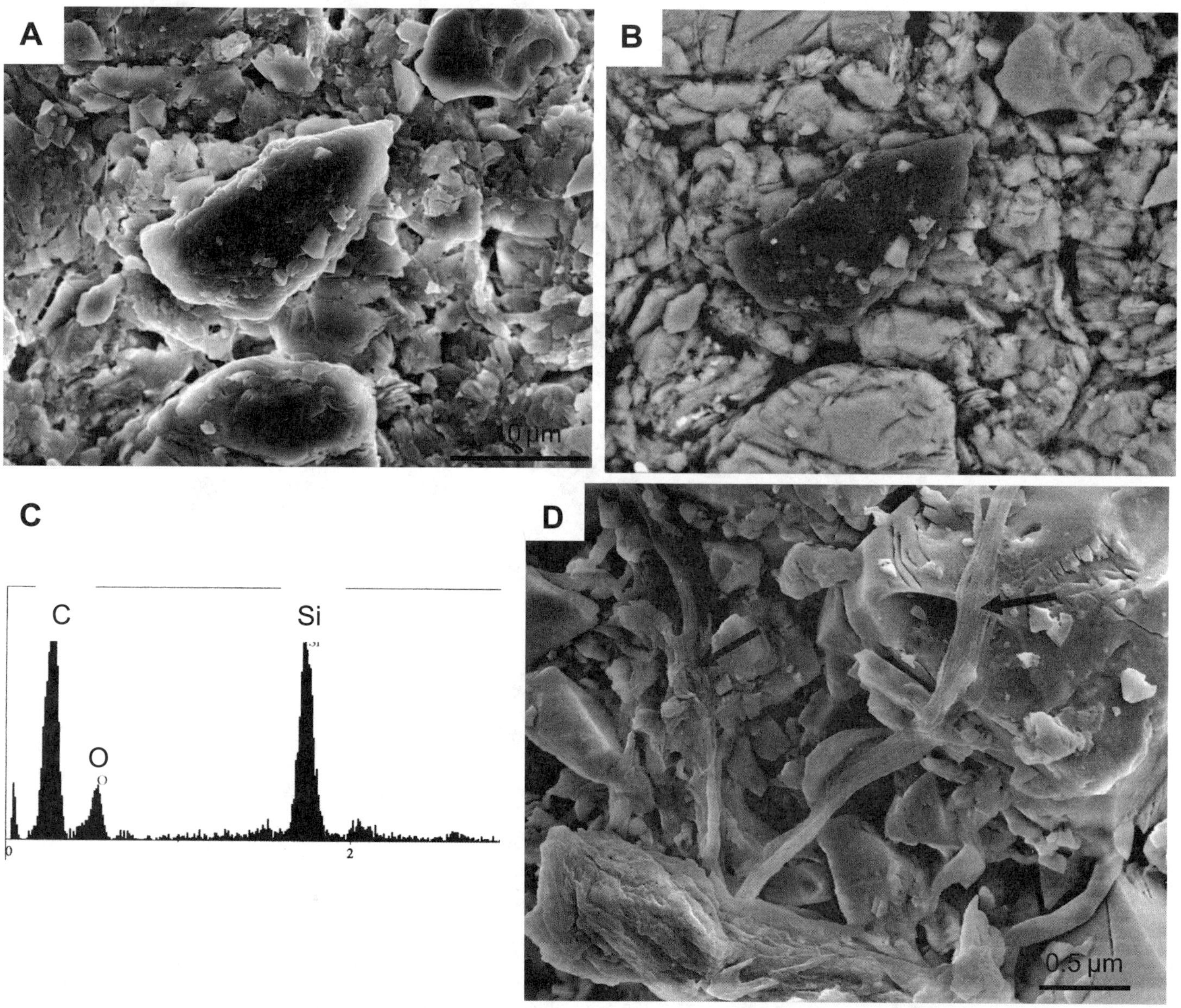

Figure 4. (A, B) SEM photomicrographs in secondary (A) and backscattered (B) electron mode of a subangular, detrital, carbonaceous particle in Layer 4. (C) EDX (electron diffraction X-ray spectrometry) spectrum of the large carbonaceous particle in the center of (A). (D) Detrital, fibrous, carbonaceous particle at the base of Layer 6 with degraded, striated filaments (arrows) attached to it that are similar to those in Figure 7.

millimeter-scale soft sediment deformation. The lowermost part of this layer is grain-supported, but the texture becomes matrix-supported in the midlayer. Clast sizes increase toward the top, which is marked by 1 mm-sized, angular fragments of euhedral protominerals having heterogeneous compositions and partially resembling the pumice fragments at the top of Layer 5. Layer 7 is overlain by another dark layer of microcrystalline quartz, replacing fine volcanic dust (a few µm) that constitutes Layer 8 (resembling Layer 6). It, in turn, is overlain by light-colored, flat-laminated, relatively coarser (30–300 µm) material (Layer 9).

The detailed micropaleontological studies were made in Layers 3–6.

Geochemistry

Orberger et al. (this volume) note that although the geochemical compositions of the light and dark-colored layers in this chert are similar, there are some differences as a result of the higher concentration of hydromuscovite (replacing protominerals, such as Ti-bearing mica, K-feldspar, amphiboles, glass shards, and higher contents of accessory minerals, monazite, and zircons) in the light-colored layers. Vanadium (associated with the Ti-oxides), Sr, and Rb (associated with hydromuscovite) are enriched in the light layers, as are Zr, Y, W, Nb, Ta, Ba, and HREE (heavy rare earth elements). LREE (light rare earth element) compositions of both dark and light layers are the same.

Bulk carbon contents for individual layers in this rock are generally low, 0.01%–0.05%, but nuclear microprobe spot measurements made on different volcanoclastic protominerals show wide variations from 0.007 to 0.859 atomic % in all the layers, irrespective of their grain size (Rouchon et al., 2004a, 2004b, 2005). Nitrogen contents measured on the same particles were also variable (0.013–0.125 at %), as were C/N ratios (0.08–20.95). Rouchon et al. (2004a, 2004b, 2005) found that there was no correlation in distribution between the N and the C, apart from their common association with hydromuscovite pseudomorphs. It is likely that the N occurs as NH_4^+ in the K^+ lattice sites of the hydromuscovite, and the C occurs dispersed as an organic phase or possibly as HCO_3^- in the (OH^-) site of the hydromuscovite structure. The SEM observations confirm the presence of an organic carbon component associated with the hydromuscovite (see below).

Carbon and Nitrogen Isotope Analyses by Stepped Combustion

The stepped combustion studies show that, for all the samples analyzed, there is a low temperature peak of carbon release at temperatures between 200 °C and 600 °C, with a maximum at 300–400 °C (Fig. 5). In the crushed samples this peak represents the main carbon release. At higher temperatures, the amount of C released is much smaller, and there is a gradual increase of carbon yield with increasing temperature. The uncrushed sample shows, however, a different release pattern (Fig. 5F): most of the carbon is released at high temperature with a peak at 1100 °C. Carbon content shows a positive correlation with sample weight (Fig. 5G). Note that although the measured amounts are positively correlated with the sample weight, the concentrations, in contrast, are negatively correlated. This can be explained mathematically, as follows. Amounts of the intrinsic carbon (C_{int}) are proportional to the sample weight (P). Amounts of the contamination carbon (C_{cont}) are proportional to the sample surface area or $P^{2/3}$. The total concentration (C_{int}+C_{org})/P is proportional to ($P+P^{2/3}$)/P or $1+P^{-1/2}$, as illustrated in Figure 5G.

In general, the isotopic compositions of the carbon released at low temperatures and at high temperatures are similar and range from –22‰ to –30‰ (Figs. 5A–5F). The bulk compositions range from –25.9‰ to –27.8‰ (Table 1).

Raman and High-Resolution TEM Investigations of the Kerogen

The TEM micrographs of Figure 6A show that the carbonaceous matter consists of small (nanometer-sized), more or less parallel-oriented, polyaromatic layers that are typical for mature kerogen (cf. Boulmier et al., 1982; Oberlin, 1989). The Raman spectrum (Fig. 6B) consists of two peaks at 1346 nm (D1 defect band) and 1586 nm ($E_{2\,g}$ band); the ratio D1/G indicates moderately mature kerogen (Beyssac et al., 2002a, 2002b),found in Layers 3–6.

Microfossils

In Layers 3, 4, 5, and 6 we have identified a number of microstructures that we interpret as representing fossil microorganisms and colonies or biofilms thereof. They include: (1) filaments with a modal diameter of 0.3 µm and lengths of up to more than 40 µm (they are never observed in their entirety); (2) coccoidal structures having two modal sizes of 0.45–0.5 µm and 0.75–0.8 µm; (3) rod-shaped structures ~0.8 µm in length and 0.4 µm in width; and (4) a film-like substance that embeds the other microstructures, as well as coating and sometimes completely embedding underlying sediment particles. We interpret these structures as fossilized filamentous, coccoidal, and rod-shaped microorganisms, respectively, and the film as extracellular polymeric substances (EPS) (see Discussion, below). Associations of these structures are interpreted as colonies or biofilms. These observations are summarized for each layer in Table 2.

Filaments

The best-preserved filaments retain a cylindrical form throughout their length and have diameters ranging from 0.3 to 0.45 µm (Fig. 7). The filaments are striated lengthwise (Figs. 7A and 7C), a feature that may be related to a fundamental characteristic of the outer cell wall of a particular species (a similar

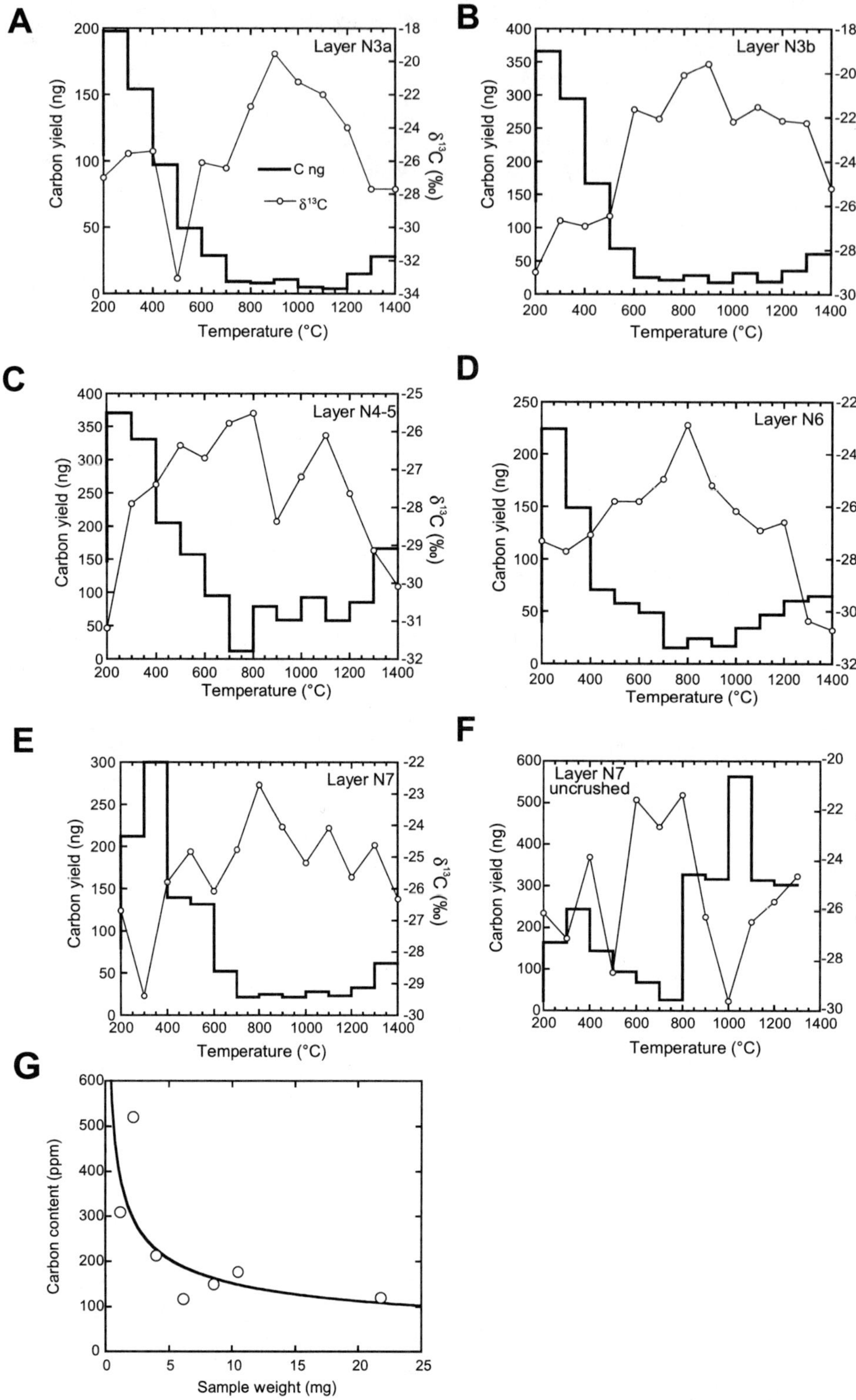

Figure 5. (A–F) Carbon content and isotopic compositions from the stepped combustion analyses of Layers 3–7. (A–E) are the results from crushed samples: (A) lower part of Layer 3; (B) upper part of Layer 3; (C) Layers 4+5; (D) Layer 6; (E) Layer 7. (F) Results for the uncrushed sample of Layer 7. (G) Carbon content as a function of sample weight. [*N.B.* although the measured amounts are positively correlated with the sample weight, the concentrations, in contrast, are negatively correlated. This can be explained mathematically. Amounts of the intrinsic carbon (Cint) are proportional to the sample weight (P). Amounts of the intrinsic carbon (C_{int}) are proportional to the sample weight (P). Amounts of the contamination carbon (C_{cont}) are proportional to the sample surface area or $P^{2/3}$. The total concentration $(C_{int}+C_{org})/P$ is proportional to $(P+P^{2/3})/P$ or $1+P^{-1/2}$].

TABLE 1. BULK DATA FOR THE CARBON CONTENT AND CARBON ISOTOPE ANALYSES OF LAYERS 3 TO 7

Layer N	Weight (mg)	Concentration (ppm)	$\delta^{13}C$ (%)
3a	6.11	117	–26.8
3b	8.50	150	–25.9
3b	1.17	309	–
4–5	10.50	177	–27.8
6	4.00	213	–27.3
7	21.80	120	–26.1
7	2.18	520	–26.2

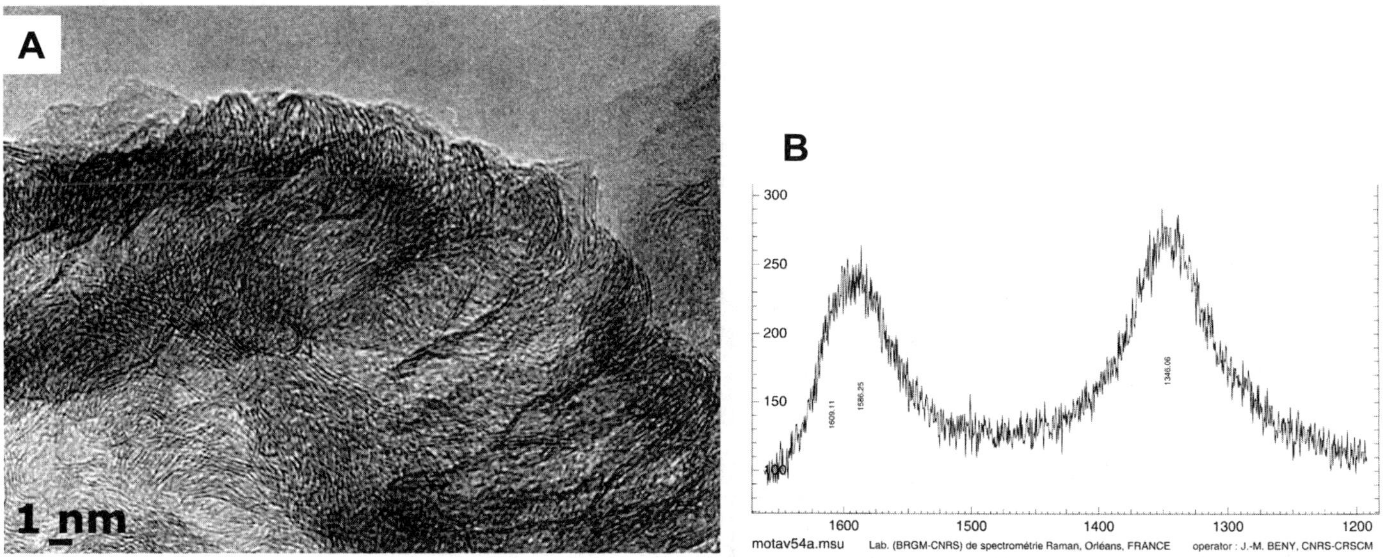

Figure 6. (A, B) High-resolution TEM photomicrograph of carbonaceous material liberated by acid maceration from Layers 5 and 6. The kerogen is characterized by small packets of short, more or less parallel-oriented polyaromatic layers, indicating some crystallographic ordering, although still far from the graphite stage. Such a structure is similar to that found in mature kerogens from biological precursors found in rock-parent of oil (Boulmier et al., 1982). (B) The ordering of the kerogen is confirmed by Raman spectroscopy that shows both the graphite peak at 1586 as well as the amorphous peak of carbon at 1346.

TABLE 2. SUMMARY OF THE MICROFOSSIL CHARACTERISTICS FOR EACH LAYER OF THE KITTY'S GAP CHERT STUDIED

Sediment layer	Filaments diam. µm range mode	Comment	Coccoids diam. µm Range	Coccoids diam. µm Mode	Coccoids SD	Comment	Rods	Colonies/biofilm	EPS
6 (dark) midlayer		Rare (1)	(1) 0.35–0.65 (2) 0.7–0.8	0.5 0.77	0.1 0.05	Relatively well preserved	None	"Carpet" colonies >100s individuals 20–30 µm	Yes
6 (dark) basal layer	0.26–0.7 0.35	Common striated surfaces, poor preservation	(1) 0.35–0.65 (2) 0.7–0.8	0.4–0.5 0.75	0.06–0.07 0.06–0.07	Poorly preserved dispersed groups Chains <10 cells dislocated cells	Rare (2) 0.8 × 0.4 µm	Biofilm of filaments, coccoids, rare rods, EPS "glued" particles	Copious smooth
5 (light)	0.2–0.5 0.35	Relatively common striated surfaces variable preservation	(1) 0.35–0.65 (2) 0.7–0.8	0.4–0.43 0.75	0.04–0.07 0.04	Moderately preserved dispersed groups Chains dislocated cells	None	Loose biofilm of filaments coccoids, EPS, "glued" particles	Smooth to granular
4 (light)	None		(1) 0.35–0.65 (2) 0.7–0.8	0.4–0.5 0.73	0.05–0.14 0–0.08	Moderately preserved dispersed groups chains dislocated cells		Loose colonies	Yes
3 (translucent)	0.1 (1 filam.)	Badly degraded fragment	(1) 0.35–0.65 (2) 0.7–0.8	0.44–0.51 0.68–0.87	0.04–0.1 0.05–0.08	Moderately preserved dispersed groups chains dislocated cells	Rare (2) 0.8 × 0.4 µm	"carpet" colonies >100s individuals 20–30 µm	Yes

Note: SD—refers to Standard Deviation; EPS—refers to Extracellular Polymeric Substances.

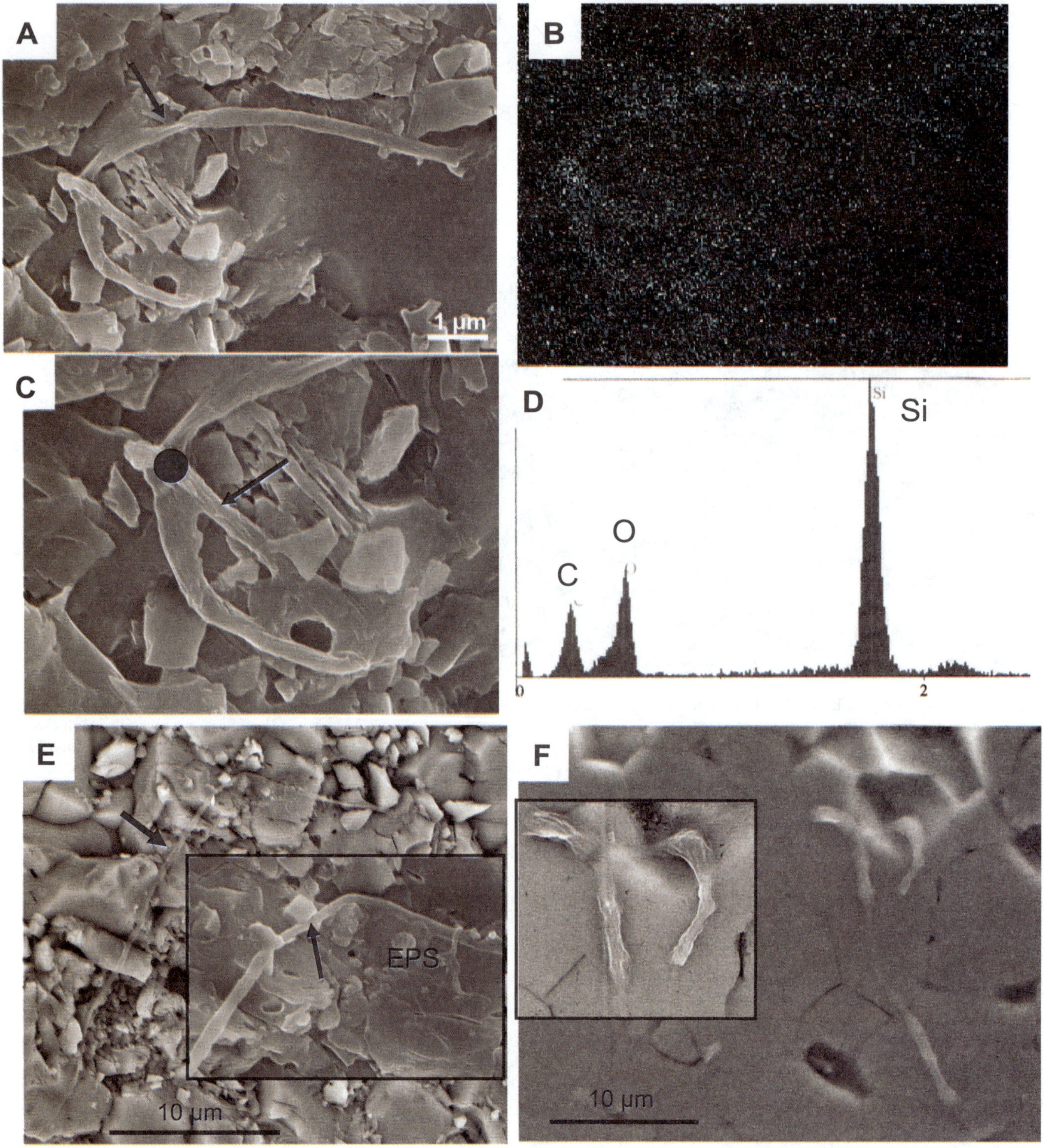

Figure 7. Filamentous microfossils from top of Layer 5. (A) Secondary electron photomicrograph of a fragment of a filament (note collapsed section of tubular structure—arrow). (B) Carbon map of same structure. (C) Detail of filament showing length-parallel striations (arrow). (D) EDX (electron diffraction X-ray spectrometry) spot analysis of filament (analysis made at black spot in [C]). (E) Detail of a long portion (>30 µm) of a filament (arrow) that has retained its tubular shape for its entire length. (Inset) Secondary electron photomicrograph detail of the striated filament. Note (i) adjacent small rectangular particle that apparently grew contemporaneously with the filament and (ii) EPS (extracellular polymeric substances) associated with filament. (F) Secondary electron photomicrograph of filaments cutting across boundaries of quartz grains in a lightly etched thin-section surface. (Inset) Detail of collapsed appearance of filaments.

phenomenon has been observed on filamentous bacteria from a hot spring environment, S. Cady, 2002, personal commun.) and/or to the degradation and shrinkage process after cell death. As a result of this characteristic, highly degraded filaments appear as frayed, fibrous strands, some of which may only be 0.1 µm in diameter (Fig. 8). All the filaments present a sinuous morphology. EDX spot measurements (Fig. 7D) document C, O, and Si, whereas carbon mapping shows that they are distinctly carbonaceous (Fig. 7B). The filaments are associated with patches of polymer and very often with coccoidal structures exhibiting features attributed to cellular division (see below) (Fig. 8). Filaments were observed in the surface of the slightly etched thin section, where they are completely embedded in quartz crystals (Fig. 7F). Filaments are common at the base of Layer 6 and relatively common at the top of Layer 5. Rare fragments of filament occur in Layers 3 and the main part of Layer 6.

Coccoidal Structures

Coccoidal structures are numerous in all layers but occur in different types of associations (colonies) in different layers (Table 2). The size/shape statistical analyses of >2000 individuals from 20 colonies showed that the coccoidal structures fall into two main size categories: one group with modal diameters between 0.4 and 0.5 µm (standard deviation [SD] 0.04–0.1) and the second with modal diameters mainly between 0.73 and 0.77 µm (SD 0–0.07) (Figs. 9A, 9C, and 10D). Most of the coccoids are surrounded by a halo of EPS (Figs. 9B, 9C, and 10A). Many of them are attached to each other either by compromised boundaries or by a neck of EPS, as if in cell division (Figs. 9 and 10). Occasionally the two parts of the compromised cells exhibit lateral dislocation, where one of the attached cells is turgid in shape but the other cell is deflated and presents an irregular surface texture (Fig. 10D). Deflation of the coccoids produces a rough, wrinkled outer surface texture, indicating that the structures were hollow (Figs. 10A, 10D, and 10E).

The coccoids occur in colonies of varying numbers of individuals. Loose associations consist of groups of a few coccoids (five to some tens) embedded in EPS that are irregularly dispersed on the substrate but occur within distances of a few µm from each other. Frequently, the coccoids in the loose colonies may form chains consisting of 3 to, at most, up to 10 individuals (Figs. 10A, 10C, and 10E). The most spectacular coccoidal colonies consist of dense associations numbering many hundreds of individuals, all embedded in EPS. We term them "carpet" colonies (Figs. 9A and 11). Such colonies have irregular but relatively distinct outlines and may be some tens of µm in size. EDX spot analyses and carbon mapping clearly demonstrate the carbonaceous nature of the coccoids and the colonies they form (Fig. 11). Layer 3 and the main part of Layer 6 (but not its base) contain exclusively "carpet" colonies. The looser colonies occur in Layers 4 and 5 as well as at the base of Layer 6, where, with the exception of Layer 4, they are commonly associated with the filaments described above (Figs. 8 and 10C).

Although the distribution of the loose and "carpet" colonies of coccoidal structures is heterogeneous within each particular layer, they are clearly concentrated around the edges of the hydromuscovite-replaced volcanic protoliths (Fig. 11C).

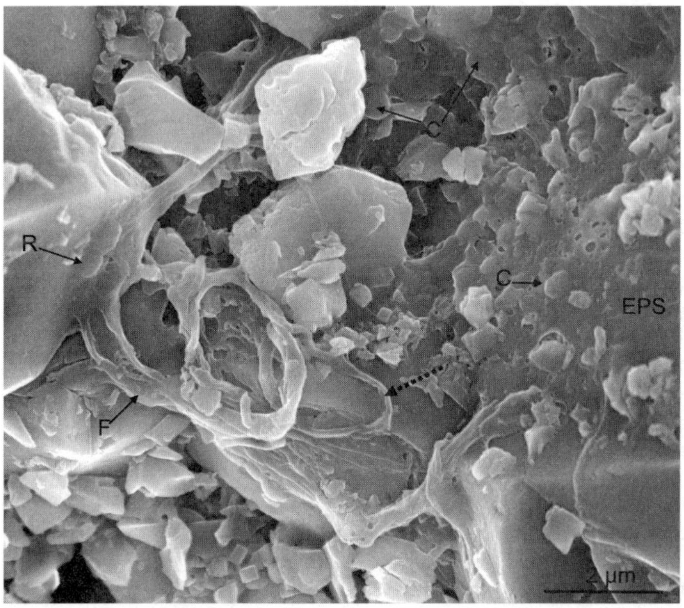

Figure 8. Secondary electron photomicrograph of the biofilm formed at the base of Layer 6. It consists of filaments (F), coccoids (C), rod-shaped microorganisms (R), and EPS (extracellular polymeric substances). The filaments are highly degraded, having broken down into fine strands in places (dotted arrow). The rod-shaped microorganisms show cell division, as do most of the coccoids. The EPS appears as a thin film that coats the sediment particles and is locally perforated by holes.

Rod-Shaped Structures

Rod-shaped structures ~0.4 µm in width and 0.8 µm in length occur rarely in this volcanoclastic sediment (Figs. 8 and 10F). Isolated individuals were observed in close association with coccoids in Layers 3, 5, and 6. The coccoid-filament association at the base of Layer 6 contains one example of two rods attached to each other by a compromised boundary at their extremity (Fig. 10F). EDX spot measurements document relatively high concentrations of C associated with the rods.

Exopolymeric Substances (EPS)

EPS occurs in two modes: (1) as an intimate extension of the coccoidal microstructures, coating them like a halo and linking dividing cells (Figs. 9B, 9C, 10A, 10C, and 10E), and (2) as a laterally extensive material that coats or embeds the microbial colonies (Figs. 8, 9A, and 11) and the surrounding mineral particles (Figs. 8 and 12B–12F). The laterally extensive EPS may have a smooth (Fig. 8) or a granular surface (Figs. 12E and 12F), which often exhibits holes and/or a cracked texture (Figs. 8, 9A, and 10A). EPS is ubiquitous throughout the layers.

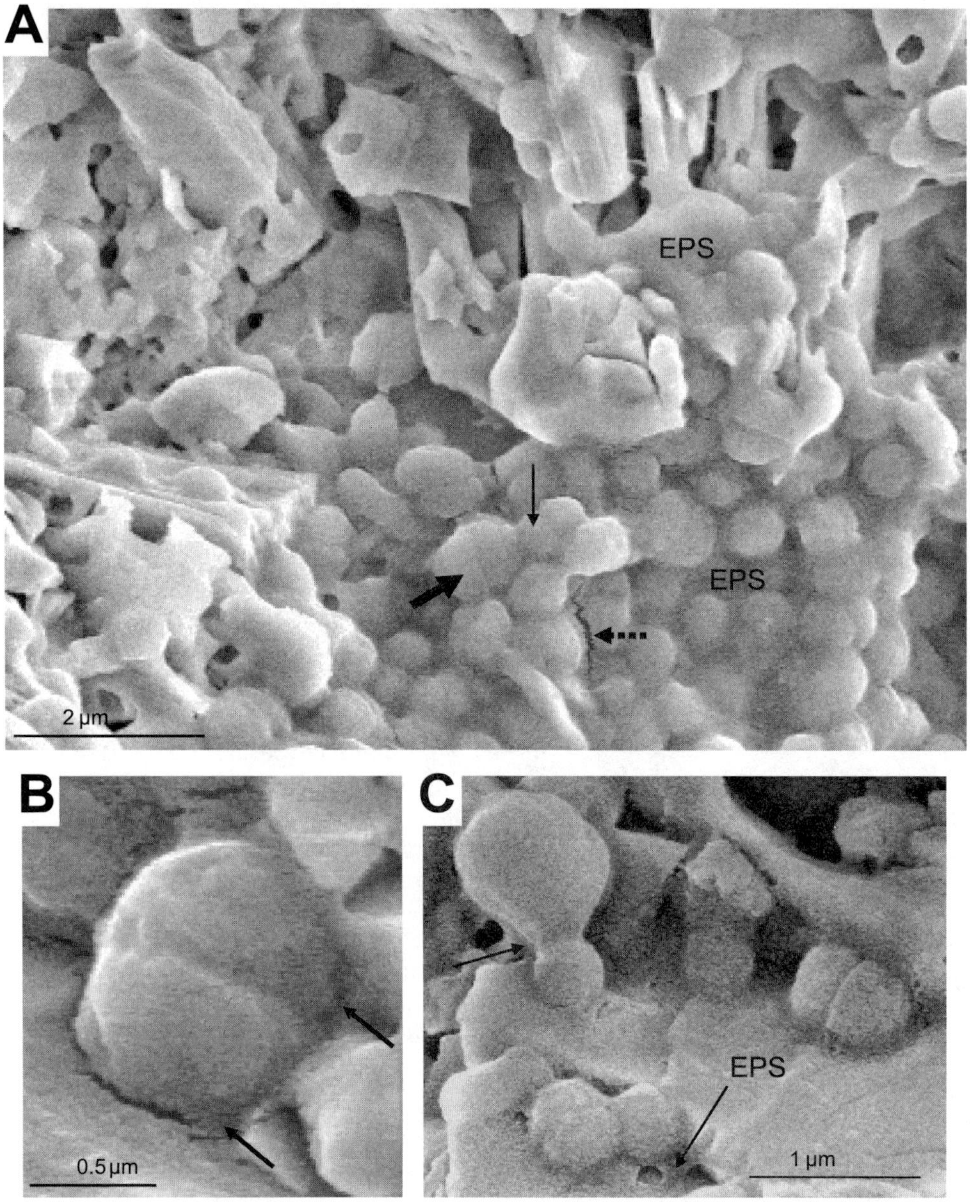

Figure 9. Secondary electron photomicrograph of coccoidal microfossils. (A) Part of a "carpet" colony of coccoids (Layer 3) at edge of a large, silicified, hydromuscovite-replaced, volcanic particle (see Fig. 11A). The hydromuscovite is visible as foliated minerals at the top of the micrograph. Note the EPS (extracellular polymeric substances) coating both the minerals and the coccoids. Two species of coccoids having two distinct modal sizes coexist in this colony: larger coccoids (large arrow) have a modal size of 0.75 µm, whereas smaller ones are 0.45 µm (small arrow). The dotted arrow points to a desiccation crack in the EPS. (B) Detail of a dividing coccoid from Layer 5. Initial separation of two parts of dividing cell, typical of cell division by fission, is clearly visible. Note collapsed, cell-bound EPS around edge of microorganisms (arrows). (C) Two species of dividing coccoids having different modal sizes from Layer 4. Note cell-bound EPS surrounding cells.

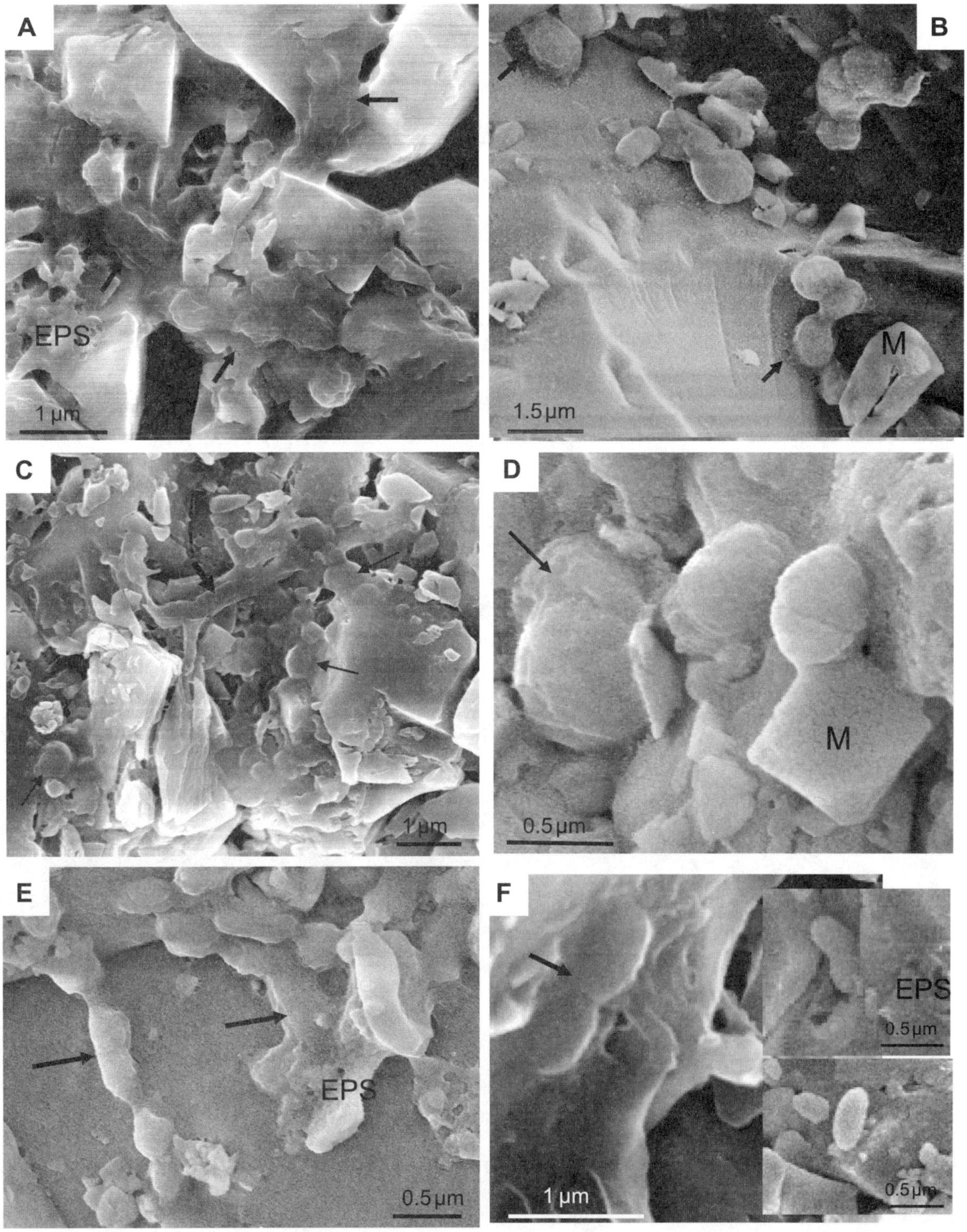

Figure 10. Secondary electron photomicrographs of coccoidal and rod-shaped microfossils in Layers 4, 5, and 6. (A) Chains of collapsed and deflated coccoids, coated in cell-bound EPS (extracellular polymeric substances; arrows) from the base of Layer 6. Note wrinkled texture on surface of these fossilized microorganisms, indicating they were dead (cell collapse and lysis) before fossilization. (B) Dividing coccoids with cell-bound EPS from base of Layer 6. M—mineral fragment. (C) Consortium consisting of a degraded filament (thick arrow) and many coccoids (thin arrows) from top of Layer 5. (D) Coccoids of two sizes in cell division in a pumice fragment. Note that one of larger coccoids has collapsed (indicating cell death) while other half still exhibits a vital turgid aspect. The mineral (M) obviously grew in coexistence with coccoids, since it appears to be attached to a small, dividing coccoid. (E) Highly degraded coccoid chains (arrows) from the base of Layer 6. (F) Rod-shaped organisms (*N.B.* structures resembling rod-shaped microorganisms are very rare in these sediments). Main micrograph shows a dividing rod from biofilm in Figure 8, the other two occur in Layer 3.

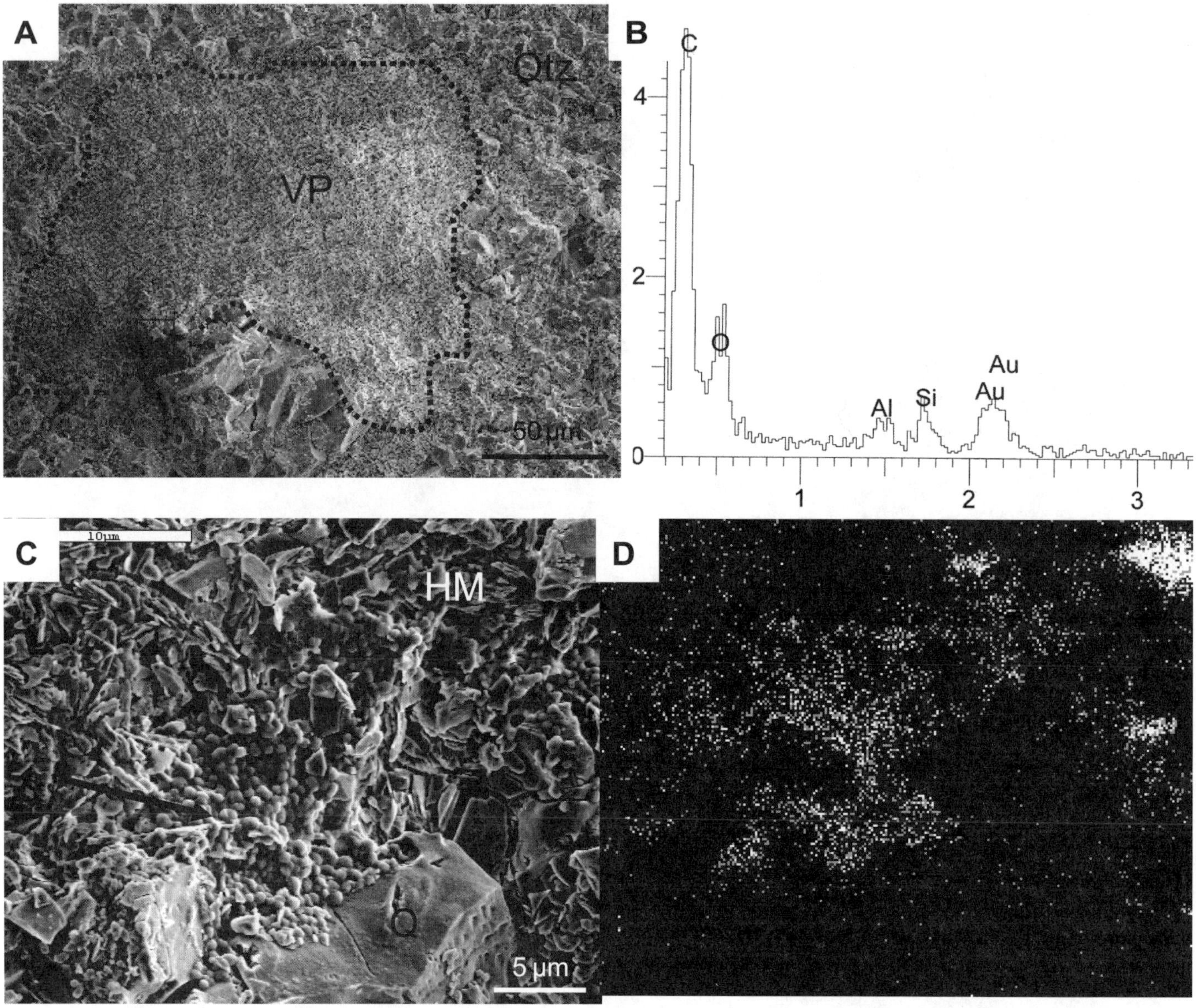

Figure 11. "Carpet" colonies in Layer 3. (A) Low magnification secondary electron photomicrograph of one of the volcanic particles (VP) in Layer 3 (dotted outline) (cf. Figs. 2E and 3A). Numerous "carpet" colonies occur around its edge. Rectangle represents location of "carpet" colony shown in (C) and in Figure 9A. (B) EDX (electron diffraction X-ray spectrometry) spot analysis of center of colony showing strong carbon concentration. (C) Secondary electron photomicrograph of a "carpet" colony (arrow) (see also detail in Fig. 9A) located at edge of hydromuscovite (HM)-replaced volcanic particle and quartz (Q) matrix. (D) EDX carbon map of colony showing that carbon is restricted to colony.

Colonies and Biofilms

As noted above, the filaments, coccoids, and rods occur with other cells of the same morphology and size and sometimes with different types of cells, in associations embedded in EPS that represent microbial colonies (Figs. 8, 9A, 10A, 10C, and 10E). The composition and style (i.e., loose or carpet colony, biofilm) of the colonies in each sediment layer are different. Layer 3 is characterized by coccoidal carpet colonies; loose colonies of small coccoid associations and chains of coccoids occur in Layer 4; at the top of Layer 5 and at the base of Layer 6, a consortium of filaments, coccoids, and (rarely) rods form biofilms; and the main part of Layer 6 consists of coccoid carpet colonies. The biofilm at the base of Layer 6 is moderately well developed and laterally extensive. It is associated with thick quantities of EPS that cement also the sediment particles (Figs. 12B–12F). It coincides with the thin brownish layer, observed in thin section, that conformably coats the top of Layer 5, including the pumice fragments (Fig. 12A). The biofilm in Layer 5 is more diffuse than that at the base of Layer 6.

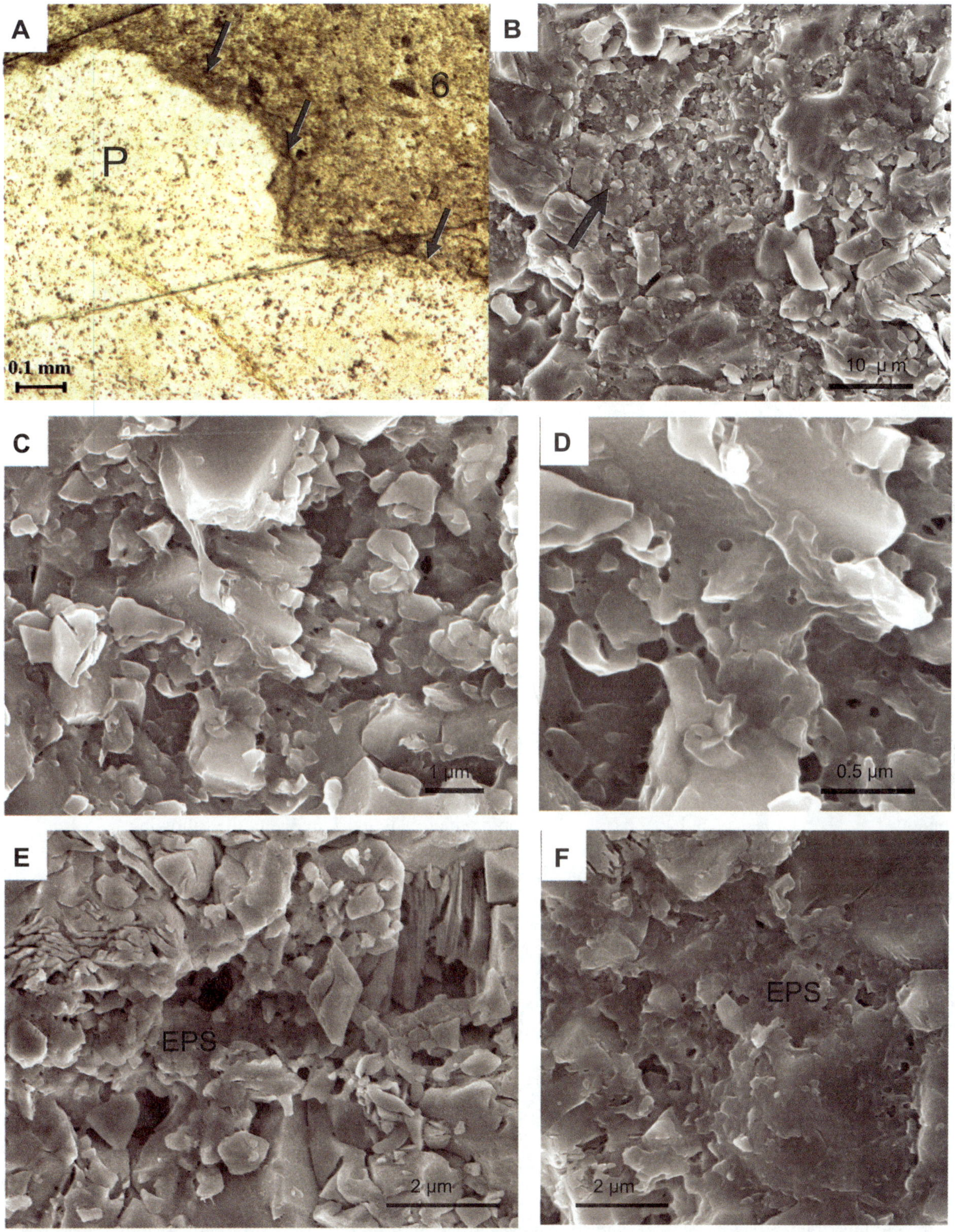

Figure 12. Extracellular polymeric substances (EPS). (A) Thin-section photomicrograph of top of layer 5, including a pumice fragment (P) and base of Layer 6. Note faint brown line at base of Layer 6 (arrows) that is conformable with underlying sediment surface and probably represents a biofilm (cf. Fig. 8). (B–D) Secondary electron photomicrographs of the layer corresponding to the brown line in (A). (B) Small zones of EPS–cemented, cryptocrystalline quartz (arrow) are associated with consortia of microfossils, such as that shown in Figure 8. (C, D) Detail of fine EPS cemented cryptocrystalline quartz particles. Note holes in EPS (D). (E, F) Granular EPS cementing sediment particles together in main part of Layer 6.

DISCUSSION

Sedimentological Environment

de Vries (2004) has demonstrated that the overall sedimentary environment for the "Kitty's Gap Chert" is a channel that was gradually filled up with sediment, termed a channel-and-flat environment. Within this broad environmental context, each layer in the sediment samples that we have studied reflects slightly different and alternating conditions related to different energetic levels of the water transporting the sediment. The light gray layers containing relatively coarse clasts were deposited in a higher energy regime than the dark gray layers consisting of dust-sized material. The source of the sediment must have been local, as the detrital particles, including those of carbonaceous composition, are extremely angular, indicating that they have not been transported far. Moreover, the filamentous carbonaceous particles always have degraded filaments attached to them (Fig. 4D) that are morphologically identical to those formed in situ in the biofilms of Layer 5 and at the base of Layer 6 (see discussion below regarding the endogeneity of the microfossils), suggesting that they formed in a similar, nearby environment.

Layer 3 stands out even in the hand specimen because of its multilayered, translucent appearance (Fig. 2E). It is the most highly silicified of the sedimentary layers studied, and most of the hydromuscovite-replaced volcanic clasts have themselves been almost completely replaced by silica. The fact that it consists of the same protoliths as Layer 2 and that it is separated from Layer 4 by a bedding-parallel chalcedony vein suggests it was subjected to a postdepositional, early diagenetic, secondary silicification event (see section below on the timing of microbial silicification).

The pumice fragments embedded in Layer 5 (Fig. 2D) probably represent ash fall deposits from a distant volcanic eruption. They show a sharp visual and compositional distinction between the translucent, orange-colored, hydromuscovite (with localized quartz replacement) upper part and the opaque, whitish, mainly quartz lower part that corresponds to the level of embedding in the top of Layer 5 (Fig. 2D). The quartz replacement of the lower part of the pumice fragments may indicate a higher saturation in silica in the sediment pore waters. The sharp distinction between the relatively coarse-grained Layer 5 and the very fine-grained Layer 6 clearly represents a significant change in the energy conditions under which the respective layers were deposited, from higher to very low energy conditions.

Biogenic or Abiogenic Origin of the Microstructures

We interpret the filamentous, coccoidal, and rod-shaped structures as microbial filaments, coccoids, and rods, respectively. Their associations are interpreted as microbial colonies, whereas the multispecies associations at the top of Layer 5 and base of Layer 6 are interpreted as microbial biofilms. Our interpretations are based on the following evaluations.

Environmental Habitat

The volcanoclastic sediments deposited in this channel-and-flat environment provided a range of microenvironments that could have hosted life, including the surfaces of the detrital particles (both exposed on the sediment surface and within the seawater-filled pore spaces of the soft sediments), as well as the boundary between Layers 5 and 6 that represents a relatively stable sediment surface.

Morphological Characteristics

The microstructures described above exhibit a range of morphological characteristics that is typical of the individual and colonial aspects of prokaryote microorganisms but not of organisms such as cyanobacteria (Westall, 1999, 2003; Madigan et al., 2000).

Filaments. The size and shape of the filaments are the same as those of modern microorganisms (Madigan et al., 2000; Handley, 2004). Moreover, their sinuosity signifies that they were formed of a material that could be plastically deformed. Partial deflation of their tubular structure suggests that they were hollow when fossilized. This implies that the filaments were already dead and had experienced cell collapse due to lysis and loss of their internal cytoplasm before fossilization. The fragmented nature of most of the filaments also supports this interpretation. The striation along the lengths of the filaments is intriguing: as noted above, similar features have been observed on other microbial filaments in hot spring environments. Although not as common as the coccoidal structures, the filaments are relatively common in Layer 5 and at the base of Layer 6 and occur in association with colonies of coccoidal structures, rare rods, and copious amounts of polymer.

Coccoids. The coccoidal structures are interpreted as being microbial in origin due to the following.

1. The consistency of their size and shape with those of modern coccoidal microorganisms (Madigan et al., 2000) (*N.B.* the two different sizes of coccoids probably represent two different species).

2. The regularity of size and shape of the individuals within each colonial association and the slight differences between different colonies in the same sedimentary layer.

3. Evidence for both living and death processes in the same colony (typical in any living microbial community). The living aspects are demonstrated by: (i) cell division; various stages of cell division (simple diploid fission) were observed ranging from the initial elongation of the cell, formation of a dividing membrane, initial separation of the two halves, almost complete separation with attachment by a "neck" of cell-bound polymer, to complete separation; (ii) chain formation; and (iii) the dislocation of dividing cells, sometimes with better preservation of one half than the other. The death aspects are documented by cell collapse resulting in an outer, wrinkled surface.

4. Association of many individuals with the same morphological characteristics in colonies that may be loosely or densely packed.

5. Association with different types of microorganisms in multispecies colonies or consortia.

6. Association with large amounts of polymer or EPS.

Rods. The size and shape of the rod-shaped organisms are typical for modern microbial rod-shaped organisms (Madigan et al., 2000), as is the evidence for cell division and their association with other structures (coccoids, filaments, and polymer) that are interpreted to be of microbial origin.

EPS. The variability of morphological characteristics of the polymer film (smooth to granular; with holes or cracked; fine films coating sediment particles or thick films embedding monospecies or multispecies colonies; as well as sediment particles) resembles that of modern extracellular polymeric substances (Westall et al., 2000 and references therein). The two types of EPS observed (cell-bound EPS directly linked to the outer walls of the coccoidal cells and EPS that links individual microorganisms in a colony or biofilm) are well known in modern microbial communities.

Colonies and biofilms. The association of many individuals together represents a colony, whether densely packed, as in the "carpet" colonies in Layers 3 and 7, or loosely packed as in the colonies in Layers 4 and 5. At the top of Layer 5 and the base of Layer 6, the totality of all these features, i.e., individuals of different morphologies within an EPS–rich layer that also incorporates sediment particles at a sediment surface, is a typical characteristic of biofilm formation. Moreover, the fact that different types of colonies and biofilms are related to different sedimentary layers supports the interpretation of formation of colonies and biofilms by different species of microorganisms in different microenvironments.

Biogeochemical Signatures

The carbonaceous composition and carbon isotope signal ($\delta^{13}C$ values of around –26‰ to –30‰) are those that would be expected from microorganisms (see discussion of organic carbon content and carbon isotopes below). The high-resolution TEM study showed that, although mature, the kerogens are far from having reached the graphite stage. The structure of the carbon platelets clearly differs from that of pyrocarbons formed by an abiogenic process, such as hydrocarbon decomposition on hot substrates, but is similar to that found in mature kerogens from biological precursors found in the rock-parent of oil (Oberlin, 1989; Skrzypczak et al., 2004).

Environmental Interactions

The coccoidal colonies and multispecies biofilms observed in the "Kitty's Gap Chert" interacted with their immediate environment in a number of ways. In the first place, the coccoidal colonies are distinctly distributed around the edges of the volcanic particles, indicating their specificity for a particular type of substrate. The indications of alteration at the edges of the volcanic clasts, as observed by optical microscopy and analyzed geochemically, coupled with the presence of microbes on the clast surfaces, together suggest that at least part of the alteration process may have been caused by the metabolic activities of the microbes. Thus, part of the supersaturation of silica in the pore waters is probably due to microbial activity, including changes in the pH.

Secondly, the formation of biofilms at the top of Layer 5 and, especially, at the base of Layer 6 greatly influenced the binding of the sediment particles and the stability of the sediment surface. Particles are coated by and at times completely engulfed by EPS that acts as a cement.

Lastly, the presence of the biogenic organic matter on the volcanic particles within the pore spaces of the sediment, as well as at the top of and immediately above a sediment surface, promoted the very early lithification (silicification) process of the whole sediment. The dissolved silica was readily complexed to the organic matter. Functional groups within it and on it acted as initial points of precipitation that led to further polymerization and, finally, contributed to the silicification of the whole sedimentary packet.

We conclude, on the basis of the totality of the microenvironmental, structural, and biogeochemical information and substrate interactions, that the filaments, coccoids, rods, and polymer represent with a high degree of probability the silicified remains of microorganisms, their colonies, and biofilms.

Abiogenic Alternatives

There is at present much debate regarding the origin of the carbonaceous matter in early Archean sediments (see also the discussion on the isotopic signal below) (Brasier et al., 2002; van Zuilen et al., 2002). Metamorphic carbon or keratite (Yushkin, 2000) and laboratory-produced carbon-silica precipitates (Garcia-Ruiz et al., 2003) can mimic certain microbial morphologies. A potential source of abiogenic carbon in the "Kitty's Gap Chert" is the hydrothermal vein that conducted silica-rich fluids into the sediments. However, although such precipitates can imitate individual morphologies, they cannot reproduce the totality of colony or biofilm characteristics exhibited by the mono- and multispecies colonies and biofilms preserved in these silicified sediments, i.e. the regularity in size and shape, the evidence of cell vitality (division), and cell death (collapsed cells), the flexibility of the filamentous forms, the formation of multispecies colonies, the association of EPS (both cell-bound and non–cell-bound), the formation of different types of colonies and biofilms in different microenvironments, the intimate nature of the interactions between the biofilms and their substrate with incorporation of sediment particles, and the carbon isotope signature. We, therefore, exclude the hypothesis that the microstructures described above could be abiogenic artifacts, such as silica spheres or filaments coated with carbon or abiogenic, carbonaceous structures that were subsequently silicified.

In Situ Formation or Later Microbial Contamination

It could be argued that the microorganisms in the "Kitty's Gap Chert" could have infiltrated the rock at a later stage and become fossilized in situ. Westall and Folk (2003) demonstrated such a situation for the 3.8-Ga-old Isua banded iron formations, in which recent endolithic fungi and cyanobacteria invaded fractures in the rock and were subsequently silicified. Great pains were taken with the "Kitty's Gap Chert" samples to exclude laboratory contamination of the preparations for SEM observation by thorough rinsing in milliQ water and flaming with alcohol during preparation. However, carbon contamination evidently occurred through the handling process of some subsamples, as indicated by the stepped combustion release of carbon at low temperatures in the sample (see discussion below). At the same time, however, the stepped combustion analyses also demonstrate the presence of carbon that is clearly intrinsic to the sample and that is released at high temperatures during stepped combustion. The HR-TEM and Raman spectroscopic evidence for the presence of mature kerogen (Beyssac et al., 2002a, 2002b) in these sediments corroborates the stepped combustion results.

There are a number of lines of evidence demonstrating that the microfossils described above are in situ and that they formed at the same time as the rock. (1) We observe different types of microbial colonies and biofilms in association with different layers representing different microenvironments, which would be difficult to explain if they resulted from field or laboratory contamination. (2) There is also an intimate interrelationship between the biofilms at the base of Layer 6, in Layer 5, and the sediment particulate substrate, with EPS incorporating sediment particles as well as microfossils, which argues for coexistence. (3) The colonies of coccoids are distributed around the edges of the volcanic clasts (Fig. 11) and are not indiscriminately disseminated in the sample, as would be expected if they were contaminants. Their close association with the surfaces of the particles indicates colonization before the silicification and cementation of the rock. (4) Filaments attached to some of the carbonaceous detrital clasts are identical to those observed in association with the biofilm (Figs. 4 and 7). They must have been formed contemporaneously and be very locally derived, otherwise they would have been completely degraded before deposition. Such transported filaments are difficult to explain by contamination. (5) The variability in the level of cellular preservation before silicification also argues for the fact that these organisms must be intrinsic to the rock, for example, the fact that the filaments and coccoids in the biofilm at the base of Layer 6 are more poorly preserved than the colonies of coccoids in the underlying and overlying layers. (6) The microfossils were observed in an etched, cut-rock surface and not on a fracture surface. They were also observed on the surface of a polished thin section, where filaments can be seen traversing the grain boundaries between quartz crystals (Fig. 7F), demonstrating that they must have formed in situ. (7) The maturity of the kerogen (probably biogenic according to the HR-TEM studies) in the "Kitty's Gap Chert" is consistent with the metamorphic grade of the rock (prehnite-pumpellyite), suggesting contemporaneous formation.

We conclude that the microfossils occurring in the "Kitty's Gap Chert" are intrinsic to the rock and are not later infiltrations and that they predated the silicification of the rock.

Microbial Preservation and Timing of Silicification

The microorganisms and their colonies and biofilms have been preserved by silicification, as indicated by the EDX spot and mapping analyses. The process of mineralization must have been relatively rapid because otherwise these delicate, organic structures would not have been preserved. Experimental studies (Westall et al., 1995; Toporski et al., 2002; Orange, 2004) as well as field studies in silica hot springs (de Ronde et al., 2002; Handley, 2004) have shown that the fossilization process can start very rapidly, within hours, and can continue for some days to weeks. The microorganisms in the "Kitty's Gap Chert" exhibit varying degrees of preservation prior to fossilization. A living colony of microorganisms always consists of both living (dividing) and dead (lysed) individuals. In fact, dead cells and EPS are actually more susceptible to fossilization than living cells because of the greater availability of functional groups for complexing mineral species in solution (Westall et al., 2000). Most of the fossilized colonies observed in this sample comprise both living and dead cells. The multispecies biofilm at the base of Layer 6 (fine volcanic dust layer), however, exhibits a higher degree of degradation before fossilization than any of the colonies in the other layers.

Not only are the microorganisms silicified but so is the whole sedimentary sequence. The fact that the microorganisms are preserved at all is evidence that the overall process of silicification was very rapid. The complexation of the silica ions by the organic surfaces of both the EPS as well as the microorganisms must have played an important role in the silicification of this sediment deposit. The silica probably had multiple origins, including (1) silica from the devitrification and diagenetic alteration of the volcanic material, which enriched the pore waters (microbial metabolic activity probably contributed to the process), (2) silica from hydrothermal fluids, and (3) silica in the seawater (*N.B.* until the appearance of microorganisms with siliceous frustules in the Phanerozoic, silica concentrations in the seawater were considerably higher) (see also Orberger et al., this volume). The difference in the state of conservation of microorganisms from different layers, i.e., the relatively well-preserved coccoidal colonies in Layers 3, 4, 5, and 6, as opposed to the relatively poorly preserved biofilm at the base of Layer 6, may be related to differences in SiO_2 concentrations between sediment pore waters and seawater, with the latter being more diluted. The biofilm at the base of Layer 6 may, therefore, have been fossilized at a slower rate than the colonies in the other layers, thus allowing more time for degradation.

The sediments at Kitty's Gap must have been subjected to a number of silicification episodes that were closely spaced in time: initial silicification of the organic matter and probably penecontemporaneous precipitation of a siliceous cement in the sediments, followed by cross-cutting and bedding-parallel silica veining.

Carbon and Carbon Isotopes as Signatures of Microbial Activity

Carbon Content

Most of the carbon in the "Kitty's Gap Chert" is released at low temperatures from the crushed samples and a smaller amount is released at higher temperatures. As mentioned above (Results section), the uncrushed sample shows, however, a different release pattern: most of the carbon is released at high temperatures with release peaking at 1100 °C. The observation for the uncrushed sample can be explained as a result of the presence of two carbon components: a surface (laboratory) contamination component that is released at low temperatures and a second component that is intrinsic to the sample, released at high temperatures. In the crushed samples, the carbon that is intrinsic to the sample is more easily exposed to oxygen during stepped combustion and, thus, burns at lower temperatures, whereas in the uncrushed sample, oxygen does not come into direct contact with the intrinsic carbon until the temperature is high enough to crack or to melt the sample.

With respect to the concentration of carbon in the sample, we find that the larger the sample aliquot, the lower the measured total carbon concentration is. This implies that contamination is proportional to the sample surface area. The data allow a rough estimation of ~100 ppm for the concentration of the carbon intrinsic to the samples. For the uncrushed sample the amount of intrinsic carbon is 84 ppm.

We can, thus, conclude:

1. The samples contain both laboratory contamination and intrinsic carbon.
2. The combustion temperatures of the two carbon components are similar (200–600 °C).
3. The amount of carbon contamination is comparable with the surface area of each of the subsamples.

The generally low concentrations of carbon in the sample are related to two factors: the low biomass and the silicification of the microorganisms. In terms of weight, the average carbon content in a bacterial cell is $\sim 3.54 \times 10^{-13}$ g. This translates into an even smaller amount of carbon in an individual microfossil because of the effects of degradation and loss of the organic carbon compounds during the mineralization of the cell. Although the numbers of individuals in a colony may reach many hundreds and colonies may be relatively common in the layers of the Kitty's Gap sediments, the overall weight of carbon in a particular layer is diluted by the volcanic components and, especially, the diluting and fossilizing silica. Thus, in a sample with variable microenvironments ranging from pore water spaces containing colonies of microorganisms attached to particle surfaces to sediment surfaces hosting biofilms or mats, there will be heterogeneously distributed pockets with higher concentrations of carbon, although the overall bulk carbon content may be low. In situ analyses of carbon would be the ideal method of analysis and, indeed, would have been successfully used on larger, younger microfossils (e.g., House et al., 2000). However, the ability to measure carbon and carbon isotopes in submicrometer structures is still in the developmental stages.

Carbon Isotopic Compositions

The isotopic compositions of the various layers show that the $\delta^{13}C$ values for the contaminating carbon and the carbon intrinsic to the sample are similar, varying from –22‰ to –30‰ in the combustion steps, with bulk compositions between –25.9‰ and –27.8 ‰. Thus, although in most cases we were not able to distinguish between the contamination and sample carbon, there is strong evidence that the samples contain carbon intrinsic to the sample that has isotopic compositions between –26‰ and –28‰. This result, coupled with the relatively low combustion temperature observed for the carbon (<600 °C), suggests that the intrinsic carbon can be identified with the presence of the microfossils. The HR-TEM observations of probable biogenic kerogen also support this hypothesis.

An alternative explanation for the carbon isotope values is abiogenic fractionation of carbon by Fischer-Tropsch synthesis in hydrothermal systems, since this process can produce an isotopic composition similar to that produced by microbial fractionation (Holm and Charlou, 2001). Although hydrothermal activity was associated with the early diagenetic silicification processes occurring at Kitty's Gap, the overwhelming evidence for microbial colonization in each of the sediment layers investigated suggests that, in this case, the isotope composition of the intrinsic carbon is probably due to microbial fractionation.

Microbial Ecosystems in the Early Archean, Near-Shore, Volcanoclastic Sediments

Subsurface Pore Space: Chemolithotrophic Colonies

The mineralogical substrate on which these microorganisms flourished is volcanic and, according to the protomineralogy, of rhyodacitic composition. The microbial colonies occur around the edges of the pseudomorphosed clasts in Layers 3, 4, and 5, form a biofilm at the top of Layer 5 and at the base of Layer 6, and are mixed in with the fine volcanic dust in Layer 6. The juxtaposition of the colonies with the volcanic material suggests a close relationship between the substrate and microbial metabolism. The coccoidal microfossils may represent at least chemolithotrophic microorganisms that obtained their source of energy from electron donors and acceptors present or extracted from the surfaces of the volcanic particles that they colonized. Their carbon could have come from CO_2 that was abundant in the early Archean environment (Kasting 1993), from organic molecules synthesized on Earth (Nakashima and Shiota, 2001; Russell and Arndt, 2005), or from organic carbon of extraterrestrial origin (Brack et al., 2001; Maurette et al., 2001). In fact, volcanic materials provide a ready source of nutrients for microbes, and studies of recent volcanic glass and pillow basalt surfaces have shown that they are invariably infested by microbes (Ross and Fischer, 1986; Thorseth et al., 1992). The rhyolitic protoliths in the Kitty's Gap sediments would have had fewer nutrients than mafic basalts, but they would still have provided an important source of macronutrients, such as K, as well as minor Fe and some Mg. It is possible that they could have extracted further Mg and Ca from the seawater. Moreover, given the fact that chemolithotrophic microorganisms excrete enzymes to break down the surfaces that they colonize in order to liberate H_2 as an electron donor for their metabolic processes (Brantley et al., 2001), microbial activity may have played a major role in the early diagenetic alteration of the volcanic particles and the liberation of silica.

Sediment Surface: Photosynthesizing Mats?

The boundary between Layers 5 and 6 probably represents a strong change in the energetic regime and a short period of nondeposition. It is at the top of Layer 5 and at the base of Layer 6 that a multispecies biofilm containing filamentous and coccoidal microorganisms formed. It could be hypothesized that some of the microorganisms in the biofilm, such as the filaments, may have been photosynthetic microorganisms, as the biofilm formed conformably on a sediment surface in the photic zone (the overall environment of deposition is a channel-and-flat environment, possibly influenced by tidal activity (de Vries, 2004). The development of photosynthesis (oxygenic or anoxygenic) by the early-mid–Archean has previously been hypothesized (e.g., Walter, 1983; Byerly et al., 1986; Walsh, 1992, 2004; Hofmann et al., 1999; Westall, 2003, 2004; Westall et al., 2001, 2004; Tice and Lowe, 2004). If this were the case for the filaments in the "Kitty's Gap Chert," the photosynthesizing microorganisms must have been anoxygenic, as geochemical studies of the chert (Orberger et al., this volume) indicate that the environment at the time of deposition was anoxic.

Early-Mid–Archean Seawater Temperatures

Knauth and Lowe (2003) noted that the early Archean seawater was probably very warm, >50 °C, if not 80 °C. The microorganisms inhabiting the channel-and-flat sediments at Kitty's Gap may therefore have been mesophilic to thermophilic. de Vries (2004) noted evidence for hydrothermal activity during or immediately after sedimentation. Thus, a contemporaneous influence of warm fluids from hydrothermal venting is a possibility for the Kitty's Gap locality. These warm fluids would have "bathed" the microbial communities and were also a source of the silica in the very early diagenetic silicification processes. The enrichment of Cu and Zn in the darker, finer-grained layers may be a tracer of such hydrothermal fluids (Orberger et al., this volume).

Implication for the Search for Past Traces of Life on Mars

The early Archean volcanoclastic sediments represent suitable analogues for the study of Noachian-age (4.5 to ca. 3.5 Ga) volcanic material on Mars (Westall, 2003, 2004, 2005b; Westall and Walsh, 2003). These sediments were formed at the end of the period when Mars still had liquid water at its surface and in the type of near-shore environment (at the edge of a small basin) that was probably relatively common on Mars. The common occurrence of fossilized colonies of probable chemolithotrophic microorganisms on the surfaces of volcanic clasts in pore spaces and the occurrence of biofilms and mats formed by possible anoxygenic photosynthesizers on the sediment surfaces in the photic zone suggest that a similar situation could have been present on Mars, if life ever arose there. Martian volcanic material could, therefore, be suitable for past life studies.

A recent study of volcanic rock samples concluded that such materials are not suitable for light microscope investigations of microbial biofilms (Walsh, 2004). However, our investigation has shown that volcanic material may be rich in microfossils but, given the submicrometer size range of the organisms and the small sizes of their colonies, highly sophisticated methods of investigation, such as high-resolution scanning electron microscopy, are necessary for their observation (see also Westall, 2005a). The Kitty's Gap microfossils are invisible in thin section because they are intermixed with or embedded by microcrystalline quartz that has the same submicrometer size as the individual organisms themselves (Figs. 12C–12E). Although the submicrometer-sized quartz particles have a definite, angular morphology that distinguishes them from the rounded coccoids in the SEM photomicrographs (Fig. 8), the resolution capabilities in optical microscopy are insufficient to distinguish the shapes of such small particles, whatever their origin. On the other hand, our observations show that the alteration of the volcanic particles in the Kitty's Gap sediments, visible in thin section, may be a telltale indication of microbial activity.

Our study, however, underlines the difficulties involved in making an in situ search for potential microfossil traces in martian rocks: (1) In the first place, sophisticated instrumentation was necessary in order to observe and analyze the signatures of the fossilized microorganisms, although thick accumulations of well-developed microbial biofilms and mats (stromatolites) formed by probable photosynthetic microorganisms can be observed with this method (Walsh, 1992, 2004; Walsh and Lowe, 1999; Tice and Lowe, 2004). Such instrumentation is, at the moment, not miniaturizable for in situ research by a lander/rover on Mars. (2) Secondly, although the microbial colonies are commonly distributed throughout the volcanoclastic sediment, the amount of biogenic organic carbon present in the bulk sediment (or even simply in the individual layers) is very small (0.01%). This means that it would be undetectable with the instrumentation presently designed for future landers/rovers. (3)

Furthermore, it may be difficult to interpret any carbon isotope signal measured in situ. These considerations argue for the return of samples from Mars in order to make a reliable search for life investigation.

CONCLUSIONS

The multidisciplinary approach that we have used, combining detailed environmental, sedimentological, and biogeochemical studies with a high-resolution SEM investigation of microfossils, has proven to be essential for understanding the ancient microcosms present in a 3.446-Ga, volcanoclastic, near-shore, channel-and-flat environment from the Kitty's Gap locality in the Coppin Gap greenstone belt, and our observations have provided a wealth of new information regarding the early Archean microbial habitats. Each millimeter- to centimeter-thick sediment layer formed under slightly different energetic conditions and provided different microenvironments for potential microbial colonization.

The amount of morphological detail made available by the use of the HR-SEMs, together with the environmental and the biogeochemical data, allowed the microstructures to be interpreted, with a high degree of probability, as representing fossilized microorganisms, colonies, and biofilms. Densely packed, "carpet"-like colonies of coccoids having two modal sizes around 0.44 and 0.78 µm colonized the surfaces of the volcanic particles in the pore spaces of Layer 3 and in the fine, volcanic dust in Layer 6; looser colonies of coccoids, many forming chains, occurred on the surfaces of the volcanic shards and mineral debris in the pore spaces of Layers 4 and 5 (modal sizes of 0.4–0.5 and 0.73–0.75 µm). They probably represent chemolithotrophic microorganisms. Biofilms of filaments (width mode 0.35 µm, lengths up to 40 µm), coccoids (modal sizes of 0.4–0.5 and 0.75–0.8 µm), rare rods (0.8 µm in length), and copious quantities of EPS formed at a sediment surface in the photic zone representing a brief period of cessation in sedimentation. The filaments in the biofilms may have been anoxygenic, photosynthetic microorganisms. All the microorganisms may have been mesothermophiles, if the ambient seawater temperatures were as high as estimated (Knauth and Lowe, 2003).

Silicification of the colonies and biofilms was rapid, with the silica originating from supersaturated pore waters owing to the possibly microbially mediated devitrification of the volcanic particles, as well as from hydrothermal and seawater sources. Silicification of the organic matter contributed to silicification of the sediments.

These volcanoclastic sediments with their microfossil communities are an excellent analogue for Noachian-aged rocks on Mars that formed in shallow-water environments. The sophistication of the methods necessary for successful identification of microbial remains precludes in situ search on Mars at current instrumental detection limits, but the knowledge and experience gained about these very ancient terrestrial rocks will be invaluable for the future study of samples returned from Mars.

ACKNOWLEDGMENTS

FW acknowledges the NRC (National Research Council, Washington), JSC (Johnson Space Center, Houston), LPI (Lunar and Planetary Science Institute, Houston), CNRS (Centre National de la Recherche Scientifique, Paris), and CNES (Centre National d'Etudes Spatiales, Paris) for funding made available over the 4 years that these investigations were made. The following people are acknowledged for their technical help: A. Richard (HR-SEM, Orléans) and T. Cacciaguerra and C. Clinard (HR-TEM and Raman spectrometer). VR thanks the French Ministry of Foreign Affairs for supporting his stay at the Planetary Science Laboratory at Osaka University. D. Prieur, M. Walter, and U. Reimold are thanked for their critical comments.

REFERENCES CITED

Allwood, A.C., Walter, M.R., van Kranendonk, M.J., and Kamber, B.S., 2005a, 3.43 Ga stromatolites, rocky shorelines and a carbonate platform: Strelley Pool Chert, Pilbara Craton, Western Australia: NASA Astrobiology Institute Biennial Meeting: Astrobiology, v. 5, no. 2, p. 184.

Allwood, A.C., Walter, M.R., and van Kranendonk, M.J., 2005b, Stromatolite facies of the 3.43 Ga Strelley Pool Chert: Pilbara Craton, Western Australia: NASA Astrobiology Institute Biennial Meeting: Astrobiology, v. 5, no. 2, p. 184.

Beyssac, O., Rouzaud, J.N., Goffe, B., Brunet, F., and Chopin, C., 2002a, Graphitization in a high-pressure, low-temperature metamorphic gradient: A Raman microspectrometry and HRTEM study: Contributions to Mineralogy and Petrology, v. 143, p. 19–31.

Beyssac, O., Goffe, B., Chopin, C., and Rouzaud, J.N., 2002b, Raman spectra of carbonaceous materials in metasediments: A new geothermometer: Journal of Metamorphic Geology, v. 20, p. 859–872, doi: 10.1046/j.1525-1314.2002.00408.x.

Boulmier, J.L., Oberlin, A., Rouzaud, J.N., and Villey, M., 1982, Natural organic matters and carbonaceous materials: A preferential field of application for transmission electron microscopy, *in* O'Hare, A.M.F., ed., Scanning Electron Microscopy: Chicago, SEM Inc., p. 1523–1538.

Brack, A., Barbier, B., Boillot, F., and Chabin, A., 2001, Extraterrestrial organic molecules and the origin of life, *in* Nakashima, S., et al., eds., Geochemistry and the Origin of Life: Tokyo, Universal Academy Press, p. 17–38.

Brantley, S.L., Lierman, L., Bau, M., and Wu, S., 2001, Uptake of trace metals and rare elements from hornblende by a soil bacterium: Geomicrobiology Journal, v. 18, p. 37–61, doi: 10.1080/01490450151079770.

Brasier, M.D., Green, O.R., Jephcoat, A.P., Kleppe, A.K., van Kranendonk, M., Lindsay, J.F., Steele, A., and Grassineau, N., 2002, Questioning the evidence for Earth's oldest fossils: Nature, v. 416, p. 76–81, doi: 10.1038/416076a.

Buick, R., 1990, Microfossil recognition in Archean rocks: An appraisal of spheroids and filaments from a 3500 m.y. old chert-barite unit at North Pole, Western Australia: Palaios, v. 5, p. 441–459.

Byerly, G.R., Walsh, M.M., and Lowe, D.L., 1986, Stromatolites from the 3300–3500 Myr Swaziland Supergroup, Barberton Mountain Land, South Africa: Nature, v. 319, p. 489–491, doi: 10.1038/319489a0.

Cady, S.L., and Farmer, J.D., 1996, Fossilization processes in siliceous thermal springs: Trends in preservation along the thermal gradient, *in* Bock, G.R., et al., eds., Evolution of hydrothermal ecosystems on Earth (and Mars): Chichester, John Wiley, Ciba Symposium 202, p. 150–173.

Cady, S.L., Farmer, J.D., Grotzinger, J.P., Schopf, W.J., and Steele, A., 2003, Morphological biosignatures and the search for life on Mars: Astrobiology, v. 3, p. 351–368, doi: 10.1089/153110703769016442.

Characklis, W.G., and Wilderer, P.A., eds., 1989, Structure and function of biofilms—Dahlem Workshop report: Chichester, Wiley, 387 p.

de Ronde, C.E.J., Stoffers, P., Garbe-Schönberg, D., Christenson, B.W., Jones, B., Manconi, R., Browne, P.R.L., Hissmann, K., Botz, R., Davy, B.W., Schmitt, M., and Battershill, C.N., 2002, Discovery of active hydrothermal venting in Lake Taupo, New Zealand: Journal of Volcanology and Geothermal Research, v. 115, p. 257–275.

de Vries, S.T., 2004, Early Archean sedimentary basins: Depositional environment and hydrothermal systems: Examples from the Barberton and Coppin Gap greenstone belts: Geologica Ultraiectina: Utrecht, University of Utrecht, 159 p.

DiMarco, M.J., and Lowe, D.R., 1989, Stratigraphy and sedimentology of and early Archean felsic volcanic sequence, eastern Pilbara Block, Western Australia, with special reference to the Duffer Formation and implications for crustal evolution: Precambrian Research, v. 44, p. 147–169, doi: 10.1016/0301-9268(89)90080-6.

Eriksson, K.A., Krapez, B., and Fralick, P.W., 1994, Sedimentology of Archean greenstone belts: Signatures of tectonic evolution: Earth Science Reviews, v. 37, p. 1–88, doi: 10.1016/0012-8252(94)90025-6.

Fedo, C.M., 2000, Setting and origin for problematic rocks from the >3.7 Ga Isua Greenstone Belt, southern west Greenland: Earth's oldest coarse sediments: Precambrian Research, v. 101, p. 69–78, doi: 10.1016/S0301-9268(99)00100-X.

Furnes, H., Banerjee, N.R., Muehlenbachs, K., and Kotninen, A., 2005, Preservation of biosignatures in metaglassy volcanic rocks from the Jormua ophiolite complex, Finland: Precambrian Research, v. 136, p. 125–137, doi: 10.1016/j.precamres.2004.09.009.

Garcia-Ruiz, J.M., Hyde, S.T., Carnerup, A.M., Christy, A.G., van Krankendonk, M.J., and Welham, N.J., 2003, Self-assembled silica-carbonate structures and detection of ancient microfossils: Science, v. 302, p. 1194–1197, doi: 10.1126/science.1090163.

Handley, K.M., 2004, *In situ* experiments on the growth and textural development of subaerial microstromatolites, Champagne Pool, Waiotapu, New Zealand [M.S. thesis]: Auckland, University of Auckland, 111 p.

Hickman, A.H., 1983, Geology of the Pilbara Block and its environs, Western Australia: Geological Survey Bulletin, v. 127, 268 p.

Hickman, A.H., 1990, Excursion no. 5: Pilbara and Hammersley Basin, *in* Ho, S.E., et al., eds., Excursion guidebook, Third International Archean Symposium, Perth, 1990: University of Western Australia, Geology Department External Services, v. 21, p. 1–58.

Hofmann, H.J., Gray, K., Hickman, A.H., and Thorpe, R.I., 1999, Origin of 3.45 Ga coniform stromatolites in Warrawoona Group, Western Australia: Geological Society of America Bulletin, v. 111, p. 1256–1262.

Holm, N.G., and Charlou, J.L., 2001, Initial indicators of abiogenic formation of hydrocarbons in the Rainbow ultramafic hydrothermal system, Mid-Atlantic Ridge: Earth and Planetary Science Letters, v. 191, p. 1–8, doi: 10.1016/S0012-821X(01)00397-1.

House, C.H., Schopf, J.W., McKeegan, K.D., Coath, C.D., Harrison, T.M., and Stetter, K.O., 2000, Carbon isotopic composition of Precambrian microfossils: Geology, v. 28, p. 707–710.

Kasting, J.F., 1993, Earth's early atmosphere: Science, v. 259, p. 920–926.

Kisch, H.J., and Nijman, W., 2004, Metamorphic grade from the K-micas in the metasediments from the Pilbara and Barberton greenstone belts, *in* Reimold W.U., and Hofmann, A., eds., Field Forum on Processes on the Early Earth, Kaapvaal Craton, South Africa: Johannesburg, University of Witwatersrand, p. 47–48.

Knauth, L.P., and Lowe, D.R., 2003, High Archean climatic temperature inferred from oxygen isotope geochemistry of cherts in the 3.5 Ga Swaziland Supergroup, South Africa: Geological Society of America Bulletin, v. 115, p. 566–580, doi: 10.1130/0016-7606(2003)115<0566:HACTIF>2.0.CO;2.

Krumbein, W.E., Paterson, D.M., and Stal, L.J., eds., 1994, Biostabilization of sediments: Oldenburg, Biblioteks und Informationssystem der Carl von Ossietzky Universität, 526 p.

Levitt, G.S., and Krumbein, W.E., 2003, Is there and adequate terminology of biofilms and microbial mats?, *in* Krumbein, W.E., et al., eds., Fossil and Recent Biofilms: Amsterdam, Kluwer, p. 333–341.

Lowe, D.R., 1994, Abiological origin of described stromatolites older than 3.2 Ga: Geology, v. 22, p. 287–290.

Madigan, M.T., Martinko, J.M., and Parker, J., eds., 2000, Brock: Biology of Microorganisms: Upper Saddle River, New Jersey, Prentice Hall, 991 p.

Maurette, M., Matrajt, G., Gounelle, M., Engrand, C., and Duprat, J., 2001, La matière extraterrestre primitive et les mystères de nos origines, *in* Gargaud, M., et al., eds., Éléments d'exobiologie: L'environnement de la Terre primitive et l'origine de la vie: Bordeaux, Presses University of Bordeaux, p. 99–127.

McGregor, V.R., and Mason, B., 1977, Petrogenesis and geochemistry of metabasaltic and metasedimentary enclaves in the Amitsoq gneisses: West Greenland: American Mineralogist, v. 62, p. 887–904.

McKay, D.S., Gibson, E.K., Thomas-Keprta, K., Vali, H., Romanek, C.S., Clemett, S.J., Chillier, X.D.F., Maechling, C.R., and Zare, R.N., 1996, Search for past life on Mars: Possible relic biogenic activity in Martian meteorite ALH84001: Science, v. 273, p. 924–930.

Monty, C.L.V., Westall, F., and van der Gaast, S.T., 1991, Diagenesis of siliceous particles in subantarctic sediments, ODP Leg 114, Hole 699A: Possible microbial mediation, *in* Ciesielski, P.F., et al., eds., Proceedings ODP Scientific Results, 114: College Station, Texas, Ocean Drilling Program, p. 685–710.

Nakashima, S., and Shiota, D., 2001, Organic-inorganic interactions and the origin and evolution of life, *in* Nakashima, S., et al., eds., Geochemistry and the Origin of Life: Tokyo, Universal Academy Press, p. 135–177.

Nijman, W., Willigers, B.J.A., and Krikke, A., 1999, Tensile and compressive growth structures: Relationships between sedimentation, deformation and granite intrusion in the Archean Coppin Gap greenstone belt, Eastern Pilbara, Western Australia: Precambrian Research, v. 95, p. 277–302, doi: 10.1016/S0301-9268(99)00013-3.

Oberlin, A., 1989, High resolution TEM studies of carbonization and graphitization, *in* Thrower, P.A., ed., Chemistry and Physics of Carbon: New York, Marcel Dekker, v. 22, p. 1–143.

Orange, F., 2004, Expériences de silicification de bactéries extrêmophiles: Application à la recherche de signatures de vie dans les roches primitives et extra-terrestres [M.S. thesis]: Orléans, University of Orléans, 33 p.

Orberger, B., Rouchon, V., Westall, F., de Vries, S.T., Pinti, D.L., Wagner, C., Wirth, R., and Hashizume, K., 2006, this volume, Microfacies and origin of some Archean cherts (Pilbara, Australia), *in* Reimold, W.U., and Gibson, R.L., Processes on the early Earth: Geological Society of America Special Paper 405, 10.1130/2006.2405(08).

Pflug, H.D., 1979, Archean fossil finds resembling yeasts: Geologie und Palaeontologie, v. 13, p. 1–8.

Pflug, H.D., 2001, Earliest organic evolution, essay to the memory of Bartholomew Nagy: Precambrian Research, v. 106, p. 79–92, doi: 10.1016/S0301-9268(00)00126-1.

Pflug, H.D., and Jaeschke-Boyer, H., 1979, Combined structural and chemical analysis of 3,800-Myr-old microfossils: Nature, v. 280, p. 483–486, doi: 10.1038/280483a0.

Robbins, E.I., 1987, *Appelella ferrifera*, a possible new iron-coated microfossil in the Isua Iron-Formation, southwestern Greenland, *in* Appel, P.W.U., et al., eds., Precambrian Iron Formations: Athens, Theophrastes, p. 141–154.

Rosing, M.T., 1999, 13C depleted carbon microparticles in >3700 Ma seafloor sedimentary rocks from West Greenland: Science, v. 283, p. 674–676, doi: 10.1126/science.283.5402.674.

Ross, K.A., and Fischer, R.V., 1986, Biogenic grooving on glass shards: Geology, v. 14, p. 571–573, doi: 10.1130/0091-7613(1986)14<571:BGOGS>2.0.CO;2.

Rouchon, V., and Orberger, B., Westall, F., Gallien, J.-P., and Pinti, D.L., 2004a, Archean protominerals and their evolution during chert formation: Kittys Gap, Pilbara Complex: Goldschmidt Conference: Abstracts, Geochimica et Cosmochimica Acta, v. 68 (11S), p. A-751.

Rouchon, V., Sugihara A., Orberger, B., Pinti, D.L., Hashizume, K., Gallien, J.P., Westall, F., 2004b, Using nuclear probe and mass spectrometry to constrain the nitrogen composition of an early Archean chert: The Mass Spectrometry Society of Japan Abstracts, Yamagata, 24–26 November 2004.

Rouchon, V., Pinti, D.L., Gallien, J.P., Orberger, B., Daudin, L., and Westall, F., 2005, NRA analyses of N and C in hydromuscovite aggregates from a 3.5 Ga chert from Kittys Gap, Pilbara, Australia: Nuclear Methods and Instruments in Physics Research, v. B231, p. 536–540.

Rouzaud, J.N., and Clinard, C., 2002, Quantitative high resolution transmission electron microscopy: A promising tool for carbon materials characterization: Fuel Processing Technology, v. 77–78, p. 229–235, doi: 10.1016/S0378-3820(02)00053-X.

Russell, M., and Arndt, N., 2005, Geodynamic and metabolic cycles in the Hadean: Biogeosciences, v. 2, p. 97–111.

Schopf, J.W., 1993, Microfossils of the early Archean Apex Chert: New evidence of the antiquity of life: Science, v. 260, p. 640–646.

Schopf, J.W., and Packer, B.M., 1987, Early Archean (3.3-billion- to 3.5-billion-year-old) microfossils from Warrawoona Group, Australia: Science, v. 237, p. 70–73.

Schopf, J.W., and Walter, M.R., 1983, Archean microfossils: New evidence of ancient microbes, *in* Schopf, J.W., ed., Earth's Earliest Biosphere: Princeton, Princeton University Press, p. 214–239.

Schopf, J.W., Kudryavtsvev, A.B., Agresti, D.G., Wdowiak, T.J., and Czaja, A.D., 2002, Laser-Raman imagery of Earth's earliest fossils: Nature, v. 416, p. 73–76, doi: 10.1038/416073a.

Skrzypczak, A., Derenne, S., Robert, F., Binet, L., Gourier, D., Rouzaud, J.-N., and Clinard, C., 2004, Characterization of the organic matter in an Archean chert (Warrawoona, Australia): Lunar and Planetary Science Conference Abstracts XXXV, no. 1241.

Stal, L.J., 1994, Microbial mats: Ecophysiological interactions related to biogenic sediment stabilization, *in* Krumbein, W.E., et al., eds., Biostabilization of Sediments: Oldenburg, Biblioteks und Informationssystem der Carl von Ossietzky Universität, p. 41–54.

Thorseth, I.H., Furnes, H., and Heldal, M., 1992, The importance of microbiological activity in the alteration of natural basaltic glass: Geochimica et Cosmochimica Acta, v. 56, p. 845–850, doi: 10.1016/0016-7037(92)90104-Q.

Tice, M., and Lowe, D.R., 2004, Photosynthetic microbial mats in the 3,416-Myr-old ocean: Nature, v. 431, p. 549–552, doi: 10.1038/nature02888.

Toporski, J.K.W., Westall, F., Thomas-Keprta, K.A., Steele, A., and Mckay, D.S., 2002, The simulated silicification of bacteria—New clues to the modes and timing of bacterial preservation and implications for the search for extraterrestrial microfossils: Astrobiology, v. 2, p. 1–26, doi: 10.1089/153110702753621312.

Van Kranendonk, M.J., Hickman, A.H., Smithies, R.H., Nelson, D.R., and Pike, G., 2002, Geology and tectonic evolution of the Archean North Pilbara Terrain, Pilbara Craton, Western Australia: Economic Geology and the Bulletin of the Society of Economic Geologists, v. 97, p. 695–732.

van Zuilen, M., Lepland, A., and Arrhenius, G., 2002, Reassessing the evidence for the earliest traces of life: Nature, v. 418, p. 627–630, doi: 10.1038/nature00934.

van Zuilen, M., Lepland, A., Teranes, J., Finarelli, J., Wahlen, M., and Arrhenius, G., 2003, Graphite and carbonates in the 3.8 Ga old Isua Supracrustal Belt, southern West Greenland: Precambrian Research, v. 126, p. 331–348, doi: 10.1016/S0301-9268(03)00103-7.

Verchovsky, A.B., Sephton, M.A., Wright, I.P., and Pillinger, C.T., 2002, Separation of planetary noble gas carrier from bulk carbon in enstatite chondrites during stepped combustion: Earth and Planetary Science Letters, v. 199, p. 243–255, doi: 10.1016/S0012-821X(02)00592-7.

Walsh, M.M., 1992, Microfossils and possible microfossils from the early Archean Onverwacht Group, Barberton Mountain Land, South Africa: Precambrian Research, v. 54, p. 271–293, doi: 10.1016/0301-9268(92)90074-X.

Walsh, M.M., 2004, Evaluation of early Archean volcanoclastic and volcanic flow rocks as possible sites for carbonaceous fossil microbes: Astrobiology, v. 4, p. 429–437, doi: 10.1089/ast.2004.4.429.

Walsh, M.M., and Lowe, D.R., 1999, Modes of accumulation of carbonaceous matter in the early Archaean: A petrographic and geochemical study of carbonaceous cherts from the Swaziland Supergroup, *in* Lowe, D.R., and Byerly, G.R., eds., Geologic evolution of the Barberton greenstone belt, South Africa: Geological Society of America Special Paper 329, p. 115–132.

Walter, M.R., 1983, Archean stromatolites: Evidence of the Earth's earliest benthos, *in* Schopf, J.W., ed., Earth's Earliest Biosphere: Princeton, Princeton University Press, p. 187–213.

Westall, F., 1997, The influence of cell wall composition on the fossilization of bacteria and the implications for the search for early life forms, *in* Cosmovici, C.B., et al., eds., Astronomical and Biochemical Origins and the Search for Life in the Universe: Bologna, Editrici Compositori, p. 491–504.

Westall, F., 1999, The nature of fossil bacteria: Journal of Geophysical Research, Planets, v. 104, p. 16,437–16,451, doi: 10.1029/1998JE900051.

Westall, F., 2003, Stephen Jay Gould, les procaryotes et leur évolution dans le contexte géologique: Palevol, v. 2, p. 485–501, doi: 10.1016/j.crpv.2003.09.009.

Westall, F., 2004, Early life on earth: The ancient fossil record, *in* Ehrenfreund, P., et al., eds., Astrobiology: Future Perspectives: Dordrecht, Kluwer, p. 287–316.

Westall, F., 2005a, Life on the early Earth: A sedimentary view: Science, v. 308, p. 366–367, doi: 10.1126/science.1107227.

Westall, F., 2005b, The geological context for the origin of life and the mineral signatures of fossil life, *in* Martin, H., et al., eds., The Early Earth and the Origin of Life: Berlin, Springer, p. 195–219.

Westall, F., and Folk, R.L., 2003, Exogenous carbonaceous microstructures in early Archean cherts and BIFs from the Isua greenstone belt: Implications for the search for life in ancient rocks: Precambrian Research, v. 126, p. 313–330, doi: 10.1016/S0301-9268(03)00102-5.

Westall, F., and Rincé, Y., 1994, Biofilms, microbial mats and microbe-particle interactions: Electron microscope observations from diatomaceous sediments: Sedimentology, v. 41, p. 147–162.

Westall, F., and Walsh, M.M., 2003, Fossil biofilms and the search for life on Mars, *in* Krumbein, W.E., et al., eds., Fossil and Recent Biofilms: Amsterdam, Kluwer, p. 447–465.

Westall, F., Boni, L., and Guerzoni, M.E., 1995, The experimental silicification of microbes: Palaeontology, v. 38, p. 495–528.

Westall, F., Steele, A., Toporski, J., Walsh, M., Allen, C., Guidry, S., Gibson, E., Mckay, D., and Chafetz, H., 2000, Polymeric substances and biofilms as biomarkers in terrestrial materials: Implications for extraterrestrial samples: Journal of Geophysical Research, Planets, v. 105, p. 24,511–24,527, doi: 10.1029/2000JE001250.

Westall, F., De Wit, M.J., Dann, J., Van Der Gaast, S., de Ronde, C., and Gerneke, D., 2001, Early Archean fossil bacteria and biofilms in hydrothermally influenced, shallow water sediments, Barberton greenstone belt, South Africa: Precambrian Research, v. 106, p. 93–116, doi: 10.1016/S0301-9268(00)00127-3.

Westall, F., Orberger, O., Rouchon, V., Rouzaud, J.-N., and Wright, I., 2004, On the identification of early Archaean microfossils in cherts from Barberton and the Pilbara, *in* Reimold, U.W., and Hofmann, A., comp., Abstract volume, Field Forum on Processes on the Early Earth, Kaapvaal craton, 4–9 July, 2004: Johannesburg, University of the Witwatersrand, p. 94–97.

Williams, I.R., 1990, Geology of the Muccan: Western Australia Geological Survey, Geological Series Explanatory Notes, scale 1:100,000, 39 p.

Wright, I.P., and Pillinger, C.T., 1989, Carbon isotopic analysis of small samples by use of stepped-heating extraction and static mass spectrometry, *in* Shanks, W.C., and Criss, R.E., eds., New Frontiers in Stable Isotopic Research: Laser Proves, Ion Probes, and Small-Sample Analysis: U.S. Geological Survey Bulletin, v. 7890, p. 9–34.

Yushkin, N.P., 2000, Biomineral homologies, organismobiosis, and the problem of biomarkers, *in* Hoover, R.B., ed., Instruments, Methods, and Missions for Astrobiology III, Proceedings of SPIE, the International Society for Optical Engineering, 4137, p. 22–35.

Manuscript Accepted by the Society 4 October 2005

Printed in the USA

Geological Society of America
Special Paper 405
2006

Microfacies and origin of some Archean cherts (Pilbara, Australia)

Beate Orberger[†]
Département des Sciences de la Terre, UMR IDES 8148, Interaction et Dynamique des Environnements de Surface, Université Paris Sud, Bât. 504, 91405 Orsay Cedex, France
Virgile Rouchon
Département des Sciences de la Terre, UMR IDES 8148, Interaction et Dynamique des Environnements de Surface, Université Paris Sud, Bât. 504, 91405 Orsay Cedex, France and
Department of Earth and Space Sciences, Graduate School of Science, Osaka University, Toyonaka, Osaka 560-0043, Japan
Frances Westall
Centre de Biophysique Moléculaire, CNRS, Rue Charles Sadron, 45071 Orléans cedex 2, France
Sjoukje T. de Vries
Faculty of Geosciences, Utrecht University, Budapestlaan 4, 3584 CD Utrecht, The Netherlands
Daniele L. Pinti
Département des Sciences de la Terre et de l'Atmosphère, Université du Québec à Montréal et GEOTOP-UQAM-McGill, C.P. 8888, Succ. Centre-Ville, Montréal, QC H3C 3P8, Canada
Christiane Wagner
Laboratoire de Pétrologie-PMMP, CNRS-FR32, Université Paris VI, 4 Place Jussieu, 75251 Paris cedex, France
Richard Wirth
GeoForschungsZentrum Potsdam, Department 4, Telegrafenberg, D-14482 Potsdam, Germany
Ko Hashizume
Department of Earth and Space Sciences, Graduate School of Science, Osaka University, Toyonaka, Osaka 560-0043, Japan

ABSTRACT

Four cherts sampled in the East Pilbara craton (Western Australia) at Marble Bar (Towers Formation), North Pole Dome (Dresser and Apex Basalt Formation), and Kittys Gap (Panorama Formation) were studied for micro- and nanomineralogy and geochemistry to determine their protoliths and to provide new insights on the physico-chemical and biological conditions of their depositional environments. The Marble Bar chert was formed at the interface with a basaltic rock. Hydrothermal fluids leached major and trace elements from the basalt and silicified the protolith of this chert. The elements Fe, Mn, Si, Ca, Mg, REE, Au, Pd, Cr, and Ni precipitated as a microbanded iron formation (BIF) under reducing and alkaline conditions. The chert is composed of magnetite, carbonates, and quartz and forms a stromatolite-like structure. Later oxidizing fluids replaced magnetite and carbonates with Fe-Mn oxyhydroxides. They show vermicular microtextures and filamentous nanotextures. Each filament is composed of euhedral nanoscopic hematite. These oxides contain several thousands of ppm of N and C, and measured C/N ratios are similar to those observed in organic matter preserved in marine sediments, thus suggesting an organic activity.

[†]E-mail: orberger@geol.u-psud.fr.

Orberger, B., Rouchon, V., Westall, F., de Vries, S.T., Pinti, D.L., Wagner, C., Wirth, R., and Hashizume, K., 2006, Microfacies and origin of some Archean cherts (Pilbara, Australia), *in* Reimold, W.U., and Gibson, R.L., Processes on the Early Earth: Geological Society of America Special Paper 405, p. 133–156, doi: 10.1130/2006.2405(08). For permission to copy, contact editing@geosociety.org.

Two black cherts from hydrothermal dykes of the North Pole Dome are interpreted as having had a black shale precursor, based on the REE (rare earth elements) and trace metal characteristics. These rocks were probably entrained into the dykes and hydrothermally overprinted. Although these two cherts had the same history, the physico-chemical conditions differed during their formation. The chert from the chert-barite unit of the Dresser Formation was formed under reducing and alkaline conditions. This is clearly indicated by clusters of nanosulfide spherules replacing precursor minerals; weblike Fe-sulfides intergrown with sphalerite; As-pyrite and vaesite; and the presence of carbonates. The black chert from the Apex Basalt Formation was formed under oxidizing conditions, as indicated by clusters of nanospherules of Fe-oxides and a negative Ce anomaly.

A black and white laminated chert from Kittys Gap was formed in a shallow marine to subaerial environment, by silicification of a rhyodacitic volcaniclastic rock. This process was accompanied by the development of microbial mats on the sediment surfaces and the formation of microbial colonies around precursor K-feldspars, Ti-bearing biotites, amphiboles, and ghost spherulites. The environment was slightly oxidizing, as indicated by the negative Ce anomaly and the presence of Ti-oxides. The presence of K-bearing phyllosilicates rather than K-feldspars indicates that the environment was also slightly acidic. Elevated Cu and Zn contents in the black laminae point to a limited influence from hydrothermal fluids. The silica probably originated mainly from alteration of the minerals of the volcaniclastic rock due to diagenetic alteration by seawater.

Keywords: Archaean sediments, chert, Banded Iron Formations, iron oxyhydroxides, microbial activity.

INTRODUCTION

Archean cherts are at the center of much debate because they are considered to host the first traces of early life, either as physical (e.g., stromatolites, microbial mats, bacteriomorph structures) or chemical signatures (N, C, and S) interpreted as being derived from biological activity (Schidlowski et al., 1983; Byerly et al., 1986; Walsh, 1992; Beaumont and Robert, 1999; Westall et al., 2001, this volume; Pinti et al., 2001b). These siliceous rocks can have different origins depending on their environment of formation. In modern and Paleozoic settings (Grenne and Slack, 2003a, 2003b), silica can precipitate at mid-ocean ridges (MOR) (Halbach et al., 2002), in shallow littoral subaquatic (Tarasov et al., 1999), or in lacustrine environments (Eugster, 1967; Schubel and Simonson, 1990; Umeda, 2003; Renaut et al., 2002). In the latter case, they are known as siliceous sinter deposits and occur in hot springs related to geothermal activity. Fluid temperatures at the surface are generally below 100 °C, but reservoir temperatures vary from 175 to 350 °C (Fournier and Rowe, 1966). Modern hydrothermal environments contain microorganisms (mostly prokaryotes) that can be rapidly mineralized and, thus, have a high preservation potential in the geological record (Cady and Farmer, 1996). Cherts can also form by replacement of preexisting magmatic (volcanic or plutonic) or sedimentary rocks, by silica metasomatism (de Wit et al., 1982; Pirajno and Grey, 2002; Orberger et al., 2005a), or they can be of biogenic-organic origin (such as diatomites, radiolarites; e.g., Beauchamp and Baud, 2002). The latter circumstances are not considered to be relevant in the Archean.

The analysis of biogenicity in Archean cherts needs an integrated, multidisciplinary approach to analyze depositional conditions and subsequent diagenetic or metamorphic transformations, which can modify the primary biogenic and abiogenic signatures (Farmer and Des Marais, 1999; Grotzinger and Knoll, 1995). In this paper, we report the results of analysis of four Archean chert samples from the Pilbara craton: a chert from the Marble Bar, Towers Formation; two cherts from the North Pole Dome, belonging to the Dresser and Apex Basalt Formation, and a chert from the Kittys Gap locality of the Coppin Belt, Panorama Formation (Fig. 1). We used micromineralogy and REE–geochemistry to reconstruct the depositional environments for these rocks and their physico-chemical conditions during silicification.

GEOLOGICAL SETTINGS

The four cherts belong to the Warrawoona Group of the East Pilbara Greenstone Belt, Western Australia (Fig. 1). The Marble Bar chert (sample A458; UTM AMG50, 782100E, 7654600N) belongs to the A-section of the Towers Formation (Kato and Nakamura, 2003; Fig. 1). Following the sampling scheme of Kato and Nakamura (2003), the sample was obtained from a horizon 2251–2308 cm from the bottom of the section. The A-section is the thickest (45 m thick) stratum in this forma-

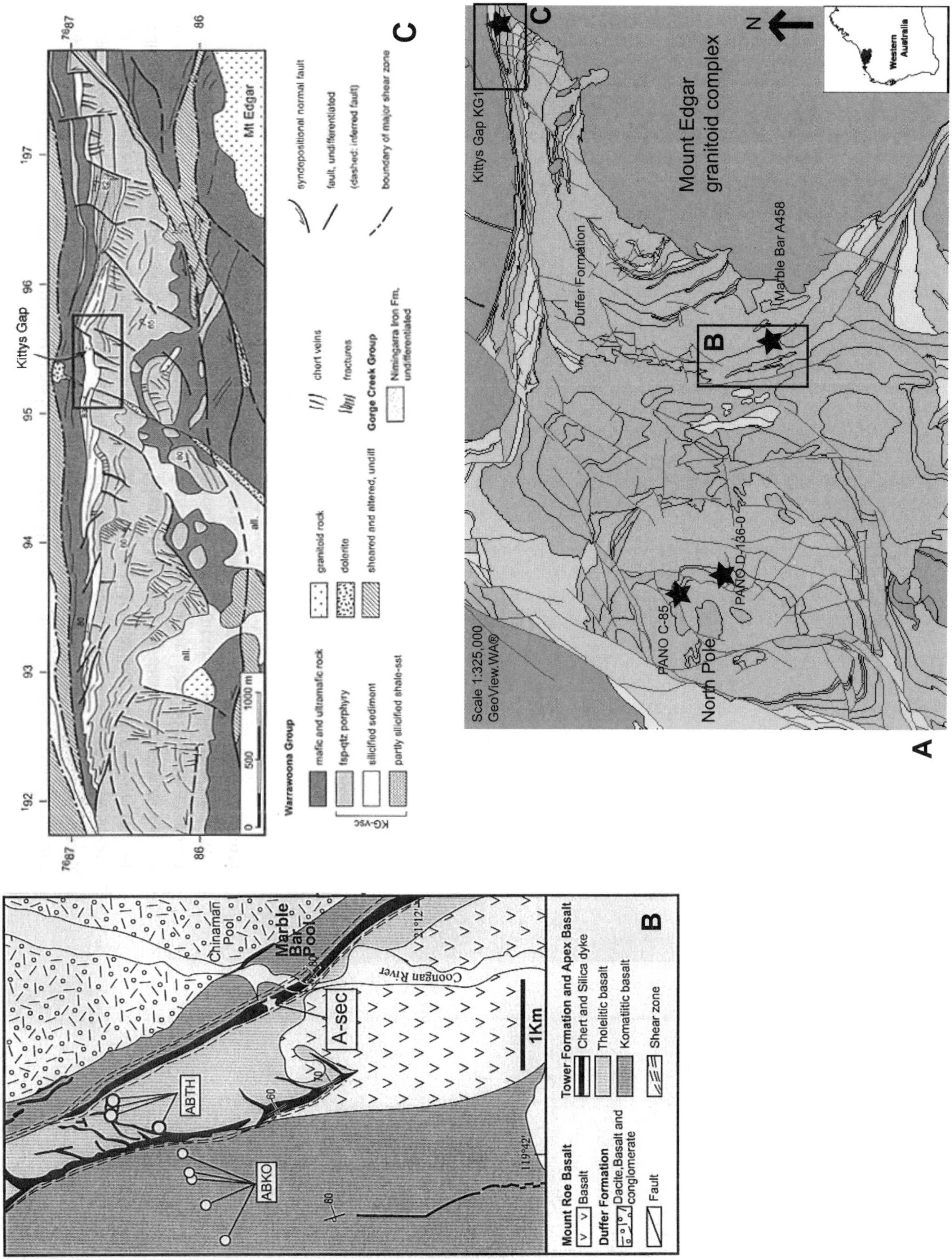

Figure 1. Simplified geological map of a part of Western Australia with the sampling positions of the studied cherts PANO D-136–0 and PANO C-85, A458 and KG-1. Map source (modified): Department of Industry and Resources of Western Australia, http://www.doir.wa.gov.au/aboutus/geoview_launch.asp.

tion. It is composed of highly ferruginous cherts directly overlying Fe-rich, K-poor, tholeiitic basalts having MORB–type (mid-oceanic-ridge basalt) geochemical patterns (Kato and Nakamura, 2003). The estimated depositional age of the Marble Bar Chert Unit is between 3474 and 3433 Ma (Thorpe et al., 1992; van Kranendonk et al., 2001). The interpretation of the depositional environment of the Marble Bar belt is controversial. Suggestions range from a volcanic- or near-arc setting, based on the chemical composition of the tholeiitic and calc-alkaline basalts (Barley, 1993; Kitajima et al., 2001), to an oceanic plateau setting, based on the association of tholeiites and komatiites (Condie, 1997). The cherts from the A-section have been recently interpreted as in situ precipitates from a high-temperature hydrothermal solution emanating from a MOR (Kato and Nakamura, 2003).

Chert sample PANO D-136–0 (UTM AMG50 754300E, 7662100N; previously named PANO D-136 *in* Pinti et al., 2001b) represents the North Pole Dome locality (Van Kranendonk et al., 2001) and belongs to the Talga-Talga Subgroup, Dresser Formation, Chert-Barite Unit (Fig. 1). This unit has an estimated depositional age of 3490 Ma (Thorpe et al., 1992). The Dresser Formation (Towers Formation in the former stratigraphical scheme of the area, Van Kranendonk et al., 2001) is composed of five horizons of stromatolitic chert-barite horizons, interbedded with pillow basalts. The sample studied here is a black chert from one of the numerous silica dykes that were emplaced within and immediately above listric normal growth faults that were active during chert deposition (Y. Isozaki, 1997, personal commun.; Van Kranendonk et al., 2001). These silica dykes have been recently interpreted as hydrothermal veins (Isozaki et al., 1999; Nijman et al., 1998; van Kranendonk et al., 2001; Ueno et al., 2004) and are possibly the remains of white hydrothermal smokers (Nijman et al., 1998). The silica would have been precipitated from hydrothermal fluids during a lacco-caldera formation (Nijman et al., 1998; van Kranendonk et al., 2001).

Chert sample PANO C-85 (UTM AMG50 751900E, 7667100N) is a zoned black and white chert from a vein that has been stratigraphically assigned to the 3465-Ma-old Salgash Subgroup, Apex Basalt Formation, Apex Chert Unit (Y. Ueno, 2004, personal commun.). Pinti et al. (2001a) erroneously related it to the volcaniclastic, felsic rock-rich Panorama Formation. The Apex Basalt Formation is dominated by pillow basalts, which are largely tholeiitic. A number of mainly black and white, well-laminated chert horizons are interbedded with the volcanics. Numerous chert dykes cut across the underlying stratigraphy and terminate at the level of the chert horizons (Lindsay et al., 2003).

The Kittys Gap locality (UTM AMG50 195200E, 7687200N) belongs to the Coppin Gap greenstone belt, north of the Mount Edgard Complex (Figs. 1A and 1C). The Coppin Gap is a weakly metamorphosed stratigraphic sequence of basalt and felsic volcanics and siliceous sedimentary rocks, offset by normal faults (Nijman et al., 1998; Van Kranendonk et al., 2001; de Vries, 2004). The felsic rocks, grading from rhyolitic to dacitic toward the top, belong to the Panorama Formation (de Vries, 2004). The chert from Kittys Gap, sample KG 1, comes from the 3446 Ma Panorama Formation, Kittys Gap Chert lower unit (de Vries, 2004). The outcrop is composed of a small-scale, heteroclitic, ripple-laminated chert.

ANALYTICAL METHODS

Polished thin sections were studied with a reflected and transmitted light optical microscope and by scanning electron microscopy (SEM, Philips XL 30) at the University of Paris Sud, France. Semiquantitative chemical analyses were performed at 20–30 kV using an EDX-PGT Ge-detector installed on the SEM. Some samples were etched for 5 s in 10 times diluted HNO_3 and later observed at 5 kV and analyzed at 10kV with field emission gun scanning electron microscopy (FEG-SEM of the Hitachi S 3400 type at the University of Orléans, France).

Major and trace element whole rock analyses were carried out at the SARM-CRPG (Centre de Recherches Pétrographiques et Géochimiques) of Nancy, France. The samples were fused with $LiBO_2$, dissolved in HNO_3, analyzed by inductive coupled plasma atomic emission spectrometry (ICP-AES: major and minor elements) and inductive coupled plasma mass spectrometry (ICP-MS: trace elements), and calibrated with international standards. Selenium was determined by Atomic Absorption Spectroscopy (AAS). Sulfur and CO_2 were analyzed using infrared absorptiometry. Organic carbon was obtained through calcination at 1400 °C, after destruction of the carbonates. The FeO contents were analyzed by volumetry. For details on the precision and detection limits, see the appendix. Trace elements in the Fe-Mn–oxides, carbonates, sulfides, and hydromuscovites were analyzed by electron microprobe (Cameca SX50®) at the Centre de Microanalyses de Paris (CAMPARIS) Center, Université Pierre et Marie Curie of Paris, France. Counting time for each element was 10 s. Standards were diopside (Mg, ca), orthoclase (K, Al), pyrophanite (Mn, Ti), Cr_2O_3 (Cr), Fe_2O_3 (Fe), ZnS (Zn), NiO (Ni), Co metal (Co), and Cu metal (Cu). Detection limits are at the hundred ppm level. Representative analyses are shown in Tables 1A–1D. Focused-ion beam (FIB) transmission electron microscopy was performed at the GeoForschungsZentrum (GFZ) in Potsdam (Germany). The method applied is described in Wirth (2004).

Results

Marble Bar Chert–Towers Formation (A458)

Mineralogy. Sample A458 is a composite chert consisting of a dark gray part (zone I), a microbanded iron formation a few millimeters thick (micro–BIF, zone II), and a light gray part (zone III, Fig. 2A). Transmitted light microscopy indicates that, prior to silicification, the precursor rock of zone I had a porphyritic or porphyroclastic texture, typical of volcanic or volcaniclastic rocks. Euhedral to subhedral phenocrysts were replaced by microcrystalline quartz. These replacement structures sometimes contain euhedral Fe-sulfides showing a Fe-oxide rim (Fig. 2B). Zone II is composed of alternating, millimeter-wide

TABLE 1. REPRESENTATIVE ELECTRON MICROPROBE ANALYSES OF Fe-Mn OXYHYDROXIDES IN MARBLE BAR CHERT A458; Fe AND Ni SULFIDES IN BLACK CHERT PANO D-136–0; CARBONATES IN BLACK CHERT PANO D-136–0; AND HYDROMUSCOVITE IN KITTYS GAP CHERT KG 1

(A) Fe-Mn oxyhydroxides: A458

wt%	*in BIF*				in Qz vein	
Mg0	1.82	1.82	1.64	3.88	1.72	1.67
Al_2O_3	0.02	0.02	0.03	0.00	0.05	0.02
SiO_2	4.75	4.54	3.84	1.26	4.33	5.60
Cr2O3	0.03	0.06	0.00	0.00	0.00	0.00
CuO	0.08	0.00	0.02	0.19	0.10	0.00
ZnO	0.14	0.00	0.05	0.00	0.00	0.00
TiO_2	0.04	0.01	0.00	0.00	0.00	0.00
MnO	0.63	0.55	14.57	45.20	4.79	6.14
FeO	72.50	72.15	56.94	13.52	64.36	63.31
NiO	0.02	0.00	0.10	0.03	0.07	0.13
K_2O	0.00	0.01	0.08	0.59	0.00	0.04
CaO	0.13	0.05	0.36	0.48	0.30	0.25
Total	80.16	79.20	77.62	65.15	75.72	77.18

(B) Fe- and Ni-Sulfides: D-136–0

at%	Pyrite									Vaesite
S	66.57	64.02	66.68	65.61	67.14	66.85	66.48	66.23	66.39	67.26
Fe	33.30	28.72	32.11	30.63	31.65	32.33	32.46	32.81	32.67	0.74
Ni	0.00	0.41	0.66	0.90	0.64	0.28	0.09	0.26	0.28	30.72
Cu	0.03	0.05	0.14	0.00	0.10	0.06	0.01	0.01	0.05	0.01
Zn	0.00	0.04	0.01	0.01	0.02	0.00	0.03	0.01	0.03	0.05
As	0.00	0.15	0.24	0.80	0.30	0.35	0.86	0.57	0.47	0.52
Se	0.00	0.45	0.04	0.14	0.03	0.01	0.00	0.01	0.01	0.61
Pb	0.09	6.18	0.12	1.91	0.12	0.12	0.08	0.10	0.10	0.09
Total	99.99	100.02	100.00	100.00	100.00	100.00	100.00	100.00	100.00	100.00

(C) Carbonates: D-136–0

wt%	Dolomite		Calcite		
CaO	27.18	27.32	53.14	53.00	47.71
MgO	14.95	15.07	0.33	0.35	6.45
FeO	7.45	7.70	0.06	0.14	0.31
MnO	1.63	1.37	0.57	0.46	0.51
P205	0.33	0.35	0.69	0.82	0.56
F	0.10	0.00	0.09	0.00	0.00
	51.65	51.85	54.95	54.84	55.60

(D) Hydromuscovite: KG

wt%				Average of 78 analyses
SiO_2	45.11	45.62	46.34	47.42
Al_2O_3	35.50	34.81	34.50	35.56
K_2O	6.44	10.09	9.32	9.12
MgO	0.48	0.63	1.44	0.65
Na_2O	0.37	0.46	0.36	0.38
Fe_2O_3	3.07	0.41	0.61	0.34
TiO_2	0.12	0.12	0.33	0.16
CaO	0.05	0.03	0.06	0.06
Cl	0.01	0.01	0.00	0.01
MnO	0.03	0.00	0.00	0.01
Total	91.17	92.16	92.97	93.72

Notes: Data in wt%. Electronmicroprobe analyses, Centre Camparis, Université Paris VI. Abbreviations: BIF—Banded Iron Formation; Qz—quartz.

iron-manganese oxyhydroxide (FeMnOH) and quartz bands (microbanded iron formations, micro–BIFs hereafter; Fig. 2A). The iron-rich bands are formed of skeletal goethite aggregates and euhedral rhombic-shaped phenocrysts of Fe-Mn oxyhydroxides (Fig. 2C). The phenocrysts have an inner euhedral core of vermicular textures and a heterogeneous composition of alternating bright and dark microbands. A euhedral overgrowth zone, containing traces of As and Ca is separated from the inner crystal by K-Al silicates (Fig. 2D). The focused ion beam transmission electron microscopy (FIB-TEM) investigations indicate that each microband is composed of nanoscopic filaments, each of them consisting of euhedral hematite crystals. The quartz bands between the FeMnOH rims contain numerous euhedral Fe-Mg oxide (magnesian-bearing magnetite or magnesioferrite) inclusions (Fig. 2E). Xenotime grains (Fig. 2F), rich in heavy rare earth elements (HREE) and Au-Pd-Fe-Cu-Ni alloys, occur interstitially with the FeMnOH. Zone III has a fine-grained homogeneous texture and is composed of microcrystalline quartz crosscut by millimeter-wide quartz veins, which contain FeMnOH with similar textures to those observed in the micro–BIFs and Ca-Mg carbonates which are partly or completely dissolved and replaced by FeMnOH (Fig. 2G). The microcrystalline quartz

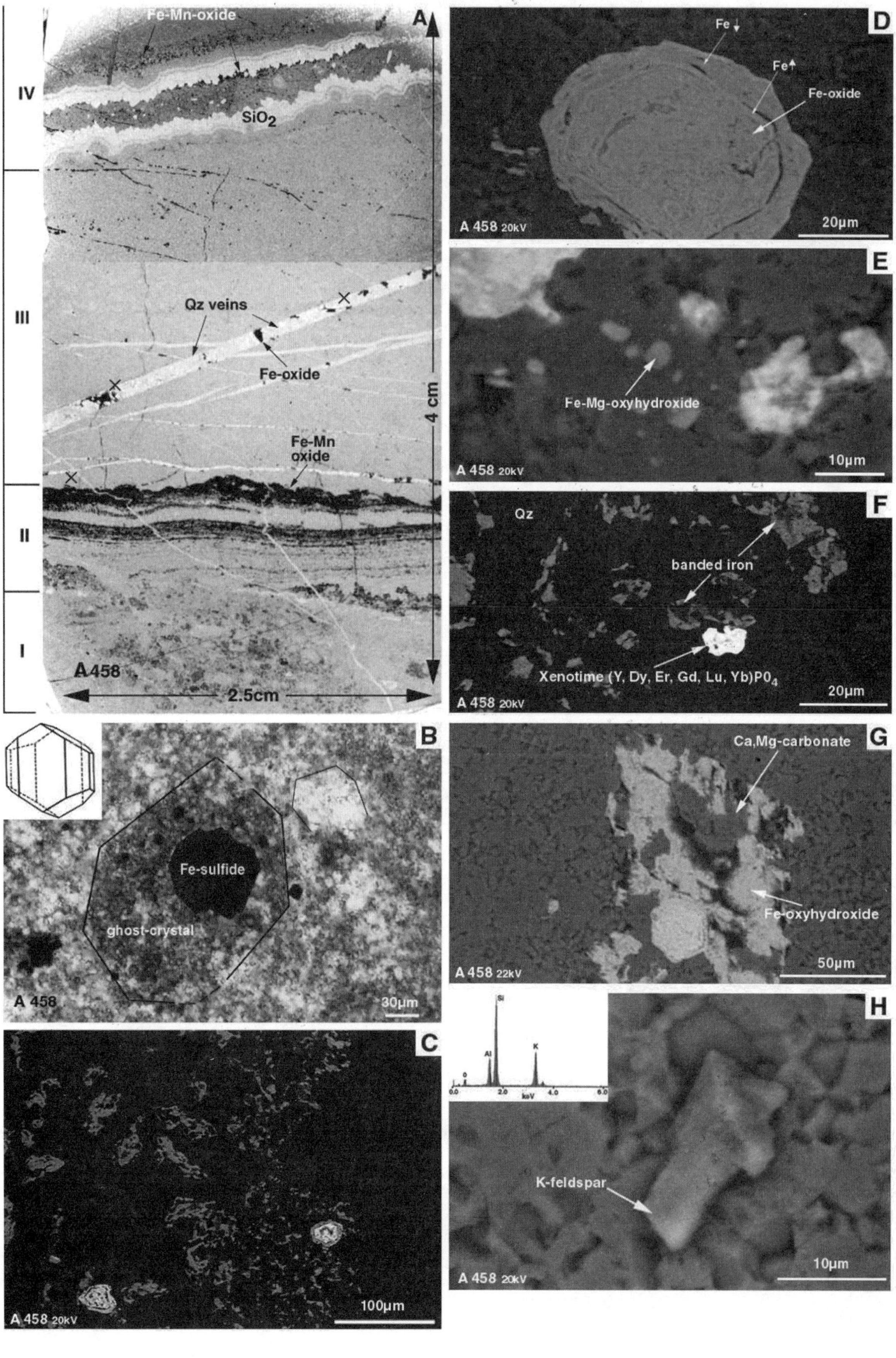
A
Fe-Mn-oxide
IV
SiO2
III
Qz veins
Fe-oxide
4 cm
Fe-Mn oxide
II
I
A 458
2.5cm
B
Fe-sulfide
ghost-crystal
A 458
30µm
C
A 458 20kV
100µm
D
Fe↓
Fe↑
Fe-oxide
A 458 20kV
20µm
E
Fe-Mg-oxyhydroxide
A 458 20kV
10µm
F
Qz
banded iron
Xenotime (Y, Dy, Er, Gd, Lu, Yb)P04
A 458 20kV
20µm
G
Ca,Mg-carbonate
Fe-oxyhydroxide
A 458 22kV
50µm
H
Si
Al
K
keV
K-feldspar
A 458 20kV
10µm

matrix of zone III contains micrometer-sized K-Al silicates (Fig. 2H), rarely barite and Ba-bearing mica, and occasionally, Fe-sulfides with similar textures to the FeMnOH.

Electron microprobe analyses and high-resolution SEM combined with Energy Dispersive X-ray Analysis (EDX) on these highly heterogeneous and porous samples show that the FeOH contains traces of MgO (~2 wt%), MnO (0.36–1.1 wt%), CaO (500–1700 ppm), and NiO (<400 ppm). Mn-oxyhydroxides contain 3–6 wt% FeO, NiO (<500 ppm), MgO (4 wt%), CaO, and K_2O (9000 ppm for both these elements). Locally, traces of Cu, Zn, and Ti (<200 ppm) were detected in Fe and Mn minerals. Microprobe analyses give totals <100 wt%, which can be related to volatile contents but also to the high porosity of these phases. It is noteworthy that the FeMnOH minerals do not have stoichiometric compositions (Table 1A).

Geochemistry

Based on their optical characteristics, chert sections of different color (zones I, II, III, Fig. 2A) were separated and analyzed by ICP-MS (Table 2). The three parts differ in chemical composition. The Fe-oxyhydroxide rich (FOH) micro–BIFs are about twice as enriched in Fe_{tot}, corresponding to Fe_2O_3 (Table 2) and MnO as the porphyritic part, and slightly richer in MgO. The micro–BIFs have K contents close to the detection limit of 0.05 wt%, which is related to the K-Al silicates intergrown with the FOH. The loss on ignition of 2.29 wt% reflects molecular and adsorbed water and other volatiles in FOH. The micro–BIFs are also richer in divalent iron (0.11 wt% FeO) compared to 0.055 wt% in the porphyritic part. The presence of divalent iron might be related to magnetite inclusions in the quartz between the FOH bands. The porphyritic part and the micro–BIFs have similar trace element contents of As, Co, Cr, Ni, Pb, Rb, and U (Table 2). The micro–BIFs are about twice as rich in Ba, Zn, and Y, whereas the porphyritic part is richer in Sb, Sr, Zr, and V. The chert precipitate (zone III, Fig. 2A) is depleted in all trace elements, except for Ge, Nb, Ta, and Zr (Table 2). Bulk sulfur content is low (0.085 wt%) and is related to the rare Fe-sulfides and tiny barite crystals in the chert precipitate. The Se/S ratio is 1.2×10^{-4} (Se: 100 ppb). The C1 normalized (Evensen et al., 1978) REE element patterns of the porphyritic part and the micro–BIFs are similar. They are enriched in HREE (heavy rare earth elements) and characterized by a positive Eu anomaly. The porphyritic part has slightly lower REE contents compared to those of the micro–BIFs. Zone III is essentially composed of microquartz and micro–K-Al silicates and has comparatively the lowest REE contents, with most REE giving values below the detection limit (Table 2).

Figure 2. (A) Thin-section scan of chert A458 showing the silicified precursor basalt part (zone I), and, at the top, the iron-manganese oxide precipitation (zone II) in contact with the silica precipitate that is crosscut by secondary quartz veins, which also contain Fe-Mn oxides. (B) Transmitted not polarized optical microscopic photograph: precursor mineral, now composed of microcrystalline quartz. Euhedral Fe-sulfide crystals are precipitated preferentially in these precursor relicts. (C) Scanning electron microscopy-backscattered secondary electrons (SEM-BSE) image: Skeletal Fe-Mn oxide containing euhedral Fe-Mn oxides; enlargements are shown in Figure 2D. (D) SEM-BSE image: euhedral Fe-oxides, with internal vermicular structure of heterogeneous chemical composition. The interface between the overgrowth and the internal crystal is made up of K-Al silicate. (E) SEM-BSE image: euhedral Fe-Mg-oxide (magnetite or magnesioferrite) inclusions in quartz interstitial to the Fe-Mn oxyhydroxide bands. (F) SEM-BSE image: HREE–rich (heavy rare earth element) xenotime in Fe-Mn oxyhydroxide bands. (G) SEM-BSE image: Ca-Mg carbonate replaced by Fe-Mn oxyhydroxides. (H) SEM-BSE image: subhedral K-feldspar (compare Energy Dispersive Spectroscopy [EDS] spectrum in inset) intergrown with microquartz

NORTH POLE CHERT–BARITE UNIT–DRESSER FORMATION (PANO D-136–0)

Mineralogy

The PANO D-136–0 sample is a compositionally homogeneous black chert. Microscopic crosscutting veins are filled with quartz. Pull-apart structures indicate extensional tectonic deformation during syn- or post-chert formation. Transmitted light microscopy shows that, prior to silicification, the rock had a brecciated texture (Fig. 3A). White and dark, subhedral to euhedral former crystal shapes are filled with microcrystalline quartz (Fig. 3B). Abundant organic matter occurs dispersed as filaments in the chert matrix or as oval inclusions in silica (Figs. 3C and 3D). The chert matrix hosts numerous K-Al-silicate crystals (5 μm) that are in some cases palagonite, according to FEG-SEM (Fig. 3E). Euhedral dolomite and calcite (Figs. 3F and 3G; Table 1C), Sm-Yb-Cl bearing apatite, xenotime, anhedral florencite, native gold, and Ti-oxides were also found (Figs. 3H and 3I). Spherulitic Fe-sulfides (<2 μm) occur as clusters in carbonates (Fig. 3F), where they replace subhedral to euhedral minerals and occur along the grain boundaries of microcrystalline quartz (Figs. 3J, 3K, and 3M). The Fe-sulfide nanospherules contain internal compartmentalized voids (Fig. 3L). Weblike Fe-sulfides are intergrown with sphalerite that hosts arsenopyrite inclusions (Figs. 3N and 3O). Galena occurs as inclusions in Fe-sulfides. The weblike Fe-sulfides contain trace amounts of Pb (0.03–7.37 at%), Ni (0.03–6.57 at%), As (0.17–1.55 at%), Cu (0.2 at%), Zn (up to 0.069 at%), and Se (up to 0.51 at%). Vaesite (Ni_2S) contains, on average, 0.67 at% Se, 0.575 at% As, 0.81 at% Fe, up to 0.05 at% Zn, and 0.02 at% Cu (Table 1B). The carbonates include Mg-calcite containing ~0.5 wt% MnO and 7000–8200 ppm P_2O_5 and dolomite containing ~7 wt% Fe, 1.5 wt% Mn, and ~3500 ppm P_2O_5 (Table 1C).

Geochemistry

This black chert has a low total iron content (Fe_{tot} as Fe_2O_3: 0.27 wt%) with ~50% divalent Fe (0.13 wt%, Table 2). Sulfur and Se are 0.14 wt% and 0.67 ppm, respectively, giving a Se/S ratio of 4.8×10^{-4}. Selenium is incorporated in Fe- and Ni-

TABLE 2. BULK ROCK GEOCHEMISTRY OF CHERTS AND CHERT PARTS

Lithology	A section Tower Formation				Kittys Gap		Dresser Formation		
protolith	A 458 Bulk	A 458 I porphyric chert basalt	A 458 II BIF (FOH) BIF (Mag) precipitate	A 458 III white chert precipitate	KG 5 white chert	KG 6 black chert	D-136–0 black chert black shale	PANO C 85a white chert precipitate	PANO C 85b black chert black shale
					(pyroclastic sediments)				
SiO_2 (wt%)	94.26	91.89	83.91	99.21	92.36	96.28	98.79	99.4	96.26
Al_2O_3	0.19	0.42	0.16	<L.D.	5.51	2.63	0.27	<L.D.	0.24
Fe_2O_3	3.52	5.89	12.49	0.18	1.59	0.78	0.24	<L.D.	0.43
MnO	0.16	0.34	0.64	<L.D.	<L.D.	<L.D.	<L.D.	<L.D.	<L.D.
MgO	0.24	0.29	0.39	<L.D.	<L.D.	<L.D.	<L.D.	<L.D.	<L.D.
CaO	<L.D.	<L.D.	<L.D.	<L.D.	<L.D.	<L.D.	<L.D.	<L.D.	0.21
Na_2O	<L.D.	<L.D.	<L.D.	<L.D.	<L.D.	<L.D.	<L.D.	<L.D.	<L.D.
K_2O	<L.D.	<L.D.	0.05	<L.D.	1.59	0.78	0.05	<L.D.	<L.D.
TiO_2	<L.D.	<L.D.	<L.D.	<L.D.	0.28	0.09	<L.D.	<L.D.	<L.D.
P2O5	<L.D.	<L.D.	<L.D.	<L.D.	<L.D.	<L.D.	<L.D.	<L.D.	0.13
PF	0.79	1.58	2.29	<L.D.	<L.D.	<L.D.	0.41	0.39	1.38
Total	99.16	100.41	99.93	99.3	99.74	99.78	99.76	99.79	98.65
FeO	0.1	0.055	0.11	n.a.	n.a.	n.a.	0.13	n.a.	n.a.
S	0.085	n.a.	n.a.	n.a.	n.a.	n.a.	0.14	n.a.	0.39
CO_2	0.15	n.a.	n.a.	n.a.	n.a.	n.a.	0.58	n.a.	1.63
Corg	0.01	n.a.	n.a.	n.a.	n.a.	n.a.	0.14	n.a.	n.a.
H_2O+	0.56	n.a.	n.a.	n.a.	n.a.	n.a.	n.a.	n.a.	n.a.
H_2O-	0.05	n.a.	n.a.	n.a.	n.a.	n.a.	n.a.	n.a.	n.a.
Se (ppm)	0.1	n.a.	n.a.	n.a.	n.a.	n.a.	0.67	n.a.	n.a.
As	51.2	6.75	5.67	2.21	1.29	1.647	14.2	16.59	36.63
Ba	52.9	35.9	56.1	26.1	61.04	47.75	40.1	40.82	127.7
Be	<L.D.	<L.D.	<L.D.	<L.D.	<L.D.	<L.D.	<L.D.	<L.D.	<L.D.
Bi	<L.D.	<L.D.	<L.D.	<L.D.	<L.D.	<L.D.	0.12	0.25	0.46
Cd	<L.D.	<L.D.	<L.D.	<L.D.	<L.D.	<L.D.	1.07	<L.D.	1.95
Ce	0.86	1.03	0.56	0.29	14.6	12.95	1.15	0.58	6.65
CO_2	2.95	2.89	3.37	0.86	0.34	0.55	14	0.33	3.3
Cr	<L.D.	9.8	9.1	<L.D.	14.25	14.79	7.6	<L.D.	8.90
Cs	<L.D.	<L.D.	<L.D.	<L.D.	0.971	0.56	0.35	<L.D.	0.51
Cu	12.6	<L.D.	<L.D.	<L.D.	8.03	69.36	11.4	<L.D.	20.71
Dy	0.19	0.16	0.33	0.03	1.46	0.97	0.09	0.07	1.03
Er	0.15	0.15	0.23	0.02	0.87	0.52	0.05	0.04	0.57
Eu	0.05	0.05	0.08	<L.D.	0.49	0.38	0.03	0.02	0.28
Ga	0.4	0.99	0.56	0.34	9.01	4.33	0.46	0.23	0.8
Gd	0.15	0.13	0.2	<L.D.	1.44	1.08	0.09	0.06	1.01
Ge	1.26	1.16	1.08	1.18	2.08	2.09	0.85	0.89	0.91
Hf	<L.D.	0.07	<L.D.	0.04	1.65	1.01	0.1	<L.D.	0.12
Ho	0.04	0.04	0.08	<L.D.	0.29	0.18	0.02	0.02	0.22
In	<L.D.	0.26	0.26	0.15	<L.D.	<L.D.	0.06	0.18	0.31
La	0.45	0.52	0.26	0.10	7.37	6.27	0.57	0.35	4.47
Lu	0.03	0.04	0.08	<L.D.	0.13	0.07	0.01	0.01	0.07
Mo	0.93	<L.D.	<L.D.	<L.D.	<L.D.	<L.D.	0.77	78.91	1569
Nb	<L.D.	0.27	0.11	0.13	3.73	1.51	0.17	<L.D.	0.28
Nd	0.4	0.45	0.26	<L.D.	6.84	6.09	0.51	0.27	3.62
Ni	15.3	23.4	25.4	<L.D.	<L.D.	7.42	42	37.81	577.9
Pb	2.44	2.67	2.26	<L.D.	4.08	5.6	10	10.50	16.911
Pr	0.1	0.12	0.07	<L.D.	1.79	1.56	0.14	0.06	0.89
Rb	0.68	0.64	0.65	<L.D.	30.31	15.46	2.16	<L.D.	1.58
Sb	2.31	2.68	1.58	1.06	2.31	2.257	<L.D.	<L.D.	15.34
Sm	0.1	0.11	0.09	<L.D.	1.46	1.255	0.12	0.04	0.79
Sn	0.94	0.71	<L.D.	<L.D.	2.57	9.48	<L.D.	0.97	2.2
Sr	3.08	5.9	0.6	<L.D.	11.32	7.12	<L.D.	<L.D.	13.17
Ta	<L.D.	0.03	<L.D.	0.02	0.35	0.15	0.02	<L.D.	0.02
Tb	0.03	0.02	0.04	<L.D.	0.23	0.17	0.02	<L.D.	0.16
Th	<L.D.	0.11	<L.D.	<L.D.	2.19	1.37	0.32	<L.D.	0.28
Tm	0.02	0.03	0.05	<L.D.	0.14	0.07	0.01	0.01	0.08
U	0.07	0.13	0.1	<L.D.	0.57	0.40	0.13	0.35	15.81
V	1.87	7.2	4.6	<L.D.	42.89	23.69	3.5	<L.D.	20.41
W	0.11	<L.D.	<L.D.	<L.D.	2.36	1.34	0.07	0.16	0.84
Y	1.72	1.47	2.9	0.19	9.02	5.33	0.47	0.51	8.55
Yb	0.18	0.24	0.43	0.02	0.89	0.48	0.04	0.05	0.47
Zn	13.2	19.4	40.5	<L.D.	<L.D.	49.69	373	<L.D.	80.69
Zr	1.25	2.94	1.64	1.83	68.81	56.94	3.89	2.22	5.63
LaN/YbN	1.749	1.51	0.43	3.05	5.84	9.22	9.06	4.78	6.65

Notes: Analyses Service d'Analyse des Roches et Minéraux–Centre de Recherches Pétrographiques et Géochimiques (SARM–CRPG), Nancy, France. Abbreviations: BIF—Banded Iron Formation; FOH—Iron OxyHydroxides; L.D.—lower limit of detection; Mag—magnetite; n.a.—not analyzed. Major element data in wt%, trace element data in ppm.

sulfides (Table 1B). The presence of CO_2 (0.58 wt%) is related to Mg-Fe-Mn carbonates (Figs. 3F and 3G), and the organic carbon (0.14 wt%) is related to the organic matter embedded in microcrystalline quartz (Fig. 3D). High Zn contents of 373 ppm reflect the presence of sphalerite. Trace metals, such as As, Ni, Cu, Bi, Mo, Pb, and Cd, are incorporated in the Fe-Ni-Zn-sulfide structures (Table 1B). The C1-normalized REE pattern is slightly fractionated and depleted in HREE compared to LREE (light rare earth elements). A negative Eu and a positive Tm anomaly has been also noted.

NORTH POLE APEX BASALT FORMATION CHERT (PANO C-85)

Mineralogy

Chert PANO C-85 comprises a white and a black zone, each of which have a distinct mineralogy (Fig. 4A). Thin section observations reveal that a diffuse microcrystalline quartz vein system overprinted a previously formed chert. The white zone represents a part of the quartz vein system that contains mostly microquartz (Fig. 4B). Microscopic pull-apart structures, filled with relatively coarser-grained quartz, indicate that these veins were formed during an early extensional event. The black part has a complex mineralogical composition. Micrometer-sized Fe-oxides are abundant, and they exhibit different morphologies. (1) There are clusters of crystallites that pseudomorph mineral shapes that existed prior to silicification (Fig. 4C). The crystallites occur either as micro-ring–shaped structures or as spherules (Figs. 4D–4F). The spherules and ring-shaped structures are associated with K-Al silicates (K-feldspar?). (2) There are crystallites that are some tenths of a micrometer in size or occur as large subhedral to euhedral Fe-oxides showing heterogeneous agglomerated internal textures (Fig. 4G). These Fe-oxides contain $CaSO_4$ (possibly $+H_2O$) with traces of Cr. EDX analyses show that subhedral Fe-oxides contain V and As. (3) There are poikilotopic Fe-oxides of a heterogeneous composition (Fig. 4H) that contain inclusions of xenotime with traces of REE and W (Fig. 4I), as well as barite (Fig. 4J). In the black zone, micrometer-sized chalcopyrite occurs in interstitial and microcrystalline quartz. Furthermore, two types of poorly crystallized phases were observed: porous (with round holes), poorly crystallized, Si-rich phases, containing high amounts of Ca, Ba, and S, possibly indicating the presence of gypsum (anhydrite) and barite (Figs. 4K and 4L); and multielement (K-Al-Si-Ti-V-P-As-S-Sr-Ca-Pb-Ba) amorphous phases that sometimes include Fe-oxide in exsolution (Fig. 4M). Subhedral K-feldspar is rare.

Geochemistry

The white and black zones of sample PANO C-85 were analyzed separately (Table 2). The black zone has similar Al_2O_3 and Fe_2O_3 contents to those of the black chert PANO D-136–0. This zone hosts the highest P_2O_5 contents, at 0.13 wt%. Phosphorous occurs either as phosphates or in poorly crystallized multielement phases (Fig. 4I). The sulfur content of this black zone is twice that measured in the black chert PANO D-136–0. It is related to the presence of Ca-sulfates (gypsum or anhydrite) or barite (128 ppm Ba), rather than sulfides. The CO_2 content is three times higher than that in the white chert zone. It reflects the presence of carbonates. The most common base metals are Mo and Ni (at 1570 ppm and 578 ppm, respectively), which are more abundant than in the black chert PANO D-136–0, by a factor of ten.

The white zone is depleted in all elements compared to the black zone. However, As, Ba, Ni, and Pb abundances are as high as in the black chert PANO D-136–0. The REE patterns of the black and white zones are similar. They are LREE enriched (Fig. 5). However, the white zone is REE depleted by a factor of 10, in comparison to the black zone. Both zones show a slight negative Ce anomaly, which is more pronounced in the black than in the white part (Fig. 5A).

KITTYS GAP CHERT–PANORAMA FORMATION (KG 1)

Mineralogy

Sample KG 1 is composed of submillimeter- to millimeter-thick black and white laminae (Fig. 6A). Sedimentary structures indicative of formation in a tidal environment include channels, varying and bimodal current directions, and flaser-linsen bedding (de Vries, 2004). The major components of sample KG 1 are microcrystalline quartz (85–90 vol%) and hydromuscovite. The white and black laminae show different textures and grain size that were determined prior to silicification; however no difference in the respective mineralogies has been observed. The

Figure 3. Microphotographs of chert PANO D-136–0. (A) Thin-section scan. (B) Plane polarized light microphotograph showing the brecciated texture of the precursor black shale; dark precursor minerals are now composed of microquartz and sulfide clusters. (C) Organic matter dispersed in the microquartz matrix. (D) Field Emission Gun Scanning Electron Microscopy (FEG-SEM) photograph: NO_3-etched microquartz grain with organic matter. (E) Scanning electron microscopy-backscattered secondary electrons (SEM-BSE) image: subhedral K-Al silicate, probably K-feldspar, in cryptocrystalline quartz. (F) Plane polarized light microphotograph showing opaque mineral clusters (Fe-sulfides) on Mg-carbonate. (G) SEM-BSE image euhedral, zoned Mg-carbonate; the central zone contains traces of Cr and Mn. (H) SEM-BSE image: euhedral, HREE–bearing (heavy rare earth element) apatite in quartz matrix. (I) Florencite in quartz matrix. (J) Plane polarized light and reflected light microphotographs: spherulitic sulfide clusters replace euhedral precursor mineral. (K) Plane polarized light microphotograph: spherulitic sulfide clusters composed of nanospherules. (L) SEM-BSE image showing details of sulfide clusters. They are composed of Fe-sulfides with internal void structures. (M) Fe-sulfide in the form of alignments of spherules. (N) Massive Fe-sulfides intergrown with arsenopyrite and sphalerite. (O) Fe-sulfide with alternating more massive and weblike zones.

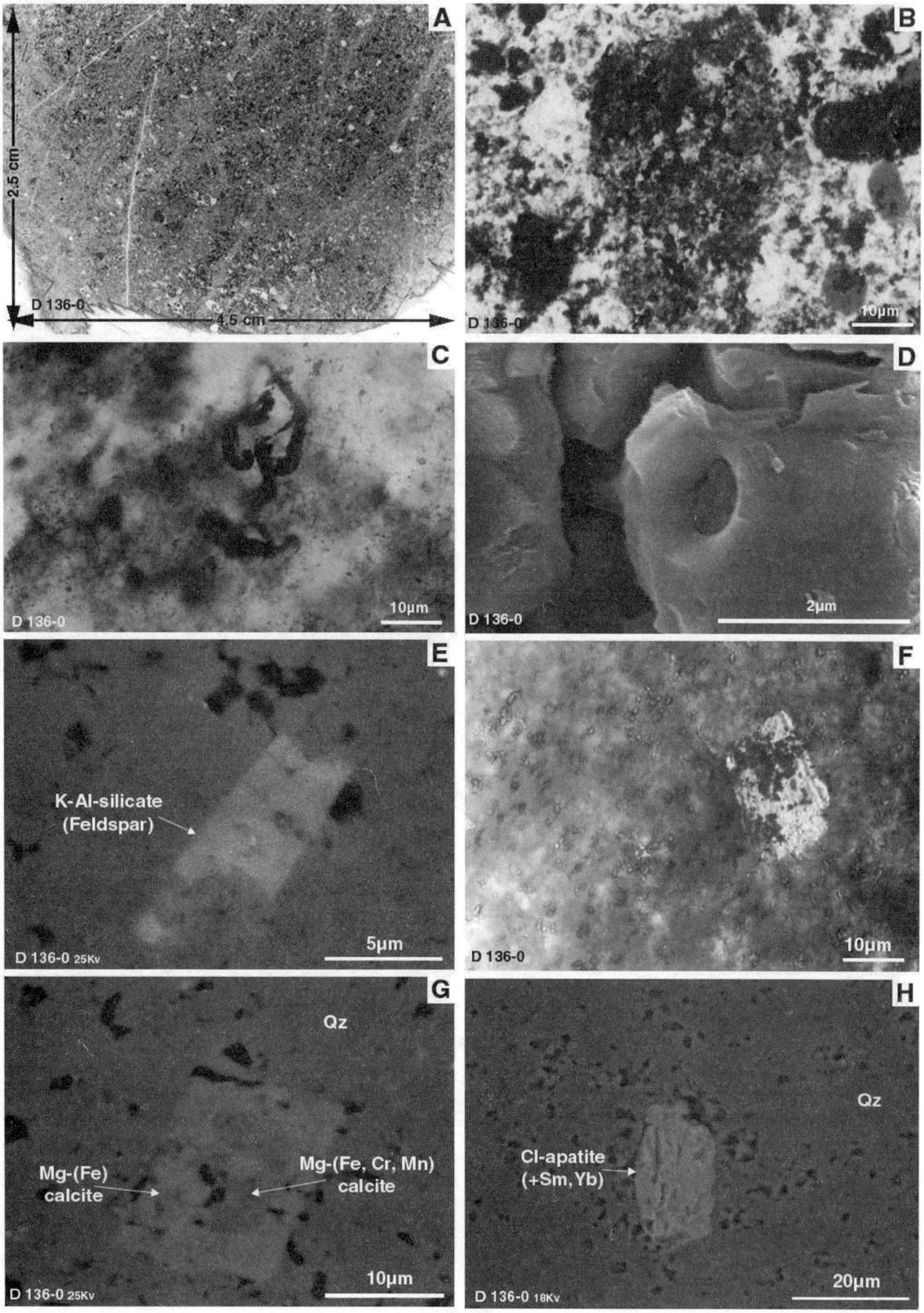
A
2.5 cm
4.5 cm
D 136-0
B
D 136-0
10µm
C
D 136-0
10µm
D
D 136-0
2µm
E
K-Al-silicate
(Feldspar)
D 136-0 25Kv
5µm
F
D 136-0
10µm
G
Qz
Mg-(Fe)
calcite
Mg-(Fe, Cr, Mn)
calcite
D 136-0 25Kv
10µm
H
Qz
Cl-apatite
(+Sm,Yb)
D 136-0 18Kv
20µm

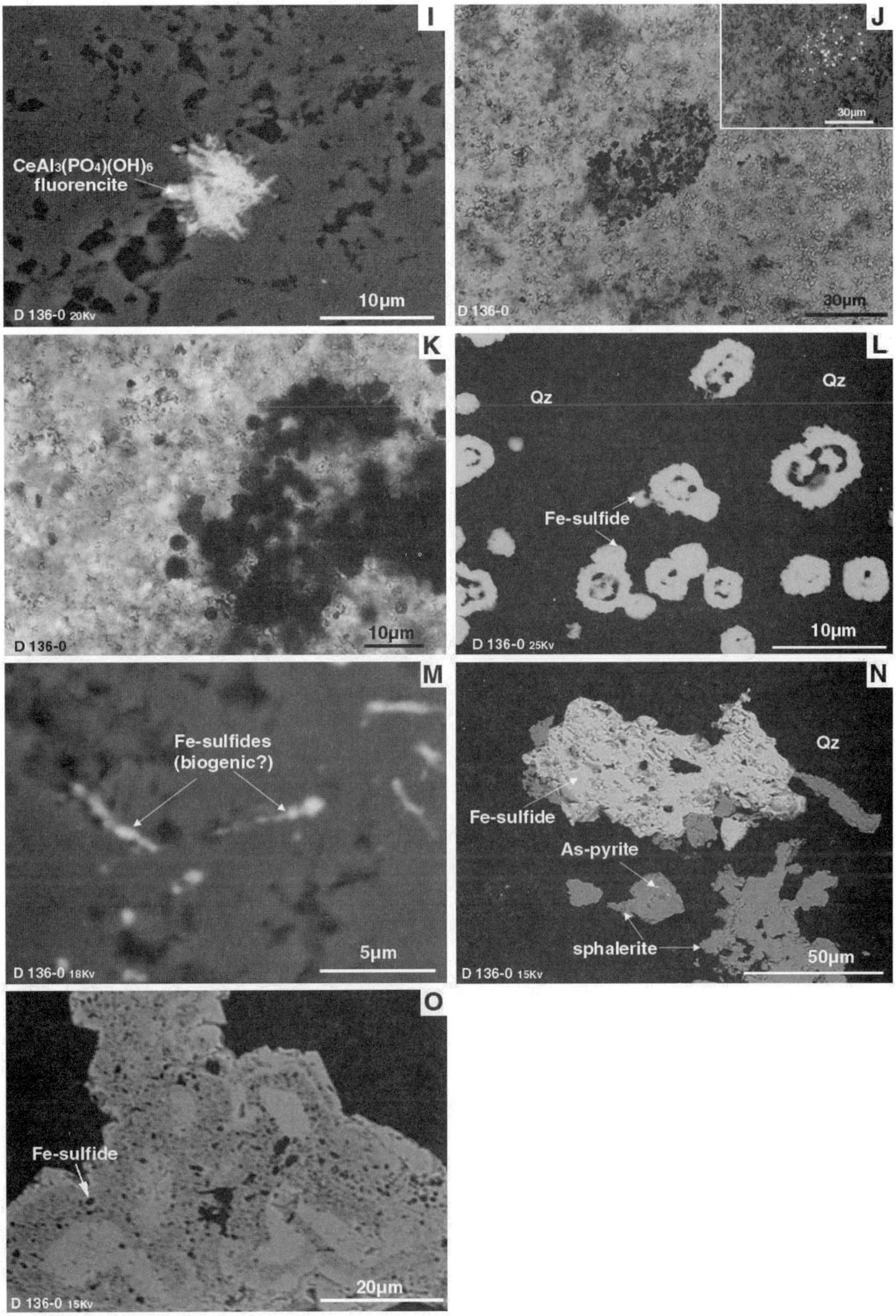
I
CeAl3(PO4)(OH)6
fluorencite
10μm
D 136-0 20Kv
J
30μm
D 136-0
30μm
K
10μm
D 136-0
L
Qz
Qz
Fe-sulfide
10μm
D 136-0 25Kv
M
Fe-sulfides
(biogenic?)
5μm
D 136-0 18Kv
N
Qz
Fe-sulfide
As-pyrite
sphalerite
50μm
D 136-0 15Kv
O
Fe-sulfide
20μm
D 136-0 15Kv

black laminae are fine grained and contain fewer porphyroclasts, which are now ghost clasts, than the white laminae (Fig. 6B). Three types of ghost clasts have been distinguished: (1) euhedral to subhedral crystal shapes, replaced by hydromuscovite and outlined by nanometer-sized Ti-oxide spherules (Fig. 6C); (2) euhedral crystals (100–300 µm) replaced by hydromuscovite (Fig. 6D); and (3) euhedral ghost crystals with aggregates of hydromuscovite and Ti-oxides intergrown with quartz (Fig. 6E). These ghost crystals are interpreted to be either Ti-bearing mica, K-feldspar, or amphibole. Anhedral aggregates of hydromuscovite and minor quartz, outlined by Ti-oxide spherules, could represent glass shards or mineral debris (Fig. 6F). Accessory monazite, florencite, and zoned and corraded zircons are more abundant in the white laminae (Figs. 6G and 6H). Poorly crystallized phases composed of Zr-Ti-REE and U are rare (Fig. 6D). Organic matter occurs as clusters around the volcaniclasts or as delicate mats on laminae surfaces (Fig. 6I) and also as detrital particles (Westall et al., this volume). Hydromuscovite has variable chemistry, containing traces of MgO (0.38–1.44 wt%), Na_2O (0.22–0.79 wt%), Fe_2O_3 (=Fe_{tot}: 0.02–3.07 wt%), TiO_2 (0–0.52 wt%), CaO (0.03–0.11 wt%), Cl (up to 0.03 wt%), and MnO (up to 0.5 wt%; Table 1D). KG 1–6 lamina was analyzed for N and C by nuclear reaction analysis (Rouchon et al., 2005), because it contains a silicified microbial mat at the interface with the KG 5 lamina. Hydromuscovite contains N contents between 0.013 and 0.112 at% in KG 1–5 lamina and between 0.054 and 0.087 at% in KG 1–6 lamina. Carbon amounts range from 0.01 to 0.60 at% in lamina KG 1–5 and from 0.1 to 0.5 at% in KG 1–6 lamina. The C/N atomic ratio ranges from 0.08 to 20.9 and varies with the C concentration (Rouchon et al., 2005).

Geochemistry

The black and the white chert laminae in sample KG 1 are compositionally different. The white laminae have twice the abundances of Al_2O_3, Fe_2O_3, and K_2O in the black part, owing to a higher abundance of hydromuscovite. Three times higher Ti contents indicate also that Ti-oxides are more common in the white than in the black laminae. The poorly crystalline, multielement phases are a second Ti carrier (Fig. 6D). Only the trace elements, such as V (Ti-oxides), Rb, and Sr (hydromuscovite) related to the above-mentioned minerals, are enriched in the white laminae. Copper, Zn, Ni, and Sn are enriched in the black laminae, whereas the white laminae are relatively richer in Zr, Y, W, Nb, Ta, and Ba. The REE patterns show a slight fractionation of LREE compared to the HREE. Black and white laminae have the same LREE abundances, but white laminae are richer in HREE (Fig. 5A).

DISCUSSION

Most of the Archean cherts are considered to be derived from silicification of either orthochems, carbonaceous and volcaniclastic sediments (Lowe and Worrell, 1999; Walsh and Lowe, 1999; Lowe 1999), or mafic and ultramafic rocks (de Wit et al., 1982; Appel et al., 2001; Myers, 2001). Cherts in banded iron formations are generally attributed to chemical precipitation (Bolhar et al., 2004, Bau and Dulski, 1996) or to a combination of exhalation and chemical sedimentation (Gross, 1980; Lydon, 1984; Kimberley, 1989).

Many of the replacement cherts show sedimentary structures, such as fine laminations, wave-rippled strata, carbonaceous matter influx prior to silicification (e.g., cherts from Hoggenoeg and Kromberg Formations, Barberton Greenstone Belt, South Africa: Lowe and Worrell, 1999; de Vries, 2004; the Kittys Gap chert from the Coppin belt and North Pole Chert–Barite Unit, Pilbara Craton, Western Australia; Nijman et al., 1998; de Vries, 2004; Dunlop, 1978; Barley, 1993; Dunlop and Buick, 1981). In the Pilbara craton, the depositional environments are mainly interpreted as low-energy marine coastal-lagoonal submarine to subaerial or partly evaporitic environments. Depositional depths are estimated at less than 100 m (Buick and Barnes, 1984; Lowe and Worrell, 1999; de Vries, 2004). In contrast, deep-water settings and high-temperature hydrothermal environments similar to modern MOR settings were suggested for the Marble Bar Greenstone belt (Kato and Nakamura, 2003) and the North Pole Chert–Barite Unit (Kitajima et al., 2001). As sedimentary or magmatic features are well preserved, all authors agree that silicification is an early synsedimentary diagenetic and/or exhalative process, occurring at the water-sediment interface (Knauth and Lowe, 1978; Lowe and Byerly, 1999; Lowe, 1999). However, post-depositional silicification by circulating hydrothermal fluid has also been proposed (de Wit et al., 1982; Barley, 1993; Buick and Barnes, 1984; Paris et al., 1985; Duchac, 1986; Lowe and

Figure 4. Microphotographs of chert PANO C-85. (A) Scan of thin section with white and black parts of the chert. (B) Thin section of the white part showing several generations of silicification. (C) Scanning electron microscopy-backscattered secondary electrons (SEM-BSE) image: iron-oxide clusters precipitated on precursor crystal. (D) SEM-BSE image: detail of Fe-oxide cluster and the corresponding Energy Dispersive Spectroscopy (EDS) spectrum. It has to be noted that K-Al-silicate is always present. Fe-oxides show voids and spherulitic agglomeration texture. (E) SEM-BSE image: clusters of Fe-oxide with voids; in foreground, spherulitic Fe-oxide. (F) SEM-BSE image: spherulitic Fe-oxide texture. (G) SEM-BSE photograph: euhedral Fe-oxide showing spherulitic Fe-oxides at its lower rim. Its heterogeneous texture might reflect that it was formed by recrystallization of spherulitic oxides. (H) SEM-BSE image: poikilotopic Fe-oxide. The corresponding EDS spectrum shows that these oxides contain traces of V and As. (I) SEM-BSE image: xenotime crystals rich in REE (rare earth elements; see corresponding EDS spectrum) precipitated as secondary product at the edge of Fe-oxides. (J–M) SEM-BSE images: poorly crystalline, gel-like, silica-rich phases and their corresponding qualitative EDS spectra: (J) SEM-BSE image: crystals of barite in poorly crystallized silica-rich phase. (K–L) SEM-BSE images: Silica gel–like phase containing Ca-Ba and S, probably reflecting the presence of nanometer-sized crystals of barite and gypsum. (M) SEM-BSE image: silica phase with high amounts of K and Al, traces of V and Ti, as well as distinct crystals of Fe-oxides.

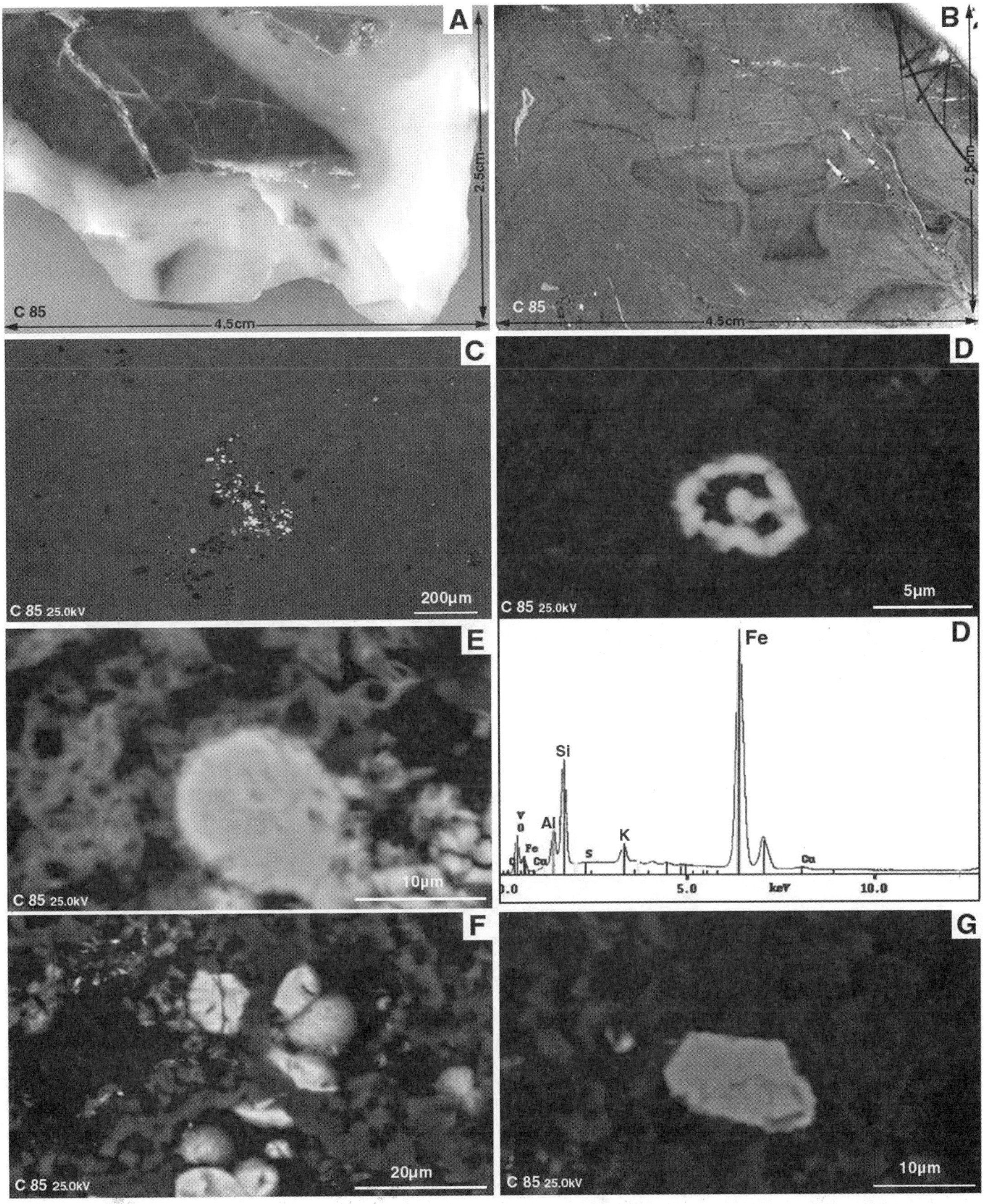
A
C 85
2.5cm
4.5cm
B
C 85
2.5cm
4.5cm
C
C 85 25.0kV
200µm
D
C 85 25.0kV
5µm
E
C 85 25.0kV
10µm
D
Fe
Si
Al
K
V
O
Fe
Cu
S
Cu
0.0
5.0
keV
10.0
F
C 85 25.0kV
20µm
G
C 85 25.0kV
10µm

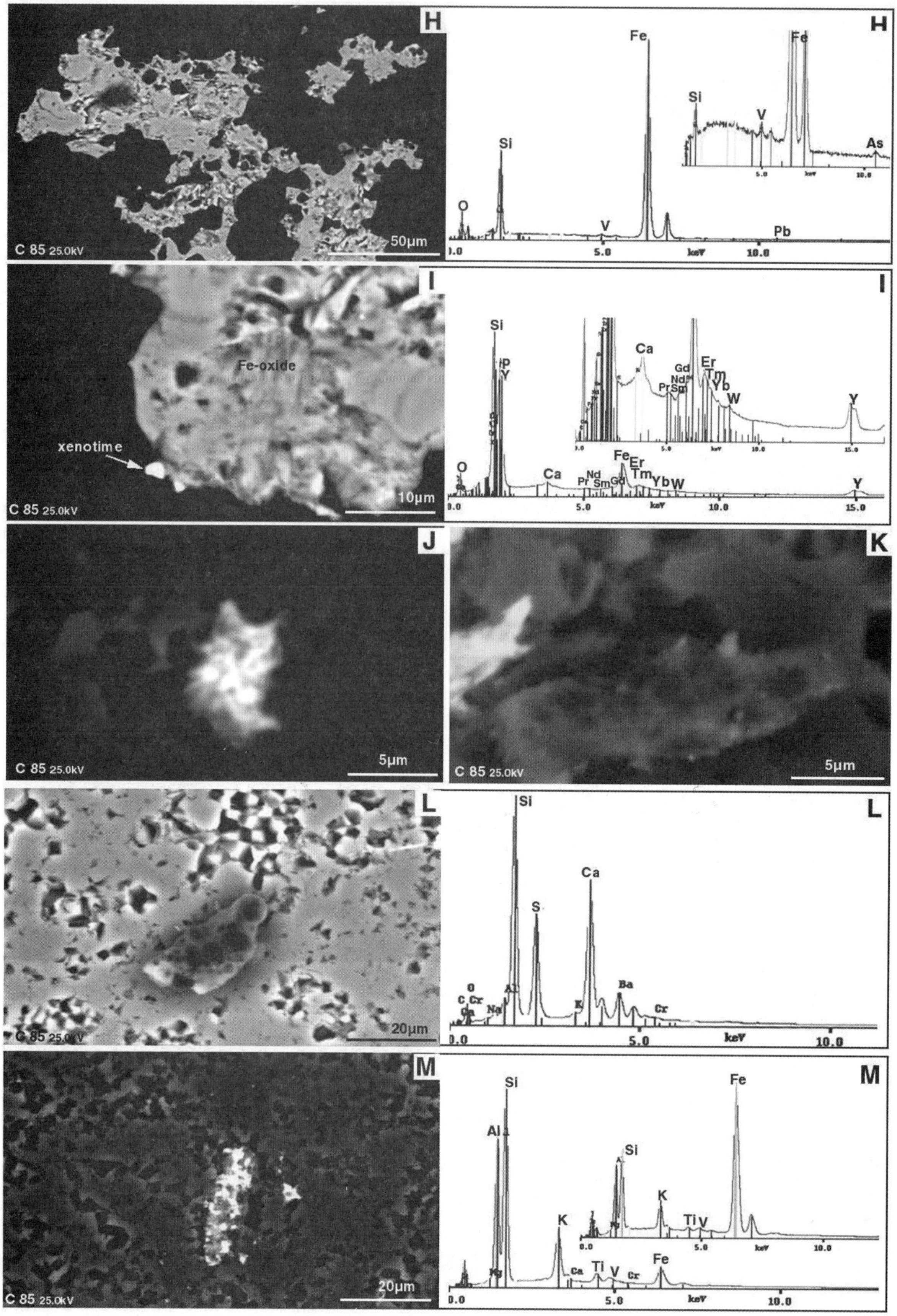
H
C 85 25.0kV
50µm
Fe
Si
V
As
O
Pb
I
Fe-oxide
xenotime
C 85 25.0kV
10µm
Si
P
Y
Ca
Gd
Nd
Er
Tm
Pr
Sm
Yb
W
O
Fe
J
C 85 25.0kV
5µm
K
C 85 25.0kV
5µm
L
20µm
Si
Ca
S
Ba
Cr
Al
Na
M
C 85 25.0kV
20µm
Si
Fe
Al
K
Ti
V
Ca
Cr

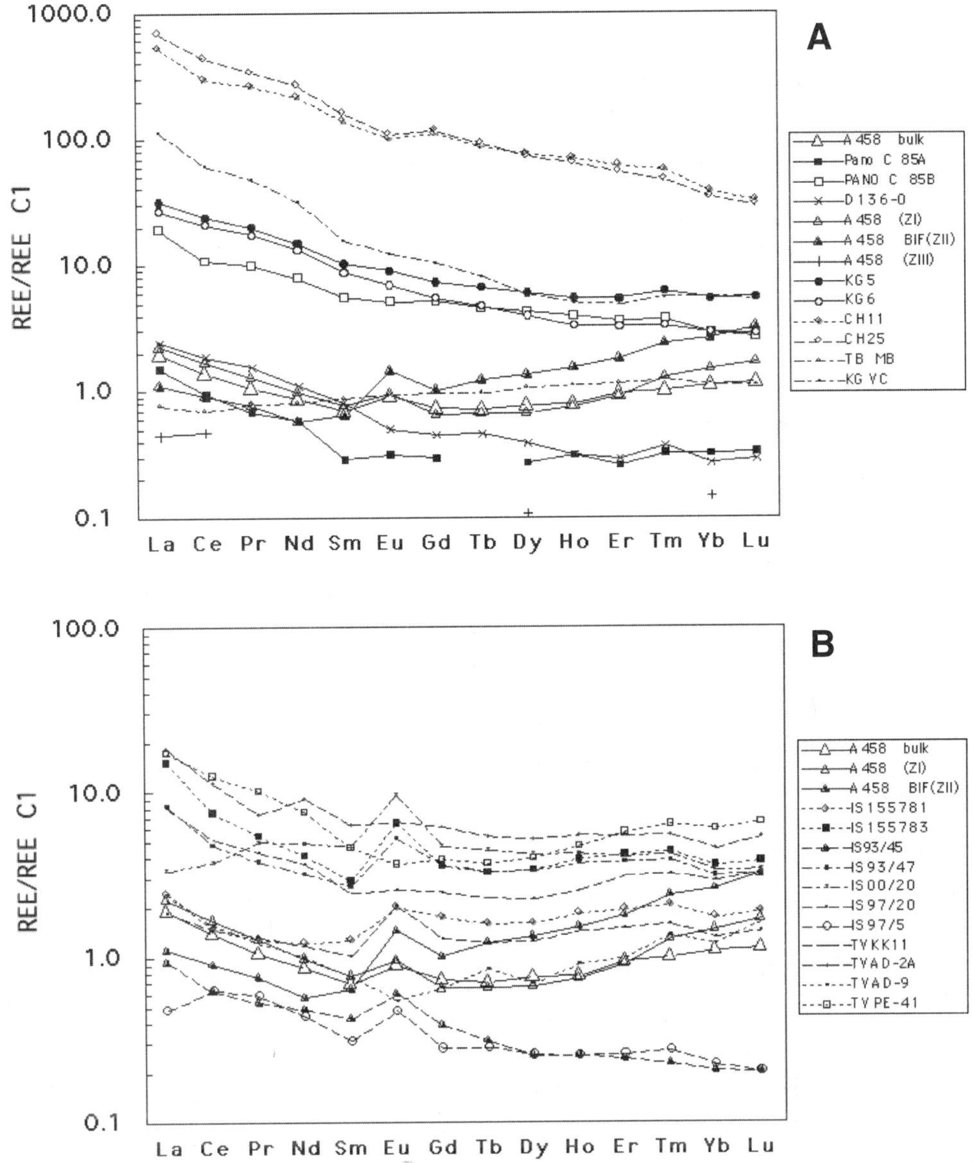

Figure 5. C1 normalized REE (rare earth element) patterns (normalization values after Evensen et al., 1978) for the studied samples from the Pilbara complex, and for comparison, for other samples. (A) A 458 bulk: A section Towers Formation, ZI (zone I)—porphyritic part (precursor: basalt); ZII (zone II)—micro–BIF (Banded Iron Formation); ZIII (zone III)—microcrystalline quartz intergrown with K-Al silicates; PANO C-85 A—white part; PANO C-85 B—black part; PANO D-136–0—black chert; KG 1–5—white chert lamina of chert KG 1; KG 1–6: black chert part of chert KG 1; TB MB—tholeiitic basalt, Marble Bar, Towers Formation (from Kato and Nakamura, 2003); KG VC—Rhyodacite, Kittys Gap (de Vries, 2004); CH11, CH25—black shales from the Nuitang Formation, South China (Lower Cambrian) (Orberger et al., 2005b). (B) Comparison of the A458 zone I and II (micro–BIF) to TV (Transvaal Supergroup): BIF Transvaal Supergroup (Bau and Dulski, 1996) and IS—BIF Isua Supracrustal Belt, Greenland (Bolhar et al., 2004).

Byerly, 1999; Hanor and Duchac, 1990). The origin of the silica has been related to either (1) hydrothermal discharge from deep-seated magmatic systems, based on combined field observations and fluid inclusion data (e.g., de Vries, 2004); (2) silica-rich, downward-circulating ocean water (Holland, 1973; Siever, 1992; de Ronde et al., 1997); (3) alteration of volcaniclastic material (Lowe and Byerly, 1999); or (4) a combination of these processes.

The geodynamic settings of the cherts studied here are well constrained, and hydrothermalism is considered, at least in part, as one of the major processes that controlled silicification (e.g., Sugitani, 1992; Van Kranendonk et al., 2001; Nijman et al., 1998), although silica-rich seawater circulation might also have been an important process during silicification of the Kittys Gap sediments (de Vries, 2004). Marble Bar chert A458 comes from

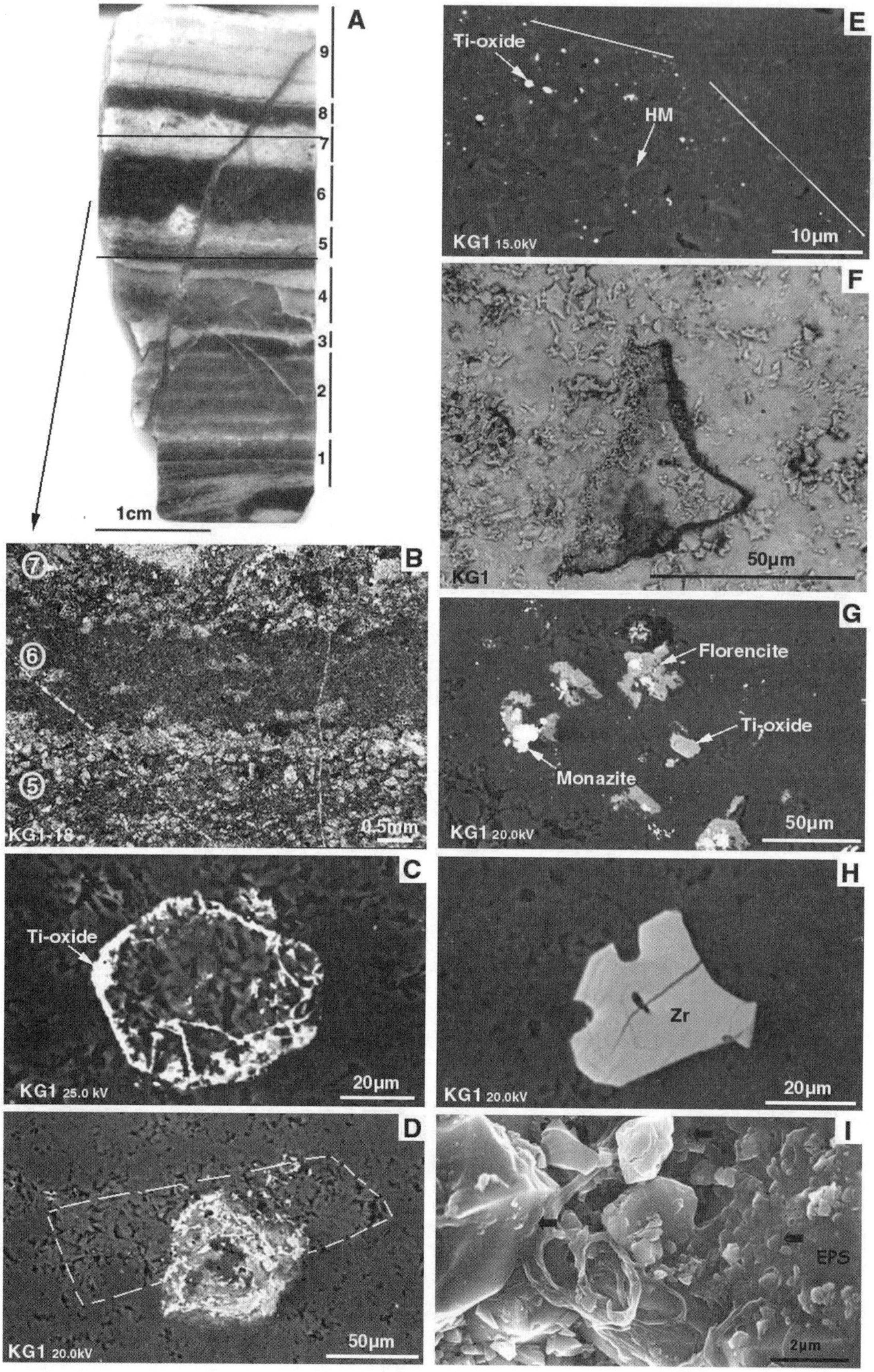
A
9
8
7
6
5
4
3
2
1
1cm
B
⑦
⑥
⑤
KG1-18
0.5mm
C
Ti-oxide
KG1 25.0 kV
20µm
D
KG1 20.0kV
50µm
E
Ti-oxide
HM
KG1 15.0kV
10µm
F
KG1
50µm
G
Florencite
Ti-oxide
Monazite
KG1 20.0kV
50µm
H
Zr
KG1 20.0kV
20µm
I
EPS
2µm

the A-section of the Towers Formation. Cherts from this section are interpreted as precipitates derived from a high-temperature hydrothermal solution emanating from a MOR (Kato and Nakamura, 2003). Mantlelike xenon anomalies observed in this chert (Pinti et al., 2001a) have been interpreted as contamination by mantle-derived pervasive fluid or, alternatively, produced by neutron capture reactions with Te, a rare element concentrated in hydrothermal-derived ores. The two chert samples PANO D-136–0 and PANO C-85 come from silica dykes of probable hydrothermal origin (Isozaki et al., 1999; Nijman et al., 1998; Ueno et al., 2004; Nijman et al., 1998). The Kittys Gap chert was deposited in a shallow water environment on top of rhyodacitic to rhyolitic rocks (de Vries, 2004). Our study shows that the physical, biological, and chemical conditions during silicification were controlled by the protolith mineralogy and chemistry and the nature of the fluid.

Hydrothermal Fluid-Basalt Interaction

In modern and Phanerozoic environments, banded iron or diffuse iron-oxide bearing cherts can be formed in coastal shallow-water settings accompanied by volcanic activity, as those of Papua–New Guinea (Tarasov et al., 1999), or at MOR, where Fe-Mn oxyhydroxides precipitate from hydrothermal plumes up to some distance from the hydrothermal vent (e.g., Grenne and Slack, 2003a). Biologically mediated precipitation of Fe was also suggested for BIFs (Nealson and Myers, 1990) and occurs in Phanerozoic shallow-water carbonate rocks (Préat et al., 2004).

Marble Bar chert A458 is partly characterized by porphyritic or volcaniclastic textures with euhedral to subhedral precursor minerals. The trace element composition of microquartz being pseudomorphous after such precursor minerals indicates that the dark quartz is rich in Fe, Mg, Ti, Ca, and Mn, whereas the bright transparent quartz is poor in mafic elements (Gallien et al., 2003). The texture, mineralogy, and mineral chemistry strongly favor a basaltic (volcaniclastic) protolith for the zone I of this chert.

Figure 6. Microphotographs of the chert from Kittys Gap (KG 1). (A) Thin-section scan showing alternating black and white chert bands with sedimentary textures. (B) Reflected light microphotograph of layers 5–7, as indicated in (A). (C) Scanning electron microscopy-backscattered secondary electrons (SEM-BSE) image: precursor mineral, probably biotite, with rims outlined by Ti-oxide; the precursor mineral is replaced by hydromuscovite and microquartz. (D) SEM-BSE image: precursor K-feldspar (as emphasized by dashed lines), now replaced by hydromuscovite in microcrystalline quartz matrix. (E) SEM-BSE image: a precursor mineral (amphibole?) (margin marked by the white line) is now composed of Ti-oxide and hydromuscovite and is intergrown with microquartz. (F) SEM-BSE image: precursor glass shard replaced by microquartz and outlined by microspherules of Ti-oxide. (G) SEM-BSE image: florencite and monazite intergrown with Ti-oxide in microquartz. (H) SEM-BSE image: Corroded zircon. (I) Field Emission Gun Scanning Electron Microscopy (FEG-SEM) image: silicified organic mats. EPS—extracellular polysaccharide.

The interface between the precursor basalt and the pure chert precipitate is marked by the precipitation of micro–BIFs, now composed of Fe-Mn oxyhydroxides and exhibiting stromatolite-like structures. These REE patterns are similar to those observed in the silicified basalt, but have higher REE abundance. This indicates a similar source and a similar fractionation process for these two zones. The presence of euhedral sulfides with textures similar to those of the later-formed vermicular Fe-oxides in the protobasaltic rock (zone I) and in the chert precipitate (zone III) indicates a reducing environment. The occurrence of Mg-carbonate relicts suggests that this environment was also alkaline. The magnetite or magnesioferrite inclusions in the silica bands intercalated with the Fe-Mn-oxyhydroxide bands indicate that the BIF was previously composed of magnetite and a reducing environment (see reaction [3] below). Metals such as Cr, Ni, Au, Pd, P, Y, and HREE coprecipitated with Fe-Mn oxides and may have been desorbed during later infiltration of low-temperature, oxidizing fluids. During this episode, magnetite was oxidized into hematite and Fe-Mn oxyhydroxides. Since the Fe-Mn oxyhydroxides contain traces of Mg and Ca, we suggest that part of these elements derived from the dissolution of carbonate precursors and were incorporated into the hydroxide structure. The following reactions illustrate the pathway to the presently observed phases. Mg-carbonate and silica formation might be the result of a classical Urey reaction.

$$CaSiO_3 + CO_2 = CaCO_3 + SiO_2. \quad (1)$$

Magnetite can form through hydrolysis or silicification, according to reactions (2) and (3).

$$3Fe_2SiO_4 + H_2SiO_4 = 2Fe_3O_4 + 4SiO_2 + H_2. \quad (2)$$

$$3Fe_2SiO_4 + 2H_2O = 2Fe_3O_4 + 3SiO_2 + 2H_2. \quad (3)$$

Sulfide formation and simultaneous magnetite formation can occur during simple hydrolysis through reaction (4):

$$6FeS + 4H_2O = Fe_3O_4 + 3FeS_2 + 4H_2. \quad (4)$$

The REE patterns of this porphyritic chert part (zone I) and the micro–BIFs differ clearly from that of the North Pole Dome black cherts and from the silicified volcaniclastic sediments of Kittys Gap. Both have a strong positive Eu anomaly and the HREE are enriched compared to LREE, leading to La_N/Yb_N ratios of 0.43 for the micro–BIFs and 1.75 for the silicified basalt of zone I (Fig. 5A, Table 2). As chert A458 is intercalated with tholeiitic basalts, a representative analysis (Kato and Nakamura, 2003) was plotted in Figure 5A. The chert is REE depleted by about five times compared to the abundances in tholeiitic basalt of the Towers Formation and differs by the presence of the positive Eu anomaly. We interpret the REE pattern of zone I and II of chert A458 as being the result of high-temperature hydrothermal fluid–tholeiitic basalt interaction, which led to leaching of Eu from plagioclases of the oceanic basaltic crust. The absence of a Ce anomaly confirms the conclusion that silicification of the basalt occurred under fO_2 conditions equal to the hematite-magnetite buffer.

For comparison, representative data of BIFs from the Isua Supracrustal Belt (often associated with mafic and ultramafic rocks as in the Archean Amitsoq gneiss complex), from the Island of Akilia (Bolhar et al., 2004), and from the Penge and Kuruman iron formations, Transvaal Supergroups, are plotted in Figure 5B (Bau and Dulski, 1996). The Isua rocks all have similar patterns showing a positive Eu anomaly but slightly lower HREE contents (La_N/Yb_N of 1.4–4.7) compared to the A458 micro–BIFs (zone II) and the basaltic silicified protolith of zone I. The Isua BIFs are all rich in mafic minerals (pyroxenes and amphiboles). However, the considerable amounts of carbonates and apatite in the BIFs might be responsible for the lower HREE composition (Bolhar et al., 2004). With the exception of the rocks from the Isua Supracrustal Belt, all other samples were interpreted as being replacement cherts and not chemical sediments (Bolhar et al., 2004). However, BIFs from the Transvaal Supergroups have been interpreted as precipitates from a mixture of seawater and hydrothermal fluids (Bau and Dulski, 1996). Figure 5B shows that the Transvaal samples have no or only slight positive Eu-anomalies, but all are enriched in HREE. This could be due to a low hydrothermal fluid contribution to seawater. The case studied here represents both a basaltic rock replacement (A458 zone I) and a chemical (possibly biochemical) precipitate (A458 zone II, micro–BIFs) at the contact with the precursor basalt. The pronounced positive Eu anomaly, thus, is interpreted to reflect intensive fluid-basalt interaction whereby Eu was leached from the basaltic plagioclase and was subsequently adsorbed onto FeMnOH surfaces during BIF precipitation. Zone III of chert A458 is a pure precipitate from a silica-rich, K-Al-S-Ba–bearing aqueous solution, likely the residual fluid left after crystallization of the micro–BIFs. With the exception of La, Ce, and Yb, all other REE were retained during micro–BIF precipitation (Table 2, Fig. 5A).

Later oxidizing fluids infiltrated along fractures and oxidized magnetite to Fe-Mn-oxyhydroxides, dissolved carbonates, and replaced them pseudomorphically by Fe-Mn oxyhydroxides. Au-Pd-Ni-Cr alloys as well as xenotime are concentrated in the BIF Zone II. Hematitized, originally magnetite-bearing BIFs occur in the Isua Supracrustal Belt, where the hematitization was related to meteoric, low-temperature, oxygenic, alkaline fluid infiltration after the metamorphic event into penetrative zones (Myers, 2001). On the other hand, hematite-goethite deposits can be primary deposits formed in modern and Phanerozoic low-temperature, submarine, hydrothermal environments (<100 °C; Uyeda, 1987; Herzig et al., 1988; Juniper and Fouquet, 1988; Grenne and Slack, 2003a). Assuming the magnetite precursor for the FeMnOH, the following reactions can be proposed:

$$2Fe_3O_4 + H_2O => 2\ Fe_2O_3 + H_2. \quad (5)$$

$$2FeO + 2H_2O => 2\ FeOOH + H_2. \quad (6)$$

Sulfide, the second precursor mineral can be transformed by:

$$2FeS_2 + Fe_3O_4 + 4O_2 => Fe_2O_3 + 2SO_4 + 4H^+. \quad (7)$$

The vermicular microscopic textures and filamentous, clustered nanotextures of the hematite might be related to microbial activity. Similar nanotextures were observed for ferrihydrite formed by the activity of Fe-oxidizing bacteria growing in near-neutral pH groundwaters (Banfield and Zhang, 2001). A further argument supporting microbial activity is the observed nonstoichiometry of the Fe-Mn oxyhydroxides (Tebo et al., 1997). However, according to FIB-TEM investigations, the nonstoichiometry may result from high nanoporosity in the nanofilaments and clusters. On the other hand, it is well known that microbial activity participates in the precipitation of Fe deposits (Grenne and Slack, 2003a). Thus, it is possible that microbes could have been active during the formation of the micro–BIFs. Mineralized filaments, similar in size and form to biofilm-forming bacteria, have been observed in Fe-smectite deposits near deep-seated hydrothermal vents. These microbial structures are similar to the features observed in the laminar overgrown zone on K-Al silicates at the interface of the Fe-oxide grains (Fig. 2C).

Nitrogen and carbon were analyzed by nuclear reactions in different mineral phases of the chert A458. Nitrogen was mainly found concentrated in the Fe-Mn oxyhydroxides in an unknown chemical form, possibly ammonium (Gallien et al., 2003). The C/N atomic ratio in the Fe-Mn oxyhydroxides ranges from 2.3 to 19.4, which is very similar to the C/N ratio measured in organic matter preserved in modern or Phanerozoic marine sediments (e.g., Rau et al., 1987; Muzuka et al., 1991) or starved marine bacteria in oceanic and MOR sediments (Takano et al., 2004). Previous N isotopic analyses of this chert (Pinti et al., 2001b; K. Hashizume, 2005, personal commun.) showed the occurrence of a main N component released at low temperature (600–800 °C) during stepped combustion. This component has a $\delta^{15}N$ value ranging from 6.7 ± 1.6 to 7.4 ± 2.4‰. New N isotopic analyses on BIF separates of the zone II show an N component released mainly at 800 °C and having a $\delta^{15}N$ of 7.6 ± 0.5‰. This isotopic ratio is identical to that of Phanerozoic marine sedimentary organic matter of +6 to +7‰ (Sweeney et al., 1978). The nitrogen isotopic composition of organic matter in modern environments is controlled by denitrification ($NO_3^- \rightarrow N_2$). During denitrification, ^{14}N returns preferentially to the atmosphere, leaving NO_3^- enriched in ^{15}N in oceanic water. As NO_3^- is the main nitrogen-bearing nutrient, marine organisms record its isotopic signature and are systematically enriched in ^{15}N by ~ +6‰ relative to the atmosphere. The N isotopic ratios measured in A458 could have recorded an Archean N cycle dominated by oxide species as that observed in Phanerozoic ages, or, alternatively, could have been added successively during postdepositional alteration (Fig. 4 *in* Pinti et al., 2001b). The first hypothesis seems not to be supported by the available data (Beaumont and Robert, 1999; Pinti et al., 2001b; Pinti and Hashizume, 2001) that suggest, to the contrary, an Archean N cycle dominated by fixation of reduced species such as NH_4^+.

The second hypothesis is linked to the timing of the oxidizing fluid entry, which has not yet been determined. It can be speculated that intensive modern weathering would have affected the entire sample and not only selective areas. However, numerous sulfides are still present. Nitrogen can be introduced in cherts by modern weathering, but this component is always released at combustion temperatures between 100 and 450 °C (Pinti et al., 2001b, 2003; van Zuilen et al., 2005), and it shows much larger positive or negative isotopic fractionations (from –13 to +20‰; Pinti et al., 2001b) than that observed in this case. The N component released at 800 °C has possibly been added later, but probably before silicification of the sample. Further investigation is needed to determine the timing of oxidation and the consequences that it has on the interpretation of the N chemical signature.

Hydrothermal Fluid–Black Shale Interaction

Black cherts have many types of occurrences as (1) thin layers within cross-bedded volcaniclastic sandstone; (2) lenticular units; (3) fine laminations, drapes, and/or flaser layers between and within cross-beds in shallow marine to lagoonal sediments (Lowe and Byerly, 1999); and (4) irregular veinlets and dykes as, for example, seen in the Msauli chert, Onverwacht Group, Barberton Greenstone Belt (Hofmann, 2004), or at the North Pole Dome (Nijman et al., 1998; de Vries, 2004). These dykes were interpreted either as representing reinjected carbonaceous sediments from the subsurface (Lowe and Byerly, 1999) or as being related to hydrothermal activity and associated with tectono-metamorphic processes (Nijman et al., 1998; Hofmann, 2004). Our black chert samples (PANO D-136–0 and C-85) were taken from dykes, suggesting a direct association with hydrothermal activity (Nijman et al., 1998). However, our studies show that the two black cherts are not simply hydrothermal silica precipitates but rather silicified brecciated sediments. The brecciated precursor texture is still visible in all three analyzed chert parts (the black of the PANO D-136–0 and the white and black parts of PANO C-85), and all parts show identical REE patterns. The PANO C-85 black chert is enriched in REE by about ten times compared to the white part. The black chert PANO D-136–0 is REE depleted by ~8 times, compared to the black part of chert PANO C-85 (Fig. 5A). The main difference between the two chert samples is the presence of a more pronounced negative Ce anomaly in the black zone of chert PANO C-85, indicating that silicification occurred under, at least, slightly oxidizing conditions (Fig. 5A). The white zone has formed later because white microquartz veins crosscut the black zone. This indicates clearly that the more pronounced negative Ce anomaly of the black zone is related to the first silicification event. It can be speculated that Fe-oxide clusters were derived from sulfides, as observed in PANO D-136–0. However, barite and gypsum grains were found interstitial to the microquartz, favoring oxidizing conditions during silicification.

The PANO D-136–0 black chert is enriched in sulfides, organic matter, and carbonates and does not show a negative Ce anomaly. It was, thus, formed under reducing and alkaline conditions. The PANO C-85 chert contains multielement, poorly crystallized, porous phases, which were not found in the PANO D-136–0 sample. These amorphous textures might be the result of rapid silicification that conserved traces of the metal-charged silica gel–like phase. Such phases might represent the precursor from which inorganic colloidal particles precipitated (Williams and Crerar, 1985). Similar structures were observed in jaspers from the Lokken ophiolite (Grenne and Slack, 2003b). One of the striking similarities between the two PANO cherts is the texture and location of micrometer-sized Fe-oxide and Fe-sulfide clusters. Compartmentalized sulfide structures, such as those observed as clusters on the precursor minerals, are also typical for sulfur bacteria in modern environments (Boetius, 2004). Light $\delta^{15}N$ values (–7.4 ± 1.0‰ and –1.3 ± 1.6‰ respectively) measured in these rocks would support the hypothesis of biological activity at the time of their formation (Pinti et al., 2001b). Preliminary FIB-TEM investigations combined with electron energy loss spectroscopy (EELS) show that nitrogen is, in part, related to nanometer-sized films around microquartz grains (Wirth and Orberger, 2005, personal commun.). This indicates that N emplacement is clearly contemporaneous with silicification and not related to a later fluid event. In situ S and Fe isotopic measurements would be necessary to clearly identify a biological influence on the formation of these minerals.

Did hydrothermal fluids cause the silicification? There are arguments in favor of and against this hypothesis. No positive Eu anomaly, which would clearly indicate a hot hydrothermal fluid contribution, was observed in the black cherts (Douville et al., 1999; Fig. 5A). However, if we compare our results with REE patterns of Archean marine shales that have not been altered by hydrothermal fluids but were deposited by seawater, we should observe a negative Eu anomaly (Condie, 1991; Grenne and Slack, 2003b). The absence of this negative Eu anomaly in our PANO cherts can, thus, be interpreted as reflecting low-temperature hydrothermal fluid alteration. Low water-rock interaction does not necessarily produce a well-pronounced positive Eu anomaly as observed in high T hydrothermal fluids (e.g., Grenne and Slack, 2003a; Orberger et al., 2005b). REE analyses of another black chert from the Dresser Formation, sampled at the root of a silica dyke and within pillowed basalts (sample Pi-47–00; Pinti et al., 2003), show a well-defined positive Eu anomaly and a REE pattern identical to those observed in MOR effluents and chimney sediments (Mills and Elderfield, 1995). A hydrothermal fluid origin is also supported by the mineralogical and metal associations characteristic for submarine hydrothermal environments found in chert PANO D-136–0 (sphalerite, galena, chalcopyrite, pyrite, and Ni-sulfides), PANO C-85 (barite, gypsum, Fe-oxides) REE data from the black shale deposit of the Lower Cambrian Nuitang Formation in South China have been plotted together with our data in Figure 5A. These shales were deposited in shelf settings and were clearly infiltrated by hydrothermal fluids as proven by sulfur isotopic compositions of the sulfides and sulfates and by fluid inclusion data (Steiner et al., 2001; Orberger et al., 2005b). They are associated with cherts and phosphorites (Horan et al., 1994; Steiner et al., 2001; Orberger et al., 2005b). Several features observed in PANO D-136–0 and C-85 are similar to metal-rich black shales (e.g., from the Upper Devonian Selwyn Basin, Yukon, Canada: Horan et al., 1994; Orberger et al., 2003; or from the Lower Cambrian of southern China: Steiner et al., 2001). Specific features of these rocks are: (1) the association of metals of contrasting geochemical behavior and typical of hydrothermal environments, such as Zn, Fe, Pb, Fe, Se, and As in PANO D-136–0; (2) Se/S ratios in the order of 1×10^{-4} found in the sulfide-rich chert PANO D-136–0, which is in the same order of magnitude as that of BIF–bearing chert A458 and typical of hydrothermal volcanogenic environments (Auclair et al., 1987; Huston et al., 1995; Orberger and Alleweldt, 1994); (3) the presence of abundant organic matter; (4) the clastic-brecciated precursor texture; and (5) the sulfide morphologies and assemblages as well as the carbonate, phosphate, and sulfate mineralogy. The REE patterns of the black cherts PANO D-136–0 and C-85 are identical to those of the Lower Cambrian black shales (Steiner et al., 2001; Orberger et al., 2005b), but the absolute concentrations are lower by a factor of ~150–200 (Fig. 5A), which is interpreted as a dilution effect produced by the silicification. These black shales, although infiltrated by hydrothermal fluids, do not show a positive Eu anomaly as would be expected and as observed in our samples. Indeed, they are even characterized by a slightly negative Eu anomaly (Orberger et al., 2005b).

The Archean, metal-rich black cherts from the Nichol Well, Roerbourne Megasequence, Pilbara, have been interpreted as black shale deposits in the vicinity of a hydrothermal vent (Sugitani et al., 2002). The Apex Basalt Formation, from which the PANO C-85 chert was obtained, contains well-preserved black shale deposits with sulfide nodules, which have been found by The Archean Biosphere Drilling Project (Nedachi, 2004). Extremely light carbon isotopic compositions up to –40‰ measured in these lithologies suggest the activities of metanothrophs in black shales (Nedachi, 2004). Interestingly, this latter feature is in agreement with light nitrogen isotopic compositions found in PANO C-85 ($\delta^{15}N$ of -1.3 ± 1.6‰) and PANO D-136–0 ($\delta^{15}N$ of -7.4 ± 1.0‰). These N isotopic signatures have been interpreted as the result of the metabolic activity of chemoautotrophic bacteria in the proximity of an Archean MOR (Pinti and Hashizume, 2001). With regard to the black shale precursor rocks, the depositional environment can, however, be constrained as a low-temperature, hydrothermal shallow-water setting.

As the black cherts investigated here were obtained from dykes, we suggest that the dyke-derived black cherts represent material that was entrained into the hydrothermal system during hydrofracturing and brecciation. In that case, two generations of organic matter would be present in these samples: (1) organic carbon (kerogenes) encapsulated in silica as relict organic matter from the black shales, and (2) organic matter that decayed during sulfurization and oxidation contemporaneous with the silicification event. The environment in the immediate vicinity of the vent could have been highly conducive to microbial activity. The Lower Cambrian mineralized black shale of the Niutang Formation exemplifies such a situation (Mao et al., 2002; Steiner et al., 2001). These black shales contain evidence for a sudden mass distribution of fossils during Cambrian explosion (the so-called Cambrian explosion; Conway Morris, 1989; Bergström, 1989). Abundant fauna and mats of sheathed sulfur bacteria were observed in adjacent sediments, whereas arthropods, sponges, and undetermined shelly remains occur in the unit itself (Steiner et al., 2001). In this example, hydrothermal seeps or vents were the mediators for promoting biocolonization of the shallow-water environments. In light of this discussion, (black) cherts with a black shale precursor must represent important targets in the search for evidence of early Archean life.

Seawater-Rhyodacitic Volcaniclastic Sediment Interactions

The 3.446 Ga Kittys Gap volcano-sedimentary deposits were deposited in a shallow-water environment where hydrothermal venting was active during or shortly after sedimentation (de Vries, 2004). Multiple silicification events produced different vein generations and silicified different parts of the sedimentary rocks. Although black and white laminae of sample KG 1 do not show a significant difference in mineralogy except for a higher abundance of hydromuscovite and Ti-oxides in the white laminae, the phyllosilicates seem to be responsible for the slightly higher HREE content compared to that found in the black laminae (Fig. 5A). LREE patterns are almost identical in the two types of laminae. When comparing the REE pattern of the black and white laminae to that of the underlying rhyodacitic rocks (KG-rhyodacite, Fig. 5A), we note that the latter show a slight negative Ce anomaly, indicating slightly oxidizing conditions during deposition. Furthermore, the rhyodacitic rocks are a factor of two higher in LREE but show identical HREE contents compared to the white laminae. It can, thus, be concluded that although Ti-oxides are abundant in the cherts and no sulfides are observed, slightly reducing conditions were likely the cause for the absence of the negative Ce anomaly. The silicifying environment was likely more acidic, resulting in preferential precipitation of K-mica over K-feldspar. These redox conditions were favorable for the growth of the microbial mats on sediment surfaces and around volcaniclastic particles, whose signatures are expressed as microfossils and $\delta^{13}C$ ratios of –22 to –30‰ (Westall, 2005, 2003; Westall et al., 2004; see also Westall et al., this volume). In certain horizons of the sediment, the fossilized microbial remains appear relatively degraded. Nitrogen found in hydromuscovite may have originated from the decay of organic matter during silicification and incorporation in the phyllosilicate structure as NH_4^+ replacing K^+. Systematic enrichment of Cu, Cr, and Zn in all laminae, compared to average rhyodacitic rocks, might indicate a slight hydrothermal influence during silicification.

The Role of Phosphates

Phosphorous is one of the most important biophile elements, and its chemical cycle is closely linked to that of oxygen, carbon, sulfur, nitrogen, and iron (Föllmi, 1990). In the Archean, phosphorous in cherts hosting possible traces of life could have originated from the decay of organic matter. Microbes could have obtained P, an important nutrient, from the bioalteration of phosphorous-containing volcanic particles. Phosphorous could also have been leached from P-bearing minerals during silicification or could have been introduced by seawater or hydrothermal fluids. The origin and significance of phosphates in Archean rocks has been a controversial subject since Mojzsis et al. (1996) interpreted the association of isotopically light carbon isotopic values of graphite and apatite in 3760 Ma rocks (interpreted as BIFs) from the Isua Supracrustal Belt as a potential biomarker. Fedo and Whitehouse (2002), on the other hand, showed that the BIF was not of sedimentary origin. Lepland et al. (2002) and van Zuilen et al. (2003) demonstrated that the apatite and graphite, respectively, had an abiogenic origin by showing that graphite was formed through thermal decomposition of siderite at 800 °C and that the REE pattern of apatite reflects a chemical precipitate.

The phosphate mineralogy of our samples can be used to reconstruct the protolith of cherts: LREE phosphates (monazite, florencite, apatite) seem to be more abundant in the rhyodacitic protoliths, as at Kittys Gap, whereas HREE phosphates (xenotimes) are more common in basaltic rocks as in the A458 zone I. HREE- (xenotime) and LREE–rich (apatite) phosphates are characteristic of black shale protoliths influenced by metal-rich exhalative fluids, as in chert PANO D-136–0. The Kittys Gap chert contains euhedral to subhedral phosphates (florencite, monazite) intergrown with the silica matrix, reflecting the chemical composition of the silicifying fluid or relict detrital minerals (zircons), whereas anhedral textures or fillings of interstitial pore space, such as florencite in the Kittys Gap chert, could be attributed also to postsilicification events. Further study of our samples that include cherts with abundant organic matter containing a higher content and a larger variety of abiogenic phosphate minerals, as well as cherts that are poor in organic matter, might provide valuable information for elucidating the origin of P-containing minerals, thus making an important contribution to the solution of the Archean phosphate controversy.

CONCLUSIONS

Micromineralogical and detailed geochemical studies of four early Archean chert samples show that characterization of their depositional environments contributes significantly to the understanding of their origins (protoliths) and of the physical, biological, and chemical conditions during silicification. The studied cherts were partially infiltrated by hydrothermal fluids. Three of the cherts had a sedimentary protolith. The black cherts PANO D-136–0 and PANO C-85 had a black shale precursor, whereas the white and black laminated chert from the Kittys Gap had a rhyodacitic volcaniclastic precursor. The PANO cherts, which were sampled from hydrothermal dykes, might represent entrained, brecciated sediments (black shales), which were silicified by hydrothermal fluids. For the Kittys Gap chert, the enriched Cu and Zn values in the black laminae might indicate a slight contribution from hydrothermal fluid infiltration. The geochemical signature of the A458 chert from the Towers Formation is a result of basalt-hydrothermal fluid interaction. With the exception of the chert PANO C-85, silicification of the studied samples occurred in all cases under reducing conditions. Mineral structures, such as Fe-sulfides infilling voids, Fe-oxides occurring as clusters, and vermicular Fe-Mn oxide structures composed of nanofilaments, which consist of aligned hematite nanoclusters, suggest that microbial activity may have influenced mineral formation. For the PANO and the Kittys Gap cherts, this microbial activity is likely contemporaneous with the silicification event. Additional arguments for the presence of life at the time of silicification are provided by C/N ratios, the $\delta^{13}C$ and $\delta^{15}N$ values, and the location of N in nanofilms around microquartz grains or incorporated in hydromuscovite structures as ammonium (Rouchon et al., 2005). The nanometer-sized, filamentous, clustered hematite in the micro–BIFs of A458 zone II is related to a postsilicification oxidizing fluid event.

APPENDIX

Details on the precision, detection limits and analytical procedures are available at http://www.crpg.cnrs-nancy.fr/SARM/index.html.

ACKNOWLEDGMENTS

DLP wishes to thank Drs. Maruyama, Isozaki, and Ueno for providing samples A458, PANO D-136-0, and C-85, together with geological information. This study was funded by the French national programs (CNRS [Centre National de la Recherche Scientifique] GDR [Groupe de Recherche]-Exobiologie, CNRS-PNP [Project National Planétologie], CNES [Centre National d'Etudes Spatiales]-Exobiology) and by the CNRS-UPS UMR-8148 IDES (Interactions and Dynamique des Environments de Surface) at Université Paris XI. We thank M. Fialin and F. Couffigal for helping with electron microprobe observation and A. Richard during FEG-SEM analysis. The comments and corrections of M. Walsh and A. Hoffmann greatly improved the paper.

REFERENCES CITED

Appel, P.W.U., Rollinson, H.R., and Touret, J.L.R., 2001, Remnants of an early Archaean (>3.75 Ga) seafloor, hydrothermal system in the Isua Greenstone Belt: Precambrian Research, v. 112, p. 27–49, doi: 10.1016/S0301-9268(01)00169-3.

Auclair, G., Fouquet, Y., and Bohn, M., 1987, Distribution of selenium in high-temperature hydrothermal sulfide deposits at 13 degrees North, East Pacific Rise: Canadian Mineralogist, v. 25, p. 577–587.

Banfield, J.F., and Zhang, H., 2001, Nanocrystals as model systems for pressure-induced structural phase transitions, *in* Banfield J.F., and Navrotsky, A., eds., Nanoparticles and the environment: Reviews in Mineralogy and Geochemistry, Volume 44: Washington, D.C., Mineralogical Society of America, p. 1–51.

Barley, M.E., 1993, Volcanic, sedimentary and tectonostratigraphy environments of the ~3.46 Ga Warrawoona Megasequence: A review: Precambrian Research, v. 60, p. 47–67, doi: 10.1016/0301-9268(93)90044-3.

Bau, M., and Dulski, P., 1996, Distribution of yttrium and rare-earth elements in the Pengeand Kuruman iron formations, Transvaal Supergroup, South Africa: Precambrian Research, v. 79, p. 37–55, doi: 10.1016/0301-9268(95)00087-9.

Beauchamp, B., and Baud, A., 2002, Growth and demise of Permian biogenic chert along northwest Pangea: Evidence for end-Permian collapse of thermohaline circulation: Palaeogeography, Palaeoclimatology, Palaeoecology, v. 184, p. 37–63, doi: 10.1016/S0031-0182(02)00245-6.

Beaumont, V., and Robert, F., 1999, Nitrogen isotope ratios of kerogens in Precambrian cherts: A record of the evolution of atmosphere chemistry?: Precambrian Research, v. 96, p. 63–82, doi: 10.1016/S0301-9268(99)00005-4.

Bergström, J., 1989, Metazoan evolution around the Precambrian-Cambrian transition, *in* Simonetta, A.M., and Conway Morris, S., eds., The early evolution of Metazoa and the significance of problematic taxa: Cambridge, UK, Cambridge University Press, p. 25–34.

Boetius, A., 2004, Linking Biosphere to Geosphere: Anoxic microbial habitats at continental margins: 2nd Swiss Geoscience meeting, Lausanne, Switzerland: Abstracts, v. 1, p. 91.

Bolhar, R., Kamber, B.S., Moorbath, S., Fedo, C.M., and Whitehouse, M.J., 2004, Characterisation of early Archaean chemical sediments by trace element signatures: Earth and Planetary Science Letters, v. 222, p. 43–60.

Brasier, M.D., Green, O.R., Jephcoat, A.P., Kleepe, A.K., Van Kranendonk, M.J., Lindsay, F., Steele, A., and Grassineau, N.V., 2002, Questioning the evidence for Earth's oldest fossils: Nature, v. 416, p. 76–81.

Buick, R., and Barnes, K.R., 1984, Cherts in the Warrawoona Group: Early Archaean silicified sediments deposited in shallow-water environments, *in* Muhling, J.R., Groves, D.I., and Blake, T.S., eds., Archean and Proterozoic basins of the Pilbara, Western Australia: Geology Department (Key Centre) and University Extension, The University of Western Australia Publication 9, p. 37–53.

Byerly, G.R., Kröner, A., Lowe, D.R., Todt, W., and Walsh, M.M., 1986, Stromatolites from the 3,300–3,500 Myr Swaziland Supergroup, Barberton Mountain Land, South Africa: Nature, v. 319, p. 489–491, doi: 10.1038/319489a0.

Cady, S.L., and Farmer, J.D., 1996, Fossilization processes in siliceous thermal springs: Trends in preservation along thermal gradients, *in* Brock, G.R., and Goode, J.A., eds., Evolution of Hydrothermal Systems on Earth (and Mars?): New York, Wiley, p. 150–170.

Condie, K.C., 1991, Another look at rare-earth elements in shales: Geochimica et Cosmochimica Acta, v. 55, p. 2527–2531, doi: 10.1016/0016-7037(91)90370-K.

Condie, K.C., 1997, Sources of Proterozoic mafic dyke swarms: Constraints from Th/Ta and La/Yb ratios: Precambrian Research, v. 81, p. 3–14, doi: 10.1016/S0301-9268(96)00020-4.

Conway Morris, S., 1989, The Burgess shale faunas and the Cambrian explosion: Science, v. 246, p. 339–346.

de Ronde, C.E.J., Channer, D.M., Faure, K., Bray, C.J., and Spooner, T.C., 1997, Fluid chemistry of Archean seafloor hydrothermal vents: Implications for the composition of circa 3.2 Ga seawater: Geochimica et Cosmochimica Acta, v. 61, p. 4025–4042.

de Vries, S.T., 2004, Early Archaean sedimentary basins: Depositional environment and hydrothermal systems: Geologica Utraiectina, v. 244, 160 p.

de Wit, M., Hart, R., Martin, A., and Abbott, P., 1982, Archean abiogenic and probable biogenic structures associated with mineralized hydrothermal vent systems and regional metasomatism, with implications for greenstone belt studies: Economic Geology and the Bulletin of the Society of Economic Geologists, v. 77, p. 1783–1802.

Douville, E., Bienvenu, P., Charlou, J.L., Donval, J.P., Fouquet, Y., Appriou, P., and Gamo, T., 1999, Yttrium and rare earth elements in fluids from various deep-sea hydrothermal systems: Geochimica et Cosmochimica Acta, v. 63, p. 627–643, doi: 10.1016/S0016-7037(99)00024-1.

Duchac, K.C., 1986, Metasomatic alteration of a komatiic sequence into chert [M.S. thesis]: Baton Rouge, Louisiana State University, 240 p.

Dunlop, J.S.R., 1978, Shallow-water sedimentation at North Pole, Pilbara Block, Western Australia: Geology, Department and University Extension Publication, v. 2, p. 390–438.

Dunlop, J.S.R., and Buick, R., 1981, Archaean epiclastic sediments derived from mafic volcanics, North Pole, Pilbara, Western Australia: Geological Society of Australia Special Publications, v. 7, p. 225–233.

Eugster, H.P., 1967, Hydrous sodium silicates from Lake Magadi, Kenya: Precursors of bedded cherts: Science, v. 157, p. 1177–1180.

Evensen, N.M., Hamilton, P.J., and O'Nions, R.K., 1978, Rare earth abundances in chondritic meteorites: Geochimica et Cosmochimica Acta, v. 42, p. 1199–1212, doi: 10.1016/0016-7037(78)90114-X.

Farmer, J.D., and Des Marais, D.J., 1999, Exploring for a record of ancient Martian life: Journal of Geophysical Research, v. 104, p. 26,977–26,995.

Fedo, C.M., and Whitehouse, M.J., 2002, Metasomatic origin of quartz-pyroxene rock, Akilia, Greenland, and implications for Earth's earliest life: Science, v. 296, p. 1448–1452, doi: 10.1126/science.1070336.

Föllmi, K.B., 1990, Condensation and phosphogenesis: Example of the Helvetic mid-Cretaceous (northern Tethyan margin), *in* Notholt, A.J.G, and Jarvis, I., eds., Phosphorite Research and Development: Geological Society [London] Special Publication 52, p. 237–252.

Fournier, R.O., and Rowe, J.J., 1966, Estimation of underground temperatures from the silica content of water from hot-springs and wet-steam wells: American Journal of Science, v. 264, p. 685–697.

Gallien, J.P., Orberger, B., Pinti, D.L., Wagner, C., Fialin, M., Daudin, L., and Hashizume, K., 2003, Mineralogy and geochemistry of an Archaean chert: In quest of N-sites: Geochimica et Cosmochimica Acta, v. 67, p. A115.

Grenne, T., and Slack, J.F., 2003a, Bedded jaspers of the Ordovician Lokken ophiolite, Norway: Seafloor deposition and diagenetic maturation of hydrothermal plume-derived silica-iron gels: Mineralium Deposita, v. 38, p. 625–639, doi: 10.1007/s00126-003-0346-3.

Grenne, T., and Slack, J.F., 2003b, Paleozoic and Mesozoic silica-rich seawater: Evidence from hematitic chert (jasper) deposits: Geology, v. 31, p. 319–322, doi: 10.1130/0091-7613(2003)031<0319:PAMSRS>2.0.CO;2.

Gross, G.A., 1980, A classification of iron formations based on depositional environments: Canadian Mineralogist, v. 18, p. 171–187.

Grotzinger, J.P., and Knoll, A.H., 1995, Anomalous carbonate precipitates in the Precambrian the key to the Permian?: Palaios, v. 10, p. 578–596.

Halbach, M., Halbach, P., and Lueders, V., 2002, Sulfide-impregnated and pure silica precipitates of hydrothermal origin from the central Indian Ocean: Chemical Geology, v. 182, p. 357–375, doi: 10.1016/S0009-2541(01)00323-0.

Hanor, J.S., and Duchac, K., 1990, Isovolumetric silicification of early Archaean komatiites: Geochemical mass balances and constraints on origin: The Journal of Geology, v. 98, p. 863–877.

Herzig, P.M., Becker, K.P., Stoffers, P., Bäcker, H., and Blum, N., 1988, Hydrothermal silica chimney fields in the Galapagos spreading center at 86°W: Earth and Planetary Science Letters, v. 89, p. 261–272, doi: 10.1016/0012-821X(88)90115-X.

Hofmann, A., 2004, Sedimentary versus hydrothermal processes for the origin of carbonaceous cherts in the Barberton Greenstone belt, *in* Reimold, W.U., and Hofmann, A., eds., Field Forum: Processes on the Early Earth, Kaapvaal Craton, South Africa: Johannesburg, Abstracts, p. 35–37.

Holland, H.D., 1973, The oceans: A possible source of iron in iron-formations: Economic Geology and the Bulletin of the Society of Economic Geologists, v. 68, p. 1169–1172.

Horan, M.F., Morgan, J.W., Grauch, R.I., Coveney, J.R.M., Murowchick, J.B., and Hulbert, L.J., 1994, Rhenium and osmium isotopes in black shales and Ni-Mo-PGE-rich sulfide layers, Yukon Territory, Canada, and

Hunan and Guizhou provinces, China: Geochimica et Cosmochimica Acta, v. 58, p. 257–265.

Huston, D.L., Sie, S.H., and Suter, G.F., 1995, Selenium and its importance to the study of ore genesis: The theoretical basis and its application to volcanic-hosted massive sulfide deposits using pixeprobe analysis: Nuclear Instruments & Methods in Physics Research, Section B, Beam Interactions with Materials and Atoms, v. 104, p. 476–480, doi: 10.1016/0168-583X(95)00462-9.

Isozaki, Y., Kabashima, T., Ueno, Y., Kitajima, K., Maruyama, S., Kato, Y., and Terabayashi, M., 1999, Early Archean mid-ocean ridge rocks and early life in the Pilbara Craton, W. Australia, [abs.]: Eos (Transactions, American Geophysical Union), v. 78, Supplement, p. 399.

Juniper, S.K., and Fouquet, Y., 1988, Filamentous iron-silica deposits from modern and ancient hydrothermal sites: Canadian Mineralogist, v. 26, p. 859–869.

Kato, Y., and Nakamura, K., 2003, Origin and global tectonic significance of early Archean cherts from the Marble Bar greenstone belt, Pilbara Craton, Western Australia: Precambrian Research, v. 125, p. 191–243, doi: 10.1016/S0301-9268(03)00043-3.

Kimberley, M.M., 1989, Exhalative origins of iron formations: Ore Geology Reviews, v. 5, p. 130–145, doi: 10.1016/0169-1368(89)90003-6.

Kitajima, K., Maruyama, S., Utsunomiya, S., and Liou, J.G., 2001, Seafloor hydrothermal alteration at an Archaean mid-ocean ridge: Journal of Metamorphic Geology, v. 19, p. 583–597, doi: 10.1046/j.0263-4929.2001.00330.x.

Knauth, L.P., and Lowe, D.R., 1978, Oxygen isotope geochemistry of cherts from the Onverwacht Group (3.4 billion years), Transvaal, South Africa, with implications for secular variations in the isotopic composition of cherts: Earth and Planetary Science Letters, v. 41, p. 209–222, doi: 10.1016/0012-821X(78)90011-0.

Lepland, A., Van Zuilen, M., and Arrhenius, G., 2002, Origin and biologic significance of graphite and apatite in early Archean supracrustal rocks from Isua belt and Akilia association: Geochimica et Cosmochimica Acta, v. 66, p. A448.

Lindsay, J.F., McKay, D.S., and Allen, C.C., 2003, Earth's earliest biosphere—A proposal to develop a collection of curated Archean geologic reference materials: Astrobiology, v. 3, p. 739–754, doi: 10.1089/153110703322736060.

Lowe, D.R., 1999, Petrology and sedimentology of cherts and related silicified sedimentary rocks in the Swaziland Supergroup, *in* Lowe, D.R., and Byerly, G.R., eds., Geologic evolution of the Barberton Greenstone Belt, South Africa: Geological Society of America Special Paper 329, p. 83–114.

Lowe, D.R., and Byerly, G.R., 1999, Geologic evolution of the Barberton Greenstone Belt, South Africa: Geological Society of America Special Paper 329, 324 p.

Lowe, D.R., and Worrell, G.F., 1999, Sedimentology, mineralogy, and implications of silicified evaporites in the Kromberg Formation, Barberton Greenstone Belt, South Africa, *in* Lowe, D.R., and Byerly, G.R., eds., Geologic evolution of the Barberton Greenstone Belt, South Africa: Geological Society of America Special Paper 329, p. 167–187.

Lydon, J.W., 1984, Volcanogenic massive sulphide deposits, I, A descriptive model, *in* Roberts, R.G., and Shean, P.A., eds., Ore Deposit Models, Geoscience Canada Reprint Series, Volume 3, p. 145–153.

Mao, J., Lehmann, B., Du, A., Zhang, G., Ma, D., Wand, Y., Zeng, M., and Kerrich, R., 2002, Re-Os dating of polymetallic Ni-Mo-PGE-Au mineralization in Lower Cambrian black shales and its geologic signification: Economic Geology and the Bulletin of the Society of Economic Geologists, v. 47, p. 1051–1061.

Mills, R.A., and Elderfield, H., 1995, Rare earth element geochemistry of hydrothermal deposits from the active TAG Mound, 26°N Mid-Atlantic Ridge: Geochimica et Cosmochimica Acta, v. 59, p. 3511–3524, doi: 10.1016/0016-7037(95)00224-N.

Mojzsis, S.L., Arrhenius, G., and Friend, C.R.L., 1996, Evidence for life on Earth before 3,800 million years ago: Nature, v. 384, p. 55–57, doi: 10.1038/384055a0.

Muzuka, A.N.N., Macko, S.A., and Pedersen, T.F., 1991, Stable carbon and nitrogen isotope compositions of organic matter from Site 724 and Site 725, Oman margin, *in* Prell, W.L., and Niitsuma, N., eds., Initial Report ODP Leg 117: Proceedings of the Ocean Drilling Program, Scientific Results, v. 117, p. 571–586.

Myers, J.S., 2001, Protoliths of the 3.8–3.7 Ga Isua greenstone belt: West Greenland: Precambrian Research, v. 105, p. 129–141, doi: 10.1016/S0301-9268(00)00108-X.

Nealson, K.H., and Myers, C.R., 1990, Iron reduction by bacteria: A potential role in the genesis of banded iron formations: American Journal of Science, v. 290, p. 35–45.

Nedachi, M., 2004, New discoveries from the Archean Biosphere Drilling Project (ABDP) [abs.]: Eos (Transactions, American Geophysical Union), v. 85, Supplement, B32B-03.

Nijman, W., de Bruijne, K.C.H., and Valkering, M.E., 1998, Growth fault control of early Archaean cherts, barite mounds and chert-barite veins, North Pole Dome, Eastern Pilbara, Western Australia: Precambrian Research, v. 88, p. 25–52, doi: 10.1016/S0301-9268(97)00062-4.

Orberger, B., and Alleweldt, J., 1994, Contribution to the petrogenesis of the platinum palladium-bearing black dunites of the Acoje block, Zambales ophiolite, Philippines; selenium and sulfur-contents, clinopyroxenes, initial weathering: Journal of Southeast Asian Earth Sciences, v. 9, p. 229–239, doi: 10.1016/0743-9547(94)90030-2.

Orberger, B., Pasava, J., Gallien, J.-P., Daudin, L., and Trocellier, P., 2003, Se, As, Mo, Ag, Cd, In, Sb, Pt, Au, Tl, Re traces in biogenic and abiogenic sulfides from black shales (Selwyn Basin, Yukon territories, Canada): A nuclear microprobe study: Nuclear Instruments & Methods in Physics Research, Section B, Beam Interactions with Materials and Atoms, v. 210, p. 441–448, doi: 10.1016/S0168-583X(03)01073-5.

Orberger, B., Gallien, J.P., Pinti, D.L., Fialin, M., Daudin, L., Gröcke, D.R., and Pasava, J., 2005a, Nitrogen and carbon partitioning in diagenetic and hydrothermal minerals from Paleozoic black shales (Selwyn Basin, Yukon Territories, Canada): Chemical Geology, v. 218, p. 249–264.

Orberger, B., Wagner, C., Vymalanova, A., Pasava, J., Kribek, B., and Gallien, J.P., 2005b, Rare metal sequestration and mobility in mineralized black shales from the Zunyi region, South China, *in* Mao, J., Bierlein, F.P., eds., Mineral Deposit Research, meeting the global challenge: Springer Verlag, v. 1, p. 167–171.

Paris, I., Stanistreet, I.G., and Hughes, M.J., 1985, Cherts of the Barberton Greenstone Belt interpreted as products of submarine exhalative activity: The Journal of Geology, v. 93, p. 111–130.

Pinti, D.L., and Hashizume, K., 2001, N-15-depleted nitrogen in early Archean kerogens: Clues on ancient marine chemosynthetic-based ecosystems? A comment to Beaumont, V., and Robert, F., 1999, Precambrian Research v. 96, p. 62–82: Precambrian Research, v. 105, p. 85–88, doi: 10.1016/S0301-9268(00)00100-5.

Pinti, D.L., Matsuda, J., and Maruyama, S., 2001a, Anomalous xenon in Archean cherts from Pilbara Craton, Western Australia: Chemical Geology, v. 175, p. 387–395, doi: 10.1016/S0009-2541(00)00331-4.

Pinti, D.L., Hashizume, K., and Matsuda, J., 2001b, Nitrogen and argon signatures in 3.8 to 2.8 Ga metasediments: Clues on the chemical state of the Archean ocean and the deep biosphere: Geochimica et Cosmochimica Acta, v. 65, p. 2301–2315, doi: 10.1016/S0016-7037(01)00590-7.

Pinti, D.L., Hashizume, K., Philippot, P., Foriel, J., and Rey, P., 2003, Nitrogen quest in Archean metasediments of Pilbara, Australia: Geochimica et Cosmochimica Acta, v. 67, p. A379.

Pirajno, F., and Grey, K., 2002, Chert in the Palaeoproterozoic Bartle Member, Killara Formation, Yerrida Basin, Western Australia: A rift-related playa lake and thermal spring environment?: Precambrian Research, v. 113, p. 169–192, doi: 10.1016/S0301-9268(01)00196-6.

Préat, A., Mamet, B., de Jong, J., Durlet, C., Morano, S., and Matielli, N., 2004, Origine de la pigmentation de l'Ammonitico Rosso d'Italie du nord (Jurassique moyen, region de Verone): Biosédimentologie; diagenèse et géochimie isotopique, *in* Camoin, G., and Gautret, P., eds., Microbialite and microbial communities in sedimentary systems: Paris, Workshop, 6–9 September, Publications de l'Association Sédimentologique Français, v. 46, p. 79–82.

Rau, G.H., Arthur, M.A., and Dean, W.E., 1987, 15N/14N variations in Cretaceous Atlantic sedimentary sequences: Implication for past changes in marine nitrogen biogeochemistry: Earth and Planetary Science Letters, v. 82, p. 269–279, doi: 10.1016/0012-821X(87)90201-9.

Renaut, R.W., Jones, B., Tiercelin, J.J., and Tarits, C., 2002, Sublacustrine precipitation of hydrothermal silica in rift lakes: Evidence from Lake Baringo, central Kenya Rift Valley: Sedimentary Geology, v. 148, p. 235–257, doi: 10.1016/S0037-0738(01)00220-2.

Rouchon, V., Pinti, D.L., Gallien, J.P., Orberger, B., Daudin, L., and Westall, F., 2005, NRA analyses of N and C in hydromuscovite from a 3.5 Ga chert from Kitty's Gap, Pilbara, Australia: Nuclear Instruments and Methods in Physics Research, v. B231, p. 536–540.

Schidlowski, M., Hayes, J.M., and Kaplan, I.R., 1983, Isotopic inferences of ancient biochemistries: Carbon, sulfur, hydrogen, and nitrogen, *in* Schopf, W.J., ed., Earth's Earliest Biosphere: Princeton, Princeton University Press, p. 149–186.

Schubel, K.A., and Simonson, B.M., 1990, Petrography and diagenesis of cherts from Lake Magadi, Kenya: Journal of Sedimentary Petrology, v. 60, p. 761–776.

Siever, R., 1992, The silica cycle in the Precambrian: Geochimica et Cosmochimica Acta, v. 56, p. 3265–3272, doi: 10.1016/0016-7037(92)90303-Z.

Steiner, M., Wallis, E., Erdtmann, B.D., Yuanlong, Z., and Yang, R., 2001, Submarine-hydrothermal exhalative ore layers in black shales from South China and associated fossils—Insights into a Lower Cambrian facies and bio-evolution: Paleogeography, Paleoclimatology, Paleoecology, v. 169, p. 165–191, doi: 10.1016/S0031-0182(01)00208-5.

Sugitani, K., 1992, Geochemical characteristics of Archean cherts and other sedimentary rocks in the Pilbara Block, Western Australia: Evidence for Archean seawater enriched in hydrothermally derived iron and silica: Precambrian Research, v. 57, p. 21–47, doi: 10.1016/0301-9268(92)90093-4.

Sugitani, K., Yamamoto, K., Wada, H., Binu-Lal, S.S., and Yoneshige, M., 2002, Geochemistry of Archean carbonaceous cherts deposited at immature island-arc setting in the Pilbara Block, Western Australia: Sedimentary Geology, v. 151, p. 45–66, doi: 10.1016/S0037-0738(01)00230-5.

Sweeney, R.E., Liu, K.K., and Kaplan, I.R., 1978, Oceanic nitrogen isotopes and their uses in determining the source of sedimentary nitrogen: Department of Scientific Industrial Resources New Zealand Bulletin, v. 220, p. 9–26.

Takano, Y., Kobayashi, K., Yamanaka, T., Marumo, K., and Urabe, T., 2004, Amino acids in the 308[deg]C deep-sea hydrothermal system of the Suiyo Seamount, Izu-Bonin Arc, Pacific Ocean: Earth and Planetary Science Letters, v. 219, p. 147–153, doi: 10.1016/S0012-821X(03)00699-X.

Tarasov, V.G., Gebruk, A.V., Shulkin, V.M., Kamenev, V.M., Fadeev, V.I., Kosmynin, V.N., Malakhov, V.V., Starynin, D.A., and Obzhirov, A.I., 1999, Effect of shallow-water hydrothermal venting on the biota of Matupi Harbour (Rabaul Caldera, New Britain Island, Papua New Guinea): Continental Shelf Research, v. 19, p. 79–116, doi: 10.1016/S0278-4343(98)00073-9.

Tebo, B.M., Ghiorse, W.C., van Waasbergen, L.G., Siering, P.L., and Caspi, R., 1997, Bacterially mediated mineral formation: Insights into manganese (II)-oxidation from molecular genetic and biochemical studies, *in* Banfield, J.F., and Nealson, K.H., eds., Geomicrobiology: Interactions between microbes and minerals, Reviews in Mineralogy, Volume 35: Washington, D.C., Mineralogical Society of America, p. 225–260.

Thorpe, R.I., Hickman, A.H., Davis, D.W., Mortensen, J.K., and Trendall, A.F., 1992, U-Pb zircon geochronology of Archean felsic units in the Marble Bar Region, Pilbara Craton, Western Australia: Precambrian Research, v. 56, p. 169–189, doi: 10.1016/0301-9268(92)90100-3.

Ueno, Y., Yoshioka, H., Maruyama, S., and Isozaki, Y., 2004, Carbon isotopes and petrography of kerogens in similar to 3.5-Ga hydrothermal silica dikes in the North Pole area, Western Australia: Geochimica et Cosmochimica Acta, v. 68, p. 573–589, doi: 10.1016/S0016-7037(03)00462-9.

Umeda, M., 2003, Precipitation of silica and formation of chert-mudstone-peat association in Miocene coastal environments at the opening of the Sea of Japan: Sedimentary Geology, v. 161, p. 249–268, doi: 10.1016/S0037-0738(03)00117-9.

Uyeda, S., 1987, Active hydrothermal mounds in the Okinawa back arc trough [abs.]: Eos (Transactions, American Geophysical Union), v. 68, p. 737.

Van Kranendonk, M.J., Hickman, A.H., Williams, I.R., and Nijman, W., 2001, Archaean geology of the East Pilbara granite-greenstone terrane, Western Australia—A field guide: Perth, Western Australia, Geological Survey of Western Australia, 134 p.

van Zuilen, M.A., Lepland, A., Terranes, J., Finarelli, J., Wahlen, M., and Arrhenius, G., 2003, Graphite and carbonates in the 3.8 Ga old Isua Supracrustal Belt, southern West Greenland: Precambrian Research, v. 126, p. 331–348.

Walsh, M., 1992, Microfossils and possible microfossils from the early Archean Onverwacht group, Barberton Mountain Land, South Africa: Precambrian Research, v. 54, p. 271–293, doi: 10.1016/0301-9268(92)90074-X.

Walsh, M.M., and Lowe, D.R., 1999, Modes of accumulation of carbonaceous matter in the early Archean: A petrographic and geochemical study of the carbonaceous cherts of the Swaziland Supergroup, *in* Lowe, D.R., and Byerly, G.R., eds., Geologic evolution of the Barberton Greenstone Belt, South Africa:: Geological Society of America Special Paper 329, p. 115–132.

Westall, F., 2005, The geological context for the origin of life and the mineral signatures of fossil life, *in* Gardaud, M., Barbier, B., Martin, H., and Reisse, G., eds., Lectures in Astrobiology, Vol. I: Berlin, Springer-Verlag (in press).

Westall, F., de Wit, M.J., Dann, J., van der Gaast, S., de Ronde, C.E.J., and Gerneke, D., 2001, Early Archean fossil bacteria and biofilms in hydrothermally influenced sediments from the Barberton greenstone belt, South Africa: Precambrian Research, v. 106, p. 93–116, doi: 10.1016/S0301-9268(00)00127-3.

Westall, F., Orberger, B., Rouchon, V., Rouzaud, J.N., and Wright, I., 2004, On the identification of early Archaean microfossils in cherts from Barberton and the Pilbara, *in* Reimold, W.U., and Hofmann, A., compilers, Field Forum: Processes on the early Earth, Kaapvaal Craton, South Africa: Abstracts: Johannesburg, p. 94–97.

Westall, F., de Vries, S.T., Nijman, W., Rouchon, V., Orberger, B., Pearson, V., Watson, J., Verchovsky, A., Wright, I., Rouzaud, J.-N., Marchesini, D., and Severine, A., 2006, this volume, The 3.466 Ga Kitty's Gap Chert, an early Archean microbial ecosystem, *in* Reimold, W.U., and Gibson, R.L., Processes on the early Earth: Geological Society of America Special Paper 405, doi: 10.1130/2006.2405(07).

Williams, L.A., and Crerar, D.A., 1985, Silica diagenesis, II, General mechanism: Journal of Sedimentary Petrology, v. 55, p. 312–321.

Wirth, R., 2004, Focused Ion Beam (FIB): A novel technology for advanced application of micro- and nanoanalysis in geosciences and applied mineralogy: European Journal of Mineralogy, v. 16, p. 863–876, doi: 10.1127/0935-1221/2004/0016-0863.

Manuscript Accepted by the Society 4 October 2005

Printed in the USA

Geological Society of America
Special Paper 405
2006

Contrasting source components of clastic metasedimentary rocks in the lowermost formations of the Barberton greenstone belt

Annika Dziggel
Institute of Mineralogy and Economic Geology, RWTH Aachen, Wuellnerstr. 2, 52056 Aachen, Germany
Gary Stevens
Department of Geology, University of Stellenbosch, Private Bag X1, Matieland 7602, South Africa
Marc Poujol
Department of Earth Sciences, Memorial University of Newfoundland, 300 Prince Phillip Drive, St. John's, Canada A1B 3X5
Richard A. Armstrong
Research School of Earth Science, Australian National University, Canberra ACT 0200, Australia

ABSTRACT

Rare occurrences of clastic metasedimentary rocks within the lowermost stratigraphic units of the Barberton greenstone belt represent one of the few lines of evidence that argue for the existence of some component of evolved granitoid crust prior to formation of the lower greenstone belt sequence. This study has examined the field occurrence, petrography, geochemistry, and geochronology of these unique sedimentary rocks in order to place constraints on their provenance and tectonic setting. On the basis of their field occurrence and mineralogy, two types can be distinguished. The most common type is a medium-grained, planar-bedded, and compositionally banded rock displaying a broadly mafic geochemical affinity. It locally displays graded bedding and occurs as thin, discontinuous layers within the predominantly mafic and ultramafic sequences. The second type is a pink, medium- to coarse-grained meta-arkose that locally contains well-preserved sedimentary structures such as trough cross-bedding and that occurs in close proximity to the previous rock type. Most of the planar-bedded rocks are depleted in incompatible elements such as Rb and Ba but contain high concentrations of CaO, Fe_2O_3, and MgO and some of the transitional trace elements, including Cr, Ni, and Co. Samples of the cross-bedded meta-arkoses are enriched in K-feldspar and contain up to 4.3 wt% K_2O. U-Pb dates of rounded detrital zircons from this unit range between ca. 3521 and 3540 Ma, indicating that at least two protoliths for these sediments predate the formation of the bulk of the greenstone belt. A minimum age of 3431 ± 11 Ma for the formation of the sediments is indicated by the age of a trondhjemitic gneiss that is intrusive into one of the greenstone remnants investigated. Combined major and trace element geochemistry indicates at least two different sources for the sediments, including reworked continental crust as well as mafic to ultramafic rock types. The major and trace element geochemistry of the sediments is highly variable, and this variability is strata-bound. This is interpreted to reflect the variety of rock compositions in the sources as well as different

Dziggel, A., Stevens, G., Poujol, M., and Armstrong, R.A., 2006, Contrasting source components of clastic metasedimentary rocks in the lowermost formations of the Barberton greenstone belt, *in* Reimold, W.U., and Gibson, R.L., Processes on the Early Earth: Geological Society of America Special Paper 405, p. 157–172, doi: 10.1130/2006.2405(09). For permission to copy, contact editing@geosociety.org.

degrees of chemical weathering in the source regions. Consequently, the data are best explained by a model involving contemporaneous deposition of relatively mature, continentally derived sediments and sediments derived from volcanic/volcaniclastic rocks from syn-depositional volcanic centers in an active tectonic environment such as an island arc.

Keywords: Barberton greenstone belt, clastic metasedimentary rocks, U-Pb geochronology, geochemistry.

INTRODUCTION

Clastic sedimentary successions are an important component of Archean greenstone belts. They may provide crucial information on their provenance, the tectonic setting during sedimentation, as well as palaeoclimatic and weathering processes that prevailed during erosion, transport, and deposition (e.g., McLennan and Taylor, 1984; Condie, 1993; Eriksson et al., 1997; Thurston and Kozhevnikov, 2000). In the Barberton greenstone belt of South Africa, terrigenous clastic sedimentation mainly took place during a major accretionary episode, in which several thousand meters of greywackes, shales, and sandstones of the ca. 3260–3220 Ma Fig Tree and Moodies Groups were deposited in several isolated basins (e.g., Lowe, 1994, 1999; Heubeck and Lowe, 1994; Lowe and Byerly, 1999). During this event, which eventually resulted in the amalgamation of the northern and southern terranes along the Saddleback-Inyoka fault system (arc and trench blocks I and II, De Ronde and De Wit, 1994), crustal thickening was associated with voluminous TTG (tonalite-trondhjemite-granodiorite) magmatism, as well as high-P, medium-T (650–700 °C; 8–11 kbar) metamorphism of older granitoids and associated greenstone belt lithologies in the mid- and lower crust (e.g., Stevens et al., 2002; Dziggel et al., 2002; Diener et al., 2005). This paper examines previously unrecognized clastic metasedimentary rocks from a much more elusive period of greenstone belt history prior to 3430 Ma. During this time, thin layers of clastic sediments were deposited in association with mafic to ultramafic volcanic units that most likely represent early oceanic crustal components (e.g., De Wit et al., 1983, 1987a; De Wit, 2004). These units as well as the intercalated sediments were subjected to high-grade metamorphism and pervasive deformation during terrane amalgamation (Dziggel et al., 2002; Stevens et al., 2002). Transposition of lithological boundaries, as well as metamorphic reaction and recrystallization has obscured much of the primary sedimentological structural record within these rocks. This study focuses on the textural evidence that has been preserved in unique low strain domains as well as on the major and trace element geochemisty in an attempt to constrain the composition and age of the source rocks and the tectonic setting of sedimentation.

GEOLOGICAL SETTING

The granitoid-dominated terrain immediately to the south of the Barberton greenstone belt (Fig. 1) exposes some of the oldest rocks of the Barberton Mountain Land. The terrain is dominated by ca. 3445 Ma trondhjemitic gneisses that are considered to have formed during an early period of compressional tectonics (D_1; e.g., De Wit et al., 1983). The trondhjemites intruded into and are, generally, in tectonic contact with the lowermost formations of the Onverwacht Group, namely the Sandspruit and Theespruit Formations (Fig. 1; Viljoen and Viljoen, 1969a, 1969b; De Wit et al., 1983). These formations are, in turn, separated from the bulk of the Barberton greenstone belt by a major shear zone, the Komati Fault (Fig. 1; De Wit et al., 1983, 1987a; De Ronde and De Wit, 1994). The trondhjemitic gneisses enclose numerous highly tectonized greenstone remnants that record peak P-T conditions of up to 650–700 °C and 8–11 kbar that were attained during the main compressional episode at ca. 3230 Ma (D_2; Dziggel et al., 2002). The post-peak juxtaposition of these high-grade rocks against the low-grade greenschist facies rocks of the Barberton greenstone belt has been interpreted to be a result of the orogenic collapse of the belt during or shortly after the D_2-deformation (Kisters et al., 2003). This was followed by a late transtensional or transpressional reactivation at ca. 3100 Ma, during which large sheets of potassic granites intruded along major thrust zones (D_3; e.g., Westraat et al., 2005; Belcher and Kisters, this volume). In the study area (Fig. 1), the greenstone remnants comprise mafic and ultramafic schists, minor intercalations of banded iron formation and chert, and locally developed felsic agglomerates and tuffs, as well as a clastic metasedimentary unit that has previously been mapped as calc-silicate rocks (C.R. Anhaeusser, 2000, personal commun.). The original classification of the greenstone remnants as belonging to either the Sandspruit Formation or the overlying Theespruit Formation is mainly based on the presence of felsic volcanic/volcaniclastic rocks (Theespruit Formation) and the predominance of ultramafic rocks and minor chemical sediments (Sandspruit Formation), although their regional distribution has also been considered for purposes of classification (Viljoen and Viljoen, 1969a, 1969b; Anhaeusser et al., 1983a). The greenstone rem-

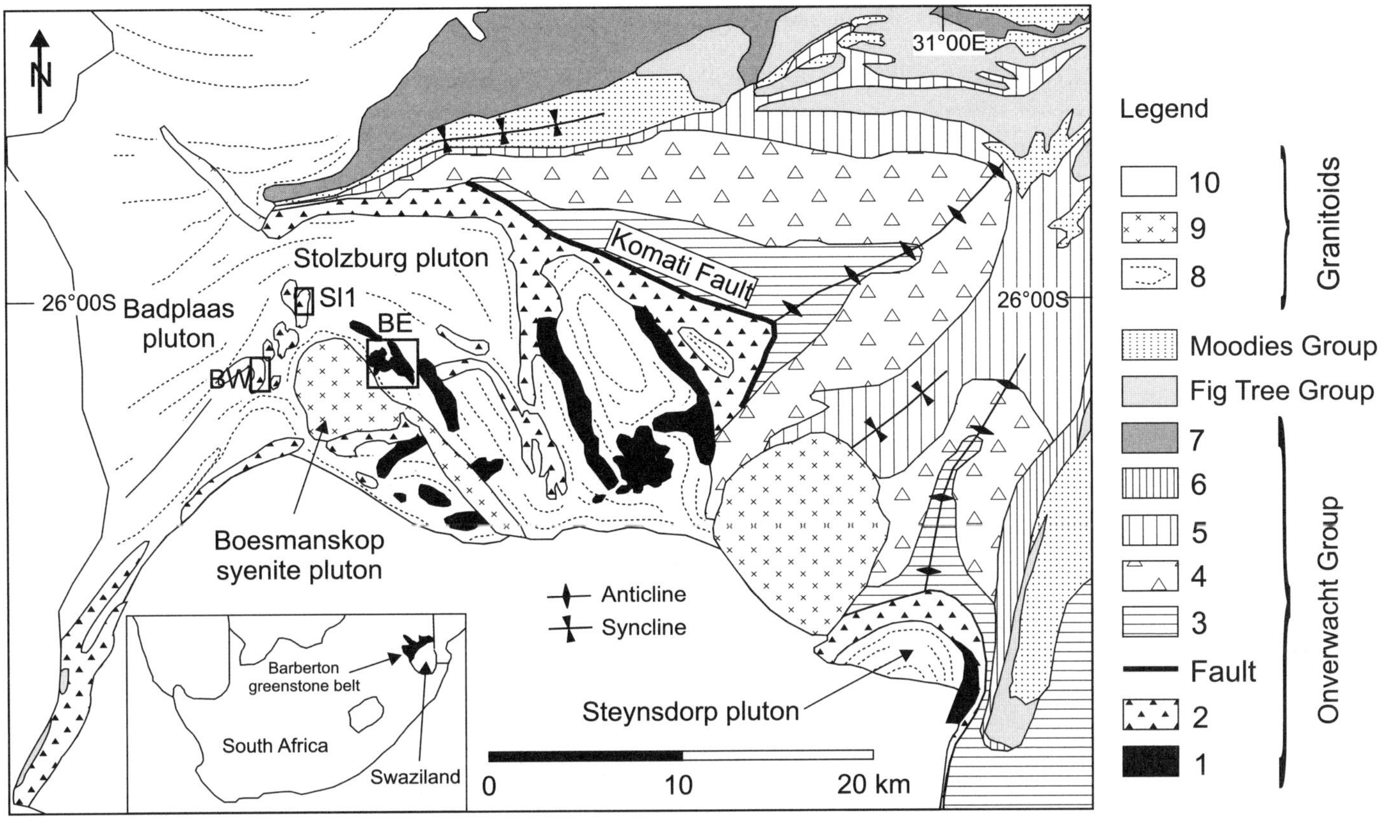

Figure 1. Geological map of the southern part of the Barberton greenstone belt and adjacent granitoid terrain (modified after Anhaeusser et al., 1983a; Kröner et al., 1996). Inset shows location of the Barberton greenstone belt in southern Africa. The Onverwacht Group in the southern part of the greenstone belt comprises: (1) Sandspruit Formation, (2) Theespruit Formation, (3) Komati Formation, (4) Hoggenoeg Formation, (5) Kromberg Formation, (6) Zwartkoppie/Mendon Formation; Onverwacht Group rocks in the north (7) are undifferentiated. The granitoid rocks have been subdivided into: (8) tonalite-trondhjemite-granodiorite plutons (including structural trend lines), (9) late potassic granites, and (10) the ca. 3100 Ma Mpuluzi and Heerenveen Batholiths. Rectangles show the areas investigated in this study.

nants investigated in this study are situated to the north, east, and west of the ca. 3107 Ma Boesmanskop syeno-granite complex and have been assigned to both the Theespruit and Sandspruit Formations (Fig. 1; Anhaeusser and Robb, 1980; Anhaeusser et al., 1983a, 1983b; Kamo and Davis, 1994). For consistency with previously published data, they will be referred to as greenstone remnants SL1, BW, and BE (cf. Dziggel et al., 2002).

FIELD OCCURRENCE AND PETROLOGY OF CLASTIC METASEDIMENTARY ROCKS

Clastic metasedimentary rocks form a distinct but minor component in many of the greenstone remnants in the study area, where they occur as discontinuous, foliation-parallel layers that vary in thickness between ~0.5 and 8 m. Although metamorphic assemblages have replaced their primary mineralogy, original sedimentary features are locally well preserved. Two texturally and mineralogically different varieties can be distinguished. The most common type is a fine- to medium-grained, compositionally banded and commonly boudinaged rock of mafic geochemical affinity, within which foliation-parallel compositional banding is well developed. Within low-strain domains, sedimentary graded bedding is locally preserved, and the compositional banding is, thus, interpreted to represent primary bedding. The second variety is medium to coarse grained and has a light pink color. This variety is a K-feldspar–rich impure arkose that is characterized by a cm-scale compositional banding both parallel to and, in low-strain domains, at an angle to the regional foliation. The compositional banding in this rock type commonly has a wavy appearance. These features are interpreted as primary bedding and trough cross-bedding. These two rock types will be referred to as "mafic" and "arkosic" facies, respectively.

Mafic Facies

The planar-bedded and compositionally banded mafic facies occurs in all of the greenstone remnants investigated. In the greenstone remnants to the east of the Boesmanskop syeno-granite complex, it forms part of a north-south–trending chain of greenstone remnants that are aligned along the contact between the Badplaas and Stolzburg plutons (Fig. 1). In greenstone remnant SL1 (Fig. 2), the subvertically dipping and north-south–striking units are pervasively foliated and are dominated by mafic and ultramafic schists. Mineral lineations are absent. Thin and discontinuous layers of garnet-bearing banded iron formation and finely banded clastic sedimentary rocks are interlayered with the mafic and ultramafic units and are generally less than half a meter thick. The clastic metasediments are characterized by a strong compositional banding that parallels the pervasive foliation in the amphibolites. The banding is defined by green, clinopyroxene-dominated layers that alternate with white, quartz-feldspar–rich layers (Figs. 3A and 3B). Thin, up to 1 cm thick, banding-parallel layers of amphibolite are common within this unit. Idioblastic grossular-almandine garnet up to 2 cm in diameter occurs within the pyroxene-rich layers (Fig. 3B). The peak-metamorphic mineral assemblage for all samples is clinopyroxene + quartz + plagioclase + garnet + K-feldspar ± hornblende. Additional retrograde minerals are common and include epidote, sphene, albite, and, more rarely, actinolite. The planar-bedded mafic facies of the clastic metasediments in greenstone remnant BE (Fig. 4) is very similar to that exposed elsewhere in the terrain. In contrast to the greenstone remnants to the west of the Boesmanskop syeno-granite complex, the layers are generally thicker (up to several m) and are commonly boudinaged. They are closely associated and interlayered with the arkosic facies in some places. The well-equilibrated peak assemblages in this rock type comprise diopside, plagioclase, quartz, K-feldspar, and hornblende. Locally, poikiloblastic garnet up to ~1 cm in diameter is observed. Retrogression in these rocks is indicated by a locally extensive replacement of the peak assemblages by epidote and sphene.

Arkosic Facies

To date, the only known occurrence of arkoses in the lowermost formations of the Onverwacht Group is located in greenstone remnant BE immediately east of the Boesmanskop syeno-granite complex (Dziggel et al., 2002). Mapping of this greenstone remnant indicates that it is much larger than originally proposed by Anhaeusser et al. (1983a). It is unique among the greenstone remnants in that it contains, in addition to the predominantly mafic and ultramafic schists, a well-developed unit of clastic metasedimentary rocks, which locally reaches a

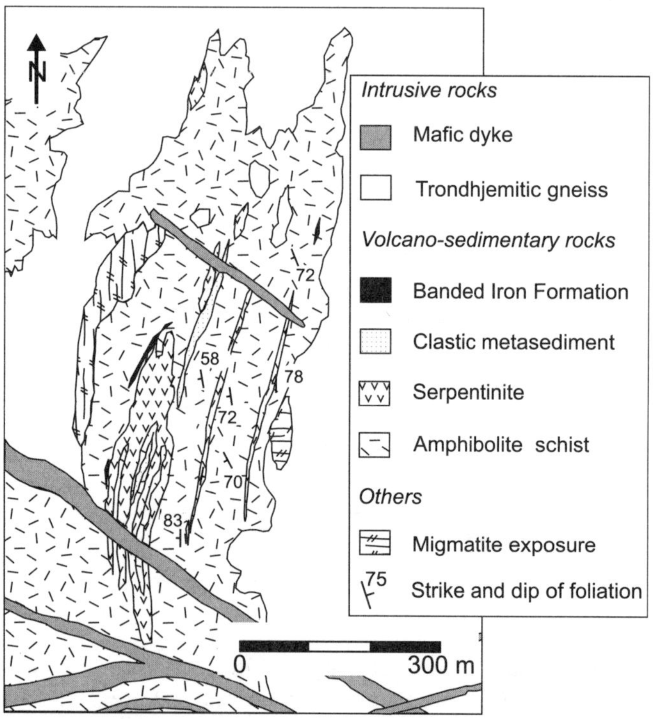

Figure 2. Simplified geological map of greenstone remnant SL1 (modified after Anhaeusser, 2000, personal commun.). The greenstone remnant is located on the farm Batavia 151 JT, from 26°00′15″ S 030°39′89″ E (south) to 25°29′76″ S 030°40′05″ E (north).

Figure 3. Field occurrence of the compositionally banded mafic facies of the clastic metasedimentary rocks in greenstone remnant SL1. (A) Finely laminated clinopyroxene-dominated layer with white stringers consisting of quartz and plagioclase; (B) idioblastic grossular-almandine garnet (Grt) in a medium-grained matrix consisting of clinopyroxene, quartz, plagioclase, K-feldspar, hornblende, epidote, and sphene. The scale (5-Rand coin in the center of the image) has a diameter of 2.5 cm.

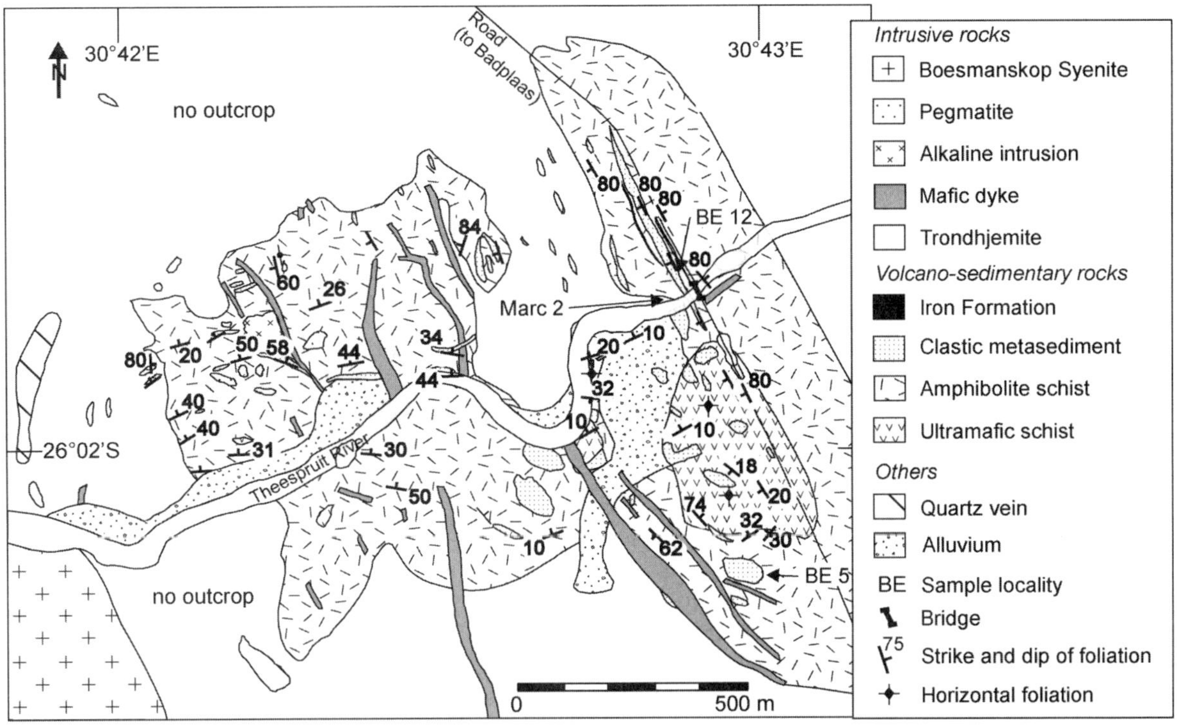

Figure 4. Geological map of greenstone remnant BE, situated on the farm Nederland 151 JT, east of the Boesmanskop Syenite pluton.

thickness of up to 8 m. Where exposed, the contacts between greenstone remnant BE and the surrounding trondhjemitic gneisses are intrusive. Additionally, and in striking contrast to the generally subvertical fabrics in and around other greenstone remnants in the study area, greenstone remnant BE has a rather complex structure, and a large proportion of its bedding-parallel fabrics, particularly in its western and southern parts, dip at low to moderate angles to the north and northwest (Fig. 4). Steeply dipping, northwest-southeast–trending fabrics dominate only in the northeastern sector.

The pink, medium- to coarse-grained, quartz- and K-feldspar–rich rocks are exceptionally well preserved and contain varying proportions (~2–10 vol%) of mafic minerals such as pyroxene. Exposures of relatively undeformed, medium- to coarse-grained meta-arkose have only been identified in one locality (locality BE 5, Fig. 4). In this low-strain domain, the meta-arkose is characterized by well-preserved bedding and trough cross-beds and even records grain-size and compositional variations between different beds, despite the high-grade metamorphic overprint (Figs. 5A and 5B). More commonly, the meta-arkose has been strongly deformed and has a schistose appearance. At locality BE 12 (Fig. 4), it occurs in a narrow, northwest-southeast–trending and subvertically dipping sequence, where it is in contact with the mafic facies. The meta-arkose consists of variable amounts of quartz, microcline, plagioclase, diopside, and muscovite. Retrogression of the peak metamorphic minerals is minor and is marked by the replacement of muscovite by chlorite and the growth of fibrous actinolite needles and small (~50 μm) epidote grains at mineral triple points. In low-strain domains, the arkosic facies contains granoblastic-polygonal textures, indicating that the coarse grain size may, to some degree, be a consequence of metamorphic recrystallization. K-feldspar is believed to constitute a detrital component of these rocks, as the well-preserved sedimentary structures clearly confirm the coarse clastic nature of the sediments. This is inconsistent with K_2O introduction via a clay mineral or white mica component. In addition to this, the Al_2SiO_5 excess that would occur with metamorphism of these minerals to K-feldspar–bearing assemblages is not evident in these sediments.

GEOCHEMISTRY

Methodology

Samples for chemical analyses were powdered using a jaw crusher and swing mill. Fused and pressed powder pellets were analyzed for major and trace element contents by standard X-ray fluorescence techniques utilizing the Philips PW 1400 X-ray Fluorescence Spectrometer in the School of Geosciences, University of the Witwatersrand, Johannesburg. The analytical precision is considered to be better than 3% of the measured value for major elements and ~5%–10% of the measured value for trace elements.

Figure 5. Field occurrence of the arkosic facies in greenstone remnant BE (locality BE 5). (A) Boulder of a cross-bedded meta-arkose. The white arrows trace cross-laminations, the black arrows indicate other possible cross-bedding in the rock. (B) Compositional variations interpreted to represent primary sedimentary bedding.

In unmetamorphosed sedimentary sequences, the major element composition of clastic rocks reflects their primary mineralogy, which in turn is sensitive to the intensity of chemical weathering during erosion and transport (e.g., Taylor and McLennan, 1985; McLennan et al., 1993). Because rocks in the study area experienced high-grade metamorphism, standard sedimentary-petrographic techniques cannot be applied, so that the characterization of the sediments has to be carried out on the basis of geochemical data. The results of XRF analysis of representative samples from the arkosic and mafic facies are presented in Table 1. The 16 samples show a wide compositional range in terms of their major and trace element concentrations and record 51.5–85.4 wt% SiO_2, 4.8–17.0 wt% Al_2O_3, 0.7–9.4 wt% Fe_2O_3, 0.3–10.7 wt% MgO, 1.0–18.9 wt% CaO, and 0.0–4.3 wt% K_2O. Except for one sample, the loss on ignition (LOI) is small (0.0–2.0 wt%), reflecting the relatively anhydrous mineralogy and the absence of carbonate phases such as calcite and dolomite.

TABLE 1. WHOLE-ROCK MAJOR (WT%) AND TRACE ELEMENT (PPM) COMPOSITION OF THE METASEDIMENTS FROM THREE GREENSTONE REMNANTS INVESTIGATED IN THIS STUDY

Sample	SL1–6	SL1–8	SL1–8a	SL1–9	BE1	BE4*	BE5*	BE8	BE9	BE10*	BE11	BE12*	BE13	BW211	BW214	BW219
SiO_2	64.06	55.87	71.02	55.77	57.51	81.25	62.7	55.88	53.7	85.43	77.19	61.62	54.6	54.02	51.54	69.29
TiO_2	0.26	0.54	0.3	0.29	0.82	0.2	0.22	0.46	0.47	0.24	0.22	0.74	0.85	0.53	0.9	0.26
Al_2O_3	9.4	4.86	10.57	10.89	11.9	6.23	11.96	6.33	7.35	7.69	5.68	13.93	12.51	13.03	17.05	11.4
Fe_2O_3	6.09	9.48	4.21	7.57	5.67	0.98	3.23	5.36	5.59	0.77	2.62	7.66	7.6	7.92	7.02	5.4
MnO	0.25	0.24	0.15	0.17	0.41	0.14	0.53	0.72	0.7	0.05	0.31	0.29	0.32	0.26	0.16	0.16
MgO	3.72	8.96	1.99	8.15	7.38	0.86	5.73	10.78	10.72	0.3	4.02	3.39	5.24	5.03	4.1	1.39
CaO	11.49	17.6	9.84	11.24	10.7	2.13	10.81	18.92	18.02	1.07	7.32	5	14.72	16.66	12.02	11.34
Na_2O	3.46	1.66	2.18	2.72	3.55	1.85	0.91	1.18	2.72	0.57	1.31	1.57	3.08	1.65	4.9	0
K_2O	0.55	0.49	0.24	1.34	1.21	3.41	3.75	0.2	0.4	4.23	1.04	4.3	0.65	0.15	0.46	0
P_2O_5	0.07	0.02	0.04	0	0.04	0.07	0.02	0.02	0.02	0.08	0.04	0.07	0.07	0.05	0.06	0.08
LOI	1.04	0.78	0.04	1.4	1.15	3.01	0.44	0.5	0.9	0.62	0.67	1.75	0.92	1.19	1.99	1.23
Total	100.39	100.5	100.49	99.56	100.34	100.13	100.31	100.35	100.59	101.05	100.43	100.33	100.56	100.49	100.22	100.57
Rb	34	44	22	63	79	104	59	29	35	64	89	124	41	21	20	8
Sr	331	147	206	232	346	181	63	129	222	32	113	68	333	166	224	169
Y	10	10	7	11	31	10	5	14	20	3	7	15	23	14	19	8
Zr	28	36	35	28	76	64	62	40	43	72	19	47	77	28	37	15
Nb	7	10	8	9	9	8	9	8	8	8	7	7	10	5	7	7
Co	48	65	34	51	74	11	9	39	50	9	22	48	57	43	57	27
Ni	464	372	307	420	560	26	9	295	418	11	187	156	295	165	217	163
Cu	2	2	2	2	2	9	2	2	2	2	9	97	38	12	941	40
Zn	56	86	56	64	140	705	28	104	99	31	51	97	91	57	1296	35
V	109	197	116	118	274	20	15	142	149	21	62	330	304	237	264	152
Cr	1729	1373	1559	1802	3356	30	23	1742	2007	26	677	160	1662	518	662	719
Ba	122	110	66	160	186	1829	1057	53	72	824	157	103	201	127	30	15
CIA[†]	25	12	32	29	30	37	32	14	16	50	25	46	27	28	35	35

Notes: Analyses were conducted by standard X-ray fluorescence techniques at the School of Geosciences, University of the Witwatersrand, Johannesburg.
*Indicates arkosic samples.
[†]CIA values were calculated on the basis of molar proportions (cf. Fedo et al., 1995).

Major Elements

Mafic Facies

Major element variations are displayed in Figures 6 and 7. The planar-bedded, compositionally banded rocks record high concentrations of CaO, Fe_2O_3, and MgO (Figs. 6A–6C). These elements show significant positive correlations and record correlation coefficients of 0.79 for the element pair CaO-MgO, 0.67 for CaO-FeO, and 0.51 for FeO-MgO. Compared to typical mafic and ultramafic rocks in the study area, a large proportion of the samples that are macroscopically classified as mafic (based on the abundance of typical mafic minerals such as clinopyroxene and epidote) are enriched in SiO_2 and relatively depleted in MgO and $Fe_2O_{3(tot)}$ (Figs. 6B and 6C). The inverse correlation of these elements is attributed to SiO_2 dilution. The CaO content of the samples investigated records strong variations, and a number of samples are enriched in CaO relative to typical mafic and ultramafic rocks from the study area. Compared to the arkosic facies, these samples are depleted in SiO_2, and their K_2O concentrations are lower than 1 wt%. Similarly, the samples record K_2O-depletion and CaO-enrichment relative to the average composition of Fig Tree Group greywackes (Toulkeridis et al., 1999). The heterogeneous geochemical signatures of the mafic facies are substantially different from the typically homogeneous compositions of single beds of mature and well-sorted clastic sedimentary rocks. This is apparent on a plot of FeO + MgO versus $K_2O/(K_2O + Na_2O)$ (Fig. 7), which shows that the planar-bedded and compositionally banded facies, including all samples from greenstone xenolith BW and SL1, have highly variable iron and magnesium concentrations. Furthermore, many samples are substantially enriched in iron and magnesium contents, but commonly have lower $K_2O/(K_2O + Na_2O)$ ratios than typical greywackes.

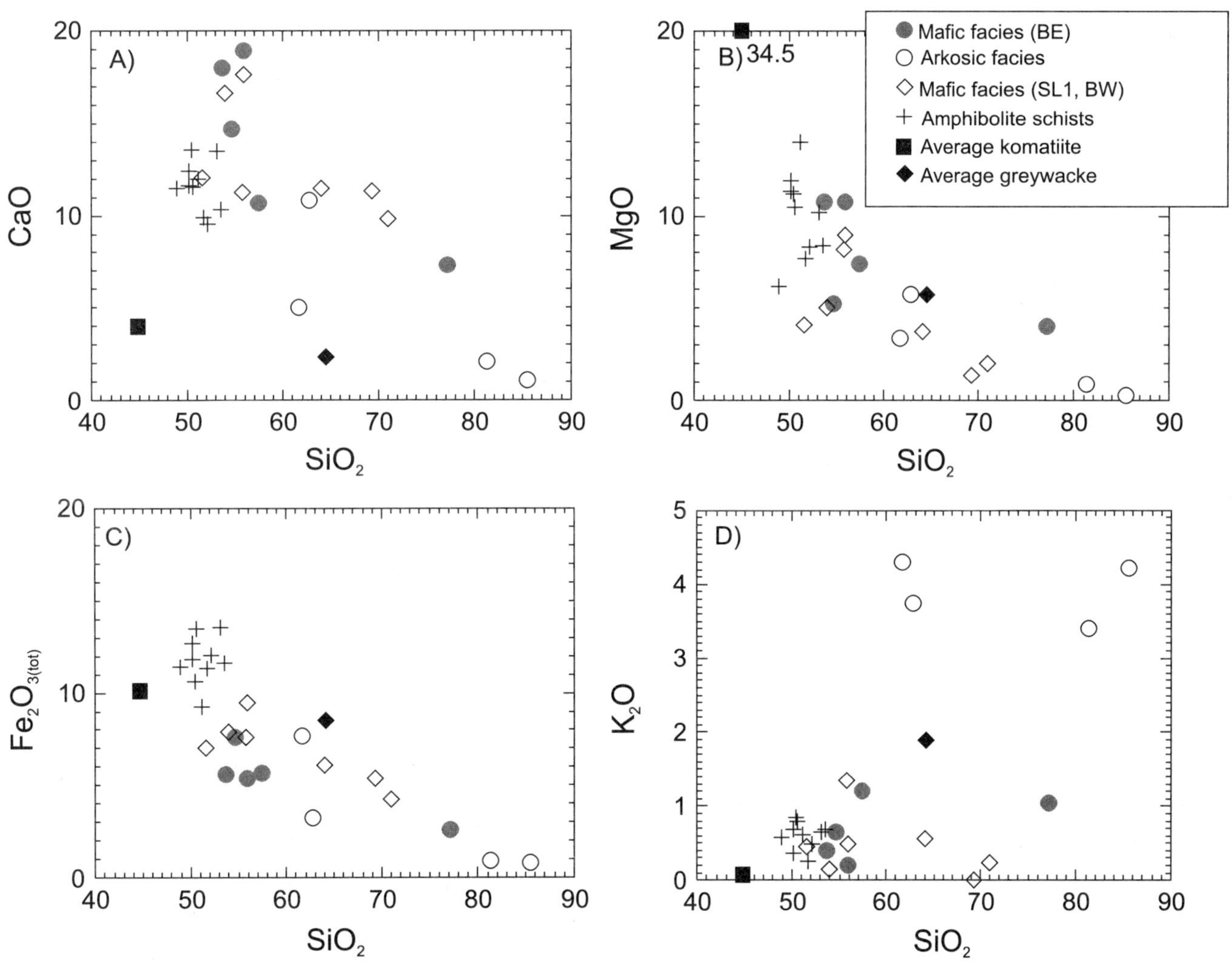

Figure 6. Harker-type diagrams showing the geochemical variations in the clastic metasedimentary rocks compared to amphibolite schists from the study area (Robb, 1981) and average geochemistries of komatiites (data from Lahaye et al., 1995) and Fig Tree Group greywackes (data from Toulkeridis et al., 1999).

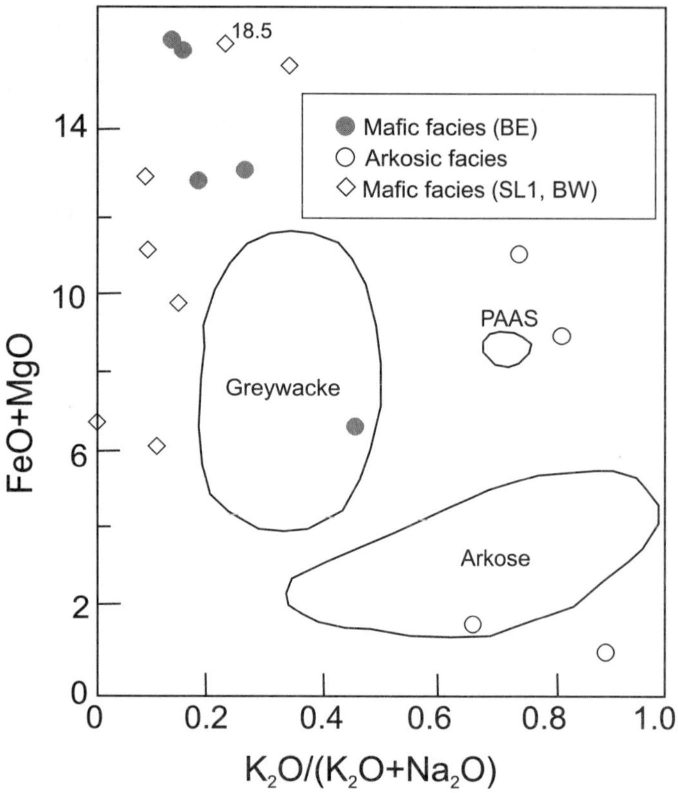

Figure 7. Plot of FeO + MgO versus $K_2O/(K_2O + Na_2O)$ (wt%; after Chacko et al., 1992) comparing the geochemistry of the clastic metasedimentary rocks with typical greywackes, arkoses, and PAAS (Average Post-Archean Australian Shale). The different fields were defined using data of Taylor and McLennan (1985) as well as Pettijohn et al. (1987).

Arkosic Facies

The cross-bedded meta-arkoses record the highest SiO_2 concentrations, are low in Na_2O, and contain up to 4.3 wt% K_2O (Table 1). Despite the large variation in the major element concentrations in the whole data set, the element pairs K_2O and SiO_2 (Fig. 6D) as well as SiO_2 and Al_2O_3 ($r = -0.4$, not shown) do not exhibit strong correlations. Fe_2O_3 and MgO concentrations range between 0.7 and 7.6 wt% and 0.8–5.7 wt%, respectively, indicating variable input of (ultra-) mafic detritus during sedimentation (Figs. 6B and 6C). In contrast to the mafic facies, samples from the arkoses record high $K_2O/(K_2O + Na_2O)$ ratios (Fig. 7). Due to the high content in Fe_2O_3 + MgO, two of the samples are compositionally similar to PAAS (Average Post-Archean Average Australian Shale; Taylor and McLennan, 1985), whereas the remaining samples fall into or near the field for arkoses.

Trace Elements

Mafic Facies

A large proportion of samples macroscopically classified as "mafic" display very high values for some of the "ferromagnesian" transition trace elements (Fig. 8). These samples have Cr

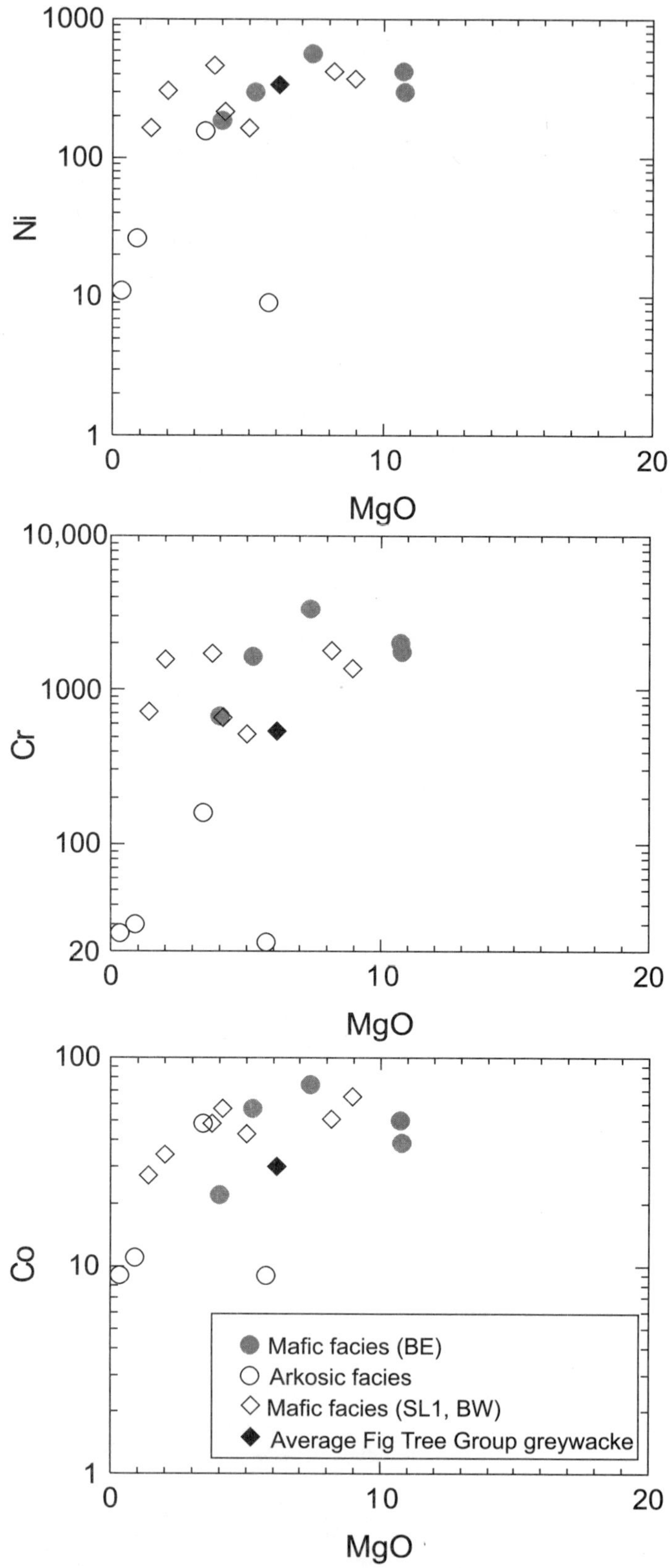

Figure 8. Plot of (A) Ni, (B) Cr, and (C) Co against MgO illustrating the high abundance of these elements in the mafic facies of the clastic metasedimentary rocks. For comparison, the average Fig Tree Group shale composition has been included.

and Ni concentrations of up to 3356 ppm and 560 ppm, respectively. The elements exhibit a strong positive correlation (r = 0.95) with Cr/Ni ratios ranging between 3 and 6. Cr, Ni, and also Co co-vary with MgO and Fe_2O_3, indicating a significant proportion of mafic and ultramafic material in their provenance (e.g., Garver et al., 1996). Most samples are highly enriched in these elements compared to average Fig Tree Group greywackes (Fig. 8). Large ion lithophile element (LILE) concentrations are variable and range from 8 to 124 ppm Rb, 129–346 ppm Sr, and 15–186 ppm Ba. Rb and Ba are positively correlated with K_2O, whereas Sr shows a marked positive correlation with Na_2O.

Arkosic Facies

The arkosic facies is characterized by comparatively low concentrations of most of the trace elements investigated in this study. Compared to the mafic facies, samples previously assigned to the arkosic facies have low Cr and Ni concentrations (≤30 ppm; Table 1). Sample BE 12, due to the high Fe_2O_3 + MgO content, is relatively enriched in Cr, Ni, and Co. Owing to the high content of K-feldspar in these rocks, the arkoses are enriched in LILE and record high, though varying, Rb and Ba concentrations, ranging from 59 to 124 ppm and 103–1829 ppm, respectively. Sr is relatively depleted compared to the mafic facies (63–181 ppm), which is interpreted to indicate a lower plagioclase content in the arkoses. Except for two samples (BE1 and BE13), the samples are relatively enriched in Zr (62–72 ppm) compared to the compositionally banded rocks (15–47 ppm) but have relatively low Zr concentrations compared to typical granitoid rocks from the Barberton greenstone belt that record Zr concentrations of 100–300 ppm (Kleinhanns et al., 2003). This is interpreted to be linked to heavy mineral sorting during sedimentation.

GEOCHRONOLOGY

In order to obtain constraints on the timing of sedimentation of the clastic metasedimentary rocks as well as to determine the age of possible source rocks, the cross-bedded arkosic facies as well as a trondhjemitic gneiss intrusive into greenstone remnant BE were chosen for U-Pb TIMS (thermal ionization mass spectrometry) and ion microprobe dating. A detailed description of the analytical procedures used in the generation of the geochronological data presented in this study has been given by Dziggel et al. (2002).

Zircons extracted from the cross-bedded meta-arkose at sample locality BE 5 have a wide range of shapes and colors and vary in size between ~100 and 300 μm. The predominant zircon type is dark purple and long-prismatic, with subrounded terminations that are consistent with limited mechanical transport and abrasion (Fig. 9). Other zircon types include small purple and yellowish-brown grains, which are generally more rounded than the long-prismatic zircons. All zircon types are concentrically and compositionally zoned, consistent with a primary igneous origin. Combined backscattered and cathodoluminescence imaging shows that the investigated grains are characterized by different degrees of fracturing, abundant mineral inclusions such as apatite and feldspar, as well as a thin, uranium-poor overgrowth, which is interpreted to be of metamorphic origin. For the present study, a total of 14 single grains as well as 27 spots on 25 different grains were analyzed by TIMS and SHRIMP (sensitive high-resolution ion microprobe) dating techniques, respectively (Fig. 10). Most of the zircons analyzed suffered variable Pb loss (Table 2), and the general trend of the data indicates nonrecent Pb loss at ca. 1000 Ma, which might be related to the Namaquan (ca. 1060–1030 Ma) or Kibaran (ca. 1200 Ma) orogenies (e.g., Robb et al., 1999). Only 5 analyses are concordant within error and imply a range of source rock ages for the cross-bedded arkoses. The two youngest concordant dates give an average $^{207}Pb/^{206}Pb$ date of 3521.5 ± 5.6 Ma, which is considered as the best estimate for the maximum age of the sedimentation of the arkose. The remaining dates range between ca. 3525 and 3540 Ma. The Th/U ratios for these zircons vary between 0.13 and 0.75 but do not record any systematic trend with increasing (or decreasing) age. Similarly, the different mineral shapes described above do not correlate with any particular $^{207}Pb/^{206}Pb$ date.

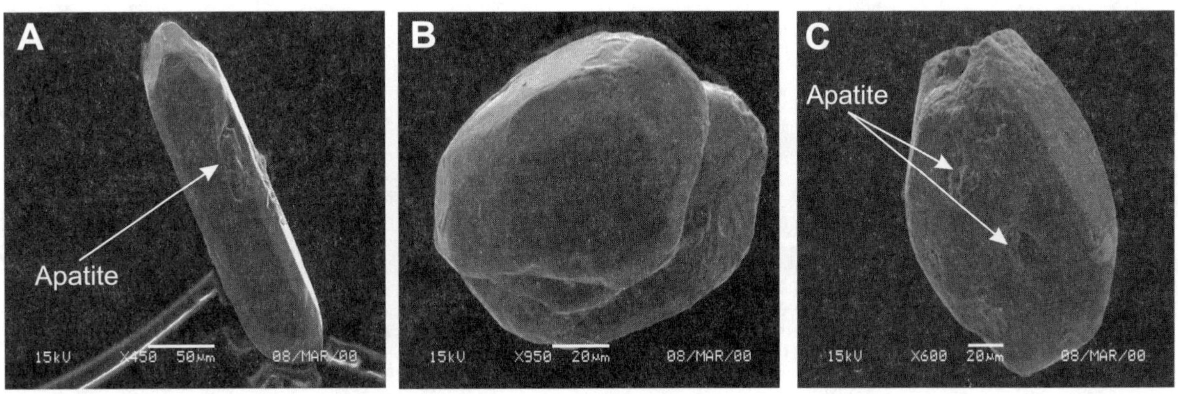

Figure 9. SEM (scanning electron microscope) backscattered electron images showing zircon morphologies in sample BE5. Mineral inclusions are apatite.

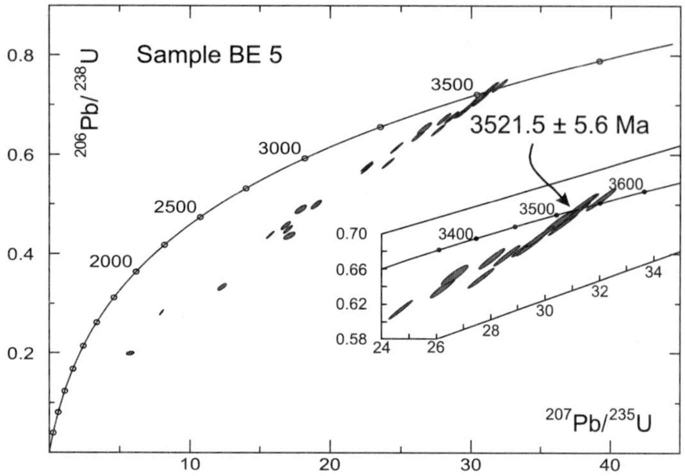

Figure 10. SHRIMP (sensitive high-resolution ion microprobe) U-Pb Concordia diagram for clastic metasediments (sample BE 5). Inset shows data that are 90% concordant or more. Error ellipses are at 2 sigma.

Zircons extracted from a foliated trondhjemitic gneiss intrusive into greenstone remnant BE and presumably part of the ca. 3445 Ma (Dziggel et al., 2002) Stolzburg pluton (sample locality Marc 2, Fig. 4) are predominantly purple to yellowish-brown. The most abundant type is long-prismatic. Five out of the six grains analyzed define a discordia with an upper intercept date of 3431 ± 11 Ma and a lower intercept date of 71 ± 35 Ma (Fig. 11). Although slightly younger, the age of 3431 ± 11 Ma is indistinguishable, within error, from the age estimates of 3445 ± 4 Ma and 3459 + 35/–23 Ma for the Stolzburg pluton (Kröner et al., 1991; Kamo and Davis, 1994; Table 3). Another explanation, however, may be that the age of ca. 3431 Ma reflects a slightly younger pulse of magmatism in this area. If the latter is assumed, the sedimentation of the arkosic facies is bracketed between 3521 ± 5.6 Ma (the age of the youngest concordant zircon in the arkose) and 3431 ± 11 Ma (the age of intrusion of this trondhjemite).

DISCUSSION

Element Mobility, Source-Rock Weathering, and Provenance

Because of the predominance of TTG rocks and mafic and ultramafic volcanic sequences in granite-greenstone terrains, Archean clastic sedimentary rocks are commonly interpreted to have been derived from bimodal rock suites (e.g., McLennan and Taylor, 1984). The highly variable geochemical character of the clastic metasediments, as indicated by the great diversity in their major and trace element concentrations, does not favor a generally high degree of chemical weathering in their source region(s). Many of the element variations observed in the clastic metasediments may, thus, be interpreted to reflect a mixed rock derived from both an (ultra-) mafic and a granitic source. However, because minor geochemical modifications cannot easily be detected in the clastic sedimentary rocks, it is important to critically analyze the field and geochemical data and to evaluate the effects of secondary processes on element mobility such as source rock weathering, hydrothermal alteration, and metamorphism.

TABLE 2. SHRIMP U-Th-Pb DATA FOR SAMPLE BE5

						Radiogenic ratios							Ages (in Ma)				
Grain spot	U (ppm)	Th (ppm)	Th/U	$^{206}Pb_c$ (%)	$^{206}Pb^*$ (ppm)	$^{207}Pb/^{206}Pb$	± (%)	$^{207}Pb/^{235}U$	± (%)	$^{206}Pb/^{238}U$	± (%)	Error Corr.	$^{206}Pb/^{238}U$	±	$^{207}Pb/^{206}Pb$	±	Disc. (%)
1.1	163	78	0.50	0.05	95.8	0.31018	0.29	29.32	1.5	0.6856	1.4	.981	3366	±38	3521.8	±4.5	4
1.2	91	9	0.10	0.52	26.2	0.2673	1.2	12.29	2.0	0.3335	1.6	.790	1855	±26	3290	±19	44
2.1	237	88	0.39	N.D.	142	0.31029	0.28	29.82	1.4	0.6970	1.4	.981	3409	±37	3522.4	±4.3	3
3.1	512	183	0.37	0.25	220	0.2762	0.81	19.00	1.6	0.4990	1.4	.859	2609	±29	3341	±13	22
4.1	396	85	0.22	0.09	209	0.29176	0.24	24.69	1.4	0.6138	1.4	.985	3085	±33	3427	±3.7	10
5.1	298	155	0.54	0.05	185	0.31176	0.19	31.10	1.4	0.7235	1.4	.990	3509	±37	3529.6	±3.0	1
5.2	250	147	0.61	0.02	157	0.31073	0.21	31.40	1.4	0.733	1.4	.989	3545	±38	3524.6	±3.2	-1
6.1	591	289	0.51	0.49	233	0.2673	0.75	16.83	1.6	0.4567	1.4	.878	2425	±28	3290	±12	26
7.1	600	237	0.41	0.02	297	0.28410	0.19	22.59	1.4	0.5766	1.3	.990	2935	±32	3385.6	±2.9	13
8.1	400	307	0.79	0.08	231	0.3027	0.38	28.03	1.4	0.6716	1.4	.963	3312	±35	3484.3	±5.9	5
9.1	246	178	0.75	0.02	156	0.31433	0.30	32.05	1.4	0.739	1.4	.978	3568	±38	3542.3	±4.7	-1
10.1	248	32	0.13	0.01	156	0.31012	0.21	31.34	1.4	0.733	1.4	.989	3545	±38	3521.5	±3.2	-1
11.1	460	124	0.28	0.58	195	0.2651	1.1	17.89	1.8	0.4894	1.4	.783	2568	±30	3277	±18	22
12.1	213	68	0.33	0.03	131	0.3101	0.41	30.48	1.4	0.7129	1.4	.959	3469	±37	3521.3	±6.3	1
13.1	236	54	0.24	0.16	129	0.2979	0.42	26.22	1.4	0.6383	1.4	.957	3182	±35	3459.6	±6.5	8
14.1	384	90	0.24	0.02	189	0.28649	0.19	22.61	1.4	0.5723	1.4	.990	2917	±32	3398.6	±3.0	14
15.1	317	128	0.42	0.15	184	0.30713	0.29	28.59	1.4	0.6751	1.4	.979	3326	±36	3506.6	±4.4	5
16.1	514	306	0.61	0.23	194	0.26060	0.30	15.73	1.4	0.4379	1.3	.976	2341	±26	3250.4	±4.7	28
17.1	823	140	0.18	0.42	200	0.20478	0.34	7.96	1.4	0.2819	1.4	.972	1601	±20	2864.7	±5.5	44
18.1	245	96	0.40	0.09	123	0.29971	0.25	24.09	1.4	0.5829	1.4	.983	2961	±33	3468.7	±3.9	15
19.1	925	169	0.19	2.05	162	0.2092	3.6	5.75	3.8	0.1992	1.4	.353	1171	±14	2900	±58	60
20.1	504	149	0.30	0.04	195	0.2744	0.52	16.99	1.5	0.4491	1.4	.933	2391	±27	3331.6	±8.2	28
21.1	339	178	0.54	0.06	207	0.31285	0.17	30.65	1.4	0.7105	1.4	.992	3460	±36	3535.0	±2.7	2
22.1	216	123	0.59	0.18	128	0.31097	0.28	29.46	1.4	0.6870	1.4	.979	3371	±36	3525.8	±4.4	4
23.1	696	274	0.41	N.D.	261	0.2841	1.3	17.09	1.9	0.4362	1.3	.718	2334	±26	3386	±20	31
24.1	172	71	0.42	0.12	96.3	0.30850	0.29	27.63	1.4	0.6496	1.4	.980	3227	±36	3513.5	±4.4	8
25.1	305	129	0.44	0.05	171	0.2962	0.59	26.66	1.5	0.6530	1.4	.918	3240	±35	3450.3	±9.1	6

Notes: Errors listed are 1σ; Pb_c and Pb* indicate the common and radiogenic portions, respectively. Error in standard calibration was 0.63% (not included in above errors but required when comparing data from different mounts). Common Pb corrected using measured ^{204}Pb. SHRIMP—sensitive high-resolution ion microprobe.

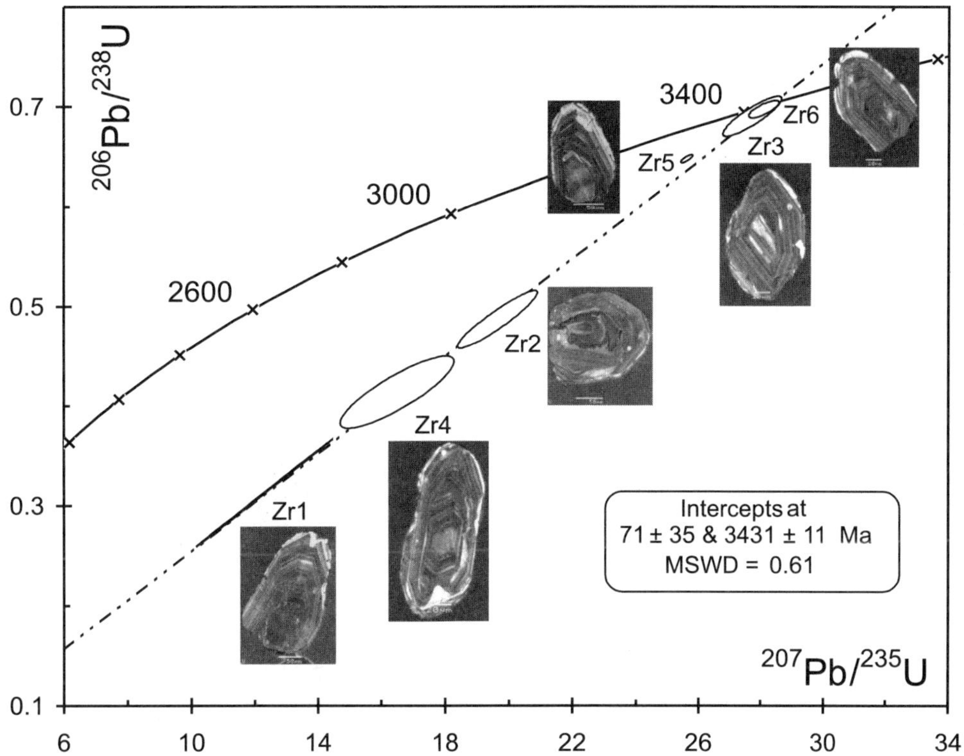

Figure 11. U-Pb Concordia diagram for sample Marc 2. Error ellipses are at 2 sigma. MSWD—mean standard weighted deviates.

TABLE 3. U-Pb DATA FOR SAMPLE MARC 2

Sample, grain	Weight (μg)	U (ppm)	Pb (ppm)	$^{206}Pb/^{204}Pb$	Radiogenic Ratios $^{206}Pb/^{238}U$	±	$^{207}Pb/^{235}U$	±	$^{207}Pb/^{206}Pb$	±	$^{208}Pb/^{206}Pb$	Apparent age (in Ma) $^{207}Pb/^{206}Pb$	Cor. Coef.	Disc. (%)
Zr 1, p, t	6	197	79	348	0.3118	14.1	12.357	14.1	0.2874	0.2	0.077	3404	0.99	49
Zr 2, d, p, b	6	303	207	354	0.4859	4.8	19.697	5.2	0.2940	1.8	0.149	3439	0.93	26
Zr 3, p, t	5	80	53	294	0.6902	2.3	27.728	2.8	0.2914	1.3	0.090	3425	0.88	1
Zr 4, p, t	3	122	81	233	0.4128	7.0	16.550	8.8	0.2907	4.5	0.299	3422	0.86	35
Zr 5, p, t	6	113	77	294	0.6478	0.5	25.719	0.6	0.2879	0.2	0.109	3407	0.92	5
Zr 6, p, t	4	50	23	259	0.6972	1.0	28.102	1.4	0.2923	0.8	0.177	3430	0.78	1

Notes: Errors are listed at 2σ. Abbreviations: p—pink, t—translucent, b—brown; d—dark.

A large-scale postdepositional redistribution of elements during hydrothermal alteration and metamorphism is considered to be unlikely, because the surrounding mafic, ultramafic, and granitoid rocks do not record elevated K_2O and SiO_2 concentrations (Dziggel, 2002). Further, many metasomatized mafic and ultramafic rock types in the Barberton greenstone terrain that are interpreted to have been silicified during early seafloor alteration have distinctly different field and geochemical characteristics (e.g., Duchac and Hanor, 1987). These rocks, which are commonly overlain by cherts, do record sharp contacts with the overlying sediments, but the intensity of alteration decreases down-section. In contrast, both the arkosic and mafic facies of the clastic metasedimentary rocks described here record sharp upper and lower contacts with their country rocks, despite their minor thickness. Strongly silicified komatiites record SiO_2 concentrations of more than 80 wt%, and, consequently, most of the mobile major and trace elements are strongly depleted. The majority of samples from the mafic facies have SiO_2 concentrations lower than 60 wt%, thus arguing against intense hydrothermal alteration. Similarly, the high SiO_2 and K_2O concentrations in samples from the arkosic facies are consistent with a clastic input of quartz and K-feldspar in these rocks. Other factors mitigating against a metasomatic origin during early seafloor alteration are the presence of unaltered amphibolitic layers within the mafic facies, the continuity of the clastic metasedimentary rocks along strike, as well as the fact that no crosscutting zones of Si and/or K alteration have been observed in either the metamafic rocks associated with the sediments or the granitoid rocks bounding the greenstone remnants within which the sediments occur. Thus, although limited chemical modifications caused by metasomatism cannot be ruled out, the negative correlation between SiO_2 and most major and trace elements (cf. Fig. 6 and Table 1) most probably indicates varying input of felsic detritus derived from a granitoid source.

Despite the apparent low volatile content and the absence of carbonate phases in the samples investigated, the high CaO concentrations in some of the samples from the mafic facies may potentially indicate a calcareous precursor for this type of sediment, i.e., a carbonated basalt or a marl-like component, from which CO_2 was removed during prograde metamorphism via reactions of the following type: $CaMg(CO_3)_2 + SiO_2 = CaMgSiO_4 + 2\ CO_2$. For the very same reason, a chemical evaluation of the degree of source rock weathering by means of the Chemical Index of Alteration ($CIA = [Al_2O_3/(Al_2O_3 + CaO^* + Na_2O + K_2O)] \times 100$; Nesbitt and Young, 1982) may be misleading, because all CaO, part of which may have been originally bound in carbonate phases, is now incorporated into metamorphic silicate phases. The calculated CIA values (Table 1) of samples from the mafic facies range between 12 and 40, which, assuming that no carbonates were present in the original sediment, would argue against intense chemical weathering in the source region. The Al_2O_3-(CaO^* + Na_2O)-K_2O diagram (Fig. 12; Nesbitt and Young, 1984) illustrates the geochemical signatures of the mafic and arkosic facies compared to average compositions of typical igneous rock types from the study area, as well as Fig Tree Group greywackes. Also shown are Archean shales from the Belingwe greenstone belt believed to be derived from submarine mafic lavas (Hofmann et al., 2003). These data were included because they share many geochemical characteristics with the mafic facies. As for the Zeederbergs shales (Hofmann et al., 2003), most samples of the mafic facies have CIA values typical for unweathered mafic rock types. The unweathered state of rocks from the mafic facies implies that the source rocks were probably submerged and not exposed to subaerial weathering. Samples highly enriched in CaO plot close to the CaO + Na_2O apex, possibly indicating that carbonate was present in the original sediment. In contrast, samples from the arkosic facies have slightly higher CIA values between 37 and 50. As expected, the samples plot away from the (CaO + Na_2O)-Al_2O_3 boundary. Two of the samples are, as a first approximation, chemically similar to potassic granitoid rock types, whereas the other two plot significantly below the plagioclase–K-feldspar join. Therefore, a simple mixing between, for example, mafic and/or ultramafic rocks and granitoid rocks of tonalitic and trondhjemitic compositions would not explain the high K_2O concentrations in the arkosic facies. Typical granitoid rocks of tonalitic and trondhjemitic compositions in the Barberton Mountain Land have potassium concentrations in the order of ~1–2 wt% (e.g., Kleinhanns et al., 2003). Clearly, most granitic rock types have K_2O concentrations smaller than ~3 wt% (e.g., Wilson, 1989). As almost all K_2O in the clastic metasedimentary rocks is located in feldspar, its relative abundance most probably indicates that K-feldspar was resistant to chemical weathering (relative to ferromagnesian minerals) and that transport distances were short enough for K-feldspar to withstand chemical weathering. Granitoid rocks of the TTG series only contain traces of K-feldspar; the main potassium-bearing phase in these rocks is biotite. It thus seems likely that the cross-bedded arkoses were derived from a weathered granitoid source that already contained significant amounts of K-feldspar (cf. Fig. 12). K_2O-rich granites that predate the Onverwacht Group and that would, thus, be potentially suitable as a source for the clastic metasediments are not exposed in the Barberton terrain but have been reported as clasts in Moodies Group conglomerates in the northern part of the belt (Kröner and Compston, 1988).

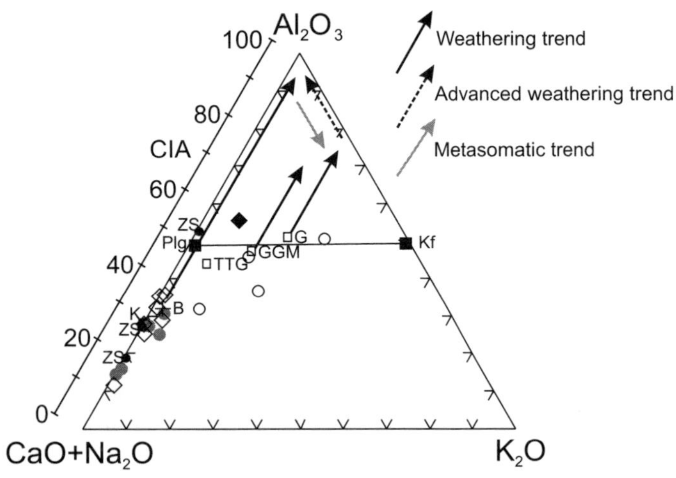

Figure 12. Al_2O_3-(CaO^* + Na_2O)-K_2O diagram for the mafic and arkosic facies (Nesbitt and Young, 1984). CaO* represents silicate fraction only; compositions are given as molar proportions. Symbols as in Figure 6. B—average amphibolite schist (Robb, 1981); K—average komatiite (Lahaye et al., 1995); TTG and GGM (granodiorite-granite-monzogranite) are from Kleinhanns et al. (2003); g (average granite), Plg (plagioclase), and Kf (K-feldspar) have been adopted from Nesbitt and Young (1984). ZS (Zeederbergs shales) are after Hofmann et al. (2003).

It is obvious from the previous discussion that much of the geochemical record of the clastic metasedimentary rocks may be modeled by mixing variable proportions of weathered felsic and unweathered mafic detritus. However, the high Cr and Ni concentrations in some of the samples indicate the presence of an ultramafic component in the source region. Elevated concentrations for these elements are a well-known phenomenon in many Archean clastic sedimentary rocks, including shales and greywackes of the Fig Tree and Moodies Groups (e.g., McLennan and Taylor, 1984; Toulkeridis et al., 1999). The reason for the high Ni and Cr concentrations, however, has not been established unequivocally. Possible enrichment scenarios include the greater abundance of mafic and ultramafic components in Archean greenstone belts (e.g., McLennan and Taylor, 1984; Condie, 1993), originally higher values for these elements in siliceous volcaniclastic rocks (e.g., Gibbs et al., 1986), as well as enrichment processes during weathering and sedimentation (Chacko et al., 1992). Typical Cr/Ni ratios in ultramafic rocks are around 1.6. Similar ratios have also been recorded in greywackes of the Fig Tree Group (e.g., Toulkeridis et al., 1999). However, the absolute Cr and Ni concentrations in these greywackes (~540 ppm Cr and 336 ppm Ni) are much lower than in typical ultramafic rocks. Rocks of the mafic facies contain up to 3356 ppm Cr and 560 ppm Ni and record Cr/Ni ratios ranging between 3 and 6. These high ratios have been interpreted to reflect significant geochemical fractionation of the two elements during weathering and/or sediment transport (Garver et al., 1996). The Cr concentrations are well within the compositional range for komatiites (e.g., Condie, 1993), whereas the Ni concentrations are much lower than would be expected in an ultramafic rock of this type. Thus, their abundance in the mafic facies of the clastic metasedimentary rocks may reflect an originally ultramafic source that underwent significant geochemical fractionation of these elements. A comparison of the Cr and Ni concentrations and resulting ratios of the mafic facies with ultramafic rock types in the Barberton greenstone belt (e.g., Byerly, 1999; Vennemann and Smith, 1999), however, allows for an alternative interpretation. These studies have shown that the Cr/Ni ratios in komatiitic rocks of the Kromberg and Mendon Formations are closely related to a partial accumulation of olivine and chromite. For example, pyroxene spinifex rocks of the Mendon Formation have Cr and Ni concentrations of up to 1181 and 505 ppm, respectively, similar to rocks of the mafic facies. The ratios range between 2.9 and 5.1. Olivine spinifex rocks and, particularly, olivine cumulates have much higher Ni concentrations (715–2800 ppm), and the ratios are much smaller (0.4–1.8). Thus, the high Cr/Ni ratios in rocks of the mafic facies do not necessarily require element fractionation during sediment transport or weathering. These elements may have been derived from an ultramafic rock that did not contain olivine.

Tectonic Setting and Age of Source

Many previous studies have suggested an active convergent margin setting for the Theespruit Formation (e.g., Armstrong et al., 1990; De Ronde and De Wit, 1994). However, the exact stratigraphic position and age relationships of the Sandspruit and Theespruit Formations relative to the overlying sequences of the Swaziland Supergroup remain poorly understood, because the existing geochronological data indicate felsic igneous activity both pre- and postdating the overlying Komati Formation. For example, Armstrong et al. (1990) dated felsic volcaniclastic rocks from the type locality of the Theespruit Formation, which yielded a date of 3453 ± 6 Ma, and concluded that the bulk of the Theespruit Formation might, in fact, be younger than the mafic and ultramafic rocks of the overlying Komati Formation, which has been dated at 3482 ± 5 Ma. In contrast, Kröner et al. (1996) showed that the Theespruit Formation in the Steynsdorp area (Fig. 1) is ~100 m.y. older than other Onverwacht Group rocks from the southern part of the Barberton greenstone belt, yielding U-Pb dates of ca. 3545 Ma. This age, however, is similar to an age estimate of ca. 3538 Ma for wedges of tectonically emplaced tonalitic gneisses, as well as an older and presumably inherited zircon population from felsic volcaniclastic rocks from the type locality of the Theespruit Formation (Armstrong et al., 1990; Kamo and Davis, 1994). Consequently, it was suggested that the felsic units dated by Kröner et al. (1996) may also represent tectonically emplaced older continental crust rather than syn-volcanic felsic igneous rocks (De Ronde and De Wit, 1994). The Theespruit Formation has, therefore, been interpreted as an active convergent margin sequence, with which the oceanic crust of the overlying Komati Formation was tectonically juxtaposed within 50 m.y. of its formation (De Wit, 1991; De Ronde and De Wit, 1994). Kröner et al. (1996), however, favored a setting similar to an oceanic plateau due to the lack of andesitic compositions and trenchlike sediments.

The identification of at least partly continentally derived clastic sedimentary rocks within the Sandspruit and Theespruit Formations sheds new light on these issues. The geochronological data show that the deposition of the arkosic facies now exposed to the east of the Boesmanskop Syenite pluton must have occurred between ca. 3521 Ma and 3431 Ma, contemporaneously with the erosion of spatially associated older, and presumably K_2O-rich, granitoid rocks. A granitoid protolith has also been inferred for felsic metavolcaniclastic rocks of the Theespruit Formation (Kohler and Anhaeusser, 2002), which record strongly fractionated REE (rare earth element) patterns and a marked negative Eu anomaly. Attempts to date detrital zircons extracted from the mafic facies from the western side of the Boesmanskop syeno-granite complex were not as successful due to strong postdepositional isotopic resetting and recrystallization of zircon in those rocks (Dziggel et al., 2005). However, a core of one recrystallized zircon grain yielded an age of 3560 ± 45 Ma, also pointing to a source predating the bulk of the Barberton greenstone belt. Older continental crust therefore not only existed prior to the intrusion of the trondhjemite plutons, but was, in all likelihood, also exposed during the earliest tectono-magmatic episode recorded in the Barberton greenstone belt (cf. De Wit et al., 1983). We therefore suggest that the Theespruit and Sandspruit Formations represent a rock suite that formed in a tectonically and magmatically active setting, perhaps comparable to modern convergent continental margins or island arcs. After the deposition of the clastic sediments, the southern part of the Barberton greenstone belt was exposed to intense calc-alkaline magmatic activity, which resulted in the formation of the ca. 3445 Ma TTG plutons and their high-level equivalents in the Hooggenoeg Formation (e.g., De Wit et al., 1987b). A subduction-related scenario is also supported by geochemical and experimental studies on the genesis of typical Archean TTGs, either through dehydration melting of a basaltic precursor such as amphibolite or eclogite (e.g., Rapp et al., 1991; Rapp, 1997; Wyllie et al., 1997) or wet partial melting of the peridotitic mantle wedge underlying the volcanic arc (Kleinhanns et al., 2003). At this time, the trondhjemite plutons presumably intruded at relatively shallow crustal levels, and the associated volcaniclastic rocks were deposited on top of the spatially closely associated mafic and ultramafic sequences of the Komati Formation (Armstrong et al., 1990).

Depositional Environments and Source Components

Although the detrital zircon data provide information about the age spectra in the continental source of the clastic sediments, the contributions of fine-grained felsic volcanic rocks or zircon-poor mafic and ultramafic units remain unrecorded. A number of stratigraphic, sedimentological, and geochemical features, however, suggest that the clastic metasedimentary rocks of both the Sandspruit and Theespruit Formations were deposited over a relatively short period of time, contemporaneous with mafic and ultramafic volcanic/volcaniclastic rocks in an active tectonic environment. Abundant pillow lavas in less deformed sections in these formations indicate that volcanism and, thus, presumably also sedimentation occurred under mostly subaqueous (and probably submarine) conditions. This is further substantiated by the low CIA values for the mafic facies, which are consistent with a submerged (ultra-) mafic source that was not exposed to subaerial weathering. Alternatively, the (ultra-) mafic detritus could represent volcaniclastic material that was derived from local volcanic centers. Thin intercalations of aluminous metapelites within the Theespruit Formation (e.g., De Wit et al., 1983) indicate the presence of a continental hinterland that was exposed to subaerial weathering. This extrabasinal detritus also points to some depositional breaks within the sequence, but these layers are extremely rare and minor in thickness. The rare occurrence of Algoma-type banded iron formation (Dziggel, 2002) and other chemical sediments further indicates a hydrothermal input into the basin, perhaps at localized hydrothermal vents (e.g., Huston and Logan, 2004). In contrast, the relatively mature and cross-stratified arkosic facies within the Sandspruit Formation suggests a deposition of continentally derived detritus that was reworked by strong, persistent currents. Moderate degrees of subaerial weathering, as indicated by the high content of K-feldspar in these rocks (cf. Fig. 12), were probably important for the granitoid-derived portions of these sediments, whereas the mixing with and deposition of the mafic detritus did not generally require additional weathering and erosion. The low CIA values (37–50; cf. Fig. 12) calculated for the arkosic facies are, thus, interpreted to reflect mixed values derived from a moderately weathered granitic source and unweathered (ultra-) mafic rock types. Moderate degrees of subaerial weathering in the granitic source of the arkoses are in accordance with an active tectonic environment in the provenance area, but to date no such relict ancient continental crust has been identified.

CONCLUSIONS

Petrological and geochemical data imply that the clastic metasedimentary rocks in the granitoid-dominated terrain to the south of the Barberton greenstone belt were derived from a variety of sources, including moderately weathered, highly evolved continental crust combined with varying proportions of mafic and ultramafic rocks. U-Pb dating of detrital zircons from the arkosic facies correlative with the Sandspruit Formation range between ca. 3540 and 3520 Ma, suggesting the presence of an older continental crust predating the Barberton greenstone belt. Based on the mineralogy and high K_2O concentrations, this early crust was presumably more potassium-rich than typical TTG granitoid rocks surrounding the belt. Although similar in their field occurrence and petrology to other, more fine-grained, planar-bedded mafic facies, these relatively mature coarse-clastic metasediments seem to be restricted to the Sandspruit Formation. K-feldspar and quartz had their sources in granitic rocks that are no longer preserved, with igneous ages between 3540 and 3520 Ma. The geochemical signatures of both types of sediment are in accordance with a contemporaneous deposition of continentally derived sediments and sediments derived from (ultra-) mafic volcanic/volcaniclastic rocks in an active tectonic environment such as an island arc.

ACKNOWLEDGMENTS

This work formed part of a Ph.D. dissertation that was supported by National Research Foundation funding to Gary Stevens and funding from the Economic Geology Research Institute at the University of the Witwatersrand, Johannesburg. Carl Anhaeusser is thanked for introducing us to the geology of the study area as well as for sharing his original maps and expertise on the Barberton greenstone belt. Annika Dziggel would also like to acknowledge logistical support as well as bursary funding from the University of the Witwatersrand. Ken Eriksson and Axel Hofmann are thanked for constructive and thorough reviews of this paper.

REFERENCES CITED

Anhaeusser, C.R., and Robb, L.J., 1980, Regional and detailed field and geochemical studies of Archaean trondhjemitic gneisses, migmatites and greenstone xenoliths in the southern part of the Barberton Mountain Land, South Africa: Precambrian Research, v. 11, p. 373–397, doi: 10.1016/0301-9268(80)90073-X.

Anhaeusser, C.R., Robb, L.J., and Viljoen, M.J., 1983a, Notes on the provisional geological map of the Barberton greenstone belt and surrounding granitic terrane, eastern Transvaal and Swaziland (1:250,000 color map): Geological Society of South Africa Special Publications 9, p. 221–223.

Anhaeusser, C.R., Robb, L.J., and Barton, J.M., Jr., 1983b, Mineralogy, petrology and origin of the Boesmanskop syeno-granite complex, Barberton Mountain Land, South Africa: Geological Society of South Africa Special Publication 9, p. 169–183.

Armstrong, R.A., Compston, W., De Wit, M.J., and Williams, I.S., 1990, The stratigraphy of the 3.5–3.2 Ga Barberton greenstone belt revisited; a single zircon ion microprobe study: Earth and Planetary Science Letters, v. 101, p. 90–106, doi: 10.1016/0012-821X(90)90127-J.

Belcher, R.W., and Kisters, A.F.M., 2006, this volume, Syntectonic emplacement and deformation of the Heerenveen batholith: Conjectures on the structural setting of the 3.1 Ga granite magmatism in the Barberton granite-greenstone terrain, South Africa, *in* Reimold, W.U., and Gibson, R.L., Processes on the early Earth: Geological Society of America Special Paper 405, doi: 10.1130/2006.2405(12).

Byerly, G.R., 1999, Komatiites of the Mendon Formation: Late-stage ultramafic volcanism in the Barberton greenstone belt, *in* Lowe, D.R., and Byerly, G.R., eds., Geologic Evolution of the Barberton Greenstone Belt, South Africa: Geological Society of America Special Paper 329, p. 189–211.

Chacko, T., Kumar, G.R., Meen, J.K., and Rogers, J.W., 1992, Geochemistry of high-grade supracrustal rocks from the Kerala Khondalite belt and adjacent massive charnokites: South India: Precambrian Research, v. 55, p. 469–489, doi: 10.1016/0301-9268(92)90040-U.

Condie, K.C., 1993, Chemical composition and evolution of the upper continental crust: Contrasting results from surface samples and shales: Chemical Geology, v. 104, p. 1–37, doi: 10.1016/0009-2541(93)90140-E.

De Ronde, C.E.J., and De Wit, M.J., 1994, Tectonic history of the Barberton greenstone belt, South Africa; 490 million years of Archaean crustal evolution: Tectonics, v. 13, p. 983–1005, doi: 10.1029/94TC00353.

De Wit, M.J., 1991, Archaean greenstone belt tectonism and basin development: Some insights from the Barberton and Pietersburg greenstone belts, Kaapvaal Craton, South Africa: Journal of African Earth Sciences, v. 13, p. 45–63, doi: 10.1016/0899-5362(91)90043-X.

De Wit, M.J., 2004, Archaean greenstone belts do contain fragments of ophiolites: *in* Kusky, T.M., ed., Precambrian Ophiolites and Related Rocks, Developments in Precambrian Geology, v. 13, p. 599–613.

De Wit, M.J., Fripp, R.E.P., and Stanistreet, I.G., 1983, Tectonic and stratigraphic implications of new field observations along the southern part of the Barberton greenstone belt: Geological Society of South Africa Special Publication 9, p. 21–30.

De Wit, M.J., Hart, R.A., and Hart, R.J., 1987a, The Jamestown ophiolite complex, Barberton mountain belt; a section through 3.5 Ga oceanic crust: Journal of African Earth Sciences, v. 6, p. 681–730, doi: 10.1016/0899-5362(87)90007-8.

De Wit, M.J., Armstrong, R.A., Hart, R.J., and Wilson, A.H., 1987b, Felsic igneous rocks within the 3.3- to 3.5-Ga Barberton greenstone belt: High crustal level equivalents of the surrounding tonalite-trondhjemite terrain, emplaced during thrusting: Tectonics, v. 5, p. 529–549.

Diener, J.F.A., Stevens, G., Kisters, A.F.M., and Poujol, M., 2005, Metamorphism and exhumation of the basal part of the Barberton greenstone belt, South Africa: Constraining the rates of Mesoarchaean tectonism: Precambrian Research, v. 143, p. 87–112.

Duchac, K.C., and Hanor, J.S., 1987, Origin and timing of the metasomatic silicification of an early Archean komatiite sequence, Barberton Mountain Land, South Africa: Precambrian Research, v. 37, p. 125–146, doi: 10.1016/0301-9268(87)90075-1.

Dziggel, A., 2002, The petrogenesis of 'lower' Onverwacht Group clastic metasediments and related metavolcanic rocks in the southern part of the Barberton Mountain Land, South Africa [Ph.D. thesis]: Johannesburg, University of the Witwatersrand, 230 p.

Dziggel, A., Stevens, G., Poujol, M., Anhaeusser, C.R., and Armstrong, R.A., 2002, Metamorphism of the granite-greenstone terrane south of the Barberton greenstone belt, South Africa: An insight into the tectono-thermal evolution of the lower portions of the Onverwacht Group: Precambrian Research, v. 114, p. 221–247, doi: 10.1016/S0301-9268(01)00225-X.

Dziggel, A., Armstrong, R.A., Stevens, G., and Nasdala, L., 2005, Growth of zircon and titanite during metamorphism in the granitoid-gneiss terrain south of the Barberton greenstone belt, South Africa: Mineralogical Magazine, v. 69, p. 1021–1038.

Eriksson, K.A., Krapez, B., and Fralick, P.W., 1997, Sedimentological aspects, *in* De Wit, M.J., and Ashwal, L.D., eds., Greenstone belts: Oxford Monographs on Geology and Geophysics 35, p. 33–54.

Fedo, C.M., Nesbitt, H.W., and Young, G.M., 1995, Unraveling the effects of potassium metasomatism in sedimentary rocks and paleosols, with implications for paleoweathering conditions and provenance: Geology, v. 23, p. 921–924, doi: 10.1130/0091-7613(1995)023<0921:UTEOPM>2.3.CO;2.

Garver, J.I., Royce, P.R., and Smick, T.A., 1996, Chromium and nickel in shale of the Taconic foreland: A case study for the provenance of fine-grained sediments with an ultramafic source: Journal of Sedimentary Research, v. 66, p. 100–106.

Gibbs, A.K., Montgommary, C.W., O'Day, P.A., and Erslev, E.A., 1986, The Archaean-Proterozoic transition: Evidence from the geochemistry of metasedimentary rocks of Guyana and Montana: Geochimica et Cosmochimica Acta, v. 50, p. 2125–2141.

Heubeck, C., and Lowe, D.R., 1994, Depositional and tectonic setting of the Archaean Moodies Group, Barberton greenstone belt, South Africa: Precambrian Research, v. 68, p. 257–290, doi: 10.1016/0301-9268(94)90033-7.

Hofmann, A., Bolhar, R., Dirks, P., and Jelsma, H., 2003, The geochemistry of Archaean shales derived from a mafic volcanic sequence, Belingwe greenstone belt, Zimbabwe: Provenance, source area unroofing and submarine versus subaerial weathering: Geochimica et Cosmochimica Acta, v. 67, p. 421–440, doi: 10.1016/S0016-7037(02)01086-4.

Huston, D.L., and Logan, G.A., 2004, Barite, BIFs and bugs: Evidence for the evolution of the Earth's early hydrosphere: Earth and Planetary Science Letters, v. 220, p. 41–55, doi: 10.1016/S0012-821X(04)00034-2.

Kamo, S.L., and Davis, D.W., 1994, Reassessment of Archaean crustal development in the Barberton Mountain Land, South Africa, based on U-Pb dating: Tectonics, v. 13, p. 167–192, doi: 10.1029/93TC02254.

Kisters, A.F.M., Stevens, G., Dziggel, A., and Armstrong, R.A., 2003, Extensional detachment faulting at the base of the Barberton greenstone belt: Evidence for a 3.2 Ga orogenic collapse: Precambrian Research, v. 127, p. 355–378, doi: 10.1016/j.precamres.2003.08.002.

Kleinhanns, I.C., Kramers, J.D., and Kamber, B.S., 2003, Importance of water for Archaean granitoid petrology: A comparative study of TTG and potassic granitoids from Barberton Mountain Land, South Africa: Contributions to Mineralogy and Petrology, v. 145, p. 377–389, doi: 10.1007/s00410-003-0459-9.

Kohler, E.A., and Anhaeusser, C.R., 2002, Geology and geodynamic setting of Archaean silicic metavolcaniclastic rocks of the Bien Venue Formation, Fig Tree Group, northeast Barberton greenstone belt, South Africa: Precambrian Research, v. 116, p. 199–235, doi: 10.1016/S0301-9268(02)00021-9.

Kröner, A., and Compston, W., 1988, Ion microprobe ages of zircons from early Archaean granite pebbles and greywacke, Barberton greenstone belt, southern Africa: Precambrian Research, v. 38, p. 367–380, doi: 10.1016/0301-9268(88)90034-4.

Kröner, A., Byerly, G.R., and Lowe, D.R., 1991, Chronology of early Archaean granite-greenstone evolution in the Barberton mountain land, southern Africa, based on precise dating by single zircon evaporation: Earth and Planetary Science Letters, v. 103, p. 41–54, doi: 10.1016/0012-821X(91)90148-B.

Kröner, A., Hegner, E., Wendt, J.I., and Byerly, G.R., 1996, The oldest part of the Barberton granitoid-greenstone terrain, South Africa, evidence for crust formation between 3.5 and 3.7 Ga: Precambrian Research, v. 78, p. 105–124, doi: 10.1016/0301-9268(95)00072-0.

Lahaye, Y., Arndt, N., Byerly, G.R., Chauvel, C., Fourcade, S., and Gruau, G., 1995, The influence of alteration on the trace-element and Sm-Nd isotopic compositions of komatiites: Chemical Geology, v. 126, p. 43–64, doi: 10.1016/0009-2541(95)00102-1.

Lowe, D.R., 1994, Accretionary history of the Archaean Barberton greenstone belt (3.55–3.22 Ga), South Africa: Geology, v. 22, p. 1099–1102, doi: 10.1130/0091-7613(1994)022<1099:AHOTAB>2.3.CO;2.

Lowe, D.R., 1999, Geologic evolution of the Barberton greenstone belt and vicinity, *in* Lowe, D.R., and Byerly, G.R., eds., Geologic Evolution of the Barberton Greenstone Belt, South Africa: Geological Society of America Special Paper 329, p. 287–312.

Lowe, D.R., and Byerly, G.R., 1999, Stratigraphy of the west-central part of the Barberton greenstone belt, South Africa, *in* Lowe, D.R., and Byerly, G.R., eds., Geologic Evolution of the Barberton Greenstone Belt, South Africa: Geological Society of America Special Paper 329, p. 1–36.

McLennan, S.M., and Taylor, S.R., 1984, Archaean sedimentary rocks and their relation to the composition of the Archaean continental crust, *in* Kröner, A., Hanson, G.N., and Goodwin, A.M., eds., Archaean geochemistry: The origin and evolution of the Archaean continental crust: Berlin, Heidelberg, Springer, 286 p.

McLennan, S.M., Hemming, S., McDaniel, D.K., and Hanson, G.N., 1993, Geochemical approaches to sedimentation, provenance, and tectonics: Geological Society of America Special Paper 284, p. 21–40.

Nesbitt, H.W., and Young, G.M., 1982, Early Proterozoic climates and plate motions inferred from major element chemistry of lutites: Nature, v. 299, p. 715–717, doi: 10.1038/299715a0.

Nesbitt, H.W., and Young, G.M., 1984, Prediction of some weathering trends of plutonic and volcanic rocks based on thermodynamic and kinetic considerations: Geochimica et Cosmochimica Acta, v. 48, p. 1523–1534, doi: 10.1016/0016-7037(84)90408-3.

Pettijohn, F.J., Potter, P.E., and Siever, R., 1987, Sand and Sandstone, 2nd ed.: New York, Springer, 553 p.

Rapp, R.P., 1997, Heterogeneous source regions for Archaean granitoids: Experimental and geochemical evidence, *in* De Wit, M.J., and Ashwal, L.D., eds., Greenstone Belts: Oxford Monographs on Geology and Geophysics 35, p. 267–279.

Rapp, R.P., Watson, E.B., and Miller, C.F., 1991, Partial melting of amphibolite/eclogite and the origin of Archaean trondhjemites and tonalites: Precambrian Research, v. 51, p. 1–25, doi: 10.1016/0301-9268(91)90092-O.

Robb, L.J., 1981, The geological and geochemical evolution of tonalite-trondhjemite gneisses and migmatites in the Barberton region, Eastern Transvaal [Ph.D. thesis]: Johannesburg, University of Witwatersrand, 342 p.

Robb, L.J., Armstrong, R.A., and Waters, D.J., 1999, The history of granulite-facies metamorphism and crustal growth from single zircon U-Pb geochronology: Namaqualand, South Africa: Journal of Petrology, v. 40, p. 1747–1770, doi: 10.1093/petrology/40.12.1747.

Stevens, G., Droop, G.T.R., Armstrong, R.A., and Anhaeusser, C.R., 2002, The Schapenburg schist belt: Amphibolite-facies metamorphism in Fig Tree Group metasediments as a consequence of c. 3.23 Ga terrane accretion in the Barberton greenstone belt, South Africa: South African Journal of Geology, v. 105, p. 271–284, doi: 10.2113/1050271.

Taylor, S.R., and McLennan, S.M., 1985, The continental crust: Its composition and evolution: Oxford, Blackwell, 312 p.

Thurston, P.C., and Kozhevnikov, V.N., 2000, An Archean quartz arenite-andesite association in the eastern Baltic Shield, Russia: Implications for assemblage types and shield history: Precambrian Research, v. 101, p. 313–340, doi: 10.1016/S0301-9268(99)00093-5.

Toulkeridis, T., Clauer, N., Kröner, A., Reimer, T.O., and Todt, W., 1999, Characterisation, provenance, and tectonic setting of Fig Tree greywackes from the Archaean Barberton greenstone belt, South Africa: Sedimentary Geology, v. 124, p. 113–129, doi: 10.1016/S0037-0738(98)00123-7.

Vennemann, T.W., and Smith, H.S., 1999, Geochemistry of mafic and ultramafic rocks in the Kromberg Formation in its type section, Barberton greenstone belt, South Africa, *in* Lowe, D.R., and Byerly, G.R., eds., Geologic Evolution of the Barberton Greenstone Belt, South Africa: Geological Society of America Special Paper 329, p. 133–149.

Viljoen, M.J., and Viljoen, R.P., 1969a, An introduction to the geology of the Barberton granite-greenstone terrain: Geological Society of South Africa Special Publication 2, p. 9–28.

Viljoen, M.J., and Viljoen, R.P., 1969b, The geology and geochemistry of the lower ultramafic unit of the Onverwacht Group and a proposed new class of igneous rocks: Geological Society of South Africa Special Publication 2, p. 55–86.

Westraat, J.D., Kisters, A.F.M., Poujol, M., and Stevens, G., 2005, Transcurrent shearing, granite sheeting and the incremental construction of the tabular 3.1 Ga Mpuluzi batholith, Barberton granite-greenstone terrain, South Africa: Journal of the Geological Society of London, v. 162, p. 373–388, doi: 10.1144/0016-764904-026.

Wilson, M., 1989, Igneous petrogenesis: A global tectonic approach: London, Chapman and Hall, 466 p.

Wyllie, P.J., Wolf, M.B., and van der Laan, S.R., 1997, Conditions for formation of tonalites and trondhjemites: Magmatic sources and products, *in* De Wit, M.J., and Ashwal, L.D., eds., Greenstone Belts: Oxford Monographs on Geology and Geophysics 35, p. 256–266.

Manuscript Accepted by the Society 4 October 2005

Printed in the USA

Geological Society of America
Special Paper 405
2006

The significance of subduction-related accretionary complexes in early Earth processes

John W. Shervais
Department of Geology, Utah State University, Logan, Utah 84322-4505, USA

ABSTRACT

The transition from Hadean-style convective overturn, driven by heating from below, and Phanerozoic-style plate tectonics, driven by the sinking of cool lithospheric slabs, was a major turning point in the thermal evolution of the Earth. Prior to this transition the formation of stable, long-lived crust was rare; after this transition, the formation and amalgamation of continental crust became a central theme of tectonic processes. Evidence for Phanerozoic-style plate tectonics includes (1) the formation of oceanic crust at mid-ocean ridge spreading centers, (2) the formation of island arc volcanic and plutonic complexes, and (3) the formation of accretionary mélange complexes during the subduction of oceanic crust. Rock assemblages characteristic of oceanic crust and island arc suites may also be formed during Hadean-style convection, making the recognition of accretionary mélange complexes the most reliable indicator of Phanerozoic-style tectonics. Accretionary mélange complexes result from the juxtaposition of oceanic crust and pelagic sediments in a matrix of arc-derived constituents and may be recognized even after high-grade metamorphic events.

At least three potential Archean mélange complexes have been recognized to date: the Dongwanzi ophiolite mélange of the North China craton, the Schreiber-Hemlo mélange of the Superior province, and the Farmington Canyon complex of the Wyoming province; it has also been proposed that the Isua supracrustals of the Greenland craton represent an accretionary complex, but this proposal has been disputed. Each of these terranes contains lithotectonic elements that could not have formed in a single tectonic setting but are now intimately intermixed. Oceanic components include mafic and ultramafic volcanic rocks, gabbros, harzburgite tectonites, and pelagic sediments (chert); arc components are represented by the mélange matrix, consisting of felsic gneiss with a greywacke composition. These studies of Archean crust show that recognition of accretionary mélange complexes may be a reliable indicator of Phanerozoic-style plate tectonics and that the initiation of recognizable plate-tectonic processes occurred by 3.0 Ga, which coincides with the oldest Phanerozoic-style ophiolite assemblages. This transition may have started as early as 3.8 Ga, which coincides with the end of the terminal lunar bombardment.

Keywords: Archean, Hadean, mélange, ophiolite, early Earth.

Shervais, J.W., 2006, The significance of subduction-related accretionary complexes in early Earth processes, *in* Reimold, W.U., and Gibson, R.L., Processes on the Early Earth: Geological Society of America Special Paper 405, p. 173–192, doi: 10.1130/2006.2405(10). For permission to copy, contact editing@geosociety.org.

INTRODUCTION

One of the central questions of plate tectonics is when the transition occurred between post-magma ocean Hadean thermal regimes, characterized by high heat flows, chaotic convection, and generalized partial melting of the mantle, and plate-tectonic processes that are recognizably similar to those that occur today. This transition represents a fundamental change in the thermal structure of the planet, from one controlled by the upwelling of mantle plumes to one controlled by the sinking of cold lithospheric plates. Thermal modeling (e.g., Zhong et al., 2000; Bercovici, 2003) shows that convective heat transfer driven by heating from below results in hot material rising in columns, which is thought to be the dominant mode of Hadean heat transfer (Fig. 1). In contrast, Phanerozoic-style plate tectonics is driven by the gravitational sinking of cool, dense lithospheric plates (Carlson, 1981; Hager and O'Connell, 1981; Jurdy and Stefanick, 1991; Anderson, 2001).

The onset of Phanerozoic-style plate tectonics may be marked by the preservation of one or more distinct assemblages: (1) oceanic crust, as represented by ophiolite assemblages or by eclogites that represent subducted oceanic crust, (2) island arc volcanic or plutonic series, or (3) accretionary complexes that contain fragments of oceanic crust and lithosphere in a greywacke or serpentinite matrix mélange. Each of these associations may be preserved in Archean crustal assemblages but not all are equally diagnostic.

The correlation of greenstone belts with oceanic crust requires that the observed stratigraphic sequences represent thin slices of crust ± mantle lithosphere that are repeated by thrust faulting and stacked onto felsic continental crust (e.g., de Wit et al., 1992; de Wit, 1998, 2004; Kusky and Polat, 1999; Hofmann and Kusky, 2004). Evidence for such thin-skin tectonics is controversial in most greenstone belts, and many detailed studies still support autochthonous relations (e.g., intracontinental rifts or ensialic back-arc basins; Lowe, 1994; Bickle et al., 1994; Corcoran and Dostal, 2001; Prendergast, 2004; Mueller et al., 2005). Oceanic crust may also be preserved as eclogite xenoliths brought to the surface by kimberlites or other alkaline lamprophyres. Eclogite xenoliths with chemical and isotopic signatures that require formation and alteration as oceanic crust are well established (e.g., MacGregor and Manton, 1986; Shervais et al., 1988; Neal et al., 1990; Shirey et al., 2001), but similar oceanic crust could form during Hadean-style thermal convection and may not be evidence for Phanerozoic-style plate tectonics.

Archean tonalite-trondhjemite-granodiorite (TTG) gneiss complexes have been correlated with island arc plutonic series formed by slab melting (Drummond and Defant, 1990; Rapp et al., 2003), and it has been proposed that Archean greenstone belts may represent essentially coeval island arc volcanism (e.g., Parman and Grove, 2004; Parman et al., 2001, 2004; Polat and Kerrich, 2001, 2002, 2004). The progression within greenstone belts from ultramafic and mafic volcanism to more felsic, dacitic, and rhyolitic volcanism is now well established and correlates well with current knowledge of island arc suites from the western Pacific (e.g., Hawkins et al., 1984). High-resolution single crystal zircon dating of ash beds intercalated with the greenstone lavas and comparable dates for TTG gneiss complexes show that the TTG suites are generally coeval to slightly younger than the greenstone belts, suggesting a related or consanguineous relationship (e.g., Kröner and Todt, 1988; Kröner et al., 1991; Armstrong et al., 1990; Kamo and Davis, 1994). In each craton, however, there seems to be a

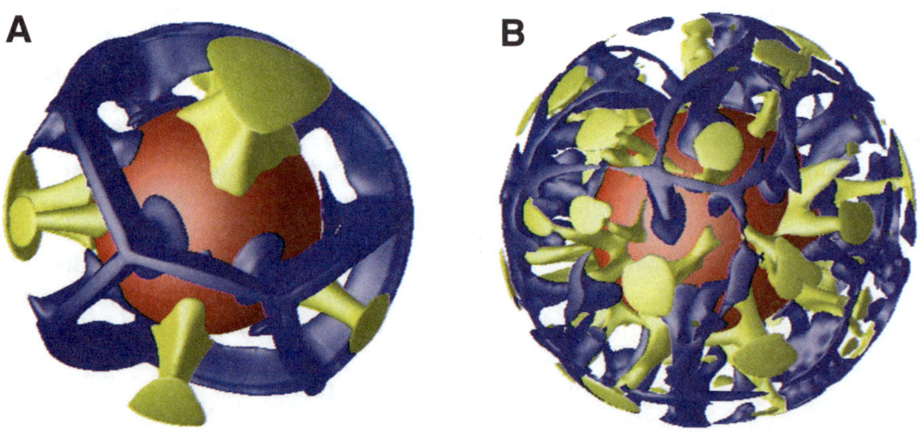

Figure 1. Numerical model of thermal convection driven by heating from below (from Zhong et al., 2000). Red—core, yellow—rising plumes of hot mantle, heated from below by core, blue—sheetlike and columnar zones of convergence and sinking of cooler return flow into mantle. Note that mantle upwelling occurs in plumelike columns, while sinking occurs largely along vertical zones of symmetric convergence. Convergence at triple junctions leads to accumulation of excess lithosphere and sinking in columns. (A) Layered viscosity structure and (B) layered viscosity coupled with temperature dependent viscosity; see Zhong et al., 2000. Reproduced/modified by permission of American Geophysical Union.

nucleus of older granitic gneiss that predates the adjacent greenstones (Ridley et al., 1997).

These correlations are not universally accepted, however (e.g., Arndt et al., 1997a, 1998; Arndt, 2004; Hamilton, 1998), and it is possible for both greenstone belts and TTG suites to form during Hadean convective overturn, prior to the onset of Phanerozoic-style asymmetric subduction. As seen in Figure 1, even during thermal convection driven by heating from below, the sinking of previously created protocrust along symmetric zones of convergence is required. Remelting of this protocrust as it sinks back into the mantle would generate melts similar to the TTG suite, so formation in an island arc setting is not required.

In contrast, subduction zone accretionary complexes are clearly the result of plate-tectonic processes that mimic those we see today and may represent the best evidence for modern-style, asymmetric subduction of lithospheric plates (Kusky and Polat, 1999; Shervais, 2004). In the following sections I develop this thesis in some detail. But first we need to examine Archean stratigraphy and the implications of thermal modeling for Hadean and early Archean tectonics.

ARCHEAN STRATIGRAPHY

The application of plate-tectonic models to the Archean has become more common in recent years, although the correlations are not exact (e.g., de Wit et al., 1992; de Wit, 1998, 2004; Kusky and Polat, 1999; Dirks and Jelsma, 1998; Connelly and Ryan, 1996; Calvert and Ludden, 1999). Much of this debate hinges on the stratigraphy of Archean greenstone belts, their internal structure, and their relationship to the surrounding TTG gneisses.

The oldest intact rock sequences on Earth have been dated at around 3.6–4.03 Ga (DeRonde and de Wit, 1994; Nutman et al., 1993, 1997, 2004; Kusky, 1998; Bowring et al., 1990; Bowring and Williams, 1999), slightly older than the end of the terminal impact cataclysm on the moon at 3.85 Ga (Ryder, 2000). Oxygen isotope analyses and SHRIMP (sensitive high-resolution ion microprobe) dating of detrital zircons from the Jack Hills and Mount Narryer metaquartzites imply that felsic crust reacted with liquid water as early as 4.3 Ga to 4.4 Ga (Wilde et al., 2001; Mojzsis et al., 2001; Peck et al., 2001). This interpretation is tempered by the occurrence of zircons with similar ages in lunar regolith that are clearly not derived from terrestrial-style continental crust (Meyer et al., 1996). Archean TTG suites are compositionally similar to hydrous melts of subducted eclogites (Drummond and Defant, 1990; Martin, 1994; Ridley et al., 1997; Barth et al., 2002; Rapp et al., 2003). However, questions still arise about the processes that formed this felsic crust: was Phanerozoic-style plate tectonics involved, or did they form by complex processes related to rising mantle plumes (Wyman and Kerrich, 2002) or sinking of eclogitic protocrust (Zegers and van Keken, 2001)?

The stratigraphy of older Archean terranes is subject to some controversy. In the classic view, greenstone belts, comprising refractory mafic and ultramafic lava flows and tuffs intercalated with more felsic lavas and shallow marine sedimentary sequences, are deposited unconformably on top of preexisting felsic gneiss complexes (e.g., Kröner, 1985; Bickle et al., 1994; Nisbet and Fowler, 2004; Prendergast, 2004). More recent work suggests two alternatives. The first proposes that greenstone belts represent oceanic crust that was thrust over the TTG gneiss complexes during continental collision events (e.g., Kusky and Kidd, 1992; Kusky, 1998; de Wit, 1998, 2004; Wyman and Kerrich, 2002; Polat and Kerrich, 2001, 2002). These workers interpret TTG assemblages as the product of subduction zone magmatism and the greenstone belts as products of back-arc basin or oceanic plume volcanism. The second alternative proposes that greenstone belts are everywhere slightly older than the TTG suites that intrude them and that both the greenstone belts and their approximately coeval, intrusive TTG suites formed in the same suprasubduction zone setting (Drummond and Defant, 1990; Grove et al., 1999; Parman et al., 2001, 2004; Parman and Grove, 2004; Corcoran et al., 2004).

Clear evidence for plate-tectonic processes typically focuses on finding rock associations that resemble modern oceanic crust or ophiolites or evidence for their subducted remains. Ophiolites are relatively common in the Neoproterozoic (e.g., Stern et al., 2004), but although many greenstone belts have assemblages that have been interpreted as ophiolites, true ophiolite assemblages are rare in the Proterozoic (e.g., 1.95 Ga Jormua ophiolite, Finland; Kontinen, 1987; Peltonen and Kontinen, 2004; 1.73 Ga Payson ophiolite, Arizona; Dann, 1991, 2004; 1.9 Ga Birch Lake assemblage [Flin-Flon belt], Trans-Hudson orogen, Canada; Wyman, 1999; ≈1.8 Ga Buckhorn Creek ophiolite, Colorado; Cavosie and Selverstone, 2003) and almost unknown in the Archean (2.5 Ga Dongwanzi ophiolite; Kusky et al., 2001; Li et al., 2002; 2.8 Ga North Karelian greenstone belt, Russia; Shchipansky et al., 2004; 3.0 Ga Olondo greenstone belt, Tuva; Puchtel, 2004). Other greenstone belts seem not to represent ophiolites, e.g., several terranes of the 2.7 Ga Slave craton (primitive island arc, back-arc basin, and rifted margin terranes; Corcoran et al., 2004; Corcoran and Dostal, 2001; Mueller et al., 2005) and the 3.4 Ga Jamestown "ophiolite" of the Barberton greenstone belt (de Wit et al., 1987; De Ronde and de Wit, 1994) where the sheeted dikes are ≈100 m.y. younger than the volcanic rocks they intrude (Kamo and Davis, 1994).

It is now recognized, however, that most Phanerozoic ophiolites do not represent true oceanic crust formed at mid-ocean ridge spreading centers. Structurally intact Phanerozoic ophiolites form above nascent or reconfigured subduction zones in response to sinking of the subducting slab and extension of the overlying fore-arc and are commonly associated with boninite suite volcanism (e.g., Pearce et al., 1984; Shervais and Kimbrough, 1985; Robinson and Malpas, 1990; Stern and Bloomer, 1992; Pearce, 2003; Shervais, 1990, 2001; Shervais et al., 2004, 2005a). Thus, while ophiolites may be indicative of Phanerozoic-style plate tectonics, they cannot be correlated with the formation of oceanic crust at mid-ocean ridges. True oceanic crust formed at mid-ocean ridge spreading centers is almost invariably destroyed during subduction, although fragments of such crust may be preserved in accretionary complexes (Shervais and Kimbrough, 1987; MacPherson et al., 1990) or recycled as eclogites (Shervais et al., 1988; Helmstaedt and Schulze, 1989). The occurrence of boninitic lavas in several late Archean to early Proterozoic terranes, however, suggests that similar suprasubduction zone processes were active by ca. 2.8 Ga (Polat and Kerrich, 2004; Shchipansky et al., 2004).

Ancient subduction processes may be inferred from eclogite xenoliths with gabbroic or basaltic protoliths and isotopic compositions that reflect ancient seawater alteration (e.g., MacGregor and Manton, 1986; Shervais et al., 1988; Shirey et al., 2001). The chemical and isotopic compositions of these eclogites demonstrate a supracrustal origin as oceanic crust, while their current mineralogy demonstrates that they have been subject to burial deep within the upper mantle. The question remains, however, whether this deep recycling into the mantle occurred in response to Phanerozoic-style asymmetric subduction or during Hadean-style symmetric convergence and subduction, as discussed in the following section.

HADEAN AND ARCHEAN (?) CRUSTAL PROCESSES

The processes that controlled crustal formation during the Hadean (pre–3.8 Ga; Cloud, 1972) and the Archean (≈3.8–2.5 Ga) remain an area of active debate. It is now generally acknowledged that the earliest Hadean was characterized by core formation and partial melting of the outer mantle (Walter and Trønnes, 2004). The terrestrial magma ocean phase (like the corresponding lunar magma ocean) was probably short-lived and frozen by ≈4.44 Ga, a period we can refer to informally as the Tartarean, after Tartarus (ταρταροω), in Greek mythology the level below Hades where the Titans were held captive. After the Tartarean magma ocean phase, heat flow would have remained high throughout the Hadean and much of the Archean, driving vigorous thermal convection of the mantle by heating from below and forming a large number of small crustal plates (Bickle, 1978; Campbell and Jarvis, 1984; Pollack, 1997). In the following section we examine the implications of this process for Hadean and early Archean tectonics.

Thermal Convection Driven From Below

Physical and numerical models of thermal convection are characterized by plumelike upwellings that bring heat to the surface (Zhong et al., 2000; Bercovici, 2003). These thermal plumes are consistent with models for Archean komatiite formation that suggest oceanic plateaux as modern analogues (Desrochers et al., 1993; Arndt, 2004; Arndt et al., 1997a, 1997b; Wyman and Kerrich, 2002), since these features are thought to represent plume volcanism in an intraoceanic setting (Richards et al., 1989; Campbell et al., 1989; Hill, 1991). If the melting that forms komatiites occurs under dry conditions, mantle potential temperatures of ≈1500 to 1700 °C are implied: hotter than modern plume heads and consistent with an Archean mantle that was hotter than today's, thermally buoyant, and actively convecting (Nisbet et al., 1993; Nisbet and Fowler, 2004).

In contrast, these same physical and numerical models of thermal convection are characterized by the sinking of cooler lithospheric plates along vertical zones of convergence, with material from both sides of the convergence zone sinking back into the hot asthenosphere symmetrically (Fig. 1). Modern, asymmetric convergence, which is driven by the sinking of cold lithospheric plates, is not observed (Zhong et al., 2000; Bercovici, 2003) and may not have occurred in the Hadean or early Archean.

A speculative model for the formation of greenstone belts at the sites of plumelike upwellings from the mantle and their preservation at zones of convergence, along with TTG suite granitoids formed by partial melting of the sinking mafic crust, is shown in Figure 2. This model is based on the pure thermal convection model presented in Figure 1, with hot plumes rising from the core-mantle boundary and melting to form a proto-oceanic crust that is thicker than modern oceanic crust and resembles oceanic plateau crust (e.g., Arndt et al., 1997a). Consistent with the numerical modeling shown in Figure 1, sinking of this proto-oceanic crust at zones of convergence is essentially vertical and symmetric along convection cell walls (Fig. 2).

This model has some interesting implications. First, geometric constraints require that the thick proto-oceanic plateau formed above the rising plume must spread laterally as it moves radially away from the rising plume, in order to preserve crustal volume as the surface area increases (Fig. 2). This spreading will be viscous in the lower crust and brittle in the uppermost crust, where normal faulting should be common. As a result, the preserved crustal thickness should be much thinner than the primary thickness of the plateau (<15% of the original thickness for spreading over two radii of the original plateau). Second, convergence at triple junctions where three cells meet will focus this radial flow back into a single descending "anti-plume"; converging proto-oceanic crust may tend to be preserved here, especially if the position of this triple junction is unstable (Fig. 2). This will preserve the proto-oceanic crust (greenstone belts?) as small, deltoidal blocks, sitting above the triple junction and subject to compressive stress during convergence. Proto-oceanic crust that does sink (i.e., most of it) will be strongly deformed by this convergence and partially melt to form TTG suite magmas that will rise and intrude the overlying greenstone belt. The residue of this melting will be eclogite similar to that formed by subduction processes today.

The implication of this thermal convection model is that even though mafic crust resembling modern oceanic crust, or oceanic plateaux, may have formed, there is no assurance that this crust was destroyed along Phanerozoic-style, asymmetric subduction zones. Sinking of mafic and ultramafic crust along vertical convergence zones would still recycle crustal material back into the mantle, preserving the chemical and isotopic traces of this process and preserving remnants of this mafic

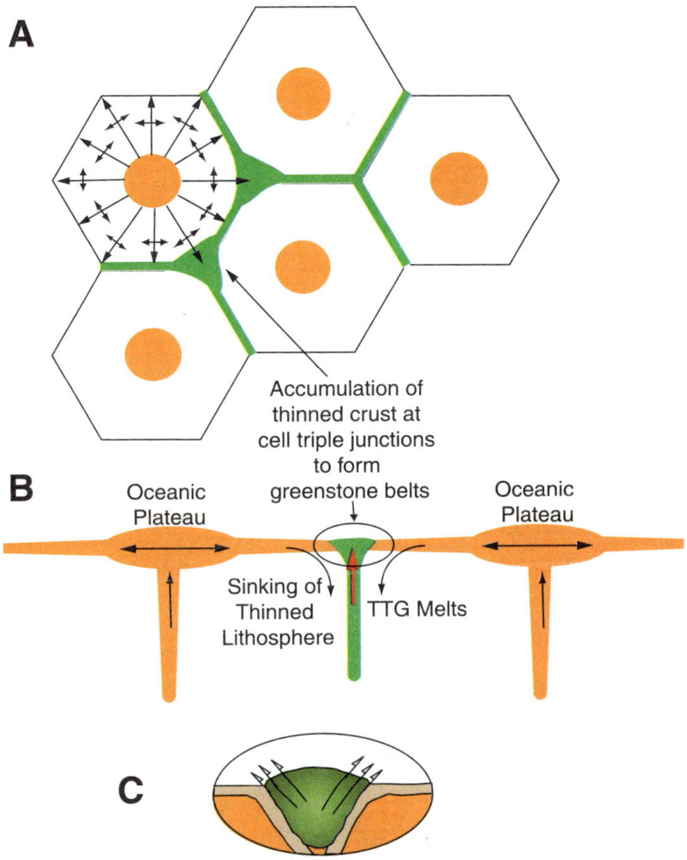

Figure 2. Schematic diagram for origin of greenstone belts and trondhjemite-tonalite-granodiorite (TTG) suites consistent with thermal convection driven by heating from below, showing plan view (A) and cross section (B). Greenstone belts form at upwelling plumes of mantle and spread radially from the plume center toward zones of symmetric convergence and sinking. Greenstone belts most commonly preserved at triple junctions of intersecting convergent zones (detail in inset C). TTG suites represent partial melts of the sinking mafic crust, which typically pools at triple junctions of intersecting convergent zones, intruding the overlying greenstone belts (red arrow).

crust as eclogite. So evidence for the existence and recycling of oceanic-like crust does not conclusively show that modern-style plate tectonics was operative.

The preservation of oceanic crust at symmetric convergence zones is likely to involve significant deformation and thrusting, forming schüppenzones of imbricate thrust slices of oceanic crust that stack older crust onto younger. These thrust complexes would not resemble modern accretionary complexes, however, because there would be no emergent volcanic arc to provide the sediments that dominate modern complexes.

Convection Driven by Sinking Slabs

In contrast, it has been known for some time that Phanerozoic-style plate tectonics is driven by the sinking of cool, dense lithospheric plates, with a small component of push as the plates slide off the uplifted oceanic ridge system—all under the influence of gravity (Carlson, 1981; Hager and O'Connell, 1981; Jurdy and Stefanick, 1991; Anderson, 2001). Oceanic spreading ridges form by the passive upwelling of asthenosphere into the breach created by plate motion, and their only contribution to plate motion is the relief created by the thermal buoyancy of the asthenosphere, which drives the "ridge push" component of plate motion (Carlson, 1981; Hager and O'Connell, 1981; Jurdy and Stefanick, 1991; Anderson, 2001). Oceanic crust formed by this process is thinner than crust created by thermal plumes, but it does not thin by lateral spreading. Island arcs form by hydrous melting of the mantle wedge that forms above the sinking slab.

It has been proposed that komatiites formed by hydrous melting in Archean subduction zones, much like boninites in modern arc settings (Grove et al., 1999; Parman et al., 2001, 2004; Parman and Grove, 2004). This proposal is supported by stratigraphy of Archean greenstone belts, which are dominated by more evolved lavas (basalts, andesites, dacites, rhyolites) and tuffs (commonly silicified to cherts; Lowe, 1994) that resemble primitive arc assemblages in the western Pacific (e.g., Hawkins et al., 1984). Similarly, TTG associations, which dominate the felsic gneiss terranes of Archean cratons, have been correlated with adakitic suites in modern arc terranes, formed by the melting of down-going oceanic crust (Drummond and Defant, 1990; Polat and Kerrich, 2001, 2002). If correct, these models imply that Archean mantle may not have been superheated significantly compared to modern plumes (e.g., Grove and Parman, 2004; Valley et al., 2002). However, komatiites lack negative Nb anomalies and other trace element characteristics of modern arc lavas, so a subduction zone origin is far from clear (e.g., Storey et al., 1991; Arndt, 2004).

PLATE TECTONICS AND ACCRETIONARY PROCESSES

One reliable indicator of Phanerozoic-style plate tectonics is the occurrence of an accretionary prism that includes sediments eroded from the upper, arc-volcanic plate and oceanic crust (mafic volcanic rocks, ultramafic tectonites, and abyssal sediments) from the lower plate (Fig. 3). Such a prism is the clear result of asymmetric subduction, with one plate subducting beneath another, more buoyant plate; subsequent ridge subduction may also be important in modifying the primary assemblage (Shervais et al., 2004, 2005a; Kusky et al., 2004a). In the following discussion we examine first the type example of a Phanerozoic accretionary complex, followed by a well-documented example of a similar rock assemblage overprinted by a major collisional orogeny. Finally, we examine several examples of potential Archean accretionary complexes and their correlation to these well-documented Phanerozoic examples.

It should be noted that not all convergent margins are accretionary. Nonaccreting margins (von Huene and Scholl, 1991) are characterized by subduction erosion that exposes fore-arc basement in the hanging wall of the trench. These mar-

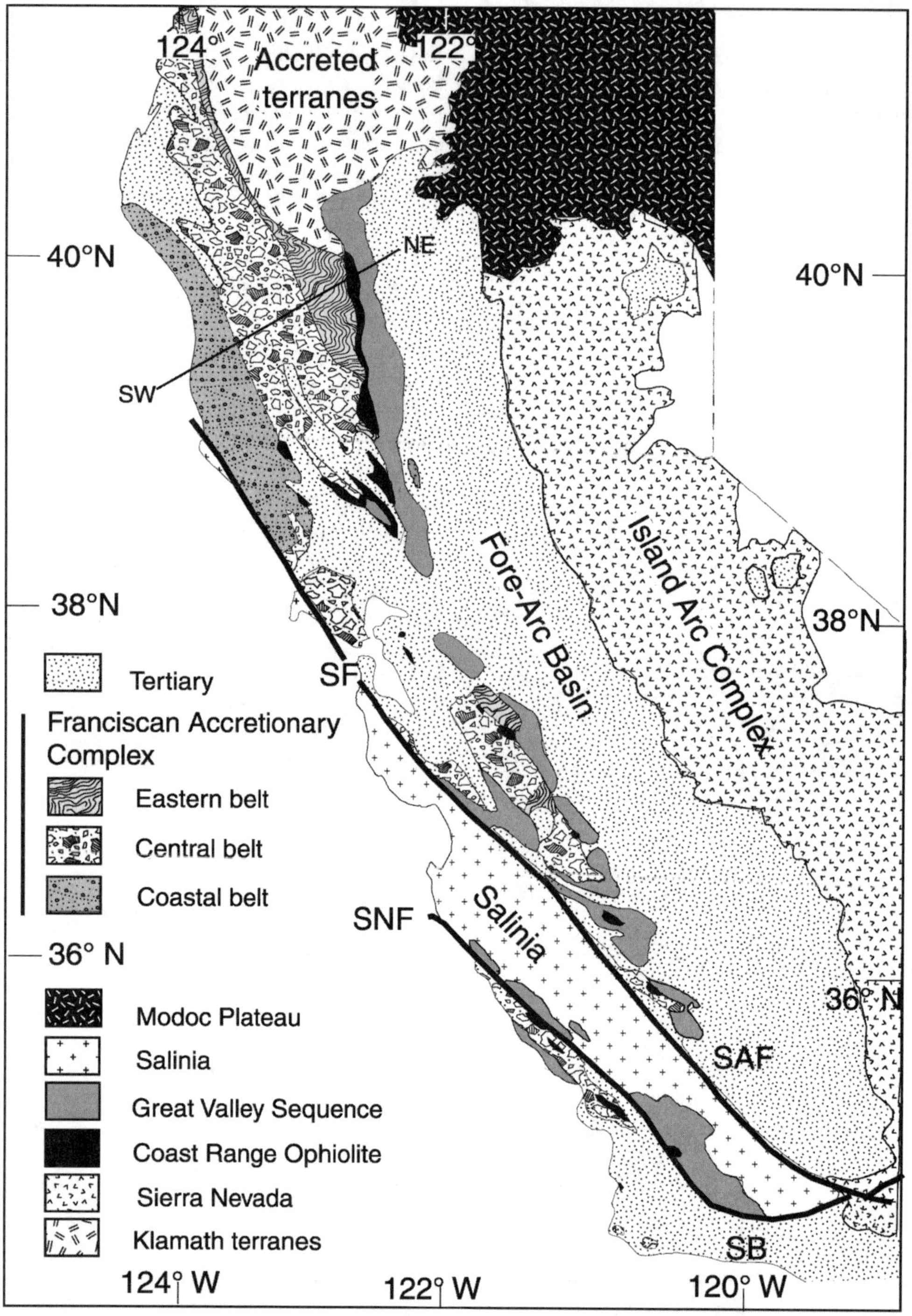

Figure 3. Geologic sketch map showing the distribution of the Franciscan assemblage in California. Eastern, Central, and Coastal belts after Bailey et al. (1964), Blake et al. (1985), and Ernst (1993). The coeval island arc complex and fore-arc basin are labeled, along with accreted terranes of the Klamath Mountains. Arc (Salinia) and Franciscan rocks are repeated west of the San Andreas (SAF) and Sur-Nacimiento (SNF) fault zones. SF—San Francisco, SB—Santa Barbara. Line of section in Figure 4 labeled southwest-northeast.

gins are typically sediment-starved because there are few emergent volcanoes (intraoceanic arcs) or because rivers carry most of the sediment into the back-arc region (continental arcs); what sediment does enter the trench is largely subducted (von Huene and Scholl, 1991). The Mariana-Bonin arc is the type example of a nonaccreting intraoceanic margin, while the Andean arc is the best example of a nonaccreting continental arc margin.

Phanerozoic Accretionary Complexes

Phanerozoic accretionary complexes include those that are currently active and fossil systems that are both well preserved and well exposed. Currently or recently active systems include the Aleutian trench system (Fruehn et al., 1999), the Java-Sumatra trench system (Hamilton, 1979; Moore and Karig, 1980), the Antilles fore-arc (Moore et al., 1982), off-shore exposures of the Makran (Kukowski et al., 2001), and the Mariana fore-arc (Fryer et al., 2000). Fossil systems include the Lichi mélange in Taiwan (Page and Suppe, 1981), the Chugach-Kodiac mélange in Alaska (Byrne, 1984; Kusky et al., 1997a, 1997b; Kusky et al., 2004a), onshore exposures of the Makran (Platt et al., 1985), and the Franciscan complex of California (Bailey et al., 1964; Cowan, 1978).

The Franciscan complex of California is the classic example of an accretionary complex formed during the subduction of oceanic lithosphere beneath an active arc terrane (Bailey et al., 1964; Hsü, 1968; Ernst, 1993; Wakabayashi, 1999). It encompasses most of the Coast Ranges of central and northern California, cropping out over an area some 700 km long and up to 200 km wide, east of the San Andreas fault; it also crops out over a small area west of the Sur-Nacimiento fault (Fig. 3). The Franciscan complex consists of three tectonic belts, each with its own distinct character and lithologies: the Eastern belt, the Central belt, and the Coastal belt (Bailey et al., 1964). The distribution of these belts is shown in Figure 3, and their relationships in northern California are shown in cross section in Figure 4.

The Eastern belt comprises coherent sheets of blueschist-facies greywacke and basalt up to hundreds of meters thick transposed along low-angle thrust faults, as well as exotic blocks in a chaotic mélange with a metagreywacke matrix (Bailey et al., 1964; Ernst, 1993; Blake and Jones, 1974; Blake and Wentworth, 1999). Eastern belt metasediments have early to mid-Cretaceous depositional ages with mid- to late Cretaceous metamorphic ages (Blake et al., 1982). This entire assemblage is juxtaposed structurally below the mid-Jurassic Coast Range ophiolite and its cover of fore-arc sedimentary rocks

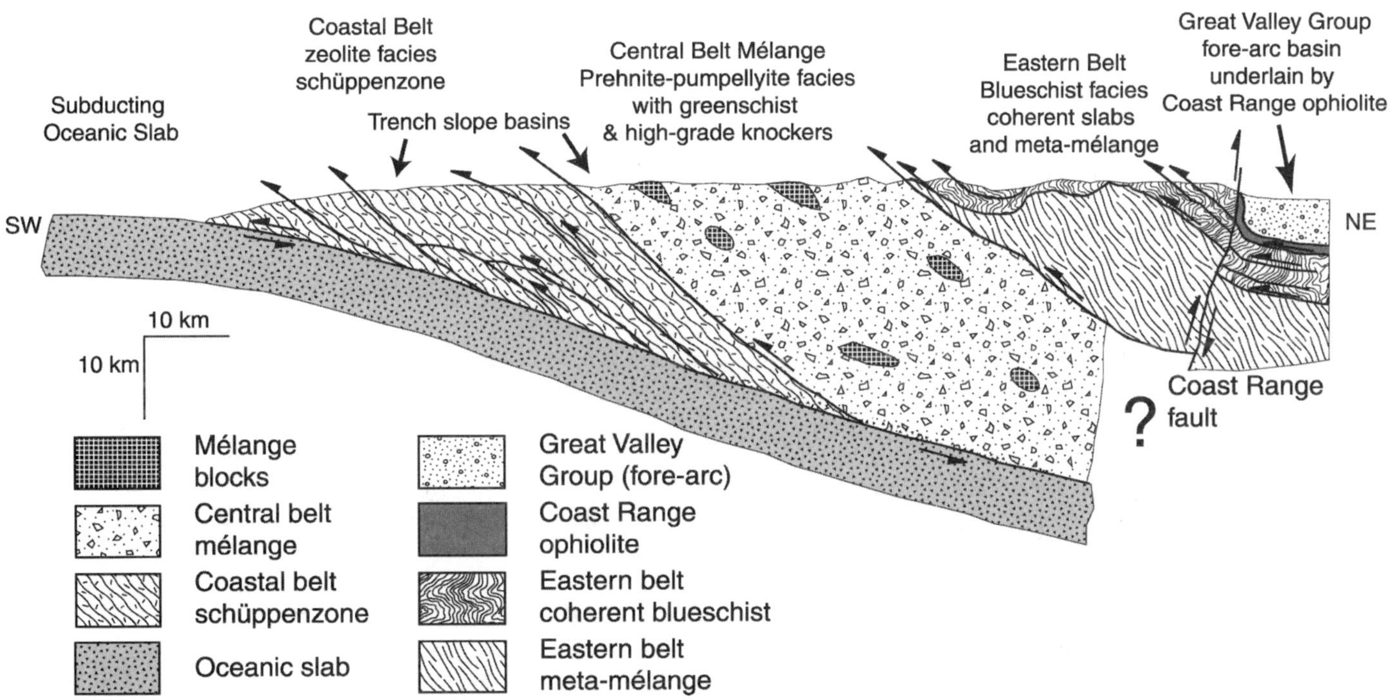

Figure 4. Schematic diagram of Franciscan assemblage accretionary complex, illustrating Phanerozoic-style plate tectonics. Coastal belt represents a schüppenzone of zeolite facies greywackes and shales, largely in coherent tracts but including broken formation. The Central belt of the Franciscan represents the classic mélange terrane, with blocks of low-grade (prehnite-pumpellyite to greenschist facies) greywacke, chert, basalt, diabase, and peridotite, and blocks of high-grade (garnet amphibolite-eclogite-blueschist facies) basalt and greywacke, in a matrix of scaly clay/microgreywacke (or more rarely, serpentinite). Eastern belt of the Franciscan consists of imbricate thrust sheets with coherent tracts of blueschist facies greywacke, schist, and basalt, overlying blueschist facies metamélange. Eastern belt is overlain at depth by Coast Range ophiolite (fore-arc basement) and Great Valley Group (fore-arc basin sediments), but these units are currently juxtaposed along the Neogene Coast Range fault, a high-angle reverse fault. Franciscan relations after Blake et al. (1985); trench side of Coastal belt (which is truncated by the San Andreas fault system) reconstructed from seismic profiles of Japan trench (von Huene and Scholl, 1991).

(Blake et al., 1982; Blake and Jones, 1974; Shervais, 2001; Shervais et al., 2004, 2005a).

The Coastal belt, which lies outboard of the Central and Eastern belts, comprises a schüppenzone of thin, west-verging thrust slices that generally young to the west (Blake et al., 1985). The easternmost thrust slice (Yager terrane) consists of coherent greywacke strata; more westerly terranes are partially to largely disrupted broken formations of arkose, mudstone, conglomerate, and, in the Point Delgado subterrane, basaltic pillow lava and diabase (Blake et al., 1985). The Coastal belt ranges in age from mid-Cretaceous to Miocene (in the King Range terrane; McLaughlin et al., 1982). This belt, which represents the youngest part of the accretionary complex, is characterized by modes that are richer in K-feldspar than older greywackes of the complex, reflecting the unroofing of plutonic rocks in the eastern Sierra arc terrane (Bailey et al., 1964).

The Franciscan central belt forms the classic mélange terrane that is commonly associated with the Franciscan accretionary complex. The central belt contains blocks and knockers of greywacke, greenstone, serpentinite, chert, limestone, blueschist, eclogite, and garnet amphibolite (Bailey et al., 1964; Blake and Jones, 1974; Blake and Wentworth, 1999), as well as large tracts of arkosic wackes interpreted as slope basin deposits (Becker and Cloos, 1985). The mélange matrix is a finely comminuted microgreywacke siltstone/shale that has been strongly sheared to produce the classic argille scagliose texture (Hsü, 1966; Blake and Jones, 1974). Detailed studies of the mélange matrix show that sedimentary processes (olistostromes) are important in some shale-matrix mélanges (Cowan, 1978, 1985), in addition to tectonic disruption by ductile flow (Cloos, 1982).

Serpentinite matrix mélange, which may represent accreted fracture zones, is less common but may represent the primary transport medium for high-grade blocks (Bailey et al., 1964; Coleman, 2000; Hopson and Pessagno, 2005). Serpentinite mélange includes serpentinite broken formation (which consists of massive blocks of serpentinized peridotite enclosed in a matrix of serpentinite schist) as well as true mélange, which consists of exotic blocks of high-grade metamorphic rocks (Figs. 5A and 5B) or greenstone/chert (Figs. 5C and 5D) in a serpentinite schist matrix (Huot and Maury, 2002; Shervais et al., 2005b; Hopson and Pessagno, 2005).

High-grade metamorphic blocks include blueschist, eclogite, amphibolite, and garnet amphibolite (Bailey et al., 1964; Moore and Blake, 1989; Moore, 1984). Many of these blocks are polymetamorphic and preserve evidence of an early high-temperature, medium-pressure assemblage (amphibolite, garnet amphibolite) overprinted by eclogite or blueschist facies assemblages (e.g., Moore and Blake, 1989). These blocks commonly preserve a rind of lower-grade minerals (actinolite, chlorite) with compositions that suggest equilibrium with serpentinite, even when found in shale matrix mélange, implying primary exhumation in serpentinite mélange before being remixed into shale matrix mélange. The rinds are typically polished, striated, and gouged, showing that the blocks formed rigid inclusions within a viscous matrix. Though rare, these high-grade blocks clearly document tectonic mixing of exotic components into matrix components that formed at much lower temperatures and pressures.

The most common tectonic block in most shale-matrix mélange is greywacke. In a few examples, this greywacke sits depositionally on chert-basalt, but more commonly it occurs alone in large, slablike knockers. These greywacke blocks represent either less disrupted material that was parental to the matrix or in some cases slope basin deposits that were deposited on a previously deformed substrate (Cowan, 1978; Becker and Cloos, 1985).

Chert and greenstone are both common knocker lithologies in the Franciscan assemblage. Chert knockers are typically slablike due to their bedding, while the greenstones may include pillow lava, sheet flows, hyaloclastites, or diabase sill complexes (Figs. 5E and 5F). One of the more common knocker assemblages is composite chert-greenstone blocks (Figs. 5C and 5D). These blocks consist of banded cherts deposited on pillowed or massive submarine lavas or intruded by sills of oceanic basalt composition (e.g., Murchey, 1984; Shervais and Kimbrough, 1987; Huot and Maury, 2002). Banded cherts may form decameter-scale lenses intruded by diabase sills (e.g., Aliso Canyon; Shervais and Kimbrough, 1987) or extensive terranes of chert + basalt stacked along thrust faults (e.g., Marin Headlands; Murchey, 1984; Karl, 1984). In parts of the Franciscan composite basalt-chert blocks are stacked in thin thrust sheets (Isozaki and Blake, 1994); similar relations are found in the Japanese accretionary complex (Isozaki et al., 1990, Isozaki, 1997).

Accretionary complexes may be intruded by igneous rocks during ridge collision/subduction events, further complicating their interpretation. This is clearly seen in south-central Alaska, where ridge subduction occurred in the Paleogene (Bradley et al., 2003; Kusky et al., 2003, 2004) and in the Franciscan complex in response to migration of the Farallon triple junction (Johnson and O'Neil, 1984; Cole and Basu, 1995). Some felsic magmas are derived from direct melting of the hot subducting slab, which forms adakites (Sr-rich granites that resemble the Archean TTG suite; Drummond and Defant, 1990). Other magmas include calc-alkaline basalt, trachybasalt, andesite, dacite, and rhyolite (Johnson and O'Neil, 1984; Cole and Basu, 1995).

Effects of Collisional Metamorphism on Accretionary Complexes

What would this assemblage look like if it were metamorphosed to upper amphibolite facies conditions during a continent-continent or arc-continent collision? We have a good model for this process in the Eastern Blue Ridge province of the southern Appalachians, which is separated from the Western Blue Ridge by the Haysville-Fries fault system (Williams and Hatcher, 1983; Adams et al., 1995). The Eastern Blue Ridge province is distinct from amphibolite to granulite facies gneisses of the Western Blue Ridge province (which reflect Grenville-age

Figure 5. Franciscan mélange field photos. (A) Serpentinite matrix mélange near Figueroa Mountain, California. Knockers of diabase and basalt (larger blocks) and high-grade metamorphics (smaller blocks and groups of blocks) are circled for clarity. (B) Knocker of blueschist in serpentinite matrix mélange near Figueroa Mountain, ~30 m across. (C) Large composite knocker of banded chert (top) and diabase (bottom) in shale matrix mélange, Aliso Canyon, California. (D) Banded ribbon chert from composite chert-diabase knocker, Aliso Canyon, California. (E) Smaller (3 m) block of basalt in sheared volcanic matrix, Paskenta, California. (F) Smaller (≈1 m) block of harzburgite in sheared serpentinite-matrix mélange, Paskenta, California.

metamorphism) and more closely resembles rocks of the Inner Piedmont terrane in western South Carolina (Williams and Hatcher, 1983); together the Eastern Blue Ridge province and the Inner Piedmont belt comprise the Jefferson terrane. The Jefferson terrane forms part of the infrastructure of the Alleghenian orogeny, which represents collision of the passive eastern margin of Laurentia with exotic arc terranes and the cratonic margin of Gondwana (Secor et al., 1986; Adams et al., 1995; Adams and Trupe, 1997; Shervais et al., 2003; Dennis et al., 2000, 2004).

The Jefferson terrane consists of biotite-quartz-feldspar gneisses, migmatitic in part, with extensive zones of pegmatite intrusion. Within this groundmass of paragneiss are blocks and slabs of exotic lithologies, including amphibolite, garnet amphibolite, anthophyllite dunite, and eclogite (Ash metamorphic suite, Alligator Back formation; Adams et al., 1995; Adams and Trupe, 1997). Two aspects of this terrane support its origin within an accretionary complex. First is the characteristic association of "oceanic" lithologies (metabasalts, metaserpentinites) within paragneisses derived from immature clastic sediments. In particular, the volumetric dominance of the paragneiss and the irregular distribution of the mafic and ultramafic rocks as blocks and slabs within these paragneisses recall the distribution of exotic blocks in the Franciscan mélange. Second, the occurrence of large slabs of metabasalt with relict eclogite assemblages is consistent with a subduction zone origin, despite later overprinting by amphibolite facies assemblages and retrogression of the primary eclogite phase assemblage (e.g., Willard and Adams, 1994). These slabs resemble garnet amphibolite in the field but preserve relict eclogite assemblages (garnet-omphacite) in the midst of hornblende-rich overprinting. Whole-rock geochemical data and paleo–pressure-temperature work show that these eclogites represent oceanic crust (mid-ocean ridge basalt [MORB]) that was metamorphosed under high pressure–low temperature conditions in a subduction zone long before continental collision (Willard and Adams, 1994; Adams et al., 1995).

Eclogite and high-pressure granulite facies rocks in the central Carolina terrane represent MORB dikes emplaced into the basement of the Carolina arc terrane where they were subject to high-pressure, medium-temperature metamorphism during an arc-arc collision (Dennis et al., 2000, 2004; Shervais et al., 2003). These rocks are not strictly analogous to the Blue Ridge eclogites but did suffer a similar amphibolite facies retrograde overprint during a collisional orogeny similar to the Alleghenian orogeny (Shervais et al., 2003). In the Carolina eclogites, garnet preserves extreme zoning that can be mapped chemically with the microprobe to establish which garnet composition was in equilibrium with which assemblage. Sodic pyroxene (omphacite) has been completely replaced by diopside-albite symplectites that reflect increased temperature and lowered pressure during thermal equilibration of the subducted lithosphere (Shervais et al., 2003). These diopside-albite symplectites are characteristic of retrograde eclogite assemblages and can be recombined using image analysis techniques to recover the composition of the primary omphacite. The resulting compositions can be used to calculate the pressure-temperature history of the former eclogites, despite amphibolite and even granulite facies overprints (Shervais et al., 2003).

Archean Accretionary Complexes

Recognition of Archean accretionary complexes is complicated by the protracted history of metamorphism and tectonism that commonly characterizes rock assemblages formed on active plate boundaries (e.g., Kusky and Polat, 1999). These rocks are more likely to be overprinted by both high-pressure and high-temperature metamorphic events than rocks formed in continental interiors or even arc volcanic suites formed above the same subduction zone (being on the upper plate, volcanic arcs rarely subduct and are commonly preserved even in major continental collision zones; e.g., the Kohistan-Ladakh arc terrane in the western Himalaya; Shah and Shervais, 1999). Nonetheless, by recognizing the effects of these later tectonic and metamorphic events, it is possible to interpret the original history of these rock assemblages in terms of modern plate-tectonic theory.

Possibly our best evidence for Archean plate tectonics at this time is found in three potential accretionary mélange complexes: the 2.5 Ga Dongwanzi ophiolite mélange in the north China craton (Kusky et al., 2001, 2004b; Li et al., 2002), the 2.7 Ga Schreiber-Hemlo greenstone belt in the Superior province of Canada (Polat et al., 1998; Polat and Kerrich, 1999), and the 2.6 Ga Farmington Canyon complex in north-central Utah (Shervais, 2004). The 3.8 Ga Isua Supergroup of Greenland (Komiya et al., 1999, 2004) may provide a fourth example that is significantly older, but its interpretation as an accretionary complex has been disputed (Myers, 2001, 2004).

Dongwanzi Ophiolite Mélange

The Dongwanzi ophiolite mélange crops out in the Zunhua structural belt, a continuation of the late Archean Central orogenic belt of the North China craton (Li et al., 2002). This late Archean mobile belt consists of gneiss, greenstone belts, and mafic/ultramafic complexes that represent in part an ophiolitic mélange (Kusky et al., 2001, 2004b; Li et al., 2002; Huson et al., 2004; Huang et al., 2004). The mélange resembles the Eastern Blue Ridge province in many respects, consisting of mafic-ultramafic plutonic rocks (cumulates and harzburgite tectonites with podiform chromite), dike complex, pillow lava, and banded iron formations (≈chert) in a mélange matrix of foliated and sheared biotite gneiss (Li et al., 2002; Kusky et al., 2001, 2004b). More than 1000 ultramafic boudins have been recognized in the Zunhua belt (Li et al., 2002; Kusky et al., 2004b), an association that is difficult to reconcile with the biotite gneiss country rock except as blocks in mélange.

Within or adjacent to this mélange is the Dongwanzi ophiolite, the best example to date of a true Archean ophiolite (Kusky et al., 2001, 2004b). This ophiolite fragment is ≈50 km long and 10 km wide. It is located near the eastern margin of the Zunhua structural belt, along the leading edge of the orogen (Kusky et al., 2001). To the west are found greenstones and TTG rocks that may represent the associated volcanic arc complex, along with retrograde eclogites and high-pressure granulites of the Hengshan belt (Li et al., 2002). The Dongwanzi ophiolite mélange is interpreted as an accretionary prism formed along an active continental margin that was subsequently trapped during collision of the western and eastern blocks of the North China craton and metamorphosed to amphibolite facies.

Schreiber-Hemlo Greenstone Belt

The late Archean (≈2670–2750 Ma) Schreiber-Hemlo greenstone belt of the Superior province crops out along the north shore of Lake Superior in southern Ontario, Canada, over an area ~90 km long and 10–20 km wide (Polat et al., 1998, 1999; Wyman et al., 2002). Metamorphic grade ranges from lower greenschist facies in the western (Schreiber) area to lower amphibolite facies in the eastern (Hemlo) region (Pan and Fleet, 1993). Volcanic rocks include oceanic plateaux assemblages (tholeiitic basalts, komatiites) and oceanic island arc assemblages (basalts, andesites, dacites with arc tholeiite or calc-alkaline compositions). Sedimentary rocks are siliciclastic turbidites interpreted as trench deposits (Polat et al., 1998; Polat and Kerrich, 1999). All of these rock assemblages have been intruded by arc-derived tonalite-trondhjemite-granodiorite plutons that are only slightly younger than the host rocks.

As described by Polat and coworkers (Polat et al., 1998; Polat and Kerrich, 1999, 2001, 2002, 2004), the oceanic plateau basalts, island arc volcanics, and siliciclastic trench turbidites were tectonically juxtaposed before and during intrusion of the TTG suite. All of the subduction-related igneous and sedimentary rocks are characterized by LREE–enrichment (light rare earth elements) and by distinct negative Ti, Zr, and Nb anomalies on primitive-mantle normalized spider diagrams, whereas the oceanic plateau basalts have relatively flat REE (rare earth element) patterns and lack negative anomalies in the primitive-mantle normalized high-field strength elements (Polat et al., 1998; Polat and Kerrich, 1999, 2001, 2002, 2004).

For the most part these distinct assemblages appear to form a schüppenzone of large, km-scale fault-bounded blocks, with little mixing between adjacent regimes. Locally, blocks of material exotic to the surrounding matrix are recognized, including arc rhyolites in a matrix of sheared ocean plateau tholeiite and blocks of gabbro, tonalite, and volcanic rock in a matrix of sheared and foliated shale or wacke (Polat et al., 1998; Polat and Kerrich, 1999). In some cases these rocks may represent boudinaged dikes and not true exotic mélange blocks; in others their exotic nature seems more clear. Native blocks are also recognized, including komatiite, gabbro, or chert in sheared ocean plateau tholeiite, forming a broken formation. Exotic and native blocks range in size from a few meters to decameters.

Polat and coworkers (Polat et al., 1998, 1999; Polat and Kerrich, 1999, 2001, 2002) interpret these relationships to represent formation (and deformation) within an accretionary complex in a convergent margin, subduction zone setting. They recognize that much of the deformation observed probably formed along the contacts of adjacent thrust sheets and by the progressive layer-parallel extension of more rigid horizons within the sheared matrix.

Despite the apparent occurrence of both mélange and broken formations, there are some significant differences between the Schrieber-Hemlo greenstone belt and Phanerozoic accretionary complexes. The primary difference is the predominance of arc-derived igneous lithologies: modern accretionary complexes are dominated by siliciclastic sediments derived by erosion from the adjacent island arc, but tectonic blocks of arc volcanic or plutonic rocks are rare to nonexistent (Bailey et al., 1964; MacPherson et al., 1990; Shervais and Kimbrough, 1987). Igneous rocks within Phanerozoic accretionary complexes are almost all derived from the subducting oceanic plate, along with blocks of chert and limestone that are exotic to the trench depositional setting. Another difference is the dominance of large, fault-bounded blocks of volcanic and plutonic rocks derived from the subducting ocean crust as well as from the arc: in modern accretionary complexes these rocks are subordinate to the more common shale or greywacke matrix. Large basalt-chert terranes may occur (e.g., Marin Headlands terrane, California), but these are relatively rare. Since many komatiite sequences are now interpreted as primitive arc volcanics, it seems possible that rather than an accretionary complex, the Schreiber-Hemlo greenstone belt may represent an island arc terrane deformed during a continent-continent collision, comparable to the Kohistan arc terrane in northern Pakistan (Coward et al., 1982; Shah and Shervais, 1999). Alternatively, it may represent a nonaccreting convergent margin similar to the Mariana-Bonin fore-arc, which exposes older arc basement in the hanging wall of the trench (Hawkins et al., 1984).

The Farmington Canyon Complex

The late Archean (≈2.6 Ga) Farmington Canyon complex of northern Utah, which lies on the southwestern margin of the Wyoming province, has been interpreted to represent a passive margin sedimentary wedge that was metamorphosed to amphibolite facies conditions in the middle Proterozoic (Bryant 1988a, 1988b; Nelson et al., 2002). The Archean age of this complex (≈2.6 to possibly 3.0 Ga) is based on inheritance in zircon U/Pb systematics, old Nd model ages, and high Sr initial ratios in orthogneiss (Hedge et al., 1983). The complex was intruded by granitic gneiss and metamorphosed ca. 1.8 Ga (Hedge et al., 1983). Nelson et al. (2002) date the complex at ca. 1800 Ma using the monazite Th-Pb microprobe technique; they suggest the older ages may reflect the age of the sediment provenance and not its formation age. However, the monazite ages are coincident with the metamorphism and granite intrusion dated by Hedge et al. (1983) and may simply reflect mid-Proterozoic metamorphism.

The southern Farmington Canyon complex consists largely of migmatitic quartzo-feldspathic gneiss (Fig. 6A) containing isolated blocks and slabs of (1) mafic metavolcanics (amphibolite, pyroxene amphibolite, garnet amphibolite), (2) ultramafic metavolcanics (now serpentine-anthophyllite), and (3) quartzite. The migmatitic gneiss has a chemical composition similar to greywacke (Bryant, 1988a, 1988b), consistent with its formation from a Franciscan-style greywacke mélange matrix. There is some variability in its appearance, and some areas may represent coherent slope basin deposits of more arkosic wackes, analogous to the Cambria slab in the Franciscan complex. The gneisses and amphibolites are strongly deformed and may be

Figure 6. Farmington Canyon complex mélange field photos. (A) Migmatitic felsic gneiss, interpreted to represent metamorphosed matrix of greywacke matrix mélange. (B) Large knocker of banded chert, cut by amphibolite dikes, in Farmington Canyon; compare banding with Franciscan ribbon chert in Figure 4. (C) Close-up of banding in metachert, near Ogden, Utah. (D) Composite knocker of amphibolite (upper part of outcrop) and metachert (lower part of outcrop), Centerville Canyon; white line marks approximate contact. (E) Garnet amphibolite with large garnets surrounded by plagioclase-hornblende rims formed in response to decreased pressure or elevated temperature, Sessions Mountains, Utah. (F) Composite knocker of komatiite (under hammer), amphibolite (halfway up slope), and chert (top of slope) in Farmington Canyon complex near Ogden, Utah.

mylonitic in places; these rocks are crosscut by undeformed granite pegmatites with sharp discordant contacts that represent much younger anorogenic melts (Shervais, 2004).

Quartzite in the southern Farmington Canyon complex may represent metachert rather than an epiclastic quartzite (Shervais, 2004). These quartzites lack clastic sedimentary structures, and bedding is suggested by color variations and micaceous horizons (commonly the Cr-mica fuchsite) that mimic bedding features in chert (Figs. 6B and 6C). A banded chert protolith is suggested for some outcrops based on the intercalation of quartzite beds 5–20 cm thick with schist layers 1–2 cm thick (Fig. 6B). Chert and amphibolite typically occur in the same block (Fig. 6D), suggesting an oceanic crust sediment-basement contact or sills that intrude oceanic sediments off axis (e.g., Shervais and Kimbrough, 1987; Figs. 5C and 5D).

Amphibolite lenses in the Farmington Canyon complex range up to 200 m long and 50 m wide and are similar in size to volcanic blocks in the Franciscan assemblage (e.g., Bailey et al., 1964). Mineral assemblages include plagioclase and hornblende ± clinopyroxene, garnet, and quartz. Garnet porphyroblasts up to 4 cm across in amphibolite are commonly mantled by a low-pressure, high-temperature symplectite of plagioclase and hornblende (±pyroxene) that implies a clockwise pressure-temperature-time path (Fig. 6E). The common association of amphibolite with "quartzite" (metachert) in composite bodies is similar to the basalt-chert assemblage in the Franciscan complex (Figs. 6B and 6C).

Ultramafic lenses in the Farmington Canyon complex consist of serpentine, pyroxenes, anthophyllite, and other mafic minerals and are commonly associated with amphibolite or hornblende-plagioclase gneiss and quartzite (Yonkee et al., 2000). The largest lenses, up to 80 m × 20 m, lie in a matrix of felsic layered gneiss in contact with lenses of metachert (Fig. 6F). The association of ultramafic rocks with metachert implies that the ultramafics represent metavolcanic assemblages (e.g., komatiite) rather than the mantle tectonites that are common in the Franciscan complex.

The combination of old Nd model ages and Archean inherited zircon components implies that this accretionary complex formed on the southwestern margin of the Wyoming province in conjunction with a coeval continental margin arc that may be represented in part by the northern portion of the Farmington Canyon complex (orthogneiss, migmatite, pegmatite). This continental margin arc apparently collided with the Santaquin arc in the mid-Proterozoic (e.g., Nelson et al., 2002).

The Isua Supracrustals

The Isua supracrustals (Bridgwater et al., 1976) represent some of the oldest supracrustal rocks preserved on earth, with U/Pb zircon dates of 3.71–3.85 Ga (Bridgwater et al., 1976; Nutman et al., 1993, 1997, 2001, 2004). Detailed mapping demonstrates a range in metamorphic grade from epidote amphibolite in the southern zone to greenschist/amphibolite transition in the northern zone. Rocks of the northern zone are not only lower in metamorphic grade but are generally less deformed as well (Komiya et al., 1999, 2004; Nutman et al., 1996; Myers, 2001, 2004).

Komiya et al. (1999) interpret the lower half of the northern zone as mélange that consists of angular to subround blocks of greenstone (including well-preserved to flattened pillow lavas) in a matrix of sheared mudstone. The upper half of the northern unit comprises several thin thrust sheets of basalt (pillow lava ± hyaloclastite) overlain by metachert or banded iron formation. The southern unit is a schüppenzone of thin thrust sheets that forms duplex structures along layer parallel faults (Komiya et al., 1999). The thrust sheets generally comprise pillow basalt and chert overlain by turbidites of mafic wackes, arkose, mudstones, calcareous sandy shale, and minor conglomerate (Fedo, 2000). The southern and northern units are separated by a middle unit that consists dominantly of chert with overlying turbidite, similar to the sediments that rest on pillow basalt in the southern unit (Komiya et al., 1999).

In contrast, Myers (2001, 2004) interprets the entire sequence as a schüppenzone of mafic volcanics and chert. He suggests that the "mafic wackes, arkose, and mudstones" of Komiya et al. (1999) are in fact sheared volcanics and that the carbonates are metasomatic in origin, not sediments (Rosing et al., 1996). In addition, Nutman et al. (1996) have shown that rocks within the supracrustal belt span an age range of ≈100 Ma. If this interpretation is correct, then the Isua greenstones cannot be interpreted as a Phanerozoic-style accretionary complex; it does, however, resemble what we might expect to form during Hadean-style symmetric convergence.

Basalts in the Isua greenstone belt include both MORB and ocean island basalt (OIB)–like compositions. The Isua MORB is enriched in FeO compared to modern MORB and was derived from a fractionated mantle (light rare earth element/heavy rare earth element rations < 1 [heavy rare earth elements]) with a potential temperature ~1480 °C (Komiya et al., 2004). The Isua OIB has a flat REE pattern, suggesting derivation from a primitive source. The occurrence of true oceanic basalts in the Isua greenstone belt, rather than the arclike volcanics characteristic of other greenstone belts, is consistent with modern settings in which true oceanic basalts are found in accretionary complexes, not ophiolites, which represent arc volcanism. It is also consistent with Hadean-style convergence.

DISCUSSION

The recognition of subduction-related accretionary complexes hinges largely on our ability to identify assemblages of rocks that could not have formed together in a single tectonic or depositional setting and must represent tectonic mixtures of widely divergent rock types. The juxtaposition of discrete blocks of oceanic lithologies that could not form on a passive margin with a mélange matrix derived from proximal, arc-dominated sources comprises first-order evidence in support of the accretionary complex model. The occurrence of chert blocks in microgreywacke mélange matrix also seems to require physical mixing of rocks from distinct depositional settings and may be proof enough for an accretionary origin. Confirmation of this interpretation requires the discovery of relict high-pressure assemblages that document formation in a collisional or subduction environment.

Hadean tectonics was likely dominated by upwelling thermal plumes and symmetric sinking of opposing lithospheric plates, consistent with models of thermal convection driven by heating from below (e.g., Fig. 1). Geometric constraints require that any crust generated above these rising plumes (presumably thick, plateaulike accumulations of mantle-derived basalts and komatiites; Arndt, 1994, 2004; Arndt et al., 1997a) must be thinned rapidly by lateral spreading as the crust expands radially from the plume-centered plateau (Fig. 2). This continual thinning of the crust will promote rapid cooling and thickening of the underlying mantle lithosphere, which will also thin by lateral spreading as it expands radially from the plateau. The crust generated above the rising plumes may accumulate and be preserved at zones of symmetric convergence, especially at triple-plate junctions where excess crust is forced into a descending "anti-plume" (Fig. 2).

The occurrence of subduction-related accretionary complexes among Archean rocks ranging in age from 2.5 Ga to 2.7 Ga, as discussed above, suggests that Phanerozoic-style plate tectonics was operational by the late Archean. If the Isua rocks are indeed a mélange, then such tectonics may have commenced as early as 3.8 Ga. This conclusion is reinforced by the occurrence of Phanerozoic-style ophiolites with boninitic volcanism in rocks ranging in age from 3.0 to 2.8 Ga (Puchtel, 2004; Shchipansky et al., 2004). This is consistent with the proposal that Archean greenstone belts represent primitive island arc terranes (Grove et al., 1999; Parman et al., 2001, 2004; Parman and Grove, 2004).

The time period between 3.8 Ga and 3.0 Ga appears to represent a transition from Hadean-style tectonics to late Archean Phanerozoic-style plate tectonics. There are no clear examples of Phanerozoic-style accretionary complexes or ophiolites older than 3.0 Ga, but some mid-Archean greenstone belts resemble primitive island arc terranes (e.g., Grove et al., 1999). One scenario for how this transition may have occurred is shown in Figure 7: greenstone-granite terranes accumulate at triple junctions between convergent plates (Fig. 7A); waning plume volcanism on one plate allows the adjacent plate to expand as the convergence zone shifts toward the waning plume (Fig. 7B); finally, the expanding plate begins to sink under the hotter, more buoyant shrinking plate to form an asymmetric convergent boundary (Fig. 7C). The expanding plate will be denser because it is farther from its parent plume and because thinning of the overlying crust on the expanding plate will enhance cooling and thickening of its subjacent mantle lithosphere. The onset of asymmetric convergence will induce passive upwelling of the asthenosphere along linear rifts, transforming the radially expanding plumes into ridge-centered plumes on oceanic spreading centers.

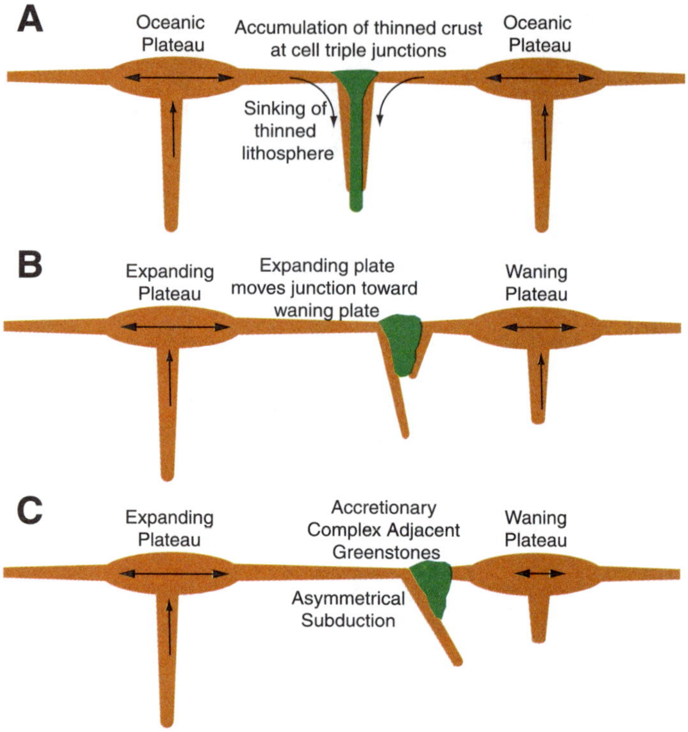

Figure 7. Schematic model for transition from Hadean-style tectonics to late Archean Phanerozoic-style plate tectonics: (A) Greenstone-granite terranes accumulate at triple junctions between convergent plates; (B) waning plume volcanism on one plate allows the adjacent plate to expand as the convergence zone shifts toward the waning plume; (C) finally, the expanding plate begins to sink under the hotter, more buoyant shrinking plate to form an asymmetrical convergent boundary. The expanding plate (left side) is denser because it is farther from its parent plume and because thinning of the overlying crust on the expanding plate enhances cooling and thickening of its subjacent mantle lithosphere. The onset of asymmetrical convergence (C) will induce passive upwelling of the asthenosphere along linear rifts, transforming the radially expanding plumes into ridge-centered plumes on oceanic spreading centers.

The accumulating evidence for Phanerozoic-style plate tectonics during the late Archean implies that cooling of Earth's mantle during the Hadean, subsequent to the Tartarean magma ocean and core formation, was relatively efficient and rapid. This efficient heat loss may have been driven in part by basin-forming impacts during the terminal bombardment around 3.9–3.8 Ga. The resulting decrease in geothermal gradient and development of a stable thermal boundary layer on the surface allowed the accumulation of sufficient liquid water to cool oceanic lithosphere hydrothermally and promote Phanerozoic-style asymmetric subduction, passive upwelling of asthenosphere, and the formation of accretionary subduction complexes that mixed arc-derived clastic sediments with exotic fragments of oceanic crust.

The loss of heat in response to this continuous thinning of the thermal boundary layer during vigorous convection was apparently rapid enough to cool Earth's surface below 100 °C by ca. 3.8 Ga when we first observe pillow lava and water-laid sediments (e.g., Komiya et al., 1999; Bolhar et al., 2004) and possibly sooner. Wilde et al. (2001) and Mojzsis et al. (2001) have argued that the oxygen isotopic compositions in some primordial detrital zircons require liquid water soon after planetary accretion, as early as 4.4 Ga. Valley et al. (2002) use these same data to argue for a cool early Earth with "continental" type crust by 4.4 Ga; however, they also suggest formation in a "tectonic environment similar to Iceland today" (Valley et al., 2002, p. 352). Since Iceland may represent our best analogue for plume-related oceanic plateaux in the Hadean, this tectonic setting is consistent with Hadean crustal processes discussed earlier. And studies of zircons from lunar regolith breccias show that granitic rocks need not be abundant for zircon to be preserved and concentrated (Meyer et al., 1996).

CONCLUSIONS

Phanerozoic-style plate tectonics may be recognized in Proterozoic and Archean rock assemblages by the occurrence of (1) oceanic crust or ophiolites, (2) island arc volcanic or plutonic suites, or (3) accretionary complexes formed in asymmetric subduction zones. However, rock assemblages similar to those formed at mid-ocean ridges or in island arcs may also form during Hadean-style convective overturn. In contrast, subduction zone accretionary complexes result from the physical juxtaposition of rock assemblages formed in different tectonic settings, making them unique to Phanerozoic-style plate tectonics. Recognition of these assemblages in the Archean rock record is thus *prima facie* evidence for Phanerozoic-style plate tectonics. This evidence is reinforced by the occurrence of boninites in the 2.8 Ga North Karelian ophiolites (Shchipansky et al., 2004) and by Phanerozoic-style island arc assemblages in the 2.7 Ga Slave province (Corcoran et al., 2004).

The increasing evidence for Archean subduction accretionary complexes implies a fundamental change in the thermal structure of the planet between 3.0 and 3.8 Ga that resulted in a mantle potential temperature only slightly higher than that observed today. If the Isua Supergroup is an accretionary complex and marks the onset of rigid lithosphere like today's, the transition from Hadean-style thermal convection driven by heating from below to Phanerozoic-style tectonics driven by cooling and sinking of lithospheric slabs appears to be coincident with the end of the late meteorite bombardment at ca. 3.8 Ga (Ryder, 2000). An older transition may be suggested by oxygen isotopes in primordial zircons, but evidence for a fractionated mantle at 4.4 Ga is debated (e.g., Amelin et al., 1999; Peck et al., 2001). The presence of liquid water on Earth's surface does not require lower mantle potential temperatures, but it does limit how high those temperatures may have been.

ACKNOWLEDGMENTS

This research was supported by National Science Foundation (NSF) grants EAR0337334 (Archean mélange) and EAR9805159 (high-pressure metamorphism in the southern Appalachians). Additional funding was provided by NSF (Coast Ranges of California), the United States Geological Survey (USGS) EdMap program (California and South Carolina), and SCUREF (South Carolina Universities Research Foundation) Task 170 (basement subcrop in the southern Appalachians). Discussions with the participants and leaders of the Field Forum on Processes on the Early Earth helped clarify my thinking on Archean processes in general and greenstone belts in particular, and detailed reviews by N.T. Arndt and T.M. Kusky challenged many of my arguments and forced me to state my case more clearly as a result. Nonetheless, the final conclusions are my own responsibility.

REFERENCES CITED

Adams, M.G., and Trupe, C.H., 1997, Conditions and timing of metamorphism in the Blue Ridge thrust complex, northwestern North Carolina and eastern Tennessee, *in* Stewart, K.G., Adams, M.G., and Trupe, C.H., eds., Paleozoic Structure, Metamorphism, and Tectonics of the Blue Ridge of Western North Carolina: Durham, North Carolina, Duke University, Carolina Geological Society 1997 Field Trip Guidebook, p. 33–48.

Adams, M.G., Stewart, K.G., Trupe, C.H., and Willard, R.A., 1995, Tectonic significance of high-pressure metamorphic rocks and dextral strike-slip faulting in the southern Appalachians, *in* Hibbard, J., van Staal, C., Cawood, P., and Colman-Sadd, S., eds., New Perspectives in the Appalachian-Caledonian Orogen: Geological Association of Canada Special Paper 41, p. 21–42.

Amelin, Y., Lee, D.C., Halliday, A.N., and Pidgeon, R.T., 1999, Nature of the Earth's earliest crust from hafnium isotopes in single detrital zircons: Nature, v. 399, p. 252–255, doi: 10.1038/20426.

Anderson, D., 2001, Top-down tectonics?: Science, v. 293, p. 2016–2018, doi: 10.1126/science.1065448.

Armstrong, R.A., Compston, W., de Witt, M.J., and Williams, I.S., 1990, The stratigraphy of the 3.5–3.2 Ga Barberton Greenstone Belt revisited: A single zircon ion microprobe study: Earth and Planetary Science Letters, v. 101, p. 90–106, doi: 10.1016/0012–821X(90)90127-J.

Arndt, N.T., 1994, Archean komatiites, *in* Condie, K.C., ed., Archean Crustal Evolution: Amsterdam, Elsevier, p. 11–44.

Arndt, N.T., 2004, Komatiites, kimberlites and boninites: Journal of Geophysical Research, v. 108, no. B6, p. 2293, doi: 10.1029/2002JB002157.

Arndt, N.T., Albarede, F., and Nisbet, E.G., 1997a, Mafic and ultramafic magmatism, *in* de Witt, M., and Ashwal, L.D., eds., Greenstone Belts: Oxford Monographs in Geology and Geophysics 35, p. 233–254.

Arndt, N.T., Kerr, A.C., and Tarney, J., 1997b, Dynamic melting in plume heads: The formation of Gorgona komatiites and basalts: Earth and Planetary Science Letters, v. 146, p. 289–301, doi: 10.1016/S0012–821X(96)00219–1.

Arndt, N.T., Albarède, F., Cheadle, M., Ginibre, C., Herzberg, C., Jenner, G., Chauvel, C., and Lahaye, Y., 1998, Were komatiites wet?: Geology, v. 26, p. 739–742, doi: 10.1130/0091–7613(1998)026<0739:WKW>2.3.CO;2.

Bailey, E.H., Irwin, W.P., and Jones, D.L., 1964, Franciscan and related rocks, and their significance in the geology of western California: Bulletin 183—California: Division of Mines and Geology, v. 183, p. 1–177.

Barth, M.G., Foley, S.F., and Horn, I., 2002, Partial melting in Archean subduction zones: Constraints from experimentally determined trace element partition coefficients between eclogitic minerals and tonalitic melts under upper mantle conditions: Precambrian Research, v. 113, no. 3-4, p. 323–340, doi: 10.1016/S0301–9268(01)00216–9.

Becker, D.G., and Cloos, M., 1985, Mélange diapirs into the Cambria Slab; a Franciscan trench slope basin near Cambria, California: The Journal of Geology, v. 93, no. 2, p. 101–110.

Bercovici, D., 2003, The generation of plate tectonics from mantle convection: Earth and Planetary Science Letters, v. 205, p. 107–121, doi: 10.1016/S0012–821X(02)01009–9.

Bickle, M.J., 1978, Heat loss from the Earth: A constraint on Archaean tectonics from the relation between geothermal gradients and the rate of heat production: Earth and Planetary Science Letters, v. 40, p. 301–315, doi: 10.1016/0012–821X(78)90155–3.

Bickle, M.J., Nisbet, E.G., and Martin, A., 1994, Archean greenstone belts are not oceanic crust: The Journal of Geology, v. 102, p. 121–138.

Blake, M.C., Jr., and Jones, D.L., 1974, Origin of Franciscan mélanges in northern California, *in* Modern and Ancient Geosynclinal Sedimentation; Problems of palinspastic restoration: Tulsa, Oklahoma, Society of Economic Paleontologists and Mineralogists Special Publication 19, p. 345–357.

Blake, M.C., Jr., and Wentworth, C.M., 1999, Structure and metamorphism of the Franciscan Complex, Mt. Hamilton area, Northern California: International Geology Review, v. 41, no. 5, p. 417–424.

Blake, M.C., Jr., Jayko, A.S., and Howell, D.G., 1982, Sedimentation, metamorphism and tectonic accretion of the Franciscan assemblage of Northern California, *in* Leggett, J., ed., Trench-forearc geology; sedimentation and tectonics on modern and ancient active plate margins: Geological Society [London] Special Publication 10, p. 433–438.

Blake, M.C., Jr., Jayko, A.S., and McLaughlin, R.J., 1985, Tectonostratigraphic terranes of the Northern Coast Ranges, California, *in* Howell, D.G., ed., Tectonostratigraphic terranes of the Circum-Pacific Region: Houston, Texas, Circum-Pacific Council for Energy and Mineral Resources, p. 159–171.

Bolhar, R., Kamber, B.S., Moorbath, S., Fedo, C.M., and Whitehouse, M.J., 2004, Characterisation of early Archean chemical sediments by trace element signatures: Earth and Planetary Science Letters, v. 222, p. 43–60, doi: 10.1016/j.epsl.2004.02.016.

Bowring, S.A., and Williams, I.A., 1999, Priscoan (4.00–4.03 Ga) orthogneisses from northwestern Canada: Contributions to Mineralogy and Petrology, v. 134, p. 3–16, doi: 10.1007/s004100050465.

Bowring, S.A., Housh, T.B., and Isachsen, C.E., 1990, The Acasta gneisses: Remnant of Earth's early crust, *in* Newsome, H., and Jones, J., eds., Origin of the Earth: Oxford, UK, Oxford University Press, p. 319–343.

Bradley, D.C., Kusky, T.M., Haeussler, P., Rowley, D.C., Goldfarb, R., and Nelson, S., 2003, Geologic signature of early ridge subduction in the accretionary wedge, forearc basin, and magmatic arc of south-central Alaska, *in* Sisson, V.B., Roeske, S., and Pavlis, T.L., eds., Geology of a transpressional orogen developed during a ridge-trench interaction along the North Pacific Margin: Geological Society of America Special Paper 371, p. 19–50.

Bridgwater, D., Keto, L., McGregor, V.R., and Myers, J.S., 1976, Archaean gneiss complex of Greenland, *in* Escher, A., and Watt, W.S., eds., Geology of Greenland: Copenhagen, Grønlands Geologiske Undersøgelse, p. 18–75.

Bryant, B., 1988a, Geology of the Farmington Canyon Complex, Wasatch Mountains, Utah: U.S. Geological Survey Professional Paper 1476, 54 p.

Bryant, B., 1988b, Evolution and early Proterozoic history of the margin of the Archean continent in Utah, *in* Ernst, W.G., ed., Metamorphism and crustal evolution of the Western United States, Rubey Volume 7: Englewood Cliffs, New Jersey, Prentice-Hall, p 431–445.

Byrne, T., 1984, Early deformation in mélange terranes of the Ghost Rocks formation, Kodiak Islands, Alaska, *in* Raymond, L.A., ed., Mélanges: Their nature, origin, and significance: Geological Society of America Special Paper 198, p. 21–51.

Calvert, A.J., and Ludden, J.N., 1999, Archean continental assembly in the southeastern Superior province of Canada: Tectonics, v. 18, no. 3, p. 412–429, doi: 10.1029/1999TC900006.

Campbell, I.H., and Jarvis, G.T., 1984, Mantle convection and early crustal evolution: Precambrian Research, v. 26, p. 15–56, doi: 10.1016/0301–9268(84)90016–0.

Campbell, I.H., Hill, R.E.T., and Griffiths, R.W., 1989, Melting in an Archaean mantle plume: Heads it's basalts, tails it's komatiites: Nature, v. 339, p. 697–699, doi: 10.1038/339697a0.

Carlson, R.L., 1981, Boundary forces and plate tectonics: Geophysical Research Letters, v. 8, p. 958–961.

Cavosie, A., and Selverstone, J., 2003, Early Proterozoic oceanic crust in the northern Colorado Front Range: Implications for crustal growth and initiation of basement faults: Tectonics, v. 22, p. 1–23.

Cloos, M., 1982, Flow mélanges; numerical modeling and geologic constraints on their origin in the Franciscan subduction complex, California: Geological Society of America Bulletin, v. 93, no. 4, p. 330–344, doi: 10.1130/0016–7606(1982)93<330:FMNMAG>2.0.CO;2.

Cloud, P., 1972, A working model of the primitive Earth: American Journal of Science, v. 272, no. 6, p. 537–548.

Cole, R.B., and Basu, A.R., 1995, Nd-Sr isotopic geochemistry and tectonics of ridge subduction and middle Cenozoic volcanism in western California: Geological Society of America Bulletin, v. 107, no. 2, p. 167–179, doi: 10.1130/0016–7606(1995)107<0167:NSIGAT>2.3.CO;2.

Coleman, R.G., 2000, Prospecting for ophiolites along the California continental margin, *in* Dilek, Y., Moores, E., Elthon, D., and Nicolas, A., eds., Ophiolites and oceanic crust: Geological Society of America Special Paper 349, p. 351–364.

Connelly, J.N., and Ryan, B., 1996, Late Archean evolution of the Nain Province, Nain, Labrador: Imprint of a collision: Canadian Journal of Earth Sciences, v. 33, no. 9, p. 1325–1342.

Corcoran, P.L., and Dostal, J., 2001, Development of an ancient back-arc basin overlying continental crust: The Archean Peltier Formation, Northwest Territories, Canada: The Journal of Geology, v. 109, p. 329–348, doi: 10.1086/319976.

Corcoran, P.L., Mueller, W.U., and Kusky, T.M., 2004, Inferred ophiolites in the Archean Slave Craton, *in* Kusky, T.M., ed., Precambrian Ophiolites and Related Rocks, Developments in Precambrian Geology, v. 13: Amsterdam, Elsevier, p. 363–404.

Cowan, D.S., 1978, Origin of blueschist-bearing chaotic rocks in the Franciscan Complex, San Simeon, California: Geological Society of America Bulletin, v. 89, no. 9, p. 1415–1423, doi: 10.1130/0016–7606(1978)89<1415:OOBCRI>2.0.CO;2.

Cowan, D.S., 1985, Structural styles in Mesozoic and Cenozoic mélanges in the western Cordillera of North America: Geological Society of America Bulletin, v. 96, p. 451–462, doi: 10.1130/0016–7606(1985)96<451:SSIMAC>2.0.CO;2.

Coward, M.P., Jan, M.Q., Rex, D., Tarney, J., Thirwall, M., and Windley, B.F., 1982, Geotectonic framework of Himalaya of northern Pakistan: Journal of the Geological Society of London, v. 139, p. 299–308.

Dann, J.C., 1991, Early Proterozoic ophiolite, central Arizona: Geology, v. 19, p. 590–593, doi: 10.1130/0091–7613(1991)019<0590:EPOCA>2.3.CO;2.

Dann, J.C., 2004, The 1.73 Ga Payson ophiolite, Arizona, USA, *in* Kusky, T.M., ed., Precambrian Ophiolites and Related Rocks, Developments in Precambrian Geology, v. 13: Amsterdam, Elsevier, p. 73–94.

De Ronde, C.E.J., and de Wit, M.J., 1994, Tectonic history of the Barberton greenstone belt, South Africa: 490 million years of Archean crustal evolution: Tectonics, v. 13, no. 4, p. 983–1005, doi: 10.1029/94TC00353.

de Wit, M.J., 1998, On Archean granites, greenstones, cratons and tectonics: Does the evidence demand a verdict?: Precambrian Research, v. 91, no. 1-2, p. 181–226, doi: 10.1016/S0301–9268(98)00043–6.

de Wit, M.J., 2004, Archean greenstone belts do contain fragments of ophiolites, *in* Kusky, T.M., ed., Precambrian Ophiolites and Related Rocks, Developments in Precambrian Geology, v. 13: Amsterdam, Elsevier, p. 599–614.

de Wit, M.J., Hart, R.J., and Hart, R.A., 1987, The Jamestown ophiolite complex, Barberton mountain belt: A section through 3.5 Ga ocean crust: Journal of African Earth Sciences, v. 6, p. 681–730, doi: 10.1016/0899–5362(87)90007–8.

de Wit, M.J., Roering, C., Hart, R.J., Armstrong, R.A., de Ronde, C.E.J., Green, R.W.E., Tredoux, M., Peberdy, E., and Hart, R.A., 1992, Formation of an Archaean continent: Nature, v. 357, p. 553–562, doi: 10.1038/357553a0.

Dennis, A.J., Shervais, J.W., and Secor, D.T., Jr., 2000, Newberry, South Carolina Eclogite: Structural setting and style of occurrence, *in* Wyeth, A., ed., A Compendium of Field Trips of South Carolina Geology: Columbia, South Carolina Department of Natural Resources, Geological Survey, p. 29–38.

Dennis, A.J., Shervais, J.W., Mauldin, J., and Maher, H.D., Jr., 2004, Petrology and geochemistry of Neoproterozoic volcanic arc terranes beneath the Atlantic Coastal Plain, Savannah River Site, South Carolina: Geological Society of America Bulletin, v. 116, p. 572–593, doi: 10.1130/B25240.1.

Desrochers, J.-P., Hubert, C., Ludden, J.N., and Pilote, P., 1993, Accretion of Archean oceanic plateau fragments in the Abitibi Greenstone Belt, Canada: Geology, v. 21, p. 451–454, doi: 10.1130/0091–7613(1993)021 <0451:AOAOPF>2.3.CO;2.

Dirks, P.H.G.M., and Jelsma, H.A., 1998, Horizontal accretion and stabilization of the Archean Zimbabwe Craton: Geology, v. 26, p. 11–14, doi: 10.1130/0091–7613(1998)026<0011:HAASOT>2.3.CO;2.

Drummond, M., and Defant, M., 1990, A model for trondhjemite-tonalite-dacite genesis and crustal growth via slab melting: Archean to modern comparison: Journal of Geophysical Research, v. 95, no. B13, p. 21,503–21,521.

Ernst, W.G., 1993, Metamorphism of Franciscan tectonostratigraphic assemblage, Pacheco Pass area, east-central Diablo Range, California Coast Ranges: Geological Society of America Bulletin, v. 105, no. 5, p. 618–636, doi: 10.1130/0016–7606(1993)105<0618:MOFTAP>2.3.CO;2.

Fedo, C.M., 2000, Setting and origin for problematic rocks from the >3.7 Ga Isua greenstone belt, southern West Greenland, Earth's oldest coarse clastic sediments: Precambrian Research, v. 101, p. 69–78.

Fruehn, J., von Huene, R., and Fisher, M.A., 1999, Accretion in the wake of terrane collision; the Neogene accretionary wedge off Kenai Peninsula, Alaska: Tectonics, v. 18, no. 2, p. 263–277.

Fryer, P., Lockwood, J.P., Becker, N., Phipps, S., and Todd, C.S., 2000, Significance of serpentine mud volcanism in convergent margins, *in* Dilek, Y., Moores, E.M., Don, M.E., and Adolphe, N., eds., Ophiolites and oceanic crust: New insights from field studies and the Ocean Drilling Program: Geological Society of America Special Paper 349, p. 35–51.

Grove, T.L., and Parman, S.W., 2004, Thermal evolution of the Earth as recorded by komatiites: Earth and Planetary Science Letters, v. 219, p. 173–187, doi: 10.1016/S0012–821X(04)00002–0.

Grove, T.L., Parman, S.W., and Dann, J.C., 1999, Conditions of magma generation for Archean komatiites from the Barberton Mountainland, South Africa, *in* Fei, Y., Bertka, C.M., and Mysen, B.O., eds., Mantle petrology: Field observations and high-pressure experimentation; A tribute to Francis R. (Joe) Boyd: Geochemical Society Special Publication 6, p. 155–167.

Hager, B., and O'Connell, R., 1981, A simple global model of plate dynamics and mantle convection: Journal of Geophysical Research, v. 86, p. 4843–4867.

Hamilton, W.B., 1979, Tectonics of the Indonesian region: U.S. Geological Survey Professional Paper 1078, 345 p.

Hamilton, W.B., 1998, Archean magmatism and deformation were not products of plate tectonics: Precambrian Research, v. 91, no. 1-2, p. 143–179, doi: 10.1016/S0301–9268(98)00042–4.

Hawkins, J.W., Bloomer, S.H., Evans, C.A., and Melchior, J.T., 1984, Evolution of intra-oceanic arc-trench systems, *in* Carlson, R.L., and Kobayashi, K., eds., Geodynamics of back-arc regions: Tectonophysics, p. 175–205.

Hedge, C.E., Sacey, J.S., and Bryant, B., 1983, Geochronology of the Farmington Canyon complex, Wasatch Mountains, Utah, *in* Miller, D.M., ed., Tectonic and stratigraphic studies of the eastern Great Basin: Geological Society of America Memoir 157, p. 37–44.

Helmstaedt, H., and Schulze, D.J., 1989, Southern African kimberlites and their mantle sample: Implications for Archean tectonics and lithosphere evolution, *in* Boyd, F.R., and Meyer, H.O.A., eds., Kimberlites and Related Rocks: Carleton, Blackwell, p. 358–368.

Hill, R.I., 1991, Starting plumes and continental breakup: Earth and Planetary Science Letters, v. 104, p. 398–416, doi: 10.1016/0012–821X(91) 90218–7.

Hofmann, A., and Kusky, T., 2004, The Belingwe Greenstone Belt: Ensialic or Oceanic?, *in* Kusky, T.M., ed., Precambrian Ophiolites and Related Rocks, Developments in Precambrian Geology, v. 13: Amsterdam, Elsevier, p. 487–538.

Hopson, C.A., and Pessagno, E.A., 2005, Tehama-Colusa Serpentinite mélange: A remnant of Franciscan Jurassic oceanic lithosphere: Northern California, International Geology Review, v. 47, p. 65–100.

Hsü, K.J., 1966, Franciscan rocks of the Santa Lucia range, California, and the Argille scagliose of the Apennines, Italy; a comparison in style of deformation: Geological Society of America Special Paper 87, p. 210–211.

Hsü, K.J., 1968, Principles of mélanges and their bearing on the Franciscan-Knoxville paradox: Geological Society of America Bulletin, v. 79, no. 8, p. 1063–1074.

Huang, X., Jianghai, L., Kusky, T.M., and Chen, Z., 2004, Microstructures of the Zunhua 2.50 Ga Podiform Chromite, North China Craton and implications for the deformation and rheology of the Archean oceanic lithospheric mantle, *in* Kusky, T.M., ed., Precambrian Ophiolites and Related Rocks, Developments in Precambrian Geology, v. 13: Amsterdam, Elsevier, p. 321–338.

Huot, F., and Maury, R.C., 2002, The Round Mountain serpentinite mélange, northern Coast Ranges of California: An association of backarc and arc-related tectonic units: Geological Society of America Bulletin, v. 114, no. 1, p. 109–123.

Huson, R., Kusky, T.M., and Li, J.H., 2004, Geochemical and petrographic characteristics of the Central Belt of the Archean Dongwanzi Ophiolite Complex, *in* Kusky, T.M., ed., Precambrian Ophiolites and Related Rocks, Developments in Precambrian Geology, v. 13: Amsterdam, Elsevier, p. 283–322.

Isozaki, Y., 1997, Jurassic accretion tectonics of Japan: The Island Arc, v. 6, p. 25–51.

Isozaki, Y., and Blake, M.C., 1994, Biostratigraphic constraints on formation and timing of accretion in a subduction complex, an example from the Franciscan complex in northern California: The Journal of Geology, v. 102, p. 283–296.

Isozaki, Y., Maruyama, S., and Furuoka, F., 1990, Accreted oceanic materials in Japan: Tectonophysics, v. 181, p. 179–205, doi: 10.1016/0040–1951(90) 90016–2.

Johnson, C.M., and O'Neil, J.R., 1984, Triple junction magmatism: A geochemical study of Neogene volcanic rocks in western California: Earth and Planetary Science Letters, v. 71, p. 241–262, doi: 10.1016/0012–821X(84)90090–6.

Jurdy, D., and Stefanick, M., 1991, The forces driving the plates: Constraints from kinematics and stress observations: Philosophical Transactions of the Royal Society of London, ser. A, v. 337, p. 127–138.

Kamo, S.L., and Davis, D.W., 1994, Reassessment of Archean crustal development in the Barberton Mountain Land, South Africa, based on U-Pb dating: Tectonics, v. 13, no. 1, p. 167–192, doi: 10.1029/93TC02254.

Karl, S.M., 1984, Sedimentologic, diagenetic, and geochemical analysis of upper Mesozoic ribbon cherts from the Franciscan assemblage at the Marin Headlands, California, *in* Blake, C., Jr., ed., Franciscan geology of Northern California: Field Trip Guidebook 43: Los Angeles, Pacific Section, Society of Economic Paleontologists and Mineralogists, p. 71–88.

Komiya, T., Maruyama, S., Masuda, T., Nohda, S., Hayashi, M., and Okamoto, K., 1999, Plate tectonics at 3.8–3.7 Ga: Field evidence from the Isua accretionary complex, Southern West Greenland: The Journal of Geology, v. 107, p. 515–554, doi: 10.1086/314371.

Komiya, T., Maruyama, S., Hirata, T., Yurimoto, H., and Nohda, S., 2004, Geochemistry of the oldest MORB and OIB in the Isua supracrustal belt, southern west Greenland: Implications for the composition and temperature of early Archean upper mantle: The Island Arc, v. 13, p. 47, doi: 10.1111/j.1440–1738.2003.00416.x.

Kontinen, A., 1987, An early Proterozoic ophiolite: The Jourma mafic-ultramafic complex, northeastern Finland: Precambrian Research, v. 35, p. 313–341, doi: 10.1016/0301–9268(87)90061–1.

Kröner, A., 1985, Evolution of the Archean continental crust: Annual Review of Earth and Planetary Sciences, v. 13, p. 49–74, doi: 10.1146/annurev.ea.13.050185.000405.

Kröner, A., and Todt, W., 1988, Single zircon dating constraining the maximum age of the Barberton Greenstone Belt, Southern Africa: Journal of Geophysical Research, v. 93, no. b12, p. 15,329–15,337.

Kröner, A., Byerly, G.R., and Lowe, D.R., 1991, Chronology of early Archaean granite-greenstone evolution in the Barberton Mountain Land, South Africa, based on precise dating by single zircon evaporation: Earth and Planetary Science Letters, v. 103, p. 41–54, doi: 10.1016/0012–821X(91)90148-B.

Kukowski, N., Schillhorn, T., Huhn, K., von Rad, U., Husen, S., and Flueh, E.R., 2001, Morphotectonics and mechanics of the central Makran accretionary wedge off Pakistan: Marine Geology, v. 173, p. 1–19, doi: 10.1016/S0025–3227(00)00167–5.

Kusky, T.M., 1998, Tectonic setting and terrane accretion of the Archean Zimbabwe craton: Geology, v. 26, p. 163–166, doi: 10.1130/0091–7613(1998)026<0163:TSATAO>2.3.CO;2.

Kusky, T.M., and Kidd, W.S.F., 1992, Remnants of an Archean oceanic plateau, Belingwe greenstone belt, Zimbabwe: Geology, v. 20, p. 43–46, doi: 10.1130/0091–7613(1992)020<0043:ROAAOP>2.3.CO;2.

Kusky, T.M., and Polat, A., 1999, Growth of granite-greenstone terranes at convergent margins, and stabilization of Archean cratons: Tectonophysics, v. 305, p. 43–73, doi: 10.1016/S0040–1951(99)00014–1.

Kusky, T.M., Bradley, D.C., Haeussler, P., and Karl, S., 1997a, Controls on accretion of flysch and mélange belts at convergent margins: Evidence from the Chugach Bay thrust and Iceworm mélange, Chugach Terrane, Alaska: Tectonics, v. 16, p. 855–878, doi: 10.1029/97TC02780.

Kusky, T.M., Bradley, D.C., and Haeussler, P., 1997b, Progressive deformation of the Chugach accretionary complex, Alaska, during a Paleogene ridge-trench encounter: Journal of Structural Geology, v. 19, p. 139–157, doi: 10.1016/S0191–8141(96)00084–3.

Kusky, T.M., Li, J.-H., and Tucker, R.D., 2001, The Archean Dongwanzi ophiolite complex, North China craton: 2.505-billion-year-old oceanic crust and mantle: Science, v. 292, no. 5519, p. 1142–1145, doi: 10.1126/science.1059426.

Kusky, T.M., Bradley, D.C., Donley, D.T., Rowley, D., and Haeussler, P., 2003, Controls on intrusion of near-trench magmas of the Sanak-Baranof belt, Alaska, during Paleogene ridge subduction, and consequences for forearc evolution, *in* Sisson, V.B., Roeske, S., and Pavlis, T.L., eds., Geology of a transpressional orogen developed during a ridge-trench interaction along the North Pacific Margin: Geological Society of America Special Paper 371, p. 269–292.

Kusky, T.M., Ganley, R., Lytwyn, J., and Polat, A., 2004a, The Resurrection Peninsula Ophiolite, Mélange and Accreted Flysch Belts of Southern Alaska as an Analog for Trench-Forearc Systems in Precambrian Orogens, *in* Kusky, T.M., ed., Precambrian Ophiolites and Related Rocks, Developments in Precambrian Geology, v. 13: Amsterdam, Elsevier, p. 627–674.

Kusky, T.M., Jainghai, L., Glass, A., and Huang, X.N., 2004b, Origin and Emplacement of Archean Ophiolites of the Central Orogenic Belt, North China Craton, *in* Kusky, T.M., ed., Precambrian Ophiolites and Related Rocks, Developments in Precambrian Geology, v. 13: Amsterdam, Elsevier, p. 223–274.

Li, J.H., Kusky, T.M., and Huang, X., 2002, Archean podiform chromitites and mantle tectonites in ophiolitic mélange, North China Craton; a record of early oceanic mantle processes: GSA Today, v. 12, no. 7, p. 4–11, doi: 10.1130/1052–5173(2002)012<0004:APCAMT>2.0.CO;2.

Lowe, D.R., 1994, Accretionary history of the Archean Barberton Greenstone Belt (3.55–3.22 Ga), southern Africa: Geology, v. 22, p. 1099–1102, doi: 10.1130/0091–7613(1994)022<1099:AHOTAB>2.3.CO;2.

MacGregor, I.D., and Manton, W.I., 1986, Roberts Victor eclogites: Ancient oceanic crust: Journal of Geophysical Research, v. 91, p. 14,063–14,079.

MacPherson, G.J., Phipps, S.P., and Grossman, J.N., 1990, Diverse sources for igneous blocks in Franciscan mélanges, California Coast Ranges: The Journal of Geology, v. 98, no. 6, p. 845–862.

Martin, H., 1994, The Archean grey gneisses and the genesis of continental crust: *in* Condie, K.C., ed., Archean Crustal Evolution: Amsterdam, Elsevier, v. 11, p. 205–259.

McLaughlin, R.J., Blake, M.C., Jr., and Jones, D.L., 1982, Franciscan mélange in central northern California, Geological Survey research 1982: U.S. Geological Survey Professional Paper P1275, 70 p.

Meyer, C., Williams, I.S., and Compston, W., 1996, Uranium-lead ages for lunar zircons; evidence for a prolonged period of granophyre formation from 4.32 to 3.88 Ga: Meteoritics, v. 31, no. 3, p. 370–387.

Mojzsis, S.J., Harrison, T.M., and Pidgeon, R.T., 2001, Oxygen-isotope evidence from ancient zircons for liquid water at the Earth's surface 4,300 Myr ago: Nature, v. 409, p. 178–182, doi: 10.1038/35051557.

Moore, D.E., 1984, Metamorphic history of a high-grade blueschist exotic block from the Franciscan Complex, California: Journal of Petrology, v. 25, no. 1, p. 126–150.

Moore, D.E., and Blake, M.C., Jr., 1989, New evidence for polyphase metamorphism of glaucophane schist and eclogite exotic blocks in the Franciscan Complex, California and Oregon: Journal of Metamorphic Geology, v. 7, no. 2, p. 211–228.

Moore, G.F., and Karig, D.E., 1980, Structural geology of Nias Island, Indonesia: Implications for subduction zone tectonics: American Journal of Science, v. 280, p. 193–223.

Moore, J.C., Biju, D.B., Bergen, J.A., Blackinton, G., Claypool, G.E., Cowan, D.S., Duennebier, F., Guerra, R.T., Hemleben, C.H.J., Hussong, D., Marlow, M.S., Natland, J.H., Pudsey, C.J., Renz, G.W., Tardy-M, Willis, M.E., Wilson, D., and Wright, A.A., 1982, Offscraping and underthrusting of sediment at the deformation front of the Barbados Ridge; Deep Sea Drilling Project Leg 78A: Geological Society of America Bulletin, v. 93, p. 1065–1077.

Mueller, W.U., Corcoran, P.L., and Pickett, C., 2005, Mesoarchean Continental Breakup: Evolution and Inferences from the >2.8 Ga Slave Craton–Cover Succession, Canada: The Journal of Geology, v. 113, p. 23–45, doi: 10.1086/425967.

Murchey, B.L., 1984, Biostratigraphy and lithostratigraphy of chert in the Franciscan Complex, Marin Headlands, California, *in* Blake, C., Jr., ed., Franciscan geology of Northern California: Field Trip Guidebook 43, Pacific Section, SEPM: Los Angeles, California, Society of Economic Paleontologists and Mineralogists, Pacific Section, p. 51–70.

Myers, J.S., 2001, Protoliths of the 3.8–3.7 Ga Isua greenstone belt: West Greenland: Precambrian Research, v. 105, p. 129–141, doi: 10.1016/S0301–9268(00)00108-X.

Myers, J.S., 2004, Isua illusions: Illusive tectonic, sedimentary, volcanic and organic features of the >3.8 – >3.7 Ga Isua greenstone belt, south-west Greenland, *in* Eriksson, P., Altermann, W., Nelson, D., Mueller, W., and Catuneau, O., eds., The Precambrian Earth: Tempos and Events: Amsterdam, Elsevier, p. 66–74.

Neal, C.R., Taylor, L.A., Davidson, J.P., Holden, P., Halliday, A.N., Paces, J.B., Clayton, R.N., and Mayeda, T.K., 1990, Eclogites with oceanic crustal and mantle signatures from the Bellsbank kimberlite, South Africa, Part 2: Sr, Nd, and O isotope chemistry: Earth and Planetary Science Letters, v. 99, p. 362–379, doi: 10.1016/0012–821X(90)90140-S.

Nelson, S.T., Harris, R.A., Dorais, M.J., Heizler, M., Constenius, K., and Barnett, D.A., 2002, Basement complexes in the Wasatch Fault, Utah, provide new limits on crustal accretion: Geology, v. 30, p. 831–834, doi: 10.1130/0091–7613(2002)030<0831:BCITWF>2.0.CO;2.

Nisbet, E.G., and Fowler, M., 2004, The origin and earliest history of the Earth, *in* Holland, H., and Turekian, K., eds., Treatise of Geochemistry, Volume 8, Biogeochemistry: Amsterdam, Elsevier, p. 1–39.

Nisbet, E.G., Cheadle, M.J., Arndt, N.T., and Bickle, M.J., 1993, Constraining the potential temperature of the Archaean mantle: A review of the evidence from komatiites: Lithos, v. 30, p. 291–307, doi: 10.1016/0024–4937(93)90042-B.

Nutman, A.P., Friend, C.R.L., Kinny, P.D., and McGregor, V.R., 1993, Anatomy of an early Archean gneiss complex: 3900 to 3600 Ma crustal evolution in southern West Greenland: Geology, v. 21, p. 415–418, doi: 10.1130/0091–7613(1993)021<0415:AOAEAG>2.3.CO;2.

Nutman, A.P., McGregor, V.R., Friend, C.R.L., Bennett, V.C., and Kinny, P.D., 1996, The Itsaq Gneiss Complex of southern West Greenland; the world's most extensive record of early crustal evolution (3900–3600 Ma): Precambrian Research, v. 78, p. 1–39.

Nutman, A.P., Mojzsis, S.J., Friend, C.R.L., and McGregor, V.R., 1997, Recognition of >3850 Ma waterlaid sediments in West Greenland and their significance for the early Archean Earth: Geochimica et Cosmochimica Acta, v. 61, p. 2475–2484.

Nutman, A.P., Friend, C.R.L., and Bennett, V.C., 2001, Review of the oldest (4400–3600 Ma) geological record: Glimpses of the beginning: Episodes, v. 24, p. 93–101.

Nutman, A.P., Friend, C.R.L., Bennett, V.C., and McGregor, V.R., 2004, Dating of the Ameralik dyke swarms of the Nuuk district, southern West Greenland: Mafic intrusion events starting from c. 3510 Ma: Journal of the Geological Society of London, v. 161, p. 421–430.

Page, B.M., and Suppe, J., 1981, The Pliocene Lichi mélange of Taiwan; its plate-tectonic and olistostromal origin: American Journal of Science, v. 281, no. 3, p. 193–227.

Pan, Y., and Fleet, M.E., 1993, Polymetamorphism in the Archean Hemlo-Heron Bay greenstone belt, Superior Province; P-T variations and implications for tectonic evolution: Canadian Journal of Earth Sciences, v. 30, p. 985–996.

Parman, S.W., and Grove, T.L., 2004, Petrology and geochemistry of Barberton komatiites and basaltic komatiites: Evidence of Archean fore-arc magmatism, *in* Kusky, T.M., ed., Precambrian Ophiolites and Related Rocks, Developments in Precambrian Geology, v. 13: Amsterdam, Elsevier, p. 539–565.

Parman, S.W., Grove, T.L., and Dann, J.C., 2001, The production of Barberton komatiites in an Archean subduction zone: Geophysical Research Letters, v. 28, no. 13, p. 2513–2516, doi: 10.1029/2000GL012713.

Parman, S.W., Grove, T.L., Dann, J.C., and de Wit, M.J., 2004, A subduction origin for komatiite and cratonic lithospheric mantle: South African Journal of Geology, v. 107, no. 1-2, p. 107–118.

Pearce, J.A., 2003, Supra-subduction zone ophiolites; the search for modern analogues, *in* Dilek, Y., and Newcomb, S., eds., Ophiolite concept and the evolution of geological thought: Geological Society of America Special Paper 373, p. 269–293.

Pearce, J.A., Lippard, S.J., and Roberts, S., 1984, Characteristics and tectonic significance of supra-subduction zone ophiolites, *in* Kokelaar, B.P., and Howells, M.F., eds., Marginal basin geology: Geological Society [London] Special Publication 16, p. 74–94.

Peck, W.H., Valley, J.W., Wilde, S.A., and Graham, C.M., 2001, Oxygen isotope ratios and rare earth elements in 3.3 to 4.4 Ga zircons: Ion microprobe evidence for high 18O continental crust and oceans in the early Archean: Geochimica et Cosmochimica Acta, v. 65, no. 22, p. 4215–4229, doi: 10.1016/S0016–7037(01)00711–6.

Peltonen, P., and Kontinen, A., 2004, The Jormua Ophiolite: A mafic-ultramafic complex from an ancient ocean-continent transition zone, *in* Kusky, T.M., ed., Precambrian Ophiolites and Related Rocks, Developments in Precambrian Geology, v. 13: Amsterdam, Elsevier, p. 35–72.

Platt, J.P., 1986, Dynamics of orogenic wedges and the uplift of high-pressure metamorphic rocks: Geological Society of America Bulletin, v. 97, no. 9, p. 1037–1053, doi: 10.1130/0016–7606(1986)97<1037:DOOWAT>2.0.CO;2.

Platt, J.P., Leggett, J.K., Young, J., Raza, H., and Alam, S., 1985, Large-scale sediment underplating in the Makran accretionary prism: Southwest Pakistan: Geology, v. 13, p. 507–511.

Polat, A., and Kerrich, R., 1999, Formation of an Archean tectonic mélange in the Schreiber-Hemlo greenstone belt, Superior Province, Canada: Implications for Archean subduction-accretion process: Tectonics, v. 18, p. 733–755, doi: 10.1029/1999TC900032.

Polat, A., and Kerrich, R., 2001, Magnesian andesites, Nb-enriched basalt-andesites, and adakites from late-Archean 2.7 Ga Wawa greenstone belts, Superior Province, Canada: Implications for late Archean subduction zone petrologenetic processes: Contributions to Mineralogy and Petrology, v. 141, no. 1, p. 36–52.

Polat, A., and Kerrich, R., 2002, Nd-isotope systematics of ~2.7 Ga adakites, magnesian andesites, and arc basalts, Superior Province: Evidence for shallow crustal recycling at Archean subduction zones: Earth and Planetary Science Letters, v. 202, no. 2, p. 345–360, doi: 10.1016/S0012–821X(02)00806–3.

Polat, A., and Kerrich, R., 2004, Precambrian arc associations: Boninites, adakites, magnesian andesites, and Nb-enriched basalts, *in* Kusky, T.M., ed., Precambrian Ophiolites and Related Rocks, Developments in Precambrian Geology, v. 13: Amsterdam, Elsevier, p. 567–598.

Polat, A., Kerrich, R., and Wyman, D.A., 1998, The late Archean Schreiber-Hemlo and White River–Dayohessarah greenstone belts, Superior Province: collages of oceanic plateaus, oceanic arcs, and subduction-accretion complexes: Tectonophysics, v. 289, p. 295–326, doi: 10.1016/S0040–1951(98)00002-X.

Polat, A., Kerrich, R., and Wyman, D.A., 1999, Geochemical diversity in oceanic komatites and basalts from the late Archean Wawa greenstone belt, Superior Province, Canada: Trace element and Nd isotopic evidence for a heterogeneous mantle: Precambrian Research, v. 94, p. 139–173, doi: 10.1016/S0301–9268(98)00110–7.

Pollack, H.N., 1997, Thermal Characteristics of the Archaean, *in* deWit, M., and Ashwal, L.D., eds., Greenstone Belts: Oxford Monographs in Geology and Geophysics, v. 35, p. 223–232.

Prendergast, M.D., 2004, The Bulawayan Supergroup: A late Archaean passive margin-related large igneous province in the Zimbabwe craton: Journal of Geological Society of London, v. 161, p. 431–445.

Puchtel, I.S., 2004, 3.0 Ga Olondo Greenstone Belt in the Aldan Shield, E. Siberia, *in* Kusky, T.M., ed., Precambrian Ophiolites and Related Rocks, Developments in Precambrian Geology, v. 13: Amsterdam, Elsevier, p. 405–424.

Rapp, R.P., Shimizu, N., and Norman, M.D., 2003, Growth of early continental crust by partial melting of eclogite: Nature, v. 425, p. 605–609, doi: 10.1038/nature02031.

Richards, M., Duncan, R., and Courtillot, V., 1989, Flood basalts and hot-spot tracks: Plume heads and tails: Science, v. 246, p. 103–107.

Ridley, J.R., Vearncombe, J.R., and Jelsma, H.A., 1997, Relations between greenstone belts and associated granitoids, *in* deWit, M., and Ashwal, L.D., eds., Greenstone Belts: Oxford Monographs in Geology and Geophysics, v. 35, p. 376–397.

Robinson, P.T., and Malpas, J., 1990, The Troodos ophiolite: New perspective on its origin and emplacement, *in* Malpas, J., Moores, E., Panayiotou, A., and Xenophontos, C., eds., Ophiolites; oceanic crustal analogues: Proceedings of the symposium "Troodos 1987," Nicosia, Cyprus, Geological Survey Department, p. 13–26.

Rosing, M.T., Rose, N.M., Bridgwater, D., and Thomsen, H.S., 1996, Earliest part of Earth's stratigraphic record: A reappraisal of the >3.7 Ga Isua (Greenland) supracrustal sequence: Geology, v. 24, p. 43–46, doi: 10.1130/0091–7613(1996)024<0043:EPOESS>2.3.CO;2.

Ryder, G., 2000, Glass beads tell a tale of lunar bombardment: Science, v. 287, p. 1768–1769.

Secor, D.T., Snoke, A.W., Bramlett, K.W., Costello, O.P., and Kimbrell, O.P., 1986, Character of the Alleghanian orogeny in the southern Appalachians, Part I, Alleghanian deformation in the eastern Piedmont of South Carolina: Geological Society of America Bulletin, v. 97, p. 1314–1328.

Shah, M.T., and Shervais, J.W., 1999, The Dir-Utror metavolcanic sequence, Kohistan arc terrane, northern Pakistan: Journal of Asian Earth Sciences, v. 17, no. 4, p. 459–475, doi: 10.1016/S1367–9120(99)00009–7.

Shchipansky, A.A., Samsonov, A.V., Bibikova, E.V., Babarina, I.I., Konilov, A.N., Krylov, K.A., Slabunov, A.I., and Bogina, M.M., 2004, 2.8 Ga Boninite-Hosting partial suprasubduction zone ophiolite sequences from the North Karelian Greenstone Belt, NE Baltic Shield, Russia, *in* Kusky, T.M., ed., Precambrian Ophiolites and Related Rocks, Developments in Precambrian Geology, v. 13: Amsterdam, Elsevier, p. 425–486.

Shervais, J.W., 1990, Island arc and ocean crust ophiolites: Contrasts in the petrology, geochemistry, and tectonic style of ophiolite assemblages in the California Coast Ranges, *in* Malpas, J., Moores, E., Panayiotou, A., and Xenophontos, C., eds., Ophiolites: Oceanic Crustal Analogues: Nicosia, Cyprus, Geological Survey Department, Proceedings of the Symposium Troodos 1987, p. 507–520.

Shervais, J.W., 2001, Birth, death, and resurrection: The life cycle of supra-subduction zone ophiolites: Geochemistry, Geophysics, Geosystems, v. 2, no. 2000GC000080, 20,925 words, 8 fig., 3 tables.

Shervais, J.W., 2004, Is the southern Farmington Canyon complex a late Archean/early Proterozoic accretionary complex? *in* Reimold, W.U., and Hofmann, A., eds., Abstract volume, Field Forum on Processes on the Early Earth, Kaapvaal Craton, Geological Society of South Africa, Johannesburg, SA, 4–9 July 2004, p. 88–89.

Shervais, J.W., and Kimbrough, D.L., 1985, Geochemical evidence for the tectonic setting of the Coast Range ophiolite; a composite island arc-oceanic crust terrane in western California: Geology, v. 13, no. 1, p. 35–38, doi: 10.1130/0091-7613(1985)13<35:GEFTTS>2.0.CO;2.

Shervais, J.W., and Kimbrough, D.L., 1987, Alkaline and transitional subalkaline metabasalts in the Franciscan complex mélange, *in* Morris, E.M., and Pasteris, J.D., eds., Mantle metasomatism and alkaline magmatism: Geological Society of America Special Paper 215, p. 167–182.

Shervais, J.W., Taylor, L.A., Lugmair, G.W., Clayton, R.N., Mayeda, T.K., and Korotev, R.L., 1988, Early Proterozoic oceanic crust and the evolution of sub-continental mantle: Eclogites and related rocks from southern Africa: Geological Society of America Bulletin, v. 100, p. 411–423, doi: 10.1130/0016-7606(1988)100<0411:EPOCAT>2.3.CO;2.

Shervais, J.W., Dennis, A.J., McGee, J.J., and Secor, D.T., 2003, Deep in the heart of Dixie: Pre-Alleghanian eclogite and HP granulite metamorphism in the Carolina terrane, South Carolina, USA: Journal of Metamorphic Geology, v. 21, no. 1, p. 65–80, doi: 10.1046/j.1525-1314.2003.00416.x.

Shervais, J.W., Kimbrough, D.L., Renne, P.R., Hanan, B.B., Murchey, B., Snow, C.A., Zoglman-Schuman, M.M., and Beaman, J., 2004, Multi-stage origin of the Coast Range Ophiolite, California: Implications for the life cycle of supra-subduction zone ophiolites: International Geology Review, v. 46, no. 4, p. 289–315.

Shervais, J.W., Murchey, B., Kimbrough, D.L., Renne, P., and Hanan, B.B., 2005a, Radioisotopic and biostratigraphic age relations in the Coast Range Ophiolite, Northern California: Implications for the tectonic evolution of the Western Cordillera: Geological Society of America Bulletin, v. 117, no. 5/6, p. 633–653.

Shervais, J.W., Kolesar, P., and Andreasen, K., 2005b, A field and chemical study of serpentinization—Stonyford, California: Chemical flux and mass balance: International Geology Review, v. 47, p. 1–23.

Shirey, S.B., Gurney, J.J., Pearson, D.G., Harris, J.W., Wiechert, U., Carlson, R.W., Richardson, S.H., and Menzies, A., 2001, Archean emplacement of eclogitic components into the lithospheric mantle during formation of the Kaapvaal craton: Geophysical Research Letters, v. 28, no. 13, p. 2509–2512, doi: 10.1029/2000GL012589.

Stern, R.J., and Bloomer, S.H., 1992, Subduction zone infancy; examples from the Eocene Izu-Bonin-Mariana and Jurassic California arcs: Geological Society of America Bulletin, v. 104, no. 12, p. 1621–1636, doi: 10.1130/0016-7606(1992)104<1621:SZIEFT>2.3.CO;2.

Stern, R.J., Johnson, P.R., Kroner, A., and Yibas, B., 2004, Neoproterozoic ophiolites of the Arabian Shield: Field Relations and Structure, *in* Kusky, T.M., ed., Precambrian Ophiolites and Related Rocks, Developments in Precambrian Geology, v. 13: Amsterdam, Elsevier, p. 95–128.

Storey, M., Mahoney, J.J., Kroenke, L.W., and Saunders, A.D., 1991, Are oceanic plateaus sites of komatiite formation?: Geology, v. 19, p. 376–379, doi: 10.1130/0091-7613(1991)019<0376:AOPSOK>2.3.CO;2.

Valley, J.W., Peck, W.H., King, E.M., and Wilde, S.A., 2002, A cool early Earth: Geology, v. 30, p. 351–354, doi: 10.1130/0091-7613(2002)030<0351:ACEE>2.0.CO;2.

von Huene, R., and Scholl, D.W., 1991, Observations at convergent margins concerning sediment subduction, subduction erosion, and the growth of continental crust: Reviews of Geophysics, v. 29, p. 279–316.

Wakabayashi, J., 1999, The Franciscan; California's classic subduction complex, *in* Moores, E.M, Sloan, D., and Stout, D., eds., Classic Cordilleran concepts; a view from California: Geological Society America Special Paper 338, p. 111–121.

Walter, M.J., and Trønnes, R.G., 2004, Early Earth differentiation: Earth and Planetary Science Letters, v. 225, p. 253–269, doi: 10.1016/j.epsl.2004.07.008.

Wilde, S.A., Valley, J.W., Peck, W.H., and Graham, C.M., 2001, Evidence from detrital zircons for the existence of continental crust and oceans on the Earth 4.4 Gyr ago: Nature, v. 409, p. 175–178, doi: 10.1038/35051550.

Willard, R.A., and Adams, M.G., 1994, Newly discovered eclogite in the southern Appalachian orogen, northwestern North Carolina: Earth and Planetary Science Letters, v. 123, p. 61–70, doi: 10.1016/0012-821X(94)90257-7.

Williams, H., and Hatcher, R.D., Jr., 1983, Appalachian suspect terranes, *in* Hatcher, R.D., Jr., Williams, H., and Zietz, I., eds., Contributions to the tectonics and geophysics of mountain chains: Geological Society of America Memoir 158, p. 33–53.

Wyman, D.A., 1999, Paleoproterozoic boninites in an ophiolite-like setting, Trans-Hudson orogen, Canada: Geology, v. 27, p. 455–458, doi: 10.1130/0091-7613(1999)027<0455:PBIAOL>2.3.CO;2.

Wyman, D.A., and Kerrich, R., 2002, Formation of Archean continental lithospheric roots: The role of mantle plumes: Geology, v. 30, p. 543–546.

Wyman, D.A., Kerrich, R., and Polat, A., 2002, Assembly of Archean cratonic mantle lithosphere and crust: Plume-arc interaction in the Abitibi-Wawa subduction-accretion complex: Precambrian Research, v. 115, p. 37–62, doi: 10.1016/S0301-9268(02)00005-0.

Yonkee, W.A., Willis, G.C., and Doelling, H.H., 2000, Petrology of Precambrian rocks of the Farmington Canyon complex, Antelope Island, Utah, *in* King, J.K., and Willis, G.C., eds., The geology of Antelope Island, Davis County, Utah: Utah Geological and Mineral Survey, Miscellaneous Publication MP-00-1, 163 p.

Zegers, T.E., and van Keken, P.E., 2001, Middle Archean continent formation by crustal delamination: Geology, v. 29, p. 1083–1086, doi: 10.1130/0091-7613(2001)029<1083:MACFBC>2.0.CO;2.

Zhong, S., Zuber, M., Moresi, L., and Gurnis, M., 2000, Role of temperature-dependent viscosity and surface plates in spherical shell models of mantle convection: Journal of Geophysical Research, v. 105, p. 11,063–11,082, doi: 10.1029/2000JB900003.

Manuscript Accepted by the Society 4 October 2005

Printed in the USA

Geological Society of America
Special Paper 405
2006

A reevaluation of Archean intracratonic terrane boundaries on the Kaapvaal Craton, South Africa: Collisional suture zones?

Carl R. Anhaeusser[†]
Economic Geology Research Institute, School of Geosciences, University of the Witwatersrand, Johannesburg, South Africa

ABSTRACT

Increasing support for the adoption of plate-tectonic processes to account for Archean geotectonic events has led to the reevaluation of selected terranes in southern Africa in order to determine the viability of plate tectonics as a mechanism for terrane accretion and the growth of continents. Geotectonic reconstructions indicate that a number of Archean ultramafic complexes and associated volcanic rocks are developed in three linear zones on the Kaapvaal Craton that may represent the fossil traces of Archean sutures or oceanic crustal collisional zones. The ultramafic complexes consist primarily of massive and schistose serpentinitic and pyroxenitic rocks derived from dunite, harzburgite, lherzolite, pyroxenite (ortho- and clinopyroxenite), and lesser gabbroic bodies believed to have been derived from high-Mg, komatiitic-type parent magmas. The ultramafic complexes have previously been interpreted to have formed as a result of differentiation processes and were emplaced either magmatically or tectonically into the early Archean crust. An alternative view suggests that the ultramafic bodies may be likened to oceanic crust or "layered series" assemblages of Phanerozoic ophiolites. Serpentinized ultramafic components of ophiolites are, in turn, commonly encountered in orogenic belts where they mark the positions of suture zones.

The three areas where extraordinary developments of serpentinized ultramafic rocks are to be found on the Kaapvaal Craton include: (1) the northern flank of the Barberton greenstone belt; (2) the area south of the Murchison greenstone belt; and (3) the region extending across the southern portion of the Johannesburg Dome. The age of the ultramafic bodies has not been determined directly, but intrusive granitoid rocks suggest they are mostly older than ca. 3200 Ma and in some cases may be as old as ca. 3600 Ma. In each of the three regions discussed the proposed suture zones mark the position of collisional convergent plate boundaries, now represented by intracratonic terrane boundaries. The suture zones separate protocontinental crustal blocks that formed as a result of the amalgamation of multicomponent, continental basement rocks, which initially developed as volcanic arcs and which are today preserved on the Kaapvaal Craton as Archean granite-greenstone complexes ranging in age between ca. 3600 and 2700 Ma. The suggested linear suture zones, which have not yet been examined in detail geophysically, are shown to be part of more extensive intracratonic linear features, such as the Thabazimbi-Murchison and Barberton Lineaments, and are believed to represent the only physical expression of terrane boundaries on the Kaapvaal Craton.

[†]E-mail: anhaeusserc@geosciences.wits.ac.za.

Anhaeusser, C.R., 2006, A reevaluation of Archean intracratonic terrane boundaries on the Kaapvaal Craton, South Africa: Collisional suture zones?, in Reimold, W.U., and Gibson, R.L., Processes on the Early Earth: Geological Society of America Special Paper 405, p. 193–210, doi: 10.1130/2006.2405(11).

Keywords: Archean suture zones, intracratonic terrane boundaries, Kaapvaal Craton, South Africa.

INTRODUCTION

Ultramafic and related rocks are of widely differing types and occur in a broad range of geological and geotectonic environments (Wyllie, 1967). Some occur in variably sized stratiform intrusions like the Bushveld, Great Dyke, Stillwater, Skaergaard, and Muskox igneous complexes (Eales and Cawthorn, 1996; Irvine and Smith, 1967; McCallum, 1996; McBirney, 1996; Wilson, 1996), while others occur as alkalic ring complexes or kimberlite intrusions (Gold, 1967; Dawson, 1967). Close scrutiny worldwide, however, reveals that most ultramafic rocks not associated with igneous complexes and Archean greenstone belts occur as remnants of oceanic crust and upper mantle and are restricted to suture zones in Proterozoic and Paleozoic orogenic belts as a result of plate-tectonic processes (Burke et al., 1977). These include: (1) the Alpine-type peridotite-serpentinite bodies distributed along deformed mountain chains (e.g., the Zagros Crush Zone and the Indus Suture occurring along the Alpine-Himalayan continental collision zone); (2) ancient and modern island arcs (Thayer, 1967; Dewey, 1969, 1977; Hamilton, 1970); and (3) concentrically zoned dunite-peridotite intrusive complexes like those of southeastern Alaska and the Ural Mountains of Russia (Irvine, 1967; Taylor, 1967). Most Alpine-type ultramafic-mafic associations are regarded as being synonymous with ophiolite suites, the latter considered to be slices of oceanic crust and mantle tectonically emplaced in orogenic belts. Some of the best-known ophiolite occurrences include those of Troodos in Cyprus (generally regarded as the classic type example; Moores and Vine, 1971; Robertson and Xenophontos, 1993), the ophiolites of the Oman Tethys (Semail Ophiolite along the coastline of the Gulf of Oman; Lippard et al., 1986), and the Appalachian ophiolites of Newfoundland (Smith, 1958; Dewey and Bird, 1971).

Ophiolite suites were described by Dewey and Bird (1970, 1971) as having been generated at oceanic ridges and by slow spreading in marginal basins behind and within island-arc complexes. They showed that many ophiolites in orogenic belts occur in a variety of tectonic settings, including those found as severely deformed and serpentinized bodies in suture zones, in sutured mélange terranes, and as thrust wedges. Terrane accretion has also been considered a viable mechanism of ophiolite emplacement, with ophiolites commonly occurring at terrane margins (Coleman, 1971; Kearey and Vine, 1995).

Because serpentinized ultramafic rocks are a common feature of most early Archean greenstone belts (Arndt and Nisbet, 1982; Nisbet, 1987; Arndt et al., 1997) there have been attempts to equate these rock associations with Phanerozoic ophiolite complexes (De Wit et al., 1987; Fripp and Jones, 1997). However, adherence to the Geological Society of America Penrose Conference definition of an ophiolite (Anonymous, 1972) shows several important differences (one of these being the complete absence of mafic sheeted dyke complexes in Archean terranes), making direct comparison untenable. Applying the Penrose ophiolite definition in its strict sense to Archean or even Proterozoic terranes has thus proved to be controversial (Bickle et al., 1995; Hamilton, 1998; McCall, 2003). Recently, in part to overcome this difficulty, Dewey (2003) referred to the Penrose ophiolites as *ophiolites sensu stricto* and reserved the term *ophiolites sensu lato* for other heterogeneous remnants of ocean floors now embedded in the continents. Şengör and Natal'in (2004) went a step further, restricting the term *ophiolite* to the more or less complete suite and called its various fragments formed by processes incorporating them into continental crust *ophirags*.

Thus, if the "strict definition" of an ophiolite suite were to be disregarded it could be argued that some Archean greenstone belts may contain fragmented ophiolites (the ophirags of Şengör and Natal'in, 2004). These Archean ophirags would, in turn, represent dismembered pieces of oceanic lithosphere produced as prototypes to present-day, plate-tectonic–related oceanic crust. This would add further strength to theoretical and actualistic models that are increasingly pointing to a Mesozoic to Archean tectonic regime not fundamentally different from that of today (Talbot, 1973; Dewey and Spall, 1975; Windley, 1976; Lowe, 1994, 1999).

Increasing support for plate-tectonic processes to account for Archean geotectonic events has led to the reevaluation of selected terranes in southern Africa in order to determine the viability of such processes as a mechanism for Archean terrane accretion and continental growth.

Evidence suggests that magmatic arc formation began on the Kaapvaal Craton ca. 3600 Ma (Poujol et al., 2003; Eglington and Armstrong, 2004) and continued episodically at intervals of between 100 and 200 m.y., resulting in the construction of discrete microcontinents that eventually coalesced to form the Kaapvaal Craton (De Wit et al., 1992; Poujol et al., 2003).

In an attempt to verify the existence of at least some of the possible boundary zones between the various terranes postulated by De Wit et al. (1992), this paper examines the nature and distribution of various mafic and ultramafic rocks found on the Kaapvaal Craton. These distinctive, yet relatively sparsely distributed rock types may, it is contended, represent ophirag remnants, which could mark the traces of collisional zones between ancient protocontinental blocks. Three separate localities on the Kaapvaal Craton have been identified that display exceptional developments of serpentinized ultramafic rocks and may signal possible suture zones formed during the late Paleoarchean to Mesoarchean (ca. 3500–2700 Ma). These include ultramafic rocks distributed linearly along: (1) the northern flank of the Barberton greenstone belt; (2) the terrane south of the Murchison greenstone belt; and (3) the southern sector of the Johannesburg Dome (see Fig. 1 for localities).

BARBERTON GREENSTONE BELT

Ultramafic and mafic rocks are extensively developed in the Barberton greenstone belt and comprise large volcanic outpourings of komatiite, komatiitic basalt, and high-Mg tholeiite as well as intrusive or tectonically emplaced layered ultramafic complexes (Viljoen and Viljoen, 1969, 1970; Anhaeusser, 1972, 1976a, 1985, 2001; Wuth, 1980; Viljoen et al., 1983). The ultramafic-mafic lava successions have been likened to modern oceanic plateau assemblages (Cloete, 1999; Dann, 2000; Anhaeusser, 2001), whereas the layered ultramafic complexes have been regarded as sill-like intrusive bodies formed contemporaneously with the lavas in shallow-seated magma chambers where they underwent differentiation from komatiitic parent magmas (Viljoen and Viljoen, 1970; Anhaeusser, 1985). It has also been recognized that many of the layered complexes appear to be fault-bounded and may have been tectonically emplaced (Anhaeusser, 1985, 2001).

De Wit et al. (1987) and De Wit and Tredoux (1988) were of the opinion that the ultramafic to mafic rocks of the Barberton greenstone belt formed a pseudostratigraphy comparable to that of Phanerozoic ophiolites, which they named the Jamestown Ophiolite Complex in deference to the pioneering geological endeavors of Hall (1918) and Visser (1956). Although objections to the ophiolite analogy have been raised (Hamilton, 1998, 2003; McCall, 2003) there may be merit in the broader application of the ophiolite/ophirag concept mentioned earlier. One significant difference with respect to the Barberton terrane (and indeed Archean terranes worldwide) remains the complete absence of sheeted dykes, which have been considered to be a fundamental component of ophiolites *sensu stricto*. For example McCall (2003), in comparing Archean and Phanerozoic tectonics, emphasized that sheeted dykes are uncharacteristic of Archean eruptive suites. Some examples have, however, been reported from a ca. 2500 Ma ophiolite complex in China (Kusky et al., 2001) and from the Archean Yilgarn Craton in Western Australia (Fripp and Jones, 1997), but these are regarded as contentious (Zhai et al., 2002; McCall, 2003). In view of the ophirag suggestion of Şengör and Natal'in (2004) outlined earlier, the absence of the sheeted-dyke component of ophiolites in Archean terranes may be a consequence of their dismemberment, or more likely, their position within the orogenic edifice. Some additional factors, not yet fully understood, may also have played a role. Archean magmas are, for example, generally believed to have formed in high-heat flow environments (e.g., melting in plumes; Arndt et al., 1997). The eruptive environment of these dominantly low viscosity komatiitic and high-Mg tholeiitic magmas may have favored fissure-style eruptions rather than feeder dyke intrusions that are generally rare in Archean terranes (Arndt, 1999).

Anhaeusser (1985) recorded the presence of 27 ultramafic complexes in the Barberton greenstone belt, most of which show pronounced magmatic differentiation and the common development of cyclically repeated layering consisting dominantly of dunite, orthopyroxenite, and harzburgite and volumetrically subordinate websterite and anorthositic gabbro-norite units. Although Archean ultramafic complexes are known from around the world, most are considerably less magnesian than the Barberton examples, which appear to have developed from parental komatiite magmas containing ~28% MgO. It has been estimated that dunite, harzburgite, and lherzolite constitute as much as 80% by volume of some of the complexes, the remainder consisting of ortho- and clinopyroxenites and minor gabbro-norite phases (Wuth, 1980; Anhaeusser, 1985; Rodel, 1993). Most of the ultramafic rocks in these complexes have been extensively serpentinized and steatized and have also been altered to various talc ± tremolite ± chlorite ± carbonate schists. The complexes display variable effects of heterogeneous strain with both massive and schistose rock assemblages being represented in places, in different parts of the same body.

Anhaeusser (1969, 2002) showed that many of the Barberton ultramafic complexes occur along the northern flank of the greenstone belt (Fig. 2A) and proposed a model reconstruction suggesting that they may once have formed a continuous, northeast-southwest–trending, sheetlike intrusion, or a series of related, but probably discontinuous sill-like bodies traceable over a distance of ~110 km (Fig. 2B). Because younger cover sequences obscure the northeastern and southwestern extensions of the Barberton greenstone belt it is not known whether the ultramafic bodies continue beyond the exposed limits.

Barberton Granitoid History

The model reconstruction referred to above involves the "removal" of the intrusive granitoid bodies that forced their way, balloonlike, into the northern Barberton terrane and which were, in turn, responsible for the prising apart of the supracrustal greenstones to form such prominent features as the northwest-trending Jamestown and Nelshoogte schist belts (Fig. 2A). The Kaap Valley tonalite pluton, dated at 3227 ± 1 Ma (Kamo and Davis, 1994), and the Nelshoogte trondhjemite pluton, dated at 3236 ± 1 Ma (De Ronde and Kamo, 2000), are regarded as responsible for the structural disturbance and dismemberment of the ultramafic complexes in the northwestern sector of the Barberton belt (Anhaeusser, 1966, 1969, 2002). The actual ages of the ultramafic complexes have not yet been determined directly because of the unsuitability of the rocks for dating. Visser (1956) was of the opinion that the ultramafic rocks were post-Moodies in age. Moodies Group deposition is bracketed between ca. 3225 Ma (Kröner et al., 1991) and the ca. 3079 Ma age determined by Heubeck et al. (1993) for the Salisbury Kop pluton, which intrudes folded Moodies Group rocks. However, field relationships combined with evidence

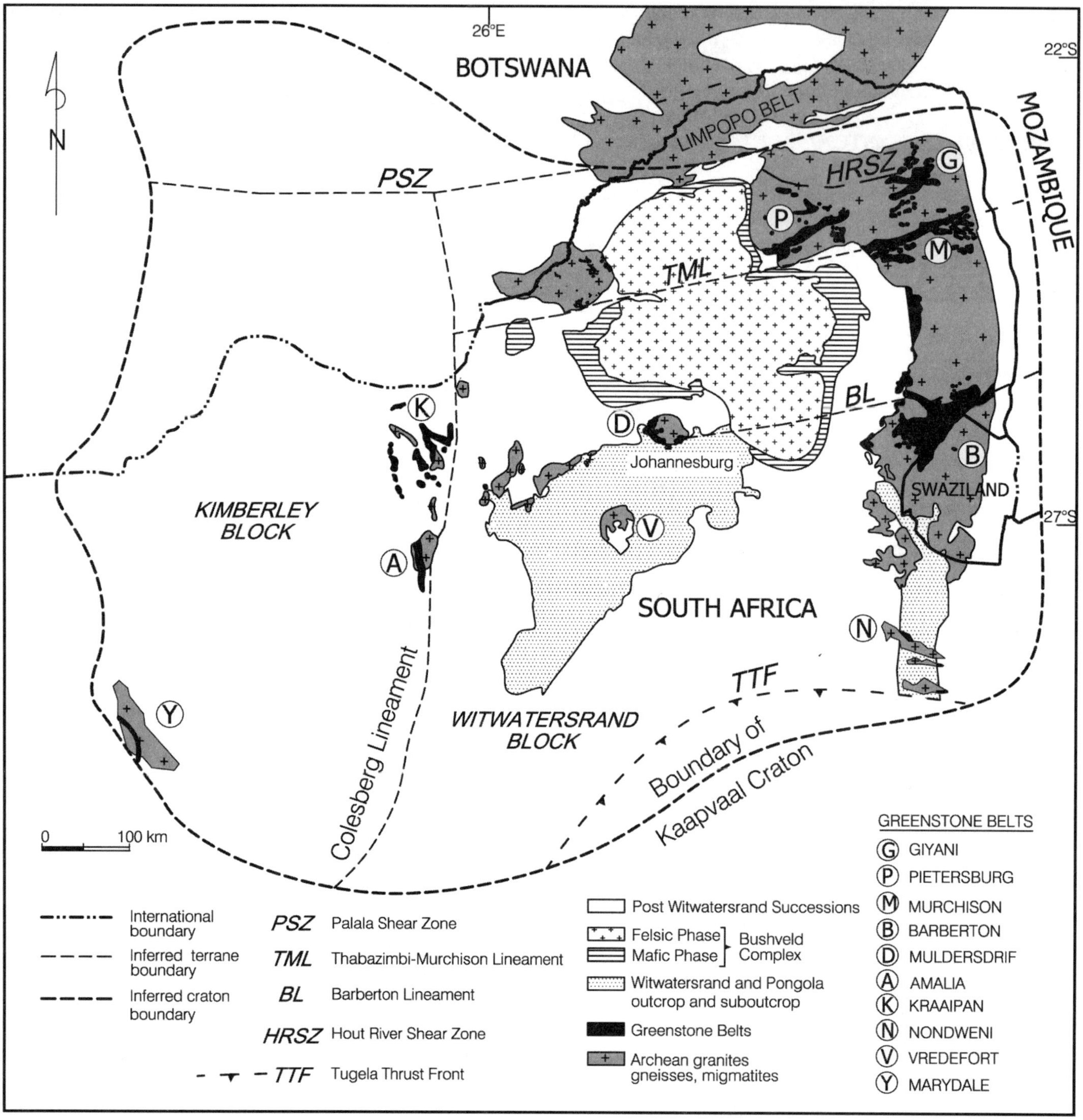

Figure 1. Simplified regional geological map showing the main linear structural features that have been described on the Kaapvaal Craton and the locations of the ultramafic-mafic rocks defining the positions of the postulated Archean suture zones proposed in the paper.

suggesting a minimum age of ca. 3226 Ma for the ultramafic complexes in the Weltevreden area southwest of Barberton (De Ronde and Kamo, 2000) places the age of the ultramafic rocks considerably older than the Moodies sedimentary sequence. Further confirmation of an older age for the ultramafic complexes relates to the previously discussed ca. 3236–3227 Ma ages determined for the Nelshoogte and Kaap Valley plutons, which intrude and deform the ultramafic rocks.

In the northeast, the Stentor pluton embays into the northern sector of the Barberton greenstone belt and, like its counterparts to the west, had a structurally disruptive influence on the ultramafic bodies in the vicinity (Fig. 2). Kohler (1994)

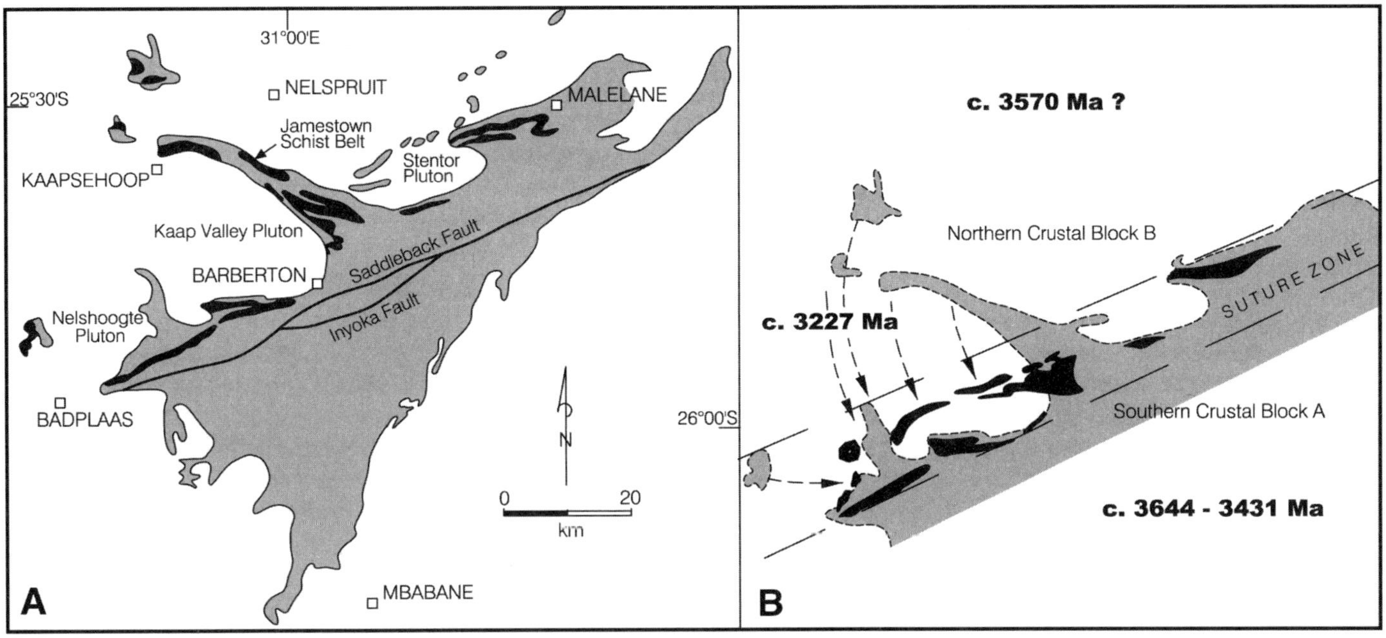

Figure 2. (A) Simplified map showing the present distribution of layered ultramafic complexes (black) along the northern flank of the Barberton greenstone belt (gray shaded). The location of the possible terrane boundary envisaged by previous workers coincides with the position of the Saddleback and Inyoka Faults (see text for details). The granitoid plutons that fragment and deform the ultramafic complexes were emplaced at ca. 3227 Ma. (B) Schematic reconstruction, minus the influence of the Kaap Valley, Nelshoogte, and Stentor granitoid plutons, showing the alignment of the ultramafic complexes along a possible suture or convergence zone (as proposed in this paper) separating a southern crustal block (ca. 3644–3431 Ma) from a northern block (possibly ca. 3570 Ma; see text).

regarded the Stentor pluton as a phase of the Nelspruit batholith, but geochemical, petrological, and geochronological investigations have shown that this pluton is more complex than either the Kaap Valley or Nelshoogte plutons further to the west. Granitoid ages reported from the Stentor pluton range from 3347 +67/–60 Ma to 3250 ± 30 Ma (Tegtmeyer and Kröner, 1987). Ages ranging from 3107 ± 5 Ma to 3082 ± 3 Ma were also recorded from this pluton by Kamo and Davis (1994). The ca. 3250 Ma age of Tegtmeyer and Kröner (1987) closely coincides with ages obtained for the Kaap Valley and Nelshoogte plutons, whereas the younger age of ca. 3107 Ma coincides with the 3105 ± 2 Ma age established for the Nelspruit batholith by Kamo and Davis (1994). The oldest age of ca. 3347 Ma also closely corresponds, within error, to the age of 3303 ± 6.4 Ma recorded by Kamo and Davis (1994) for tonalitic xenoliths found in porphyritic granodiorites of the Nelspruit batholith north of the Sabie River near Hazyview (Robb, 1977).

The view that the terrane north of the Barberton greenstone belt contains remnants of older granitic crust was first recorded by Visser (1956), who described granite pebbles in Moodies Group conglomerates in the Eureka syncline northeast of Barberton (Sheba Siding vicinity). These authors noted that "the granite pebbles represent a parent body which has hitherto not been recognized in outcrop in the area" (Visser, 1956, p. 78). The deformed granitoid pebbles, which have a distinctive graphic or granophyric texture (Anhaeusser, 1966), occur predominantly along the northern flank of the syncline (Anhaeusser, 1969, 1976b), suggesting their source was largely from the north, the terrane now occupied by the ca. 3105 Ma Nelspruit batholith. Van Niekerk and Burger (1978) established that the granite boulders in the conglomerates were derived from granitoid terranes older than the ca. 3227 Ma Kaap Valley pluton. This was later confirmed by Kröner and Compston (1988), who undertook ion microprobe analyses on zircons from the granitoid clasts, establishing their age at between 3570 ± 6 Ma and 3518 ± 11 Ma. Thus, it appears that the granitoid terrane north of the Barberton greenstone belt formed as a result of successive plutonic events that began coalescing into a microcontinent in late Eoarchean times (ca. 3600 Ma), but which was subsequently transformed in Paleoarchean to early Mesoarchean times (ca. 3500–3000 Ma) by later crust-forming granitoid emplacement episodes that culminated with the emplacement of the Nelspruit batholith at ca. 3105 Ma. During approximately the same period, similar crust-forming events were taking place elsewhere on the evolving Kaapvaal Craton, which led to the amalgamation of the granitoid terrane that presently occupies the regions south and southeast of the Barberton greenstone belt in the Komati River valley and in Swaziland. This southern terrane, including the Barberton greenstone belt, has been viewed as a compound terrane formed through long-term magmatic activity (ca. 3644–3431 Ma) and late-stage tectonic accretion (Compston and Kröner, 1988; Lowe et al., 1989; De Wit et al., 1992; Kröner et al., 1992, 1996; De Ronde and De Wit, 1994). Lowe (1994, 1999) divided the southern region into four major blocks or protocontinents, each comprising an ultramafic-mafic oceanic volcanic stage followed by a felsic tonalite-trondhjemite-granodiorite (TTG) magmatic stage that had characteristics akin to subduction-related volcanic arcs. More recently, Kisters et al. (2003) have suggested that the granitoids and enclosed metavolcanic and metasedimentary remnants that occur in the southern Barberton region constitute an exhumed ancient core complex containing peak metamorphic assemblages indicative of burial to mid- to lower-crustal levels.

Barberton Tectonic History

A number of attempts have been made to describe the tectonic history of the Barberton granite-greenstone terrane. Regional mapping led to the recognition of numerous isoclinal folds, which are commonly bounded by major, generally steeply dipping faults that partitioned or segmented the greenstone belt into a number of preferentially developed synclinoria (Visser, 1956; Anhaeusser et al., 1968, 1969; Anhaeusser, 1984). Stratigraphic and sedimentological studies led to the recognition of different facies of Fig Tree and Moodies sedimentary rocks, the present boundary between the two sedimentary realms (Northern and Southern Facies) coinciding with a major strike-slip fault system (the Saddleback-Inyoka fault zone, Fig. 2A) occurring in the northern half of the Barberton greenstone belt (Visser, 1956; Heinrichs and Reimer, 1977; Eriksson, 1980; Lowe et al., 1999; Heubeck and Lowe, 1999). Subsequent workers came to regard this transitional zone as representing an accretionary convergent domain involving arc- and trench-related processes and culminating in arc amalgamation and suturing (De Wit, 1991; De Ronde and De Wit, 1994; De Ronde and Kamo, 2000; Lowe, 1999).

While the author is in agreement that some form of collisional tectonic process played a role in juxtaposing the northern and southern facies sediments in the Barberton greenstone belt, the position of the suture remains debatable. In this contribution it is argued that the zone occupied by the layered ultramafic complexes is more appropriate as the locus of suturing or oceanic crustal collision. The convergence of two ancient protocontinental crustal blocks, one from the north and the other from the south, as described earlier, would have counterparts similar to obducted oceanic crust (ophiolite suites) comparable to those occurring in Phanerozoic orogenic suture zones and sutured mélange terranes. The collisional process is likely to have been complex, involving a variety of possible plate, ocean, and continental combinations, as outlined by Dewey and Bird (1970, 1971). Evidence from the Barberton greenstone belt, provided by De Ronde and De Wit (1994) and Lowe (1999), suggests that the granite-greenstone terrane represents a compound crustal block constructed of a number of smaller crustal segments, which amalgamated over a time span ranging over 330–490 m.y. A dramatically shorter time frame of ~3 m.y. was later outlined by De Ronde and Kamo (2000) for what they described as an arc-arc D_2 collisional event in the Weltevreden area southwest of Barberton, which occurred between 3229 and 3227 Ma.

Massive serpentinized ultramafic rocks also occur in the Msauli-Havelock mine areas in the southeastern part of the Barberton greenstone belt, adjacent to the Swaziland border (Büttner, 1984; Barton, 1986). These ultramafic rocks, which are hosts to major chrysotile asbestos deposits, occur in a linear, faulted zone that can be traced for ~15 km and may also reflect a zone of collisional suturing within the greenstone belt of smaller crustal segments of the type outlined above by Lowe (1999).

Precise dating techniques now clearly identify an older southern crustal domain (>3445 Ma) from a younger (ca. 3227 Ma) northern domain (Kamo and Davis, 1994). It is suggested that the ultramafic-mafic (ophirag) wedge separating these two crustal blocks reflects a consuming margin along and into which were emplaced TTG granitoids at ca. 3227 Ma following subduction-related arclike volcanism. Figure 3 indicates schematically the possible nature of the collisional interface, with the Fig Tree and Moodies sediments as well as earlier calc-alkaline volcanic rocks (Upper Onverwacht lithologies) occurring to the south and representing products of the subduction event. This could imply that the poorly time-constrained ultramafic-mafic zone in the north as well as the Onverwacht type-locality komatiitic assemblages on the southern side of the Barberton greenstone belt may be part of the same oceanic crust and lithosphere upon which the later arclike events of volcanism and magmatism developed.

MURCHISON GREENSTONE BELT

A number of ultramafic-mafic bodies south of the Murchison greenstone belt (Fig. 4) were first recorded by Hall (1912) in the Ofcolaco-Mica-Phalaborwa region. They include massive bodies of serpentinite, amphibolite, and gabbro as well as schistose

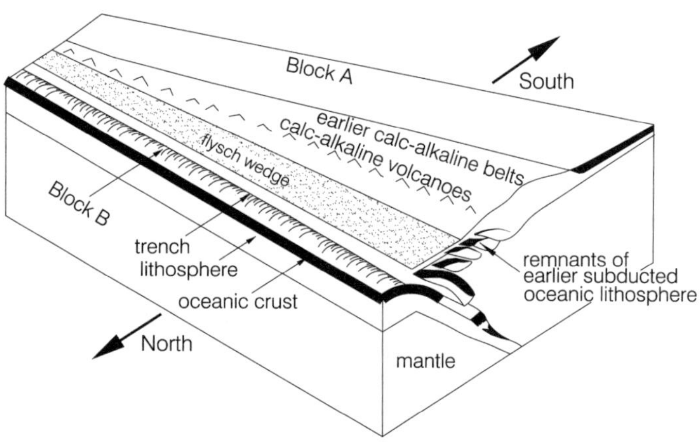

Figure 3. Schematic block diagram (after Dewey and Bird, 1971) showing the collisional interface or suture zone along which the oceanic crust and ultramafic complexes (ophirags) were emplaced on the northern flank of the Barberton greenstone belt. Fig Tree and Moodies sediments (the flysch wedge) and calc-alkaline, arclike volcanics (Upper Onverwacht Group) occupy the area to the south on amalgamated crustal block A. Oceanic crust subducted along the trench is believed to have generated tonalite-trondhjemite-granodiorite (TTG) magmas, which led to the intrusion of the Kaap Valley, Nelshoogte, and Stentor granitoid plutons (Fig. 2A).

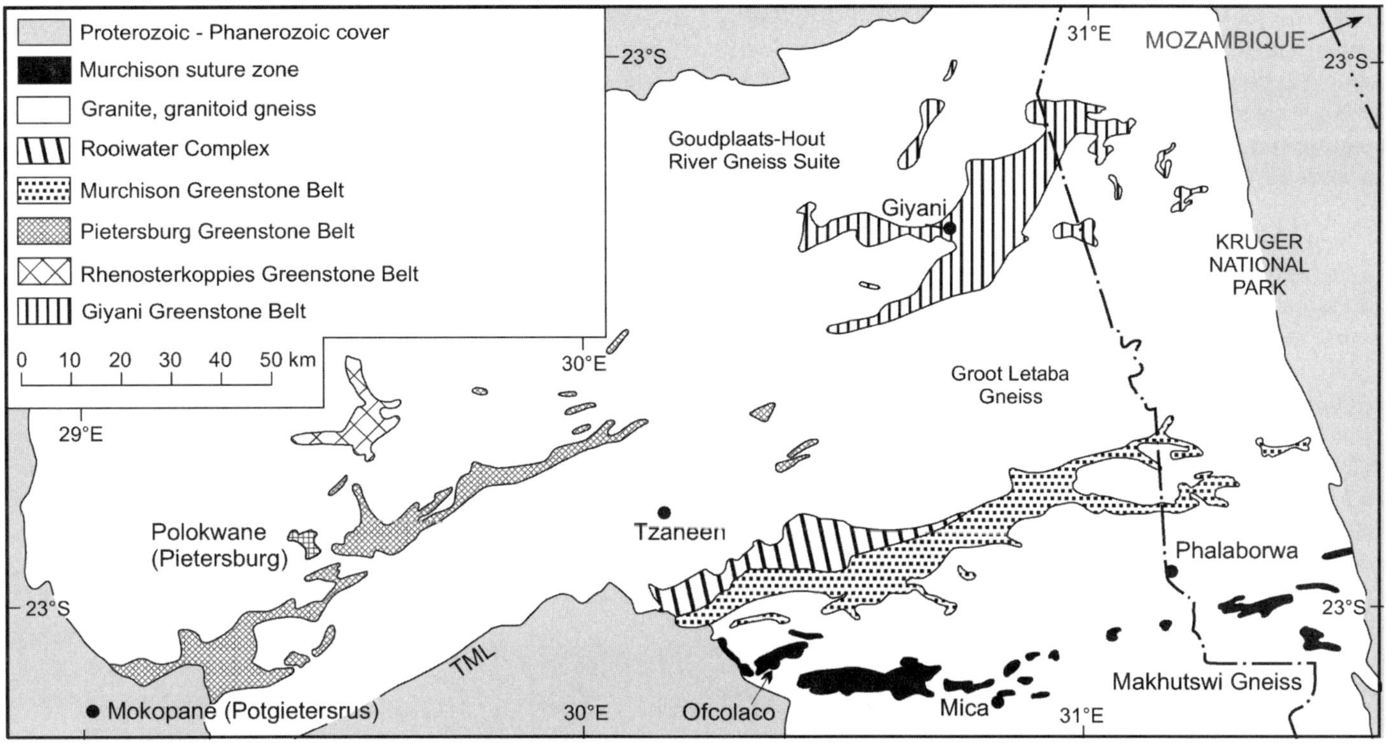

Figure 4. Simplified geological map of the northeastern sector of the Kaapvaal Craton showing the location of the Pietersburg, Giyani (Sutherland), and Murchison greenstone belts and surrounding granitoid basement. The distribution of numerous serpentinized ultramafic-mafic bodies, intruded by granites and pegmatites and discussed in this paper, is shown in a zone south of the Murchison greenstone belt in the vicinity of Ofcolaco, Mica, and Phalaborwa. The Thabazimbi-Murchison Lineament (TML) is shown extending west-southwest along the northeast-southwest projection of the Murchison greenstone belt.

rocks, including hornblende, talc, and chlorite schists. Exposures are generally poor, but these rocks can be traced sporadically from beneath the Neoarchean Wolkberg Group formations of the Transvaal Drakensberg Escarpment in the west toward Phalaborwa and the Kruger National Park in the east—a distance of ~120 km (see 1:250,000 map sheets, 2430 Pilgrim's Rest; Walraven, 1986; and 2330 Tzaneen; Brandl, 1985). The largest bodies occur in the west near Ofcolaco and north of the Makhutswi River east of Alsace. A few of the larger bodies also occur in the east in the Kruger National Park, north of the Olifants River (Walraven, 1986). Between these localities, from just west of Mica to the region south of Phalaborwa, the bodies have been dismembered by intrusive granitoid rocks, including numerous pegmatites of the Selati Line–Olifants River Mica Field (Hall, 1912; Robb and Robb, 1986).

Reconnaissance investigations by the author suggest that the ultramafic occurrences south of the Murchison greenstone belt (Fig. 4) are lithologically similar to the layered ultramafic complexes described above in the Barberton area and those discussed later in the Johannesburg Dome. Their age has not been determined precisely, but they are intruded by granitoid rocks ranging in age from 2700 to 3330 Ma (Poujol et al., 2003), including U-Pb zircon results pointing to a minimum age of 2850 Ma for the pegmatites (Poujol and Robb, 1999).

Together with the Barberton ultramafic-mafic complexes and those on the Johannesburg Dome, the occurrences of serpentinized ultramafic rocks in the Ofcolaco-Mica-Phalaborwa region constitute the largest development of this rock association on the Kaapvaal Craton. These rocks, like their counterparts described earlier, may represent ophiolite-type (ophirag) remnants dispersed along a suture (or oceanic collisional zone) separating two older crustal remnants located north and south of the Murchison greenstone belt.

In the north and northwest are vestiges of the ca. 3282–3333 Ma (Kröner et al., 2000; Brandl and Kröner, 1993) Goudplaats-Hout River Gneiss Suite. These gneisses are located northwest of the Pietersburg and Giyani (Sutherland) greenstone belts (Fig. 4) and south of the Southern Marginal Zone of the Limpopo Belt (i.e., south of the Hout River Shear Zone—HRSZ, Fig. 1). They are pervaded by younger granitoid rocks, emplaced at around 2900 Ma, which were formed during a major magmatic phase involving dehydration melting of amphibolite and biotite-gneiss protoliths (Du Toit et al., 1983). The granitoid rocks in the terrane further to the northeast, occurring between the Murchison and Pietersburg-Giyani greenstone belts, are grouped together under the term Groot Letaba Gneiss (Fig. 4). These gneisses are virtually identical to the Goudplaats-Hout River gneisses and have only recently been distinguished on geochronological grounds. The oldest gneiss from the area south of the Giyani greenstone belt yielded a single zircon Pb-Pb age of 3171 ± 6 Ma (Brandl and Kröner, 1993). Younger ages (ca. 2853–2886 Ma) were obtained from tonalitic and trondhjemitic gneisses and discordant leucogneisses east of Polokwane (Pietersburg) and near the northeastern end of the Pietersburg greenstone belt (Brandl and Kröner, 1993; Kröner et al., 2000).

South of the Murchison greenstone belt more gneisses occur, among which is the Makhutswi Gneiss that extends for some 50 km southward before merging with the Klaserie Gneiss. The Klaserie Gneiss, in turn, adjoins the granitoid terrane that extends to the Nelspruit region north of the Barberton greenstone belt described earlier. The Makhutswi Gneiss has tonalitic to granodioritic compositions and has yielded an age of 3228 ± 12 Ma (Poujol et al., 1996) close to the Murchison greenstone belt. Younger intrusive granitoids in the Makhutswi terrane vary between 3112 ± 5 and 3078 ± 6 Ma (Brandl and Kröner, 1993). Rocks of approximately similar age to those reported near the eastern end of the Pietersburg greenstone belt occur south of the Murchison greenstone belt and include pegmatites and undeformed porphyritic granite yielding ages of 2848 ± 58 and 2820 ± 38 Ma, respectively (Poujol and Robb, 1999; Poujol, 2001). The pegmatites in the Mica area, ~30 km south of the Murchison belt, intrude the Makhutswi tonalitic and trondhjemitic gneisses as well as numerous greenstone remnants comprising mainly serpentinites, talc and talc-chlorite schists, and amphibolites (Hall, 1912; Walraven, 1986; Robb and Robb, 1986). The Union Mine pegmatite, dated at 2912 ± 2.6 Ma (Kruger et al., 1998), and an additional pegmatite from the Olifants River Mica Field, dated at 2848 ± 58 Ma (Poujol and Robb, 1999), provide a minimum age for the ultramafic-mafic remnants, here postulated to represent oceanic crustal rocks.

Murchison Tectonic History

The geotectonic history of the northeastern sector of the Kaapvaal Craton is relatively poorly understood, due in part to the restricted exposure (Viljoen et al., 1993) and the paucity of detailed geological and geochronological data (Poujol et al., 2003). Exceptions include the Murchison and Giyani greenstone belts, which give rise to low ranges of hills that have been exploited for gold, antimony, copper-zinc, titaniferous magnetite, and pegmatite mineralization (Hall, 1912; Van Eeden et al., 1939; Viljoen et al., 1978; Vearncombe et al., 1992; Ward and Wilson, 1998). The concept of Archean plate tectonics in the Murchison area was first mooted by Vearncombe (1991), who maintained that the Murchison greenstone belt formed part of an exhumed volcanic island arc sequence. As modern island arcs are developed above or close to subduction zones, this suggests that the Murchison belt may once have formed in a collisional oceanic tectonic regime. Vearncombe (1991) did not identify a suture zone nor was any detailed subduction configuration provided, but he maintained that the Rooiwater Complex (a layered mafic igneous body with lithologies similar to the Bushveld Complex; Fig. 4) represented a subvolcanic magma chamber overlain by calc-alkaline felsic volcanic rocks of the Rubbervale Formation. Subsequent U-Pb isotopic studies have shown, however, that the Rubbervale Formation (2971 ± 10 Ma) is at least 200 m.y. older than the Rooiwater Complex (2740 ± 4 Ma; Poujol et al., 1996), suggesting that the latter complex is unrelated to the island arc evolutionary model proposed by Vearncombe (1991).

If a volcanic arc model is, however, accepted for the Murchison region as a whole, the suggestion offered here is that two protocontinental blocks (the northern and southern gneiss terranes described earlier) converged in post-Rubbervale Formation times (ca. 2971 Ma) entrapping part of the Murchison volcanic arc as well as some of the mafic-ultramafic oceanic crust in a suture zone to the south of the main Murchison greenstone belt. This "ophiolite-like" oceanic crust was then dismembered and transformed into a variety of mafic and ultramafic schists and massive serpentinite and amphibolite remnants (ophirags) by pegmatites intruded along the Olifants River Mica Field between ca. 2848 and 2912 Ma.

JOHANNESBURG DOME

Studies undertaken on the Johannesburg Dome have led to the recognition of a variety of Archean granitoid rocks (Anhaeusser, 1973a) as well as a suite of mafic and ultramafic rocks analogous to the komatiitic lavas and layered ultramafic complexes encountered in the Barberton greenstone belt. The mafic and ultramafic rocks are exposed intermittently within a WNW-ESE–trending zone across the southern portion of the Johannesburg Dome, extending from beyond Muldersdrif in the west (Anhaeusser, 1977, 1978, 1992) to the Edenvale-Modderfontein area in the east (Anhaeusser, 1973a, 2004; Chaumba, 1992), a distance of ~50 km (Fig. 5). With the exception of the Barberton greenstone belt and the Ofcolaco-Mica area described earlier, the mafic-ultramafic rocks on the Johannesburg Dome represent the largest development of serpentinized ultramafic rocks on the Kaapvaal Craton.

The principal rock types in the ultramafic complexes in the Johannesburg Dome include serpentinized dunite, harzburgite, lherzolite, ortho- and clinopyroxenite, and gabbro (Anhaeusser, 2004). Pillow basalts and ultramafic lava sequences of komatiitic affinity have also been reported from localities in the Muldersdrif area (Anhaeusser, 1977, 1978). In places, the serpentinites contain significant quantities of Cr (up to 5335 ppm) and Ni (up to 2240 ppm), which may account for the presence of detrital chromites and high nickel values found in the conglomerates and shales of the Witwatersrand Supergroup to the south, dated at ca. 2970–2780 Ma (Robb and Robb, 1998).

The mafic-ultramafic assemblages on the Johannesburg Dome have previously been regarded as remnants of ancient greenstone belts, but the absence of other typical greenstone lithologies (viz., chert, banded iron formation, jaspilite, felsic

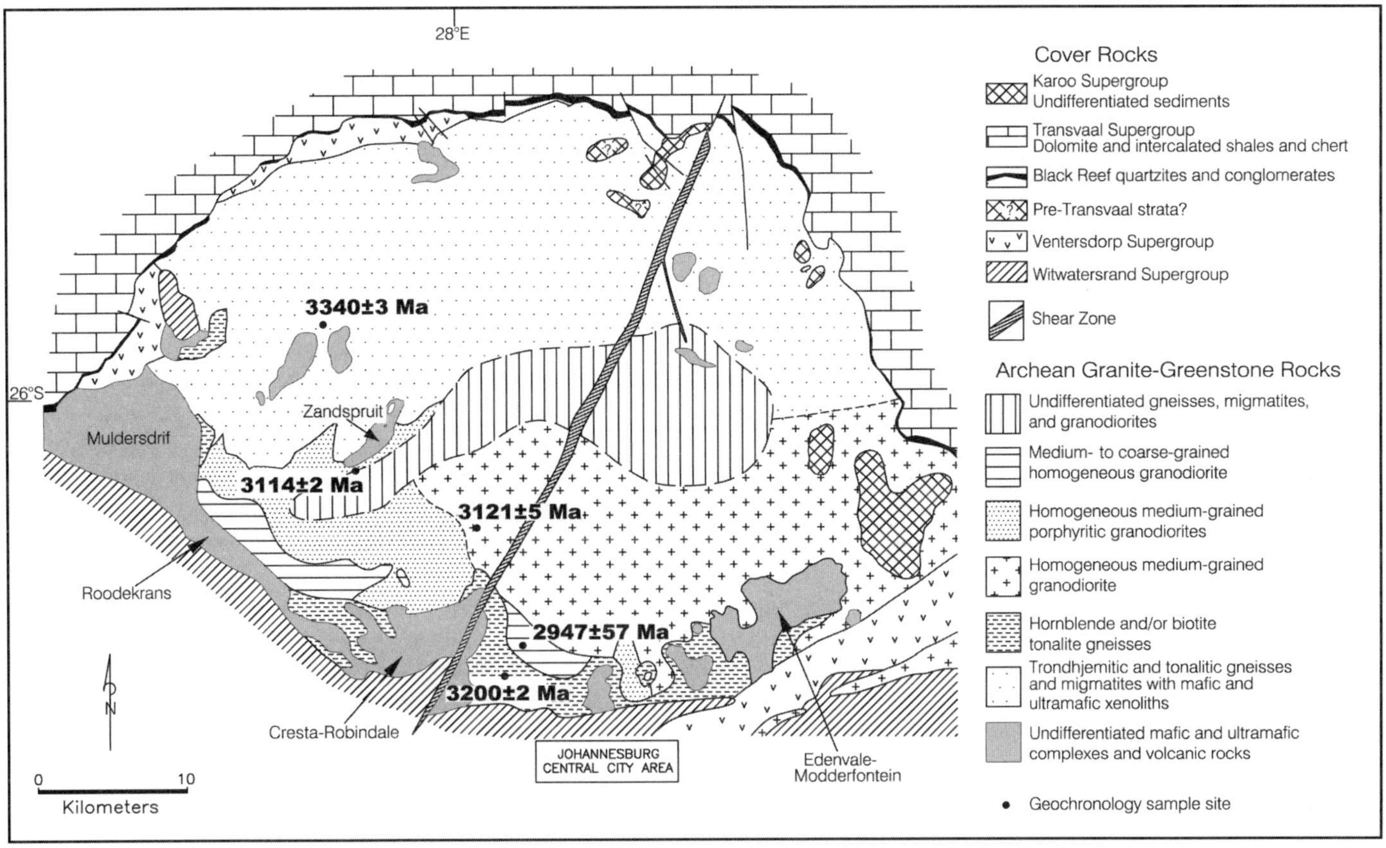

Figure 5. Simplified geological map of the Johannesburg Dome showing the distribution and ages of the various granitoid rocks and the location of Archean ultramafic-mafic complexes that define the position of a possible WNW-ESE–trending, linear suture zone along the southern margin of the dome (after Anhaeusser, 1973a).

volcanoclastic rocks) suggests that these rocks may be more closely allied to upper mantle or oceanic crustal rocks.

Progressive urban development in recent years north of Johannesburg, Roodepoort, and Edenvale has resulted in the destruction of most of the key areas of greenstone exposure on the Johannesburg Dome. The best-preserved, most continuous exposures of mafic-ultramafic rocks occur in the west where the layered successions of the Muldersdrif Ultramafic Complex (Anhaeusser, 1978) form a series of low, WNW-ESE–trending ridges and valleys (Fig. 6). The complex comprises a number of cyclic units consisting of serpentinized dunite, harzburgite, wehrlite, lherzolite, and Mg-rich ortho- and clinopyroxenite cumulate rocks. The cyclic units are terminated by thin (<1-m-wide) medium- to fine-grained metagabbroic layers, which also provide marker horizons delineating chrysotile asbestos fiber development. The rocks of the layered complex have been extensively deformed with superimposed fold events having produced a complex outcrop pattern of interference fold structures (Anhaeusser, 1976a, 1986).

Southeast of the Muldersdrif Ultramafic Complex in the Roodekrans-Ruimsig area, rocks of the Roodekrans Ultramafic Complex form a layered succession largely deformed and altered to talc, talc-chlorite, and talc-tremolite schists. Massive serpentinite layers and pods are present in places and a prominent northeast-trending, pre-Witwatersrand Supergroup cleavage is developed oblique to the layering and to the earlier-formed schistosity, testifying to a complex tectonic history for the rocks in the region (Anhaeusser, 2004).

Ultramafic rock exposures also occur ~5 km northeast of Roodekrans where they occur as remnants of the Zandspruit Ultramafic Complex that crops out discontinuously over a distance of ~4 km (Anhaeusser, 1992). The succession, consisting of alternating, cyclically repetitive, cumulate layers of serpentinized dunite, harzburgite, and metapyroxenite, dips to the southeast at shallow angles (10°–30°) and is surrounded by intrusive porphyritic granodiorites.

In the northwestern suburbs of Johannesburg, massive and schistose ultramafic-mafic rocks, referred to by Anhaeusser (2004) as the Cresta-Robindale complexes, are sporadically exposed over a distance of 10 km. These bodies, consisting predominantly of serpentinized and metamorphosed dunite, harzburgite, ortho- and clinopyroxenite, and minor gabbro, are intruded by hornblende-biotite tonalitic gneiss. Likewise, in the northeastern suburbs on the southeastern edge of the Johannesburg Dome, the Edenvale-Modderfontein Complex (Chaumba, 1992; Anhaeusser, 2004) consists predominantly of east-west–

Figure 6. Oblique aerial photograph, looking west, of part of the Muldersdrif Ultramafic Complex in the southwestern sector of the Johannesburg Dome. The WNW-ESE–trending ridges seen in the photograph consist predominantly of folded, sheared, and altered serpentinized dunite, harzburgite, and pyroxenite.

and northeast-southwest–trending massive and schistose serpentinite (including talc-chlorite and talc-carbonate schists) and pyroxenite layers that are extensively uralitized and steatized (tremolite-actinolite schists). Cyclical layering is again evident, with upward of 14 cyclical units having been identified by Chaumba (1992). Trondhjemitic gneisses and late-stage pegmatites intrude the ultramafic-mafic rocks, but exposures are limited due to the advancement of housing and industrial development in the area.

Johannesburg Dome Geochronology

Recent geochronological investigations have contributed to a fuller understanding of the complex geological and tectonic evolution of events in the Johannesburg Dome. U-Pb zircon dating has yielded ages of 3340 ± 3.3 Ma for trondhjemitic gneisses on the northern half of the dome and an age of 3200 ± 2 Ma for tonalitic gneisses on the southern rim (Poujol and Anhaeusser, 2001). Ages ranging between 3121.2 ± 5 and 2947 ± 57 Ma were determined for K-enriched granodiorites and pegmatites in the south-central parts of the dome (Barton et al., 1999; Poujol and Anhaeusser, 2001). Sediments of the Witwatersrand Basin were deposited unconformably on this basement after ca. 3074 Ma, the age of the upper lava sequence of the Dominion Group (Armstrong et al., 1991), which underlies the Witwatersrand Supergroup ~120 km southwest of the Johannesburg Dome.

The ages of the ultramafic-mafic rocks in the dome have not yet been determined directly because of their unsuitability for dating purposes, but their relationships with intrusive granitoids on the dome provide important age constraints. In the west, the Muldersdrif, Roodekrans, and Zandspruit complexes are intruded by porphyritic granodiorites dated at ca. 3114 Ma (Poujol and Anhaeusser, 2001). The Roodekrans Complex is also partly overlain by ca. 2914 Ma (or older) sediments of the West Rand Group of the Witwatersrand Supergroup (Robb and Robb, 1998). In the Cresta-Robindale area the ultramafic complexes are intruded by hornblende-biotite gneisses that have yielded U-Pb zircon ages of ca. 3200 Ma (Anhaeusser and Burger, 1982; Poujol and Anhaeusser, 2001). Similarly, the Edenvale-Modderfontein Complex has been intruded by biotite-trondhjemite gneisses, which are petrologically and geochemically identical to the ca. 3340 Ma gneisses dated on the northwestern side of the dome by Poujol and Anhaeusser (2001). Thus, although there are no direct ages for the ultramafic-mafic rocks of the Johannesburg Dome, there is evidence that they at least predate the early stages of Witwatersrand deposition (at ca. 2914 Ma) and that they also clearly predate the ca. 3200 Ma tonalitic gneisses. They may equally well be older than the ca. 3340 Ma trondhjemitic granitoids in the northwest (Fig. 5), but no direct contact relationships with these rocks have been recorded.

Johannesburg Dome Tectonic History

Petrological and geochemical studies have demonstrated that the TTG granitoids on the dome have characteristics similar to Archean TTG granitoids interpreted as having evolved in a tectono-magmatic setting analogous to that found in modern-day volcanic-arc environments (Anhaeusser, 1999). As discussed elsewhere in this contribution, Phanerozoic plate-tectonic models involving collisional tectonics commonly show suture zones defined by the preservation of slivers of oceanic lithosphere (ophiolites and ophirags), which formerly separated continental masses (Kearey and Vine, 1995). A reassessment of the Archean basement in the Johannesburg Dome therefore suggests that the ultramafic-mafic exposures described above may represent the locus of collision of two protocontinental blocks, thereby constituting a Paleoarchean suture zone. A possible scenario for the Johannesburg Dome, in pre-Witwatersrand times, envisages an ancient trondhjemitic gneiss-migmatite terrane (ca. 3340 Ma) in the north colliding with another crustal fragment (now hidden beneath the Witwatersrand Basin) in the south. The collision event may correspond with the emplacement, at ca. 3200 Ma, of hornblende tonalite gneisses along the southern margin of the dome, which resulted in the fragmentation of the oceanic lithosphere trapped between the colliding microplates. Subsequent plutonism due to partial melting of the crust of the microcontinent and possibly resulting from overthickening during or after collision (as described by Dewey and Kidd, 1974) thereafter produced the ca. 3121–2947 Ma intrusive granodioritic to granitic rocks on the dome. Further support for this crust-mantle melting interaction, and hence a significant orogenic event, may be linked to a suite of foliated, Mg-tholeiitic basaltic to calc-alkaline lamprophyric dykes emplaced on the Johannesburg Dome at ca. 3120 Ma (Prevec et al., 2004).

The crustal block buried beneath the Witwatersrand Basin probably manifests itself in the ~45-km-wide Archean crystalline basement core of the Vredefort Dome exposed some 120 km SSW of the Johannesburg Dome. Extreme uplift, associated with the Vredefort impact event at 2023 ± 4 Ma (Kamo et al., 1996), has exposed polydeformed high-grade tonalite-trondhjemite-granodiorite (TTG) gneisses, migmatites, and late-tectonic intrusive granitoids metamorphosed to upper amphibolite- to granulite-facies conditions (Hart et al., 1999; Gibson and Reimold, 2001; Lana, 2003; Lana et al., 2004). Isotopic studies carried out on various lithologies in the core of the Vredefort Dome have been summarized by Gibson and Reimold (2001) and Lana (2003). These and more recent results indicate that metamorphism took place at ca. 3100 Ma and that this event postdated most of the TTG–greenstone lithologies identified on the dome. Upper intercept ages obtained from zircons found in these rocks yielded values ranging between 3310 and 3425 Ma (Kamo et al., 1996; Hart et al., 1999). Lana (2003) also reported ca. 3400 Ma zircons from mafic, xenolith-rich enderbitic gneisses. These findings suggest that the Archean basement at Vredefort experienced a protracted and complicated Mesoarchean evolutionary history, much like that found in the other crystalline basement terranes on the Kaapvaal Craton discussed previously.

It is envisaged that the northern and southern blocks outlined above migrated toward one another as a result of oceanic plate consumption and eventually led to continent/continent collision as envisaged by Dewey and Bird (Fig. 15 *in* Dewey and Bird, 1970). The trench zone between the colliding protocontinental blocks was ultimately reduced to a narrow suture from which oceanic crust and accompanying sediments were extruded. Admittedly, this reconstruction is based on circumstantial evidence, but there does not appear to be a readily apparent way to confirm or negate this argument given the lack of exposure.

DISCUSSION

The events that took place not only in the vicinity of the Johannesburg Dome but also north of Barberton and south of Murchison occurred some 3000 m.y. ago or more. Tectonic and other changes including erosion have largely destroyed or removed whatever confirmative evidence that may once have existed (with the possible exception of the Barberton greenstone belt). If the suture concept is correct, the serpentinized ultramafic-mafic rocks preserved in these terrane boundary zones remain as the only acceptable testimony to Archean oceanic plate consumption.

Witwatersrand Basin and Plate Tectonics

Numerous models have been suggested for the origin and evolution of the Witwatersrand Basin located in the central portion of the Kaapvaal Craton (Fig. 1). Prior to the emergence of the theory of plate tectonics in the 1960s, the models employed to explain the genesis of the Witwatersrand Supergroup leaned heavily on taphrogenic or rift-related basin models (Brock and Pretorius, 1964a, 1964b) together with refinements involving intracratonic alluvial plain and lacustrine models (Pretorius, 1976; Myers et al., 1990). Van Biljon (1980) was the first to employ plate-tectonic theory to explain the origin of the Witwatersrand Basin, concluding that it represented an embayment along a subduction zone between two Archean minicontinents. Later, a cratonic foreland model developed in a continental back-arc plate-tectonic setting was proposed (Burke et al., 1985, 1986; Winter, 1987). This was followed by Stanistreet and McCarthy (1991) who introduced the concept of a Wilson cycle for the tectonic evolution of the Witwatersrand triad, which includes the underlying Dominion and overlying Ventersdorp successions. Despite the application of plate-tectonic theory, none of the studies mentioned above provided any details pertaining to the nature of the pre-Witwatersrand basement geology and the influence this may have had on the eventual location and evolution of the Witwatersrand Basin.

Reconstructions of the crustal architecture of the Kaapvaal Craton have led to general agreement among researchers that terrane amalgamation was accomplished by processes akin to but different in scale from those responsible for present-day global plate-tectonic activity. The model that has gained the most support for early Archean crustal accretion suggests that a mosaic of protocontinental subdomains consisting predominantly of nucleated granite-greenstone terranes was welded together by processes similar to those identified as being responsible for Phanerozoic and present-day plate-tectonic activity (Anhaeusser, 1973b, 1999; De Wit et al., 1992; McCourt, 1995; Barley, 1997; Brandl and De Wit, 1997; Myers and Swagers, 1997; Stott, 1997). While the concept of terrane amalgamation in the manner described may be appealing, problems remain in realistically determining the terrane boundaries. Undue speculation as to the shape, size, isotopic characteristics, and boundary relationships of unsubstantiated crustal fragments, some considered exotic or allochthonous and others that may once have formed part of a coherent geological province, suggests that further confirmatory work needs to be undertaken (Mogk et al., 1992).

A number of major crustal features have been identified on the Kaapvaal Craton, many of which display protracted periods of reactivation in post-Archean times (Fig. 1). Some of the larger lineaments include the Palala Shear Zone (Du Plessis and Walraven, 1990; McCourt, 1995), the Hout River Shear Zone (Smit et al., 1992), the Thabazimbi-Murchison Lineament (Du Plessis, 1991; Good and De Wit, 1997), and the Barberton Lineament (Fripp et al., 1980; Van Biljon, 1980), the latter being coincident with the King Fault of Stanistreet et al. (1986). In addition, three major east-northeast–trending, left-lateral wrench-fault systems are located on the northern margin of the Witwatersrand Basin. These strike-slip systems, which came into existence during Witwatersrand times, appear to be of craton-wide scale and are thought to extend eastward at least as far as the Barberton greenstone belt (Van Biljon, 1980; Stanistreet et al., 1986; fig. 1 of Stanistreet and McCarthy, 1991; Fig. 7 herein).

Other lineaments with north-south trends exist elsewhere on the craton (e.g., Colesberg; Corner et al., 1990; Doucouré and De Wit, 2002; Schmitz et al., 2004; Fig. 1 herein) but whether or not all, or some, of these extensive linear features initially constituted Archean terrane boundaries remains debatable. De Wit and Tinker (2004) recently interpreted deep seismic reflection profiles across the Kaapvaal Craton and claim to have recognized four crustal panels (>200 km by 10–20 km) of pre-Ventersdorp age resembling a tectonically stacked series of crustal slivers extending from west to east across the Kimberley and Witwatersrand blocks (Fig. 1). These blocks, also defined on geochronological grounds (Schmitz et al., 2004), are separated by the north-south–trending Colesberg Magnetic Lineament (Corner et al., 1990). This magnetic lineament, it has been conceded by De Wit and Tinker (2004), is not a feature that can be easily linked to an underlying suture in the lower crust and mantle. If a suture zone existed at all it is suggested that it was positioned to the west of the craton boundary, dipping to the west, and indicating east-directed tectonic accretion of the Amalia-Kraaipan terrane across a Witwatersrand basement panel or block.

One clearly identifiable crustal domain boundary is that of the Tugela Terrane of KwaZulu-Natal (Thomas, 1989). Matthews (1972) recognized and interpreted rocks of oceanic affinity, which were thrust northward (obducted) onto the southern flank of the Kaapvaal Craton, as an allochthonous ophiolite complex. Large, dismembered serpentinite bodies (ophirags) occur within and along this zone, which was initially referred to as the Natal Thrust Belt and Natal Nappe Complex by Matthews (1972) and Matthews and Charlesworth (1981). In Figure 1 it is shown as the Tugela Thrust Front. Unlike the inferred suture zones on the Kaapvaal Craton, the Tugela Terrane boundary collision zone separates ca. 3000 Ma Archean cratonic rocks from lithologies of Mesoproterozoic age (ca. 1200–1150 Ma; McCourt et al., 2000). This southern margin of the Kaapvaal Craton has been imaged by seismic profiling (De Wit and Tinker, 2004) and is consistent with the existing geological interpretation originally proposed by Matthews (1972).

Only two of the lineaments on the Kaapvaal Craton (viz., Thabazimbi-Murchison and Barberton) have discernable characteristics linking them to possible intracratonic suturing. In the case of the Thabazimbi-Murchison Lineament the eastern sector exposed in the Archean granite-greenstone basement comprises the highly deformed Murchison greenstone belt, which has been likened to a volcanic arc (Vearncombe, 1991). The Murchison collision zone, as contended, includes both the greenstone belt and the adjacent ultramafic-mafic rocks to the south. The zone of tectonic shortening, which includes the folded, faulted, and sheared granite-greenstones, is ~35–40 km wide. The Mhlapitsi Fold Belt, which represents the westward extension of the earlier-formed Murchison collision zone, is believed to have developed in pre-Bushveld times (Button, 1973). This folded zone, which has been shortened by upward of 35% (Potgieter, 1992), was probably reactivated during the emplacement (at ca. 2060 Ma) of the Bushveld Complex. The full extent of the Thabazimbi-Murchison Lineament, which includes the Mhlapitsi Fold Belt, projects for over 500 km in an ENE-WSW direction across the Kaapvaal Craton and delineates a major intracratonic terrane boundary (McCourt and Vearncombe, 1987; Du Plessis, 1991; Good and De Wit, 1997), the deep structure of which has been supported by seismic anisotropy studies (Vinnik et al., 1995).

The Barberton Lineament was shown by Van Biljon (1980) to strike approximately N60°E in the Barberton region, and it was suggested that if this trend was projected to the southwest beneath the younger Transvaal and Karoo Supergroup cover sequences, it would reemerge in the vicinity of the Witwatersrand Basin (Fig. 7). As was shown by Stanistreet et al. (1986) and Stanistreet and McCarthy (1991), three major, pre-Transvaal, east-northeast–trending wrench fault systems are known to have influenced the northern edge of the Witwatersrand Basin in the Johannesburg vicinity (viz., the Kromdraai, Rietfontein, and Sugarbush fault systems), and these authors presented evidence suggesting that one or more of these fault systems may merge with those in the Barberton region (e.g., the King Fault; Fig. 7). Additional support for a possible geotectonic link between the Johannesburg and Barberton terranes includes the presence of dunite-peridotite-pyroxenite-gabbro rocks beneath the younger cover sequences in the Delmas, Bethal and Carolina areas. These rocks, encountered in borehole intersections,

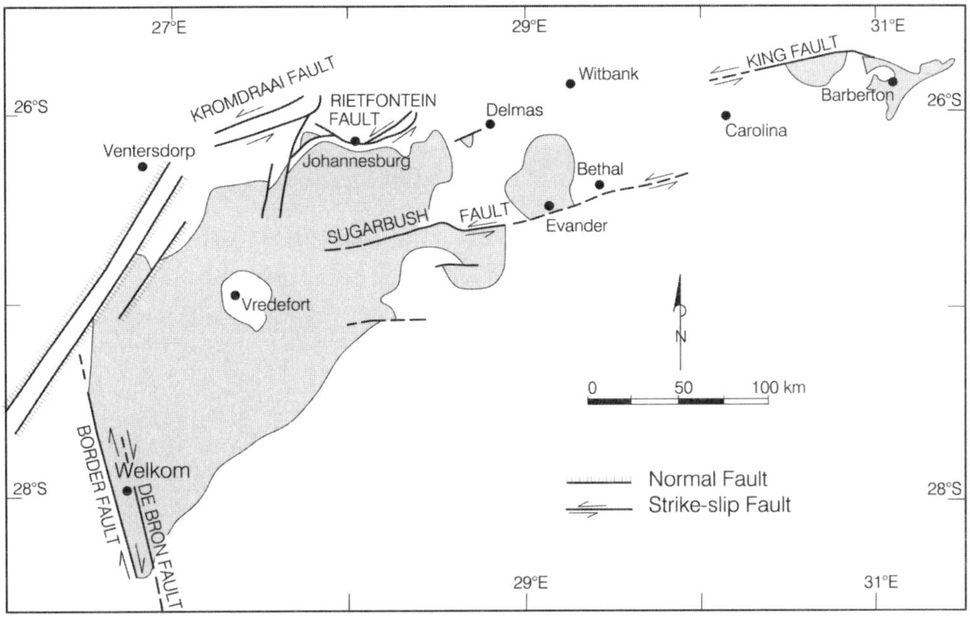

Figure 7. Simplified geological map showing the extent of the Witwatersrand Basin in the central part of the Kaapvaal Craton (gray shaded area south of Johannesburg). Some faults are shown, including the Rietfontein and Sugarbush fault systems, which may extend eastward beneath younger cover toward the Barberton greenstone belt (see text). Diagram modified after Stanistreet et al. (1986) and Stanistreet and McCarthy (1991).

were correlated with Swaziland Supergroup rocks occurring in the Barberton greenstone belt (Ferraz, 1989). If this interpretation is correct it could be argued that the ultramafic-mafic complexes in the Barberton and Johannesburg Dome areas (and the intervening covered areas) may represent components of the same craton-wide suture or terrane boundary.

Equally, it could be argued that the major Rietfontein Fault boundary structure immediately adjacent to the northern margin of the Witwatersrand Basin in the Central Rand area (Myers et al., 1990) owes its position to inherent crustal weaknesses along a pre-Witwatersrand terrane boundary or suture zone. The northern margin of the Witwatersrand Basin may also have been determined or influenced by this terrane boundary. Likewise, the Thabazimbi-Murchison tectonic fault zone could conceivably represent reactivation along a long-lived structure that may once have constituted a suture zone separating Archean crustal blocks developed north and south of the Murchison greenstone belt.

CONCLUDING REMARKS

Proposals have been made suggesting that the Kaapvaal Craton resulted from the amalgamation of a number of Archean protocontinental crustal blocks and that the nucleation of these ancient blocks was accomplished by processes akin to modern-day plate-tectonic activity. Geological and geochemical evidence continues to mount from Archean terranes suggesting that greenstone belt and TTG–granitoid evolution took place in environments not far removed from those seen today in volcanic arc environments (see papers in Kröner, 1981; Condie, 1994; De Wit and Ashwal, 1997). While comparisons between Archean and Phanerozoic geotectonic processes have become increasingly popular, there are differences that still require explanation. Issues such as higher heat flow during the Archean, the absence of high-pressure blueschist facies rocks, and "paired metamorphism," so characteristic of island arcs, together with the absence in Archean terranes of ophiolites (more specifically those not accompanied by sheeted dyke systems, as per the Penrose ophiolite definition) and tectonic mélanges, have been used to dispute direct comparisons. Some would even argue that there is no justification for any analogy between Archean and Phanerozoic tectonics (Hamilton, 1998, 2003; McCall, 2003). The debate continues!

In this contribution the position adopted is one of acceptance of Archean plate-tectonic activity, albeit that this may have taken a modified form from that which is advocated in the theory of uniformitarianism, or the age-old maxim stating that "the present is the key to the past." In an attempt to offer an explanation for the unique development of the massive serpentinite and pyroxenite bodies found in only a few restricted localities on the Kaapvaal Craton, the concept of Archean sutures emerges as a viable possibility for their origin. Their clustering in linear arrays in the Barberton, Murchison, and Johannesburg regions suggests that they may be relics of upper mantle or oceanic crust trapped in Archean plate-tectonic convergent zones separating protocontinental sialic nuclei. Acceptance of plate tectonics during the Archean also acknowledges the presence of ophiolites at this stage in earth history. However, the author only favors the modified form of the ophiolite definition being applied to the Archean (viz., the incompletely developed or dismembered ophiolites referred to by Şengör and Natal'in [2004] as ophirags).

How does one test or verify the existence of fossil suture zones extending as far back in time to the Archean? How also does one attempt to determine the geometric orientation of the subduction zone? At present there is little or no available geophysical profiling across the Kaapvaal Craton ultramafic-mafic zones highlighted in this paper. The great age of these rocks, coupled with the multiplicity of geological and geotectonic events that have occurred on the craton over the past 3 billion years or more, suggest that evidence for the presence of collisional terrane boundaries might have been destroyed as a consequence of the welding process accompanying the intrusion and fragmentation of the suture zone by later granitoid rocks. If, as has been suggested by De Wit et al. (1992), the Kaapvaal Craton is made up of a mosaic of subdomains welded together over a time span extending from ca. 3700–2700 Ma, then the possibility exists that at least some of these terrane boundaries have an identifiable physical presence, as has been proposed in this paper, thereby reducing the degree of speculation concerning their whereabouts. Archean subdomains, defined using lithological and geochronological criteria are recognizable in the Barberton granite-greenstone terrane (Lowe, 1999). By contrast, the scheme outlined by De Wit et al. (1992), whereby the Kaapvaal Craton is envisaged as having been formed by the amalgamation of upward of 12 subdomains, is largely conjectural and has not been adequately proven either in the field or on geophysical grounds.

One final consideration emerging from the suture zone suggestion proposed in this paper concerns the anomalous distribution and amount of gold on the Kaapvaal Craton (particularly that deposited in the Witwatersrand Basin, the outline of which is shown in Fig. 1). It has been calculated that production from the Witwatersrand goldfields, from commencement of mining up to 2002, has amounted to 49,332 tons of gold (Handley, 2004). Potentially, a further 40,000 tons of gold remains to be exploited. This gold occurs in mines located in a "Golden Arc" occupying an area ~8000 km^2 in size out of a total area of ~100,000 km^2 estimated for the greater Witwatersrand Basin.

The origin and source of this gold has been the subject of debate for over 100 yr (Robb and Robb, 1998), and no final solution has yet emerged. Gold production from the Witwatersrand goldfields has mainly been along the northern, northwestern, and western perimeter of the basin, leading to one suggestion that the source may have been from a crescent-shaped, juvenile volcanic arc developed around the northern and western rim of the Kaapvaal Craton (Poujol et al., 2003). However, sedimentological evidence favors the view that the gold may have been derived from a source much closer to the present-day basin edge, leading to the possible link between the Witwatersrand gold mineralization and the suture zones or terrane boundaries postulated in this paper. Likewise, it may not be fortuitous that the main distribution of gold mines and the principal production of gold from the Barberton goldfield has been from localities adjacent to the suture zone suggested in Figure 2B.

If the terrane boundary concept has any validity, the protocontinental collisional events outlined in this paper may have been preceded by ocean-floor consumption and volcanic arc development. Hydrothermal exhalative activity along such zones could have introduced auriferous solutions into the volcanic arc and basin edge formations, which were subsequently eroded and introduced into the fluvial systems as detrital particles that were deposited in the Witwatersrand Basin (Hutchinson and Viljoen, 1988).

ACKNOWLEDGMENTS

This paper has been improved following the suggestions of the reviewers, Dr. Steve Prevec (Rhodes University) and Dr. Marthinus Cloete (Council for Geoscience, Pretoria). Professor Roger Gibson (University of the Witwatersrand) also made useful recommendations to improve the manuscript. Drafting assistance was rendered by Lynda Whitfield and Lynnette Greyling.

REFERENCES CITED

Anhaeusser, C.R., 1966, Facets of the granitic assemblage on the northwest flank of the Barberton Mountain Land: Economic Geology Research Unit, University of the Witwatersrand, Johannesburg, Information Circular 32, 27 p.

Anhaeusser, C.R., 1969, The stratigraphy, structure and gold mineralization of the Jamestown and Sheba Hills area of the Barberton Mountain Land [Ph.D. thesis]: Johannesburg, University of the Witwatersrand, 322 p.

Anhaeusser, C.R., 1972, The geology of the Jamestown Hills area of the Barberton Mountain Land, South Africa: Geological Society of South Africa Transactions, v. 75, p. 225–263.

Anhaeusser, C.R., 1973a, The geology and geochemistry of the Archaean granites and gneisses of the Johannesburg-Pretoria Dome: Geological Society of South Africa Special Publication 3, p. 361–385.

Anhaeusser, C.R., 1973b, The evolution of the early Precambrian crust of southern Africa: Philosophical Transactions of the Royal Society of London, ser. A, v. 273, p. 359–388.

Anhaeusser, C.R., 1976a, The nature of chrysotile asbestos occurrences in southern Africa: A review: Economic Geology and the Bulletin of the Society of Economic Geologists, v. 71, p. 96–116.

Anhaeusser, C.R., 1976b, The geology of the Sheba Hills area of the Barberton Mountain Land, South Africa: Geological Society of South Africa Transactions, v. 79, p. 253–280.

Anhaeusser, C.R., 1977, Geological and geochemical investigations of the Roodekrans Ultramafic Complex and surrounding Archaean volcanic rocks, Krugersdorp District: Geological Society of South Africa Transactions, v. 80, p. 17–28.

Anhaeusser, C.R., 1978, The geology and geochemistry of the Muldersdrif Ultramafic Complex, Krugersdorp District: Geological Society of South Africa Transactions, v. 81, p. 193–203.

Anhaeusser, C.R., 1984, Structural elements of Archaean granite-greenstone terranes, as exemplified by the Barberton Mountain Land, South Africa, *in* Kröner, A., and Greiling, R., eds., Precambrian tectonics illustrated: Stuttgart, E. Schweizerbart'sche Verlagsbuchhandlung (Nägele u. Obermiller), p. 57–78.

Anhaeusser, C.R., 1985, Archaean layered ultramafic complexes in the Barberton Mountain Land, South Africa: Geological Association of Canada Special Paper 28, p. 281–301.

Anhaeusser, C.R., 1986, The geological setting of chrysotile asbestos occurrences in southern Africa, *in* Anhaeusser, C.R., and Maske, S., eds., Mineral deposits of southern Africa: Johannesburg, Geological Society of South Africa, v. 1, p. 359–375.

Anhaeusser, C.R., 1992, Archaean granite-greenstone relationships on the farm Zandspruit 191 IQ, North Riding area, Johannesburg Dome: South African Journal of Geology, v. 94, p. 94–101.

Anhaeusser, C.R., 1999, Archaean crustal evolution of the central Kaapvaal Craton, South Africa: Evidence from the Johannesburg Dome: South African Journal of Geology, v. 102, p. 303–322.

Anhaeusser, C.R., 2001, The anatomy of an extrusive-intrusive Archaean mafic-ultramafic sequence: The Nelshoogte Schist Belt and Stolzburg Layered Ultramafic Complex, Barberton greenstone belt, South Africa: South African Journal of Geology, v. 104, p. 167–204, doi: 10.2113/1040167.

Anhaeusser, C.R., 2002, Ultramafic complexes along the northern flank of the Barberton greenstone belt, South Africa: Remnants of oceanic lithosphere along an Archaean suture zone?: Extended Abstract, 11th Quadrennial IAGOD Symposium and Geocongress 2002, Windhoek, Geological Survey of Namibia, 1 pdf.

Anhaeusser, C.R., 2004, Palaeoarchaean to Mesoproterozoic (ca. 3500–1000 Ma) ultramafic and mafic intrusions of the Kaapvaal Craton and neighbouring metamorphic belts: A review: Johannesburg, Economic Geology Research Institute, University of the Witwatersrand, Information Circular 384, 63 p.

Anhaeusser, C.R., and Burger, A.J., 1982, An interpretation of U-Pb zircon ages for Archaean tonalitic gneisses from the Johannesburg-Pretoria granite dome: Geological Society of South Africa Transactions, v. 85, p. 111–116.

Anhaeusser, C.R., Roering, C., Viljoen, M.J., and Viljoen, R.P., 1968, The Barberton Mountain Land: A model of the elements and evolution of an Archaean fold belt: Geological Society of South Africa Transactions Annexure, v. 71, p. 225–253.

Anhaeusser, C.R., Mason, R., Viljoen, M.J., and Viljoen, R.P., 1969, A reappraisal of some aspects of Precambrian shield geology: Geological Society of America Bulletin, v. 80, p. 2175–2200.

Anonymous, 1972, Ophiolites: Geotimes, v. 17, p. 24–25.

Armstrong, R.A., Compston, W., Retief, E.A., Williams, I.S., and Welke, H.J., 1991, Zircon ion microprobe studies bearing on the age and evolution of the Witwatersrand triad: Precambrian Research, v. 53, p. 243–266, doi: 10.1016/0301-9268(91)90074-K.

Arndt, N.T., 1999, Why was flood volcanism on submerged continental platforms so common in the Precambrian?: Precambrian Research, v. 97, p. 155–164, doi: 10.1016/S0301-9268(99)00030-3.

Arndt, N.T., and Nisbet, E.G., eds., 1982, Komatiites: London, George Allen and Unwin, 526 p.

Arndt, N.T., Albarède, F., and Nisbet, E.G., 1997, Mafic and ultramafic magmatism, *in* de Wit, M.J., and Ashwal, L.D., eds., Greenstone belts: Oxford, Clarendon Press, p. 233–254.

Barley, M.E., 1997, The Pilbara Craton, *in* De Wit, M.J., and Ashwal, L.D., eds., Greenstone belts: Oxford, Clarendon Press, p. 657–664.

Barton, C.M., 1986, The Havelock asbestos deposit in Swaziland, Barberton greenstone belt, *in* Anhaeusser, C.R., and Maske, S., eds., Mineral deposits of southern Africa, Johannesburg, Geological Society of South Africa, v. 1, p. 395–407.

Barton, J.M., Jr., Barton, E.S., and Kröner, A., 1999, Age and isotopic evidence for the origin of the Archaean granitoid intrusives of the Johannesburg Dome, South Africa: Journal of African Earth Sciences, v. 28, p. 693–702, doi: 10.1016/S0899-5362(99)00039-1.

Bickle, M.J., Nisbet, E.G., and Martin, A., 1995, Archaean greenstone belts are not oceanic crust: The Journal of Geology, v. 102, p. 121–138.

Brandl, G., compiler, 1985, Geological Series map (2330 Tzaneen): Pretoria, Geological Survey of South Africa, scale 1:250,000.

Brandl, G., and De Wit, M.J., 1997, The Kaapvaal Craton, *in* De Wit, M.J., and Ashwal, L.D., eds., Greenstone belts: Oxford, Clarendon Press, p. 581–607.

Brandl, G., and Kröner, A., 1993, Preliminary results of single zircon studies from various Archaean rocks of the northeastern Transvaal: Mbabane, Swaziland, Extended Abstract, 16th International Colloquium of African Geology, p. 54–56.

Brock, B.B., and Pretorius, D.A., 1964a, An introduction to the stratigraphy and structure of the Rand Goldfields, *in* Haughton, S.H., ed., The geology of some ore deposits in southern Africa: The Geological Society of South Africa, v. I, p. 25–61.

Brock, B.B., and Pretorius, D.A., 1964b, Rand basin sedimentation and tectonics, *in* Haughton, S.H., ed., The geology of some ore deposits in southern Africa: The Geological Society of South Africa, v. I, p. 549–599.

Burke, K., Dewey, J.F., and Kidd, W.S.F., 1977, World distribution of sutures: The sites of former oceans: Tectonophysics, v. 40, p. 69–99, doi: 10.1016/0040-1951(77)90030-0.

Burke, K., Kidd, W.S.F., and Kusky, T.M., 1985, Is the Ventersdorp rift system of southern Africa related to a continental collision between the Kaapvaal and Zimbabwe Cratons at 2.64 Ga ago?: Tectonophysics, v. 115, p. 1–24, doi: 10.1016/0040-1951(85)90096-4.

Burke, K., Kidd, W.S.F., and Kusky, T.M., 1986, Archean foreland basin tectonics in the Witwatersrand, South Africa: Tectonics, v. 5, p. 439–456.

Büttner, W., 1984, The serpentinites and related rocks of the Msauli asbestos deposit in the Archaean Barberton greenstone belt, South Africa: Johannesburg, Economic Geology Research Unit, University of the Witwatersrand, Information Circular No. 168, 32 p.

Button, A., 1973, A regional study of the stratigraphy and development of the Transvaal Basin in the eastern and northern Transvaal: [Ph.D. thesis]: Johannesburg, University of the Witwatersrand, 352 p.

Chaumba, J.B., 1992, The geology and geochemistry of part of the Edenvale granite-greenstone terrane, Johannesburg-Pretoria granite dome: [B.S. dissertation]: Johannesburg, University of the Witwatersrand, 62 p.

Cloete, M., 1999, Aspects of volcanism and metamorphism of the Onverwacht Group lavas in the southwestern portion of the Barberton greenstone belt: Geological Survey of South Africa Memoir 84, 232 p.

Coleman, R.G., 1971, Plate tectonic emplacement of upper mantle peridotites along continental edges: Journal of Geophysical Research, v. 76, p. 1212–1222.

Compston, W., and Kröner, A., 1988, Multiple zircon growth within early Archean tonalitic gneiss from the Ancient Gneiss Complex, Swaziland: Earth and Planetary Science Letters, v. 87, p. 13–28, doi: 10.1016/0012-821X(88)90061-1.

Condie, K.C., ed., 1994, Archean crustal evolution: Amsterdam, Elsevier, 528 p.

Corner, B., Durrheim, R.J., and Nicolaysen, L.O., 1990, Relationships between the Vredefort structure and the Witwatersrand Basin within the tectonic framework of the Kaapvaal Craton as interpreted from regional gravity and aeromagnetic data: Tectonophysics, v. 171, p. 49–61, doi: 10.1016/0040-1951(90)90089-Q.

Dann, J.C., 2000, The 3.5 Ga Komati Formation, Barberton greenstone belt, South Africa, Part I: New maps and magmatic architecture: South African Journal of Geology, v. 103, p. 47–68, doi: 10.2113/103.1.47.

Dawson, J.B., 1967, A review of the geology of kimberlite, *in* Wyllie, P.J., ed., Ultramafic and related rocks: New York, John Wiley and Sons, p. 241–251.

De Ronde, C.E.J., and De Wit, M.J., 1994, Tectonic history of the Barberton greenstone belt, South Africa: 490 million years of Archean crustal evolution: Tectonics, v. 13, p. 983–1005, doi: 10.1029/94TC00353.

De Ronde, C.E.J., and Kamo, S., 2000, An Archaean arc-arc collisional event: A short-lived (ca. 3 m.y.) episode, Weltevreden area, Barberton greenstone belt, South Africa: Journal of African Earth Sciences, v. 30, p. 219–248, doi: 10.1016/S0899-5362(00)00017-8.

Dewey, J.F., 1969, Evolution of the Appalachian/Caledonian orogen: Nature, v. 221, p. 124–129.

Dewey, J.F., 1977, Suture zone complexities: A review: Tectonophysics, v. 40, p. 53–67, doi: 10.1016/0040-1951(77)90029-4.

Dewey, J.F., 2003, Ophiolites and lost oceans: Rifts, ridges, arcs, and/or scrapings?, *in* Dilek, Y., and Newcomb, S., eds., Ophiolite concept and the evolution of geological thought: Geological Society of America Special Paper 373, p. 153–158.

Dewey, J.F., and Bird, J.M., 1970, Mountain belts and the new global tectonics: Journal of Geophysical Research, v. 75, p. 2625–2647.

Dewey, J.F., and Bird, J.M., 1971, Origin and emplacement of the ophiolite suite: Appalachian ophiolites in Newfoundland: Journal of Geophysical Research, v. 76, p. 3179–3206.

Dewey, J.F., and Kidd, W.S.F., 1974, Continental collisions in the Appalachian-Caledonian belt: Variations related to complete and incomplete suturing: Geology, v. 2, p. 543–546, doi: 10.1130/0091-7613(1974)2<543:CCITAO>2.0.CO;2.

Dewey, J.F., and Spall, H., 1975, Pre-Mesozoic plate tectonics: Geology, v. 3, p. 422–424, doi: 10.1130/0091-7613(1975)3<422:PPTHFB>2.0.CO;2.

De Wit, M.J., 1991, Archaean greenstone belt tectonism and basin development: Some insights from the Barberton and Pietersburg greenstone belts, Kaapvaal Craton, South Africa: Journal of African Earth Sciences, v. 13, p. 45–63, doi: 10.1016/0899-5362(91)90043-X.

De Wit, M.J., and Ashwal, L.D., eds., 1997, Greenstone belts: Oxford, Clarendon Press, 809 p.

De Wit, M.J., and Tinker, M., 2004, Crustal structures across the central Kaapvaal Craton from deep-seismic reflection data: South African Journal of Geology, v. 107, p. 185–206, doi: 10.2113/107.1-2.185.

De Wit, M.J., and Tredoux, M., 1988, PGE in the 3.5 Ga Jamestown Ophiolite Complex, Barberton greenstone belt, with implications for PGE distribution in simatic lithosphere, *in* Prichard, H.M., Potts, P.J., Bowles, J.F.W., and Cribbs, S.J., eds., Geo-Platinum 87: London, Elsevier Applied Science, p. 319–341.

De Wit, M.J., Hart, R.A., and Hart, R.J., 1987, The Jamestown Ophiolite Complex, Barberton mountain belt: A section through 3.5 Ga oceanic crust: Journal of African Earth Sciences, v. 6, p. 681–730, doi: 10.1016/0899-5362(87)90007-8.

De Wit, M.J., Roering, C., Hart, R.J., Armstrong, R.A., De Ronde, C.E.J., Green, R.W.E., Tredoux, M., Peberdy, E., and Hart, R.A., 1992, Formation of an Archaean continent: Nature, v. 357, p. 553–562, doi: 10.1038/357553a0.

Doucouré, C.M., and De Wit, M.J., 2002, Temporal variation in rigidity and mechanical behaviour of old thick continental lithosphere: South African Journal of Geology, v. 105, p. 39–50, doi: 10.2113/1050039.

Du Plessis, C.P., 1991, Tectonics along the Thabazimbi-Murchison Lineament: [Ph.D. thesis]: Johannesburg, University of the Witwatersrand, 307 p.

Du Plessis, C.P., and Walraven, F., 1990, The tectonic setting of the Bushveld Complex in southern Africa, Part 1, Structural deformation and distribution: Tectonophysics, v. 179, p. 305–319, doi: 10.1016/0040-1951(90)90296-K.

Du Toit, M.C., Van Reenen, D.D., and Roering, C., 1983, Some aspects of the geology, structure and metamorphism of the Southern Marginal Zone of the Limpopo Metamorphic Complex: Geological Society of South Africa Special Publication 8, p. 121–142.

Eales, H.V., and Cawthorn, R.G., 1996, The Bushveld Complex, *in* Cawthorn, R.G., ed., Layered intrusions: Amsterdam, Elsevier, 531 p.

Eglington, B.M., and Armstrong, R.A., 2004, The Kaapvaal Craton and adjacent orogens, southern Africa: A geochronological database and overview of the geological development of the craton: South African Journal of Geology, v. 107, p. 13–32, doi: 10.2113/107.1-2.13.

Eriksson, K.A., 1980, Transitional sedimentation styles in the Moodies and Fig Tree Groups, Barberton Mountain Land, South Africa: Evidence favouring an Archaean continental margin: Precambrian Research, v. 12, p. 141–160, doi: 10.1016/0301-9268(80)90027-3.

Ferraz, M.F.F., 1989, The nature of the Archaean basement in the provenance areas of the East Rand and Evander goldfields: [M.S. thesis]: Johannesburg, University of the Witwatersrand, 133 p.

Fripp, R.E.P., and Jones, M.G., 1997, Sheeted intrusions and peridotite-gabbro assemblages in the Yilgarn Craton, Western Australia: Elements of Archaean ophiolites, *in* De Wit, M.J., and Ashwal, L.D., eds., Greenstone belts: Oxford, Clarendon Press, p. 422–437.

Fripp, R.E.P., Van Nierop, D.A., Callow, M.J., Lilly, P.A., and Du Plessis, L.U., 1980, Deformation in part of the Archaean Kaapvaal Craton, South Africa: Precambrian Research, v. 13, p. 241–251, doi: 10.1016/0301-9268(80)90006-6.

Gibson, R.L., and Reimold, W.U., 2001, The Vredefort Impact Structure, South Africa—The scientific evidence and a two-day excursion guide: Council for Geoscience, South Africa, Memoir 92, 111 p.

Gold, D.P., 1967, Alkaline ultrabasic rocks in the Montreal area, Quebec, *in* Wyllie, P.J., ed., Ultramafic and related rocks: New York, John Wiley and Sons, p. 288–302.

Good, N., and De Wit, M.J., 1997, The Thabazimbi-Murchison lineament of the Kaapvaal Craton, South Africa: 2700 Ma of episodic deformation: Journal of the Geological Society of London, v. 154, p. 93–97.

Hall, A.L., 1912, The geology of the Murchison Range and district: Geological Survey of South Africa Memoir 6, 184 p.

Hall, A.L., 1918, The geology of the Barberton gold mining district: Geological Survey of South Africa Memoir 9, 347 p.

Hamilton, W.B., 1970, The Uralides and the motion of the Russian and Siberian platforms: Geological Society of America Bulletin, v. 81, p. 2553–2576.

Hamilton, W.B., 1998, Archaean magmatism and deformation were not products of plate tectonics: Precambrian Research, v. 91, p. 143–179, doi: 10.1016/S0301-9268(98)00042-4.

Hamilton, W.B., 2003, An alternative earth: GSA Today, v. 13, no. 11, p. 4–12.

Handley, J.R.F., 2004, Historic overview of the Witwatersrand goldfields: Howick, South Africa, Handley, 224 p.

Hart, R.J., Moser, D., and Andreoli, M., 1999, Archaean age for the granulite facies metamorphism near the center of the Vredefort structure, South Africa: Geology, v. 27, p. 1091–1094, doi: 10.1130/0091-7613(1999)027<1091:AAFTGF>2.3.CO;2.

Heinrichs, T.K., and Reimer, T.O., 1977, A sedimentary barite deposit from the Archean Fig Tree Group of the Barberton Mountain Land (South Africa): Economic Geology and the Bulletin of the Society of Economic Geologists, v. 72, p. 1426–1441.

Heubeck, C., and Lowe, D.R., 1999, Sedimentary petrography and provenance of the Archean Moodies Group, Barberton greenstone belt, *in* Lowe, D.R., and Byerly, G.R., eds., Geologic evolution of the Barberton greenstone belt, South Africa: Geological Society of America Special Paper 329, p. 259–286.

Heubeck, C., Wendt, J.I., Toulkeridis, T., Kröner, A., and Lowe, D.R., 1993, Timing of deformation of the Archaean Barberton greenstone belt, South Africa: Constraints from zircon dating of the Salisbury Kop pluton: South African Journal of Geology, v. 96, p. 1–8.

Hutchinson, R.W., and Viljoen, R.P., 1988, Re-evaluation of gold source in Witwatersrand ores: South African Journal of Geology, v. 91, p. 157–173.

Irvine, T.N., 1967, The Duke Island ultramafic complex, southeastern Alaska, *in* Wyllie, P.J., ed., Ultramafic and related rocks: New York, John Wiley & Sons, p. 84–97.

Irvine, T.N., and Smith, C.H., 1967, The ultramafic rocks of the Muskox intrusion, Northwest Territories, Canada, *in* Wyllie, P.J., ed., Ultramafic and related rocks: New York, John Wiley & Sons, p. 38–49.

Kamo, S., and Davis, D.W., 1994, Reassessment of Archaean crustal development in the Barberton Mountain Land, South Africa based on U-Pb dating: Tectonics, v. 13, p. 167–192, doi: 10.1029/93TC02254.

Kamo, S.L., Reimold, W.U., Krogh, T.E., and Colliston, W.P., 1996, A 2.023 Ga age for the Vredefort impact event and a first report of shock metamorphosed zircons in pseudotachylite breccias and granophyre: Earth and Planetary Science Letters, v. 144, p. 369–388, doi: 10.1016/S0012-821X(96)00180-X.

Kearey, P., and Vine, F.J., 1995, Global tectonics: Oxford, Blackwell Science, 302 p.

Kisters, A.F.M., Stevens, G., Dziggel, A., and Armstrong, R.A., 2003, Extensional detachment faulting and core-complex formation in the southern

Barberton granite-greenstone terrain, South Africa: Evidence for a 3.2 Ga orogenic collapse: Precambrian Research, v. 127, p. 355–378, doi: 10.1016/j.precamres.2003.08.002.

Kohler, E.A., 1994, The geology of the Archaean granitoid-greenstone terrane in the vicinity of Three Sisters, Barberton greenstone belt: [Ph.D. thesis]: Johannesburg, University of the Witwatersrand, 309 p.

Kröner, A., ed., 1981, Precambrian plate tectonics: Amsterdam, Elsevier, 781 p.

Kröner, A., and Compston, W., 1988, Ion microprobe ages of zircons from early Archaean granite pebbles and greywacke, Barberton greenstone belt, southern Africa: Precambrian Research, v. 38, p. 367–380, doi: 10.1016/0301-9268(88)90034-4.

Kröner, A., Byerly, G.R., and Lowe, D.R., 1991, Chronology of early Archean granite-greenstone evolution in the Barberton Mountain Land, South Africa, based on precise dating by single grain zircon evaporation: Earth and Planetary Science Letters, v. 103, p. 41–54, doi: 10.1016/0012-821X(91)90148-B.

Kröner, A., Hegner, E., Byerly, G.R., and Lowe, D.R., 1992, Possible terrane identification in the early Archean Barberton greenstone belt, South Africa, using single zircon geochronology: Eos (Transactions, American Geophysical Union), v. 73, p. 616.

Kröner, A., Hegner, E., Wendt, J.I., and Byerly, G.R., 1996, The oldest part of the Barberton granitoid-greenstone terrain, South Africa: Evidence for crust formation at 3.5 and 3.7 Ga: Precambrian Research, v. 78, p. 105–124, doi: 10.1016/0301-9268(95)00072-0.

Kröner, A., Jaeckel, P., and Brandl, G., 2000, Single zircon ages for felsic to intermediate rocks from the Pietersburg and Giyani greenstone belts and bordering granitoid orthogneisses, northern Kaapvaal Craton, South Africa: Journal of African Earth Sciences, v. 30, p. 773–793, doi: 10.1016/S0899-5362(00)00052-X.

Kruger, F.J., Kamber, B.S., and Harris, P.H., 1998, Isotopic peculiarities of an Archaean pegmatite (Union Mine, Mica, South Africa): Geochemical and geochronological implications: Precambrian Research, v. 91, p. 253–267, doi: 10.1016/S0301-9268(98)00052-7.

Kusky, T.M., Jiang-Hai, Li, and Tucker, R.D., 2001, The Archean Dongwanzi ophiolite complex, North China Craton: 2,505 billion year old crust and mantle: Science, v. 292, p. 1142–1145.

Lana, C., 2003, Structural and petrogenetic studies related to the geological evolution of the Archaean basement complex of the Vredefort Dome, South Africa: [Ph.D. thesis]: Johannesburg, University of the Witwatersrand, 328 p.

Lana, C., Reimold, W.U., Gibson, R.L., Koeberl, C., and Siegesmund, S., 2004, Nature of the Archean midcrust in the core of the Vredefort Dome, Central Kaapvaal Craton, South Africa: Geochimica et Cosmochimica Acta, v. 68, p. 623–642, doi: 10.1016/S0016-7037(03)00447-2.

Lippard, S.J., Shelton, A.W., and Gass, I.G., 1986, The ophiolite of northern Oman: Geological Society of London Memoir 11, 178 p.

Lowe, D.R., 1994, Accretionary history of the Archean Barberton greenstone belt (3.55–3.22 Ga) southern Africa: Geology, v. 22, p. 1099–1102, doi: 10.1130/0091-7613(1994)022<1099:AHOTAB>2.3.CO;2.

Lowe, D.R., 1999, Geologic evolution of the Barberton greenstone belt and vicinity, *in* Lowe, D.R., and Byerly, G.R., eds., Geologic evolution of the Barberton greenstone belt, South Africa: Geological Society of America Special Paper 329, p. 287–312.

Lowe, D.R., Byerly, G.R., and Kröner, A., 1989, Relationship between episodic dacitic volcanism and tonalitic plutonism in the Archean Barberton greenstone belt (BGB), South Africa: Eos (Transactions, American Geophysical Union), v. 70, p. 139.

Lowe, D.R., Byerly, G.R., and Heubeck, C., 1999, Structural divisions and development of the west-central part of the Barberton greenstone belt, *in* Lowe, D.R., and Byerly, G.R., eds., Geologic evolution of the Barberton greenstone belt, South Africa: Geological Society of America Special Paper 329, p. 37–82.

Matthews, P.E., 1972, Possible Precambrian obduction and plate tectonics in southeast Africa: Nature, v. 240, p. 37–39, doi: 10.1038/240037a0.

Matthews, P.E., and Charlesworth, E.G., 1981, A geological review of the northern margin of the Proterozoic Namaqua-Natal mobile belt in Natal: Guidebook to post-congress excursion to Natal: Pretoria, Geological Society of South Africa Geocongress, 83 p.

McBirney, A.R., 1996, The Skaergaard intrusion, *in* Cawthorn, R.G., ed., Layered intrusions: Amsterdam, Elsevier, p. 147–180.

McCall, G.J.H., 2003, A critique of the analogy between Archaean and Phanerozoic tectonics based on regional mapping of the Mesozoic-Cenozoic plate convergent zone in the Makran, Iran: Precambrian Research, v. 127, p. 5–17, doi: 10.1016/S0301-9268(03)00178-5.

McCallum, I.S., 1996, The Stillwater Complex, *in* Cawthorn, R.G., ed., Layered intrusions: Amsterdam, Elsevier, p. 441–483.

McCourt, S., 1995, The crustal architecture of the Kaapvaal crustal block South Africa, between 3.5 and 2.0 Ga: Mineralium Deposita, v. 30, p. 89–97, doi: 10.1007/BF00189338.

McCourt, S., and Vearncombe, J.R., 1987, Shear zones bounding the Central Zone of the Limpopo Mobile Belt, Southern Africa: Journal of Structural Geology, v. 9, p. 127–137, doi: 10.1016/0191-8141(87)90021-6.

McCourt, S., Armstrong, R.A., and Johnston, S.T., 2000, New U-Pb SHRIMP ages from the Tugela terrane: Implications for the evolution of the Natal belt [abs.]: Durban, IGCP 418, 4th International Field Meeting, p. 26.

Mogk, D.W., Mueller, P.A., and Wooden, J.L., 1992, The nature of Archean terrane boundaries: An example from the northern Wyoming Province: Precambrian Research, v. 55, p. 155–168, doi: 10.1016/0301-9268(92)90020-O.

Moores, E.M., and Vine, F.J., 1971, Troodos Massif, Cyprus and other ophiolites as oceanic crust: Evaluation and implications: Royal Society of London Philosophical Transactions, ser. A, v. 268, p. 443–466.

Myers, J.S., and Swagers, C., 1997, The Yilgarn Craton, *in* De Wit, M.J., and Ashwal, L.D., eds., Greenstone belts: Oxford, Clarendon Press, p. 640–656.

Myers, R.E., McCarthy, T.S., and Stanistreet, I.G., 1990, A tectono-sedimentary reconstruction of the development and evolution of the Witwatersrand Basin, with particular emphasis on the Central Rand Group: South African Journal of Geology, v. 93, p. 180–201.

Nisbet, E.G., 1987, The young Earth—an introduction to Archaean geology: Boston, Allen and Unwin, 402 p.

Potgieter, G.J., 1992, Tektonisme langs die noordoostelike rand van die Bosveldkompleks, Suid-Afrika: [Ph.D. thesis]: Pretoria, University of Pretoria, 214 p.

Poujol, M., 2001, U-Pb isotopic evidence for episodic granitoid emplacement in the Murchison greenstone belt, South Africa: Journal of African Earth Sciences, v. 33, p. 155–163, doi: 10.1016/S0899-5362(01)90096-X.

Poujol, M., and Anhaeusser, C.R., 2001, The Johannesburg Dome, South Africa: New single zircon U-Pb isotopic evidence for early Archaean granite-greenstone development within the central Kaapvaal Craton: Precambrian Research, v. 108, p. 139–157, doi: 10.1016/S0301-9268(00)00161-3.

Poujol, M., and Robb, L.J., 1999, New U-Pb zircon ages on gneisses and pegmatites from south of the Murchison greenstone belt, South Africa: South African Journal of Geology, v. 102, p. 93–97.

Poujol, M., Robb, L.J., Respaut, J.P., and Anhaeusser, C.R., 1996, 3.07–2.97 Ga greenstone belt formation in the northeastern Kaapvaal Craton: Implications for the origin of the Witwatersrand Basin: Economic Geology and the Bulletin of the Society of Economic Geologists, v. 91, p. 1455–1461.

Poujol, M., Robb, L.J., Anhaeusser, C.R., and Gericke, B., 2003, A review of the geochronological constraints on the evolution of the Kaapvaal Craton, South Africa: Precambrian Research, v. 127, p. 181–213, doi: 10.1016/S0301-9268(03)00187-6.

Pretorius, D.A., 1976, The nature of the Witwatersrand gold-uranium deposits, *in* Wolf, K.H., ed., Handbook of strata-bound and stratiform ore deposits: Amsterdam, Elsevier, v. 7, p. 29–88.

Prevec, S.A., Anhaeusser, C.R., and Poujol, M., 2004, Evidence for Archaean lamprophyre from the Kaapvaal Craton, South Africa: South African Journal of Science, v. 100, p. 549–555.

Robb, L.J., 1977, The geology and geochemistry of the Archaean granite-greenstone terrane between Nelspruit and Bushbuckridge, Eastern Transvaal: [M.S. thesis]: Johannesburg, University of the Witwatersrand, 190 p.

Robb, L.J., and Robb, V.M., 1986, Archaean pegmatite deposits in the north-eastern Transvaal, *in* Anhaeusser, C.R., and Maske, S., eds., Mineral deposits of southern Africa: Johannesburg, Geological Society of South Africa, v. 1, p. 437–449.

Robb, L.J., and Robb, V.M., 1998, Gold in the Witwatersrand Basin, *in* Wilson, M.G.C., and Anhaeusser, C.R., eds., The mineral resources of South Africa: Council for Geoscience, South Africa, Handbook 16, p. 294–349.

Robertson, A.H.F., and Xenophontos, C., 1993, Development of concepts concerning the Troodos ophiolite and adjacent units in Cyprus, *in* Prichard, H.M., Alabaster, T., Harris, N.B.W., and Neary, C.R., eds., Magmatic processes and plate tectonics: Geological Society of London Special Publication 76, p. 85–119.

Rodel, J.E., 1993, The petrography and geochemistry of the Stolzburg and Rosentuin layered ultramafic complexes, Barberton Mountain Land, eastern Transvaal [M.S. thesis]: Johannesburg, University of the Witwatersrand, 232 p.

Schmitz, M.D., Bowring, S.A., De Wit, M.J., and Gartz, V., 2004, Subduction and terrane collision stabilize the western Kaapvaal Craton tectosphere 2.9 billion years ago: Earth and Planetary Science Letters, v. 222, p. 363–376, doi: 10.1016/j.epsl.2004.03.036.

Şengör, A.M.C., and Natal'in, B.A., 2004, Phanerozoic analogues of Archaean oceanic basement fragments: Altaid ophiolites and ophirags, *in* Kusky, T.M., ed., Precambrian ophiolites and related rocks: Developments in Precambrian Geology: Amsterdam, Elsevier, v. 13, p. 675–726.

Smit, C.A., Roering, C., and Van Reenen, D.D., 1992, The structural framework of the southern margin of the Limpopo Belt, South Africa: Precambrian Research, v. 55, p. 51–67, doi: 10.1016/0301-9268(92)90014-F.

Smith, C.H., 1958, Bay of Islands igneous complex, western Newfoundland: Canadian Geological Survey Memoir 290, 132 p.

Stanistreet, I.G., and McCarthy, T.S., 1991, Changing tectono-sedimentary scenarios relevant to the development of the late Archaean Witwatersrand Basin: Journal of African Earth Sciences, v. 13, p. 65–81, doi: 10.1016/0899-5362(91)90044-Y.

Stanistreet, I.G., McCarthy, T.S., Charlesworth, E.G., Myers, R.E., and Armstrong, R.A., 1986, Pre-Transvaal wrench tectonics along the northern margin of the Witwatersrand Basin, South Africa: Tectonophysics, v. 131, p. 53–74, doi: 10.1016/0040-1951(86)90267-2.

Stott, G.M., 1997, The Superior Province, Canada, *in* De Wit, M.J., and Ashwal, L.D., eds., Greenstone belts: Oxford, Clarendon Press, p. 480–507.

Talbot, C.J., 1973, A plate tectonic model for the Archaean crust: Philosophical Transactions of the Royal Society of London, v. A273, p. 413–428.

Taylor, H.P., 1967, The zoned ultramafic complexes of southeastern Alaska, *in* Wyllie, P.J., ed., Ultramafic and related rocks: New York, John Wiley & Sons, p. 97–121.

Tegtmeyer, A.R., and Kröner, A., 1987, U-Pb zircon ages bearing on the nature of early Archean greenstone belt evolution Barberton Mountain Land southern Africa: Precambrian Research, v. 36, p. 1–20, doi: 10.1016/0301-9268(87)90014-3.

Thayer, T.P., 1967, Chemical and structural relations of ultramafic and feldspathic rocks in Alpine intrusive complexes, *in* Wyllie, P.J., ed., Ultramafic and related rocks: New York, John Wiley & Sons, p. 222–239.

Thomas, R.J., 1989, A tale of two tectonic terranes: South African Journal of Geology, v. 92, p. 306–321.

Van Biljon, W.J., 1980, Plate tectonics and the origin of the Witwatersrand Basin: Proceedings of the Fifth Quadrenial IAGOD Symposium, v. 1, Stuttgart, E. Schweizerbart'sche Verlagsbuchandlung, p. 217–226.

Van Eeden, O.R., Partridge, F.C., Kent, L.E., and Brandt, J.W., 1939, The mineral deposits of the Murchison Range east of Leydsdorp: Geological Survey of South Africa Memoir 36, 172 p.

Van Niekerk, C.B., and Burger, A.J., 1978, The age of the Moodies conglomerate boulders: Geological Society of South Africa Special Publication 4, p. 99–106.

Vearncombe, J.R., 1991, A possible Archaean island arc in the Murchison belt, Kaapvaal Craton, South Africa: Journal of African Earth Sciences, v. 13, p. 299–304, doi: 10.1016/0899-5362(91)90093-E.

Vearncombe, J.R., Barton, J.M., Cheshire, P.E., De Beer, J.H., Stettler, E.H., and Brandl, G., 1992, Geology, geophysics and mineralization of the Murchison schist belt, Rooiwater Complex and surrounding granitoids: Geological Survey of South Africa Memoir 81, 139 p.

Viljoen, M.J., and Viljoen, R.P., 1969, The geology and geochemistry of the Lower Ultramafic Unit of the Onverwacht Group and a proposed new class of igneous rock: Geological Society of South Africa Special Publication 2, p. 55–85.

Viljoen, M.J., Van Vuuren, C.J.J., Pearton, T.N., Minnitt, R.C.A., Muff, R., and Cilliers, P., 1978, The regional geological setting of mineralization in the Murchison Range with particular reference to antimony: Geological Society of South Africa Special Publication 4, p. 55–76.

Viljoen, M.J., Viljoen, R.P., Smith, H.S., and Erlank, A.J., 1983, Geological, textural and geochemical features of komatiitic flows from the Komati Formation: Geological Society of South Africa Special Publication 9, p. 1–20.

Viljoen, M.J., Coward, D., and Wedepohl, C., 1993, Geological mapping of the granite-greenstone terrane between the Murchison and Sutherland ranges, Transvaal, using SPOT data: South African Journal of Science, v. 89, p. 436–441.

Viljoen, R.P., and Viljoen, M.J., 1970, The geology and geochemistry of the layered ultramafic bodies of the Kaapmuiden area, Barberton Mountain Land: Geological Society of South Africa Special Publication 1, p. 661–688.

Vinnik, L.P., Green, R.W.E., and Nicolaysen, L.O., 1995, Recent deformations on the deep continental roots in southern Africa: Nature, v. 375, p. 50–52, doi: 10.1038/375050a0.

Visser, D.J.L., compiler, 1956, The geology of the Barberton area: Geological Survey of South Africa Special Publication 15, 253 p.

Walraven, F., compiler, 1986, Geological Series map (2430 Pilgrim's Rest): Pretoria, Geological Survey of South Africa, scale 1:250,000.

Ward, J.H.W., and Wilson, M.G.C., 1998, Gold outside the Witwatersrand Basin, *in* Wilson, M.G.C., and Anhaeusser, C.R., eds., The mineral resources of South Africa: Council for Geoscience, South Africa, Handbook 16, p. 350–386.

Wilson, A.H., 1996, The Great Dyke of Zimbabwe, *in* Cawthorn, R.G., ed., Layered intrusions: Amsterdam, Elsevier, p. 365–402.

Windley, B.F., 1976, New tectonic models for the evolution of Archaean continents and oceans, *in* Windley, B.F., ed., The early history of the Earth: London, John Wiley & Sons, p. 105–111.

Winter, H., de la R., 1987, A cratonic foreland model for Witwatersrand Basin development in a continental back-arc, plate-tectonic setting: South African Journal of Geology v. 90, p. 409–427.

Wuth, M., 1980, The geology and mineralization potential of the Oorschot-Weltevreden schist belt, south-west of Barberton—Eastern Transvaal [M.S. thesis]: Johannesburg, University of the Witwatersrand, 185 p.

Wyllie, P.J., 1967, Ultramafic and ultrabasic rocks, *in* Wyllie, P.J., ed., Ultramafic and related rocks: New York, John Wiley & Sons, p. 1–7.

Zhai, M., Zhao, G., Zhang, Q., Kusky, T.M., and Li, J.-H., 2002, Is the Dongwanzi Complex an Archean ophiolite?: Science, v. 295, p. 923, doi: 10.1126/science.295.5557.923a.

MANUSCRIPT ACCEPTED BY THE SOCIETY 4 OCTOBER 2005

Printed in the USA

Geological Society of America
Special Paper 405
2006

Syntectonic emplacement and deformation of the Heerenveen batholith: Conjectures on the structural setting of the 3.1 Ga granite magmatism in the Barberton granite-greenstone terrain, South Africa

Richard W. Belcher
Alexander F.M. Kisters†
Department of Geology, University of Stellenbosch, Private Bag X1, Matieland, 7602, Republic of South Africa

ABSTRACT

The Heerenveen batholith in the Mesoarchean Barberton granite-greenstone terrain is a horizontal, tabular, and compositionally composite granite body belonging to the 3.1 Ga magmatic suite of granodiorites, monzogranites, and syenogranites, the granodiorite-monzogranite-syenogranite (GMS) suite, that has intruded large parts of the Kaapvaal Craton. In this paper, we present structural data of magmatic and penetrative solid-state fabrics of the Heerenveen batholith that was hitherto considered to be undeformed. The orientation and kinematics of conjugate sets of synmagmatic shear zones along the margins and within the Heerenveen batholith, the presence of regionally persistent magmatic and solid-state fabrics, and the progressive deformation of dikes and sills related to the pluton all point to the syntectonic emplacement of the granites during bulk coaxial, northwest-southeast subhorizontal crustal shortening. Both the timing as well as the strain recorded by the Heerenveen batholith and other plutons of the 3.1 Ga GMS suite in the Barberton granite-greenstone terrain indicate an emplacement during the late-stage D3 convergent tectonics that is responsible for the northeast-southwest–trending main structural grain within the Barberton greenstone belt. One of the characteristic features of the D3 deformation in the granite-gneiss terrain is the very pronounced localization of the bulk strain into melt-bearing structures such as synmagmatic shear zones and granite sheets. Variations in the D3 strain intensity recorded in wide parts of the Barberton greenstone belt may be related to this strain localization, reflecting the proximity or absence of syntectonic intrusions. The late-stage transtensional and/or extensional deformation of the granite-greenstone terrain shortly follows the intrusion of the extensive tabular granites. The transition from convergent to divergent tectonics may be related to the gravitational collapse of the crust above the upper crustal melt sheets. We speculate that the setting of the GMS suite in the Barberton granite-greenstone terrain may be a contractional continental back arc, related to collisional plate-margin processes in the central parts of the Kaapvaal Craton.

Keywords: Kaapvaal Craton, Barberton granite-greenstone terrain, granite emplacement, Archean geology.

†Corresponding author e-mail: akisters@sun.ac.za.

Belcher, R.W., and Kisters, A.F.M., 2006, Syntectonic emplacement and deformation of the Heerenveen batholith: Conjectures on the structural setting of the 3.1 Ga granite magmatism in the Barberton granite-greenstone terrain, South Africa, *in* Reimold, W.U., and Gibson, R.L., Processes on the Early Earth: Geological Society of America Special Paper 405, p. 211–231, doi: 10.1130/2006.2405(12). For permission to copy, contact editing@geosociety.org.

INTRODUCTION

The first regional-scale cratonization of the central-eastern parts of the Kaapvaal Craton in southern Africa at ca. 3.1 Ga coincided with an episode of short-lived but voluminous potassic plutonism (e.g., Hunter, 1973; Anhaeusser and Robb, 1983; Robb et al., 1983; De Wit et al., 1992; Kamo and Davis, 1994). This plutonic event gave rise to granites, monzogranites and syenogranites, the so-called GMS (granodiorite-monzogranite-syenogranite) suite, that record for the first time the large-scale generation of granite magmas *sensu stricto,* indicating the widespread melting of preexisting continental crust mainly made up of tonalites, trondhjemites, and granodiorites of earlier plutonic episodes (Anhaeusser, 1973; De Wit et al., 1992; Poujol and Anhaeusser, 2001). Extensive midcrustal plutonism and associated high-grade metamorphism at 3.1 Ga is recorded, for example, in the Vredefort dome and also parts of the Johannesburg dome in the central parts of the Kaapvaal Craton (Fig. 1) (Hart et al., 1999; Moser et al., 2001; Poujol and Anhaeusser, 2001; Lana et al., 2003, 2004). The best-exposed and probably most extensively studied examples of the GMS suite, however, occur in the Barberton granite-greenstone terrain in the eastern parts of the Kaapvaal Craton (Figs. 1 and 2). Here, the 3.1 Ga granites form large batholiths covering an area of over 20,000 km^2 (Anhaeusser et al., 1981). The composite and commonly tabular granitoid bodies were emplaced at relatively shallow crustal levels (Anhaeusser et al., 1983; Robb et al., 1983) and are interpreted to represent the near-surface equivalents of the midcrustal plutonism and melting recorded in, for example, the Vredefort dome (Hart et al., 1999; Lana et al., 2004).

The origin of this craton-wide, short-lived plutonic event is as enigmatic as the tectonic setting of the plutonism and the accompanying crustal consolidation (De Wit et al., 1992; De Ronde and De Wit, 1994; Hart et al., 1999; Moser et al., 2001; Lana et al., 2003). The midcrustal granite plutonism and high-grade metamorphism at 3.1 Ga in the Vredefort dome are tentatively linked to subduction-related, accretionary processes along an active continental margin (Hart et al., 1999; Lana et al., 2004). Moser et al. (2001) also speculated that the craton-wide plutonism was the expression of the coupling between the Kaapvaal crust and its mantle root at that time, which led to the subsequent stabilization of the central-eastern

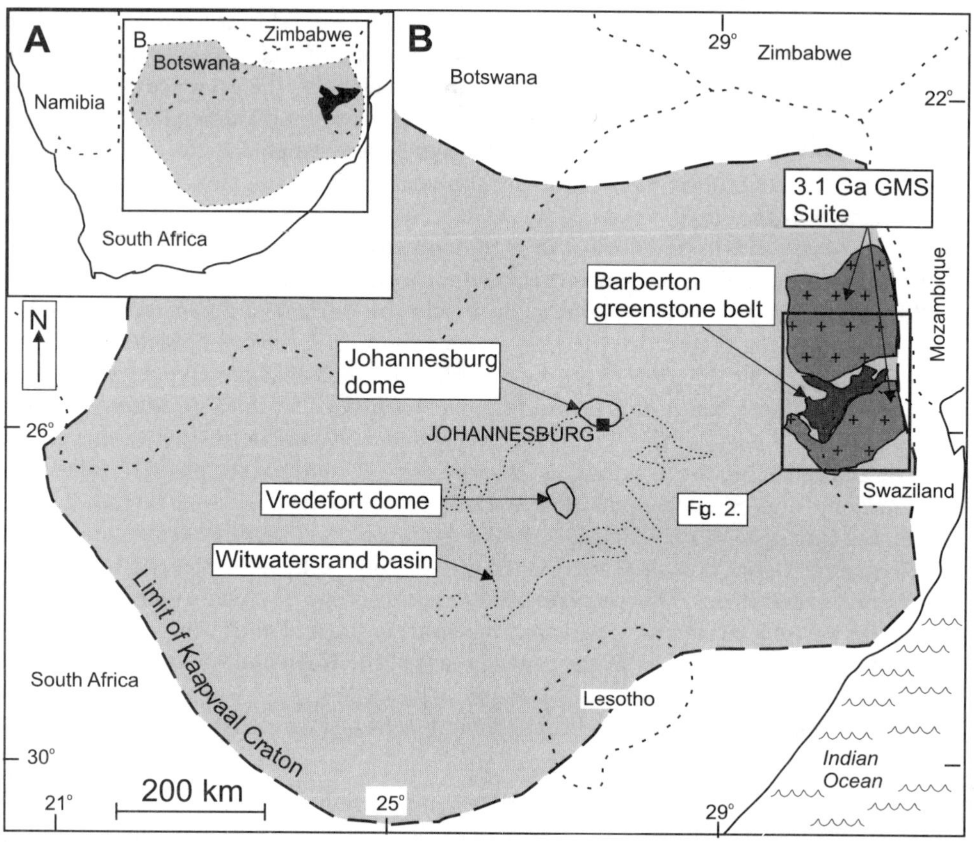

Figure 1. Sketch map of the Kaapvaal Craton in southern Africa. (A) Extent of the Kaapvaal Craton (gray) and location of the Barberton greenstone belt (black) in southern African. (B) Simplified map showing the location of the Barberton greenstone belt and extent of the granodiorite-monzogranite-syenogranite (GMS) suite in the eastern Kaapvaal Craton as well as other localities discussed in the text. Modified from Poujol et al. (2003).

Kaapvaal Craton. Unlike the midcrustal, intensely deformed granite-gneisses and migmatites in the central Kaapvaal Craton, 3.1 Ga granite batholiths in the Barberton region have commonly been described to be largely undeformed, so that earlier studies suggested an anorogenic setting for the granite plutonism (Hunter, 1973; Anhaeusser and Robb, 1983; Robb et al., 1983). Alternatively, the lack of a regional deformation event recorded in the shallow-crustal rocks was also interpreted to indicate a partitioning of the bulk strain during the 3.1 Ga convergent tectonics into the ductile, partially molten midcrustal levels while the brittle, upper crust remained unaffected (Moser et al., 2001). The studies by, for example, De Wit et al. (1992), De Ronde et al. (1992), De Ronde and De Wit (1994), and Kamo and Davis (1994) proposed an extensional, rift-type or a transtensional setting for the GMS suite in the Barberton granite-greenstone terrain. This is mainly based on the alkaline nature of the GMS suite and the dikelike geometry of some volumetrically minor but petrologically and geochemically distinct GMS suite intrusions that are syenitic and syenogranitic in composition. Hence, to date there is no general consensus as to the setting and tectonic environment into which the GMS suite was intruded.

In this contribution, we present the first comprehensive fabric data set from one of the GMS suite batholiths, the Heerenveen batholith in the southern Barberton granite-gneiss terrain, that forms one of the smaller batholiths of the GMS suite (Fig. 2). The understanding of the formation and significance of magmatic and solid-state fabrics in granitic plutons has greatly improved over the last ten to fifteen years (e.g., Paterson et al., 1989, 1998; Tobisch and Williams, 1998; Vernon et al., 2004), and granite plutons have increasingly been used for the structural analysis of terrains. The study of granite fabrics is particularly useful in polyphase deformed terrains, where a distinction between individual tectonic events may be difficult in the complexly deformed country rocks. In these cases, syn- and late-kinematic granites can be used as strain and kinematic markers that, depending on the timing of emplacement, record the last deformation of the terrain, thereby indicating the tectonic evolution of the particular orogenic phase (e.g., Benn et al., 1998, 2001; Gleizes et al., 1998).

In this paper we emphasize the synkinematic emplacement of the Heerenveen batholith during regional deformation and describe the syn- and post-emplacement deformation of the composite batholith. Regional mapping of the Heerenveen batholith has identified the presence of pervasively developed magmatic and solid-state fabrics as well as synmagmatic shear zones within and along marginal parts of the batholith. The timing of emplacement of the 3.1 Ga (by inference) Heerenveen batholith coincides with the last phase of regional deformation recorded in supracrustal rocks of the adjoining Barberton greenstone belt (late D3/D4, after De Ronde and De Wit, 1994; Kamo and Davis, 1994; De Ronde and Kamo, 2000). The aim of this paper is to use the fabric and kinematic data from the Heerenveen batholith to constrain the hitherto poorly defined late-stage D3/D4 deformation of the granite-greenstone terrain as well as the regional structural setting of the GMS suite plutonism in this part of the Kaapvaal Craton.

GEOLOGICAL SETTING

The Barberton greenstone belt and its surrounding granite-gneiss terrain in the eastern parts of the Kaapvaal Craton (Figs. 1 and 2) were assembled through several tectonomagmatic episodes between ca. 3.5 and 3.1 Ga. The main phase of crustal amalgamation and terrane accretion was associated with extensive TTG (tonalite-trondhjemite-granodiorite) plutonism and high-grade metamorphism in the surrounding granite-gneiss terrain and occurred at 3.23–3.22 Ga (Armstrong et al., 1990; De Ronde and De Wit, 1994; Kamo and Davis, 1994; Lowe, 1994; Dziggel et al., 2002). This progressive deformation event (D2 and early D3, after De Ronde and De Wit, 1994; Kamo and Davis, 1994; De Ronde and Kamo, 2000) is responsible for the main northeast-southwest–trending structural grain of the low-grade metamorphic Barberton greenstone belt. The intrusion of the GMS suite marks a massive plutonic event, manifested in the emplacement of several tens of thousands of cubic kilometers of granitic magmas in the Barberton granite-greenstone terrain at 3.1 Ga. Existing age data indicate that the emplacement of the large and composite plutons of the GMS suite occurred within 10–15 m.y. between 3100 and 3115 Ma, peaking at 3105–3110 Ma (Kamo and Davis, 1994; De Ronde and Kamo, 2000; Westraat et al., 2005). The timing of the GMS plutonism overlaps with the late D3 and D4 events described by De Ronde and De Wit (1994) for the Barberton greenstone belt, marking the transition from crustal convergence (late D3, between 3126 ± 21 and 3084 ± 18 Ma; De Ronde et al., 1992) to transtensional tectonics (D4, younger than 3084 ± 18 Ma), the latter deformation event being tentatively related to the collapse stage of the granite-greenstone terrain (De Ronde and De Wit, 1994; Kamo and Davis, 1994; De Ronde and Kamo, 2000). Most studies regard the GMS suite plutonism as part of this event, also heralding the onset of extensional tectonics on the Kaapvaal Craton manifested by the bimodal, rift-type volcanism and sedimentation of the Dominion Group at 3074 ± 6 Ma (Armstrong et al., 1991). Regional, penetrative deformation phases subsequent to the 3.1 Ga plutonism and the late D3 and D4 deformation events are not recorded in the Barberton granite-greenstone terrain.

Granitoids of the GMS suite are heterogeneous, ranging in composition from trondhjemite, granodiorite, and monzogranite to syenogranite and syenite. Individual plutons may cover several thousand square kilometers, the largest being the Nelspruit batholith in the north and the Mpuluzi and Pigg's Peak batholiths to the south and southeast of the Barberton greenstone belt, respectively (Fig. 2). Regional mapping studies have shown that most of the GMS suite batholiths represent subhorizontal, sheetlike intrusions with thicknesses between only 500 and 1000 m (Hunter, 1973; Anhaeusser and Robb, 1983; Robb

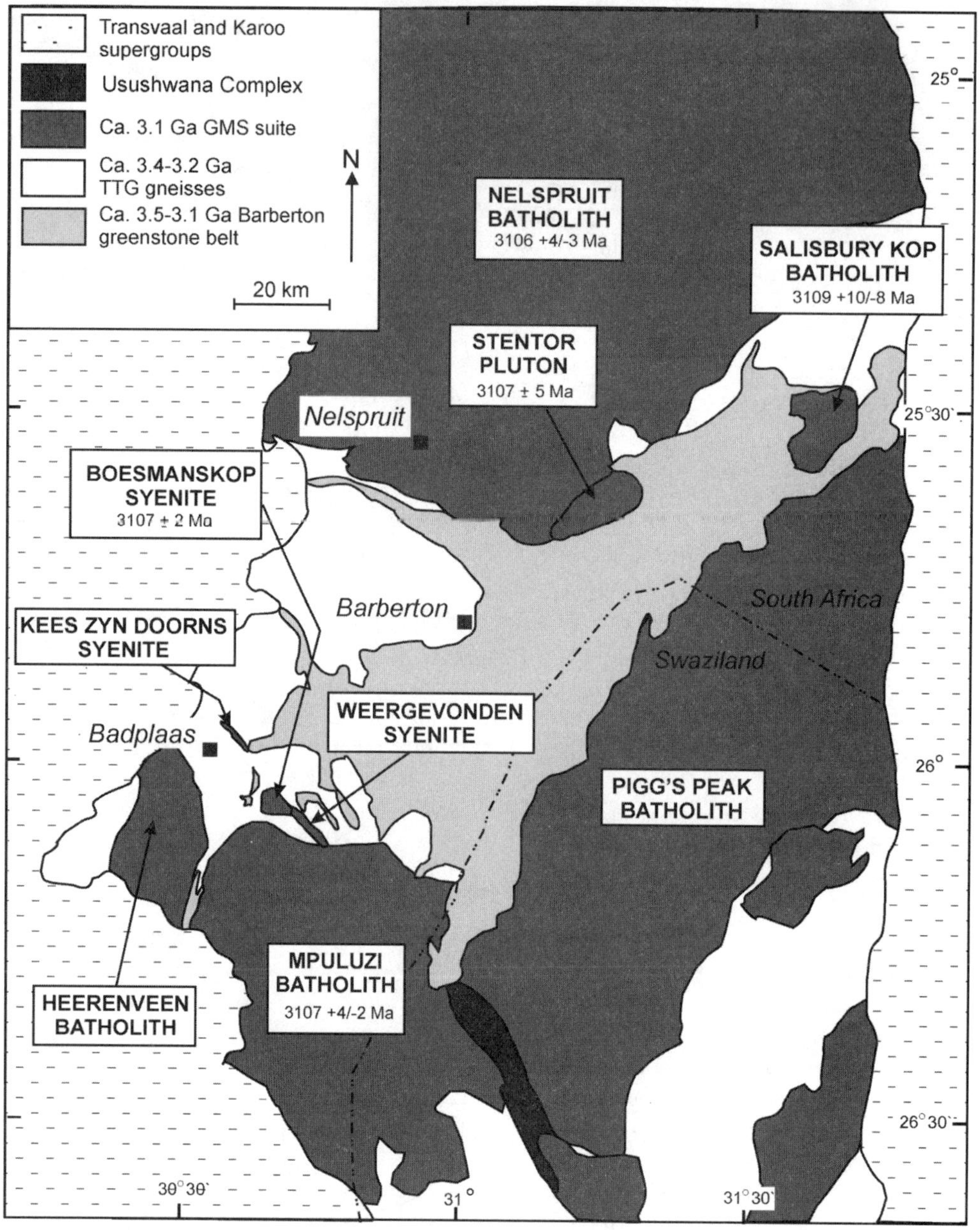

Figure 2. Distribution and ages of the volumetrically extensive, ca. 3.1 Ga batholiths of the granodiorite-monzogranite-syenogranite (GMS) suite that are intrusive into the older (>3.2 Ga) tonalite-trondhjemite-granodiorite (TTG) basement and greenstone lithologies of the Barberton greenstone belt. Modified from Anhaeusser et al. (1981); ages are from Kamo and Davis (1994).

et al., 1983). Locally, the tabular granites can be shown to be underlain by granitic dike complexes that have been interpreted as the feeders for the overlying granite sheets (Hunter, 1973; Anhaeusser and Robb, 1983; Westraat et al., 2005). Smaller plug- and dikelike intrusions of syenitic and syenogranitic compositions, such as the Boesmanskop, Weergevonden, and Kees Zyn Doorns intrusions (Fig. 2), are spatially and temporally closely associated with the large batholiths (Anhaeusser, 1980; Anhaeusser and Robb, 1983; Kamo and Davis, 1994).

THE HEERENVEEN BATHOLITH

The Heerenveen batholith is an elongate, NNE-SSW–trending body. It is exposed over a length of 30 km and a width of ~15 km (Fig. 3). It is concealed by rocks of the early Proterozoic Transvaal Supergroup in the north and of the Phanerozoic Karoo Supergroup in the south (Figs. 2 and 3). The composite batholith is intrusive into banded TTG gneisses and amphibolite-facies greenstone lithologies, including the Schapen-

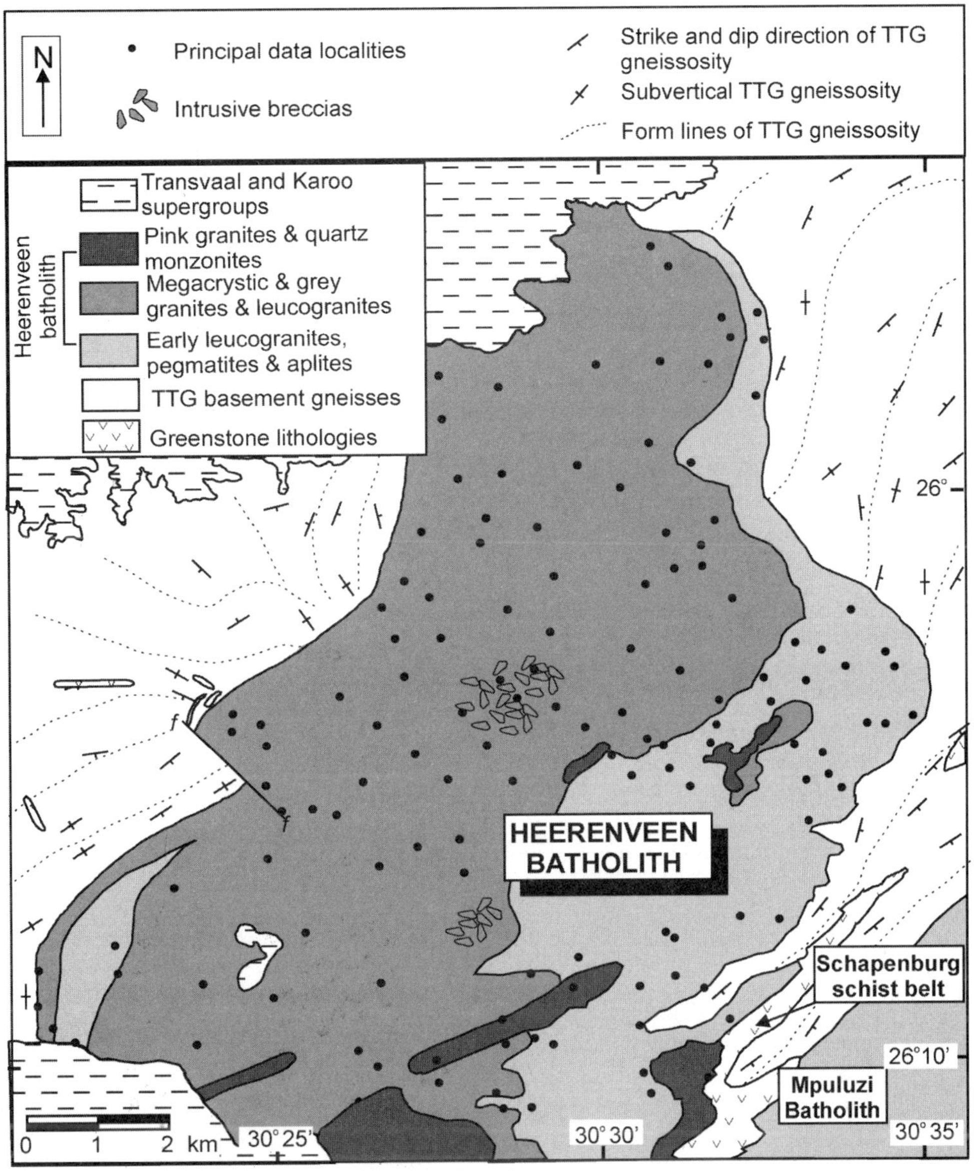

Figure 3. Geological map showing the distribution and extent of the various granitic intrusive phases that constitute the Heerenveen batholith. Note the asymmetric distribution and general northeast-southwest trend of individual granitic phases. Dots indicate the location of prominent pavements in the Heerenveen batholith, where detailed structural observations were made; TTG—tonalite-trond-hjemite-granodiorite. Geological boundaries of the Schapenburg schist belt and Transvaal and Karoo supergroups after Anhaeusser et al. (1981).

burg schist belt along its southeastern margin. The floor of the pluton is well exposed along the eastern margin of the Heerenveen batholith, where the Theespruit River and its tributaries deeply incise through the granite into the underlying TTG-greenstone basement. Significantly, this lower contact can be followed for ~15 km at approximately the same elevation, indicating the subhorizontal and flat geometry of the floor in this part of the Heerenveen batholith. Although the roof rocks are nowhere exposed, the flat floor of the Heerenveen batholith points to an overall tabular geometry of the pluton, as is typically described for most of the other batholiths of the GMS suite, such as the Nelspruit batholith (Robb et al., 1983) and the Mpuluzi and Pigg's Peak batholiths (Hunter, 1973; Anhaeusser and Robb, 1983; Westraat et al., 2005).

The Heerenveen batholith is heterogeneous and at least six petrographically and texturally distinct main granitoid phases can be distinguished in the field (Table 1). Geochemically, the Heerenveen batholith can be subdivided into two types: a medium-potassic and a high-potassic series, which are both calc-alkaline and metaluminous to slightly peraluminous. Com-

TABLE 1. SUMMARY OF THE DIFFERENT GRANITIC PHASES OF THE HEERENVEEN BATHOLITH AND THEIR MUTUAL INTRUSIVE RELATIONSHIPS

Phases	Appearance	Occurrence	Intrusion relationships
Pink granites	Medium- to fine-grained, light pink to pink	NE-SW elongated bodies (composed of subvertical dykes) spatially associated with the synmagmatic shear zones.	Crosscuts megacrystic granite, leucogranite, and sheeted granite complex. Some phases posttectonic. Youngest of the intrusive phases.
Quartz monzonites	Medium-grained, deep pink, speckled with black hornblende	Discrete NE-SW elongated body within synmagmatic shear zone, and equidimensional bodies in the south of batholith.	Co-magmatic with pink granites.
Grey granites	Medium-grained homogeneous gray	Limited outcrops in south-central parts of the batholith associated with the leucogranites.	Age relationship unknown, although always spatially associated with the megacrystic and leucogranites.
Leucogranites	Medium- to fine-grained, pinkish-white to greyish	NE- and N-trending elongated outcrops (composed of NE-SW to N-S subvertical dykes, respectively).	Co-magmatic with megacrystic granites. Crosscuts the underlying sheeted granites.
Megacrystic granite	Coarse-grained, white to greyish-white, with K-feldspar megacrysts (pinkish to white)	The dominant volumetric phase occurring across the northern and central parts of the batholith.	Co-magmatic with leucogranites. Crosscuts the underlying sheeted granites.
Sheeted granite complex	Fine-grained, white to gray, including pegmatites and aplites	Intrudes as shallow dipping sheets along the gneissosity of the basement. Increase in volume at structurally higher levels.	Oldest of the intrusive phases. Intrudes into basement.

positions range from tonalite and trondhjemite to granodiorite and granite (Anhaeusser and Robb, 1983; Yearron, 2003). The main mineral components of both types are quartz, plagioclase, and microcline. Accessory minerals include biotite, muscovite together with chlorite, zircon, apatite, and opaque minerals. A distinct hornblende-bearing phase is confined to the southern parts of the batholith. Crosscutting relationships have been used to establish the relative chronology between the different phases (Table 1), but there are also clearly multiple intrusive relationships recorded within individual granitic phases.

In detail, the Heerenveen batholith is a sheeted granitoid, made up of subhorizontal and subvertical granitic sheets. The earliest phases of the Heerenveen batholith intrude the banded TTG basement gneisses in a lit-par-lit fashion as foliation-parallel, shallow-southeasterly dipping sills and subordinate dikes in the eastern parts of the batholith. These phases have also been recognized by Anhaeusser and Robb (1983), who described the rocks as the marginal, migmatitic phase of the Heerenveen batholith. The sill complexes of granites, aplites, and pegmatites contain, in places, a solid-state gneissosity, parallel to the enveloping basement gneisses, which possibly suggests a reactivation of the earlier fabrics in the banded TTG gneisses. The bulk of the Heerenveen batholith (~75% of the exposed surface area) consists of two main phases, namely (1) a megacrystic microcline granite, which represents the coarse and very coarse porphyritic granite described by Anhaeusser and Robb (1983); and (2) a fine- to medium-grained homogeneous leucocratic granite, equivalent to the homogeneous hood-type granite of Anhaeusser and Robb (1983). The granites form mainly sheetlike bodies, and multiple intrusive relationships are common. Along the western margin of the batholith, the granites intrude as laterally continuous subvertical sheets, ranging in thickness from <1 m to >10 m. This sheeted western contact is between 100 m and 1 km wide, grading into more homogeneous granites toward the east. Westraat et al. (2005) have recently described similar multiple intrusive relationships between and within the commonly sheetlike phases of the Mpuluzi batholith to the immediate east (Fig. 2), testifying to the incremental assembly of the large GMS suite batholiths through multiple granite sheeting. The last major phase of granites occurs as a series of fine-grained granites that generally intruded as discrete northeast-southwest–trending linear bodies (Table 1). The internal architecture of the bodies is composed of cm- to m-scale subvertical sheets intruding into earlier granites. In areas of prominent topography, these granites can be seen to also form laterally extensive subhorizontal sheets at higher structural levels within the Heerenveen batholith. The following descriptions deal with the main and late-stage phases of the Heerenveen batholith that constitute the bulk of the batholith.

Structures and Kinematics

The Heerenveen batholith contains penetrative gneissose fabrics, and only discrete zones within the pluton and minor late-stage leucogranites are devoid of macroscopic fabrics. Both magmatic and solid-state fabrics can be distinguished that are subparallel to each other throughout the batholith. However, the fabric intensities vary considerably. The description of fabric intensities presented here is qualitative, based on factors such as the degree of dynamic recrystallization of the rocks, the degree of grain-shape preferred orientation of minerals, aspect ratios of, for example, quartz ribbons, and the presence of intrafolial folds and transposition fabrics. Based on this, protomylonitic and mylonitic solid-state foliations delineate three linear high-strain belts within and along the western margin of the batholith. Intrusive relationships and the development of solid-state fabrics indicate that these high-strain zones mark the presence of synmagmatic shear zones. Details of the orientation, distribution, and intensity of fabrics in the Heerenveen batholith, together with intrusive relationships and deformation of intrusive phases and country rocks, are presented below.

Magmatic Fabrics

The preferred alignment of microcline phenocrysts in the megacrystic phase of the batholith is evidence for the presence of magmatic fabrics in large parts of the Heerenveen batholith.

Other common igneous fabrics such as a compositional magmatic layering or schlieren textures (e.g., Paterson et al., 1989; Miller and Paterson, 1994) are rare, and individual phases of the Heerenveen batholith appear homogeneous in outcrop (Anhaeusser and Robb, 1983). Microcline megacrysts reach lengths of up to 8 cm. They may occur as closely clustered aggregates, forming up to 20% of the megacrystic granite, but also as rather sporadic and isolated phenocrysts set within the medium-grained groundmass of the granite. A macroscopic magmatic zonation is defined by the inclusion of small quartz and plagioclase crystals along growth zones of microcline phenocrysts. Several crystal habits can be distinguished, including rhomb-shaped, equidimensional phenocrysts and tabular crystals with three unequal axes (Figs. 4A–4C). The latter crystals define a regionally consistent east-northeast–trending magmatic foliation in the southern parts of the Heerenveen batholith, which assumes a more north-northeast trend in the northern parts of the pluton (Figs. 4D–4F). Due to the mainly flat outcrops, it was not possible to accurately measure the dip of the foliation defined by the tabular megacrysts or the plunge of the long axis of the phenocrysts; only the strike of the foliation was mapped (Fig. 5). However, the dip of the foliation appeared to be steep at all localities where dips could be estimated.

Solid-State Fabrics

Most rocks of the Heerenveen batholith contain a solid-state fabric, resulting in the generally gneissose texture of the granites. The solid-state gneissosity is parallel to the magmatic foliation in the megacrystic granite (Fig. 6). Fabric intensities are highly variable, ranging from almost undeformed granites to protomylonitic and mylonitic textures. In low strain domains, the weak gneissosity is defined by elongated quartz grains. Quartz shows undulose extinction and subgrain formation, resulting in "chessboard"-like patterns. The recrystallization of larger, interstitial quartz grains between feldspars into smaller quartz-grain aggregates results in the grain-shape preferred orientation of quartz that imparts the macroscopic fabric to the weakly deformed granite phases. Feldspars are mainly euhedral and show a random orientation, but both plagioclase and K-feldspar can be seen to have undergone marginal recrystallization in thin section (Fig. 7A). The grain boundaries of the large, euhedral microcline phenocrysts in the megacrystic granite are also highly irregular and dynamically recrystallized in response to grain-boundary migration (Fig. 7B). The primary magmatic origin of the larger porphyroclasts is, however, clearly preserved, and plagioclase, for instance, commonly shows An-rich cores replaced by fine-grained sericite, whereas the An-poor rims appear unaltered (Fig. 7C). Myrmekites of quartz and plagioclase replacing the margins of the phenocrysts are common (Fig. 7D).

For the most part, the Heerenveen batholith contains a well-developed solid-state gneissosity that is most prominently expressed by mm- to cm-sized, positively weathering, elongate quartz and quartz-grain aggregates. Aspect ratios of the quartz discs range, on the mainly horizontal surfaces, from 2:1–5 to 10:1, and, at higher fabric intensities, quartz ribbons with aspect ratios well over 10:1 may result (Fig. 7E). The quartz ribbons consist of recrystallized polygonal quartz aggregates that wrap around feldspars of the groundmass as well as microcline megacrysts. Plagioclase and microcline of the groundmass have undergone almost pervasive recrystallization and grain-size reduction, and relict feldspars form mantled porphyroclasts (Figs. 7F and 8A). Where present, the alignment of biotite and muscovite accentuates the gneissic fabric defined by the grain-shape preferred orientation of quartz and quartz-feldspar aggregates. A stretching lineation defined by elongated quartz and quartz-feldspar aggregates is associated with the solid-state foliation. The lineation has a consistent southwesterly plunge at shallow to moderate angles across the whole of the batholith (Fig. 9).

Synmagmatic Shear Zones

Protomylonitic and mylonitic fabrics occur in three distinct, several-hundred-meter-wide belts within the Heerenveen batholith and along its western margin (Figs. 6 and 9). The high-strain zones show a compositional banding and a penetrative gneissic foliation defined by quartz ribbons that alternate with finely recrystallized quartz-feldspar aggregates (Fig. 7E). Quartz ribbons show, on horizontal sections, aspect ratios of between 10:1 and 20:1. Microcline megacrysts form rounded, centimeter-sized, mantled porphyroclasts, resulting in augen-textured gneisses (Figs. 7A, 7E, and 7F), while large feldspars from pegmatites are commonly fragmented along microfaults, resulting in bookshelf-type structures (Fig. 8A).

The protomylonitic fabrics can be traced for several kilometers and up to 20 km along strike (Fig. 6). The high-strain fabrics are subparallel to the magmatic foliation and solid-state gneissosity of the Heerenveen batholith, and a gradual increase in the intensity of the foliation into the shear zones is recorded. In detail, the shear fabrics undulate but define ENE-WSW trends on a regional scale, with subvertical to steep southerly dips. Stretching lineations are defined by rodded quartz and quartz-feldspar aggregates that show shallow to moderate southwest plunges, parallel to the lineation developed across most of the Heerenveen batholith. Notable exceptions from this general east-northeast trend of high-strain zones are two northerly trending segments along the western margin of the Heerenveen batholith (Fig. 6). The northerly trending parts of the western boundary of the Heerenveen batholith show subvertical, down-dip stretching lineations (Figs. 8B and 9).

Shear sense indicators are common and include widespread S-C fabrics (Figs. 8D and 8E) together with σ-clasts (Figs. 7F and 8A) and brittle microfaults in tiled K-feldspar megacrysts (Fig. 8A). The east-northeast–trending shear zones consistently show a component of dextral strike-slip. In contrast, shear-sense indicators along the northerly trending shear zones consistently show a component of sinistral strike-slip (Fig. 8F). Notably, shear sense indicators are only observed in horizontal

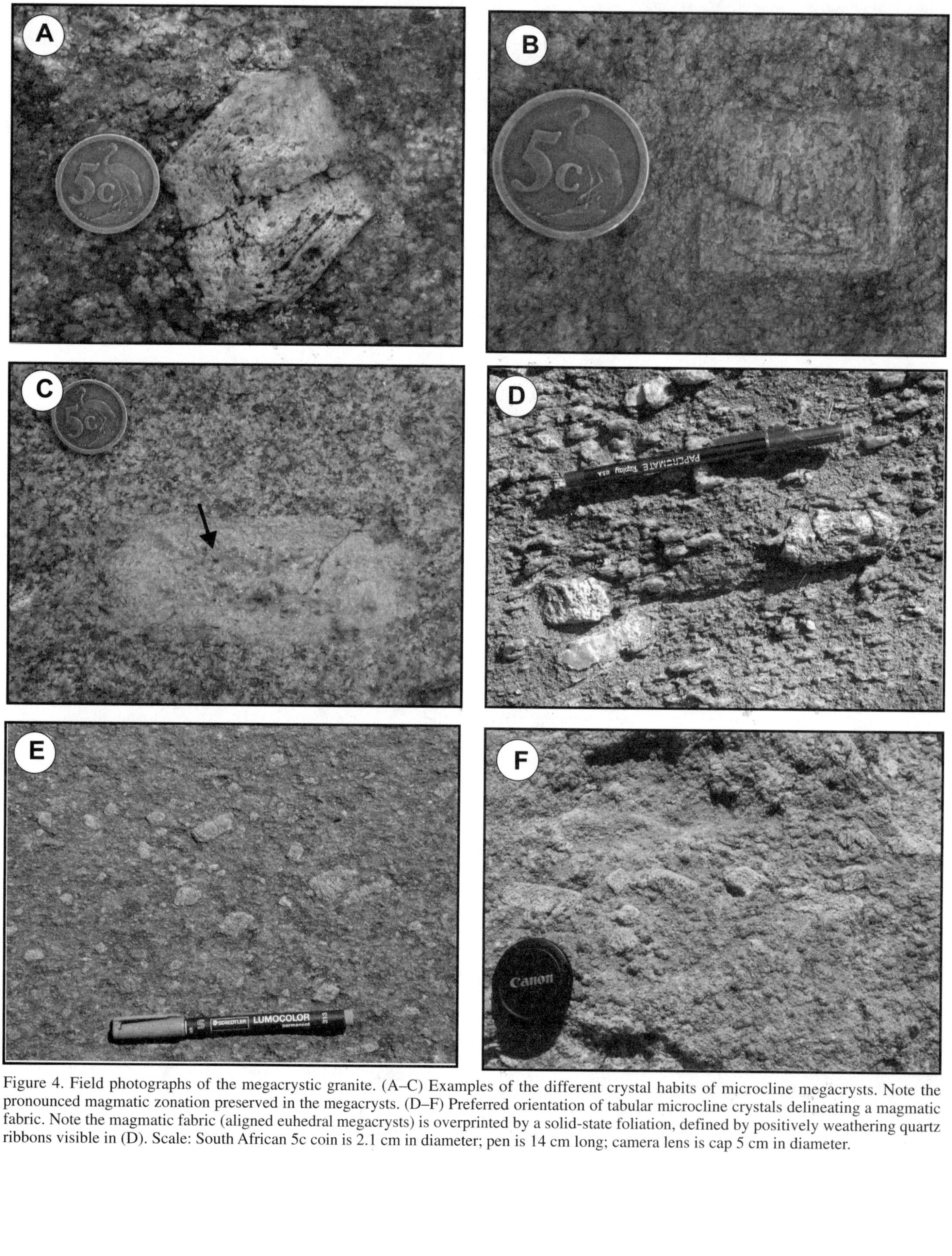

Figure 4. Field photographs of the megacrystic granite. (A–C) Examples of the different crystal habits of microcline megacrysts. Note the pronounced magmatic zonation preserved in the megacrysts. (D–F) Preferred orientation of tabular microcline crystals delineating a magmatic fabric. Note the magmatic fabric (aligned euhedral megacrysts) is overprinted by a solid-state foliation, defined by positively weathering quartz ribbons visible in (D). Scale: South African 5c coin is 2.1 cm in diameter; pen is 14 cm long; camera lens is cap 5 cm in diameter.

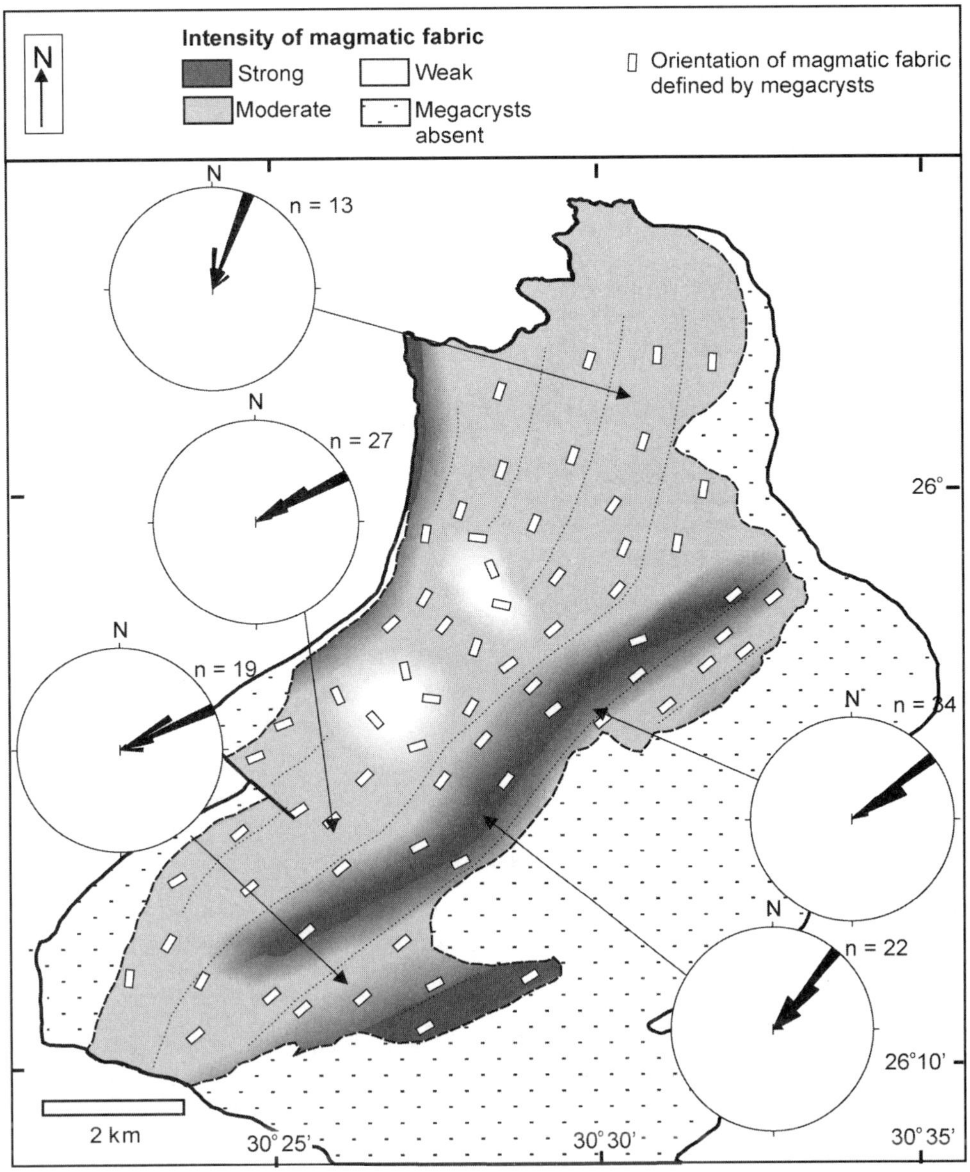

Figure 5. Trend of the magmatic foliation across the Heerenveen batholith as defined by the preferred orientation of euhedral tabular microcline megacrysts in the megacrystic granite. Thick black line denotes the extent of the exposed batholith (see Fig. 3). Note the randomly oriented megacrysts in the center of the batholith indicating the absence of a magmatic foliation. Rose diagrams show the preferred orientation of tabular feldspars for different areas within the batholith (n—number of readings).

sections, perpendicular to the subvertical stretching lineations. The subvertical stretching combined with the strike-slip movement suggests that the northerly trending shear zones are sinistral transpressive shear zones. Provided that the northerly trending transpressive shear zones and the east-northeast–trending, dextral strike-slip shear zones formed at the same time, the high-strain zones within and along the western margin of the Heerenveen batholith form a conjugate shear zone pattern (Fig. 6).

The occurrence of high-strain fabrics closely corresponds to zones of extensive granite sheeting characterized by dike-in-dike intrusive relationships between a multitude of granitic, pegmatitic, and aplitic phases (Figs. 10A and 10B). Individual granitic sheets vary in thickness from <1 m to >10 m and show steep, mainly southeasterly dips and east-northeast to northeast trends parallel to the solid-state gneissosity and the shear-zone boundaries. Intrusive relationships can still be discerned, and intrusive breccias are well preserved, particularly at lower fabric intensities (Fig. 10C). Angular fragments of the megacrystic granite contain magmatic and solid-state fabrics that are parallel to the solid-state fabric in the intrusive fine-grained leucogranitic sheets. Notably, the solid-state fabric intensity in many of the late-stage intrusive sheets may be considerably

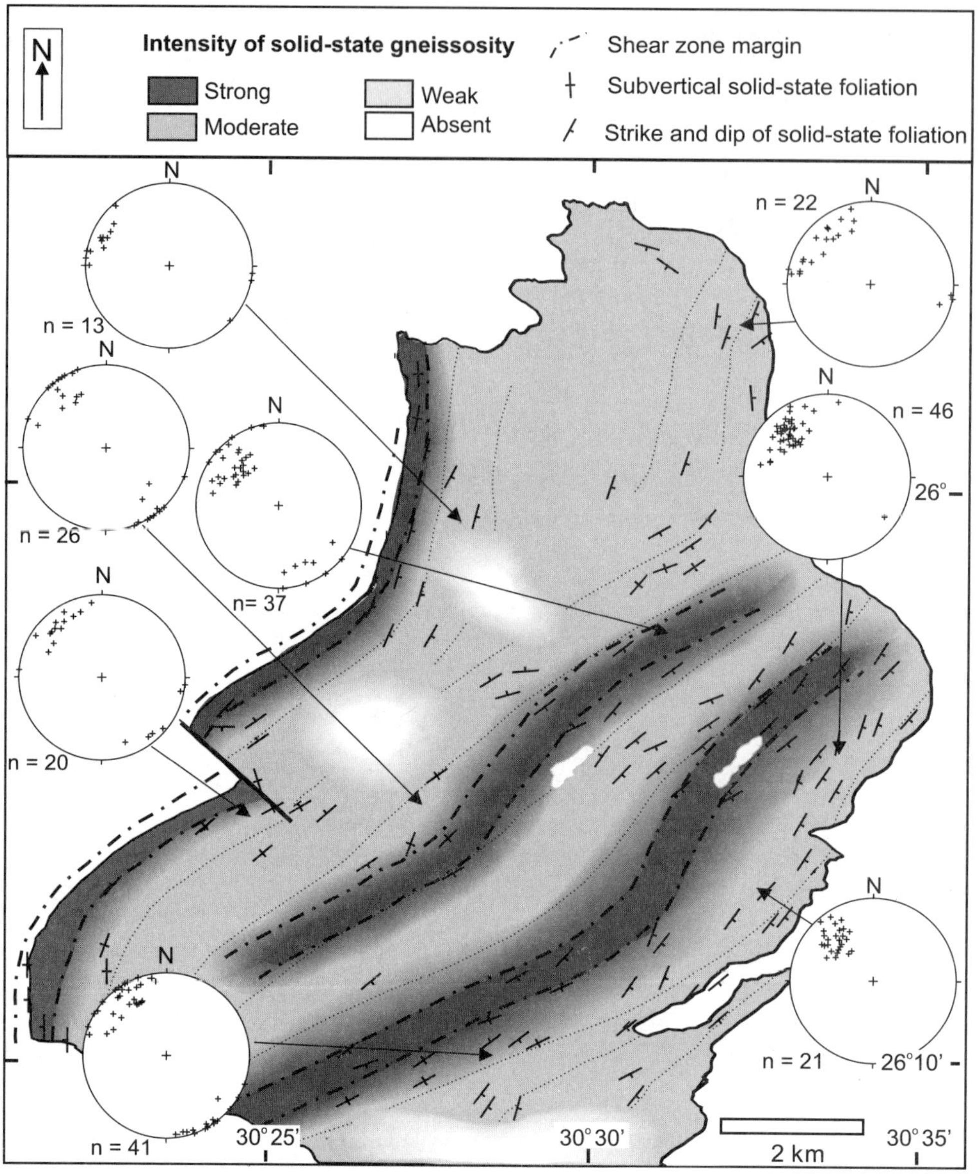

Figure 6. Trend of the solid-state foliation across the Heerenveen batholith. Shading reflects the intensity of the foliation as defined by, for example, the degree of recystallization and grain-shape preferred orientation of minerals, aspect ratios of quartz ribbons or presence of transposition fabrics. Stereonets are equal area lower hemisphere projections showing poles to the solid-state gneissosities for the respective domains (n—number of readings).

higher compared to that in the wall rocks that they intrude (Figs. 10D and 10E). This strain localization into intrusive granitoids together with the close spatial relationship, on a regional scale, between high-strain fabrics and granitic sheeting suggests synmagmatic deformation along the shear zones. The positive feedback between the presence of melts and deformation is also illustrated by the abrupt along-strike termination of the two prominent, east-northeast–trending shear zones close to the eastern margin of the Heerenveen batholith (Fig. 6). There is no evidence of the several-hundred-meter-wide shear zones outside the Heerenveen batholith, in the older TTG country-rock gneisses.

The intrusion of late-stage leucogranites (*sensu lato*) that are widespread in the central, high-lying portions of the Heerenveen batholith is an important time marker in the emplacement history of the batholith. These granitic phases are devoid of macroscopic fabrics and truncate the high-strain fabrics and compositional banding of the synmagmatic shear zones (Figs. 6 and 10F). The post-kinematic emplacement of the latest phases of the Heerenveen batholith is clear evidence that plutonic activity in the area and progressive granitic sheeting that led to the assembly of the composite batholith outlasted regional deformation.

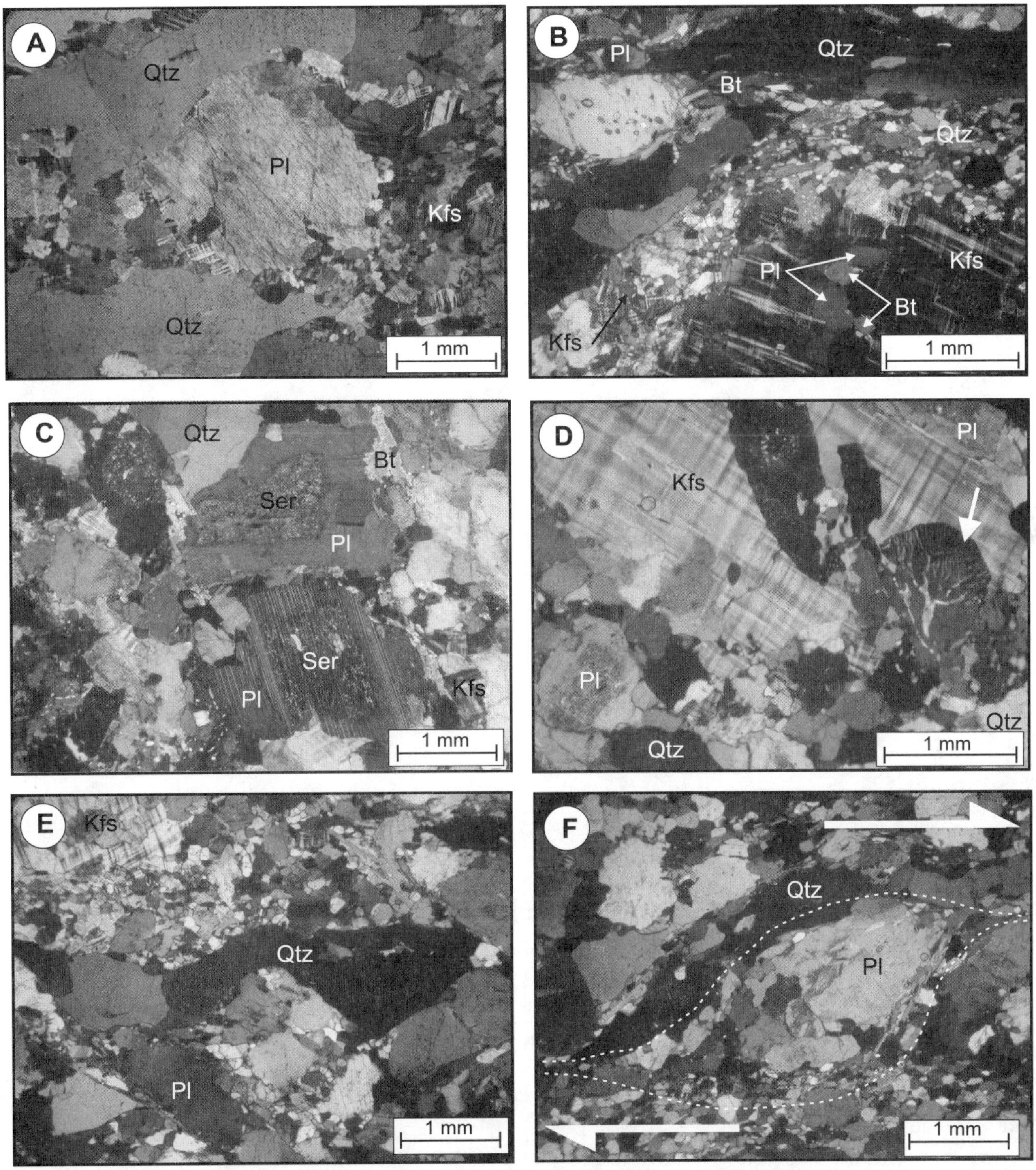

Figure 7. Photomicrographs of solid-state deformation textures within the granitic phases of the Heerenveen batholith (all images taken with crossed polarizers). (A) Elongated quartz grains wrapping around marginally recrystallized plagioclase clasts. Note the pervasive recrystallization of K-feldspar in the right-hand part of the photo. (B) Large microcline crystal (bottom) undergoing marginal recrystallization. Quartz ribbons and biotite laths impart a strong fabric. (C) Evidence for the magmatic origin of plagioclase as seen by the preferential alteration of the cores to sericite due to the difference in the anorthite content between the cores and the rims. (D) Myrmekite development in K-feldspar (arrowed). The myrmekites either form small bloblike (as shown) or linear structures, predominantly parallel or subparallel to the solid-state gneissosity. (E) Quartz ribbons commonly often have aspect ratios of >10:1 and define the strong foliation on both a micro- and macroscopic scale. Note the almost complete recrystallization of K-feldspar in the upper part of the photo. (F) Polygonal recrystallized quartz wrapping around a plagioclase porphyroclast showing dextral sense of shear. Mineral symbols from Kretz (1983).

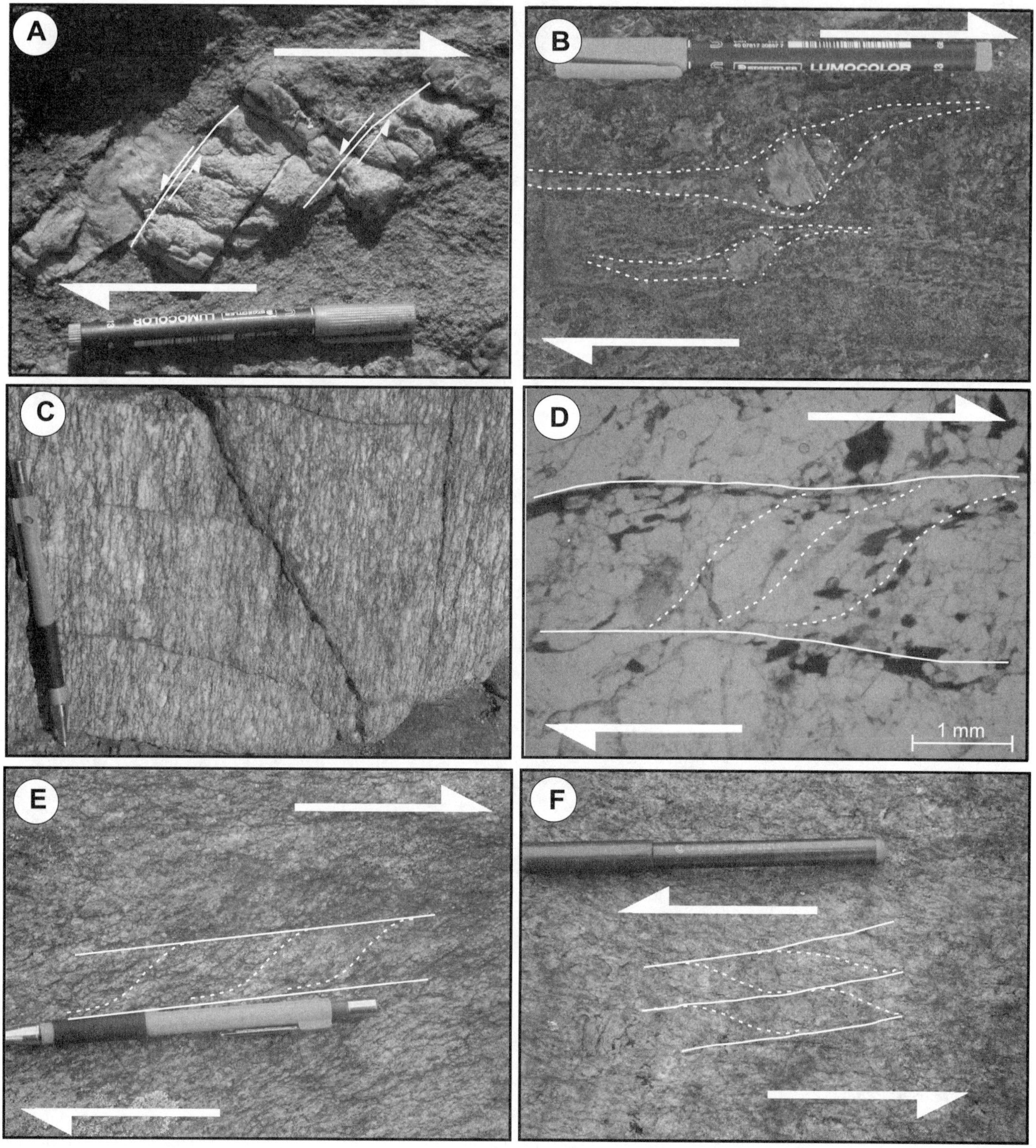

Figure 8. Kinematic indicators and stretching lineations from synmagmatic shear zones within the Heerenveen batholith. (A) Domino-type K-feldspar from syntectonic pegmatites within a northeast-trending shear zone in the southeastern parts of the Heerenveen batholith. The shear sense indicated by the segmentation of the feldspar is ambiguous, but dextral strike-slip kinematics are suggested by common S-C fabrics in the vicinity. (B) σ-type K-feldspar porphyroclasts indicating dextral sense of shear. (C) Subvertical stretching lineation defined by stretched quartz and quartz-feldspar aggregates (from the north-trending, sinistral segments of the western marginal shear zone). (D) and (E) Well-developed S-C fabrics on a mm to cm scale identifying dextral sense of shear. (F) Centimeter-scale S-C fabric in leucogranite along the north-trending shear zone identifying sinistral sense of shear. Pens are 14 cm long.

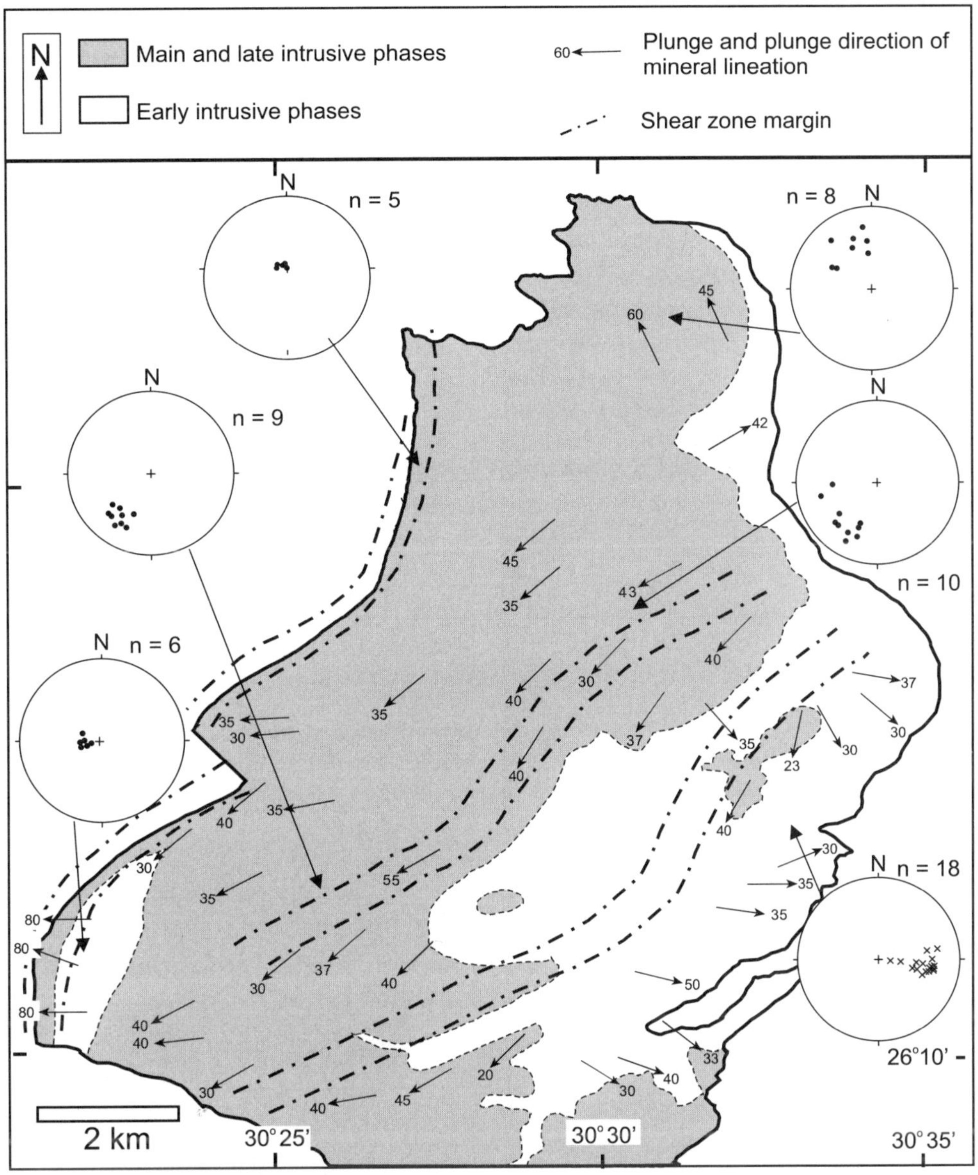

Figure 9. Plunge of mineral stretching lineations in the Heerenveen batholith. The thick black line denotes the extent of the batholith and shading shows the location of main and later granitic phases that make up the bulk of the Heerenveen batholith. Southwesterly plunging lineations dominate across most of the batholith, but subvertical stretching lineations characterize the northern and southern sectors of the western margin in the sinistral transpressive shear zones. Southeasterly plunges are restricted to the earliest, sill-like phases of the Heerenveen batholith along its eastern margin. Lower hemisphere equal area projections.

Deformation of Dikes and Sills Related to the Heerenveen Batholith

Granitic, aplitic, and pegmatitic dikes and sills related to the Heerenveen batholith are found throughout the batholith and also in the country rocks, up to within ~500 m of the main body of the batholith. Many of these dikes are clearly related to the last intrusive phases of the Heerenveen batholith, as they crosscut all major granitic phases and appear undeformed. Numerous dikes and sills, however, are folded and boudinaged, containing pervasive solid-state fabrics (Fig. 11A). The folding and boudinage of intrusive granitic and pegmatitic dikes and sills are particularly common and well developed in the amphibolite-facies supracrustal schists and metavolcanic rocks of the Schapenburg schist belt (Stevens et al., 2002) (Fig. 3). Folds of pegmatitic and granitic dikes in the Schapenburg schist belt range in wavelength from centimeters to several tens of meters (Figs. 11B and 11C). The axial surfaces are steep and

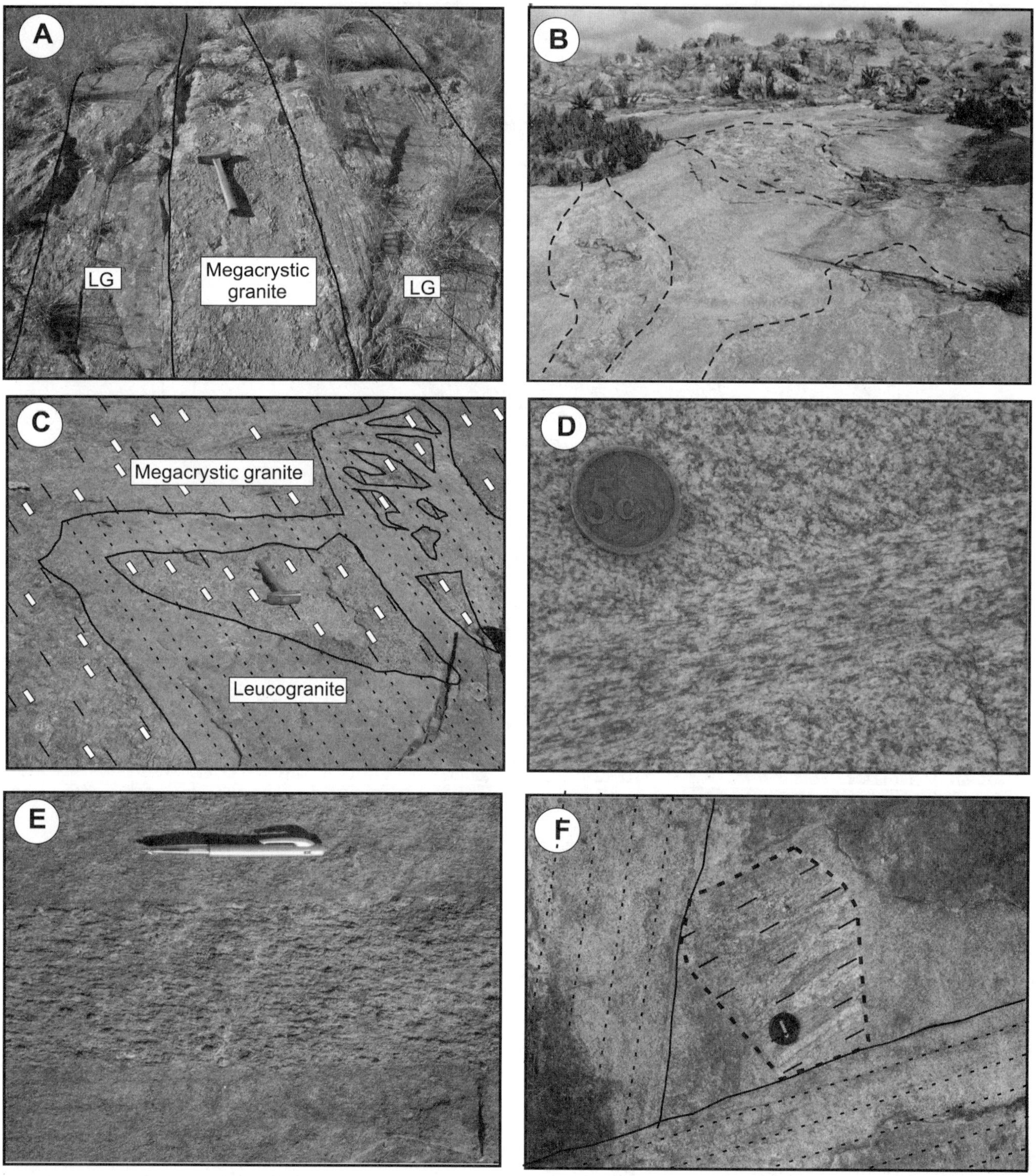

Figure 10. Intrusive relationships in the Heerenveen batholith. (A) Dike-in-dike intrusion of subvertical leucogranite and megacrystic granite sheets on a meter scale. Note the pervasive, dike-parallel fabrics in the leucogranites (LG) as well as in the megacrystic granite. Hammer is 37 cm long. (B) Intrusive brecciation of earlier, coarse-grained, granitic phases by fine-grained granitic sheets. These breccias may form bodies that can extend for several square kilometers (Fig. 3). Outcrop width is ~6 m. (C) Intrusive breccia showing the megacrystic granite containing a pervasive magmatic (rectangles) and solid-state foliation (by longer dashes) intruded by leucogranite that also contains a solid-state foliation (short dashes) subparallel to both the foliations in the megacrystic granite. (D–E) Later centimeter-scale dikes intruding earlier phases of the batholith. The pervasive development of solid-state fabrics in the dikes compared to their wall rocks is interpreted to point to strain localization into the intrusive dikes. (F) Late-tectonic granites containing a weak solid-state fabric (short dashes) truncating a banded gneissosity developed in early syntectonic granites (long dashes). Camera lens cap is 5 cm in diameter.

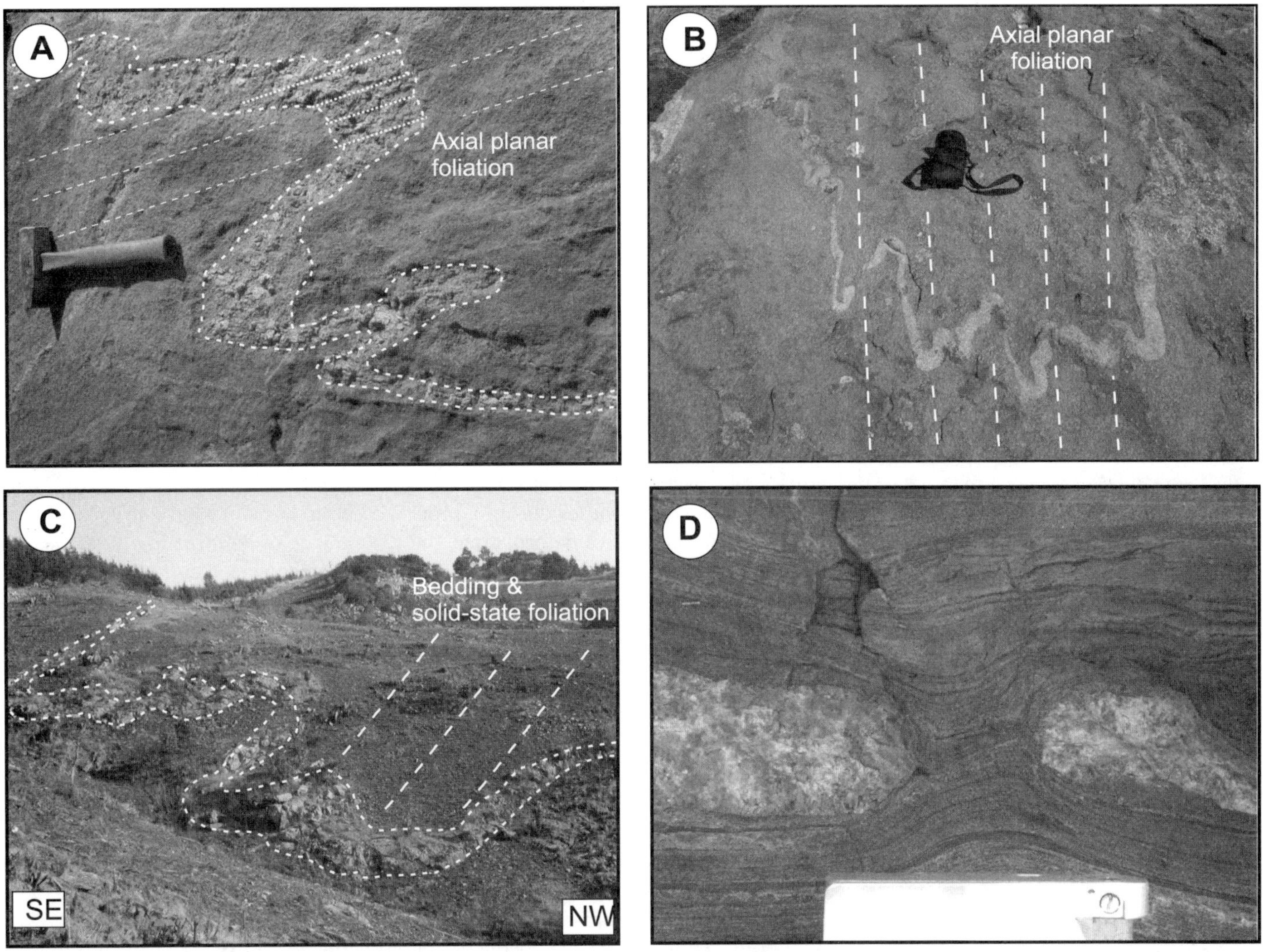

Figure 11. (A–B) Folded pegmatite dikes within leucogranites of the Heerenveen batholith from the western (A) and southeastern (B) margins of the pluton. A strong northeast-trending, subvertical foliation in the pegmatite dikes is defined by the grain-shape preferred orientation of quartz and feldspar. The foliation in the pegmatites is parallel to the solid-state gneissosity developed in the surrounding granite, which, in turn, is axial planar to the folds in the pegmatite. Scale: hammer is 37 cm long and camera bag is 16 cm long. (C) Decameter-scale folding of a pegmatite dike intrusive into the Schapenburg schist belt (field of view ~150 m across). A northeast-trending, subvertical foliation is developed within the pegmatite, which is parallel to the bedding/amphibolite–grade gneissosity in the surrounding amphibolites and graywackes of the Schapenburg schist belt. Both foliations show an axial planar orientation with regard to the folds developed in the pegmatite dike. (D) Horizontal view of a bedding-parallel granitic sill related to the Heerenveen batholith displaying layer-parallel boudinage in northeast-southwest–trending, steeply southeast-dipping metasediments of the Schapenburg schist belt. Boudinage is observed on both horizontal and vertical surfaces, indicating a subhorizontal northwest-southeast–directed shortening strain during boudinage. Compass length is 9 cm.

trend northeast to east-northeast and folds plunge steeply to the southeast and east. A penetrative axial-planar foliation is defined by large quartz ribbons and the grain-shape preferred orientation of quartz-feldspar aggregates. This axial planar foliation is parallel to the solid-state gneissosity in the Heerenveen batholith and the bedding and foliation of the Schapenburg schist belt. The interlimb angles of the mainly symmetrical folds vary from open to tight and isoclinal, even within the same outcrop. The tight- to isoclinal fold shapes indicate a minimum of 50%–60% shortening. The variations in fold shape and interlimb angles may be interpreted to reflect the different timing of emplacement of the granitic dikes during progressive deformation.

Boudinage is common in northeast-trending, bedding-parallel sills and dikes that cut the bedding of the supracrustal succession at low angles (Fig. 11D). The boudin necks are filled with quartz or quartz-feldspar stringers, and the amphibolite-facies mineral parageneses of the supracrustal rocks in boudin necks show no evidence of a retrogression (Stevens et al., 2002), indicating that deformation has occurred at moderate to high temperatures. The boudinage of the intrusive granitoids is observed in plan view as well as in vertical sections. Thus, the chocolate-tablet boudinage outlined above indicates a significant component of northwest-southeast subhorizontal layer-normal flattening. This is consistent with the upright, northeast-trending symmetrical folds of crosscutting granitic sills and dikes.

Granitic and pegmatitic dikes in the TTG basement surrounding the Heerenveen batholith are abundant and show mainly northeast to east-northeast trends and subvertical dips. The width of dikes ranges from merely several centimeters to 10 m. Folding and/or boudinage is only rarely observed, but most dikes contain a penetrative solid-state foliation parallel to the dike margins. The solid-state foliation is confined to the dikes, while the country-rock gneisses appear unaffected. This suggests that the deformation of the dikes occurred during and/or shortly after their emplacement, reflecting strain localization into the comparatively hot and rheologically weaker intrusions relative to the country rocks (e.g., Zulauf and Helferich, 1997). Due to the flat nature of the outcrops, lineations could not be measured. S-C fabrics are common in the northeast- to east-northeast–trending dikes and indicate a component of dextral strike-slip shear along the dikes, similar to the kinematics along the subparallel shear zones within the Heerenveen batholith.

Conditions of Solid-State Deformation

The pervasive recrystallization of feldspars, the preservation of biotite, and, locally, hornblende, and the general absence of any widespread retrogression of minerals that define the gneissic fabric point to moderate to high temperatures of deformation under at least lower amphibolite-facies conditions (>500 °C, Tullis and Yund, 1987). Similarly, country-rock gneisses and greenstone lithologies show unretrogressed amphibolite-facies mineral parageneses, and wall-rock deformation is ductile as exemplified by the rotation of the gneissosity of the wall rocks into parallelism with synmagmatic shear zones along the western margin of the Heerenveen batholith (Fig. 3). Peak-metamorphic amphibolite-facies conditions in the supracrustal greenstone lithologies of the Schapenburg schist belt and the surrounding TTG basement gneisses, however, were reached already during the main tectonometamorphic event at ca. 3230 Ma (Stevens et al., 2002; Dziggel et al., 2002), some 120 m.y. prior to the intrusion of the GMS suite. The intrusion level of the batholiths of the GMS suite corresponds to the present exposure level of the Barberton greenstone belt, the bulk of which has only undergone lower-greenschist facies metamorphism (T < 350 °C; Xie et al., 1997). Thus, the moderate- to high-temperature solid-state fabrics in the 3.1 Ga Heerenveen batholith, as well as in the immediate country rocks, most likely reflect deformation shortly after the emplacement and crystallization of the batholith while the rocks were cooling.

DISCUSSION

Syntectonic Emplacement of the GMS Suite

In this study, we have presented evidence for the synkinematic emplacement of the Heerenveen batholith in order to use the structural inventory of the batholith to constrain the regional strain field during granite magmatism and to allow speculation about the tectonic environment of the GMS suite plutonism at 3.1 Ga. Evidence for the syntectonic emplacement of the Heerenveen batholith includes: (1) the close spatial correlation between high-strain zones and corridors of multiple granite sheeting, together with the strain localization into dikes that intrude basement gneisses outside the confines of the Heerenveen batholith; (2) the highly different degrees of folding and/or boudinage of sills and dikes in the batholith and in the country rocks, suggesting an emplacement of the different phases during a protracted period of time and during progressive deformation; and (3) the lack of deformation in late-stage leucogranite phases as well as undeformed crosscutting granite and pegmatite dikes, illustrating that the GMS suite plutonism outlasted regional deformation that is otherwise recorded for the bulk of the batholith by the penetrative fabric development.

The dominant set of east-northeast–trending shear zones that show predominantly a component of dextral strike-slip shear, together with the northerly trending, sinistral transpressive shear zones, define a conjugate pattern of synmagmatic shear zones in the batholith (Fig. 12A). The high-strain zones are separated by several kilometer-wide, elongate domains of lower fabric intensity. The orientation and kinematics of the conjugate shear zones point to a subhorizontal northwest-southeast shortening strain. This bulk shortening strain is also recorded by the chocolate-tablet boudinage and symmetrical folding of granite dikes and sills in the batholith and in the country rocks (Fig. 12B). Batholith-wide magmatic and, in particular, solid-state foliations are oriented at high angles to this shortening strain. The penetrative gneissosity is axial planar to folds and parallel to the plane of chocolate-tablet boudinage of sills and dikes, which underlines that the gneissosity approximates the orientation of the XY plane ($X \geq Y \geq Z$) of the finite strain ellipsoid in large parts of the batholith.

The northwest-southeast–directed, subhorizontal shortening and shear zone kinematics agree very well with the findings of Jackson and Robertson (1983) and Westraat et al. (2005), who undertook structural studies to the immediate east, in various parts of the large Mpuluzi batholith. Jackson and Robertson (1983) document northwest-southeast–shortening strains of up to 80% in the northeast-trending Motjane schist belt, in the eastern parts of the Mpuluzi batholith (Fig. 12C). Westraat et al. (2005) record similar flattening strains of up to 60% and associated dextral strike-slip kinematics along the 2–3 km wide, protomylonitic western margin of the Mpuluzi batholith. Both studies emphasize the syntectonic emplacement of the granites and the close correlation between high-strain fabrics and zones of multi-

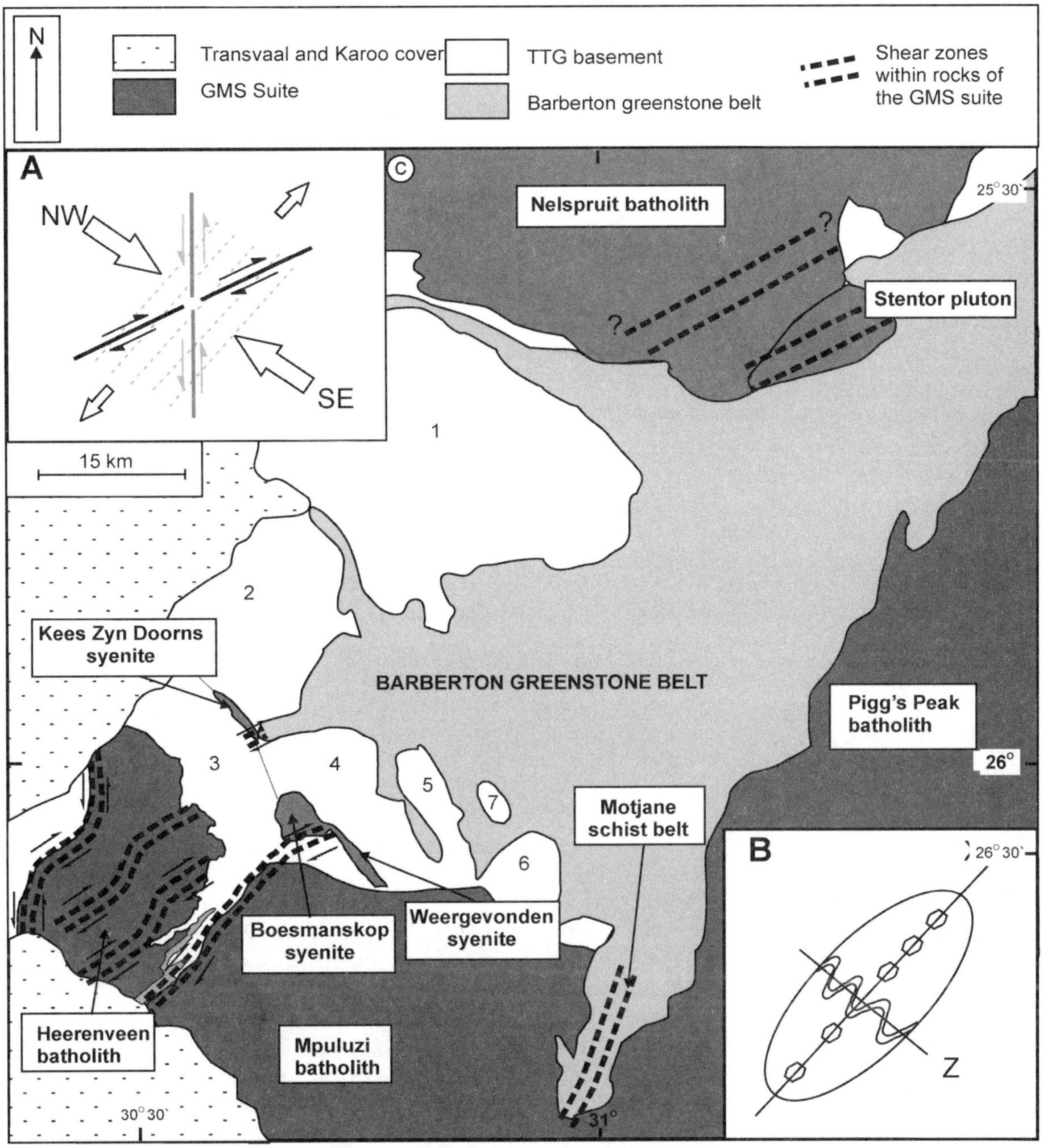

Figure 12. (A) Regional synopsis showing the occurrence of well-developed magmatic and high-strain solid-state fabrics in rocks of the GMS (granodiorite-monzogranite-syenogranite) suite in the Barberton granite-greenstone terrain, outlining the presence of synmagmatic shear zones; Numbers (1–7) refer to older TTG (tonalite-trondhjemite-granodiorite) plutons: 1—Kaap Valley, 2—Nelshoogte, 3—Badplaas, 4—Stolzburg, 5—Theespruit, 6—Dalmein, and 7—Doornhoek (Anhaeusser et al., 1981; Westraat et al., 2005; and this study). (B) Synoptic sketch, illustrating the orientation and kinematics of the 3.1 Ga synmagmatic shear zones and magmatic and solid-state fabrics. The north-trending sinistral and east-northeast–trending dextral shear zones define a conjugate shear zone pattern that, together with the pluton-wide northeast- to east-northeast–trending magmatic and solid-state fabrics, implies their formation during northwest-southeast–oriented subhorizontal shortening. (C) Geometric relationship between batholith gneissosity and folding and boudinage in sills and dykes in the surrounding wallrocks. The penetrative gneissosity in the batholiths is axial planar to folds and parallel to the plane of chocolate-tablet boudinage of sills and dikes. This underlines the subvertical, northeast-trending orientation of the XY plane of the finite strain ellipsoid during the emplacement of the Heerenveen batholith ($X \geq Y \geq Z$). Z—the short axis of the finite strain ellipsoid.

ple granite sheeting. Outside the northeast-trending gneiss belts, the bulk of the Mpuluzi batholith appears largely undeformed or contains only weak magmatic and solid-state fabrics (e.g., Anhaeusser and Robb, 1983; Westraat et al., 2005).

The northwest-southeast shortening strain during the emplacement of the GMS suite is also consistent with the orientation of the small, but distinctly dikelike, northwest-southeast–trending intrusions of the Kees Zyn Doorns and Weergevonden syenogranites, the orientations of which correspond to extensional fractures (Fig. 12C). The partitioning of the bulk strain into melt-bearing zones can be demonstrated even for these small and seemingly isolated intrusions. For example, the Kees Zyn Doorns syenite contains an up-to-200-m-wide, east-northeast–trending zone of pervasively developed S-C fabrics along its southern tip (Anhaeusser and Robb, 1983). The high-strain fabrics are confined to the narrow syenite body. The orientation and the dextral-strike slip kinematics recorded by the S-C fabrics in the protomylonitic gneisses of the Kees Zyn Doorns syenite form part of the regional-scale pattern of conjugate, synmagmatic, ca. 3.1 Ga shear zones that characterizes the GMS suite batholiths in the southern Barberton granite-gneiss terrain (Fig. 12C).

The conjugate shear zone pattern documented here corresponds closely to the observations by Grujic and Mancktelow (1998), who described the formation of shear zones in analogue models and field examples of melt-bearing shear zones during coaxial shortening. In their analogue models, sets of conjugate shear zones nucleate during bulk coaxial shortening around weak inclusions, i.e., melt patches of natural systems. The obtuse angle between the shear zones faces the maximum shortening direction. In natural examples studied (e.g., Pawley et al., 2002), similar conjugate shear zone patterns were observed around melt patches and leucosomes, although Grujic and Mancktelow (1998) noted that there was no dominant sense of vorticity along the shear zones.

In summary, the orientation and kinematics of synmagmatic shear zones, regionally consistent northeasterly foliation trends, the northwest-southeast trends of dikelike intrusions of the GMS suite, and the progressive deformation of intrusive dikes and sills all point to an emplacement of the GMS suite during northwest-southeast subhorizontal crustal contraction.

Structural Setting of Plutonism in the Barberton Granite-Greenstone Terrain

Most regional studies relate the emplacement of the GMS suite to the D4 transtensional or extensional phase that has also been described as the collapse stage of the Barberton greenstone belt and its surrounding granite-gneiss terrain (De Ronde et al., 1992; De Wit et al., 1992; De Ronde and De Wit, 1994; Kamo and Davis, 1994; De Ronde and Kamo, 2000). Existing age data place the onset of this collapse stage younger than 3084 ± 18 Ma (De Ronde et al., 1992). The peak of the GMS suite plutonism occurred, however, at 3105–3110 Ma (Kamo and Davis, 1994). This overlaps with the late-D3 (3126 ± 21 Ma to 3084 ± 18 Ma; De Ronde et al., 1992) convergent tectonism described for the Barberton greenstone belt, during which the regional structural grain of northeast-southwest–trending folds and northeast-trending thrusts in the supracrustal belt was steepened and tightened (De Ronde and De Wit, 1994). The fabric development and strains illustrated here for the GMS suite batholiths clearly corroborate the late-D3 timing of the bulk of the GMS suite plutonism during northwest-southeast crustal shortening.

The waning stages of the GMS suite magmatism coincide with the transition to the D4 transtensional or extensional tectonism and collapse stage documented by De Ronde and De Wit (1994) for the northern parts of the granite-greenstone terrain (De Ronde and De Wit, 1994). Recent work in the southern parts of the Barberton granite-greenstone terrain, however, has suggested that the extensional collapse and exhumation of mid- and deep-crustal basement gneisses occurred already at ca. 3225 Ma (Kisters et al., 2003), shortly following the main D2 collisional tectonic event in the region (De Ronde and De Wit, 1994; Kamo and Davis, 1994; Dziggel et al., 2002). The timing of this exhumation and core complex formation is based on metamorphic ages of exhumation fabrics (Kisters et al., 2003; Diener et al., 2005), the onset of deposition and sedimentological evolution of the molasse-type Moodies Group (Lowe et al., 1999), and the age of the post-tectonic 3216 ± 1 Ma Dalmein Pluton (Kamo and Davis, 1994). It is particularly the age of the post-tectonic Dalmein Pluton that places an upper time bracket on the main phase of tectonism in the southern part of the greenstone belt. The lack of post-D2, ca. 3.22 Ga deformational events along the southern granite-greenstone contacts is in conflict with field and geochronological data from the northern parts of the belt that provide clear evidence for tectonism at ca. 3.1 Ga (De Ronde and De Wit, 1994). Lithologies of the greenstone belt show penetrative high-temperature ductile and brittle-ductile deformation fabrics confined to a <1 km wide zone along the southern margin of the GMS suite batholiths (e.g., Tomkinson and King, 1991; De Ronde et al., 1992; Harris et al., 1995). Available age data indicate intrusion of the Nelspruit batholith at 3106 +4/–3 Ma and an intrusive age of 3107 ± 5 Ma for the Stentor pluton (Kamo and Davis, 1994), and a clear time marker for this deformation in the supracrustal belt is provided by up to several-tens-of-meters-wide pegmatite dikes related to the Nelspruit batholith that are pervasively mylonitized. Similarly, regionally developed belts of penetrative gneissosities are documented from the Nelspruit batholith (Robb et al., 1983), and the smaller Stentor pluton is, in most parts, strongly gneissose (Anhaeusser et al., 1981; Kohler, 1994).

The results presented in this study may offer a key to the seemingly contradicting evidence for the timing of the late-stage tectonism throughout the Barberton greenstone belt. One of the most pertinent characteristics of the D3 strain is the high degree of strain localization into zones of syntectonic granite sheeting and magma emplacement. In the southern Barberton greenstone belt, the large 3.1 Ga batholiths only intrude to within 10–15 km of the granite-greenstone contact; this includes the areas around the Nelshoogte, Badplaas, Stolzburg, and Theespruit plutons (Fig. 12C). Most parts of the southern granite-greenstone contacts are, thus, largely unaffected by deformation and heat advection related to the 3.1 Ga granites. The Mpuluzi batholith is only in contact with the southernmost extremity of the Barberton greenstone belt along the Motjane schist belt (Fig. 12C). It is this region where Jackson and Robertson (1983) described the pervasive refoliation and refolding of greenstone lithologies that are intruded by multiple sheets of the Mpuluzi batholith. Similarly, 3.1 Ga deformation in the northern parts of the greenstone belt is most prominent where the supracrustals are intruded by or in close contact with rocks related to the Nelspruit batholith or the Stentor pluton.

This spatial and temporal relationship leads us to postulate that the seemingly diachronous and heterogeneous distribution of late-stage deformation in and around the Barberton greenstone belt is the result of the pronounced strain localization of the bulk D3 strain into syntectonic intrusions. The strain intensity and timing of the late-stage deformation is largely determined by the regional-scale geometry and distribution of synmagmatic shear zones and melt sheets related to the GMS suite in the granite-gneiss terrain and the greenstone belt. These results may also have significance on the localization of the main gold mines in the Barberton greenstone belt, given the close spatial relationship between gold mineralization and the late-stage D3 deformation in the belt (e.g., De Ronde et al., 1992).

Regional Considerations

The 3.1 Ga granite plutonism on the Kaapvaal Craton is a short-lived, craton-wide event, and the causes of the widespread magmatism and associated high-grade metamorphism recorded in mid- and lower crustal sections of, for example, the Vredefort dome are somewhat enigmatic (De Wit et al., 1992; Hart et al., 1999; Poujol et al., 2003). A collisional plate margin setting was postulated by Moser et al. (2001) to account for the deep-crustal burial of metasediments and concomitant magmatism in the Vredefort dome at 3.1 Ga. Moser et al. (2001) also suggested that the lithospheric strain during the 3.1 Ga convergent tectonics did not affect the upper, brittle crust but was rather partitioned into the hotter, ductile lower crust. The results presented in this study agree with a convergent setting for the GMS suite plutonism. The pervasive deformation of the Heerenveen batholith and other batholiths of the GMS suite also demonstrates that the lithospheric strain did, indeed, penetrate into the upper crust. Much of the bulk strain, however, was effectively partitioned into zones of melt ascent and/or emplacement. If a collisional plate margin is assumed in the central Kaapvaal Craton in the area of the Vredefort dome, the Barberton granite-greenstone terrain may correspond to a contractional continental back-arc region. The source of the craton-wide granite plutonism and whether a crust-mantle coupling occurred at that time, as proposed by Moser et al. (2001), must remain speculative.

Less speculative is the assertion that the GMS suite plutonism has most likely had a significant influence on the post-collisional evolution of the Kaapvaal crust. In the Barberton granite-greenstone terrain, the voluminous and laterally extensive granite sheets, the bulk of which intruded over a short period of time of only 5–10 m.y., must have represented an almost continuous low-viscosity melt layer within the crust. If the midcrustal section of the Vredefort dome is representative for larger parts of the Kaapvaal Craton, then substantial parts of the Kaapvaal crust were significantly weakened at this time (Moser et al., 2001). In the Barberton region, the bulk of the GMS suite plutonism is shortly followed by transtensional and/or extensional tectonics. This may suggest that the transition from convergent to extensional tectonics may not be related to a change in far-field stresses but rather reflect gravitational instabilities and gravity-driven flow induced by the presence of regional-scale melt layers in the crust (e.g., Vanderhaeghe and Teyssier, 2001). This would correspond to the D4 collapse stage of the Barberton granite-greenstone terrain described by De Ronde and De Wit (1994). However, significant tectonic thickening cannot be demonstrated for the granite-greenstone terrain at 3.1 Ga, and the gravitational potential energy that drives the exhumation of mid- and deep-crustal levels during lateral flow was likely very limited. The highest exhumation rates occur in regions that were significantly thickened during the collisional stage corresponding to areas of largest gravitational potential energy (Vanderhaeghe and Teyssier, 2001). High post-collisional exhumation rates are indicated, for example, for the rocks of the Vredefort dome, where the 3074 ± 6 Ma Dominion Group unconformably overlies basement rocks that underwent granulite-facies metamorphism at 3.1 Ga (Hart et al., 1999), suggesting exhumation rates of well in excess of 1 mm/year.

ACKNOWLEDGMENTS

This material is based upon work supported by a grant from the South African National Research Foundation (grant no. NRF 2053186) awarded to AFMK. We greatly appreciate the cooperation of farmers and residents in the Badplaas region for their hospitality and access to their lands. Carl Anhaeusser is thanked for his introduction to the GMS suite rocks and insightful discussions. Many ideas presented here were developed during stimulating discussions in the field and at the Department of Geology, University of Stellenbosch, with colleagues working in the Barberton area. U. Riller and C. Anhaeusser are thanked for their detailed reviews of the manuscript.

REFERENCES CITED

Anhaeusser, C.R., 1973, The evolution of the early Precambrian crust of southern Africa: Philosophical Transactions of the Royal Society of London, v. A273, p. 359–388.

Anhaeusser, C.R., 1980, A geological investigation of the Archaean granite-greenstone terrane south of the Boesmanskop syenite pluton, Barberton

Mountain Land: Transactions of the Geological Society of South Africa, v. 83, p. 93–106.

Anhaeusser, C.R., and Robb, L.J., 1983, Geological and geochemical characteristics of the Heerenveen and Mpuluzi batholiths south of the Barberton greenstone belt and preliminary thoughts on their petrogenesis, *in* Anhaeusser, C.R., ed., Contributions to the Geology of the Barberton Mountain Land: Special Publication of the Geological Society of South Africa, v. 9, p. 131–152.

Anhaeusser, C.R., Robb, L.J., and Viljoen, M.J., 1981, Provisional geological map of the Barberton greenstone belt and surrounding granitic terrane, eastern Transvaal and Swaziland: Johannesburg, Geological Society of South Africa, scale 1:250,000, 1 sheet.

Anhaeusser, C.R., Robb, L.J., and Barton, J.M., Jr., 1983, Mineralogy, petrology and origin of the Boesmanskop Syeno-granite complex, Barberton Mountain Land, South Africa, *in* Anhaeusser, C.R., ed., Contributions to the Geology of the Barberton Mountain Land: Special Publication of the Geological Society of South Africa, v. 9, p. 169–184.

Armstrong, R.A., Compston, W., De Wit, M.J., and Williams, I.S., 1990, The stratigraphy of the 3.5–3.2 Ga Barberton greenstone belt revisited: A single zircon ion microprobe study: Earth and Planetary Science Letters, v. 101, p. 90–105, doi: 10.1016/0012-821X(90)90127-J.

Armstrong, R.A., Compston, W., Retief, E.A., Williams, I.S., and Welke, H.J., 1991, Zircon ion microprobe studies bearing on the age and evolution of the Witwatersrand triad: Precambrian Research, v. 53, p. 243–266, doi: 10.1016/0301-9268(91)90074-K.

Benn, K., Ham, N.M., Pignotta, G.S., and Bleeker, W., 1998, Emplacement and deformation of granites during transpression: Magnetic fabrics of the Archean Sparrow pluton, Slave Province, Canada: Journal of Structural Geology, v. 20, p. 1247–1259, doi: 10.1016/S0191-8141(98)00065-0.

Benn, K., Paterson, S.R., Lund, S.P., Pignotta, G.S., and Kruse, S., 2001, Magmatic fabrics in batholiths as markers of regional strains and plate kinematics: Example of the Cretaceous Mt. Stuart batholith: Physics and Chemistry of the Earth: Part A: Solid Earth and Geodesy, v. 26, p. 343–354, doi: 10.1016/S1464-1895(01)00064-3.

De Ronde, C.E.J., and De Wit, M.J., 1994, Tectonic history of the Barberton greenstone belt, South Africa: 490 million years of Archean crustal evolution: Tectonics, v. 13, p. 983–1005, doi: 10.1029/94TC00353.

De Ronde, C.E.J., and Kamo, S.L., 2000, An Archæan arc-arc collisional event: A short-lived (ca 3 Myr) episode, Weltevreden area, Barberton greenstone belt, South Africa: Journal of African Earth Sciences, v. 30, p. 219–248, doi: 10.1016/S0899-5362(00)00017-8.

De Ronde, C.E.J., Spooner, E.T.C., De Wit, M.J., and Bray, C.J., 1992, Shear zone related, Au-quartz vein deposits in the Barberton greenstone belt, South Africa: Field and petrographic characteristics, fluid properties and light stable isotope geochemistry: Economic Geology and the Bulletin of the Society of Economic Geologists, v. 87, p. 366–402.

De Wit, M.J., Roering, C., Hart, R.J., Armstrong, R.A., De Ronde, C.E.J., Green, R.W.E., Tredoux, M., Perberdy, E., and Hart, R.A., 1992, Formation of an Archean continent: Nature, v. 357, p. 553–562, doi: 10.1038/357553a0.

Diener, J., Stevens, G., Kisters, A.F.M., and Poujol, M., 2005, Metamorphism and exhumation of the basal parts of the Barberton greenstone belt, South Africa: Constraining the rates of mid-Archean tectonism: Precambrian Research, v. 143, p. 87–112.

Dziggel, A., Stevens, G., Poujol, M., Anhaeusser, C.R., and Armstrong, R.A., 2002, Metamorphism of the granite-greenstone terrane south of the Barberton greenstone belt, South Africa: An insight in the tectonothermal evolution of the 'lower' portions of the Onverwacht Group: Precambrian Research, v. 114, p. 221–247, doi: 10.1016/S0301-9268(01)00225-X.

Gleizes, G., Leblanc, D., Santana, V., Olivier, P., and Bouchez, J.L., 1998, Sigmoidal structures featuring dextral shear during emplacement of the Hercynian granite complex of Cauterets-Panticosa (Pyrenées): Journal of Structural Geology, v. 20, p. 1229–1245, doi: 10.1016/S0191-8141(98)00060-1.

Grujic, D., and Mancktelow, N.S., 1998, Melt-bearing shear zones: Analogue experiments and comparison with examples from southern Madagascar: Journal of Structural Geology, v. 20, p. 673–680, doi: 10.1016/S0191-8141(98)00006-6.

Harris, P.D., Robb, L.J., and Tomkinson, M.J., 1995, The nature and structural setting of rare-element pegmatites along the northern flank of the Barberton greenstone belt, South Africa: South African Journal of Geology, v. 98, p. 82–94.

Hart, R.J., Moser, D.E., and Andreoli, M., 1999, Archean age for the granulite facies metamorphism near the centre of the Vredefort Structure, South Africa: Geology, v. 27, p. 1091–1094, doi: 10.1130/0091-7613(1999)027<1091:AAFTGF>2.3.CO;2.

Hunter, D.R., 1973, The granitic rocks of the Precambrian in Swaziland: Special Publication of the Geological Society of South Africa, v. 3, p. 131–145.

Jackson, M.P.A., and Robertson, D.I., 1983, Regional implications of early-Precambrian strains in the Onverwacht Group adjacent to the Lochiel Granite, north-west Swaziland, *in* Anhaeusser, C.R., ed., Contributions to the Geology of the Barberton Mountain Land: Special Publication of the Geological Society of South Africa, v. 9, p. 45–62.

Kamo, S.L., and Davis, D.W., 1994, Reassessment of Archean crustal development in the Barberton Mountain Land, South Africa, based on U-Pb dating: Tectonics, v. 13, p. 167–192, doi: 10.1029/93TC02254.

Kisters, A.F.M., Stevens, G., Dziggel, A., and Armstrong, R.A., 2003, Extensional detachment faulting and core-complex formation in the southern Barberton granite-greenstone terrain, South Africa: Evidence for a 3.2 Ga orogenic collapse: Precambrian Research, v. 127, p. 355–378, doi: 10.1016/j.precamres.2003.08.002.

Kohler, E.A., 1994, The geology of the Archaean granitoid-greenstone terrane in the vicinity of three Sisters, Barberton greenstone belt [Ph.D. thesis]: Johannesburg, University of Witwatersrand, 309 p.

Kretz, R., 1983, Symbols for rock-forming minerals: The American Mineralogist, v. 68, p. 277–279.

Lana, C., Gibson, R.L., Kisters, A.F.M., and Reimold, W.U., 2003, Archean crustal structure of the Kaapvaal Craton, South Africa—Evidence from the Vredefort dome: Earth and Planetary Science Letters, v. 206, p. 133–144, doi: 10.1016/S0012-821X(02)01086-5.

Lana, C., Reimold, W.U., Gibson, R.L., Koeberl, C., and Siegesmund, S., 2004, Nature of the Archean midcrust in the core of the Vredefort Dome, Central Kaapvaal Craton, South Africa: Geochimica et Cosmochimica Acta, v. 68, p. 623–642, doi: 10.1016/S0016-7037(03)00447-2.

Lowe, D.R., 1994, Accretionary history of the Archean Barberton Greenstone Belt (3.55–3.22 Ga), southern Africa: Geology, v. 22, p. 1099–1102, doi: 10.1130/0091-7613(1994)022<1099:AHOTAB>2.3.CO;2.

Lowe, D.R., Byerly, G.R., and Heubeck, C., 1999, Structural divisions and development of the west-central part of the Barberton Greenstone Belt, *in* Lowe, G.R., and Byerly, D.R., eds., Geological evolution of the Barberton Greenstone Belt, South Africa: Geological Society of America, Special Paper 392, p. 37–83.

Miller, R.B., and Paterson, S.R., 1994, The transition from magmatic to high-temperature solid-state deformation: Implications from the Mount Stuart batholith, Washington: Journal of Structural Geology, v. 16, p. 853–865, doi: 10.1016/0191-8141(94)90150-3.

Moser, D.E., Flower, R.M., and Hart, R.J., 2001, Birth of the Kaapvaal tectonosphere 3.08 billion years ago: Science, v. 291, p. 465–468, doi: 10.1126/science.291.5503.465.

Paterson, S.R., Vernon, R.H., and Tobisch, O.T., 1989, A review of criteria for the identification of magmatic and tectonic foliations in granitoids: Journal of Structural Geology, v. 11, p. 349–363, doi: 10.1016/0191-8141(89)90074-6.

Paterson, S.R., Fowler, T.K., Jr., Schmidt, K.L., Yoshinobu, A.S., Yuan, E.S., and Miller, R.B., 1998, Interpreting magmatic fabric patterns in plutons: Lithos, v. 44, p. 53–82, doi: 10.1016/S0024-4937(98)00022-X.

Pawley, M.J., Collins, W.J., and Van Kranendonk, M.J., 2002, Origin of fine-scale sheeted granites by incremental injection of magma into active shear zones: Examples from the Pilbara Craton, NE Australia: Lithos, v. 61, p. 127–139, doi: 10.1016/S0024-4937(02)00076-2.

Poujol, M., and Anhaeusser, C.R., 2001, The Johannesburg Dome, South Africa: New single zircon U-Pb isotopic evidence for the early Archaean granite-greenstone development within the central Kaapvaal Craton: Precambrian Research, v. 108, p. 139–157, doi: 10.1016/S0301-9268(00)00161-3.

Poujol, M., Robb, L.J., Anhaeusser, C.R., and Gericke, B., 2003, A review of the geochronological constraints on the evolution of the Kaapvaal Craton, South Africa: Precambrian Research, v. 127, p. 181–213, doi: 10.1016/S0301-9268(03)00187-6.

Robb, L.J., Anhaeusser, C.R., and Van Nierop, D.A., 1983, The recognition of the Nelspruit batholith north of the Barberton greenstone belt and its significance in terms of Archaean crustal evolution: Special Publication of the Geological Society of South Africa, v. 9, p. 117–130.

Stevens, G., Droop, G.T.R., Armstrong, R.A., and Anhaeusser, C.R., 2002, Amphibolite facies metamorphism in the Schapenburg schist belt: A record of mid-crustal response to ~3.23 Ga terrane accretion in the Barberton greenstone belt: South African Journal of Geology, v. 105, p. 271–284, doi: 10.2113/1050271.

Tobisch, O.T., and Williams, Q., 1998, Use of microgranitoid enclaves as solid state strain markers in deformed granitic rock: An evaluation: Journal of Structural Geology, v. 20, p. 727–743, doi: 10.1016/S0191-8141(98)00009-1.

Tomkinson, M.J., and King, V.J., 1991, The tectonics of the Barberton greenstone belt: An overview, *in* Ashwal, L.D., ed., Two Cratons and an Orogen: Johannesburg, University of Witwatersrand, Excursion Guidebook, IGCP Project 280, p. 69–83.

Tullis, J., and Yund, R.A., 1987, Transition of cataclasite flow to dislocation creep of feldspar: Mechanisms and microstructures: Geology, v. 15, p. 606–609, doi: 10.1130/0091-7613(1987)15<606:TFCFTD>2.0.CO;2.

Vanderhaeghe, O., and Teyssier, C., 2001, Partial melting and flow of orogens: Tectonophysics, v. 342, p. 451–472, doi: 10.1016/S0040-1951(01)00175-5.

Vernon, R.H., Johnson, S.E., and Melis, E.A., 2004, Emplacement-related microstructures in the margin of a pluton: The San José tonalite, Baja California, México: Journal of Structural Geology, v. 26, p. 1867–1884, doi: 10.1016/j.jsg.2004.02.007.

Westraat, J.D., Kisters, A.F.M., Poujol, M., and Stevens, G., 2005, Transcurrent shearing, granite sheeting and the incremental construction of the tabular 3.1 Ga Mpuluzi batholith, Barberton granite-greenstone terrain, South Africa: Journal of the Geological Society of London, v. 162, p. 373–388, doi: 10.1144/0016-764904-026.

Xie, X., Byerly, G.R., and Ferrel, R.E., Jr., 1997, IIb trioctahedral chlorite from the Barberton greenstone belt: Crystal structure and rock composition constraints with implications to geothermometry: Contributions to Mineralogy and Petrology, v. 126, p. 275–291, doi: 10.1007/s004100050250.

Yearron, L.M., 2003, Archean granite petrogenesis and the implications for the evolution of the Barberton Mountain Land, South Africa [Ph.D. thesis]: London, UK, Kingston University, 315 p.

Zulauf, G., and Helferich, S., 1997, Strain and strain rate in a synkinematic trondhjemitic dike: Evidence for melt-induced strain softening during shearing (Bohemian Massif, Czech Republic): Journal of Structural Geology, v. 19, p. 639–652, doi: 10.1016/S0191-8141(96)00100-9.

Manuscript Accepted by the Society 4 October 2005

Printed in the USA

Geological Society of America
Special Paper 405
2006

SHRIMP zircon age constraints on Mesoarchean crustal development in the Vredefort dome, central Kaapvaal Craton, South Africa

Richard A. Armstrong†
Research School of Earth Sciences, The Australian National University, Canberra 0200, A.C.T., Australia
Cristiano Lana‡
Impact Cratering Research Group, School of Geosciences, University of the Witwatersrand, Private Bag 3, 2050 Wits, Johannesburg, South Africa
Wolf Uwe Reimold§
Impact Cratering Research Group, School of Geosciences, University of the Witwatersrand, Private Bag 3, 2050 Wits, Johannesburg, South Africa, and
Museum für Naturkunde (Mineralogy), Humboldt University, Invalidenstrasse 43, 10115 Berlin, Germany
Roger L. Gibson
Impact Cratering Research Group, School of Geosciences, University of the Witwatersrand, Private Bag 3, P.O. Wits 2050, Johannesburg, South Africa

ABSTRACT

SHRIMP (sensitive high-resolution ion microprobe) U-Pb zircon dating of the Archean basement rocks in the core of the Vredefort dome, South Africa, is integrated with results of lithological and structural mapping and geochemical data to assess the tectonic evolution of the central parts of the Kaapvaal Craton. Recent tectonic models have envisaged the rocks in the core of the dome as part of a much larger, >3200 Ma, Archean cratonic shield (Kaapvaal Shield) that attained stability during partial melting of its early crust and emplacement of voluminous, Mesoarchean, crustally derived granitoid bodies between 3200 and 3100 Ma. Our studies have indicated, however, that the oldest sialic rocks (tonalite-trondhjemite-granodiorite—TTG) in the core of the dome were only formed between 3120 and 3100 Ma, probably in an island arc setting, and that they were accreted onto the craton shortly thereafter. Tectonic accretion of the TTG rocks led to crustal thickening, partial melting, and emplacement of 3100–3080 Ma granites and granodiorites at midcrustal levels under high-grade metamorphic conditions, slightly later than the main pulse of granitoid magmatism in the Kaapvaal Shield. This event was followed almost immediately by collapse of the thickened crust, leading to rapid exhumation of the midcrustal rocks prior to the deposition of rift-related volcanic and sedimentary supracrustal strata at ca. 3074 Ma. The Mesoarchean tectonomagmatic activity in the core of the dome was terminated with emplacement of 3068 Ma aplite dikes.

Keywords: Kaapvaal Craton, Vredefort dome, Archean tectonics, zircon, SHRIMP.

†E-mail: richard.armstrong@anu.edu.au.
‡Present Address: Impacts and Astromaterials Research Centre, Imperial College London, Royal School of Mines, South Kensington Campus, SW7 2AZ, UK.
§E-mail: uwe.reimold@museum.hu-berlin.de.

Armstrong, R.A., Lana, C., Reimold, W.U., and Gibson, R.L., 2006, SHRIMP zircon age constraints on Mesoarchean crustal development in the Vredefort dome, central Kaapvaal Craton, South Africa, *in* Reimold, W.U., and Gibson, R.L., Processes on the Early Earth: Geological Society of America Special Paper 405, p. 233–253, doi: 10.1130/2006.2405(13). For permission to copy, contact editing@geosociety.org.

INTRODUCTION

While the Kaapvaal Craton of southern Africa (Fig. 1) does not contain the oldest rocks on Earth, it contains one of the most comprehensive Archean geological records from ca. 3.6 Ga onward (Hunter, 1974; de Wit et al., 1992; Kröner and Layer, 1992; Kröner et al., 1996; Poujol et al., 2003). In recent years, the volume of geochronological data for the craton has grown considerably and has allowed further refinement of geodynamic models for its evolution (see reviews in Poujol et al., 2003; Eglington and Armstrong, 2004). In their original synthesis, de Wit et al. (1992) proposed a two-stage evolution for the craton, with the initial cratonic shield (Kaapvaal Shield) having grown through collisional tectonics and vertical magmatic accretion between 3600 and 3200 Ma, after which lateral accretion along the margins of the Shield dominated.

The results of more recent field mapping and geochemical studies and U-Pb single zircon and monazite dating are consistent with the de Wit et al. (1992) model for the eastern parts of the craton (Armstrong et al., 1990; Kamo and Davis, 1994; de Ronde and de Wit, 1994; Lowe, 1994, 1999; Kröner et al., 1996), with the geochronological data having allowed the identification of distinct microcontinental blocks within the Barberton region (Fig. 1) (e.g., Lowe, 1994, 1999). Between 3200 and 3100 Ma, however, local-scale tectonomagmatic processes appear to have been replaced by regional-scale partial melting of the preexisting >3200 Ma tonalite-trondhjemite-granodiorite (TTG) suites and emplacement of high-level granite-granodiorite-monzogranite (GGM) batholiths throughout the Kaapvaal Shield (Robb et al., 1991; de Wit et al., 1992; de Wit, 1998; Moser et al., 2001; Poujol et al., 2003). This regional anatectic/magmatic event, the last major episode of vertical differentiation in the Kaapvaal Shield, has been interpreted by some workers as having been responsible for the establishment of a thick, stable lithosphere that was capable of supporting the large Neoarchean to Paleoproterozoic volcano-sedimentary basins that developed subsequently on the craton (Robb et al., 1991; de Wit et al., 1992; Kamo and Davis, 1994; de Wit, 1998; Moser et al., 2001; Poujol and Anhaeusser, 2001; Poujol et al., 2003). Although the exact cause of the regional GGM event

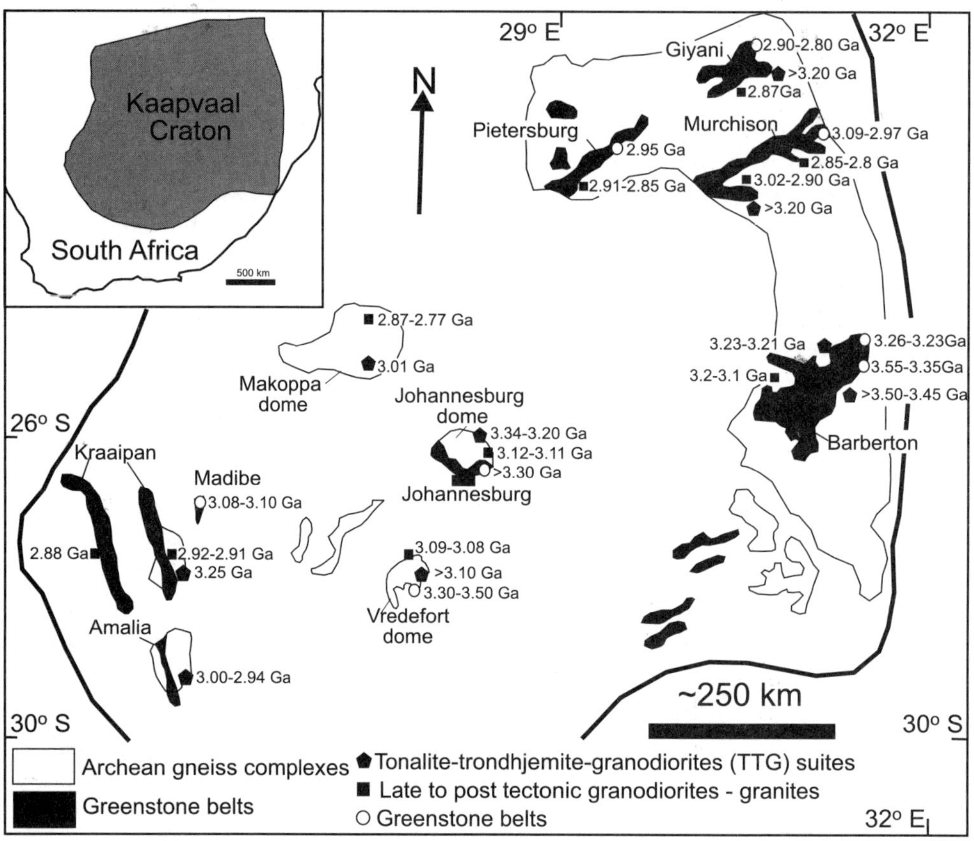

Figure 1. Schematic map of the Kaapvaal Craton, based on geological, aeromagnetic, and gravity data (after de Wit et al., 1992), showing the outcrop locations of the Archean greenstones and gneiss complexes. The most recently published U-Pb isotope dilution thermal ionization mass spectrometry (ID-TIMS) and SHRIMP (sensitive high-resolution ion microprobe) ages of the greenstones, tonalite-trondhjemite-granodiorite (TTG) suites, and granodiorite-granite-monzogranite (GGM) suites are indicated. Geochronological data from each sector of the craton has been compiled by Poujol et al. (2003).

remains elusive, several workers have suggested that it incorporated upper mantle melting and substantial reconfiguration of the subcrustal lithosphere beneath the craton (Moser et al., 2001; Flowers et al., 2003; Prevec et al., 2004).

In the central parts of the Kaapvaal Craton, Hart et al. (1999), Moser et al. (2001), Flowers et al. (2003), and Poujol et al. (2003) have suggested a similar Mesoarchean evolutionary history in the Johannesburg and Vredefort domes to that proposed by de Wit et al. (1992), with a >3.2 Ga TTG–greenstone crust that experienced high-grade metamorphism and GGM magmatism at ca. 3.1 Ga. In this paper, we present new SHRIMP (sensitive high-resolution ion microprobe) zircon U-Pb geochronological results from the Vredefort dome that indicate that emplacement of the oldest sialic rocks (TTG suite) in the core of the dome occurred only after the assembly of the Kaapvaal Shield at 3200 Ma and that the GGM event was also younger than the one that affected the Shield. This necessitates a revised geodynamic model for the evolution of these rocks.

LOCAL GEOLOGY

The ~80-km-wide Vredefort dome, located ~120 km southwest of Johannesburg, is the eroded remnant of the central uplift of the originally 250–300-km-diameter, 2023 ± 4 Ma (Kamo et al., 1996) Vredefort impact structure (for more details, refer to review by Gibson and Reimold, 2001). The dome consists of an ~40-km-wide core of Mesoarchean crystalline basement, which is surrounded by a 15–20-km-wide collar of Neoarchean to Paleoproterozoic supracrustal strata (Fig. 2). These supracrustal rocks, which range in age from 3.07 to ca. 2.1 Ga (Armstrong et al., 1991), belong to (from the oldest to the youngest) the Dominion Group and the Witwatersrand, Ventersdorp, and Transvaal supergroups.

The core of the Vredefort dome exposes a complex amphibolite- to granulite-facies migmatite terrane, consisting dominantly of stromatic migmatites and syntectonic granitoids and volumetrically minor mafic granulites, amphibolites, and nonmigmatitic ortho- and paragneisses (Fig. 2) (Stepto, 1979, 1990; Lana et al., 2003a, 2004; Lana, 2004). The migmatites, restitic gneisses, and granitoids record a complex Archean tectonometamorphic history involving four ductile deformation events (D1 to D4) and widespread melting (Lana et al., 2003a, 2003b, 2003c). Published U-Pb single zircon and monazite data from various rock types in the central parts of the core of the dome indicate that the high-grade metamorphism occurred ~3.1 b.y. ago (Kamo et al., 1996; Hart et al., 1999; Moser et al., 2001; Flowers et al., 2003). Analyses of two multigrain fractions of shocked and highly cracked zircon from a postmetamorphic dolerite dike gave discordant, minimum $^{207}Pb/^{206}Pb$ ages of ca. 2.57 Ga, data used by Hart et al. (1999) to indicate the minimum age for the Archean metamorphic and deformation events. Recently, Moser et al. (2001) and Flowers et al. (2003) obtained U-Pb single zircon ages between 3.13 and 3.08 Ga for several synmetamorphic dikes of granite, tonalite, and charnockite. The results presented here form part of an integrated field, geochemical, and geochronological study that has produced the first coherent lithological and structural map of the core of the dome (Lana et al., 2003a, 2004; Fig. 2). They have allowed the identification of four main periods of tectonomagmatic activity in these rocks.

The oldest event recorded in the dome involved submarine to subaerial ultramafic-mafic volcanic activity with eruption of likely >3.4 Ga komatiites, komatiitic basalts, and tholeiites. This was followed by or interspersed with deposition of pelites, graywackes, and banded ironstones (Stepto, 1979, 1990; Hart et al., 1981, 1990; Minnitt et al., 1994; Lana et al., 2003b, Lana, 2004). The komatiitic lavas are characterized by massive and pillowed basalt lava flows displaying vesicles, variolites, and spinifex textures. The lavas alternate with thin layers of chert and banded iron formation (Minnitt et al., 1994; Lana et al., 2003b). Field, petrological, and geochemical data from these sequences indicate that the mafic and ultramafic rocks might have a similar origin to the komatiites and komatiitic basalts in the Johannesburg dome and the Barberton Greenstone belt (Lana et al., 2003b). The best outcrops in the Vredefort dome occur in an inlier in the southeastern core (Fig. 2).

Over much of the core of the dome, the volcano-sedimentary sequence forms volumetrically minor rafts within large, heterogeneous tonalite-trondhjemite-granodiorite intrusive bodies (Stepto, 1990; Flowers et al., 2003; Lana et al., 2004). The earliest recognizable intrusion is a biotite-rich trondhjemitic gneiss (sample ABBG of this study), which is found as xenoliths within the main trondhjemitic gneiss (sample ABG of this study) and which contains evidence of an early S1 fabric that has been largely transposed by D2 (Lana et al., 2004; Lana, 2004). Both xenoliths and host gneisses are found in scattered outcrops in the northeastern sector of the core of the dome (Fig. 2). These rocks were intruded by a porphyritic granodiorite in the western parts of the core of the dome (Fig. 2) (Lana et al., 2004). Emplacement of this granodiorite was accompanied by significant deformation, involving the development of a dominant subhorizontal S2 fabric in the trondhjemitic gneisses. S2 has, however, been rotated to the vertical in the outer parts of the core during the impact-induced doming event (Fig. 2; Lana et al., 2003c).

Emplacement of leucosomes and granitoids with a wide range of compositions occurred during the peak amphibolite- to granulite-facies metamorphic event in the TTG and supracrustal rocks (Lana et al., 2004; Lana, 2004). This was accompanied by the development of northwest-trending peak-metamorphic S3 high-strain zones characterized by both coaxial and noncoaxial strains (Fig. 2) during a northeast-southwest–directed compressional event (Lana et al., 2003a, 2004; Lana, 2004). Field and geochemical evidence (Lana et al., 2004) indicates that the leucosomes that intruded the upper levels of the migmatite (amphibolite-facies zone) are compositionally similar to highly evolved GGM rocks that are found elsewhere in the craton as upper crustal sheeted dike and batholithic granitic bodies (Anhaeusser, 1999; Poujol and Anhaeusser, 2001; Poujol

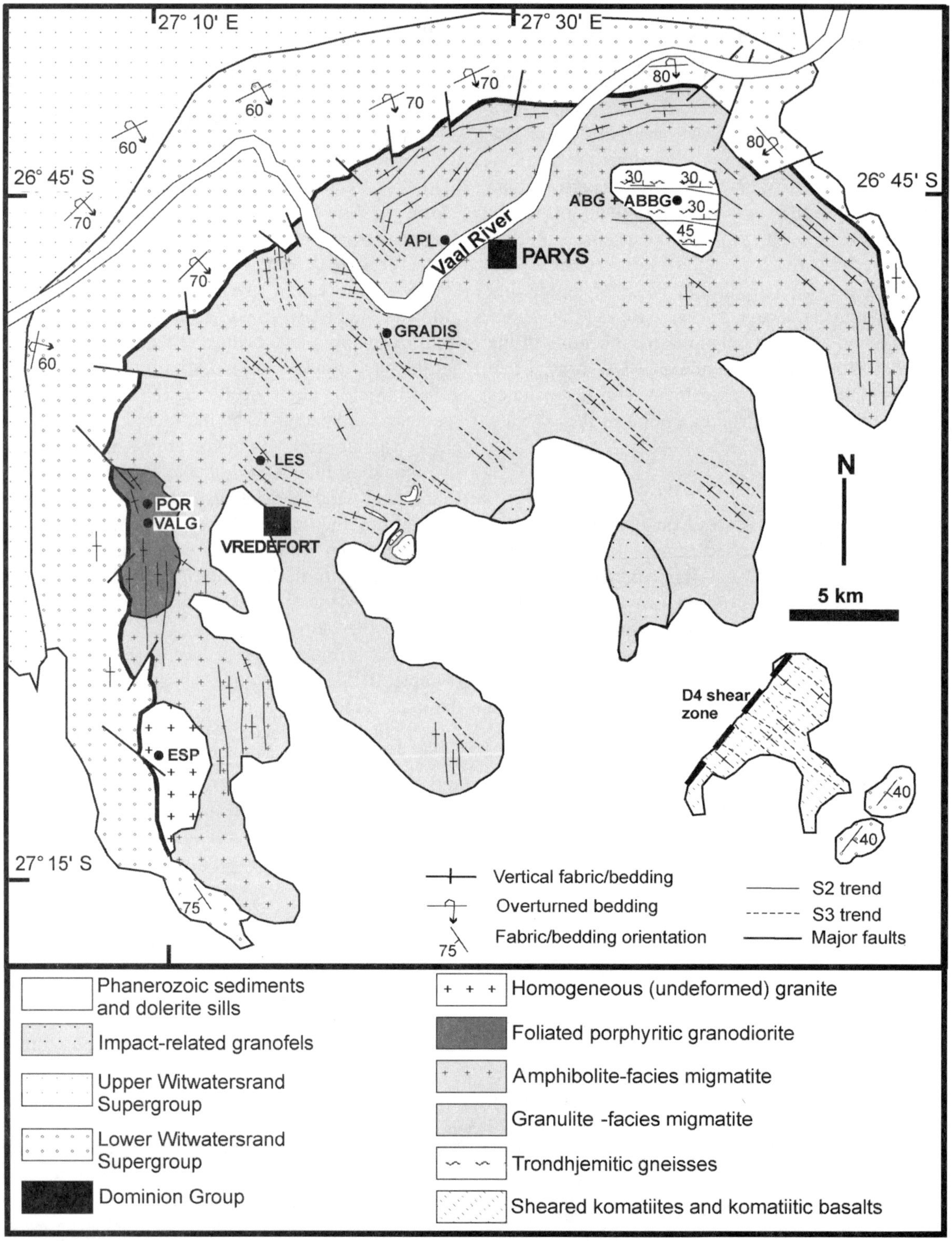

Figure 2. Simplified geological map of the Vredefort dome showing the main lithologies and structures of the Archean gneiss complex in core and strata of inner collar of the dome. Black dots indicate the localities of samples collected for SHRIMP (sensitive high-resolution ion microprobe) analysis. Sample names refer to original sample set collected by Lana (2004).

et al., 2003). Leucosomes in the granulite-facies zone range from enderbitic to charnockitic in composition (Lana et al., 2004; Lana, 2004). The charnockites have similar compositions to, and probably share a common source with, the granitic leucosomes in the amphibolite-facies zone (Lana et al., 2004). The enderbites are relatively more mafic and are characterized by significant amounts of xenoliths of mafic granulite (e.g., remnants of greenstones; Lana et al., 2004; Lana, 2004).

Collapse of the thickened crust along northeast-trending, subhorizontal, extensional D4 mylonitic shear zones followed immediately after the attainment of peak metamorphic conditions (Colliston and Reimold, 1990; Lana et al., 2003b). The collapse led to juxtaposition of the greenschist-facies greenstone sequence in the southeastern sector of the dome with granulite-facies migmatites in the central parts of the dome (Lana et al., 2003b) and may have been accompanied by emplacement of several aplite dikes that intrude into or crosscut minor mylonitic shear zones (Lana et al., 2004; Lana, 2004).

SAMPLING AND METHODOLOGY

Zircons from the nonmigmatitic gneisses, migmatites, and granites exposed in the core of the Vredefort dome were separated for SHRIMP U-Pb dating. The host trondhjemitic gneiss (ABG) and the xenoliths of biotite-rich gneiss (ABBG) were collected in the northeastern sector of the dome (Fig. 2), whereas the syn-D2 porphyritic granodiorite (POR) and trondhjemitic melanosome (VALG) originate from the western sector of the dome (Fig. 2). Samples of the enderbitic leucosome (LES) and magnetite-rich granodiorite (GRADIS), all representing partial melts formed during D3 (Lana et al., 2003a, 2004), were collected in the amphibolite-granulite facies transition zone, some 5 km northeast of Vredefort town (Fig. 2). One sample of an undeformed aplite dike (APL), which crosscuts S4 and which should, thus, provide a minimum age for the polyphase tectonometamorphic events in the dome, was collected in the Kudu dimension stone quarry ~3 km north of Parys (Fig. 2).

Mineral separates were prepared from 7 to 10 kg rock samples that were crushed and pulverized at the School of Geosciences, University of the Witwatersrand, Johannesburg. Zircons were separated using a Wilfley table, heavy liquids (bromoform and methylene iodide), and a Frantz isodynamic separator. SHRIMP II and SHRIMP RG U-Th-Pb analyses were performed at the Research School of Earth Sciences at the Australian National University, Canberra. Zircons were handpicked under a binocular microscope and mounted in epoxy, together with the Research School of Earth Sciences zircon standards FC1 and SL13. All grains were then polished to half their thickness to expose internal structures. Preparation for the SHRIMP analyses included extensive use of transmitted and reflected light microphotography, together with SEM (scanning electron microscope) cathodoluminescence imaging, to decipher the growth complexities in the zircon grains and to target the best areas for in situ analysis. For the zircon age calibration, the Pb/U ratios were normalized relative to a value of 0.1859 for the $^{206}Pb^{*}/^{238}U$ ratio of the FC1 reference zircons, equivalent to an age of 1099 Ma (Paces and Miller, 1993). Uranium and thorium concentrations were determined relative to the SL13 standard. SHRIMP analyses comprised 6 repeated scans through the species $^{196}Zr_2O$, ^{204}Pb (common Pb), background ^{206}Pb, ^{207}Pb, ^{208}Pb, ^{238}U, ^{248}ThO, and ^{254}UO.

The data were reduced in a manner similar to that described by Williams (1998, and references therein), using the SQUID Excel Macro of Ludwig (2000a). Uncertainties in the Pb/U calibration from the various analytical sessions are included in the computed ages (e.g., the concordia ages) by SQUID but are not included in the individual analyses as reported in the data tables (Tables 1–7; uncertainties in the calibrations are, however, given in the tables). Corrections for common Pb were made using the appropriate model values of Stacey and Kramers (1975).

Uncertainties given for individual analyses (ratios and ages) are at the 1σ level unless stated otherwise; however, uncertainties in the calculated weighted mean, intercept, or concordia ages are reported at 95% confidence limits. Concordia plots and age calculations were carried out using Isoplot/Ex (Ludwig, 2000b) or SQUID.

RESULTS

Nonmigmatitic Gneisses

A sample of the dominant trondhjemitic gneiss (ABG) in the dome contains light brown, generally subhedral zircons. Many grains are highly fractured, with some radial cracks linked to volume expansion in areas surrounding (metamict) high-U zones (e.g., Lee and Tromp, 1995) and also display sets of parallel, near-planar fractures as a result of the 2.02 Ga impact (see also Kamo et al., 1996; Gibson et al., 1997). Resorption features and reworking of terminations indicate metamorphic modification. The zircons have the oscillatory zoning typical of crystallization in felsic magma. Many grains also preserve core structures (Fig. 3A). These are best identified through cathodoluminescence imaging, but the overall poor (dark) luminescence of these grains makes unraveling the more subtle internal structures difficult and sometimes impossible. For the analysis of these complex zircons from this and other samples, SHRIMP spots were targeted in the cores, the zoned magmatic grains, and, where spatial resolution allowed, in the reworked or metamorphic rims. Least altered areas were targeted, but given the overall poor state of preservation of the zircons from these rocks, this was not always possible.

The SHRIMP U-Pb analyses of zircons from the trondhjemitic gneiss (ABG) are largely discordant, with the data scattering about a discordia trend that appears to be dominated by Mesoproterozoic Pb loss (Fig. 4; Table 1), although the scatter of data does not necessarily indicate a unique trend. Even the more

TABLE 1. RESULTS OF THE SHRIMP U-Pb ANALYSES FOR ZIRCONS FROM SAMPLE ABG

Grain spot	% $^{206}Pb_c$	U (ppm)	Th (ppm)	$^{232}Th/^{238}U$	$^{206}Pb^*$ (ppm)	$^{206}Pb/^{238}U$ age (Ma)[†]	$^{207}Pb/^{206}Pb$ age (Ma)[†]	% Disc.	$^{207}Pb^*/^{206}Pb^{*\dagger}$	± (%)	$^{207}Pb^*/^{235}U^{\dagger}$	± (%)	$^{206}Pb^*/^{238}U$	± (%)	Err. Corr.
1.1m	0.15	523	187	0.37	215	2524 ± 15	2937.3 ± 5.5	14	0.21416	0.34	14.15	0.80	0.4793	0.73	.907
2.1c	0.13	327	216	0.68	106	2059 ± 13	2680.3 ± 5.7	23	0.18300	0.34	9.493	0.83	0.3762	0.76	.911
3.1m	0.43	456	76	0.17	189	2532 ± 15	2924.7 ± 4.7	13	0.21249	0.29	14.10	0.79	0.4811	0.74	.929
4.1m	0.57	202	41	0.21	99.4	2910 ± 22	3055.6 ± 6.3	5	0.23051	0.39	18.14	1.0	0.5706	0.92	.921
5.1m	0.04	260	274	1.09	140	3139 ± 20	3110.2 ± 4.5	−1	0.23853	0.28	20.64	0.86	0.6274	0.82	.944
5.2m	0.30	248	221	0.92	129	3038 ± 32	3110.1 ± 7.5	2	0.2385	0.47	19.79	1.4	0.6019	1.3	.941
6.1m	1.64	324	169	0.54	147	2688 ± 17	3034.8 ± 7.0	11	0.22753	0.44	16.23	0.90	0.5173	0.79	.876
6.2c	3.88	21	1	0.05	10.8	2908 ± 44	3131 ± 19	7	0.2417	1.2	19.00	2.1	0.5700	1.7	.817
7.1r	0.08	208	40	0.20	109	3076 ± 20	3092.6 ± 5.2	1	0.23592	0.33	19.89	0.89	0.6115	0.83	.929
7.2m	0.03	535	259	0.50	254	2835 ± 29	3046.0 ± 3.1	7	0.22913	0.19	17.45	1.3	0.5523	1.3	.989
8.1m	0.04	790	542	0.71	278	2210 ± 13	2815.7 ± 7.6	21	0.19872	0.46	11.206	0.85	0.4090	0.71	.837
9.1m	0.15	274	142	0.54	140	2997 ± 19	3101.3 ± 4.6	3	0.23721	0.29	19.36	0.83	0.5920	0.78	.937
9.2m	0.04	415	223	0.56	220	3099 ± 32	3111.6 ± 3.2	0	0.23873	0.20	20.32	1.3	0.6173	1.3	.988
9.3c	0.23	239	108	0.47	123	3027 ± 32	3089.9 ± 4.8	2	0.23552	0.30	19.46	1.4	0.5992	1.3	.975
10.1r	9.71	496	129	0.27	137	1647 ± 20	2682 ± 100	39	0.183	6.1	7.36	6.3	0.2912	1.4	.219
11.1m	0.00	268	25	0.10	138	3021 ± 22	3076.4 ± 4.5	2	0.23352	0.28	19.25	0.97	0.5978	0.93	.957
11.2m	0.05	276	54	0.20	138	2958 ± 31	3097.5 ± 4.1	5	0.23664	0.26	19.00	1.3	0.5822	1.3	.981
11.3m	0.50	167	83	0.51	85.3	3001 ± 32	3080.1 ± 6.2	3	0.23406	0.39	19.13	1.4	0.5928	1.3	.961
12.1r	0.97	475	156	0.34	220	2759 ± 18	3023.1 ± 4.8	9	0.22588	0.30	16.63	0.85	0.5341	0.80	.935
14.1m	0.23	442	201	0.47	166	2332 ± 30	2996.6 ± 4.3	22	0.22219	0.27	13.35	1.5	0.4357	1.5	.985
15.1m	0.00	493	388	0.81	256	3050 ± 31	3116.0 ± 3.1	2	0.23941	0.19	19.97	1.3	0.6051	1.3	.989
16.1m	0.96	150	140	0.97	78.4	3035 ± 34	3101.4 ± 7.5	2	0.2372	0.47	19.67	1.5	0.6014	1.4	.948
17.1m	0.43	739	307	0.43	274	2308 ± 25	2797.2 ± 4.1	17	0.19648	0.25	11.66	1.3	0.4305	1.3	.981
18.1m	0.01	733	179	0.25	351	2857 ± 29	3031.2 ± 2.9	6	0.22702	0.18	17.45	1.3	0.5576	1.3	.990
20.1m	0.06	418	294	0.73	205	2915 ± 30	3064.2 ± 3.9	5	0.23175	0.24	18.27	1.3	0.5717	1.3	.983
21.1c	13.87	812	324	0.41	200	1421 ± 17	2673 ± 39	47	0.1822	2.4	6.20	2.7	0.2467	1.4	.500
21.2m	1.26	566	259	0.47	226	2440 ± 26	2900.8 ± 5.5	16	0.20938	0.34	13.28	1.3	0.4601	1.3	.967
22.1m	1.32	69	45	0.66	32.9	2804 ± 34	3008 ± 30	7	0.2238	1.9	16.81	2.4	0.5449	1.5	.630
22.2m	0.32	1567	68	0.04	343	1459 ± 17	2169.8 ± 4.7	33	0.13544	0.27	4.744	1.3	0.2540	1.3	.978
23.1r	0.11	328	216	0.68	159	2871 ± 30	3055.3 ± 5.2	6	0.23047	0.33	17.83	1.3	0.5612	1.3	.970
24.1r	0.05	490	185	0.39	218	2684 ± 30	2970.8 ± 3.6	10	0.21865	0.22	15.57	1.4	0.5165	1.4	.987
25.1m	0.27	513	178	0.36	236	2760 ± 29	3006.8 ± 3.9	8	0.22360	0.25	16.48	1.3	0.5344	1.3	.982
25.2m	0.46	607	106	0.18	186	1961 ± 22	2739 ± 29	28	0.1897	1.8	9.30	2.2	0.3556	1.3	.591
26.1m	0.27	219	49	0.23	77.9	2227 ± 25	2887 ± 13	23	0.2076	0.82	11.81	1.6	0.4126	1.3	.850
26.2m	0.32	124	49	0.41	61.0	2913 ± 33	3095.9 ± 9.5	6	0.2364	0.60	18.62	1.5	0.5712	1.4	.919

Notes: Errors are 1σ. Pb_c and Pb* indicate the common and radiogenic portions, respectively. Error in the standard calibration was 0.37% (1σ). Spot location and zircon characterization: m—magmatic zircon, c—core, r—structural rim, overgrowth, or embayment.

[†]Common Pb corrected using the measured ^{204}Pb.

TABLE 2. RESULTS OF THE SHRIMP U-Pb ANALYSES FOR ZIRCONS FROM SAMPLE ABBG

Grain spot	% $^{206}Pb_c$	U (ppm)	Th (ppm)	$^{232}Th/^{238}U$	$^{206}Pb^*$ (ppm)	$^{206}Pb/^{238}U$ age (Ma)[†]	$^{207}Pb/^{206}Pb$ age (Ma)[†]	% Disc.	$^{207}Pb^*/^{206}Pb^{*\dagger}$	± (%)	$^{207}Pb^*/^{235}U^{\dagger}$	± (%)	$^{206}Pb^*/^{238}U$	± (%)	Err. Corr.
1.1r	0.33	74	24	0.33	37.6	2997 ± 27	3078 ± 13	3	0.2338	0.80	19.08	1.4	0.5918	1.1	.821
1.2c	0.58	347	87	0.26	158	2723 ± 17	3008.2 ± 5.6	9	0.22379	0.35	16.22	0.86	0.5257	0.78	.914
1.3c	0.06	316	103	0.34	167	3089 ± 19	3113.0 ± 4.1	1	0.23896	0.26	20.25	0.83	0.6147	0.79	.951
1.4r	1.12	58	19	0.34	30.9	3104 ± 25	3082 ± 15	−1	0.23437	0.92	19.98	1.4	0.6184	1.0	.744
1.5c	0.39	254	78	0.32	134.6	3090 ± 17.5	3113.8 ± 5.5	1	0.23907	0.35	20.27	0.8	0.6149	0.71	.899
2.1r	0.15	486	30	0.06	208	2602 ± 16	2926.5 ± 4.0	11	0.21274	0.25	14.59	0.78	0.4973	0.74	.949
3.2r	2.16	631	87	0.14	262	2493 ± 16	2770 ± 31	10	0.1932	1.9	12.58	2.0	0.4722	0.78	.382
4.1r	0.00	336	31	0.09	174	3037 ± 19	3076.6 ± 4.1	1	0.23356	0.25	19.38	0.83	0.6017	0.79	.952
4.2c	0.07	204	97	0.49	106	3048 ± 21	3106.3 ± 5.8	2	0.23795	0.36	19.84	0.94	0.6046	0.87	.924
5.1r	0.35	674	317	0.49	284	2563 ± 15	2846.2 ± 4.2	10	0.20247	0.26	13.63	0.77	0.4882	0.72	.941
6.1c	0.33	482	213	0.46	160	2102 ± 13	2718.5 ± 7.7	23	0.18728	0.46	9.958	0.88	0.3856	0.75	.850
7.1c	0.21	385	252	0.68	174	2725 ± 17	3016.1 ± 5.5	10	0.22489	0.34	16.31	0.85	0.5261	0.78	.916
8.1c	0.12	137	102	0.77	69.5	2986 ± 23	3067.5 ± 6.7	3	0.23223	0.42	18.86	1.1	0.5891	0.96	.917
8.2c	0.19	190	147	0.80	96.7	3000 ± 21.7	3082.5 ± 4.9	3	0.23442	0.31	19.16	1.0	0.5927	0.91	.947
9.1c	0.35	242	117	0.50	113	2785 ± 19	3085.7 ± 7.3	10	0.2349	0.46	17.50	0.96	0.5403	0.85	.881
10.1c	0.25	141	72	0.53	62.8	2692 ± 20	2999.3 ± 6.8	10	0.22256	0.42	15.91	1.0	0.5183	0.92	.909
11.1r	0.20	1202	43	0.04	340	1830 ± 11	2588.6 ± 4.8	29	0.17318	0.29	7.838	0.75	0.3283	0.69	.922
12.1c	0.09	363	142	0.40	155	2602 ± 17	2988.4 ± 4.6	13	0.22105	0.29	15.16	0.84	0.4973	0.79	.940
12.2r	0.17	374	48	0.13	177	2824 ± 18	3038.1 ± 4.3	7	0.22801	0.27	17.28	0.83	0.5496	0.79	.945
13.1r	0.35	1090	49	0.05	269	1621 ± 10	2120 ± 24	24	0.1317	1.4	5.190	1.5	0.2859	0.71	.464
14.1r	0.39	896	47	0.05	311	2177 ± 16	2621.4 ± 5.9	17	0.17662	0.36	9.784	0.93	0.4018	0.86	.923
14.2c	0.67	272	147	0.56	130	2826 ± 19	3043.1 ± 6.6	7	0.22871	0.41	17.35	0.94	0.5501	0.84	.897
15.1r	0.78	1797	59	0.03	347	1299.3 ± 8.1	1893.4 ± 7.6	31	0.11586	0.42	3.567	0.81	0.2233	0.69	.854
15.2c	0.58	260	106	0.42	99.6	2364 ± 17	3021.0 ± 7.0	22	0.22559	0.44	13.78	0.95	0.4430	0.85	.888
16.1r	0.22	502	353	0.73	225.8	2709 ± 14.4	2969.2 ± 3.9	9	0.21843	0.24	15.73	0.7	0.5223	0.65	.936
17.1c	0.26	224	108	0.50	117.8	3075 ± 17.7	3120.2 ± 4.9	1	0.24003	0.31	20.23	0.8	0.6113	0.72	.921
17.2r	0.34	989	64	0.07	279.7	1829 ± 9.8	2413.6 ± 6.1	24	0.15608	0.36	7.058	0.7	0.3280	0.62	.865
18.1c	0.08	143	102	0.74	69.9	2911 ± 18.6	3096.5 ± 5.6	6	0.23649	0.35	18.61	0.9	0.5707	0.79	.914
19.1c	0.63	124	64	0.53	61.8	2925 ± 19.5	3064 ± 10	5	0.23179	0.63	18.35	1.0	0.5742	0.83	.794
19.2r	0.56	488	32	0.07	193.3	2432 ± 13.2	2857.9 ± 6.9	15	0.20393	0.42	12.88	0.8	0.4582	0.65	.837

Notes: Errors are 1σ. Pb_c and Pb* indicate the common and radiogenic portions, respectively. Error in the standard calibration was 0.37% (1σ). Spot location and zircon characterization: m—magmatic zircon, c—core, r—structural rim, overgrowth, or embayment.

[†]Common Pb corrected using the measured ^{204}Pb.

TABLE 3. RESULTS OF THE SHRIMP U-Pb ANALYSES FOR ZIRCONS FROM SAMPLE POR

Grain spot	% ^{206}Pb$_c$	U (ppm)	Th (ppm)	^{232}Th/^{238}U	^{206}Pb* (ppm)	^{206}Pb/^{238}U age (Ma)†	^{207}Pb/^{206}Pb age (Ma)†	% Disc.	^{207}Pb*/^{206}Pb*†	± (%)	^{207}Pb*/^{235}U†	± (%)	^{206}Pb*/^{238}U	± (%)	Err. Corr.
1.1m	0.05	186	20	0.11	100.0	3140 ± 21	3122.9 ± 5.5	–1	0.24045	0.34	20.80	0.91	0.6275	0.84	.926
2.1r	0.22	191	40	0.21	98.9	3042 ± 20	3113.4 ± 6.2	2	0.23901	0.39	19.87	0.92	0.6031	0.84	.905
3.1r	0.04	162	31	0.20	84.9	3065 ± 21	3098.9 ± 5.8	1	0.23685	0.36	19.88	0.93	0.6088	0.86	.920
4.1m	0.00	96	113	1.22	52.2	3169 ± 36	3132.0 ± 9.9	–1	0.2418	0.62	21.17	1.5	0.6349	1.4	.916
4.2r	0.04	215	22	0.10	106	2918 ± 19	3069.2 ± 6.3	5	0.23247	0.40	18.35	0.90	0.5724	0.81	.898
5.1m	0.05	31	19	0.63	16.7	3122 ± 36	3119 ± 21	0	0.2398	1.3	20.60	2.0	0.6231	1.5	.744
5.2m	0.02	167	41	0.25	89.4	3116 ± 21	3133.3 ± 5.6	1	0.24202	0.35	20.74	0.93	0.6215	0.86	.926
6.1m	0.03	165	32	0.20	89.9	3165 ± 22	3120.3 ± 5.6	–1	0.24006	0.35	20.98	0.96	0.6338	0.90	.931
7.1m	0.05	107	55	0.53	54.7	3013 ± 23	3135.5 ± 7.3	4	0.2424	0.46	19.91	1.1	0.5958	0.95	.901
8.1r	0.49	270	108	0.41	91.0	2121 ± 14	2779.2 ± 7.5	24	0.19434	0.46	10.439	0.91	0.3896	0.79	.865
8.2m	0.14	46	23	0.53	19.3	2579 ± 25	2882 ± 14	10	0.2070	0.88	14.04	1.5	0.4920	1.2	.807
9.1m	1.20	41	21	0.53	20.8	2960 ± 31	3098 ± 16	4	0.2368	0.99	19.02	1.6	0.5827	1.3	.797
9.2r	0.00	161	50	0.32	84.8	3086 ± 21	3102.4 ± 5.7	1	0.23737	0.36	20.10	0.93	0.6140	0.86	.924
9.3r	0.08	217	41	0.20	112	3027 ± 19	3108.5 ± 5.3	3	0.23829	0.33	19.69	0.87	0.5994	0.80	.922
10.1r	0.09	149	37	0.26	75.9	2997 ± 21	3088.8 ± 6.2	3	0.23535	0.39	19.21	0.96	0.5919	0.87	.912
10.2r	0.07	406	90	0.23	171	2575 ± 14	2984.9 ± 4.8	14	0.22059	0.29	14.93	0.73	0.4909	0.67	.914
12.1m	0.02	146	37	0.26	78.4	3133 ± 22	3115.7 ± 5.9	–1	0.23936	0.37	20.65	0.98	0.6257	0.90	.925
12.2r	0.18	207	49	0.24	107	3024 ± 20	3105.5 ± 6.7	3	0.23788	0.41	19.63	0.92	0.5985	0.82	.892
13.1m	0.00	46	29	0.65	25.9	3226 ± 31	3126 ± 10	–3	0.2410	0.64	21.58	1.4	0.6494	1.2	.887
13.2m	0.06	143	33	0.24	76.8	3130 ± 22	3125.4 ± 6.0	0	0.24081	0.38	20.76	0.96	0.6252	0.88	.919
13.3r	0.21	222	39	0.18	114	3016 ± 19	3071.0 ± 5.5	2	0.23279	0.33	19.14	0.87	0.5966	0.80	.919
14.1c	0.27	228	59	0.27	121	3094 ± 18	3130.6 ± 6.0	1	0.24168	0.36	20.52	0.83	0.6159	0.74	.893
15.1c	0.33	316	105	0.34	140	2668 ± 15	3014.6 ± 5.8	11	0.22476	0.34	15.88	0.78	0.5126	0.69	.886
16.1r	0.14	194	46	0.24	99.2	3010 ± 19	3110.4 ± 5.6	3	0.23859	0.35	19.57	0.87	0.5950	0.80	.917
17.1r	0.05	180	45	0.26	94.5	3074 ± 19	3107.3 ± 5.0	1	0.23811	0.31	20.06	0.86	0.6110	0.80	.931
18.1r	0.02	169	54	0.33	90.0	3102 ± 20	3103.3 ± 5.2	0	0.23750	0.32	20.23	0.89	0.6179	0.82	.931
18.2c	0.06	63	35	0.58	34.1	3156 ± 30	3109.7 ± 8.6	–1	0.2385	0.54	20.77	1.3	0.6317	1.2	.911
19.1c	0.55	72	42	0.61	38.1	3088 ± 28	3111 ± 11	1	0.2388	0.68	20.22	1.3	0.6146	1.1	.846
20.1r	0.08	209	55	0.27	98.6	2818 ± 17	3023.9 ± 5.2	7	0.22600	0.33	17.09	0.83	0.5484	0.76	.919
20.2c	0.26	86	42	0.50	41.5	2860 ± 24	2998.3 ± 9.4	5	0.2225	0.57	17.12	1.2	0.5583	1.1	.875
21.1r	0.11	195	56	0.30	95.0	2895 ± 19	3038.8 ± 5.6	5	0.22813	0.35	17.83	0.87	0.5669	0.80	.916
22.r1	0.22	400	252	0.65	162	2488 ± 17	3025.0 ± 4.9	18	0.22620	0.30	14.69	0.87	0.4710	0.82	.936
23.1r	0.12	180	41	0.24	93.9	3051 ± 20	3113.1 ± 5.6	2	0.23900	0.35	19.95	0.89	0.6054	0.82	.919
24.1r	0.14	192	44	0.24	96.7	2968 ± 19	3079.9 ± 5.7	4	0.23408	0.35	18.87	0.88	0.5848	0.81	.916
25.1c	0.02	197	40	0.21	104	3094 ± 20	3130.4 ± 5.1	1	0.24159	0.32	20.52	0.87	0.6160	0.81	.930

Notes: Errors are 1σ. Pb$_c$ and Pb* indicate the common and radiogenic portions, respectively. Error in the standard calibration was 0.37% (1σ). Spot location and zircon characterization: m—magmatic zircon, c—core, r—structural rim, overgrowth, or embayment.

†Common Pb corrected using the measured ^{204}Pb.

TABLE 4. RESULTS OF THE SHRIMP U-Pb ANALYSES FOR ZIRCONS FROM SAMPLE VALG

Grain spot	% ^{206}Pb$_c$	U (ppm)	Th (ppm)	^{232}Th/^{238}U	^{206}Pb* (ppm)	^{206}Pb/^{238}U age (Ma)†	^{207}Pb/^{206}Pb age (Ma)†	% Disc.	^{207}Pb*/^{206}Pb*†	± (%)	^{207}Pb*/^{235}U†	± (%)	^{206}Pb*/^{238}U	± (%)	Err. Corr.
1.1m	0.19	1032	103	0.10	324	2007 ± 13	2717.8 ± 3.7	26	0.18723	0.22	9.426	0.80	0.3652	0.77	.961
1.2c	0.00	89	112	1.30	45.7	3016 ± 26	3092.3 ± 7.1	3	0.2359	0.4	19.40	1.2	0.5965	1.1	.923
2.1m	3.52	497	389	0.81	140.6	1767 ± 13	2591 ± 15	32	0.1739	0.87	7.542	1.2	0.3154	0.82	.678
3.1c	0.00	63	45	0.74	32.8	3046 ± 28	3085.2 ± 8.6	1	0.2348	0.5	19.56	1.3	0.6040	1.2	.909
3.2c	0.46	71	53	0.77	37.4	3072 ± 24	3094.3 ± 8.2	1	0.2362	0.52	19.88	1.1	0.6105	1.0	.889
4.1c	0.22	57	54	0.98	30.0	3088 ± 28	3108.3 ± 9.8	1	0.2383	0.61	20.19	1.3	0.6146	1.1	.881
4.2r	0.03	171	59	0.36	87.6	3010 ± 22	3082.9 ± 5.3	2	0.23450	0.33	19.24	0.96	0.5951	0.90	.938
4.3m	0.16	870	57	0.07	282	2056 ± 14	2711 ± 13	24	0.1865	0.82	9.66	1.1	0.3757	0.78	.689
4.4m	0.14	184	81	0.45	87.8	2841 ± 21	3059.5 ± 7.6	7	0.2311	0.48	17.65	1.0	0.5539	0.91	.886
5.1m	0.58	740	231	0.32	215	1869 ± 13	2660.4 ± 9.2	30	0.1809	0.55	8.385	0.96	0.3363	0.78	.815
5.2c	2.07	110	78	0.73	57.3	3007 ± 25	3120 ± 30	4	0.2400	1.9	19.67	2.2	0.5943	1.0	.475
6.1m	0.00	88	62	0.73	27.2	1979 ± 25	2084 ± 13	5	0.12897	0.75	6.39	1.7	0.3593	1.5	.892
6.2m	0.07	146	78	0.55	39.4	1760 ± 13	2075.4 ± 9.3	15	0.12835	0.53	5.555	1.0	0.3139	0.86	.853
7.1c	0.11	97	33	0.35	50.3	3038 ± 25	3111.3 ± 7.2	2	0.2387	0.45	19.81	1.1	0.6019	1.0	.914
7.2m	0.17	149	33	0.23	67.7	2735 ± 20	3019.6 ± 6.5	9	0.22542	0.40	16.42	1.0	0.5284	0.92	.915
7.3m	0.38	158	39	0.25	72.1	2734 ± 18	3017.6 ± 6.5	9	0.22511	0.41	16.39	0.89	0.5281	0.80	.890
7.4c	0.05	112	34	0.32	59.6	3112 ± 29	3103.5 ± 5.5	0	0.23752	0.34	20.32	1.2	0.6206	1.2	.959
8.1c	0.21	300	120	0.41	157	3057 ± 20	3122.8 ± 4.6	2	0.24048	0.29	20.12	0.88	0.6068	0.83	.943
9.1c	0.25	326	74	0.23	154	2820 ± 19	3141.2 ± 4.6	10	0.24328	0.29	18.40	0.88	0.5487	0.83	.945
10.1c	0.06	152	73	0.50	81.8	3139 ± 23	3154.1 ± 6.9	0	0.2452	0.44	21.21	1.0	0.6274	0.93	.905
11.1m	0.52	119	24	0.21	55.1	2775 ± 23	3042 ± 10	9	0.2285	0.64	16.95	1.2	0.5379	1.0	.843
12.1m	0.06	677	47	0.07	245	2267 ± 14	2819.0 ± 4.5	20	0.19913	0.28	11.571	0.79	0.4215	0.74	.936
13.1m	0.43	80	47	0.61	39.7	2941 ± 29	3076 ± 12	4	0.2335	0.72	18.61	1.4	0.5781	1.2	.858
14.1c	0.31	150	270	1.86	81.0	3133 ± 23	3132.2 ± 7.7	0	0.2419	0.49	20.87	1.1	0.6258	0.94	.889
15.1c	0.04	108	134	1.28	59.1	3173 ± 26	3107.9 ± 8.6	–2	0.2382	0.54	20.88	1.2	0.6359	1.0	.889
15.2m	0.16	1118	30	0.03	374	2116 ± 13	2906.4 ± 4.1	27	0.21011	0.25	11.254	0.77	0.3885	0.72	.944
16.1m	0.20	326	27	0.09	132	2478 ± 17	2913.9 ± 7.8	15	0.2111	0.48	13.64	0.94	0.4687	0.81	.859
16.2c	3.30	35	20	0.59	13.8	2356 ± 44	2928 ± 38	20	0.2129	2.3	12.95	3.2	0.4411	2.2	.688
17.2c	0.56	81	24	0.30	39.8	2892 ± 40	3058.6 ± 7.2	5	0.2309	0.45	18.03	1.8	0.5663	1.7	.967

Notes: Errors are 1σ. Pb$_c$ and Pb* indicate the common and radiogenic portions, respectively. Error in the standard calibration was 0.37% (1σ). Spot location and zircon characterization: m—magmatic zircon, c—core, r—structural rim, overgrowth, or embayment.

†Common Pb corrected using the measured ^{204}Pb.

TABLE 5. RESULTS OF THE SHRIMP U-Pb ANALYSES FOR ZIRCONS FROM SAMPLE LES

Grain spot	% $^{206}Pb_c$	U (ppm)	Th (ppm)	$^{232}Th/^{238}U$	$^{206}Pb^*$ (ppm)	$^{206}Pb/^{238}U$ age (Ma)†	$^{207}Pb/^{206}Pb$ age (Ma)†	% Disc.	$^{207}Pb^*/^{206}Pb^*$†	± (%)	$^{207}Pb^*/^{235}U$†	± (%)	$^{206}Pb^*/^{238}U$	± (%)	Err. Corr.
3.1m	0.44	146	93	0.66	75.2	3011 ± 37	3101.0 ± 6.2	3	0.23715	0.39	19.46	1.6	0.5952	1.6	.970
3.2c	0.82	40	25	0.63	23.1	3266 ± 47	3356 ± 13	3	0.2788	0.86	25.36	2.0	0.660	1.8	.905
4.1c	0.75	22	9	0.42	12.6	3247 ± 51	3365 ± 13	3	0.2803	0.84	25.31	2.2	0.655	2.0	.920
5.1c	0.56	41	17	0.43	24.9	3404 ± 47	3434 ± 13	1	0.2930	0.86	28.10	2.0	0.696	1.8	.898
6.1c	0.11	86	65	0.77	51.2	3375 ± 42	3396.5 ± 7.0	1	0.2861	0.45	27.14	1.7	0.688	1.6	.962
7.1m	0.52	173	111	0.67	87.5	2971 ± 37	3081.3 ± 7.6	4	0.2342	0.48	18.91	1.6	0.5854	1.6	.957
8.1c	0.45	43	13	0.32	26.2	3462 ± 47	3422.9 ± 8.9	−1	0.2910	0.57	28.52	1.8	0.711	1.7	.950
9.1m	0.22	59	57	1.01	31.7	3132 ± 45	3093.2 ± 8.4	−1	0.2360	0.53	20.36	1.9	0.626	1.8	.960
10.1m	0.23	51	43	0.87	26.2	3029 ± 42	3110.4 ± 9.0	3	0.2386	0.56	19.72	1.8	0.600	1.7	.951
11.1m	0.33	41	37	0.94	21.4	3064 ± 43	3095 ± 11	1	0.2363	0.66	19.83	1.9	0.608	1.7	.935
12.1m	0.17	118	72	0.63	64.1	3153 ± 38	3084.7 ± 8.9	−2	0.2348	0.56	20.42	1.6	0.6309	1.5	.940
13.1c	0.27	66	35	0.55	38.4	3306 ± 53	3416 ± 15	3	0.2897	0.95	26.76	2.3	0.670	2.1	.907
14.1e	0.20	149	1	0.01	82.9	3208 ± 39	3200.1 ± 5.7	0	0.25243	0.36	22.44	1.6	0.6447	1.5	.973
14.2c	0.59	32	14	0.46	19.1	3406 ± 48	3396 ± 10	0	0.2860	0.67	27.45	1.9	0.696	1.8	.939
15.1m	0.47	29	24	0.84	15.3	3086 ± 45	3117 ± 13	1	0.2396	0.81	20.28	2.0	0.614	1.8	.916
15.2c	0.45	45	28	0.64	25.5	3280 ± 44	3360.6 ± 8.9	2	0.2796	0.57	25.57	1.8	0.663	1.7	.948
16.1c	0.53	37	11	0.31	23.1	3468 ± 47	3434 ± 17	−1	0.2931	1.1	28.79	2.1	0.713	1.8	.850

Notes: Errors are 1σ. Pb_c and Pb* indicate the common and radiogenic portions, respectively. Error in the standard calibration was 0.37% (1σ). Spot location and zircon characterization: m—magmatic zircon, c—core, r—structural rim, overgrowth, or embayment.
†Common Pb corrected using the measured ^{204}Pb.

TABLE 6. RESULTS OF THE SHRIMP U-Pb ANALYSES FOR ZIRCONS FROM SAMPLE GRADIS

Grain spot	% $^{206}Pb_c$	U (ppm)	Th (ppm)	$^{232}Th/^{238}U$	$^{206}Pb^*$ (ppm)	$^{206}Pb/^{238}U$ age (Ma)†	$^{207}Pb/^{206}Pb$ age (Ma)†	% Disc.	$^{207}Pb^*/^{206}Pb^*$†	± (%)	$^{207}Pb^*/^{235}U$†	± (%)	$^{206}Pb^*/^{238}U$	± (%)	Err. Corr.
1.1r	0.13	60	51	0.87	29.2	2899 ± 26	3053.6 ± 9.3	5	0.2302	0.58	18.02	1.3	0.5678	1.1	.887
1.2c	—	57	52	0.95	31.5	3216 ± 29	3117.8 ± 8.9	−3	0.2397	0.56	21.38	1.3	0.6468	1.1	.899
2.1r	0.91	49	27	0.57	26.0	3066 ± 30	3076 ± 12	0	0.2337	0.75	19.60	1.5	0.6090	1.2	.851
2.2c	0.27	199	127	0.66	95.1	2848 ± 18	3014.6 ± 6.2	6	0.22475	0.37	17.21	0.85	0.5555	0.76	.893
3.1r	0.20	70	49	0.73	34.3	2911 ± 25	3047.0 ± 8.7	4	0.2293	0.54	18.04	1.2	0.5708	1.1	.891
3.2c	0.16	111	61	0.57	59.8	3127 ± 24	3089.7 ± 6.8	−1	0.23552	0.42	20.27	1.1	0.6242	0.96	.914
4.1r	0.24	72	52	0.73	37.8	3051 ± 26	3049 ± 12	0	0.2296	0.75	19.16	1.3	0.6052	1.1	.814
4.2r	0.71	96	68	0.73	46.3	2852 ± 22	3051 ± 12	7	0.2300	0.71	17.64	1.2	0.5564	0.95	.784
5.1r	0.68	255	324	1.31	87.7	2156 ± 17	2893.3 ± 6.9	25	0.20854	0.42	11.42	1.0	0.3972	0.94	.912
6.1c	0.46	74	47	0.66	39.0	3086 ± 26	3090 ± 10	0	0.2356	0.63	19.94	1.2	0.6139	1.1	.860
7.1c	0.18	209	146	0.72	98.5	2810 ± 19	2946.8 ± 5.4	5	0.21546	0.33	16.23	0.92	0.5464	0.86	.932
7.2r	0.43	61	32	0.54	32.3	3083 ± 57	3064 ± 10	−1	0.2318	0.62	19.59	2.4	0.613	2.3	.966
8.1c	0.41	82	29	0.36	39.4	2848 ± 24	3027.1 ± 9.0	6	0.2265	0.55	17.34	1.2	0.5554	1.0	.879
9.1c	0.05	37	29	0.81	18.1	2933 ± 33	3054 ± 13	4	0.2303	0.79	18.30	1.6	0.5763	1.4	.874
10.1c	1.39	240	244	1.05	92.1	2344 ± 17	2988.7 ± 8.2	22	0.2214	0.50	13.37	1.0	0.4384	0.86	.861
11.1c	0.39	205	280	1.41	86.1	2554 ± 18	3002.3 ± 6.2	15	0.22305	0.38	14.95	0.96	0.4862	0.88	.915
11.2r	1.37	49	42	0.88	24.6	2925 ± 31	2964 ± 36	1	0.2180	2.2	17.23	2.6	0.5741	1.4	.524
12.1r	0.12	66	33	0.52	35.1	3096 ± 27	3085.4 ± 8.1	0	0.2349	0.51	19.97	1.2	0.6166	1.1	.908
13.1c	1.74	40	34	0.87	21.2	3067 ± 28	3098 ± 25	1	0.2367	1.6	19.88	1.9	0.6092	1.1	.584
13.2r	0.73	316	298	0.97	140	2657.4 ± 9.3	3000 ± 12	13	0.2227	0.72	15.67	0.83	0.5102	0.43	.511
13.3r	0.46	46	34	0.77	23.0	2956 ± 25	3030 ± 14	2	0.2269	0.84	18.20	1.4	0.5819	1.1	.784
14.1c	0.08	66	37	0.58	34.7	3092 ± 28	3100.7 ± 8.4	0	0.2371	0.53	20.12	1.3	0.6154	1.1	.908
15.1r	4.06	18	29	1.66	8.40	2687 ± 36	2779 ± 61	3	0.1943	3.7	13.86	4.0	0.5172	1.6	.405
15.2r	5.88	18	22	1.27	8.59	2721 ± 38	2747 ± 86	1	0.1906	5.2	13.80	5.5	0.5252	1.7	.312
15.3c	—	97	71	0.76	50.9	3081 ± 19	3085.5 ± 7.8	0	0.2349	0.49	19.84	0.92	0.6128	0.78	.848
16.1r	0.05	53	27	0.52	28.2	3096 ± 30	3081.5 ± 9.0	0	0.2343	0.56	19.91	1.3	0.6164	1.2	.907
17.1r	0.34	119	59	0.51	62.0	3047 ± 21	3080.7 ± 7.9	1	0.2342	0.48	19.51	1.0	0.6044	0.88	.873
18.1r	0.00	79	38	0.50	41.1	3067 ± 25	3068 ± 11	0	0.2323	0.67	19.51	1.2	0.6093	1.0	.840
18.2r	0.06	87	79	0.94	45.9	3079 ± 24	3099.2 ± 7.1	1	0.2369	0.44	20.00	1.1	0.6123	10	.913
19.1r	0.20	60	44	0.75	30.2	2950 ± 28	3051.6 ± 9.7	3	0.2300	0.60	18.40	1.3	0.5803	1.2	.887
20.1r	0.09	61	34	0.59	32.5	3122 ± 29	3081.9 ± 9.0	−1	0.2344	0.56	20.13	1.3	0.6229	1.2	.900
21.1c	0.79	274	227	0.86	116	2561 ± 15	2987 ± 10	14	0.2210	0.58	14.85	0.95	0.4878	0.71	.744
22.1r	0.04	208	175	0.87	96.6	2781 ± 17	3083.5 ± 6.9	10	0.2346	0.43	17.44	0.89	0.5393	0.77	.871

Notes: Errors are 1σ. Pb_c and Pb* indicate the common and radiogenic portions, respectively. Error in the standard calibration was 0.37% (1σ). Spot location and zircon characterization: m—magmatic zircon, c—core, r—structural rim, overgrowth, or embayment.
†Common Pb corrected using the measured ^{204}Pb.

concordant points show a range in apparent $^{207}Pb/^{206}Pb$ dates (Table 1), which can be interpreted either as real differences in ages (i.e., a heterogeneous population of magmatic and inherited zircons) or the result of early Pb loss. The latter could result from metamorphism shortly after crystallization of the trondhjemite protolith or from the Paleoproterozoic impact event. Given these complexities, the determination of an emplacement age is difficult and subjective. The favored interpretation is to calculate an upper intercept age of 3113.8 ± 5.5 Ma (MSWD [mean square of weighted deviates] = 0.91; probability = 0.50) from the nine most concordant points of the zoned, magmatic population (Fig. 4). Although grain cores were targeted for dating in anticipation that these would be significantly older, this is not necessarily the case, and in most instances these cores are not discernibly older than

TABLE 7. RESULTS OF THE SHRIMP U-Pb ANALYSES FOR ZIRCONS FROM SAMPLE APL

Grain spot	% $^{206}Pb_c$	U (ppm)	Th (ppm)	$^{232}Th/^{238}U$	$^{206}Pb^*$ (ppm)	$^{206}Pb/^{238}U$ age (Ma)†	$^{207}Pb/^{206}Pb$ age (Ma)†	% Disc.	$^{207}Pb^*/^{206}Pb^*$†	± (%)	$^{207}Pb^*/^{235}U$†	± (%)	$^{206}Pb^*/^{238}U$	± (%)	Err. Corr.
1.1m	0.32	353	174	0.51	174	2912 ± 13	3016.1 ± 7.3	3	0.2249	0.46	17.71	0.70	0.5710	0.54	.760
1.2m	4.30	1283	749	0.60	218	1117.0 ± 8.3	2263 ± 71	51	0.1429	4.1	3.73	4.2	0.1892	0.81	.193
2.1m	0.00	112	64	0.59	55.2	2921 ± 32	3006.0 ± 7.8	3	0.2235	0.49	17.67	1.4	0.5734	1.4	.942
3.1m	0.04	184	114	0.64	96.3	3066 ± 18	3067.2 ± 6.1	0	0.23219	0.38	19.50	0.81	0.6090	0.72	.885
3.2m	0.03	156	90	0.59	79.3	2995 ± 20	3068.2 ± 4.4	2	0.23234	0.28	18.95	0.88	0.5914	0.83	.949
4.1m	0.03	91	59	0.68	47.2	3048 ± 22	3070.2 ± 7.6	1	0.2326	0.48	19.39	1.0	0.6045	0.91	.885
4.2m	0.08	64	35	0.57	33.2	3049 ± 24	3045.4 ± 6.8	0	0.22905	0.42	19.10	1.1	0.6047	0.99	.919
4.3m	0.03	73	45	0.64	35.8	2925 ± 23	3023.9 ± 6.7	3	0.22600	0.42	17.89	1.1	0.5742	0.96	.917
5.1m	5.66	236	52	0.23	86.8	2190 ± 15	2925 ± 62	25	0.2125	3.8	11.85	3.9	0.4045	0.79	.203
5.2m	0.26	488	255	0.54	226	2767 ± 11	2975.5 ± 5.1	7	0.21929	0.32	16.210	0.58	0.5361	0.49	.842
6.1m	0.30	141	12	0.09	68.6	2885 ± 20	3009.2 ± 6.8	4	0.22393	0.42	17.43	0.96	0.5645	0.87	.898
7.1m	0.08	169	63	0.39	84.7	2958 ± 20	3017.1 ± 4.8	2	0.22504	0.30	18.06	0.90	0.5822	0.85	.944
8.1m	0.18	96	38	0.41	43.3	2717 ± 23	2947.1 ± 7.9	8	0.2155	0.49	15.57	1.2	0.5242	1.0	.906
10.1m	0.07	136	108	0.82	63.3	2782 ± 19	3008.3 ± 5.1	8	0.22382	0.32	16.65	0.88	0.5396	0.82	.933
11.1m	0.64	117	78	0.69	52.7	2714 ± 19	3009.0 ± 8.8	10	0.2239	0.55	16.16	1.0	0.5235	0.85	.838
12.1m	2.10	59	42	0.74	24.8	2537 ± 32	2836 ± 34	11	0.2012	2.1	13.38	2.6	0.4821	1.5	.590
13.1m	0.02	226	81	0.37	105	2785 ± 17	3021.9 ± 3.8	8	0.22572	0.24	16.82	0.80	0.5403	0.77	.955
14.1m	1.23	75	45	0.61	36.3	2840 ± 23	3019 ± 18	6	0.2253	1.1	17.20	1.5	0.5536	10	.665
14.2c	2.38	102	58	0.59	47.0	2715 ± 27	3251 ± 34	16	0.2607	2.2	18.82	2.5	0.5237	1.2	.490
15.1m	0.43	84	46	0.57	38.8	2769 ± 22	2950 ± 17	6	0.2158	1.0	15.96	1.4	0.5365	0.98	.688

Notes: Errors are 1σ. Pb_c and Pb* indicate the common and radiogenic portions, respectively. Error in the standard calibration was 0.37% (1σ). Spot location and zircon characterization: m—magmatic zircon, c—core, r—structural rim, overgrowth, or embayment.

†Common Pb corrected using the measured ^{204}Pb.

the magmatic population (e.g., analyses 6.1, 6.2, and 9.3; Table 1), although discordance and early Pb loss could obscure this relationship. Where possible, analyses of small overgrowths of what appear to be metamorphic zircon or recrystallized magmatic growth zones were analyzed, and although these few analyses do hint at an age only slightly younger than that calculated for the magmatic phase, they are too discordant and scattered to provide any accurate dates (Fig. 4). One analysis (7.1) within error of concordia has a $^{207}Pb/^{206}Pb$ age of 3093 ± 10 Ma (2σ).

The zircons separated from a biotite-rich gneiss (ABBG) xenolith are brown, clear to translucent, anhedral, highly fractured, and show signs of resorption as well as shock deformation. Concentric, oscillatory, magmatic zoning is generally better preserved in the lower-U cores of individual grains, which grade into more highly altered and higher-U margins (Fig. 3B). The population as a whole is highly discordant and follows much the same pattern as that described for the zircons from the host trondhjemitic gneiss (ABG). Rare small tips or overgrowths of very high-U, low-Th/U zircon (e.g., analyses 13.1 and 15.1; Table 2) suggest growth of a later metamorphic phase, but these analyses are, unfortunately, extremely discordant, and no useful geochronological information could be obtained from them apart from minimum $^{207}Pb/^{206}Pb$ ages. A minimum age of 2120 ± 24 Ma (1σ) calculated for analysis 13.1 does, however, seem to indicate that this metamorphic growth predated the Vredefort impact event (dated at 2023 ± 4 Ma by Kamo et al., 1996).

The discordant analyses tend to scatter along a splayed discordia generated by the effects of multistage Pb loss, and as a result, calculation of an age or ages is limited to consideration of the few more concordant points from grains 1, 4, and 17 (Table 2). These grains are useful in that the cathodoluminescence images show clear core/rim structures (Figs. 3C and 3D), and multiple analyses within the different domains on two of the grains (an analysis of the rim of grain 17 was discordant) provide both the least discordant data from the whole data set and also suggest a significant age difference between the two generations of zircon. A weighted mean $^{207}Pb/^{206}Pb$ age of 3113.8 ± 4.8 Ma is calculated from the four least discordant core analyses (1.3, 1.5, 4.2, and 17.1). Similarly, the weighted mean $^{207}Pb/^{206}Pb$ age of 3077.1 ± 7.3 Ma from three analyses (1.1, 1.4, and 4.1) sited within the dark rims or embayments suggests metamorphic growth at this time. The apparent age of the cores is indistinguishable from that calculated for the host trondhjemite gneiss (ABG), and the dates calculated from the limited data sets for the rims of both samples are also similar. However, the scarcity of concordant or near-concordant analyses does not allow the effects of early Pb loss to be satisfactorily evaluated (Fig. 5), so that these results must nevertheless be considered minimum age estimates.

Syn-D2 Porphyritic Granodiorite

Zircons in the porphyritic granodiorite (POR) are light brown and clear, and although most grains are subhedral with evidence of metamorphic rounding and resorption, some original euhedral forms are preserved. All zircons are fractured, with the most intense fracturing related to shock deformation. Some grains are also traversed by fluid inclusion tracks. Cathodoluminescence imagery shows that all grains have cores of various shapes, sizes, and states of preservation (e.g., Fig. 3E). Cores generally comprise concentric, oscillatory-zoned zircon surrounded by more strongly zoned growth with higher U contents. In some cases the zoning in these central areas is truncated by the later growth, but the growth zones can also appear to be continuous, although with changing compositions. Most grains are mantled or embayed by very thin overgrowths of zircon that are often difficult to discern, even with cathodoluminescence imaging (CL) imaging.

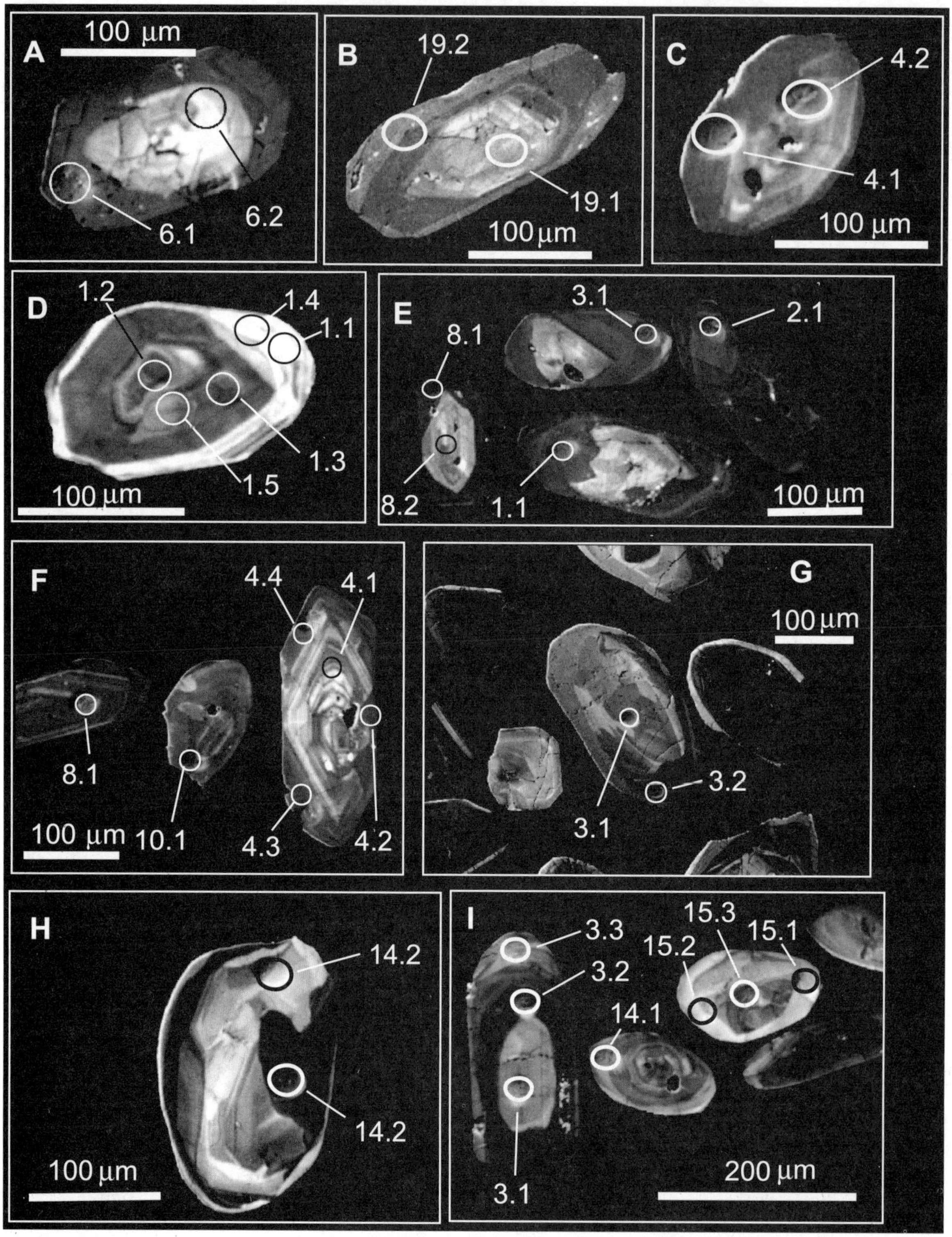

Figure 3. Cathodoluminescence (CL) images of zircons from the various basement lithologies, showing the internal structures (e.g., cores, rims) and specific features discussed in the text. The ellipses mark the location of the SHRIMP (sensitive high-resolution ion microprobe) spots analyzed, with the numbers referring to the data in the relevant tables. (A) Zircon from sample ABG showing a ca. 3.13 Ga core mantled by dark-CL zircon. (B) Grain #19 from sample ABBG showing a typical oscillatory zoned central area surrounded by higher U zircon. (C) Grain #4 from sample ABBG showing a zoned core with a dark zircon rim that embays and cuts across the zoning of the older core. (D) Grain #6 from sample ABBG showing a bright rim on a zoned ca. 3.21 Ga core. (E) Typical zircons from sample POR showing cores and rims. (F) Core and rim structures from sample VALG. (G) General internal structure of zircons from sample LES. (H) Ca. 3.4 Ga core embayed by low Th/U, 3.2 Ga zircon from sample LES. (I) Selected grains from sample GRADIS showing cores and rims.

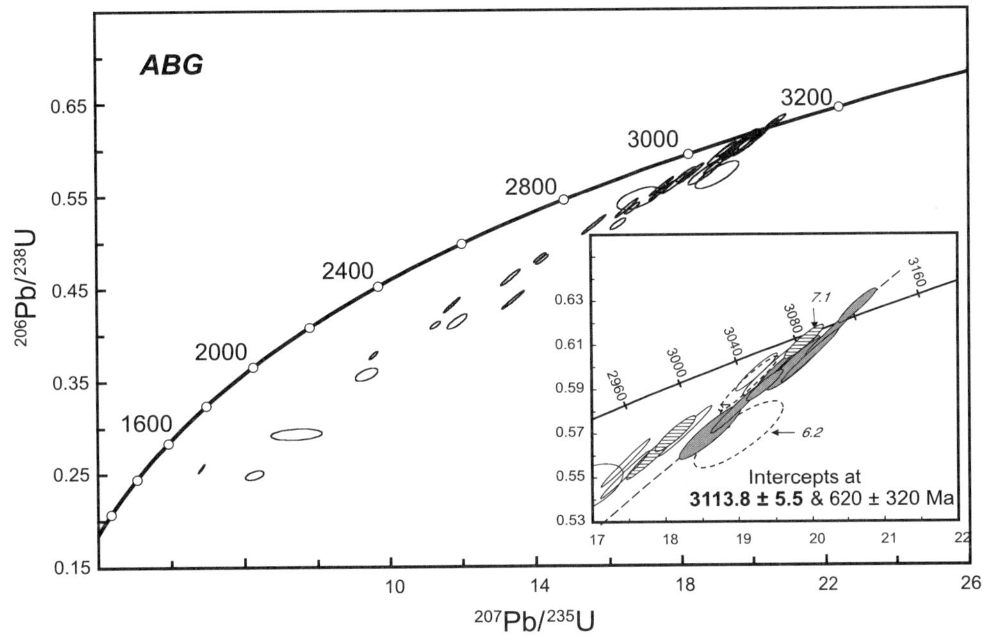

Figure 4. Concordia diagram showing all the SHRIMP (sensitive high-resolution ion microprobe) zircon U-Pb data for zircons from the trondhjemitic gneiss sample ABG. The upper intercept age of 3113.8 ± 5.5 Ma indicated in the inset concordia plot is calculated from the nine filled error ellipses representing the magmatic population. The striped error ellipses are of small (metamorphic?) overgrowths. The dashed error ellipses represent analyses of cores.

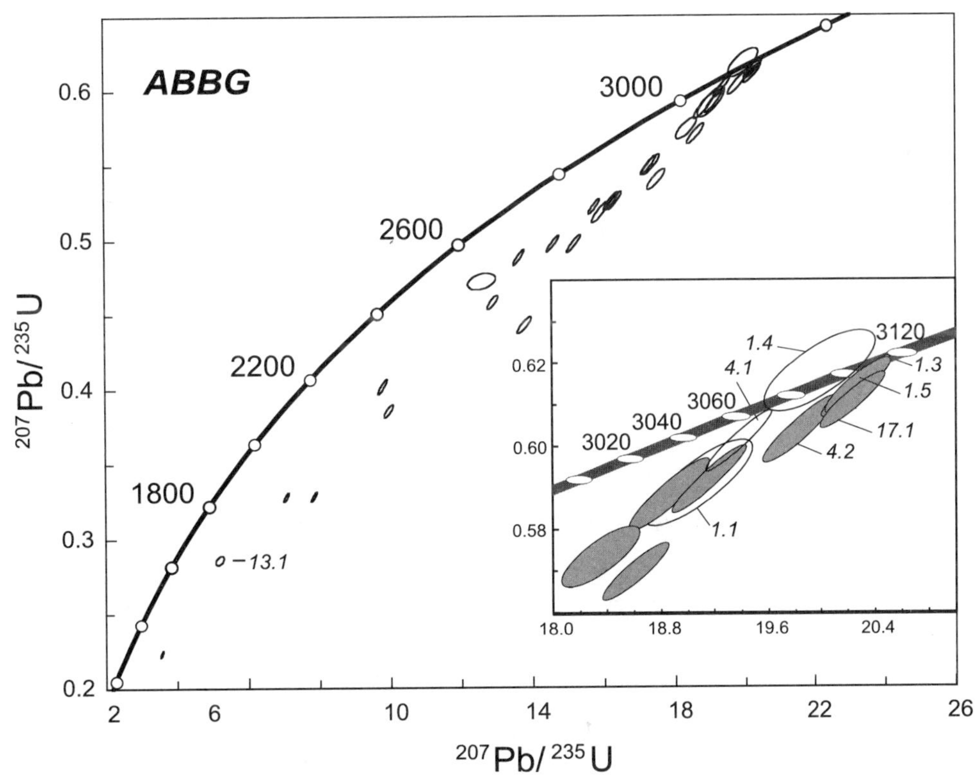

Figure 5. Concordia diagram showing the SHRIMP (sensitive high-resolution ion microprobe) U-Pb analyses of zircons from the trondhjemitic gneiss xenolith (sample ABBG). Filled error ellipses as shown in the inset concordia plot represent data from the cores. The age of 3113.8 ± 4.8 Ma is calculated from the most concordant analyses (1.3, 4.2, 1.5, and 17.1). Unfilled error ellipses represent data for the rims.

Many of the data points plot on or near the concordia but with some spread between ca. 3130 Ma and ca. 3080 Ma, and this is clearly not a single age population. The data in Figure 6 are plotted according to the zircon characterization determined from the CL images. Analyses of the cores (defined here as the central sector-zoned zircon and the surrounding dark-CL growth) and analyses of the small overgrowths representing the youngest phase of growth are plotted as open and filled ellipses, respectively. The ten more concordant data of these overgrowths combine to give a weighted mean $^{207}Pb/^{206}Pb$ age of 3105 ± 5 Ma (MSWD = 1.5; probability = 0.14). Although these younger overgrowths are volumetrically relatively small, they are interpreted to give the best estimate of the age of this porphyritic granodiorite. Many of the analyses of cores are concordant and show a small spread in ages about a mean of ca. 3125 Ma (Fig. 6 inset). These inherited cores provide an indication of the age of the preexisting crust in the region.

Migmatite

Melanosomes

The zircons from the trondhjemitic melanosome (VALG) are highly variable in shape and internal structure. The absence of shock deformation features in these grains may reflect lower peak shock pressures in rocks further from the center of the dome than the other samples or heterogeneous local shock conditions (Gibson, 2002; Gibson et al., 2002; Gibson and Reimold, 2005). Oscillatory compositional zoning is present in at least parts of many grains, which suggests a magmatic origin. Most grains have substantial cores, many of which are structurally discordant with respect to the overgrowths of zircon (Fig. 3F), and are interpreted as inherited zircons. Overall, the analyses show a great range in discordance, but most data do follow a discordia with an upper intercept date of 3111 ± 14 Ma (Fig. 7) but with some significant scatter (MSWD = 15; probability = 0.00). If the more discordant analyses plus those

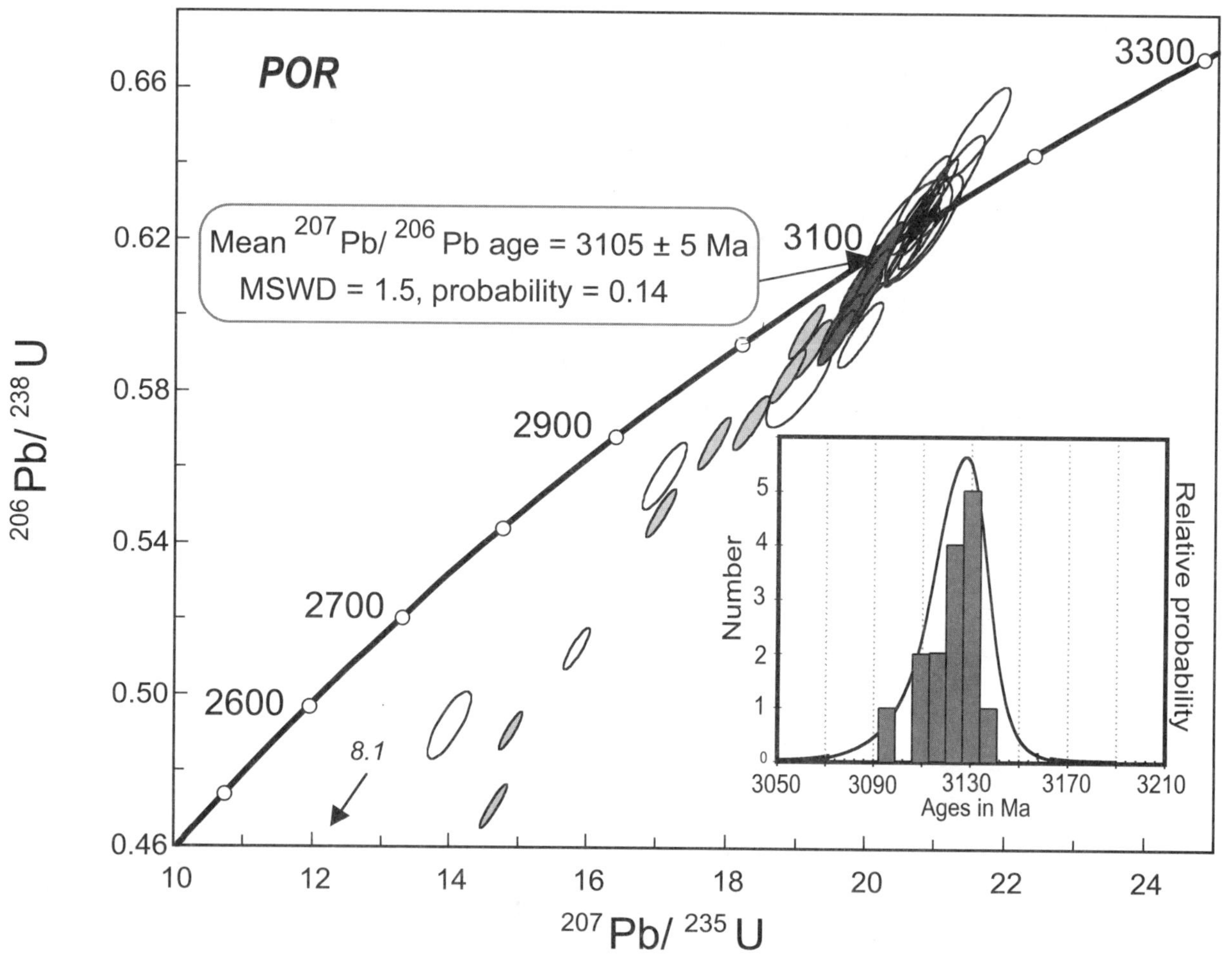

Figure 6. Concordia diagram showing SHRIMP (sensitive high-resolution ion microprobe) U-Pb data for zircons from the porphyritic granodiorite sample (POR). Unfilled error ellipses represent analyses of cores and the ca. 3125 Ma inherited zircon component that dominates this rock. Filled error ellipses represent data from the youngest overgrowth phase, with the darker coloring indicating those analyses that combine to yield a $^{207}Pb/^{206}Pb$ age of 3105 ± 5 Ma. The inset shows a cumulative probability plot for the age distribution of cores (only those that are <5% discordant); MSWD—mean square of weighted deviates.

analyses clearly sited in older, inherited cores are ignored, the data are regressed to give a more well-constrained upper intercept age of 3106.5 ± 8.4 Ma (n = 15, MSWD = 0.88, probability = 0.57; Fig. 7 inset). The calculated lower intercept indicates Mesoproterozoic Pb loss is dominant. The apparently inherited magmatic cores analyzed (8.1, 9.1, 10.1, and 14.1; Fig. 7; Table 4) give ages ranging from not much greater than the host (3107 Ma) to the oldest concordant age of 3154 ± 7 Ma. Two analyses of a rim on a rounded grain (analyses 6.1 and 6.2; Table 4) appear to provide evidence for at least a minor component of zircon growth during the Paleoproterozoic, with an upper intercept age calculated on these two points alone suggesting a date of ca. 2088 Ma. This is somewhat older than the age of the Vredefort impact event, but it is possible that the analyses incorporated a small component of an older core.

Leucosomes

Zircons in the enderbitic leucosome (LES) collected some 5 km northeast of Vredefort town (Fig. 2) are brown, generally opaque, anhedral, up to 500 μm in length, strongly resorbed or rounded, and strongly fractured. Internally, they are also extremely complex, with virtually all grains comprising an inherited core and a broad, inconspicuously zoned overgrowth interpreted to represent the magmatic zircon from the enderbitic leucosome (Figs. 3G and 3H). Rare embayments of dark-CL zircon invade some cores and appear to predate the overgrowths (analysis 14.1; Fig. 3H; Table 5). The extremely low Th/U ratio of 0.01 measured for this zircon indicates a metamorphic origin (Williams and Claesson, 1987).

The U-Pb analyses of the cores show variable discordance, and even the concordant analyses show a small spread of apparent $^{207}Pb/^{206}Pb$ ages (Table 5), possibly a consequence of early Pb loss, but equally plausibly reflecting real heterogeneity in this inherited component (Fig. 8). A weighted mean $^{207}Pb/^{206}Pb$ age for the most concordant group of 7 core analyses gives 3410 ± 14 Ma (MSWD = 2.2), but with a low probability-of-fit (0.043) reflecting the scatter of data (Fig. 8). Similarly, the data for the overgrowths give a weighted mean $^{207}Pb/^{206}Pb$ age of 3094.5 ± 9.1 Ma (MSWD = 1.8; probability = 0.087). A single analysis of a dark-CL embayment (analysis 14.1; Fig. 3; Table 5) gives a concordant, intermediate age of 3200.1 ± 5.7 Ma, the significance of which is uncertain (Fig. 8). However, it could record a hitherto unrecognized metamorphic event in the basement. This enderbite contains abundant mafic granulite xenoliths, and it is these rocks that presumably were the source for the 3410 Ma inherited cores.

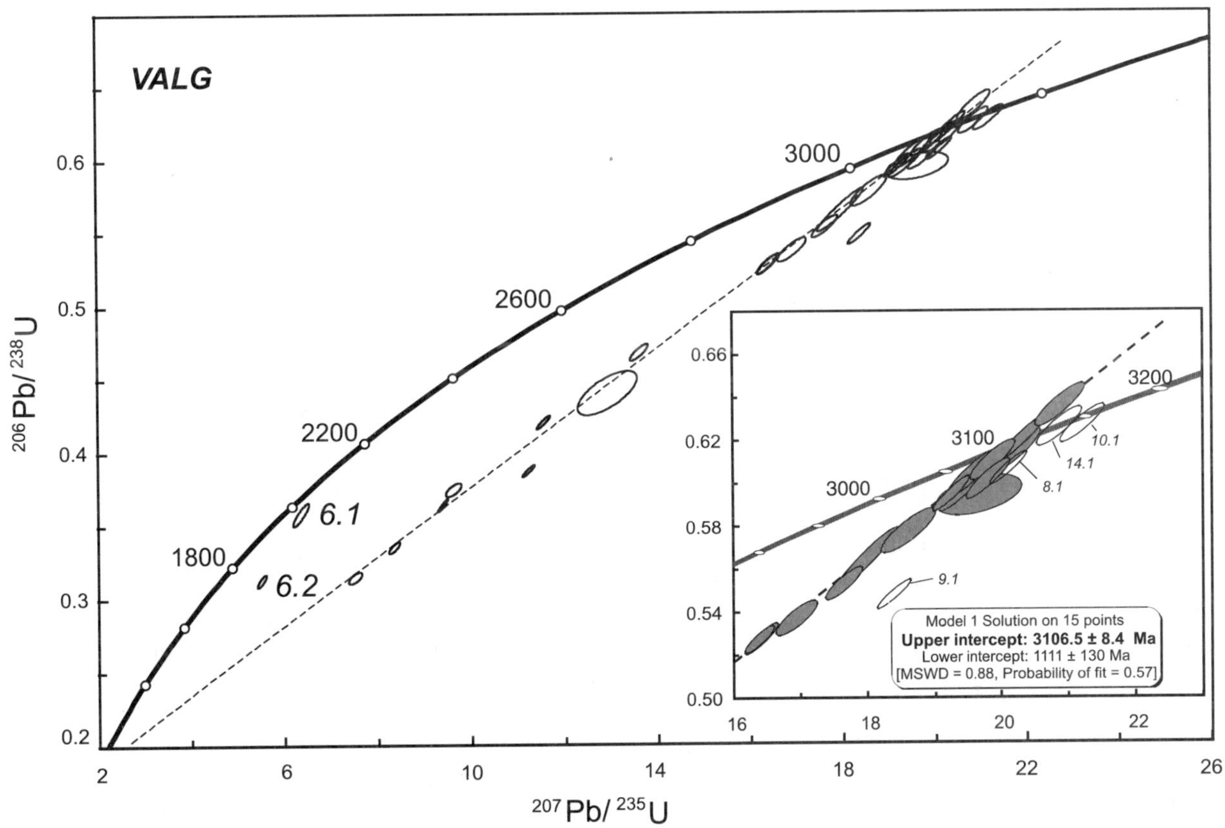

Figure 7. Concordia diagram showing SHRIMP (sensitive high-resolution ion microprobe) U-Pb data for zircons from a trondhjemitic melanosome sample (VALG). With some obvious exceptions, most data scatter about the regressed discordia line shown in the main concordia plot, but the best age is obtained from the data for the magmatic zircons shown in the inset. The unfilled error ellipses represent analyses of inherited cores; MSWD—mean square of weighted deviates.

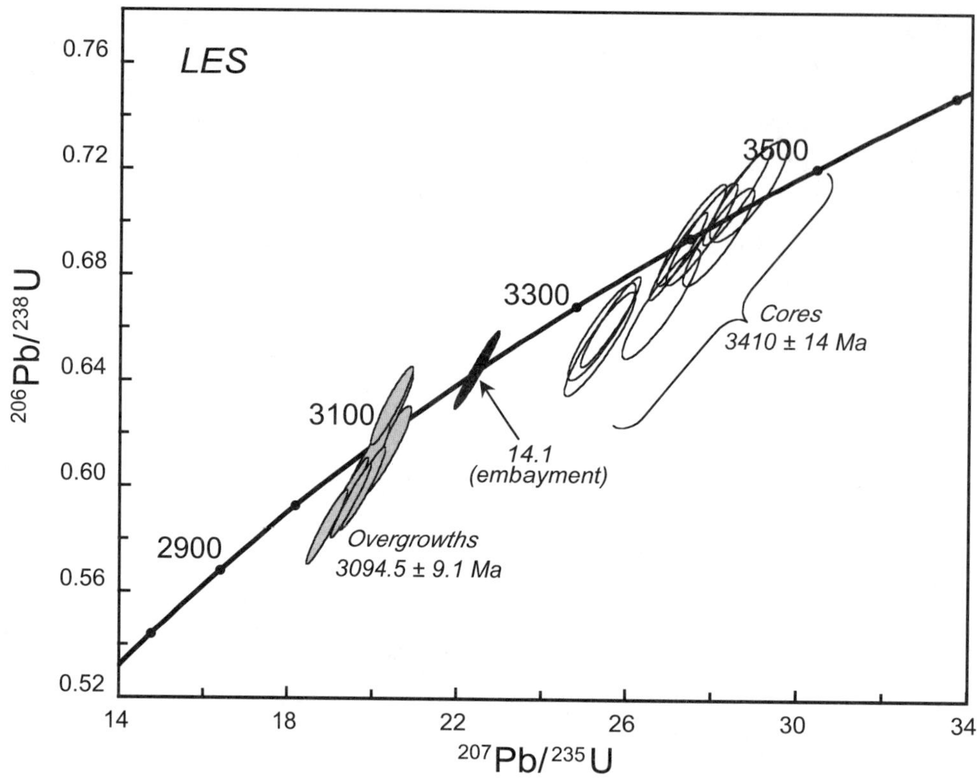

Figure 8. Concordia diagram showing U-Pb data for zircons from the enderbitic leucosome sample (LES). Unfilled error ellipses represent data for cores, gray error ellipses represent overgrowths, and the single analysis of an embayment is shown as a black error ellipse.

Syn-D3 Magnetite-Rich Granodiorite

Zircons in this granodiorite (GRADIS) are euhedral to anhedral, with pronounced metamorphic rounding of pyramidal terminations. As with many other samples from the Vredefort dome, the zircons are highly fractured, with some radial cracks linked to volume expansion in areas surrounding (metamict) high-U zones. Many grains show typical (e.g., Kamo et al., 1996; Gibson et al., 1997) near-planar shock fracture patterns. Cathodoluminescence imaging shows almost all zircons comprise a central, relatively strongly zoned "core," which is mantled by weaker, broad zones of zircon growth. Characterizing the central zones as exotic cores or simply as an early phase of magmatic growth is difficult as in some cases the change from central to outer areas is abrupt and in other instances it is transitional (Fig. 3I). Elevated Th/U ratios in both types would be consistent with a magmatic origin for both structural types, and it is assumed they have a common origin. A number of the zircons have very thin bright-CL rims or spots within the outer parts of the grains that appear to be metamorphic overgrowths plus reworking of the rims. Only one of these bright-CL rims (on Grain #15; see Fig. 3I) was thick enough to access with the SHRIMP, and it has an extremely low U content of ~18 ppm and a significantly younger apparent age of about ca. 2769 Ma.

Overall, the SHRIMP analyses show a complex Pb loss and age pattern (Fig. 9). Some data cluster along the concordia, giving a concordia age of 3083.7 ± 8.6 Ma from the 14 analyses shown in Figure 9. Many other points are highly discordant and follow a multibranched pattern of Pb loss.

A cautionary note here is that if the analyses are treated separately according to their petrographic characterization as "core" or "overgrowth" (as interpreted from the CL images), two separate $^{207}Pb/^{206}Pb$ ages can be determined from the concordant analyses. The cores give a $^{207}Pb/^{206}Pb$ age of 3092 ± 6.2 Ma (n = 7; MSWD = 0.62; probability = 0.72) and the rims an age of 3074 ± 10 Ma (n = 7; MSWD = 1.3; probability = 0.25). It is hardly possible to know whether this data treatment is valid or not, given the apparent complications of early Pb loss and the difficulty in characterizing these structures unambiguously. It is possible that the rims have preferentially suffered the early Pb loss discussed above or that the analyses of the rims included the small bright-CL reworked zones, thus producing a mixed, apparently younger, date. If the latter scenario is correct, then the date of 3092 ± 6.2 Ma cited above would be the most accurate estimate of the time of crystallization of this granodiorite.

The more discordant data trend toward a Mesoproterozoic lower intercept. A number of analyses (11.2, 15.1, and 15.2; Fig. 9; Table 6) plot along a seemingly different, shallow, discordia trend and clearly show the effects of either the 2.02 Ga

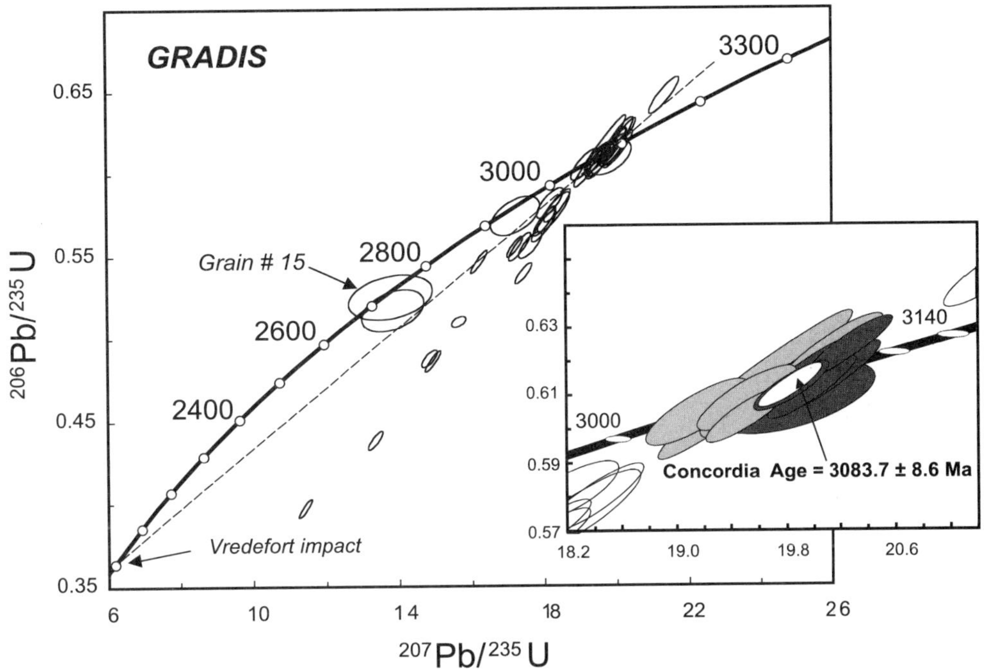

Figure 9. Concordia diagram and inset showing the SHRIMP (sensitive high-resolution ion microprobe) U-Pb zircon data for zircons from the granodiorite sample (GRADIS). The dashed line in the main concordia plot represents a reference discordia line assuming a lower intercept age of 2.02 Ga (the time of the Vredefort impact event). The two overlapping analyses from grain # 15 are shown. In the inset diagram the more concordant data for the "cores" are plotted as filled, dark-gray error ellipses, the "overgrowths" or "rims" as filled, light-gray ellipses, and the filled, white error ellipse represents the concordia age plotted with 95% uncertainty.

impact or some intermediate but early metamorphic event (Fig. 9). The latter scenario appears to be more probable given the difference between the observed Pb-loss trend and the theoretical 2.02 Ga discordia shown in Figure 9.

Aplite

Zircons extracted from aplites, pegmatites, and other highly evolved rocks can have unusual compositions (e.g., extreme U contents and discordance), which makes them difficult to date accurately. Inheritance is also a common problem. The zircons from the undeformed aplite sample (APL) analyzed in this study are anhedral or fragmental, with very rare euhedral forms. Many show dense shock fracture patterns. The dominant zircon type has a platey form and is thus generally fragmented (presumably in the sample crushing) and has some internal zoning (but this is generally weak and not continuous) and high-U margins. A number of large (>250 μm), anhedral platey fragments that are relatively clear and uncomplicated are also present and probably represent broken fragments of the magmatic zircon described above. These unusual grains are interpreted to represent the magmatic population of this aplite. A small number of possibly inherited cores were found.

The U-Pb analyses are generally discordant, and it is not possible to define a single Pb-loss trend (Fig. 10). Apart from analysis 14.2, which was of an inherited core, all analyses were sited within zircons interpreted to represent the magmatic population. The only concordant or near-concordant data came from analyses within two zircons from the clear, anhedral population. A weighted mean $^{207}Pb/^{206}Pb$ age calculated from three analyses on grains #3 and #4 gives 3068.3 ± 6.3 Ma (MSWD = 0.046, probability = 0.96). Repeat analyses on different areas of grain #4 (analyses 4.1–4.3; Table 7), however, did not reproduce this age, and it is clear that this particular grain must either be zoned in terms of age or, more likely, must have suffered early Pb loss. The single analysis of a core (#14.2) was discordant and gave a minimum $^{207}Pb/^{206}Pb$ age of 3251 ± 34 Ma (1σ) (Table 7).

If our interpretation of these zircons as magmatic zircons that crystallized in the aplite is correct, then this date provides a firm age constraint on the aplite. This is important as it provides a lower age constraint on the timing of deformation of the basement granitoids, which must have been complete by ca. 3068 Ma.

DISCUSSION

U-Pb Discordance

This study highlights a number of factors that can complicate how U-Pb data are deciphered and dates are determined from metamorphic terranes that have been subjected to multiple

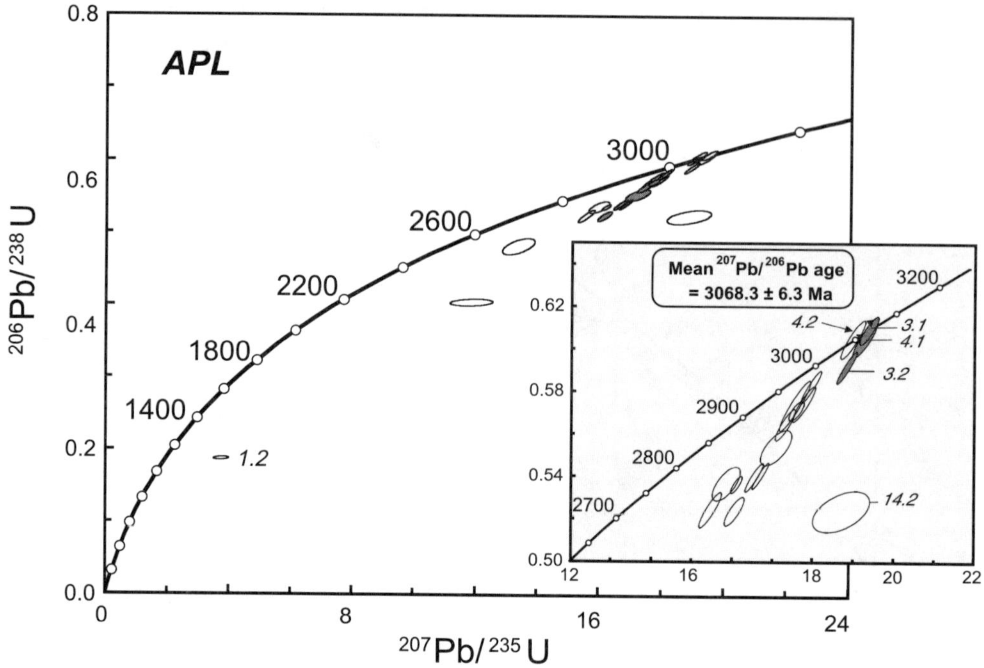

Figure 10. Concordia diagram showing isotopic data for zircons from the aplite sample (APL), with the inset showing the more concordant data. The weighted mean $^{207}Pb/^{206}Pb$ age of 3068.3 ± 6.3 Ma (MSWD [mean square of weighted deviates] = 0.046, probability = 0.96) shown is calculated from the three analyses represented by the filled error ellipses 3.1, 3.2, and 4.1.

igneous and/or metamorphic events. The first is the very complex nature of almost all the zircon populations. Even with detailed microphotography and cathodoluminescence imaging, it was difficult to definitively categorize many of the zircon types (e.g., identify whether apparent cores were older, inherited grains or simply an early stage of crystallization in continuously evolving conditions). Some cores that were clearly inherited (e.g., fragments of grains surrounded by overgrowths of magmatic zircon) gave apparent ages that are sometimes indistinguishable from those of the overgrowths, despite using the small-scale sampling of the SHRIMP technique (at the 20–30 μm scale used in this study).

A second problem is that many of the zircons display sets of parallel, near-planar fractures as a result of the 2.02 Ga impact (Kamo et al., 1996; Gibson et al., 1997). This, as well as the highly altered and/or metamict nature of many of the grains (Reimold et al., 2002), sometimes made selection of sites for analysis difficult but critical, as these altered areas have invariably suffered severe Pb loss. Apart from this more recent Pb loss, the zircons from the basement lithologies in the core of the dome appear to have lost Pb shortly after their crystallization and during one or more Proterozoic events. Results of SHRIMP U-Pb analysis of Archean rocks from the western parts of the craton (e.g., Poujol et al., 2002) and those presented here suggest that early Pb loss seems to be a common feature regionally. The early Pb loss in the basement rocks of the Vredefort dome must have occurred within a few million years of crystallization of the parent rocks and could, plausibly, be related to polyphase intrusions of TTG magma into the crust and to the ensuing high-grade metamorphic conditions as well as to the formation of syntectonic granites and granodiorites emplaced during the metamorphism.

Pb loss during the Proterozoic has previously been related to the ca. 2020 Ma impact event and the ca. 1100–1300 Ma Kibaran/Grenville orogeny (Kamo et al., 1996). The impact event has been dated at 2023 ± 4 Ma based on a concordant age obtained from unshocked zircons from impact-related pseudotachylitic breccia and Vredefort Granophyre (impact melt rock; Kamo et al., 1996). Kamo et al. (1996) observed that the Pb loss seems to be more prominent in populations of highly shocked zircons. Our SHRIMP analyses do not indicate significant Pb loss during the impact event (or shortly thereafter). This may be partly because the samples analyzed here did not contain significant volumes of pseudotachylitic breccia and, therefore, have had less chance of thermally induced Pb loss associated with the impact event, and also because severely shocked zircons were generally avoided where possible. The dominant Pb loss indicated by this study seems to have occurred during the Mesoproterozoic, possibly coincident with the Kibaran-Namaquan orogeny, and also associated with intrusion in the dome of several laterally extensive gabbroic sheets of the same age (Reimold et al., 2000). These sheets also seem to have induced significant hydrothermal alteration and resetting of the Ar isotopic system not only in the core of the Vredefort dome

(Reimold et al., 1990; Spray et al., 1995), but also throughout the Witwatersrand basin (e.g., Reimold et al., 1990; Trieloff et al., 1994; Friese et al., 2003). An important observation from this study is that even within a population of zircons from a single sample, there is not necessarily a single Pb-loss pattern, and certainly the Pb-loss pattern for the region is variable. Clearly, an appreciation for the appropriate Pb-loss model is needed for interpretation of intercept ages that are calculated from the Archean basement complex from discordant data and especially from multigrain analyses or small data sets. In the case of the Vredefort dome, several previous studies (e.g., Kamo et al., 1996; Moser, 1997; Hart et al., 1999) have made use of results calculated from upper intercept dates assuming Pb loss to be exclusively associated with the impact event at 2.02 Ma (i.e., the assumed lower intercept age). In some instances this might be appropriate (e.g., for highly shock-metamorphosed zircon), but in other cases this can lead to artificially high apparent upper intercept dates. Significantly, the Archean rocks in the western parts of the craton also record evidence of Pb loss during the Kibaran/Namaquan orogeny (Poujol et al., 2002), although these samples were derived from an area substantially closer to the locus of geological activity at this time along the southern and western edges of the craton.

Timing of Magmatic and Deformation Events in the Core of the Vredefort Dome

Previous models for the evolution of the rocks exposed in the core of the Vredefort dome (e.g., Moser et al., 2001; Flowers et al., 2003) have favored a two-stage process in which an initial, >3.3 Ga, TTG component underwent partial melting between 3.13 and 3.08 Ga to produce a suite of trondhjemitic gneisses and tonalites, granites, and quartz syenites. They have suggested that partial melting of the older crust occurred during a 3.107 Ga (Hart et al., 1999) high-grade metamorphic event and correlated this with the craton-wide magmatic event that led to the formation of the voluminous granodiorite-granite-monzogranite (GGM) suites in the upper crust of the Kaapvaal Shield (see review in Belcher and Kisters, this volume). In these models, little attention has been given to the deformation structures found in the rocks in the dome or their tectonic significance, although large-scale tectonic inferences have been made.

As part of their model, Moser et al. (2001) maintained that at least 40% of the lower crust supposedly exposed in the central core of the Vredefort dome melted to form a 10-km-thick granitic layer that is now exposed in the outer part of the core of the dome. However, new structural and geochemical data (Lana et al., 2003a, 2004) have indicated that the core of the dome, while being compositionally heterogeneous on an outcrop scale due to its migmatitic and polydeformed character, shows no definitive gross kilometer-scale layering consistent with such a model. Lana et al. (2004) also indicated that the section exposed in the dome comprises the same TTG suite, with only metamorphic grade increasing toward the center, and that voluminous granitic/charnockitic rocks intruded throughout the core of the dome during peak-metamorphic conditions. The SHRIMP data presented here provide further refinements of the geological model proposed by Lana et al. (2004), with 3 magmatic suites and associated tectonic features identifiable.

The first suite of sialic rocks is represented by the trondhjemitic gneisses, including the melanosomes in the amphibolite-facies zone (Lana et al., 2004), which intruded the greenstone sequence before or during D1 and the porphyritic granodiorite that was emplaced during the D2 deformation event (Lana et al., 2003a, 2003c, 2004). Both the xenolith ABBG and its host, the trondhjemitic gneiss ABG, gave ages of ca. 3114 Ma. These age data are indistinguishable from the 3106 ± 8 Ma ages for trondhjemitic melanosomes in the migmatites (sample VALG) and support geochemical interpretations that link the melanosomes to the nonmigmatized TTG suite (Lana et al., 2004). The 3105 ± 5 Ma age obtained for zircons extracted from the porphyritic granodiorite (POR) also shows that its emplacement occurred at this time or only slightly later than the trondhjemitic gneisses. Thus, the data suggest multiple intrusions of TTG magma within a very restricted period around 3.11 Ga. The ages of D1 and D2 can also be similarly bracketed as both events occurred during emplacement of the TTG granitoids. An approximate age of 3105 Ma is suggested for D2, given that this event occurred during the emplacement of the porphyritic granitoid.

The second suite of sialic rocks, represented by the evolved granitic (charnockitic) leucosomes and granitic to granodioritic bodies, intruded during a northeast-southwest compressional event under low-P, high-T metamorphic conditions. The age of this tectonomagmatic event is indicated by zircons in the magnetite-rich granodiorite (GRADIS), which yielded ages of ca. 3090 Ma. The enderbitic leucosomes (LES), which host a large volume of mafic granulite xenoliths (Kamo et al., 1996; Lana et al., 2004; Lana 2004), yielded two distinct ages of 3410 ± 14 Ma and 3095 ± 9 Ma. The 3095 ± 9 Ma age obtained from zircon rims may confirm the fact that most of the leucosomes and syntectonic granitoids crystallized between 3100 and 3080 Ma. This age interval is broadly consistent with the ages of granitic and quartz syenite intrusions in the dome dated by Moser et al. (2001) and Flowers et al. (2003). The SHRIMP results from the GRADIS and LES samples also suggest that peak metamorphic conditions were reached between 3100 and 3080 Ma, which is within error of or slightly younger than the 3107 ± 9 Ma age previously proposed by Hart et al. (1999) based on a single concordant U-Pb monazite age. The ca. 3410 Ma age obtained from zircon cores in the enderbite is broadly consistent with the 3400–3500 Ma Rb-Sr and Th-Pb whole-rock ages previously reported by Hart et al. (1981) and with inherited zircons from elsewhere in the central part of the Kaapvaal Craton (Armstrong, et al., 1991). These age data from zircon cores confirm previous interpretations that the oldest greenstone rocks in the core of the dome are similar in age to those exposed in the central and eastern parts of the craton (Hart et al., 1981; Hart et al., 1999; Lana et al., 2003b) and that the enderbites were derived by partial melting of greenstone remnants (Lana et al., 2004).

Although the TTG rocks and the syntectonic granitoids and leucosomes were emplaced during distinct deformation events, it is worth noting that the differences in age between these rocks are within uncertainty. The age results and the geochemical data for these lithologies (see Lana et al., 2004, for discussion) do not exclude the possibility that the TTG suite and the more granitic leucosomes and granitoids are part of a single, continuous, magmatic episode in which the leucosomes represent more fractionated melts from a parental trondhjemitic magma (e.g., Condie and Hunter, 1976).

The youngest suite of rocks analyzed is represented by the volumetrically minor aplite dikes. These dikes intruded parallel to subparallel to the D4 mylonitic shear zones and were not deformed by the D4 mylonitic deformation. The 3068 ± 6 Ma age obtained for the aplite sample is identical to the age established for the deposition of the Dominion Group volcano-sedimentary sequence (3074 ± 6 Ma; Armstrong et al., 1991; Robb et al., 1991) that unconformably overlies the crystalline basement rocks in the core of the dome. It has important implications for the tectonic evolution of the central parts of the Kaapvaal Craton (see below).

Tectonic Implications

Existing models for the Archean crustal evolution in the Kaapvaal Craton suggest formation of a >3.2 Ga cratonic nucleus (the Kaapvaal Shield) during repeated magmatic accretion and tectonic amalgamation of small island arc terranes (de Wit et al., 1992; de Ronde and de Wit, 1994; Kamo and Davis, 1994; Lowe, 1994, 1999; Poujol et al., 2003). New geochronological results from the central parts of the craton (Poujol and Anhaeusser, 2001) suggest that the Kaapvaal Shield extended at least as far west as the Johannesburg dome (Fig. 1) where tonalitic and trondhjemitic gneisses older than 3.2 Ga and GGM rocks ranging in age from 3.11 and 3.13 Ga are exposed (Poujol et al., 2003). In contrast, the crystalline rocks in the northern parts of the craton (Murchison, Pietersburg, and Giyani greenstone belts; Fig. 1) have been interpreted as the product of a series of juvenile volcanic arcs that were successively accreted along a convergent plate boundary between 3.1 and 2.8 Ga (Poujol et al., 2003). More recently, 3.3–3.2 Ga inherited zircons and Nd model ages for gneisses in the Kraaipan belt exposed in the western parts of the craton (Fig. 1) have been used by Schmitz et al. (2004) to suggest that the craton comprises two distinct fragments—a younger Kimberley Block in the west and an older Witwatersrand Block in the east—that collided at ca. 2.9 Ga along a north-south–trending suture immediately east of the Kraaipan-Amalia belts (Fig. 1).

Integral to the models of Poujol et al. (2003) and Schmitz et al. (2004) is that the assembly of the Kaapvaal Shield (or Witwatersrand Block of Schmitz et al., 2004), which also included the crust now exposed in the Vredefort dome, occurred prior to 3.2 Ga and that it was subsequently affected by the regional-scale partial melting of the 3.2 Ga old TTG rocks and emplacement of kilometer-scale GGM suite batholiths at >3.1 Ga. This "final granite bloom" (Bleeker, 2002) removed minimum melt fractions, fluids, and heat-producing elements from the lower crust, cooling it and increasing its rigidity and likely enhancing its coupling to the mantle lithosphere. However, while field, geochemical, and geochronological data indicate the presence of oceanic crust as old as ca. 3.3–3.4 Ga in the Vredefort dome (Reimold et al., 1988; Hart et al., 1981, 1990, 1999; Tredoux et al., 1999; Lana et al., 2003b; Lana, 2004), our study has found no direct evidence of significant TTG magmatism prior to 3.15 Ga. This indicates that the rocks exposed in the core of the dome are too young to have been part of the (>3.2 Ga) Kaapvaal Shield. The geochemical composition of the TTG suites in the core of the dome (e.g., Lana et al., 2004; Lana, 2004) is also consistent with them having been formed during partial melting of hydrated oceanic crust (e.g., Skjerlie and Johnson, 1993; Singh and Johannes, 1996) rather than from remelting of older TTG crust.

The TTG rocks in the core of the Vredefort dome could have formed either in an intraoceanic environment (i.e., island arc or thrust stack; de Wit, 1998; Lowe, 1994, 1999), in similar fashion to other greenstone-TTG associations in the central and western parts of the craton (e.g., Anhaeusser, 1971, 1973, 1999; de Wit et al., 1992; de Ronde and de Wit, 1994; Kamo and Davis, 1994; Lowe, 1994, 1999; Poujol et al., 2003), or in a magmatic arc along the western margin of the Kaapvaal Shield. The paucity of >3.15 Ga inherited zircons in the TTG rocks in the dome is consistent with an oceanic arc, with a relatively thin crustal substrate (e.g., Mason and McDonald, 1978; Whalen, 1985) rather than a continent-margin magmatic arc environment, in which a thick underlying crust would have facilitated formation of voluminous granitoid batholiths (Windley, 1973, 1982). Whatever the cause, our data suggest that the idea of a monolithic Kaapvaal Shield (extending from Barberton to west of the Vredefort dome) that was created exclusively before 3.2 Ga needs revision and that addition of juvenile crustal material to its western margin might have occurred regularly between 3.2 Ga and the collision of the Kimberley Block. Any such terranes are now largely hidden beneath thick younger supracrustal sequences (Fig. 1).

A similar distinction may also be drawn between the formation of the GGM rocks in the dome versus those in the Shield. Although there may be some overlap in the ages, the regional "final granite bloom" (Bleeker, 2002) in the Shield appears to be 10–20 m.y. older than the GGM intrusions in the dome (see reviews in Poujol et al., 2003; Belcher and Kisters, this volume). In terms of tectonic environment, the granitoids in the dome are associated with a strong, high-grade, subvertical northwest-trending S3 fabric that, so far, has not been identified in the Johannesburg dome, but which seems to be parallel to subparallel to the regional foliation in the relatively younger granite-greenstone rocks of the Amalia/Kraaipan terrane in the western part of the craton (Poujol and Anhaeusser, 2001, 2003; Hirner, 2003). In contrast, the 3.1 Ga granite batholiths in the Barberton region appear to have been emplaced within regional northeast-trending transpressional strike-slip zones (Belcher and Kisters, this volume).

The emplacement of the 3100–3080 Ma granitoids and the development of the D3 northwest-trending crustal shear zones in the dome indicate that high-grade metamorphic conditions were achieved during a northeast-southwest compressional event (Stevens et al., 1997; Lana et al., 2003a). The structural features and the anticlockwise P-T path derived from the greenstone fragments during the high-grade metamorphic event (Stevens et al., 1997) indicate that D3 was intimately linked to crustal thickening. Accretion of the Vredefort rocks onto the Kaapvaal Shield could have occurred during this thickening of the crust at 3100–3080 Ma, which is less than 10–25 m.y. prior to the outpouring of the 3074 ± 6 Ma (Armstrong et al., 1991) Dominion Group lavas in the central parts of the craton.

The fact that the orogenic climax between 3100 and 3080 Ma is only slightly older than Dominion Group volcanism implies that exhumation of the midcrustal levels exposed in the core of the dome must have been particularly rapid. The most effective way to achieve this would be by extensional unroofing of the high-grade rocks shortly after the D3 tectonic and magmatic thickening (e.g., Dewey, 1988; Wong and Gans, 2003; Kisters et al., 2003). Accommodation of the deformation related to extensional unroofing might have occurred principally along a >500-m-wide northeast-trending S4 mylonitic shear zone exposed along the northern margin of the greenstone inlier in the southeastern sector of the dome (Fig. 2) and several smaller mylonitic shear zones that occur in the northern sector of the core of the dome (Lana et al., 2003b). Geothermobarometry results from peak metamorphic assemblages suggest at least 5 km of vertical displacement along the southeastern D4 shear zone located between the upper amphibolite-facies gneisses and greenschist-facies greenstones (Lana et al., 2003b). Other evidence linking D4 with extensional collapse of thickened crust is the peak to post-peak metamorphic assemblages in the S4 fabric, which indicate that D4 was initiated under high-grade metamorphic conditions and that shearing progressed under conditions of declining temperature (Lana et al., 2003b). These observations are consistent with the tectonic setting for the deposition of the Dominion Group lavas in an extensional-intracontinental basin (e.g., Robb et al., 1991; Jackson, 1992) that developed in the collapsing hanging wall of the evolving extensional detachment.

CONCLUSIONS

Results of U-Pb SHRIMP analysis of Archean crystalline basement rocks exposed in the core of the Vredefort dome suggest a comparatively short-lived, three-stage magmatic and tectonic evolution. Sialic crust formation commenced with multiple intrusions of 3120–3100 Ma TTG rocks during D1 and D2. The chemical compositions of the TTG rocks (Lana et al., 2004) together with the geochronological data suggest that they were probably derived from partial melting of hydrated oceanic crust in an island arc setting. The TTG and greenstone sequences were later partially melted and intruded by 3100–3080 Ma leucosomes and other granitoids during upper amphibolite- to granulite-facies regional metamorphism that coincided with northeast-southwest D3 compression. D3 and the high-grade metamorphic event are related here to tectonic collision of the Vredefort island arc terrane with the >3200 Ma Kaapvaal Shield to the east. The final Mesoarchean structural event recorded in the core of the dome was associated with the collapse of the thickened crust along northeast-trending, normal dip-slip, peak-to-retrograde, subhorizontal mylonitic shear zones (D4). The collapse was probably associated with or was shortly followed by the emplacement of 3068 ± 6 Ma aplite dikes. This extension may be linked to the rifting that accompanied the outpouring of the 3074 ± 6 Ma Dominion Group volcanic sequence that marked the start of the development of a succession of large sedimentary and volcanic basins on the craton over the next thousand million years.

ACKNOWLEDGMENTS

We would like to thank Carl Anhaeusser for fruitful discussions on the geological evolution of the central parts of the Kaapvaal Craton. Reviews of earlier drafts of the manuscript by Sandra Kamo, Stephen Prevec, and Christian Koeberl are gratefully acknowledged. RAA would like to thank the staff of the Electron Microscopy Unit (EMU) at the Australian National University (ANU) for access to their excellent scanning electron microscope (SEM) facilities. Financial support from the Research Council of the University of the Witwatersrand is gratefully acknowledged. University of the Witwatersrand Impact Cratering Research Group Contribution No. 74.

REFERENCES CITED

Anhaeusser, C.R., 1971, The Barberton Mountain land, South Africa—A guide to the understanding of the Archaean geology of western Australia: Geological Society of Australia Special Publication, v. 3, p. 103–120.

Anhaeusser, C.R., 1973, The geology and geochemistry of the Archaean granites and gneisses of the Johannesburg-Pretoria Dome: Geological Society of South Africa Special Publication, v. 9, p. 361–385.

Anhaeusser, C.R., 1999, Archaean crustal evolution of the central Kaapvaal Craton, South Africa: Evidence from the Johannesburg Dome: South African Journal of Geology, v. 102, p. 303–322.

Armstrong, R.A., Compston, W., de Wit, M.J., and Williams, I.S., 1990, The stratigraphy of the 3.5–3.2 Ga Barberton greenstone belt revisited: A single zircon ion microprobe study: Earth and Planetary Science Letters, v. 101, p. 90–106, doi: 10.1016/0012-821X(90)90127-J.

Armstrong, R.A., Compston, W., Retief, E.A., Williams, L.S., and Welke, H.J., 1991, Zircon ion microprobe studies bearing on the age and evolution of the Witwatersrand Basin: Precambrian Research, v. 53, p. 243–266, doi: 10.1016/0301-9268(91)90074-K.

Belcher, R.W., and Kisters, A.F.M., 2006, this volume, Syntectonic emplacement and deformation of the Heerenveen batholith: Conjectures on the structural setting of the 3.1 Ga granite magmatism in the Barberton granite-greenstone terrain, South Africa, *in* Reimold, W.U., and Gibson, R.L., eds., Processes on the Early Earth: Geological Society of America Special Paper 405, 10.1130/2006.2405(13).

Bleeker, W., 2002, Archean tectonics: A review with illustrations from the Slave craton, *in* Fowler, C.M.R., Ebinger, C.J., and Hawkesworth, C.J., eds., The Early Earth: Physical, Chemical and Biological Development, Geological Society [London] Special Publications 199, p. 151–181.

Colliston, W.P., and Reimold, W.U., 1990, Structural studies in the Vredefort Dome: Preliminary interpretations of results on the southern portion of the structure: Johannesburg, Economic Geology Research Unit Information Circular 229, University of the Witwatersrand, 42 p.

Condie, K.C., and Hunter, D.R., 1976, Trace element geochemistry of Archaean granitic rocks from the Barberton region, South Africa: Earth and Planetary Science Letters, v. 29, p. 389–400, doi: 10.1016/0012-821X(76)90144-8.

de Ronde, C.E.J., and de Wit, M.J., 1994, Tectonic history of the Barberton greenstone belt, South Africa: 490 million years of Archaean crustal evolution: Tectonics, v. 13, p. 983–1005, doi: 10.1029/94TC00353.

Dewey, J.F., 1988, Extensional collapse of orogens: Tectonics, v. 7, p. 1123–1139.

de Wit, M.J., 1998, On Archaean granites, greenstones, cratons and tectonics: Does the evidence demand a verdict?: Precambrian Research, v. 91, p. 181–226, doi: 10.1016/S0301-9268(98)00043-6.

de Wit, M.J., Roering, C., Hart, R.J., Armstrong, R.A., de Ronde, C.E.J., Green, R.W.E., Tredoux, M., Pederdy, E., and Hart, R.A., 1992, Formation of an Archaean continent: Nature, v. 357, p. 553–562, doi: 10.1038/357553a0.

Eglington, B.M., and Armstrong, R.A., 2004, The Kaapvaal craton and adjacent orogens, southern Africa: A geochronological database and overview of the geological development of the craton: South African Journal of Geology, v. 107, p. 13–32, doi: 10.2113/107.1-2.13.

Flowers, R.M., Moser, D.E., and Hart, R.J., 2003, Evolution of the amphibolite-granulite facies transition exposed by the Vredefort impact structure, Kaapvaal Craton, South Africa: The Journal of Geology, v. 111, p. 455–470, doi: 10.1086/375282.

Friese, A.E.W., Reimold, W.U., and Layer, P.W., 2003, ^{40}Ar-^{39}Ar dating and structural information on tectonite-bearing faults in the Witwatersrand Basin: Evidence for multi-stage, tectonothermal activity in the Central Kaapvaal Craton: South African Journal of Geology, v. 106, p. 41–70, doi: 10.2113/1060041.

Gibson, R.L., 2002, Impact-induced melting in Archaean granulites in the Vredefort Dome, South Africa. I.: Anatexis of metapelitic granulites: Journal of Metamorphic Geology, v. 20, p. 57–70, doi: 10.1046/j.0263-4929.2001.00358.x.

Gibson, R.L., and Reimold, W.U., 2001, The Vredefort Impact Structure, South Africa: The scientific evidence and a two-day excursion guide: Pretoria, South Africa, Council for Geoscience Memoir 92, 111 p.

Gibson, R.L., and Reimold, W.U., 2005, Shock pressure distribution in the Vredefort impact structure, South Africa, *in* Kenkmann, T., Hörz, F., and Deutsch, A., eds., Large Meteorite Impacts III: Geological Society of America Special Paper 384, p. 329–349.

Gibson, R.L., Armstrong, R.A., and Reimold, W.U., 1997, The age and thermal evolution of the Vredefort Impact Structure: A single-grain U-Pb zircon study: Geochimica et Cosmochimica Acta, v. 61, p. 1531–1540, doi: 10.1016/S0016-7037(97)00013-6.

Gibson, R.L., Reimold, W.U., Ashley, A.J., and Koeberl, C., 2002, Metamorphism on the Moon: A terrestrial analogue in the Vredefort Dome, South Africa: Geology, v. 30, p. 475–478, doi: 10.1130/0091-7613(2002)030<0475:MOTMAT>2.0.CO;2.

Hart, R.J., Welke, H.J., and Nicolaysen, L.O., 1981, Geochronology of the deep profile through the Archaean Basement of Vredefort, with implications for early crustal evolution: Journal of Geophysical Research, v. 86, p. 10,663–10,680.

Hart, R.J., Andreoli, M.A.G., Tredoux, M., and de Wit, M.J., 1990, Geochemistry across an exposed section of Archaean crust at Vredefort, South Africa: With implications for mid-crustal discontinuities: Chemical Geology, v. 82, p. 21–50, doi: 10.1016/0009-2541(90)90072-F.

Hart, R.J., Moser, D., and Andreoli, M.A.G., 1999, Archaean age for the granulite facies metamorphism near the center of the Vredefort structure, South Africa: Geology, v. 27, p. 1091–1094, doi: 10.1130/0091-7613(1999)027<1091:AAFTGF>2.3.CO;2.

Hirner, A., 2003, Geology and gold mineralization of the Madibe Greenstone Belt, eastern parts of the Kraaipan terrane, Kaapvaal Craton, South Africa [Ph.D. thesis]: Johannesburg, University of the Witwatersrand, 502 p.

Hunter, D.R., 1974, Crustal development in the Kaapvaal Craton, the Archean: Precambrian Research, v. 1, p. 259–294, doi: 10.1016/0301-9268(74)90002-3.

Jackson, M.C., 1992, Geochemistry and metamorphic petrology of Dominion Group metavolcanics in the Vredefort area, South Africa: Johannesburg, South Africa, Economic Geology Research Unit, Information Circular 256, University of the Witwatersrand, 36 p.

Kamo, S.L., and Davis, D.W., 1994, Geochronology and tectonic evolution of the Barberton greenstone belt: Tectonics, v. 13, p. 167–192, doi: 10.1029/93TC02254.

Kamo, S.L., Reimold, W.U., Krogh, T.E., and Colliston, W.P., 1996, A 2.023 Ga age for the Vredefort impact event and a first report of shock metamorphosed zircons in pseudotachylitic breccia and granophyre: Earth and Planetary Science Letters, v. 144, p. 369–387, doi: 10.1016/S0012-821X(96)00180-X.

Kisters, A.F.M., Stevens, G., Dziggel, A., and Armstrong, R.A., 2003, Extensional detachment faulting and core-complex formation in the southern Barberton granite-greenstone terrain, South Africa: Evidence for a 3.2 Ga orogenic collapse: Precambrian Research, v. 127, p. 355–378, doi: 10.1016/j.precamres.2003.08.002.

Kröner, A., and Layer, P.W., 1992, Crust formation and plate motion in the early Archean: Science, v. 256, p. 1405–1411.

Kröner, A., Hegner, E., Wendt, J.I., and Byerly, G.R., 1996, The oldest part of the Barberton granitoid-greenstone terrain South Africa: Evidence for crust formation between 3.5 and 3.7 Ga: Precambrian Research, v. 78, p. 105–124, doi: 10.1016/0301-9268(95)00072-0.

Lana, C., 2004, Structural and petrogenetic studies related to the geological evolution of the Archaean basement complex of the Vredefort dome, South Africa [Ph.D. thesis]: Johannesburg, University of the Witwatersrand, 238 p.

Lana, C., Gibson, R.L., Kisters, A., and Reimold, W.U., 2003a, Archean crustal structure of the Kaapvaal craton, South Africa—Evidence from the Vredefort dome: Earth and Planetary Science Letters, v. 206, p. 133–144, doi: 10.1016/S0012-821X(02)01086-5.

Lana, C., Gibson, R.L., Reimold, W.U., and Minnitt, R.C.A., 2003b, Geology and geochemistry of a granite-greenstone association in the southeastern Vredefort dome, South Africa: South African Journal of Geology, v. 106, p. 291–314, doi: 10.2113/106.4.291.

Lana, C., Gibson, R.L., and Reimold, W.U., 2003c, Impact tectonics in the core of the Vredefort dome: Implications for formation of central uplifts in large impact structures: Meteoritics & Planetary Science, v. 38, p. 1093–1107.

Lana, C., Reimold, W.U., Gibson, R.L., Koeberl, C., and Siegesmund, S., 2004, Nature of the Archean mid-crust in the core of the Vredefort Dome, Central Kaapvaal Craton, South Africa: Geochimica et Cosmochimica Acta, v. 68, p. 623–642, doi: 10.1016/S0016-7037(03)00447-2.

Lee, J.K.W., and Tromp, J., 1995, Self-induced fracture generation in zircon: Journal of Geophysical Research, v. 100, B9, p. 17,753–17,770, doi: 10.1029/95JB01682.

Lowe, D.R., 1994, Accretionary history of the Archean Barberton Greenstone Belt (3.55–3.22 Ga), southern Africa: Geology, v. 22, p. 1099–1102, doi: 10.1130/0091-7613(1994)022<1099:AHOTAB>2.3.CO;2.

Lowe, D.R., 1999, Geological evolution of the Barberton greenstone belt and vicinity, *in* Lowe, D.R., and Byerly, G.R., eds., Geologic evolution of the Barberton greenstone belt, South Africa: Geological Society of America Special Paper 329, p. 287–312.

Ludwig, K.R., 2000a, SQUID: 1.00, A User's Manual: Berkeley, California, Berkeley Geochronology Center Special Publication 2, 17 p.

Ludwig, K.R., 2000b, Isoplot/Ex: A geochronological toolkit for Microsoft Excel: Berkeley, California, Berkeley Geochronology Center Special Publication 1a, 53 p.

Mason, D.R., and McDonald, J.A., 1978, Intrusive rocks and porphyry copper occurrences of the Papua New Guinea–Solomon Islands region: A reconnaissance study: Economic Geology and the Bulletin of the Society of Economic Geologists, v. 73, p. 857–877.

Minnitt, R.C.A., Reimold, W.U., and Colliston, W.P., 1994, The geology of the Greenlands Greenstone Complex and selected granitoid terranes in the south-eastern quadrant of the Vredefort Dome: Johannesburg, South Africa, Economic Geology Research Unit, Information Circular, 281, University of Witwatersrand, 46 p.

Moser, D.E., 1997, Dating the shock wave and thermal imprint of the giant Vredefort impact, South Africa: Geology, v. 25, p. 7–10, doi: 10.1130/0091-7613(1997)025<0007:DTSWAT>2.3.CO;2.

Moser, D.E., Flowers, R.M., and Hart, R.J., 2001, Birth of the Kaapvaal Tectosphere 3.08 billion years ago: Science, v. 291, p. 465–468, doi: 10.1126/science.291.5503.465.

Paces, J.B., and Miller, J.D., 1993, Precise U-Pb ages of Duluth Complex and related mafic intrusions, northeastern Minnesota: Geochronological insights to physical, petrogenetic, paleomagnetic and tectonomagmatic processes associated with the 1.1 Ga midcontinent rift system: Journal of Geophysical Research, v. 98B, p. 13,997–14,013.

Poujol, M., and Anhaeusser, C.R., 2001, The Johannesburg Dome, South Africa: New single U-Pb isotopic evidence for early Archaean granite-greenstone development within the central Kaapvaal craton: Precambrian Research, v. 108, p. 139–157, doi: 10.1016/S0301-9268(00)00161-3.

Poujol, M., Anhaeusser, C.R., and Armstrong, R.A., 2002, Episodic granitoid emplacement in the Archaean Amalia-Kraaipan terrane, South Africa: Confirmation from single zircon U-Pb geochronology: Journal of African Earth Sciences, v. 35, p. 147–161.

Poujol, M., Robb, L., Anhaeusser, C.R., and Gericke, B., 2003, Geochronological constraints on the evolution of the Kaapvaal craton, South Africa: Precambrian Research, v. 127, p. 181–213, doi: 10.1016/S0301-9268(03)00187-6.

Prevec, S.A., Anhaeusser, C.R., and Poujol, M., 2004, Origin and evolution of late mafic dykes in an Archaean gneissic assemblage, Kaapvaal Craton, South Africa: South African Journal of Science, v. 100, p. 549–555.

Reimold, W.U., Colliston, W.P., and Robertson, A.S., 1988, Chronological and structural work in the Vredefort Dome: Geocongress '88: Durban, Biannual Congress of the Geological Society of South Africa, p. 501–504.

Reimold, W.U., Jessberger, E.K., and Stephan, T., 1990, Ar-Ar dating of pseudotachylite from the Vredefort Impact Structure, South Africa—A progress report: Tectonophysics, v. 171, p. 139–152, doi: 10.1016/0040-1951(90)90095-P.

Reimold, W.U., Pybus, G.Q.F., Kruger, F.J., Layer, P.W., and Koeberl, C., 2000, The Anna's Rust Sheet and related gabbroic intrusions in the Vredefort Dome—Part of a widespread Kibaran magmatic event on the Kaapvaal craton and beyond?: Journal of African Earth Sciences, v. 31, p. 499–521, doi: 10.1016/S0899-5362(00)80004-4.

Reimold, W.U., Leroux, H., and Gibson, R.L., 2002, Shocked and thermally metamorphosed zircon from the Vredefort impact structure, South Africa: A transmission electron microscopic study: European Journal of Mineralogy, v. 14, p. 859–868, doi: 10.1127/0935-1221/2002/0014-0859.

Robb, L.J., Davis, D.W., and Kamo, S.L., 1991, Chronological framework for the Witwatersrand Basin and environs: Towards a time-constrained depositional model: South African Journal of Geology, v. 94, p. 86–95.

Schmitz, M.D., Bowring, S.A., de Wit, M.J., and Gartz, V., 2004, Subduction and terrane collision stabilize the western Kaapvaal craton tectosphere 2.9 billion years ago: Earth and Planetary Science Letters, v. 222, p. 363–376, doi: 10.1016/j.epsl.2004.03.036.

Singh, J., and Johannes, W., 1996, The hydration melting of tonalities, Part II, composition of melts and solids: Contributions to Mineralogy and Petrology, v. 125, p. 26–44, doi: 10.1007/s004100050204.

Skjerlie, K.P., and Johnson, A.D., 1993, Vapour-absence melting from 10–20 kbar of crustal rocks that contain multiple hydrous phases: Implications for anatexis in the deep to very deep continental crust and active continental margins: Journal of Petrology, v. 37, p. 661–691.

Spray, J.G., Kelley, S.P., and Reimold, W.U., 1995, Laser-probe ^{40}Ar-^{39}Ar dating of pseudotachylites and the age of the Vredefort impact event: Meteoritics & Planetary Science, v. 30, p. 335–343.

Stacey, J.S., and Kramers, J.D., 1975, Approximation of terrestrial lead isotope evolution by a two-stage model: Earth and Planetary Science Letters, v. 26, p. 207–221, doi: 10.1016/0012-821X(75)90088-6.

Stepto, D., 1979, A geological and geophysical study of the central portion of the Vredefort Dome structure [Ph.D. thesis]: Johannesburg, University of the Witwatersrand, 378 p.

Stepto, D., 1990, The geology and gravity field in the central core of the Vredefort structure: Tectonophysics, v. 171, p. 75–103, doi: 10.1016/0040-1951(90)90091-L.

Stevens, G., Gibson, R.L., and Droop, G.T.R., 1997, Mid-crustal granulite facies metamorphism in the Central Kaapvaal craton: The Bushveld Complex connection: Precambrian Research, v. 82, p. 113–132, doi: 10.1016/S0301-9268(96)00043-5.

Tredoux, M., Hart, R.J., Carlson, R.W., and Shirey, S.B., 1999, Ultramafic rocks at the center of the Vredefort Structure: Further evidence for the crust-on-edge model: Geology, v. 27, p. 923–926, doi: 10.1130/0091-7613(1999)027<0923:URATCO>2.3.CO;2.

Trieloff, M., Reimold, W.U., Kunz, J., Boer, R.H., and Jessberger, E.K., 1994, Ar-Ar thermochronology of pseudotachylites at the Ventersdorp Contact Reef, Witwatersrand Basin: South African Journal of Geology, v. 97, p. 365–384.

Whalen, J.B., 1985, Geochemistry of an island-arc plutonic suite: the Uasilau–Yau Yau intrusive complex, New Britain, P.N.G.: Journal of Petrology, v. 26, p. 603–632.

Williams, I.S., 1998, U-Th-Pb geochronology by ion microprobe, *in* McKibben, M.A., Shanks, W.C., and Ridley, W.I., eds., Application of microanalytical techniques to understanding mineralizing processes: Reviews in Economic Geology, v. 7, p. 1–35.

Williams, I.S., and Claesson, S., 1987, Isotopic evidence for the Precambrian provenance and Caledonian metamorphism of high-grade paragneisses from the Seve Nappes, Scandinavian Caledonides, II, Ion microprobe zircon U-Th-Pb: Contributions to Mineralogy and Petrology, v. 97, p. 205–217, doi: 10.1007/BF00371240.

Windley, B.F., 1973, Crustal development in the Precambrian: Royal Society of London Philosophical Transactions, v. 273, p. 321–341.

Windley, B.F., 1982, The evolving continents: Chichester, John Wiley and Sons, 399 p.

Wong, M., and Gans, P.B., 2003, Tectonic implications of early Miocene extensional unroofing of the Sierra Mazatan metamorphic core complex, Sonora, Mexico: Geology, v. 31, p. 953–956, doi: 10.1130/G19843.1.

Manuscript Accepted by the Society 4 October 2005

Geological Society of America
Special Paper 405
2006

Physical volcanology and compositions of the basaltic lavas in the Archean Nzuse Group, White Mfolozi inlier, South Africa

Allan H. Wilson†
Claire E. Grant
School of Geological Sciences, University of KwaZulu-Natal, Durban 4041, South Africa

ABSTRACT

The Mesoarchean Pongola Supergroup is a thick volcano-sedimentary succession that is unique in the early record of Earth's history because of the nature and composition of the lava flows that, in addition to basalts, have a high proportion of intermediate and acid compositions. The succession contrasts with other contemporaneous volcanic deposits in southern Africa and in other parts of the world that gave rise to greenstone belts by having a low state of deformation and a uniformly low grade of metamorphism.

The White Mfolozi inlier is a deeply eroded river section that exposes spectacular structures and textures of two sequences of volcanic rocks, the lower Nhlebela Formation and the upper Agatha Formation. The Nhlebela Formation contains pillowed successions, peperite-like breccias, and extensive fragmental deposits formed as a result of hydroclastic eruption. The shallow water hydromagmatic deposits developed into amygdaloidal pahoehoe units higher in the succession as the lava eruptions raised the ground level. The Agatha Formation, studied in two traverses, commenced with the formation of a hydroclastic tuff formed by the emplacement of a high-Mg basalt magma interacting with unconsolidated sediment. Subaqueous eruptions rapidly progressed into subaerial deposits consisting initially of aa lavas that then developed into pahoehoe lavas higher in the succession.

The pahoehoe lavas in the Agatha Formation preserve spectacular ropy surfaces. The internal structures of the flow lobes and distribution of vesicles indicate that lava inflation was the dominant mechanism of emplacement. Therefore important processes that are observed in modern and recent subaerial and continental basalts also operated in the mid-Archean.

Changes in structures and styles of the pahoehoe lava flow lobes can be linked to magma compositions. Compositions include basalts, Fe-rich basalts, siliceous Mg basalts, andesites, and dacites. Trace elements identify at least four flows made up of 75 flow lobes that make up the 210-m-thick Agatha Formation. Geochemical tectonic discrimination diagrams fail to resolve the tectonic setting because of crustal contamination, but field observations indicate it is likely that the depositional setting in this area was a subsiding continental margin.

Keywords: Archean volcanism, pahoehoe, flow lobes geochemistry.

†E-mail: wilsona@ukzn.ac.za.

Wilson, A.H., and Grant, C.E., 2006, Physical volcanology and compositions of the basaltic lavas in the Archean Nzuse Group, White Mfolozi inlier, South Africa, *in* Reimold, W.U., and Gibson, R.L., Processes on the Early Earth: Geological Society of America Special Paper 405, p. 255–289, doi: 10.1130/2006.2405(14). For permission to copy, contact editing@geosociety.org.

INTRODUCTION

Volcanic rocks of the Archean (older than 2.5 Ga) are an integral part of granite-greenstone belt terranes that are usually deformed by a combination of early thrusting events and later stage folding (de Wit and Ashwal, 1997). The Pongola Supergroup is essentially undeformed and consistently of low metamorphic grade and therefore is unlike a greenstone belt in the normal sense, but it contains rock components similar to those of greenstone belts. Similar to most Archean greenstone belts the Pongola Supergroup comprises a basal sequence of volcanic rocks overlain by a thick sedimentary succession. Most greenstone belts are Archean in age and the lavas contained in them were produced by a diverse set of physical and chemical processes operating in a variety of paleoenvironments. The interpretation of these environments is seldom straightforward. In addition, alteration and later tectonic modification may have partly or completely obscured the original textures and compositions. The excellent preservation of many of the sequences in the Pongola Supergroup provides an opportunity to study continuous and well-preserved volcanic rocks that are seldom available in greenstone belts of similar age.

The Pongola Supergroup, located close to the southeastern margin of the Kaapvaal Craton (Fig. 1, inset A), is a mid-Archean volcano-sedimentary succession that formed during two distinct depositional phases. The older Nsuze Group is predominantly volcanic, while dominantly sedimentary volcanic rocks form the younger Mozaan Group. The Pongola Supergroup is exposed where the Phanerozoic cover sequences have been eroded to reveal a number of inliers found throughout southeastern Mpumalanga, southern Swaziland, and KwaZulu-Natal of South Africa (Fig. 1, inset A). A general time interval between 3100 Ma and 2840 Ma has been postulated for the deposition of the Pongola Supergroup (Hegner et al., 1984).

The Pongola Supergroup is regionally subdivided into a Northern Region (southeastern Mpumalanga and Swaziland) and a Central and Southern Region in KwaZulu-Natal (Fig. 1). The subdivision is based on distinct facies variations relating to

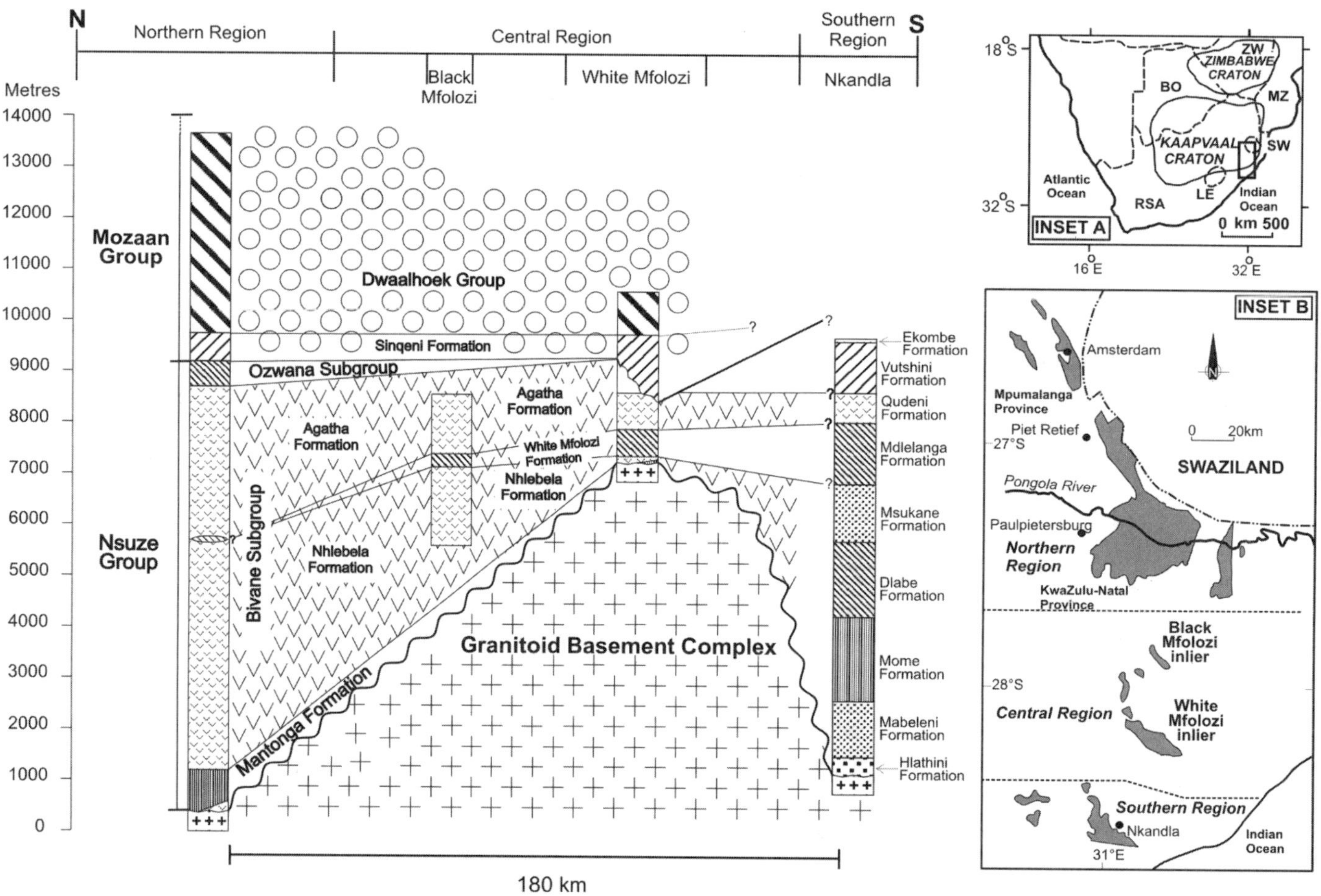

Figure 1. Locality of the study area and schematic diagram of the stratigraphy of the Pongola Supergroup. (Inset A) Locality in relation to the Kaapvaal and Zimbabwe Cratons. Also shown are the international boundaries for Swaziland (SW), Zimbabwe (ZW), Mozambique (MZ), Lesotho (LE), Botswana (BO), and South Africa (RSA). (Inset B) Regional subdivision of the Pongola Supergroup into Northern, Central, and Southern Regions. Stratigraphy based on SACS (1980), Beukes and Cairncross (1991), Armstrong et al. (1982), Groenewald (1984), Hatfield (1990), Preston (1987), Gold (1993), and Linström (1987).

major depositional environments and the nature of the pre-Pongola basement. Volcanic rocks in the Nsuze Group predominate in the Northern Region, whereas sedimentary rocks predominate in the Southern Region. Unlike most Paleo- and Mesoarchean volcanic sequences, which are generally regarded as representing ocean basin volcanism, the Nsuze Group has abundant evidence of shallow water or subaerial environments of deposition. The precise tectonic setting of the Nsuze Group remains a matter of debate. Hypotheses have included continental extension, resulting in the formation of rift-controlled basins (Burke et al., 1985; Crow et al., 1989; Hegner et al., 1984; Matthews, 1990); nonorogenic continental magmatism due to plumes beneath the Kaapvaal Craton (Armstrong, 1980; Armstrong et al., 1986); and volcanic arc setting due to the occurrence of andesites and geochemical signatures (Tankard et al., 1982). The dominant deposition environments of the Mozaan are shallow marine shelf facies and fluvial braidplan (Beukes and Cairncross, 1991). Shoreline deposits only account for a minor component of the Mozaan Group sediments, and a tide-dominated environment is identified relating to an intertidal sequence (von Brunn and Mason, 1977).

The White Mfolozi inlier of the Central Region of the Pongola Supergroup is exposed in the deep gorge of the White Mfolozi River, providing an almost unbroken stratigraphic record over an exposed distance of nearly 12 km. In this section the basal contact is observed for the Nsuze Group passing upward into the sediments of the Mozaan Group. It is surrounded by Phanerozoic sedimentary sequences of the Devonian Natal Supergroup and the Mesozoic Karoo Supergroup. The lavas of the Nsuze Group in this area are probably some of the best preserved in the Archean geological record. In this paper we present observations of the nature and depositional environment of the volcanic rocks as well as chemical compositions for well-preserved sequences in the White Mfolozi River valley.

RECENT AND ARCHEAN BASALTIC LAVA FLOWS

Basaltic lava (used here in the generic sense to include also the more evolved rock types such as basaltic andesite and andesite) occurs in three distinct eruptive environments on Earth: deep water, relatively shallow water and littoral, and subaerial. Each of these environments produces distinct and common textures and flow structures. Superimposed on these are additional controls relating to proximity to the source, nature of the substrate, composition of the magmas, and temperature. Large volumes of fast-moving lava may be channelized where diminished heat loss causes it to remain in the liquid state (Kilburn, 2000). Initially the lava forms a channel or a tube with a crust or a surface chill, which, if pulled apart, allows further lava flows to develop in an open channel. Channel flow in lava fields is relatively short lived, and the greater proportion of lavas that constitute most subaerial basalts and continental flood basalts were emplaced by the inflation mechanism (Hon et al., 1994; Self et al., 1997, 1998; Walker, 1991). The inflation process extends over a period of time during which lava lobes expanded, thickened, and coalesced as the advancing sheet flows were fed through internal pathways.

Hydromagmatic eruptions include pillow lavas and hydroclastic breccias. Pillow lava is one of the most common features of subaqueous lava flows with its distinctive spherical or subspherical forms. It can form in abyssal and in shallow-water regimes, including shorelines, rivers, and lakes. The occurrence, size, and form of pillows are commonly related to effusion rates with large pillows forming at the slowest extrusion rates. Magma composition and viscosity also influence pillow size. Hyaloclastites and pillow breccias can form more readily in shallow water because of greater fragmentation steam pressure at decreased hydrostatic pressure (Staudigel and Schmincke, 1984). Pillows associated with abundant hyaloclastite and breccias can also form on steep slopes, usually in shallow water (Batiza and White, 2000).

Hydroclastic volcanic rocks are those that have interacted with water during their eruption and can form extensive fragmental deposits. Hyaloclastite is glassy debris related to nonexplosive shattering of lava surfaces as a result of rapid quenching by aqueous chilling (Batiza and White, 2000) and occurs in all depositional environments from subaerial to abyssal submarine. However, depth of water and nature of the volcanic topography appear to exert an influence in the formation of hyaloclastite and related breccias (Batiza and White, 2000). Mixtures of hyaloclastite, hydroclastic breccia, and fragmented and coherent lava flows can form in single eruptions as a result of fragmentation and integration of the lava. Remnants of pillows and fragments of pillows commonly make up an important part of the breccias.

Most basalts in greenstone belts formed as submarine deposits (Sylvester et al., 1997) and exhibit a range of textural types that include massive (sheet), pillowed, and brecciated flow units (Dimroth et al., 1985), probably reflecting the proximity to vent zones and effusion rates. Pillowed and brecciated lava flows are commonly associated with massive flow units that may reflect relative distances from the volcanic vent. Brecciated basalts include hyaloclastites formed by quenching and contraction fracturing as a result of the interaction of lava with water. Others formed as a result of hydraulic fracturing or by continued movement of the partly solidified lava (Cousineau and Dimroth, 1982). Symmetrical distribution, both laterally and vertically, of massive, pillowed, and brecciated textural lava types is observed in some cases (Dimroth et al., 1978, 1985; Sylvester et al., 1997). Massive flows may have formed in the proximity of volcanic vents, whereas pillows and then breccias formed in more distal zones. The tops and bases of flow units are sometimes marked by pillows or breccia zones. Massive units have also been attributed to high effusion rates, and pillows and breccias result from lower emplacement rates. The interaction of lava with wet sediment to produce peperites is an important environmental indicator but relatively few examples are reported for Archean successions. It has also been pointed out (Skilling et al., 2002) that in ancient deposits it may be difficult to distinguish fragments derived from this process from those of tectonic origin.

Pahoehoe lava with its characteristic smooth, bulbous, or ropy surfaces and characteristic distributions of vesicles is one of the most common types of subaerial basaltic volcanism, but there are few examples of this type described for the Archean. Basalt successions in many greenstone belts range in thickness from hundreds to even thousands of meters but estimates of thickness from a single center are usually exaggerated because lateral interdigitation from adjacent volcanic centers may not have been recognized (Sylvester et al., 1997). Further complications arise from tectonic juxtaposition of different volcanic packages of supracrustal sequences (Thurston and Chivers, 1990). Archean basalt flow sequences are commonly incomplete because of poor exposure or tectonic disturbance, and therefore valuable contextual information that allows interpretations of the environment of eruption is commonly limited.

The sequences of basaltic rock in the White Mfolozi River valley are continuously exposed over wide distances and therefore their description and interpretation is of value in the debate on Archean volcanism. The low degree of deformation and metamorphism of these rocks and lack of structural duplication allow lava lobes and flows to be evaluated and thicknesses to be determined with confidence.

AGE AND REGIONAL GEOLOGY OF THE PONGOLA SUPERGROUP

Pre-Pongola Basement

The Mpuluzi granitoid sheet dated at 3107 ± 4 Ma (U-Pb single zircon; Kamo et al., 1990) forms the basement to the Pongola Supergroup, which in the Northern Region is also the type section of the Nsuze Group (Fig. 1). The most recent age determination (U-Pb on zircons) for pre-Pongola lithologies is 3147 ± 13 Ma for a gneissic granite near Piet Retief (Griffin, 2002). These ages therefore represent the maximum age of the Pongola Supergroup, and the Archean basement rocks in this area probably formed the original depositional platform for the Nsuze Group (Hunter and Wilson, 1988). In the White Mfolozi inlier a nonconformable contact between granite-gneiss basement and the Nsuze Group sedimentary sequence is exposed in the White Mfolozi River (Cole, 1994; Matthews, 1967; observations from this work). In the Southern Region of the Nkandlha area (Fig. 1), the relationship of the Nsuze Group to older units has been obscured by strong regional deformation (Groenewald, 1984).

The contact between the pre-Pongola basement and the Nsuze Group in the area east of Paulpietersburg (Fig. 1, inset B) is marked by a well-developed quartz-sericite paleosaprolite 2–10 m thick, which grades into the underlying granite (Matthews and Scharrer, 1968). A distinct lack of feldspars and abundant micaceous minerals distinguish it from the granite. The limited thickness but considerable lateral extent of the paleosaprolite is evidence for extremely low topographical relief prior to the deposition of the Pongola Supergroup (Matthews and Scharrer, 1968). A similar, but less mature paleosaprolite is also observed in the White Mfolozi inlier.

Nsuze Group

The Nsuze Group (SACS, 1980) in the Northern Region and type area comprises basic to acid volcanic rocks, with minor intercalated lenses of clastic sediments. The proportion of sediments within the Group increases toward the south. Named after the Nsuze River, the outcrop of the Nsuze Group trends north-south and is present as far as Nkandlha in the Southern Region, a distance of ~180 km (Fig. 1, inset B).

The Nsuze Group can be divided into several lithological units assigned subgroup or formation status (SACS, 1980) broadly based on the dominance of volcanic or sedimentary rocks. The thickness of the Nsuze Group is greatest in the southern sector of the Northern Region (Fig. 1) where it is nearly 9000 m thick and thins dramatically to the north and south. In the southern Central Region of the White Mfolozi inlier the thickness is ~1000 m. South of the White Mfolozi inlier in the Southern Region, the Nsuze Group thickens again to close to 7000 m but is different to the northern region in that it is dominated by sediments (Fig. 1), making up nine formations (Groenewald, 1984).

In the Northern Region and type area, the Nsuze Group is divided into two subgroups: the dominantly volcanic, thick Bivane Subgroup (8000 m) and the thin upper Ozwana Subgroup (300–500 m), consisting of both primary and reworked pyroclastics and volcaniclastic sediments. The Bivane Subgroup is made up of four formations: the basal Mantonga Formation, consisting of predominantly clastic sediments with minor pyroclastic and lava lenses; the lower dominantly volcanic Nhelebela Formation; the middle sedimentary White Mfolozi Formation (nonexistent in some areas in the Northern Region); and the upper dominantly volcanic Agatha Formation (Fig. 1). The Bivane Subgroup has a marked facies variation to the south with the development of a major sedimentary package of the White Mfolozi Formation and marked thinning of the two predominantly volcanic formations (Fig. 1).

Dating of the Nsuze lavas from the Northern Region has yielded the following results: Pb-Pb bulk zircon 3090 ± 90 Ma (Burger and Coertze, 1973); Pb-Pb single zircon evaporation 2978 ± 2 Ma (Walraven and Pape, 1994); U-Pb bulk zircon 2940 ± 14 Ma (Hegner et al., 1984) and the same sample redated (Hegner et al., 1994), using U-Pb single zircon chemistry, resulted in an age of 2985 ± 1 Ma. The oldest lava unit of the Nsuze Group has a U-Pb zircon age of 2980 ± 10 Ma (Griffin, 2002). A rhyolitic unit found near the top of the Nsuze Group has a U-Pb zircon age of 2968 ± 6 Ma (Griffin, 2002). Both samples of the latter study are from the Northern Region.

Mozaan Group

The dominantly sedimentary Mozaan Group is preserved in the Northern and Central Regions, but it has not been recognized in the Southern Region in the vicinity of Nkandlha (Fig. 1, inset B). Pb-Pb whole rock analysis of the ferruginous shale of the Sinqeni Formation in the Amsterdam area gave an age of 2860 ± 26 Ma (Walraven and Pape, 1994). Recent age determination studies on an andesite lava flow found within the Tobolsk Member, a thin volcanic unit at the top of the Mozaan Group (Griffin, 2002), give a significantly older U-Pb zircon age of 2954 ± 9 Ma. This indicates a considerably shorter depositional period for the Pongola Supergroup than previously suggested. Nine lithological formations have been identified for the Mozaan Group in the Northern Region (Beukes and Cairncross, 1991; Gold, 1993), which include conglomerates, quartzites, argillites, shales, and minor lavas and volcaniclastic rocks toward the top of the sequence.

THE STRATIGRAPHY OF THE NSUZE GROUP WITH PARTICULAR EMPHASIS ON THE WHITE MFOLOZI INLIER

The Bivane Subgroup comprises the dominantly volcanic lower Nhlebela and the upper Agatha Formations, separated by the sedimentary White Mfolozi Formation in the southern facies. The Bivane Subgroup thins from the maximum thickness of 7500 m in the type area to ~1200 m at the White Mfolozi inlier over a distance of ~120 km. The sections studied are shown in Figures 2 and 3.

The volcanic rocks of the Nhlebela Formation overlie sediments of the Mantonga Formation (Fig. 1), which, in turn, unconformably overlie granitic basement rocks. In the White Mfolozi River section the contact with the granites is exposed in two areas, referred to as the northern and southern localities (Fig. 2). In the northern locality the Mantonga Formation is 20 m thick and is composed of well-banded quartz arenites with a coarse-grained arkosic layer at the base formed from juvenile gravels of the underlying granite (Figs. 4A and 4B). A conglomerate layer containing well-rounded black quartz pebbles is variably developed at the contact, which dips at 12° to the east. The underlying granite is strongly altered and kaolinized to a depth of 1–5 m and represents a gravel paleosol that predated the deposition of the sediments. In the southern exposure of the contact of the Mantonga Formation and the granite basement, there is a progressive thickening of the sedimentary unit from 5 to 60 m within a series of fault blocks up to several hundred meters wide. These may represent deposition in variable topography controlled by local block faulting that became reactivated at a later time. At a number of localities normal strike faulting has caused a repetition of the sequence where the Mantonga Formation is in contact with the granite.

In the type area of the Northern Region (Fig. 1) the volcanic rocks of the Nhlebela Formation are predominantly andesite and basaltic andesite with minor basalt (Armstrong et al., 1982; Hatfield, 1990). Porphyritic and nonporphyritic varieties of dacite and rhyolite are present. Thin rhyolite and dacite flows are found in the upper parts of the sequence and comprise a variety of volcaniclastic rocks, including airfall tuffs as well as possibly high-temperature pyroclastic flows (Manley, 1992). Lavas of similar siliceous compositions are not found in the Nhlebela Formation in the White Mfolozi inlier.

In the Northern Region the andesites and basaltic andesites of the Nhlebela Formation are commonly highly porphyritic with large plagioclase phenocrysts. This contrasts with the White Mfolozi inlier where lavas of similar composition are not porphyritic. Basalts in both the Northern and Central Regions are generally amygdaloidal, aphanitic, and lack phenocrysts (Armstrong et al., 1982). The Nhlebela Formation in the White Mfolozi inlier is estimated to be 130–165 m thick and conformably overlies the Mantonga Formation. Dykes and sills of similar composition to the lavas intrude the sequence in this area and were most likely contemporaneous with the lavas.

The Nhlebela Formation in the White Mfolozi inlier is overlain by sediments of the 480-m-thick White Mfolozi Formation (Armstrong et al., 1982; Matthews, 1967), which is also the type section. Basal rock types include diamictites and quartzites with trough, planar, and herringbone cross-bedding. These grade into a series of upward-fining sequences comprising alternating quartzite, siltstone, and shale, with calcareous arenites at the top of each cycle (Beukes and Lowe, 1989; Matthews, 1967). The overlying succession comprises dolomitic limestones, quartzitic dolomites, calcareous siltstones, and a stromatolitic unit 4.5 m thick (Mason and von Brunn, 1977; von Brunn and Mason, 1977). Capping this sequence is a series of quartzite units containing cross-beds, shale rip-up clasts, gritstone lags, and a carbonate-cemented coarse-grained arenite that grades into a nonstromatolitic dolomite and calcareous quartzite. The uppermost unit of the White Mfolozi Formation is a quartzite with rip-up mud clasts, poorly sorted grits, and sands at the base, grading upward into a well-sorted quartzite with well-developed trough and planar cross-beds.

The Agatha Formation overlies the White Mfolozi Formation and represents a resumption of volcanic activity. In the Northern Region it is dominated by mafic and felsic lavas in the lower section overlain by felsic volcaniclastics (Armstrong et al., 1982; Matthews, 1967; Preston, 1987) and is ~3000 m thick. The lavas are generally amygdaloidal andesite with minor components of basaltic andesite and basalt as well as an acid phase at the base of the formation represented by rhyolitic and dacitic flow units. In the White Mfolozi inlier the Agatha Formation is ~210 m thick and comprises amygdaloidal, nonporphyritic basalts, and basaltic andesites. Rhyolites are not present in the Central Region.

The Ozwana Subgroup overlies the Agatha Formation in the Northern Region and is a highly variable (0–50 m thick) volcaniclastic-sedimentary succession comprising argillites (volcanogenic sediments) and air-fall tuffs, minor arenites, and

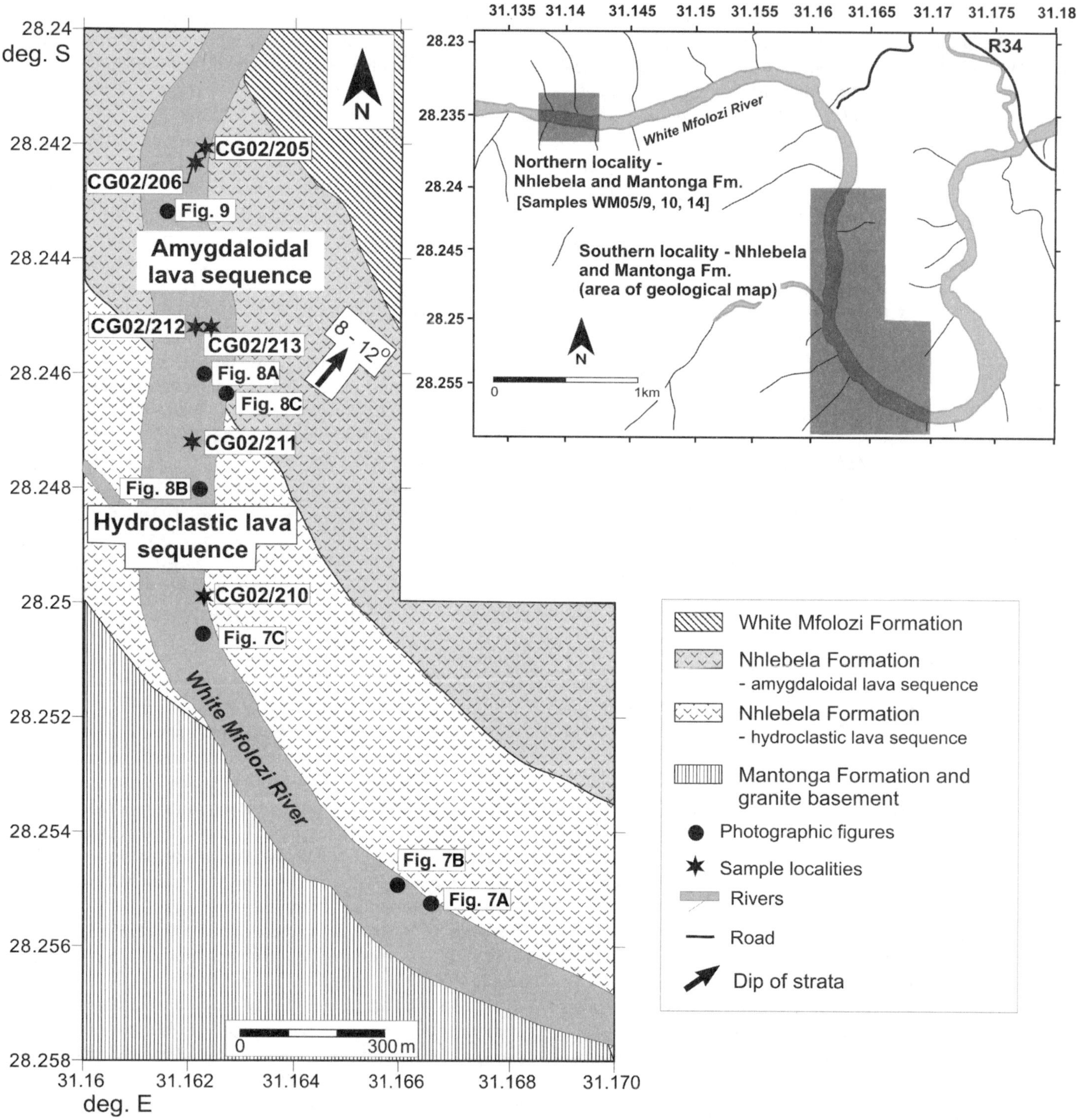

Figure 2. Southern and northern localities of the Nhlebela Formation (shaded areas on small-scale map). Large-scale map is of the southern locality showing the distributions of the underlying basement and Mantonga Formation and the lower hydroclastic and upper amygdaloidal sequence. Positions of photographic figures and sample localities are shown.

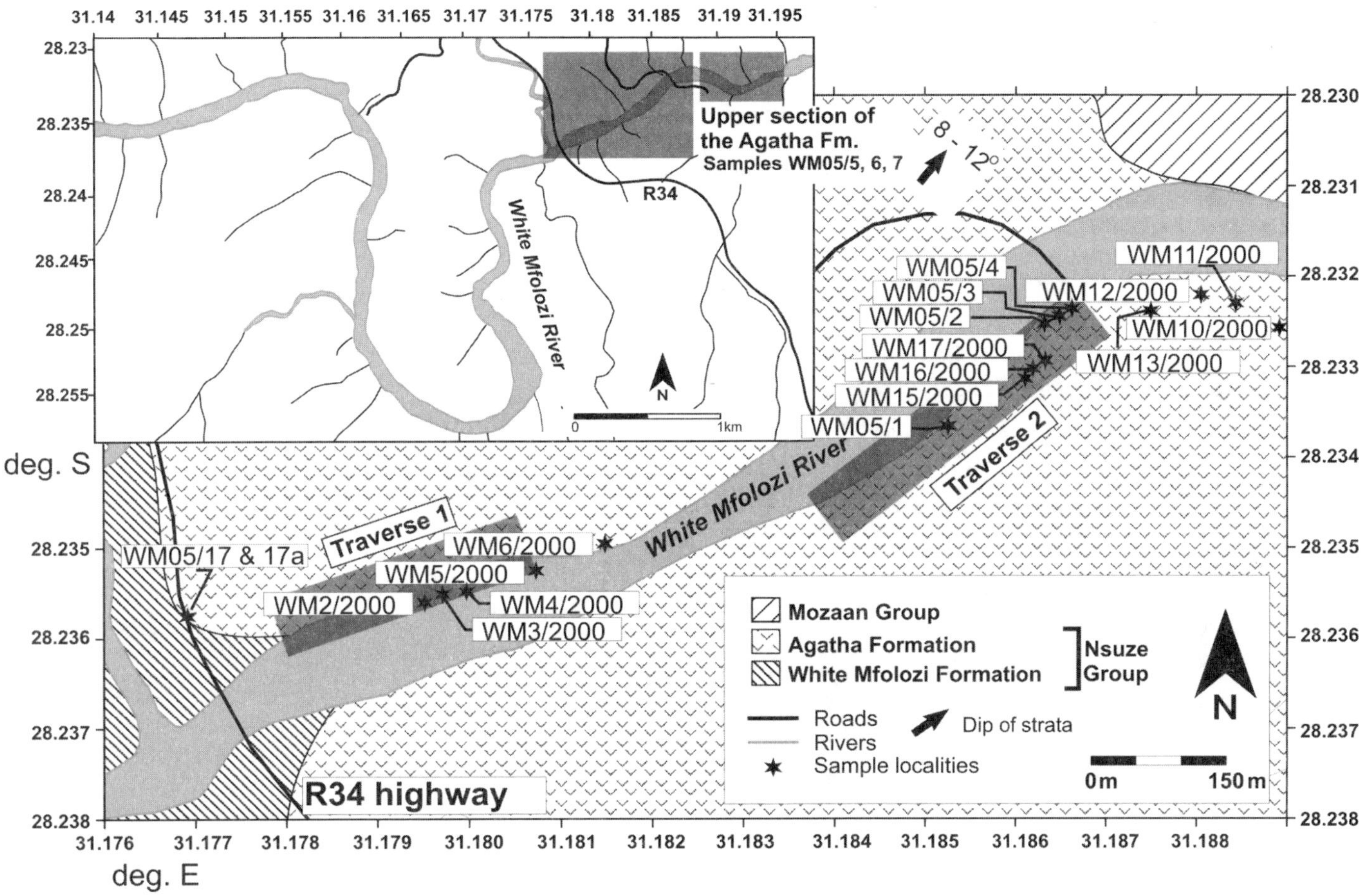

Figure 3. Locality of the Agatha Formation. Localities of sample sites in Traverses 1 and 2 studied in detail (sections in Fig. 10) are shown on the large-scale map and highlighted as the western block on the small-scale map. The highlighted block on the east represents the upper section of the Agatha Formation.

acid lava units (Gold, 1993). Epiclastic volcanic breccia and accretionary lapilli lenses are also found within the volcanogenic sediments (Armstrong et al., 1982). While development of the Ozwana Subgroup is reported in some parts of the White Mfolozi inlier as a sequence of ferruginous shales and minor quartzites (SACS, 1980), it is highly irregular in form and is absent in the well-exposed White Mfolozi River section.

The contact between the Nsuze and Mozaan Groups represents the stratigraphic level above which (with one exception relating to the topmost unit of the Mozaan Group) volcanic rocks (including pyroclastics or volcanogenic sediments) are absent. In the White Mfolozi River section, quartz sandstones of the basal formation of the Mozaan Group directly overlie the basaltic rocks of the Nsuze Group (Agatha Formation) at a well-exposed contact. A marked angular unconformity is present with the lava flow surfaces dipping at 18°–22° to the northeast (the same as the average dip of the White Mfolozi Formation), in contrast to the dip of the dominant bedding planes of the basal Mozaan sediments at 8°–12° to the east. Contact with the upper 3–5 m of the lava is marked by a highly foliated and ferruginized talc layer. The tectonic foliation becomes progressively less developed over 30 m into the underlying lavas. In some areas the dip of the lava flows has increased to over 70° to the north. The foliation and deformation immediately below the contact with the Mozaan Group quartz sandstones is interpreted in this study to be the result of considerable deformation arising from a low angle thrust at the contact.

DEPOSITIONAL ENVIRONMENT OF THE SEDIMENTS OF THE NSUZE GROUP

The basal Mantonga Formation in the Northern Region is interpreted as a distal braided stream environment (Watchorn and Armstrong, 1980). Rapid lateral migration of channels resulted in the deposition of tabular sandstones, with limited mudstone layers. In contrast, the Mantonga Formation exposed in the Central Region of the White Mfolozi inlier represents a near-littoral environment, with deposition occurring in a transgressive sea (Matthews, 1967). The shoreline must therefore have been situated between the two regions. The lack of feldspathic material in the quartz wackes indicates that the detritus was derived from the weathering of the paleosaprolite. The

Figure 4. Basal contact of the Mantonga Formation with the granitoid basement. (A) Banded quartz sandstone and a pebble conglomerate overlying partly decomposed granite gravel. (B) Detail of the basal conglomerate and the decomposed granite gravel.

source rocks for the subordinate arkosic arenites were unweathered granitic basement (Matthews, 1967; Watchorn and Armstrong, 1980). The paleocurrent directions vary from northeast to southeast, and the higher proportion of conglomerates at the base of the Mantonga Formation represents a gradual degradation of the source area with time (Armstrong et al., 1982).

The White Mfolozi Formation in the Central Region has been interpreted to represent a prograding tidal environment (Beukes and Lowe, 1989; Preston, 1987; von Brunn and Mason, 1977). The lack of development of the White Mfolozi Formation in the Northern Region indicates that the shoreline transgressed between the two regions. This is supported by the development of the White Mfolozi Formation east of Piet Retief and is considered to have formed as a result of fluvial processes (Hatfield, 1990).

In the Northern Region the uppermost unit of the Nsuze Group, the Ozwana Subgroup, marks the shoreline as both fluvial and marine deposits are present. In the Central and Southern Regions, the lithologies of the Ozwana Subgroup, where they occur, are representative of tidal to distal shelf shallow-water marine deposits. Therefore the subaerial depositional environment deduced for the lavas of the Agatha Formation, as demonstrated in the following sections, reverted back to shallow-water marine conditions, presumably by subsidence following the termination of the volcanic phase.

VOLCANOLOGY OF THE WHITE MFOLOZI INLIER

Overview of the Lava Types in the Nhlebela and Agatha Formations

In the northern locality of the Nhlebela Formation (Fig. 2) volcanic rocks overlie sediments of the Mantonga Formation, which in turn rest unconformably on basement granites. In the southern locality the volcanic rocks overlie both the sediments and the granites as a series of off-lapping relationships controlled by pre- or syn-depositional block faults.

The upper volcanic Agatha Formation overlies quartz sandstones of the White Mfolozi Formation with a possible slight angular unconformity and is a very well preserved sequence of mainly subaerial lavas. Two detailed traverses (Fig. 3) through the riverside outcrop of the White Mfolozi River gorge provide an excellent section through this formation. Traverse 1 is along the northern bank of the White Mfolozi River and covers a horizontal distance of 285 m in an east-northeastern direction, approximately parallel to the average dip direction. Traverse 2 is on the southern bank of the White Mfolozi River over a distance of 375 m in a northeastern direction, 15° off the average dip direction of the strata. Areas not included in the traverses have rubbly outcrop or poor exposure. The uppermost section of the Agatha Formation is exposed to the east of Traverse 2 (Fig. 3) over a distance of ~500 m but continuity is hampered by boulders from the steep cliffs of the overlying Mozaan Group sandstones. This section is shown in the inset to Figure 3 and is referred to as the upper section of the Agatha Formation.

Important features and diagnostic lava textures and structures will be described in the sections below. Characteristics of the lavas can be broadly subdivided into categories relating to subaqueous and subaerial emplacement. Volcanic rocks formed by subaqueous emplacement include pillow lavas, hydroclastic breccias, and tuffs. Volcanic rocks that have been erupted subaerially are represented by pahoehoe, aa, and blocky lavas. These different types relate not only to the eruptive environment but also to compositions, temperature, rate of effusion, and nature and slope of the substrate. In line with the definitions of Self et al. (1998), a lobe describes the smallest coherent package of lava with flows being built up from lobes. A compound flow is made up of many similar-sized lobes and is the result of a single outpouring of lava. Flow fields are the result of several flows becoming juxtaposed. In the subaqueous environment pillow lavas are closely related to pahoehoe in both eruption mechanism and morphology (Self et al., 1998).

Many of the lavas contain vesicles (cavities formed from

exsolved gas bubbles). It is likely that most if not all the vesicles were partly or completely filled with secondary minerals. In some cases this material has weathered and leached away to reveal the original cavity. To avoid unnecessary description and to more closely reflect the original state of the lavas, the terms vesicles and amygdales will be used synonymously.

THE NHLEBELA FORMATION

Contact of the Volcanic Rocks with Sediments

The lowermost unit of the Nhlebela Formation in the northern locality is a relatively poorly exposed massive basalt of variable thickness and morphology. The contact with the underlying sediment is not exposed, but the lava is generally highly fractured and associated with zones of fragmentation in which angular clasts of lava are mixed with sediment as well as fine-grained reworked fragments of lava. The jigsaw fit of many of the clasts (Fig. 5A) indicates in situ fragmentation similar to that which characterizes peperite. Peperite is caused by the disintegration of magma followed by mingling with unconsolidated sediment (White et al., 2000). A high proportion of the sediment appears to be spalled fragments of lava that have been reworked in a subaqueous environment and formed a wet substrate to further lava emplacement. Other exposures have similar mixtures of sediment and lava and commonly occur as compositionally variable layers (in terms of sediment and mafic lava fragments) in attitudes that vary from parallel to the lava contact to steeply dipping (Fig. 5B), truncated by the overlying lava.

In spite of the extensive strike length of the contact of the lavas with the underlying Mantonga Formation in the southern locality (~500 m), it is exposed in only a few localities. Where the contact is exposed it is sharp, and there is no evidence of lava mixing with underlying sediment. Peperite-type textures with jigsaw fits of lava fragments are common some meters above the sediment contact, indicating interaction with wet endogenous sediment. Rounded sand grains have been observed in this material, but it is mostly spalled glass derived from the lavas.

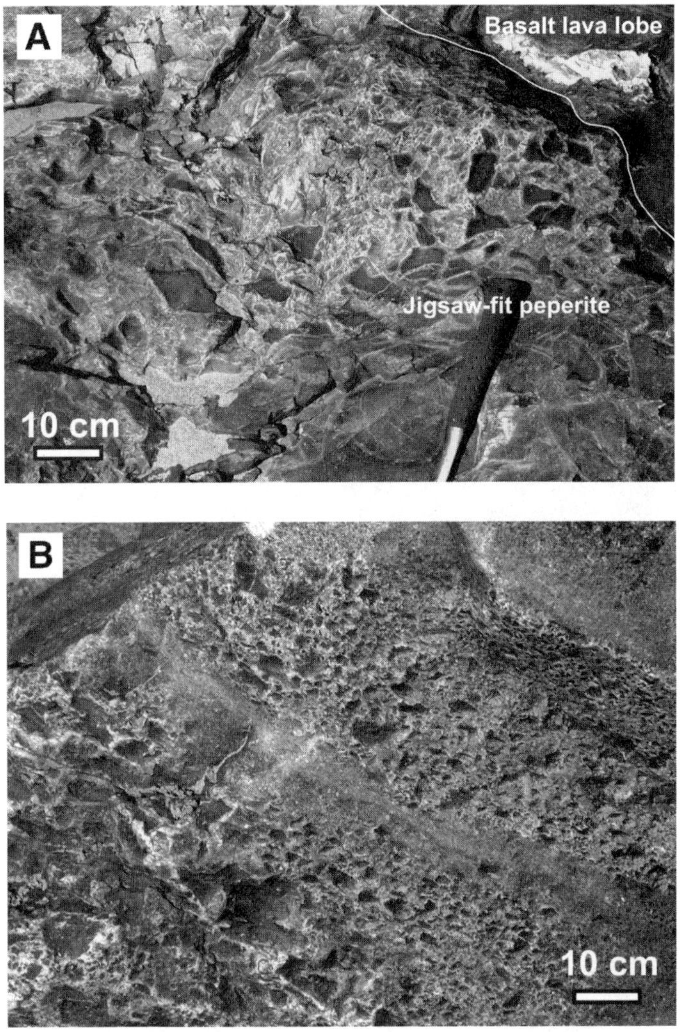

Figure 5. Hydroclastic breccia at the base of the Nhlebela Formation. (A) Fragmented jigsaw-fit peperite formed by emplacement of andesitic lava into a wet and mainly endogenous sediment. (B) Beds of reworked juvenile angular lava fragments and quartz clastic sedimentary material.

Subaqueous Lavas

Pillow lavas are extensively developed near the base of the Nhlebela Formation in both the northern and southern localities. In the northern locality the pillow lavas form a unit ~20–30 m thick and overlie the basal massive lava with a variable contact reflecting local topography (Fig. 6A). The pillows are generally bulbous, 1–2 m in diameter (Fig. 6B), and tend to form distinct layers with pillows of different average size interlayered with elongate sheetlike bodies (Fig. 6C) 0.3–1 m in thickness. Relatively large pillows tend to overlie the sheet lava. The pillows are commonly molded into one another and have extensive development of vesicles. Interpillow hyaloclastite is well developed. Both pillowed and the sheetlike units contain hyaloclastite zones, unlike the peperite-like fragmental material derived from reaction with wet sediment. Pillow chill zones range from 5–10 cm in thickness, and discontinuous multiple rinds (up to three) are observed in some pillows.

The size and shape of pillow forms would have been dependent mainly on the lava viscosity, with secondary controls relating to the slope and lava supply rate (Walker, 1992). Irregular sheet flows as thin as 4 cm have been observed to develop on steeper slopes in Pliocene deposits in the Azores and have been termed para-pillows (Walker, 1992). Lateral gradation from massive lava flows to pillows in the Rouyn-Noranda area have also been attributed to a slowing of the lava supply rate (Dimroth et al., 1978).

The southern locality of the Nhlebela Formation (Fig. 2) is divided into a lower predominately hydroclastic sequence

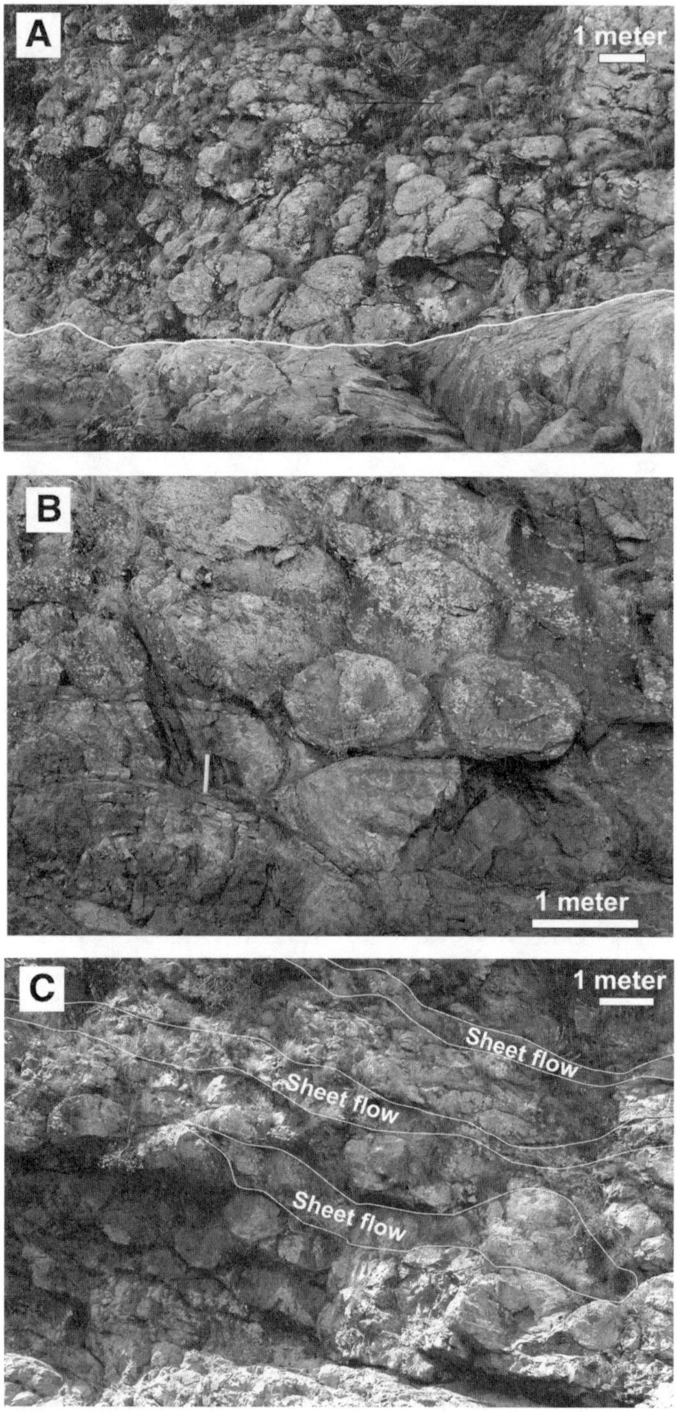

Figure 6. Pillow lavas in the northern locality of the Nhlebela Formation. (A) Well-developed pillows overlying a massive lava unit. (B) Large pillows showing well-developed multiple rims, interpillow hyaloclastite, and amygdaloidal centers. Some have radial fractures. (C) Interspersed layers of pillows and sheet flows.

(40–75 m thick) and an upper highly vesicular lava sequence (65–90 m thick). In the present context hydroclastic rocks are regarded as lava that has been fragmented in the presence of water. The hydroclastic sequence is highly variable comprising pillows, pillow fragments, breccias, and narrow lava sheets (Fig. 7A). The pillows and lava sheets are interlayered with brecciated zones comprising angular fragments ranging in size from 1–10 cm. The pillows have small spherical vesicles distributed throughout and have a strongly developed radial fracture pattern toward the margins (Fig. 7B). Pillows tend to be better developed toward the base of the sequence and can be up to 5 m in length. The pillows have thick (up to 10 cm), glassy devitrified skins with highly vesicular interiors. Multiple rinds and spalling off of chill margins are observed, which resulted from penetration of water beneath the first skin (Walker, 1992). The proportion of fragmental zones to pillows increases upward but is also very irregular. The lava sheets together with the interlayered breccia zones are similar to para-pillows (Walker, 1992) described for the northern exposure.

Passive emplacement of lavas into bodies of water requires moderate to low effusion rates, a shallow gradient, and minimum disruption in the water itself, i.e., waves (Carr and Jones, 2001). The formation of pillows represents the lowest effusion rate of lava (Gregg and Fink, 1995). Increasing slope and effusion rates result in the pillows becoming larger and flatter, eventually forming lobate flows or toes to lobate flows. The association of pillows, irregular flow lobes, and breccias in the southern area may indicate variable eruption rates and highly irregular local gradients.

The fragmental zones are commonly several meters thick and occur as distinct zones sometimes grading into elongate lava lobes with chill margins. In other cases they are highly irregular bodies up to 10 m in length. In general, the proportion of pillows to fragmental zones increases upward in the sequence, and thick zones of angular fragments of variable size commonly form a matrix or a substrate to the pillows (Fig. 7C).

A sporadically developed matrix-supported heterolithic breccia (Fig. 8A) 2–5 m in thickness is located in several areas at the same horizon close to the top of the hydroclastic unit. It comprises angular to subrounded clasts of basaltic glass, granitic, chert, and quartzite fragments and both highly altered and unaltered lava fragments ranging in size from 5–20 cm. The lack of emplacement structures and the composition of the breccia indicate that it is a lahar or a volcanic debris flow. Lahars indicate the presence of water to cause fluidization of the debris and often follow preexisting valleys, which may explain the relatively limited extent of the exposure (Fisher and Schmincke, 1984).

Transition to Vesicular Lavas

The pillows grade into flow lobes with thick chill margins and rounded surfaces (Fig. 8B) with sporadic development of marginal breccia. The transition from the lower hydroclastic sequence to the vesicular lavas is marked by a zone of several

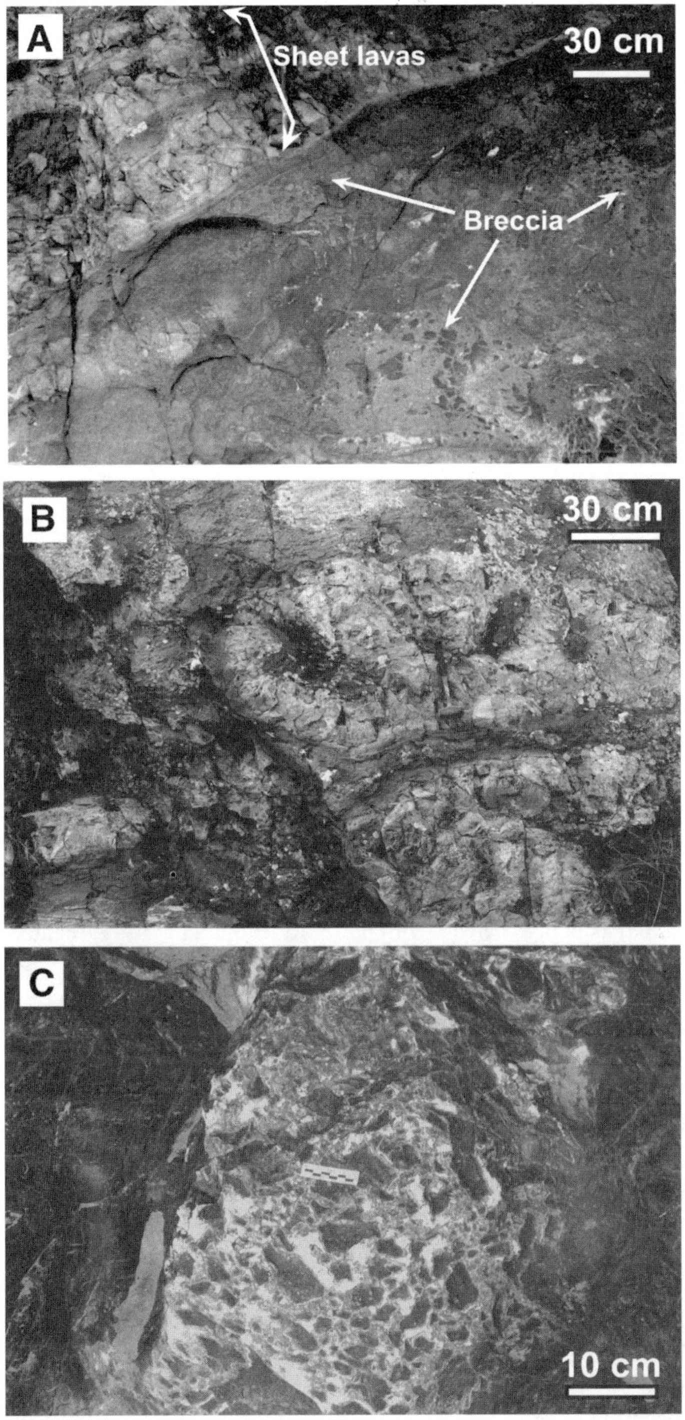

Figure 7. Hydroclastic breccias and pillow lavas in the southern locality of the Nhlebela Formation. (A) Sheet flows and breccia in contact with an andesite lava lobe. (B) Pillow lavas with well-developed radial fractures, hyaloclastite, and breccias. (C) Hydroclastic angular breccia in contact with nonbrecciated lava lobes.

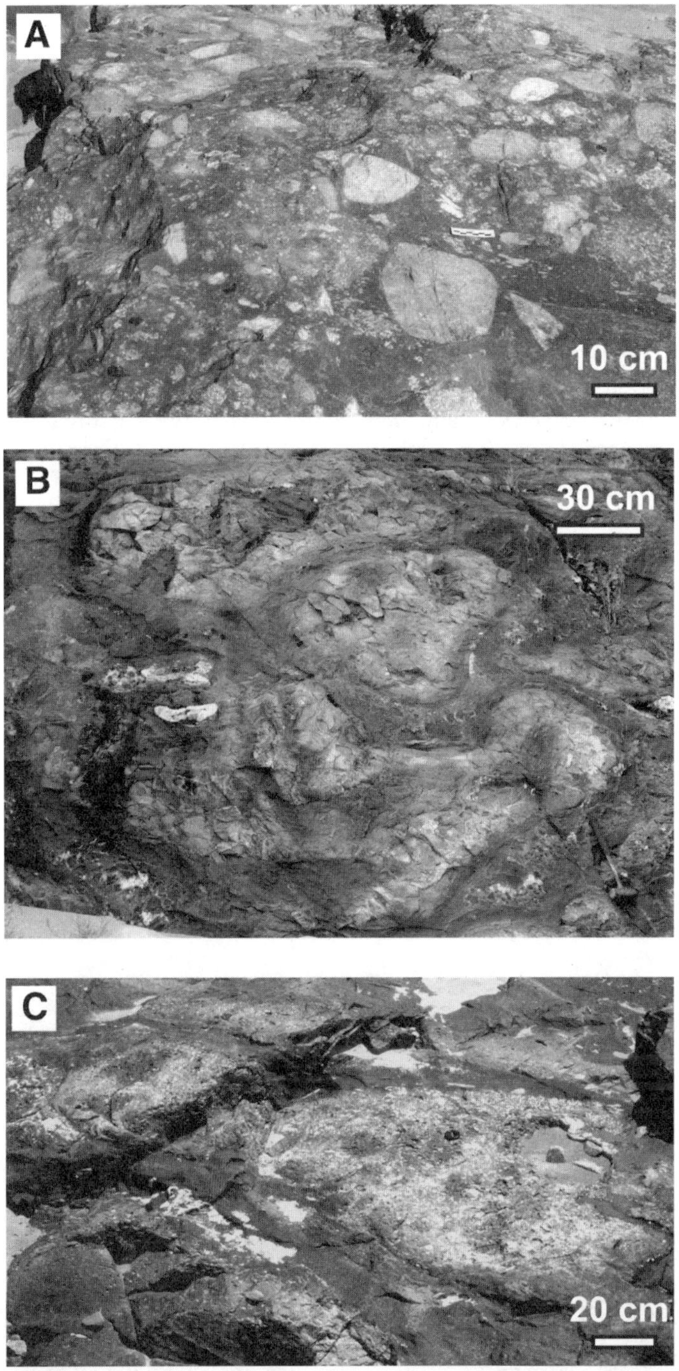

Figure 8. Rock types and structures in the southern locality of the Nhlebela Formation. (A) Heterolithic laharic breccia near the top of the hydroclastic sequence with a wide variety of rounded and angular fragments. (B) Irregularly shaped, highly vesicular lava lobes interspersed with angular hydroclastic fragments. (C) Large highly vesicular pillows with thick chill margins at the top of the hydroclastic sequence.

meters of large (3–5 m), close-packed pillows with no interpillow material (Fig. 8C). Well-developed chilled margins are developed on the pillows with only minor development of hyaloclastite. These pillows are characteristically highly vesicular.

The amygdaloidal lavas of the upper part of the Nhlebela Formation are characterized by laterally extensive flow lobes with highly vesicular flow tops and dense interiors (Fig. 9). These commonly have a lower chill zone containing small spherical vesicles. The average thickness of the flow lobes is 5–7 m, and they range in size from 2–10 m. The flow lobes have a lower chill zone containing small spherical vesicles and rare pipe vesicles with vertical jointing. In the upper part of the flows the vesicles increase in size and density and occur as spherical or large irregular gas blisters. Thicker flow lobes have larger and more irregular vesicles and gas blisters (Fig. 9). Flow surfaces tend to be smooth and undulating, classifying the lavas as pahoehoe lobes. The consistency and continuity of the flow lobes indicate emplacement onto a relatively flat eruption surface. The dense middle zone does not exhibit columnar jointing.

A significant proportion of the amygdaloidal lava sequence comprises sills (estimated to be ~30%) that are largely conformable with the flows and have variable and sporadic development of chill margins. Compositionally the sills are similar to that of the flows and are most likely an intrusive phase contemporaneous with the volcanism.

AGATHA FORMATION

The locations of Traverses 1 and 2 in the Agatha Formation and positions of samples taken are shown in Figure 3. The details of the sections through the flow sequences in the two traverses together with typical field views of the flow lobes and positions of photographs are shown in Figure 10. Distinct characteristics and styles of volcanism in the Agatha Formation allow subdivision (together with approximate thicknesses) from the base upward, as follows: the lower fragmental lavas (25 m); pahoehoe lavas of Traverse 1 (60 m); the pahoehoe lavas of Traverse 2 (55 m); the sequence of pahoehoe lavas east of Traverse 2 (30 m); and the uppermost sequence in the Agatha Formation (40 m).

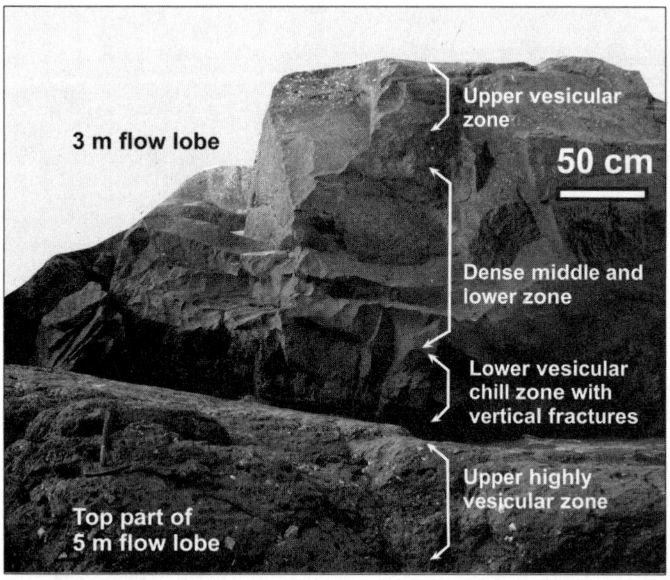

Figure 9. Pahoehoe lava flow lobes in the amygdaloidal lava sequence in the southern locality of the Nhlebela Formation. The upper and lower zones are highly vesicular, and the centers are dense and nonvesicular.

Contact with the Sediments

The basal contact of the Agatha Formation with the underlying sedimentary sequence is a zone of fragmental clasts of lava and highly rounded sand grains. In some cases the fragmental zone is only a few centimeters in thickness and separates massive lava and quartz sandstone. In a well-preserved outcrop on R34 highway (samples WM05/17 and WM05/17a; Fig. 3), the zone is several meters in thickness and comprises clasts of lava randomly dispersed within the sediment in a series of layers with differing grain sizes and compositions (relating to the proportions of sediment and mafic clasts) (Fig. 11). Juvenile volcanic clasts range in size from <3 mm to 3 cm and are typically elongated and highly irregular in form with tapered ends. Some are amygdaloidal, and rare accretionary and armored lapilli are present. The larger fragments have an imbricated arrangement indicating reworking and redistribution in an aqueous environment, although there are no distinct sedimentary structures, such as cross-bedding. The sedimentary matrix is medium to fine grained and is made up largely of rounded quartz grains together with angular and splintery quartz grains. This is interpreted as being derived from partly to completely disaggregated sediment, possibly in an explosive environment. Other fragments comprise more felsic material (possibly dacitic in composition) containing quartz crystals that cannot be linked with any other known rock type in the area.

These rocks have abnormally high concentrations of Cr (727 ppm) and Ni (558 ppm). Separated fragments of the mafic clasts have 51.7 wt% SiO_2, 10.8 wt% MgO, 0.24 wt% Cr_2O_3, and 0.18 wt% NiO (Table 1), indicating a precursor high-Mg basalt that has not been identified as a lava.

The compositions, textures, overall structure, and rock-type associations of this unit are remarkably similar to basaltic pyroclastic rocks formed as the first volcanic rocks to overlie the sedimentary succession by the onset of flood volcanism in the lower Paleogene Kangerlussuaq Basin, East Greenland (Ukstins et al., 2003). They are interpreted as subaerially deposited hydroclastic basaltic tuffs.

The Lower Fragmental Lavas of Traverse 1

The basal flow lobes (units 1–4 in Fig. 10C) are poorly defined, highly fragmented, and are attributed to hydroclastic interaction of lava and water. They are most likely associated

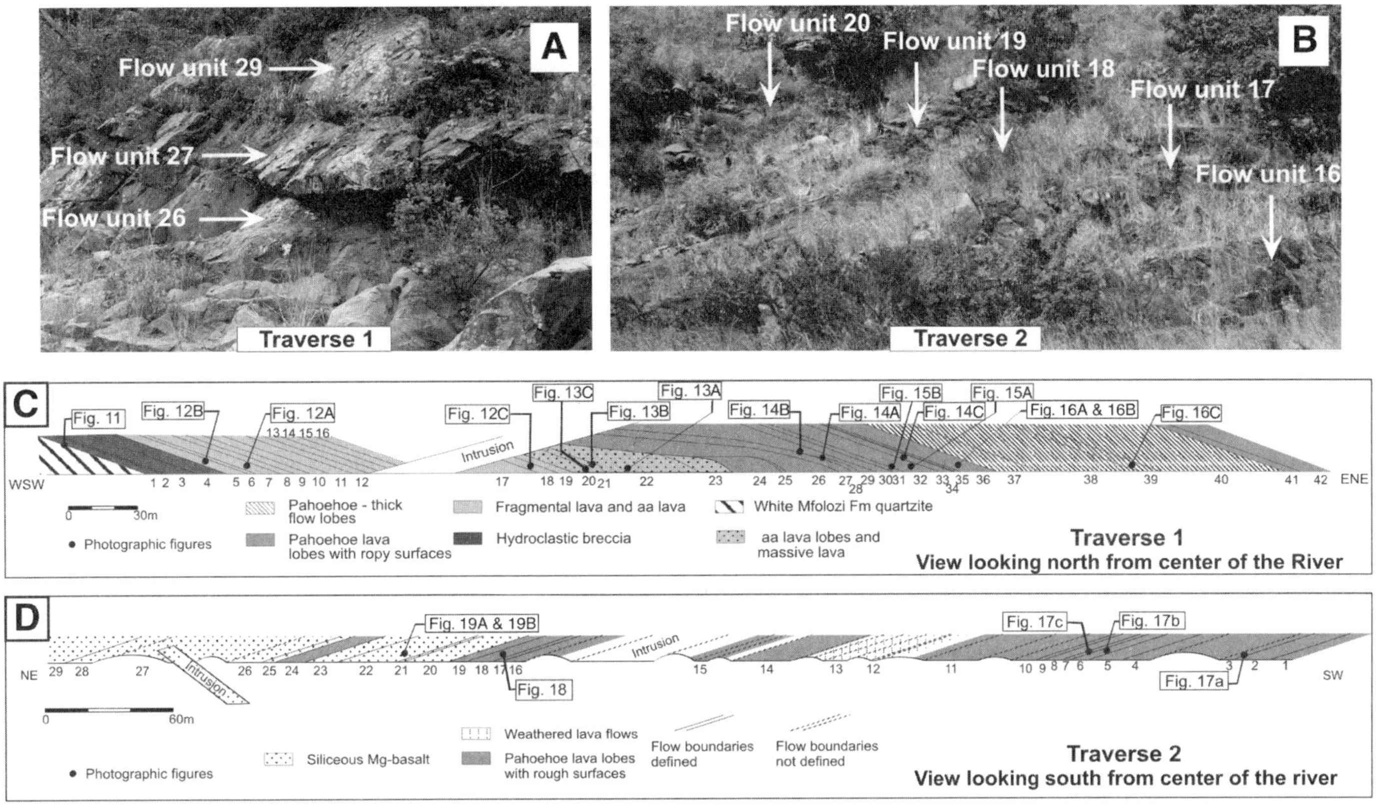

Figure 10. Diagrammatic sections and field views of the lava lobes studied in Traverses 1 and 2 in the Agatha Formation. The positions of the traverses are shown in Figure 3. Labels and positions refer to figure numbers of field photographs. The individual lava lobes are numbered. (A) Flow lobes of units 26–29 from Traverse 1, looking toward the north. (B) Flow lobes of units 16–20 in Traverse 2, looking toward the south. (C) View looking north showing the lower section of the Agatha Formation and the lava lobes encountered. (D) View looking south showing the lava lobes encountered. The siliceous Mg-rich basalt occurs on the northeastern section. It is associated with a dyke intrusion of similar composition.

with the hydromagmatic basaltic tuff overlying the sedimentary unit. There is no suggestion of pillow formation. Some of the fragments appear to have been leached, and the altered margins were the result of reaction with water. The overlying sequence of lava flow lobes (units 5–23) is 4–10 m thick, highly irregular in form, and is made up of lava fragments and breccia together with massive or nonfragmental components. The flow lobes exhibit typical aa lava structure with massive interiors and rough fragmented flow tops (Cas and Wright, 1987). Blocks of vesicular lava range in size from 2 cm to over 1 m in length. Fragments of vesicular lava and chill margins are recognizable, and magma squeezed between the blocks appears as lava ribbons (Fig. 12A). Other features that are typical of aa lavas include columnar jointing in the massive interiors of the flow units and irregular gas blisters.

In this sequence, some aa flow lobes are entirely fragmental but others have a sharp contact with the massive parts of the lobes (Figs. 12B and 12C) with no evidence of chill margins. The fragments include crenulated and ropy pahoehoe surfaces, smooth chills, and compound welded fragments from previously formed fragmental flows (Fig. 13A). Lava flowage continued below the surface of fragmental flows in places, and lava

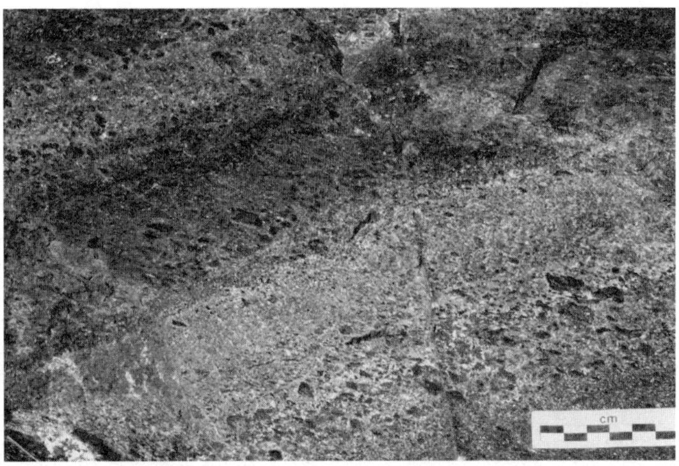

Figure 11. Hydroclastic tuff marked as sample position WM05/17 on the R34 highway in Figure 3. The tuff has been reworked to give a strongly imbricated fabric of juvenile angular mafic fragments and quartz sand grains. The mafic fragments are high-Mg basalt.

TABLE 1. CHEMICAL ANALYSES OF THE VOLCANIC ROCKS OF THE WHITE MFOLOZI INLIER

	Nhlebela Formation Northern and Southern Loacilities									Agatha Formation West Section				
	CG02/205	CG02/206	CG02/210	CG02/211	CG02/212	CG02/213	WM05/9	WM05/10	WM05/14	WM2/200C	WM3/2000	WM4/2000	WM5/2000	WM6/2000
	Flow int	Flow int	Chill	Chill	Flow int	Flow int	chill	dyke	dyke chill	Flow int	Flow int	Flow int	Flow int	Flow int
Major Elements wt%														
SiO_2	61.06	64.87	52.73	65.18	59.82	63.24	58.61	55.50	59.91	52.80	47.85	49.19	50.19	49.13
Al_2O_3	13.88	12.76	12.89	14.11	14.50	13.81	14.82	14.40	14.06	14.70	15.21	15.81	16.37	16.23
FeO*	9.80	10.33	14.61	8.27	9.98	9.45	9.52	12.21	9.48	14.22	16.01	12.58	12.83	12.27
MnO	0.13	0.12	0.22	0.10	0.14	0.12	0.16	0.19	0.14	0.17	0.22	0.19	0.20	0.24
MgO	2.24	3.77	4.21	1.76	3.71	2.47	4.25	5.24	3.00	4.32	4.66	3.64	4.96	4.27
CaO	5.97	2.52	8.66	2.44	4.86	3.18	6.38	7.90	7.04	8.16	9.87	10.87	8.22	10.62
Na_2O	4.92	3.59	1.97	2.82	4.52	4.21	2.73	3.69	2.71	1.95	3.29	4.50	4.39	3.75
K_2O	0.81	0.90	1.59	3.94	1.22	2.39	2.34	0.05	2.47	1.19	0.18	0.33	0.40	1.18
TiO_2	0.85	0.79	2.26	0.84	0.89	0.86	0.96	0.84	1.00	1.61	1.66	1.71	1.59	1.54
P_2O_5	0.16	0.13	0.35	0.16	0.16	0.14	0.20	0.13	0.21	0.29	0.30	0.30	0.28	0.26
Total#	99.82	99.78	99.49	99.61	99.81	99.87	99.98	100.15	100.02	99.39	99.22	99.09	99.40	99.47
Trace Elements ppm														
Rb	30	19	61	136	42	36	49	1	75	36	3	6	6	37
Zr	228	215	178	232	238	228	188	133	203	194	186	193	187	176
Y	32	31	57	30	33	30	41	28	45	38	34	34	33	29
Ni	68	67	43	58	73	69	30	111	16	89	88	81	107	88
Cr	17	16	32	19	15	16	6	64	3	39	47	45	149	112
Nb	16.23	14.52	15.17	16.29	16.43	16.38	14.46	7.91	16.48	9.22	8.80	9.16	8.85	8.86
La	37.17	33.87	30.85	35.58	32.41	30.56	36.60	20.42	41.68	24.98	14.94	18.85	17.6	15.29
Ce	73.37	65.91	66.43	67.87	67.07	58.66	73.09	39.55	82.37	48.18	32.12	39.12	38.19	30.51
Pr	8.29	7.60	8.35	7.53	7.96	6.52	8.44	4.50	9.41	6.23	4.55	5.18	5.32	4.14
Nd	28.54	27.31	33.30	25.18	28.94	22.82	30.70	16.41	33.75	23.60	18.64	20.06	20.70	17.12
Sm	6.24	5.98	7.97	5.72	6.16	5.35	6.72	3.66	7.22	5.84	4.98	4.96	5.28	4.14
Eu	1.57	1.47	2.04	1.28	1.51	1.27	1.55	0.93	1.62	2.03	1.06	0.89	2.94	1.16
Gd	6.46	6.30	8.91	5.84	6.30	5.54	7.09	4.47	7.64	7.39	6.59	6.41	6.77	4.98
Tb	0.96	0.93	1.47	0.89	0.95	0.85	1.11	0.73	1.19	1.05	0.98	0.95	0.97	0.72
Dy	5.49	5.34	9.18	5.21	5.48	5.05	6.75	4.58	7.19	5.79	5.38	5.27	5.14	4.10
Ho	1.09	1.06	1.92	1.04	1.09	1.03	1.41	0.96	1.51	1.11	1.05	1.04	0.99	0.87
Er	2.97	2.89	5.42	2.86	3.04	2.84	4.00	2.64	4.28	3.02	2.92	2.87	2.72	2.54
Tm	0.45	0.43	0.84	0.43	0.46	0.43	0.62	0.40	0.66	0.42	0.40	0.40	0.38	0.36
Yb	2.73	2.65	5.23	2.56	2.85	2.67	3.84	2.44	4.14	2.51	2.49	2.50	2.38	2.35
Lu	0.41	0.40	0.77	0.38	0.42	0.39	0.54	0.36	0.57	0.35	0.34	0.35	0.33	0.34
Th	12.62	11.87	6.49	12.97	13.49	12.93	8.82	6.33	10.33	2.23	2.21	2.29	2.70	2.62
U	3.24	3.05	1.87	3.30	3.44	3.27	1.97	1.56	2.38	0.34	0.34	0.35	0.44	0.44

(continued)

TABLE 1. CHEMICAL ANALYSES OF THE VOLCANIC ROCKS OF THE WHITE MFOLOZI INLIER (*continued*)

	Agatha Formation East Section											
	WM10/2000 Chill	WM11/2000 Chill	WM12/2000 Chill	WM13/2000 Chill	WM15/2000 Chill	WM16/2000 Chill	WM17/2000 Chill	WM05/1 Chill	WM05/2 Chill	WM05/3 Chill	WM05/4 dyke	WM05/5 Chill
Major Elements wt%												
SiO_2	49.11	54.14	55.41	54.16	51.58	53.76	52.03	53.73	54.62	54.91	52.26	55.08
Al_2O_3	15.44	15.02	14.72	14.94	15.64	14.17	14.31	14.79	14.59	14.83	14.72	16.89
FeO*	15.23	11.08	11.16	10.96	14.23	11.18	11.11	13.32	11.00	10.83	11.04	11.02
MnO	0.23	0.18	0.17	0.17	0.21	0.17	0.20	0.15	0.16	0.17	0.21	0.25
MgO	4.65	5.79	6.26	6.45	4.88	6.30	6.50	3.50	6.75	6.07	6.68	4.49
CaO	10.35	8.77	8.01	8.09	7.04	9.25	10.65	8.81	8.38	9.26	10.66	7.29
Na_2O	3.35	3.65	3.14	2.95	3.83	2.24	2.35	2.54	2.34	2.08	2.51	0.61
K_2O	0.02	0.02	0.01	0.58	0.36	1.06	0.83	1.06	0.99	0.85	0.76	2.92
TiO_2	1.15	0.89	0.87	0.82	1.20	0.88	1.08	1.55	0.89	0.91	1.08	1.04
P_2O_5	0.19	0.17	0.16	0.16	0.21	0.17	0.18	0.28	0.16	0.18	0.18	0.18
Total#	99.69	99.68	99.90	99.28	99.17	99.15	99.23	99.74	99.88	100.08	100.10	99.78
Trace Elements ppm												
Rb	1	1	1	10	4	11	19	27	12	12	19	85
Zr	126	123	120	122	142	129	104	119	92	90	106	122
Y	26	23	21	23	26	24	29	37	23	23	29	22
Ni	147	101	106	108	112	117	55	116	126	116	70	103
Cr	139	212	216	233	146	243	238	125	250	217	328	69
Nb	5.74	5.44	5.27	5.42	6.64	5.66	6.99	10.16	6.27	6.62	7.71	6.98
La	9.04	10.85	12.19	12.67	11.47	13.30	9.50	23.92	15.94	16.55	10.68	15.00
Ce	18.82	23.10	25.00	26.24	25.59	27.37	20.45	52.09	34.06	35.61	24.15	32.00
Pr	2.53	2.95	3.17	3.30	3.38	3.44	2.72	6.65	4.26	4.43	3.25	3.89
Nd	10.43	11.72	12.49	13.05	13.67	13.54	11.42	27.32	17.02	17.76	13.93	15.49
Sm	2.81	2.83	2.88	3.15	3.49	3.24	3.08	6.53	3.94	4.07	3.61	4.15
Eu	0.73	0.93	1.09	1.10	1.04	1.17	1.04	2.07	1.31	1.36	1.15	1.34
Gd	3.89	3.67	3.68	3.87	4.48	4.02	4.14	7.09	4.24	4.39	4.06	3.94
Tb	0.64	0.56	0.56	0.58	0.70	0.60	0.66	1.11	0.66	0.69	0.71	0.62
Dy	3.91	3.31	3.22	3.37	4.08	3.49	4.08	6.52	3.92	4.07	4.47	3.73
Ho	0.81	0.67	0.64	0.66	0.81	0.70	0.86	1.28	0.79	0.82	0.95	0.77
Er	2.29	1.88	1.79	1.88	2.26	1.96	2.48	3.36	2.13	2.21	2.62	2.16
Tm	0.32	0.26	0.25	0.27	0.32	0.27	0.36	0.48	0.32	0.33	0.41	0.32
Yb	2.04	1.66	1.58	1.68	2.00	1.73	2.37	2.87	1.98	2.02	2.64	2.10
Lu	0.29	0.23	0.22	0.23	0.27	0.23	0.35	0.37	0.27	0.28	0.40	0.31
Th	1.41	1.04	1.00	1.05	1.57	1.07	1.41	1.79	1.05	1.10	1.36	2.51
U	0.29	0.19	0.23	0.18	0.24	0.18	0.31	0.32	0.20	0.20	0.33	0.60

(*continued*)

TABLE 1. CHEMICAL ANALYSES OF THE VOLCANIC ROCKS OF THE WHITE MFOLOZI INLIER (*continued*)

	Agatha Formation				
	Upper Section			Hydroclastic Tuff	
	WM05/6 Chill	WM05/7 Chill	WM05/8 Chill	WM05/17 Tuff	WM05/17a Mafic clast
Major Elements wt%					
SiO_2	53.73	52.09	54.57	58.61	51.64
Al_2O_3	15.89	16.23	14.80	14.82	14.18
FeO*	11.06	11.57	10.93	9.52	18.17
MnO	0.21	0.26	0.18	0.16	0.14
MgO	5.53	5.80	6.08	4.25	10.78
CaO	8.11	8.93	8.61	6.38	1.26
Na_2O	3.80	3.55	2.81	2.73	1.41
K_2O	0.29	0.36	0.87	2.34	0.00
TiO_2	0.99	1.02	0.89	0.96	1.57
P_2O_5	0.16	0.17	0.17	0.20	0.07
Total#	99.77	99.98	99.91	99.98	99.21
Trace Elements ppm					
Rb	9	12	17	1	
Zr	120	121	94	48	
Y	19	18	22	9	
Ni	105	106	112	559	1565
Cr	72	73	213	728	2580
Nb	6.19	6.29	6.17	3.30	
La	17.10	17.42	15.37	3.71	
Ce	35.09	37.83	32.45	8.11	
Pr	4.36	4.43	4.07	1.07	
Nd	17.33	16.53	16.48	4.77	
Sm	3.98	3.98	3.73	1.32	
Eu	1.31	1.21	1.21	0.41	
Gd	4.02	3.86	4.06	1.64	
Tb	0.57	0.52	0.64	0.27	
Dy	3.21	2.89	3.79	1.63	
Ho	0.66	0.60	0.77	0.30	
Er	1.90	1.81	2.09	0.79	
Tm	0.28	0.28	0.31	0.12	
Yb	1.97	1.84	1.91	0.72	
Lu	0.30	0.28	0.27	0.10	
Th	2.12	2.25	1.06	2.47	
U	0.56	0.58	0.19	0.82	

Note: Flow int—interior of flow lobe; Total#—anhydrous total.
*FeO—total Fe as FeO

Figure 12. Lower aa volcanic section of Traverse 1 in the Agatha Formation (see Fig. 10 for positions). (A) Ribbons of lava in fragments of aa lava. (B) Contact of aa lava and massive vesicular lava. (C). Boundary between aa and massive lava.

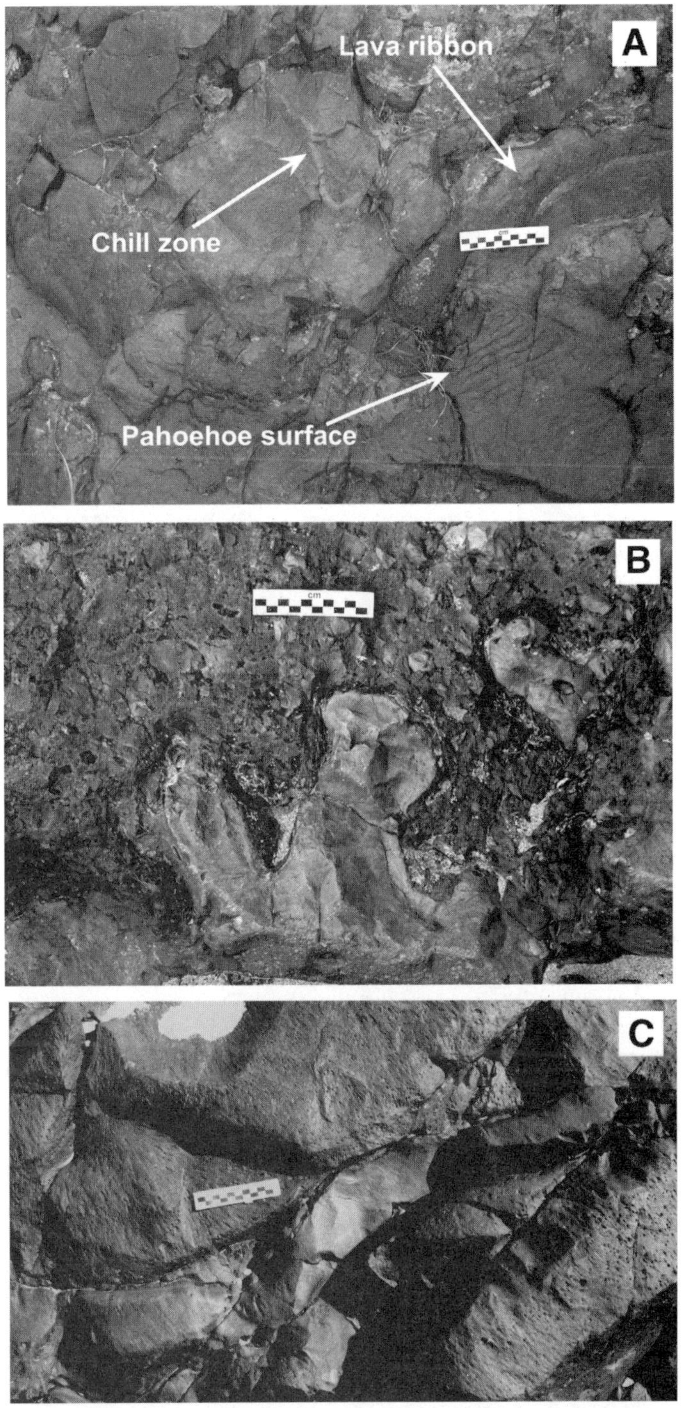

Figure 13. Lower aa volcanic section of Traverse 1 in the Agatha Formation (see Fig. 10 for positions). (A) Fragments of lava with chills, vesicles, and pahoehoe surfaces. (B) Lava diapirs at base of aa flow lobe. (C) Narrow feeder dyke breaking through fragmental and amygdaloidal lava.

penetrated upward between the fragments (Fig. 13B). Penetration of solidified fragmental lavas also took place through narrow dykes (Fig. 13C) that worked around the aa fragments.

The highly irregular form of the flow surface of unit 20 (Fig. 10C) suggests that the lava flowed over a relatively steep surface and thereby experienced higher rates of shear, allowing the development of aa lavas. Other critical controls on the development of aa lava include effusion rates, effective distance from the source, temperature, composition, crystallinity, and volatile content (Cashman et al., 1999). Observations of aa lava indicate that it tends to form early in the eruptive sequencer when effusion rates are high (Macdonald, 1953). The development of the aa lava sequence in the lower part of the Agatha Formation may therefore also be indicative of high effusion rates.

Pahoehoe Lava Flows—Traverse 1

The first occurrence in the stratigraphic succession of a well-developed pahoehoe lava lobe is in flow unit 20 (Fig. 10). This is, however, not taken as the top of the fragmental lava zone as several fragmental units overlie it and indicate a change in eruption style from aa to pahoehoe. In flow unit 22 there are angular fragments with crenulated and ropy pahoehoe surfaces (Fig. 13A), indicating the prior formation of pahoehoe lava. Flow 19 is massive lava overlying a fragmental flow 18 with a sharp contact. The change from aa to pahoehoe may be due to a drop in effusion rate from the erupting volcanic center or may represent the interaction of flow lobes of different flows.

The transition from fragmental lavas to pahoehoe lobes with ropy surfaces is not clearly defined, but a crenulated (not ropy) surface and a well-developed chill zone with cooling joints, typical of pahoehoe, is evident for unit 24 (Figs. 14A and 14B). The surface of this 5-m-thick lobe is underlain by a jointed zone 15-cm-thick chill in turn overlying a highly vesicular zone in which many of the vesicles are elongated in the plane of the flow lobe. Several zones of vesicles with distinctive forms and distributions provide evidence for the emplacement mechanism of the flow lobe. The presence of highly vesicular zones within the lower flow lobes (units 24 and 25 in Fig. 10) and the dominance of small, regular, spheroidal vesicles are typical of pahoehoe lava, which may contain 20%–50% vesicles. In some cases an accumulation of fine fragmental material is present at the boundary between two flow lobes (Fig. 14B), but in most cases the contact is knife sharp. The structure of these flow lobes with a well-defined upper crust, core, and basal crust, the thicknesses of these zones, as well as the nature and distribution of vesicles and fractures are identical to pahoehoe lava formed by flow inflation in the Columbia River Basalts (Self et al., 1997).

The overlying flow lobes are clearly pahoehoe with beautiful ropy surfaces as seen on Flow unit 33 (Fig. 14C). Intersecting ropy surfaces of flow lobes are common together with clefts resulting from rupturing of the chill zone (Fig. 15A) and breakouts from the molten interiors of the flow lobes. While flow

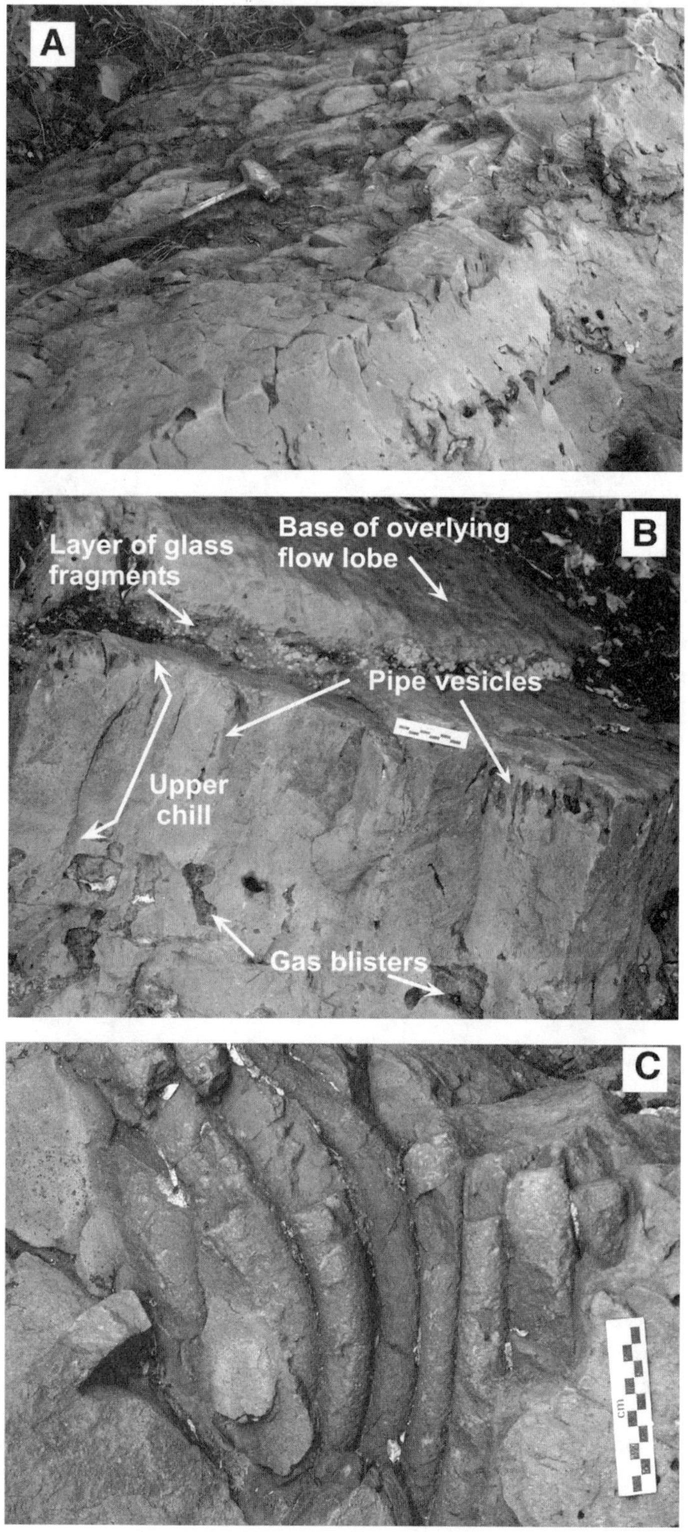

Figure 14. Pahoehoe lava succession in Traverse 1 in the Agatha Formation (see Fig. 10 for positions). (A) Crenulated pahoehoe surface with jointed upper crust and chill zone underlain by large coalesced vesicles. (B) Top of pahoehoe lava lobe with pipe vesicles in the upper chill underlain by large gas blisters. The overlying flow is separated by a narrow zone of (originally) spalled glass fragments. (C) Ropy lava surface on pahoehoe flow lobe.

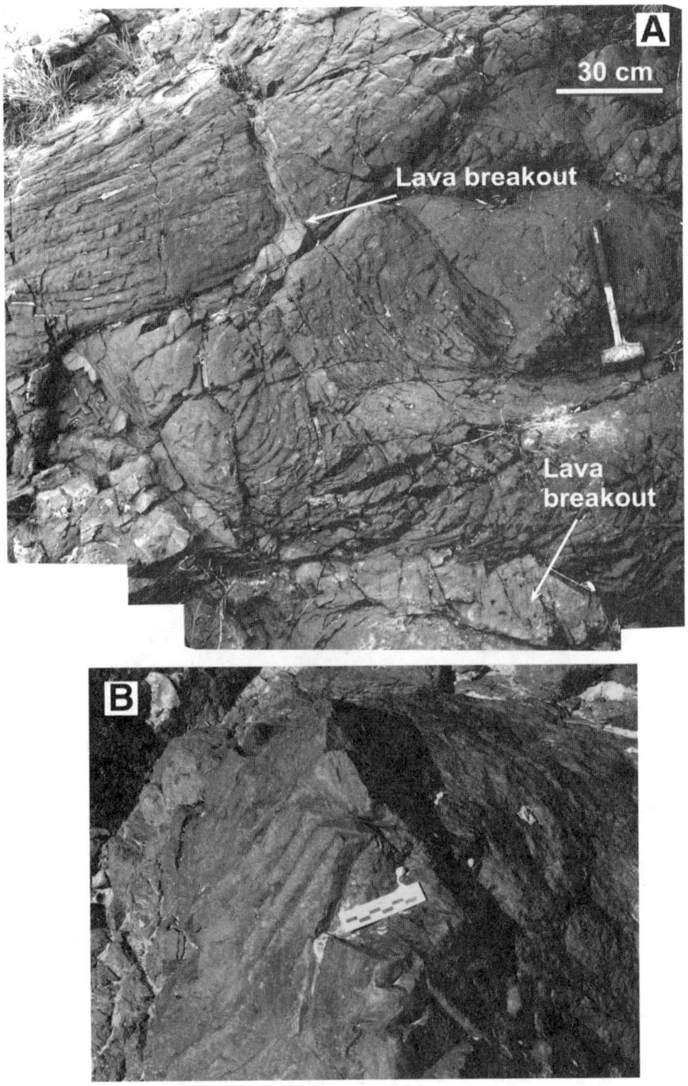

Figure 15. Pahoehoe lava succession in Traverse 1 in the Agatha Formation (see Fig. 10 for positions). (A) Intersection of ropy flow surfaces and vesicular lava breakouts in pull-apart structures. (B) Dislodged block of lava with a ropy surface formed as a tumulus.

directions can be determined by using the convex trains of the ropes, the extent of the exposed surfaces is not sufficient for paleoflow evaluation, as some pahoehoe ropes may have formed due to localized eddies on the flow surface (Cas and Wright, 1987). In the same local area (close to unit 33), hummocky blocks are similar to tumuli described from many recent pahoehoe fields. In one example (Fig. 15B) the in situ block has a well-developed ropy surface that has been displaced at a marked angle to the enclosing flow lobes and exposes a cross section through the upper part of the lobe.

The lava flows develop a compound structure upward from unit 24 (Fig. 10C). The complexity of the relationships between the lava lobes increases from this point to unit 36. This may indicate a decrease in effusion rate, a decrease in viscosity, and possibly a combination of both factors. The surface structures of the underlying flow are commonly cast into the base of the overlying lava lobes (Fig. 16A). The surface textures are remarkably well preserved, which suggests that even if the effusion rate from the vent decreased, it did not reduce sufficiently to expose flow surfaces for sufficiently long periods of time to allow weathering to take effect (Cas and Wright, 1987). However, in this regard climate plays a significant role with pahoehoe surfaces being well preserved in arid areas even after prolonged exposure (J. Marsh, 2005, personal commun.). The top surface of the lobe has a rupture (Fig. 16B), filled in by the overlying flow, the base of which has a very well pronounced jointed chill.

The thickness of the pahoehoe flow lobes varies from <1 m to 4 m. Two flow units (units 26 and 32 in Fig. 10E) are ~8–10 m thick. The thicker flow lobes in general have vesicles in their interiors, which are elongated in the plane of the flow lobe. Ropy pahoehoe surfaces are the dominant forms in units 24–36 of Traverse 1. The thickness of the lava lobes tends to increase higher in the succession (units 37–40 in Fig. 10C) with some being greater than 20 m. While the quality of the exposures deteriorates in this region, the lobes are generally dense with small and relatively few vesicles and strongly developed columnar jointing. A smooth glassy lava toe overlying a pahoehoe surface is preserved in unit 39 (Figs. 10C and 15C). The thick overlying flow is vesicle free at the base and has well-developed columnar joints perpendicular to the flow surface. Large vesicles occur toward the upper surface of the flow lobe.

Pahoehoe Lavas—Lower Sequence of Traverse 2

Traverse 2, on the south bank of the White Mfolozi River, represents a section approximately midway through the Agatha Formation (Figs. 3 and 10D). The flow lobes are pahoehoe. Fragmental and aa lava are not present. Pahoehoe flow lobes are evident in flow units 1–19 (Fig. 10D) but do not have extensive development of ropy surfaces seen in the lower flows of Traverse 1. Flow lobes are 3–10 m thick (Fig. 10B), and contacts tend to be smooth and planar. Small spherical vesicles occur close to the base of the flow lobes, and large vesicles and gas blisters, commonly elongated in the plane of the flow, occur toward the top.

The flow tops of the pahoehoe lava lobes are generally marked by a sudden increase in the content of the vesicles (Fig. 17A) and typically contain two distinct size populations of vesicles. Small, spherical, densely packed vesicles are less than 3 mm in diameter. Large gas blisters 5–20 cm in the long axis are concentrated below the base of the upper chill zones. There is a decrease in vesicle size upward toward the flow surface. Small spherical vesicles dominate the flow surface. Concentrations of small vesicles are so great on certain flow surfaces that they must have been frothy prior to solidification. Some flow surfaces higher in the sequence from unit 5 resemble ripply pahoehoe (Kilburn, 1990) or rough and toothpaste pahoehoe (Fig. 17B; Rowland and Walker, 1988). Rather poorly developed ropy sur-

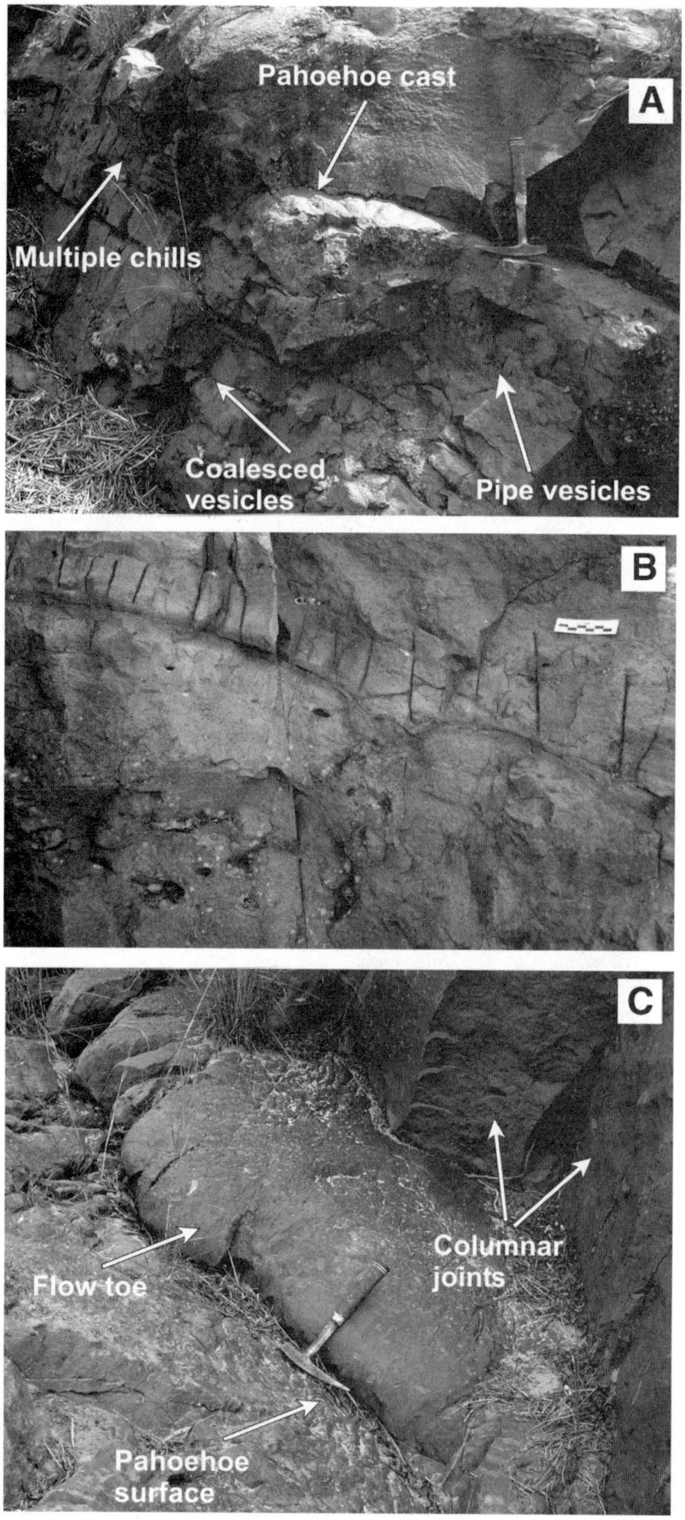

Figure 16. Pahoehoe lava succession in Traverse 1 in the Agatha Formation (see Fig. 10 for positions). (A) Base of lava lobe molded onto a ropy pahoehoe surface. The flow lobe has multiple chills and pipe vesicles. (B) Jointed lower chill with cooling joints molded into cleft caused by a pull-apart of the underlying chill. (C) Lava toe over pahoehoe surface and overlain by glassy massive flow with well-developed vertical columnar joints.

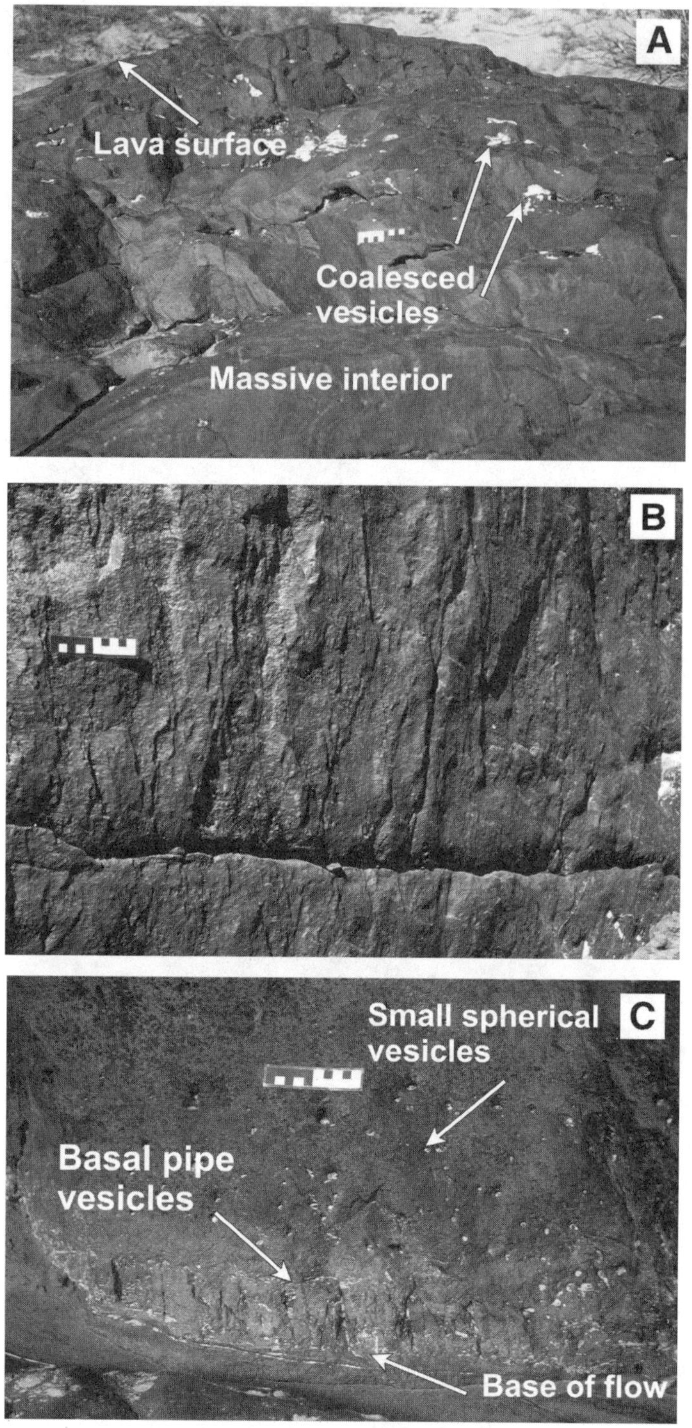

Figure 17. Pahoehoe lava succession in Traverse 2 in the Agatha Formation (see Fig. 10 for positions). (A) Massive interior of flow lobes with large gas blisters trapped in the upper zone below the chill. (B) Rough pahoehoe surface texture resulting from increased viscosity. (C) Small pipe vesicles at the base of the flow lobe and small spherical vesicles in the interior.

faces are observed in isolated occurrences within the rough or ripply pahoehoe and toothpaste lava and can be seen on the surfaces in ~10% of the pahoehoe-type lava flow units in this section. These may have been squeeze-ups that pushed through the surface of the tearing rough/ripply pahoehoe and toothpaste lava flow tops. Small irregularly shaped vesicles (1–3 mm diameter) occur in the thin crust. These features suggest an increase in viscosity and the onset of aa lava development. Where the bases of the flows are exposed the lower portion is commonly vesicular. These vesicles are small and spherical (<2 mm diameter), and small pipe vesicles are ~2 mm wide and 2–6 cm long (Fig. 17C).

Flow 18 (Fig. 10D) exhibits large vesicles that are not present in any of the other flow lobes. The vesicles are large: 15–20 cm wide and 10–15 cm high (Fig. 18). The tops of the vesicles are convex while the bottom surface is concave. Migration trails are evident at the widest points of the vesicles. These migration trails are generally fractures and thin elongated cavities that formed perpendicular to the surface of the lava flow and are interpreted as large gas bubbles that rose through the liquid. Similar structures, termed domed vesicles, are observed in the Cohasset Flow of the Grande Ronde Basalt in south-central Washington, United States (McMillan et al., 1987). These domed vesicles are thought to have formed by relatively few nuclei that experienced exceptional growth by accumulation of smaller gas bubbles in a relatively viscous lava.

Pahoehoe Lavas—Upper Sequence of Traverse 2

Units 20–29 in Traverse 2 and the upper section of the Agatha Formation (Fig. 10D) have well-developed flow lobes with planar flow contacts. They have marked conchoidal fracture and undulatory flow surfaces with a rugged texture that resembles rough or toothpaste pahoehoe rather than ropy surfaces (poorly developed ropy surfaces are present in only very rare cases). On Hawaii, rough pahoehoe surface forms from lava slightly more viscous than that forming ropy pahoehoe (Rowland and Walker, 1988). Toothpaste lava forms from even more viscous lava. The surface of toothpaste lava is covered with spines and longitudinal grooves and ridges. While a relatively high viscosity is required to form toothpaste lava, the dynamic property that distinguishes it from forming aa lava is the slow flow rate of the lava.

Figure 18. Pahoehoe lava succession in Traverse 2 in the Agatha Formation (see Fig. 10 for positions). Coalesced domed vesicles with migration trails.

In this area thin elongate and rounded flow lobes up to 30 cm in thickness or diameter are molded into each other (Fig. 19A). In many cases the lobes taper to give lava toes. The surfaces of lobes that are longer than 10 m are characteristically highly planar. Compound chills derived from each lobe are well developed (Fig. 19B). Thicker lobes in which vesicles are developed in the upper portions overlie the rounded and bulbous flow lobes. The bulbous lobes resemble pillows in cross section but lack other diagnostic features of pillows such as radial features, hyaloclastite, and hydroclastic breccias. These structures are particularly well developed in flow units 21, 22, and 25 (Fig. 10D).

The characteristics of units 20–29 in Traverse 2 may be attributed to a change in eruptive mechanism, effusion rates, temperature, and nature of the substrate. However, magma composition, which strongly influences viscosity, could also have been a major contributing factor, and it will be shown that these rocks are more siliceous (52–55 wt% SiO_2) compared with that of the underlying lavas (48%–53%). While these flows can clearly be distinguished in the excellent outcrop, there is also a dykelike feature of similar siliceous composition that crosscuts the underlying flows and appears to be a feeder to the flows (Fig. 10).

Upper Section of the Agatha Formation

The uppermost section of the Agatha Formation (Fig. 3) has well-defined flow lobes and flow toes. The thickness of the flow lobes is 3–5 m, and flow lobes are characterized by well-defined chills 2–5 cm thick. The surface contacts between the flow lobes are characteristically planar. Flow lobes are bulbous in section and mold onto the planar surface of the underlying lobes (Fig. 20A). These lavas are highly vesicular, with very well developed pipe and cylindrical vesicles (Fig. 20B) and therefore contrast with flow lobes in the underlying sequence. Characteristics of the vesicles are described in the following section.

An interesting structural feature is observed in the flow lobes on approaching the contact with the overlying Mozaan Group. The angle of the flow contacts increases progressively toward the contact over a vertical distance of ~30 m. In some cases the dip is as much as 70° to the north. The pipe vesicles in these cases are reoriented close to the horizontal as a result of drag deformation resulting from low-angle thrusting of the Mozaan Group on the contact with the Nzuse lavas.

Vesicle Types and Distribution in the Agatha Formation

Vesicle types and distributions are important because they provide evidence of emplacement and flow mechanisms as well as rates of effusion of the lavas and the cooling history. A wide

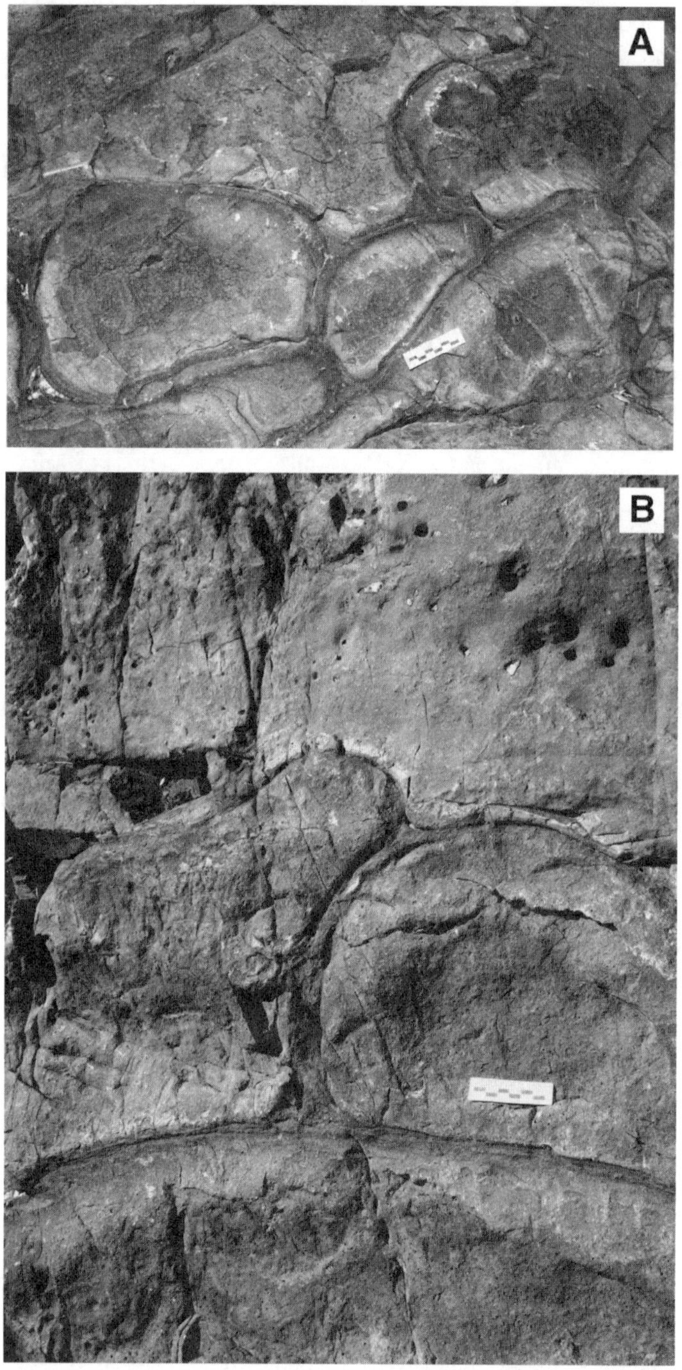

Figure 19. Pahoehoe lava in the upper part of Traverse 2 in the Agatha Formation (see Fig. 10 for positions). (A) Molded pahoehoe flow lobes overlain by a laterally extensive flow lobe. These lobes have concentric rather than radial structures and are not pillows. (B) Pahoehoe lobes with thin chills overlain by an extensive vesicular flow lobe.

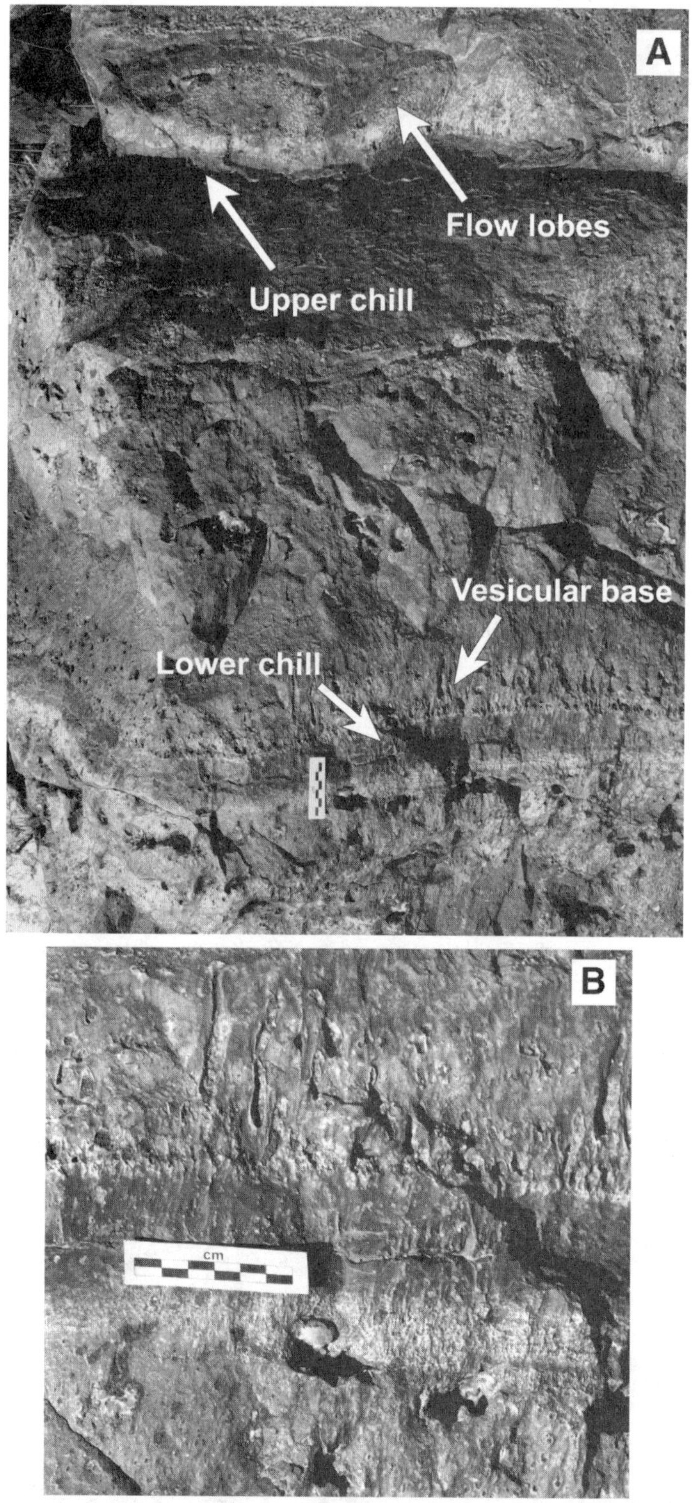

Figure 20. Highly vesicular flow lobes in the uppermost section of the Agatha formation (see Fig. 3). (A) Flow lobe with lower chill and pipe vesicles and a planar upper surface overlain by flow toes. (B) Detail of pipe vesicles above the lower chill margin.

variety of vesicle types are observed in the flows of the Agatha Formation. Three of the main types are described below and relate to other macrofeatures previously described.

One of the best-preserved sections giving a three-dimensional view is near the base of the pahoehoe section in Traverse 1 in flow unit 29 (Figs. 14A and 14B). The flow lobe consists of an upper chill with well-developed subvertical fractures below a crenulated pahoehoe surface (Fig. 21). The chill is 30 cm thick and has cylindrical or pipe vesicles in the upper 10–15 cm. It is most unusual for cylindrical vesicles to occur in upper chill zones. Gas blisters up to 20 cm in length are developed below the chill and extend for 50–70 cm into the flow. In the interior the gas blisters decrease in abundance but increase in size to nearly 40 cm in length. The midzone is massive and relatively free of vesicles except for a narrow band 10–15 cm wide. The lower part of the flow has small vesicles developed above a narrow chill. There are no pipe vesicles at the base of this flow, but the overlying flow of similar structure and thickness has sparsely developed pipe vesicles (Fig. 21). Columnar joints are present in the lower part of the flow, extend for 50–100 cm, and terminate at a level marked by a layer of relatively coarse spherical vesicles.

Flow unit 38 in Traverse 1 (Fig. 10C) is ~20 m thick with large vesicles (but no large gas blisters) developed in the upper portions. The large vesicles are highly elongated in the plane of the flow and tend to be concentrated in specific horizons. Bands of smaller spherical to slightly elongate vesicles form trains and are also concentrated in the flow plane.

In contrast to the sparse development of pipe vesicles in the lower parts of the Agatha Formation, in the uppermost section of the Agatha Formation the flow lobes are characterized by extremely well formed pipe and cylindrical vesicles, commonly as multiple zones. In a typical example (Fig. 22A) the lower chill margin has a high percentage (estimated at ~30%–40%) of very small vesicles (<0.5 mm in diameter) indicating that the magma was degassing at the time of the emplacement. This example has 4–5 zones of pipe and cylindrical vesicles separated by narrow selvages of vesicle-deficient lava. Immediately above the chill margin and in contact with it, a zone of stubby vesicles is present

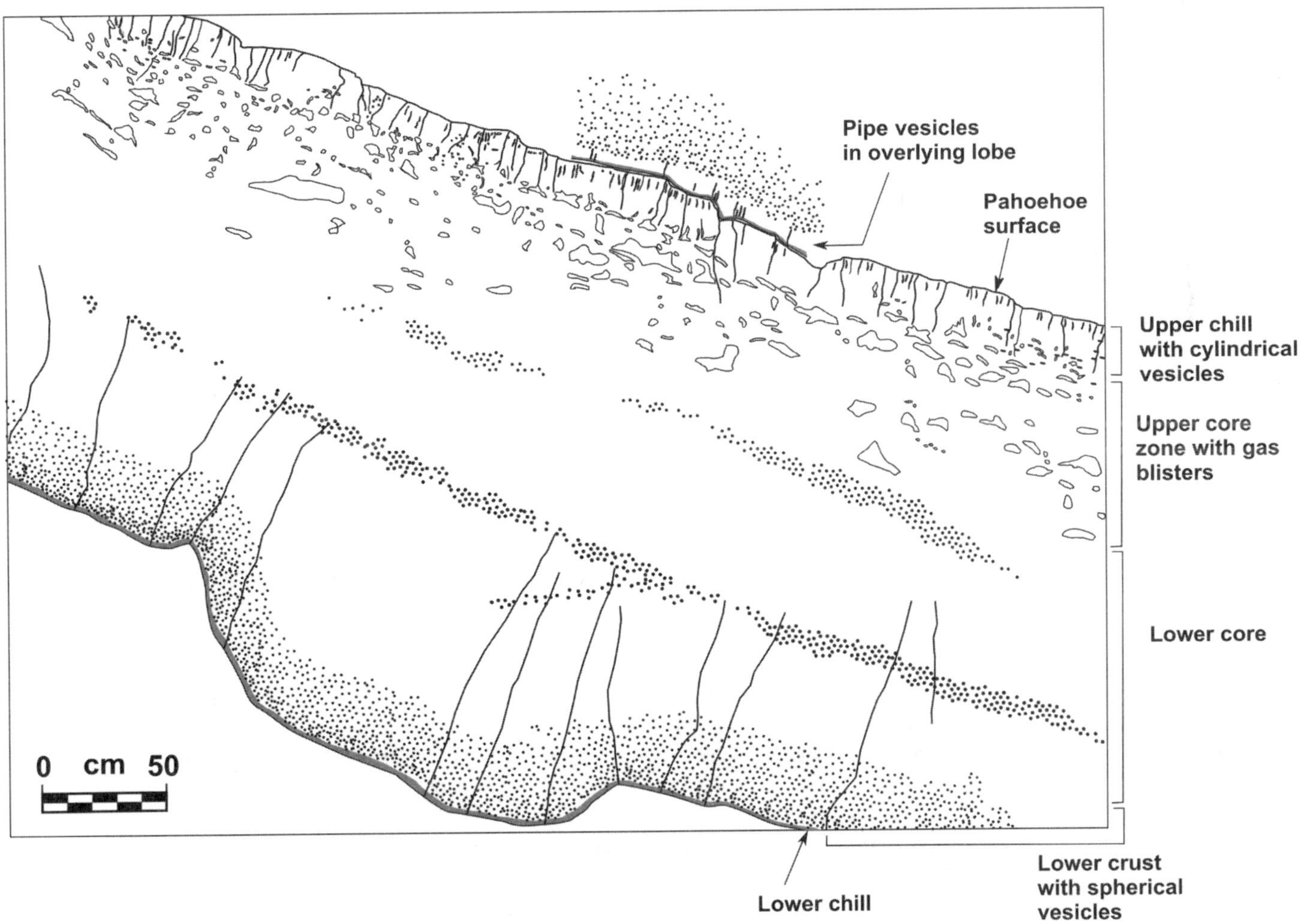

Figure 21. Details of vesicle types and distributions in flow unit 29 (see also Fig. 14). The structure of the flow lobe represented by an upper chill, coarse vesicular zone, dense core, and lower chill is consistent with formation by lava flow inflation.

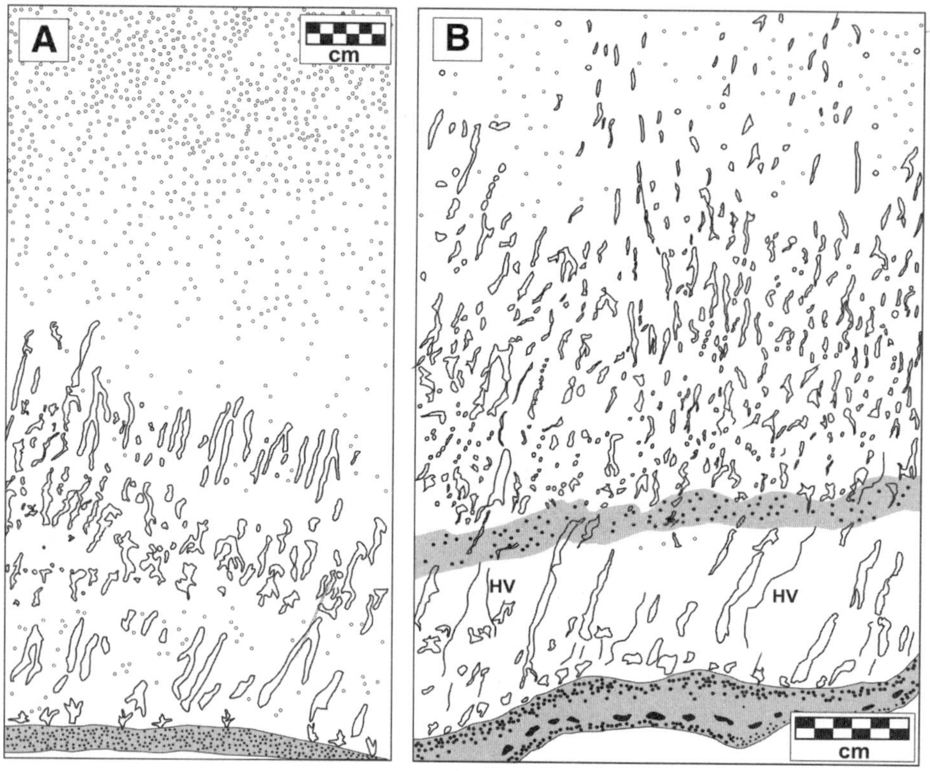

Figure 22. Details of vesicle types and distributions in flow lobes in the upper section of the Agatha Formation. (A) Multiple layers of inclined pipe vesicles. (B) Multiple chills and layers of pipe vesicles and healed vesicle pathways (HV).

and tends to fan upward. Above the vesicle-poor layer is a population of cylindrical vesicles. It is common for two or even three separate vesicles to merge into a single cylinder. The inclined angle of the pipes and cylinders indicates continued movement of the solidifying lava shortly after the formation of the vesicles. The interior of the lobe has a high abundance of spherical vesicles 1–2 mm in diameter, which increase in abundance away from the basal chill zone. The small spherical vesicles are depleted in the zones of the pipe and cylindrical vesicles, indicating coalescence and subsequent movement of larger gas bubbles. Some of the boundaries between the various layers of pipe and cylindrical vesicles are noticeably finer grained and appear to have acted as barriers to further upward migration of the vesicles.

A variant of this type has several vesicle populations separated by very fine-grained chill zones (Fig. 22B). The lower chill 3 cm thick has itself a layered distribution of small spherical vesicles on the margins and larger irregular vesicles in the center. A zone of inclined pipe vesicles as well as a set of parallel lineations (not fractures) that may have been previous vesicle channels that later closed up, overlie the lower chill. A very fine-grained layer, appearing as a second chill zone, overlies the cylindrical vesicles again and contains small spherical vesicles. Short cylindrical vesicles (which are not inclined) overlie the fine-grained layer, and these grade into small spherical bubbles in the interior of the flow.

COMPOSITIONS OF THE LAVAS

Samples were selected for chemical analyses from the areas described, and major and trace element data are given in Table 1. Localities of the samples are shown in Figures 2 and 3. The major element compositions presented here will focus on differences and similarities between the groups and to the extent to which they allow further understanding of the textures and mode of occurrence of the lavas. The analyses are given in Table 1, and details of methodology are given in Appendix 1.

Major element chemical analyses are shown for selected oxides plotted against SiO_2 and MgO (Fig. 23). The alkali-SiO_2 classification diagram of volcanic rocks (Le Maitre, 1989) indicates that all compositions lie in the subalkaline field over the range from basalt to dacite. Rocks of the Nhlebela Formation are andesite to dacite and are clearly separated from the basalts and basaltic andesites of the Agatha Formation. The siliceous lavas of the Agatha Formation identified in Traverse 2 (Fig. 10D) are clearly distinguishable from the underlying lavas in their lower Al_2O_3, FeO, and TiO_2 contents. Three analyses of lavas from the upper section of the Agatha Formation (Fig. 3) plot differently than either of the other two groups. For this group of three, Al_2O_3 contents are similar to lavas of Traverse 1, but FeO and TiO_2 contents place them with the

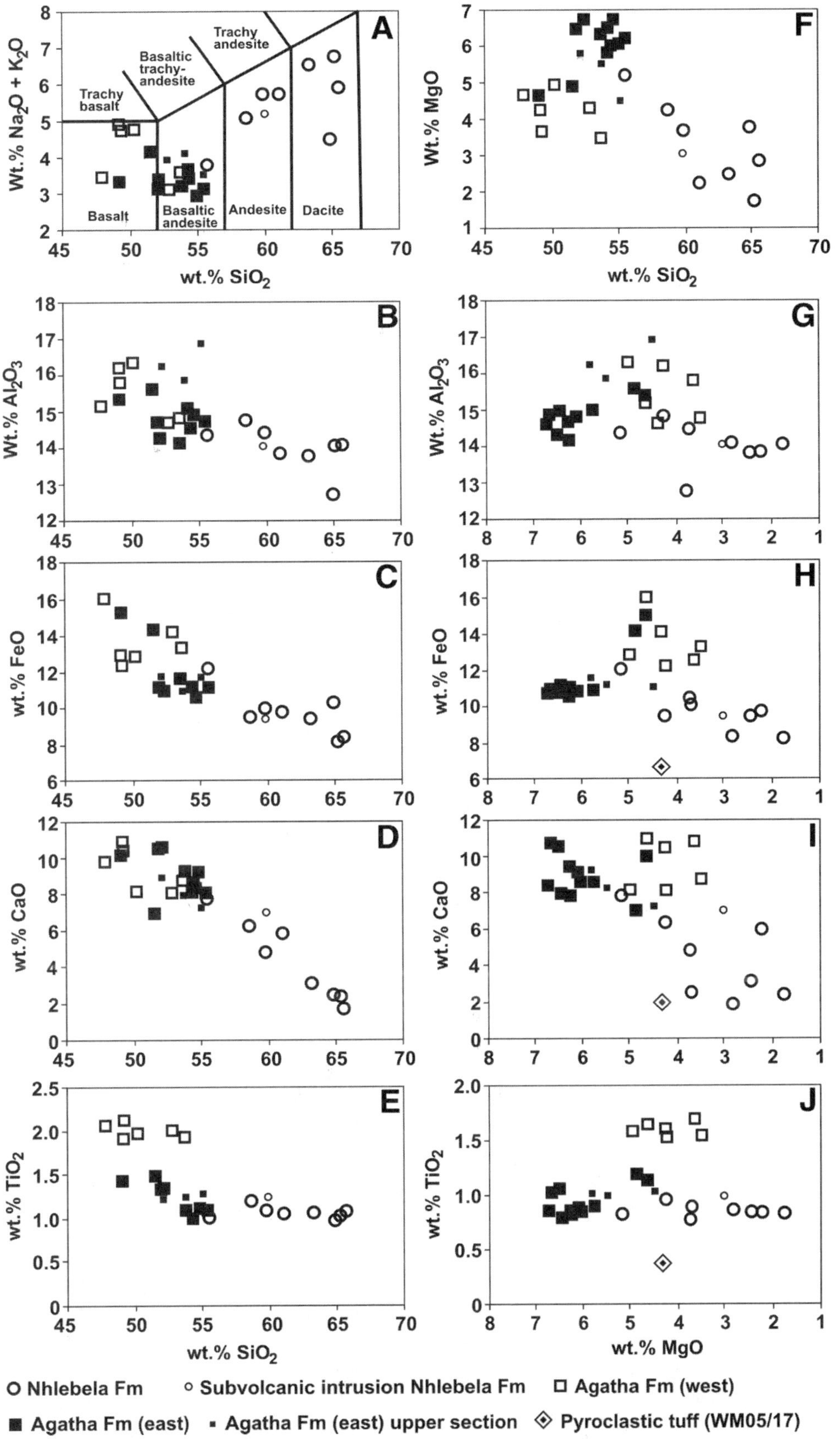

Figure 23. Major element variation diagrams for selected oxides plotted against wt% SiO_2 (A–E) and wt% MgO (F–J). Locations of samples are seen in Figures 2 and 3.

more siliceous samples of Traverse 2. While it is not possible to specifically evaluate the compositional controls on the textures observed, the similarities and differences indicate that compositions played a role in the vesicularity and viscosities of the lavas.

Major element oxides plotted against MgO indicate a more complex relationship. The rocks from the Nhlebela Formation have, as expected, the lowest MgO contents, and the three sets of data from the Agatha Formation are also separated from each other. The siliceous lavas from Traverse 2 (Fig. 3) have the highest MgO contents in the Agatha Formation. The three samples from the upper section of the Agatha formation have MgO contents intermediate to the other two groups.

DISCUSSION OF THE VOLCANOLOGY OF THE WHITE MFOLOZI INLIER

Physical characteristics of lava flows provide important evidence of the styles and environments of eruption. Excellent preservation of extensive flows is generally rare for the Archean, and observations are commonly limited to two-dimensional views of restricted outcrops. In this study we have described a wide range of textures and structures, allowing deductions to be drawn regarding the emplacement of these lavas and comparisons to be made with well-understood recent volcanism.

Overview of the Volcanic Rocks

There are both similarities and differences in the two formations of volcanic rocks of the White Mfolozi inlier. Both show progression from basal subaqueous volcanism progressing to subaerial eruption. Prior to the onset of volcanic activity of the Nhlebela Formation, the depositional environment was possibly relatively deep shoreline facies that progressed into a fluvial system. Pillow structures and hydroclastic breccia define the lower volcanic units. The upper amygdaloidal lava flows show no interaction with water, and structures are typical of subaerial eruptions. Volcanism was terminated for a substantial period of time, resulting in the development of the sedimentary White Mfolozi Formation in a littoral depositional environment.

The environment of the initial outpourings of lava of the Agatha Formation was shallow water or wet subaerial, possibly a lagoon or shoreline as proposed for the uppermost White Mfolozi Formation (Matthews, 1967). The volcanism terminated the shallow-water sedimentary processes of the White Mfolozi Formation due to extensive and relatively rapid building of the lava pile, which probably raised the ground level above the water line. The lava flows of the Agatha Formation in this area contrast with the extensive pyroclastic eruptions in the Northern Region (Fig. 1; Armstrong et al., 1982), indicating a regional change in the style of volcanism and also, most likely, compositions of the lavas.

Interaction with Sediments and Peperite Formation

Clear evidence of direct interaction between wet underlying sediments and the lavas is not observed at the base of either of the two formations. However, both show the involvement of wet sediments in the earliest eruption stages. Near the base of the Nhlebela Formation juvenile close-packed clasts with jigsaw fits indicate in situ fragmentation of lava in a wet, endogenous sediment, resulting in the formation of peperite (White et al., 2000; Skilling et al., 2002; Doyle, 2000). The presence of rounded quartz grains in some of these occurrences is indicative of sediments being involved, but the greater proportion of the host material is lava fragments that have been altered by water interaction. The endogenous sediment is therefore derived from the lavas and is of similar composition to the juvenile clasts with incorporation of some quartz grains from disaggregation of preexisting sediments. It has also been pointed out (Skilling et al., 2002) that in ancient deposits it may be difficult to distinguish fragments derived from tectonic processes from those generated by wet sediment interaction. In the case of the Nhlebela Formation, the unequivocal evidence of hydroclastic eruptives mitigate strongly against a tectonic origin for the fragments. The peperite observed in the Nhlebela Formation is similar in appearance and clast-matrix composition to that described in the Neoproterozoic Schakalsberg Complex in Namibia (Mueller, 2003).

The bedded deposits of mixed sediment-derived sand grains and magma clasts at the base of the Agatha Formation are indicative of hydromagmatism. Juvenile lava fragments may have been derived from pyroclastic eruptions that incorporated unconsolidated sediments (Ukstins et al., 2003) or even from preexisting peperite, which was subsequently disintegrated and reworked in an aqueous environment. The abnormally high Cr and Ni concentrations of the rock also indicate that a more ultramafic magma may initially have erupted but no trace of it has been found to date.

Hydromagmatic Eruptions

The basal units of the Nhlebela and Agatha Formations were emplaced subaqueously. The nature of the underlying sediments (cross-bedded quartz sandstones) indicates that in both cases the environment was relatively shallow water, which would have had an important influence on the type of eruption (White et al., 2003). Hydroclastic volcanic rocks are present in the basal sequences of both formations but are of different types. In the southern outcrop of the Nhlebela Formation there is a close association of pillows, discontinuous flow lobes, and fragmental rocks. In the northern outcrop the pillowed sequence is clearly separated from the fragmental zones. Significant fragmentation associated with pillow inflation is not common (Moore et al., 1973; White et al., 2003), and fragmentation in this deposit would most likely have been controlled by the relatively high gas content, low confining pressure, and high viscosity. Considering the variable thickness of the underlying Mantonga Formation, the depth of water may have varied quite considerably. The combined dispersal medium would have been magmatic gas and vaporized external water. The variable sizes of pillows and association of pods or sheets can be related to the nature and slope of the substrate. Multiple chill selvages appear almost identical to features of Eocene pillows in New Zealand (Walker, 1992), explained by contraction during implosion at shallow-water levels (Kawachi and Pringle, 1988). The thick pillow rinds are characteristic of large pillows that have taken several hours to form (Solomon, 1969). The narrow sheets (para-pillows) associated with the pillow and breccias are interpreted as pillows that failed to inflate due to either steepness of the slope or inadequate lava supply (Walker, 1992).

The basal sequence of the Agatha Formation does not have pillows, and the fragmented zones attributed to hydromagmatism are less extensive than those in the Nhlebela Formation. The progression from subaqueous to subaerial eruption occurred relatively rapidly, indicating a very shallow shoreline or even a fluvial environment.

Aa/Blocky Lavas and the Transition to Pahoehoe Lavas

Aa lavas form when effusion rates are high and crustal development is disrupted, while at the same time cooling and subsequent crystallization changes the rheological properties of the magma (Cashman et al., 1999). In most cases pahoehoe lava flows evolve into aa lava some distance from the vent. The rapid transport of lavas through open channels inhibits the formation of vesicles (Rowland and Walker, 1990). The rheological controls on aa formation as compared with pahoehoe are as follows: in the former, the chilled zones and crusts fail continuously, whereas crusts of pahoehoe fail episodically (Kilburn, 1993). The transition from aa to pahoehoe, as observed in Traverse 1 in the Agatha Formation, could be explained by decreasing effusion rates (Cashman et al., 1999) in the same flow field or by the interaction of eruptions from two volcanic centers with the initially higher effusion type rapidly declining. Blocky lava forms by a similar mechanism but differs from aa lava in that the fragments tend to be smooth and angular with fragments ranging in size from tens of cm to meters (Kilburn, 2000).

Pahoehoe Flows and Vesicle Characteristics in the Agatha Formation

Pahoehoe lava flows are common in every basalt province but their documentation and description in Archean occurrences is sparse, even in comprehensive review articles (Sylvester et al., 1997). This may partly be due to poor preservation, but it may also reflect the relatively rare occurrence of subaerial volcanism at that time. Deformation may also have obscured many of the subtle textural features. Most, if not all, pahoehoe flows (the subaerial equivalent of pillow lavas) are made up of compound flow lobes that become inflated and thickened by continued delivery of lava through channels and pathways in the cores of the flow lobes (Hon et al., 1994). Thin flow lobes, initially tens of centimeters thick, may eventually inflate to sheets many meters thick. The delivery of lava to the flow front is at significantly slower rates than that of the lava within the flow lobes. This unifying and integrated model for subaerial lava emplacement explains many of the textures and structures seen in the flows (Self et al., 1997, 1998).

Many of the morphological features described in the Agatha Formation are consistent with the lava inflation model (Self et al., 1997), which therefore indicates that similar processes that operate in the formation of current and recent basalt lava flows also took place in the Archean. A narrow flow front arrived with molten lava contained within a stretchable viscoelastic skin (Fig. 16C) that contorted, stretched, and rolled to give billowy or ropy chilled surfaces (Fig. 14C; Fink and Fletcher, 1978). Gas bubbles were trapped within the skin. Continued inflow of lava caused the lobe to inflate, lifting the upper skin. Bubbles rose from the molten core, combined and coalesced and were trapped beneath the thickening upper crust as horizontal vesicular zones (Fig. 14B). At the same time the high rate of cooling and stress caused by continued motion of the crust resulted in the development of closely spaced regular joints. Growth of the lower crust was slower and commonly produced pipe vesicles. As the interior of the lobe cooled it became devoid of vesicles and developed more widely spaced and longer joints. Opening of cooling fractures caused clefts, and failure of the viscoelastic skin allowed magma from the core to break out (Fig. 15A). In some cases these clefts were filled from the overlying lava lobe (Fig. 16B). Continued movement and localized inflation caused blocks of crust to be displaced as tumuli (Fig. 15B; Self et al., 1998).

Different types of pahoehoe (originally described on Hawaii) are distinguished on the basis of vesicle distribution (Wilmoth and Walker, 1993). Flows classed as P-type have relatively low vesicle contents (largely degassed) and are characterized by pipe vesicles and formed under inflation conditions. In contrast S-type flows have a higher proportion of spherical vesicle and no pipe vesicles and formed under conditions of minimal inflation. P-type lava developed as a result of relatively long residence times in inflating lava tubes (Wilmoth and Walker, 1993), which allowed extensive bubble growth and overall loss of gas. P-type lavas develop most readily on shallow slopes where the lava undergoes slow movement and the bubbles migrate as a result of the surface tension within the bubbles acting against the deformation forces (Wilmoth and Walker, 1993). This is supported by the commonly curved and nonvertical attitude of many pipe vesicles (Fig. 22). Multiple vesicle layers (Fig. 22) are most likely to have originated by repeated lava emplacement relating to inflation of the lava lobe, causing rejuvenation of pipe growth.

The three zones identified in lava flows in Hawaii that formed under inflation conditions are present in the amygdaloidal basalts of the Nhlebela Formation and in most of the subaerial flow lobes in the Agatha Formation. The zones are (Fig. 21): a basal vesicle zone, a middle vesicle deficient or massive zone, and an upper vesicle-rich zone (Aubele et al., 1988; Cashman and Kauahikaua, 1997). A vesicle-poor crust or chill margin often marks the extremities of the lava flow units. In general, the thinner the flow, the higher the vesicle content within the middle zone as a result of the more rapid solidification, but there is less of a tendency to form coalesced vesicles. Small spherical vesicles are found within the middle zone in minor proportions, most often toward the top of the zone. The distribution of vesicles as observed for Hawaii flow lobes (Cashman and Kauahikaua, 1997) is remarkably similar to those observed in Traverse 1. The presence of squeeze-ups (Fig. 15A) and lava ribbons suggests that fresh lava was being injected beneath the pahoehoe crust.

Ripply pahoehoe (Kilburn, 1990) and rough and toothpaste pahoehoe surfaces (Fig. 17B; Rowland and Walker, 1988) are thought to have formed by irregular tearing of the exposed crust of the lava flow (Kilburn, 1990). These types may represent a transitional phase between pahoehoe and aa, particularly if the lava was slightly more viscous than that which formed ropy pahoehoe (Rowland and Walker, 1988). Similar textures observed in Traverse 2 are located in the region corresponding to the compositional change to more siliceous lava and may, at least to some extent, reflect increased viscosity. The clear differences in eruptive style and distribution and types of vesicles between the lower and middle sections of the Agatha Formation and those in the upper section can at least in part be attributed to compositional differences.

TRACE ELEMENT CHARACTERIZATION AND CONSTRAINTS ON PETROGENESIS

Trace elements have been selected to represent the compatible and incompatible groups (Fig. 24). Trends for compatible elements such as Cr and Ni reflect high-level fractionation processes while the incompatible elements (such as U, Th, and Nb) are an indicator of the source magmas or of contamination.

Quite different geochemical patterns are indicated for Cr and Ni versus SiO_2 and MgO. The three groups of samples again remain distinct with the more silica-rich rocks of the Nhlebela Formation generally most depleted in Ni and Cr. While there is considerable overlap of Ni contents for all rocks from the Agatha Formation, Cr reveals distinct groupings relating to different parts of the succession. The highest Cr contents are from the siliceous Mg-rich rocks of the Agatha Formation in Traverse 2. Basalts from the western occurrence of the Agatha Formation in Traverse 1 overlap with those from the top of the succession in the far east.

The incompatible trace elements Nb, Th, and U are strongly enriched in rocks of the Nhlebela Formation. The relative enrichment in Nb for samples from the Nhlebela Formation is a factor of 2 to 3 times greater than that for the Agatha Formation, whereas relative enrichment for U and Th is a factor of 6 to 7 times greater. This indicates that, in the broadest sense, the magmas from the two formations cannot be linked by fractionation and are also unlikely to have had a common source. There is a small but consistent difference in the contents of Nb and Th for the sections of the Agatha Formation represented by Traverses 1 and 2, with the siliceous Mg-rich samples having the lowest concentrations. The concentrations of these trace elements for the uppermost samples from the Agatha Formation in the far east (uppermost) overlaps with those from the base of the Formation in Traverse 1.

Chondrite-normalized (Sun and McDonough, 1989) rare-earth element (REE) patterns (Fig. 25) also reveal distinct signatures for the groups of lavas. The Nhlebela Formation has the highest concentrations of REE and also the steepest pattern (average [La/Lu]n 7.9) and pronounced negative Eu anomaly (Eu* average –10.5). Distinct patterns and REE ratios separate the rocks of the Agatha Formation into four groups. The siliceous high-Mg rich rocks of Traverse 2 have the lowest total REE concentrations and also the flattest pattern (average [La/Lu]n 4.8) and small negative Eu anomaly (Eu* average –2.2). The lowermost rocks of Traverse 1 have moderately steep patterns ([La/Lu]n 5.7) and small negative Eu anomaly (Eu* average –2.2). Samples of the Agatha Formation from Traverse 2 have similar-shaped patterns to other rocks of the formation but with essentially no Eu anomaly (average [La/Lu]n 5.9 and average Eu* –0.6). A further group of compositions for samples located above the siliceous Mg-rich lavas (far-east Agatha Formation) with distinctly flat heavy–REE patterns is indicative of a further magma variant. The four groups of samples from the Agatha Formation, identified on the basis of the trace element geochemistry, may therefore be indicative of four flows making up the succession. The marked negative Eu anomalies for the andesites and dacites of the Nhlebela Formation, indicative of plagioclase fractionation, are not accompanied by the presence of feldspar phenocrysts and contrast with the highly porphyritic andesites in the Nsuze group in the Northern Region.

Basalts occurring in greenstone belts are commonly described and characterized on the basis of chondrite-normalized ([La/Sm]n) as this ratio varies with major element composition (Sylvester et al., 1997). The compositions of the volcanic rocks from the White Mfolozi inlier are compared to fields of greenstone basalts (Fig. 26A). The Nhlebela lavas are andesites and dacites and therefore are not appropriately discriminated in this diagram, but lie in a compositional field between the Agatha basalts and the granitoid basement rocks. The wide range of FeO contents (Fig. 26B) may indicate extreme fractionation, and some of the basalts from the Agatha Formation lie in the field of Archean iron-rich tholeiites (Arndt et al., 1977). On the basis of [La/Sm]n and FeO contents, the rocks of the Nhlebela Formation lie outside the field of normal greenstone belt basalts, and the compositions are consistent with crustal contamination from the underlying basement rocks. The clasts of the hydroclastic tuff at the base of the Agatha Formation are high-Mg basalts.

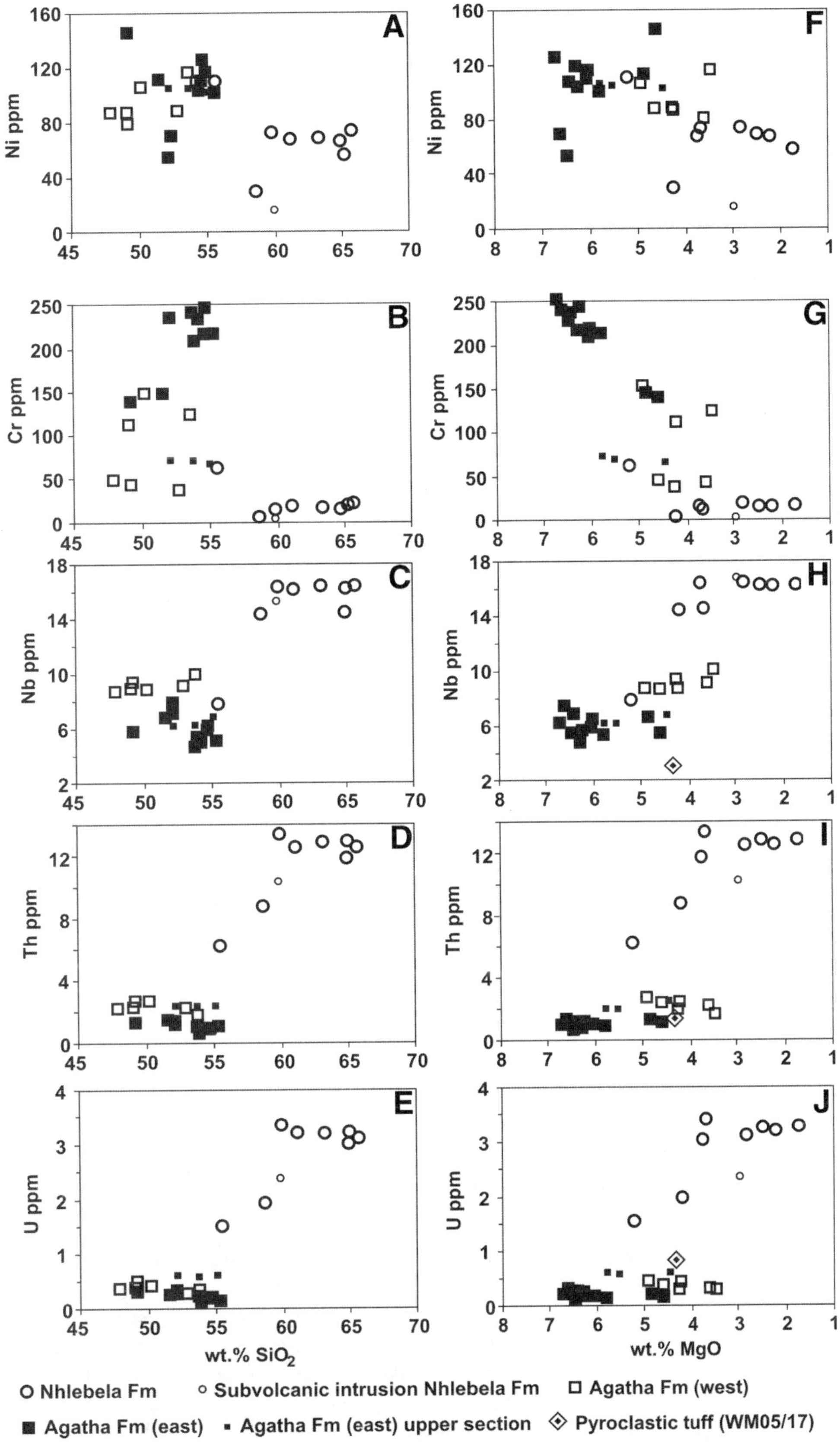

Figure 24. Trace element variation diagrams for selected compatible and incompatible elements plotted against wt% SiO_2 (A–E) and wt% MgO (F–J). Locations of samples are seen in Figures 2 and 3.

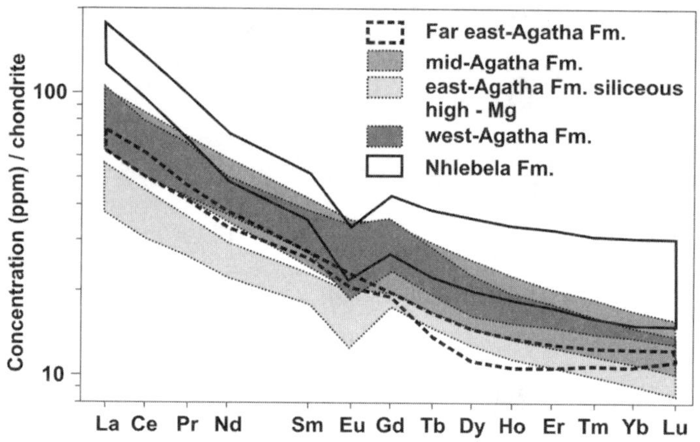

Figure 25. Chondrite normalized rare earth patterns for volcanic rocks from the Nhlebela and Agatha Formations.

Trace element discrimination diagrams employing immobile incompatible trace elements can be useful in identifying tectonic settings and also possible contamination. The Hf-Ta-Th diagram (Wood, 1980) is effective in identifying volcanic arc environments as well as the effects of crustal contamination on various basalt types. The diagram in Figure 27A shows the compositions of the White Mfolozi inlier as well as a suite of samples from the granitoid gneisses, which are representative of the underlying basement of the Pongola Supergroup (Hunter and Wilson, 1988). The volcanic rocks lie on a distinct linear trend parallel to that for the basement rocks (not plotted for tectonic discrimination purposes) within the calc-alkaline volcanic arc field. The role of crustal contamination is therefore difficult to ascertain. A limitation of this diagram (Rollinson, 1993) is that Th enrichment can occur as a result of removal of Hf and Ta in crystallizing mineral phases.

The La-Y-Nd discrimination diagram (Cabanis and Lecolle, 1989) after Rollinson (1993) discriminates between volcanic arc, oceanic, and continental basalts. Analyses of the White Mfolozi volcanic rocks (Fig. 27B; and the basement granitoids shown to assess possible contamination) straddle the boundary between the continental and calc-alkaline volcanic arc fields. The distribution is consistent with contamination of continental basalts from the basement granitoids.

The four distinct compositional types of basalts in the Agatha Formation are reflected in the distinct styles of volcanism and textures. These occur over clearly defined stratigraphic intervals and indicate that the formation was made up of four compound flows. The eruptive centers may have been geographically separated but this cannot be ascertained from the limited exposures in the river section.

SUMMARY AND CONCLUSIONS

Volcanic rocks from the earliest part of Earth's history displaying exceptionally well preserved primary structures are rare. The deeply eroded White Mfolozi inlier exposes such a succession of basalts, basaltic andesites, and dacites of the Nsuze Group of the Pongola Supergroup. The two volcanic sequences, the lower Nhlebela Formation and the upper Agatha Formation, are separated by the fluvial, lagoonal, and shoreline sediments of the White Mfolozi Formation. The compositions of the lavas range from mainly andesite-dacite in the Nhlebela Formation to basalt and basaltic andesites in the Agatha Formation. The combination of distinct compositional types together with a variety of depositional and physical environments has given rise to a wide range of primary volcanic textures and structures.

In both formations volcanic activity commenced with shallow-water hydromagmatic eruptions. In the Nhlebela Formation these gave rise to a variety of fragmental rock types. Jigsaw-fit peperites formed by the emplacement of lava into wet sediment, derived mainly from spalled volcanic glass and minor clastic quartz grains. Lava lobes and sheet flows are interspersed with large, highly vesicular pillows. The basal deposit of the Agatha Formation is a hydroclastic tuff made of quartz sand grains and juvenile clasts of high-Mg basalt (10 wt% MgO) overlain by hydroclastic basalt breccias. The lower hydromagmatic eruptions are overlain in both formations by subaerially emplaced lavas. Well-developed aa and blocky lavas with massive amygdaloidal interiors dominate the lower part of the Agatha Formation. This is overlain by a beautifully preserved succession of pahoehoe lavas, with ropy surfaces and flow crenulations. Chill zones, joint patterns, and distributions and types of vesicles are consistent with these flow lobes having formed by the inflation mechanism, widely accepted as the dominant process for extensive development of recent continental flood basalts. There is no evidence of extensive chanalization, indicating that rates of lava effusion and mechanisms of emplacement were similar in the Archean to recent and modern eruptions in similar environments. The transition from subaqueous to subaerial emplacement in both formations is indicative of rising ground elevation due to the volcanic activity. Change in eruptive style from aa to pahoehoe lavas higher in the sequence in the Agatha Formation may be indicative of a reduction of effusion rates after the first outpourings.

A compositional change to more siliceous and Mg-rich basalts higher in the Agatha Formation is accompanied by distinct styles of pahoehoe lavas. Small domed and bulbous lobes overlain and molded into extensive sheets with planar upper surfaces characterize the flow lobes. The surface textures are ripply or rough as the result of tearing of the relatively high viscosity lava. Multiple chill zones with well-developed pipe vesicles are common. The varieties of lava types in the Agatha Formation have distinct chemical signatures, but at present there are insufficient data to clearly investigate the interaction of flow lobes of different compositions. At least four compound flows are indicated to be responsible for the development of the volcanic pile.

Trace elements clearly distinguish lava compositions from the Nhlebela and Agatha Formations. The compound flows of the Agatha Formation overlap the field of strongly enriched Archean basalts on the MgO-[La/Sm]n diagram and extend into

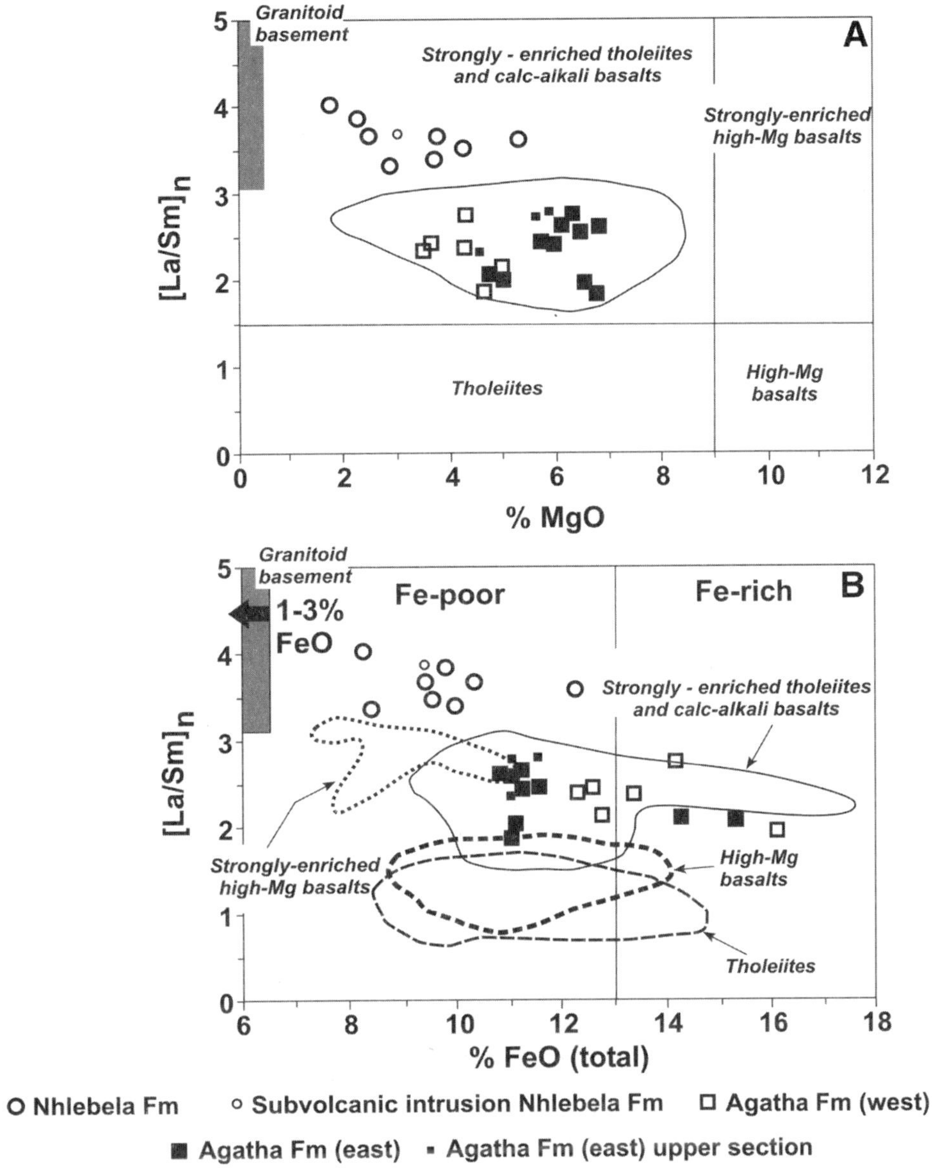

Figure 26. Subdivision of Archean basalt compositions after Sylvester et al. (1997). Fields relating to the basement granitoids are also shown in the toned block. (A) The White Mfolozi basalts are enriched in relation to [La/Sm]n and lie in the same field as other Archean basalts in this class (outlined). (B) The White Mfolozi basalts overlap with fields of Archean strongly enriched tholeiites and calc-alkali basalts and extend into the field of Fe-rich varieties. Plots of the compositions of the andesites and dacites of the Nhlebela Formation are shown for comparison.

the field of Fe-rich tholeiites. The discrimination diagrams Hf-Th-Ta and Y-La-Nb place these lavas in the calc-alkali volcanic arc field overlapping into continental tholeiites, but these indicators may have been severely affected by contamination from the underlying basement granitoids and are not considered to provide a reliable indicator of tectonic setting.

A previously unreported observation for the uppermost lavas of the Agatha Formation is the steepening of the dip and increasing deformation on approaching the base of the Mozaan quartzites. This is attributed to a major thrust on the contact between the Nsuze and Mozaan Groups and questions the widely held acceptance of a primary angular unconformity at this contact.

At least 75 flow lobes are present in the Agatha Formation, the greater proportion of which formed in a subaerial environment, indicating the presence of a well-established continental basement at this time. The interplay of subaerial and shallow-water volcanism, together with shallow to moderately deep sedimentary depositional environments, indicates a subsiding continental margin setting for the Nsuze Group in this area.

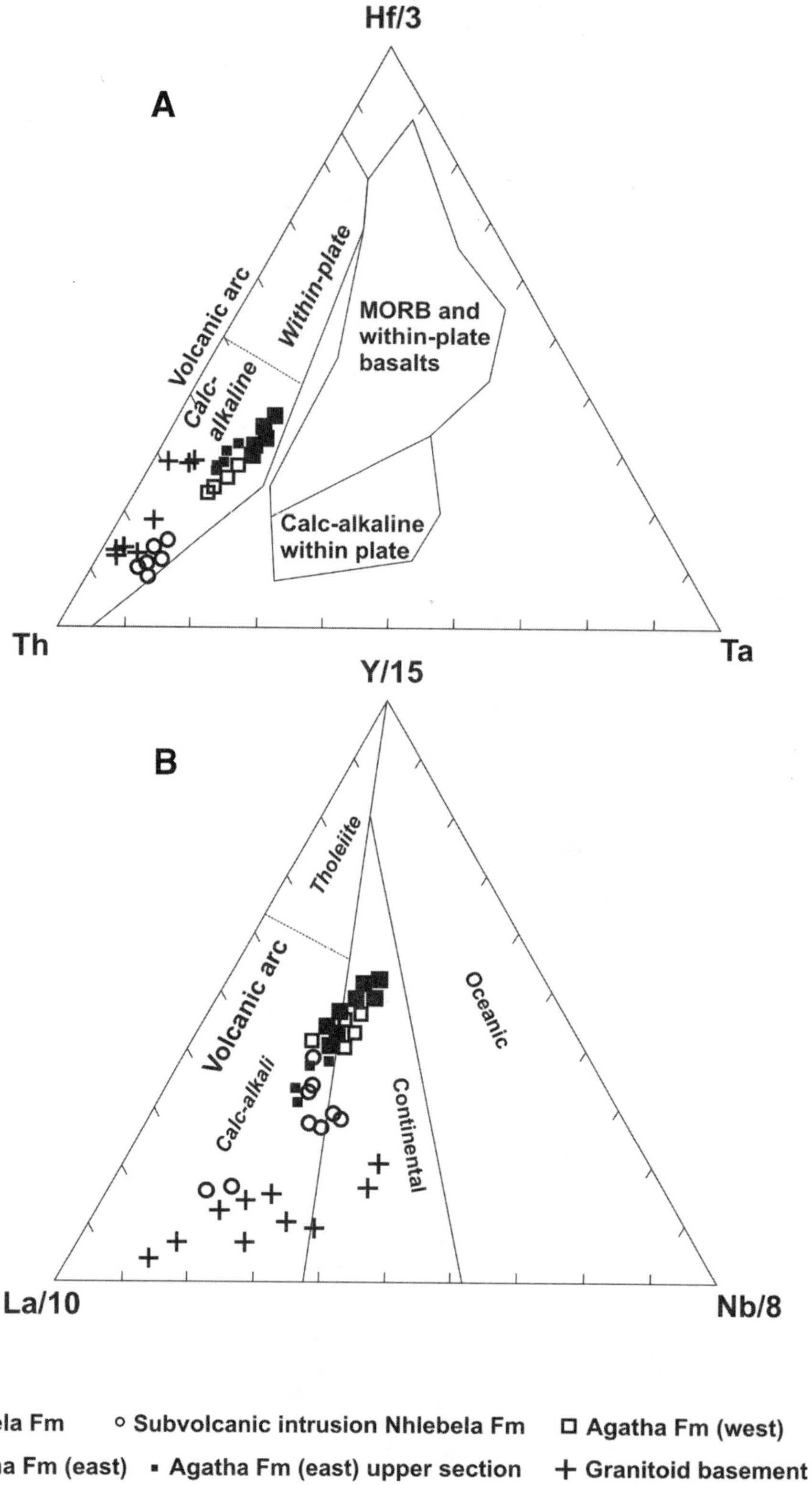

Figure 27. Trace element discrimination diagrams for the volcanic rocks of the White Mfolozi inlier. Also plotted are compositions of basement granitoids in the area. (A) Plots in the Hf-Th-Ta diagram (Wood, 1980) indicate a calc-alkaline volcanic arc environment but these compositions are likely to have been influenced by contamination and high degrees of fractionation. (B) Y-La-Nb (Cabanis and Lecolle, 1989) for basalts after Rollinson (1993) also indicates a calc-alkali volcanic arc but these compositions could also have been modified by basement contamination

APPENDIX

Analytical Methods Used

Large samples (1 kg) were collected and carefully prepared to exclude weathered or altered material, as well as amygdales. The samples were crushed in a carbon steel mill and major elements were determined by XRF (X-ray fluorescence) and trace elements using a Perkin-Elmer 6100 ICP-MS (inductively-coupled mass spectrometer) according to standard analytical methods at the University of KwaZulu-Natal (Wilson, 2003). XRF calibrations were against certified standard rock materials. Errors in major elements were determined to be lower than 1% of the true values for all elements. Dissolutions for ICP-MS were carried out using the Anton-Paar microwave digestion system, and calibrations were against certified standard solutions. Standard reference materials BCR-1 and BHVO-1 (U.S. Geological Survey certified rock materials) were run with all sample batches. Agreement for all trace elements was better than 10% and in most cases better than 5%.

ACKNOWLEDGMENTS

AHW and CEG acknowledge the University of KwaZulu-Natal Research fund and the National Research Foundation (NRF) for financial support of this project, including an M.S. student bursary to CEG. Guy Charlesworth and Goonie Marsh are thanked for detailed and critical reviews that greatly improved the manuscript.

REFERENCES CITED

Armstrong, N.V., 1980, Geology and geochemistry of the Nsuze Group in northern Natal and southeastern Transvaal [Ph.D. thesis]: Pietermaritzburg, University of Natal, 386 p.

Armstrong, N.V., Hunter, D.R., and Wilson, A.H., 1982, Stratigraphy and petrology of the Archaean Nsuze Group, northern Natal and southeastern Transvaal: Precambrian Research, v. 19, p. 75–107, doi: 10.1016/0301-9268(82)90020-1.

Armstrong, N.V., Wilson, A.H., and Hunter, D.R., 1986, The Nsuze Group, Pongola Sequence, South Africa: Geochemical evidence for Archaean volcanism in a continental setting: Precambrian Research, v. 34, p. 175–203, doi: 10.1016/0301-9268(86)90056-2.

Arndt, N.T., Naldrett, A.J., and Pyke, D.R., 1977, Komatiitic and iron-rich tholeiitic lavas of Munro Township, northeastern Ontario: Journal of Petrology, v. 18, p. 319–369.

Aubele, J.C., Crumpler, L.S., and Elston, W.E., 1988, Vesicle zonation and vertical structure of basalt flows: Journal of Volcanology and Geothermal Research, v. 35, p. 349–374, doi: 10.1016/0377-0273(88)90028-5.

Batiza, R., and White, J.D.L., 2000, Submarine lavas and hyaloclastite, *in* Sigurdson, H., ed., Encyclopedia of Volcanoes: San Diego, Academic Press, p. 361–381.

Beukes, N.J., and Cairncross, B., 1991, A lithostratigraphic-sedimentological reference profile for the late Archaean Mozaan Group, Pongola Sequence: Application to sequence stratigraphy and correlation with the Witwatersrand Supergroup: South African Journal of Geology, v. 94, p. 44–69.

Beukes, N.J., and Lowe, D.R., 1989, Environmental control on diverse stromatolite morphologies in the 3000 Myr Pongola Supergroup, South Africa: Sedimentology, v. 36, p. 383–397.

Burger, A.J., and Coertze, F.J., 1973, Radiometric age measurements on rocks from southern Africa to the end of 1971: Bulletin of the Geological Survey of South Africa, v. 58, p. 1–46.

Burke, K., Kidd, W.S.F., and Kusky, T.M., 1985, The Pongola structure of the southeastern Africa: The world's oldest preserved rift?: Journal of Geodynamics, v. 2, p. 35–49, doi: 10.1016/0264-3707(85)90031-6.

Cabanis, B., and Lecolle, M., 1989, Le diagramme La/10-Y/15-Nb/8: Un outil pour la discrimination des series volcaniques et la mise en evidence des processus de melange et ou de contamination crustale: C.R.: Academic Science Series II, v. 309, p. 2023–2029.

Carr, P.F., and Jones, B.G., 2001, The influence of palaeoenvironment and lava flux on the emplacement of submarine, near-shore late Permian basalt lavas, Sydney Basin (Australia): Journal of Volcanology and Geothermal Research, v. 112, p. 247–266, doi: 10.1016/S0377-0273(01)00244-X.

Cas, R.A.F., and Wright, J.V., 1987, Volcanic successions: Modern and Ancient: London, Allen and Unwin, 528 p.

Cashman, K.V., and Kauahikaua, J.P., 1997, Re-evaluation of vesicle distributions in basaltic flows: Geology, v. 25, p. 419–422, doi: 10.1130/0091-7613(1997)025<0419:ROVDIB>2.3.CO;2.

Cashman, K.V., Thornber, C., and Kauahikaua, J.P., 1999, Cooling and crystallization of lava in open channels and the transition from pahoehoe lava to aa: Bulletin of Volcanology, v. 61, p. 306–323, doi: 10.1007/s004450050299.

Cole, E.G., 1994, Lithostratigraphic and depositional environment of the Archaean Nsuze Group, Pongola Supergroup [M.S. thesis]: Johannesburg, Rand Afrikaans University, 166 p.

Cousineau, P., and Dimroth, E., 1982, Interpretation of the relations between massive, pillowed, and brecciated facies in an Archean submarine andesite–Amulet andesite, Rouyn-Noranda, Canada: Journal of Volcanology and Geothermal Research, v. 13, p. 83–102, doi: 10.1016/0377-0273(82)90021-X.

Crow, C., Condie, K.C., Hunter, D.R., and Wilson, A.H., 1989, Geochemistry of volcanic rocks from the Nsuze Group, South Africa: Arc-like volcanics in a 3.0 Ga intracratonic rift: South African Journal of Earth Science, v. 9, p. 589–597, doi: 10.1016/0899-5362(89)90043-2.

de Wit, M.J., and Ashwal, L.D., 1997, Convergence towards divergent models of greenstone belts, *in* de Wit, M.J., and Ashwal, L.D., eds., Greenstone belts: Oxford, Clarendon Press, p. ix–xvii.

Dimroth, E., Cousineau, P., Leduc, M., and Sanschagrin, Y., 1978, Structure and organization of Archean subaqueous basalt flows, Rouyn-Noranda area, Quebec, Canada: Canadian Journal of Earth Sciences, v. 15, p. 902–918.

Dimroth, E., Imreh, L., Cousineau, P., Leduc, M., and Sanschagrin, Y., 1985, Paleogeographic analysis of mafic submarine flows and use in exploration for massive sulphide deposits, *in* Ayres, L.D., Thurston, P.C., Card, K.D., and Weber, W., eds., Evolution of Archean Supracrustal Sequences: Geological Association of Canada Special Paper 28, p. 203–222.

Doyle, M.G., 2000, Clast shape and textural associations in peperite as a guide to hydromagmatic interactions: Upper Permian basaltic and basaltic andesite examples from Kiama, Australia: Australian Journal of Earth Sciences, v. 47, p. 167–177, doi: 10.1046/j.1440-0952.2000.00773.x.

Fink, J.H., and Fletcher, R.C., 1978, Ropy pahoehoe: Surface folding of a viscous fluid: Journal of Volcanology and Geothermal Research, v. 4, p. 151–170, doi: 10.1016/0377-0273(78)90034-3.

Fisher, R.V., and Schmincke, H.-U., 1984, Pyroclastic rocks: Heidelberg, Springer-Verlag, 472 p.

Gold, D.J.C., 1993, The geological evolution of the Pongola basin, southeastern Kaapvaal Craton [Ph.D. thesis]: Pietermaritzburg, University of Natal, 168 p.

Gregg, T.K.P., and Fink, J.H., 1995, Quantification of submarine lava-flow morphology through analog experiments: Geology, v. 23, p. 73–76, doi: 10.1130/0091-7613(1995)023<0073:QOSLFM>2.3.CO;2.

Griffin, K.R., 2002, Magmatic and Tectonic Development of the Pongola Supergroup and post-Pongola Granitoids, Kaapvaal Craton, South Africa [M.S. thesis]: University of Michigan, 58 p.

Groenewald, P.B., 1984, The lithostratigraphy and petrogenesis of the Nsuze Group, northwest of Nkandla, Natal [M.S. thesis]: Pietermaritzburg, University of Natal, 333 p.

Hatfield, R.A., 1990, The geology and structure of the Pongola Sequence east of Piet Retief, southeastern Transvaal [M.S. thesis]: Pietermaritzburg, University of Natal, 109 p.

Hegner, E., Kröner, A., and Hoffman, A.W., 1984, Age and isotope geochemistry of the Archaean Pongola and Usushwana suites in Swaziland, southern Africa: A case for crustal contamination of the mantle-derived magma: Earth and Planetary Science Letters, v. 70, p. 267–279, doi: 10.1016/0012-821X(84)90011-6.

Hegner, E., Kröner, A., and Hunt, P., 1994, A precise U-Pb zircon age for the Archaean Pongola Supergroup volcanics in Swaziland: Journal of African Earth Sciences, v. 18, p. 339–341, doi: 10.1016/0899-5362(94)90072-8.

Hon, K., Kauahikaua, J., Delinger, R., and MacKay, K., 1994, Emplacement and inflation of pahoehoe sheet flows: Observations and measurements of active lava flows on Kilauea Volcano, Hawaii: Geological Society of America Bulletin, v. 106, p. 351–370, doi: 10.1130/0016-7606(1994)106<0351:EAIOPS>2.3.CO;2.

Hunter, D.R., and Wilson, A.H., 1988, A continuous record of Archaean evolution from 3.5 Ga to 2.6 Ga in Swaziland and northern Natal: South African Journal of Geology, v. 91, p. 57–74.

Kamo, S.L., Davis, D.W., and de Witt, M.J., 1990, U-Pb geochronology of Archaean plutonism in the Barberton region, S. Africa: 800 Ma of crustal evolution: Abstract of the 7th International Conference on Geochronology, Cosmochronology and Isotope Geology: Canberra, Geological society of Australia, p. 53.

Kawachi, Y., and Pringle, I.J., 1988, Mutiple-rind structure in pillow lava as an indicator of shallow water: Bulletin of Volcanology, v. 50, p. 161–168, doi: 10.1007/BF01079680.

Kilburn, C.R.J., 1990, Surface of aa flow-fields on Mount Etna, Sicily: Morphology, rheology, crystallization and scaling phenomena, *in* Fink, J.H., ed., Lava flows and domes: Emplacement mechanisms and hazard implications: IAVCEI (International Association of Volcanology and Chemistry of the Earth's Interior), Proceedings in Volcanology 2: Berlin, Springer-Verlag, p. 129–156.

Kilburn, C.R.J., 1993, Lava crusts, aa flow lengthening and the pahoehoe-aa transition, *in* Kilburn, C.R.J., ed., Active Lavas: London, University College London Press, p. 263–280.

Kilburn, C.R.J., 2000, Lava flows and flow fields, *in* Sigurdson, H., ed., Encyclopedia of Volcanoes: San Diego, Academic Press, p. 291–305.

Le Maitre, R.W., 1989, A Classification of Igneous Rocks and Glossary of Terms: Oxford, Blackwell Scientific, 193 p.

Linström, W., 1987, The geology of the Dundee area: Expl. of sheet 2830: Pretoria, Geological Survey of South Africa, 52 p.

Macdonald, G.A., 1953, Pahoehoe, aa and block lavas: American Journal of Science, v. 251, p. 169–191.

Manley, C.R., 1992, Extended cooling and viscous flow of large, hot rhyolite lavas: Implications of numerical modelling results: Journal of Volcanology and Geothermal Research, v. 53, p. 27–46, doi: 10.1016/0377-0273(92)90072-L.

Mason, T.R., and von Brunn, V., 1977, 3 G-yr old stromatolites from South Africa: Nature, v. 266, p. 47–49, doi: 10.1038/266047a0.

Matthews, P.E., 1967, The Pre-Karroo Formation of the White Mfolozi Inlier, Northern Natal: Transactions of the Geological Society of South Africa, v. 70, p. 39–63.

Matthews, P.E., 1990, A plate tectonic model for the late Archaean Pongola Supergroup in Southeastern Africa, *in* Sychanthavong, S.P.H., ed., Crustal evolution and orogeny: New Delhi, Oxford Publishers, p. 41–72.

Matthews, P.E., and Scharrer, R.H., 1968, A graded unconformity at the base of the early Precambrian Pongola System: Transactions of the Geological Society of South Africa, v. 71, p. 257–272.

McMillan, K., Cross, R.W., and Long, P.E., 1987, Two-stage vesiculation in the Cohassett flow of the Grande Ronde basalt, south-central Washington: Geology, v. 15, p. 809–812, doi: 10.1130/0091-7613(1987)15<809:TVITCF>2.0.CO;2.

Moore, J.G., Phillips, R.L., Grigg, R.W., Peterson, D.W., and Swanson, D.A., 1973, Flow of lava into the sea 1969–1971, Kilauea Volcano, Hawaii: Geological Society of America Bulletin, v. 84, p. 537–546, doi: 10.1130/0016-7606(1973)84<537:FOLITS>2.0.CO;2.

Mueller, W.U., 2003, A subaqueous eruption model for shallow-water, small volume eruptions: Evidence from two Precambrian examples, *in* White, J.D.L., Smellie, J.L., and Clague, D.A., eds., Explosive Subaqueous Volcanism: Washington, D.C., American Geophysical Union, p. 189–203.

Preston, V.A., 1987, The geology and geochemistry of the Mpongoza inlier, northern Natal [M.S. thesis]: Pietermaritzburg, University of Natal, 282 p.

Rollinson, H., 1993, Using Geochemical Data: New York, John Wiley and Sons, 352 p.

Rowland, S.K., and Walker, G.P.L., 1988, Mafic-crystal distributions, viscosities and lava structures of some Hawaiian lava flows: Journal of Volcanology and Geothermal Research, v. 35, p. 55–66, doi: 10.1016/0377-0273(88)90005-4.

Rowland, S.K., and Walker, G.P.L., 1990, Pahoehoe and aa lava in Hawaii: Volumetric flow rate controls the lava structure: Bulletin of Volcanology, v. 52, p. 615–641, doi: 10.1007/BF00301212.

Self, S., Thordarson, T., and Keszthelyi, L., 1997, Emplacement of continental flood basalt lava flows, *in* Mahoney, J.J., and Coffin, M.F., eds., Large Igneous Provinces: Washington, D.C., American Geophysical Union, p. 381–410.

Self, S., Keszthelyi, L., and Thordarson, T., 1998, The importance of pahoehoe: Annual Reviews in Earth and Planetary Science, v. 26, p. 81–110, doi: 10.1146/annurev.earth.26.1.81.

Skilling, I.P., White, J.D.L., and McPhie, J., 2002, Peperite: A review of magma-sediment mingling: Journal of Volcanology and Geothermal Research, v. 114, p. 1–17, doi: 10.1016/S0377-0273(01)00278-5.

Solomon, M., 1969, The nature and possible origin of pillow lavas and hyaloclastite breccias of King Island, Australia: Quarterly Journal of the Geological Society of London, v. 124, p. 159–169.

South African Commission for Stratigraphy (SACS), 1980, Lithostratigraphy of the Republic of South Africa, South West Africa/Namibia, and the Republics of Bophuthatswana, Transkei and Venda, *in* Kent, L.E., ed., Stratigraphy of South Africa, Part 1: Pretoria, Geological Survey of South Africa, p. 690.

Staudigel, H., and Schmincke, H.-U., 1984, The Pliocene seamount series of La Palma/Canary Island: Journal of Geophysical Research, v. 89, p. 11,195–11,215.

Sun, S.-s., and McDonough, W.F., 1989, Chemical and isotopic systematics of oceanic basalts: Implications for mantle composition and processes, *in* Saunders, A.D., and Norry, M.J., eds., Magmatism in the Ocean Basins: London, Geological Society, p. 313–345.

Sylvester, P.J., Harper, G.D., Byerly, P.W., and Thurston, P.C., 1997, Volcanic aspects, *in* de Wit, M.J., and Ashwal, L.D., eds., Greenstone Belts: Oxford, Clarendon Press, p. 55–90.

Tankard, A.J., Jackson, M.P.A., Eriksson, K.A., Hobday, D.K., Hunter, D.R., and Minter, W.E.L., 1982, Crustal evolution of Southern Africa: New York, Springer-Verlag, 523 p.

Thurston, P.C., and Chivers, K.M., 1990, Secular evolution in greenstone sequence development emphasizing Superior Province, Canada: Precambrian Research, v. 46, p. 21–58, doi: 10.1016/0301-9268(90)90065-X.

Ukstins, P., Larsen, M., and Lesher, C.E., 2003, The transition from sedimentation to flood volcanism in the Kangerlussuaq Basin, East Greenland: Basaltic pyroclastic volcanism during initial Palaeogene continental breakup: Journal of the Geological Society, v. 160, p. 759–772.

von Brunn, V., and Mason, T.R., 1977, Siliciclastic-carbonate tidal deposits from the 3000 Ma Pongola Supergroup, South Africa: Sedimentary Geology, v. 18, p. 245–255, doi: 10.1016/0037-0738(77)90014-8.

Walker, G.P.L., 1991, Structure, and origin by injection under surface crust, of tumuli, 'lava rises,' 'lava-rise pits,' and 'lava inflation clefts' in Hawaii: Bulletin of Volcanology, v. 53, p. 546–558, doi: 10.1007/BF00298155.

Walker, G.P.L., 1992, Morphometric study of pillow-size spectrum among pillow lavas: Bulletin of Volcanology, v. 54, p. 459–474, doi: 10.1007/BF00301392.

Walraven, F., and Pape, J., 1994, Pb-Pb whole-rock ages of the Pongola Supergroup and the Usushwana Complex, South Africa: Journal of African Earth Sciences, v. 18, p. 297–308, doi: 10.1016/0899-5362(94)90069-8.

Watchorn, M.B., and Armstrong, N.V., 1980, Contemporaneous sedimentation and volcanism at the base of the Precambrian Nsuze Group, South Africa: Transactions of the Geological Society of South Africa, v. 83, p. 231–238.

White, J.D.L., McPhie, J., and Skilling, I., 2000, Peperite: A useful genetic term: Bulletin of Volcanology, v. 62, p. 65–66, doi: 10.1007/s004450050293.

White, J.D.L., Smellie, J.L., and Clague, D.A., 2003, Introduction: A deductive outline and topical overview of subaqueous explosive volcanism, *in* White, J.D.L., Smellie, J.L., and Clague, D.A., eds., Explosive Subaqueous Volcanism: Washington, D.C., American Geophysical Union, p. 1–23.

Wilmoth, R.A., and Walker, G.P.L., 1993, P-type and S-type pahoehoe: A study of vesicle distribution patterns in Hawaiian lava flows: Journal of Volcanology and Geothermal Research, v. 55, p. 129–142, doi: 10.1016/0377-0273(93)90094-8.

Wilson, A.H., 2003, A new class of silica enriched, highly depleted komatiites in the southern Kaapvaal Craton, South Africa: Precambrian Research, v. 127, p. 125–141, doi: 10.1016/S0301-9268(03)00184-0.

Wood, D.A., 1980, The application of a Th-Hf-Ta diagram to problems of tectonomagmatic classification and to establishing the nature of crustal contamination of basaltic lavas of the British Tertiary volcanic province: Earth and Planetary Science Letters, v. 50, p. 11–30, doi: 10.1016/0012-821X(80)90116-8.

Manuscript Accepted by the Society 4 October 2005

Printed in the USA.

Geological Society of America
Special Paper 405
2006

The structural-metamorphic evolution of the northern margin of the Zimbabwe Craton and the adjacent Zambezi belt in northeastern Zimbabwe

Paul H.G.M. Dirks
School of Geosciences, University of the Witwatersrand, P Bag 3, Wits 2050, Johannesburg, South Africa
Hielke A. Jelsma
Strategy and Targeting, De Beers Exploration Division, P.O. Box 82232, Southdale 2135, South Africa and *School of Geosciences, University of the Witwatersrand, P Bag 3, Wits 2050, Johannesburg, South Africa*

ABSTRACT

The Archean Zimbabwe Craton in northeastern Zimbabwe is bounded along its northern margin by the east-trending Zambezi belt, which has been subdivided from south to north into the Archean Migmatitic Gneiss Terrain, overlain by Proterozoic gneiss units of the Marginal Gneiss Terrain and the Allochthonous Gneiss Terrain. The transition from the Zimbabwe Craton into the Migmatitic Gneiss Terrain occurs over a 10–15-km-wide zone across which the metamorphic grade increases from 3 to 4 Kbar and 500 °C in the northern part of the craton, to 6–7 Kbar and 700 °C in the Migmatitic Gneiss Terrain. This P-T (Pressure-Temperature) gradient was established in the late Archean and reflects exhumation of the Migmatitic Gneiss Terrain as a result of D_1 sinistral transpression due to oblique collision between the Zimbabwe Craton and an unknown rock mass to the north. D_1 structures are absent within the Marginal and Allochthonous Gneiss Terrains.

The craton margin and Migmatitic Gneiss Terrain were subsequently intruded and crosscut by the 2.58 Ga Great Dyke. Between 1.1 and 0.5 Ga renewed tectonism occurred along the craton margin, forming the Zambezi belt. Deformation fabrics linked to this are mainly preserved in the Marginal and Allochthonous Gneiss Terrains and include 0.85–1.05 Ga high-P granulite-facies fabrics (D_2) extensively overprinted by D_3 fabrics that formed ~0.80 b.y. ago. D_3 fabrics preserve mylonitic characteristics across the Marginal Gneiss Terrain, which is part of a 3–5-km-thick crustal-scale, shallowly north- to northeast-dipping shear zone that can be traced along the full length of the Zambezi belt north of the Zimbabwe Craton. This shear zone accommodated extension and exhumation of rocks in the Marginal Gneiss Terrain, the Migmatitic Gneiss Terrain, and the craton margin. A renewed burial and exhumation cycle affected the rocks of the Marginal Gneiss Terrain, the Migmatitic Gneiss Terrain, and the craton margin ca. 0.53 Ga, as units of the Allochthonous Gneiss Terrain overthrust the margin of the Zimbabwe Craton.

Keywords: Zimbabwe Craton, Zambezi belt, tectonics, Pan-African extension, transpression, low-angle detachment, Zimbabwe.

Dirks, P.H.G.M., and Jelsma, H.A., 2006, The structural-metamorphic evolution of the northern margin of the Zimbabwe Craton and the adjacent Zambezi belt in northeastern Zimbabwe, *in* Reimold, W.U., and Gibson, R.L., Processes on the Early Earth: Geological Society of America Special Paper 405, p. 291–313, doi: 10.1130/2006.2405(15). For permission to copy, contact editing@geosociety.org.

INTRODUCTION

The contacts between Archean cratons and adjacent Archean gneiss terrains involve either sharp or gradational structural-metamorphic transition zones (e.g., De Wit and Ashwal, 1997). In terms of rock types and age, the gneiss terrains are commonly comparable to the cratons, and they have been variably interpreted as: (1) lateral equivalents of the craton that have been involved in subsequent orogenic events; (2) deep-crustal equivalents of the craton that have been exhumed in (oblique) crustal sections; or (3) unrelated terranes that have been accreted against the craton in a convergent plate margin-type setting (Shackleton, 1995; De Wit and Ashwal, 1997; Percival et al., 1997). Examples have been documented in the Superior Province (e.g., Stott, 1997), Yilgarn Block (e.g., Myers, 1987), Antarctica (e.g., Sims et al., 1994), and the Kaapvaal Craton (e.g., Brandl and De Wit, 1997).

The Zimbabwe Craton in southern Africa is an example of an Archean craton that is bounded on all sides by gneiss terrains including the Limpopo belt to the south, the Magondi belt to the west, the Mozambique belt to the east, and Zambezi belt to the north (Fig. 1; Jelsma and Dirks, 2002; Dirks et al., 2003). Of these, only the contact between the craton and the Limpopo belt is reasonably well documented (e.g., Treloar et al., 1992; Rollinson, 1993), while all others are poorly understood or remain undefined as is the case with the eastern contact of the craton. In this paper, we provide a description of the sequence of deformational events that occurred along the northern contact of the Zimbabwe Craton and the adjacent Zambezi belt, with a view to elucidate its tectonic evolution.

The Zimbabwe Craton is characterized by granite-greenstone terrains that formed during a time span of at least 900 m.y. (3.5–2.6 Ga; Jelsma and Dirks, 2002). The greenstone belts consist of (ultra-)mafic to felsic volcanic units and coarse clastic sedimentary rocks intruded by granitoid plutons. They were generally metamorphosed at low pressures and low to moderate temperatures (Stagman, 1978; Jelsma and Dirks, 2002).

Along the northern margin of the craton, the Zambezi belt stretches for over 400 km (Fig. 1). The Zambezi belt is part of a transcontinental, Pan-African, orogenic system that includes from west to east, the Damara belt, the Lufillian Arc, the Zambezi belt, and the Mozambique belt, and it separates the Congo Craton to the north from the Kalahari Craton to the south (Fig. 1; e.g., Dirks et al., 2003; Hanson, 2003). The Zambezi belt was defined in central Zambia as an east-southeast–trending orogenic zone characterized by Neoproterozoic volcano-sedimentary sequences of the Katangan Supergroup, which truncate Mesoproterozoic gneiss of the Irumide belt to the northeast and the Choma-Kaloma block to the southwest (e.g., Coward and Daly, 1984; Hanson et al., 1994; De Waele, 2005; Fig. 1). It extends eastward, across poorly mapped domains in eastern Zambia and northwestern Mozambique, and merges with the Mozambique belt across a triple junction characterized by continuous structural trends and metamorphic isograds, indicating that both belts formed part of a related Proterozoic orogenic system (e.g., Barton et al., 1991; Goscombe et al., 2000; Dirks et al., 2003).

The Zambezi rift valley along the center of the Zambezi belt divides the belt into two parts. To the north, reworked Proterozoic gneiss terrains occur that characteristically record 900–1100 Ma and 500–600 Ma events (e.g., Goscombe et al., 2000; Johnson and Oliver, 2000; Hanson, 2003; De Waele, 2005). South of the rift, gneiss and metasedimentary sequences occur along with reworked Archean crust, and 800–900 Ma and 500–600 Ma events dominate (e.g., Stocklmayer, 1980; Barton et al., 1991; Dirks et al., 1999; Vinyu et al., 2001; Fig. 1).

In this paper we present structural data from a 70-km-long and 5–15-km-wide transect across the contact zone between the Zimbabwe Craton and the Zambezi belt (Fig. 2). The scope of this transect is to distinguish Archean and Proterozoic tectono-metamorphic events and better constrain the following: (1) the kinematic nature and extent of late Archean deformation events along the northern margin of the Zimbabwe Craton and within the adjacent Zambezi belt, and (2) the tectonic significance of Neoproterozoic to early Paleozoic events recorded in the Zambezi belt in northeastern Zimbabwe and the impact of these events on the craton margin.

REGIONAL SETTING

The Zimbabwe Craton in southern Africa (Fig. 1) has an areal extent of 260,000 km^2 and is characterized by a greenstone-granite (*sensu lato*) ratio of ~1:8. It has a long history of crustal growth that started in the early Archean (ca. 3.5 Ga) and became a stable crustal segment at around 2.6 Ga after the emplacement of late- to post-tectonic, sheetlike, monzogranite batholiths (Chilimanzi Suite; Becker et al., 2000), shortly followed by the Great Dyke (2.58 Ga; Armstrong and Wilson, 2000; Wingate, 2000). The majority of rocks (~90%) formed between 2.9 and 2.6 Ga and include volcanic and sedimentary units of the Belingwean (2.9 Ga), Lower Bulawayan (2.8 Ga), Upper Bulawayan, and Shamvaian (2.7–2.6 Ga) greenstone sequences and voluminous granite plutons that have been assigned to the Chingezi (2.9 Ga), Sesombi, and Wedza (2.7–2.6 Ga) suites (Stagman, 1978; Wilson et al., 1995).

The Zambezi belt occurs to the north and northeast of the Zimbabwe Craton (Fig. 1) and has been divided into three domains (Barton et al., 1991; Dirks et al., 1998). From south to north these are (Fig. 2): (1) a domain comprising high-grade Archean gneiss overprinted during post-Archean orogenic events and referred to as the Migmatitic Gneiss Terrain (MigGT; Barton et al., 1991); (2) the Marginal Gneiss Terrain (MarGT; Barton et al., 1991), which consists of a high-grade, Proterozoic, paragneiss sequence, tectonically interleaved with orthogneiss units and intruded by 800 Ma, syntectonic granite sheets. The Marginal Gneiss Terrain experienced intense deformation and metamorphism during Pan-African orogenic events (Barton et al., 1991, 1993; Carney et al., 1991) and is com-

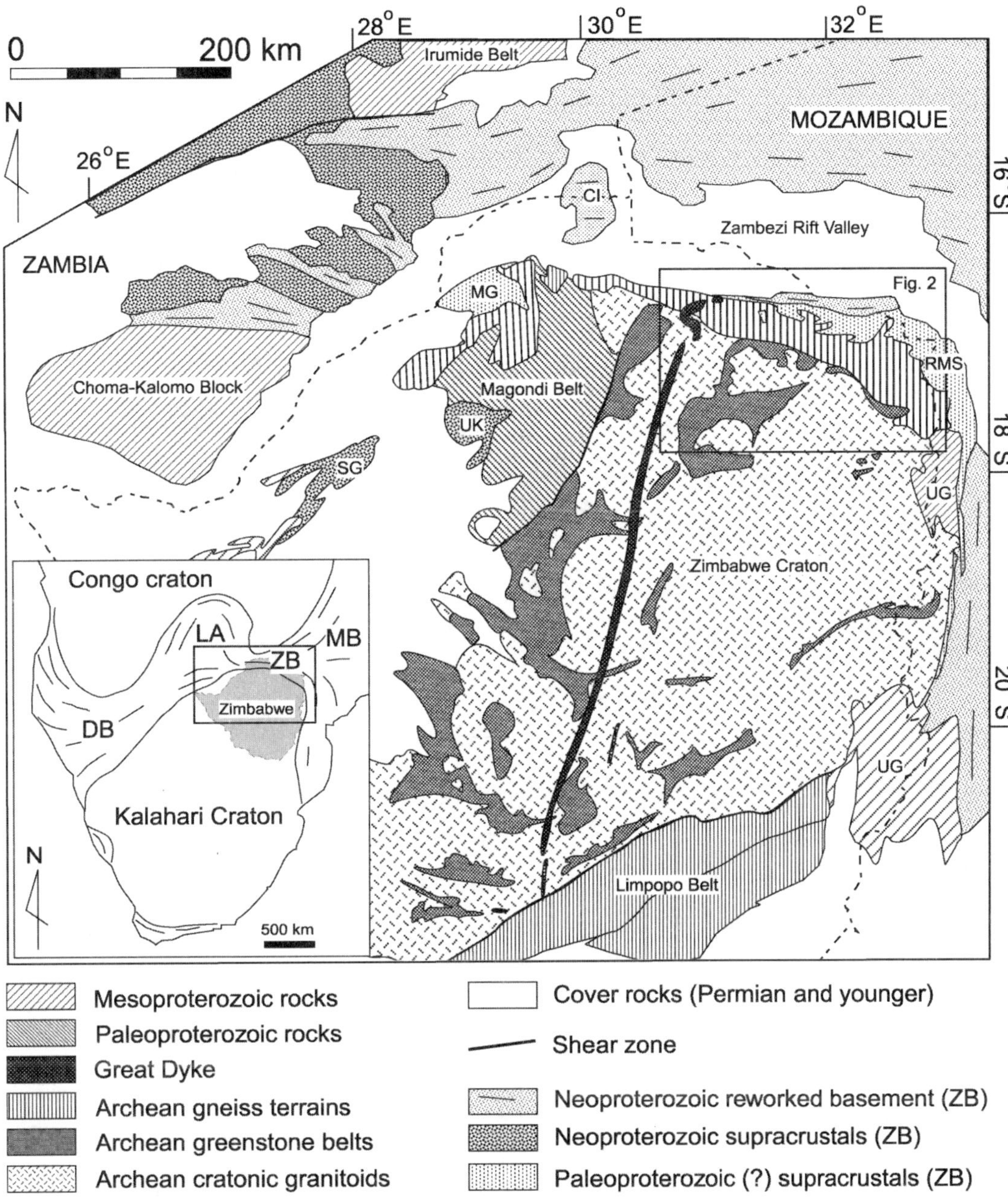

Figure 1. Geological map of the Zimbabwe Craton and Zambezi belt (after Dirks et al., 2002 and Hargrove et al., 2003). CI—Chewore Inliers, MG—Makuti Group, UK—Urungwe Klippe, SG—Sijarira Group, RMS—Rushinga Metamorphic Suite, UG—Umkondo Group. Inset: Pan-African, transcontinental mobile belts in southern Africa. DB—Damara belt, LA—Lufilian Arc, ZB—Zambezi belt, MB—Mozambique belt.

plexly infolded with the underlying Migmatitic Gneiss Terrain (Fig. 2; Barton et al., 1991, 1993; Dirks et al., 1998); and (3) the Allochthonous Gneiss Terrain, which consists of felsic and mafic gneiss units that experienced intense deformation and metamorphism during 800 Ma and 500 Ma events (Fig. 2; Barton et al., 1991; Mueller, 2004).

The northern limit of the Zimbabwe Craton is generally taken as the northern contact of a near-continuous stretch of east-west–trending greenstone belts that include (from east to west) the Makaha, Makati, Dindi, and Mount Darwin greenstone belts (Fig. 2; Stagman, 1978; Barton et al., 1991). This contact is referred to as the Pfunzi Boundary Slide (Fig. 2; Barton et al., 1991) and approximately coincides with the northernmost extent of post-tectonic granite batholiths belonging to the Chilimanzi Suite.

The structural-metamorphic transition from the Zimbabwe Craton into the Zambezi belt is gradational and occurs within a 10–15-km-wide zone, in this paper referred to as the craton mar-

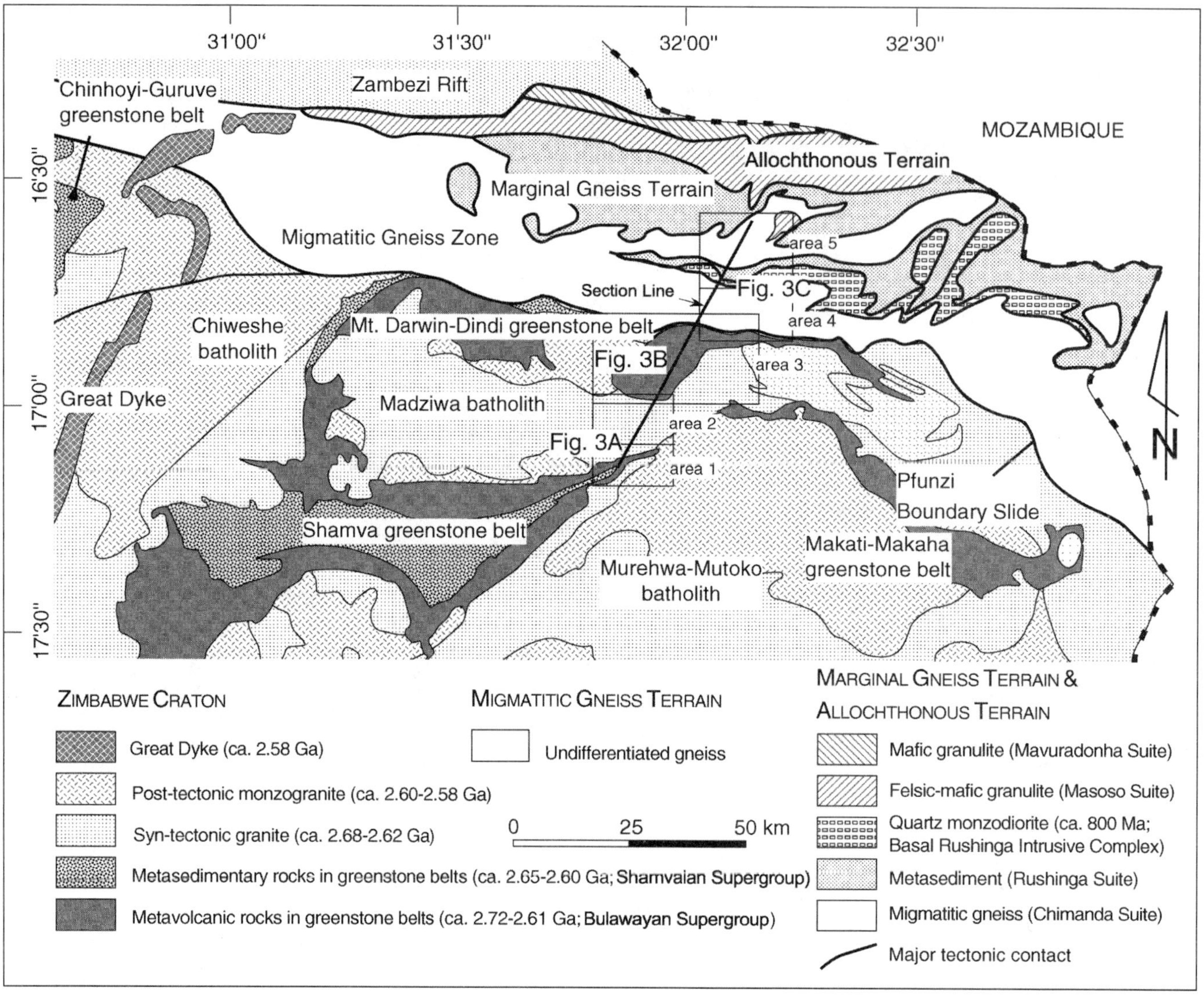

Figure 2. Tectono-stratigraphic map of the contact between the Zimbabwe Craton and the Zambezi belt in northeastern Zimbabwe, delineating the extent of the main lithotectonic domains as well as the relationship with the Great Dyke. The positions of the transect detailed in Figures 3A–3C and areas 1–5 are indicated.

gin, which exposes granite-greenstone units that display progressively higher metamorphic conditions as the Zambezi belt is approached. From south to north across the craton margin, the metamorphic grade increases from 3 to 4 Kbar and 500 °C, typical for greenstone sequences in the northern part of the craton (Jelsma and Dirks, 2000, 2002), to 6–7 Kbar and 700 °C typical for gneiss units in the Zambezi belt (Barton et al., 1991; Carney et al., 1991; Dirks et al., 1998).

North of the study area, within the Zambezi rift valley, several gneiss inliers that are younger than 1400 Ma occur within the Zambezi belt. These inliers experienced extensive tectonism around 940–1070 Ma and again around 530 Ma, but they preserve no evidence for 800 Ma events as observed in the study area (Fig. 1; Oliver et al., 1998; Johnson and Oliver, 2000; Goscombe et al., 2000; Hanson, 2003). The relationships between these inliers and the terrains discussed in this paper are poorly established, due to lack of continuous outcrop.

Zimbabwe Craton and Craton Margin

The northern Zimbabwe Craton consists of greenstone belts separated by granite-gneiss terrains (Fig. 2). Away from the craton margin, metamorphic grades in the greenstone belts are low (Barton et al., 1991; Jelsma and Dirks, 2002), and greenstone belts such as the Shamva-Bindura belt located 30–50 km south of the craton margin (Fig. 2) are not directly affected by orogenic events external to the craton, except for discrete later faults (Jelsma, 1993; Jelsma and Dirks, 2000).

Greenstone belts along the craton margin near the contact with the Migmatitic Gneiss Terrain are intensely sheared, dismembered, and metamorphosed (Johnson, 1968; Stocklmayer, 1978; Barton et al., 1991; Hofmann et al., 2002a). These greenstone belts as well as the southern margin of the Migmatitic Gneiss Terrain are intruded by ca. 2.6-Ga-old Chilimanzi-type monzogranite plutons (Murehwa-Mutoko, Chiweshe, and Madziwa batholiths, Fig. 2; Hofmann et al., 2002b; Jelsma and Dirks 2002). The plutons are massive with weak or moderate foliations developed near and parallel to the contact with the Migmatitic Gneiss Terrain (Leitner and Phaup, 1974; Bache et al., 1984; Barton et al., 1991; Becker et al., 2000). The contact between the craton and the Migmatitic Gneiss Terrain is intruded and crossed by the 2575 Ma (Armstrong and Wilson, 2000; Wingate, 2000; Table 1) Great Dyke (Fig. 2), which at this point is unmetamorphosed and unaffected by any of the tectonic events that characterize the craton margin.

Migmatitic Gneiss Terrain

The Migmatitic Gneiss Terrain is exposed to the north of the craton margin over a strike length of at least 400 km and a width of 10–40 km (Figs. 1 and 2). The Migmatitic Gneiss Terrain preserves strongly retrogressed high-grade gneiss units that are of Archean origin, including mafic volcanic and sedimentary units, which probably originated as greenstone sequences (Barton et al., 1991; Hofmann et al., 2002a). The dominant gneissic layering in this zone strikes west-northwest, parallel to

TABLE 1. COMPILATION OF RELEVANT AGE DATA FOR THE STUDY AREA

Stratigraphic unit	Lithology†	Position		Age	Error	Interpretation	Technique
		Long.	Lat.	(Ma)	(Ma)		
Zimbabwe craton margin and Great Dyke							
Great Dyke	Quartz gabbro (1)	30.458	−18.060	2571	9	Crystallization age	SHRIMP, zircon
Great Dyke	Quartz gabbro (1)	30.458	−18.060	2573	2	Crystallization age	SHRIMP, zircon
Great Dyke	Websterite (2)	30.050	−19.630	2579	3	Crystallization age	SHRIMP, zircon
Great Dyke	Gabbro (3)	30.421	−18.061	2586	16	Crystallization age	Sm-Nd, mineral–whole rock
Great Dyke	Gabbro (3)	30.421	−18.061	2587	8	Crystallization age	U-Pb, single rutile, TIMS
Arcturus Formation	Amphibolite (4)	31.718	−17.296	2572	44	Metamorphic age (cooling age)	Ar-Ar, hornblende
Arcturus Formation	Amphibolite (4)	31.579	−17.387	2606	8	Metamorphic age (cooling age)	TIMS, U/Pb garnet
Arcturus Formation	Amphibolite (4)	31.579	−17.387	2621	60	Metamorphic age (cooling age)	Ar-Ar, hornblende
Black Cat porphyry	Andesitic porphyry (5)	31.583	−17.340	2672	12	Crystallization age	U-Pb, zircon fractions, TIMS
Chilimanzi Suite	Granite (6)	32.831	−17.252	2570	0.2	Crystallization age	Pb-Pb, single zircon, Kober
Chilimanzi Suite	Granite (5)	31.820	−17.290	2601	14	Crystallization age	U-Pb, zircon fractions, TIMS
Maparu Formation	Water-laid tuff (7)	31.526	−17.200	2697	9	Crystallization age	Pb-Pb, single zircon, SHRIMP
Shamvaian Supergroup	Dacitic porphyry (5)	31.571	−17.286	2641	65	Crystallization age	Pb-Pb, whole rock
Shamvaian Supergroup	Dacitic tuff (8)	31.580	−17.325	2666	36	Crystallization age	Pb-Pb, single zircon, SHRIMP
Migmatitic Gneiss Terrain							
Mudzi Suite	Granitic gneiss (6)	32.831	−17.252	2593	0.6	Crystallization age	Pb-Pb, single zircon, Kober
Mudzi Suite	Granodiorite gneiss (6)	32.831	−17.252	2601	0.6	Crystallization age	Pb-Pb, single zircon, Kober
Pfungwe Suite	Hbl, amphibolite (9)	32.035	−16.824	491	2	Metamorphic age (cooling age)	Ar-Ar, hornblende
Pfungwe Suite	Hbl, amphibolite (9)	32.043	−16.722	508	3	Metamorphic age (cooling age)	Ar-Ar, hornblende
Pfungwe Suite	Tonalitic orthogneiss (10)	32.035	−16.751	2621	9	Crystallization age	U-Pb, single zircon, TIMS
Pfungwe Suite	Granitic leucogneiss (10)	32.032	−16.827	2623	1	Crystallization age	U-Pb, single zircon, TIMS
Pfungwe Suite	Granite (10)	32.052	−16.738	2625	11	Crystallization age	U-Pb, single zircon, TIMS
Marginal Gneiss Terrain and Basal Rushinga Intrusive Complex (BRIC)							
BRIC	Granodioritic gneiss (9)	32.039	−16.726	521	71	Metamorphic age	TIMS, U-Pb, zircon
BRIC	Granodioritic gneiss (9)	32.039	−16.726	805	11	Crystallization age	TIMS, U-Pb, zircon
Chironga Gneiss	Granitic gneiss (11)	31.698	−16.496	795	2.1	Crystallization age	TIMS, U-Pb, zircon
Allochtonous Gneiss Terrain							
Masoso Suite	Banded mafic gneiss (9)	32.143	−16.519	525	3	Metamorphic age (cooling age)	Ar-Ar, hornblende
Masoso Suite	Banded mafic gneiss (9)	32.143	−16.519	534		Metamorphic age	TIMS, U-Pb, zircon
Masoso Suite	Leucomigmatite (9)	32.144	−16.521	535		Metamorphic age (cooling age)	TIMS, U-Pb, titanite
Masoso Suite	Leucomigmatite (9)	32.144	−16.521	535		Metamorphic age	TIMS, U-Pb, zircon
Masoso Suite	Corona-textured monzodiorite (14)	32.191	−16.618	854	1	Crystallization age	Pb-Pb, single zircon, Kober
Masoso Suite	Leucomigmatite (9)	32.144	15.274	870	1	Crystallization/metamorphic age	TIMS, U-Pb, zircon
Masoso Suite	Garnet-bearing granitic gneiss (11)	31.523	−16.439	870	5	Crystallization age	TIMS, U-Pb, zircon
Masoso Suite	Porphyritic granodioritic gneiss (12)	31.848	−16.468	1022	7	Crystallization age	U-Pb, single zircon, TIMS
Masoso Suite	Tonalitic gneiss (12)	31.795	−16.401	1028	28	Crystallization age	U-Pb, single zircon, TIMS
Masoso Suite	Granitic gneiss (11)	31.846	−16.458	1052	14	Crystallization age	TIMS, U-Pb, zircon
Masoso Suite	Pegmatite (11)	31.844	−16.460	1059	19	Crystallization age	TIMS, U-Pb, zircon
Masoso Suite	Metagabbro (13)	32.187	−16.622	849	2	Crystallization age	TIMS, U-Pb, zircon
Mavuradonha Suite	Anorthosite (12)	31.795	−16.389	502	6	Metamorphic age (cooling age)	U-Pb, single rutile TIMS
Mavuradonha Suite	Amphibolite (11)	31.748	−16.445	530		Metamorphic age (cooling age)	TIMS, U-Pb, titanite
Mavuradonha Suite	Metagabbro (12)	31.779	−16.405	538	2	Metamorphic age	U-Pb, single zircon, TIMS
Mavuradonha Suite	Pegmatite vein (12)	31.754	−16.395	540	19	Crystallization/metamorphic age	U-Pb, single zircon, TIMS
Mavuradonha Suite	Pegmatite vein (12)	31.716	−16.428	545	3	Crystallization age	U-Pb, single zircon, TIMS
Mavuradonha Suite	Anorthosite (12)	31.795	−16.389	546	9	Metamorphic age (cooling age)	Sm-Nd; gnt-WR isochron
Mavuradonha Suite	Amphibolite (11)	31.738	−16.441	550		Metamorphic age (cooling age)	TIMS, U-Pb, titanite
Mavuradonha Suite	Metagabbro (12)	31.779	−16.405	862	8	Crystallization age	SHRIMP, zircon

Note: Rb-Sr whole-rock data have not been included. TIMS—tertiary ion mass spectrometry; SHRIMP—sensitive high-resolution ion microprobe.

†Numbers in parentheses correspond to the following: (1) Wingate (2000); (2) Armstrong and Wilson (2000); (3) Mukasa et al. (1998); (4) Van Dijk and Kater (1996); (5) Jelsma et al. (1996); (6) Hofmann et al. (2002a); (7) Wilson et al. (1995); (8) Nesbitt et al. (2000); (9) Vinyu et al. (1999); (10) Vinyu et al. (2001); (11) Hargrove et al. (2003); (12) Mueller (2004); (13) Mariga (1999); (14) Dirks et al. (2000).

the craton margin, and is transected by the Great Dyke. Gneiss units of the Migmatitic Gneiss Terrain have yielded U-Pb zircon ages for syntectonic granitoids of 2560–2620 Ma (Vinyu et al., 2001; Hofmann et al., 2002a; Table 1), suggesting that the formation of the penetrative gneissic layering is synchronous with the development of the younger granite-greenstone terrains in the craton (Dirks et al., 1998; Jelsma and Dirks, 2002).

Barton et al. (1991) postulated two Archean orogenic events for the Migmatitic Gneiss Terrain. Evidence for an earlier event is preserved in isolated pods and sheets of felsic and mafic granulite with complicated metamorphic textures and relict fold interference structures. These rocks were exhumed during a late Archean event that Barton et al. (1991) correlate with 2.6–2.7 Ga tectonism within the adjacent granite-greenstone terrains.

Marginal Gneiss Terrain

The Marginal Gneiss Terrain (Barton et al., 1991) of the Zambezi belt is exposed between the Migmatitic Gneiss Terrain to the south and the Allochthonous Gneiss Terrain and Mesozoic Zambezi rift basin to the north (Fig. 2). The Marginal Gneiss Terrain can be subdivided into a number of mostly shear-zone–bounded units of varying age (Barton et al., 1991; Vinyu et al., 1999; Dirks et al., 2000), which include medium-grade metasedimentary and metavolcanic gneiss units (Rushinga Metamorphic Suite) infolded with reworked orthogneiss fragments of Paleoproterozoic age (Dirks et al., 2000) and intruded by voluminous, peraluminous granite sheets of Neoproterozoic age (700–800 Ma, Basal Rushinga Intrusive Complex; Barton et al., 1991; Vinyu et al., 1999; Hanson, 2003; Table 1). To the north, the Marginal Gneiss Terrain is in tectonic contact with high-grade, quartzo-feldspathic and mafic gneiss units of Neo- and Mesoproterozoic age (Vinyu et al., 1999) that contain granulite and eclogite lenses (Masoso Suite of the Allochthonous Gneiss Terrain; Barton et al., 1991; Carney et al., 1991). These allochthonous gneiss units are themselves tectonically overlain by mafic to ultramafic gneiss belonging to a deformed, 860-Ma-old, layered igneous complex (Mavuradonha Suite of the Allochthonous Gneiss Terrain; Mueller, 2004; Table 1, Fig. 2).

The age of the sediments of the Rushinga Metamorphic Suite is unconstrained. The unit was first linked to the Umkondo Group exposed along the edge of the craton in southeastern Zimbabwe (e.g., Johnson, 1968; Fig. 1) but is now more commonly interpreted as an 800 Ma rift-fill related to the Katangan Sequence in Zambia (e.g., Bache et al., 1984; Hanson et al., 1994; Hanson, 2003). The latter correlation hinges on the interpretation of evenly grained, massive, laminated, quartzo-feldspathic units, common throughout the Rushinga Metamorphic Suite and dated at 795 ± 2 Ma (TIMS [Tertiary Ion Mass Spectrometry], zircon; Hargrove et al., 2003), as felsic metavolcanic rocks (e.g., Munyanyiwa et al., 1997). This interpretation is uncertain, because the quartzo-feldspathic units are invariably mylonitic in nature and display ample evidence of mesoscopic isoclinal folding, transposition, extensive grain-size reduction, formation of ribbon grains, fabric asymmetries, and post-tectonic annealing and grain growth (e.g., Dirks et al., 1999; Sithole, 2002). It is therefore possible that the units represent intrusive granite sheets, equivalent to the Basal Rushinga Intrusive Complex (BRIC) with which they share geochemical similarities (e.g., Barton et al., 1991; Hargrove et al., 2003), that intruded the Rushinga Metamorphic Suite and were intensely sheared during Pan-African events. If so, the Rushinga Metamorphic Suite must be older than 800 Ma, which is consistent with a limited number of provenance dates for zircons from metapelite units in the Rushinga Metamorphic Suite, which are all in excess of 2100 Ma (Pb-Pb Kober ages; Dirks et al., 2000). We therefore suggest that the Rushinga Metamorphic Suite may be related to the sedimentary sequences of the Magondi Group, common along the western margin of the craton (Fig. 1).

STRUCTURAL HISTORY

Detailed structural observations are presented from five adjacent areas selected along a 70-km-long transect (Fig. 2). From south to north these areas are: Area 1—Craton: Shamva greenstone belt; Area 2—Craton: Madziwa batholith; Area 3—Craton margin: Dindi greenstone belt; Area 4—Migmatitic Gneiss Terrain; Area 5—Marginal Gneiss Terrain (Figs. 2 and 3).

Deformation structures are grouped into 4 major events: D_1 to D_4. In this paper, all Archean events are referred to as D_1, all 850–1100 Ma compressional events are grouped as D_2, 750–850 Ma extensional events are collectively termed D_3, and Pan-African 500–600 Ma events are grouped as D_4. Each of these four deformation events comprises a range of structures that resulted from dynamic processes, which involved folding, transposition, and refolding, generally within a non-coaxial flow regime. This resulted in the progressive development of a range of foliations and fold geometries, which we will present as temporally discrete events (e.g., D_{1a}, D_{1b}), even though some spatially distinct fabrics and fold geometries probably developed synchronously in separate parts of a partitioned deformation system (see Dirks et al., 1998). In many places deformation geometries and fabrics that are closely related share a common lineation direction (mineral lineation, intersection lineation, or fold axis).

The metamorphic history in the different areas is complex and characterized by multiple overprints. A full discussion of this is beyond the scope of this paper and will be presented elsewhere. For each area only a brief summary of the general metamorphic conditions is given, and a summary of all events is presented in Tables 1 and 2.

Area 1: Craton–Shamva Greenstone Belt

The northeastern part of the Shamva greenstone belt (SGB; Fig. 3A) has been divided into two lithotectonic blocks that are separated by a prominent, 50–100-m-wide, east-northeast–trending, D_1 layer-parallel shear zone (Jelsma and Dirks, 2000;

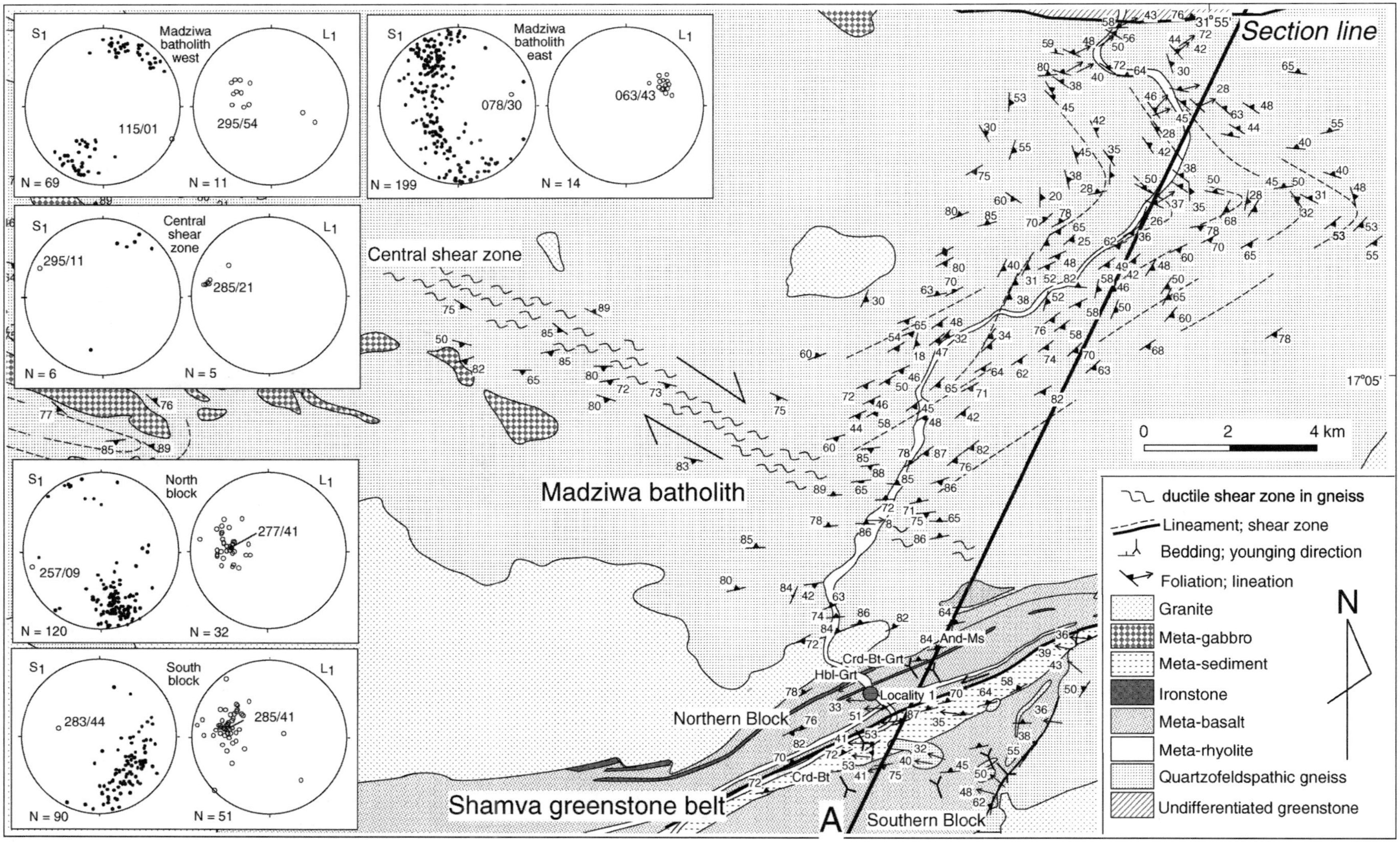

Figure 3. *(this and following two pages)* (A) Geological map of the northeastern part of the Shamva greenstone belt (area 1) and the Madziwa batholith (area 2). Shown are the main lithologies, structures, lithotectonic domains, prominent shear zones, a locality for strain analysis (Loc. 1), and some indicative metamorphic mineral assemblages in pelitic units. The insets show lower hemisphere equal area stereoplots for poles to S_1 foliation planes (left) and for L_1 lineation directions (right). Structural data is based on detailed field studies by former Bachelor of Science Honors students of the University of Zimbabwe, by Stidolph (1977) and authors' data.

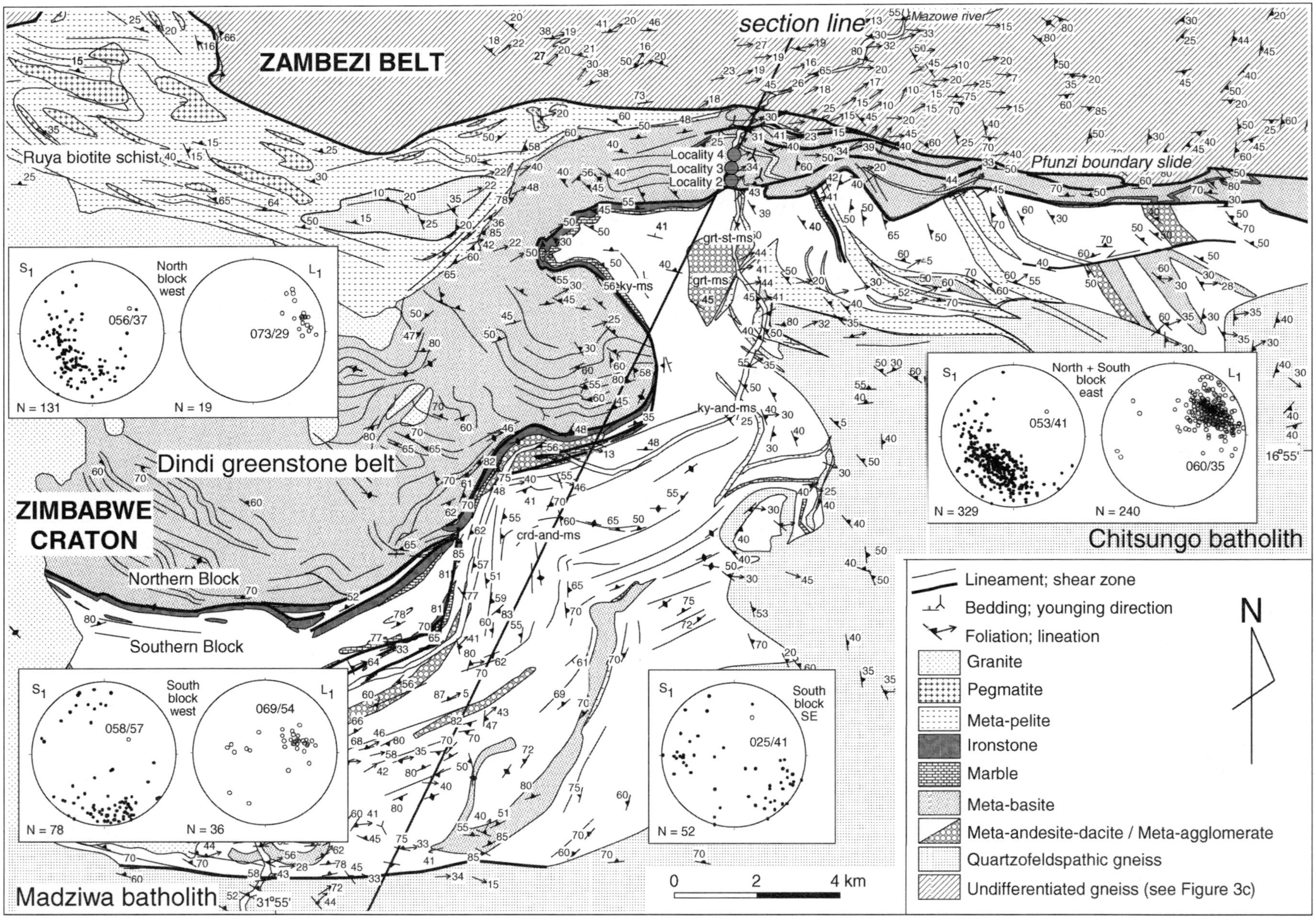

Figure 3. *(continued)* (B) Geological map of the Dindi greenstone belt (area 3). Shown are the main lithologies, structures, lithotectonic domains, prominent shear zones, localities for strain analysis (Locations 2–4), and some indicative metamorphic mineral assemblages in pelitic units. The insets show lower hemisphere equal area stereoplots for poles to S_1 foliations (left) and for L_1 lineations (right). Structural data is based on detailed field studies by former B.S. Honors students of the University of Zimbabwe, Bache et al. (1984), Barton et al. (1991), and authors' data. Mineral abbreviations used in the figure are as follows: Crd—cordierite, Grt—garnet, Bt—biotite, Hbl—hornblende, Ms—muscovite, St—staurolite, Ky—Kyanite, Sit—Sillimanite, And—andaulusite.

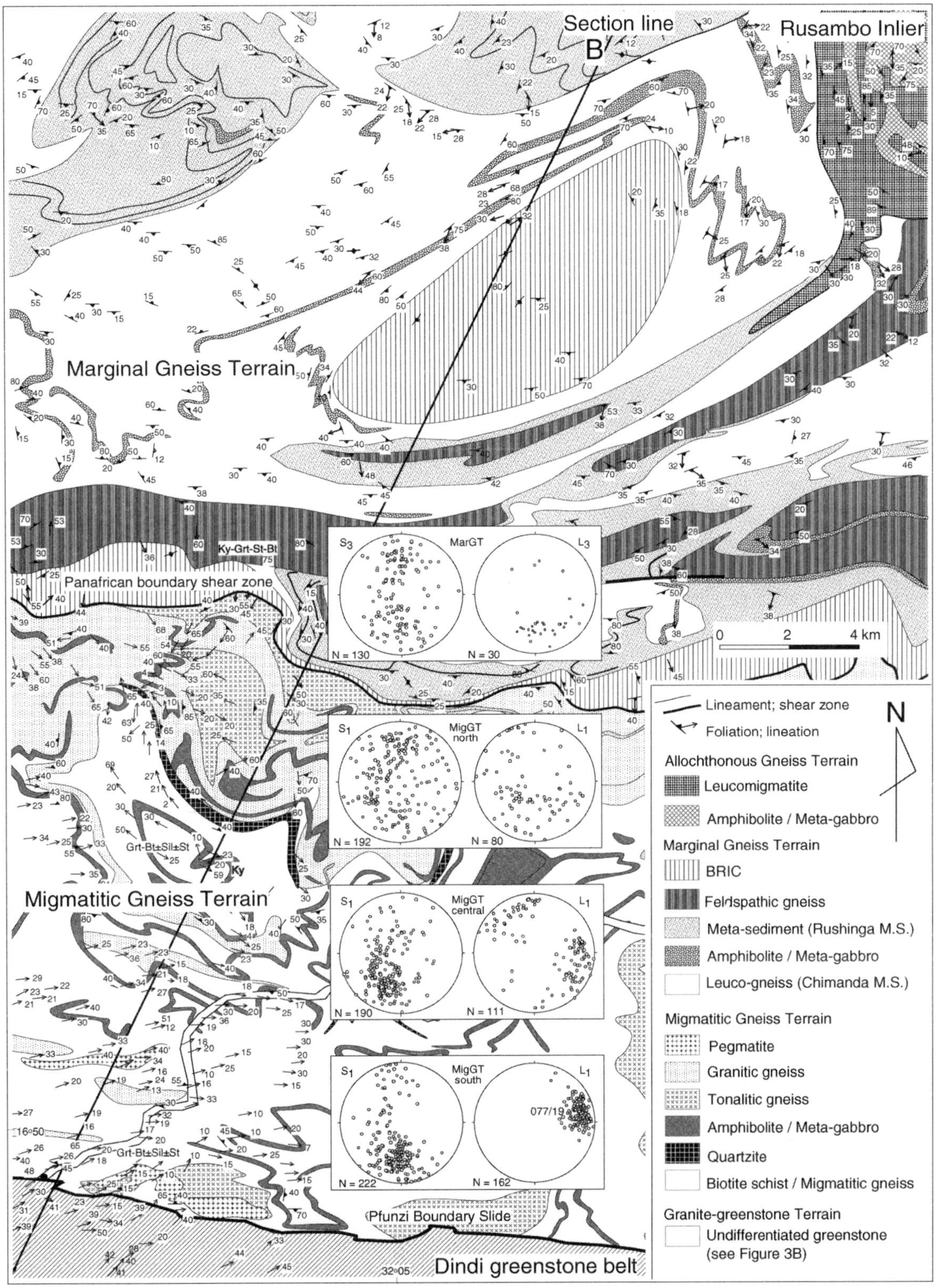

Figure 3. *(continued)* (C) Geological map of part of the Migmatitic Gneiss Terrain (area 4) and Marginal Gneiss Terrain (area 5). Shown are the main lithologies, structures, lithotectonic domains, prominent shear zones, and some indicative metamorphic mineral assemblages in pelitic units. The insets show lower hemisphere equal area stereoplots for poles to S_1 foliations (left) and for L_1 lineations (right) for the Migmatitic Gneiss Terrain (MigGT) and for poles to S_2 foliations (left) and L_2 lineations (right) for the Marginal Gneiss Terrain (MarGT). BRIC—Basal Rushinga Intrusive Complex. Structural data based on detailed field studies by former BSc Honors students from the University of Zimbabwe, Barton et al. (1991), and authors' data.

TABLE 2. SUMMARY DIAGRAM OF THE INTRUSIVE AND DEFORMATIONAL EVENTS RECORDED IN THE ROCKS ALONG TRANSECT A–B SHOWN IN FIGURE 2

Rock formation and deformation events	Craton		Craton margin	Migmatitic GT		Marginal GT	Allochthonous GT		Def. phase
	Shamva GB	Madziwa batholith	Dindi GB	Southern part	Northern part		Masoso Suite	Mavuradonha Suite	
1: Formation of mafic-felsic greenstone units	2700 Ma (7)		?						
2: Deposition of sedimentary greenstone units	2640–2660 Ma (7)		?						
3: Emplacement of syntectonic granitoids		>2600 Ma							
4: Isoclinal folding and reversals in younging, layer-parallel shear, thrust stacking, and upright folding around a constant, ENE-WSW–trending, L1 mineral lineation	L1 = 285/41, NE-trending, sinistral reverse shears. Strain: flattening	L1 = 295/54 (SW) and 063/43 (NE), NW-trending dextral reverse shears.	L1 = 069/54 (S) and 073/29 (N), NE-trending sinistral reverse shears. F1a and F1b co-linear folding. Strain: constrictional						D1a,b
5: Crenulation and reclined folding of the shear zone fabrics			Reclined F1c folding interpreted as fault-bend folds						D1c
6: Emplacement of late-tectonic Chilimanzi Granites, solid state diapirism, and contact metamorphism in greenstone sequences	2570–2600 Ma (5, 6) T = 600 °C, P = 3.3 Kbar at 2606 Ma (4)		T = 570 °C, P = 4.8 Kbar						
7: Development of retrograde, slickensided strike-slip faults with Au, located mostly along earlier shear zones	present	present	L1d lineation plunges NW along steep, W-trending sinistral strike-slip zones						D1d
8: Great Dyke emplacement, 2575 Ma (1, 2, 3)	present	present	present	present	present				
9: Formation of the Rushinga Metamorphic Suite						Possibly >2200 Ma (14)			
10: Emplacement of tonalitic to granodioritic units in the Zambezi belt							1060–1020 and 870 Ma (9, 11, 12)		D2 (?)
11: Emplacement of gabbroic units								850–865 Ma (12, 13)	D2

(*continued*)

TABLE 2. SUMMARY DIAGRAM OF THE INTRUSIVE AND DEFORMATIONAL EVENTS RECORDED IN THE ROCKS ALONG TRANSECT A–B SHOWN IN FIGURE 2 (*continued*)

Rock formation and deformation events	Craton		Craton margin	Migmatitic GT		Marginal GT	Allochthonous GT		Def. phase
	Shamva GB	Madziwa batholith	Dindi GB	Southern part	Northern part		Masoso Suite	Mavuradonha Suite	
12: Burial, high-P metamorphism, and exhumation of high-P gneiss, presumably due to S-directed thrusting.			Loading?	Loading?	Loading?	D2 thrust stacking of Rushinga Suite and basement gneiss units (?). No fabrics	T = 850 °C, P = 15 Kbar at 850–870 Ma (11, 13)	?	D2
13: Development of a penetrative shear fabric with a top-to-the-N sense of movement and associated with several generations of tight to isoclinal folding					Folding of the L1 lineation around F3a, tight recumbent folds	Development of a pervasive mylonitic foliation, S3. L3 plunges moderately S, but varies. Shearing associated with 2–3 generations of tight to isoclinal F3 folds; T = 650 °C, P = 6–7 Kbar	Development of a pervasive mylonitic foliation S3, in at least the S part of the Masoso suite. L3 trends N to NE; T = 650 °C, P = 6–7 Kbar.	?	D3a
14: Emplacement of high-K, monzogranite sheets of the Basal Rushinga Intrusive Complex (BRIC)						805 Ma (11); Emplacement into D3 shear zones during development of S3			D3a
15: Development of domal, open upright folds			Exhumation from below the D3 extensional shear zone?	Exhumation from below the D3 extensional shear zone?	Folding of the L1 lineation and F3a folds by open upright, F3b folds with NE-trending axial planes.	Large-scale doming of basement units and BRIC into overlying gneiss. Domes have a NE-elongation direction, approx. parallel to L3			D3b
16: S-directed thrusting			Loading and static recrystallization of earlier fabrics; T = 650 °C, P = 5.8 Kbar (could be due to a D2 overprint)	Loading and static recrystallization of earlier fabrics; T = 660 °C, P = 6–7 Kbar (could be due to a D2 or D3 overprint)	Loading and static recrystallization of earlier fabrics. Cooling below 500 °C around 500 Ma (9)	Loading and static recrystallization of earlier fabrics; leucosome-filled shear bands; T = 680 °C, P = 7–8 Kbar	Loading and static recrystallization of earlier fabrics; T = 680 °C, P = 7–8 Kbar around 535 Ma (9)	Development of a thrust at the base of the Mavuradonha Suite around 530 Ma (12)	D4

Note: Numbers in parentheses correspond to those used in Table 1: (1) Wingate (2000); (2) Armstrong and Wilson (2000); (3) Mukasa et al. (1998); (4) Van Dijk and Kater (1996); (5) Jelsma et al. (1996); (6) Hofmann et al. (2002a); (7) Wilson et al. (1995); (8) Nesbitt et al. (2000); (9) Vinyu et al. (1999); (10) Vinyu et al. (2001); (11) Hargrove et al. (2003); (12) Mueller (2004); (13) Mariga (1999); (14) Dirks et al. (2000). GB—Greenstone Belt; Def. phase—Deformation phase.

Hofmann et al., 2002b; Fig. 3A). This shear zone comprises hornblende-biotite–bearing, volcano-sedimentary schist units and truncates the trends of primary layering in footwall and hanging-wall lithologies, which constitute the south and north blocks, respectively.

In the northern hanging-wall block, the stratigraphy comprises a locally preserved, basal sliver of calc-alkaline meta-andesite and an overlying pile of (up to 6 km thick) pillowed and massive tholeiitic metabasalt. These metabasaltic rocks contain several intercalations of metarhyolitic volcanic and volcaniclastic metasedimentary rocks that have been dated at 2697 ± 9 Ma (U/Pb single zircon; Wilson et al., 1995). The contact between the metabasalt unit and the Madziwa batholith to the north is intrusive. The southern footwall block comprises tholeiitic and komatiitic metavolcanic units, intercalated with ironstone, chert, pelite, and limestone horizons, which are overlain by clastic metasediment of the Shamvaian Supergroup (Jelsma, 1993; Hofmann et al., 2002b).

The lithologies of the northern hanging-wall block are overturned to the northwest and structurally overlie the younger Shamvaian metasedimentary sequence. Directly south of the shear zone contact, the Shamvaian metasediment units are folded into a 500-m-wide antiformal imbricate stack above thrust surfaces within the metasediment units. Further south, in the greenstone units of the southern footwall block, a planar northward younging sequence is preserved (Fig. 3A; Jelsma and Dirks, 2000).

Lithological contacts in both blocks are invariably sheared along zones characterized by intense S/L fabrics, disharmonic folding, and S-C fabrics that formed during a first stage of regional deformation, D_1. Consequently, both the footwall and hanging-wall blocks include internally stacked, mafic-felsic greenstone stratigraphies. An associated regional foliation, S_1, is composite in nature; i.e., it is a transposition fabric preserving intrafolial folds.

In both lithotectonic blocks, a penetrative L_1 fabric (lineations defined by mica and amphibole alignment, X-axes of strain markers, and boudin necks) and F_1 folds in the Upper Bulawayan and Shamvaian lithologies plunge at shallow to moderate angles to the west (285/41° for the footwall block and 277/41° for the hanging-wall block; Fig. 3A). The sense of shear on these D_1 shear zones, as indicated by asymmetric porphyroclasts and S-C fabrics, indicates oblique (sinistral), westward movement of the northern hanging-wall block. Deformed pillows in pillow basalt along the Mazowe River (locality 1; Fig. 3A, Table 3) indicate high-strain intensities (d = 9.7) and uniaxial flattening strains (k = 0.3).

Some garnet-bearing amphibolite units from close to the contact with the Madziwa batholith preserve post-kinematic garnet blasts that have been linked to contact metamorphic growth associated with the emplacement of the Madziwa batholith (Jelsma, 1993; Jelsma and Dirks, 2000), indicating that thermal effects outlived D_1 shearing. The mid- to upper-amphibolite facies metamorphic conditions experienced during D_1 in the northeastern sector of the Shamva belt were reported by Jelsma (1993) as 3.3 ± 1.0 Kbar and 600 ± 25 °C and have been dated at 2606 ± 4 Ma (TIMS U-Pb garnet; Van Dijk and Kater, 1996; Tables 1 and 2).

Area 2: Craton–Madziwa Batholith

The Madziwa batholith (MB, Fig. 3A) comprises foliated or banded, hornblende-bearing tonalitic gneiss, intercalated with lenses of fine-grained amphibolite and intruded by massive to foliated monzogranite and leucogranodiorite (Stidolph, 1977), which also intrude the Shamva greenstone belt. Locally within the tonalitic gneiss, lenses of calc-silicate and sillimanite-biotite–bearing metapelite are preserved. In the central part of the Madziwa batholith is a series of km-scale lenses of amphibolitized gabbronorite belonging to the Madziwa Intrusive Complex. The only direct dates for the batholith are Rb-Sr dates of 2614 and 2597 Ma for the tonalite gneiss (Birch, 1985; Vinyu, 1995).

D_1 in the Madziwa batholith is associated with a penetrative gneissic layering, S_1, defined by oriented mica or amphibole grains separating quartz-feldspar–rich domains. S_1 is pervasive throughout the batholith, but associated fold geometries vary in orientation across a major shear zone that transects the central part of the batholith (Fig. 3A; Stidolph, 1977). This northwest-trending shear zone dips moderately to the southwest and comprises a 1–2-km-wide mylonite zone characterized by a shallowly west-plunging mineral lineation, L_1, and fabric asymmetries consistent with a dextral, top-to-the-west, normal sense of movement.

To the southwest of the central shear zone, S_1 is axial planar to km-scale, tight to isoclinal folds defined by aligned lenses of the Madziwa Intrusive Complex. These folds plunge moderately westward, with fold axes parallel to the pervasive mineral lineation L_1 (295/54°) that is also parallel to L_1 in the adjacent Shamva greenstone belt (Fig. 3A). The gneissic layering is locally refolded by D_{1b} upright folds, which also predominantly plunge to the west. To the northeast of the central shear zone, the L_1 mineral lineation plunges moderately east-northeast (063/43°) and parallels L_1 in the adjacent Dindi greenstone belt (see below). The gneissic layering in this part of the batholith is folded around a 10-km-scale antiformal fold structure with an easterly plunge (078/30°), parallel to L_1.

Area 3: Dindi Greenstone Belt

The Dindi greenstone belt (DGB, Fig. 3B) is the northeastern extension of the Shamva–Mount Darwin–Dindi greenstone belts that envelop the Madziwa batholith (Fig. 2). The greenstone belt is bound to the north by the Migmatitic Gneiss Terrain and to the east and south by tonalitic-granitic gneiss of the Chitsungo and Madziwa batholiths, respectively (Fig. 2). The Dindi greenstone belt is intruded along its western margin by monzogranite plutons that resemble ca. 2.6 Ga Chilimanzi Suite granite intrusions in other parts of the craton in terms of

TABLE 3. STRAIN ANALYSIS RESULTS FOR PILLOW BASALTS FROM FOUR LOCALITIES ALONG THE MAZOWE RIVER IN THE SHAMVA AND DINDI GREENSTONE BELTS

Locality	S_1	L_1	Axial ratios		d	k
1 (Shamva)	330/75	290/55	X:Y = 3.92	Y:Z = 10.3	9.7	0.3
2 (Dindi)	048/41	051/35	X:Y = 8.8	Y:Z = 1.7	7.7	11.0
3 (Dindi)	345/63	039/52	X:Y = 15.2	Y:Z = 2.1	14.2	2.9
4 (Dindi)	058/40	056/39	X:Y = 21.5	Y:Z = 9.4	22.1	2.5

Note: Localities are marked on Figure 3. Axial ratios are given as harmonic means, strain intensities, d, as $\sqrt{\{(R_{XY}-1)^2 + (R_{YZ}-1)^2\}}$ and strain types, k, as $(R_{XY}-1) / (R_{YZ}-1)$ (Flinn, 1962; Ramsay and Huber, 1983).

modal mineralogy, texture, and field appearance (Bache et al., 1984). The greenstone belt can be divided into two major lithotectonic packages separated by a prominent, moderately north- to northwest-dipping, D_1 shear zone that runs along the center of the Dindi greenstone belt and truncates primary layering in both the footwall and hanging-wall rocks (Fig. 3B).

The southern block in the shear zone footwall (Fig. 3B) is predominantly felsic in composition and comprises steeply dipping metapelite and meta-andesitic to metadacitic schist overlain by marble and banded-iron stone horizons. The northern block in the shear zone hanging wall (Fig. 3B) is predominantly mafic in composition and consists of a volcanic pile of pillowed and massive metabasalt with intercalated mafic to ultramafic schist units and rare felsic volcanic horizons. The northern block is structurally overlain to the north by a sequence of biotite-rich schist (Ruya Biotite Schist; Fig. 3B) that has been equated with Shamvaian sediment found further south in the Shamva greenstone belt (Bache et al., 1984; Barton et al., 1991). Gabbro sills and granodiorite dikes intruded the greenstone belt at all structural levels. Younging directions in the two blocks are generally to the north as indicated by graded bedding in tuffaceous horizons and pillow structures in metabasalt. Contacts between the Dindi greenstone belt lithologies and either the Chitsungo batholith to the east or the Madziwa batholith to the south are generally concordant (lit-par-lit intrusive and sheared). Foliations and lineations in the batholiths are concordant and continuous with those of the Dindi greenstone belt (Fig. 3B).

D_1 is the dominant deformational event in the Dindi greenstone belt and at least four generations of D_1 structures can be recognized, two of which (D_{1a} and D_{1b}) are closely related and linked by a common lineation direction, L_1. The dominant foliation in the Dindi greenstone belt, S_{1a}, locally intensifies along 50–200-m-wide, anastomosing, D_{1b} shear zones (hornblende-, muscovite-, talc-, or graphite schist and intensely deformed or mylonitic marble, chert, or ironstone horizons), which represent both stratigraphic and structural discontinuities in the original stratigraphic pile. The shear zones are characterized by high-strain deformation features such as a penetrative S/L fabric, S-C fabrics, ribbon grains, low-angle foliation truncations, intrafolial folds, and spiral inclusion trails in garnet porphyroblasts. The movement sense on the shear zones is predominantly oblique reverse (sinistral), east-northeast over west-southwest. Across the D_{1b} shear zones, lithological units are distributed as a series of stacked lenses of internally less strongly deformed rock that define duplexlike arrangements at a variety of scales (Fig. 3B).

S_{1b} is a regional schistosity that originated as a crenulation cleavage overprinting S_{1a}. S_{1a} and S_{1b} typically dip northward and are associated with colinear fabric elements that include: (1) an L_1 mineral lineation defined by mica and amphibole alignments or quartz and feldspar rodding; (2) F_{1a} and F_{1b} fold axes; (3) boudin necks; and (4) prolate pillow- and other strain markers. The composite linear fabric is the most characteristic strain feature and developed in an identical orientation during all stages of D_1 deformation. It is collectively referred to as L_1 and plunges at shallow to moderate angles to the northeast (Fig. 3B).

L_1 lineations and associated structural elements (S_{1a}, S_{1b}) have been refolded by F_{1c} folds (Fig. 4A). F_{1c} folds are map-scale southwest-verging, inclined folds with moderately to steeply plunging fold axes and open to closed interlimb angles. Although F_{1c} folds refold all earlier structures, many F_{1c} folds cannot be traced across major D_{1b} shear zones (including the central D_{1b} shear zone), suggesting that F_{1c} folding is an integral part of the D_{1b} shearing process. F_{1c} fold geometries can be explained as large-scale fault-bend faults and antiformal imbricate stacks (for more detailed examples see Dirks et al., 2002). In fold hinge regions, F_{1c} folds are associated with a regional northwest-southeast striking axial planar crenulation cleavage, S_{1c} (Fig. 4A)

Pillow basalt strain markers at different locations along the Mazowe River (Fig. 3B) indicate that D_{1a-b} strain in this part of the Dindi greenstone belt was constrictional. Finite strains are high and increase from south to north, toward the contact with the Migmatitic Gneiss Terrain (Table 3).

The contact between the Dindi greenstone belt and the Migmatitic Gneiss Terrain is inclined at an angle of ~40–50°N, in apparent structural concordance with the regionally developed S_{1a} foliation. Although Barton et al. (1991) mapped the contact as a discrete shear zone (the Pfunzi Boundary Slide; Fig. 3B), field data show that the contact is gradational in terms of D_{1a-b} structures and metamorphic grade, both on a local and regional scale. The actual northern contact of the Dindi greenstone belt consists of a 20-m-wide talc-tremolite-chlorite schist zone with a poorly defined northwest-plunging mineral lineation, which is oblique to L_1 in quartzo-feldspathic gneiss of the Migmatitic Gneiss Terrain to the north and metabasalt units of the Dindi greenstone belt to the south. Similar greenschist facies (ultra-)mafic schist bands occur as discrete, anastomosing shear horizons within the Dindi greenstone belt and have been grouped as D_{1d} structures. They bound relatively massive sigmoidal lenses of amphibolite-grade metabasalt or metagabbro and locally display gold mineralization. The close association with gold mineralization allows the D_{1d} shear zones to be correlated with a regional system of late-Archean strike-slip zones common to many greenstone belts in the craton (e.g., Dirks and Jelsma, 1998).

Synkinematic garnet, staurolite, andalusite, and kyanite have been reported from sheared $S_{1a/b}$ metapelite horizons in

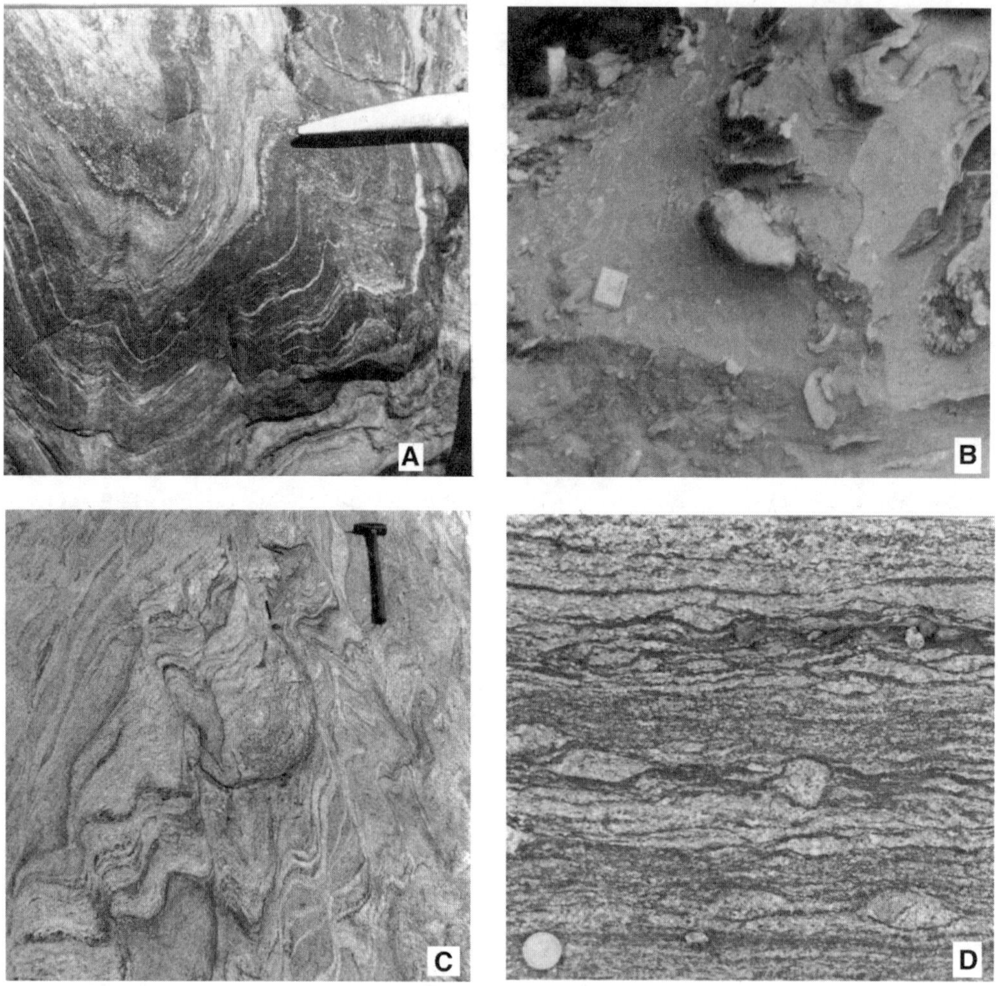

Figure 4. Structural elements in the different areas along transect A–B (Fig. 2) across the contact between the Zimbabwe Craton and the Zambezi belt. (A) Tightly folded hornblende-plagioclase schist and S_{1c} crenulation of an earlier S_{1b} fabric, Dindi greenstone belt (DGB; U.S.943375). (B) Plan view of isoclinal intrafolial and rootless folds in biotite gneiss, Migmatitic Gneiss Terrain (MigGT). The folds are defined by boudinaged quartz and pegmatite veins and are colinear with a mineral lineation, L_1 (U.S.969388). A matchbox to the left of the picture is shown as scale. (C) D_3 leucosome-filled shear bands and associated D_3 folding in the Migmatitic Gneiss Terrain (U.S.981451). (D) Sheared monzonite belonging to the Basal Rushinga Intrusive Complex along the contact between the Migmatitic Gneiss Terrain (MigGT) and Marginal Gneiss Terrain (MarGT) indicative of a top-to-the-north movement sense (U.S.977507).

the central Dindi belt (Barton et al., 1991), while sillimanite parallel to L_1 is present in metapelite layers in the Migmatitic Gneiss Terrain (Barton et al., 1991; Dirks et al., 1998) within 1 km north of the contact with the Dindi belt. This suggests that during D_1, metamorphic conditions were close to 4.8 Kbar and 570 °C (i.e., close to the aluminosilicate triple point of Holland and Powell, 1990). A younger, post-kinematic replacement assemblage in these shear zones (garnet overgrowths, unoriented chlorite and muscovite, unoriented kyanite) has been estimated at 650 °C and 5.8 Kbar (P. Dirks, 2000, personal commun.) and reflects the effects of a later, static loading event in the area (D_2 or D_4; Table 2, see below).

Area 4: Migmatitic Gneiss Terrain

The Migmatitic Gneiss Terrain (MigGT, Fig. 3C) consists of an assemblage of gneiss units of partly sedimentary (~30%) and partly igneous (~70%) origin. The most structurally complex unit in this terrain occurs in the western and central parts of the Migmatitic Gneiss Terrain and comprises biotite-hornblende gneiss, biotite schist, granitic to trondhjemitic gneiss and pegmatite (leucosome) with layers and inclusions of amphibolite, quartzite, calc-silicate, and hypersthene-bearing granulite. The biotite schist is sedimentary in origin as suggested by the presence of aluminosilicate and rounded (i.e., detrital) zircon grains and resembles the metasedimentary units of the Dindi greenstone belt. The gneiss units have been intruded by and are infolded with sheetlike bodies of trondhjemitic and tonalitic gneiss dated at 2623 Ma (Pb-Pb zircon; Vinyu et al., 2001; Table 1) and structurally later megacrystic granitic gneiss and lens-shaped bodies of intrusive leucogneiss with a similar age (Vinyu et al., 2001; Table 1). Farther east, the Migmatitic Gneiss Terrain is dominated by a more homogenous mass of tonalitic to granitic orthogneiss containing enclaves of mafic and feldspathic granulite. These lithologies are intruded by several generations of mafic dikes and sills (some of which are garnetiferous) comprising intensely deformed (foliated and lineated) amphibolite and relatively massive corona-textured gabbro lenses.

Structural elements within the Migmatitic Gneiss Terrain originated during Archean D_1 and Proterozoic D_3 events (Tables 1 and 2). Archean D_1 structures are best preserved (Figs. 4B and 4C) and include a penetrative gneissic layering (S_{1a}), a mineral lineation (L_1), and two generations of folds (F_{1a}, F_{1b}), which can be correlated with D_1 structures in the Dindi greenstone belt. Both generations of folds have fold axes parallel to the regional mineral lineation, L_1.

L_1 is a mineral or stretching lineation defined by mica and hornblende, quartz and feldspar rodding, elongated leucosome pockets, and pegmatite veins. In F_{1a} or F_{1b} fold hinges, the linear fabric is dominant over the planar fabric (L-tectonite, Fig. 4B). F_1 fold axes, boudin necks, and L_1 are colinear and generally plunge at shallow angles to the east-northeast (Fig. 3C). The distribution of poles to S_{1a} defines a great circle perpendicular to the trend and plunge of the linear fabrics, indicating folding during D_1.

A second generation of structures, here collectively called D_3 (note that the annotation D_3 is used to ensure that the deformation schemes in the Migmatitic Gneiss Terrain and the adjacent Marginal Gneiss Terrain correlate; evidence for the regional D_2 events has not been found in the Migmatitic Gneiss Terrain; Table 2) is only developed in a 5–10-km-wide zone in the northern part of the Migmatitic Gneiss Terrain, to the south of and parallel to the contact with the Marginal Gneiss Terrain. In the absence of a penetrative D_3 fabric, the presence of D_3 structures is most clearly demonstrated by the dispersion of the linear D_1 fabric (Fig. 3C). D_3 structures include two generations of colinear folds with northeast-plunging fold axes (F_{3a} and F_{3b}), both of which refold the earlier D_1 structures. F_{3a} folds comprise outcrop-scale to map-scale, tight recumbent folds, while the later F_{3b} folds are large-scale, gentle-open upright folds with northeast-striking axial planes. Neither of these fold generations is associated with a penetrative axial planar cleavage. Overprinting fold geometries produced type 2 and type 3 (e.g., Ramsay and Huber, 1983) fold interference patterns. Locally, partial melt pockets and leucosome-filled shear bands have developed along the axial planar surfaces of D_3 folds (Fig. 4C; see also Dirks et al., 1998).

D_1 and post-D_1 textural relationships are best preserved in garnetiferous, metapelitic, and metabasic inclusions in the Migmatitic Gneiss Terrain. Metapelitic rafts characteristically contain an S_{1a} foliation defined by mm-scale differentiated layering of alternating quartzofeldspathic and biotite-rich domains. Remnant sillimanite oriented parallel to L_1 occurs together with inclusion-rich garnet. The S_{1a}/L_1 fabric is strongly recrystallized and overgrown by unstrained biotite, kyanite, rutile, and inclusion-free garnet that formed as a result of post-D_1 events at around 660 °C and 6–7 Kbar (Dirks et al., 1998). Even close to the contact with the Marginal Gneiss Terrain, where macroscopic D_3 structures are prominent, the post-D_1 assemblages remain poorly oriented and do not define a clear foliation trace along the axial planes of F_3 folds. This suggests that the assemblages formed late during, and probably postdate, the D_3 events (Dirks et al., 1998).

Area 5: Marginal Gneiss Terrain

The section of the Marginal Gneiss Terrain (MarGT, Fig. 3C) investigated in this study includes leucocratic and biotite gneiss of the Chimanda Metamorphic Suite (CMS) and paragneiss and migmatite belonging to the Rushinga Metamorphic Suite (RMS), which comprise aluminous mica schist, quartzite, and massive feldspathic and quartzo-feldspathic gneiss. To the north and northeast, these rocks are in tectonic contact with gneiss of the Masoso Suite within the Allochthonous Gneiss Terrain (Figs. 2 and 3C), which contains abundant 850–1050 Ma (Mueller, 2004) granodioritic to tonalitic rocks. These envelop corona-textured metagabbroic gneiss units that preserve high-P granulite assemblages (Carney et al., 1991; Barton et al., 1991; Hargrove et al., 2003; Mueller, 2004). To explain the presence of these 850–1050 Ma intrusive rocks and the associated high-P metamorphic events to the north of the Marginal Gneiss Terrain, we invoke a D_2 crustal thickening event linked to magma emplacement, burial, and subsequent exhumation of the high-P rocks (Table 2). D_2 deformation presumably resulted from S-directed thrusting; however, we have not observed any penetrative fabrics in the Marginal Gneiss Terrain that can be linked to D_2 (contrary to what has been reported by Barton et al., 1991). Instead, fabrics found in the Marginal Gneiss Terrain are consistent with top-to-the north- to northeast-directed shearing and the concomitant rise of basement diapirs (see also Talbot, 1987; Dirks et al., 1998). These events have been collectively called D_3 (Table 2).

Near the transect, structures that dominate the rocks of the Marginal Gneiss Terrain include a prominent S_3 foliation that parallels the primary compositional layering in the metasedimentary units and an associated mineral lineation, L_3, which is variable in orientation but generally plunges southward. The contact between the Marginal Gneiss Terrain and the Migmatitic Gneiss Terrain dips moderately south and is defined by a mylonitic metapelite horizon intruded by a 20–100-m-wide, sheetlike intrusive body of quartz monzonite that belongs to the ca. 830–800 Ma (Barton et al., 1991; Vinyu et al., 2001) Basal Rushinga Intrusive Complex (BRIC; Fig. 3C). Outcrops in the Basal Rushinga Intrusive Complex preserve a well-developed mylonitic S_3-L_3 fabric, but the nature and strength of this fabric varies along strike. Locally, cm-scale feldspar porphyroclasts are preserved with excellent monoclinic fabric asymmetries and S-C fabrics consistent with north-directed thrusting (Fig. 4D). In many places the strength of the fabric is more intense and porphyroclasts have recrystallized into feldspar ribbons, resulting in an even-grained, laminated quartzo-feldspathic gneiss with few fabric asymmetries.

The Basal Rushinga Intrusive Complex is overlain by garnet-biotite-kyanite–bearing metapelite horizons of the Rushinga Metamorphic Suite interlayered with quartzofeldspathic gneiss units of the Chimanda Metamorphic Suite. In the metapelite, S_3 is defined by oriented biotite and is locally enhanced by segregations of leucocratic material; kyanite defines L_3. Well-developed S-C fabrics and spiral inclusion patterns in garnet indicate north-directed thrust movement. The S_3/L_3 fabric is associated with a series of inclined isoclinal folds with F_{3a} fold axes that parallel L_3.

Further north, the Basal Rushinga Intrusive Complex is exposed within a large domal structure that consists of a quartz-monzonite core and a granitic rim. This domal structure represents an F_{3b} fold with a pronounced northeastern elongation, but no axial planar foliation has developed. Around the dome, S_3/L_3 fabrics are reoriented, causing a scatter of L_3 in this area (Fig. 3C).

F_{3b} domal structures cored by the Basal Rushinga Intrusive Complex or the Migmatitic Gneiss Terrain are common in the Marginal Gneiss Terrain, and D_3 fabrics commonly weaken toward the cores of larger domes (Talbot, 1987; Barton et al., 1991; Dirks et al., 1998). The long axes of most domal F_{3b} fold structures trend in a northeasterly direction, and many are asymmetric in shape, verging to the northeast. On a regional scale, the L_3 lineation generally plunges to the southwest or northeast (Barton et al., 1991; Dirks et al., 1998), parallel to the long axes of the domes.

To the northeast of the transect, the Marginal Gneiss Terrain is infolded with a high-grade gneiss unit consisting of granulite and variably deformed monzonite, gabbro, and pyroxenite intrusions belonging to the Masoso Suite (the Rusambo Inlier; Fig. 3C). This unit has been variably interpreted as the top of a basement gneiss dome (Talbot, 1987) or a klippe related to south-directed thrusting of deep-crustal material (Barton et al., 1991; Mariga, 1999). The contacts of the unit are intensely sheared with well-developed D_3 fabrics containing tailed objects (delta-clasts) consistent with a top-to-the-northeast normal sense of shear.

D_4 structures in the Marginal Gneiss Terrain are relatively rare and include minor upright folding and the development of leucosome-filled crenulation bands (Dirks et al., 1998). D_4 events are associated with pervasive static recrystallization and post-tectonic grain growth (Dirks et al., 1998). Major regional structures related to D_4 are preserved to the north of the study area. The most notable of these occurs at the base of the Mavuradonha Complex (Fig. 2), where a 100–200-m-wide, moderately north-dipping mylonitic amphibolite horizon occurs that has retrogressed hanging-wall metagabbro containing high-P assemblages (Mueller, 2004). The amphibolitic mylonite zone is associated with a north-plunging lineation and south-directed thrust movement (Mueller, 2004), which is opposite to the D_3 shear sense preserved in the underlying gneiss units of the Masoso Suite and Marginal Gneiss Terrain (Hargrove et al., 2003; Mueller, 2004).

P-T conditions for D_3 and D_4 events near the transect have been estimated at 650 °C and 6–7 Kbar and 650–700 °C and 7–8 Kbar respectively (Carney et al., 1991; Dirks et al., 1998). Going from west to east within the Marginal Gneiss Terrain, the dominant aluminosilicate in metapelitic rocks changes from kyanite to sillimanite, and the transition between these two minerals more or less occurs within the area shown in Figure 3C (Barton et al., 1991). Kyanite occurs, both as a texturally early phase oriented in S_3 and locally replaced by fibrolitic sillimanite and as a texturally late phase in syn-D_4 leucosome pockets that transect the dominant S_3 foliation (Dirks et al., 1998).

DISCUSSION

Structural Evolution of the Zimbabwe Craton–Zambezi Belt Transition Zone

D_1 Events

Deformation in the Zimbabwe Craton along the craton margin and along the southern margin of the Migmatitic Gneiss Terrain is characterized by D_1 structures that include several overprinting S_1 transposition fabrics related to a series of colinear folding events (Table 2). The fabrics share a relatively constant east- or west-plunging mineral lineation L_1 (Fig. 3) that parallels the several generations of F_1 fold axes, intersection lineations, and boudin necks, as well as deformed strain markers such as pillow structures in metabasalt. The colinearity of the structures and high k-values for strain markers (Table 3) indicate intense deformation in a constrictional environment. This type of deformation has also been noted in other sections of the Bindura-Shamva greenstone belt, where it has been linked with penetrative deformation predating the emplacement of ballooning diapirs such as the Chinamora batholith at 2.6 Ga (Becker et al., 2000; Jelsma and Dirks, 2000, 2002).

Although L_1 is relatively constant over large areas, the intensity of related deformation geometries varies as a function of metamorphic grade. In the lower-grade central parts of the Shamva greenstone belt, D_1 strain is mainly partitioned into discrete shear zones that anastomose around low-strain domains with relatively pristine primary features (bedding, pillows; Jelsma and Dirks, 2000). In this low-grade section of the belt (P = 1.5–3 Kbar; Jelsma, 1993), large segments of the stratigraphy appear to be relatively undisturbed, and shear zones typically parallel primary layering. Shear sense indicators (S-C fabrics, shear bands, rotated objects) are consistent with an east-over-west, sinistral sense of movement on east- to north-northeast–trending shears and an east-over-west, dextral sense of movement on north- to northwest-trending shears. The former, northeast-trending shear zones dominate in the area. Jelsma and Dirks (2000) related these shear zones to the emplacement of the Madziwa batholith within the core of a large, sheath-shaped, nappe structure that overrode sections of the Shamva greenstone belt in an oblique, transpressional manner (Fig. 5A). Internal deformation in this nappe was intense and concordant with deformation in the adjacent greenstone belts.

The higher metamorphic pressures (4–6 Kbar) in the Dindi greenstone belt indicate that the craton margin exposes a deeper section of Archean crust. L_1 is still the dominant geometrical feature, but the general strain intensity is higher than within the Shamva belt (Fig. 3B; Table 2). The strain type is more constrictional (higher k-values; Table 3), and unequivocal evidence of primary bedding is rarely preserved. Most fabrics resulted from transposition involving several stages of progressive D_1 folding. High-strain zones form an anastomosing network that envelops and juxtaposes a great variety of different rock types. They locally preserve good shear sense indicators (S-C fabrics, shear bands, rotated objects, inclusion trails) that consistently indicate an east-over-west, sinistral, oblique thrusting sense of movement. On a km scale, stratigraphic repetitions are common across the shear zones, and duplex geometries between roof and floor thrusts can be identified (Figs. 3B and 5).

The Migmatitic Gneiss Terrain to the north of the craton margin preserves the highest metamorphic pressures linked to Archean events (6–8 Kbar and possibly higher), exposing an Archean midcrustal section. The contact between the Dindi greenstone belt and the Migmatitic Gneiss Terrain is gradational, and structures within the Migmatitic Gneiss Terrain are continuous with structures in the Dindi greenstone belt. Finite strains throughout the Dindi belt and the Migmatitic Gneiss Terrain are high and independent of metamorphic grade. There is no evidence for a major thrust zone at the contact, as reported by Barton et al. (1991) who described a "sudden" change in the plunge direction of the mineral lineation and an inversion of metamorphic grade. The contact is reactivated as a minor, retrograde shear zone, probably during the late Archean.

Within the Migmatitic Gneiss Terrain, L_1 remains the most prominent fabric element, and complicated fold/boudin structures occur within one dominant, straight foliation that trends east-west. This foliation originated as a high-grade gneissic layering characterized by intense transposition fabrics. Primary layering is not preserved, and the intensely folded nature of relatively late and, therefore, relatively less disturbed, mafic dikes bears witness to the extremely high strains accommodated within the Migmatitic Gneiss Terrain (Fig. 3C). With respect to D_1, few unequivocal shear sense indicators occur, probably reflecting the high metamorphic grade during which deformation occurred (Passchier and Trouw, 1996) and the effects of recrystallization and annealing during subsequent metamorphic overprints (Dirks et al., 1998). However, the characteristic L_1 lineation in the Migmatitic Gneiss Terrain can be traced to the east (east of the Makaha greenstone belt, Fig. 2; Johnson, 1968; Dirks et al., 1998; Hofmann et al., 2002a) and in the craton lithologies to the south (Bindura-Shamva greenstone belt; Jelsma and Dirks, 2000) where a wide variety of kinematic indicators consistently show a top-to-the-west, sinistral thrust movement. Therefore, the east-west–trending Migmatitic Gneiss Terrain is interpreted as an oblique section through a crustal scale, transpressional shear zone that accommodated sinistral thrusting.

Age of D_1

D_1 fabrics in the craton and Migmatitic Gneiss Terrain can be traced westward where they are transected by the Great Dyke, dated at 2575 Ma (Mukasa et al., 1998; Armstrong and Wilson, 2000; Wingate, 2000). This age puts a definite minimum age limit on D_1 deformation. In northeastern Zimbabwe, direct ages for D_1 are scarce. In the Migmatitic Gneiss Terrain, Vinyu et al. (2001) dated trondhjemitic and tonalitic gneiss with a strong S_1/L_1 foliation at 2625 ± 11 Ma and 2621 ± 9 Ma, respectively, while an intrusive leucogneiss was dated at 2623 ± 1 Ma (all single zircon U-Pb ages). The leucogneiss contains xenoliths and schlieren of strongly foliated and lineated biotite gneiss and generally occurs as large elongated lens-shaped bodies. In many of these units megacrysts and augen define a mineral elongation lineation, indicating a syn-D_1 timing of intrusion. In the Madziwa batholith, a syntectonic tonalitic gneiss with a moderately developed S_1/L_1 fabric has yielded Rb-Sr ages of 2597 and 2614 Ma with error bars of several hundred m.y. (Birch, 1985; Vinyu, 1995). In the eastern part of the Shamva greenstone belt mineral ages on post-kinematic garnet (Pb-Pb) and amphibole (Ar-Ar) blasts are between 2580 and 2620 Ma (Van Dijk and Kater, 1996), indicating that mineral growth driven by the thermal effects of diapirism and doming outlived D_1 events (Dirks and Jelsma, 1998). Along the northeastern margin of the craton, Hofmann et al. (2002a) report Pb-Pb Kober ages of 2601–2593 Ma for magmatic zircon grains from syntectonic granite intrusions that contain S_1/L_1 fabrics and an age of 2570 Ma for a post-tectonic Chilimanzi-type granite intrusion (Table 1).

D_2 and D_3 Events

Coward and Daly (1984), Daly (1986), Barton et al. (1991), and Hanson (2003) interpreted the units of the Marginal Gneiss Terrain to be tectonically stacked and thrust onto the Migmatitic Gneiss Terrain and craton margin at 0.8 Ga. This interpretation relies on large-scale reasoning and was put forward to explain the exhumation of high-P granulite lenses in rocks of the Masoso Suite of the Allochthonous Gneiss Terrain north of the Marginal Gneiss Terrain but was never substantiated with kinematic data. Barton et al. (1991) suggested that the tectonic units were subsequently folded, inverted, and backthrust during later stages of this deformation event, simultaneous with the emplacement of the Basal Rushinga Intrusive Complex. Dirks et al. (1998) showed that within the Marginal

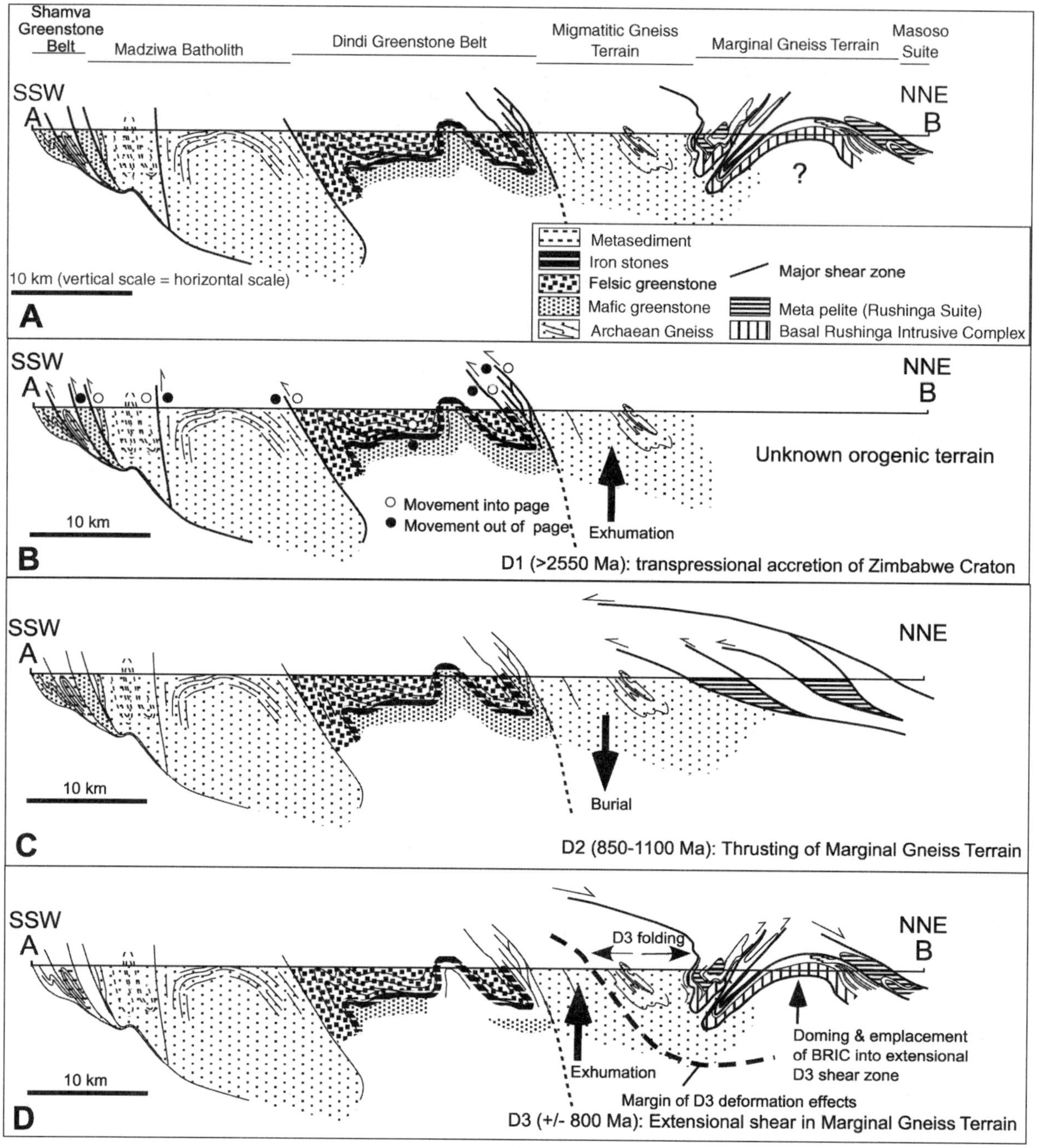

Figure 5. Structural profile (south-southwest–north-northeast) along transect A–B (Fig. 2). Structures have been extrapolated at depth. The major structural elements developed during D_1, D_2, and D_3 are shown. (A) Section as it appears currently. (B) Section outlining the late-Archean, D_1 events that led to accretion of the craton and exhumation of the Migmatitic Gneiss Terrain. (C) D_2, 850–1100 Ma events probably resulted in the exhumation of high-P rocks across a thrust stack. The geometries associated with these events are inferred. (D) D_3, ca. 800 Ma events were associated with the development of a large north-dipping extensional shear zone that comprises the gneiss units of the Marginal Gneiss Terrain as well as parts of the Allochthonous Gneiss Terrain (Masoso Suite). Basement gneiss domes and granitic sheets belonging to the BRIC (Basal Rushinga Intrusive Complex) were emplaced into the base of the D_3 shear zone and caused extensive refolding of the mylonitic fabrics. The limit of D_3 folding events within the Migmatitic Gneiss Terrain is shown with a dashed line.

Gneiss Terrain in northeastern Zimbabwe, fabrics are extensional, and there is no direct kinematic evidence to support the S-directed thrusting events postulated by Barton et al. (1991). Therefore, exhumation of the high-P granulite units could be the result of a more complex sequence of events that may have included an early stage of thrusting followed by transposition, extension, and the diapiric rise of basement domes.

The gneiss of the Marginal Gneiss Terrain is complexly infolded with rocks of the Allochthonous Gneiss Terrain (Masoso Suite; Figs. 2 and 3C). The latter contains pockets of low D_3-strain, where D_2 manifests itself through remnant fabrics that can be linked to high-P assemblages (Carney et al., 1991). These remnant fabrics attest to a phase of crustal thickening that probably resulted in a tectonic stack of para- and orthogneiss units that make up the rocks of the Marginal Gneiss Terrain and Masoso Suite (Barton et al., 1991). It appears that D_2 fabrics in these units were all but obliterated during D_3 as a result of regional transposition accompanied by extensive metamorphic reworking, partial melting, and the emplacement of numerous high-K intrusions (the Basal Rushinga Intrusive Complex; Figs. 3C and 5).

Our research suggests that the dominant geometries preserved in the Marginal Gneiss Terrain of the Zambezi belt formed during D_3. They comprise a complex array of progressive folding and shearing events that are mostly colinear and that can be linked by a common south-southwest– to northeast-plunging lineation direction developed in a pervasive shear foliation that is characterized by a top-to-the-northeast sense of movement. The pervasive, mylonitic D_3 fabric has been folded around north- to east-northeast–trending F_{3a} folds that parallel L_3 and bend around syntectonic domal structures (see also Talbot, 1987 and Dirks et al., 1998; Fig. 5). These domes probably originated from the diapiric or solid state emplacement of the Migmatitic Gneiss Terrain into the base of the Marginal Gneiss Terrain (Talbot, 1973, 1987; Barton et al., 1991; Dirks et al., 1998; Fig. 5D) in a process not dissimilar to that described for the formation of core complexes, albeit on a smaller scale.

The laterally extensive sheets of sheared monzogranite (the Basal Rushinga Intrusive Complex; Fig. 3C) near the base of the Marginal Gneiss Terrain have a high alkali content with abundant sodic amphibole or pyroxene and trace element patterns that are consistent with a within-plate tectonic setting involving attenuated continental lithosphere (Barton et al., 1991; Hargrove et al., 2003). This is consistent with the observed intrusive relationships for the Basal Rushinga Intrusive Complex, which suggest syn-D_3 emplacement, i.e., emplacement into the extensional D_3 shear zones within the Marginal Gneiss Terrain.

On a regional scale, the contact between the Migmatitic Gneiss Terrain and the Marginal Gneiss Terrain is complexly folded, with inliers of older gneiss occurring throughout the Marginal Gneiss Terrain (Talbot, 1973; Bache et al., 1984; Barton et al., 1991; Figs. 3C and 5). In spite of outcrop-scale complexities, the overall attitude of the contact zone (i.e., the envelope) is a relatively flat-lying surface with a gentle dip to the north or northeast. The overall shear sense on this surface was normal dip-slip, and the surface can be interpreted as the lower boundary of a crustal-scale low-angle detachment zone that accommodated crustal extension during the Neoproterozoic.

Age of D_2 and D_3

D_2 crustal thickening events have been related to the high-P metamorphism recorded in metagabbro units that occur as lenses within the highly strained quartzofeldspathic gneiss units of the Masoso Suite, infolded with the Marginal Gneiss Terrain in the northern part of the study area (Figs. 2 and 3C). Felsic to intermediate rocks of the Masoso Suite in northern Zimbabwe have yielded crystallization ages of 1020 Ma to 1060 Ma (Table 1; Mueller, 2004; Hargrove et al., 2003). The felsic units preserve an intense tectonic fabric and envelop high-P granulites, which are mainly corona-textured metagabbro units. Direct ages for such units include an age for the Mavuradonha Metagabbro Complex of 862 ± 8 Ma (SHRIMP [sensitive high-resolution ion microprobe] zircon age; Mueller, 2004), and a monzodiorite associated with metagabbro in the Rusambo Klippe (Fig. 3C), which has yielded an igneous age of 854 Ma (Pb-Pb Kober zircon ages; Dirks et al., 2000). A lower age limit for the high-P events is reported by Mariga (1999), who obtained a crystallization age of 849 ± 2 Ma (TIMS, zircon) from a metagabbro unit in the Rusambo Inlier that intrudes and crosscuts corona-textured metagabbro. The only direct age for the high-grade, granulite facies, D_2 events in the Masoso Suite is an age of 870 ± 1 Ma (TIMS, zircon; Vinyu et al., 1999) for a leucosome pocket associated with high-P metagabbro lenses (Table 1).

The ages indicate extended igneous activity between 1060 Ma and 860 Ma (Table 1) during or prior to the D_2 compressional events that resulted in deep burial of the rocks. High-P metamorphism may have culminated in the period 850–870 Ma but may have occurred before that time. To the northwest of the study area in the Chewore Inliers (Fig. 1) in the central Zambezi valley, Oliver et al. (1998) describe an ophiolite complex with an igneous age of 1393 Ma (U-Pb SHRIMP, zircon), preserving eclogite facies assemblages of unknown age. In this area, 1080 Ma granitic orthogneiss experienced a high-T, low-P metamorphic overprint associated with a shortening event that could have lasted for an extended period between 945 and 1070 Ma (Goscombe et al., 2000) and that may be linked to the D_2 events in the study area.

The composite D_3 shearing event is best constrained with ages reported for the Basal Rushinga Intrusive Complex, which was probably emplaced in a syn-D_3 extensional, intracontinental environment (Barton et al., 1991; Dirks et al., 1998). Vinyu et al. (2001) reported a single zircon age of 804 ± 10 Ma (TIMS) for this unit. The age of the chemically related and sheared Chironga Gneiss (Barton et al., 1991) has been reported as 795 ± 2 Ma (Hargrove et al., 2003). Dirks et al. (1999) also reported 850–750 Ma zircon ages (Kober, Pb-Pb) from chemically similar granite sheets that intruded into a low-angle crustal-scale shear zone along the southern margin of the Zambezi belt in northwestern Zimbabwe. Other relevant ages for D_3 include an 803 Ma Ar-Ar plateau age for amphibole from a shear zone in the Masoso Suite (Hargrove et al., 2003) to the northwest of the area of investigation. Shear zones in this area are related to a shallowly north-plunging L_3 lineation and a normal sense of movement as confirmed by Hargrove et al. (2003) and Mueller (2004). D_3 events appear to be limited to the contact zone with the deformed Zimbabwe Craton and have not been reported from the Chewore Inliers further north (Goscombe et al., 2000).

D_4 Events

Most structural-metamorphic textures in the craton margin, the Migmatitic Gneiss Terrain, and the Marginal Gneiss Terrain are largely statically recrystallized, which happened contemporaneous with Pan-African tectonic loading referred to as D_4. In the Marginal Gneiss Terrain, D_4 recrystallization is related to partial migmatization and the formation of kyanite-bearing leucosome pockets and pegmatite dikes, concentrated along crenulation cleavage planes (Dirks et al., 1998). In the Migmatitic Gneiss Terrain and the craton margin, late-stage recrystallization, textural annealing, and the growth of nonoriented muscovite and chlorite are common features that have been linked to a syn-D_4 metamorphic overprint.

To explain the syn-D_4 mid-amphibolite facies metamorphic overprint affecting the Marginal Gneiss Terrain, the Migmatitic Gneiss Terrain, and the craton margin, we suggest that the area was buried below a stack of south-verging, Pan-African thrust sheets. The lowermost of such thrusts has been described from the base of the Mavuradonha Complex (Mueller, 2004) within the Allochthonous Gneiss Terrain (Barton et al., 1991; Dirks et al., 1998; Mueller, 2004).

Age of D_4

Early Paleozoic, ca. 500 Ma, K-Ar and Rb-Sr ages from the Zambezi belt have been widely reported (e.g., Vail and Snelling, 1971; Cahen et al., 1984; Hahn and Steiner, 1990; Barton et al., 1991) and indicate that the belt was thermally active around that time. These ages are consistent with the age of late syntectonic granite veins (ca. 535 Ma, 555 Ma; single zircon, U-Pb), reported by Vinyu et al. (1999) and Mueller (2004), and zircon ages from a late tectonic pegmatite (550 Ma), Ar-Ar hornblende cooling ages (525–490 Ma), and sphene ages (ca. 530 Ma) reported by Hargrove et al. (2003) and Hanson et al. (1997, 1998), as well as metamorphic zircon ages (554 Ma; single zircon, U-Pb) and garnet-WR (Whole Rock), Sm-Nd isochron ages (546 Ma) from the Mavurahdonha Complex (Mueller, 2004). From these data it appears that the rocks in the study area experienced a thermal overprint associated with minor deformation, which culminated at ca. 530 Ma. To the northwest, in the Chewore inliers, Pan African D_4 events are pronounced and linked to southwest-directed thrusting and crustal overthickening associated with a mid-amphibolite facies metamorphic event at 530 Ma (7–9 Kbar, 590–720 °C; Goscombe et al., 2000). Mueller (2004) suggests that some of the high-P assemblages preserved in the Mavuradonha complex could have originated around 530 Ma, which opens up the possibility that evidence for more than one high-P granulite or eclogite event is preserved in the area.

SUMMARY AND CONCLUSIONS

The transition from the Archean Zimbabwe Craton into the Zambezi belt is gradational and occurs over a 10–15-km-wide zone across which the metamorphic grade increases from 3 to 4 Kbar and 500 °C, typical for greenstone sequences in the northern part of the craton (Jelsma and Dirks, 2000, 2002) to 6–7 Kbar and 700 °C typical for Archean gneiss in the Zambezi belt (Barton et al., 1991; Carney et al., 1991; Dirks et al., 1998). This P-T gradient was established in the late Archean and reflects exhumation of the Migmatitic Gneiss Terrain. The craton margin and adjacent Migmatitic Gneiss Terrain were subsequently intruded and crosscut by the 2.58 Ga Great Dyke.

Archean structures in the craton and the adjacent Migmatitic Gneiss Terrain represent a continuum and are characterized by a penetrative, east- to northeast-trending lineation. In the craton, strain is partitioned along an anastomosing network of mostly east- to northeast-trending and north-dipping shear zones that envelop low strain domains. The sense of shear on such shear zones is consistently sinistral, with a northeast-over-southwest thrusting sense of movement. Northwest-trending shear zones are rare, but one such shear zone occurs within the Madziwa batholith where it accommodated dextral normal faulting. The shear zones accommodated accretion of the greenstone belts and gneiss terrains of the northern parts of the Zimbabwe Craton (Jelsma and Dirks, 2002) and occurred within a transpressional environment, characterized by sinistral translation. As the craton margin is traversed toward the Zambezi belt and metamorphic grade increases, D_1 strains become higher and more homogeneously distributed, but the lineation direction and associated shear sense remain constant. Discrete shear zones are lacking in the Migmatitic Gneiss Terrain, which is characterized by L-tectonite, with lineation directions identical to those in the adjacent greenstone belts. By inference, the Migmatitic Gneiss Terrain also experienced noncoaxial flow within a constrictional, sinistral transpressional environment, some time between 2580 and 2620 Ma (Tables 1 and 2). This deformation event resulted in exhumation of midcrustal granulite and may have resulted from oblique collision of the Zimbabwe Craton with an unknown craton to the north.

During Neoproterozoic and early Paleozoic times (1.1–0.5 Ga), renewed tectonism occurred along the Zambezi belt. Deformation fabrics linked to these events are mainly preserved in the Marginal and Allochthonous Gneiss Terrains, while the craton margin and the Migmatitic Gneiss Terrain resisted penetrative deformational overprints, instead experiencing two stages of burial and exhumation due to tectonic loading and unloading of the Archean crust (Fig. 5).

The sections of the Zambezi belt exposed along the northern margin of the Zimbabwe Craton only represent a small part of the belt (Fig. 1). The history in this area is probably unique when compared to other parts of the Zambezi belt, as suggested by the contrasting histories described for the Chewore Inliers and areas north of the Zambezi Rift (e.g., Johnson and Oliver, 2000; Goscombe et al., 2000; Dirks et al., 2003; Hanson, 2003; De Waele, 2005). Characteristic features for the Zambezi belt along the northern margin of the craton include the presence of reworked Archean crust, the preponderance of high-K, 800 Ma granite sheets with geochemical characteristics consistent with a within-plate, extensional setting (Barton et al., 1991; Munyanyiwa et al., 1997; Hanson, 2003; Hargrove et al., 2003), as well as a shallowly north- to northeast-dipping, highly sheared gneiss sequence that records a top-to-the-north, D_3 sense of movement (Dirks et al., 1998; Dirks and Sithole, 1999; Sithole, 2002). The lower boundary of these sheared gneiss units is the shallowly north- to northeast-dipping contact zone between the Migmatitic Gneiss Terrain and the Marginal Gneiss Terrain (Dirks et al., 1998). The upper contact is represented by the moderately north-dipping base of the Mavurahdonha Complex (Mueller, 2004). We interpret the entire, 3–5-km-thick lithostratigraphic pile represented by the highly sheared and folded gneiss units of the Marginal Gneiss Terrain and the Masoso Suite as a shallow-dipping extensional shear zone (D_3) or low-angle detachment zone that was active between 750 and 850 Ma and that allowed exhumation of lower crustal rocks from below the detachment, resulting in the numerous gneiss domes within the Marginal Gneiss Terrain (e.g., Dirks et al., 1998; Talbot, 1987; Fig. 5). The extensional shear zone can be traced along the entire northern margin of the Zimbabwe Craton and is represented in northwestern Zimbabwe by the mylonitic rocks of the Makuti Group (Fig. 1; Dirks et al., 1999; Dirks and Sithole, 1999; Sithole, 2002). Within this shear zone, the relationship between the Marginal Gneiss Terrain and rocks of the Masoso Suite remains unclear. Although Barton et al. (1991) suggested that the Masoso Suite was thrust onto the Marginal Gneiss Terrain, contacts between the two units are complexly folded and commonly characterized by D_3 shear fabrics recording a top-to-the-north sense of movement.

Within the D_3 extensional shear zone represented by the Marginal Gneiss Terrain and rocks of the Masoso Suite, high-P granulite facies assemblages have been reported from 850 to 870 Ma metagabbro units (e.g., Carney et al., 1991; Mariga, 1999; Dirks et al., 2000) that escaped D_3 shearing. High-P granulite facies metamorphism in the Masoso Suite culminated between 850 and 870 Ma (e.g., Hargrove et al., 2003) but may have occurred before that time and could be related to a long-lived compressional episode, D_2, that is also associated with the emplacement of tonalitic to granitic units in the Masoso Suite between 1060 Ma and 860 Ma. This D_2 compressional event may have resulted in partial exhumation of rocks of the Masoso Suite, even though the regional structures (i.e., thrusts) responsible for such events have not been preserved (Fig. 5). In the Chewore Inliers to the northwest (Fig. 1), a 1390 Ma ophiolite and volcanic arc complex preserving eclogite facies assemblages (Johnson and Oliver, 2000) and 1080 Ma granitic orthogneiss units experienced a high-T, low-P metamorphic overprint associated with a shortening event that could have lasted for an extended period between 945 and 1070 Ma (Goscombe et al., 2000) and may have extended before that time, when a Kibaran ocean was closed between the Congo and Zimbabwe cratons (Johnson and Oliver, 2000; De Waele, 2005). In the Chewore Inliers, shortening structures that can be correlated with D_2 in the Marginal Gneiss Terrain are intact, and the rocks were not affected by a pervasive extensional, D_3 overprint.

It is not clear whether D_2 thrusting, if it occurred, resulted in loading of the margin of the Zimbabwe Craton. K-Ar and Rb-Sr ages from the northern part of the craton do not provide evidence for such loading events, but instead record an extensive 500–550 Ma overprint (Vail and Snelling, 1971; Cahen et al., 1984; Hahn and Steiner, 1990).

Around 865 Ma, at the time that D_2, high-P metamorphism was still affecting rocks in the Masoso Suite, the layered mafic-ultramafic complex of the Mavuradonha Complex was emplaced (Mueller, 2004). The base of this complex is now represented by a 200-m-wide amphibolitic thrust horizon across which the Mavurahdonha Complex was transported southward on top of the Marginal Gneiss Terrain and Masoso Suite, sometime between 550 and 530 Ma (Mueller, 2004). Further south, within the Marginal Gneiss Terrain and the Migmatitic Gneiss Terrain, no evidence for similar 500–550 Ma shear zones is found. The rocks did, however, experience a major metamorphic overprint at that time (Barton et al., 1991; Vinyu et al., 1999; Hargrove et al., 2003), reaching conditions of 650–700 °C and 6–8 Kbar in the Marginal Gneiss Terrain and 600–650 °C and 5–6 Kbar in the central Dindi belt. This suggests that D_4 thrusting extended south onto the craton, causing significant loading of the craton margin. Erosion remnants of such thrusts may be preserved in western Zimbabwe in the Urungwe Klippe (Fig. 1; Shackleton et al., 1966).

ACKNOWLEDGMENTS

This research was financially supported by grants from the Stichting Dr. Schürmannfonds (grants no. 1995-2003/13), the Research Board of Zimbabwe (3YYH-10/3784), the Volkswagen Stiftung, Masasa Mines, and Prospecting Ventures. The paper was originally submitted in December 1999 to Precambrian Research, where it was reviewed by M. De Wit, J. Barton, D. van Reenen, A. Smit, and K. Kusky, who made numerous comments for improvements. However, before the reviews were received, things had changed significantly in Zimbabwe, and the manuscript disappeared on the pile of things to do. Although the paper is now appearing elsewhere, we would like to thank the reviewers and the editors of Precambrian Research for all their help and useful input. In addition we would like to thank R. Gibson for an additional detailed review of the updated manuscript. Much of the data presented in this paper was collected over an 8 year period with Honors students from the University of Zimbabwe. These were: S. Malenga, M. Wray, and K. Danda in the Madziwa batholith; W. Mtisi, C. Benhura, C. Kambewa, and L. Changara in the Dindi greenstone belt; and W. Mtisi and C. Benhura in the Migmatitic and Marginal Gneiss Terrains. Our thanks also go to Hubert Munyanyiwa and Michael Vinyu, who helped in the compilation of much of the original data, but who departed this world much too early. We remember you fondly and with great respect.

REFERENCES CITED

Armstrong, R., and Wilson, A.H., 2000, A SHRIMP U-Pb study of zircons from the layered sequence of the Great Dyke, Zimbabwe, and a granitoid anatectic dyke: Earth and Planetary Science Letters, v. 180, p. 1–12, doi: 10.1016/S0012-821X(00)00162-X.

Bache, J.J., Dallas, S., and Milian, J.F., 1984, A preliminary report of the country around Centenary and Mount Darwin: Zimbabwe Geological Survey Annals, v. 15, p. 34–58.

Barton, C.M., Carney, J.N., Crow, M.J., Dunkley, P.N., and Simango, S., 1991, The geology of the country around Rushinga and Nyamapanda: Zimbabwe Geological Survey Bulletin, v. 92, 220 p.

Barton, C.M., Carney, J.N., Crow, M.J., Evans, J.A., and Simango, S., 1993, Geology and structural framework of the Zambezi belt, NE Zimbabwe, *in* Findlay, R.H., Unrug, R., Banks, M.R., and Veevers, J.J., eds., Gondwana Eight: Rotterdam, Balkema, p. 55–67.

Becker, J., Siegesmund, S., and Jelsma, H.A., 2000, The Chinamora Batholith, Zimbabwe: Structure and emplacement-related magnetic rock fabric: Journal of Structural Geology, v. 22, p. 1837–1854.

Birch, J.G., 1985, The petrology of the Madziwa Mafic Intrusion, Zimbabwe [Ph.D. thesis]: London, Imperial College, University of London, 220 p.

Brandl, G., and De Wit, M.J., 1997, The Kaapvaal Craton, *in* De Wit, M.J., and Ashwal, L.D., eds., Greenstone Belts: Oxford, Oxford University Press, p. 581–607.

Cahen, L., Snelling, N.J., Delhal, J., and Vail, J.R., 1984, The geochronology and evolution of Africa: Oxford, Clarendon Press, 512 p.

Carney, J.N., Treloar, P.J., Barton, C.M., Crow, M.J., Evans, J.A., and Simango, S., 1991, Deep-crustal granulites with migmatitic and mylonitic fabrics from the Zambezi belt, northeastern Zimbabwe: Journal of Metamorphic Geology, v. 9, p. 461–479.

Coward, M.P., and Daly, M.C., 1984, Crustal lineaments and shear zones in Africa: Their relationship to plate movements: Precambrian Research, v. 24, p. 27–45, doi: 10.1016/0301-9268(84)90068-8.

Daly, M.C., 1986, Crustal shear zones and thrust belts: Their geometry and continuity in Central Africa: Philosophical Transactions of the Royal Society of London, ser. A, v. 317, p. 111–128.

De Waele, B., 2005, The Proterozoic geological history of the Irumide belt, Zambia [Ph.D. thesis]: Perth, Australia, Curtin University of Technology, 516 p.

De Wit, M.J., and Ashwal, L.D., 1997, Greenstone belts: Oxford, Oxford University Press, 810 p.

Dirks, P.H.G.M., and Jelsma, H.A., 1998, Horizontal accretion and stabilization of the Zimbabwe Craton: Geology, v. 26, p. 11–14, doi: 10.1130/0091-7613(1998)026<0011:HAASOT>2.3.CO;2.

Dirks, P.H.G.M., and Sithole, T.A., 1999, Eclogites in the Makuti Gneisses, Zimbabwe; implications for the tectonic evolution of the Zambezi belt in southern Africa: Journal of Metamorphic Geology, v. 17, p. 593–612, doi: 10.1046/j.1525-1314.1999.00215.x.

Dirks, P.H.G.M., Jelsma, H.A., Vinyu, M.L., and Munyanyiwa, H., 1998, The structural history of the Zambezi belt in NE Zimbabwe: Evidence for crustal extension during the early Pan African: South African Journal of Geology, v. 101, p. 1–16.

Dirks, P.H.G.M., Kröner, A., Jelsma, H.A., Sithole, T.A., and Vinyu, M.L., 1999, Pb-Pb zircon dates from the Makuti gneisses; evidence for a crustal-scale Pan African shear zone in the Zambezi belt, NW Zimbabwe: Journal of African Earth Sciences, v. 28, p. 427–442, doi: 10.1016/S0899-5362(99)00013-5.

Dirks, P.H.G.M., Kröner, A., Jelsma, H.A., Maneya, C., and Jamal, D.L., 2000, Pb-Pb zircon ages and a tectonic framework for the Zambezi belt, Zimbabwe, 18th Colloquium of African Geology, Graz: Journal of African Earth Sciences, v. 30, p. 23–24.

Dirks, P.H.G.M., Jelsma, H.A., and Hofmann, A., 2002, Accretion of an Archean greenstone belt in the Midlands of Zimbabwe: Journal of Structural Geology, v. 24, p. 1707–1727, doi: 10.1016/S0191-8141(02)00002-0.

Dirks, P.H.G.M., Jelsma, H.A., and Blenkinsop, T.G., 2003, The Geology of Africa, *in* de Vivo, B., Stüwe, K., and Grasemann, B., eds., Encyclopedia of Life Support Systems: Oxford, UK, EOLSS Publishers Co. Ltd., February 2003, http:www.eolss.net.

Flinn, D., 1962, On folding during three dimensional progressive deformation: Quarterly Journal of the Geological Society of London, v. 118, p. 385–428.

Goscombe, B., Armstrong, R., and Barton, J.M., 2000, Geology of the Chewore inliers, Zimbabwe: Constraining the Mesoproterozoic to Paleozoic evolution of the Zambezi belt: Journal of African Earth Sciences, v. 30, p. 589–627, doi: 10.1016/S0899-5362(00)00041-5.

Hahn, L., and Steiner, L., 1990, Geology and mineral prospecting in the Makonde and Guruve Districts, Zimbabwe: Hannover, Technical Cooperation Project Rep. 84.2170.3, 296 p.

Hanson, R.E., 2003, Proterozoic geochronology and tectonic evolution of southern Africa, *in* Yoshida, M., Windley, B., and Dasgupta, S., eds., Proterozoic East Gondwana: Supercontinent assembly and break-up: Geological Society [London] Special Publication 206, p. 428–463.

Hanson, R.E., Wilson, T.J., and Munyanyiwa, H., 1994, Geologic evolution of the Neoproterozoic Zambezi orogenic belt in Zambia: Journal of African Earth Sciences, v. 18, p. 135–150, doi: 10.1016/0899-5362(94)90026-4.

Hanson, R.E., Vinyu, M.L., Jelsma, H.A., Martin, M.W., Bowring, S.A., Krol, M.A., and Hodges, K.V., 1997, Geochronological constraints on the tectonic evolution of the 870–530 Ma Zambezi orogenic belt, NE Zimbabwe: Geological Society of America Abstracts with Programs, v. 29, p. A468.

Hanson, R.E., Hargrove, U.S., Martin, M.W., Bowring, S.A., Krol, M.A., Hodges, K.V., Munyanyiwa, H., and Blenkinsop, T.G., 1998, New geochronological constraints on the tectonic evolution of the Pan-African Zambezi belt, south-central Africa: Journal of African Earth Sciences, v. 27, p. 104–105.

Hargrove, U.S., Hanson, R.E., Martin, M.W., Blenkinsop, T.G., Bowring, S.A., Walker, N., and Munyanyiwa, H., 2003, Tectonic evolution of the Zambezi orogenic belt: Geochronological, structural and petrological constraints from northern Zimbabwe: Precambrian Research, v. 123, p. 159–186, doi: 10.1016/S0301-9268(03)00066-4.

Hofmann, A., Yagoutz, O., Kröner, A., Dirks, P.H.G.M., and Jelsma, H.A., 2002a, The Chirwa dome: Granite emplacement during late Archean thrusting along the northeastern margin of the Zimbabwe Craton: South African Journal of Geology, v. 105, p. 285–300, doi: 10.2113/1050285.

Hofmann, A., Dirks, P.H.G.M., and Jelsma, H.A., 2002b, Late Archean clastic sedimentary rocks (Shamvaian Group) of the Zimbabwe craton: First observations from the Bindura-Shamva greenstone belt: Canadian Journal of Earth Sciences, v. 39, p. 1689–1708, doi: 10.1139/e02-074.

Holland, T.J.B., and Powell, R., 1990, An enlarged and updated internally consistent thermodynamic data set with uncertainties and correlations: The system K_2O-Na_2O-CaO-MgO-MnO-FeO-Fe_2O_3-Al_2O_3-TiO_2-SiO_2-C-H_2-O_2: Journal of Metamorphic Geology, v. 8, p. 89–124.

Jelsma, H.A., 1993, Granites and greenstones in northern Zimbabwe: Tectonothermal evolution and source regions [Ph.D. thesis]: Amsterdam, Free University, 198 p.

Jelsma, H.A., and Dirks, P.H.G.M., 2000, Tectonic evolution of a greenstone sequence in northern Zimbabwe: Sequential early stacking and pluton diapirism: Tectonics, v. 19, p. 135–152, doi: 10.1029/1999TC900058.

Jelsma, H.A., and Dirks, P.H.G.M., 2002, NeoArchean tectonic evolution of the Zimbabwe Craton, *in* Fowler, C.M.R., Ebinger, C., and Hawkesworth, C.J., eds., The Early Earth: Physical, Chemical and Biological Development: Geological Society [London] Special Publication 199, p. 183–211.

Jelsma, H.A., Vinyu, M.L., Valbracht, P.J., Davies, G.R., Wijbrans, J.R., Verdurmen, E.A.T., 1996, Contraints on Archean crustal evolution of the Zimbabwe craton: A U-Pb zircon, Sm-Nd and Pb-Pb whole-rock isotope study: Contributions to Mineralogy and Petrology, v. 124, p. 55–70.

Johnson, R.L., 1968, Structural history of the western front of the Mozambique belt in northeast Southern Rhodesia: Geological Society of America Bulletin, v. 79, p. 513–526.

Johnson, S.P., and Oliver, G.J.H., 2000, Mesoproterozoic oceanic subduction, island arc formation and the initiation of back-arc spreading in the Kibaran belt of central southern Africa: Evidence from the ophiolite terrain, Chewore Inliers, northern Zimbabwe: Precambrian Research, v. 103, p. 125–146, doi: 10.1016/S0301-9268(00)00075-9.

Leitner, E.G., and Phaup, A.E., 1974, The geology of the country around Mt. Darwin: Geological Survey of Zimbabwe Bulletin, v. 73, 151 p.

Mariga, J., 1999, Structural and geochronological evolution of deep-crustal granulites, supracrustal rocks, and deformed plutons in the Zambezi orogenic belt, Rusambo Mission area, NE Zimbabwe [M.S. thesis]: Fort Worth, Texas, Texas Christian University, 110 p.

Mueller, M., 2004, The Mavuradonha Layered Complex: Neoproterozoic emplacement and Pan-African granulite-facies metamorphism in the Zambezi Allochthonous Terrane of the Mt. Darwin Area, Zambezi belt, NE-Zimbabwe [Ph.D. thesis]: Mainz, Germany, Johannes Gutenberg University, 181 p.

Mukasa, S.B., Wilson, A.H., and Carlson, R.W., 1998, A multielement geochronologic study of the Great Dyke, Zimbabwe: Significance of the robust and reset ages: Earth and Planetary Science Letters, v. 164, p. 353–369, doi: 10.1016/S0012-821X(98)00228-3.

Munyanyiwa, H., Hanson, R.E., Blenkinsop, T.G., and Treloar, P.J., 1997, Geochemistry of amphibolites and quartzofeldspathic gneisses in the Pan-African Zambezi belt, NW Zimbabwe: Evidence for bimodal magmatism in a continental rift setting: Precambrian Research, v. 81, p. 179–196, doi: 10.1016/S0301-9268(96)00034-4.

Myers, J.S., 1987, High-grade terrains in and around the Yilgarn Block of Western Australia, *in* Park, R.G., and Tarney, J., eds., Evolution of the Lewisian and Comparable Precambrian High Grade Terrains: Geological Society [London] Special Publication 27, p. 277–284.

Nesbitt, R.W., Fanning, C.M., Wilson, J.F., Jelsma, H.A., and Horstwood, M.S.A., 2000, Zircon U-Pb geochronology of the Zimbabwean Craton; untangling the stratigraphy of greenstone belts [abs.]: Geoscience 2000, 17–20 April.

Oliver, G.J.H., Johnson, S.P., Williams, I.S., and Herd, D.A., 1998, Relict 1.4 Ga oceanic crust in the Zambezi Valley, northern Zimbabwe: Evidence for Mesoproterozoic supercontinent fragmentation: Geology, v. 26, p. 571–573, doi: 10.1130/0091-7613(1998)026<0571:RGOCIT>2.3.CO;2.

Passchier, C.W., and Trouw, R.A.J., 1996, Microtectonics: Berlin, Heidelberg, Springer-Verlag, 289 p.

Percival, J., Roering, C., van Reenen, D., and Smit, C.A., 1997, Tectonic evolution of associated greenstone belts and high-grade terrains, *in* De Wit, M.J., and Ashwal, L.D., eds., Greenstone Belts: Oxford, Oxford University Press, p. 398–420.

Ramsay, J.G., and Huber, M.I., 1983, The Techniques of Modern Structural Geology, 1: Strain Analysis: London, Academic Press, 307 p.

Rollinson, H.R., 1993, A terrane interpretation of the Archean Limpopo belt: Geological Magazine, v. 130, p. 755–765.

Shackleton, R.M., 1995, Tectonic evolution of greenstone belts, *in* Coward, M.P., and Ries, A.C., eds., Early Precambrian processes: Geological Society [London] Special Publication 95, p. 53–65.

Shackleton, R.M., Vail, J.R., and Wood, D.S., 1966, Preliminary report on the origin and significance of the Urungwe Klippe, Rhodesia: Annual Report, Research Institute of African Geology, University of Leeds, v. 10, p. 10–13.

Sims, J.P., Dirks, P.H.G.M., Carson, C.J., and Wilson, C.J.L., 1994, Structural evolution of the Rauer Group, East Antarctica; mafic dykes as passive markers in a composite Proterozoic terrain: Antarctic Science, v. 6, p. 379–394.

Sithole, T., 2002, The tectonic significance of the Makuti Group, NW Zimbabwe [Ph.D. thesis]: Harare, University of Zimbabwe, 243 p.

Stagman, J.G., 1978, An outline of the geology of Rhodesia: Geological Survey of Zimbabwe Bulletin, v. 80, 126 p.

Stidolph, P.A., 1977, The geology of the country around Shamva: Geological Survey of Zimbabwe Bulletin, v. 78, 249 p.

Stocklmayer, V.R., 1978, The geology of the Inyanga North-Makaha area: Geological Survey of Zimbabwe Bulletin, v. 79, 173 p.

Stott, G.M., 1997, The Superior Province, Canada, *in* De Wit M.J., and Ashwal, L.D., eds., Greenstone Belts: Oxford, Oxford University Press, p. 480–507.

Talbot, C.J., 1973, The geology at the eastern end of the Zambezi orogenic belt, Rhodesia: Transactions of the Geological Society of South Africa, v. 76, p. 113–133.

Talbot, C.J., 1987, Strains and vorticity beneath a tabular batholith in the Zambezi belt: NE Zimbabwe: Tectonophysics, v. 138, p. 121–158, doi: 10.1016/0040-1951(87)90036-9.

Treloar, P.J., Coward, M.P., and Harris, N.B.W., 1992, Himalayan-Tibetan analogies for the evolution of the Zimbabwe Craton and Limpopo belt: Precambrian Research, v. 55, p. 571–587, doi: 10.1016/0301-9268(92)90046-Q.

Vail, J.R., and Snelling, N.T., 1971, Isotopic age measurements for the Zambezi orogenic belt and the Urungwe Klippe, Rhodesia: Geologische Rundschau, v. 60, p. 619–630, doi: 10.1007/BF02000472.

Van Dijk, E., and Kater, L., 1996, Geology and metamorphic evolution of the Shamva greenstone belt, NE Zimbabwe: Amsterdam, Internal report, Free University of Amsterdam, 52 p.

Vinyu, M.L., 1995, The geochronology and geochemistry of the post-orogenic granites within the Shamva-Harare greenstone belt [Ph.D. thesis]: Zimbabwe, University of Harare, 177 p.

Vinyu, M.L., Martin, M.W., Bowring, S., Hanson, R., Jelsma, H.A., and Dirks, P.H.G.M., 1999, U-Pb and ^{40}Ar/^{39}Ar geochronological constraints on the tectonic evolution of the easternmost part of the Zambezi Orogenic belt, northeastern Zimbabwe: Precambrian Research, v. 98, p. 67–82, doi: 10.1016/S0301-9268(99)00039-X.

Vinyu, M.L., Martin, M.W., Bowring, S., Hanson, R., Jelsma, H.A., and Dirks, P.H.G.M., 2001, U-Pb zircon ages from a craton margin, Archean orogenic belt in northern Zimbabwe: Journal of African Earth Sciences, v. 32, p. 103–114, doi: 10.1016/S0899-5362(01)90021-1.

Wilson, J.F., Nesbitt, R.W., and Fanning, C.M., 1995, Zircon geochronology of Archean felsic sequences in the Zimbabwe Craton: A revision of greenstone stratigraphy and a model for crustal growth, *in* Coward, M.P., and Ries, A.C., eds., Early Precambrian processes: Geological Society [London] Special Publication 95, p. 109–126.

Wingate, M.T.D., 2000, Ion microprobe U-Pb zircon and baddeleyite ages for the Great Dyke and its satellite dykes, Zimbabwe: South African Journal of Geology, v. 103, p. 74–80, doi: 10.2113/103.1.74.

Manuscript Accepted by the Society 4 October 2005

Printed in the USA

Geological Society of America
Special Paper 405
2006

The Limpopo Belt: A result of Archean to Proterozoic, Turkic-type orogenesis?

J.M. Barton Jr.[†]
Department of Geology, University of Johannesburg (formerly RAU), P.O. Box 524, Auckland Park 2006, South Africa
R. Klemd[‡]
Department of Geology, University of Johannesburg (formerly RAU), P.O. Box 524, Auckland Park 2006, South Africa and *Mineralogical Institute, University of Würzburg, Am Hubland, 97074 Würzburg, Germany*
A. Zeh[§]
Mineralogical Institute, University of Würzburg, Am Hubland, 97074 Würzburg, Germany

ABSTRACT

Evidence is presented relating to the tectonic evolution of the Limpopo Belt of southern Africa. It is demonstrated that this evolution was protracted, lasting at least 700 m.y. between ca. 2.7 and ca. 2.04 Ga, and comprised the complex assembly of terranes along shear zones through processes involving compression, extension, and strike-slip strain (transpression and transtension). It is argued that this evolution is more akin to Turkic-type craton building than to Himalayan- or Alpine-type collision between the Kaapvaal and Zimbabwe cratons at either ca. 2.7 Ga or ca. 2.04 Ga, to which most earlier workers have subscribed. The steeply dipping shear zones separating terranes are tens of kilometers wide with gradational contacts, as opposed to narrow (<1 km) shear zones with relatively sharp contacts that characterize the Himalayas and Alps. Ductile deformation with both subvertical and horizontal transport directions is common in the Turkic-type orogenesis rather than thin skin, thrust tectonics. The shear zones are interpreted to be artifacts of long-lived subduction zones. Relics of calc-alkaline (I-type) magmatic arcs that formed both on continental and oceanic crust are identified and are related to subduction-accretion of the various terranes. In contrast, S-type magmatism (melts derived from sedimentary protoliths) characterizes deformation after continent-continent collision in the Alps and Himalayas. Individual terranes are characterized by distinct P-T-time (pressure-temperature) paths that show they experienced different tectonic histories under high-grade amphibolite- to granulite-facies conditions instead of lower-grade, greenschist- to amphibolite-grade conditions, which are more common in the Alps and Himalayas. Individual terranes also have distinct metallogenic signatures.

Keywords: Limpopo Belt, tectonics, terranes, P-T-time paths, Turkic orogenesis.

[†]E-mail: Jay.Barton@telkomsa.net.
[‡]E-mail: reiner.klemd@mail.uni-wuerzburg.de.
[§]E-mail: armin.zeh@mail.uni-wuerzburg.de.

Barton, J.M., Jr., Klemd, R., and Zeh, A., 2006, The Limpopo Belt: A result of Archean to Proterozoic, Turkic-type orogenesis?, *in* Reimold, W.U., and Gibson, R.L., Processes on the Early Earth: Geological Society of America Special Paper 405, p. 315–332, doi: 10.1130/2006.2405(16). For permission to copy, contact editing@geosociety.org.

INTRODUCTION

The Altaids of central Asia have been interpreted by Şengör and Natal'in (1996) as a wide complex of terranes formed by a subduction-accretion assemblage of crustal elements onto the Russian and Angara cratons over a period of several hundred m.y., lasting from the Vendian to the Permian (Table 1). Individual terranes evolved geographically independent of one another but were tectonically linked along one or more long-lived subduction zones, along which the strain changed among compression, transtension, or transpression and extension several times. The Russian and Angara (Siberian) cratons each increased in size by this mechanism before they were finally juxtaposed in their present orientation. The individual terranes comprising the Altaids therefore interacted with one another in diverse ways so that accretion in one area might have involved the welding of pieces of continental crust, while in others it might have involved the joining of oceanic magmatic arcs with continental nuclei. Some terranes may even be joined on a smaller scale along sutures similar to those in Alpine and Himalayan orogens, and individual terranes might also contain evidence for more than one period of tectonism (Şengör and Natal'in, 1996). Şengör and Natal'in (1996) argue that this type of continent building, termed by them *Turkic-type,* involves magmatic arcs on both continental and oceanic crust, bulk shortening of the subduction-accretion complexes (steeply dipping, ductile tectonics), and often high-grade metamorphism and was probably also the primary mechanism for forming Precambrian cratons. It is exemplified by the accretion process presently active along the western margin of North and South America.

The individual terranes within the Altaids are pieces of continental and, to a lesser extent, oceanic crust (ophiolites) that are bounded by steeply dipping shear zones, tens of kilometers wide with gradational contacts and that comprise rocks with: (1) different geological histories as outlined by isotopic signatures including ages in the absence of fossils; (2) distinct metamorphic P-T-time (pressure-temperature) path(s) that may be related to distinct tectonic environments; (3) unique mineral deposits, and (4) lithological units that cannot be traced across shear zones from one crustal element to another along their boundaries (criteria from Jones et al., 1983). The magmatic arcs on continental crust are recognized by the linear orientation of calc-alkaline granitoid intrusions, whose compositions may be directly related to subduction magmatic processes. The bulk shortening is recognized by relatively steeply plunging, ductile deformation in shear zones occurring at up to high-grade conditions and high-grade metamorphic rocks with upper-amphibolite to granulite-facies metamorphic assemblages.

These characteristics are in marked contrast to those of Alpine- and Himalayan-type orogenies (Table 1). In the Western Alps (the portion of the Alps that most modelers of the Limpopo Belt use as an analogue; Şengör and Natal'in, 1996; Stampfli et al., 2002; www-sst.unil.ch/research/alpes/cross_section.htm), the African Plate was thrust over the European Plate, resulting in tectonic polarity of northward-directed thrust sheets and crustal thickening through thrust stacking. Uplift during tecton-

TABLE 1. COMPARISONS OF CHARACTERISTICS OF ALPINE, HIMALAYAN, AND TURKIC OROGENS WITH THOSE OF THE LIMPOPO BELT

Characteristics	Alpine orogen	Himalayan orogen	Turkic orogen	Limpopo belt
Lateral subduction zone links among terranes before joining	No	No	Yes	Yes
Duration	Two episodes, one during the Cretaceous and the second ca. 40 m.y. ago	One episode beginning ca. 50 Ma and continuing until today	ca. 400 m.y. from ca. 650 to ca. 245 Ma	ca. 700 m.y. from ca. 2.7 to ca. 2.0 Ga
Granitoid magmatism associated with tectonism	Uncommon	Common in backthrust belt	Common	Common
I-type, calc-alkaline granitoid magmatism	Common, early before craton-craton collision	Common, early before craton-craton collision	Common, dominant	Common, dominant
S-type granitoid magmatism	Common, after plate-plate collision	Common, after plate-plate collision	Minor	None recorded
Ophiolite complexes recognized	Yes	Yes	Yes	No
Terranes involved	2, Africa and Europe in the Western Alps	2, Asia and India	Many	>10
Width of major boundaries between terranes	<1 km	<1 km	>10 km	>20 km
Horizontal tectonics	Dominant	Dominant on the Indian Plate	Occurs	Occurs
Craton-craton overthrustng	Yes	No	Locally occurring	No
Syntectonic, foreland sedimentary basins	Yes	Yes	Uncommon	None recognized
Uplift rates after peak metamorphism	Variable but generally ~10 mm/year	Variable but generally ~10 mm/year	Generally <1 mm/year	All determined <1 mm/year (e.g., 0.3 mm/year)

Note: For sources of information, see text

ism was variable but during peak deformation it was on the order of 10 mm per year. Major foreland sedimentary basins formed syntectonically, both to the north and to the south. Magmatism was minor and exhumation of granulite-facies metamorphic rocks was restricted to the Ivrea Zone, the tectonically upturned, lower portion of the African Plate. The suture between the African and European Plates is situated at the base of the Ivrea Zone and is a few hundred meters wide. Alpine tectonism was rapid, occurring over a few tens of million years. Terranes were not involved in the tectonism of the Western Alps but are involved to the east.

In the Himalayas proper (Table 1 herein, Molnar and Tapponnier, 1975; Tapponnier et al., 1982; Searle et al., 1987; Şengör and Natal'in, 1996), tectonism began in the Eocene and is still going on. No overthrusting of cratons is involved, and the suture between the Asian and Indian Plates is narrow, a few hundred meters wide. The Indian Plate is being shortened by the southward stacking of thrusts, and the Asian Plate is being deformed by escape tectonism along wrench faults. Present uplift rates in the Himalayas are variable, but rates of 10 mm per year are not uncommon. Syntectonic sedimentary foreland basins occur both to the north and to the south. Granitic rocks in the Himalayas proper are predominantly S-type, formed from melting of sediments with large water contents. Calc-alkaline magmatism occurred on the Asian Plate as a result of subduction of oceanic crust beneath it prior to collision with the Indian Plate; i.e., in this instance calc-alkaline magmatism is pretectonic. Exposure of high-grade metamorphic rocks is uncommon, although it occurs in the east (e.g., Liu and Zhong, 1997).

The Limpopo Belt, although smaller, has long been described as either an Alpine- or a Himalayan-type orogenic belt caused by the oblique collision of the Zimbabwe and Kaapvaal cratons (e.g., Coward, 1976; Fairhead and Scovell, 1976; Coward and Fairhead, 1980; Fripp, 1982; Light, 1982; van Reenen et al., 1987, 1992; Roering et al., 1992; Treloar et al., 1992; Smit and van Reenen, 1997). This collision was interpreted to have occurred either at ca. 2.7 Ga or at ca. 2.04 Ga (e.g., Van Breemen and Dodson, 1972; Barton and van Reenen, 1992a; Barton et al., 1994; Kamber et al., 1995b; Barton and Sergeev, 1997; Jaeckel et al., 1997; Holzer et al., 1998, 1999). In these models, the continent-continent collision involved the thrusting of the Kaapvaal craton over the Zimbabwe craton, resulting in thickened continental crust. Only rocks belonging to the colliding cratons were involved; i.e., by implication the Limpopo Belt should contain no terranes other than the Zimbabwe and Kaapvaal cratons. Also in these models, the isostatic reaction to this crustal thickening was rapid uplift and erosion or "pop-up" (rapid, isothermal decompression; e.g., van Reenen et al., 1987; Roering et al., 1992), and the uplifted material was thrust outward onto the colliding cratons. The abundant granulite-facies metamorphic rocks presently exposed at surface supposedly formed at the bottom of the thickened crust and were uplifted and exposed during the final stages of isostatic adjustment. Despite the proposed rapid uplift, no foreland sedimentary basins related to it have been identified to the north or south. Sediments of the ca. 1.85 Ga Waterberg and Soutpansberg groups unconformably overlie the high-grade rocks of the Limpopo Belt (e.g., Watkeys, 1979, 1984; Barker, 1983; Watkeys et al., 1983; Cheney et al., 1990; Barton et al., 2003).

Alternative models for the Limpopo Belt involving terranes besides the Zimbabwe and Kaapvaal cratons have been proposed but not widely accepted. These recognized that the dominant strike-slip motions along the Palala-Zoetfontein and Triangle shear zones (Fig. 1) are incompatible with a continent-continent collision. Barton and Key (1981) argued, largely on the basis of available geochronological data, that the Central Zone (see next section for discussion of terminology) of the Limpopo Belt was exotic, unrelated to the adjacent cratons but incorporated between them along these strike-slip shear zones. McCourt and Vearncombe (1987, 1992) modeled an exotic Central Zone as an indenter, thrust from the east along these shear zones onto a combined Zimbabwe-Kaapvaal craton. Holzer et al. (1998, 1999) modeled the strike-slip motion along these shear zones as a necessary result of oblique collision between the Zimbabwe and Kaapvaal cratons, resulting from southeastward thrusting of the Magondi orogen to the west.

Several observations, however, many of which have been known for decades, are incompatible with a single craton-craton collision scenario. In this paper, we review these observations and suggest that the creation of the Limpopo Belt might be better interpreted in terms of Turkic-type rather than Alpine- or Himalayan-type orogenesis.

CRUSTAL ELEMENTS OF THE LIMPOPO BELT

As originally defined (Cox et al., 1965; Mason, 1973), the Limpopo Belt (Fig. 1) consists of three crustal elements: a Central Zone characterized by a unique succession of leucocratic, granitic, and high-grade, commonly migmatitic, quartzo-feldspathic rocks, and supracrustal lithologies (quartzite and calc-silicate, carbonate, metapelitic, and para-amphibolitic rocks) separated by the wide Palala-Zoetfontein and Triangle shear zones from two marginal zones. The Southern Marginal Zone consists of high-grade, intensely deformed and metamorphosed, tonalitic, and trondhjemitic rocks with minor greenstones and was interpreted as reworked rocks of the Kaapvaal craton (e.g., Coward et al., 1976; van Reenen et al., 1987, 1992), whereas the Northern Marginal Zone is dominated by enderbite and enderbitic orthogneisses with very minor greenstones and a few granitic rocks (e.g., Mason, 1973; James, 1975; van Reenen et al., 1987, 1992; Kamber and Biino, 1995; Jelsma and Dirks, 2002). The rocks of the marginal zones are themselves thrust, respectively, southward and northward, onto the respective cratons along >10 km wide, steeply dipping shear zones, the Hout River and Umlali shear zones, respectively. All four shear zones mentioned above have complex strain histories and gradational contacts (see discussion below and Table 1). The three zones of the Limpopo belt were thought to have been

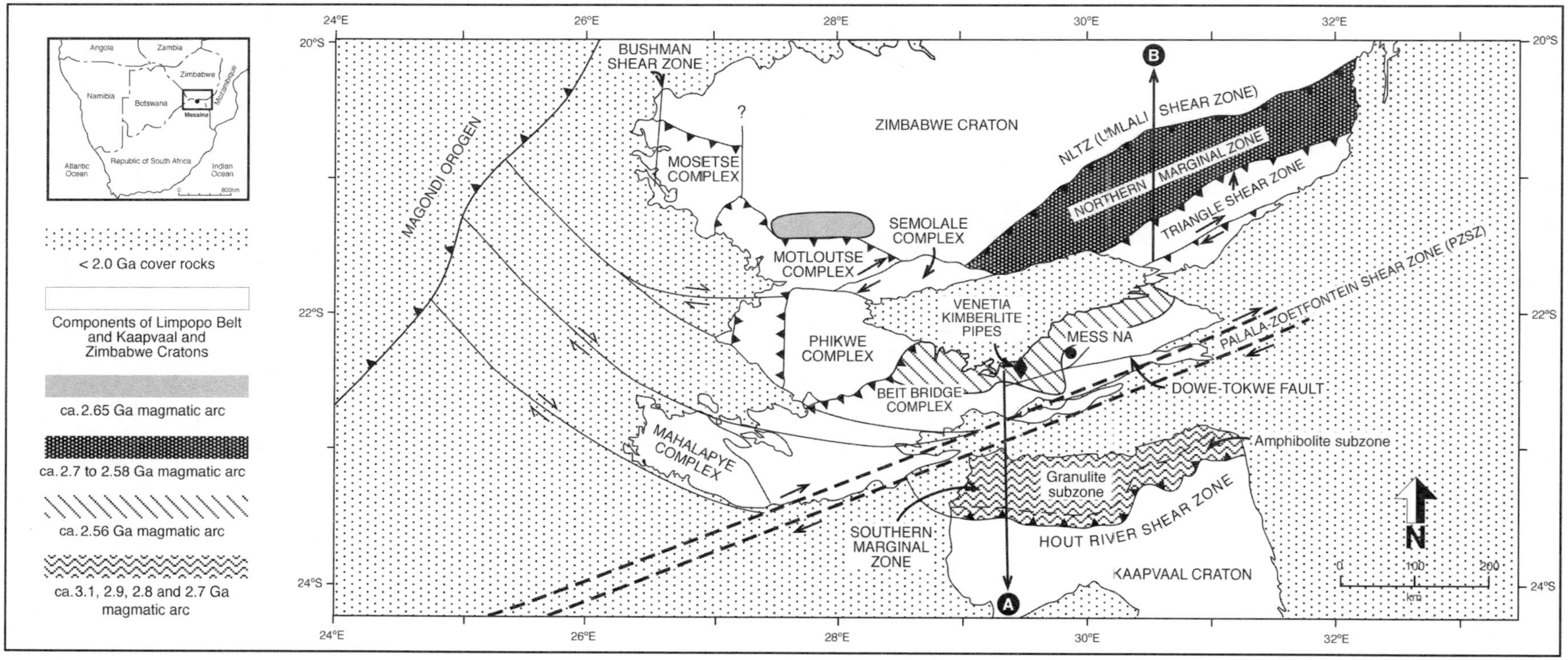

Figure 1. Geological map of the Limpopo Belt showing the terranes comprising it, rocks belonging to magmatic arcs and the Magondi orogen (modified from Aldiss, 1991; Carney et al., 1994; Barton and Pretorius, 1998; Holzer et al., 1999; Barton et al., 2003). NLTZ—northern Limpopo transition zone.

formed under granulite-facies conditions and subsequently overprinted under retrograde conditions (e.g., van Reenen et al., 1987, 1992). Their lateral extent to the east and west is poorly defined because they are covered by younger sedimentary and volcanic rocks.

Subsequent mapping, primarily in Botswana, delineated other shear zone–bounded crustal elements that had originally been considered part of the western Central Zone (e.g., Hickman and Wakefield, 1975; Aldiss, 1991; Brandl, 1992; McCourt et al., 2004). In addition, what was traditionally called the Central Zone proper was subdivided by Aldiss (1991) into (1) the Mahalapye Complex, which is dominated by ca. 2.02 Ga to ca. 2.0 Ga igneous rocks with minor metasedimentary rocks (Hisada and Miyano, 1996; Hisada and Paya, 1997; McCourt and Armstrong, 1998; Holzer et al., 1999; Chavagnac et al., 2001; Hisada et al., 2004); (2) the Phikwe Complex dominated by Archean, hornblende-bearing, tonalitic and trondhjemitic gneisses and igneous rocks (e.g., Brandl, 1992); and (3) the Beit Bridge Complex, with the infra- and supracrustal lithologies originally ascribed to the Central Zone. The granitic quartzofeldspathic rocks of the Beit Bridge Complex are now recognized as Archean, ca. 3.2 Ga orthogneisses (Boryta, 1988; Boryta and Condie, 1990; Barton and Pretorius, 1998; Barton et al., 2003; Barnett, 2005), whereas the supracrustal paragneisses may be divided into two, unconformity bounded successions, one ≥3.1 Ga and the other ≤2.66 Ga in age (e.g., Eriksson et al., 1988; Brandl, 1992; Barton et al., 2003; Buick et al., 2003). Quartzite from within the older unconformity-bounded succession contain zircons with U-Pb ages of ca. 3.2 Ga and ca. 3.8 Ga, suggesting that an unrecognized, very old protolith must exist within the Beit Bridge Complex (W. Compston, 1985, personal commun. And W. Compston and J.M. Barton Jr., 1985, personal commun.; Armstrong et al., 1987; Barton and Sergeev, 1997). The contact between the Phikwe and Beit Bridge complexes is clearly indicated on geological maps by changes in lithologies and roughly follows the valley of the Limpopo River (e.g., Aldiss, 1991; Brandl, 1992). However, because of poor exposure, the nature of this contact is uncertain.

Although the major shear zones separating the various crustal elements of the Limpopo Belt are easily observable on air photographs, satellite images, and geophysical surveys (e.g., Cox et al., 1965; Bahnemann, 1972; Mason, 1973), they are for the most part obscured by younger sedimentary rocks of middle Proterozoic age (Waterberg and Soutpansberg Groups) and Mesozoic (Karoo Supergroup) sediments and deep soils. Where exposed (Fig. 1; Table 2), they are tens of kilometers wide, anastomize, have long and complicated strain histories, and contain rocks of various metamorphic grades (see references below). As such, they are not technically shear zones with a coherent fabric formed in response to deformation in a single stress field but instead are wide zones of deformation in which reactivation in response to different stress fields has occurred. However, because the names of the "shear zones" are entrenched in the literature, the incorrect term "shear zone" is retained for them in this paper to avoid confusion.

TABLE 2. SUMMARY OF CHARACTERISTICS OF THE PRIMARY SHEAR ZONES WITHIN THE LIMPOPO BELT

Shear zones/characteristic	Hout River shear zone	Palala-Zoetfontein shear zone including the straightening zone of Bahnemann (1972)	Triangle shear zone or North Marginal transition zone	Umali shear zone or North Limpopo thrust zone
Orientation	Arcuate from west-northwest in the west to east-northeast in the east	Linear east-northeast	Linear east-northeast	Linear northeast
Separating	Kaapvaal Craton from Southern Marginal Zone	Separating Kaapvaal and Southern Marginal Zone from the Beit Bridge and Mahalapye Complexes	Separating the Beit Bridge Complex from the Northern Marginal Zone	Separating the Northern Marginal Zone from the Zimbabwe Craton
Orientation	Dipping steeply northward	Dipping both steeply northward and steeply southward	Dipping steeply southward	Dipping steeply southward
Width	>10 km	>22 km	~35 km	~5 km but deformation continues northward into the Zimbabwe craton
Displacement	Southwestward thrusting	Southward thrusting followed by right and left lateral strike-slip motion and then by normal faulting	Northward thrusting followed by right lateral strike-slip motion	Northward thrusting
Time(s) of movement	ca. 2.71 to ca. 2.67 Ga	ca. 2.05 to ca. 1.9 Ga; Post-Triassic	ca. 2.0 Ga	ca. 2.6 Ga
Primary lithologies involved	Calc-alkaline, quartzofeldspathic rocks of tonalitic to granodioritic compositions	Primarily calc-alkaline, quartzofeldspathic rocks of tonalitic, enderbitic, and granitic compositions with minor metasedimentary and gabbroic rocks	Calc-alkaline, quartzofeldspathic gneisses and minor metasedimentary rocks	Calc-alkaline enderbitic to charnoenderbitic rocks with minor greenstone lithologies
Tectonics	Ductile	Ductile and brittle	Ductile	Ductile

Note: For sources of data, see text.

It is impossible to precisely demarcate the edges of these shear zones because strain along them was heterogeneous and decreased gradually outward over long distances. In any given shear zone, both brittle-deformed rocks (cataclasites and sometimes pseudotachylitic breccia; Brandl and Reimold, 1990) and ductile-deformed rocks (mylonites and sometimes steeply dipping sheath folds with straight gneisses) may be found (Table 1, see references below). However, because of limited exposure, only specific areas of any of them have been documented in detail. This limited documentation introduces uncertainty regarding sweeping regional pronouncements about the deformation histories illustrated in these possibly unrepresentative areas.

The arcuate, west-northwest– to east-northeast–trending Hout River and the northeast-trending Umlali steeply dipping shear zones have complicated strain histories, which are dominated by compression. The Hout River shear zone contains north-dipping shear foliations, whereas shear foliations in the Umlali shear zone dip to the south. For detailed descriptions of the Hout River shear zone that separates the Southern Marginal Zone from the Kaapvaal craton, see De Wit et al. (1992), Roering et al. (1992), Smit et al. (1992), Smit and van Reenen (1997), and Kreissig et al. (2001). For similar information for the Umlali shear zone or Northern Limpopo thrust zone separating the Northern Marginal Zone from the Zimbabwe craton, see Blenkinsop et al. (1995), Mkweli et al. (1995), and Holzer et al. (1998, 1999). In these shear zones, steeply plunging mineral elongation lineations predominate, and kinematic indicators in these zones show thrusting southwestward and northward, respectively, onto the adjacent cratons. Subhorizontal fabrics are minor. Geochronologic studies of the Hout River shear zone indicate that it was active between ca. 2.71 Ga and ca. 2.6 Ga (Kreissig et al., 2001), whereas motion on the Umlali shear zone occurred at or before ca. 2.6 Ga (Mkweli et al., 1995).

The east-northeast–trending, steeply southward dipping Triangle shear zone (Fig. 1, terminology of Kamber et al., 1995b) or Northern Marginal Transitional Zone (terminology of Holzer et al., 1999) is a deformation zone at least 35 km wide, in which a poorly constrained, early northward-thrusting event is overprinted by right-lateral transpression or transtension at ca. 2.0 Ga.

In the east-northeast–trending Palala-Zoetfontein shear zone exposed within the Koedoesrand window (McCourt, 1993; Schaller et al., 1999), rocks with specific strain senses are localized, and hence, the shear zone may be subdivided into domains that have different deformation histories at different metamorphic grades during deformation ranging from granulite to probably greenschist facies (e.g., McCourt, 1983; McCourt and Vearncombe, 1987, 1992; Brandl and Reimold, 1990; Broekhuizen et al., 1995; Schaller et al., 1999). For example, following the interpretations of McCourt (1983) and Schaller et al. (1999), in the ~4-km-wide southern domain, rocks belonging to the ca. 1.7 Ga to ca. 2.0 Ga Palala granite (e.g., Barton and McCourt, 1983; Walraven et al., 1990), the compositionally unique Koeodesrand sediments of uncertain age and ca. 2.04 Ga mafic rocks of the Bushveld Complex are all cataclastically deformed and subsequently mylonitized, indicating variations in P-T conditions during strain. Shear planes dip steeply to the south. The southern domain is terminated on the south across the Abbottspoort shear zone by undeformed igneous rocks of the Bushveld Complex. The >10-km-wide central domain is composed of lenses of >2.6 Ga paragneisses of uncertain origin occurring within calc-alkaline felsic, mafic, and enderbitic mylonites and ultra-mylonites. Shear planes dip steeply to the north with subhorizontal mineral elongation lineations. In the ~2.5-km-wide northern domain, calc-alkaline, quartzofeldspathic gneisses, retrogressed calc-alkaline, <2.1 Ga charnockitic rocks, and intermediate to felsic, calc-alkaline mylonites surround layers of calc-silicate rocks, compositionally similar to those within the Beit Bridge Complex to the north. On steeply northward dipping shear planes, older, steeply dipping mineral elongation lineations associated with southward transport are overprinted by subhorizontal, amphibolite to greenschist facies mineral elongation lineations associated with kinematic indicators showing primarily right-lateral, with lesser left-lateral, strain. The northern boundary of the northern domain is defined as the Melinda fault, which offsets Mesozoic Karoo rocks, showing it was active at some point more recently than the Triassic and, hence, may be a young feature superimposed on the Palala-Zoetfontein shear zone. Deformation associated with the Palala-Zoetfontein shear zone extends at least 5 km further to the north into the Beit Bridge Complex in Bahnemann's (1972) straightening zone. Steeply northward-dipping shear planes show both subvertical and subhorizontal mineral elongation lineations in predominantly calc-alkaline, quartzofeldspathic rocks and minor paragneisses, including Proterozoic magnetite quartzite. Geochronologic studies (Schaller et al., 1999) indicate that the main movement of the Palala-Zoetfontein shear zone occurred between ca. 2.05 Ga and ca. 1.9 Ga.

In the west, the Palala-Zoetfontein and Triangle shear zones and their interpreted extensions have been interpreted to be truncated by the ca. 2.0 Ga to ca. 1.8 Ga Magondi orogen (Fig. 1, e.g., Hutchins and Reeves, 1980; Meixner and Peart, 1984; Carney et al., 1994; Holzer et al., 1998, 1999), and in some cases, movements along them control younger sedimentation and mafic volcanism, e.g., the rocks of the ca. 1.85 Ga Waterberg/Soutpansberg/Palapye Groups and those of the Permian to Triassic, Karoo Supergroup (e.g., Barker, 1983; Brandl, 1992).

CRUSTAL STRUCTURE BENEATH THE LIMPOPO BELT

Within the Limpopo Belt, three types of evidence exist with regard to deep crustal and upper lithospheric structure: (1) seismic, gravity, and resistivity geophysical data, (2) xenoliths from the Venetia kimberlite pipes, and (3) initial $^{207}Pb/^{206}Pb$ ratios of originally igneous and sedimentary rocks, which are related to μ values ($^{238}U/^{206}Pb$). These data exist only for the eastern portion of the Belt incorporating the Northern and Southern Marginal Zones and the Beit Bridge and Phikwe complexes.

Seismic, Gravity, and Resistivity Data

Refraction seismic data from a traverse from the Zimbabwe craton through the Northern Marginal Zone in Zimbabwe (Fig. 1, Traverse B) indicate that the Moho there shallows from ~40 km to 30 km going southward (Stuart and Zengeni, 1987). Similar data from a traverse from Pietersburg (now Polokwane) to Harare less precisely indicate the Moho but show a distinct, vertical, traveltime discontinuity at the position of the Palala-Zoetfontein shear zone (Durrheim et al., 1992). These data from Durrheim et al. (1992) also defined the deep crustal and upper mantle seismic structures used to produce the lithologic column under the Venetia kimberlite pipes (see next section). Data from a vibroseis traverse between Pont Drift and south of Pietersburg in South Africa (Fig. 1, Traverse A) come from insufficient depth (10 seconds two-way traveltime) to resolve the Moho under the Beit Bridge Complex, Southern Marginal Zone, and Kaapvaal craton (e.g., De Beer and Stettler, 1992; Durrheim et al., 1992; Roering et al., 1992). They do, however, indicate a reflector beneath the Beit Bridge Complex at ~10 km depth, underlain by a homogeneous zone with few if any reflectors. The reflection seismic data also do not resolve the Palala-Zoetfontein shear zone directly, indicating that it is a vertical to steeply subvertical feature.

On a more regional scale, Nguuri et al. (2001) interpreted refraction seismic data across the eastern Limpopo Belt to indicate an average depth to Moho of ~37 km beneath the Northern Marginal Zone and of ~40–42 km beneath the Southern Marginal Zone and the Kaapvaal craton. Data are ambiguous for rocks beneath the Beit Bridge Complex but suggest that the crust must be in excess of ~40–45 km thick, with possible discontinuities at ~35–30 km and between ~40 and 53 km. These data also indicate a dense lower crust beneath the Beit Bridge Complex. Deep seismic data (Fouch et al., 2004) indicate that the lithospheric keel beneath the Zimbabwe craton, Northern Marginal Zone, and Beit Bridge Complex extends to ~225–250 km, whereas the keel under the Kaapvaal craton and Southern Marginal Zone extends deeper to at least ~250–300 km. A discontinuity is indicated between these portions of crust at the position of the Palala-Zoetfontein Shear Zone.

Regional gravity data are interpreted to indicate that a zone of higher-density rocks extends from the Northern Marginal Zone southward under the Beit Bridge Complex, with its top at the same depth, ~10 km, indicated by seismic data (Fairhead and Scovell, 1976; Coward and Fairhead, 1980; De Beer and Stettler, 1992). No lower limit to the higher density rocks was modeled. Resistivity data also indicate a lithologic break at ~10 km depth beneath the Beit Bridge Complex (Van Zyl, 1978). The gravity data also show a distinct discontinuity, tens of km wide, at the position of the Palala-Zoetfontein shear zone (De Beer and Stettler, 1992).

Data from Crustal and Lithospheric Xenoliths

Crustal and upper mantle xenoliths brought to the surface in the ca. 530 Ma Venetia kimberlite pipes were studied by Pretorius (1996), Barton and Pretorius (1998), and Pretorius and Barton (2003a, 2003b). Lithologic and chemical compositions along with mineral-equilibrium pressure data and measured and calculated p-wave velocity data were combined with the seismic and gravity data (De Beer and Stettler, 1992; Durrheim et al., 1992) to construct a lithological section beneath the Venetia kimberlite pipes (Fig. 2). This approach of comparing geophysical data that reflect present structures with lithologies brought to the surface ~530 m.y. ago assumes that the crustal structure has not changed since the Venetia kimberlite pipes were emplaced. This assumption seems probable because this area is presently in isostatic equilibrium (De Beer and Stettler, 1992), and analysis of the kimberlite structures within the Venetia pipes suggests at most only a couple of kilometers of crust could have been eroded from the top since emplacement (Kurszlaukis and Barnett, 2003).

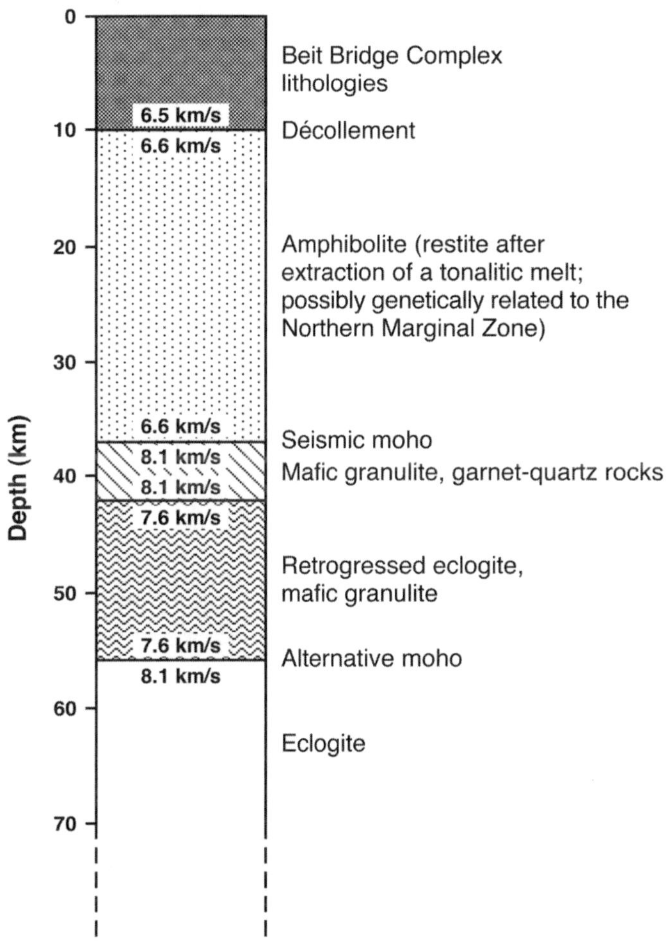

Figure 2. Lithological section beneath the ca. 530 Ma Venetia kimberlite pipes in the Beit Bridge Terrane constructed from pressure and P-wave velocity data and compositions of xenoliths from Venetia kimberlite (Pretorius, 1996; Pretorius and Barton, 2003a, 2003b).

Data from xenoliths extracted from beneath the Venetia kimberlite pipes indicate that lithologies of the Beit Bridge Complex extend only to a depth of ~10 km, where they terminate against amphibolite-facies, metabasitic rocks. This contact, also indicated by the seismic, gravity, and resistivity data, was interpreted as a décollement separating the supracrustal rocks from mafic rocks forming the lower crust (Barton and Pretorius, 1998). The mafic rocks, which extend to a depth of ~37 km, were interpreted as restites from which tonalitic magma has been extracted by partial melting and these restites correspond to the higher-density material indicated by the gravity and seismic data. These geophysical data, along with chemical data, were used to link the mafic rocks (mostly amphibolite) to the enderbitic rocks of the Northern Marginal Zone, suggesting that these lithologies may be genetically related (Pretorius, 1996; Pretorius and Barton 2003a) and, hence, might be the roots and top of an oceanic magmatic arc (see below, Fig. 3). However, the Triangle shear zone separates the Beit Bridge Complex from the Northern Marginal Zone, making this interpretation less tenable. The amphibolite is texturally homogeneous, which may explain the lack of reflectors observed in the vibroseis data.

Below ~37 km, p-wave velocities jump to ~8.1 km/s, caused by the increased density of garnet-quartz rocks and mafic granulite. The mafic granulite is chemically similar to the overlying amphibolite, suggesting that the seismic discontinuity is a metamorphic isograd. At ~42 km, p-wave velocities drop to ~7.6 km/s in mafic granulite and retrogressed eclogite that is again compositionally similar to the overlying amphibolite. P-wave velocities then increase again to 8.1 km/s at ~56 km depth where eclogite is encountered. Pretorius (1996) and Pretorius and Barton (2003a) suggested that this latter jump in p-wave velocities may represent the bottom of the continental crust, i.e., the lithological Moho. However, eclogite could also be a mafic component of continental crust metamorphosed to eclogite facies conditions, and so the base of the continental crust could be deeper. The various lithologic breaks between ~37 and 56 km may be those indicated in some of the regional refraction seismic data of Nguuri et al. (2001).

Pb-Isotopic Data

Large $^{207}Pb/^{204}Pb$ ratios compared to $^{206}Pb/^{204}Pb$ ratios (Fig. 4) characterize most igneous and meta-igneous rocks of the Beit Bridge and Phikwe Complexes and in the Northern Marginal Zone and Zimbabwe craton, suggesting that these complexes were derived from a source with a long-lived, high μ value (data from Barton et al., 1983; Taylor et al., 1991; J.M. Barton Jr., 1995, personal commun.; Barton, 1996; Kreissig et al., 2000). These large Pb isotopic ratios are in marked contrast to those for rocks exposed in the Southern Marginal Zone and the Kaapvaal craton, which were derived from sources with μ values close to bulk Earth (Fig. 4). Hence, the rocks of the Zimbabwe craton, Northern Marginal Zone, and Beit Bridge and Phikwe Complexes may be derived from the same mantle source, but they cannot be genetically related in an easy way to those of the Kaapvaal craton and Southern Marginal Zone.

A Cross Section Across the Limpopo Belt

Taking these geophysical and lithological data together, it is possible to create a schematic north-to-south cross section (Fig. 3) across the Limpopo Belt along the line of the vibroseis survey from south of Pietersburg on the Kaapvaal craton through the Southern Marginal Zone and Beit Bridge Complex to Pont Drift (De Beer and Stettler, 1992; Durrheim et al., 1992), and then northward along the seismic traverse of Stuart and Zengeni (1987) and Durrheim et al. (1992) through the Northern Marginal Zone and onto the Zimbabwe craton (Traverses A and B, Fig. 1). This cross section clearly shows the Palala-Zoetfontein shear zone as a wide crustal break, separating the Kaapvaal craton with the Southern Marginal Zone thrust upon it from the Beit Bridge Complex, and the Zimbabwe craton with the Northern Marginal Zone thrust upon it. The Beit Bridge Complex is shown as a, 10-km-thick veneer of rocks overlying mafic rocks that are separated from the Northern Marginal Zone by the Triangle shear zone. The cross section, as shown in Figure 3, is in

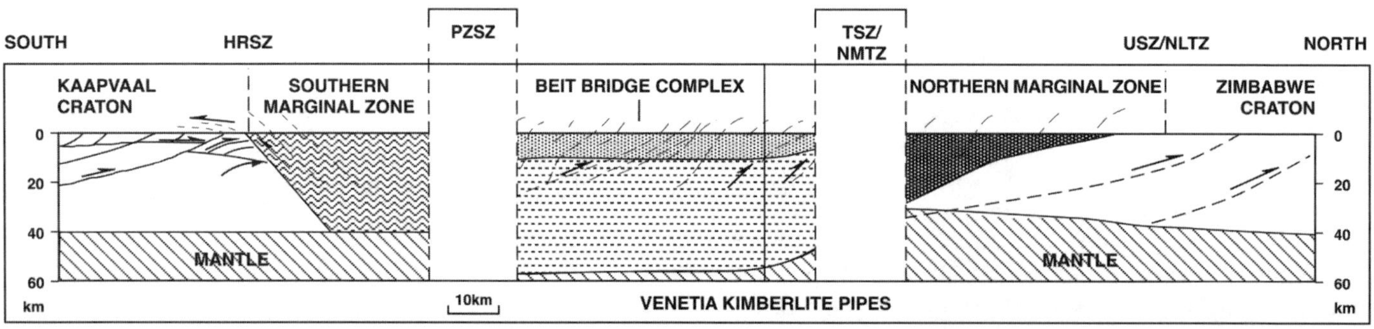

Figure 3. Cross section through the Limpopo Belt based on seismic and gravity data (modified from DeBeer and Stettler, 1992; Roering et al., 1992) and xenolith compositions from the Venetia kimberlite pipes (Pretorius, 1996; Pretorius and Barton, 2003a, 2003b). Note that the widths of the Palala-Zoetfontein (PZSZ) and Triangle (TSZ/NMTZ) shear zones are minimum estimates, and note that some authors show widths in excess of 35 km. HRSZ—Hout River shear zone, USZ/NLTZ—Umlali shear zone/northern Limpopo transition zone.

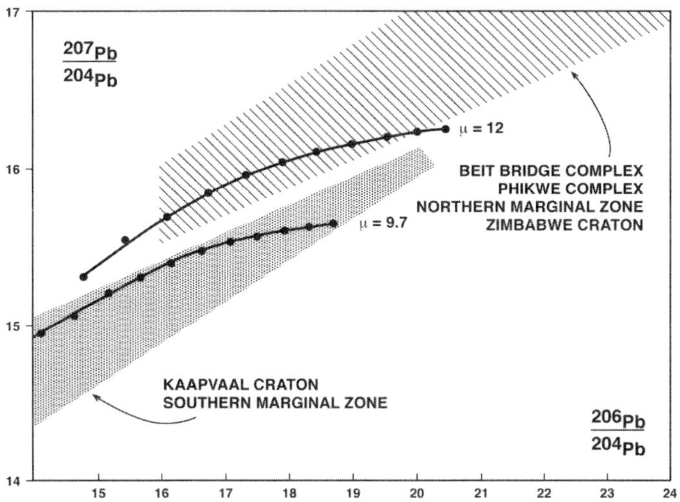

Figure 4. Plot of $^{207}Pb/^{204}Pb$ versus $^{206}Pb/^{204}Pb$ data for rocks from different terranes in the Limpopo Belt and the Zimbabwe and Kaapvaal cratons (Barton et al., 1983; Taylor et al., 1991; Barton, 1995; and Barton, 2005, personal commun.; Kreissig et al., 2000). These data indicate that the rocks of the Kaapvaal craton and the Southern Marginal Zone are genetically different from those of the Beit Bridge Terrane, Phikwe Complex, Northern Marginal Zone and Zimbabwe craton and that the latter were derived from a source with a very long-lived, high μ composition.

marked contrast to that proposed by Roering et al. (1992, Fig. 2 herein), which was drawn to fit the "pop-up" model for the Limpopo Belt (e.g., van Reenen et al., 1987).

MAGMATIC ARCS

Calc-alkaline, tonalitic, granodioritic, and granitic lithologies, which are usually the result of melts formed and emplaced in magmatic arcs on continental crust in modern tectonic regimes, are extensive and are volumetrically significant within the Limpopo Belt, where they form orthogneiss complexes in linear bands (Fig. 1). Their presence implies that significant amounts of oceanic crust were subducted, dehydrated, and sometimes partially melted under continental crust during formation of the Limpopo Belt prior to ca. 2.0 Ga. The distribution of these rocks indicates where the subduction zones were located adjacent to them at the edge of the crustal blocks. However, ophiolites, blueschist-facies rocks, or extensive volumes of basaltic rocks with MORB (mid-oceanic-ridge basalt) signatures have not been identified within the Limpopo Belt, although one small occurrence of podiform chromite, often associated with sutures in continent-continent collisions, has been recognized within the Southern Marginal Zone (Smit, 1984). The lack of these subduction-related indicators could be used to argue that consumption of significant amounts of oceanic crust did not occur during formation of the Limpopo Belt and the Kalahari craton (see discussion in Light, 1982) or that this oceanic crust was completely subducted by a so-far unknown process.

The most obvious magmatic arc in the Limpopo Belt comprises the ca. 2.75 Ga to ca. 2.58 Ga charnoenderbitic to charnockitic igneous rocks making up ~90% of the Northern Marginal Zone (e.g., Berger et al., 1995; Kramers et al., 2001). As mentioned above, these rocks formed by partial melting of amphibolite and higher-grade rocks of an appropriate composition, similar to those preserved at a depth of >10 km beneath the Beit Bridge Complex (Pretorius, 1996; Pretorius and Barton, 2003a). McCourt et al. (2004) have also defined a ca. 2.65 Ga, granitic Francistown magmatic arc complex, probably related to subduction during the accretion of the Motloutse Complex onto the Zimbabwe craton. Magmatism in the Northern Marginal Zone and Motloutse Complex may be tectonically related to the emplacement of the upper greenstone succession in the Zimbabwe craton (Jelsma and Dirks, 2002).

Other examples of magmatic arcs within the Limpopo Belt are the ca. 2.63 to ca. 2.57 Ga tonalitic to granodioritic to granitic igneous rocks and orthogneisses within the Beit Bridge Complex, including the Bulai batholith and Singelele gneiss (e.g., Boryta, 1988; Boryta and Condie, 1990; Barton et al., 1994, 2003; Kröner et al., 1999; Boshoff et al., 2004; Barnett, 2005). These rocks are characterized by high U and Th contents that assist in their identification and in the delineation of their areal extent using radiometric remote sensing techniques (e.g., Andreoli et al., 2003; Barton et al., 2003; Barnett, 2005; Council for Geosciences, South Africa, 2003, personal commun.). In the Southern Marginal Zone and adjacent Kaapvaal craton, ca. 3.1–2.9 Ga tonalitic rocks (e.g., the Goud Plaats and Bavianskloof gneisses) and ca. 2.8 and 2.7 Ga granitic to granodioritic igneous rocks and orthogneisses (e.g., the Turfloop batholith and Matok pluton, respectively) probably indicate other magmatic arcs, although, with the exception of the Turfloop batholith, their limits have not been adequately defined (e.g., Barton et al., 1992; Brandl and Kröner, 1993; Henderson et al., 2000).

PRESSURE-TEMPERATURE-TIME PATHS

Pressure-temperature (P-T) paths throughout the Beit Bridge Complex and Southern Marginal Zone were initially interpreted to be parallel (identical), with significant isothermal decompression in retrograde segments followed by nearly isobaric cooling (e.g., van Reenen, 1983; van Reenen et al., 1987, 1988, 1990; Hisada and Miyano, 1996; Holzer et al., 1998; 1999; Kröner et al., 1999). It was thought that this was the result of a single orogenic event at ca. 2.7 Ga (e.g., van Reenen et al., 1987, 1990; McCourt and Vearncombe, 1987, 1992; Roering et al., 1992). Furthermore, it was suggested that all units within the Limpopo Belt underwent granulite-facies metamorphism, as supported by the results of petrologic studies, mostly from within a few tens of kilometers of Messina (now renamed Musina) within the Beit Bridge Complex (e.g., Windley et al., 1984; Droop, 1989; Perchuk et al., 2000) but also from the Mahalapye Complex (Hisada and Paya, 1997; Chavagnac et al., 2001), the Phikwe Complex (Tsunogae et al., 1989; Hisada and Miyano, 1996), the Northern Marginal Zone (Tsunogae et al., 1992; Kamber and Biino, 1995; Rollinson and Blenkinsop, 1995), and the Southern Marginal Zone (van Reenen, 1986; Stevens and van Reenen, 1992; Perchuk et al., 2000). These observations led to the conclusion that the high-grade metamorphic rocks of the Central and Marginal Zones of the Limpopo Belt were rapidly exhumed by "pop-up"–style tectonics (e.g., van Reenen et al., 1987; Roering et al., 1992; Holzer et al., 1998). Perchuk et al. (2004) presented a variation on this theme, whereby the pop-up was replaced by diapirism due to density differences among the rocks involved.

With the employment of quantitative phase diagrams such as pseudosections (e.g., Zeh et al., 2004b, 2005a, 2005b), variations in P-T paths of the Beit Bridge Complex in the Messina and Venetia areas have become apparent (Fig. 5A). Within the Beit Bridge Complex itself, different P-T-time paths are found within a few kilometers of one another in the Messina area and also in the area around the Venetia kimberlite pipes (Fig. 5A; compare and contrast results by, e.g., Droop, 1989; Klemd et al., 2003; Sachau, 2003; Zeh et al., 2003, 2004a, 2004b, 2005a, 2005b; van Reenen et al., 2004). One such example involves the P-T paths for the amphibolite-facies Endora and Venetia klippen near the Venetia kimberlite pipes that are separated by less than 1 km and the P-T path of the granulite-facies rocks of the Alldays area ~20 km to the west (Fig. 5A). The P-T vectors of most of these paths are similar to that obtained from the granulites of the Messina area but are shifted to lower pressure and temperature conditions. However, none of these paths displays significant isothermal decompression followed by isobaric cooling. Instead, they display simultaneous cooling and decompression after peak metamorphism and, thus, do not correlate with the paths from the Southern Marginal Zone, the Phikwe Complex, or the Mahalapye Complex (Fig. 5C; e.g., van Reenen, 1983, 1986; van Reenen et al., 1988; Stevens and van Reenen, 1992; Tsunogae et al., 1992; Kamber and Biino, 1995; Hisada and Miyano, 1996; Hisada and Paya, 1997; Hisada et al., 2000; Perchuk et al., 2000). Therefore, no single P-T path is characteristic of the Limpopo Belt as a whole.

When geochronological results are added to these P-T paths, further variations may be noted. In the Southern Marginal Zone, high-grade metamorphism occurred at or before ca. 2.7 Ga, as it is constrained by the intrusion at ca. 2.67 Ga of the Matok pluton through the retrograde, ortho-amphibole isograd (van Reenen, 1986; Barton et al., 1992). Much later, the Southern Marginal Zone was overprinted by greenschist- and amphibolite-facies metamorphism between ca. 2.0 and 1.9 Ga, as shown by Rb-Sr ages for biotite (Barton and van Reenen, 1992b) and U-Pb ages for apatite (Berger and Braun, 1997). At the same time, the Beit Bridge Complex of the Central Zone was affected by a major high-grade tectono-metamorphic event dated at ca. 2.04 Ga (e.g., Barton et al., 1994, 2003; Kamber et al., 1995a, 1995b; Jaeckel et al., 1997; Holzer et al., 1998, 1999; Kröner et al., 1999; Buick et al., 2003; Boshoff et al., 2004), with no evidence for a similar metamorphic event at ca. 2.7 Ga. Furthermore, no evidence for a ca. 2.7 Ga event was found in either the

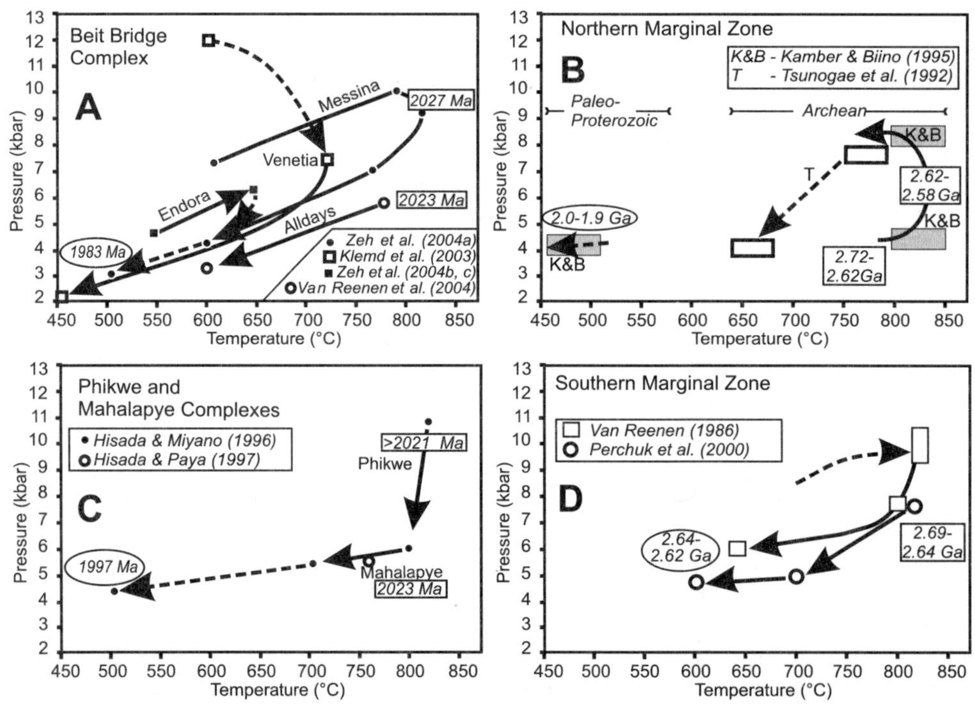

Figure 5. Contrasting pressure-temperature-time paths for various localities within the Limpopo Belt. For sources of the different age determinations, see discussion in text.

Mahalapye Complex (Holzer et al., 1999; Chavagnac et al., 2001) or the Phikwe Complex (Hickman and Wakefield, 1975; Holzer et al., 1999). A U-Pb age for apatite of 1983 ± 14 Ma (Holzer et al., 1998) and a Rb-Sr biotite age of 1970 ± 50 Ma (Barton et al., 1983) indicate that the Central Zone finally cooled below ~500 °C during the Paleoproterozoic.

In addition to the differences described above, the rocks of the Beit Bridge Complex show evidence of two high-grade metamorphic events, whereas those from other complexes do not (Barton et al., 1990). At the famous Sand River outcrop (Fripp, 1981, 1982, 1983; Buhlmann, 2003), migmatitic rocks of the Sand River Gneisses formed at >3.2 Ga and then were intruded by ca. 3.0 Ga mafic dykes under brittle conditions, as is shown by the local preservation of angular xenoliths of migmatite within them. Subsequently, these dykes underwent ductile deformation and granulite-facies metamorphism, resulting in partial melting of all other lithologies in the area. This granulite-facies overprint occurred at ca. 2.04 Ga (e.g., Barton et al., 1994; Jaeckel et al., 1997; Boshoff et al., 2004; Smit et al., 2004), much later than the intrusion of the Bulai batholith (~20 km to the west), which formed part of a ca. 2.57 Ga magmatic arc (see above). So far, there is no petrological, structural, nor geochronological evidence suggesting that the intrusion of the Bulai batholith was accompanied by an Archean high-grade regional metamorphic event (e.g., Barton and Sergeev, 1997; Jaeckel et al., 1997). Pb stepwise leaching ages of ca. 2.57 Ga, as reported by Holzer et al. (1998), were obtained from metamorphic garnet and sillimanite from xenoliths within the batholith. These ages, which are identical to that of the enclosing orthogneiss, are not indicative for a high-grade regional event at ca. 2.57 Ga but are more likely to reflect contact metamorphism. In fact, no isotopic or structural evidence for ca. 2.57 Ga metamorphism and deformation has been identified within the nearby Sand River Gneisses. The Sand River Gneisses commonly contain ca. 3.2 Ga zircons that only show metamorphic overgrowth at ca. 2.0 Ga, not at ca. 2.57 Ga (Barton and Sergeev, 1997). Furthermore, all structures can be correlated through their relationships with dated leucosomes with either the ca. 3.2 Ga or ca. 2.0 Ga events (e.g., Fripp, 1981, 1982, 1983; Barton et al., 1990; Jaeckel et al., 1997).

P-T paths for metapelitic rocks in the Messina area of the Beit Bridge Complex (Fig. 5A) indicate granulite-facies conditions were achieved along a clockwise metamorphic P-T path from 600 °C at 7.0 kbar to 780 °C at 9–10 kbar (the pressure peak) to 820 °C at 8 kbar (the thermal peak), followed by a decompression-cooling path to 600 °C at 4.0 kbar (Zeh et al., 2003, 2004b). A similar retrograde decompression-cooling path but with lower P conditions from 780 °C at 5.7 kbar (thermal peak) to 600 °C at 3.3 kbar was inferred by van Reenen et al. (2004) for high-grade gneisses from the central part of the Beit Bridge Complex northwest of Alldays. The peak P-T conditions derived by van Reenen et al. (2004) are similar to those inferred by Hisada and Paya (1997) for rocks from the Mahalapye Complex further southwest (>750 °C at ~5 kbar).

In contrast to the areas around Messina, northwest of Alldays and the Mahalapye Complex, metapelitic rocks of the Venetia klippe ~15 km north-northeast of Alldays and ~80 km west of Messina only underwent amphibolite-facies metamorphism, which is, so far, a unique feature within the Limpopo Belt (Klemd et al., 2003). Furthermore, two distinct P-T paths are distinguished, indicating that the area surrounding the Venetia kimberlite pipes forms part of a klippen complex. Rocks of the Venetia nappe provide evidence for a prograde decrease from 600 °C at 13 kbar (pressure peak) to 720 °C at 7–8 kbar (thermal peak), followed by retrogression to 350 °C at 2 kbar (Klemd et al., 2003). In contrast, metapelitic rocks and garnet-grunerite amphibolites from the Endora klippe, 1 km north of the Venetia klippe, provide evidence for a prograde P-T increase from 540 °C at 4.5 kbar to 650 °C at 6.5 kbar (metamorphic peak) followed by a P-T decrease (Zeh et al., 2004a, 2005a, 2005b). As already described, most P-T paths for rocks of the Beit Bridge Complex are similar but display varying pressure and temperature conditions as a consequence of horizontal tectonics. The only exception, so far, is the P-T path obtained by Klemd et al. (2003) from rocks from the Venetia klippe, which begins at much higher pressure. Its significance is not yet clear.

Despite their different early histories, P-T-time paths for rocks in the Beit Bridge, Phikwe, and Mahalapye complexes (Fig. 5) are similar, suggesting that the last deformation in them is related to a single orogenic event beginning at ca. 2.04 Ga that lasted no more than 30–60 m.y. (e.g., Hisada and Miyano, 1996; Zeh et al., 2004a, 2004b). The thermal peaks in rocks located ~10 km northwest of Messina occurred at ca. 2.03 Ga (Jaeckel et al., 1997; Holzer et al., 1998), in the area northwest of Alldays at ca. 2.02 Ga (Boshoff et al., 2004), in the Mahalapye Complex at ca. 2.02 Ga (Hisada and Miyano, 1996; Holzer et al., 1999), in the Phikwe Complex at ca. 2.08 Ga (Holzer et al., 1999), and in the Venetia klippe at ca. 2.02 Ga (Barton et al., 2003; Klemd et al., 2003). These ages are essentially the same within experimental error.

From a combined petrological, geochronological, and structural database, Zeh et al. (2004a, 2004b, 2005a, 2005b) concluded that the prograde evolution of the Beit Bridge Complex resulted from successive stacking due to horizontal nappe tectonics at ca. 2.04 Ga. Cooling from the thermal peak of this event is bracketed by a zircon age of 2027 ± 6 Ma (Jaeckel et al., 1997) and a U-Pb apatite age of 1983 ± 14 Ma (Holzer et al., 1998) for rocks from the Messina area. If it is assumed that the spherical zircon dated by Jaeckel et al. (1997) grew close to the thermal peak at 820 °C and 8 kbar and a closure temperature of ~500 °C is taken for the U-Pb system in apatite (e.g., Dahl, 1997), a linear cooling rate for these rocks of around 7 °C/Ma is indicated. Furthermore, a linear exhumation rate of ~0.3 mm/year is gained by assuming that the lithostatic pressure gradient was 270 bar/km and exhumation to 4 kbar at 500 °C. This slow exhumation rate and the retrograde P-T path involving simultaneous cooling and decompression argue against the pop-up model. The slow exhumation rate, mainly maintained by erosion (Zeh et al., 2004b), could, however, explain the general lack of reverse, dip-slip displacement fabrics associated with the ca. 2.04 Ga event.

In the Southern Marginal Zone, P-T paths (Fig. 5D) indicate peak metamorphic conditions of 820 °C at >9 kbar followed by decompression to 800 °C at 8.6–7.2 kbar and nearly isobaric cooling to 650–625 °C at >6 kbar (van Reenen, 1983, 1986; Stevens and van Reenen, 1992). Perchuk et al. (2000) derived a slightly different retrograde path, characterized by decompression and cooling from 870 to 850 °C at 8.2–7.2 kbar (thermal peak) to 700 °C at 5 kbar followed by isobaric cooling to 600 °C at 5–4.5 kbar. Kreissig et al. (2001) argued that this evolution occurred between ca. 2.69 and ca. 2.64 Ga.

Tsunogae et al. (1992, Fig. 5B herein) showed that rocks from the eastern portion of the Northern Marginal Zone experienced peak metamorphic conditions of 790–740 °C at 7.9–7.5 kbar followed by retrogression to 660–630 °C at 4.5–3.8 kbar. In contrast, Kamber and Biino (1995) inferred an anticlockwise P-T path from 850 to 800 °C at 4.5 kbar to 850–800 °C at 8.5 kbar, which took place at ca. 2.6 Ga, more recent than in the Southern Marginal Zone (Berger et al., 1995; Kamber and Biino, 1995; Mkweli et al., 1995).

The P-T paths described above for the Limpopo Belt are typical of many granulite-facies terranes worldwide (e.g., Harley, 1989). The anticlockwise P-T path of the Northern Marginal Zone involves isobaric cooling, whereas the clockwise P-T paths for most areas in the Southern Marginal Zone as well as the Phikwe and Mahalapye complexes involve a period of almost isothermal decompression followed by isobaric cooling. However, the slope of the clockwise P-T path for the Beit Bridge Complex indicates prograde simultaneous burial and heating during crustal thickening followed by contemporaneous decompression and retrograde cooling. The P-T paths of the Southern Marginal Zone and the Beit Bridge Complex are the result of tectonic thickening at different times, ca. 2.7 and ca. 2.04 Ga, respectively, followed by exhumation, which was dominated either by erosion (Beit Bridge Complex) or tectonic denudation (Southern Marginal Zone, Phikwe and Mahalapye complexes), producing a steeper-sloped decompression P-T path (e.g., Thompson and England, 1984). The prograde P-T paths of the Beit Bridge Complex and the Northern Marginal Zone are consistent with burial along a moderate geothermal gradient (intermediate P/T metamorphism) characteristic of continental collisions and, thus, in clear contrast to Alpine- or Himalayan-type metamorphism. The latter usually is associated with a steep geothermal gradient (high P/T metamorphism), which typically occurs in subduction zones (e.g., Ernst, 1988; O'Brien et al., 2001).

However, the most compelling difference between the tectonic evolution of most parts of the Limpopo Belt and Alpine- or Himalayan-type settings lies in the different post-peak metamorphic P-T evolutions. The Alpine or Himalayan types are characterized by a period of fast, almost isothermal, decompression (Ernst, 1988; Liu and Zhong, 1997; O'Brien et al., 2001), which is usually controlled by pure shear tectonics (e.g., Thompson and England, 1984; Ruppel et al., 1988). However, such tectonism is not supported by the retrograde P-T paths inferred from rocks of the Northern Marginal Zone and the Beit Bridge Complex, which either reflect isobaric cooling or contemporaneous cooling and decompression.

METALLOGENIC SIGNATURES

Although not particularly known as a major mineral-producing area, different crustal elements within the Limpopo Belt are characterized by different types and styles of metallogenesis, which formed prior to the final assembly of the Limpopo Belt at ca. 2.0 Ga. The Beit Bridge Complex was a significant source of alluvial corundum and also produced significant amounts of graphite, magnesite, and asbestos (Söhnge, 1945; Söhnge et al., 1948). It has also been a significant source and is a potential future source of magnetite iron ore (e.g., Söhnge, 1945; Du Plessis et al., 1997) derived from either hydrothermal alteration or weathering of banded iron formation that may be coeval with the ca. 2.6 Ga banded iron formation in the Transvaal succession (e.g., Walraven and Martini, 1995; Martin et al., 1998). The Cu and diamond mineralization in the Beit Bridge Complex is younger than ca. 2.0 Ga (e.g., Söhnge, 1945; Barton et al., 2003). The Phikwe Complex is unique in the Belt by containing a world-class Ni-Cu-Co deposit at Selebi-Phikwe with numerous minor occurrences around it (e.g., Van Breemen and Dodson, 1972; Gordon, 1973; Hickman and Wakefield, 1975; Wakefield, 1976; Lear, 1979; Gallon, 1986; Brown, 1987; Barton and van Reenen, 1992b). This mineralization appears to be associated with intrusion of one or more ultramafic sills that were subsequently subjected to intense ductile deformation at ca. 2.04 Ga (e.g., Wakefield, 1976; Barton and van Reenen, 1992b). Gold mineralization occurs in sheared rocks of the Southern Marginal Zone at Bochum (Wilson and Anhaeusser, 1998) and the Northern Marginal Zone at Renco Mine (e.g., Blenkinsop and Frei, 1996; Kisters et al., 1998; Blenkinsop et al., 2004). In the Bochum occurrence, mineralization is post–peak metamorphism (<2.7 Ga; Wilson and Anhaeusser, 1998), but in the case of Renco Mine, it apparently occurred during peak metamorphism (ca. 2.7–2.6 Ga) and was remobilized at ca. 2.0 Ga (e.g., Blenkinsop and Frei, 1996; Kisters et al., 1998; Blenkinsop et al., 2004).

ARE THE CRUSTAL ELEMENTS OF THE LIMPOPO BELT TERRANES?

From the evidence described above, the crustal elements recognized in the Limpopo Belt have all the characteristics of terranes (Jones et al., 1983). They are shear-zone bounded and have distinct geological histories as indicated by geochronology and P-T-time paths and different mineralization. They also contain lithologies that cannot be traced across the shear zones. Therefore, application of the word "complex" should be changed to the use of the word "terrane": e.g., the Beit Bridge Terrane.

ASSEMBLY OF THE LIMPOPO BELT

At this stage, knowledge is incomplete of the order and tectonic style in which the various terranes comprising the Limpopo Belt were assembled. Nevertheless, it is known that this assembly involved wide, steeply dipping, ductile shear zones and magmatic arcs over an extended period of time. The Mosetsi and Motloutse Terranes apparently were accreted onto the Zimbabwe craton at ca. 2.6 Ga (McCourt et al., 2004), and the Northern Marginal Zone apparently was accreted onto the Zimbabwe craton by ca. 2.58 Ga (Blenkinsop et al., 1995; Mkweli et al., 1995; Holzer et al., 1998, 1999). The Southern Marginal Zone had been accreted onto the Kaapvaal craton by ca. 2.7 Ga (see discussion above). The Beit Bridge Terrane, including a ca. 2.6 Ga magmatic arc complex, was accreted onto the Northern Marginal Zone along the Triangle shear zone at ca. 2.04 Ga and juxtaposed by transpression or transtension along the Palala-Zoetfontein shear zone with the Kaapvaal craton and Southern Marginal Zone at about the same time (e.g., Holzer et al., 1998, 1999). It is obvious, therefore, that the assembly of the terranes of the Limpopo Belt and Kalahari craton took place over a time span of ~700 m.y. and did not occur as a result of a simple and relatively rapid collision between the Zimbabwe and Kaapvaal cratons at either ca. 2.7 Ga or 2.04 Ga. The transtensional or transpressional stress field that existed along the Palala-Zoetfontein shear zone between ca. 2.04 and ca. 1.97 Ga could have produced regions of extension within the various terranes that it bounded and the Kaapvaal craton, allowing for the emplacement of the igneous Bushveld and Phalaborwa complexes at this same time (age constraints in Eriksson, 1984; Kruger et al., 1987; Walraven et al., 1990; Reischmann, 1995; Mapeo et al., 2004). It also seems unrealistic to think that the formation of the Limpopo Belt may be divorced from the evolution of the Magondi orogen that immediately followed it in the west and the formation of the Palapye, Waterberg, and Soutpansberg sedimentary basins on top of it shortly afterward.

CONCLUSIONS

Because the terranes comprising the Limpopo Belt have different tectono-metamorphic histories and were assembled independently onto the Zimbabwe and Kaapvaal cratons over a period of ~700 m.y. ending at ca. 2.04 Ga, it is possible and even probable that they evolved at widely spaced positions on the Earth's surface, as was the case with the Altaids (Şengör and Natal'in, 1996). The wide shear zones with gradational contacts separating them would be artifacts of major subduction zones connecting the terranes. This possibility is supported by the presence of magmatic arcs of different ages within the belt and the fact that the shear zones separating the terranes were reactivated at different times with different shear senses and at different metamorphic grades that could have resulted from either different crustal depths or different geothermal gradients. In addition, P-T-time paths for different terranes clearly reflect different tectonic regimes, even though activity within them sometimes overlapped in time. Because the Palala-Zoetfontein shear zone separates terranes with distinct Pb isotopic signatures, it probably represents the locus of final juxtaposition of the Zimbabwe craton with its associated terranes with the Kaapvaal craton and its associated terranes. Therefore, the assembly of the Limpopo Belt between ca. 2.7 Ga and 2.04 Ga was not similar to the formation of the Alps or the Himalayas. Instead it more closely resembles Turkic-type continental growth.

It is obvious from the above discussion of the currently available database that substantial further work is required in many parts of the Limpopo Belt to fully understand its tectonic evolution. Any new results will provide additional tests for the proposed Turkic-type model for the evolution of the Limpopo Belt.

ACKNOWLEDGMENTS

Inspiration for writing this paper came from a field trip to the Limpopo Belt in 2003 with Mike Searle, during which he took great pains to explain to JMB that he saw no Himalayan analogues there and from JMB listening to a series of lectures given by Kevin Burke, John Dewey, and Celâl Şengör in 2002 and again in 2003 at the University of the Witwatersrand, where the implications of transtension, transpression, and Turkic orogenesis were explained. Suddenly all the inconsistencies of applying an Alpine- or Himalayan-type model to the evolution of the Limpopo Belt became clear and a means of reconciling the inconsistencies was apparent. We also wish to thank all of our colleagues whose work we quote here for their contribution to the present understanding of the evolution of the Limpopo Belt. It is easier to construct models with limited data than to do so when all of the data are available and must be accommodated. We thank Paul Dirks, Roger Gibson, and Uwe Reimold for their critical and constructive reviews of an earlier version of this manuscript.

REFERENCES CITED

Aldiss, D.T., 1991, The Motloutse Complexes and the Zimbabwe Craton/Limpopo Belt transition in Botswana: Precambrian Research, v. 50, p. 89–109, doi: 10.1016/0301-9268(91)90049-G.

Andreoli, M.A.G., Coetzee, H., Hat, R., Ashwal, L.D., Huizenga, J.M., Smit, A., Brandl, G., and Grantham, G., 2003, The Erlank paradox: High-U granulites, low-U amphibolite facies gneisses in the mobile belts of southern Africa, *in* Cuney, M., ed., Proceedings Volume of the International Conference on Uranium Geochemistry: France, University of Nancy, p. 47–50.

Armstrong, R., Compston, W., Dodson, M.H., Kröner, A., and Williams, I.S., 1987, Archaean crustal history in southern Africa: The Australian National University Research School of Earth Sciences Annual Report, v. 1987, p. 62–64.

Bahnemann, K.P., 1972, A review of the structure, the stratigraphy and the metamorphism of the basement rocks in the Messina district, northern Transvaal: [Ph.D. thesis]: Pretoria, South Africa, University of Pretoria, 156 p.

Barker, O.B., 1983, A proposed geotectonic model for the Soutpansberg Group within the Limpopo Mobile Belt, South Africa: Geological Society of South Africa Special Publication 8, p. 181–190.

Barnett, M., 2005, The geology of the Farm Gotha, Beit Bridge Terrane, South Africa: [M.S. thesis]: Johannesburg, South Africa, University of Johannesburg, 101 p., plus appendices.

Barton, J.M., Jr., 1996, The Messina Layered Intrusion, Limpopo belt, South Africa: An example of *in situ* contamination of an Archean anorthosite complex by continental crust: Precambrian Research, v. 78, p. 139–150, doi: 10.1016/0301-9268(95)00074-7.

Barton, J.M., Jr., and Key, R.M., 1981, The tectonic development of the Limpopo Mobile Belt, and the evolution of the Archaean cratons of southern Africa, *in* Kröner, A., ed., Precambrian Plate Tectonics: Amsterdam, Elsevier, p. 185–212.

Barton, J.M., Jr., and McCourt, S., 1983, Rb-Sr age for the Palala granite, Limpopo Mobile Belt: Geological Society of South Africa Special Publication 8, p. 45–46.

Barton, J.M., Jr., and Pretorius, W., 1998, Crustal xenoliths in Venetia kimberlite pipes indicate a *décollement* at ~10 km beneath the Central Zone of the Limpopo Belt, South Africa: South African Journal of Geology, v. 101, p. 323–327.

Barton, J.M., Jr., and Sergeev, S., 1997, High-precision U-Pb analyses of single grains of zircon from quartzite in the Beit Bridge Group yield a discordia: South African Journal of Geology, v. 100, p. 37–42.

Barton, J.M., Jr., and van Reenen, D.D., 1992a, When was the Limpopo Orogeny?: Precambrian Research, v. 55, p. 7–16, doi: 10.1016/0301-9268(92)90010-L.

Barton, J.M., Jr., and van Reenen, D.D., 1992b, The significance of Rb-Sr ages of biotite and phlogopite for the thermal history of the Central and Southern Marginal Zones of the Limpopo Belt of southern Africa and the adjacent portions of the Kaapvaal Craton: Precambrian Research, v. 55, p. 17–32, doi: 10.1016/0301-9268(92)90011-C.

Barton, J.M., Jr., Ryan, B., and Fripp, R.E.P., 1983, Rb-Sr and U-Th-Pb isotopic studies of the Sand River Gneisses, Central Zone, Limpopo Mobile Belt: Geological Society of South Africa Special Publication 8, p. 9–18.

Barton, J.M., Jr., van Reenen, D.D., and Roering, C., 1990, The significance of 3000 Ma granulite facies mafic dikes in the Central Zone of the Limpopo Belt, southern Africa: Precambrian Research, v. 48, p. 299–308, doi: 10.1016/0301-9268(90)90014-H.

Barton, J.M., Jr., Doig, R., Smith, C.B., Bohlender, F., and van Reenen, D.D., 1992, Isotopic and REE characteristics of the intrusive charnoenderbite and enderbite geographically associated with the Matok Pluton, Limpopo Belt, southern Africa: Precambrian Research, v. 55, p. 451–467, doi: 10.1016/0301-9268(92)90039-Q.

Barton, J.M., Jr., Holzer, L., Kamber, B., Doig, R., Kramers, J.D., and Nyfeler, D., 1994, Discrete metamorphic events in the Limpopo belt, southern Africa: Implications for the application of P-T paths in complex metamorphic terranes: Geology, v. 22, p. 1035–1038, doi: 10.1130/0091-7613(1994)022<1035:DMEITL>2.3.CO;2.

Barton, J.M., Jr., Barnett, W.P., Barton, E.S., Barnett, M., Doorgapershad, A., Twiggs, C., Klemd, R., Martin, J., Mellonig, L., and Zenglein, R., 2003, The geology of the area surrounding the Venetia kimberlite pipes, Limpopo Belt, South Africa: A complex interplay of nappe tectonics and granitoid magmatism: South African Journal of Geology, v. 106, p. 109–128, doi: 10.2113/106.2-3.109.

Berger, M., and Braun, I., 1997, Pb-Pb dating of apatite by a stepwise dissolution technique: Chemical Geology, v. 143, p. 17–42, doi: 10.1016/S0009-2541(97)00097-1.

Berger, M., Kramers, J.D., and Nägler, T.F., 1995, Geochemistry and geochronology of charnoenderbitic rocks in the Northern Marginal Zone of the Limpopo Belt, southern Africa, and genetic models: Schweizerische Mineralogische und Petrographische Mitteilungen, v. 75, p. 17–42.

Blenkinsop, T.G., and Frei, R., 1996, Archean and Proterozoic mineralization and tectonics at the Renco Mine (Northern Marginal Zone, Limpopo Belt, Zimbabwe): Economic Geology and the Bulletin of the Society of Economic Geologists, v. 91, p. 1225–1238.

Blenkinsop, T.G., Mkweli, S., Rollinson, H.R., Fedo, C.M., Paya, B.K., Kamber, B., Kramers, J.D., and Berger, M., 1995, The Northern Limpopo thrust zone (NLTZ): The northern boundary of the Limpopo belt in Zimbabwe and Botswana: Johannesburg, Extended Abstracts, Geocongress 1995, Geological Society of South Africa, p. 174–177.

Blenkinsop, T.G., Kröner, A., and Chiwara, V., 2004, Single stage, late Archean exhumation of granulites in the Northern Marginal Zone, Limpopo Belt, Zimbabwe, and relevance to gold mineralization at Renco Mine: South African Journal of Geology, v. 107, p. 377–396, doi: 10.2113/107.3.377.

Boryta, M.D., 1988, Geochemistry and origin of igneous rocks from the Archean Beit Bridge Complex, Limpopo Belt, South Africa: [M.S. thesis]: Socorro, New Mexico Institute of Technology, 111 p.

Boryta, M., and Condie, K.C., 1990, Geochemistry and origin of the Archaean Beit Bridge complex, Limpopo Belt, South Africa: Journal of the Geological Society of London, v. 147, p. 229–239.

Boshoff, R., van Reenen, D.D., Smit, C.A., Armstrong, R.A., Perchuk, L.L., and Kramers, J.D., 2004, New data linking major fold types in the Central Zone of the Limpopo Complex, to distinct tectono-metamorphic events: Johannesburg, South Africa, Extended Abstracts, Geoscience Africa, University of the Witwatersrand, p. 76–77.

Brandl, G., 1992, Geological Map of the Limpopo Belt and Environs: Geological Survey of South Africa, scale 1:500,000.

Brandl, G., and Kröner, A., 1993, Preliminary results of single zircon studies from various Archaean rocks of the northeastern Transvaal: Swaziland, Extended Abstracts, 16th Colloquium of African Geology, p. 54–56.

Brandl, G., and Reimold, W.U., 1990, The structural setting and deformation associated with pseudotachylite occurrences in the Palala Shear Belt and Sand River gneiss, Northern Transvaal: Tectonophysics, v. 171, p. 201–220, doi: 10.1016/0040-1951(90)90099-T.

Broekhuizen, A., McCourt, S., and Snyman, C.P., 1995, Structural evolution of the Koedoesrand hills and surrounding area, North-West Transvaal, South Africa: Johannesburg, Extended Abstracts, Geocongress 1995, Geological Society of South Africa, p. 280–282.

Brown, P.J., 1987, Petrogenesis of Ni-Cu ore bodies, their host rocks and country rocks at Selebi-Phikwe, eastern Botswana: [Ph.D. thesis]: UK, University of Southampton, 370 p.

Buhlmann, S., 2003, Geological and mineralogical investigations of granulite-facies rocks in the Central Zone of the Limpopo Belt (Messina Nature Reserve, South Africa): [Diploma thesis]: University of Würzburg, Germany, 99 p., plus appendices.

Buick, I.S., Williams, I.S., Gibson, R.L., Cartwright, I., and Miller, J.A., 2003, Carbon and U-Pb evidence for a Palaeoproterozoic crustal component in the Central Zone of the Limpopo Belts, South Africa: Journal of the Geological Society of London, v. 160, p. 601–612.

Carney, J.N., Aldiss, D.T., and Lock, N.P., 1994, The Geology of Botswana: Bulletin of the Geological Survey of Botswana, v. 37, 113 p.

Chavagnac, V., Kramers, J.D., Nägler, T., and Holzer, L., 2001, The behavior of Nd and Pb isotopes during 2.0 Ga migmatisation in paragneisses of the Central Zone of the Limpopo Belt (South Africa and Botswana): Precambrian Research, v. 112, p. 51–86, doi: 10.1016/S0301-9268(01)00170-X.

Cheney, E.C., Barton, J.M., Jr., and Brandl, G., 1990, Extent and age of the Soutpansberg sequences of southern Africa: South African Journal of Geology, v. 93, p. 664–675.

Coward, M.P., 1976, Archean deformation patterns in southern Africa: Philosophical Transactions of the Royal Society of London, ser. A, v. 283, p. 313–331.

Coward, M.P., and Fairhead, J.D., 1980, Gravity and structural evidence for the deep structure of the Limpopo belt, southern Africa: Tectonophysics, v. 68, p. 31–43, doi: 10.1016/0040-1951(80)90007-4.

Coward, M.P., James, P.R., and Wright, L., 1976, Northern margin of the Limpopo mobile belt, southern Africa: Geological Society of America Bulletin, v. 87, p. 601–611, doi: 10.1130/0016-7606(1976)87<601:NMOTLM>2.0.CO;2.

Cox, K.G., Johnson, R.L., Monkman, L.J., Vail, J.R., and Wood, D.N., 1965, The geology of the Nuanetsi Igneous Province: Philosophical Transactions of the Geological Society of London, ser. A, v. 257, p. 71–218.

Dahl, P.S., 1997, A crystal chemical basis for Pb retention and fission track annealing systematics in U-bearing minerals, with implications for geochronology: Earth and Planetary Science Letters, v. 150, p. 277–290, doi: 10.1016/S0012-821X(97)00108-8.

De Beer, J.H., and Stettler, E.H., 1992, The deep structure of the Limpopo Belt from geophysical studies: Precambrian Research, v. 55, p. 173–186, doi: 10.1016/0301-9268(92)90022-G.

De Wit, M.J., van Reenen, D.D., and Roering, C., 1992, Geologic observations across a tectono-metamorphic boundary in the Babangu area, Giyani (Sutherland) Greenstone Belt, South Africa: Precambrian Research, v. 55, p. 111–122, doi: 10.1016/0301-9268(92)90018-J.

Droop, G.T.R., 1989, Reaction history of garnet-sapphirine granulites and conditions of Archaean high-pressure metamorphism in the Central Limpopo Mobile Belt, Zimbabwe: Journal of Metamorphic Geology, v. 7, p. 383–403.

Du Plessis, G., Jonck, G.J., and Kruger, R., 1997, Potential low-grade iron ore deposits in metamorphosed banded iron formations, Northern Province, South Africa: Mineralium Deposita, v. 32, p. 362–370, doi: 10.1007/s001260050102.

Durrheim, R.J., Barker, W.H., and Green, R.W.E., 1992, Seismic studies of the Limpopo Belt: Precambrian Research, v. 55, p. 187–200, doi: 10.1016/0301-9268(92)90023-H.

Eriksson, K.A., Kidd, W.S., and Krapez, B., 1988, Basin analysis in regionally metamorphosed and deformed early Archaean terrains: Examples from southern Africa and Western Australia, *in* Kleinspehn, K.L., and Poala, C., eds., Frontiers in Sedimentary Geology, New Perspectives in Basin Analysis: New York, Springer-Verlag, p. 329–335.

Eriksson, S.C., 1984, Age of carbonatites and phoscorite magmatism of the Phalaborwa Complex (South Africa): Isotope Geoscience, v. 2, p. 291–299.

Ernst, W.G., 1988, Tectonic history of subduction zones inferred from retrograde blueschist P-T paths: Geology, v. 16, p. 1081–1084, doi: 10.1130/0091-7613(1988)016<1081:THOSZI>2.3.CO;2.

Fairhead, J.D., and Scovell, P., 1976, Gravity study of the Limpopo belt, southern Africa: UK, 20th Annual Report of the President of the Institute of African Geology, University of Leeds, p. 31–35.

Fouch, M.J., James, D.E., VanDecar, J.C., van der Lee, S., and Kaapvaal Seismic Group, 2004, Mantle seismic structure beneath the Kaapvaal and Zimbabwe Cratons: South African Journal of Geology, v. 107, p. 33–44.

Fripp, R.E.P., 1981, The ancient Sand River Gneisses, Limpopo Mobile Belt, South Africa: Geological Society of Australia Special Publication 7, p. 329–335.

Fripp, R.E.P., 1982, The Precambrian geology of the area around the Sand River near Messina, northern Transvaal: [Ph.D. thesis]: Johannesburg, South Africa, University of the Witwatersrand, 251 p.

Fripp, R.E.P., 1983, The Precambrian geology of the area around the Sand River near Messina, Central Zone, Limpopo Mobile Belt: Geological Society of South Africa Special Publication 8, p. 89–102.

Gallon, M.L., 1986, Structural re-interpretation of the Selebi-Phikwe nickel-copper sulphide deposits, eastern Botswana, *in* Anhaeusser, C.R., and Maske, S., eds., Mineral Deposits of Southern Africa: Johannesburg, Geological Society of South Africa, p. 1663–1669.

Gordon, P.S.L., 1973, The Selebi-Pikwe nickel-copper deposits, Botswana, *in* Lister, L.A., ed., Symposium on Granites, Gneisses and Related Rocks: Geological Society of South Africa Special Publication 3, p. 167–188.

Harley, S.L., 1989, The origin of granulites: A metamorphic perspective: Geological Magazine, v. 126, p. 215–247.

Henderson, D.R., Long, L.E., and Barton, J.M., Jr., 2000, Isotopic ages and chemical and isotopic composition of the Archaean Turfloop batholith, Pietersburg, granite-greenstone terrane, Kaapvaal Craton, South Africa: South African Journal of Geology, v. 103, p. 38–46, doi: 10.2113/103.1.38.

Hickman, M.H., and Wakefield, J., 1975, Tectonic implications of new geochronologic data from the Limpopo belt at Pikwe, Botswana, southern Africa: Geological Society of America Bulletin, v. 86, p. 1468–1472, doi: 10.1130/0016-7606(1975)86<1468:TIONGD>2.0.CO;2.

Hisada, K., and Miyano, T., 1996, Petrology and microthermometry of aluminous rocks of the Botswanan Limpopo Central Zone: Evidence or isothermal decompression and isobaric cooling: Journal of Metamorphic Geology, v. 14, p. 183–197, doi: 10.1046/j.1525-1314.1996.05857.x.

Hisada, K., and Paya, B.K., 1997, Phase analysis of migmatites in the Mahalapye plutonic Block, *in* Dirks, P.H.G.H., and Jelsma, H.A., eds., Abstracts of the Conference on Intraplate Magmatism and Tectonics in Southern Africa: Harare, University of Zimbabwe, p. 15.

Hisada, K., Paya, B.K., and van Reenen, D.D., 2000, Partial melting of politic rocks in the Mahalapye plutonic block Botswana: Implications for a 2.0 Ga, low P, high T migmatization in the Limpopo granulite complex: Petrology, v. 8, p. 19–37.

Hisada, K., Perchuk, L.L., Gerya, T.V., van Reenen, D.D., and Paya, B.K., 2004, Fluid infiltration and LP-HT event in the Mahalapye Complex of the Limpopo high-grade terrain, eastern Botswana: Extended Abstracts, Geoscience Africa: Johannesburg, South Africa, University of the Witwatersrand, p. 276–277.

Holzer, L., Frei, R., Barton, J.M., Jr., and Kramers, J.D., 1998, Unraveling the record of successive high grade events in the Central Zone of the Limpopo Belt using Pb single phase dating of metamorphic minerals: Precambrian Research, v. 87, p. 87–115, doi: 10.1016/S0301-9268(97)00058-2.

Holzer, L., Barton, J.M., Paya, B.K., and Kramers, J.D., 1999, Tectonothermal history of the western part of the Limpopo Belt: Tectonic models and new perspectives: Journal of African Earth Sciences, v. 28, p. 383–402, doi: 10.1016/S0899-5362(99)00011-1.

Hutchins, D.G., and Reeves, C.V., 1980, Regional geophysical exploration of the Kalahari in Botswana: Tectonophysics, v. 69, p. 201–220, doi: 10.1016/0040-1951(80)90211-5.

Jaeckel, P., Kröner, A., Kamo, S.L., Brandl, G., and Wendt, J.I., 1997, Late Archean to early Proterozoic granitoid magmatism and high-grade metamorphism in the central Limpopo belt, South Africa: Journal of the Geological Society of London, v. 154, p. 25–44.

James, P.R., 1975, A deformation study across the northern margin of the Limpopo belt, Rhodesia: [Ph.D. thesis]: UK, University of Leeds, 303 p.

Jelsma, H.A., and Dirks, P.H.G.M., 2002, Neoarchean tectonic evolution of the Zimbabwe Craton, *in* Fowler, C.M.R., Ebinger, C.B., and Hawkesworth, C.J., eds., The Early Earth: Physical, Chemical and Biological Development: The Geological Society [London] Special Publication 199, p. 183–211.

Jones, D.L., Howell, D.G., Coney, P.J., and Monger, J.W.H., 1983, Recognition, character, and analysis of tectonostratigraphic terranes in western North America, *in* Hashimoto, M., and Uyeda, S., eds., Accretion Tectonics in the Circum-Pacific Regions: Tokyo, Japan, Terra Scientific Publishing Company, p. 21–35.

Kamber, B., and Biino, G.G., 1995, The evolution of high T–low P granulites in the Northern Marginal Zone sensu stricto, Limpopo Belt, Zimbabwe—The case for petrography: Schweitzerische Mineralogische Petrologische Mitteilung, v. 75, p. 427–454.

Kamber, B., Blenkinsop, T.G., Villa, I.M., and Dahl, P.S., 1995a, Proterozoic transpressive deformation in the Northern Marginal Zone, Limpopo Belt, Zimbabwe: The Journal of Geology, v. 103, p. 493–508.

Kamber, B., Kramers, J.D., Napier, R., Cliff, R.A., and Rollinson, H.R., 1995b, The Triangle Shearzone, Zimbabwe, revisited: New data document an important event at 2.0 Ga in the Limpopo Belt: Precambrian Research, v. 70, p. 191–213, doi: 10.1016/0301-9268(94)00039-T.

Kisters, A.F.M., Kolb, J., and Meyer, F.M., 1998, Gold mineralization in high grade metamorphic shear zones of the Renco mine, Southern Zimbabwe: Economic Geology and the Bulletin of the Society of Economic Geologists, v. 93, p. 587–601.

Klemd, R., Martin, J., Schmidt, A., and Barton, J.M., Jr., 2003, P-T path constraints from calc-silicate metapelitic rocks east of the Venetia kimberlite pipes, Central Zone, Limpopo Belt, South Africa: Have these rocks seen granulite facies conditions?: South African Journal of Geology, v. 106, p. 129–148, doi: 10.2113/106.2-3.129.

Kramers, J.D., Kreissig, K., and Jones, M.Q.W., 2001, Crustal heat production and style of metamorphism: A comparison between two Archean high grade provinces in the Limpopo Belt, southern Africa: Precambrian Research, v. 112, p. 149–163, doi: 10.1016/S0301-9268(01)00173-5.

Kreissig, K., Nägler, T.F., Kramers, J.D., van Reenen, D.D., and Smit, C.A., 2000, An isotopic and geochemical study of the northern Kaapvaal Craton and the Southern Marginal Zone of the Limpopo Belt: Are they juxtaposed terranes?: Lithos, v. 50, p. 1–25, doi: 10.1016/S0024-4937(99)00037-7.

Kreissig, K., Holzer, L., Frei, R., Villa, I.M., Kramers, J.D., Kröner, A., Smit, C.A., and van Reenen, D.D., 2001, Geochronology of the Hout River Shear Zone and the metamorphism in the Southern Marginal Zone of the Limpopo Belt, Southern Africa: Precambrian Research, v. 109, p. 145–173, doi: 10.1016/S0301-9268(01)00147-4.

Kröner, A., Jaeckel, P., Brandl, G., Nemchin, A.A., and Pidgeon, R.T., 1999, Single zircon ages for granitoid gneisses in the Central Zone of the Limpopo Belt, Southern Africa and geodynamic significance: Precambrian Research, v. 93, p. 299–337, doi: 10.1016/S0301-9268(98)00102-8.

Kruger, F.J., Cawthorn, R.G., and Walsh, K., 1987, Strontium isotopic evidence against magma addition in the Upper Zone of the Bushveld Complex: Earth and Planetary Science Letters, v. 84, p. 51–58, doi: 10.1016/0012-821X(87)90175-0.

Kurszlaukis, S., and Barnett, W.P., 2003, Volcanological and structural aspects of the Venetia kimberlite cluster—A case study of South African kimberlite maar-diatreme volcanoes: South African Journal of Geology, v. 106, p. 165–192, doi: 10.2113/106.2-3.165.

Lear, P.A., 1979, The ore mineralogy of the Phikwe and Selebi nickel-copper deposits, Botswana: Geological Society of South Africa Special Publication 5, p. 117–132.

Light, M.P.R., 1982, The Limpopo Mobile Belt: A result of continental collision: Tectonics, v. 1, p. 325–342.

Liu, Y., and Zhong, D., 1997, Petrology of high-pressure granulites from the eastern Himalayan syntaxis: Journal of Metamorphic Geology, v. 15, p. 451–466, doi: 10.1111/j.1525-1314.1997.00033.x.

Mapeo, R.B.M., Kampunzu, A.B., Ramokate, L.V., Corfu, F., and Key, R.M., 2004, Bushveld-age magmatism in southeastern Botswana: Evidence from U-Pb zircon and titanite geochronology, the Moshaneng Complex: South African Journal of Geology, v. 107, p. 219–232, doi: 10.2113/107.1-2.219.

Martin, D.McB., Clendenin, C.W., Krapez, B., and McNaughton, N.J., 1998, Tectonic and geochronological constraints on late Archean and Palaeoproterozoic stratigraphic correlations within and between the Kaapvaal and Pilbara cratons: Journal of the Geological Society of London, v. 155, p. 311–322.

Mason, R., 1973, The Limpopo mobile belt—Southern Africa: Philosophical Transactions of the Royal Society of London, ser. A, v. 273, p. 463–485.

McCourt, S., 1983, Archaean lithologies of the Koedoesrand area, north-west Transvaal, South Africa: Geological Society of South Africa Special Publication 8, p. 113–119.

McCourt, S., 1993, Shear zones in the Limpopo Belt and northern Kaapvaal Craton: Insights into late Archaean collision tectonics: Mbabane, Swaziland, Extended Abstracts, 16th Colloquium of African Geology, p. 212–213.

McCourt, S., and Armstrong, R.A., 1998, SHRIMP U-Pb zircon geochronology of granites from the Central Zone, Limpopo Belt, southern Africa: Implications for the age of the Limpopo Orogeny: South African Journal of Geology, v. 101, p. 329–338.

McCourt, S., and Vearncombe, J.R., 1987, Shear zones bounding the central zone of the Limpopo mobile belt, southern Africa: Journal of Structural Geology, v. 9, p. 127–137, doi: 10.1016/0191-8141(87)90021-6.

McCourt, S., and Vearncombe, J.R., 1992, Shear zones of the Limpopo Belt and adjacent granitoid-greenstone terranes: Implications for late Archaean collision tectonics in southern Africa: Precambrian Research, v. 55, p. 553–570, doi: 10.1016/0301-9268(92)90045-P.

McCourt, S., Kampunzu, A.B., Bagai, Z., and Armstrong, R.A., 2004, The crustal architecture of Archaean terranes in northeastern Botswana: South African Journal of Geology, v. 107, p. 147–158, doi: 10.2113/107.1-2.147.

Meixner, H.M., and Peart, R.J., 1984, The Kalahari Drilling Project: Bulletin of the Geological Survey of Botswana, v. 27, 224 p.

Mkweli, S., Kamber, B.S., and Berger, M., 1995, Westward continuation of the craton–Limpopo Belt tectonic break in Zimbabwe and new age constraints on the timing of the thrusting: Journal of the Geological Society of London, v. 152, p. 77–83.

Molnar, P., and Tapponnier, P., 1975, Cenozoic tectonics of Asia: Effects of continental collision: Science, v. 189, p. 419–426.

Nguuri, T.K., Gore, J., James, D.E., Webb, S.J., Wright, C., Zengeni, T.G., Gwavava, O., Snoke, J.A., and Kaapvaal Seismic Group, 2001, Crustal structure beneath southern Africa and its implications for the formation of the Kaapvaal and Zimbabwe cratons: Geophysical Research Letters, v. 28, p. 2501–2504.

O'Brien, P.J., Zotov, N., Law, R., Khan, M.A., and Jan, M.Q., 2001, Coesite in Himalayan eclogite and implications for models of India-Asia collision: Geology, v. 29, p. 435–438, doi: 10.1130/0091-7613(2001)029<0435:CIHEAI>2.0.CO;2.

Perchuk, L.L., Gerya, T.V., van Reenen, D.D., Krotov, A.V., Safonov, O.G., Smit, C.A., and Shur, M.Y., 2000, Comparative petrology and metamorphic evolution of the Limpopo (South Africa) and Lapland (Fennoscandia) high-grade terrains: Mineralogy and Petrology, v. 69, p. 69–107, doi: 10.1007/s007100050019.

Perchuk, L.L., Gerya, T.V., van Reenen, D.D., and Smit, C.A., 2004, A mechanism for the late-Archean exhumation of the Southern Marginal Zone of the Limpopo high-grade terrain: Johannesburg, South Africa, Extended Abstracts, Geoscience Africa, University of the Witwatersrand, p. 513–514.

Pretorius, W., 1996, A geochemical and geophysical investigation of a suite of crustal and upper mantle nodules from Venetia kimberlite pipes, Limpopo Belt, South Africa: [M.S. thesis]: Johannesburg, South Africa, Rand Afrikaans University, 275 p.

Pretorius, W., and Barton, J.M., Jr., 2003a, Petrology and geochemistry of crustal and upper mantle xenoliths from the Venetia Diamond Mine—Evidence for Archean crustal growth and subduction: South African Journal of Geology, v. 106, p. 213–230, doi: 10.2113/106.2-3.213.

Pretorius, W., and Barton, J.M., Jr., 2003b, Measured and calculated compressional wave velocities of crustal and upper mantle rocks in the Central Zone of the Limpopo Belt, South Africa—Implications for lithospheric structure: South African Journal of Geology, v. 106, p. 205–212, doi: 10.2113/106.2-3.205.

Reischmann, T., 1995, Precise U/Pb age determination with baddeleyite (ZrO_2), a case study from the Phalaborwa igneous complex, South Africa: South African Journal of Geology, v. 98, p. 1–4.

Roering, C., van Reenen, D.D., Smit, C.A., Barton, J.M., Jr., de Beer, J.H., de Wit, M.J., Stettler, E.H., van Schalkwyk, J.F., Stevens, G., and Pretorius, S., 1992, Tectonic model for the evolution of the Limpopo Belt: Precambrian Research, v. 55, p. 539–552, doi: 10.1016/0301-9268(92)90044-O.

Rollinson, H.R., and Blenkinsop, T., 1995, The magmatic, metamorphic and tectonic evolution of the Northern Marginal Zone of the Limpopo Belt in Zimbabwe: Journal of the Geological Society of London, v. 152, p. 66–75.

Ruppel, C., Royden, L., and Hodges, K.V., 1988, Thermal modeling of extensional tectonics: Application to pressure-temperature-time history of metamorphic rocks: Tectonics, v. 7, p. 947–957.

Sachau, T., 2003, Geological and mineralogical investigations on an area near Messina (South Africa): [Diploma thesis]: University of Würzburg, Germany, 95 p., and appendices.

Schaller, M., Steiner, O., Studer, I., Holzer, L., Herwegh, M., and Kramers, J.D., 1999, Exhumation of the Limpopo Central Zone granulites and dextral continent-scale transcurrent movement at 2.0 Ga along the Palala shear zone, Transvaal: Precambrian Research, v. 96, p. 263–288, doi: 10.1016/S0301-9268(99)00015-7.

Searle, M.P., Windley, B.F., Coward, M.P., Cooper, D.J.W., Rex, D., Tingdong, Li, Xuchang, Xiao, Jan, M.Q., Thakur, V.C., and Kumar, S., 1987, The closing of Tethys and the tectonics of the Himalaya: Geological Society of America Bulletin, v. 98, p. 678–701.

Şengör, A.M.C., and Natal'in, B.A., 1996, Turkic-type orogeny and its role in the making of the continental crust: Annual Reviews of Earth and Planetary Science, v. 24, p. 263–337, doi: 10.1146/annurev.earth.24.1.263.

Smit, C.A., 1984, Petrochemistry and tectonic significance of a chromite-bearing ultramafic suite of rocks in the Limpopo metamorphic complex, South Africa: Transactions of the Geological Society of South Africa, v. 87, p. 303–314.

Smit, C.A., and van Reenen, D.D., 1997, Deep crustal shear zones, high-grade tectonites and associated alteration in the Limpopo belt, South Africa: Implications for deep crustal processes: The Journal of Geology, v. 105, p. 37–57.

Smit, C.A., Roering, C., and van Reenen, D.D., 1992, The structural framework of the southern margin of the Limpopo Belt, South Africa: Precambrian Research, v. 55, p. 51–67, doi: 10.1016/0301-9268(92)90014-F.

Smit, C.A., van Reenen, D.D., Boshoff, R., and Perchuk, L.L., 2004, The Limpopo complex of southern Africa: New insight into its evolution: Johannesburg, South Africa, Extended Abstracts, Geoscience Africa, University of the Witwatersrand, p. 605.

Söhnge, P.G., 1945, The geology of the Messina Copper Mines and surrounding country: Geological Survey of South Africa Memoir 40, 280 p.

Söhnge, P.G., Le Roux, H.D., and Nel, H.J., 1948, The geology of the country around Messina, an explanation of sheet no. 46 (Messina): Pretoria, Geological Survey of South Africa, 82 p.

Stampfli, G.M., Borel, G.D., Marchant, R., and Mosar, J., 2002, Western Alps geological constraints on western Tethyan reconstructions, *in* Rosenbaum, G., and Lister, G.S., eds., Reconstruction of the evolution of the Alpine-Himalayan orogeny: Journal of the Virtual Explorer (PDF), http://www.virtualexplorer.com.au/2002/8/stampfli/paper5.html (last accessed on June 4, 2005).

Stevens, G., and van Reenen, D.D., 1992, Partial melting and the origin of metapelitic granulites in the Southern Marginal Zone of the Limpopo Belt, South Africa: Precambrian Research, v. 55, p. 279–296, doi: 10.1016/0301-9268(92)90028-M.

Stuart, G.W., and Zengeni, T.G., 1987, Seismic crustal structure of the Limpopo mobile belt, Zimbabwe: Tectonophysics, v. 144, p. 323–335, doi: 10.1016/0040-1951(87)90300-3.

Tapponnier, P., Peltzer, G., Le Dain, A.Y., Armiyo, R., and Cobbold, P., 1982, Propagating extrusion tectonics in Asia: New insights from simple experiments with plasticine: Geology, v. 10, p. 611–616, doi: 10.1130/0091-7613(1982)10<611:PETIAN>2.0.CO;2.

Taylor, P.N., Kramers, J.D., Moorbath, S., Wilson, J.F., Orpen, J.L., and Martin, A., 1991, Pb/Pb, Sm-Nd and Rb-Sr geochronology in the Archean Craton of Zimbabwe: Chemical Geology, v. 87, p. 175–196.

Thompson, A.B., and England, P.C., 1984, Pressure-temperature-time paths of regional metamorphism II: Their influence and interpretation using mineral assemblages in metamorphic rocks: Journal of Petrology, v. 25, p. 929–955.

Treloar, P.J., Coward, M.P., and Harris, N.B.W., 1992, Himalayan-Tibetan analogies for the evolution of the Zimbabwe Craton and Limpopo Belt: Precambrian Research, v. 55, p. 571–587, doi: 10.1016/0301-9268(92)90046-Q.

Tsunogae, T., Miyano, T., and Machacha, T.P., 1989, Retrograde orthoamphiboles in metapelites near Maratele, southwest Selebi-Phikwe, Botswana: Annual Report of the Institute of Geoscience, the University of Tsukuba, Japan, v. 15, p. 259–277.

Tsunogae, T., Miyano, T., and Ridley, J., 1992, Metamorphic P-T profiles from the Zimbabwe Craton to the Limpopo Belt, Zimbabwe: Precambrian Research, v. 55, p. 259–277, doi: 10.1016/0301-9268(92)90027-L.

Van Breemen, O., and Dodson, M.H., 1972, Metamorphic chronology of the Limpopo Belt, Southern Africa: Geological Society of America Bulletin, v. 83, p. 2005–2018.

Van Reenen, D.D., 1983, Cordierite + garnet + hypersthene + biotite–bearing assemblages as a function of changing metamorphic conditions in the Southern Marginal Zone of the Limpopo Metamorphic Complex, South Africa: Geological Society of South Africa Special Publication 8, p. 143–167.

Van Reenen, D.D., 1986, Hydration of cordierite and hypersthene and a description of the retrograde orthoamphibole isograd in the Limpopo belt, South Africa: The American Mineralogist, v. 71, p. 900–915.

Van Reenen, D.D., Barton, J.M., Roering, C., Smit, C.A., and van Schalkwyk, J.F., 1987, Deep crustal response to continental collision: The Limpopo belt of southern Africa: Geology, v. 15, p. 11–14, doi: 10.1130/0091-7613(1987)15<11:DCRTCC>2.0.CO;2.

Van Reenen, D.D., Roering, C., Smit, C.A., van Schalkwyk, J.F., and Barton, J.M., Jr., 1988, Metamorphic evolution of the northern high-grade margin of the Kaapvaal Craton, South Africa: The Journal of Geology, v. 96, p. 549–560.

Van Reenen, D.D., Roering, C., Brandl, G., Smit, C.A., and Barton, J.M., Jr., 1990, The granulite facies rocks of the Limpopo belt, southern Africa, *in* Vielzeuf, D., and Vidal, Ph., eds., Granulites and Crustal Evolution: Dordrecht, The Netherlands, Kluwer, p. 257–289.

Van Reenen, D.D., Roering, C., Ashwal, L.D., and de Wit, M.J., 1992, Regional geological setting of the Limpopo Belt: Precambrian Research, v. 55, p. 1–5, doi: 10.1016/0301-9268(92)90009-D.

Van Reenen, D.D., Perchuk, L.L., Smit, C.A., Varlamov, D.A., Boshoff, R., Huizenga, J.M., and Gerya, T.V., 2004, Structural and P-T evolution of a major cross fold in the Central Zone of the Limpopo high-grade terrain, South Africa: Journal of Petrology, v. 45, p. 1413–1439, doi: 10.1093/petrology/egh028.

Van Zyl, J.S.V., 1978, The relationship between the deep electrical resistivity structure and tectonic provinces in southern Africa, Part 1, results obtained by Schlumberger soundings: Transactions of the Geological Society of South Africa, v. 81, p. 129–142.

Wakefield, J., 1976, The structural and metamorphic evolution of the Pikwe Ni-Cu sulfide deposit, Selebi-Pikwe, eastern Botswana: Economic Geology and the Bulletin of the Society of Economic Geologists, v. 71, p. 988–1005.

Walraven, F., and Martini, J., 1995, Zircon Pb-evaporation age determinations of the Oak Tree Formation, Chuniespoort Group, Transvaal sequence: Implications for Transvaal-Griqualand West basin correlations: South African Journal of Geology, v. 98, p. 58–67.

Walraven, F., Armstrong, R.A., and Kruger, F.J., 1990, A chronostratigraphic framework for the north-central Kaapvaal Craton, the Bushveld Complex and the Vredefort structure: Tectonophysics, v. 171, p. 23–48, doi: 10.1016/0040-1951(90)90088-P.

Watkeys, M.K., 1979, Explanation of the geological map of the country west of Beitbridge: Salisbury (Harare), Short Report of the Geological Survey of Rhodesia, v. 45, 96 p.

Watkeys, M.K., 1984, The Precambrian geology of the Limpopo Belt north and west of Messina [Ph.D. thesis]: Johannesburg, South Africa, University of the Witwatersrand, 49 p.

Watkeys, M.K., Light, M.P.R., and Broderick, T.J., 1983, A retrospective view of the Central Zone of the Limpopo Belt, Zimbabwe: Geological Society of South Africa Special Publication 8, p. 65–80.

Wilson, M.G.C., and Anhaeusser, C.R., 1998, The mineral resources of South Africa: Pretoria, Handbook of the Council for Geoscience, South Africa, v. 16, 740 p.

Windley, B.F., Ackermand, D., and Herd, R.K., 1984, Sapphirine/kornerupine–bearing rocks and crustal uplift history of the Limpopo belt, Southern Africa: Contributions to Mineralogy and Petrology, v. 86, p. 342–358, doi: 10.1007/BF01187139.

Zeh, A., Klemd, R., Buhlmann, S., and Barton, J.M., 2003, Application of qualitative phase diagrams for the inference of detailed P-T paths from magmatic rocks: An example from the Beit Bridge Complex (Limpopo Belt, South Africa): Geophysical Research Abstracts, v. 5, no. 05585.

Zeh, A., Klemd, R., and Barton, J.M., 2004a, Different P-T paths in the Beit Bridge Complex (Limpopo Belt, South Africa) and geological implications: Johannesburg, South Africa, Extended Abstracts, Geoscience Africa, University of the Witwatersrand, p. 724.

Zeh, A., Klemd, R., Buhlmann, S., and Barton, J.M., 2004b, Pro- and retrograde P-T evolution of granulites of the Beit Bridge Complex (Limpopo Belt, South Africa): Constraints from quantitative phase diagrams and geotectonic implications: Journal of Metamorphic Geology, v. 22, p. 79–95, doi: 10.1111/j.1525-1314.2004.00501.x.

Zeh, A., Holland, T.J.B., and Klemd, R., 2005a, Phase relationship in grunerite-garnet–bearing amphibolites in the system CFMASH, with applications to metamorphic rocks from the Central Zone of the Limpopo Belt, South Africa: Journal of Metamorphic Geology, v. 23, p. 1–17, doi: 10.1111/j.1525-1314.2005.00554.x.

Zeh, A., Klemd, R., and Barton, J.M., Jr., 2005b, Petrological evolution in the roof of the high-grade metamorphic Central Zone of the Limpopo Belt, South Africa: Geological Magazine, v. 142, p. 229–240, doi: 10.1017/S001675680500052X.

Manuscript Accepted by the Society 4 October 2005

Printed in the USA

Geological Society of America
Special Paper 405
2006

Origin of hydrothermal fluids and gold mineralization associated with the Ventersdorp Contact Reef, Witwatersrand Basin, South Africa: Constraints from S, O, and H isotopes

B. Zhao†
Department of Geology, University of Fort Hare, Private Bag X1314, Alice 5700, South Africa
L.J. Robb
EGRI, School of Geosciences, University of the Witwatersrand, Johannesburg, Private Bag X3, Wits 2050, South Africa
C. Harris
Department of Geological Sciences, University of Cape Town, Private Bag, Rondebosch 7700, South Africa
L.J. Jordaan
Council for Geoscience, Private Bag X112, Pretoria 0001, South Africa

ABSTRACT

The Ventersdorp Contact Reef, one of the major gold-bearing conglomerate horizons in the Witwatersrand Basin, occurs as a distinct horizon between the overlying Klipriviersberg Group lavas (2714 Ma) and the underlying Central Rand Group rocks (<2894 to >2714 Ma). The Ventersdorp Contact Reef has been metamorphosed and hydrothermally altered under greenschist facies conditions (290–350 °C and 0.2–0.3 GPa). A study of S, O, and H isotopes was carried out to constrain the sources of hydrothermal fluids and gold mineralization in the Ventersdorp Contact Reef. A narrow range of $\delta^{34}S$ values (–1.5‰ to +1.8‰) for authigenic pyrite, pyrrhotite, chalcopyrite, sphalerite, and galena from the Ventersdorp Contact Reef and its hanging-wall and footwall lithologies suggests a reconstitution of detrital sulfides during fluid circulation at peak metamorphic conditions. The $\delta^{18}O$ values of authigenic quartz and calcite range from +8.9‰ to +11.3‰, and the $\delta^{18}O$ and δD values of muscovite separates are between +7.2‰ and +8.4‰ and –62‰ and –31‰ respectively. The estimated $\delta^{18}O$ (+4.8‰ to +6.1‰) and δD (–27‰ to –39‰) values of the fluids in equilibrium with these minerals under the relevant P-T conditions suggest a metamorphic origin for the fluid. It is concluded that the metamorphic fluids involved were probably derived from the Basin itself, and allogenic sulfides in the Ventersdorp Contact Reef were reconstituted during fluid circulation at peak metamorphic conditions. Accordingly, gold appears to have been locally remobilized and possibly derived from preexisting placer concentrations.

Keywords: Ventersdorp Contact Reef, Witwatersrand, sulfur isotope, oxygen isotope, hydrogen isotope, hydrothermal fluid, gold mineralization.

†E-mail: bzhao@ufh.ac.za.

Zhao, B., Robb, L.J., Harris, C., and Jordaan, L.J., 2006, Origin of hydrothermal fluids and gold mineralization associated with the Ventersdorp Contact Reef, Witwatersrand Basin, South Africa: Constraints from S, O, and H isotopes, *in* Reimold, W.U., and Gibson, R.L., Processes on the Early Earth: Geological Society of America Special Paper 405, p. 333–352, doi: 10.1130/2006.2405(17). For permission to copy, contact editing@geosociety.org.

INTRODUCTION

The Ventersdorp Contact Reef is one of the best-preserved and consistently mineralized reef horizons in the Witwatersrand Basin, which is the world's largest gold deposit with dimensions of ~350 km in length and 250 km in width. Total production of gold from the Witwatersrand Basin was 43,500 tons between 1887 and 1994 (Robb and Robb, 1998). The Ventersdorp Contact Reef was first mined at the Zuurbult Mine in the West Rand goldfield in 1888 and has contributed a large portion of the annual gold production of South Africa (Viljoen and Reimold, 1994).

There has been considerable debate as to the environment of deposition of the Ventersdorp Contact Reef, although it is now generally agreed that the conglomerate layer formed in a low-relief, mature, terraced fluvial system. A variety of terrace and slope (or rises) landforms have been recognized (Henning et al., 1994; McWha, 1994) that gave rise to variability in sediment coarseness, degree of packing, reef width, and gold grade (Schweitzer et al., 1994). It is noted that gold-bearing conglomerate strata in the Witwatersrand Basin are traditionally known as "reefs." Despite the apparent relationship with sedimentary features, it has nevertheless become clear that the distribution of gold in the Witwatersrand Basin, including the Ventersdorp Contact Reef, is also structurally controlled. Jolley et al. (2004) have shown that these structural controls can be seen at scales ranging from millimeters to kilometers.

It is also widely recognized that the Ventersdorp Contact Reef is one of the most intensely metamorphosed and hydrothermally altered of all the conglomerate horizons in the Witwatersrand Basin; it occurs immediately beneath or interdigitated with the Ventersdorp Supergroup volcanics (e.g., Henckel and Schweitzer, 1994; Smits, 1994; Zhao et al., 1994). Much of the gold in the Ventersdorp Contact Reef is intimately related to alteration mineral assemblages, and there is little or no preservation of primary, detrital gold accumulation (Frimmel and Gartz, 1997). These features, together with the intimate association with basaltic volcanic rocks of the overlying Ventersdorp Supergroup, contribute to the complex paragenesis of gold mineralization in the Ventersdorp Contact Reef.

The purpose of this sulfur, oxygen, and hydrogen isotope study of different minerals in the Ventersdorp Contact Reef conglomerates and related lithologies is to characterize the isotopic and chemical composition of the fluids involved in post-depositional fluid circulation, as well as the possible source of ore constituents, specifically gold, in the Ventersdorp Contact Reef. The samples for stable isotope study were collected from Western Deep Levels (now known as Mponeng) mine in the Carletonville area of the West Wits Line, 90 km west of Johannesburg and 50 km east of Potchefstroom (Fig. 1).

REGIONAL GEOLOGICAL SETTING

The Ventersdorp Contact Reef occurs principally along the West Wits Line and has been mined as the primary reef in all the mines of this goldfield. It also occurs in the Klerksdorp, Welkom, and West Rand goldfields (Fig. 1, inset). No econom-

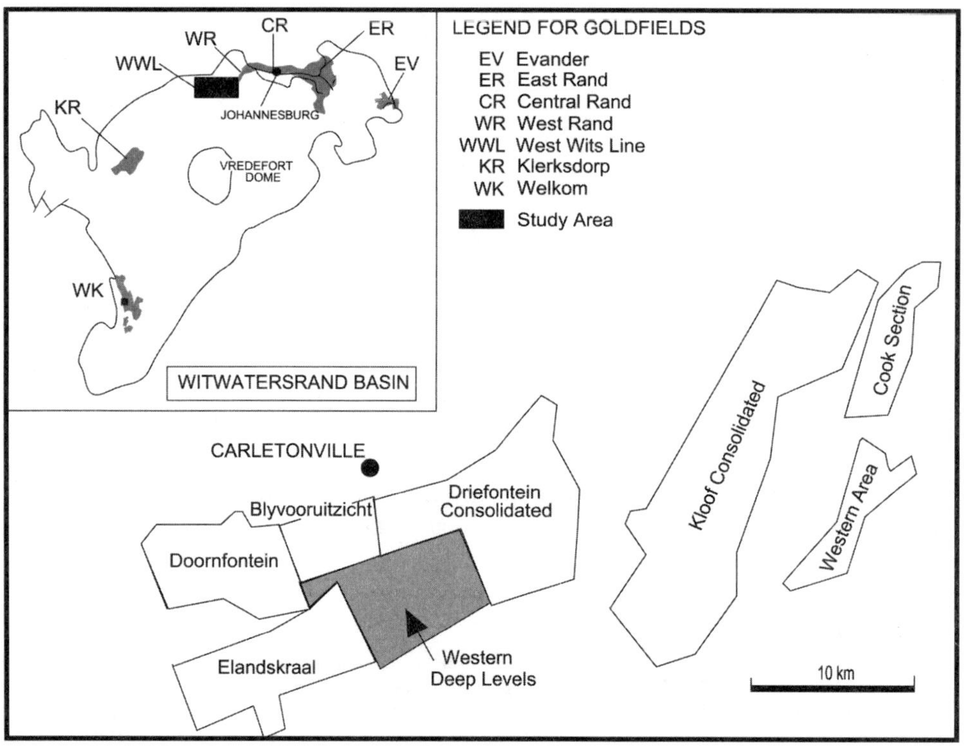

Figure 1. Locality plan showing Western Deep Levels (Mponeng) mine within the West Wits Line.

ically mineable Ventersdorp Contact Reef has been reported in the Central Rand, East Rand, and Evander goldfields.

Table 1 presents a stratigraphic column of the West Wits Line goldfield, summarized from South African Committee for Stratigraphy (SACS, 1980; SACS–Task Group for the Witwatersrand Supergroup, draft report, T.S. McCarthy, 1996, personal commun.), and the regional geological setting of the Ventersdorp Contact Reef in the West Wits Line is from Engelbrecht et al. (1986). Witwatersrand Supergroup sediments (comprising the West Rand and Central Rand Groups) were deposited on an Archean granite-greenstone basement, overlain by the Dominion Group, which is older than 3086 Ma (Robb and Meyer, 1995). The Witwatersrand sequence was terminated by deposition of the Venterspost Conglomerate Formation, the deposition of which was terminated by the extrusion of the Kliprivierburg Group basic lavas at 2714 Ma (Armstrong et al., 1991). The Ventersdorp Supergroup, together with the younger Transvaal Supergroup, represents the cover sequences that helped to preserve the Ventersdorp Contact Reef, as well as much of the underlying basin, from erosion.

The Venterspost Conglomerate Formation comprises the Ventersdorp Contact Reef conglomerates, as well as interdigitated quartzites known as Ventersdorp Quartzite. In addition, these sediments are commonly intercalated with lavas of the Westonaria Formation of the Klipriviersberg Group, such as at Kloof and Driefontein Consolidated Gold Mines in the West Wits Line (Hall, 1994). The conglomerates of the Ventersdorp Contact Reef are usually dark in color and characterized by sili-

TABLE 1. STRATIGRAPHIC COLUMN OF THE WEST WITS LINE

STRATIGRAPHY					AGE
TRANSVAAL SUPERGROUP (6 km)		Pretoria Group		Dolomite	2550–2200 Ma
		Chuniespoort Group		Carbonaceous shale, quartzite, conglomerate	
		Black Reef Quartzite Formation		Quartzite, conglomerate, shale	
		Wolkberg Group		Quartzite, arkoses, shale, basalt	
VENTERSDORP SUPERGROUP (±2 km)		Platberg Group	Rietgat Formation	Porphyritic lava, sediments	2714 Ma[1]
			Makwassie Formation	Quartz porphyries	
			Kameeldoorns Formation	Volcanic clasts	
		Klipriviersberg Group	Edenville Formation	Lavas and clasts	
			Orkney Formation	Nonamygdaloidal aphanitic lava	
			Alberton Formation	Porphyritic lava	
			Westonaria Formation	Andesitic tuff, lava	
VENTERSPOST CONGLOMERATE Formation: Ventersdorp Contact Reef (VCR)					±2714 Ma[1]
WITWATERSRAND SUPERGROUP	CENTRAL RAND GROUP	TURFFONTEIN SUBGROUP	Mondeor Formation	Quartzite and conglomerate (Deelkraal Reef, Uranium Bands [Ubs] Reef, Elansrand Reef)	2894–>2714 Ma[2]
			Elsburg Formation	Quartzite	
			Gold Estates Formation	Quartzite and conglomerate (Kloof Reef, Libanon Reef)	
			Robinson Formation	Quartzite, shale, conglomerate (Kleinfontein Reef)	
		JOHANNESBURG SUBGROUP	Booysens Formation	Quartzite, argillaceous quartzite, shale	
			Krugersdorp Formation	Quartzite, conglomerate (Bird Reef)	
			Luipaardsvlei Formation	Quartzite and conglomerate (Cobble Reef, Doorfontein Reef, Basal Grit, Large Pebble Marker, Khaki Pebble Reefs)	
			Randfontein Formation	Quartzite, conglomerate (Livingstone Reefs, Johnstone Reefs)	
			Main Formation	Quartzite, conglomerate (Main Reef Leader, Green Bar, Carbon Leader)	
			Blyvooruitzicht Formation	Quartzite, conglomerate (North Leader)	
	WEST RAND GROUP	JEPPESTOWN SUBGROUP	Roodepoort Formation	Shale, shaly quartzite	3070–2914 Ma[2]
			Crown Formation	Amygdaloidal lava	
			Florida Quartzite Formation	Quartzite, small pebble conglomerate	
		GOVERNMENT SUBGROUP	Witpoortjie Formation	Green-black, green silty shale, siltstones, argillaceous, gritty quartzite	
			Coronation Shale Formation	Black magnetic shale	
			Promise Quartzite Formation	Sericitic, argillaceous quartzite, ferruginous, magnetic shale, conglomerate (Promise Reef)	
		HOSPITAL HILL SUBGROUP	Brixton Formation	Green orthoquartzite, shale	
			Parktown Shale Formation	Shale quartzite	
Dominion Group (2–3 km): Porphyritic and andesitic volcanics and quartzite					3086–3074 Ma[2]
Archean granitic basement					>3086 Ma[2]

Note: Summarized from SACS (1980). 1—Armstrong et al. (1991); 2—Robb and Meyer (1995).

ceous, chloritic, and micaceous alteration (Zhao et al., 1994). This formation unconformably overlies the Witwatersrand Supergroup and underlies the Transvaal Supergroup.

The base of the Ventersdorp Contact Reef is marked by an angular unconformity that cuts into and eliminates progressively older Witwatersrand strata toward the east. It also represents a hiatus of several million years, during which time the Central Rand Group was tilted and eroded prior to inundation by the Ventersdorp lavas. The occurrence of the volcanic rocks of the Klipriviersberg Group makes a good case for contemporaneous sedimentation and subaerial volcanism, which is supported by the observation of tuffaceous material in the upper Ventersdorp Contact Reef and volatile-release structures in lavas immediately overlying what must have been largely unconsolidated sediment. Pillow structures observed in lavas of the Ventersdorp Contact Reef hanging wall indicate deposition in a subaqueous environment (Hall and Els, 2002). For this reason, the Ventersdorp Contact Reef is attributed stratigraphically to a distinct unit termed the Venterspost Conglomerate Formation, whose association is more strictly with the Ventersdorp Supergroup than with the Witwatersrand Supergroup, although this correlation is still somewhat contentious (Hall, 1994; Winter, 1994). A detailed study of load-induced deformation structures in the Ventersdorp Contact Reef and its lava hanging wall has, for example, concluded that the Ventersdorp Contact Reef was deposited without a major break in time over a substantial portion of the Witwatersrand Basin (Hall and Els, 2002).

All the gold mines in this goldfield occur along the southern limb of the Rand Anticline and the western limb of the Bank Anticline, so that the Ventersdorp Contact Reef dips roughly to the southwest. Two major structures, the Master Bedding and the Tear faults, occur in the northern part of the goldfield, whereas two other major structures, the Bank and West Rand faults, occur in the eastern part of the goldfield. The Bank Fault divides the goldfield into a western section including Doornfontein, Blyvooruitzicht, Elandskraal (including the former Deelkraal and Elandsrand gold mines), Western Deep Levels and Driefontein Consolidated Mines, and an eastern section including Kloof Consolidated Mines (Robb and Robb, 1998). The West Rand Fault, acting as a boundary structure, separates the West Wits Line from the West Rand goldfield to the east. Pseudotachylitic breccia–filled structures related to the Vredefort impact event at 2.02 Ga (see Fig. 1 for location of the central part of the Vredefort impact structure, the Vredefort Dome) are common, cutting across the Ventersdorp Contact Reef in the West Wits Line (e.g., Reimold et al., 1999) and possibly in other goldfields of the Witwatersrand Basin. The fault models of Myers et al. (1990) and McCarthy (1994) suggested that the West Wits Line goldfield is surrounded by rifted blocks of the Vreysrus Dome in the north and Westerdam Dome in the west. These blocks controlled the tectono-sedimentary development of the Ventersdorp Contact Reef, which is further supported by the study of Vermaakt and Chunnett (1994). The Ventersdorp Contact Reef was also affected by postdepositional tectonic activity such as dykes, tectonic fault breccia formations (Berlenbach and Roering, 1992), and faults.

Several sedimentary models have been proposed for the deposition of the various Ventersdorp Contact Reef facies, and these can be subdivided broadly into favoring either degradational processes (e.g., Thomas, 1979; Els, 1989) or aggradational processes (e.g., Minter, 1970; Chapman and Briggs, 1989; Mullins, 1993). The channel directions observed in the Ventersdorp Contact Reef trend mainly from east-northeast to west-southwest in the West Wits Line goldfield, and northwest-southeast to the east of it. Winter (1994) suggested that the Ventersdorp Contact Reef was deposited in a beach environment and represents an unconformity-bounded depositional sequence, whereas other studies have recognized a number of other environments (e.g., Hall, 1994; Germs and Schweitzer, 1994). McCarthy (1994) recognized two distinctive sedimentary environments termed (1) the Western Deep Levels type, characterized by pronounced paleorelief at its base, extensive, residual pediment deposits and gravel sheets, and small angular unconformities (<4°) in the footwall, which occurs mainly in the western section of the goldfield; and (2) the South Deep type, characterized by narrow conglomeratic zones adjacent to active faults, larger angular unconformities (10–15°) with the footwall, and rapid facies changes, which occurs mainly in the West Rand goldfield.

Major post–Ventersdorp Contact Reef thermal events in the Witwatersrand Basin have been suggested to have occurred at ca. 2500 Ma and at ca. 2300 Ma. Evidence for this has been provided by the formation of authigenic pyrite and rutile linked to fluid-circulation events associated with the episodic deposition of the Transvaal Sequence (Robb et al., 1991, 1997). However, the major hydrothermal event in the Witwatersrand Basin, and possibly over much of the Kaapvaal Craton, occurred at around 2050–2000 Ma. The major thermal events that affected the Witwatersrand Basin at that time include the emplacement of the Bushveld Complex and related igneous rocks to the north, dated by U-Pb techniques at 2055 ± 5 Ma (Walraven, 1997) and the Vredefort impact event at 2020 Ma (e.g., Spray et al., 1995; Kamo et al., 1996; Gibson and Stevens, 1997). Similar (and possibly related) ages of 1950 Ma have been reported for shales (whole rock K-Ar) by Layer et al. (1988), 1900–2050 Ma for clay separates of the Ventersdorp Contact Reef (K-Ar) by Zhao et al. (1999), and an age of 2006 Ma was reported for pseutotachylitic breccia associated with the Ventersdorp Contact Reef (^{40}Ar-^{39}Ar) by Trieloff et al. (1994). The Ventersdorp Contact Reef has been severely affected by these thermal events, in addition to the fact that it is directly overlain by the Ventersdorp lavas; and it is not surprising that it has been more pervasively metamorphosed and hydrothermally altered than most parts of the Witwatersrand sequence. The P-T conditions that affected the Ventersdorp Contact Reef have been determined at ~350 ± 50 °C and less than 0.3 GPa (e.g., Phillips et al., 1989; Wallmach and Meyer, 1990; Law, 1990; Frimmel et al., 1993; Frimmel, 1994, 1997; Phillips and Law, 1997; Boer et al., 1995; Drennan, 1997; Zhao, 1998).

MINE GEOLOGY AND VENTERSDORP CONTACT REEF ALTERATION MINERALOGY

The Regional Ventersdorp Contact Reef Working Group (1994) presented hanging-wall and footwall plans of the Ventersdorp Contact Reef. The direct hanging wall of the Ventersdorp Contact Reef generally consists of the Alberton Formation in the west, i.e., at Western Deep Levels, Elandskraal gold mine, but the Westonaria Formation occurs as the Ventersdorp Contact Reef hanging wall in the east, at Driefontein and Kloof Gold Mines (see Table 1). The footwall of the Ventersdorp Contact Reef consists of the Johannesburg Subgroup in the core of the Bank Anticline and the Turffontein Subgroup on its flanks (see Table 1).

The mineralogy of the Ventersdorp Contact Reef has been widely studied, including the work by Smits (1994) at Kloof Mine, Henckel and Schweitzer (1994) at Elandskraal (merging of Enlandsrand and Deelkraal Gold Mines), and Zhao et al. (1994) at Western Deep Levels (Mponeng) mine. The silicate mineralogy of the Ventersdorp Contact Reef is dominated by chlorite, muscovite, and quartz, and the major sulfide minerals are pyrrhotite, pyrite, chalcopyrite, galena, sphalerite, and arsenopyrite. Minor minerals include albite, calcite, siderite, epidote, sphene, pyrophyllite, talc, pentlandite, zircon, and chromite, with trace minerals including bismuthinite, tetrahedrite, stibnite, molybdenite, anatase, silver, monazite, apatite, and gold.

The characteristics and depositional environment of the Ventersdorp Contact Reef at Western Deep Levels South Mine (now part of Mponeng mine) have been described by Henning et al. (1994) and McWha (1994), whereas the nature and paragenesis of alteration has been outlined by Zhao et al. (1994) and Zhao (1998).

The detailed sedimentological and petrographic work carried out on the Ventersdorp Contact Reef at Western Deep Levels (Mponeng) Mine was the reason for selection of this location for the present investigation. The Ventersdorp Contact Reef is associated with a 1–3-m-thick zone of intense alteration that extends from the hanging-wall basalt into the footwall metasediments. Seven alteration zones were recognized on the basis of alteration mineral assemblages and coloration. Distinctive alteration zones in the Ventersdorp Contact Reef and associated lithologies can be recognized in the photographs of polished slabs (Zhao et al., 1994). Alteration involves albitization, chloritization, carbonatization, sulfidization, and muscovitization. Phyllosilicates, especially muscovite, represent the most abundant alteration mineral phases and are suitable for oxygen and hydrogen isotopic study.

At least three types of white mica were recognized optically, on the basis of paragenetic sequence and grain size. Coarse-grained (>200 μm) detrital muscovite is often kinked, banded, and replaced by chlorite. Medium-grained (100 μm to 200 μm) muscovite has a tabular texture and occurs either as overgrowths on quartz grains or as interstitial filling replacing albite and chlorite. Fine-grained muscovite (<10 μm) with chlorite, recognized by X-ray diffraction (XRD) studies, occurs in the matrix of the sediments and as alteration products in the hanging-wall lava. SEM (scanning electron microscopy) work has revealed that authigenic fine-grained muscovite coexists with authigenic chlorite and quartz, and TEM (transmission electron microscopy) studies have shown that muscovite exhibits an authigenic morphology (Zhao et al., 1999).

Ore mineralogy in the Ventersdorp Contact Reef is also characterized by a predominance of authigenic phases, and there are few detrital heavy minerals other than zircon. Round pyrite grains of possible detrital origin are rare and where present are overgrown by pyrrhotite, which is also associated with minor arsenopyrite, sphalerite, chalcopyrite, and galena. Gold in the Ventersdorp Contact Reef is invariably associated with the alteration mineral assemblage, and there is little or no textural evidence for allogenic remnants. The interrelationships between the silicate and sulfide alteration mineral assemblages in the Ventersdorp Contact Reef demonstrate a multiple paragenetic sequence of hydrothermal alteration in the reef (Zhao et al., 1994).

The orebodies are mainly comprised of conglomerates with internal quartzites that occur between two or more conglomerate units in the Ventersdorp Contact Reef. The conglomerates are auriferous, oligomictic, poorly sorted, and clast to matrix supported. The clasts of the conglomerates comprise mainly white and smoky vein quartz, with minor chert. The size of vein quartz ranges from 1 to 10 cm in diameter, whereas the chert is usually much smaller than the vein quartz. The matrix of the conglomerates is chloritic, micaceous, quartzitic, and sulfidic and usually displays a dark-gray to yellowish-green color. The conglomerates host major gold ore bodies. Two types of the quartzites occur in the Ventersdorp Contact Reef: an apple-green, fine-grained immature quartzite and a fine- to medium-grained, dark-gray mature quartzite. Gold concentrations in the quartzites are very low (<1 g/t).

Authigenic gold mineralization in the Ventersdorp Contact Reef is found associated with four distinctive mineral assemblages (Zhao, 1998): (1) gold together with authigenic sulfides (mainly pyrite); (2) a gold-chlorite assemblage, which is common; (3) a gold-bitumen assemblage; and (4) a gold-uranium assemblage. The interrelationships between the silicate and sulfide alteration mineral assemblages in the Ventersdorp Contact Reef and the paragenetic sequence were summarized by Zhao et al. (1994), which indicated multiple stages in the paragenetic sequence of hydrothermal alteration in the Ventersdorp Contact Reef.

DESCRIPTION OF THE SAMPLES

Samples for Sulfur Isotopic Analyses

A total of 14 rock samples, all containing sulfides, were collected from the hanging-wall lavas of the Klipriviersberg Group (n = 2), Ventersdorp Contact Reef conglomerates (n = 4), and footwall quartzites in the Central Rand Group (n = 8; Table 2).

TABLE 2. δ^{34}S DATA OF SULFIDES FROM THE VCR CONGLOMERATE, HANGING-WALL LAVA, AND FOOTWALL QUARTZITE

Sample number	δ^{34}S‰ Py	Po	Sph	Cpy	Ga
Lava					
R-1	0.0	–0.7	N.D.	–0.3	–1.3
B0413a	–0.6	N.D.	N.D.	N.D.	N.D.
VCR					
X0427	N.D.	0.6	N.D.	N.D.	N.D.
XR1654	N.D.	1.0	N.D.	N.D.	N.D.
XR1645–1	0.8	0.3	N.D.	N.D.	N.D.
XR1645–2	0.1	–0.4	N.D.	–0.2	N.D.
DGG quartzite					
B0384	N.D.	N.D.	0.9	N.D.	–1.5
B0376	1.8	0.3	0.9	N.D.	N.D.
B0372	N.D.	N.D.	0.4	N.D.	–0.8
X0439	0.1	–0.4	N.D.	N.D.	N.D.
XR1646	1.0	–0.2	N.D.	N.D.	N.D.
XR1643	–0.1	–0.5	N.D.	N.D.	N.D.
XR1642	1.0	–0.1	N.D.	N.D.	N.D.
XR1609	0.1	N.D.	N.D.	N.D.	N.D.

Note: VCR—Ventersdorp Contact Reef, Py—pyrite, Po—pyrrhotite, Sph—sphalerite, Cpy—chalcopyrite, Ga—galena, N.D.—not determined, DGG—dark grey alteration zone.

Mineral assemblages of sulfides in these samples are pyrite-pyrrhotite-chalcopyrite-sphalerite-galena. In total, 28 separates of single sulfides from the 14 rock samples have been analyzed, which includes authigenic pyrite (10), pyrrhotite (10), sphalerite (3), chalcopyrite (2), and galena (3).

Samples for Oxygen and Hydrogen Isotopic Analyses

Material collected from the Ventersdorp Contact Reef and related lithologies includes three types of samples: (1) individual minerals (quartz, calcite); (2) muscovite-rich mineral separates; and (3) bulk rocks. The individual mineral samples include the following.

1. Vein quartz from calcite-quartz veins, which normally cut through the Ventersdorp Contact Reef conglomerates, as well as related lithologies, e.g., hanging-wall lava of the Westonaria Formation in the Klipriviersberg Group, Ventersdorp Supergroup, and footwall quartzites in the Central Rand Group.

2. Amygdale quartz from amygdaloidal lavas in the Ventersdorp Contact Reef hanging wall. The lavas belong to the Westonaria Formation, Klipriviersberg Group, Ventersdorp Supergroup. The amygdales are up to 300 mm in diameter and consist of coarse-grained white quartz and calcite with minor amounts of chlorite and sulfides. A fluid inclusion study indicates moderate temperatures (290 °C and 350 °C) and moderate pressures (0.27 GPa) for the associated hydrothermal fluid (Zhao et al., 1994).

3. Vein calcite from quartz-calcite veins, coarse-grained in size and formed at moderate temperatures of 290 °C and 350 °C and at a moderate pressure of 0.27 GPa (Zhao et al., 1994).

Muscovite-rich samples and a chlorite-rich sample were obtained for oxygen and hydrogen isotope analysis using conventional natural water settling techniques for mineral separation, developed at the University of Strasbourg, France. The nine muscovite-rich mineral separates contain 95%–100% muscovite, with less than 5% chlorite. It is noted that chlorite may have 3 times as much H as muscovite (see discussion). The single chlorite-rich mineral separate comprised 85% chlorite with 15% muscovite. The mineral separates also contained trace quartz. These samples have been analyzed for both their oxygen and hydrogen isotopic compositions.

Bulk rocks were derived from different alteration zones and rock types. A total of fourteen bulk rocks have been analyzed and include: (1) five samples from the altered hanging-wall lava; (2) three Ventersdorp Contact Reef conglomerate samples; and (3) six quartzite samples from the altered footwall quartzites, namely the dark gray zone and the greenish gray zone. The dark gray zone represents a zone that was subjected to strong hydrothermal alteration in the footwall quartzite immediately below the Ventersdorp Contact Reef conglomerates. It is ~20–50 cm wide and comprises a mineral assemblage of chlorite-muscovite-quartz-sulfides-albite-(rutile). The greenish gray zone occurs directly below the dark gray zone and is ~20–100 cm wide. It comprises a mineral assemblage of muscovite-chlorite-quartz-pyrite-pyrrhotite-albite.

ANALYTICAL METHODS

Sulfur Isotopic Analysis

Sulfide mineral separation was carried out at the Bernard Price Institute of Geophysical Research, University of the Witwatersrand, Johannesburg, following these procedures: (1) crushing the samples to chips and further to fine grains of 125–1700 μm size; (2) heavy fluid separation using bromoform; and (3) application of a Franz Isodynamic Magnetic Separator to separate different sulfides. The samples were finally purified with a vacuum hand-picker at the Council for Geoscience of South Africa, Pretoria. Separated minerals include authigenic pyrite, pyrrhotite, chalcopyrite, sphalerite, and galena.

Sulfur isotope analyses were carried out at the Isotope Laboratory of the Council for Geoscience of South Africa, Pretoria, using conventional SO_2 extraction facilities and a Finnigan Mat 251 gas source mass spectrometer. The international reference materials NBS123 and Soufre de Lacq are used to calibrate the inhouse SO_2 bottle standard. Errors are produced from the sulfur extraction line and mass spectrometer measurement. In this study, the total two sigma error of δ^{34}S values is less than 0.4‰. All δ^{34}S values are reported in per mil relative to the Canyon Diablo Troilite.

Oxygen and Hydrogen Isotopic Analysis

All stable isotope ratios were measured using a Finnigan MAT252 mass spectrometer and are reported in the familiar δ notation where $\delta = (R_{sample}/R_{standard} - 1)*1000$ and $R = {}^{18}O/{}^{16}O$, D/H or $^{13}C/^{12}C$. Oxygen isotope data for whole-rock samples and silicate minerals were obtained by conventional methods using ClF_3 as the reagent (Borthwick and Harmon, 1982). Further details of the extraction methods for oxygen from silicates employed at the University of Cape Town were given by Vennemann and Smith (1990) and Harris and Erlank (1992). The quartz standard NBS-28 was analyzed in duplicate along with batches of eight samples, and data were normalized to the standard mean ocean water (SMOW) scale using the value of 9.64‰ recommended by Coplen et al. (1983). The average mean deviation of 12 duplicate NBS-28 quartz standard analyses made during the project period is 0.10‰.

Hydrogen was produced from ~30 mg muscovite separates and 100 mg whole-rock samples by the method of Vennemann and O'Neil (1993), with "Indiana" Zn being used to reduce the water to H_2. The amount of Zn was ~5 times that normally required to reduce the water to prevent "poisoning" of the Zn by any fluorine that might be liberated from the minerals. For hydrogen, an internal water standard ($\delta D = -9‰$) was used to calibrate the data to the SMOW scale, and the data were normalized so that Standard Light Antarctic Precipitation (SLAP) gave a value of –428‰ on the SMOW scale, as recommended by Coplen et al. (1983). Water contents of minerals and whole rocks were estimated from the voltage measured on the mass 2 collector on the mass spectrometer, as described by Vennemann and O'Neil (1993).

The oxygen isotope ratios of carbonate minerals were determined on CO_2 produced by reaction of whole-rock powders with "103%" phosphoric acid (McCrea, 1950). Samples were reacted overnight at 50 °C, and analyses are of the total carbonate present. The dominant carbonate mineral present was calcite, and hence the calcite-phosphoric acid fractionation factor (1.009) was used in the correction procedure. Data were normalized to the SMOW scale using an internal standard calibrated against NBS-19 ($\delta^{18}O = 28.64‰$).

SULFUR ISOTOPE STUDY

Previous Work

Mineralogical studies have revealed the existence of both allogenic and authigenic sulfide phases in mineralized Witwatersrand conglomerates; these include a variety of pyrite morphologies including detrital, mudball (round and porous), and crystalline pyrite, all of which may occur with carbonaceous matter in the Witwatersrand Basin (e.g., Viljoen, 1963; Robb and Meyer, 1986; Meyer et al., 1990; Robb et al., 1994, 1997). The pyrite types have formed in different environments; for example, allogenic pyrite is derived from the source area of the sulfides and associated heavy minerals such as zircon and chromite, whereas mudball pyrite is a syndepositional to diagenetic product and reflects processes taking place during burial and compaction. Authigenic pyrite as well as other sulfides are postdepositional products, which reflect much later processes of metamorphism and hydrothermal alteration. Detrital, mudball, and authigenic pyrite all coexist in the Ventersdorp Contact Reef, together with pyrrhotite and a variety of other sulfides such as chalcopyrite, arsenopyrite, galena, and sphalerite.

Sulfur isotope studies on various Witwatersrand pyrite types as well as other sulfides and carbonaceous matter have been carried out by workers such as Hoefs et al. (1968), Chukhrov et al. (1970), Hattori et al. (1983), Förster (1986), Drennan (1997), and England et al. (2002).

Published $\delta^{34}S$ values for detrital pyrite from the Witwatersrand reefs from the above authors define a narrow range from –1‰ to +5.6‰, which is much more restricted than that for sedimentary sulfides between –35‰ and +25‰ (Hoefs et al., 1968). Based on the above, most workers agree that sulfur in detrital pyrite in the Witwatersrand reefs originates from a magmatic source (Hoefs et al., 1968; Drennan, 1997). Förster (1986) also suggested that sulfur in detrital pyrite originated from subaerial volcanic smokers in the Witwatersrand Basin, a view also supported by Hutchinson and Viljoen (1988).

It is suggested that $\delta^{34}S$ values of syngenetic mudball pyrite between –0.5‰ and +4.0‰ were best attributable to anaerobic conditions at that time, which inhibited oxidation of sulfur in the sedimentary cycle (Hattori et al., 1983). This conclusion supports the models of low atmospheric oxygen during Proterozoic time as suggested by England et al. (2002).

For authigenic pyrite, the available $\delta^{34}S$ values range from –2‰ to +6.5‰ (Hoefs et al., 1968; Förster, 1986; Drennan, 1997; England et al., 2002), which is very similar to the range for detrital pyrite of –1‰ to +5.6‰ (e.g., Hoefs et al., 1968; Hattori et al., 1983; Förster, 1986; Drennan, 1997). In addition, no obvious difference in $\delta^{34}S$ values has been found between the authigenic pyrite, the pyrite associated with carbonaceous matter, and carbonaceous matter itself (Hoefs et al., 1968; Hattori et al., 1983).

Results of $^{34}S/^{32}S$ Analyses

A total of 28 S isotope determinations were carried out for authigenic grains of pyrite, pyrrhotite, sphalerite, galena, and chalcopyrite, as tabulated in Table 2. The $\delta^{34}S$ values of authigenic sulfides range from –1.5‰ to +1.8‰, with a mean around 0.0‰. In detail, sulfur isotopic compositions are between –1.5‰ and –0.8‰ for galena; between –0.7‰ and +1.0‰ for pyrrhotite; between –0.3‰ and –0.2‰ for chalcopyrite; between +0.4‰ and +0.9‰ for sphalerite; and between –0.6‰ and 1.8‰ for pyrite. The $\delta^{34}S$ values of single sulfide minerals are plotted in a histogram (Fig. 2), which indicates that the $\delta^{34}S$ values for both pyrite and pyrrhotite have almost identical distribution patterns with two modes around –0.5‰ and +0.5‰ (Figs. 2A and 2B).

Figure 3 shows $\delta^{34}S$ values of individual sulfide minerals from different host rocks. The $\delta^{34}S$ values of individual sulfides from two complete sample profiles comprising hangingwall lavas, Ventersdorp Contact Reef conglomerate, and footwall quartzites shows more or less similar distributions (Figs. 3A–3F), with no obvious difference in the distribution of $\delta^{34}S$ values for different rock types. The $\delta^{34}S$ values of sulfides show almost the same distribution in the different altered rock

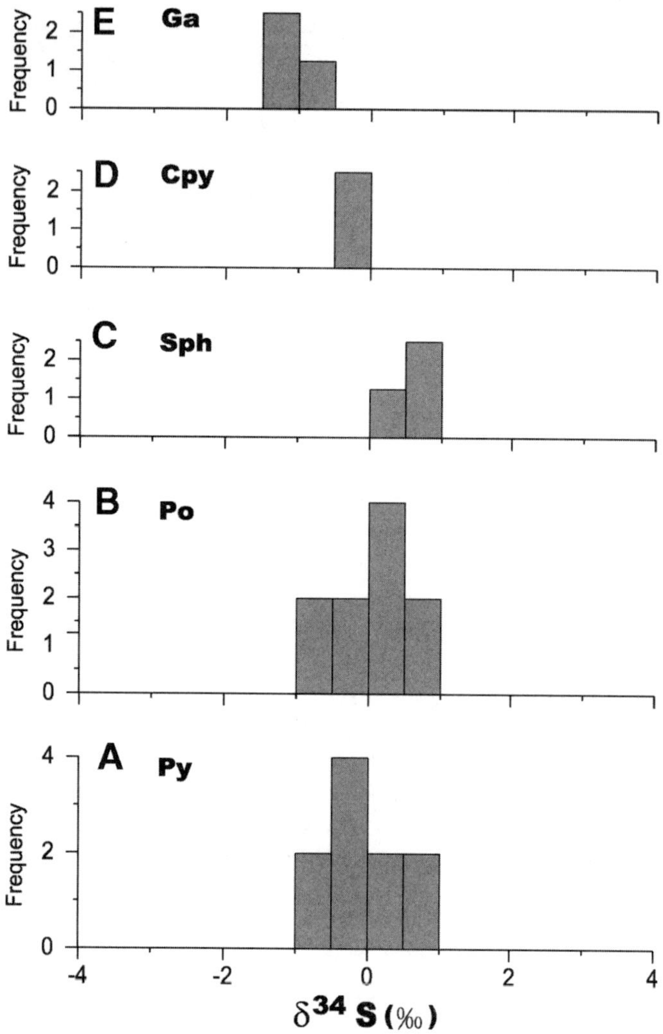

Figure 2. Histogram of $\delta^{34}S$ values in individual authigenic sulfide minerals from the Ventersdorp Contact Reef and its related rocks. Py—pyrite, Po—pyrrhotite, Sph—sphalerite, Cpy—chalcopyrite, Ga—galena.

types such as lavas, Ventersdorp Contact Reef conglomerates, and footwall quartzites (Figs. 3G–3J). Figure 3K shows the distribution of the $\delta^{34}S$ values from sphalerite for footwall quartzites only.

Interpretation of the Data

When sulfide minerals are precipitated from aqueous solutions or crystallize from sulfide liquids, small differences may occur in $\delta^{34}S$ values of cogenetic minerals due to isotopic fractionation among the solids and between the solid and liquid (Sakai, 1957; Faure, 1986). The differences in the isotopic composition of S in cogenetic sulfide minerals may reflect the temperature of isotopic equilibration. Theoretical considerations by Sakai (1968) and Bachinski (1969) based on bond strengths indicated that the ^{34}S enrichment of some common sulfide minerals should decrease in the order pyrite > sphalerite > chalcopyrite > galena. This is consistent with the results of this study: sphalerite (between +0.4‰ and +0.9‰) > chalcopyrite (between –0.3‰ and –0.2‰) > galena (between –1.5‰ and –0.8‰) (Fig. 2). However, pyrite (–0.6‰ and +1.8‰) and pyrrhotite (between –0.7‰ and +1.0‰) do not fit into the sequence. This may be attributed to the two paragenetic sequences of early pyrite-pyrrhotite and later sphalerite-chalcopyrite-galena (Zhao et al., 1994) that possibly did not form in S-isotope equilibrium.

$\delta^{34}S$ values of authigenic sulfides in the Ventersdorp Contact Reef from this study are compared to those from other studies (Hoefs et al., 1968; Hattori et al., 1983; Drennan, 1997) in Figure 4. These data have been categorized after Drennan (1997) into (1) detrital pyrite; (2) detrital pyrite associated with carbonaceous matter; (3) authigenic pyrite; (4) authigenic pyrite associated with carbonaceous matter; (5) other authigenic sulfides (pyrrhotite, sphalerite, chalcopyrite, and galena); and (6) S in carbonaceous matter.

The $\delta^{34}S$ values of detrital pyrite from the Witwatersrand reefs define a spread in $\delta^{34}S$ values from 0 to +8‰, with a roughly normal distribution with a mode of +2.8‰ (Fig. 4A; Drennan, 1997). Moreover, the $\delta^{34}S$ values of detrital pyrite associated with bitumen are in the range from –1‰ to +4‰, but they show a slightly scattered distribution pattern (Fig. 4B). No significant difference between the two distributions of $\delta^{34}S$ values for pyrite associated with carbonaceous matter and other pyrite can be detected. The pattern of $\delta^{34}S$ values for detrital pyrite (Fig. 4A) probably can be related to a magmatic source, as suggested by other workers (Hoefs et al., 1968; Förster, 1986; Drennan, 1997).

For authigenic/recrystallized pyrite, $\delta^{34}S$ values in the Ventersdorp Contact Reef from this study are similar to those measured for pyrite in other Witwatersrand conglomerates, except for a slightly narrower compositional range for the Ventersdorp Contact Reef data (Fig. 4C). The distribution of $\delta^{34}S$ values for authigenic pyrite is a mode of +0.5‰ (Fig. 4C).

Similarly, $\delta^{34}S$ values of other authigenic sulfides (pyrrhotite, sphalerite, chalcopyrite, galena) in the Ventersdorp Contact Reef indicate no difference from the data of Drennan (1997). These sulfides have lower $\delta^{34}S$ values compared to those of the authigenic pyrite and have a roughly normal distribution with a mode at –0.3‰. This indicates normal equilibrium fractionation among the sulfides, i.e., $\delta^{34}S$ values in a decrease order of pyrite > sphalerite, chalcopyrite, and galena at the above-mentioned temperature of 290–350 °C.

The $\delta^{34}S$ values of authigenic pyrite associated with carbonaceous matter are between +1.0 and +5.0‰ (Fig. 4E), similar to the data for the detrital pyrite (Fig. 4A) and for detrital pyrite associated with carbonaceous matter (Fig. 4B). Moreover, the $\delta^{34}S$ values for carbonaceous matter, representing very small grains, are between +2.0 and +4.0‰ (Fig. 4F), which is also indistinguishable from the data for the detrital pyrite (Fig. 4A) and the detrital pyrite associated with bitumen (Fig. 4B).

There is a large overlap of sulfur isotopic compositions for sulfides from the hanging-wall lavas, the Ventersdorp Contact

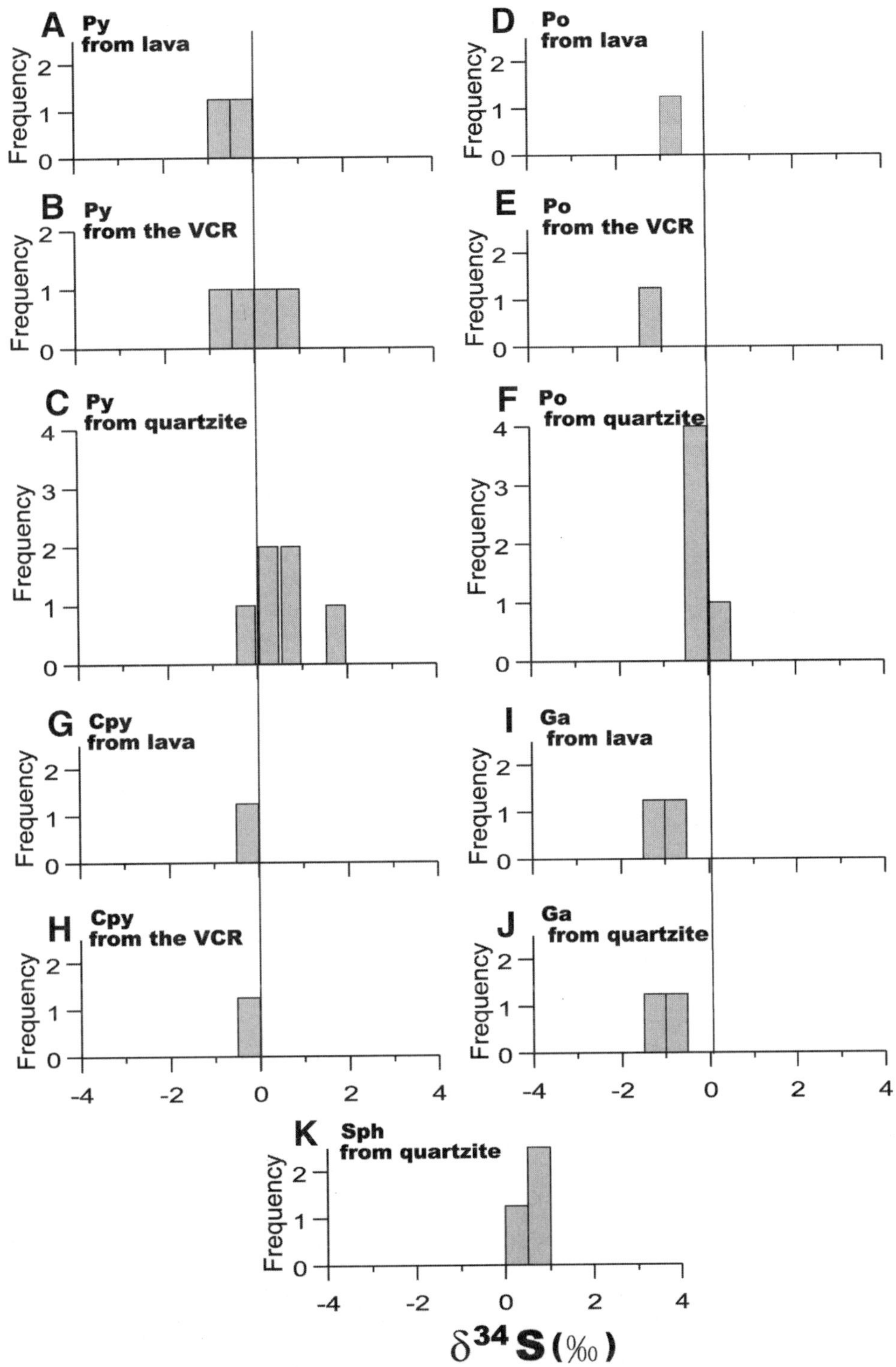

Figure 3. Histogram of δ^{34} S values in various authigenic sulfide minerals and in different host rocks from the Ventersdorp Contact Reef and its related rocks. Py—pyrite, Po—pyrrhotite, Cpy—chalcopyrite, Ga—galena, Sph—sphalerite.

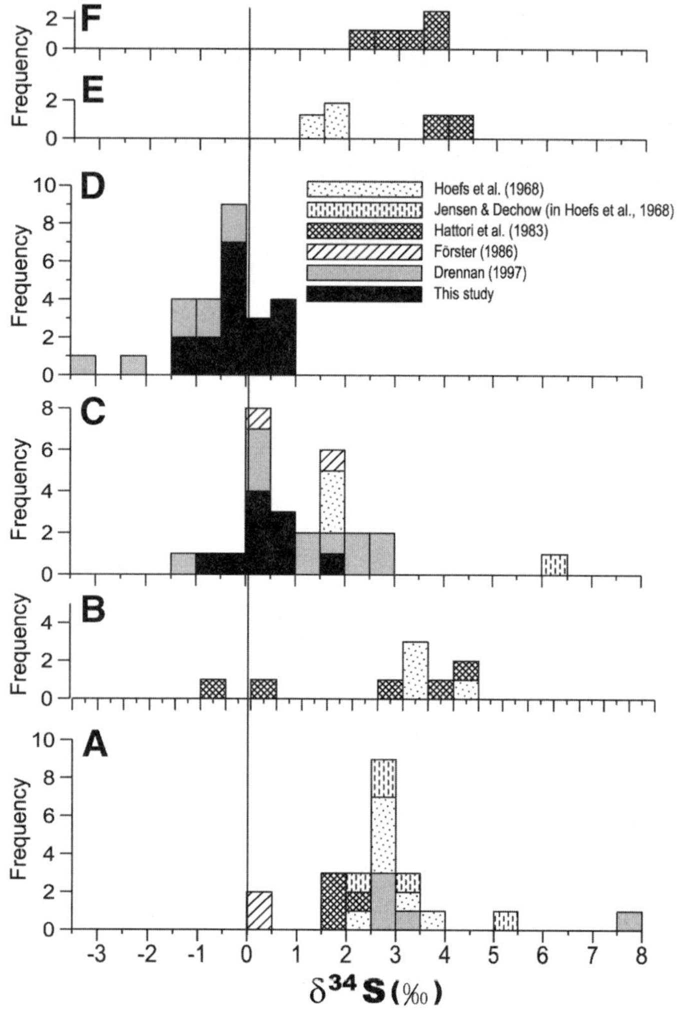

Figure 4. $\delta^{34}S$ values for sulfides from the Ventersdorp Contact Reef in this study, together with sulfur isotope data for sulfides and bitumen (carbonaceous matter) from the Witwatersrand reefs and quartz veins (after Drennan, 1997). (A) Detrital pyrite; (B) detrital pyrite associated with bitumen; (C) authigenic pyrite; (D) authigenic chalcopyrite, galena, sphalerite, and pyrrhotite; (E) authigenic pyrite associated with bitumen; and (F) bitumen.

Reef, and the footwall quartzites (Table 2, Fig. 3). There is no clear relationship between $\delta^{34}S$ values for sulfides and for their host rocks. Based on the similarity in S isotope compositions between carbon materials, detrital pyrite, and authigenic pyrite, it would seem that an influence of bacteria on the fractionation of sulfur isotopes in the Witwatersrand system is unlikely.

Two clusters of $\delta^{34}S$ values for authigenic pyrite (Fig. 4C) can be interpreted to be due to partial fractionation of the sulfur isotopes of authigenic pyrite by the newly formed chalcopyrite, galena, and sphalerite. This can be summarized as a two-stage model: (1) fractionation of detrital sulfur isotopes at a very low oxygen level, related to global anaerobic conditions; (2) partial fractionation of sulfur isotope compositions of the authigenic pyrite and related mineralization of chalcopyrite, sphalerite, and galena on a regional scale during metamorphism and/or hydrothermal activity.

The $\delta^{34}S$ values for sulfides in the Witwatersrand Basin are compared to those from two types of magmatic reservoirs as well as other distinctive types of ore deposits in Figure 5. The $\delta^{34}S$ values for detrital pyrite from the Witwatersrand Basin are very different from those for Mississippi Valley–type massive Pb-Zn deposits (–35‰ and +25‰) and for sediment-hosted massive sulfides, with three clusters between –20‰ and +15‰. The $\delta^{34}S$ values for Witwatersrand sulfides are also different from the red-bed copper mineralization in the Cheshire basin of northwestern England, which is associated with evaporite deposits with a wide range of $\delta^{34}S$ values (–1.8‰ and +16.2‰; Naylor et al., 1989). In this case, the narrow range of $\delta^{34}S$ values for detrital pyrite from the Witwatersrand Basin is consistent with a single source, whereas a wide range of $\delta^{34}S$ values is normally thought to be indicative of multiple sources or a relationship to biogenic fractionation (Ohmoto, 1986).

Magmatic sulfides (between –4.0‰ and +9.0‰) such as those of the East Pacific Rise vent (between +0.5‰ and 5‰), porphyry copper sulfides (between –3‰ and +1‰), and Cyprus-type sulfides (between –1.5‰ and +8‰) are the most likely candidates for a magmatic source of sulfur (Ohmoto, 1986). The direct source of detrital pyrite can be from erosion of the altered granites, mesothermal quartz veins, and veinlets in the granitic hinterland of the Witwatersrand Basin as suggested by Klemd and Hallbauer (1987) and Robb and Meyer (1991). A subaerial volcanic smoker source was also invoked by Förster (1986), and exhalites along the tectonically active northwestern margin of the basin, formed by shallow marine hydrothermal discharge, were suggested by Hutchinson and Viljoen (1988).

OXYGEN AND HYDROGEN ISOTOPES

Previous Work

Quartz is the most abundant mineral in the coarse clastics of the Witwatersrand Basin and includes quartz and chert pebbles, quartz sands, authigenic quartz, and vein quartz, as well as amygdale quartz from lavas. Oxygen isotope studies on these materials, together with calcite, muscovite, and chlorite, have the potential to provide valuable information on the source of the sediments, their metamorphism, and their possible mineralization history during basin evolution. Oxygen isotope studies of quartz pebbles, authigenic quartz, and postdepositional vein quartz were carried out previously in the Witwatersrand Basin by workers such as Savin and Epstein (1970), Barton et al. (1992), and Vennemann et al. (1992, 1996).

Oxygen isotope studies on white and gray quartz and chert pebbles have been carried out in the Witwatersrand Basin by Barton et al. (1992) and Vennemann et al. (1992, 1996). The samples of Barton et al. (1992) covered a large part of the Witwatersrand stratigraphy, together with the Dominion Group and

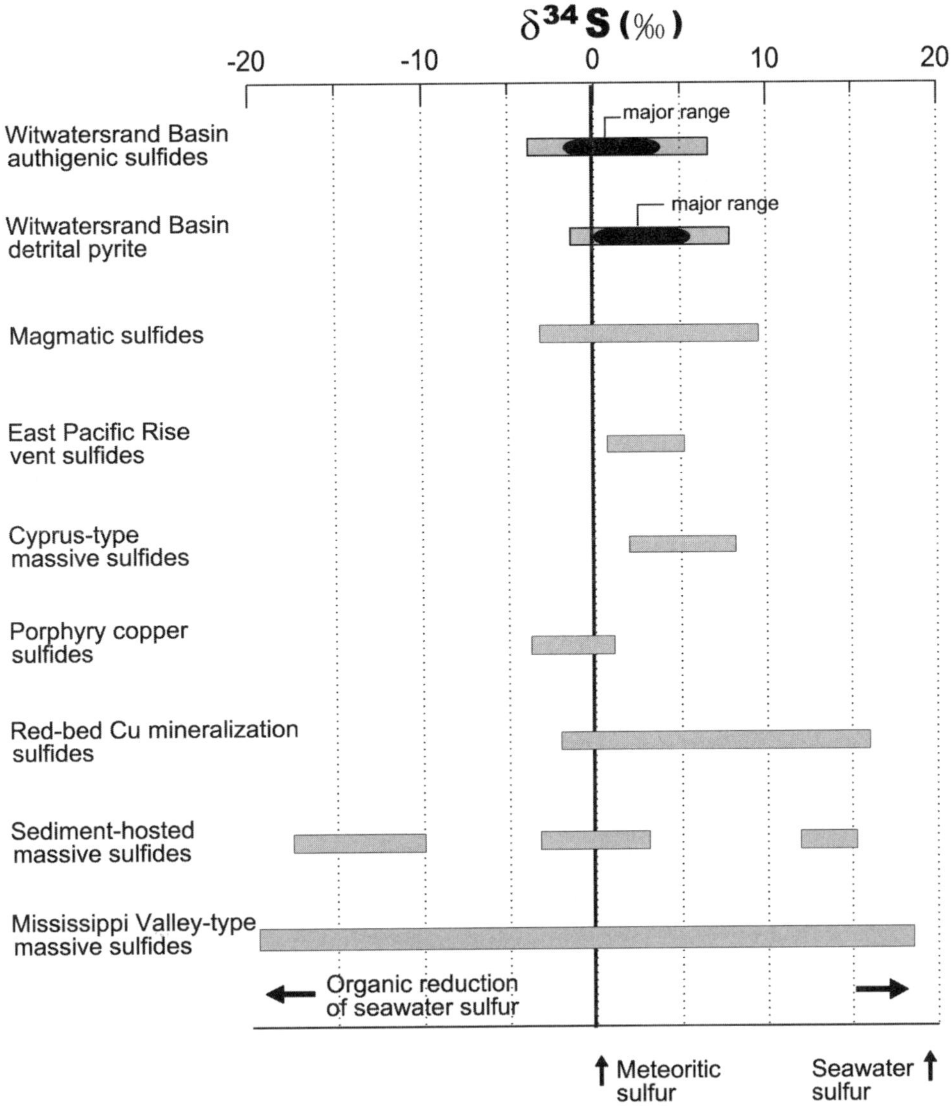

Figure 5. The δ^{34}S values for sulfides in the Witwatersrand Basin, together with the data for sulfides of typical hydrothermal and magmatic origin (compilation by Rollinson, 1993) and the Witwatersrand data from this study and other authors indicated in the text.

Transvaal Supergroup, as well as a wide geographic area, whereas those of Vennemann et al. (1992, 1996) were from several distinct reefs in the Central Rand, West Rand, and Welkom gold fields. The results of both studies are similar and indicate that the δ^{18}O values of quartz pebbles from Witwatersrand reefs occupy a δ^{18}O range from 8.8 to 14.5‰ (Vennemann et al., 1992) or 8.8‰–13.1‰ (Barton et al., 1992). Also, there is significant heterogeneity in δ^{18}O values among the quartz clasts within individual hand specimen samples (Vennemann et al., 1992) or from center to rim within individual quartz clasts (Barton et al., 1992). The absence of any textural evidence of quartz veining or overgrowth indicates that the quartz pebbles and chert clasts preserved their original oxygen isotope compositions and were not significantly recrystallized (Barton et al., 1992; Vennemann et al., 1992, 1996). The δ^{18}O values of 11 chert clasts from the Witwatersrand reefs (Barton et al., 1992) have an identical, relatively narrow range of values between 9.0‰ and 12.5‰ and are similar to the values of the quartz clasts (between 9.8‰ and 11.5‰; Vennemann et al., 1992).

A study of sand-sized quartz in the matrix of conglomerates was carried out by Vennemann et al. (1996) on the Monarch Reef and the Main Reef in the Witwatersrand Basin. They reported δ^{18}O values between 7.8‰ and 11.8‰, which is lower than the range for quartz clasts and cherts (from 9.0‰ to 12.5‰). The quartz sand population in each of these conglomerate horizons does not have δ^{18}O values in excess of 11.8‰, which suggests a different sediment source than that of some of the quartz pebbles (Vennemann et al., 1996).

In order to determine the possible source area for the Witwatersrand reefs, Barton et al. (1992) investigated a number of types of clasts from the potential source areas including (1) the Beit Bridge Group in the Central Zone of the Limpopo Belt, near Messina (between 9.5‰ and 10.7‰); (2) white hydrothermal quartz veins from the Pietersburg granite greenstone terrane (between 11.6‰ and 11.8‰); (3) gold-bearing white quartz vein, from the Pietersburg granite greenstone terrane (between 14.1‰ and 14.2‰); (4) white hydrothermal quartz veins from the Barbrook Gold Mine in the Barberton granite greenstone terrane (between 14.7‰ and 15.0‰); and (5) from the Witwatersrand Basin itself including quartzite from the Promise Formation of the West Rand Group (between 8.4‰ and 8.8‰) and from the Turffontein Subgroup of the Central Rand Group (between 9.6‰ and 10.0‰). In addition, Beattie and Barton (1992) investigated gold-bearing hydrothermal quartz veins in the northern Kaapvaal Craton and Limpopo belt and reported two populations with $\delta^{18}O$ values in ranges between 9.2‰ and 13.1‰ and between 14.0‰ and 14.4‰, respectively. The first group of $\delta^{18}O$ values almost completely overlaps those of the large clasts in the Witwatersrand Basin, and the second group is higher than those of any Witwatersrand clasts yet measured.

Vennemann et al. (1996) also compared the $\delta^{18}O$ values of quartz pebbles and quartz sands in quartz pebble conglomerates with likely known rock types from other parts of the world, including granitoids, pegmatites, paragneisses, metavolcanic rocks, and metasandstones and suggested that the compositionally homogeneous granites and related pegmatites must have been the dominant source of uranium, whereas hydrothermal lode gold veins were the dominant source for gold in the Witwatersrand.

New Results of Oxygen Isotope Analysis

A total of 36 $\delta^{18}O$ values for bulk rocks and quartz and calcite separates were obtained and are tabulated in Table 3. Bulk rock $\delta^{18}O$ values obtained in the present study range from 6.1‰ to 10.4‰ (Table 3). For different rock types, the $\delta^{18}O$ ranges are (1) 6.1‰–8.3‰ for lava samples; (2) 9.1‰–10.6‰ for conglomerates; and (3) 9.4‰–10.4‰ for quartzites (Table 3).

The $\delta^{18}O$ values for bulk rocks from this study, grouped in terms of alteration types (Zhao et al., 1994), are compared to

TABLE 3. OXYGEN ISOTOPE COMPOSITIONS FOR BULK ROCKS, MINERAL SEPARATES, VEIN AND AMYGDALE QUARTZ, AND VEIN CALCITE IN THE VCR

Sample	Type	Lithology	Zone	Description	$\delta^{18}O$‰ (SMOW)
BH020	Rock	Lava	MC	Strongly altered amygdaloidal lava	8.3
GBE-2	Rock	Lava	MC	Strongly altered amygdaloidal lava	7.4
1145A	Rock	Lava	DG	Least altered phenoclastic lava	6.1
1145B	Rock	Lava	DG	Least altered amygdaloidal lava	6.5
BHE-1	Rock	lava	DG	Least altered massive lava	6.9
B0364	Rock	VCR	RA	Ms-rich conglomerate	10.6
XR1614	Rock	VCR	RA	Light green conglomerate	9.1
XR1618	Rock	VCR	RA	Dark gray conglomerate	10.2
I0705	Rock	Qtz	DGG	Qz-Ch-Ms-sulfides	9.6
B0366	Rock	Qtz	DGG	Qz-Ms-Ch	9.8
B0415	Rock	Qtz	DGG	Qz-Ms-Ch	10.4
BH021	Rock	Qtz	GG	Qz-Ms	9.4
BHI-10	Rock	Qtz	GG	Qz-Ms	9.8
BGH14	Rock	Qtz	GG	Qz-Ms	10.0
133019	M-sep	Lava	MC	85% Ch, 15% Ms with Qz	6.1
133729	M-sep	VCR	RA	100% Ms with Qz	7.8
133730	M-sep	VCR	RA	100% Ms with minor Qz	7.2
133017	M-sep	qtz	DGG	95% Ms, 5% Ch with Qz	8.4
133731	M-sep	qtz	DGG	95% Ms, 5% Ch with Qz	8.4
133732	M-sep	qtz	DGG	100% Ms with Qz	8.5
133020	M-sep	qtz	GG	100% Ms with Qz	7.4
133734	M-sep	qtz	GG	100% Ms with Qz	7.9
133736	M-sep	qtz	GG	95% Ms, 5% Ch with Qz	8.0
133737	M-sep	qtz	AGNQ	95% Ms, 5% Ch with minor Qz	8.1
R-1	Qz-A	Lava	-	From amygdaloidal lava	10.7
XR1662	Qz-A	Lava	-	From amygdaloidal lava	10.3
XR1640	Qz-A	Lava	-	From amygdaloidal lava	10.4
BHE-6	Qz-A	Lava	-	From amygdaloidal lava	9.5
B0397	Qz-A	Lava	-	From amygdaloidal lava	10.7
X0436	Qz-V	Cc-Qz-V	-	From Qz vein with some Ms	10.3
X0451	Qz-V	Cc-Qz-V	-	From zoned Qz calcite vein	11.3
XR1669	Qz-V	Cc-Qz-V	-	From calcite Qz vein in shale	10.3
No4	Qz-V	Cc-Qz-V	-	From calcite Qz vein	9.6
B0397	Cc-V	Cc-Qz-V	-	From calcite Qz vein	9.7
X0451	Cc-V	Cc-Qz-V	-	From calcite Qz vein	9.0
XR1699	Cc-V	Cc-Qz-V	-	From calcite Qz vein	8.9

Note: VCR—Ventersdorp Contact Reef, M-sep—mineral separate, Qz—quartz, Cc—calcite, Qtz—quartzite, A—lava amygdale, V—vein, Ms—muscovite, Ch—chlorite, MC—multiple color zone in lava, DG—dark grey lava, RA—reef alteration zone, DGG—dark grey alteration zone, GG—greenish grey alteration zone, AGNQ—apple green quartzite zone.

those for Ventersdorp-age dolerite and other gabbros/basalts from the world in Figure 6. It is clear that different alteration zones for the Ventersdorp Contact Reef rocks have different O isotope compositions. The least altered lavas are more or less similar to those of the Ventersdorp-age dolerites from the Witwatersrand Basin, which cut through some quartzites of the Witwatersrand Supergroup (Harris and Watkins, 1990) and have a range from +6.1‰ to +6.9‰, similar to that for typical, unaltered gabbros/basalts on Earth (+6‰ to +8‰; cf., Faure, 1986). The Ventersdorp Contact Reef has the same whole rock $\delta^{18}O$ values as the dark gray quartzite and gray quartzite from the alteration zones. The $\delta^{18}O$ values of quartzite (Harris and Watkins, 1990) almost overlap with the quartzite from the gray quartzite zone, which is not related to any reefs and is not adjacent to dolerite dykes (Harris and Watkins, 1990).

The $\delta^{18}O$ values of muscovite and chlorite mineral separates from the Ventersdorp Contact Reef and related lithologies are plotted in Figure 7, together with those for bulk rocks and typical gabbros/basalts for comparison. This figure clearly

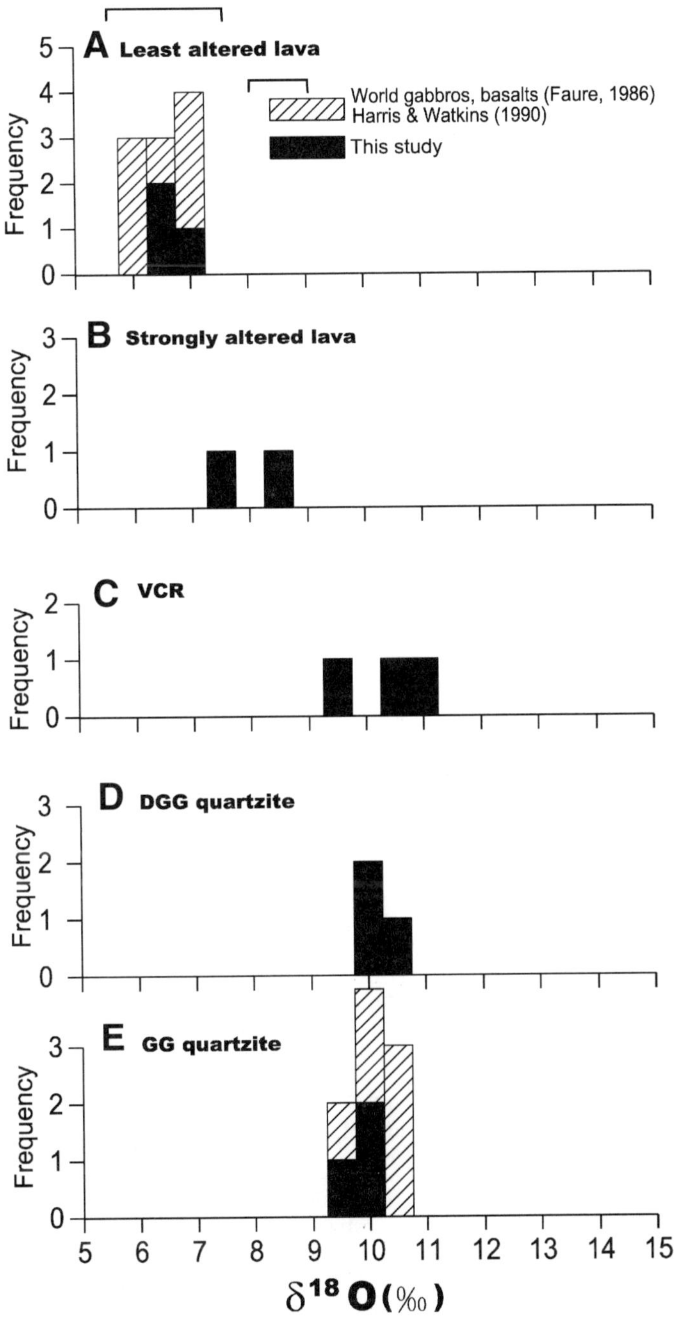

Figure 6. Histogram of $\delta^{18}O$ values (‰) for bulk rocks in the Ventersdorp Contact Reef, together with data from Harris and Watkins (1990). (A) Least altered lava; (B) strongly altered lava; (C) Ventersdorp Contact Reef (VCR) conglomerate; (D) strongly altered quartzite (DGG); (E) least altered quartzite (GG).

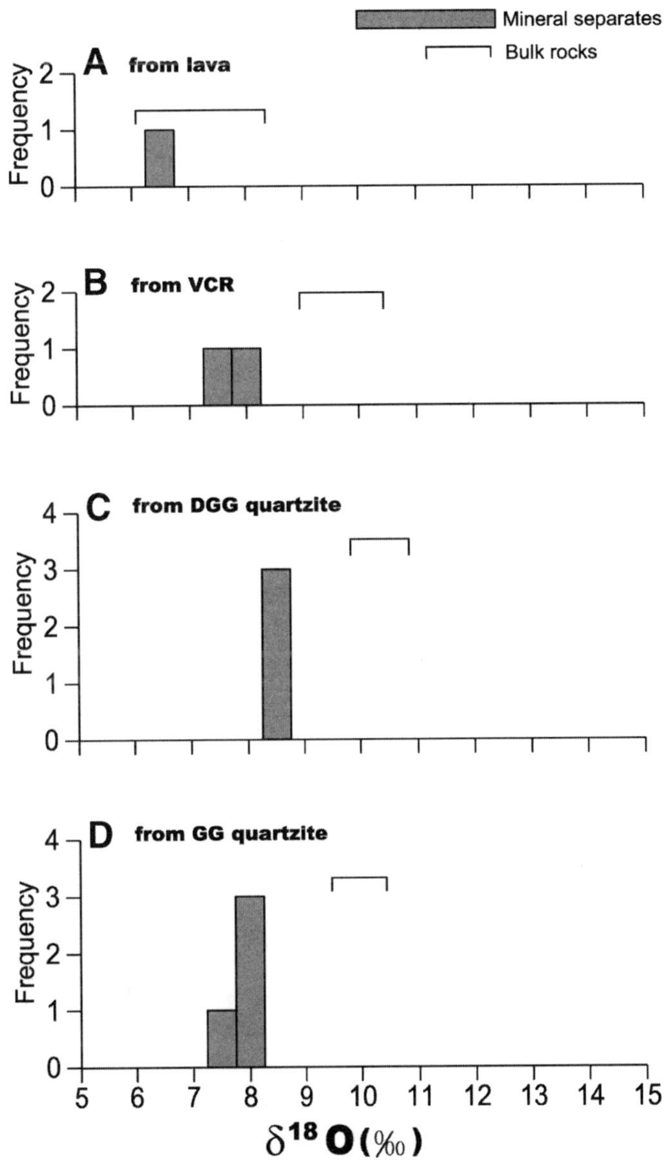

Figure 7. Histogram of $\delta^{18}O$ (‰) values for muscovite and chlorite separates from the Ventersdorp Contact Reef, together with those for the corresponding host rocks. Abbreviations as in Figure 6.

indicates a very similar distribution for the mineral separate data to that for the bulk rocks, as shown in Figure 6. All the mineral separates have lower $\delta^{18}O$ values compared to the corresponding bulk rocks, because the mineral-quartz fractionation factors will be negative at temperatures between 300 °C and 350 °C.

$\delta^{18}O$ values were measured between 9.5‰ and 10.7‰ for amygdale quartz, between 9.6‰ and 11.3‰ for vein quartz, and between 8.9‰ and 9.7‰ for vein calcite. The $\delta^{18}O$ values are plotted together with those for agates from Karoo lavas for comparison in Figure 8. The results show that the $\delta^{18}O$ values from amygdale quartz almost totally overlap with those of vein quartz, whereas the $\delta^{18}O$ values for vein calcite are partially lower than those for both amygdale and vein quartz. Moreover, $\delta^{18}O$ values for the amygdale quartz in this study are much lower than those typically found in agates.

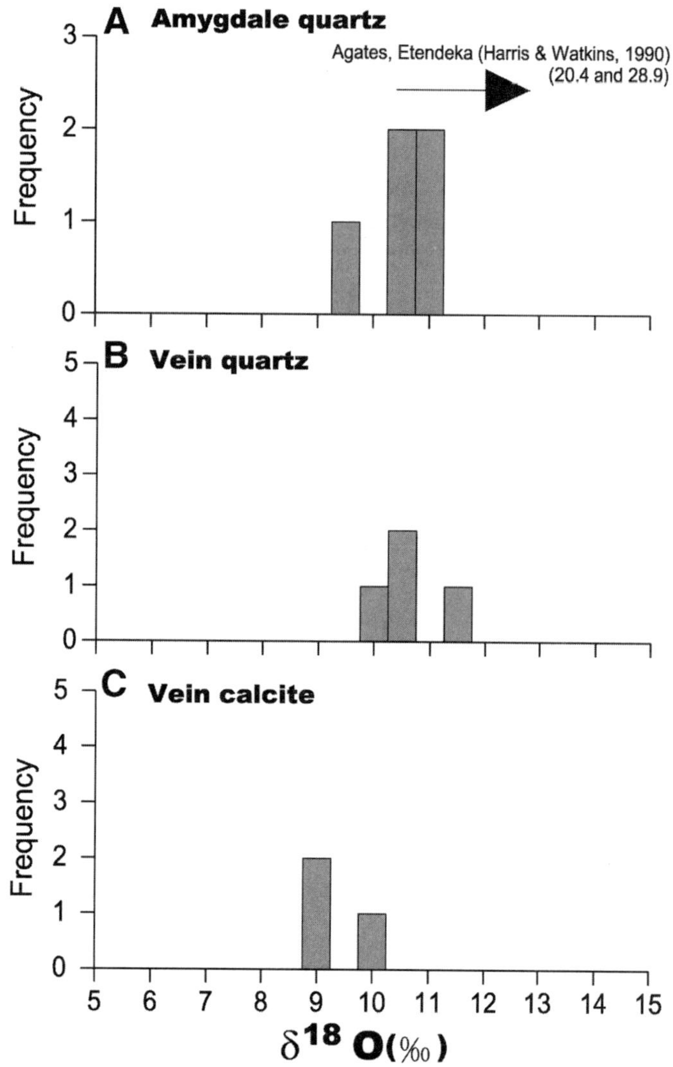

Figure 8. Histogram of $\delta^{18}O$ (‰) values for authigenic quartz (both vein and amygdale quartz) from the Ventersdorp Contact Reef, together with those for quartz pebbles from the Ventersdorp Contact Reef (Barton et al., 1992; Vennemann et al., 1996) and from other Witwatersrand reefs (excluding the Ventersdorp Contact Reef; Barton et al., 1992; Vennemann et al., 1992, 1996) and for quartz sands from the Monarch and Main reefs (Vennemann et al., 1996). The range from 20.4 to 28.9‰ is derived from Harris (1989).

Results of Oxygen and Hydrogen Isotopes in Fluids

Ten δD values were obtained for the separated chlorite-rich (1) and muscovite-rich (9) samples from hanging-wall lava, Ventersdorp Contact Reef conglomerates (Reef Alteration zone and Apple Green Quartzite), and footwall quartzite (Dark Grey quartzite zone, Greenish Grey quartzite zone). Most of the data fall between –52‰ and –62‰ (Table 4).

Calculation of Oxygen and Hydrogen Isotopes for Hydrothermal Fluids

The isotopic compositions of aqueous hydrothermal solutions can either be measured directly in fluid inclusions (e.g., Ohmoto and Rye, 1979) or indirectly calculated on the basis of laboratory calibrations of equilibria between alteration related minerals and associated fluids (Chacko et al., 2001). The latter method is used here to calculate the δD and $\delta^{18}O$ values of the fluids.

An equation of the equilibrium O-isotope fractionation for illite-water in the temperature range 200–400 °C and based on a combination of experimental, empirical and /or theoretical data was derived by Sheppard and Gilg (1996), as follows:

$$1000 \ln\alpha \text{ illite-water} = 2.39 * 10^6 * T^{-2} - 3.76. \quad (1)$$

On the basis of the structural similarity between muscovite and illite, the above equation is thought to be suitable also for the purpose of calculating the $^{18}O/^{16}O$ of fluids in equilibrium with muscovite. Based on an assumed muscovite-water equilibrium, most of the empirical and experimental data for hydrogen suggest that the fractionation is not very sensitive to temperature variation in the range 120 °C and 400 °C and can be approximated as 1000 ln α = –25‰ ± 5‰ (Sheppard and Gilg, 1996). Our mineral separates comprise muscovite, chlorite, and minor amounts of quartz. Nine of the ten mineral separates are dominated by muscovite (>95%). The nine muscovite-rich samples are used for the calculation of $^{18}O/^{16}O$ and D/H for hydrothermal fluids in the Ventersdorp Contact Reef using the equations of Sheppard and Gilg (1996). A maximum temperature of 350 °C is used for the calculation. The results show that all δD values for the hydrothermal fluids should be between –27‰ and –39‰. All $\delta^{18}O$ values for this data set fall between 4.8‰ and 6.1‰ (Table 4).

A diagram of oxygen and hydrogen isotopic compositions of the hydrothermal fluids related to alteration zones is given in Figure 9. Fields of magmatic water (Taylor, 1974) and metamorphic water (Taylor, 1974), together with the meteoric water line (Epstein et al., 1965; Epstein, 1970), are plotted for comparison. Figure 9 demonstrates that all the data of this study

TABLE 4. OXYGEN AND HYDROGEN ISOTOPE DATA FROM MINERAL SEPARATES AND FOR HYDROTHERMAL FLUIDS ESTIMATED FROM MINERAL-WATER EQUILIBRIA IN THE VCR

Sample	Lithology	Zones	Description	δD‰ (SMOW)		δ¹⁸O‰ (SMOW)		Water
				Mineral separates	Water†	Mineral separates	Water†	wt%
133019	Lava	MC	85% Ch, 15% Ms with Qz	–61	–	7.8	-	8.49
133729	VCR	RA	100% Ms with Qz	–61	–36	7.8	5.4	1.89
133730	VCR	RA	100% Ms with minor Qz	–52	–27	7.2	4.8	4.00
133017	qtz	DGG	95% Ms, 5% Ch with Qz	–58	–33	8.4	6.0	4.39
133731	qtz	DGG	95% Ms, 5% Ch with Qz	–62	–37	8.4	6.0	4.01
133732	qtz	DGG	100% Ms with Qz	–64	–39	8.5	6.1	2.95
133020	qtz	GG	100% Ms with Qz	–52	–27	7.4	5.0	4.05
133734	qtz	GG	100% Ms with Qz	–59	–34	7.9	5.5	4.26
133736	qtz	GG	95% Ms, 5% Ch with Qz	–62	–37	8	5.6	3.71

Note: VCR—Ventersdorp Contact Reef, MC—multiple color zone in lava, RA—reef alteration zone, DGG—dark grey alteration zone, GG—greenish grey alteration zone.
†Calculated values (see text).

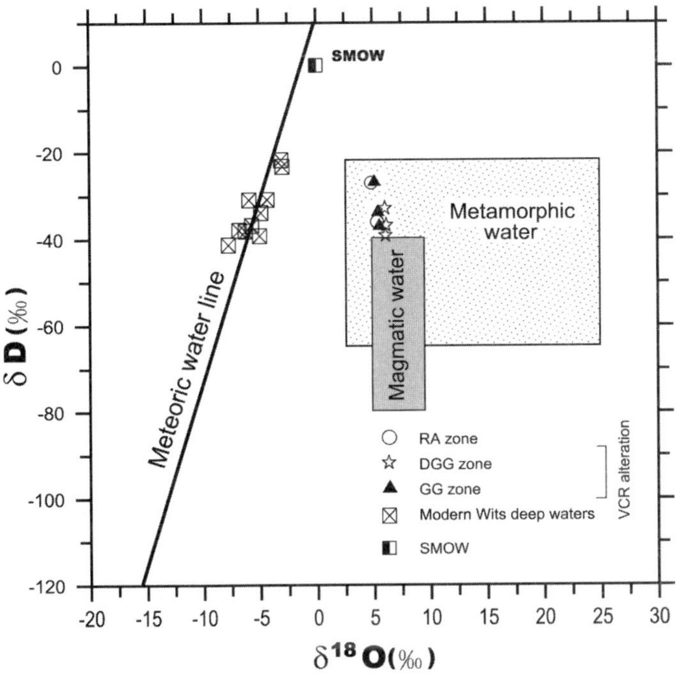

Figure 9. Plot of δD versus $\delta^{18}O$ for the compositions of hydrothermal fluids calculated from muscovite in alteration zones of the Ventersdorp Contact Reef at Western Deep Levels South Mine. The fractionation factors of oxygen and hydrogen for muscovite-water used in the calculation are from Sheppard and Gilg (1996). The data for deep waters in modern Witwatersrand gold mines (Duane et al., 1997), fields of magmatic water (Taylor, 1974) and metamorphic water (Taylor, 1974), and the meteoric water line (Epstein et al., 1965; Epstein, 1970) are also shown. RA—reef alteration zone; DGG—dark grey alteration zone; GG—greenish grey zone. SMOW—standard mean ocean water.

plot into the field of metamorphic fluids. The $\delta^{18}O$ and δD values of modern waters from several major gold mines in the Witwatersrand Basin (Duane et al., 1997) have also been added to the diagram for comparison (Fig. 9). All the samples plot along the meteoric water line, except for two data, which have been interpreted to represent meteoric water that has been subject to modifications (Duane et al., 1997).

Implications of the Oxygen and Hydrogen Isotopic Data

The isotopic data for authigenic quartz including both amygdale and vein quartz in the Ventersdorp Contact Reef, together with those of quartz pebbles from Ventersdorp Contact Reef conglomerates (Barton et al., 1992), from the Witwatersrand reefs excluding the Ventersdorp Contact Reef (Barton et al., 1992; Vennemann et al., 1992, 1996), and from quartz sands from the Monarch and Main reefs in the Witwatersrand Basin have been plotted into a histogram (Fig. 10). The results demonstrate that $\delta^{18}O$ values of authigenic amygdale and vein quartz form a cluster with a modal value of 10.5‰ (Fig. 10A). The $\delta^{18}O$ values for quartz pebbles of the Ventersdorp Contact Reef form a cluster with a modal value of 11.5‰, which shows no clear normal distribution pattern. The $\delta^{18}O$ values for quartz pebbles of the Witwatersrand reefs, excluding the Ventersdorp Contact Reef, form a cluster with a modal value of 11.0‰. The $\delta^{18}O$ values of quartz sands in the Witwatersrand Basin have a single skewed cluster with values less than 11.5‰, which suggests a different source than that of the quartz sands (in agreement with Vennemann et al., 1996).

In comparison with the $\delta^{18}O$ values of quartz pebbles in the Ventersdorp Contact Reef, quartz pebbles, and quartz sands in other reefs of the Witwatersrand Basin (Barton et al., 1992; Vennemann et al., 1992, 1996), the $\delta^{18}O$ values for authigenic quartz measured in this study are similar. The diagram of oxygen and hydrogen isotopic data clearly indicates a typical metamorphic origin for the hydrothermal waters, but no direct evidence for meteoric waters. Such fluids may well have been produced by processes of degassing, release of pore fluids from the sediments, and dewatering of early formed or detrital hydrous minerals to form muscovite and chlorite during metamorphism in the Witwatersrand Basin (Stevens et al., 1996; Drennan, 1997). Furthermore, buffering of oxygen isotopes for the hydrothermal fluids, mainly by detrital quartz materials (quartz pebbles and quartz sands), is also supported by the present results. The large spread in values obtained for the authigenic phases supports local fluid-buffering rather than exchange with a pervasively infiltrating fluid.

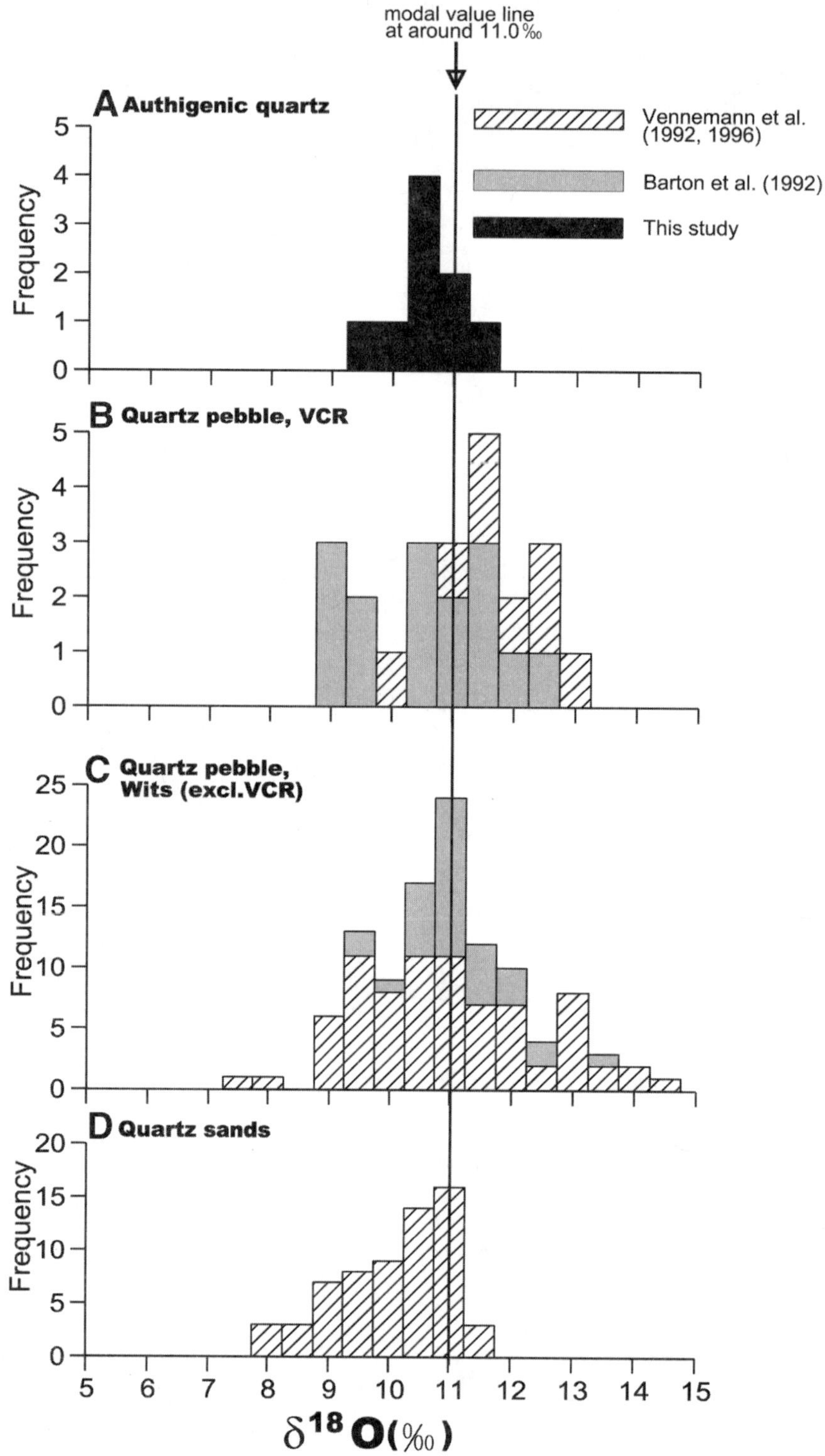

Figure 10. Histogram of d[18]O (‰) values for authigenic quartz (both vein and amygdale quartz) from the Ventersdorp Contact Reef, together with those for quartz pebbles from the Ventersdorp Contact Reef (Barton et al., 1992; Vennemann et al., 1996) and from Witwatersrand reefs (excluding the Ventersdorp Contact Reef [VCR]; Barton et al., 1992; Vennemann et al., 1992, 1996), and for quartz sands from the Monarch and Main reefs (Vennemann et al., 1996).

DISCUSSION AND CONCLUSIONS

A significant new data set of sulfur, oxygen, and hydrogen isotopic compositions for Ventersdorp Contact Reef and related lithologies and minerals has been obtained in this study (Tables 2, 3, and 4) and has been compared with the previous data for the Witwatersrand Basin (Figs. 4 and 9). $\delta^{34}S$ values for authigenic sulfides in the Ventersdorp Contact Reef have a narrow range of between −1.5‰ and +1.8‰. $\delta^{18}O$ values are in a range between +6.1‰ and +10.4‰ for bulk rocks and between +8.9‰ and 11.3‰ for authigenic quartz and calcite. $\delta^{18}O$ values and δD values for hydrothermal fluids in the Ventersdorp Contact Reef fall between +4.8‰ and +6.1‰ and between −27‰ and −39‰, respectively.

The $\delta^{34}S$ values for authigenic pyrite and for other sulfides (pyrrhotite, sphalerite, galena, chalcopyrite) in the Ventersdorp Contact Reef span two narrow ranges from −0.6‰ to +1.8‰ and from −1.5‰ to +1.0‰, respectively, which is even narrower than that determined for the detrital pyrite (−1‰ to +5.6‰). As described previously, the sulfur isotope composition for detrital pyrite was suggested to be derived from a magmatic source (Hoefs et al., 1968; Chukhrov et al., 1970; Hattori et al., 1983; Förster, 1986; Drennan, 1997). Under reduced and neutral pH physicochemical conditions, the sulfur isotope compositions of the fluids were little fractionated ($\delta^{34}S$ values <±2‰), so that the $\delta^{34}S$ values of authigenic pyrite and pyrrhotite, sphalerite, galena, and chalcopyrite are representative of the oxygen-deficient atmosphere in the Archaean Witwatersrand Basin (England et al., 2002), i.e., most of the sulfur in solution was S^{2-} of the total sulfur isotope compositions of the fluids. It is suggested that the restricted $\delta^{34}S$ values for the authigenic pyrite, pyrrhotite, sphalerite, galena, and chalcopyrite reflect a rather homogeneous source of the sulfur, which may be produced from the fractionation of sulfur derived from the detrital pyrite. Detrital pyrite and other heavy minerals are largely derived from sources related to magmatic lithologies that may include hydrothermally altered granitoids (Hallbauer and Kable, 1982; Robb and Meyer, 1986; Hallbauer, 1986), extinct volcanic smokers (Hoefs et al., 1968), or shallow marine hydrothermal discharges (Hutchinson and Viljoen, 1988)

Based on the O and D isotopic results, the hydrothermal fluids responsible for the Ventersdorp Contact Reef alteration are dominantly derived from metamorphic event(s). In addition, the S isotopic results support the view that the metamorphic fluids were derived from within the basin. Authigenic sulfides are derived by reconstitution of detrital sulfides during fluid circulation at peak metamorphic conditions. Based on the close relationship between gold and sulfide mineralization, gold might be remobilized together with authigenic pyrite, chlorite, muscovite, and albite due to metamorphism and hydrothermal alteration but within limited distance. The overall results of this study of multiple stable isotopes support the view that the gold in the Ventersdorp Contact Reef and the Witwatersrand as a whole originated from a paleoplacer that in early Proterozoic times (2.5–2.0 Ga) was subjected to one or more thermal/hydrothermal overprints.

ACKNOWLEDGMENTS

BZ would like to thank the National Research Foundation of South Africa for funding his Ph.D. study at the University of the Witwatersrand; the management of the Western Deep Levels (Mponeng) mine, especially Mike McWha, for their support and for permission to publish this paper; and Michael Meyer, Rheinisch-Westfälische Technische Hochschule (RWTH) Aachen, for his supervision at an early stage and support. We would like to express our gratitude to Uwe Reimold and Roger Gibson, editors of this special issue, for their encouragement, and to Torsten Vennemann and an anonymous reviewer for their very thorough reviews of the manuscript.

REFERENCES CITED

Armstrong, R.A., Compston, W., Retief, E.A., Williams, L.S., and Welke, H.J., 1991, Zircon ion microprobe studies bearing on the age and evolution of the Witwatersrand Triad: Precambrian Research, v. 53, p. 243–266, doi: 10.1016/0301–9268(91)90074-K.

Bachinski, D.J., 1969, Bond strength and sulfur isotopic fractionation in coexisting sulfides: Economic Geology and the Bulletin of the Society of Economic Geologists, v. 64, p. 56–65.

Barton, J.M., Jr., Wenner, D.B., and Hallbauer, D.K., 1992, Oxygen isotopic study of the nature and provenance of large quartz and chert clasts in gold-bearing conglomerates of South Africa: Geology, v. 20, p. 1123–1126, doi: 10.1130/0091–7613(1992)020<1123:OISOTN>2.3.CO;2.

Beattie, M., and Barton, J.M., 1992, Alteration and occurrence of gold mineralization in the Roodepoort Goldfield, Pietersburg granite-greenstone terrane: South African Journal of Geology, v. 95, p. 131–140.

Berlenbach, J.W., and Roering, C., 1992, Sheath-fold–like structures in pseudotachylites: Journal of Structural Geology, v. 14, p. 847–856, doi: 10.1016/0191–8141(92)90045-X.

Boer, R.H., Reimold, W.U., and Kesler, S.E., 1995, Conditions of gold remobilization in the Ventersdorp Contact Reef, Witwatersrand Basin, *in* Reimold W.U., ed., The economic significance of metamorphism and fluid movement within the Witwatersrand Basin: Johannesburg, Economic Geology Research Unit, University of the Witwatersrand, Information Circular, v. 296, p. 3–8.

Borthwick, J., and Harmon, R.S., 1982, A note regarding ClF_3 as an alternative to BrF_5 for oxygen isotope analysis: Geochimica et Cosmochimica Acta, v. 46, p. 1665–1668, doi: 10.1016/0016–7037(82)90321–0.

Chacko, T., Cole, D.R., and Harita, J., 2001, Equilibrium oxygen, hydrogen, and carbon isotope fractionation factors applicable to geologic systems, *in* Valley, J.W., and Cole, D., eds., Stable Isotope Geochemistry: Review in Mineralogy and Geochemistry, v. 43, p. 41–81.

Chapman, G.S.G., and Briggs, G.P., 1989, Geological notes on the gold bearing reefs of West Rand Consolidated Mines Ltd.: Geological Society of South Africa, Other Reefs Symposium, Abstracts, p. 29–35.

Chukhrov, F.V., Vinogradov, V.I., and Ermilova, L.P., 1970, On the isotopic sulfur composition of some Precambrian strata: Mineralium Deposita, v. 5, p. 209–222, doi: 10.1007/BF00201988.

Coplen, T.B., Kendall, C., and Hopple, J., 1983, Comparison of stable isotope reference samples: Nature, v. 302, p. 236–238.

Drennan, G.R., 1997, Fluid inclusion microthermometry of the Witwatersrand Basin [Ph.D. thesis]: Johannesburg, University of the Witwatersrand, 238 p.

Duane, M.J., Pigozzi, G., and Harris, C., 1997, Geochemistry of some deep gold mine waters from the Western portion of the Witwatersrand Basin, South Africa: Journal of African Earth Sciences, v. 24, p. 105–123, doi: 10.1016/S0899–5362(97)00030–4.

Els, B.G., 1989, Sedimentary characteristics of a bedrock placer: The VCR at Stilfontein Gold Mine: Geological Society of South Africa, Other Reefs Symposium, Abstracts, p. 87–91.

Engelbrecht, C.J., Baumbach, G.W.S., and Fletcher, P., 1986, The West Wits Line, *in* Anhaeusser, C.R., and Maske, S., eds., Mineral deposits of southern Africa: Geological Society of South Africa, v. 1, p. 599–648.

England, G.L., Rasmussen, B., Krape , B., and Groves, D.I., 2002, Palaeoenvironmental significance of rounded pyrite in siliciclastic sequences of the late Archaean Witwatersrand Basin: Oxygen-deficient atmosphere or hydrothermal evolution: Sedimentology, v. 49, p. 1133–1156, doi: 10.1046/j.1365-3091.2002.00479.x.

Epstein, S., 1970, Antarctic ice sheet: Stable isotope analysis of Byrd Station coves and interhemispheric climatic implications: Science, v. 168, p. 570–572.

Epstein, S., Sharp, R.P., and Gow, A.J., 1965, Six-year record of hydrogen and oxygen isotope variations in South Pole firn. Journal of Geophysical Research, v. 70, p. 1809–1814.

Faure, G., 1986, Principles of Isotope Geology: New York, John Wiley and Sons, 589 p.

Förster, O., 1986, Geochemische und isotopengeochemische Untersuchungen an Kohle- und Gold-führenden Konglomeraten des Proterozoikums Central Rand Group, Südafrika [Ph.D. thesis]: München, Germany, Technische Universität München, 226 p.

Frimmel, H.E., 1994, Metamorphism of Witwatersrand gold: Exploration Mining Geology, v. 3, p. 357–370.

Frimmel, H.E., 1997, Chlorite thermometry in the Witwatersrand Basin: Constraints on the Paleoproterozoic geotherm in the Kaapvaal Craton, South Africa: The Journal of Geology, v. 105, p. 601–615.

Frimmel, H.E., and Gartz, V.H., 1997, Witwatersrand gold particle chemistry matches of model of metamorphosed, hydrothermally altered place deposits: Mineralium Deposita, v. 32, p. 523–530, doi: 10.1007/s001260050119.

Frimmel, H.E., Le Roex, A.P., Knight, J., and Minter, W.E.L., 1993, A case study of the postdepositional alteration of the Witwatersrand Basal Reef gold placer: Economic Geology and the Bulletin of the Society of Economic Geologists, v. 88, p. 249–265.

Germs, G.J.B., and Schweitzer, J.K., 1994, A provisional model for the regional morphostratigraphy of the Ventersdorp Conglomerate Formation in the West Rand and Carletonville goldfields: South African Journal of Geology, v. 97, p. 279–287.

Gibson, R.L., and Stevens, G., 1997, Regional metamorphism due to anorogenic intracratonic magmatism: Johannesburg, Economic Geology Research Unit, University of the Witwatersrand, Information Circular, v. 311, 23 p.

Hall, R.C.B., 1994, Supporting evidence for the placement of the inter-reef lavas and associated sediments within the Venterspost Conglomerate Formation, Kloof Gold Mine: South Africa Journal of Geology, v. 97, p. 297–307.

Hall, R.C.B., and Els, B.G., 2002, The origin and significance of load-induced deformation structures in soft-sediment and lava at the base of the Archaean Ventersdorp Supergroup, South Africa: Journal of African Earth Sciences, v. 35, p. 135–145, doi: 10.1016/S0899–5362(01)00063-X.

Hallbauer, D.K., 1986, The mineralogy and geochemistry of Witwatersrand pyrite, gold, uranium and carbonaceous matter, *in* Anhaeusser, C.R., and Maske, S., eds., Mineral Deposits of Southern Africa: Geological Society of South Africa, v. 1, p. 731–752.

Hallbauer, D.K., and Kable, E.J.D., 1982, Fluid inclusions and trace element content of quartz and pyrite pebbles from Witwatersrand conglomerates: Their significance with respect to the genesis of primary deposits, *in* Amstutz, G.C., ed., Ore—Genesis—The state of the Art: Berlin, West Germany, Springer-Verlag, p. 742–752.

Harris, C., 1989, Oxygen-isotope zonation of agates from Karoo volcanics of the Skeleton Coast, Namibia: The American Mineralogist, v. 74, p. 476–481.

Harris, C., and Erlank, A.J., 1992, The production of large-volume, low-delta ^{18}O rhyolites during the rifting of Africa and Antarctica: The Lebombo Monocline, Southern Africa: Geochimica et Cosmochimica Acta, v. 56, p. 3561–3570, doi: 10.1016/0016–7037(92)90399–4.

Harris, C., and Watkins, T., 1990, Fluid interaction in the Witwatersrand goldfields: Oxygen isotope geochemistry of Ventersdorp age dolerite intrusions: South African Journal of Geology, v. 93, p. 611–615.

Hattori, K., Campbell, F.A., and Krouse, H.R., 1983, Sulfur isotope abundance in Aphebian clastic rocks: Implications for the coeval atmosphere: Nature, v. 302, p. 323–326, doi: 10.1038/302323a0.

Henckel, J., and Schweitzer, J.K., 1994, Chemical and mineralogical characteristics of a portion of the Ventersdorp Contact Reef at Elandsrand Gold Mine: South African Journal of Geology, v. 97, p. 332–338.

Henning, L.T., Els, B.G., and Mayer, J.J., 1994, The Ventersdorp Contact Placer at Western Deep Levels South Gold Mine—An ancient terraced fluvial system: South African Journal of Geology, v. 97, p. 308–319.

Hoefs, J., Nielsen, H., and Schidlowski, M., 1968, Sulfur isotope abundances in pyrite from the Witwatersrand conglomerates: Economic Geology and the Bulletin of the Society of Economic Geologists, v. 63, p. 975–977.

Hutchinson, R.W., and Viljoen, R.P., 1988, Re-evaluation of gold source in Witwatersrand ores: South African Journal of Geology, v. 91, p. 157–173.

Jolley, S.J., Freeman, S.R., Barnicoat, A.C., Phillips, G.M., Knipe, R.J., Pather, A., Fox, N.P.C., Strydom, D., Birch, M.T.G., Henderson, I.H.C., and Rowland, T.W., 2004, Structural controls on Witwatersrand gold mineralisation: Journal of Structural Geology, v. 26, p. 1067–1086, doi: 10.1016/j.jsg.2003.11.011.

Kamo, S.L., Reimold, W.U., Krogh, T.E., and Colliston, W.P., 1996, A 2.023 Ga age for the Vredefort event and a first report of shock metamorphosed zircon in pseudotachylite breccias and granophyre: Earth and Planetary Science Letters, v. 144, p. 369–387, doi: 10.1016/S0012–821X(96)00180-X.

Klemd, R., and Hallbauer, D.K., 1987, Hydrothermally altered peraluminous Archaean granites as a provenance model for Witwatersrand sediments: Mineral Deposita, v. 22, p. 227–235.

Law, J.D.M., 1990, Mineralogy, geochemistry and alteration of Witwatersrand meta-sediments in the Welkom area: Implications for Mineralization [M.S. thesis]: Johannesburg, University of the Witwatersrand, 198 p.

Layer, P.A., Kröner, A., McWilliams, M., and Clauer, N., 1988, Regional magnetic overprinting of Witwatersrand Supergroup sediments, South Africa: Journal of Geophysical Research, v. 93, p. 2191–2200.

McCarthy, T.S., 1994, The tectono-sedimentary evolution of the Witwatersrand Basin with special reference to its influence on the occurrence and character of the Ventersdorp Contact Reef—A review: South African Journal of Geology, v. 97, p. 247–259.

McCrea, J.M., 1950, On the isotopic chemistry of carbonates and a paleotemperature scale: The Journal of Chemical Physics, v. 18, p. 849, doi: 10.1063/1.1747785.

McWha, M., 1994, The influence of landscape on the Ventersdorp Contact Reef at Western Deep Levels South Mine: South African Journal of Geology, v. 97, p. 319–331.

Meyer, F.M., Robb, L.J., Oberthür, T., Saager, R., and Stupp, H.D., 1990, Cobalt, nickel, and gold in pyrite from primary gold deposits and Witwatersrand reefs: South African Journal of Geology, v. 93, p. 70–82.

Minter, W.E.L., 1970, Gold distribution related to the sedimentology of a Precambrian Witwatersrand conglomerate, South Africa, as outlined by using moving average analysis: Economic Geology and the Bulletin of the Society of Economic Geologists, v. 65, p. 963–969.

Mullins, M., 1993, The geology of the Ventersdorp Contact Reef at First Wesgold Mining, *in* Abstracts VCR Revisited Symposium: Geological Society of South Africa, p. 7–11.

Myers, R.E., McCarthy, T.S., and Stanistreet, I.G., 1990, A tectono-sedimentary reconstruction of the development and evolution of the Witwatersrand Basin, with particular emphasis on the Central Rand Group: South African Journal of Geology, v. 93, p. 180–201.

Naylor, H., Turner, P., Vaughan, D.J., Boyce, A.J., and Fallick, A.E., 1989, Genetic studies of red bed mineralization in the Triassic of the Cheshire basin, northwest England: Journal of Geological Society of London, v. 146, p. 685–699.

Ohmoto, H., 1986, Isotope geochemistry of ore deposits, *in* Vally, J.W., Taylor, H.P., Jr., and O'Neil, J.R., eds., Stable isotopes in high temperature geological processes: Review in Mineralogy, v. 16, p. 491–559.

Ohmoto, H., and Rye, R.O., 1979, Isotopes of sulfur and carbon, *in* Barnes, H.L., ed., Geochemistry of Hydrothermal Ore Deposits, 2nd edition: New York, John Wiley and Sons, p. 509–567.

Phillips, G.N., and Law, J.D.M., 1997, Hydrothermal origin for Witwatersrand gold: Society of Economic Geology (SEG) Newsletter, v. 31, p. 26–32.

Phillips, G.N., Myers, R.E., Law, J.D.M., Baily, A.C., Cadle, A.B., Beneke, S.D., and Guisti, L., 1989, The Witwatersrand gold fields: Part I, post-depositional history, synsediment processes, and gold distribution: Monograph of Economic Geology, v. 6, p. 585–597.

Regional VCR Working Group, 1994, Some regional characteristics of the Ventersdorp Contact Reef (VCR): South African Journal of Geology, v. 97, p. 385–386.

Reimold, W.U., Koeberl, C., Fletcher, P., Killick, A.M., and Wilson, J.D., 1999, Pseudotachylitic breccias from fault zones in the Witwatersrand Basin, South Africa: Evidence of autometasomatism and post-brecciation alteration processes: Mineralogy and Petrology, v. 66, p. 25–53, doi: 10.1007/BF01161721.

Robb, L.J., and Meyer, F.M., 1986, The nature of the Archaean basement in the hinterland of the Witwatersrand Basin: I, The Rand anticline between Randfontein and Rysmierbult: South African Journal of Geology, v. 90, p. 44–63.

Robb, L.J., and Meyer, F.M., 1991, A contribution to recent debate concerning epigenetic versus syngenetic mineralization processes in the Witwatersrand Basin: Economic Geology and the Bulletin of the Society of Economic Geologists, v. 86, p. 396–401.

Robb, L.J., and Meyer, F.M., 1995, The Witwatersrand Basin, South Africa: Geological framework and mineralization processes: Ore Geology Reviews, v. 10, p. 67–94, doi: 10.1016/0169–1368(95)00011–9.

Robb, L.J., and Robb, V.M., 1998, Gold in the Witwatersrand Basin, *in* Wilson M.G.C., and Anhaeusser, C.R., eds., The Mineral Resources of South Africa, Handbook 16: South Africa, Pretoria, Council for Geoscience, p. 294–349.

Robb, L.J., Davis, D.W., and Kamo, S.L., 1991, Chronological framework for the Witwatersrand Basin and environs: Towards a time-constrained depositional model: South African Journal of Geology, v. 94, p. 86–95.

Robb, L.J., Landais, P., Meyer, F.M., and Davis, D.W., 1994, Nodular organic matter in granites: Implications for the origin of 'kerogen' in the Witwatersrand Basin, South Africa: Exploration and Mining Geology, v. 3, p. 219–230.

Robb, L.J., Charlesworth, E.G., Drennan, G.R., Gibson, R.L., and Tongu, E.L., 1997, Tectono-metamorphic setting and paragenetic sequence of Au-U mineralisation in the Archaean Witwatersrand Basin, South Africa: Australian Journal of Earth Sciences, v. 44, p. 353–371.

Rollinson, H., 1993, Using Geochemical Data: Evaluation, Presentation, Interpretation: Essex, Longman, 352 p.

Sakai, H., 1957, Fractionation of sulfur isotopes in nature: Geochimica et Cosmochimica Acta, v. 12, p. 150–169.

Sakai, H., 1968, Isotopic properties of sulfur compounds in hydrothermal processes: Chemistry, v. 199, p. 401–413.

Savin, S.M., and Epstein, S., 1970, The oxygen isotopic composition of coarse grained sedimentary rocks and minerals: Geochimica et Cosmochimica Acta, v. 34, p. 323–329, doi: 10.1016/0016–7037(70)90109–2.

Schweitzer, J.K., Van Niekerk, A.W., Ashworth, S.G.E., and Henckel, J., 1994, Interrelationships between geomorphology, stratigraphy, and gold grade in the Ventersdorp Contact Reef, eastern portion of Elandsrand Gold Mine: South African Journal of Geology, v. 97, p. 332–338.

Sheppard, S.F.M., and Gilg, H.A., 1996, Stable isotope geochemistry of clay minerals: Clay Minerals, v. 31, p. 1–24.

Smits, G., 1994, Note on the mineralogy of the Ventersdorp Contact Reef at Kloof Gold Mine: South African Journal of Geology, v. 97, p. 357–364.

South African Committee for Stratigraphy (SACS), 1980, Stratigraphy of South Africa, Part 1 (Compiled by L.E. Kent), Lithostratigraphy of the Republic of South Africa, South West Africa, Namibia, and the Republic of Bophuthaswana, Transkei and Venda: Handbook, Pretoria, Geological Survey of South Africa, 8, 690 p.

Spray, J.G., Kelley, S.P., and Reimold, W.U., 1995, Laser-probe ^{40}Ar-^{39}Ar dating of pseudotachylites and the age of the Vredefort impact event: Meteoritics, v. 30, p. 335–343.

Stevens, G., Gibson, R.L., and Droop, G.T.R., 1996, Polyphase granulite metamorphism in the Vredefort Dome: A window into the deep Kaapvaal Craton at 2.06 Ga: Johannesburg, Economic Geological Research Unit, University of the Witwatersrand, Information Circular, v. 297, 27 p.

Taylor, H.P., Jr., 1974, The application of oxygen and hydrogen isotope studies to problems of hydrothermal alteration and ore deposition: Economic Geology and the Bulletin of the Society of Economic Geologists, v. 69, p. 843–883.

Thomas, D.E., 1979, A sedimentologically based valuation of the gold and uranium mineralization in the VCR at Orkney in the Klerksdorp Goldfield: Geological Society of South Africa, Johannesburg, Geocongress 1979, Extended Abstracts, p. 351–358.

Trieloff, M., Reimold, W.U., Kunz, J., Boer, R.H., and Jessberger, E.K., 1994, ^{40}Ar-^{39}Ar thermochronology of pseudotachylite at the Ventersdorp Contact Reef, Witwatersrand Basin: South African Journal of Geology, v. 97, p. 365–384.

Vennemann, T.W., and O'Neil, J.R., 1993, A simple and inexpensive method of hydrogen isotope and water analyses of minerals and rocks based on zinc reagent: Chemical Geology, v. 103, p. 227–234, doi: 10.1016/0009–2541(93)90303-Z.

Vennemann, T.W., and Smith, H.S., 1990, The rate and temperature of reaction of ClF_3 with silicate minerals, and their relevance to oxygen isotope analysis: Chemical Geology, v. 86, p. 83–88.

Vennemann, T.W., Kesler, S.E., and O'Neil, J.R., 1992, Stable isotope compositions of quartz and their fluid inclusions as tracers of sediment provenance: Implications for gold- and uranium-bearing quartz-pebble conglomerates: Geology, v. 20, p. 837–840, doi: 10.1130/0091–7613(1992)020<0837:SICOQP>2.3.CO;2.

Vennemann, T.W., Kesler, S.E., Frederickson, G.C., Minter, W.E.L., and Heine, R.R., 1996, Oxygen isotope sedimentology of gold- and uranium-bearing Witwatersrand and Huronian Supergroup quartz-pebble conglomerates: Economic Geology and the Bulletin of the Society of Economic Geologists, v. 91, p. 322–342.

Vermaakt, D.T., and Chunnett, I.E., 1994, Tectono-sedimentary processes which controlled the deposition of the Ventersdorp Contact Reef within the West Wits Line: Johannesburg, South African Institute of Mining and Metallurgy (SAIMM), XVth CMMI Congress, v. 3, p. 117–130.

Viljoen, M.J., and Reimold, W.U., 1994, The Ventersdorp Contact reef revisited—An introduction: South African Journal of Geology, v. 97, p. 234–237.

Viljoen, R.P., 1963, Petrographic and mineragraphic aspects of the Main Reef and Main Reef Leader of the Main-Bird Series, Witwatersrand System [M.S. thesis]: Johannesburg, University of the Witwatersrand, 193 p.

Wallmach, T., and Meyer, F.M., 1990, A petrogenetic grid for metamorphosed aluminous Witwatersrand shales: South African Journal of Geology, v. 93, p. 93–102.

Walraven, F., 1997, Geochronology of the Rooiberg Group, Transvaal Supergroup, South Africa: Johannesburg, Economic Geological Research Unit, University of the Witwatersrand, Information Circular, v. 316, 21 p.

Winter, H. de la R., 1994, The Ventersdorp Contact Reef in relation to the Witwatersrand Basin and overlying Ventersdorp volcanic rocks: South African Journal of Geology, v. 97, p. 260–278.

Zhao, B., 1998, A mineralogical and geochemical study of alteration associated with the Ventersdorp Contact Reef in the Witwatersrand Basin, South Africa [Ph.D. thesis]: Johannesburg, University of the Witwatersrand, 298 p.

Zhao, B., Meyer, F.M., Robb, L.J., and McWha, M., 1994, A preliminary study of alteration associated with the Ventersdorp Contact Reef, at Western Deep Levels South Mine, Witwatersrand Basin, South Africa: South African Journal of Geology, v. 97, p. 348–356.

Zhao, B., Robb, L.J., Zwingmann, H., Toulkeridis, T., and Meyer, F.M., 1999, K-Ar dating of micas from the Ventersdorp Contact Reef of the Witwatersrand Basin, South Africa: Timing of post-depositional alteration: Mineralogy and Petrology, v. 66, p. 149–170, doi: 10.1007/BF01161726.

Manuscript Accepted by the Society 4 October 2005

Printed in the USA

Geological Society of America
Special Paper 405
2006

The nature of hydrocarbons and related fluids in the Witwatersrand Basin, South Africa: Their role in metal redistribution

G.R. Drennan[†]
School of Geosciences, University of the Witwatersrand, P.O. Wits 2050, Johannesburg, South Africa
L.J. Robb
Economic Geology Research Institute, School of Geosciences, University of the Witwatersrand, P.O. Wits 2050, Johannesburg, South Africa

ABSTRACT

Gold and uranium mineralization within the Archean Witwatersrand Basin exhibits a close association with carbonaceous matter. In order to understand the gold and uranium mineralization associations, it is necessary to understand something about the carbonaceous matter itself as well as the fluids involved in remobilizing both hydrocarbons and mineralization within the basin.

Diagenetic maturation of primitive bacterial material released hydrocarbons that were able to migrate through the sediments of the economically important Witwatersrand Basin. Migrating hydrocarbons underwent polymerization and condensation due to the effect of ionizing radiation in the proximity of detrital uraninite during catagenesis. Isotopically heavy hydrocarbons precipitated out as bitumen seams where uraninite concentrations were high or as "fly-speck" bitumen around isolated uraninite grains, giving rise to sediment-hosted bitumen types. Lighter hydrocarbons, liberated during catagenic radiolysis, were incorporated into circulating basinal fluids. During late-catagenic/early-metagenic processes, hydrocarbon-bearing fluids migrating along aquifers (faults, bedding planes, and unconformities) precipitated insoluble carbonaceous matter, or distal/vein-related bitumen, in suitable sites. Light hydrocarbons precipitated as crosscutting veins where the vein intersected uraniferous seam bitumen, as bitumen nodules within late-stage quartz veins and as nodules and/or inclusion linings within fluid inclusions in late-stage quartz veins. Where fault zones permeated the Archean granitic basement, hydrocarbons were precipitated by radiolytic processes around high-U phases as nodular, granitoid hosted, or distal bitumen. Distal- and vein-related bitumen exhibits a higher degree of structural ordering than sediment-hosted bitumen as a result of elevated pressure-temperature conditions experienced during metagenic and fault-related processes.

Liberated light hydrocarbons in basinal fluids formed organo-urano-gold-, thio-gold-, and/or organo-urano-gold-sulphide complexes. Circulation of these hydrocarbon-rich fluids and subsequent precipitation of bitumen and mineralization provide an explanation for the apparent hydrothermal component of Witwatersrand mineralization.

[†]E-mail: drennang@geosciences.wits.ac.za.

Drennan, G.R., and Robb, L.J., 2006, The nature of hydrocarbons and related fluids in the Witwatersrand Basin, South Africa: Their role in metal redistribution, *in* Reimold, W.U., and Gibson, R.L., Processes on the early Earth: Geological Society of America Special Paper 405, p. 353–385, doi: 10.1130/2006.2405(18).

Keywords: bitumen, catagenesis, metagenesis, radiolytic polymerization, organo-urano/thio-/sulphide-gold complex.

INTRODUCTION AND BACKGROUND

Small in size but prolific in output, the late Archean Witwatersrand Basin has dominated world gold production for over a century. The prodigious gold and uranium endowment within this predominantly clastic sedimentary sequence has provoked enormous interest with respect to the nature and origin of its mineralization. Arguments ranging from placer to hydrothermal, with all shades in between, have continued for almost as long as mining has taken place. The debate has more than just economic relevance though, and several fundamental issues related to the early development of Earth's crust and atmosphere could be addressed if the metallogeny of the Witwatersrand Basin were better understood. The presence of carbonaceous matter in the gold- and uranium-bearing conglomerates of the Witwatersrand Basin is one of the more intriguing and controversial aspects of the genesis of these deposits. The occurrence of carbonaceous matter is particularly relevant to mining operations, as it is usually an indication of high grades of both gold and uranium, with values of up to 4.7 wt% gold and 11.5 wt% uranium reported from analyses of separated carbonaceous matter (Landais et al., 1990a, 1990b; Robb et al., 1997a, 1997b). Consequently, an understanding of its origins, distribution, and relationship to gold and uranium is pertinent both to the economic evaluation of individual reef horizons and to the understanding of the mineralization process. In addition, the presence of ca. 3 Ga carbonaceous matter has relevance to more fundamental issues such as the nature of organic material at this time and its subsequent transformation into primitive petroleum-type products. Unfortunately, widely differing opinions exist in the literature as to the origin of the carbonaceous material in the Witwatersrand Basin, and few attempts have, therefore, been made to utilize its occurrence, either in a predictive capacity during exploration and mining or to better understand the nature and occurrence of primitive life-forms.

Witwatersrand carbonaceous matter was originally referred to as "thucholite," an outdated term used to describe what was thought to be an authigenic Th-U-C-H-O–bearing mineral (Ellsworth, 1928). The first detailed description of Witwatersrand carbonaceous matter was made by Young (1917), who recognized three different occurrences: nodular granules; massive, compact veins; and columnar or fibrous veins. He suggested that the carbonaceous matter was unlikely to represent the in situ residue of animal or vegetable matter but a deposit from a liquid that had pervaded the sediments some time after their deposition.

The detailed ore-microscopic studies of Ramdohr (1958) and Liebenberg (1955) favored a detrital origin for uraninite grains in the conglomerates and recognized the paragenetically late aspect of carbonaceous matter, which was observed to progressively surround, transect, and, in some cases, eventually replace preexisting uraninite. It was also noted that carbonaceous matter and uraninite almost invariably occurred together and that this intimate relationship did not apply to any other mineral in the paragenesis. Early stable isotope analyses of carbonaceous matter indicated depleted $\delta^{13}C$ values typical of sedimentary organic carbon, suggesting that the material had a biogenic origin (Hoefs and Schidlowski, 1967). Support for the presence of primitive life-forms in the Witwatersrand Basin also came from the observation of circular and annular microstructures of bacterial or algal morphology preserved, ironically, not in carbonaceous matter but in detrital pyrite grains (Schidlowski, 1965). Coal petrographers likewise recognized the presence of circular and filamentous shapes within the carbonaceous matter itself and suggested that the Witwatersrand hydrocarbons contained remnants of algal and fungal organisms similar to those seen in sapropelic coals (Snyman, 1965). The in situ biogenic origin of Witwatersrand material gained popularity when Hallbauer and van Warmelo (1974) and Hallbauer (1975) published high-resolution photographs supposedly representing well-differentiated organic remains that included bacteria, algae, fungi, and lichens. This hypothesis was economically relevant, as it was suggested that the presence of organic mats in low-energy sedimentary environments could have physically trapped detrital gold and uraninite particles.

Schidlowski (1981) persistently rejected the in situ biogenic hypotheses and regarded the Witwatersrand carbonaceous matter as the solidified product of mobile hydrocarbons that had undergone radiolytic polymerization in the vicinity of detrital uraninite. He argued that in situ preservation of delicate fossil life-forms in compacted and metamorphosed sediments was unlikely and also pointed to the biological problems of having relatively advanced life-forms in existence in rocks as old as ca. 3 Ga. The hypothesis of a radiolytically polymerized origin for the Witwatersrand hydrocarbons was also supported by Zumberge et al. (1981), who suggested that the aromaticity and high concentration of free radicals was consistent with a hydrocarbon structure that had been subjected to intense radiation but that the kerogen seams contained, at the same time, coccoid and filamentous microstructures consistent with an originally biogenic (algal or bacterial) origin. Ebert et al. (1990) utilized a variety of analytical techniques (NMR [Nuclear Magnetic Resonance] and ESR [Electron Spin Resonance]) to conclude that the carbonaceous matter had not been subjected to temperatures as high as 300 °C but that palynological observations suggested that biogenic shapes, although indeterminate, were present.

Frimmel et al. (1999) analyzed crush-leach data (in association with oxygen isotope analyses of quartz and in combination with microthermometric data) for individual generations of aqueous fluid inclusions, in hydrothermal quartz from different auriferous conglomerates in the Witwatersrand Basin, in order to constrain the chemical composition, pH, temperature of formation, and the possible source of the mineralizing fluid that mobilized some of the detrital gold. Their results, particularly a lack of bromine in the fluid (indicative of seawater), the presence of organic acids, and $\delta^{18}O_{fluid}$ values around 0‰, led them to favor a largely meteoric source for the gold-mobilizing fluids. The fluids were ascribed to hydrothermal infiltration triggered by the 2020 Ma Vredefort impact (Gibson and Reimold, 1999, for a review), which created a secondary permeability in the form of a dense network of microfractures preferentially in the conglomeratic horizons of the already metamorphosed Witwatersrand sedimentary sequence (see also Reimold et al., 2005, for more on the effects of the Vredefort impact event). Frimmel et al. (1999) concluded that this fluid was different from the regional metamorphic fluid, because it was substantially more alkaline (pH 5.7–7.2). Due to the basic nature of their fluids, these authors assumed that Au was transported most likely as an $Au(HS)_2^-$ complex and that precipitation of mobilized Au was related to the precipitation of sulphides. Frimmel et al. (1999) also reported that carbonic species in fluid inclusions (Drennan et al., 1997, 1999) were related to bitumen inclusions and nodules of their study and that the bitumen was a mixture of local and extrinsic oils, with the extrinsic component being derived not from outside the basin but from within the Witwatersrand Basin. The carboxylic acids found in the fluids studied could have originated from sedimentary kerogen, in particular through hydrous pyrolysis of such kerogen during diagenesis and metamorphism (Frimmel et al., 1999). A meteoric source for the Au-mobilizing fluid precludes any magmatic derivation of the Au (Phillips et al., 1997) and associated sulphides as well as a metamorphic origin of the Au-transporting fluid (Phillips and Myers, 1988).

Zartman and Frimmel (1999) investigated Rn-generated ^{206}Pb in hydrothermal sulphide minerals and bitumen from quartz vein material crosscutting the Ventersdorp Contact Reef. They concluded that the radiogenic, mainly uranogenic, component of the Pb appeared to have been derived almost entirely from uraninite in the surrounding reef rocks. The paragenetically late chalcopyrite and pyrrhotite that was analyzed plotted on a $^{207}Pb/^{204}Pb$ versus $^{206}Pb/^{204}Pb$ diagram with nearly constant $^{207}Pb/^{204}Pb$ of 24.1–24.6, despite a huge range in $^{206}Pb/^{204}Pb$ from 60–230. A similar trend was evident, but with even higher $^{206}Pb/^{204}Pb$ ratios (up to 949) for bitumen nodules deposited on quartz crystals lining cavities in the veins (Zartman and Frimmel, 1999). Zartman and Frimmel (1999) argued that these results required the addition of virtually pure ^{206}Pb to a more normal, radiogenically enriched Pb and that this demanded the decoupling of the two U decay schemes with the acquisition of essentially pure ^{206}Pb by selectively involving one of the ^{238}U intermediate daughter isotopes. They attributed this decoupling to the leakage of ^{222}Rn from uraninite, which was then transported by aqueous fluids into the quartz veins and adsorbed onto the bitumen and late-stage sulphides. The disturbance of U-Pb isotope systems in this environment was considered to raise serious doubts as to the validity of age determinations derived from this material, in particular to those pointing to an age of ca. 2300 Ma for the carbonaceous material (Robb et al., 1994). Furthermore, Zartman and Frimmel (1999), similar to others (e.g., Burger et al., 1962; Barton and Hallbauer, 1996; Poujol et al., 1998), attributed the large variation in Pb isotopic composition of bitumen and sulphides to derivation from local sources, probably from detrital uraninite and sulphides within a given reef.

England et al. (2001) documented three forms of bitumen nodules within the Witwatersrand sediments: (1) solitary composite grains randomly distributed throughout sandstones and conglomerates ("fly-speck" carbon or random nodules), (2) solitary composite grains strung out along stratification planes (stratified nodules), and (3) groups of composite grains coalesced into stratiform aggregates (carbon seams). They observed solitary bituminous nodules in both mineralized and unmineralized units throughout the Witwatersrand Supergroup in all goldfields, indicating that no unique association of carbonaceous matter and Au-U reefs existed. England et al. (2001) confirmed that nearly all bituminous material from Witwatersrand ore deposits contained U-bearing minerals (commonly uraninite but also brannerite and galena). They identified bitumen envelopes (5–100 μm thick) isolating enclosed grains of uraninite from the surrounding mineral framework and observed that the uraninite grains were typically fragmented. Fragmentation was related to infiltration of the bitumen precursor along zones of crystallographic weakness, thereby splitting and fracturing grains. Carbon seams exhibited an internal columnar through to fibrous morphology perpendicular to bedding (England et al., 2001). They established that this texture was a postdepositional fabric, indicating that the bitumen precursor was plastic early in its development and that nodules underwent transformation from spherical to ellipsoidal structures (producing columnar or fibrous aggregates) under the influence of vertical extension (Gray et al., 1998; Parnell, 1999). Elongation of bitumen nodules was accompanied, in some seams, by development of an anastomosing network of bedding-parallel microfractures in surrounding framework minerals. Pore-space between elongated nodules was infilled by pyrophyllite, gersdorffite, florencite, brannerite, xenotime, uraninite, or gold (England et al., 2001). These observations led England et al. (2001) to conclude that Witwatersrand bitumen represented former sites of heavy mineral accumulations of placer uraninite and that bitumen nodules formed during burial, while oil migration, through primary porosity, was prevalent. This implied that carbon seams were a focus of deformation and not a consequence of it and that uraninite and polymerized hydrocarbons were present prior to structural deformation (England et al., 2001). Continual polymerization of bitumen within carbon seams resulted in the generation of a nonstructural, intracolumn secondary porosity that was subsequently infilled by secondary minerals, including gold.

Macro- and microscopic observations by Gray et al. (1998) are in agreement with those of Ramdohr (1958), Schidlowski (1981), England et al. (2001), and Spangenberg and Frimmel (1999, 2001), although they specifically described the seam carbon as a "carbonaceous mesophase" derived from a pyrobitumen representing the residue of migrated liquid hydrocarbon. Gray et al. (1998) noted that the optical anisotropy within seam "spindles" (carbon globules elongated perpendicular to bedding) was bedding perpendicular and argued that mesophase formation must have occurred during fracture opening related to a period of lateral compression. They argued that the pyrobitumen is paragenetically late and that the introduction of hydrocarbon occurred after lithification with the development of bedding-parallel fractures. They attributed observations of both carbon properties and sequencing to a pulsed hydrothermal pressure model where, as pressure rises toward lithostatic, the pressure potential gradient is from coarse- to fine-grained lithologies; the gradient is reversed once lithostatic pressure is approached and excess fluids bleed off through the anastomosing fracture network developed along the reef surface (Gray et al., 1998). They argued that this reversed pressure gradient promoted oil migration into the newly formed fracture network. Gray et al. (1998) differ from other authors (e.g., Ramdohr, 1958; Schidlowski, 1981; England et al., 2001; Spangenberg and Frimmel, 1999, 2001), however, in that they believed the uraninite was introduced at the same time as the fluid overpressuring occurred and that the uraninite precipitated on meeting the chemically reducing hydrocarbons (Barnicoat et al., 1997; Phillips and Law, 2000), which in turn stabilized the oil by polymerization. Further hotter hydrothermal flux then "coked" the bitumen to fibrous (spindle-shaped) mesophase and precipitated gold into available pore spaces created by anisotropy-controlled lateral contraction between and at the terminations of mesophase spindles (Gray et al., 1998).

Spangenberg and Frimmel (2001) used Rock-Eval pyrolysis (Spangenberg and Macko, 1998), $\delta^{13}C$ determinations of insoluble and soluble organic matter, and the chemical and isotopic composition of individual n-alkanes and acyclic isoprenoids (pristine and phytane) to investigate the origin and postdepositional history of the hydrocarbons in the Witwatersrand Basin. Their data led them to conclude that the source of hydrocarbons (Zartman and Frimmel, 1999) in the Witwatersrand reefs is best located within the Witwatersrand sediments, derived from the decomposition of biogenic organic matter buried in the clastic sediments and possibly in the shale units of the Witwatersrand Supergroup. They attributed the variability of the composition of the indigenous carbonaceous material to a variety of biomass and reducing conditions that must have existed in the Archean sedimentary environments. Their isotopic organic geochemical results did not support any model invoking the influx of large amounts of mineralizing hydrothermal fluids from a source outside of the Witwatersrand Basin (Spangenberg and Frimmel, 2001).

More recently, Frimmel (2005) argued that although most of the Witwatersrand gold appears as a precipitate within or associated with postdepositional hydrothermal phases (including carbonaceous phases) and along microfractures, all available microtextural, mineralogical, geochemical, and isotopic data indicate that this hydrothermal gold, as well as other hydrothermal phases, was derived from local mobilization of detrital particles.

Organic matter in a sedimentary basin must be studied in the context of its thermal and burial history. Phillips and Law (1994) reviewed the metamorphic history of the Witwatersrand goldfields and confirmed that a regional pressure-temperature (P-T) climax of 350 ± 50 °C and 1.5–3 kbar applied throughout the basin. Metamorphic grades are considered to have been fairly consistent basinwide, although P-T conditions increased inward toward the Vredefort Dome. Although the timing of metamorphism is difficult to constrain, reviews of isotopic age determinations in and around the basin suggest that metamorphism may have been episodic, occurring mainly in response to increments of progressive burial at ca. 2500 and 2300 Ma and then to a major thermo-tectonic perturbation at 2060–2020 Ma related to intrusion of the Bushveld Complex and the Vredefort catastrophism (Frimmel et al., 1999; Gibson and Reimold, 1999; Reimold et al., 2005; Robb and Meyer, 1995; Robb et al., 1997a, 1997b). It is during the latter episode that the basin is believed to have been subjected to its metamorphic climax.

Organic maturation in the Witwatersrand Basin was controlled by the successive increments of postdepositional burial and thermal evolution of the host rocks. The timing of organic maturation in the Witwatersrand Basin is one of the issues that remain controversial and uncertain. One view regards oil formation and migration as early, and related to, late diagenetic processes (Barnicoat et al., 1997; Phillips and Law, 2000), although this seems inconsistent with the late paragenetic appearance of these hydrocarbons. An opposing view has suggested that bitumen formation was very late, occurring at around 2300–2400 Ma (Robb et al., 1994) during deposition of the upper part of the Transvaal Supergroup (i.e., the Pretoria Group), although the U-Pb isotopic age data pointing to this link (Robb et al., 1994) now seem questionable (Zartman and Frimmel, 1999).

Witwatersrand strata contain kerogen and bitumen that resulted from the maturation of primitive organic matter. Progressive in situ maturation of primitive algal material through diagenesis and metagenesis would have produced kerogen and released light hydrocarbons (CO_2, CH_4, C_2H_6) as well as oil for migration through the basin. Evidence of organo-metallic complexation in the fluids associated with this event (Drennan, 1997; Drennan et al., 1995, 1997, 1999) is provided by high-resolution element mapping of bitumen nodules associated with pyrite and gold in fault-related quartz veins. Proton-induced X-ray emission (PIXE) micromaps show uniform distribution of metals such as Au, U, Fe, Cu, As, and Th throughout these nodules (Drennan, 1997). These metals appear to be present in elemental form (Au and U) and associated with minute sulphide phases (Fe, Cu, and As), suggesting that gold, uranium, and other metals were transported in the same fluids as the light hydrocarbons and may have been involved in the formation of a variety of chelate-type complexes.

This paper presents the results of new petrographic, chemical, and spectroscopic data for Witwatersrand carbonaceous matter that is now known to be texturally more variable than was previously thought and is contained in different host environments both within and external to the basin. These observations shed new light on the nature and origin of Witwatersrand carbonaceous matter and also contribute to an improved understanding of its relationship to gold and uranium, both within and outside the basin.

CHARACTERIZATION OF POSTDEPOSITIONAL FLUIDS IN THE WITWATERSRAND BASIN

The characteristics of postdepositional fluids affecting the Witwatersrand basin are important when considering the nature and origin of the gold and uranium mineralization, because the nature of fluids determines their metal-carrying capacity, which in turn determines the role played by these fluids in mobilizing and precipitating mineralization. An analysis of postdepositional fluids (trapped in late-stage quartz veins) in Witwatersrand sediments was undertaken by Drennan (1997) and was reported on by Drennan et al. (1995, 1997, 1999). In order to establish the relationship between these fluids, carbonaceous matter, and mineralization, it is appropriate to review some of this information.

In order to examine the nature of late-stage, postdepositional fluids affecting the Witwatersrand sediments and to assess the role played by these fluids in the origin, mobilization, and precipitation of Au and U mineralization, an extensive sampling program was undertaken of quartz veins from various stratigraphic levels throughout the basin (Fig. 1; Drennan, 1997; Drennan et al., 1999). At least three types of quartz veining were sampled: vein/"vug" infilling associated with intrusions (e.g., magmatic dikes); fault-related quartz veining; and veins exhibiting no clear relationship to major structures. The exact age relationships of these veins to the regional tectonic evolution of the basin could not always be determined (Drennan et al., 1999; Frimmel et al., 1999). The veins nevertheless

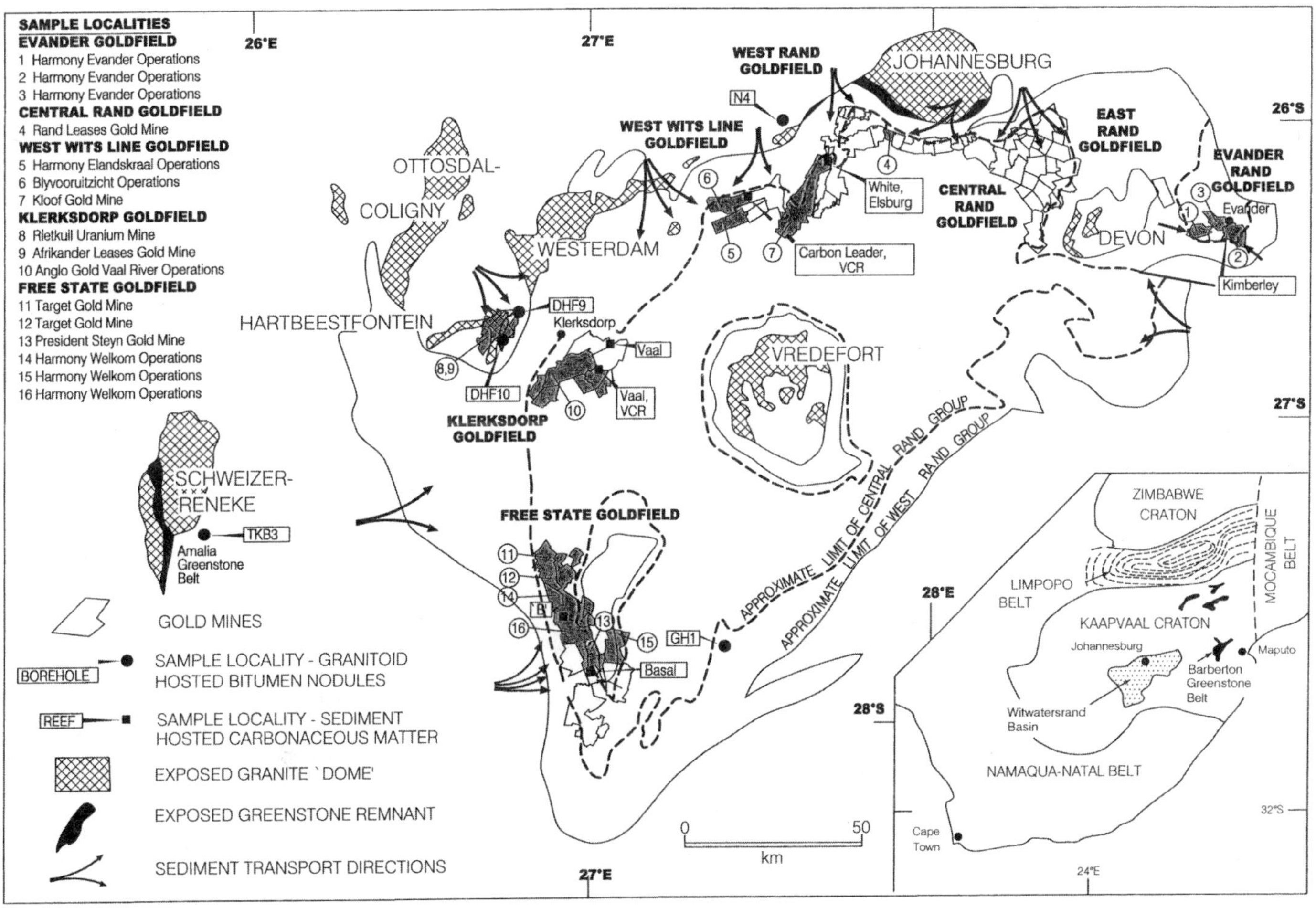

Figure 1. Simplified outline of the Witwatersrand Basin and the location of goldfields sampled for fluid inclusion and hydrocarbon studies discussed. Also given are the names of reefs sampled at these localities. VCR—Ventersdorp Contact Reef; B—B-reef; GHI, DHF9, DHF10—a borehole "name" with no specific meaning to anyone else but the mine who paid for the drilling.

record a protracted history of fluid flow, as it is evident from this study that individual veins record multiple trapping events and fluid populations. In certain cases, sampling was carried out in an environment where detailed structural analysis (Berlenbach, 1993) allowed a relative chronology of vein-forming events to be established. A summary of vein types, sample localities, and vein mineralogy is presented in Table 1.

Petrographic analysis of selected samples using transmitted and reflected light microscopy provided detailed mineralogic, textural, and recrystallization descriptions. This made it possible to identify various fluid systems and inclusion populations (summarized in Table 2) affecting the Witwatersrand sediments in postdepositional times and to suggest relative chronological constraints for inclusion populations (Drennan, 1997; Drennan et al., 1997, 1999). Paleofluids that were trapped in intracrystalline cavities within vein quartz material crosscutting the Witwatersrand sediments, associated with faulting as well as occurring with/within late-stage dyke intrusions, were exam-

TABLE 1. SAMPLE MATERIAL SUITABLE FOR MICROTHERMOMETRIC INVESTIGATIONS

Evander Goldfield	
Harmony Evander Operations (previously called Winkelhaak)	
WH3	Quartz-pyrite-chlorite vein in Hanging wall 1 quartzites
WH13	Quartz-pyrite-sphalerite vein in Inter Zone hanging wall
West Wits Line	
Kloof Gold Mine	
2729	NE thrust-related quartz vein in Ventersdorp Contact Reef
3485	Quartz infilling tension gash, movement out of basin
3753A	Shear zone in dyke; quartz fibers show NE movement
2742	Quartz-chlorite-hematite vein
Klerksdorp Goldfield	
Anglo Gold Vaal River Operations (previously called Vaal Reefs)	
VRW4/5	Quartz-carbonate vein in VCR
VRS1	Quartz below "C" Reef (possible bed-slip faulting)
VR8iii	Quartz-carbonate vein associated with intrusive
VR9	Terminated quartz crystals within vugs in intrusive
VR1A	Quartz vein below Vaal Placer
VR7A	Quartz vein crosscutting Mizpah Reef
Free State Goldfield	
President Steyn Gold Mine	
PS3	Quartz vein overlying sheared Basal Reef
PS1	Quartz vein-swarm in hanging wall waxies
Harmony Welkom Operations (previously called Free State Geduld)	
FSG6	Fault-related quartz-carbonate-pyrophyllite vein
FSG3A	Quartz-carbonate vein in hanging wall to Basal Reef: 2nd generation vein
FSG3A	Quartz-carbonate vein in hanging wall to Basal Reef: 1st generation vein
FSG3A	Quartzites adjacent to quartz-carbonate vein in HW to Basal Reef
Harmony Welkom Operations (previously called Western Holdings)	
WHD1	Quartz-carbonate-pyrite-chalcopyrite vein associated with bed-slip fault—Khaki Shales
WHD4	Quartz-pyrophyllite-pyrite vein below Basal Reef
WHD7	Quartz vein crosscutting mylonitic zone in Witpan Reef
WHD9	Quartz-galena vein in sill

Note: WH3—sample number 3 from Winkelhaak Gold Mine, now referred to as Harmony Evander Operations; 2729—sample number from Kloof Gold Mine; VRW4/5—sample numbers 4 and 5 from Vaal Reefs West Section, now referred to as Anglo Gold Vaal River Operations; VRS1—sample number 1 from Vaal Reefs South Section; VR8iii—sample number 8iii from Vaal Reefs main section; PS3—sample number 3 from President Steyn Gold Mine; FSG6—sample number 6 from Free State Geduld Gold Mine, now referred to as Harmony Welkom Operations; WHD1—sample 1 from Western Holdings Gold Mine, now referred to as Harmony Welkom Operations.

TABLE 2. SUMMARY OF FLUID INCLUSION TYPES, SYSTEMS, AND POPULATIONS WITHIN LATE-STAGE QUARTZ VEINS IN THE WITWATERSRAND BASIN

Fluid system	Inclusion population	Size (μ)	Occurrence	T_{fm} (°C)	T_h (°C)	wt% eNaCl
L_w Aqueous fluids	L_w1	<1	Earliest inclusions imparting a whispy texture, unsuitable for microthermometry			
	L_w2 H_2O rich (+ salts)	4–50	Isolated, clusters, trails	–10	170	2–15
		5–20	Crosscutting trails (see Figs. 4A and 4C)	–9 to –4	120	5–10
$L_{w\text{-}c}$	H_2O-CO_2 rich (+ salts)	4–40	Trails (see Fig. 4A)	+2 to +8	230	5–30
$L/V_{w\text{-}c\text{-}m}$	H_2O-CO_2-CH_4-N_2 rich	<20	Trails (see Fig. 4B)	–5 to 0	230–280 ($L_{w\text{-}c\text{-}m}$)	8–10
		<30	Trails	–70 to –40	–80 ($V_{w\text{-}c\text{-}m}$)	
V_m	CH_4-N_2 rich vapors	5–30	Trails (see Fig. 4D)	–40		

Note: L_w—aqueous-rich fluid inclusions, $L_{w\text{-}c}$—aqueous-carbondic-rich fluid inclusions, $L/V_{w\text{-}c\text{-}m}$—liquid or vapor-rich inclusions containing water–carbon dioxide–methane ± nitrogen and other gases, T_{fm}—final melting temperature, T_h—homogenization temperature, V_m—vapor inclusions: containing methane-nitrogen and other gases.

ined using conventional microthermometric techniques (Poty et al., 1976; Roedder, 1984; Shepherd et al., 1985). Within any given crystal, several populations of fluids were trapped, each with different molar volumes and compositions that could only be quantified by making use of additional microanalytical methods. To determine the V-X (volume-compositional) properties of individual fluid inclusions and to obtain P-T-X (pressure-temperature-compositional) and geochemical data (and, hence, determine the metal-carrying capacity of the fluids and reconstruct the P-T evolution of the fluids), independent characterization of paleofluids was obtained by subjecting selected fluid inclusion samples to Fourier transform infrared (FT-IR) spectroscopy (Fig. 2; Pironon and Barres, 1990, 1992) and laser Raman spectrometric analysis (Fig. 3; Dubessy et al., 1982, 1984; Pasteris et al., 1988; Wopenka et al., 1990).

Raman absorbance levels for H_2O, CO_2, CH_4, H_2S, N_2, and H_2 and NaCl (Fig. 3) were combined with microthermometric parameters (i.e., T_{im}, initial melting temperature; T_{fm}, final melting temperature; T_mClath, clathrate melting temperature; T_h, homogenization temperature; % vapor fill) for fluid inclusions exhibiting complex compositional characteristics (Drennan, 1997; Drennan et al., 1997, 1999). These data were used to calculate mole percentages for each of the species present in the fluids (Table 3) as well as isochores for each analyzed sample. The procedure, which converts absorbance levels into compositional data, utilizes the modified Redlich-Kwong equation of state (Bowers and Helgeson, 1983) for calculations pertaining to H_2O-CO_2–rich fluids ($L_{w\text{-}c}$, Liquid$_{water\text{-}carbondioxide}$) and the Kerrick and Jacobs (1981) equation of state for the more volatile-rich inclusions ($L/V_{w\text{-}c\text{-}m}$, Liquid/Vapor$_{water\text{-}carbondioxide\text{-}methane}$; Table 3). The Zhang and Frantz (1987) equation of state was used for salt-rich CO_2-H_2O inclusions.

Compositionally complex fluid inclusions ($L/V_{w\text{-}c\text{-}m}$ and V_m, Vapor$_{methane}$) are difficult to interpret due to a lack of suitable experimental data for such complex fluid systems and the difficulty in estimating molar volumes. However, an initial attempt was made to determine the overall compositions of some of the complex inclusions encountered in this study (Table 4). The program HELGES (J. Dubessy, 1990, CREGU [Centre de Recherches sur la Géologie des Matinères Premières Minerales et Énergetiques], Nancy, France, personal commun.), which utilizes the molar volume estimates (for fluid components) of Angus et al. (1976), was used in order to perform the overall composition calculations. For example, a 30 mol% error in H_2O composition is possible by incorrectly interpreting an inclusion with only 0.05% of its volume occupied by an aqueous phase as being 100% gas filled. This is quite possible, as gas-rich inclusions (V_m) are usually dark in color and volumes of liquid are often microscopically invisible. Molar volumes for CH_4-N_2–rich inclusions (V_m) can be obtained directly from an isoplethic V-X diagram. However, determinations of the molar volume for inclusions exhibiting critical homogenization conditions are inaccurate because the molar volume changes drastically with T_h in this critical homogenization region (Swanenberg, 1979; van den Kerkhof, 1988).

Fluid Inclusion Types

An initial petrographic study of the sample material revealed that a variety of fluid systems and inclusion populations occurred in late-stage veins affecting the Witwatersrand

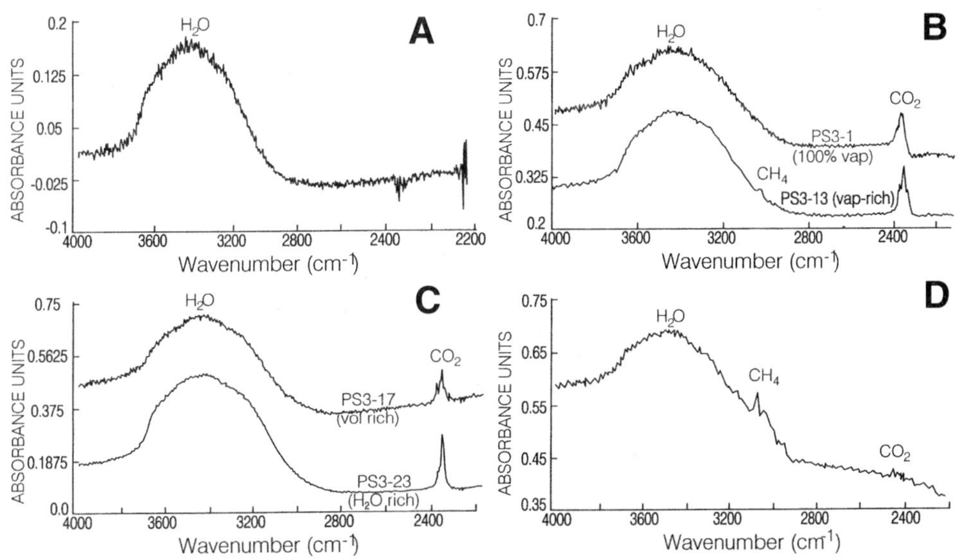

Figure 2. Examples of fourier transform infrared (FT-IR) spectra obtained for fluid inclusions of varying composition: (A) H_2O spectrum for sample VRW8B inclusion 18. (B) Spectrum indicating the presence of coexisting H_2O-CO_2 and H_2O-CO_2-CH_4 inclusions in sample PS3. (C) Variation of spectra for aqueous- and volatile-rich inclusions in sample PS3. (D) H_2O-CO_2-CH_4 spectrum for sample FSG6 (after Drennan, 1997 and Drennan et al., 1999).

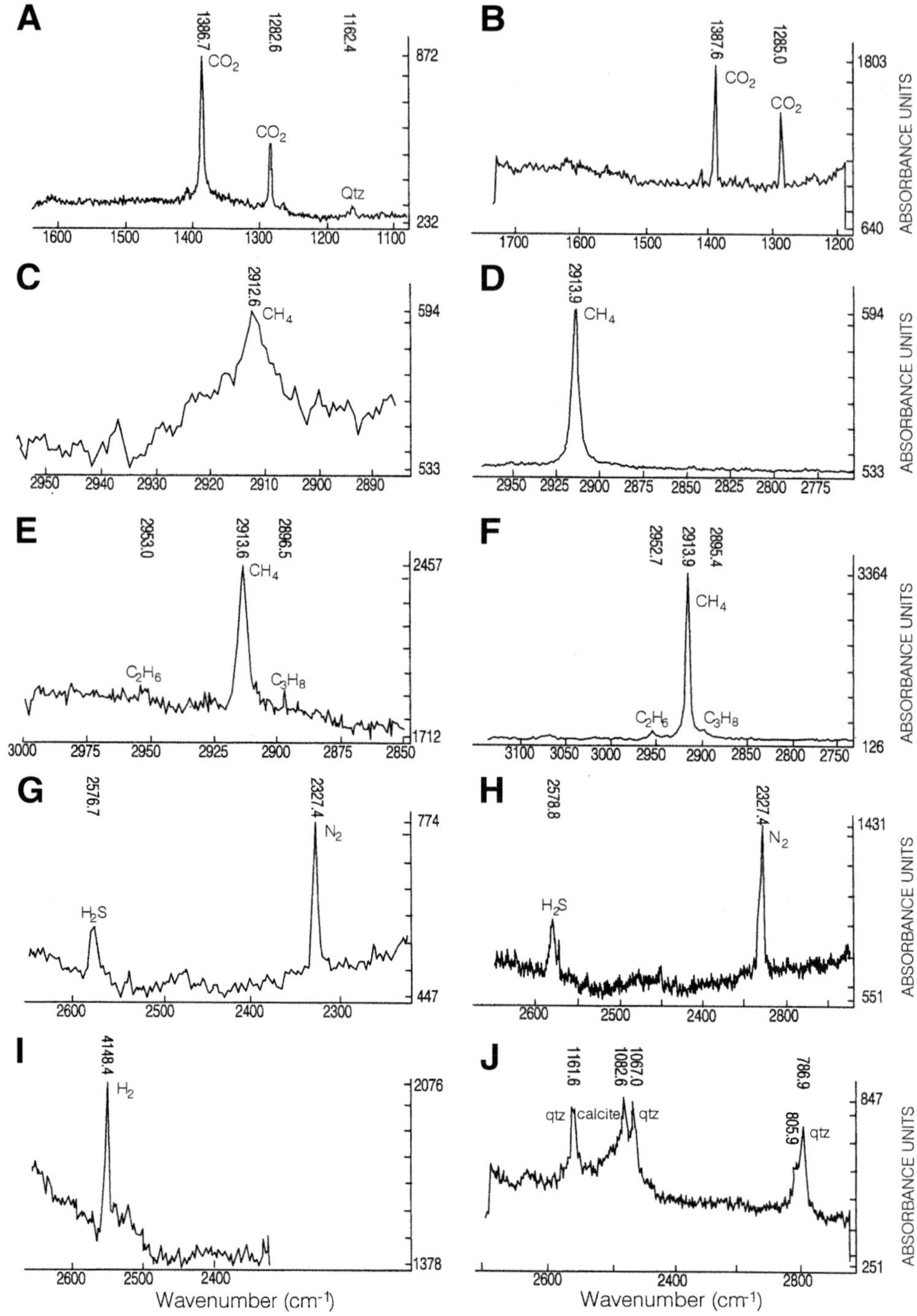

Figure 3. Examples of laser Raman spectra obtained for fluid inclusions exhibiting complex compositions: (A) CO_2 spectrum for sample VRS1/3/5. (B) CO_2 spectrum for sample PS3/1/4. (C) CH_4 spectrum for sample VRS1/3/7. (D) CH_4 spectrum for sample PS3/3/4. (E) CH_4, C_2H_6, and C_3H_8 spectra for sample FSG6/1. (F) CH_4, C_2H_6, and C_3H_8 spectra for sample 2729/4. (G) H_2S and N_2 spectra for sample 2729/x. (H) H_2S and N_2 spectra for sample 2729/4. (I) H_2 spectrum for sample 2729/x. (J) Spectrum of calcite daughter crystal, sample VRS3/1/39 (after Drennan, 1997 and Drennan et al., 1999).

TABLE 3. SUMMARY OF MICROTHERMOMETRIC DATA FOR VOLATILE/GAS-RICH (L/V_{W-C-M} AND V_M) INCLUSIONS

Sample	T_h (°C)	Density	XH_2O	XCO_2	XCH_4	XC_2H_6	XN_2	XH_2S	XH_2	wt% eNaCl	T_{trap}	Eqn*
2729/4	−99.3	0.046	0	0	0.897	0.011	0.053	0.011	0.023	-	-	K&J
2729/9	−73.6	-	0	0	0.891	0.014	0.036	0.009	0.050	-	-	K&J
VRS1/1	+178	0.810	0.815	0.171	0	0	0	0	0	5.5	460	B&H
VRS1/3	+151	1.001	0.929	0.052	0	0	0	0	0	6.3	295	B&H
VRS1/3	+145	0.974	0.849	0.042	0	0	0	0	0	29.4	530	B&H
VRS1/3	+154	1.006	0.904	0.041	0	0	0	0	0	16.5	275	B&H
FSG6/2	+151	0.837	0.090	0.001	0.003	0	0	0	0	4.6	375	Z&F
FSG6/1	−44.8	-	0	0	0.769	0.059	0.159	0.014	0	-	-	Z&F
PS3/1/4	−57.6	0.162	0	0.071	0.859	0	0.072	0.001	0	-	-	K&J
PS3/2/5	−73.1	-	0	0.094	0.794	0	0.107	0.006	0	-	-	K&J
PS3/2/1	−65.7	-	0	0.68	0.826	0	0.100	0.006	0	-	-	K&J

Note: Data from Drennan (1997) and Drennan et al. (1999). Eqn* denotes Equation of State used in calculating compositional data: K&J—Kerrick and Jacobs (1981); B&H—Bowers and Helgeson (1983); Z&F—Zhang and Frantz (1987); T_{trap}—trapping temperature (Drennan et al., 1999). L/V_{w-c-m}—liquid or vapor-rich inclusions containing water–carbon dioxide–methane ± nitrogen and other gases, T_h—homogenization temperature, V_m—vapor inclusions: containing methane-nitrogen and other gases.

TABLE 4. OVERALL COMPOSITIONS OF SELECTED FLUID INCLUSIONS DETERMINED BY MICRO-RAMAN SPECTROSCOPY

Sample	Incl.	V_{aq} (%)	H_2O	NaCl	CO_2	N_2	H_2S	CH_4	D_{incl}	V_{incl}	C%	H%	O%	N%	S%
Aqueous															
FSG6/2	25	81.20	97.50	1.46	0.77	0	0	0.27	0.84	22.45	0.35	66.21	33.44	0	0
VRS1/3	7	85	91.31	1.88	6.82	0	0	0	0.98	20.88	2.31	62.04	35.65	0	0
		90	92.90	1.91	5.19	0	0	0	1.00	20.10	1.77	63.14	35.10	0	0
		95	94.36	1.94	3.7	0	0	0	1.02	19.37	1.26	64.15	34.59	0	0
VRS1/3	4	85	90.36	4.14	5.50	0	0	0	1.01	20.98	1.91	62.84	35.25	0	0
		90	91.91	4.20	4.09	0	0	0	1.03	20.13	1.42	63.82	34.76	0	0
		95	92.94	4.26	2.80	0	0	0	1.06	19.35	0.97	64.72	34.31	0	0
VRS1/4	4	55	81.44	1.48	17.08	0	0	0	0.81	28.44	5.78	55.11	39.11	0	0
		60	83.78	1.52	14.71	0	0	0	0.83	26.88	4.98	56.71	38.31	0	0
		65	85.86	1.56	12.58	0	0	0	0.86	25.48	4.26	58.14	37.56	0	0
Volatile															
PS3/2	5	0	0	0	9.36	10.66	0.60	79.36	0.13	150.00	19.80	71.12	4.18	4.75	0.13
		0.05	30.25	0	6.87	7.39	0.42	55.05	0.18	100.48	15.38	69.92	10.93	3.67	0.10
		0.10	47.64	0	5.45	5.52	0.31	41.07	0.22	87.46	12.35	69.09	15.54	2.93	0.08

Note: Data from Drennan (1997) and Drennan et al. (1999). V_{aq}—volume of aqueous phase as a percentage, D_{incl}—density of inclusion, V_{incl}—volume of inclusion (Drennan et al., 1999).

Basin, which in turn indicated that the sediments had been affected by numerous postdepositional fluid events, evidenced by multiple crosscutting fluid inclusion trails (microfractures; see also Boer and Reimold, 2006). It was not always possible to define the inclusions as primary, pseudosecondary, or secondary inclusions, according to the criteria of Roedder (1984), because of the frequent absence of distinct growth zonation, or crystal or grain boundaries within the vein quartz material. As a consequence, inclusions that occurred in isolated three-dimensional groups or in limited planar arrays were referred to as early inclusion generations, whereas those that occurred in extensive two-dimensional trails were referred to as late inclusions. The latter are typically fracture-related, and these fracture planes are transgranular. Commonly there is a parallel to subparallel alignment of the inclusion stringers, and consecutive fracture-related inclusion stringers appear as intersecting inclusion arrays. Trails that crosscut the aligned stringers were referred to as being very late fluid inclusion generations. For most inclusions greater than 1 μm in size (>70%), more than one phase change was observed (Drennan, 1997; Drennan et al., 1997, 1999). The compositions of heterogeneous fluid populations and complex fluids were confirmed by FT-IR and laser Raman spectroscopy.

Fluid Systems

On the basis of petrographic observations and microanalytical investigations, at least four different fluid systems were distinguished. These are summarized in Table 2 and include the following:

1. Aqueous fluids (L_w), with variable salinities in 1-, 2-, and 3-phase inclusions at room temperature.
2. H_2O-CO_2–rich fluids (L_{w-c}) that occur as 2- and 3-phase inclusions at room temperature. These fluids exhibit variable liquid-to-vapor ratios.
3. H_2O-CO_2–rich fluids with variable amounts of CH_4 and N_2 that occur as monophase (liquid or vapor) inclusions, 2-phase (liquid and vapor) inclusions, and 3-phase inclusions at room temperature (L_{w-c-m} or V_{w-c-m}). These inclusions have hydrocarbons as well as daughter crystals trapped within the inclusions.
4. CH_4-N_2–rich vapors (V_m) that occur as dark, monophase inclusions at room temperature.

Fluid Inclusion Populations

The above-mentioned fluid systems are categorized into eight fluid populations.

L_w1. In many samples the earliest inclusion populations have often been destroyed by decrepitation or later overprinting and are often too small (<1 μm) and, therefore, unsuitable for microthermometric investigations. These very small inclusions impart a distinctive wispy texture to the samples.

L_w2.A subsequent population comprises aqueous fluids of varying salinity (2 wt% eNaCl to 20 wt% eNaCl), sometimes saturated with respect to salts as evidenced by the presence of daughter crystals. These inclusions are typically very small (1–5 μm) to small (5–20 μm) and are present in all the samples studied (Fig. 4A).

$L_{w\text{-}c}$. A CO_2-rich fluid, with variable CO_2/H_2O ratios, is present in a number of samples. The CO_2-rich, 2- and 3-phase inclusions are generally large (20–40 μm, Fig. 4A) and are characterized by typical "double bubbles." Some of these inclusions exhibit critical homogenization.

L_w1, L_w2, and $L_{w\text{-}c}$ inclusions generally occur in isolated three-dimensional groups or in limited planar arrays and are collectively regarded as the early fluids.

$L_{w\text{-}c\text{-}m}$ and $V_{w\text{-}c\text{-}m}$. Vapor-rich, CO_2-rich fluids with variable amounts of CH_4, C_2H_6, C_3H_8, and N_2 occur in trails that crosscut the early fluid populations (Fig. 4B) and were detected by FT-IR and Raman spectroscopy (Tables 3 and 4). Infrared absorption spectroscopy (Fig. 2) for selected volatile/gas-rich fluid inclusions ($L/V_{w\text{-}c\text{-}m}$) indicates the presence of H_2O, CO_2, and CH_4. Raman microanalysis confirmed the presence of CO_2 and CH_4 in the gaseous phases and further indicates the presence of H_2S, N_2, and H_2, as well as traces of C_2H_6 and C_3H_8, in some of these inclusions (Fig. 3). Very dark vapor coloration and prominent bubble edges often characterize these inclusions. Some of these inclusions homogenize into the liquid phase, whereas others homogenize into the vapor phase. Aqueous inclusions coexist with this generation in the same fluid inclusion trails (Fig. 4C). A common feature of these inclusions is the presence of abundant bitumen nodules as well as possible sulphide daughter crystals. Most of these inclusions also have a lining of hydrocarbon-rich material, identifiable only by Raman spectroscopy.

V_m. CH_4-N_2 vapor-rich inclusions (Fig. 4D) were also identified by Raman spectroscopy (Tables 3 and 4; Fig. 3). These inclusions are very dark in color, which makes the recognition of possible aqueous contributions to the fluid systems impossible. Possible bitumen nodules were observed in these inclusions, but their presence could also have been masked by the inclusion coloring. V_m inclusions exhibit very low (negative; Table 2) homogenization temperatures and are not, therefore, plotted on conventional microthermometric plots.

$L_{w\text{-}c\text{-}m}$ or $V_{w\text{-}c\text{-}m}$ and V_m inclusions occur in extensive two-dimensional trails and are referred to as late inclusions that postdate the earlier populations (Table 2, Figs. 4B and 4D). These inclusions are generally 10–30 μm in size and occur predominantly in fault-related samples but were also present in some samples associated with intrusions.

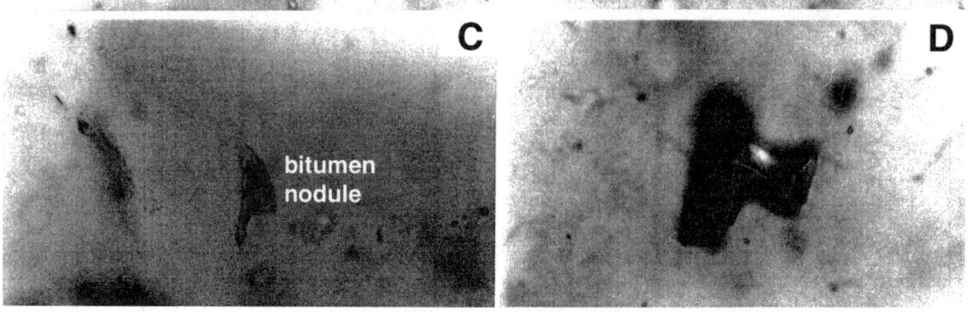

Figure 4. Examples of fluid inclusions representing the major fluid systems observed in quartz veins: (A) "Double-bubble" CO_2-rich inclusion ($L_{w\text{-}c}$) occurring together with H_2O-rich inclusion (L_w2); crosscutting this trail are later aqueous inclusion trails (L_w3), sample VRS1, width of field of view = 80 μm. (B) Coexisting H_2O-CH_4-N_2–rich ($L_{w\text{-}c\text{-}m}$) and CO_2-CH_4-C_2H_6-N_2-H_2S–rich (V_m) inclusions, sample PS3, width of field of view = 150 μm. (C) Bitumen nodule occurring within aqueous inclusion (nodules normally occupy 5% of inclusion volume but may occupy as much as 60% of the inclusion), sample VR8(iii), width of field of view = 150 μm. (D) Late volatile/gas-rich inclusion, sample FSG6, width of field of view = 100 μm (from Drennan 1997 and Drennan et al., 1999).

L_w3. A very late generation aqueous fluid occurs in most samples and appears to be more saline than the earlier aqueous inclusion generations.

Calculation of fO_2 from Fluid Inclusions

Some quartz vein–hosted fluid trails (see samples FSG6 and PS3 of Drennan, 1997; Drennan et al., 1999) contain inclusions representing aqueous-rich and volatile-rich fluids (L_{w-c} and V_m), respectively, within the C-H-O system. Information relating to these two inclusion types has been used in order to estimate fO_2 for postdepositional fluids affecting the Witwatersrand sediments.

Oxygen and sulfur fugacities can be derived from fluid inclusion data, provided that the bulk V-X properties of individual inclusions have been estimated. The compositions of inclusion fluids within the C-O-H system, as determined at room temperature, do not differ significantly from their equilibrium compositions under entrapment P-T conditions (Dubessy et al., 1987, 1989). Consequently, fO_2 can be calculated along the isochoral path from the V-X properties measured by microthermometry and Raman spectroscopy (Dubessy, 1984; Dubessy et al., 1987). Data pertaining to carbon isotope exchange between CH_4 and CO_2 in natural samples suggest that the oxidation state of fluids is controlled by the equilibrium

$$CH_4 + 2O_2 \rightarrow CO_2 + 2H_2O, \quad (1)$$

at temperatures greater than 300 °C (Sheppard, 1981; Giggenbach, 1982; Dubessy et al., 1987). At 320 °C and above, the fO_2 of carbonic fluids containing in excess of 10^{-2} mol% CH_4 is fixed around and below the Ni-NiO buffer (Dubessy et al., 1987). Thermodynamic calculations of the graphite-fluid equilibrium show that fO_2 approximates the Q-F-M (Quartz-Fayalite-Magnetite)buffer in the 400–500 °C temperature range (French, 1966; Ohmoto and Kerrick, 1977; Dubessy et al., 1987). At higher temperatures, therefore, fO_2 can be calculated using the measured V-X properties. The calculation of fugacity coefficients takes into account the concentrations of dissolved salts in terms of equivalent wt% NaCl (Dubessy et al., 1987). The volatile-rich fluid (sample PS3; V_m) was determined to be more reducing than the aqueous-rich fluid (sample FSG6; L_{w-c}; Drennan, 1997; Drennan et al., 1999). The temperature ranges mentioned here are not unlike those suggested by others working in the Witwatersrand Basin (e.g., Foya, 2002; Frimmel et al., 1999; Schreyer and Bisschoff, 1982) and are consistent with fluid inclusion thermometry. Specifically at 350 °C the range of fO_2 in the aqueous- and CH_4-rich fluids is between –31.0 and –32.1 log units.

Fluid chemistry for the Witwatersrand Basin suggests that the postdepositional C-O-H fluids affecting the sediments has oxygen fugacities between Q-F-M and Ni-NiO (Drennan et al., 1999) at temperatures ranging from ~350–500 °C and at a pH buffered by the stability of muscovite (pH around 4–5). Under these conditions in aqueous solutions, uranium is likely to occur as the $U(OH)_4$ complex (Wilde et al., 1989), and uranium solubility would amount to no more than a few tens of ppb (Meyer et al., 1991; Robb and Meyer, 1991). This accounts for the widespread preservation of detrital uraninite while at the same time not negating the possibility of limited uranium remobilization. At sites of localized higher temperatures (>400 °C), such as fault zones, remobilized uranium is likely to have been precipitated along with other carbonaceous complexes. It is also pertinent to note that hydrocarbon-bearing fluids, from which solid bitumens have precipitated in and around lags of uraninite particles, appear to have had only minimal effect on the dissolution of these grains. This view contrasts with that of Barnicoat et al. (1997) and Gray et al. (1998), who suggested that uraninite and bitumen coprecipitated from the same fluid population. Oxygen fugacity estimates at 350 °C (a realistic temperature for gold mobilization in the Witwatersrand Basin) are in the range –32.1 to –31 (Drennan et al., 1999), which is very similar to the fO_2 conditions estimated by Robb and Meyer (1991) for Witwatersrand fluids in terms of pyrite-pyrrhotite equilibrium. It is, therefore, possible that the gold-thio complex was more effective in transporting the metal than was the gold-chlorite complex.

CHARACTERIZATION OF BITUMEN IN THE WITWATERSRAND BASIN AND SUBJACENT BASEMENT

The terminology applied to carbonaceous matter is largely dependent on the extent to which its origin is understood. Mossman et al. (1993) and Nagy (1993) have simplified usage to two terms. Kerogen may be defined as the syngenetic, particulate organic substance (Mossman and Nagy, 1996) that generally reflects the residue after a mobile liquid or gaseous hydrocarbon fraction has been extracted from the rock. Kerogen also has a "random polymer-like macromolecular structure that usually consists of polycyclic aromatic hydrocarbons and is essentially immobile in rocks following deposition" (Mossman and Nagy, 1996, p. 115). In Archean sediments kerogen is reported to occur as anastomosing, semicontinuous, bedding-parallel laminae of organic matter resembling the maceral types alginate or lamalginite (Rasmussen, 2005). Bitumen, on the other hand, is a complex organic substance, containing hydrocarbons and various other molecular constituents, that was originally either liquid or solid. Its "random polymer-like" molecular structure is a product of the fact that it was once mobile as a fluid or vapor but has since solidified (Mossman and Nagy, 1996). Texturally it occurs in Archean sediments as nodules around radioactive mineral cores or as nodules and aggregates in remnant pore spaces (Rasmussen, 2005).

Kerogen is normally the most abundant organic substance in a sedimentary sequence and represents the residue after a normal cycle of organic evolution. In ore deposits, bitumens are the products of maturation and have been solidified by processes such as biodegradation, oxidation, thermal stress, or radiation-induced polymerization (Landais et al., 1993; Mossman and Nagy, 1996). Witwatersrand terminology over the past 20 years has been dominated by usage of the term kerogen because of widely held support for the in situ biogenic origin of the carbonaceous matter. Recently, the term bitumen has been applied to the Witwatersrand Basin with the recognition that mobile hydrocarbons circulated through the basin (Drennan et al., 1997, 1999; England et al., 2001; Gray et al., 1998; Mossman and Nagy, 1996; Nagy, 1993; Robb et al., 1997a, 1997b; Zartman and Frimmel, 1999).

The results of a detailed petrographic, chemical, and spectroscopic study of carbonaceous matter in the Witwatersrand Basin are presented in Robb et al. (1997b); an overview of this work is presented here. In order to obtain representative samples of carbonaceous matter, Robb et al. (1997a, 1997b) sampled a total of 7 mines across the entire basin (Fig. 1). Samples of carbonaceous matter from the "B" and Basal Reefs in the Welkom Goldfield, the Ventersdorp Contact, and Vaal Reefs in the Klerksdorp Goldfield, the Carbon Leader, Ventersdorp Contact, Elsburg and White Reefs from the Carletonville and West Rand Goldfields, and the Kimberley Reef in the Evander Goldfield were collected and then slabbed, polish-sectioned, and separated for a variety of chemical and spectroscopic analyses. In addition, bitumen nodules were separated from quartz veins cutting through the Witwatersrand sequence and from samples of rare, granitoid-hosted bitumens from a number of peraluminous granites beneath and adjacent to the Witwatersrand sediments (Robb et al., 1994).

Carbonaceous material associated with the Witwatersrand Basin occurs in a variety of locations and textural forms; these are summarized as follows.

Carbonaceous Seams

Although carbonaceous seams commonly occur at the base of a conglomeratic unit and may be underlain by shale, they are also seen to anastomose through the conglomerate (Fig. 5A), bifurcate around pebbles, and gradually disaggregate. Seams incorporate most of the allogenic mineral constituents of the conglomerate, including uraninite, pyrite, quartz, chromite, zircon, and gold. Uraninite is, however, undoubtedly the most common constituent associated with the seams, and its abundance varies from ±20 vol% to a virtually undetectable presence. Figure 5B shows part of a carbonaceous seam containing numerous fragmented and partially replaced uraninite grains. Figure 5C is a view under high magnification of a single grain of uraninite that has apparently been partially replaced by carbonaceous matter. The original outline of the now partially replaced uraninite grain is demarcated by a halo of enhanced reflectance that is believed to be the result of radiation.

Well-developed seams, particularly at the base of a conglomerate, may exhibit a fibrous aspect, which, in the past, has been attributed either to algal (or fungal) symbionts preserved as microfossils (Hallbauer, 1975) or to "fiber-vein" growths formed by precipitation of solid hydrocarbons in an extensional fracture environment (e.g., Schidlowski, 1981; England et al., 2001). Evidence from the present study clearly favors the latter suggestion. Carbonaceous seams are observed to have been precipitated along fractures representing preexisting fluid pathways. Figure 5D shows a veinlet containing sericitic material and migrated carbonaceous matter transecting a quartz pebble and merging eventually with a thicker carbonaceous seam. The bitumen within the veinlet contains no detectable uraninite fragments, whereas the thicker seam encapsulates numerous uraninite particles.

Evidence for the development of fibrous or columnar textured bitumen is presented in Figure 5E, which shows a thin seam of carbonaceous matter deposited in a fracture that transects a row of en echelon grains. The fracture appears to bisect several of the uraninite grains such that they are split laterally on either side of the dislocation. These observations are interpreted to suggest that carbonaceous matter was precipitated in a tensional fracture during passage of a hydrocarbon-bearing fluid. As the vein continued to open under extension, the fibrous texture developed at right angles to the vein walls. Solidification of bitumen was caused by the radiation flux emanating from the uraninite grains, whose deposition must have preceded the formation of the fracture. It is suggested that the uraninite represents a detrital lag that was concentrated toward the base of the conglomeratic host. The sequence of events recorded here contradicts previous suggestions that bitumen and uraninite represent coeval precipitates from a single fluid (Barnicoat et al., 1997; Gray et al., 1998). By contrast, authigenic gold (Fig. 5B) is generally observed to be paragenetically later than carbonaceous material and occurs as irregular polycrystalline aggregates on the margins of seams, as veinlets cutting through the latter or as filaments between the fibrous elements of well-developed seams.

The fibrous elements developed in this seam are unlikely to be the fossilized remnants of primitive organic life-forms. Components of carbonaceous seams that exhibit clear evidence for having replaced preexisting detrital uraninite grains or for having migrated along fractures or veins qualify texturally for designation as bitumen or pyrobitumen. The question of whether all the carbon seams can, on this basis, be inferred to be bitumen or whether some might still preserve kerogen, is, however, debatable and requires further observations.

Carbonaceous Nodules

Carbonaceous nodules are very often distributed in the matrix of the same conglomeratic units that contain well-developed carbonaceous seams but may also occur in hangingwall and footwall quartzitic sequences. The larger carbonaceous nodules are spatially related to seams and merely represent small-scale versions of the latter. Such nodules (2–3 mm diameter) may have irregular outlines with a large length-to-width aspect and

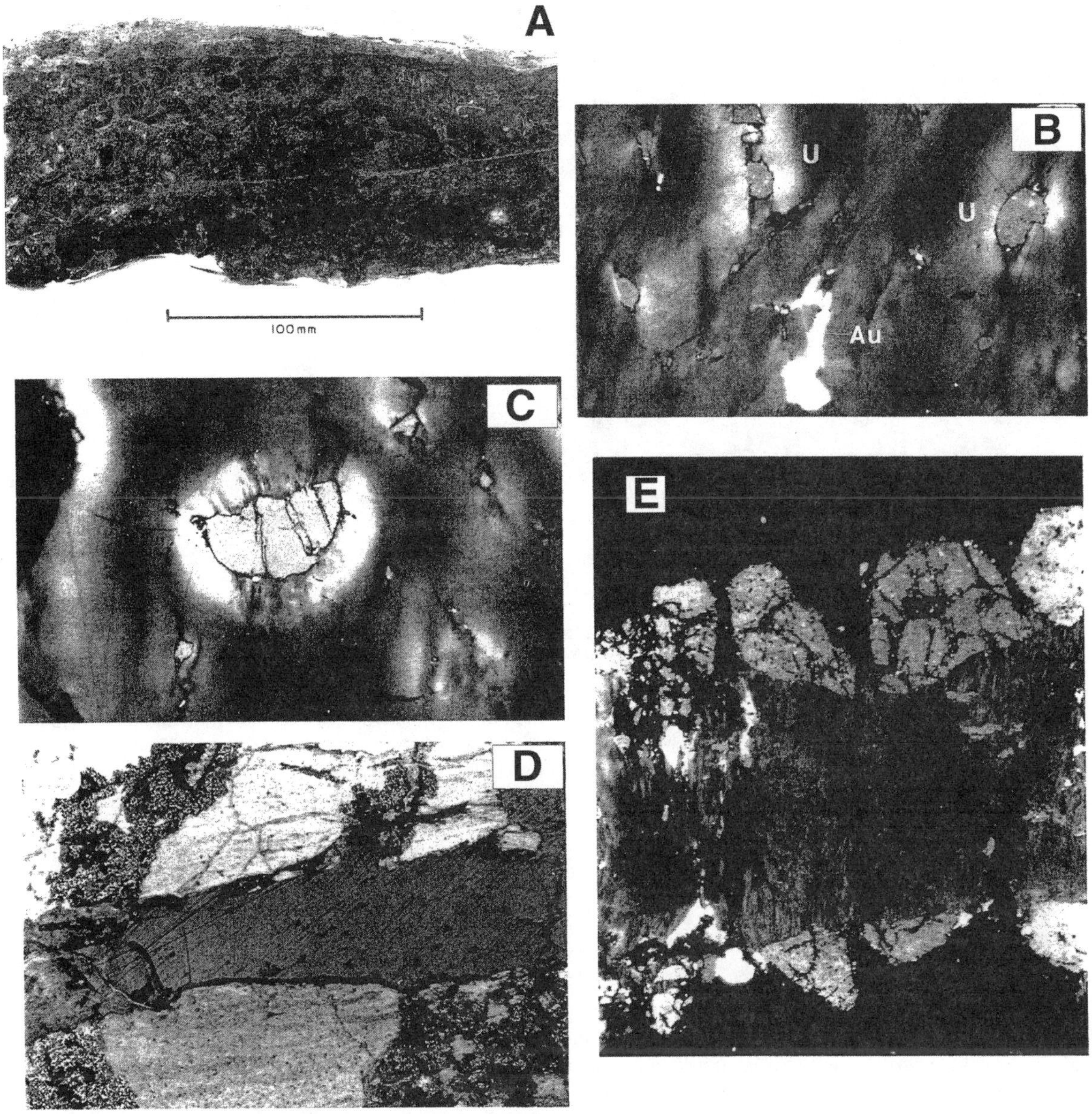

Figure 5. (A) Multiply stacked seam bitumen within Beisa Reef. (B) Authigenic gold (Au) and detrital uraninite (u) within seam bitumen (at base of "C" Reef, width of field of view = 2.0 mm). (C) Single grain of detrital uraninite undergoing fracturing and replacement by bitumen; the pleochroic halo demarcates the alpha-radiation envelope around the grain ("C" Reef, width of field of view = 0.2 mm). (D) Bituminous veinlet deposited along a preexisting bedding-parallel fracture crosscutting a quartz pebble, width of field of view = 2.5 mm. (E) Fibrous bituminous seam deposited along a bedding-parallel fracture that separates a layer of uraninite grains (photomicrograph courtesy of Dr. T. Oberthur, BGR, Hannover, Germany), width of field of view = 0.3 mm (compare Drennan, 1997; Robb et al., 1997a, 1997b).

are usually associated with numerous uraninite particles (Fig. 6A). Nodules that are not obviously related to a carbonaceous seam tend to be smaller, more circular to globular in aspect, and may not enclose visually detectable uraninite particles. This may be due either to the complete replacement of uraninite by bitumen or to a hydrocarbon fixation mechanism that is unrelated to the presence of a radiation-emitting phase. Occasionally, small nodules (~100 μm diameter) form around individual uraninite particles and represent discrete aggregates of carbonaceous matter that occur where there is little or no seam developed. In Figure 6B three uraninite grains reflect progressive increments of fragmentation replacement by bitumen; this observation lends support to the notion that some nodules might represent the products of complete replacement of preexisting uraniferous phases, a feature also observed in granitoid-hosted bitumen nodules adjacent to the basin (Robb

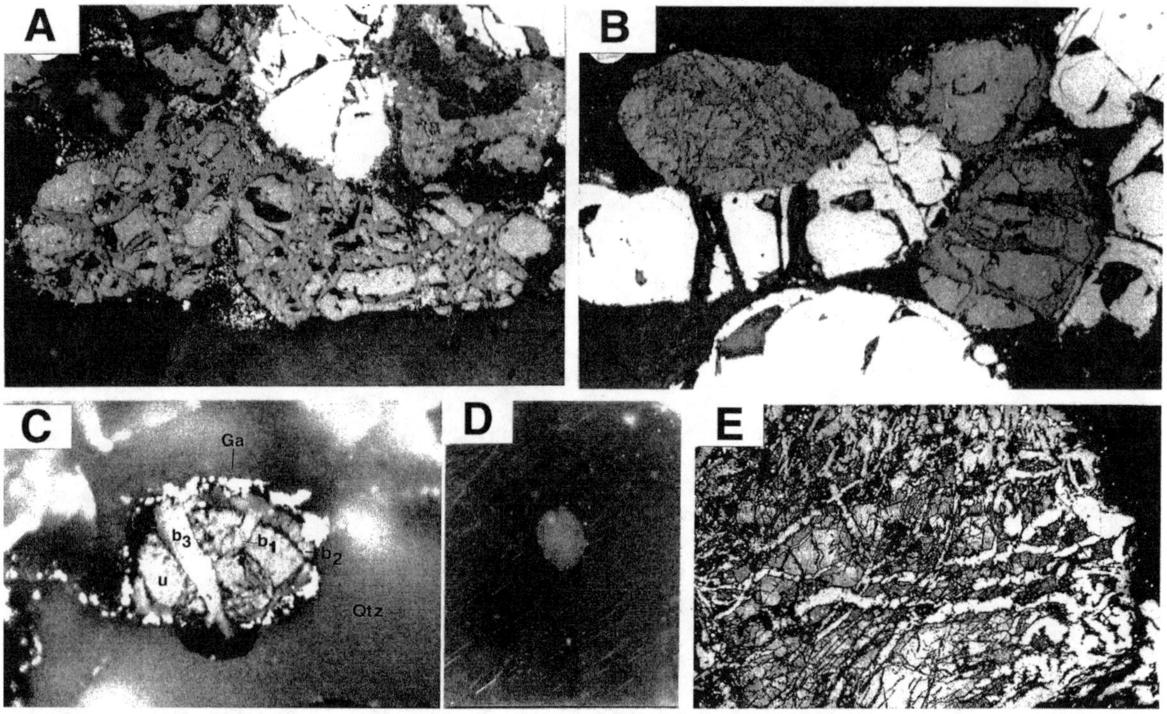

Figure 6. (A) Several uraninite grains fragmented and cemented by bitumen to form a large, elongate nodule, width of field of view = 0.6 mm. (B) Three discrete uraninite grains illustrating a progression of fragmentation and replacement by bitumen, width of field of view = 0.2 mm. (C) Multiple stages of bitumen deposition (b_1, b_2, b_3) replacing uraninite within conglomeratic matrix. Radiogenic galena (Ga) precipitates around the grain, Beisa Reef, width of field of view = 0.1 mm. (D) Bituminous nodule deposited along an annealed microfracture in a quartz pebble (microphotograph courtesy of Dr. T. Oberthur, BGR (Bundesanstalt für Geowissenschaften und Rohstoffe), Hannover, Germany), width of field of view = 0.1 mm. (E) Granitoid-hosted bitumen replacing high U-bearing phase (uranothorite), borehole GH1, width of field of view = 0.5 mm (Drennan, 1997; Robb et al., 1997a, 1997b).

et al., 1994, 1997a, 1997b). In some instances, multiple bitumen replacements can be identified. In Figure 6C a single uraninite grain is seen as being affected by three different bitumen precipitation events. In addition, small bituminous nodules are occasionally observed along the sites of annealed microfractures within quartz pebbles in the conglomerates (Oberthür, 1983; Fig. 6D).

The textures described above indicate that carbonaceous nodules do not represent the in situ remnants of microbiota but more likely the products of migrating hydrocarbons that were solidified around or replaced individual uraninite grains, as a result of radiolytically induced polymerization. As such they should be referred to as bituminous nodules. The formation of seams as opposed to nodules appears to be related to the amount of uraninite present at any point in the conglomerate and the extent to which the bitumen has invaded and replaced the uraninite.

Granitoid-Hosted Carbonaceous Nodules

Several workers have now recorded the presence of bituminous nodules in granitoids surrounding the Witwatersrand Basin (Hallbauer, 1984; Klemd and Hallbauer, 1987; Robb et al., 1992, 1994). A detailed account of their distribution and petrographic characteristics is provided in Robb et al., 1994, 1997a, 1997b. In summary, nodules or nodular stringers ranging in diameter from 0.2 to 0.8 mm commonly occur in peraluminous granitoids that contain primary uranium-thorium–bearing accessory minerals. The nodules are paragenetically late in terms of the sequence of alteration of the granitoids and exhibit progressive increments of replacement of either uraninite or uranothorite (Fig. 6E). In some cases nodules may not be associated with visually detectable uranium-bearing phases but are always associated with high concentrations of uranium (Robb et al., 1994, 1997a, 1997b). It has been suggested that the same hydrocarbon-bearing fluids that permeated the Witwatersrand sediments also passed through granitoids adjacent and subjacent to the basin (Robb et al., 1994, 1997a, 1997b). It would appear that bitumens were precipitated in these granitoids by radiolytic polymerization in much the same way as hydrocarbons were fixed within the Witwatersrand Basin.

Quartz Vein- and Fluid Inclusion-Hosted Carbonaceous Matter

A detailed study of postdepositional fluids in the Witwatersrand Basin has revealed the presence of, *inter alia,* a hydrocarbon-bearing fluid population comprising CO_2 and variable proportions of CH_4, C_2H_6, H_2S, H_2, and N_2 (Fig. 4; Drennan, 1997; Drennan et al., 1999; Robb et al., 1997a, 1997b). Inclusions bearing these fluids are occasionally coated by a thin film of solid hydrocarbon or actually contain bituminous nodules up to 10 μm in diameter (Fig. 4C; Drennan 1997; Drennan et al., 1997, 1999). Nodules have also recently been discovered in association with pyrite and gold from quartz veins that cut the conglomerates and are associated with late-stage faulting in the basin (Drennan, 1997, 1999). Quartz vein-hosted bitumen nodules either occur as discrete warty aggregates of hydrocarbon (Fig. 7A) or intimately intergrown with pyrite and occasionally gold (Figs. 7B and 7C). The presence of these bitumens clearly demonstrates that hydrocarbon circulation in the basin was not restricted to the conglomeratic horizons but also involved aqueo-carbonic solutions migrating along other conduits such as faults and shears.

A late-stage, secondary chlorite-quartz-bitumen vein is illustrated in Figures 7D and 7E. This remobilized bitumen vein (b_2) crosscuts the uraniferous bitumen (b_1; Beisa Reef, Fig. 7D) but is itself lacking any uranium distribution (see the fission-track micromap; Fig. 7E). This would suggest that where migrating aqueo-carbonic fluid channelways intersected sediment-hosted, uraniferous bitumen seams (e.g., Beisa Reef), the light hydrocarbons precipitated out as bitumen veins.

MICROANALYSIS OF WITWATERSRAND BITUMEN

Bitumen in the Witwatersrand sediments occurs as seams, often at the base of conglomeratic units, and as nodules ("fly-speck" bitumen) within the conglomerate matrix. These bituminous types are referred to as sediment-hosted bitumen (Drennan, 1997; Robb et al., 1997a, 1997b) in the remainder of this paper and equate to the "carbonaceous mesophase" described by Gray et al. (1998), derived from pyrobitumen that represents the residue of migrated liquid hydrocarbon. Released hydrocarbons (CO_2, CH_4, oil) migrated through the sediments along aquifers (unconformity surfaces, permeable facies, and/or structural discontinuities), often being trapped as fluid inclusions in secondary inclusion trails within the sediments or in late-stage quartz veins and precipitated as late-stage crosscutting bitumen veins (Drennan, 1989, 1997). It is also likely that hydrocarbon-rich fluids migrated toward the basin margin and into the adjoining granitic hinterland (Robb et al., 1992, 1994). These bitumen types are hereafter referred to as distal and vein related.

Drennan (1997) investigated the nature of the distal/vein-related bitumen and, in particular, its relation to postdepositional fluid movement through the basin. In addition to the presence of abundant hydrocarbons in the volatile phase of fluid inclusions, carbonaceous matter was also observed in three other forms: (1) optically invisible carbonaceous wall coatings to aqueous-rich fluid inclusions as well as vapor-rich inclusions; (2) rounded-to-bulbous aggregates or nodules of solid hydrocarbon within fluid inclusions; and (3) rounded-to-bulbous solid hydrocarbon nodules coprecipitated with authigenic sulphides and gold in quartz veins.

Carbonaceous wall coatings inside fluid inclusions and bitumen nodules, coprecipitated with remobilized sulphide and gold mineralization within quartz veins in the Witwatersrand Basin, were analyzed by Raman microspectroscopy (Drennan, 1997). All samples were taken from late-stage quartz veins associated with faulting. The results were compared to those obtained for seam and "fly-speck" bitumen in the Witwatersrand Basin (Landais et al., 1990a, 1990b; Robb et al., 1997a, 1997b). Authigenic bitumen nodules from quartz veins were subjected to micro–PIXE analysis to determine REE (rare earth element) concentrations, as well as the distribution of Au, U, Th, Fe, Cu, and As within these nodules (Drennan, 1997).

Spectroscopic Characteristics of Witwatersrand Bitumen

Micro-Raman analyses on quartz vein–related migrated bitumen were carried out on a multichannel X-Y Dilor spectrometer at CREGU, Nancy, France. The excitation radiation is the 514.5 nm line of an Ar laser set at low power (<10 nm for bitumen nodules and <30 nm for carbonaceous inclusion wall coatings) in order to minimize alteration effects of the organic matter. Integration times of between 5 and 20 s for 10–30 accumulations were undertaken in order to record the spectra (Fig. 8). Spectral bands were integrated and emphasis placed on the E2 g2 band attributed to C-C vibrations of aromatic carbons (located at 1575 cm^{-1} for perfect graphite; Landais et al., 1990b) and the 1340 cm^{-1} defect band.

Raman microspectroscopy provides information on the degree of organization of the hydrocarbon network. Two dominant peaks are recorded; one at around 1610 cm^{-1}, which corresponds to the C-C vibrations, and the other at around 1340 cm^{-1}, which represents the defects in the organic skeleton (Landais et al., 1990a). A third minor peak at around 1500 cm^{-1} is observed in bitumen nodules that have been coprecipitated in quartz-vein material, and this peak probably reflects the presence of methane inclusions within the nodules. Peak wave numbers and peak heights for samples analyzed are recorded in Table 5. Peak height ratios and the half height width (1/2H 1610) of the 1610 peaks are presented in Table 6. The peak ratio 1610/1340 represents a good approximation of the organization of the bitumen structure. The larger the peak ratio value, the better structurally organized the bitumen is (note: with increased structural order, bitumen becomes progressively more akin to graphite, with fewer aliphatic chains). Smaller peak height ratios imply poorly organized bitumen with abundant aliphatic chains. The half height width is calculated as an inverse measure of the aromaticity and is broadly proportional to the uranium content of the carbonaceous matter (Landais et al., 1990a).

When examining Raman spectra obtained for the different bitumen types analyzed (Fig. 8; Drennan, 1997), certain spectral differences are immediately apparent, as discussed below.

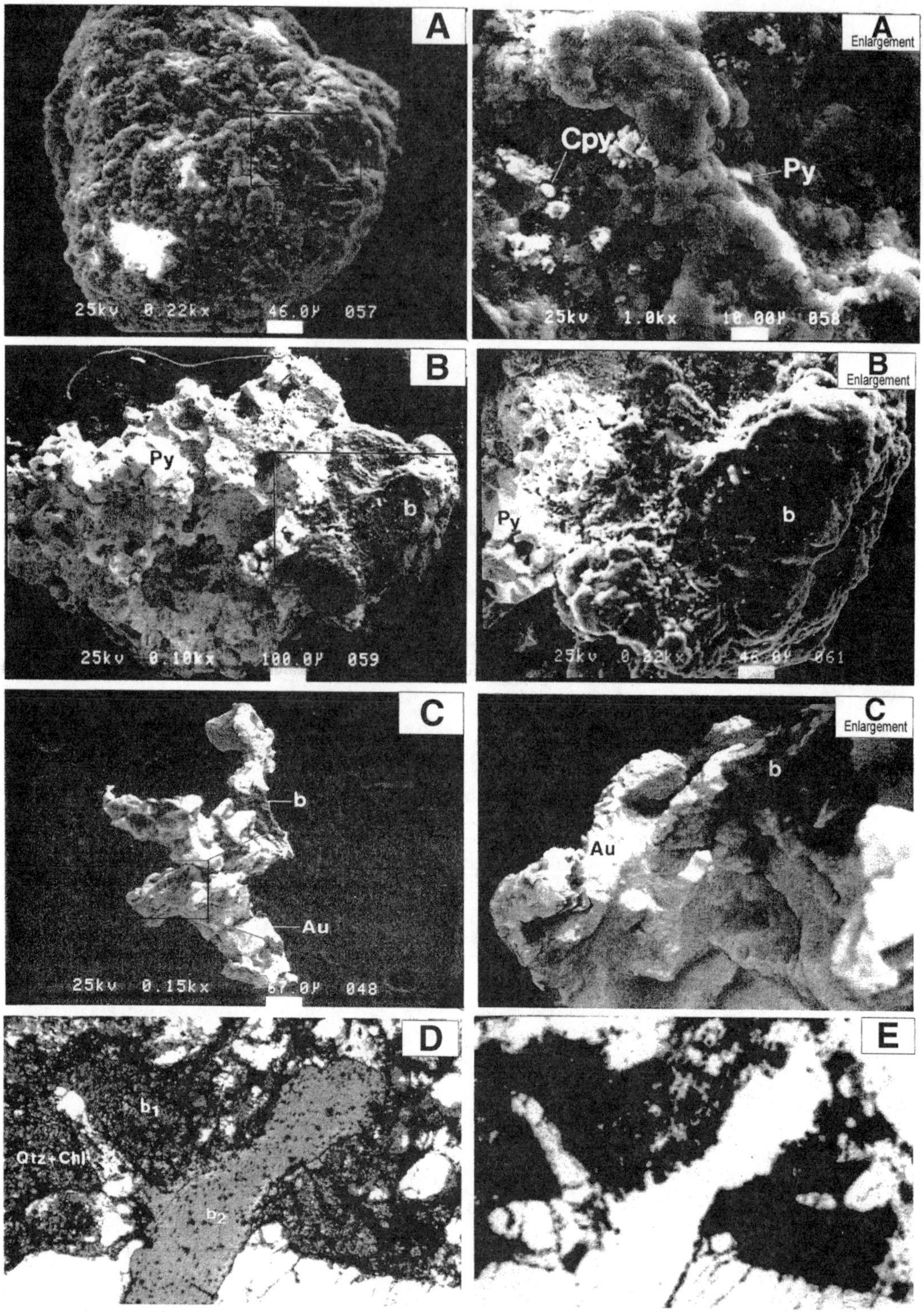

Figure 7. (A) SEM (scanning electron microscope) photomicrograph of bitumen nodule extracted by acid dissolution from quartz vein material cutting through conglomerate, sample VRW5 (Vaal Reef West 5; see Table 1 for details). Small crystals on the surface of the nodule are sulfides, scale bar = 46 µm. (B) SEM photomicrograph of bitumen nodule intergrown with pyrite extracted by acid dissolution from quartz vein material cutting through conglomerate, sample VRW5, scale bar = 100 µm. (C) SEM photomicrograph of bitumen nodule intergrown with authigenic gold extracted by acid dissolution from quartz vein material cutting through conglomerate, sample VRW5, scale bar = 100 µm. (D) Two generations of bitumen: an early bedding-parallel seam (b_1) associated with numerous uraninite fragments, cut by a later bitumen vein (b_2) that is devoid of any uranium-bearing phase, width of field of view = 2 mm. (E) Fission-track micromap of the area in Figure 7D showing the contrasting incidence of radiation-emitting phases in the 2 generations of bitumen (after Drennan, 1997; Robb et al., 1997a, 1997b).Abbreviations: Cpy—Chalcopyrite; Py—pyrite; b—bitumen; Au—gold; Qtz+Chl—quartz + chlorite vein; b_1—bitumen generation 1; b_2—bitumen generation 2.

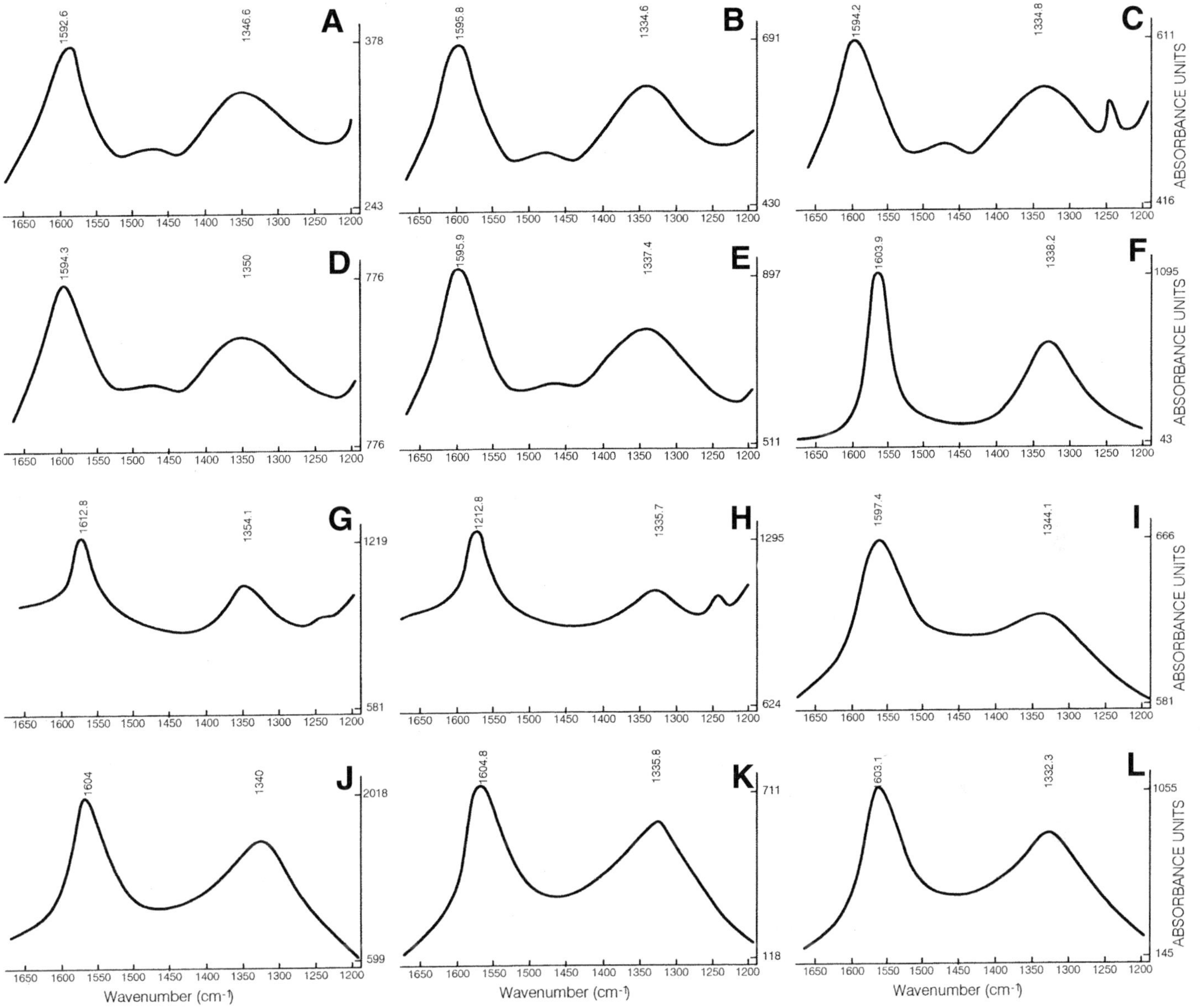

Figure 8. Examples of typical micro-Raman spectra for Witwatersrand carbonaceous matter, for Spectra (A–E): bitumen nodules in quartz veins; Spectra (F–H): hydrocarbon wall coatings to fluid inclusions; Spectrum (I): granite-hosted bitumen nodule; Spectra (J–K): seam bitumen; Spectrum (L): "fly-speck" bitumen.

TABLE 5. RAMAN SPECTRAL CHARACTERISTICS OF CARBONACEOUS MATERIAL IN QUARTZ VEINS AND FLUID INCLUSIONS

Sample	P_{1610} (cm^{-1})	h_{1610} (cm)	P_{1500} (cm^{-1})	h_{1500} (cm)	P_{1340} (cm^{-1})	h_{1340} (cm)
Nodule 1	1593	7.6	1475	1.3	1347	4.6
Nodule 2	1596	8.2	1480	1.3	1335	5.4
Pyrite-bitumen nodule 1	1594	8.1	1480	1.6	1335	5.0
Pyrite-bitumen nodule 2	1595	7.5	1470	1.2	1350	5.0
Pyrite-bitumen nodule 3	1560	8.3	1470	1.4	1337	4.5
Vapor-rich fluid inclusion	1604	10.3	1450	1.0	1338	6.1
Aqueous-rich fluid inclusion 1	1613	6.5	1480	0.6	1354	3.7
Aqueous-rich fluid inclusion 2	1613	6.8	1460	0.6	1335	3.2

Note: Data from Drennan (1997).

TABLE 6. RAMAN SPECTRAL CHARACTERISTICS OF WITWATERSRAND HYDROCARBONS

Hydrocarbon type	1610/1340	1500/1340	1610/1340	$1/2H_{1610}$
Quartz vein bitumen nodule 1	5.85	0.28	1.65	69.5
Quartz vein bitumen nodule 2	6.31	0.24	1.53	72.0
Pyrite-rich vein nodule 3	5.23	0.31	1.62	70.8
Pyrite-rich vein nodule 4	6.25	0.29	1.83	72.0
Nodule in crystallized pyrite	5.93	0.32	1.87	79.5
Carbonaceous lining of vapor-rich fluid inclusion	10.30	0.16	1.69	42.0
Carbonaceous lining of aqueous-rich fluid inclusion 1	10.83	0.16	1.76	55.7
Carbonaceous lining of aqueous-rich fluid inclusion 2	10.46	0.21	2.16	44.5
Granite bitumen nodule TKB3*	3.00	0.43	1.30	95.0
Seam bitumen (Beisa Reef)*	4.29	0.26	1.12	52.5
Seam bitumen (Vaal Reef)*	3.50	0.34	1.21	78.0
"Fly-speck" bitumen*	3.22	0.42	1.36	73.0

Notes: Data from Drennan (1997); *data from Robb et al. (1992).

Bitumen Nodules in Quartz Veins (Figs. 8A–8E)

Only spectra obtained for authigenic bitumen nodules sampled in quartz veins exhibit a minor peak at around 1500 cm^{-1}. Signal intensities for 1610 cm^{-1} range from 378 to 897. The 1610/1340 peak height ratios measure 5.23–6.31 (Table 6).

Wall Coatings to Fluid Inclusions (Figs. 8F–8H)

Vapor-rich inclusion linings (Fig. 8F) exhibit well-defined peaks around 1610 cm^{-1} and 1340 cm^{-1}, with the signal intensity of the 1610 cm^{-1} peak measuring 1095. The 1610/1340 ratio measures 10.3 (Table 6), indicating a higher degree of ordering for this carbonaceous matter compared to the nodules in quartz veins.

Aqueous-rich inclusion linings (Figs. 8G and 8H) have a minor peak around 1240 cm^{-1}. The signal intensity of the 1610 cm^{-1} peak is higher (1220–1295) than for vapor-rich inclusions. These inclusions also exhibit greater differences of peak height to background. The 1610/1340 ratios are somewhat higher than those obtained for vapor-rich inclusions (10.46–10.83; Table 6). This suggests a better-ordered structure for the bitumen in the aqueous-rich inclusions.

The above-mentioned spectral characteristics can be compared to spectra obtained for previously analyzed (Robb et al., 1994, 1995) Witwatersrand-related bitumen, as discussed below.

Granite-Hosted Nodules (Fig. 8I)

These samples exhibit a relatively well defined peak at 1610 cm^{-1}, with a less well defined, broader peak at 1340 cm^{-1}. Signal intensity of the 1610 cm^{-1} peak approaches 670. The 1610/1340 ratio measures 3.00 (Table 6), suggesting relatively poor organization of bitumen.

Seam Bitumen (Figs. 8J–8K)

Seam bitumen is characterized by peaks at 1610 cm^{-1} and 1340 cm^{-1}. Signal intensity of the 1610 cm^{-1} peak for a Beisa Reef sample (Fig. 8J) exceeds 2000, whereas for the Vaal Reef sample (Fig. 8K) the signal intensity is only slightly higher than 700. The difference of peak height to background for the Beisa Reef is greater than that for the Vaal Reef. The 1610/1340 ratios of the Beisa Reef and Vaal Reef samples, respectively, are 4.29 and 3.50 (Table 6). This suggests that the structural organization of the Beisa Reef seam bitumen is better than that of the Vaal Reef seam bitumen (i.e., Beisa Reef bitumen has fewer aliphatic chains, possibly as a result of the increased uraninite content therein).

"Fly-Speck" Bitumen (Fig. 8L)

"Fly-speck" bitumen exhibits peaks at 1610 cm^{-1} and 1340 cm^{-1}. Signal intensity of the 1610 cm^{-1} peak reaches 1055. The 1610/1340 ratio is 3.22 (Table 6), similar to that obtained for seam carbon, which suggests similar ordering for these bitumen types.

Raman spectral data for quartz vein–hosted bitumen indicate that the 1340 cm^{-1} peak, which is attributed to the presence of defects in the carbon structure, is as well developed as the 1610 cm^{-1} peak. This suggests that the analyzed bitumen generally has a poorly organized structure. The presence of a small peak at around 1500 cm^{-1} for the bitumen nodules sampled in late-stage quartz veins is indicative of additional components in the bitumen. Raman cross sections of selected pure hydrocarbons (Stephenson, 1974) indicate a methane peak in the low-frequency region at 1535 cm^{-1}. It is possible, therefore, that the observed peak for the quartz vein–hosted nodules indicates the presence of methane inclusions (either liquid or vapor) within this bitumen. This suggestion is supported by the abundance of methane in the vapor phases of fluid inclusions that occur in quartz veins from which authigenic bitumen nodules were extracted (Drennan, 1997; Robb et al., 1997a, 1997b).

Raman spectra recorded for bitumen sampled in late-stage veins (authigenic nodules and fluid inclusion wall linings) are compared to other bitumen data pertaining to Witwatersrand-related rocks (granite-hosted nodules, seam bitumen, and "fly-speck" bitumen, as well as late-stage vein bitumen). The plot showing the inverse relationship between the 1610/1500 peak intensity ratio and the 1500/1340 peak intensity ratio (Fig. 9) emphasizes distinct differences between the various bitumen types. Figure 10, which displays the intensity ratio of the two main bands versus the half height width of the 1610 cm^{-1} band, further illustrates the differences between bitumen types, whereby two trends have been identified distinguishing between distal/vein-related bitumen and sediment-hosted bitumen (Drennan, 1997; Robb et al., 1995, 1997a, 1997b).

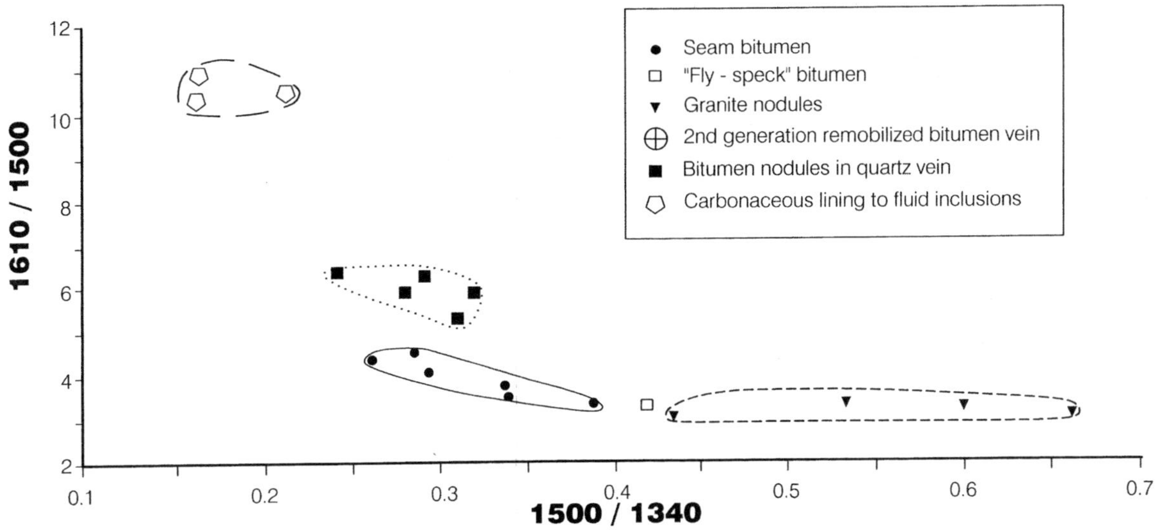

Figure 9. Plot of the 1610/1500 peak intensity ratio versus the 1500/1340 peak intensity ratio for Witwatersrand bitumen.

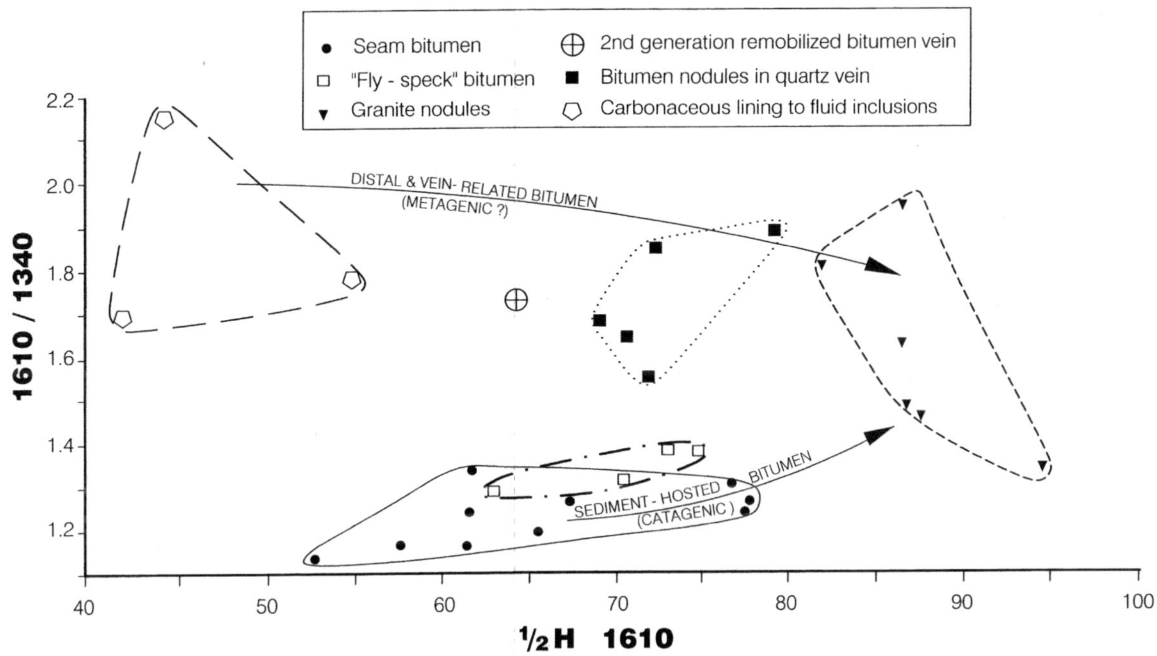

Figure 10. Plot of 1610/1340 peak intensity ratio versus half-height/width (1/2H 1610) of the 1600 cm^{-1} band for Witwatersrand bitumen.

Contrary to the meta-anthracite or graphite series, which generally exhibit a good correlation between the 1610 cm^{-1} band width and the 1610/1340 intensity ratio (Oh, 1987; Wang et al., 1989), no significant relationship is observed in the Witwatersrand seam and "fly-speck" bitumens. Distal/vein-related bitumen samples (in fluid inclusions and late-stage veining), on the other hand, exhibit a better correlation (Fig. 9). Seam bitumen and "fly-speck" nodules have very low 1610/1340 intensity ratios. Seam bitumen exhibits scattering along the X-axis (Fig. 9), which has been correlated with the uranium content of the organic matter and proximity to uraninite grains (Landais et al., 1990a). The spectral characteristics of the "fly-speck" bitumen (Fig. 9) are similar to those of the carbon seams, although the 1610/1340 ratios of the latter are somewhat higher, suggesting that the "fly-speck" nodules are less well organized than the seams. The observed variations for seam bitumen could be explained by hydrocarbon fixation due to radiolytic polymerization by beta- and gamma-radiation. The difference between seam and "fly-speck" bitumen would then suggest that the "fly-speck" carbon has been subjected to a lower degree of nuclear bombardment and radiolytic polymerization and is, therefore, less well structured than seam bitumen.

Carbonaceous wall coatings to fluid inclusions are characterized by the highest 1610/1340 intensity ratios. Bitumen nodules that coprecipitated within the same late-stage quartz veins have 1610/1340 intensity ratios intermediate with respect to ratios for fluid inclusion linings and seam/"fly-speck" samples (Fig. 9). These authigenic bitumen nodules exhibit similar characteristics to those documented for the late, second generation hydrocarbon vein (Fig. 7D; Drennan, 1989) that crosscuts an earlier seam (Fig. 9; Landais et al., 1990a). Neither the crosscutting bitumen vein nor the quartz vein nodules are associated with uraninite, although the latter nodules are characterized by enriched U (and Au) concentrations (Table 7; Fig. 11). In contrast to granite-hosted nodules, the quartz vein nodules exhibit pervasive U enrichment throughout the nodule, sometimes showing preferential concentrations toward the nodule center (Fig. 11).

Granite-hosted nodules are characterized by the largest 1610 cm^{-1} peak half height width values and exhibit 1610/1340 intensity ratios similar to those of quartz vein nodules and migrated hydrocarbon veins (Fig. 9). This implies that the granite-hosted bitumen comprises a hydrocarbon molecule that is structurally better organized than the sediment-hosted bitumen (seam and "fly-speck") but not as well organized as the carbonaceous fluid inclusion coatings, the bitumen in crosscutting veins, or the quartz vein nodules. The structure of the granite-hosted bitumen may reflect the smaller proportion of bitumen relative to replaced uraninite or uranothorite (Robb et al., 1997a, 1997b) in the granite host rock. It is interesting to note that the granite-hosted nodules also exhibit enriched Au concentrations (Table 7), although values are not as high as in the quartz vein nodules.

Compositions of Witwatersrand Bitumens

Five authigenic bitumen nodules were analyzed using the micro–PIXE facility at the Schonland Research Centre, University of the Witwatersrand (now part of iThemba Laboratories). The facility makes use of the quadrupole triplet that has been installed on the 30°W line of the EN tandem at the Centre. A 3.2 MeV proton beam was used, and the beam current varied between 500 pA and 1 mA, with a charge of 0.2 to ~2 °C, and a beam spot of ~20 × 20 μm. The proton-induced X-rays were detected using an ORTEC Si(Li) detector. An electron suppression voltage of 200 V was used. The absorber between the sample and the detector was an aluminum layer of 0.2 μm thickness. A sample matrix comprising carbon and uranium was assumed. The accumulation of data and its organization into spatial maps was achieved using a specially developed version of a multi-parameter nuclear data analysis program known as WIMPS (Witwatersrand Interactive Multi-Parameter System; Watterson et al., 1993). All PIXE spectra were analyzed using the GUPIX software package, which gives elemental concentrations and errors of limits of detection, on the basis of a calculated physical model of the PIXE process (Maxwell et al., 1989; Przybylowicz et al., 1993). The errors include the fit error, the statistical error, and also the systematic error due to the separation of peaks in the spectra. The limits of detection are calculated on the basis of three standard deviations of the background areas over a peak region. Complete nodule scans were carried out, and elemental maps were constructed. Spot scans were undertaken for areas exhibiting high Au, U, Th, Cu, and As concentrations as well as for high and low Fe concentrations. Elemental concentrations are listed in Table 7 and PIXE maps are presented in Figure 11.

Hallbauer (1986) distinguished chemical compositions of seam- and "fly-speck" carbon in the Witwatersrand sediments and argued, on the basis of differences in trace element content, that the former represented syngenetic relict algal mats, and the latter was more akin to granite-hosted nodules. Hallbauer et al. (1986) maintained that granite-hosted nodules were enriched in Na, Ca, Cr, Ni, Cu, and Co but depleted in Sr, Ba, Zn, Ag, Au, and Cd, relative to seam carbon. Robb et al. (1992, 1994, 1997a, 1997b) noted that granite-hosted nodules were markedly enriched in total REE content, relative to the host granite, exhibiting HREE (heavy rare earth element) enrichment and a negative Eu anomaly. They suggested that this REE pattern derived its signature from uranium minerals being replaced in the granite and that this should also apply to the sediment-hosted seams and nodules. There are now well-documented REE signatures for uraninite from different ore type settings (Fryer and Taylor, 1987; Pagel et al., 1987). Robb et al. (1992) also noted that REE patterns from Witwatersrand seam bitumen were similar to those obtained for granite-hosted bitumen and that the sediment-hosted bitumen derived its pattern from the uraninite with which it is intimately associated. Zumberge et al. (1978) established that a Vaal Reef bituminous seam contained higher REE abundances (by factors of 2–100 more) than average crustal rocks. They found that differences in REE distribution patterns occurred among samples from different localities within the Vaal Reef and suggested that these variations indicated differences in the absolute abundances of certain detrital minerals. The Witwatersrand bitumens, therefore, have a REE signature not unlike that of most uraninites, indicating that the carbonaceous material derives its chemical signatures from the uraniferous material it replaces.

The compositions of five nodules were determined by micro–PIXE analysis in order to examine the REE concentrations and distributions within the nodules. Full area scans (FS) were carried out in order to produce PIXE maps of selected elements (Fig. 11) and to determine a bulk REE concentration for each bitumen nodule (Table 7). Spot scans (SS) were then undertaken on areas within individual nodules exhibiting elemental enrichment to better determine the precise chemical composition of specific locations within the grains (Table 7).

TABLE 7. REE AND OTHER TRACE ELEMENT ANALYSES (DATA IN PPM) OF MIGRATED BITUMEN NODULES FROM LATE-STAGE QUARTZ VEINS, WITWATERSRAND BASIN

	Nodule BIT1			Nodule BIT2				Nodule BIT3					Nodule BIT4			Nodule BIT5			
	FS	SS high Fe	SS low Fe	FS	SS high U	SS high Th	SS high Fe	FS	SS high U	SS high Au	SS high As	SS high Fe	SS high U	SS high As	SS high Fe	FS	SS center	Granite*	Seam*
K	<1080		976	1460?	11532	15885			5756										6600
Ca	404		297	1240	4734	18368	1330?	2950	6690	4256	8944	29548				3780			
Sc																			22
Ti	<43		429		<80	280	280?			13504						170?			
Fe	7330	95851	717	4740	1347	9471	206490	1360	2054	1088	4225	55987	6261	6828	26784	340	1075		1700
Co	132		109					450	7770	814	16544	1968		5458					226
Ni	91		82					330	1420	3533	4050	2496	753	1164	359				
Cu	291	236	229	2434	1869	10366	41570	1130	84	571	200	60094	4915	5656	9982	220	326		
Zn								127	422	154	220	425							
As	249	532	186	<54	<59	<150	170	1670	12057	8520	30650	6716		8361	451				1430
Br				2489	4814	3761	1360	5450	5218	8338	6402	9106				1730	2606		
Sr				<40	<46	261		60?				214?	83?					197	
Y	2480	1981	2036	2762	7927	7232	480	620	4056	95	763	253	11391	4789	3394	5330	13294		
Zr	394?	359	327	617?	1805	1081?	550		354?	33?	103?	577	3461		638	1100	1352		
Ag									40?	3755									108
Cs				<180	1350														4
Ba				760	1721	17744			292					4311	2929	<280		1802	
La															2235			201	275
Ce	203?	439	176	164?	1642	2747?	600?	1250	810		878	4116	1309	300	185	250?	234	607	
Pr									442		78		1097						
Nd	437	162	321	590	4331	4504	130?	860	1238	66?	584	2173	2638	1196	683	850	5063	483	
Sm	196?		162	188?	1584	1724	<140		360	104			1631				2452	187	
Eu			44?	<70	609	78	<140		83?	102		704					736?	43	34
Gd	397	509	250	320	1962	1907	<260		715	82		1912	2287		295?	400?	3257		
Tb				<460	488?	1213?			327			2629	1191?					78	28
Dy			<110	<600	1283	1011			757?				4079			520?	2334		
Ho			<40	<70	<100	<100	<370		316?		308								
Yb	149?	67	80	<46	282?	124?	<100									190?	758	215	172
Au				32?	37?	33?	80?	1860	97	29167	152	408	143?	<46		<40	<40	0.09	0.3
Hg	182		7205	39?	29?	<40	40?		66?								<40		
Pb	1907	2234	2516	1750	2660	9869	1059		1675			2120	16102	6791	4016	7120	29408		
Ac								820			121?								
Th	8059	9233	4463	19064	34576	125676	7050	2390	8646	615	3852	1355	42776	37458	21351	12140	27076	2425	4470
U	25269	24786	19582	48625	148775	184511	14580	9530	7899	1751	20468	5568	156612	80185	41020	48270	160274	6040	37900

Note: Data from Drennan (1997). FS—full scan; SS—spot scan; Granite*—Granite-hosted bitumen nodule (INAA [Instrumental Neutron Activation Analysis] data from Robb et al., 1992); Seam*—Sediment-hosted Vaal Reef seam bitumen (INAA data from Robb et al., 1992); 1081?—definitely detected but percentage error in detection limit relatively high; Blank—not detected or below detection limits; REE—rare earth elements.

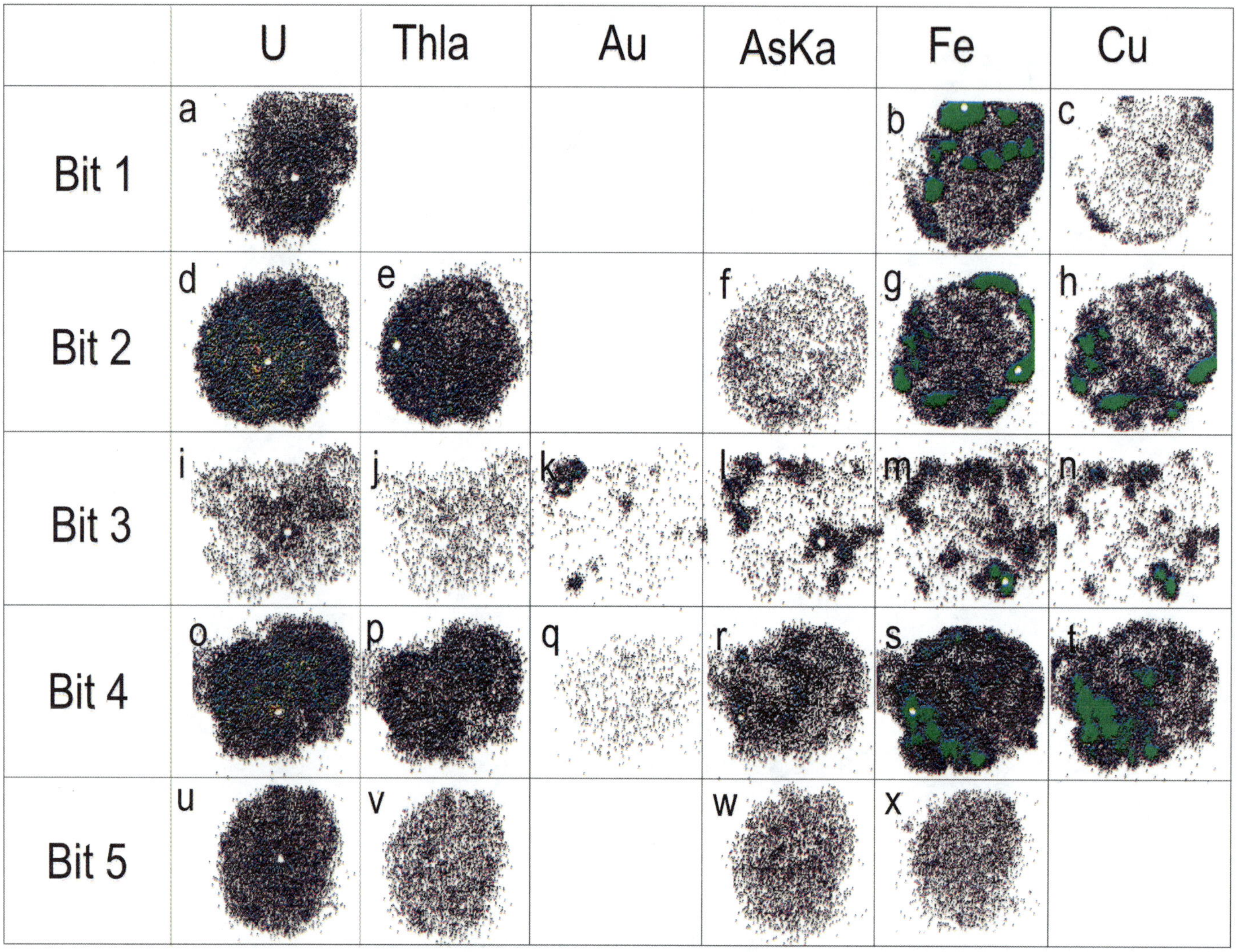

Figure 11. PIXE maps showing the distribution of uranium, thorium, gold, arsenic, iron, and copper in five bitumen nodules from late-stage quartz veins in the Witwatersrand Basin. White areas show no counts whereas green, yellow, and red areas show increasingly higher count rates. Full scan (FS) size for BIT1 and BIT2 is 0.5 × 0.5 mm, whereas for BIT3, BIT4, and BIT5 it is 0.4 × 0.4 mm. Spot scan (SS) positions are illustrated by white dots within areas of high elemental concentrations (after Drennan, 1997).

These values are compared in Table 7 to data for a granite-hosted bitumen nodule (Robb et al., 1992) and sediment-hosted seam bitumen (Zumberge et al., 1978).

Results of Proton-Induced X-Ray Emission Analysis

When examining the PIXE maps of all 5 vein nodules (Fig. 11), certain similarities are apparent.

1. U and Th do not form coatings around the edges of bitumen nodules but are pervasively distributed throughout individual nodules. Uranium concentrations are, however, sometimes higher away from grain edges (Figs. 11D and 11O). This suggests that the fluids from which the nodules were precipitated contained uranium in solution (which may have existed as urano-organo complexes) and that the uranium and hydrocarbons coprecipitated. It must be noted that no particulate grains of uraninite were observed in vein-hosted nodules, as is the case for seam- and "fly-speck" bitumen.

2. Arsenic may be pervasively distributed throughout some nodules (Figs. 11F, 11R, and 11W) but also occurs in localized concentrations (Figs. 11L and 11R). Localized concentrations of As in Figure 11L in part correlate with Au enrichment in Figure 11K, suggesting possible Au-As complexing.

3. Au occurs distributed evenly throughout some carbonaceous nodules in relatively low abundances (Figs. 11K and 11Q) as well as in localized high concentrations representing

discrete gold phases (Fig. 11K). Pervasively distributed Au indicates the possibility of organo-gold complexing and/or the coprecipitation of gold and uranium-rich hydrocarbons. Localized concentrations of Au, associated with As, Fe, and Cu, suggest that gold and sulfide minerals coprecipitated with the uranium-enriched bitumen. This may point to the existence of thio-gold and urano-organo complexes in the fluids, although this aspect requires further study to demonstrate the point unequivocally.

4. Fe and Cu vary markedly in content from nodule to nodule and occur pervasively distributed throughout the nodules (Fig. 11X) as well as in particulate form (Figs. 11B, 11C, 11G, 11H, 11M, 11N, 11S, and 11T). This again supports the concept of urano-organo-metal- and thio-metal-complexing in the same fluids. Optical investigation and SEM (scanning electron microscopy) analysis confirmed the presence of coprecipitated pyrite and chalcopyrite crystals in bitumen nodules (Fig. 7A Enlargement).

The REE and trace element concentrations analyzed by micro–PIXE (Table 7) are extremely variable within individual vein nodules and vary greatly from one nodule to another. In general, the vein nodules are characterized by total REE enrichment relative to previously analyzed Witwatersrand bitumen. Uranium and Th contents as well as Au concentrations in the vein nodules are markedly enriched compared to granite-hosted nodules reported elsewhere (Hallbauer et al., 1986; Robb et al., 1992, 1994). High Co and Ni contents are probably related to the presence of sulfides. Arsenic, Ag, and Hg may be associated with Au; however, native arsenic has been described for other hydrothermal veins associated with Ag, Co, and Ni ores (Svenek, 1987). High, localized zinc concentrations suggest the presence of sphalerite inclusions in the bitumen, and galena probably accounts for much of the Pb content. This is highly probable as sphalerite and galena form part of the sulfide suite observed within late-stage quartz veins. Localized concentrations of titanium possibly reflect inclusions of rutile in the nodules, a feature supported by the fact that rutile needles have also been identified as solid inclusions within fluid inclusions trapped in quartz veins (Drennan et al., 1997). Zirconium is generally considered an immobile element; however, hydrothermal zircons or zircon overgrowths have been described in hydrothermal veins (Johnson and McIntyre, 1983, 1986) and in mineralized shear zones (Gates, 1987) and have also been observed as overgrowths on detrital zircon grains in Witwatersrand sediments (Frimmel et al., 1999). Preliminary attempts to date these hydrothermal zircons yielded an age of ca. 2.55 Ga (Frimmel, 1994). A similar age of ca. 2.58 Ga was obtained for rutile from the West Rand Group (Robb et al., 1990). These ages point to mobilization of Ti and Zr (as well as U and Au?) during hydrothermal infiltration at this time, an event that may be related to loading of the basin during early stages of Transvaal Supergroup deposition.

Chondrite-normalized REE patterns obtained by PIXE spot scans (Drennan, 1997) of individual vein nodules appear to be complex and do not resemble the patterns for granite-hosted nodules or seam bitumen (Fig. 12A). Of the five nodules, BIT1 is the least enriched in total REEs (Fig. 12B); it does exhibit high Sm and Gd concentrations and a negative Eu anomaly. BIT2, BIT3, and BIT4 are characterized by REE enrichment (Figs. 12C–12E), whereas BIT5 has the highest REE abundances of all five grains (Fig. 12F). BIT3 and BIT5 (Figs. 12D and 12F) also exhibit negative Eu anomalies, and both nodules have elevated Pr contents. BIT3 also has very high amounts of Tb present, whereas BIT5 has very high Sm and Gd concentrations. Although BIT4 has no Eu anomaly, it is characterized by high Pr and Tb abundances (Fig. 12E). BIT2 exhibits a positive Eu anomaly and very high Tb concentrations (Fig. 12C). These highly variable REE patterns do not approximate REE patterns of uraninite or uranothorite. It cannot, therefore, be argued that the REE patterns of vein nodules derive their signatures from any preexisting uranium-rich mineral but rather that these bitumen nodules coprecipitated in quartz veins are chemically unique, having precipitated from chemically complex, hydrocarbon-rich, metalliferous fluids. It is suggested that point sources (spot scans) probably relate to actual mineral inclusions within the bitumen nodules (see Fig. 7B), whereas disseminated metal distributions relate to where metals may have been complexed with hydrocarbon ligands.

DISCUSSION

The present study describes the setting and textural characteristics of Witwatersrand bitumens as well as the nature of fluids that were implicated in the migration of these hydrocarbons through the Central Rand Group sediments and surrounds. It also presents the first direct evidence for an association between aqueo-carbonic- and hydrocarbon-bearing fluids and the migration of metals, including gold and uranium. The following sections discuss the implications of these data for Witwatersrand metallogenic models, specifically with respect to the nature and origin of carbonaceous matter and related fluids as well as their likely role in the remobilization of metals during the postdepositional history of the basin.

Bitumen or Kerogen?

The light carbon isotopic signature of carbonaceous matter in the Witwatersrand Basin clearly indicates that it is organic and biogenic in origin (Hoefs and Schidlowski, 1967; Spangenberg and Frimmel, 2001). Nevertheless, the question of the source rock for the carbonaceous material in the basin remains problematic, as fluid pathways are obscure and the distances of hydrocarbon migration are effectively unknown. Earlier suggestions that the source rock was not within the Witwatersrand Basin but in an overlying sequence such as the 2550 Ma dolomites of the Transvaal Supergroup, which contained abundant organic material now preserved as algal stromatolites, are not supported by the available evidence. Spangenberg and Frimmel (2001), for example, have demonstrated that carbona-

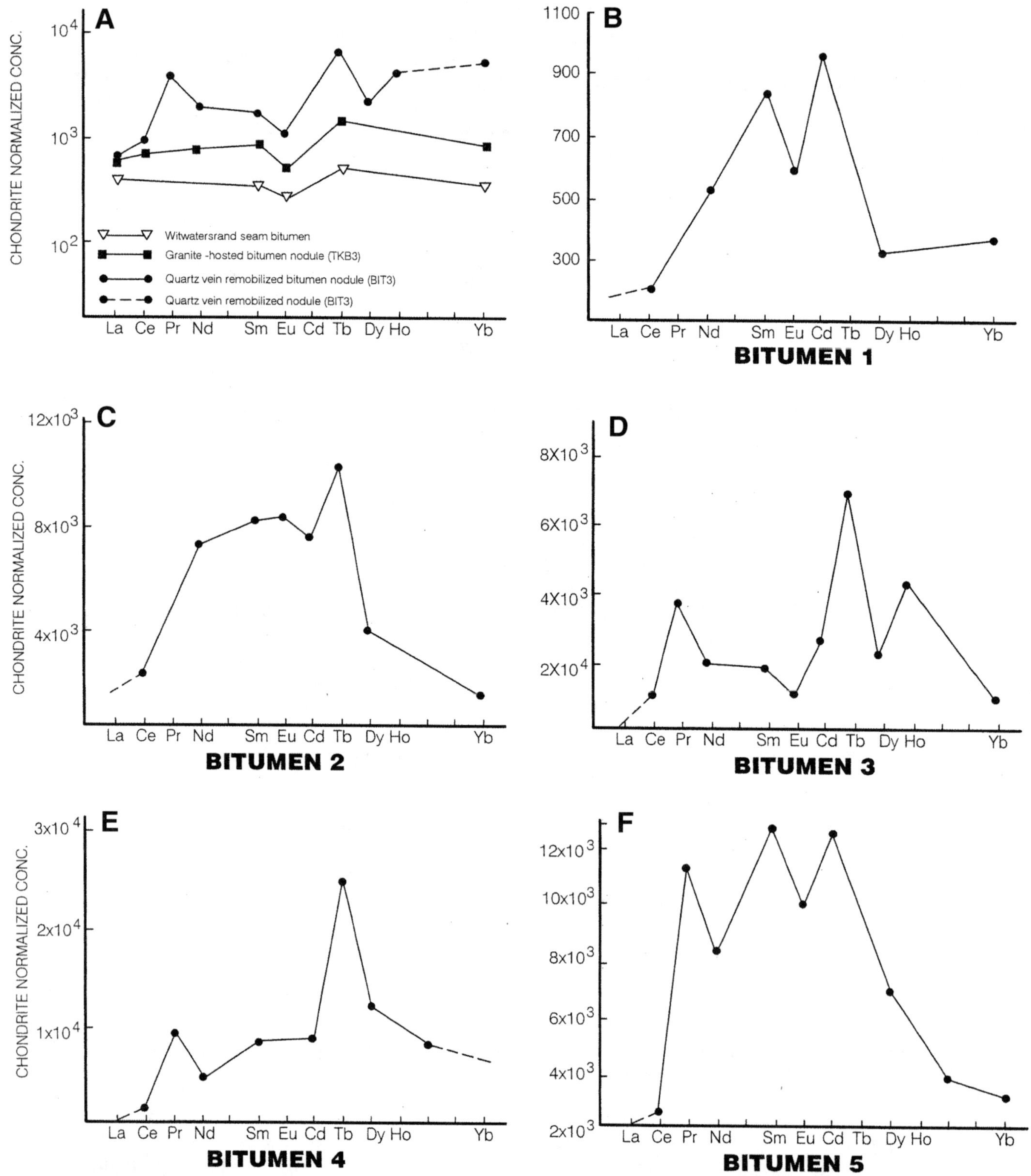

Figure 12. C1 Chondrite-normalized REE (rare-earth element) patterns for: (A) Witwatersrand sediment-hosted seam bitumen, granite-hosted bitumen nodules (Robb et al., 1990), and bitumen nodules extracted from late-stage quartz veins. (B–F) Individual bitumen nodules extracted from a late-stage quartz vein (after Drennan, 1997).

ceous matter from Transvaal dolostones have different bulk compositional and C isotopic characteristics and must be precluded as a source of Witwatersrand bitumen. They also point out that the kerogen (or insoluble organic) fraction of carbonaceous matter in the Witwatersrand basin is characterized by variable $\delta^{13}C$ values (–24 to –39‰), whereas the mobile (or extractable) fraction has a more restricted $\delta^{13}C$ signature (–25 to –27‰) that nevertheless overlaps with the bulk of the kerogen data. The more variable kerogen signature reflects primary inhomogeneities of the algal or bacterial source organisms, while the data overlap points to a genetic link between the mobile hydrocarbons and an indigenous kerogen.

It is considered likely, therefore, at least given the present database, that the source rocks were actually located within the basin and that the unconformity surfaces on which most of the major conglomeratic units were deposited represent the main sites along which prokaryotic microbiota were developed. Shales that occur mainly in the lower part (West Rand Group) of the Witwatersrand Supergroup might also represent a source, but this would have been a minimal component as the carbon content of these units is generally low.

If, as seems likely, the organic source rocks were within the Witwatersrand Basin, then both kerogen and migrated organic matter exist in the sequence. Much of the carbonaceous matter observed in the basin is bitumen and can be recognized texturally as a now-solidified migrated product. The insoluble kerogen fraction is essentially unrecognizable texturally, as are the remnants of the original source material. Since the bitumen and kerogen in the Witwatersrand sediments are so similar both texturally and in terms of reflectance, it seems likely that oil generation would have been local, as suggested by Spangenberg and Frimmel (1999). However, the question of whether the original organic source was close to the present sites of bitumen accumulation (i.e., on the unconformity surfaces upon which the conglomerates were accumulated) or in shales from lower down in the stratigraphic column remains unresolved.

Origin and Evolution of Witwatersrand Carbonaceous Matter

Sediment-hosted carbonaceous matter that is Archean in age will have been derived from the biodegraded products of prokaryotic microorganisms in the form of algae and bacteria. The duration of hydrocarbon production and the period of time subsequent to sediment deposition when the oil window is attained are highly variable and depend largely on subsidence rates. Oil production can be ongoing for any amount of time between 5 and 100 m.y., whereas the elapsed time between sedimentation and petroleum generation may vary from a few m.y. to more than 300 m.y. (Tissot and Welte, 1978; Rasmussen, 2005). In the Witwatersrand Basin, fossilized prokaryotic microorganisms would have undergone normal biodegradation and diagenesis to form Type I (algal derived) kerogen. Colonies of microbiota may have developed on the unconformity surfaces that represent hiatuses in sedimentation. Rapid sediment aggradation at these sites would have preserved and fossilized this material, although it seems doubtful that it played a significant role in the concentration of heavy minerals within the conglomerates. The basin was deposited on what had, by 2700 Ma, become a relatively stable protocontinent and was subjected to slow and protracted subsidence over the following 700 m.y. The period of maximum oil generation would, therefore, have been a long-lived event (perhaps 10^1–10^2 m.y.), and the time that elapsed between the end of sedimentation and the onset of hydrocarbon generation might have been of similar duration depending on whether the organic source rocks were in the West Rand or Central Rand Group. As in the Huronian Basin, kerogen and some bitumen may be spatially juxtaposed, indicating limited migration distances (Mossman and Nagy, 1996). Most of the organic material in the Witwatersrand Basin is intimately associated with uranium. By contrast, other uranium deposits containing significant carbonaceous material, such as those at Oklo, in Saskatchewan, in the Huronian, and in Lodeve and Arizona, contain bitumens that are both mineralized and barren with respect to uranium (Landais et al., 1990a, 1990b, 1993). This supports the view that Witwatersrand bitumens were largely solidified by radiolytic polymerization, a feature which is not always the case in other uranium deposits.

Petrographic studies reveal that carbonaceous matter in and around the Witwatersrand Basin occurs in a variety of settings and is texturally variable. In addition, it is variable in terms of chemical composition and degree of structural organization. Although highly mature, the carbonaceous matter exhibits a wider range in bulk composition than expected of unaltered, algal-derived kerogen, and this reflects modification related either to demethanation or oxidation/reduction. Many of the organic factors described above vary systematically as a function of the uranium content of the carbonaceous matter, and this is consistent with the notion that radiolysis, rather than the more typical manifestations of hydrous pyrolysis, played the major role in compositional and structural modifications (Robb et al., 1997a, 1997b).

Carbonaceous Material and Paragenetic Sequence

Many ore deposits, especially uranium and base metal sediment-hosted types, contain significant amounts of carbonaceous matter (Landais et al., 1993; Parnell, 1988, 1993). The timing of bitumen formation in relation to ore deposition is variable, and examples of pre-, syn- and post-ore hydrocarbons have been documented. In certain cases, bitumens are genetically unrelated to the process of ore formation. In many cases, however, organic matter has played a role in the precipitation of metals through redox reactions or absorption. In a few cases a close genetic link between ore and hydrocarbon suggests that organo-metallic complexes may have been implicated in metal transport in the aqueo-carbonic solutions responsible for concentration of ore components. In the Witwatersrand deposits the latter two scenarios may both have applied, as carbonaceous matter is intimately associated with gold and uranium, the deposition of which is reflected in a complex and long-lived paragenetic sequence (Robb et al., 1997a, 1997b; Frimmel, 2005).

A scheme illustrating the relative paragenetic sequence of ore formation in the Witwatersrand Basin is shown in Figure 13. The initial concentration of allogenic ore constituents, including uraninite, took place during sedimentation, which for the upper auriferous part of the sequence took place between 2900 and 2700 Ma. This would also, therefore, be the period when prokaryotic microbiota flourished in suitable environments within the basin. Postdepositional ore modification took place for at least 700 Ma after termination of sedimentation and probably occurred during at least three episodes of fluid circulation. The earliest of these would have been diagenetic to late diagenetic at 2700 Ma, followed by circulation of metamorphic fluids at ca. 2500 Ma (Frimmel and Gartz, 1997a, 1997b; Frimmel et al., 1993, 1999; Robb and Meyer, 1995; Robb et al., 1997a, 1997b). A final and very important event occurred at around 2020 Ma and was associated with meteoric fluid circulation linked to catastrophic exhumation of the central portion of the basin after the Vredefort impact event (Gibson and Reimold, 1999; Gibson et al., 1997; Reimold et al., 2005).

This scheme is different from those presented by several proponents of a purely hydrothermal origin for the gold and uranium mineralization in the Witwatersrand Basin. Barnicoat et al. (1997) and Phillips and Law (2000), for example, suggested that hydrothermal gold and uranium mineralization occurred very early in the postdepositional subsidence history of the basin (i.e., in pre-Platberg faulting times) during a single fluid fluxing event, which is contrary to studies that recognize a long-lived and episodic sequence of hydrothermal events (e.g., Drennan et al., 1999; Foya, 2002; Frimmel et al., 1999, 2005; Robb et al., 1997a, 1997b) and which is supported in the present work. This study supports the notion that there were multiple fluid events affecting the basin throughout its history. It should be noted, however, that the existence of a discrete fluid circulation event involving hydrocarbon production and fixation of bitumens at around 2380–2330 Ma (Robb et al., 1994) receive little support from more recent work that questions the time significance and integrity of U-Pb isotope ratios obtained from carbonaceous matter (Zartman and Frimmel, 1999). The present study concurs with others suggesting that hydrocarbon migration and bitumen fixation occurred prior to the final major authigenic mineral-forming event at ca. 2020 Ma, which also coincides with the attainment of peak metamorphic conditions in the basin (Stevens et al., 1997; Gibson and Reimold, 1999) and major remobilization of late sulfides and gold (Foya, 2002; Meyer et al., 1991; Reimold et al., 2005; Robb and Meyer, 1995). The presence of abundant gold associated with but postdating carbonaceous matter indicates that a substantial amount of the metal was precipitated late in the paragenetic sequence. The estimates of peak metamorphic conditions in the Witwatersrand goldfields are not inconsistent with the degree of maturation exhibited by the organic matter present in the sediments. These constraints need to be more rigorously tested, however, as the relationship between the metamorphic grades of organic material and host sediments has not been well established.

Role of Organic Fluids and Carbonaceous Matter in Remobilization and Fixation of Metals during the Mineralization Process

The role of organic matter in the genesis of ore deposits is multifaceted and complex (Saxby, 1976; Parnell, 1993). At low temperatures microorganisms such as bacteria and fungi are able to concentrate uranium and other metals from solution (Milodowski et al., 1990). Mature organic matter is also often observed as a site of preferred metal precipitation because it acts as a redox barrier to the passage of metal-charged hydrothermal solutions. In addition, however, hydrothermal solutions themselves often contain substantial amounts of organic ligands, which enhance their metal-carrying capacity by leaching and metal complexation along the fluid pathway.

It is now considered unlikely that mats of prokaryotic microorganisms growing on unconformity surfaces would have been capable of physically trapping uraninite, gold, and other heavy mineral detritus and concentrating them during sedimentation. The reasons are twofold: first, it is difficult to envisage how heavy particles would remain trapped in a high-energy depositional environment and, indeed, whether such biota existed in these environments at all. Second, uraninite is almost always the principal detrital phase that occurs within the carbonaceous seams with a notable paucity of equivalent detritus such as pyrite, zircon, or chromite. Consequently, the association between uranium and carbonaceous matter in the Witwatersrand Basin is one where primary concentrations of detrital uraninite represent the sites where bitumens were precipitated and solidified by the effects of ionizing radiation and radiolytic polymerization. By contrast, the association between gold and carbonaceous matter is probably related to the behavior of bituminous seams as redox barriers to the circulation of gold in solution. Robb and Meyer (1995) suggested that, irrespective of whether gold was in solution as a chloride- or thio-complex, a decrease in local fO_2 would have had the effect of precipitating gold out of solution. The gold-carbon association appears, therefore, to be chemical, and it points to the paragenetically late mobilization of gold in the Witwatersrand deposits.

In addition to these relatively obvious associations, there is now increasing evidence from the present study that organo-metal complexation occurred during the migration of hydrocarbon-bearing fluids in the Witwatersrand Basin. REE, Au, and U concentrations have been determined for bitumen nodules from fault-related quartz veins and are compared to concentrations obtained for granite-hosted nodules as well as some sedimentary-hosted seam bitumen. Significant concentrations of Au and U exist in bitumen nodules hosted in quartz veins compared to granite-hosted nodules. Gold concentrations in vein nodules

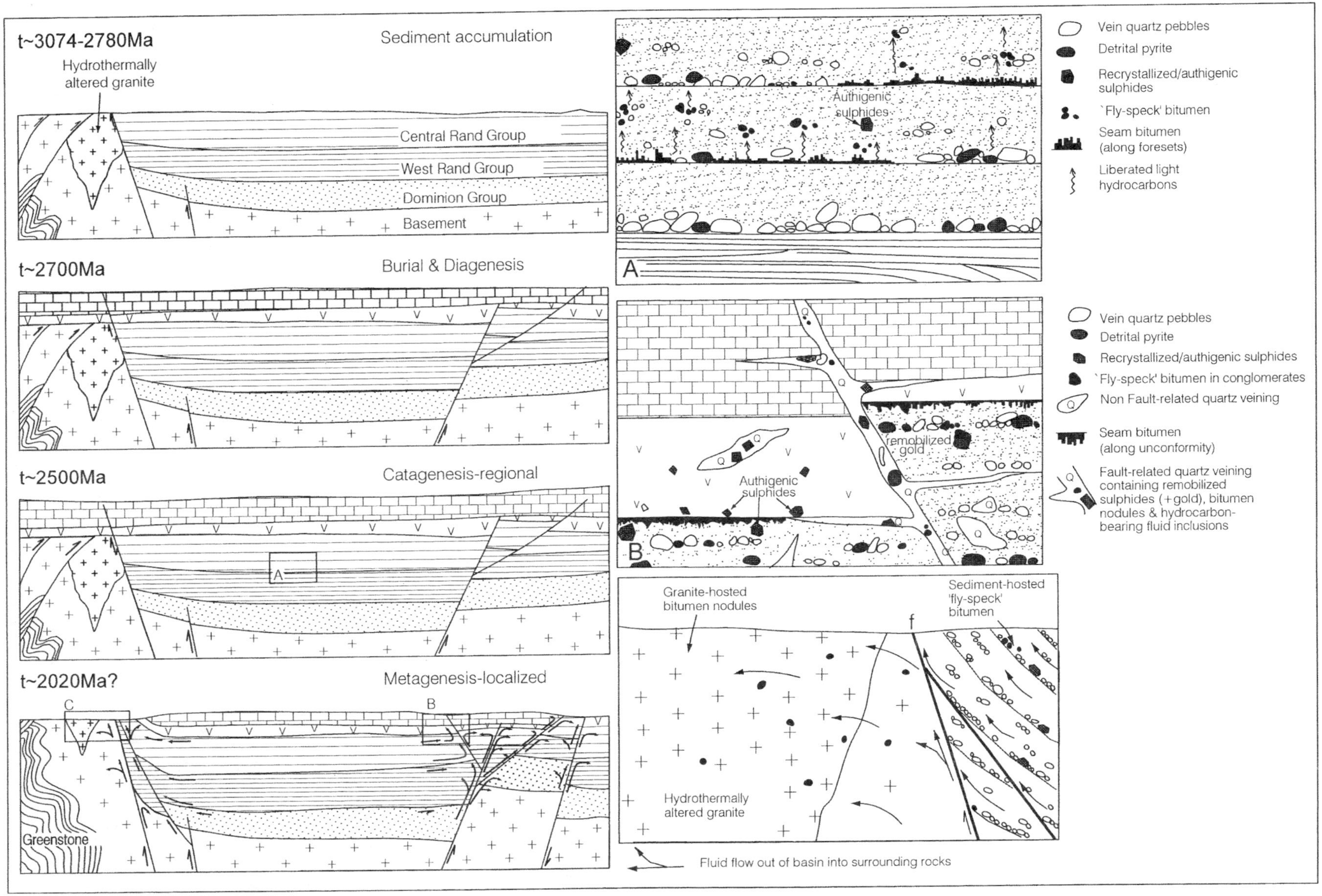

Figure 13. Diagrammatic representation of the development of the paragenetic sequence within the Witwatersrand Basin (see text for discussion; Drennan et al., 1997).

exceed those of carbon nodules in the Vaal, Basal, and Leader Reefs by approximately one to two orders of magnitude. Quartz vein nodules are depleted in U, however, when compared to reef samples. The Basal Reef, Vaal Reef, and Leader Reef nodules are roughly 1.5–3 times richer in U than quartz vein nodules.

The relation between gold and bitumen differs distinctly from that between uranium and bitumen. In sediment-hosted seam bitumen, gold occurs in veinlets and along seam edges, whereas uraninite occurs as inclusions within the bitumen. The uraninite inclusions typically form aggregates (Eakin and Gize, 1992) in which the shape of an originally larger detrital grain can be distinguished but where the constituent fragments have been moved apart (Figs. 5B and 5C). The inclusions exhibit varying degrees of replacement by hydrocarbons that are characteristically highly pleochroic adjacent to the fragments (Fig. 5C). The pleochroic halos demarcate the alpha-radiation envelope around individual uraninite grains (Schidlowski, 1981).

Precipitation of mobile hydrocarbons around preexisting uraninite as a result of radiation-induced polymerization processes was accompanied by simultaneous dissolution of uraninite (Eakin and Gize, 1992) and the generation of evolved fluids that likely contained soluble uranyl complexes despite their relatively reduced character. It is possible that liquid hydrocarbons could have developed oxygenated functional groups by development and regeneration of oxygenated functional groups during contact with oxidizing U-mineralizing fluids (Landais et al., 1987), thus enabling urano-organic complexes to form. It has been argued that such "preconcentration" is essential for the subsequent precipitation of uranium from weak mineralizing solutions (Kochenov et al., 1977).

The structural (or organizational) characteristics of the Witwatersrand sediment-hosted bitumen are clearly related to uranium abundance and proximity to individual uraninite grains. However, no such relationship exists with respect to gold content. Gold appears to have been precipitated in and around bitumen as a result of fluid-hydrocarbon interaction, and it appears that the more reduced the bitumen, the higher the gold content (Landais et al., 1990a, 1990b). Carboxyl groups and uranic acids contained in organic compounds could have stabilized colloidal gold and/or gold organic polymeric substances during relatively low temperature transport by water (Zumberge et al., 1978) from a primary source (detrital gold) to the sediment-hosted bitumen.

Within the granite-hosted nodules, the uranium distribution is associated with relic grains of uraninite that have been partially replaced by the hydrocarbons (Drennan, 1989; Robb et al., 1990). Hydrocarbon-bearing fluids originating from within the Witwatersrand Basin during a period of liquid hydrocarbon expulsion percolated along aquifers into the adjacent granites. Hydrocarbon fixation occurred around accessory uraninite and/or uranothorite, replacing these phases by radiolytic polymerization. Gold distribution within granite-hosted nodules is not known.

Experimental studies by Nakashima et al. (1984) showed that under diagenetic or hydrothermal conditions, natural organic matter could form complexes with uranyl species in two different ways: (1) at T = 45–250 °C, fixation of uranyl species by organic matter occurred without any reduction; this process did not proceed through a classical ion exchange mechanism but resulted in the formation of stable organo-metallic combinations; (2) at T = 120–400 °C, reduction of dissolved uranyl species resulted in the precipitation of uranium and stoichiometric liberation of H^+ in the solution medium, resulting in a dehydrogenation of the organic matter. This dehydrogenation was attributed to (1) an oxidation of alcohol functional groups into aldehyde or ketone functions accompanied by a simultaneous reduction of uranyl species, and (2) a dehydrogenation of aliphatic moieties induced by the uranium species. Molecular hydrogen produced during this process was subsequently consumed during continued reduction of uranyl species. Rouzaud et al. (1980) established that reduction of U^{6+} complexed to mineral-bound U^{4+} ions could also have been achieved by interaction with organic matter at T > 500 °C. Berger (1974), on the other hand, showed that reduction of hexavalent uranium ions occurred by reduced sulfur species.

Bitumen nodules that occur in fault-related quartz veining exhibit different uranium and gold distributions compared to sediment- and granite-hosted bitumen samples. Uranium is pervasively distributed throughout the vein nodules but may exhibit higher concentrations toward grain centers (Fig. 11). Gold occurs pervasively distributed throughout the nodules but is also locally concentrated. This suggests that uranium, gold, and hydrocarbons were transported in the same hydrothermal solution, possibly as organo-urano-gold entities and/or as thio-gold complexes that were coprecipitated with uranium in solution. Since fluid inclusion microthermometry for these veins suggests entrapment temperatures in excess of 250 °C (Drennan, 1997; Drennan et al., 1997; see also Boer and Reimold, 2006), it is suggested that these complexes formed as a result of reduction and dehydrogenation, similar to processes discussed by Nakashima et al. (1984).

Within the fault-related quartz veins, uranium- and gold-enriched bitumen nodules coprecipitated with euhedral, authigenic pyrite of cubic-striated form, as well as particulate gold (Figs. 7B and 7C). According to Murowichick and Barnes (1987), cubic-striated pyrite only forms at temperatures in excess of 350 °C (and up to 450 °C). Microcrystals of cubic pyrite and chalcopyrite (identified by SEM) occur within the carbonaceous matter, and globular inclusions of bitumen are also intimately associated with gold. In addition, As, Fe, and Cu are all locally concentrated within nodules and pervasively distributed throughout the bitumen (Fig. 11). This supports the suggestion that fluids migrating through the fault systems were capable of transporting sulfur, gold, uranium, and hydrocarbons. Fluid inclusion microthermometry suggests entrapment temperatures of between 300 and 400 °C for these fluids (Drennan, 1997; Drennan et al., 1999; Boer and Reimold, 2006; and references in these works). It is, therefore, argued that within fault zones localized high T (and P) conditions were experienced. Under these conditions it is suggested that organo-urano-gold-sulfide complexes (or organo-urano-gold and thio-gold complexes in the same fluid) underwent reduction and dehydrogenation, resulting in the coprecipitation of sulfides, gold, and REE–enriched bitumen nodules, as well as the entrapment of hydrocarbon-rich fluid inclusions.

CONCLUSIONS

A distinction can be made between sediment-hosted and distal/vein-related bitumen. Sediment-hosted bitumen occurs in seams and as "fly-speck" nodules within the sediment matrix. This organic material is heterogeneous, and its chemical structure is a function of proximity to uraninite grains and uranium content. Radiolytic polymerization during catagenesis within the Witwatersrand Basin resulted in the precipitation of heavy hydrocarbons around detrital uraninite (giving rise to reef- and "fly-speck" bitumen) and the liberation of light hydrocarbons (CH_4, C_2H_6, C_3H_8) into circulating basinal fluids that resulted in the eventual precipitation of distal/vein-related bitumen.

Hydrocarbons were identified within CO_2- and CH_4-rich inclusions ($L_{w\text{-}c\text{-}m}$ and V_m) and occur either as optically invisible coatings on inclusion walls or as globular bitumen nodules within the inclusions or both. These vein-related bitumen nodules are characterized by significant Au, U, Fe, Cu, As, Pb, and Ni concentrations as well as other trace elements (Drennan, 1997; Drennan et al., 1997). Uranium is pervasively distributed throughout the nodules, whereas gold occurs pervasively distributed (but in lower concentrations) throughout the nodules as well as in localized high concentrations (Drennan et al., 1995). Iron, Pb, Ni, and As occur in particular form (associated with minute sulphide grains) and pervasively distributed throughout the nodules. These nodules also exhibit a close association with authigenic sulphides and gold sampled from the same fault-related quartz veins (Drennan et al., 1995, 1997). This suggests that U, Au, and other metals were transported in the same solutions as light hydrocarbons.

It is therefore believed that urano-organo–bearing fluids were able to scavenge gold and sulfides from the sediments forming stable organic complexes, which migrated through the sediments along aquifers during subsequent metagenic processes. Where these migrating fluid channelways intersected sediment-hosted, uraniferous bitumen seams, the light hydrocarbons precipitated out as bituminous veins. Localized high P-T conditions experienced in late-stage fault-related quartz veining resulted in reduction of the uranyl species and dehydrogenation of the carbonaceous matter, causing the precipitation of uranium-enriched bitumen nodules together with gold and sulfides. The remaining fluids trapped as fluid inclusions in intercrystalline cavities within the quartz veins are characterized by abundant hydrocarbons in the vapor phases, carbonaceous coatings on inclusion walls, and solid, globular carbonaceous precipitates. As a result of the higher P-T conditions associated with faulting, these migrated bitumen/hydrocarbon phases all exhibit a higher degree of structural ordering than sediment-hosted bitumen.

Where the enriched hydrocarbon-bearing fluids percolated along conduits into granites adjacent to the Witwatersrand Basin, hydrocarbon fixation occurred around, and replaced, accessory uraninite and uranothorite as a result of radiolytic polymerization, giving rise to distal bitumen precipitation. As the ratio of bitumen to uraninite in granites is low (Robb et al., 1992, 1994), the combined effects of polymerization resulted in a more organized hydrocarbon structure than that observed in the conglomeratic reefs.

Both sediment-hosted and distal/vein-related bitumen types are seen to be paragenetically late. It is suggested that sediment-hosted bitumen was fixed during catagenic processes and that distal/vein-related bitumen is the product of subsequent metagenesis. However, striking similarities in terms of petrographic characteristics, composition, radiogenic isotope characteristics, and in the intimate relationships between bitumen and uranium phases associated with significant gold concentrations suggest that the sediment-hosted and distal/vein-related bitumen are genetically linked. U-Pb isotope analysis indicates that the granite- and sediment-hosted bitumen was fixed some 400 m.y. after sedimentation stopped. As the bitumen exhibits no sign of metamorphic alteration (Ebert et al., 1990), it is further suggested that hydrocarbon fixation postdates the period of peak metamorphism and may be considerably younger than initially thought.

ACKNOWLEDGMENTS

The data presented here formed the basis of G.R. Drennan's Ph.D. thesis. The research was sponsored by Anglo American Corporation. The authors are grateful to M.-C. Boiron, M. Cathelineau, and J. Dubessy (Centre de Recherches sur la Géologie des Matières Premières Minerales et Énergétiques, Nancy, France), P. Landais (Centre Nationale Recherches de Science, Nancy, France), and W.J. Przybylowicz (Schonland Research Centre [now part of iThemba Laboratories], University of the Witwatersrand, South Africa), who provided valuable input to the study. D. Du Toit, L. Whitfield, and R. Drennan are thanked for drafting assistance. J. Law and K. Kenyon made valuable comments on an earlier version of the manuscript.

REFERENCES CITED

Angus, S., Armstrong, B., and de Reuck, K.M., 1976, International Thermodynamic Tables of the Fluid State, Carbon Dioxide: New York, Pergamon Press, 67 p.

Barnicoat, A.C., Henderson, I.H.K., Knipe, R.J., Yardley, B.W.D., Napier, R.W., Fox, N.P.C., Kenyon, A.K., Muntingh, D.J., Strydom, D., Winkler, K.S., Lawrence, S.R., and Cornford, C., 1997, Hydrothermal gold mineralization in the Witwatersrand Basin: Nature, v. 386, p. 820–824, doi: 10.1038/386820a0.

Barton, E.S., and Hallbauer, D.K., 1996, Trace-element and U-Pb isotopic compositions of pyrite types in the Proterozoic Black Reef, Transvaal

Sequence, South Africa: Implications on genesis and age: Chemical Geology, v. 133, p. 173–199, doi: 10.1016/S0009-2541(96)00075-7.

Berger, A.I., 1974, The role of organic matter in the accumulation of uranium, *in* Formation of uranium ore deposits: Vienna, Austria, International Atomic Energy Agency, p. 99–123.

Berlenbach, J.W., 1993, The mechanics of bedding-parallel faulting associated with the VCR on the Kloof Gold Mine [Ph.D. thesis]: Johannesburg, South Africa, Rand Afrikaans University, 160 p.

Boer, R.H., and Reimold, W.U., 2006, Conditions of gold remobilization in the Ventersdorp Contact Reef, Witwatersrand Basin, South Africa, *in* Reimold, W.U., and Gibson, R.L., Processes on the early Earth: Geological Society of America Special Paper 405, p. XXX–XXX.

Bowers, T.S., and Helgeson, H.C., 1983, Calculation of the thermodynamic and geochemical consequences of nonideal mixing in the system H_2O-CO_2-NaCl on phase relations in geologic systems: Equations of state for H_2O-CO_2-NaCl fluids at high pressures and temperatures: Geochimica et Cosmochimica Acta, v. 47, p. 1247–1275, doi: 10.1016/0016-7037(83)90066-2.

Burger, A.J., Nicolaysen, L.O., and de Villiers, J.W.L., 1962, Lead isotopic compositions of galenas from the Witwatersrand and Orange Free State, and their relation to the Witwatersrand and Dominion Reef uraninites: Geochimica et Cosmochimica Acta, v. 26, p. 25–59, doi: 10.1016/0016-7037(62)90004-2.

Drennan, G.R., 1989, The nature of the Archaean basement in the hinterland to the Welkom Goldfield [M.S. dissertation]: Johannesburg, South Africa, University of the Witwatersrand, 185 p.

Drennan, G.R., 1997, Fluid Inclusion Microthermometry of the Witwatersrand Basin [Ph.D. dissert.]: Johannesburg, South Africa, University of the Witwatersrand, 228 p.

Drennan, G.R., Cathelineau, M., Boiron, M.C., Landais, P., and Robb, L.J., 1995, Characteristics of postdepositional fluids in the Witwatersrand Basin, with emphasis on hydrocarbon- bearing fluids: Symposium on the economic significance of metamorphism and fluid movement within the Witwatersrand Basin, *in* Proceedings of the Symposium on the economic significance on metamorphism and fluid movement within the Witwatersrand Basin, Carletonville: Carletonville, Western Transvaal Branch of the Geological Society of South Africa, 26 October 1995, p. 9–15.

Drennan, G.R., Cathelineau, M., Boiron, M.-C., Landais, P., and Robb, L.J., 1997, Post-depositional fluids and their relationship to mineral paragenesis in the Witwatersrand Basin, South Africa, *in* Hendry, J.P., Garey, P.F., Parnell, J., Ruffell, A.H., and Warden, R.H., eds., GEOFLUIDS II 1997, The Queen's University of Belfast, 10–14 March: Belfast, Ireland, Contributions to the Second International Conference on Fluid Evolution, Migration and Integration in Sedimentary Basins and Orogenic Belts, p. 436–439.

Drennan, G.R., Boiron, M.-C., Cathelineau, M., and Robb, L.J., 1999, Characteristics of post-depositional fluids in the Witwatersrand Basin: Mineralogy and Petrology, v. 66, p. 83–109, doi: 10.1007/BF01161723.

Dubessy, J., 1984, Simulations des équilibres chimiques dans le système C-O-H: Consequences mèthodologiques pour les inclusions fluides: Bulletin de Minéralogie, v. 107, p. 155–168.

Dubessy, J., Audeoud, D., Wilkins, R., and Kosztolanyi, C., 1982, The use of the Raman microprobe MOLE in determination of the electrolytes dissolved in the aqueous phase of fluid inclusions: Chemical Geology, v. 37, p. 137–150, doi: 10.1016/0009-2541(82)90073-0.

Dubessy, J., Beny, C., Guilhaumou, N., Dhamelincourt, P., and Poty, B., 1984, Applications of the MOLE Raman microprobe to the study of fluid inclusions in minerals: Journal de Physique, Colloque C2, suppl. 2/45, p. 811–814.

Dubessy, J., Ramboz, C., Nguyen-Trung, C., Cathelineau, M., Charoy, B., Cuney, M., Leroy, J., Poty, B., and Weisbrod, A., 1987, Physical and chemical controls (fO_2, T, pH) of the opposite behaviour of U and Sn-W as exemplified by hydrothermal deposits in France and Great Britain, and solubility data: Bulletin de Minéralogie, v. 110, p. 261–281.

Dubessy, J., Poty, B., and Ramboz, C., 1989, Advances in C-O-H-N-S fluid geochemistry based on micro-Raman spectrometric analysis of fluid inclusions: European Journal of Mineralogy, v. 1, p. 517–534.

Eakin, P.A., and Gize, A.P., 1992, Reflected-light microscopy of uraniferous bitumens: Mineralogical Magazine, v. 56, p. 85–99.

Ebert, L.B., Robbins, E.I., Rose, K.D., Katstrup, R.V., Scanlon, J.C., Gebhard, L.A., and Garcia, A.R., 1990, Chemistry and palynology of carbon seams and associated rocks from Witwatersrand Goldfields, South Africa: Ore Geology Reviews, v. 5, p. 423–444, doi: 10.1016/0169-1368(90)90045-O.

Ellsworth, H.V., 1928, Thucholite and uraninite from the Wallingford Mine near Buckingham, Quebec: The American Mineralogist, v. 13, p. 442–448.

England, G.L., Rasmussen, B., Krape , B., and Groves, D.I., 2001, The origin of uraninite, bitumen nodules, and carbon seams in Witwatersrand gold-uranium-pyrite ore deposits, based on a Permo-Triassic analogue: Economic Geology and the Bulletin of the Society of Economic Geologists, v. 96, p. 1907–1920.

Foya, S.N., 2002, Mineralogical and geochemical controls on gold distribution in the Kimberley Reefs, South Africa [Ph.D. thesis]: Johannesburg, South Africa, University of the Witwatersrand, 292 p.

French, B.M., 1966, Some geological implications of equilibrium between graphite and a C-H-O gas phase at high temperatures and pressures: Reviews of Geophysics, v. 4, p. 223–253.

Frimmel, H.E., 1994, Metamorphism of Witwatersrand gold: Exploration Geochemistry, v. 3, p. 357–370.

Frimmel, H.E., 2005, Archaean atmospheric evolution: Evidence from the Witwatersrand goldfields, South Africa: Earth-Science Reviews, v. 70, p. 1–46, doi: 10.1016/j.earscirev.2004.10.003.

Frimmel, H.E., and Gartz, V.H., 1997a, Polyphase fluid infiltration in the Witwatersrand Basin, South Africa, and its bearing on the distribution of the gold, *in* Hendry, J.P., Carey, P., Parnell, J., Ruffell, A., and Worden, R., eds., GEOFLUIDS II 1997, The Queen's University of Belfast, 10–14 March 1977: Belfast, Ireland, Contributions to the Second International Conference on Fluid Evolution, Migration and Integration in Sedimentary Basins and Orogenic Belts, p. 444–447.

Frimmel, H.E., and Gartz, V.H., 1997b, Witwatersrand gold particle chemistry matches model of metamorphosed, hydrothermally altered placer deposits: Mineralia Deposita, v. 32, p. 523–530, doi: 10.1007/s001260050119.

Frimmel, H.E., Leroux, A.P., Knight, J., and Minter, W.E.L., 1993, A case study of postdepositional alteration of the Witwatersrand Basal reef gold placer: Economic Geology and the Bulletin of the Society of Economic Geologists, v. 88, p. 249–265.

Frimmel, H.E., Hallbauer, D.K., and Gartz, V.H., 1999, Gold mobilizing fluids in the Witwatersrand Basin—Composition and possible source: Mineralogy and Petrology, v. 66, p. 55–81, doi: 10.1007/BF01161722.

Frimmel, H.E., Groves, G.I., Kirk, J., Ruiz, J., Chesley, J., and Minter, W.E.L., 2005, The formation and preservation of the Witwatersrand goldfields, the largest gold province in the world, *in* Hendenquist, J.W., Thompson, J.F.H., Goldfarb, R.J., and Richards, J.P., eds., Economic Geology, 100th Anniversary Volume: Littleton, Society of Economic Geologists, v. 100, p. 769–797.

Fryer, B.J., and Taylor, R.P., 1987, Rare-earth element distributions in uraninites: Implication for ore genesis: Chemical Geology, v. 63, p. 101–108, doi: 10.1016/0009-2541(87)90077-5.

Gates, A.E., 1987, Shear zones: Origin and mineralization [abs.]: Geological Society of America Abstracts with Program, no. 19, p. 84–85.

Gibson, R.L., and Reimold, W.U., 1999, The significance of the Vredefort Dome for the thermal and structural evolution of the Witwatersrand Basin, South Africa: Mineralogy and Petrology, v. 66, p. 5–23, doi: 10.1007/BF01161720.

Gibson, R.L., Armstrong, R.A., and Reimold, W.U., 1997, The age and thermal evolution of the Vredefort Impact Structure: A single grain U-Pb zircon study: Geochimica et Cosmochimica Acta, v. 61, p. 1531–1540, doi: 10.1016/S0016-7037(97)00013-6.

Giggenbach, W.F., 1982, Carbon 13 exchange between CO_2 and CH_4 under geothermal conditions: Geochimica et Cosmochimica Acta, v. 46, p. 159–165, doi: 10.1016/0016-7037(82)90243-5.

Gray, G.J., Lawrence, R., Kenyon, K., and Cornford, C., 1998, Nature and origin of "carbon" in the Archean Witwatersrand Basin, South Africa: Journal of the Geological Society of London, v. 155, p. 39–59.

Hallbauer, D.K., 1975, The plant origin of the Witwatersrand carbon: Mineral Sciences and Engineering, v. 7, p. 111–131.

Hallbauer, D.K., 1984, Archean granitic sources for the detrital mineral assemblage in Witwatersrand conglomerates, *in* Extended Abstract, Geocongress 1984: Potchefstoom, Biennial Conference of the Geological Society of South Africa, p. 53–56.

Hallbauer, D.K., 1986, The mineralogy and geochemistry of Witwatersrand pyrite, gold, uranium and carbonaceous matter, *in* Anhaeusser, C.R., and Maske, S., eds., Mineral Deposits of Southern Africa: Johannesburg, The Geological Society of South Africa, v. 1, p. 731–732.

Hallbauer, D.K., and Van Warmelo, T., 1974, Fossilized plants in thucholite from Precambrian rocks from the Witwatersrand, South Africa: Precambrian Research, v. 1, p. 199–212, doi: 10.1016/0301-9268(74)90010-2.

Hallbauer, D.K., Klemd, R., and Von Gehlen, K., 1986, A provenance model for the Witwatersrand gold and uranium mineralization and its implications in the recognition of gold distribution patterns in reefs, *in* Extended Abstracts, Geocongress 1986: Johannesburg, Biennial Congress of the Geological Society of South Africa, p. 133–137.

Hoefs, J., and Schidlowski, M., 1967, Carbon isotope composition of carbonaceous matter from the Precambrian of the Witwatersrand System: Science, v. 155, p. 1096–1097.

Johnson, K.M., and McIntyre, D.H., 1983, Disseminated gold-silver deposit in a rhyolite dome at the Sunbeam Mine, Custer County, Idaho: Geological Society of America Abstracts with Program, v. 15, p. 324–325.

Johnson, K.M., and McIntyre, D.H., 1986, Epithermal gold-silver mineralization related to volcanic subsidence in the Custer graben, Custer County, Idaho: Journal of Geochemical Exploration, v. 25, p. 247–248, doi: 10.1016/0375-6742(86)90042-7.

Kerrick, D.M., and Jacobs, G.K., 1981, A modified Redich-Kwang equation for H_2O, CO_2 and H_2O-CO_2 mixtures at elevated pressures and temperatures: American Journal of Science, v. 281, p. 735–767.

Klemd, R., and Hallbauer, D.K., 1987, Hydrothermally altered peraluminous Archean granites as a provenance model for Witwatersrand sediments: Mineralia Deposita, v. 22, p. 227–235.

Kochenov, A.V., Korolev, K.G., Dubinchuk, V.T., and Medveden, Yu.L., 1977, Experimental data on the conditions of precipitation of uranium from aqueous solutions: Geochemistry International, v. 14, p. 82–87.

Landais, P., Connan, J., Dereppe, J.M., George, E., Meunier, J.D., Monthioux, M., Pagel, M., Pironon, J., and Poty, B., 1987, Alterations of organic matter: A clue for uranium ore genesis: Uranium, v. 3, p. 307–342.

Landais, P., Dubessy, J., Robb, L.J., and Nouel, C., 1990a, Preliminary chemical analyses and Raman spectroscopy on selected samples of Witwatersrand kerogen: Economic Geology Research Unit, Information Circular, no. 222, 8 p.

Landais, P., Dubessy, J., Poty, B., and Robb, L.J., 1990b, Three examples illustrating the analysis of organic matter associated with uranium ores: Organic Geochemistry, v. 16, no. 1-3, p. 601–608, doi: 10.1016/0146-6380(90)90073-9.

Landais, P., Dubessy, J., Dereppe, J.-M., and Philip, R.P., 1993, Characterization of graphite alteration and bitumen genesis in the Cigar Lake deposit (Saskatchewan, Canada): Canadian Journal of Earth Sciences, v. 30, p. 743–753.

Liebenberg, W.R., 1955, The occurrence of gold and radioactive minerals in the Witwatersrand System, the Dominion Reef, the Ventersdorp Contact Reef: Transactions of the Geological Society of South Africa, v. 58, p. 101–227.

Maxwell, J.A., Campbell, J.L., and Teesdale, J., 1989, The Guelph PIXE software package: Nuclear Instruments and Methods, no. B43, 218 p.

Meyer, F.M., Drennan, G.R., Robb, L.J., Cathelineau, M., Dubessy, J., and Landais, P., 1991, Conditions of Au-U mineralization in Witwatersrand Reefs, *in* Pagel, M., and Leroy, J.L., eds., Source, transport and deposition of metals, Proceedings of the 25 Years Society for Geology Applied to Mineral Deposits Anniversary Meeting, Nancy: Rotterdam, Balkema, p. 681–684.

Milodowski, A.E., West, J.M., Pearce, J.M., Hyslop, E.K., Basham, I.R., and Hooker, P.J., 1990, Uranium-mineralized micro-organisms associated with uraniferous hydrocarbons in southwest Scotland: Nature, v. 347, p. 465–467, doi: 10.1038/347465a0.

Mossman, D.J., and Nagy, B., 1996, Solid bitumens: An assessment of their characteristics, genesis, and role in geological processes: Terra Nova, v. 8, no. 2, p. 114–128.

Mossman, D.J., Goodarzi, F., and Gentzis, T., 1993, Characterization of insoluble organic matter from the lower Proterozoic Huronian Supergroup, Eliot Lake, Ontario: Precambrian Research, v. 61, p. 279–293, doi: 10.1016/0301-9268(93)90117-K.

Murowichick, P., and Barnes, H.L., 1987, Pyrite morphology: The American Mineralogist, v. 72, p. 1241–1250.

Nagy, B., 1993, Kerogens and bitumens in Precambrian uraniferous ore deposits; Witwatersrand, South Africa, Elliot Lake, Canada and the natural fission reactors, Oklo, Gabon, *in* Parnell, J., Kucha, H., and Landais, P., eds., Bitumens in Ore Deposits: Heidelberg, Springer, p. 287–333.

Nakashima, S., Disnar, J.R., Perruchot, A., and Trichet, J., 1984, Experimental study of mechanisms of fixation and reduction of uranium by sedimentary organic matter under diagenetic or hydrothermal conditions: Geochimica et Cosmochimica Acta, v. 48, p. 2321–2329, doi: 10.1016/0016-7037(84)90228-X.

Oberthür, T., 1983, Metallogenetische Uberlegungen zur Bildung des Carbon Leader Reef, Carletonville Goldfield, Südafrika [Ph.D. thesis]: Germany, University of Cologne, 214 p.

Oh, J.O., 1987, Étude structurale de la graphitation naturelle (exemple des basins Sud Coréens) [Ph.D. thesis]: France, University of Orleans, 181 p.

Ohmoto, H., and Kerrick, D., 1977, Devolatilization equilibria in graphitic systems: American Journal of Science, v. 277, p. 1013–1044.

Pagel, M., Pinte, G., and Rotach-Toulhoat, N., 1987, The rare earth elements in natural uranium oxides, Monograph Series on Mineral Deposits, v. 27: Berlin-Stuttgart, Gebruder Bortraeger, p. 81–85.

Parnell, J., 1988, Metal enrichments in solid bitumens: Mineralia Deposita, v. 23, p. 191–199.

Parnell, J., 1993, Paragenesis of bitumens and ore in mineral deposits: Resource Geology, v. 15, Special Issue, p. 111–122.

Parnell, J., 1999, Petrographic evidence for emplacement of carbon into Witwatersrand conglomerates under fluid pressure: Journal of Sedimentary Research, v. 69, p. 164–170.

Pasteris, J.D., Wopenka, B., and Seitz, J.C., 1988, Practical aspects of quantitative laser Raman microprobe spectroscopy for the study of fluid inclusions: Geochimica et Cosmochimica Acta, v. 52, p. 979–988, doi: 10.1016/0016-7037(88)90253-0.

Phillips, G.N., and Law, J.D.M., 1994, Metamorphism of the Witwatersrand gold fields: A review: Ore Geology Reviews, v. 9, p. 1–31, doi: 10.1016/0169-1368(94)90017-5.

Phillips, G.N., and Law, J.D.M., 2000, Witwatersrand gold fields: Geology, genesis and exploration: Society of Economic Geologists Reviews, v. 13, p. 439–500.

Phillips, G.N., and Myers, R.E., 1988, The Witwatersrand gold fields, Part II, An origin for Witwatersrand gold during metamorphism and associated alteration: Economic Geology Monograph, No. 6, The Geology of Gold Deposits: Perspectives, v. 1988, p. 598–608.

Phillips, G.N., Law, J.D.M., and Stevens, G., 1997, Alteration, heat, and Witwatersrand gold: 111 years after Harrison and Langlaagte: South African Journal of Geology, v. 100, p. 377–392.

Pironon, J., and Barres, O., 1990, Semi-quantitative FT-IR microanalysis limits; evidence from synthetic hydrocarbon fluid inclusions in sylvite: Geochimica et Cosmochimica Acta, v. 54, p. 509–518, doi: 10.1016/0016-7037(90)90348-O.

Pironon, J., and Barres, O., 1992, Influence of brine-hydrocarbon interactions on FT-IR microspectroscopic analyses of intercrystalline liquid inclusions: Geochimica et Cosmochimica Acta, v. 56, p. 169–174, doi: 10.1016/0016-7037(92)90124-2.

Poty, B., Leroy, J., and Jachimowicz, L., 1976, A new device for measuring temperatures under the microscope: The Chaixmeca microthermometry apparatus: Bulletin de Société Française Minéralogie et Cristallographie, v. 99, p. 182–189.

Poujol, M., Robb, L.J., and Respaut, J.P., 1998, U-Pb and Pb-Pb isotopic studies relating to the origin of gold mineralization in the Evander goldfield, Witwatersrand Basin, South Africa: University of the Witwatersrand, Economic Geology Research Unit Information Circular, no. 320, 18 p.

Przybylowicz, W.J., Watterson, J.I.W., Annegarn, H.J., Connell, S.H., Fearick, R.W., Andeweg, A.H., and Sellschop, J.P.F., 1993, The new Schonland PIXE microprobe and applications to geological and archaeological samples: Nuclear Instruments and Methodology, v. 75, p. 539–543, doi: 10.1016/0168-583X(93)95711-D.

Ramdohr, P., 1958, New observations on the ores of the Witwatersrand in South Africa and their genetic significance: Transactions of the Geological Society of South Africa, Annexure no. 61, 50 p.

Rasmussen, B., 2005, Evidence for pervasive petroleum generation and migration in 3.2 and 2.63 Ga shales: Geology, v. 33, no. 6, p. 497–500, doi: 10.1130/G21316.1.

Reimold, W.U., Koeberl, C., Gibson, R.L., and Dressler, B.O., 2005, Economic mineral deposits in impact structures and their geological settings, *in* Koeberl, C., and Henkel, H., eds., Impact Tectonics, Impact Studies, v. 6: Heidelberg, Springer, p. 479–552.

Robb, L.J., and Meyer, M., 1991, A contribution to recent debate concerning epigenetic versus syngenetic mineralization processes in the Witwatersrand Basin: Economic Geology and the Bulletin of the Society of Economic Geologists, v. 86, p. 396–401.

Robb, L.J., and Meyer, F.M., 1995, The Witwatersrand Basin, South Africa; geological framework and mineralization processes: Ore Geology Reviews, v. 10, p. 67–94, doi: 10.1016/0169-1368(95)00011-9.

Robb, L.J., Davis, D.W., and Kamo, S.L., 1990, U-Pb ages on single detrital zircon grains from the Witwatersrand Basin, South Africa: Constraints on the age of sedimentation and on the evolution of granites adjacent to the Basin: The Journal of Geology, v. 98, p. 311–328.

Robb, L.J., Davis, D.W., Kamo, S.L., and Meyer, F.M., 1992, Ages of altered granite adjoining the Witwatersrand Basin with implications for the origin of gold and uranium: Nature, v. 357, p. 677–680.

Robb, L.J., Landais, P., Meyer, F.M., and Davis, D.W., 1994, Nodular organic matter in granites: Implications for the origin of "kerogen" in the Witwatersrand Basin, South Africa: Exploration and Mining Geology, v. 3, no. 3, p. 219–230.

Robb, L.J., Landais, P., and Drennan, G.R., 1995, Petrographic, chemical and spectroscopic studies of carbonaceous matter and their bearing on the metamorphic evolution of the Witwatersrand Basin: Symposium on the economic significance of metamorphism and fluid movement within the Witwatersrand Basin, Carletonville, Western Transvaal, *in* Proceedings of Symposium on the economic significance of metamorphism and fluid movement within the Witwatersrand Basin: Carletonville, Western Transvaal Branch of the Geological Society of South Africa, 26 October 1995, p. 39–43.

Robb, L.J., Landais, P., Drennan, G.R., and Dubessy, J., 1997a, Petrographic, chemical and spectroscopic data relating to the origin of carbonaceous matter in the Witwatersrand Basin, South Africa, *in* Hendry, J.P., Garey, P.F., Parnell, J., Ruffell, A.H., and Warden, R.H., eds., GEOFLUIDS II 1997: Belfast, Ireland, Contributions to the Second International Conference on Fluid Evolution, Migration and Integration in Sedimentary Basins and Orogenic Belts, 10–14 March, p. 452–455.

Robb, L.J., Landais, P., Drennan, G.R., and Dubessy, J., 1997b, Petrographic, chemical and spectroscopic data relating to the origin of carbonaceous matter in the Witwatersrand Basin, South Africa: University of the Witwatersrand, Economic Geology Research Unit Information Circular, no. 312, 34 p.

Roedder, E., 1984, Fluid Inclusions: Reviews in Mineralogy, Mineralogical Society of America, v. 12, 664 p.

Rouzaud, J.N., Oberlin, A., and Trichet, J., 1980, Interaction of uranium and organic matter in uraniferous sediments, *in* Douglas, A.J., and Maxwell, J.R., eds., Advances in Organic Geochemistry: New York, Pergamon Press, p. 505–516.

Saxby, J.D., 1976, The significance of organic matter in ore genesis, *in* Wolf, K.H., ed., Handbook of Strata-bound and Stratiform Ore Deposits, Geochemical Studies: Amsterdam, Elsevier, v. 2, p. 111–133.

Schidlowski, M., 1965, Probable life-forms from the Precambrian of the Witwatersrand System (South Africa): Nature, v. 208, p. 895–896.

Schidlowski, M., 1981, Uraniferous constituents of the Witwatersrand conglomerates: Ore-microscopic observations and implications for Witwatersrand metallogeny, *in* Armstrong, F., ed., Genesis of uranium- and gold-bearing Precambrian quartz-pebble conglomerates: U.S. Geological Survey Professional Paper 1161, p. 1–29.

Schreyer, W., and Bisschoff, A.A., 1982, Kyanite as a metamorphic mineral in Witwatersrand sediments at Johannesburg and Krugersdorp, South Africa: Transactions of the Geological Society of South Africa, v. 85, p. 211–214.

Shepherd, T.J., Rankin, A.A., and Alderton, D.H.M., 1985, A practical guide to fluid inclusion studies: Glasgow, Blackie and Son, 239 p.

Sheppard, S.M.F., 1981, Stable isotope geochemistry of fluids, *in* Rickard, D., and Wickman, F.E., eds., Chemistry and geochemistry of solutions at high temperatures and pressures: Oxford, UK, Physics and Chemistry of the Earth: Pergamon Press, v. 13 and 14, p. 419–443.

Snyman, C.P., 1965, Possible biogenetic structures in Witwatersrand thucholite: Transactions of the Geological Society of South Africa, v. 68, p. 225–233.

Spangenberg, J.E., and Frimmel, H.E., 1999, Tracking the origin of organic matter in the Witwatersrand Basin using $^{13}C/^{12}C$ analysis of individual hydrocarbons, *in* Stanley, C., ed., Mineral Deposits: Processes to Processing, Society for Geology Applied to Mineral Deposits—International Association for Geology of Ore Deposits Meeting, 22–25 August, 1999, London: Rotterdam, Balkema, p. 271–275.

Spangenberg, J.E., and Frimmel, H.E., 2001, Basin-internal derivation of hydrocarbons in the Witwatersrand Basin, South Africa: Evidence from bulk and molecular $\delta^{13}C$ data: Chemical Geology, v. 173, p. 339–355, doi: 10.1016/S0009-2541(00)00283-7.

Spangenberg, J., and Macko, S.A., 1998, Organic geochemistry of the San Vicente zinc-lead district, eastern Pucará Basin, Peru: Chemical Geology, v. 146, p. 1–23, doi: 10.1016/S0009-2541(97)00158-7.

Stephenson, D.A., 1974, Raman cross sections of selected hydrocarbons and freons: Journal of Quantitative Spectroscopy and Radiation Transfer, v. 14, p. 1291–1301, doi: 10.1016/0022-4073(74)90098-3.

Stevens, G., Boer, R.H., and Gibson, R.L., 1997, Metamorphism in the Witwatersrand Basin: A review of recent advances and some predictions on gold remobilization by metamorphic fluid flow: University of the Witwatersrand, Economic Geology Research Unit Information Circular, no. 303, 34 p.

Svenek, J., 1987, A Field Guide in Colour—Minerals: Artia, Prague, Octopus Books Ltd., 224 p.

Swanenberg, H.E.C., 1979, Phase equilibria in carbonic systems and their application to freezing studies of fluid inclusions: Contributions to Mineralogy and Petrology, v. 68, p. 303–306.

Tissot, B.P., and Welte, D.H., 1978, Petroleum formation and occurrence: A new approach to oil and gas exploration: New York, Springer-Verlag, 538 p.

van den Kerkhof, A.M., 1988, The system CO_2-CH_4-N_2 in fluid inclusions: Theoretical modeling and geological applications [Ph.D. thesis]: Amsterdam, The Netherlands, Free University of Amsterdam, 189 p.

Wang, A., Dhamelincourt, P., Dubessy, J., Guerard, D., Landais, P., and Lelaurain, M., 1989, Characterization of graphite alteration in a uranium deposit by micro-Raman spectroscopy, X-ray diffraction, transmission electron microscopy and scanning electron microscopy: Carbon, v. 27, no. 2, p. 209–218, doi: 10.1016/0008-6223(89)90125-5.

Watterson, J.J.W., Fearick, R.W., Connell, S.H., Annegarn, H.J., Przybylowicz, W.J., Andeweg, A.H., McQueen, I., and Sellschop, J.F.P., 1993, Innovative features of the Schonland proton microprobe: Nuclear Instruments and Methods, v. B77, p. 70–84.

Wilde, W.R., Bloom, M.S., and Wall, V.J., 1989, Transport and deposition of gold, uranium and platinum group elements in unconformity-related uranium deposits, *in* Keays, R.R., Ramsay, W.R.H., and Groves, D.I, eds., The geology of gold deposits: The perspective in 1988: Economic Geology Monograph 6, p. 637–650.

Wopenka, B., Pasteris, J.D., and Freeman, J.J., 1990, Analysis of individual fluid inclusions by Fourier transform infrared and Raman microspectroscopy: Geochimica et Cosmochimica Acta, v. 54, p. 519–533, doi: 10.1016/0016-7037(90)90349-P.

Young, R.B., 1917, The Banket of South African Gold Fields: London, Gurney and Jackson, 125 p.

Zartman, R.E., and Frimmel, H.E., 1999, Rn-generated ^{206}Pb in hydrothermal sulfide minerals and bitumen from the Ventersdorp Contact Reef, South Africa: Mineralogy and Petrology, v. 66, p. 171–191, doi: 10.1007/BF01161727.

Zhang, Y., and Frantz, J.D., 1987, Determination of the homogenization temperatures and densities of supercritical fluids in the system NaCl-KCl-$CaCl_2$-H_2O using synthetic fluid inclusions: Chemical Geology, v. 64, p. 335–350, doi: 10.1016/0009-2541(87)90012-X.

Zumberge, J.E., Sigleo, A.C., and Nagy, B., 1978, Molecular and elemental analyses of the carbonaceous matter in the gold and uranium bearing Vaal Reef carbon seams, Witwatersrand Sequence: Mineral Sciences and Engineering, v. 10, p. 223–246.

Zumberge, J.E., Nagy, B., and Nagy, L.A., 1981, Some aspects of the development of the Vaal Reef uranium-gold carbon seams, Witwatersrand Sequence: Organic geochemical and microbiological considerations, *in* Armstrong, F., ed., Genesis of uranium- and gold-bearing Precambrian quartz-pebble conglomerates: U.S. Geological Survey Professional Paper 1161, p. O1–O7.

Manuscript Accepted by the Society 4 October 2005

Printed in the USA

Geological Society of America
Special Paper 405
2006

Conditions of gold remobilization in the Ventersdorp Contact Reef, Witwatersrand Basin, South Africa

Rudy H. Boer[†]
Impact Cratering Research Group, School of Geosciences, University of the Witwatersrand, Private Bag 3, P.O. Wits 2050, Johannesburg, South Africa and
Ferret Mining and Environmental Services (Pty) Ltd, P.O. Box 72313, Lynnwood Ridge 0040, South Africa
Wolf Uwe Reimold[‡]
Impact Cratering Research Group, School of Geosciences, University of the Witwatersrand, Private Bag 3, 2050 Wits, Johannesburg, South Africa, and
Museum für Naturkunde (Mineralogy), Humboldt University, Invalidenstrasse 43, 10115 Berlin, Germany

ABSTRACT

The Late Archean Witwatersrand Basin of South Africa represents the world's foremost gold deposit. The basin is also the erosional remnant of the world's oldest (2.02 Ga) and largest known impact structure, the Vredefort impact structure. The debate about the origin of the Witwatersrand gold resource—as a placer deposit, through influx of gold-bearing hydrothermal solutions into the basin or mobilization of basinal fluids, or in the form of a so-called modified placer deposit—has been ongoing for about a century. Currently, most active Witwatersrand workers favor either a hydrothermal process or the modified placer theory, although the nature and timing of hydrothermal events affecting the basin remains a matter of debate.

Here, results of fluid inclusion petrographic and quadrupole mass spectrometric analyses on Ventersdorp Contact Reef conglomerate and footwall quartzite samples, as well as Vredefort impact-related pseudotachylitic breccias from the bedding-parallel fault zones at the top and bottom of the Ventersdorp Contact Reef, are reported. These analyses were obtained in an attempt to derive new information on the types of fluids that may have been involved in gold mineralization during the history of this important ore deposit. Samples from Tau Lekoa Mine, previously known as the Vaal Reefs No. 10 Shaft Mine in the Klerksdorp goldfield, and samples from the western part of Elandskraal Mine, previously known as the Elandsrand Gold Mine in the Carletonville goldfield, were analyzed. Three populations of predominantly secondary fluid inclusions are recognized. The earliest one comprises small, wispy-textured inclusions. The second involves a moderately saline fluid and the third one a CO_2-rich fluid with variable CO_2/H_2O ratios. Microthermometry indicates fluid entrapment temperatures from 220 to 370 °C at 2 kbar. Tau Lekoa fluid inclusions are enriched in CO_2 compared to the CH_4-enriched inclusions in Elandskraal samples. These differences are explained by comparatively more reducing conditions at the time of fluid entrapment at Elandskraal.

[†]E-mail: rudy@ferretmining.co.za.
[‡]E-mail: uwe.reimold@museum.hu-berlin.de.

Boer, R.H., and Reimold, W.U., 2006, Conditions of gold remobilization in the Ventersdorp Contact Reef, Witwatersrand Basin, South Africa, *in* Reimold, W.U., and Gibson, R.L., Processes on the early Earth: Geological Society of America Special Paper 405, p. 387–402, doi: 10.1130/2006.2405(19). For permission to copy, contact editing@geosociety.org.

Detailed textural studies indicate that fluid entrapment occurred at a late stage in the evolution of the Witwatersrand Basin. This is corroborated by previous argon chronological information as well as detailed mineralogical analysis of impact-related pseudotachylitic breccias, which also suggest that fluid migration occurred at the time of the Vredefort impact event. Thermodynamic calculations based on information gained from entrapped remnants of mineralizing fluids indicate that the solubility of gold in these fluids is, at 1–10 parts per billion, an order of magnitude lower than previously thought. These results cast doubt on the hydrothermal model for authigenic gold mineralization in the Witwatersrand Basin and favor a modified placer model.

Keywords: Witwatersrand, gold, remobilization, Ventersdorp Contact Reef.

INTRODUCTION

The Late Archean, oligomictic quartz-pebble conglomerates of the Witwatersrand Supergroup contain the world's largest resource of gold and second largest resource of uranium (e.g., Pretorius, 1986; Robb and Robb, 1998). While mining of these ores has been undertaken for more than a century, the origin of the gold and uranium in the Witwatersrand Basin remains controversial. Existing models for the origin of Witwatersrand gold can be divided into two groups: those models suggesting that gold and/or uranium were deposited as detritus, which may or may not have experienced minor to major postdepositional modification (paleoplacer or modified paleoplacer models), and those models that suggest that the gold was first introduced into the rocks via postdepositional hydrothermal activity (for recent reviews, see Phillips and Law, 2000; Frimmel and Minter, 2002; Frimmel, 2005a; Hayward et al., 2005 and references therein; and Drennan and Robb, 2006). With the realization during the past few decades that large-scale fluid migration is a normal, predictable phenomenon in sedimentary and metamorphic environments, the focus has shifted away from the paleoplacer model, sensu stricto, to analysis of the modified placer versus hydrothermal mineralization models for the origin of gold and uranium in the Witwatersrand Basin. A key aspect of these models is the proposed linkage between fluid movement and brittle to brittle-ductile deformation manifested in fault and shear zones, many of which appear to be subparallel to the bedding in the Witwatersrand Supergroup. One of the most prominent of these fault zones occurs in association with the uppermost reef in the Witwatersrand Supergroup, namely the Ventersdorp Contact Reef. Apart from being one of the most intensely altered reef horizons, the Ventersdorp Contact Reef is also one of the richest and largest gold resources in the Witwatersrand Supergroup. The Ventersdorp Contact Reef is marked by one of many bedding-parallel shear zones that occur within some of the auriferous and uraniferous conglomerates (see Viljoen and Reimold, 1994 and papers therein; Killick et al., 1988; Reimold et al., 1999). These and a host of other shear zones throughout the Witwatersrand stratigraphy had the potential to serve as channelways, along which hydrothermal fluids could migrate (Jolley et al., 2004).

In this paper, the nature of the fluids that are preserved in the Ventersdorp Contact Reef, as well as the associated hanging-wall and footwall rocks and pseudotachylitic breccias of the Ventersdorp Contact Reef fault zone, is characterized in order to evaluate in a quantitative manner the physico-chemical conditions of gold remobilization and to evaluate the extent of fluid-rock interaction.

GEOLOGICAL SETTING AND CHARACTERISTICS OF THE VENTERSDORP CONTACT REEF

The Ventersdorp Contact Reef is the uppermost auriferous and uraniferous conglomerate in the Late Archean (3.05–2.71 Ga; Robb et al., 1990; Armstrong et al., 1991; Kirk et al., 2001, 2002; Frimmel, 2005a) Witwatersrand Supergroup of the Witwatersrand Basin (Fig. 1). The Witwatersrand Supergroup has been divided into the lower, ~5-km-thick, West Rand Group and the Central Rand Group, which forms the upper 2 km. The Central Rand Group consists almost entirely of quartz arenites with many intercalated conglomerate bands, which were deposited under fluvial conditions and which are considered to represent a succession of superimposed and coalescing fanlike bodies (Robb and Robb, 1998). Paleocurrent vectors at the mines under discussion are unimodal and indicate source areas to the north-northwest of the basin.

A fault zone generally follows the contact between the Ventersdorp Contact Reef and the underlying footwall quartzite (see also Killick et al., 1988; Reimold et al., 1999). Locally, this fault zone cuts across the contact and runs parallel to bedding in one of the Ventersdorp Contact Reef units or even transgresses onto the contact with the hanging wall to the Ventersdorp Contact Reef. Brecciated and sheared rocks are most common along the fault itself but also form offshoots (injection veins) and occur in fault-sympathetic shear fractures, above and/or below the conglomerate (Killick et al., 1988; this study). Evidence for lateral movement is widespread along the fault zone. However, although both sinistral and dextral offset is observed, it has not been possible to determine a consistent sense of slip. Macroscopically, much of the fault zone appears to be cataclastic or mylonitic. Killick and Reimold (1990) and Reimold et al.

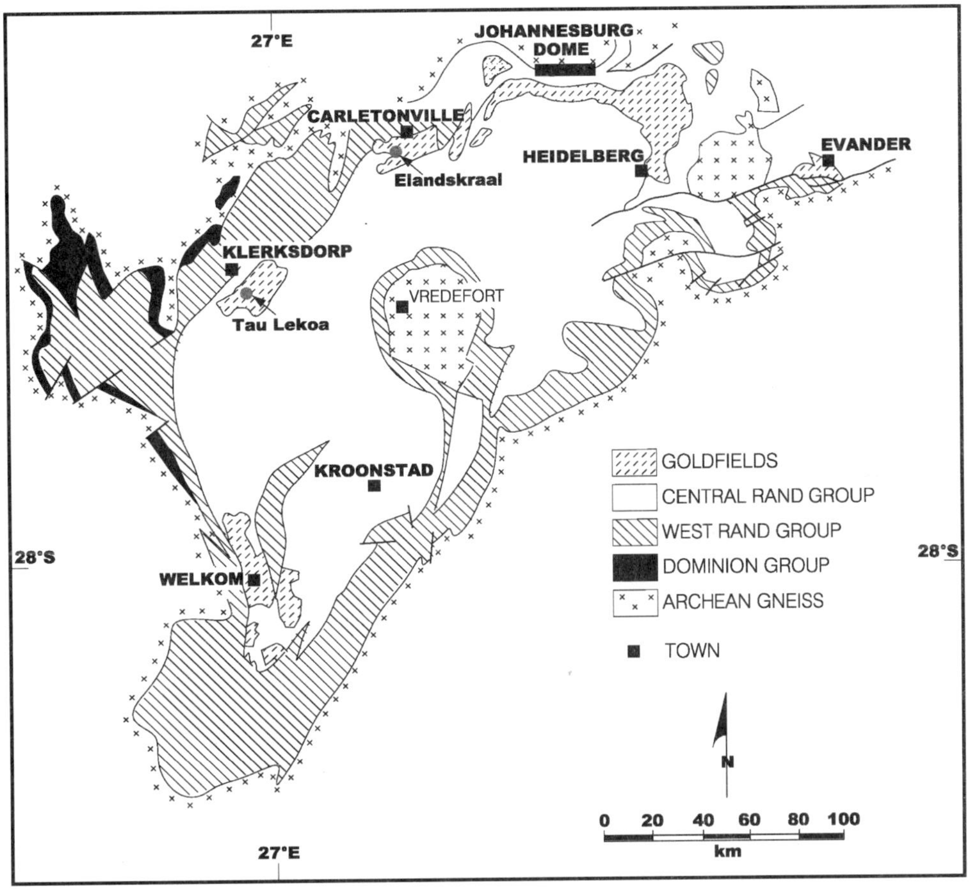

Figure 1. Locations of Tau Lekoa and Elandskraal gold mines in the northwestern and northern sectors of the Witwatersrand basin. Note the Vredefort dome, the eroded central uplift structure of the Vredefort impact structure (Gibson and Reimold, 2001), near the geographic center of the Witwatersrand Basin.

(1999) discussed the local development of pseudotachylite (friction melt) within Witwatersrand fault zones that also contain other fault rock types, including cataclasites and mylonites. The assemblages of such diverse fault breccia types are thus termed here *pseudotachylitic breccia,* by analogy with the discussion by Reimold and Gibson (2005) of multiple breccia types occurring in impact structures (as discussed later herein, it must be assumed that at least the majority of such breccia occurrences in the Witwatersrand Basin must be regarded as the result of the Vredefort impact event).

Sulfide mineralization is locally conspicuous in the fault zone. At several locations, up to 2-cm-wide hydrothermal quartz veins mark the upper and/or lower fault boundaries, either in contact with Ventersdorp Contact Reef or in the footwall quartzite. These veins contain pockets with massive sulfide mineralization containing mainly pyrrhotite with associated pyrite as well as minor chalcopyrite, sphalerite, and galena. The fault zone is cut by later joints but not by younger quartz veins. Trieloff et al. (1994) carried out ^{40}Ar-^{39}Ar dating of the pseudotachylitic breccias and obtained ca. 2000 Ma ages, which are interpreted to represent a link to the Vredefort impact event dated at 2020 ± 5 Ma (e.g., Kamo et al., 1996; for a comprehensive review of the Vredefort impact structure: Gibson and Reimold, 2001). Other argon dating of Witwatersrand fault breccias has been reported recently by Friese et al. (2003).

A second, less well developed fault zone marks the contact between the Ventersdorp Contact Reef and the Ventersdorp lavas in the hanging wall. Where fault material exists, it is generally only up to a few centimeters wide and is generally mylonitic. These mylonites are locally accompanied by hydrothermal quartz veins or quartz pockets, and, rarely, by quartz-carbonate deposits. Cataclasite and pseudotachylite do occur locally but represent only minor components of the fault rocks. Lava above this contact is consistently and strongly impregnated with siliceous veins, stringers, or nodules that exhibit locally intricate folding and deformation. The breccias along both the lower and upper contacts of the Ventersdorp Contact Reef carry locally (as pods or stringers) or over appreciable distances (decimeters to meters) quartz veins that are locally mineralized with sulfides and rarely, minor gold. So-called phyllonites (i.e., mylonitic and ultracataclastic, locally also pseudotachylitic fault rock) with appreciable hydrothermal mineralization, including gold, have also been noted along many of the other auriferous reefs in the various goldfields of the Witwatersrand basin, but particularly in the West Wits Line (Carletonville) and West Rand (Klerksdorp) goldfields (Fig. 1). Their occurrence is considered by many Witwatersrand geologists to be a strong exploration indicator for high gold grades, although this is still somewhat controversial.

Reimold et al. (1999) made a strong argument for a link between brecciation along bedding-parallel fault zones in the Witwatersrand Basin and the Vredefort impact event. Gibson and Reimold (1999) suggested impact-induced hydrothermal activity on both local and basinwide scales, and Frimmel (2005b) emphasizes that the Vredefort impact event would have been the one major permeability-generating event that affected the entire Witwatersrand Basin in post-Witwatersrand depositional times. For this reason, it must be rated rather unlikely that the 2060 Ma Bushveld Complex emplacement event would have initiated significant basinwide hydrothermal fluid activity, although the regional metamorphic peak is associated with this "just" pre-Vredefort impact magmatic phase (e.g., Gibson and Stevens, 1998; Foya, 2001). Reimold et al. (2005) recently discussed the impact-related processes that would have led to extensive hydrothermal activity in the Witwatersrand Basin.

SAMPLING AND PETROGRAPHIC ANALYSIS

Samples of the Ventersdorp Contact Reef and its footwall and hanging-wall rocks that were used in this investigation were acquired from level 82, raise 17 and level 85, raise 20 of the Elandskraal Mine (from that part of Elandskraal that was previously known as the Elandsrand Mine), situated on the West Wits Line goldfield, as well as from levels 10–50 and 1200 in the Tau Lekoa Mine (previously known as the Vaal Reefs No. 10 Shaft mine) in the Klerksdorp goldfield (Fig. 1). Some 160 samples of the various lithologies associated with the Ventersdorp Contact Reef fault zone (mainly along the lower contact of the reef) were collected along 11 traverses across the fault zone. The sample suite (Reimold et al., 1993; Reimold and Boer, 1992) comprises altered lava in the hanging wall, reef (conglomerates of the Witwatersrand Supergroup are locally referred to as "reefs" or "banket") conglomerate, fresh and altered footwall quartzite, and pseudotachylitic and cataclastic breccias and mylonites. Where pseudotachylite is found in the reef horizon, it appears to be younger than the secondary quartz veins that crosscut the Ventersdorp Contact Reef quartz-pebble conglomerate. The footwall consists of arenaceous quartzite, whereas the hanging wall comprises amygdaloidal lava. The presence of pseudotachylite and mylonite along the Ventersdorp Contact Reef horizon indicates intense postdepositional bedding-parallel shearing.

Several generations of faulting have affected the Witwatersrand and Ventersdorp supergroups (cf. also Reimold and Colliston, 1994 and references therein), with various interpretations favoring either syn- to immediately post-Witwatersrand fault activity and associated hydrothermal processes (e.g., Jolley et al., 2004), Ventersdorp tectonism and hydrothermal flux (e.g., Phillips and Law, 2000), syn-Transvaal (ca. 2.5–2.15 Ga) fluidization (Robb et al., 1997), or fluid flux at Vredefort times (Reimold et al., 2005; Hayward et al., 2005). At least one suite of quartz veins is older than the pseudotachylitic breccia and a second one is commonly found parallel to fault zones.

Sulfide mineralization is commonly more pronounced in the lower parts of the Ventersdorp Contact Reef immediately above the fault zone. Locally, it extends up to 20 cm into the reef. Higher up, but still in the reef, sulfide enrichments are found in the form of up to 2-cm-wide bands and stringers that may be aligned parallel to macroscopically visible bedding and/or fractures. Pronounced sulfide mineralization is also associated with a zone of altered lava along the fault above the reef. Sulfide aggregates are found throughout this altered zone or may be associated with quartz-filled vesicles or nodules of up to a few centimeters in diameter.

Hydrothermal phases observed and sampled for this study include: (1) quartz veins above the reef in lava; (2) quartz veins immediately above or along the lower margin of the fault zone; (3) quartz veins terminated or transected by fault rock; (4) veins in pebble conglomerate; and (5) vesicle fillings and veins in hanging-wall lava.

For Elandskraal Mine, the following sequence of structural events was recognized.

1. Mylonite formation.
2. Quartz vein 1 formation (compare Killick et al., 1988 and Killick and Reimold, 1990).
3. Pseudotachylitic breccia 1 formation, accompanied by hydrothermal alteration (quartz veins 2 as well as quartz-carbonate veins).
4. Pseudotachylitic breccia 2 and/or post-pseudotachylitic breccia 1 brittle deformation, with local hydrothermal activity (fracture fillings with secondary quartz and sulfides).

In Tau Lekoa Mine, the following scenario was encountered.

1. Quartz veins 1 emplaced (they are now folded).
2. Pseudotachylitic breccia, including mylonite, formed, with associated hydrothermal alteration.
3. Quartz veins 2 emplaced.

Several morphological types of pyrite and pyrrhotite were observed in the samples: (1) well-rounded grains that must be at least partly of detrital origin and that are commonly associated with other unequivocally detrital materials such as quartz pebbles with chromite inclusions; (2) large euhedral sulfide grains (greater than several millimeters in size); (3) finer-grained (smaller than 1 mm) euhedral sulfide crystals occurring individually or in aggregates; (4) elongated sulfide grains with well-rounded ends, which are sometimes flanked by pressure haloes formed of quartz and/or chlorite fibers; (5) very fine-grained subhedral crystallites in a matrix of pseudotachylitic breccia or altered lava; and (6) replacements of amygdale fillings in altered lava.

The majority of the fine-grained sulfide, generally with euhedral crystal shapes, occurs in fractures (usually in fractured quartz pebbles), in vesicles in the hanging-wall lava or in pseudotachylitic breccia, or disseminated through altered rocks. Sphalerite, which is relatively abundant in all sampled lithologies, occurs as a few rounded or apparently deformed grains of possible detrital origin, although most grains are clearly authigenic. Rutile is widespread, in particular in altered lava, but also in selected Ventersdorp Contact Reef and footwall quartzite specimens. Chalcopyrite and galena were observed in samples from all parts of the fault zone and in all lithologies. Pyrrhotite is more abundant in Elandskraal ore than in Tau Lekoa samples. Many studies pertaining to gold mineralization in Ventersdorp Contact Reef conglomerate have been carried out (e.g., review by Frimmel, 2005b and references therein) and have shown that the bulk of Ventersdorp Contact Reef gold occurs in the form of authigenic growths, as overgrowths on older sulfide minerals or fracture infill. These observations, *inter alia,* are the basis for the long-standing promotion of the hydrothermal and modified placer hypotheses for the genesis of Witwatersrand gold.

Lava samples are moderately vesicular, with up to 25% of a sample comprising roundish to irregularly shaped amygdales that are filled with secondary carbonate and chlorite, ± clinozoisite-quartz-sulfides. All lava samples showed evidence of metamorphic or hydrothermal effects. Most samples contain typical greenschist facies mineral assemblages (albitic feldspar, chlorite, carbonate, quartz, and iron oxides, ± actinolite-clinozoisite-pyrite). Some samples have also undergone hydrothermal alteration that formed secondary carbonate, quartz, sericite, iron oxides, and sulfides. Amygdales in the most strongly altered rocks are enriched in quartz and sulfide minerals.

BULK ROCK CHEMICAL CHARACTERISTICS

As part of a wider-ranging multidisciplinary study of Ventersdorp Contact Reef and Ventersdorp Contact Reef–related lithologies in the Elandskraal and Tau Lekoa study locations, we have carried out petrographic analysis as well as major and trace element analysis by X-ray fluorescence spectrometry (XRF) and instrumental neutron activation analysis (INAA; on 156 and 81 samples, respectively) (these results are reported by Reimold and Boer, 1992; Reimold et al., 1993; they will be discussed in detail in a separate publication). However, a summary of some of these results is pertinent to the work reported here and its discussion.

The rocks associated with the Ventersdorp Contact Reef, in general, show a high degree of sericitization that is reflected in their rather elevated K_2O and Al_2O_3 concentrations. Pseudotachylitic breccias are enriched in Al_2O_3 and K_2O and depleted in Fe_2O_3 (total Fe) and MgO, in comparison to Ventersdorp Contact Reef and lava. Petrographic observations strongly suggest that a large component of Ventersdorp Contact Reef parent material contributed to the fault breccia (see also Reimold et al., 1999). However, the lithological heterogeneity of Ventersdorp Contact Reef conglomerate and of pseudotachylitic breccia should not be underestimated; for instance, shale or quartzite inclusions in a breccia sample could strongly influence the bulk analysis of small samples measuring only a few centimeters in diameter. Potassium and Al enrichment are a general signature of the hydrothermal overprint on the fault zone, and as it is a characteristic of the fault breccia too, it could be regarded as evidence for this event either being related to the formation of the fault rock or postdating it. We favor the latter possibility, on the basis of our petrographic evidence (also results of an independent investigation of Witwatersrand fault zones by Reimold et al., 1999; Frimmel et al., 1999).

Alteration of footwall quartzite is characterized by strong depletion of Fe_2O_3 relative to reef conglomerate. However, the distribution of sulfides, especially in the footwall quartzite samples, is very heterogeneous. For example, sulfide-rich patches were observed in the sericitic matrix or in sphene-quartz-sulfide "nodules." Ti-rich sphene-quartz-sulfide "nodules" have been observed in the footwall quartzite. These observations suggest that hydrothermal activity was pervasive and affected the entire fault zone, from the upper contact zone with Ventersdorp Lava, through the conglomerate and into the footwall quartzite(also see Zhao et al., 2006 and earlier work cited therein). In recent years, a number of gold mining companies have begun to mine footwall in addition to conglomerate, as gold grades can be very high in the footwall quartzite.

On the basis of our XRF and INAA data, three basic types of vein fillings can be recognized: Type 1, which is characterized by <1 wt% Al_2O_3 and low K_2O (<0.15 wt%); Type 2, which has >7 wt% Al_2O_3, >1.5 wt% K_2O, high Sr (~75 ppm) and high Rb (>50 ppm); and Type 3, distinguished here as those veins that are folded and probably predate fault breccia formation and that are characterized by high Ca (>1 wt%), high Sr (>60 ppm), and high Cu (>400 ppm; as confirmed by the observation of massive chalcopyrite in a sample from this vein).

Chromium, Ni, As, and Sb are strongly enriched in altered lava, whereas Na is strongly depleted. One altered specimen displays Au enrichment by a factor of approximately three times the concentrations of less altered samples. The same specimen also displays the highest measured concentrations for As, Cr, and Ni. In general, altered lava is characterized by low Na and high K contents. These geochemical signatures indicate that the Ventersdorp Contact Reef rocks have experienced significant sericitization (Frimmel et al., 1999; Frimmel and Gartz, 1997).

FLUID INCLUSION PETROGRAPHY

The reason for embarking on a detailed fluid inclusion investigation was twofold: (1) to understand the relative temporal relations between the various fluid populations, and (2) to characterize, in a semiquantitative manner, the nature of the mineralizing fluids in the Ventersdorp Contact Reef.

Analytical Procedure

Fluid inclusion data were obtained by microthermometric heating and freezing, using the standard technique described by Roedder (1984). Temperatures of phase changes were obtained with a FLUID INC.–adapted U.S. Geological Survey heating/freezing system, which operated by passing cooled N_2 gas or heated air directly above or beneath the sample. The thermocouple was calibrated against synthetic fluid inclusions (Sterner and Bodnar, 1984). Reproducibility of measurements during calibrations was within 0.2 °C below 0 °C and better than ±2 °C on heating the sample. Salinities were standardized and are expressed as weight percent NaCl equivalent (wt% eNaCl) using the conversion factors of Bodnar (1993).

Results

The high density of randomly oriented fractures decorated with fluid inclusions made it difficult to determine with confidence whether a given inclusion lay along one of these fractures (secondary or pseudosecondary) or was trapped during crystal growth (primary); criteria for the recognition of inclusion types follow those of Roedder (1984). We have not attempted to classify the fluid inclusion types but rather have related groups of inclusions to various fillings of crosscutting veins. This also applies to fluid inclusions in matrix material (quartz clasts as well as groundmass quartz). In general, microscopic examination of the quartz shows many if not all of the described types of fluid inclusions "coexisting" with one another in any field of view. Although in most cases it could not be determined from simple petrographic techniques which, if any, of the many inclusions were coeval because of the abundance of fractures. CO_2-rich inclusions were conspicuously absent in the latest-stage, crosscutting, fluid inclusion trails. Thus, although the CO_2-rich inclusions coexist spatially with the aqueous inclusions, they do not appear to be coeval. This assertion is further supported by salinity and temperature data described below.

Quartzite samples show abundant triple junctions between newly grown grains (recrystallized quartz) with a wispy-textured distribution of inclusions. The wispy-textured inclusions are small (with diameters smaller than 2 µm) and well rounded. No microthermometric experiments were conducted on these small inclusions, as their small size prohibited the accumulation of well-controlled temperature data for phase changes. However, the wispy-textured inclusions indicate entrapment temperatures in excess of 350 °C.

Elandskraal Mine

Most fluid inclusions (average size ~4 µm) occur along secondary trails in the matrix of the conglomeratic reef material. No microthermometric studies were performed on the inclusions in the pebbles, as it appears irrelevant with respect to the goals of this project to study inclusions that likely are related to the geological history of the unknown pebble provenance terranes. Black carbonic daughter crystals are common in these inclusions. These fluid inclusions in the matrix exhibit dark menisci, which indicate the presence of carbonic gaseous species and are found along trails containing aqueous inclusions with variable liquid-to-vapor ratios. Some CO_2-rich inclusions lack a vapor bubble at room temperature. The high CO_2 content in these fluid inclusions correlates with a high carbon and bitumen (kerogen) content in the siliceous matrix of the sample. The quartz in this sample does not appear to be as recrystallized as the quartz closer to a pseudotachylitic breccia veinlet. No visible CO_2 was detected, but clathrates (gas hydrate structures) formed during cooling experiments. Clathrate melting temperatures vary around 5 °C but could not be duplicated with precision due to the irregular shape and relatively small size of the inclusions. Therefore, the salinities of inclusions with clathrate, in this study, were also estimated from final ice-melting temperatures.

Furthermore, the presence of CO_2-clathrate obscures the initial melting of salt-hydrates and ice, because clathrate in fluid inclusions removes pure water. Consequently, temperatures recorded as the apparent eutectic of the salt-H_2O subsystem refer to the lowest temperatures at which the liquidus assemblage, ice + CO_2-clathrate + vapor + carbonic liquid, could be recognized. The microthermometric results (Table 1) therefore provide a maximum value for the true eutectic temperature and salinity of the inclusions. Eutectic melting temperatures of around –30 °C indicate the presence of divalent cation species in the fluids, such as Ca^{2+} and Mg^{2+}. Final melting temperatures (T_m) average around –9.6 °C (13.5 wt% eNaCl), whereas the homogenization temperatures range from 165 to 180 °C (0.996 g/cm^3). The dark daughter crystals did not change shape or size during heating, and the fluid inclusions decrepitated before they could dissolve.

Large secondary inclusions occurring along trails can be divided into at least two groups. The more saline inclusions exhibit an average final melting temperature (T_m) of –11.0 °C

TABLE 1. SUMMARY OF MICROTHERMOMETRIC DATA FOR FLUID INCLUSIONS IN VARIOUS ROCK TYPES IN AND NEAR THE VENTERSDORP CONTACT REEF (VCR) IN THE ELANDSKRAAL MINE

	T_m (°C)	wt% eNaCl	T_h (°C)	Density (g/cm^3)
VCR Quartzite				
Fluid I	–11.0	15.0	190	0.99
Fluid II	–1.4	2.3	190	0.89
Fluid III			>350	
Fuchsitic footwall quartzite				
Fluid I	–9.8	13.7	180	0.99
Fluid II	–2.6	4.2	180	0.92
Altered Pseudotachylitic Breccia				
Fluid I	–2.7	4.4	140	0.96
	–6.4	9.7	140	0.99
Mineralized Quartz Vein in VCR				
Fluid I	–4.9	7.7	240	0.89
Fluid II	–10.8	14.8	115	1.05
Fluid III	–2.7	4.3	124	0.97
Quartz Matrix in Pebble Conglomerate Reef				
Fluid I	–8.1	11.8	285	0.87
Fluid II	–9.8	13.7	130	1.03

Note: T_m—melting temperature; T_h—homogenization temperature.

(Type I), indicating 15 wt% NaCl equivalent (eNaCl). The second type of inclusion contains fluid with a final melting temperature of −1.4 °C (Type II), which corresponds to a salinity of 2.3 wt% eNaCl. Homogenization temperatures (T_h) for both of these populations were unimodal at ~190 °C. Homogenization always took place into the liquid phase. Using the Brown and Lamb (1989) equation of state for H_2O-CO_2-NaCl fluids, these data were used to calculate densities of 0.99 and 0.89 g/cm^3 for the respective fluids. Temperatures at which entrapment of the fluid took place must, therefore, have been higher than 190 °C, as the homogenization temperature reflects a minimum entrapment temperature in such scenarios.

Metastable behavior was detected in some dark-colored fluid inclusions with a large vapor-to-liquid ratio (V:L = 6:4). These secondary fluid inclusions overprint the earlier fluid generation that is present as the small (<1 µm) wispy-textured inclusions.

Vein quartz occurring on the contact between pseudotachylitic breccia and green footwall quartzite is hosted in a chlorite and fuchsite matrix. Quartz fragments are cut by a multitude of aqueous inclusion trails that terminate at the grain boundaries. Daughter crystals are rare in these inclusions. Fluid inclusions show negative crystal shapes. No fluid inclusions are present in the pseudotachylitic matrix material. Two populations of low salinity inclusions are present, with final melting temperatures of −2.6 and −9.8 °C (Fluid II and Fluid I, respectively) and salinities of 4.2 and 13.7 wt% eNaCl, respectively. Both types of inclusions homogenized at 180 °C into the liquid phase, yielding densities of 0.92 and 0.99 g/cm^3, respectively.

Quartzite fragments occurring in slightly altered pseudotachylitic breccia contain trails of irregularly shaped, 2–5-µm-diameter, secondary aqueous fluid inclusions. A sharp contact is present between breccia matrix and quartzite clasts. Quartzite shows wispy-textured inclusions trapped on grain boundaries and at triple junctions of recrystallized quartz. The most important observation is the recognition of naturally decrepitated fluid cavities. Daughter crystals are present in the later secondary inclusions. No CO_2-bearing inclusions were detected in this sample. Final melting temperatures fall into a bimodal distribution with peaks at −2.7 °C (4.4 wt% eNaCl) and −6.4 °C (9.7 wt% eNaCl). The average homogenization temperature peaked at 140 °C, but some inclusions showed T_h as low as 66 °C. Fluid densities ranged from 0.96 to 0.99 g/cm^3.

A quartz vein in the Ventersdorp Contact Reef conglomerate, positioned parallel to the pseudotachylitic breccia, contained crosscutting trails of large (±15 µm diameter), irregular aqueous inclusions that are closely associated with sulfide mineralization. These inclusions behave in a very consistent manner, with final melting temperatures around −4.9 °C (Fluid I) (7.7 wt% eNaCl). Homogenization temperatures form a maximum around 240 °C, resulting in a fluid density of 0.89 g/cm^3. Daughter crystals, which are rare in these inclusions, are transparent, of cubic crystal shape, and are possibly halite (NaCl) or sylvite (KCl). A second, more saline, fluid is present with T_m −10.8 °C (Fluid II) (14.8 wt% eNaCl) and T_h 115 °C (1.05 g/cm^3). This Fluid II occurs in single large inclusions, as opposed to the former, less saline inclusions (Fluid I), which occur along trails. From petrographic considerations, it would appear as if the more saline fluid (Fluid II) was emplaced earlier than the less saline fluid (Fluid I). The relationship between the saline fluids and mineralization could not be established.

Numerous trails of secondary low salinity inclusions are present in a quartz vein that crosscuts a dolerite dyke. Fluid inclusions show unequal ratios of liquid to vapor, with some inclusions appearing empty at room temperature. No daughter crystals were observed. The average eutectic temperature of approximately −4.7 °C suggests an absence of divalent cation species. This relatively high eutectic temperature suggests an apparent eutectic due to the presence of clathrates, which incorporate the first melt fluid into the gas hydrate structure. The presence of carbonic species that form the nucleus of such hydrate structures has been confirmed by quadrupole gas analysis (see below). Final melting temperatures show a range from −2.5 to −1.0 °C (4.1–1.6 wt% eNaCl), and homogenization temperatures cluster around 136 °C (density range: 0.96–0.94 g/cm^3).

Quartz occurring in the matrix of conglomeratic reef material contains trails of low salinity secondary fluid inclusions. These inclusions occur together with earlier, irregular-shaped, randomly distributed carbonic fluid inclusions. A few of the irregularly shaped, early inclusions with final melting temperature around −10.4 to −5.7 °C (14.4–8.8 wt% eNaCl) show homogenization temperatures of 316 °C. The majority of these fluid inclusions decrepitated during heating at temperatures of ~285 °C. The existence of this peculiar type of fluid inclusion is significant and, to our knowledge, was not recognized in earlier Witwatersrand studies. We did not find it in any of the other samples from outside the reef horizon. The aqueous secondary fluid inclusions do not contain any carbonic species and have a final melting temperature of around −9.8 °C (Fluid II) with intermediate salinity (13.7 wt% eNaCl). Homogenization temperatures of these late-stage inclusions average around 130 °C.

A tourmaline-bearing quartz vein positioned immediately above the Ventersdorp Contact Reef contains secondary fluid inclusion trails in quartz, consisting of small (<5 µm), well-rounded fluid inclusions. Irregularly shaped, locally decrepitated fluid inclusions occur along trails, but are of a different type than those previously mentioned. Inclusions exhibit eutectic melting around −24.0 °C, and final melting temperatures are consistent, in a very narrow range around −10.4 °C indicating intermediate salinities (14.4 wt% eNaCl). Homogenization temperatures are consistently around 115 °C (1.05 g/cm^3).

Tau Lekoa

Footwall quartzite at Tau Lekoa Mine contains fluid inclusions showing a wispy texture, as well as large secondary fluid inclusions. A petrographic study of the temporal relations in fluid inclusions occurring in the Tau Lekoa quartzite samples provides evidence of a later stage fluid. Microthermometric results are compiled in Table 2. The wispy fluid inclusion tex-

TABLE 2. SUMMARY OF MICROTHERMOMETRIC DATA OF VARIOUS ROCK TYPES ON THE VENTERSDORP CONTACT REEF (VCR) STRATIGRAPHIC HORIZON IN THE TAU LEKOA MINE

	T_m (°C)	wt% eNaCl	T_h (°C)	Density (g/cm³)
VCR Quartzite				
Fluid I	–2.1	3.4	150	0.94
Fluid II			>350	
Pseudotachylitic Breccia				
Fluid I CO_2 rich	–59.2		16.0	0.81
Fluid II	+1.5	13.9	150–200	1.08
VCR footwall quartzite with Crosscutting Quartz Vein				
Fluid I Quartzite	–6.0	9.2	150	0.98
Fluid II Vein	–2.2	3.6	225	0.87
Pseudotachylitic Breccia Stringer in Quartzite				
Fluid I P stringer	+7.5	4.9	115–250	0.98
Fluid II Quartzite	–1.7	2.8	200	0.82

Note: T_m—melting temperature; T_h—homogenization temperature.

ture is indicative of high entrapment temperatures (>350 °C). Abundant irregularly shaped and large (diameter 10–15 µm) secondary aqueous inclusions have final melting temperatures ranging from –3.6 to +1.5 °C. These temperatures are indicative of the presence of clathrates that, in this case, proved to be CO_2 rich. Homogenization temperatures fall within a narrow range around 150 °C. No daughter crystals are present in these large inclusions. Numerous trails are composed of small dark inclusions, many of which appear empty.

Pseudotachylitic breccia contains rounded quartz fragments in a strongly altered matrix. Trails of small inclusions transect the quartz fragments. The distribution of fluid inclusions in the various quartz fragments is extremely heterogeneous, with some quartz fragments containing almost no inclusions. CO_2-H_2O inclusions with a CO_2:H_2O ratio of 60:40 are common. A slightly lowered final melting point for CO_2 to –59.2 °C reflects the presence of ~13.5 mol% CH_4 in the three-component system H_2O-CO_2-CH_4. The CO_2 homogenization temperature of +16.0 °C corresponds to a density of 0.85 g/cm³ for the CO_2, and the $T_{h(H_2O)}$ of 250 °C (Table 2) indicates a bulk density of 0.95 g/cm³ in the system H_2O-CO_2-NaCl. The salinity of the fluid as calculated from the clathrate melting temperature of 1.5 °C is 13.9 wt% eNaCl.

Abundant triple junctions are present in vein quartz from the Tau Lekoa Mine, and this quartz contains inclusion trails with decrepitated fluid inclusions. Most of these inclusions have a final melting temperature of –2.2 °C (3.6 wt% eNaCl) and a homogenization temperature of 225 °C (0.87 g/cm³). Irregularly shaped inclusions occur in quartz grains with numerous small, empty inclusions on grain boundaries. Secondary fluids in the quartzite and vein quartz are distinctly different. These two fluids are not the only fluids present. Decrepitation textures and dark empty inclusions along grain boundaries suggest that earlier fluids have escaped. Peculiar, dark (possibly of higher hydrocarbon composition) daughter crystals are present.

Veining at the contact between the altered lava in the hanging wall and the Ventersdorp Contact Reef conglomerate material consists predominantly of calcite and associated carbonates. Wispy-textured distribution of fluid inclusions with numerous additional trails and abundant triple junctions is present. Fluid inclusions in carbonaceous material are too small to study.

A pseudotachylitic breccia stringer containing quartzite clasts with wispy-textured fluid inclusions exhibits numerous crosscutting fluid inclusion trails that are terminated at grain boundaries. Chlorite fills the abundant small vesicles in the pseudotachylitic matrix. In secondary trails terminating on quartz fragment boundaries, liquid-to-gas ratios are highly variable. Clathrate melting temperatures of +7.5 °C (4.9 wt% eNaCl) indicate the presence of CO_2. Fluid densities cover a wide range of 0.9–1.05 g/cm³ for different inclusion trails. This is suggestive of different fluid populations.

Quadrupole Mass Spectrometer Gas Analysis

Quadrupole mass spectrometric analyses were carried out on small groups of small grains that were broken from doubly polished plates (~200 µm thick and less than 2 mm in size) with a total mass of less than 50 mg. These samples had already been subjected to fluid inclusion petrography. Although every effort was made to select areas from these plates that contained single fluid inclusion assemblages, most samples contained at least small amounts of other fluid inclusion assemblages because of the complex overlapping of the assemblages described above. This uncertainty was judged acceptable because the main purpose of this investigation was to determine the principal gas species present in the mineralizing system rather than distinguishing between compositions of the various fluid inclusion assemblages. Gases were released by crushing the small quartz particles at 110 °C under vacuum in a modified Nupro® valve.

Analyses of gases from fluid inclusions were made with a VG SXP600 quadrupole mass spectrometer (QMS) at the Department of Geological Sciences, University of Michigan. Gases were passed for analysis through an inlet system designed to minimize the path distance between sample and ionizing filament. Source electron energy was 50 V. A secondary electron multiplier operated at 1200 V detected ionized molecules and molecular fragments. For most samples 16 masses were monitored, allowing simultaneous real-time analysis for H_2O, CO_2, CH_4, C_2H_6, N_2, Ar, H_2S, SO_2, and HCl. Data were collected at a rate of ~100 cpm. In almost all cases, water mass response was 2–4 orders of magnitude above background.

In a quadrupole mass spectrometer four rods are aligned parallel with the best possible degree of accuracy in order to achieve optimum detection limits. Each pair of opposite rods is electrically connected, and a positive dc voltage is applied to one pair and an equal, though negative, dc voltage to the other pair. Superimposed on these dc voltages is a radiofrequency ac signal arranged in such a way that the phase of the signal applied to one pair of rods is 180 degrees different from that of the other pair. The ratio between the applied dc and ac fields is critical to successful operation of the QMS.

In almost all analyses of fluid inclusions from the Ventersdorp Contact Reef samples (Table 3), H_2O, CO_2, and CH_4 make up more than 99 mol% of the gas composition, although their relative abundances vary greatly. On a CO_2-CH_4-C_2H_6 ternary diagram, the Elandskraal and Tau Lekoa samples plot in distinctly separate fields (Fig. 2), with CO_2 not exceeding 0.60 mol% in the Elandskraal samples and CH_4 not surpassing 0.59 mol% in the Tau Lekoa samples. On a H_2O-CO_2 binary plot, the Tau Lekoa samples plot on a straight line, whereas the Elandskraal data scatter (Fig. 3). The opposite is true for a binary H_2O-CH_4 plot (Fig. 4). Higher alkanes (e.g., C_2H_6) occur at up to 0.14 mol% but do not correlate consistently with other gases. Argon, which was detected in amounts of up to 0.01 mol%, has a weak negative correlation with H_2O but not with other compounds. The average total sulfur gas content for Elandskraal and Tau Lekoa samples is ~0.02 and 0.011 mol%, respectively. The SO_2/H_2S ratio varies from 0.01 to 0.39, irrespective of locality or rock type, and a weak positive correlation is apparent between the two gases (Fig. 5). HCl is significantly enriched in the Elandskraal samples (Fig. 6), in comparison to the Tau Lekoa data. Moderate N_2 and Ar values (Fig. 7) suggest that the samples were not affected by atmospheric contamination during preparation (e.g., by air entrapment along cleavage planes).

TABLE 3. FLUID INCLUSION COMPOSITIONS FOR SAMPLES FROM ELANDSKRAAL MINE AND TAU LEKOA MINE

Sample ID	H_2O	CO_2	CH_4	C_2H_6	N_2	Ar	H_2S	SO_2	HCl
EL 17	96.63	0.015	0.170	0.146	1.089	0.027	0.023	0.008	0.072
EL 20	97.21	0.602	1.227	0.083	0.322	0.012	0.022	0.002	0.036
EL 22	95.89	0.197	2.395	0.144	0.239	0.016	0.011	0.002	0.212
EL 38	98.47	0.424	0.649	0.029	0.199	0.015	0.010	0.003	0.115
EL 47	98.87	0.132	0.727	0.037	0.063	0.007	0.020	0.000	0.133
10D02B	97.02	2.399	0.269	0.025	0.190	0.008	0.008	0.001	0.055
10D06	97.99	1.535	0.207	0.034	0.157	0.006	0.006	0.002	0.018
10D07	99.48	0.278	0.111	0.012	0.091	0.005	0.005	0.000	0.008
10D210	96.08	3.104	0.529	0.046	0.161	0.011	0.020	0.008	0.040
10D225	97.78	1.413	0.595	0.047	0.113	0.009	0.006	0.001	0.022
10G12	96.15	2.974	0.238	0.072	0.499	0.008	0.005	0.004	0.031
HO 68–203	99.58	0.103	0.007	0.012	0.195	0.003	0.003	0.001	0.022
BM 7384	98.03	1.346	0.268	0.010	0.127	0.003	0.032	0.000	0.178

Notes: Crush analyses; data in mole percent. EL—Elandskraal, 10D and 10G–Tau Lekoa, HO and BM—reference standards.

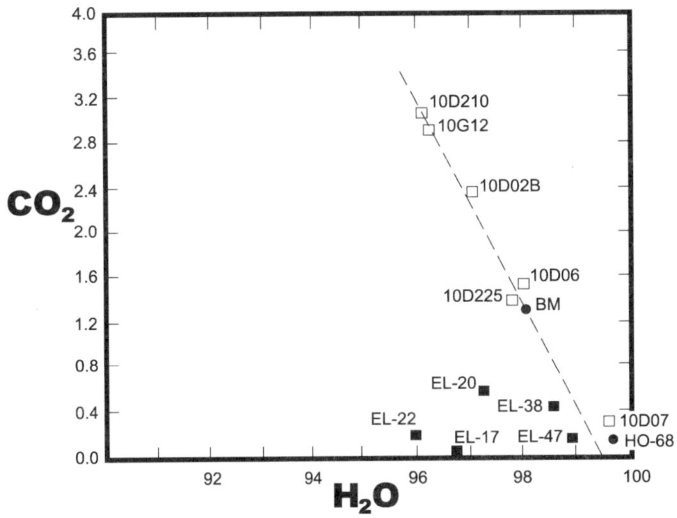

Figure 3. Binary H_2O-CO_2 diagram for Elandskraal and Tau Lekoa quadrupole mass spectrometric data; sample index: EL—Elandskraal, 10D and 10G—Tau Lekoa, HO and BM—reference standards.

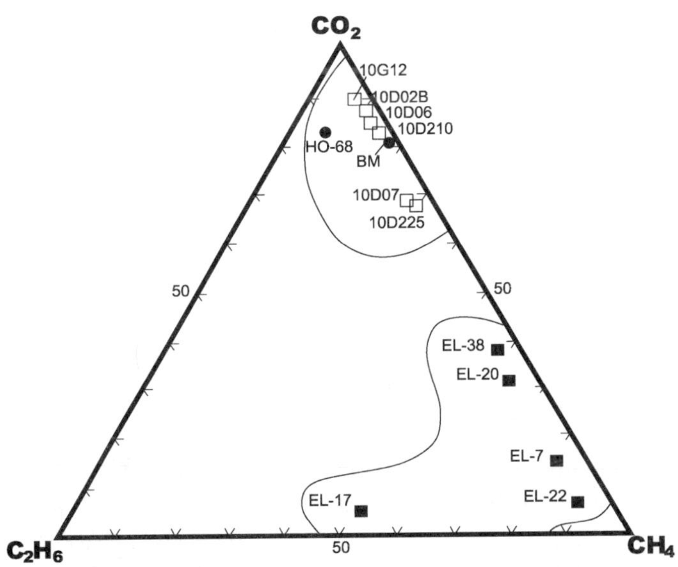

Figure 2. Ternary CO_2-CH_4-C_2H_6 diagram for Elandskraal Mine and Tau Lekoa Mine samples analyzed by quadrupole mass spectrometry; sample index: EL—Elandskraal, 10D and 10G—Tau Lekoa, HO and BM—reference standards.

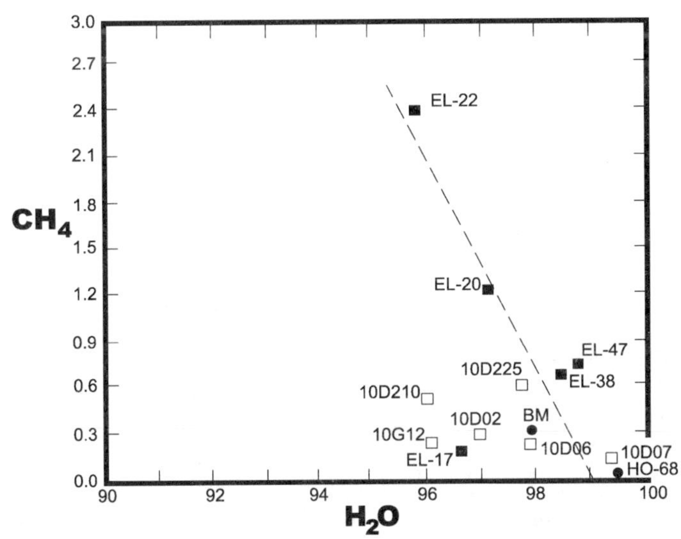

Figure 4. Binary H_2O-CH_4 diagram for Elandskraal and Tau Lekoa quadrupole mass spectrometric data; sample index: EL—Elandskraal, 10D and 10G—Tau Lekoa, HO and BM—reference standards.

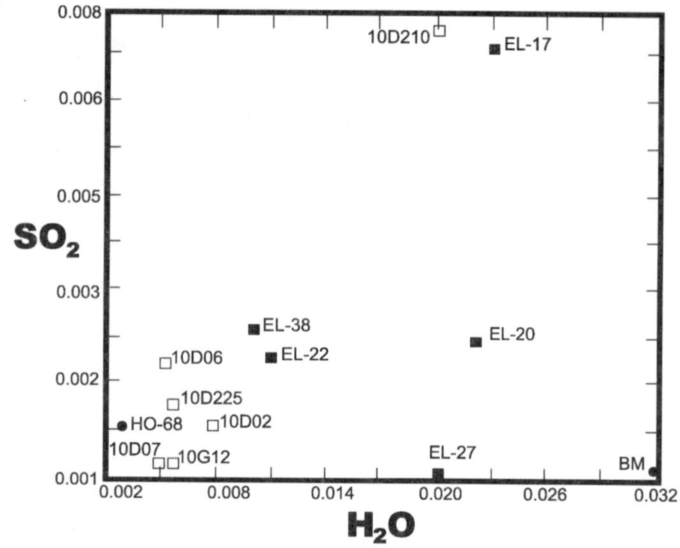

Figure 5. Binary H_2O-SO_2 diagram for Elandskraal and Tau Lekoa quadrupole mass spectrometric data; sample index: EL—Elandskraal, 10D and 10G—Tau Lekoa, HO and BM—reference standards.

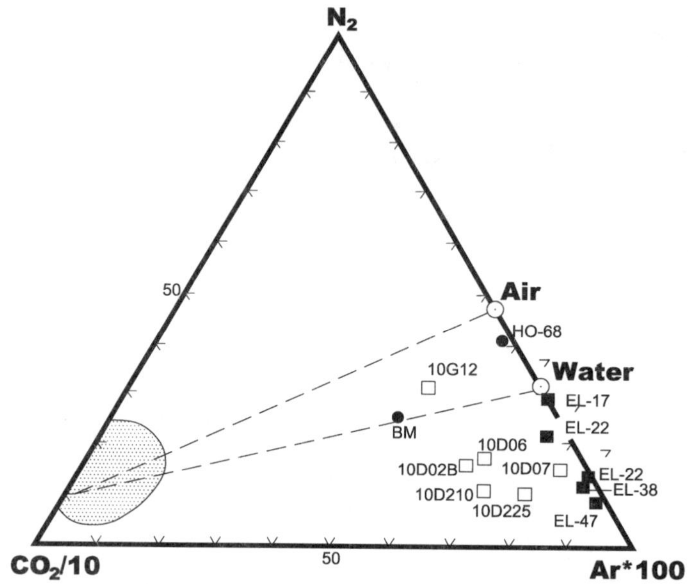

Figure 7. Ternary diagram comparing the CO_2, N_2, and Ar contents of inclusion fluids from Elandskraal and Tau Lekoa samples. The tie lines from air and water to CO_2 represent possible mixing lines between air and/or meteoric water with magmatic gas. Data for fumarolic environment after Giggenbach et al. (1990); sample index: EL—Elandskraal, 10D and 10G—Tau Lekoa, HO and BM—reference standards.

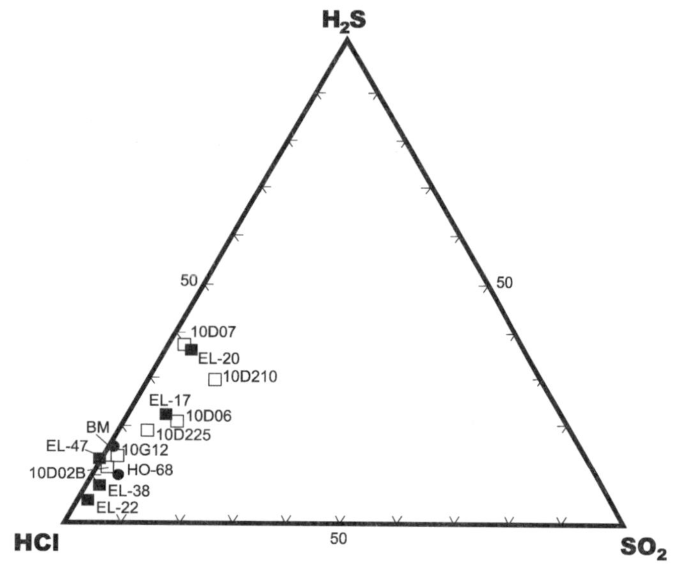

Figure 6. Ternary H_2S-SO_2-HCl diagram for Elandskraal and Tau Lekoa quadrupole mass spectrometric data; sample index: EL—Elandskraal, 10D and 10G—Tau Lekoa, HO and BM—reference standards.

DISCUSSION

Microthermometric Data

In samples from both sampling sites three dominant fluid inclusion generations have been recognized. The earliest fluid inclusions (Fluid I) are small and are found in numerous planes that form a wispy texture in the host mineral. These inclusions contain a low-salinity aqueous fluid. This population was succeeded by an aqueous carbonic fluid (Fluid II) of intermediate to high salinity (12–15 wt% eNaCl), with variable liquid-to-vapor filling ratios. It is present in a large number of samples that are closely related to the reef horizon. A third, late-stage type of inclusion generation (Fluid III) is found in trails that cut other inclusion assemblages. These inclusions contain moderately saline aqueous fluid (2–4 wt% eNaCl).

Fluids from both footwall quartzite and reef quartzites at Elandskraal and Tau Lekoa Mines have remarkably similar salinities and densities. Quartzites from both localities contain wispy textures of earlier fluid inclusions and, together with pseudotachylitic breccia specimens, are regarded as unsuitable for fluid inclusion studies. Secondary inclusions in quartz clasts in pseudotachylitic breccia from Tau Lekoa have a significantly higher CO_2 content than those from equivalent specimens from Elandskraal Mine. This observation was confirmed by quadrupole gas analysis. Similarities in secondary fluid inclusions amongst the quartzites and pseudotachylitic breccias suggest that these rock types were affected by the same post-entrapment process/es.

The homogenization temperatures presented in Tables 1 and 2 are an indication of minimum entrapment temperatures for these fluids. In order to determine the actual temperature of entrapment it is necessary to apply a pressure correction. Kyanite in quartz veins in faults that intersect the Kimberley and Monarg reefs (Central Rand Group) indicate pressures in the range of 2–2.5 kbar, which would be equivalent to a depth of ~7–9 km under lithostatic pressure (Wallmach and Meyer, 1990). This depth is considerably greater than the thickness of overlying Witwatersrand Supergroup sediments, which suggests that the kyanite formed after deposition of later volcanic and sedimentary rocks that overlie the Witwatersrand Supergroup. Because this great thickness of younger rocks was not completely deposited until about the time of intrusion of the Bushveld Complex (2.06 Ga; e.g., Walraven and Martini, 1995), this suggests that the kyanite did not form until that time. Sericite pseudomorphs after posttectonic kyanite were found in a shear zone near the top of the Central Rand Group, which, by assuming a pressure in the order of 2–2.5 kbar, places the kyanite-forming event clearly into post-Transvaal Supergroup time (Frimmel et al., 1992). The shales of the Orange Grove Quartzite Formation also contain kyanite, which according to Schreyer and Bisschoff (1982) occurs on a regional scale. T. Wallmach (Ferret Mining and Environmental Services [Pty] Ltd, 2005, personal commun.) mentioned the local occurrence of kyanite in the Ventersdorp Contact Reef fault rocks, although we did not see it in the samples used in this study. Work by Wallmach (2005, personal commun.) on the overlying Black Reef has indicated the presence of andalusite, with kyanite being absent. This observation indicates that the Ventersdorp Contact Reef situation was close to this alumosilicate phase transition and that a pressure assumption of 2–2.5 kbar for the Ventersdorp Contact Reef is reasonable. If this pressure prevailed, the carbonic fluid inclusions (Fluid II) were probably trapped at temperatures between 220 and 410 °C. Late-stage, low-salinity fluid inclusions (Fluid III) have lower homogenization temperatures and were probably trapped at lower pressures, although we have no evidence on the timing or pressure of their entrapment. The available data indicate that a temperature of 350 °C and a pressure of 2 kbar is reasonable to use in the following thermodynamic modeling.

These data are in good agreement with the microthermometric study by Killick et al. (1988) on quartz veins from near the Ventersdorp Contact Reef on Western Areas gold mine, which yielded pressure-corrected (2.5 kbar) temperatures of up to 320 °C. Entrapment temperatures obtained by Frimmel and Minter (1991; also Frimmel et al., 1993) of 200–270 °C (or 240–300 °C; Frimmel et al., 1992) for the Basal Reef in the Welkom Gold Field (Central Rand Group) are in the lower part of the range obtained in this study. In the West Rand, occurrences of biotite together with stilpnomelane have been reported, and stilpnomelane has also been identified in the Ventersdorp Contact Reef, where it is thought to have replaced K-feldspar and chlorite (Wallmach, 1991). According to Winkler (1979) and Frey (1987), parageneses including stilpnomelane and biotite are stable only during very low-grade metamorphism at temperatures less than 350 °C, which provides further support for our temperature estimates. Also the fluid inclusion study by Foya (2001) of the Kimberley Reefs (stratigraphically in the uppermost part of the Central Rand Group, not much below the Ventersdorp Contact Reef), from basinwide localities, indicated fluid entrapment temperatures of ~300 °C for pressures of 2–3 kbar.

Our primary fluid inclusion data from pseudotachylitic breccia infer formation conditions of 600–650 °C at pressures of 3–4 kbar (see also, Killick and Roering, 1998). However, the fluids studied in our pseudotachylitic breccia samples indicate much lower pressure (P), temperature (T) conditions. This is another clear indication that we are dealing with later, secondary fluids in the pseudotachylitic breccias and that the primary fluid characteristics have been largely overprinted.

All the observed secondary fluids, be it in Ventersdorp Contact Reef conglomerate, footwall quartzite, secondary quartz veins, or in pseudotachylitic breccia, were trapped at temperatures below 400 °C. The secondary fluids present in the pseudotachylitic breccias are similar to those found in the other rock types. This implies that fluid inclusions in pseudotachylitic breccia postdate the time of breccia formation and are, thus, not representative of conditions of pseudotachylitic breccia formation. ^{40}Ar-^{39}Ar dating results on some of the breccia samples used in this study (Trieloff et al., 1994) place an approximate 2000 Ma constraint on the age of major pseudotachylitic breccia formation along this fault zone. More recent chronological work by Friese et al. (2003) on samples from the Welkom goldfield in the southwestern sector of the Witwatersrand basin also supports the notion that major pseudotachylitic breccia development in the Witwatersrand Basin should be linked to the 2.02 Ga Vredefort impact event. It appears reasonable to further conclude that these low-temperature secondary fluids were formed during or right after this time, as a consequence of the Vredefort impact-induced hydrothermal event (also Foya et al., 1999; Reimold and Przybylowicz, 2000; Reimold et al., 1999; Gibson and Reimold, 1999).

Thus, in terms of the objectives for this investigation, we were able to characterize the nature of the various fluids present in the Ventersdorp Contact Reef and associated quartz-rich rocks. The data obtained strongly suggest that Upper Witwatersrand strata at Tau Lekoa and Elandskraal mines were influenced by fluid circulation during low-grade metamorphism at or just after the 2020 Ma Vredefort event.

Discussion of Quadrupole Gas Analyses

The pronounced difference in CO_2/CH_4 ratios between the Ventersdorp Contact Reef sample suites from the two study localities is noteworthy. Probable sources for CO_2, CH_4, and other higher hydrocarbons seen in these fluids (compare Table 3) include thermal breakdown of complex hydrocarbons at moderate to high temperature (>100 °C), outgassing of juvenile carbon via high-temperature hydrothermal systems, and inorganic reactions between gases and rocks at higher temperatures (>300–400 °C). The CH_4/CO_2 ratio in such a reaction is mainly determined by the oxygen fugacity. Lower fO_2 conditions, and thus a more reducing environment, would increase the CH_4/CO_2 ratio, and vice versa. This indicates that lower fO_2 conditions prevailed in the Elandskraal region compared to Tau Lekoa at the time of fluid inclusion entrapment. Naden and Shepherd (1989) suggested that reaction of a fluid with graphitic rocks to produce CH_4 can cause the fluid to separate into a liquid and a vapor phase (i.e., phase separation). In this case, H_2S will enter the vapor, promoting gold precipitation by destabilization of the gold-transporting bisulphide ligands. The data presented above are in keeping with the generation of both methane and H_2S, as is reflected in higher concentrations of these species in the Elandskraal Mine samples. However, the total concentration of sulfur species in the Ventersdorp Contact Reef rocks is low and this inhibits the gold-carrying capacity of these transporting fluids.

The positive relation between H_2O and CO_2 in Figure 3 for Tau Lekoa, as well as the positive relation between H_2O and CH_4 in Figure 4 for the Elandskraal samples, furthermore, support the likelihood that the redox state was dictated by the CH_4/CO_2 ratio. However, this does not exclude the possible incorporation of carbonic species from other sources, such as the breakdown of organic material, which could also be a reason for the data scatter on Figures 3 and 4. The lower fO_2 conditions reflected by the Elandskraal samples would clearly have favored gold precipitation. The fire-assay analyses presented in Table 4 give an indication that this was indeed the case ($Au_{Elandskraal} >> Au_{Tau\ Lekoa}$).

Hayashi and Ohmoto (1991) provided thermodynamic data that suggest that, under P-T conditions similar to those prevailing in the Ventersdorp Contact Reef environment during the time of mineralization, gold would be transported mainly as HAu(HS) in hydrothermal fluids that are in equilibrium with pyrite and at low pH. Under more alkaline conditions (pH > 5.5), Au(HS) becomes dominant (Seward, 1989; Shenberger and Barnes, 1989). Partial dissolution and subsequent precipitation of gold may have occurred due to fluctuation in $a(H_{2aq})$, such as by reaction with ferrous minerals or organic matter or by fluctuation in $a(H_2S_{aq})$. The pronounced correlation between higher H_2 and H_2S concentrations and gold assay values at Elandskraal Mine compared to Tau Lekoa seems to support the above suggestion. However, it is clear from Figure 6 that HCl abundance is greater than the total concentration of sulfur species. This important finding suggests that gold may have been transported as a chloride complex during remobilization of this element relatively late in the paragenetic sequence for Witwatersrand ores (see Robb et al., 1997; Reimold et al., 1999; Foya, 2001), rather than as a sulfur-bearing complex.

In a plot of CO_2-N_2-Ar abundances (Fig. 7), most of the fluid inclusion analyses plot away from the composition of air-saturated water, except for sample 10G12, which is a typical late-stage quartz vein. It is suggested, therefore, that the predominant fluid associated with Ventersdorp Contact Reef rocks is a secondary connate fluid that had not been in contact with the atmosphere for a considerable time prior to entrapment. This confirms that we are dealing with secondary fluids, which could be related to gold remobilization, and totally excludes any contribution from meteoric waters. Samples from porphyry-related tin-tungsten veins, Mississippi Valley–type, and micron-gold deposits, all associated with sedimentary rocks (a possible source of N_2), would plot above the water-CO_2 mixing line on a diagram similar to Figure 7 (Robert J. Bodnar, Virginia Polytechnic Institute and State University, June 2005, personal commun.). Analyses from porphyry and acid-sulfate deposits, in which magmatic gas contributions should be largest, would plot closest to the fumarole gas compositions. High Ar values are probably due to generation from the radioactive decay of potassium with subsequent incorporation of radiogenic argon into fluid inclusions.

TABLE 4. SELECTED Au FIRE-ASSAY ANALYSES

Elandskraal Mine		Tau Lekoa	
Sample ID	Au (ppm)	Sample ID	Au (ppm)
EL 17–1	8.45	10D06–1	4.30
EL 20–1	12.90	10D210–1	5.35
EL 22–1	7.00	10D225–1	1.70

Note: Results were obtained commercially.

Thermodynamic Modeling of Petrographic and Gas Analytical Data

The data discussed above provide a basis for estimating the mobility of gold in the Ventersdorp Contact Reef fluids. Pressure and temperature conditions used for these calculations are 2 kbar and 350 °C. All fluid species given in Table 3 are assumed to be in the gaseous state. The pH of the Ventersdorp Contact Reef solutions were obtained by considering the coexistence of muscovite and pyrophyllite, which are found in numerous Ventersdorp Contact Reef samples (e.g., EL15, –21, –28, 10D08, –204, –05, –217 and 10G05). Following the reaction

$$KAl_2(AlSi_3)O_{10}(OH)_2 + 3SiO_2 \leftrightarrows 1.5Al_2Si_4O_{10}(OH)_2 \quad \log K = 1.930, \tag{1}$$

we find that

$$\log K = \log aK^+ - \log aH^+. \tag{2}$$

For a salinity, as determined with microthermometry, of 14 wt% eNaCl (2.6 molal), a Na:K ratio of 4 and an activity coefficient of 0.120, the pH for the above reaction is 3.04. The neutral pH is given by the reaction

$$2H^+ + OH^- \leftrightarrows H_2O \quad \log K = 10.914. \tag{3}$$

For a pressure of 2 kbar and a temperature of 350 °C, the neutral pH is 5.1. Further proof of these acidic conditions is presented by the absence of K-feldspar in the Ventersdorp Contact Reef rocks. On this basis, as well as in terms of the reaction

$$KAl_2(AlSi_3)O_{10}(OH)_2 + 2K^+ + 6SiO_2 \leftrightarrows 3K(AlSi_3)O_8 + 2H^+ \quad \log K = -7.734, \tag{4}$$

we can calculate the upper limit of pH to ~4.97. The absence of K-feldspar dictates that the pH limit was 4.97.

In order to constrain the carbonic species in pH and log fO_2 space, the following reactions were considered:

$$\text{Graphite} + H_2O \leftrightharpoons CH_4 + O_2 \log K = -29.015. \quad (5)$$

$$\text{Graphite} + 2H_2O \leftrightharpoons CO_2 + 2H_2 \log K = +2.385. \quad (6)$$

Calculated for ideal conditions at elevated Ventersdorp Contact Reef pressures and temperatures, the oxygen fugacities ($\log fO_2$) for these reactions are –26.05 and –29.02, respectively. By calculating the activities of the various gaseous species for the Elandskraal and Tau Lekoa localities (Table 5), we are able to calculate the specific oxygen fugacities that prevailed at the time of gold remobilization at these localities.

By using these calculated activities for the Elandskraal and Tau Lekoa localities in the reactions previously calculated for ideal conditions, it is possible to calculate the oxygen and hydrogen fugacities for these specific localities. These calculations yielded oxygen fugacity values of –28.686 for the Elandskraal Mine and –26.824 for Tau Lekoa Mine. Using the reactions

$$\text{Graphite} + 2H_2O \leftrightharpoons CO_2 + 2H_2 \log K = 2.385, \quad (7)$$

and

$$CH_4 + CO_2 \leftrightharpoons 2\text{Graphite} + 2H_2O \log K = -4.191, \quad (8)$$

we were able to calculate a hydrogen fugacity ($\log fH_2$) of –1.07 for Elandskraal Mine and –3.67 for Tau Lekoa Mine, respectively.

Thus, the difference in oxygen and hydrogen fugacity between these two localities is more than two orders of magnitude. The low pH conditions, as well as the low aH_2S activities, certainly do not favor the transportation of gold by sulfur complexes. As reflected in the high HCl concentration in the gas analysis, the chloride species is more likely to have played a role in the remobilization of gold in the Ventersdorp Contact Reef.

Further Thermodynamic Modeling: Implications for Gold Deposition

From the data presented, the following information pertaining to the nature of the ore-transporting fluid and environmental conditions at the time of deposition can be obtained. Mineralization took place at a temperature of ~350 °C (range of estimated temperatures from fluid inclusion analyses and low-grade metamorphic assemblages: 220–370 °C) and a minimum pressure of 2.5 kbar (Wallmach and Meyer, 1990). Mean X_{CO_2} and X_{CH_4} concentrations in the Elandskraal fluids were 0.27 and 1.03, corresponding to activities of $aCO_2 = 0.34$ and $aCH_4 = 0.46$, respectively. Mean concentrations for Tau Lekoa samples were $X_{CO_2} = 1.95$ and $X_{CH_4} = 0.33$, corresponding to $aCO_2 = 2.44$ and $aCH_4 = 0.15$. Fluid salinities were around 2.6 molar, with a mean Na:K ratio of 4. In terms of the reaction $CO_{2(aq)} + H_2O \leftrightharpoons HCO_3^- + H^+$ ($\log K = -7.658$; calculated using SUPCRT92, Johnson et al., 1992; $\log a_K = -1.11$; ionic strength = 2.46), an acidic pH of ~3 (neutral pH at 350 °C is 5.1) was calculated for the Ventersdorp Contact Reef fluids (g_{CO_2} – 2.26; calculated from Patterson et al., 1981).

QMS gas data were utilized to determine redox conditions: fO_2 was calculated using the reactions $CH_4 + 2H_2O \leftrightharpoons 4H_2 + CO_2$ and $H_2O \leftrightharpoons H_2 + ½O_2$, which yielded values of $10^{-29.02}$ and $10^{-26.05}$ for Elandskraal and Tau Lekoa samples, respectively. QMS data yielded sulfur fugacities for Elandskraal and Tau Lekoa of $fH_2S_g = 1.099$ and 0.598, assuming $\Sigma S_{Total} = X_{H_2S} + X_{SO_2}$, which corresponds to $aH_2S_{,aq}$ of $10^{-1.5}$ and $10^{-2.6}$, respectively.

Using these fluid characteristics, inferences on the gold speciation and precipitation mechanism are possible. Solubility calculations indicate that gold in fluids circulating through the Ventersdorp Contact Reef in the Elandskraal area was an order of magnitude more soluble as a chloride complex than as a bisulfide complex. In the Tau Lekoa area, gold was equally soluble as a chloride complex and as a bisulfide complex. The calculated result is confirmed by the higher abundance of HCl in the Tau Lekoa samples compared to the Elandskraal samples. It appears that the postdepositional fluids were certainly capable of taking up appreciable amounts of gold into solution, in the order of 1–10 parts per billion. These calculated values for gold solubility are significantly lower than the speculative estimates by Phillips et al. (1990). The lower gold solubilities calculated in this study place severe constraints on the hydrothermal model for gold mineralization in the Witwatersrand basin and rather argue against large-scale gold mobilization. That late fluid activity in the Witwatersrand Basin is probably restricted to small-scale activity (though overall on a basinwide scale) and has also been discussed in detail by Hayward et al. (2005) on the basis of micropetrographic analysis of gold from all Witwatersrand goldfields. These authors conclude also that it is very likely that hydrothermal activity related to the 2020 Ma Vredefort impact event had a significant effect on gold remobilization throughout all major reefs of the Witwatersrand Basin. They stated that gold mineralization occurred under fluid-poor conditions at temperatures that did not exceed 350 °C. Furthermore, widespread circulation of mineralizing fluids within the Central Rand Group is not supported and data suggest that the Vredefort Impact Event (2.02 Ga) was the cause of late deformation, and that postimpact fluid = poor metamorphism resulted in crystallization of a significant proportion of the gold on and within mineral grains that were deformed during this event (Hayward et al., 2005).

TABLE 5. ACTIVITIES FOR THE DOMINANT GASEOUS SPECIES IN SAMPLES FROM ELANDSKRAAL AND TAU LEKOA MINES, CALCULATED FOR 2 KBAR AND 350 °C

Activity	Elandskraal Mine	Tau Lekoa
aCO_2	0.343	2.436
aCH_4	0.469	0.148
aC_2H_6	0.063	0.028
aN_2	0.128	0.047
aH_2S	0.017	0.008

Note: *a*—activity.

The precipitation mechanisms that account for gold deposition within the Ventersdorp Contact Reef are not well understood. It is possible that gold was precipitated as a result of adsorption/reduction reactions (e.g., Schoonen et al., 1992), as locally suggested by the relationship between gold content and As zonation in pyrite (Meyer, 1992). A number of precipitating mechanisms may have been active simultaneously within any reef unit.

CONCLUSIONS

The following conclusions can be drawn on the basis of these new petrographic and analytical results.

1. The samples studied contain three main fluid populations: Fluid I—this earliest fluid is characterized by small inclusions forming a wispy texture; Fluid II—a moderately saline fluid is present in all the samples studied; Fluid III—a third, CO_2-rich fluid, with variable CO_2/H_2O ratios, is present in a large number of samples. Fluids from the Ventersdorp Contact Reef quartzites, both footwall quartzite and reef quartzite at Elandskraal and Tau Lekoa Mines, are remarkably similar regarding salinities and densities. Quartzites from both localities contain wispy textures of earlier fluid inclusions.

2. Fluid inclusions are predominantly of a secondary origin. Fluid inclusion microthermometry yielded a temperature range of 220–370 °C at 2 kbar for fluid entrapment conditions. Secondary inclusions in samples from Tau Lekoa have a significantly higher CO_2 content compared to those from the Elandskraal Mine, which have, in turn, a higher CH_4 content. The variation in the CH_4/CO_2 ratio is indicative of a difference in oxygen fugacity conditions, with lower fO_2 conditions prevailing at the Elandskraal Mine.

3. The presence of higher hydrocarbon species was confirmed by analysis of relatively high concentrations of C_2H_6. Whole-rock oxygen and hydrogen isotope values indicate that the associated fluids were typical of a low-metamorphic and organic nature. Relatively low N_2 and high Ar concentrations suggest a connate fluid that had not been exposed to atmosphere for a significant part of geologic time.

4. Thermodynamic modeling for Ventersdorp Contact Reef conditions using gas analytical data yields distinctly lower oxygen and hydrogen fugacity conditions for samples from Elandskraal Mine than for samples from Tau Lekoa. The pH conditions were acidic (pH = 3.03 for samples analyzed; neutral pH = 5.1 at 2 kbar and 350 °C).

5. The activity of sulfur species in Ventersdorp Contact Reef rocks is extremely low (e.g., $aH_2S_{Elandskraal} = 0.017$ and $aH_2S_{Tau\ Lekoa} = 0.0078$). High concentrations of HCl (HCl >> sulfur species) suggest that gold was transported as a chloride-complex during remobilization.

6. The solubility of gold in the mineralizing fluids has been calculated to be in the order of 1–10 parts per billion. These values were calculated and verified using quantitative mineralogical, microthermometric, and entrapped fluid composition data, which were processed with standard thermodynamic procedures and principles.

Our textural observations relate fluid entrapment to a very late stage in the paragenetic evolution of the Ventersdorp Contact Reef. Based on the available chronological information (Trieloff et al., 1994; Friese et al., 2003) and the premise that the often massive pseudotachylitic breccia occurrences throughout much of the Witwatersrand Basin (Fletcher and Reimold, 1989; Killick and Reimold, 1990; Reimold and Colliston, 1994; Reimold and Gibson, 2005; Reimold et al., 2005) must be related to the 2.02 Ga Vredefort impact event, it appears that the low-temperature secondary fluids represent Vredefort-related fluid activity. According to Reimold et al. (2005), it is to be expected that the specific impact-triggered pressure and temperature conditions throughout a wide part of the Witwatersrand basin could have been conducive to remobilization of gold. However, the chemical characteristics of this fluid, as presented in this contribution, suggest that it may have had limited capability of mobilizing gold and thus that the ultimate source of the Ventersdorp Contact Reef gold is sedimentary mineralization.

ACKNOWLEDGMENTS

We thank Stephen Kesler for the use of the Quadrupole Mass Spectrometer analytical facilities at the Department of Geological Sciences, University of Michigan, Ann Arbor, and for his extraordinary kindness and support. Comments by Hartwig Frimmel, Laurence Robb, Gillian Drennan, and anonymous reviewers of an earlier version of this manuscript greatly improved the paper, as did the editorial input by Roger Gibson. Anglo American Prospecting Services supported us through the Ventersdorp Contact Reef Geochemical Project of 1991–1993. Lyn Whitfield assisted with drafting of the figures. This is University of the Witwatersrand Impact Cratering Research Group contribution No. 91.

REFERENCES CITED

Armstrong, R.A., Compston, W., Retief, E.A., Williams, I.S., and Welke, H.J., 1991, Zircon ion-microprobe studies bearing on the age and evolution of the Witwatersrand triad: Precambrian Research, v. 53, p. 243–266, doi: 10.1016/0301-9268(91)90074-K.

Bodnar, R.J., 1993, Revised equation and table for determining the freezing-point depressions of H_2O-NaCl solutions: Geochimica et Cosmochimica Acta, v. 57, p. 683–684, doi: 10.1016/0016-7037(93)90378-A.

Brown, P.E., and Lamb, W.M., 1989, P-V-T properties of fluids in the system H_2O-CO_2-NaCl: New graphical presentations and implications for fluid inclusion studies: Geochimica et Cosmochimica Acta, v. 53, p. 1209–1221, doi: 10.1016/0016-7037(89)90057-4.

Fletcher, P., and Reimold, W.U., 1989, Some notes and speculations on the pseudotachylites in the Witwatersrand Basin and the Vredefort Dome: South African Journal of Geology, v. 92, p. 223–234.

Foya, S.N., 2001, Mineralogical and geochemical controls on gold distribution in the Kimberley Reefs, South Africa [Ph.D. thesis]: Johannesburg, South Africa, University of the Witwatersrand, 383 p.

Foya, S.N., Reimold, W.U., Przybylowicz, W.J., and Gibson, R.L., 1999, PIXE microanalysis of gold-pyrite-pyrrhotite associations from Kimberley Reefs, Witwatersrand Basin: Nuclear Instruments and Methods in Physics, v. 158, p. 588–592, doi: 10.1016/S0168-583X(99)00364-X.

Frey, M., 1987, Very low-grade metamorphism of clastic sedimentary rocks, *in* Frey, M., ed., Low Temperature Metamorphism: Glasgow, Blackie, p. 9–58.

Friese, A.E.W., Reimold, W.U., and Layer, P.W., 2003, ^{40}Ar-^{39}Ar dating and structural information on tectonite-bearing faults in the Witwatersrand Basin: Evidence for multi-stage, tectono-thermal activity in the Central Kaapvaal Craton: South African Journal of Geology, v. 106, p. 41–70, doi: 10.2113/1060041.

Frimmel, H.E., 2005a, Archaean atmospheric evolution: Evidence from the Witwatersrand gold fields, South Africa: Earth Science Reviews, v. 70, p. 1–46, doi: 10.1016/j.earscirev.2004.10.003.

Frimmel, H.E., 2005b, Review paper: Economic Geology and the Bulletin of the Society of Economic Geologists, v. 100 (in press).

Frimmel, H.E., and Gartz, V.H., 1997, Witwatersrand gold particle chemistry matches model of metamorphosed, hydrothermally altered placer deposits: Mineralia Deposita, v. 32, p. 523–530, doi: 10.1007/s001260050119.

Frimmel, H.E., and Minter, W.E.L., 1991, The mobility of Witwatersrand gold during post-depositional alteration, *in* Pagel, M., and Leroy, J.L., eds., Extended Abstracts, Conference on Source, Transport and Deposition of Metals: Rotterdam, Balkema, p. 657–660.

Frimmel, H.E., and Minter, W.E.L., 2002, Recent developments concerning the geological history and genesis of the Witwatersrand gold deposits, South Africa, *in* Goldfarb, R.J., and Nielsen, R.L., eds., Integrated Methods for Discovery: Global Exploration in the Twenty-First Century: Littleton, Society of Economic Geologists Special Publication 9, p. 17–45.

Frimmel, H.E., Le Roux, A.P., Knight, J., and Minter, W.E.L., 1992, A case study on the Post-Depositional Alteration of the Witwatersrand Basal Reef Gold Placer: Precambrian Research Unit, University of Cape Town, Information Circular, v. 8, p. 16–33.

Frimmel, H.E., Le Roux, A.P., Knight, J., and Minter, W.E.L., 1993, A case study of the postdepositional alteration of the Witwatersrand Basal reef gold placer: Economic Geology and the Bulletin of the Society of Economic Geologists, v. 88, p. 249–265.

Frimmel, H.E., Hallbauer, D.K., and Gartz, V.H., 1999, Gold mobilizing fluids in the Witwatersrand Basin: Composition and possible sources: Mineralogy and Petrology, v. 66, p. 51–88.

Gibson, R.L., and Reimold, W.U., 1999, The significance of the Vredefort Dome for the thermal and structural evolution of the Witwatersrand Basin, South Africa: Mineralogy and Petrology, v. 66, p. 5–23, doi: 10.1007/BF01161720.

Gibson, R.L., and Reimold, W.U., 2001, The Vredefort Impact Structure, South Africa (The scientific evidence and a two-day excursion guide): Council for Geoscience, Pretoria Memoir 92, 110 p.

Gibson, R.L., and Stevens, G., 1998, Regional metamorphism due to anorogenic intracratonic magmatism, *in* Treloar, P.J., and O'Brien, P.J., eds., What Drives Metamorphism and Metamorphic reactions?: Geological Society [London] Special Publication 138, p. 121–135.

Giggenbach, W.F., Garcia, P.N., Londono, C.A., Rodrigues, V.L., Rojas, G.N., and Calcache, V.M.L., 1990, The chemistry of fumarolic vapor and thermal-spring discharge from the Nevado del Ruiz volcanic-magmatic hydrothermal system, Columbia: Journal of Volcanology and Geothermal Research, v. 42, p. 13–39, doi: 10.1016/0377-0273(90)90067-P.

Hayashi, K.I., and Ohmoto, H., 1991, Solubility of gold in NaCl- and H_2S-bearing aqueous solutions at 250–350 °C: Geochimica et Cosmochimica Acta, v. 55, p. 2111–2126, doi: 10.1016/0016-7037(91)90091-I.

Hayward, C.L., Reimold, W.U., Gibson, R.L., and Robb, L.J., 2005, Gold mineralization within the Witwatersrand Basin, South Africa: Evidence for a modified placer origin, and the role of the Vredefort impact event, *in* McDonald, I., Boyce, A.J., Butler, I.B., Herrington, R.J., and Polya, D.A., eds., Mineral Deposits and Earth Evolution: Geological Society [London] Special Publication 248, p. 31–57.

Johnson, J.W., Oelkers, E.H., and Helgeson, H.C., 1992, SUPCRT92: A software package for calculating the Standard Molal Thermodynamic Properties of minerals, gases, aqueous species, and reactions from 1 to 5000 bars and 0° to 1000 °C: Computers in Geoscience, v. 18, p. 899–947, doi: 10.1016/0098-3004(92)90029-Q.

Jolley, S.J., Freeman, S.R., Barnicoat, A.C., Phillips, G.M., Knipe, R.J., Pather, A., Fox, N.P.C., Strydon, D., Birch, M.T.G., Henderson, L.H.C., and Rowland, T.W., 2004, Structural controls on Witwatersrand gold mineralisation: Journal of Structural Geology, v. 26, p. 1067–1086, doi: 10.1016/j.jsg.2003.11.011.

Kamo, S.L., Reimold, W.U., Krogh, T.E., and Colliston, W.P., 1996, A 2.023 Ga year age for the Vredefort impact event and a first report of shock metamorphosed zircons in pseudotachylitic breccias and granophyre: Earth and Planetary Science Letters, v. 144, p. 369–387, doi: 10.1016/S0012-821X(96)00180-X.

Killick, A.M., and Reimold, W.U., 1990, Review of the pseudotachylites in and around the Vredefort Dome, South Africa: South African Journal of Geology, v. 93, p. 350–365.

Killick, A.M., and Roering, C., 1998, An estimate of the physical conditions of pseudotachylite formation in the West Rand Goldfield, Witwatersrand Basin, South Africa: Tectonophysics, v. 284, p. 247–259, doi: 10.1016/S0040-1951(97)00181-9.

Killick, A.M., Thwaites, A.M., Germs, G.J.B., and Schoch, A.E., 1988, Pseudotachylite associated with a bedding-parallel fault zone between the Witwatersrand and Ventersdorp Supergroups, South Africa: Geologische Rundschau, v. 77, p. 329–344, doi: 10.1007/BF01848694.

Kirk, J., Ruiz, J., Chesley, J., Titley, S., and Walshe, J., 2001, A detrital model for the origin of gold and sulfides in the Witwatersrand Basin based on Re-Os isotopes: Geochimica et Cosmochimica Acta, v. 65, p. 2149–2159, doi: 10.1016/S0016-7037(01)00588-9.

Kirk, J., Ruiz, J., Chesley, J., Walshe, J., and England, G., 2002, A major Archean gold and crust-forming event in the Kaapvaal Craton, South Africa: Science, v. 297, p. 1856–1858, doi: 10.1126/science.1075270.

Meyer, F.M., 1992, A review of the ore mineralogy and geochemistry of Witwatersrand Reefs with emphasis on the gold-pyrite association: Short Course Reviewing the Understanding of the Witwatersrand Basin, Department of Geology/Economic Geology Research Unit: University of the Witwatersrand, v. 2, p. 155–167.

Naden, J., and Shepherd, T.J., 1989, Role of methane and carbon dioxide in gold deposition: Nature, v. 342, p. 793–795, doi: 10.1038/342793a0.

Patterson, D.J., Ohmoto, H., and Solomon, M., 1981, Geologic setting and genesis of cassiterite-sulfide mineralization at Renison Bell, Western Tasmania: Economic Geology and the Bulletin of the Society of Economic Geologists, v. 76, p. 393–438.

Phillips, G.N., and Law, J.D.M., 2000, Witwatersrand Gold Fields: Geology, Genesis, and Exploration: Society of Economic Geologists (SEG), Reviews, v. 13, p. 439–500.

Phillips, G.N., Law, J.D.M., and Myers, R.E., 1990, The role of fluids in the evolution of the Witwatersrand Basin: South African Journal of Geology, v. 93, p. 54–69.

Pretorius, D.A., 1986, The goldfields of the Witwatersrand Basin, *in* Anhaeusser, C.R., and Maske, S., eds., Mineral Deposits of Southern Africa, v. 1: Johannesburg, Geological Society of South Africa, p. 489–493 and p. 1019–1020.

Reimold, W.U., and Boer, R.H., 1992, Final Report "Ventersdorp Contact Reef Pseudotachylite Geochemical Project" (Anglo American Prospecting Services, Johannesburg): Johannesburg, Economic Geology Research Unit, University of the Witwatersrand, 186 p.

Reimold, W.U., and Colliston, W.P., 1994, Pseudotachylites of the Vredefort Dome and the surrounding Witwatersrand Basin, South Africa, *in* Dressler, B.O., Grieve, R.A.F., and Sharpton, V.L., eds., Large Meteorite Impacts and Planetary Evolution: Geological Society of America Special Paper 293, p. 177–196.

Reimold, W.U., and Gibson, R.L., 2005, "Pseudotachylite" in large impact structures, *in* Koeberl, C., and Henkel, H., eds., Impact Tectonics, Impact Studies Series, v. 6: Berlin, Springer, p. 1–53.

Reimold, W.U., and Przybylowicz, W.J., 2000, PIXE microanalysis of concretionary pyrite from the Witwatersrand Basin, South Africa, *in* Rammelmair, D., Mederer, J., and Oberthür, Th., eds., Proceedings of the ICAM 2000 Conference, Applied Mineralogy: Rotterdam, Balkema, p. 407–410.

Reimold, W.U., Boer, R.H., and Jackson, M.C., 1993, Final report "Ventersdorp Contact Reef Pseudotachylite Geochemical Project (Extension 1992)" (to Anglo American Prospecting Services, Johannesburg): Johannesburg, Economic Geology Research Unit, University of the Witwatersrand, 133 p.

Reimold, W.U., Köberl, C., Fletcher, P., Killick, A.M., and Wilson, J.D., 1999, Pseudotachylitic breccias from fault zones in the Witwatersrand Basin, South Africa: Evidence of autometasomatism and post-brecciation alteration processes: Mineralogy and Petrology, v. 66, p. 25–53, doi: 10.1007/BF01161721.

Reimold, W.U., Gibson, R.L., Koeberl, C., and Dressler, B.O., 2005, Economic ore deposits in impact structures and their geological setting, *in* Koeberl, C., and Henkel, H., eds., Impact Tectonics, Impact Studies Series, v. 6: Berlin, Springer, p. 479–552.

Robb, L.J., and Robb, V.M., 1998, Gold in the Witwatersrand Basin, *in* Wilson, M.G.C., and Anhaeusser, C.R., eds., The Mineral Resources of South Africa: Pretoria, South Africa, Council for Geoscience, p. 294–349.

Robb, L.J., Davis, D.W., and Kamo, S.L., 1990, U-Pb ages on single detrital zircon grains from the Witwatersrand basin, South Africa: Constraints on the age of sedimentation and on the evolution of granites adjacent to the basin: The Journal of Geology, v. 98, p. 311–328.

Robb, L.J., Charlesworth, E.G., Drennan, G.R., Gibson, R.L., and Tongu, E.L., 1997, Tectono-metamorphic and paragenetic sequence of Au-U mineralization in the Archaean Witwatersrand Basin, South Africa: Australian Journal of Earth Sciences, v. 44, p. 353–371.

Roedder, E., 1984, Fluid Inclusions: Reviews in Mineralogy, v. 12, 644 p.

Schoonen, M.A.A., Fisher, N.S., and Wente, M., 1992, Gold sorption onto pyrite and goethite: A radiotracer study: Geochimica et Cosmochimica Acta, v. 56, p. 1801–1814, doi: 10.1016/0016-7037(92)90311-6.

Schreyer, W., and Bisschoff, A.A., 1982, Kyanite as a metamorphic mineral in Witwatersrand sediments at Johannesburg and Krugersdorp, South Africa: Transactions of the Geological Society of South Africa, v. 85, p. 211–214.

Seward, T.M., 1989, The hydrothermal chemistry of gold and its implications for ore formation: Boiling and conductive cooling as examples, *in* Keays, R.R., Ramsay, W.R.H., and Groves, D.I., eds., The Geology of Gold Deposits: The Perspective in 1988: Economic Geology Monographs, v. 6, p. 398–404.

Shenberger, D.M., and Barnes, H.L., 1989, Solubility of gold in aqueous sulfide solutions from 150 to 350 °C: Geochimica et Cosmochimica Acta, v. 53, p. 269–278, doi: 10.1016/0016-7037(89)90379-7.

Sterner, S.M., and Bodnar, R.J., 1984, Synthetic fluid inclusions in natural quartz I, compositional types synthesized and applications to experimental geochemistry: Geochimica et Cosmochimica Acta, v. 48, p. 2659–2668, doi: 10.1016/0016-7037(84)90314-4.

Trieloff, M., Reimold, W.U., Kunz, J., Boer, R.H., and Jessberger, E.K., 1994, ^{40}Ar-^{39}Ar thermochronology of pseudotachylites at the Ventersdorp Contact Reef, Witwatersrand Basin: South African Journal of Geology, v. 97, p. 365–384.

Viljoen, M.J., and Reimold, W.U., 1994, eds., The Ventersdorp Contact Reef: South African Journal of Geology, v. 97, p. 234–237.

Wallmach, T., 1991, Metamorphism in the Witwatersrand: Johannesburg, Extended Abstracts: 7th Annual Meeting of the Tectonics Division of the Geological Society of South Africa, v. 7, p. 15–16.

Wallmach, T., and Meyer, F.M., 1990, A petrogenetic grid for metamorphosed aluminous Witwatersrand shales: South African Journal of Geology, v. 93, p. 93–102.

Walraven, F., and Martini, J., 1995, Zircon Pb-evaporation age determinations of the Oak Tree Formation, Chuniespoort Group, Transvaal Sequence: Implications for Transvaal-Griqualand West basin correlations: South African Journal of Geology, v. 98, p. 58–67.

Winkler, H.G.F., 1979, Petrogenesis of Metamorphic Rocks, 5th Ed.: Berlin, Springer, 348 p.

Manuscript Accepted by the Society 4 October 2005

Printed in the USA

Index

N

O

P

Q

R